AF248940

DYNAMIC RESEARCH, INC.

THE DEBATE BETWEEN STIFF AND YIELDING SEATS

A NEW GENERATION OF YIELDING SEATS WITH HIGH RETENTION IN REAR CRASHES

Other SAE books of interest:

Role of the Seat in Rear Crash Safety
by David C. Viano
(Order No. R-317)

2002 SAE Active Safety Technology Collection on CD-ROM
(Order No. ACTSAFE2002)

Seat Belts: The Development of an Essential Safety Feature
by David C. Viano
(Order No. PT-92)

Air Bag Development and Performance – New Perspectives from Industry, Government and Academia
by Richard W. Kent
(Order No. PT-88)

THE DEBATE BETWEEN STIFF AND YIELDING SEATS

A NEW GENERATION OF YIELDING SEATS WITH HIGH RETENTION IN REAR CRASHES

PT-106

Edited by

David C. Viano

Published by
Society of Automotive Engineers, Inc.
400 Commonwealth Drive
Warrendale, PA 15096-0001
U.S.A.
Phone (724) 776-4841
Fax: (724) 776-5760
www.sae.org

For multiple print copies contact:

SAE Customer Service
Tel: 877-606-7323 (inside USA and Canada)
Tel: 724-776-4970 (outside USA)
Fax: 724-776-1615
Email: CustomerService@sae.org

ISBN 0-7680-1258-9
Library of Congress Catalog Card Number: 2003105036
SAE/PT-106

SAE Order No. PT-106

Printed in USA

Table of Contents

SAE AND AAAM ARTICLES ON SEATS

NHTSA STUDY ON SEATBACK PERFORMANCE IN RELATION TO FMVSS 207

Please note:
Some papers may not be included in their entirety. Appendices and other portions may have been deleted due to space limitations. Copies of entire papers may be purchased from SAE, if desired

SECTION 1:

SAFETY BELT SYSTEMS AND PERFORMANCE

CHAPTER 1:

RESOLVING THE DEBATE BETWEEN RIGID (STIFF) AND YIELDING SEATS: SEAT PERFORMANCE CRITERIA FOR REAR CRASH SAFETY

Resolving the Debate Between Rigid (Stiff) and Yielding Seats: Seat Performance Criteria for Rear Crash Safety

David C. Viano
ProBiomechanics LLC

ABSTRACT

The yielding seats of the 1960-80s had good safety performance in field crashes. A number of studies showed they provided a balance of safety over the full range of rear crashes. However, there were cases where improved occupant retention would be beneficial in limiting seatback rotation in the most severe crashes. This motivated a gradual increase in seat strength and stiffness in the 1980-90s as recliners and structures were strengthened.

Field studies showed that whiplash rates increased in the 1980-90s as seat stiffness increased with strength. This may have realized the concern of safety engineers about increasing whiplash by rigid seats. This furthered the debate between rigid (stiff) and yielding seats. A series of recent studies found a resolution to the long-standing debate. When the seat frame was strengthened, a perimeter frame was used to maintain a yielding behavior by letting the occupant displace into the seatback while the frame stayed upright. This allowed the introduction of strong, yielding seats, otherwise called high retention seats.

This study describes recent research aimed at developing specifications for seats to improve occupant protection in rear crashes. The specifications include requirements for seat responses in a Quasistatic Seat Test (QST), and rear sled tests at 16 and 36 km/h. For the QST performance, seat stiffness should be $k < 25$ kN/m, frame stiffness $j < 2.0\ ^0$/kN and seat load limit $F_L > 7.7$ kN.

For the sled tests, specifications are recommended for seat and dummy responses to assure an overall safety performance. While other "best" practices are described and evaluations with the 5% Hybrid III female dummy are recommended in the development of seats, the proposed requirements establish a minimum set of performance objectives for seats subjected to occupant loads in rear crashes. They also allow seat innovation to develop, since they are performance objectives, not design requirements.

INTRODUCTION

The work of Severy et al. (1967a,b, 1968a,b) demonstrated improvements in occupant retention with rigid seats. This started nearly a four decade debate between proponents of yielding and rigid (stiff) seats. The original studies, ones by coworkers (Blaisdell et al. 1993, Strother et al. 1994) and others (Warner et al. 1991, James et al. 1991, Prasad et al. 1997) showed the potential disbenefit with neck injuries and rigid seats. The argument for rigidizing the seat was understood, but the possible increase in whiplash in low speed crashes and injury in other situations hindered the implementation of rigid seats.

In retrospect, there were valid reasons to limit seatback rotation and there were valid reasons to maintain a yielding seat. However, no seat design provided both benefits until a new generation of yielding seats was phased into production from 1997-2002 (Viano 2002). These seats maintained the performance of the 1960-90s yielding seats and introduced a strong seat frame structure for occupant retention in severe rear crashes.

The debate was resolved by separately considering seat characteristics for strength and stiffness. Until the introduction of the new generation of yielding seats, stronger seats introduced in the 1980s-90s were also stiffer, and the two terms were treated as synonyms. The seat designs increased both characteristics as the recliner structure was strengthened without substantial variation to the seatback design. Strong structural cross-members were still used across the seatback at the hip and shoulders to increase torsional resistance and improve FMVSS 207 performance.

Designs to maintain a yielding behavior while strengthening the seat frame required an open seatback so the occupant could displace between the side frame structures. This gave distance for yielding, which lowered torso acceleration and gradually displaced the occupant forward, while the seat frame remains upright. This approach gave both a strong frame structure and a yielding stiffness.

The way to define the essential seat characteristics for occupant safety in rear crashes evolved with the development of high retention seats. They were developed and evaluated using the Quasistatic Seat Test (QST), which involved an occupant load into the seatback. This allowed measurements determining the essential seat characteristics for occupant loading in rear crashes (Viano 2002, 2003a-e):

- k Seat Stiffness (kN/m)
- j Frame Stiffness (^{0}/kN)
- F_L Seat Load Limit (kN)

There are other important seat characteristics that will be described in this article. They refine the performance of a seat for particular crash situations, but these three properties have the foremost importance.

The objective of this study is to synthesize previous work on seat design, evaluation and assessment to establish proposed performance criteria. This background and the specifications are needed for occupant retention and energy transfer in severe rear crashes and for torso loading, head-neck support and whiplash prevention in low-speed rear crashes.

The intention of this paper is to lay out a set of requirements for seats that ensure improved occupant protection in the full range of rear crashes. However, this study does not intend to address other aspects of rear crash safety, including vehicle mass, structure, EA bumpers, trailer hitch attachments and other attributes, which influence the crash pulse and limit the extent of intrusion in the most severe rear crashes. Obviously, vehicle mass is a key factor (Malliaris 1990).

NOMENCLATURE

ABTS All belts to seat

d Neck displacement, d = x – z

F Force on Occupant's Torso in kN
F_L Seat Load limit in kN
G Force on Neck in kN

i Time integer
j Slope of Seatback Angle vs Load in ^{0}/kN
k Stiffness of the Seat due to Occupant Loading in kN/m

k_R Stiffness due to Seatback Rotation in kN/m
k_S Stiffness from Trim and Suspension in kN/m

k_{HR} Stiffness of the Head Restraint in kN/m
k_N Stiffness of Neck in kN/m

m Effective Mass of the Occupant Loading the Seat in kg

m_T Mass of the Torso in kg
m_H Mass of the Head-Neck in kg

M_H H-point moment in Nm

p Length of the Seatback from the Pivot to the Top of the Head Restraint in m.
r Gap due to Seatback Rotation at the Center of Pressure of Occupant Loading in m
s Actual Head Restraint Gap w/r moving Vehicle and Head in m

Δt Delta time in s

u Drop in Head Restraint Height Due to Seatback Rotation in m
v Rearward Displacement of Head Restraint Due to Seatback Rotation in m
w Head Restraint Gap in m

x Horizontal Displacement of the Torso in m
$\dot{x}$ Velocity of the Torso in m/s
$\ddot{x}$ Acceleration of the Torso in m/s^2

x_R Torso Displacement due to Seatback Rotation
x_S Torso Displacement due to Seat Suspension Compliance

y Horizontal Displacement of the Vehicle in m
$\dot{y}$ Velocity of the Vehicle in m/s
$\ddot{y}$ Acceleration of the Vehicle in m/s^2

z Horizontal Head Displacement in m
$\dot{z}$ Head Velocity in m/s
$\ddot{z}$ Head acceleration in m/s^2

θ Seatback Angle w/r Vertical in deg

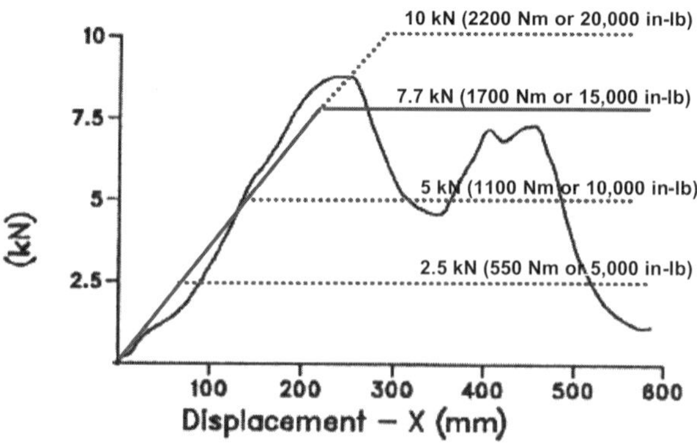

Figure 1: Example of force-deflection response of a seat loaded by a dummy in a QST. Various load limits and H-point moments are shown.

DEFINING SEAT PROPERTIES FROM QUASISTATIC SEAT TESTS

Figure 1 shows the force-deflection response from a Quasistatic Seat Test (QST) of a luxury Asian seat. The QST involves a Hybrid dummy loading the seat in a rearward direction by a hydraulic ram pressing against the lumbar joint (Viano 2002). The force is the ram load pushing the dummy into the seat and the deflection is the displacement of the ram.

The seat in this example has an initial stiffness of k = 36 kN/m for loads between F = 1-5 kN based on a linear least square analysis, where $F = k\Delta x$. The nominal limit load is F_L = 7.7 kN and occurs at about x = 215 mm of dummy displacement. The graph shows other load limits consistent with various peak H-point moments from M_H = 550-2200 Nm.

Figure 2 shows the same seat response as in Figure 1, but with different initial stiffness levels of k = 24 and 48 kN/m. These stiffnesses involve 160 and 320 mm displacement of the dummy into the seat at the limit load, respectively. This represents a 26% decrease to 49% increase in displacement at the limit load.

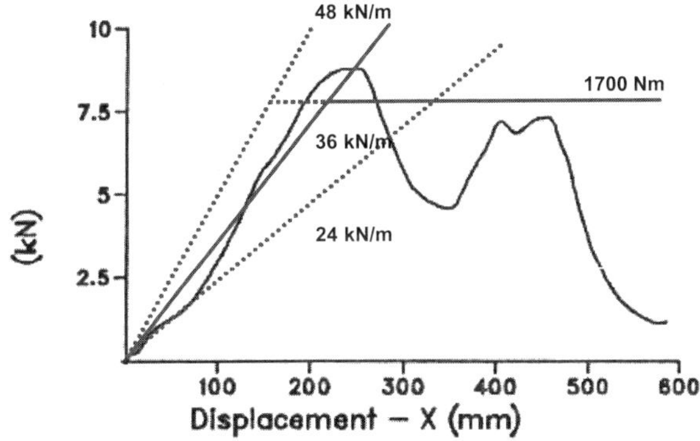

Figure 2: Typical range of frame stiffness from the early yielding seats of the 1960-80s, to strong, stiff seats of the 1990s through high retention seats.

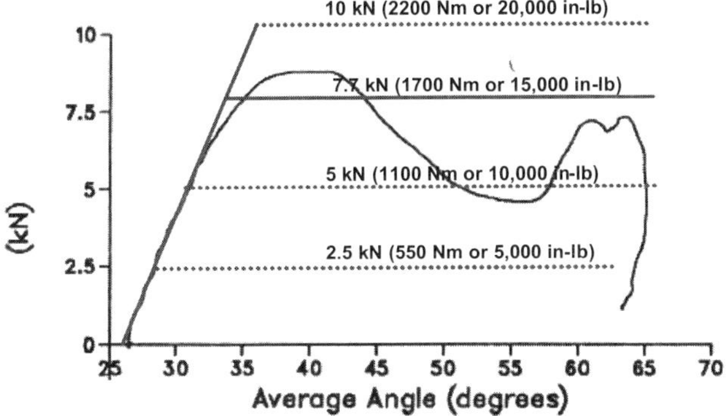

Figure 3: Force-rotation angle of the seat frame in the QST shown in Figures 1-2.

As the dummy loads the seatback, there is a combination of compression of the seat trim, suspension and rotation of the frame structure. Figure 3 shows the force-rotation response of the example seat. The nominal frame stiffness is j = 1.25 ^{0}/kN, where $\Delta\theta = jF$. While j is not really a stiffness value but rather compliance, it will be referred to as frame stiffness for convenience. This is a low value for j among seats tested (Viano 2002). However, ABTS seat can involve even lower seatback rotation. Again, variations in limit load and H-point moment are shown covering the range of traditional seats.

Figure 4 shows two additional frame stiffnesses that are typical of traditional seats. In this example, the limit load is reached at frame rotations of 9.6^0, 15.4^0 and 23.1^0, respectively. However, seats of the 1960-80s had load limits in the range of 3-6 kN, and never reached the 7.7 kN level shown in this example. They would have had a plateau at lower force levels, so the description is hypothetical in some cases.

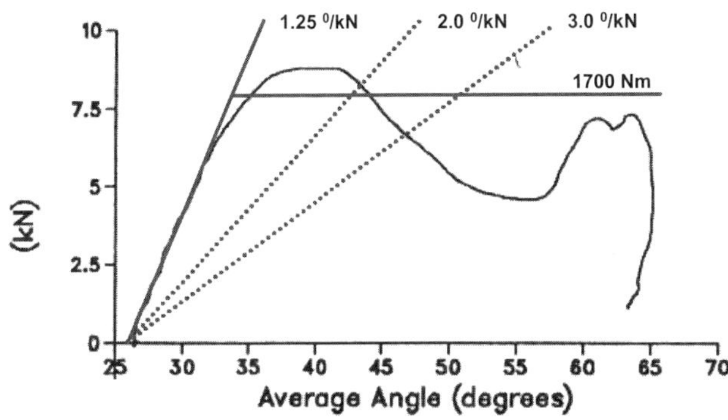

Figure 4: Typical range of initial frame stiffness from the early yielding seats of the 1960-80s, to strong, stiff seats of the 1990s through high retention seats.

The responses in Figures 1-4 define three essential characteristics or properties of a seat for occupant loading in a rear crash. They are: k seat stiffness, j frame stiffness and F_L load limit. Due to variations in force in the plateau or load limit region, there is no precise definition for a given seat. A nominal value is used referring tot the plastic region of the seat rotation response. These properties will be discussed in more detail in the following sections.

HISTORIC TRENDS IN SEAT PROPERTIES

Figure 5 shows the change in seat strength over the past four decades. It is based on seat testing by Severy (1967a,b), Strother et al. (1987, 1994) and Warner et al. (1991), and more recent QST tests by Viano (2002). While the peak force is shown, it reflects the trend in limit load, which is a somewhat lower value for each seat.

By the 1980s-90s, seats had increased in strength to 6-8 kN peak load from the lows of about 3 kN in the 1960s-70s. High retention seats further increased the strength of seats to over 14 kN, but they did so in a way that the seat stiffness was similar to the yielding seats of the 1980s-90s. Other strong seats had a comparable increase in seat stiffness as their strength increased.

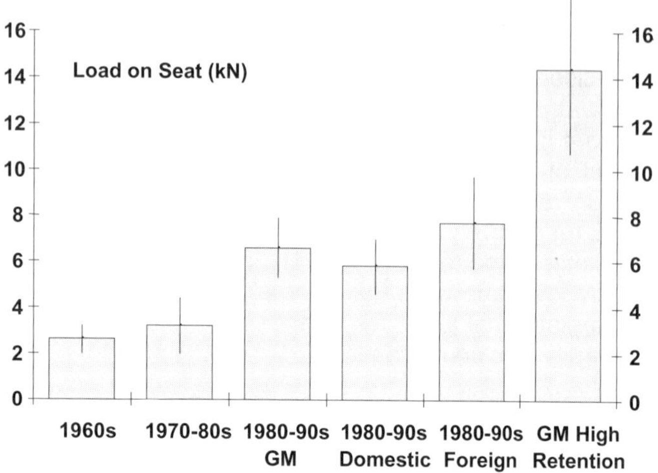

Figure 5: Trend in peak force supported by seats with an occupant or body block load toward the rear (from Viano 2003a with permission).

Figure 6 shows the range of seat responses for the yielding seats of the 1960-80s. The frame stiffness for these seats was j = 2.5-4.0 ^{0}/kN, yielding stiffness was k <25 kN/m and load limits were 3-6 kN. Some seats were designed with strong seat frames in the 1990s however stiffness increased with seat strength as the designs followed curves of constant jk = 60 ^{0}/m. This increased both the seat strength and stiffness to k = 35-50 kN/m.

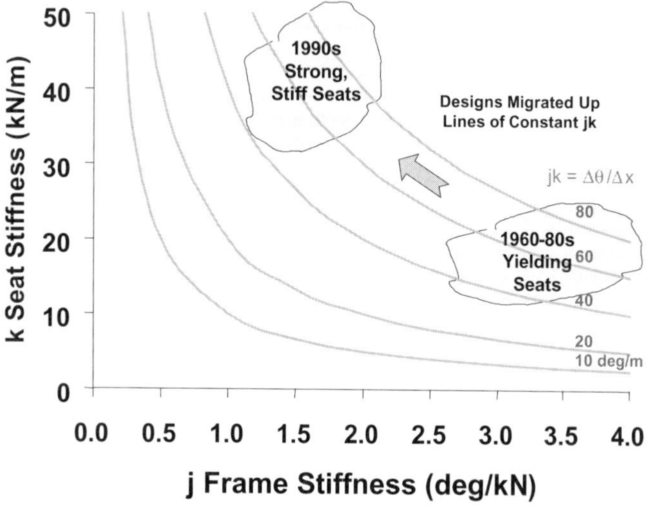

Figure 6: Shift in seat properties from the yielding seats of the 1960-80s to stronger and stiffer seats of the foreign luxury benchmark of the 1990s (Viano 2002).

Figure 7 shows the design direction for high retention seats introduced with a phase-in from 1997-2003 (Viano 2002). These seats maintained the yielding behavior of the earlier seats of 1960-80s, but introduced a much stronger seat frame to increase occupant retention in severe rear crashes.

This new generation of yielding seats involved a design philosophy that included a perimeter frame and compliant seat trim and suspension to allow a yielding stiffness with k < 25 kN/m while increasing the strength of the seat frame. These designs are in a region of the graph where no other production seat was designed. They are a combination of yielding and high retention. This emphasized the need for occupant retention in severe rear crashes and also the need for yielding in low-speed crashes to reduce chest acceleration and risks for whiplash.

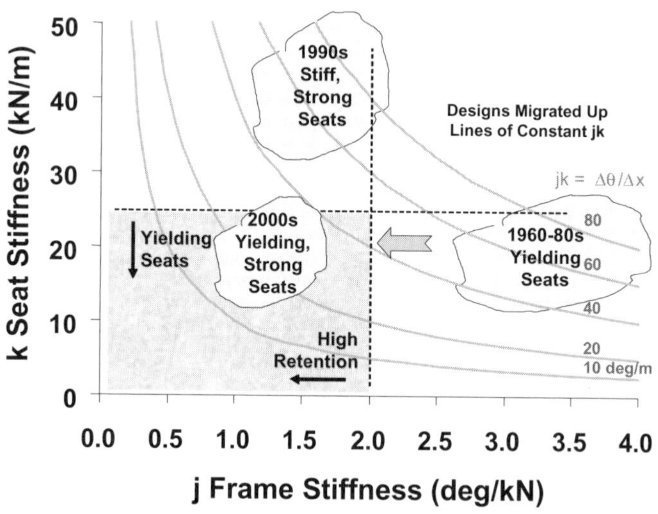

Figure 7: Seat and frame stiffness for various seats, including the new generation of high retention seats, which maintained the yielding behavior of earlier seat designs while providing high retention.

The need for a yielding seat was argued by Strother, James (1987), Warner et al. (1991), Strother et al. (1994) and Prasad et al. (1997), who provided engineering analyses and rationale for the yielding seatback in rear crashes. They argued against the rigid (stiff) seat for reasons of whiplash, out-of-position impacts and rebound (Warner et al. 1991, Benson et al. 1996). They proposed yielding seats as the best means to provide overall occupant safety in rear crashes by controlled seatback rotation that limited forces on the neck. This involved an energy absorbing restraint as the seatback yielded rearward in a crash and followed the historic trend in load-limiting and energy absorbing deformation of interior components, such as the steering system, padded dashboard and high-penetration resistant windshield.

However, researchers failed to separate the need for a strong seat from the disadvantage of a stiff seat. There are many reasons to want the benefits of a strong seat frame to limit seatback rotation in a severe rear crashes. Yet, there are ample reasons not to want stiff seats, as they increase loading on the occupant and aggravate risks for whiplash as well as other injuries.

The difference between seat strength and stiffness has been studied by Viano (2003b-e). It is clear that seats should be strong and yielding for all of the reasons of occupant protection in severe and low-speed rear crashes. However, the seat design approach offering both of these characteristics was not introduced until 1997, based on a seat patent for a perimeter frame construction (USPTO 1996). A review of the patents on seating systems has identified various yielding seatback ideas, which are consistent with the perimeter frame concept, but none were realized in production.

Figure 8 shows a different way to visualize the design direction for the new generation of yielding seats. They involve a jk < 50 ^{0}/m and load limits above F_L > 7.7 kN. Yielding seats of the 1960-80s and the strong, stiff seats of the 1990s had jk = 60-80 ^{0}/m. Ironically, these two seat types have the same initial seatback rotation in rear crashes even though the 1990s seats were substantially stronger. This is a result of the defining relationship $\Delta\theta = jk\Delta x$, where Δx is initially the forward displacement of the vehicle. High retention seats have jk = 20-35 ^{0}/m and involve about a half to a third of the initial seatback rotation as the 1960-80s yielding and 1990s strong, stiff seats (Viano 2003e).

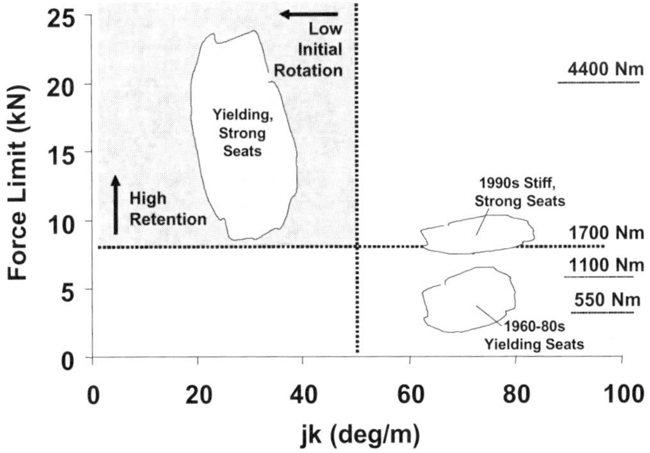

Figure 8: Seat characteristics of load limit and jk product for several seat types, including the new generation of yielding seats. Initial seatback rotation is related to jk.

The debate between rigid (stiff) and yielding seats involved reinforcing position by opposing sides in a number of litigation cases on seat performance in rear crashes. For example, the out-of-position tests by Strother et al. (1994) that demonstrated neck injury risks with rigid seats also involved an extremely stiff, rigid seat with j = 0.1 ^{0}/kN. Although the yielding seat had a low load limit, its j = 1.0 ^{0}/kN, both remarkably low j values. In part, the stiffness increased the occupant responses, and the testing by others (Saczalski et al. 1993) showed the potential benefits of rigid seats without adequately addressing potential disbenefits.

TYPICAL SEAT PROPERTIES

Seat Stiffness k, Frame Stiffness j and Load Limit F_L: Seat stiffness k and frame rotation stiffness j are determined in QSTs between occupant loads of 1 kN to 5 kN (Viano 2002, 2003a). This covers the majority of loading up to head restraint contact in rear crashes involving whiplash. The benchmark average of 14 foreign seats from high-end European and Asian manufacturers was k = 34.2 ± 6.5 kN/m with a range from k = 26-47 kN/m stiffness. The average of the five stiffest seats was k = 40.9 kN/m. The foreign benchmark was exemplified with k = 40 kN/m and j = 1.8 ^{0}/kN (Figure 9).

The 1990 C car seat (1990 Park Avenue) has a stiffness that is typical of yielding seats of the 1980-90s with an average stiffness of k = 21.1 ± 2.4 kN/m. Yield is essentially due to seatback rotation with a load limit of F_L = 5 kN and M_H = 1100 Nm.

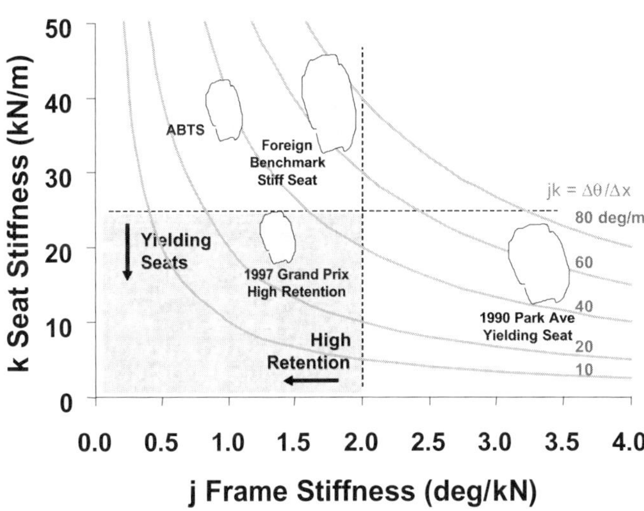

Figure 9: Seat j and k properties for the four exemplar seat types. This also defines a yielding seat with k < 25 kN/m and high retention seat with j < 2.0 ^{0}/kN.

The average stiffness of the 1997 W (Pontiac Grand Prix) and 1997 P-90 (Chevrolet Malibu), which are high retention seats, was k = 22.1 ± 5.4 kN/m. They achieved a 5 kN load at 32^0, nearly 10^0 less seatback rotation than the 1990 C car seat. In a recent study of high retention seats, Viano (2003a) found that the average stiffness of more than 20 seats was k = 25.3 ± 4.9 kN/m, which was statistically similar to the average 24.7 ± 4.9 kN/m for pre-high retention seats. High retention (HR) seats had an average peak force of 13.5

± 3.6 kN for seatback rotations below 60^0 and an average H-point moment of M_H = 2537 ± 703 Nm.

For analysis and mathematical modeling, four seat types were generically defined based on the typical values from QST testing. These seats are described in Table 1 and the following. A yielding seat from the 1960-70s was exemplified with a stiffness of k = 20 kN/m and j = 3.4 ^{0}/kN and F_L = 3 kN to simulate a moment of M_H = 660 Nm (3.0 kN with a 0.22 m moment arm). This is typical of the early vintage seats. A higher load limit of F_L = 5 kN was used to simulate yielding seats of the 1980-90s, such as the 1990 C car seat. This gave an H-point moment of M_H = 1100 Nm reflecting the higher strength.

Table 1: Seat Properties Used in Modeling.

Seat Type	k kN/m	j deg/kN	jk deg/m	Force Limit kN	H-pt Moment Nm
1960-70s Yielding Seat	20	3.4	68	3.0	660
1980-90s Yielding Seat	20	3.4	68	5.0	1100
Foreign Benchmark	40	1.8	72	7.7	1700
High Retention Seat	20	1.4	28	10.0	2200
ABTS	40	1.0	40	20.0	4400

A strong, stiff seats was modeled having a seat stiffness of k = 40 kN/m and j = 1.8 ^{0}/kN and F_L = 7.7 kN. The H-point bending moment was M_H = 1700 Nm. This seat is typical of the 1990s foreign benchmark seats. A high retention seat was modeled with k = 20 kN/m, j = 1.4 ^{0}/kN and F_L = 10 kN giving a moment of M_H = 2200 Nm. High retention (HR) seats use a strong seat frame and recliners for occupant retention in severe rear crashes to limit seatback rotation. They also have an open seatback providing compliance or yield by deformation of the seat trim and an EA pelvic strap to pocket the occupant's pelvis and lower back in a rear crash (Viano 2002, 2003a).

The fourth seat modeled was an ABTS (All Belt to Seat) seat with the stiffness of the foreign benchmark but lower frame rotation stiffness. The ABTS seat had k = 40 kN/m, j = 1.0 ^{0}/kN and F_L = 20 kN giving a moment of M_H = 4400 Nm.

Seatback Compliance and Frame Stiffness: Seat deformation includes the suspension between the side frame structures, trim and rotation of the seatback frame. Based on the measured stiffness and seatback rotation, it is possible to determine the individual stiffness components for the suspension and seatback frame, which work in series to support occupant load.

Based on the seatback angle at 5 kN occupant load, the contribution to the horizontal motion of the dummy from seatback rotation is determined using the center of occupant pressure on the seatback. The horizontal displacement due to seatback rotation is:

$$x_R = 0.22[\tan(\theta_{5kN}) - \tan(\theta_0)] \quad (1)$$

where x_R is the displacement due to seatback rotation at 22 cm above the H-point. Stiffness due to frame rotation is:

$$k_R = F / x_R = 5 / x_R \quad (2)$$

$$k_R = \Delta\theta / jx_R = 260 / j \quad (3)$$

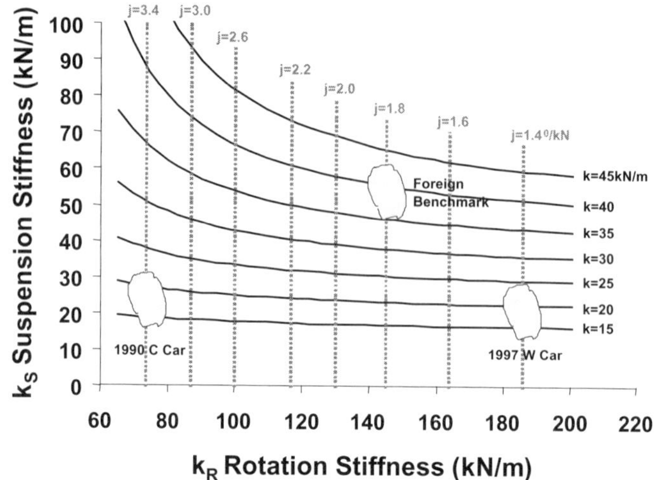

Figure 10: Relationship between seat properties j and k, and the frame stiffness and trim compliance for three of the exemplar seats.

Seat frame stiffness depends solely on the j factor. Seat suspension compliance can be determined from the measured seatback stiffness (k) and the differential displacement due to frame rotation:

$$x_S = (5/k) - x_R \quad (4)$$

so that:

$$k_S = F / x_S = 5 / x_S \quad (5)$$

$$k_S = kk_R / (k_R - k) = 260k / (260 - jk) \quad (6)$$

The trim and suspension stiffness (seat trim and compliance) is determined by both the j and k factors. Seat stiffness (k) can be verified by the relationship:

$$k = k_R k_S / (k_R + k_S) \quad (7)$$

Figure 10 shows the relationship between seat stiffness and the j and k factors. The ABTS results are off the graph since j = 1.0 ^{0}/kN.

OCCUPANT INTERACTION WITH THE SEAT

Based on the QST data on seat properties for occupant loading, it is possible to determine the early responses associated with whiplash loading of the neck. For example, using the luxury seat shown in Figure 1-4 and vehicle displacement forward, the following calculation can be made using the nominal seat properties from

QST testing. This assumes all of the seat deformation is from the forward movement of the struck vehicle and neglects the effects of torso movement, which is a delayed response in rear crashes. Assuming a vehicle displacement of 0.215 m causes the following occupant and seat responses:

$$\Delta x_i = y_i - x_i = 0.215\text{m} \tag{8}$$

$$F_i = k\Delta x_i = (36\text{kN/m})*.215 = 7.7\text{kN} \tag{9}$$

$$\Delta\theta_i = jF_i = (1.25\ ^0/\text{kN})*7.7\text{kN} = 9.6^0 \tag{10}$$

$$x_{Ri} = R\Delta\theta_i = (.22\text{m})*9.6^0 = 0.037\text{m} \tag{11}$$

$$x_{Si} = x_i - x_{Ri} = (.200 - .037) = 0.163\text{m} \tag{12}$$

A majority of the seat deflection is related to the compliance of the seat trim and suspension. Only a fraction (17%) is due to seat frame rotation.

MODELING OCCUPANT DYNAMICS IN REAR CRASHES

Vehicle Dynamics: The full analysis procedure can be found in earlier studies (Viano 2003b-d). Some aspects of the equations of motion are repeated here to give context to the analysis presented in this paper. Vehicle acceleration ($\ddot{y}_i$) is prescribed in equal time steps, so that a spreadsheet calculation can be made of the vehicle and occupant dynamics. It also can be integrated to determine velocity and displacement, which displace the seat forward.

Occupant Dynamics: Figure 11 shows the 1D model with 2D effects of seatback rotation on the gap behind the occupants head. The occupant is displaced forward in a rear crash by the motion of the vehicle seat and deformation of the seat during the impact. Force develops on the torso by the stiffness of the seat and torso displacement with respect to the vehicle:

$$F_i = k(y_i - x_i) = k\Delta x_i \tag{13}$$

Force on the occupant increases with deformation of the seat, which accelerates the occupant in proportion to the effective mass of the torso loaded by the seatback. The effective mass (m) of the Hybrid III dummy loaded by the seatback is 70% of the dummy mass, or 52.5 kg (Viano 2002). This agrees with the estimate by Severy (1968b) using an earlier vintage dummy. The mass of the head-neck is 5.5 kg, based on 4.5 kg for the head plus two-thirds of the neck mass or 1.0 kg (Backaitis, Mertz 1994), so torso mass is 47.0 kg. Torso acceleration is:

$$\ddot{x}_i = F_i / m_T = F_i / 47.0 \tag{14}$$

where Equation (14) assumes that the eventual load on the neck has a second order effect on the calculation and can be neglected, and there are no inertial effects of the seatback mass. Torso acceleration can be integrated to determine velocity and displacement of the torso.

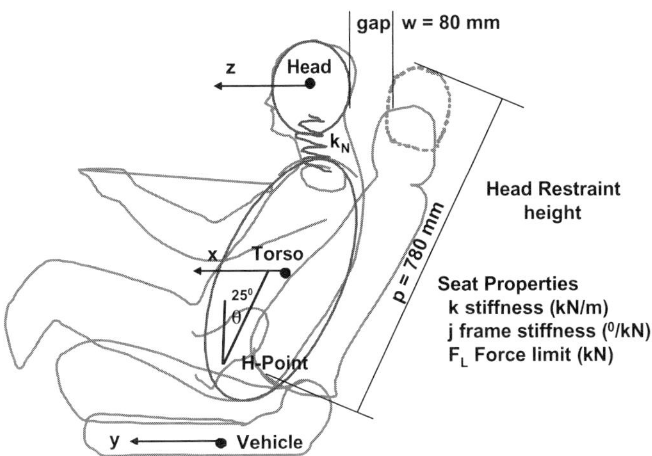

Figure 11: Two-mass representation of the torso and head with a flexible neck and the representation of the principle seat characteristics and coordinates.

Seatback Angle Change: An important issue in rear crashes is the angle change of the seatback. Torso load causes seatback rotation, which is generally linear up to the load limit. Seat stiffness is only one factor in the angle change. The seatback angle is:

$$\Delta\theta_i = \theta_i - \theta_0 = jF_i \tag{15}$$

where $\theta_0 = 25^0$ for typical seats and j is the slope of the line relating occupant load to seatback angle. Since force on the torso is related to seat stiffness, the relation can be reduced to:

$$\Delta\theta_i = jk(y_i - x_i) = jk\Delta x_i \tag{16}$$

For the exemplar seats, the change in seatback angle for occupant displacement into the stiff (foreign benchmark) seat was $\Delta\theta / \Delta x = 72\ ^0/\text{m}$. The 1990 C car and 1970s yielding seats had $\Delta\theta / \Delta x = 68\ ^0/\text{m}$. The 1997 W car and high retention seats had $\Delta\theta / \Delta x = 28\ ^0/\text{m}$, and the ABTS seat had $\Delta\theta / \Delta x = 40\ ^0/\text{m}$.

During the initial loading in a rear crash, Equation 16 reduces to $\Delta\theta_i = jky_i$ so the seatback angle change is proportional to vehicle displacement. Forces of the car crash and mass of the vehicle far exceed that of the occupant, so the occupant does not influence vehicle acceleration or displacement. Vehicle displacement directly loads the occupant in proportion to seat stiffness k and the seatback rotates proportion to jk.

For seat rotations in the constant load limit condition, the change in seatback angle was determined by the change in occupant displacement converted to a seatback angle at the height of the center of pressure.

This ensured equilibrium of motion during the load limiting phase.

Head Restraint Trajectory: As the occupant loads the seatback and it rotates, there is a downward and rearward displacement of the head restraint given by (Figure 12):

$$u_i = p[\cos(\theta_i) - \cos(\theta_0)] \tag{17}$$
$$v_i = p[\sin(\theta_i) - \sin(\theta_0)] \tag{18}$$

where p = 780 mm is used to represent the head restraint height up the seatback of modern seats. This value was used for the analysis of the 1960-70s and 1980-90s yielding seats even though the actual head restraint was lower for production seats. This allowed a direct comparison of the different seat stiffness types.

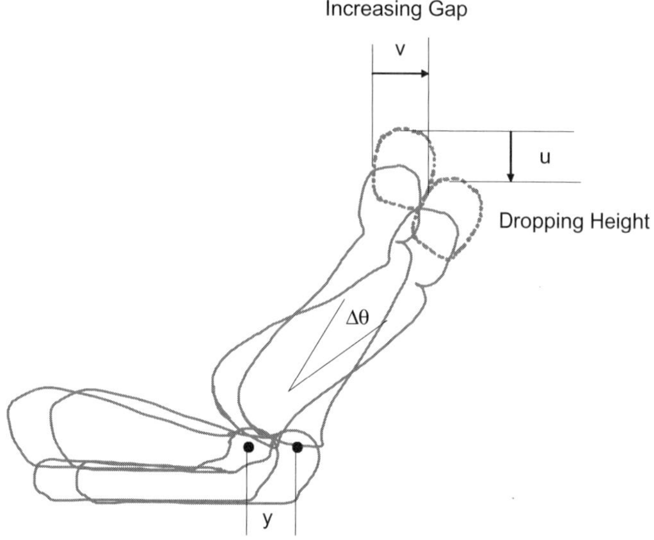

Figure 12: Seatback rotation increases the gap and drops the head restraint with respect to the seated occupant, as the seat is pushed forward in a rear crash.

The gap to the head restraint varies with the initial gap behind the head and seatback angle changes. It can be determined from the previous relationships with:

$$w_i = w_0 + v_i \tag{19}$$

where $w_o = 80$ mm is assumed typical of modern head restraint placements. This relationship does not address the effects of head-neck displacement. The actual gap between the back of occupant head and the head restraint is a function of the head and vehicle displacement, and other geometric relationships defining the head restraint position with seatback rotation. It is given by:

$$s_i = w_i - (y_i - z_i) \tag{20}$$

Head-Neck Dynamics: Force develops on the neck due to differential motion between the torso and head. It is related to the stiffness of the neck due to a combination of bending, tension and shear:

$$G_i = k_N(x_i - z_i) \tag{21}$$

Neck stiffness was k_N = 5 kN/m representing the 50% male neck extension due to bending, shear and tension (Viano 2003c-d). Force on the neck increases with the relative displacement between the head and torso, which accelerates the head in proportion to the 5.5 kg effective mass of the head and portion of the neck:

$$\ddot{z}_i = G_i / m_H = G_i / 5.5 \tag{22}$$

Head acceleration determines the velocity and displacement of the head with respect to ground, which adds to the occupant energy transfer. After head restraint contact, there is an additional force on the head in proportion to its penetration into the head restraint. A linear relationship is assumed for current head restraints. For this condition, the force on the head in Equation (9) becomes:

$$G_i = k_N(x_i - z_i) - k_{HR} s_i \tag{23}$$

Based on subsystem tests of head restraints (Viano 2002, Chapter 13), a 50 mm penetration of the back of the Hybrid III head causes a 2.8 kN load giving a stiffness of 80 kN/m; however, with the head restraint installed in the seat, the stiffness reduces to $k_{HR} = 20$ kN/m. This value was selected after comparison with model validation tests (Viano 2003c-d).

SEAT AND OCCUPANT RESPONSES IN REAR CRASH

36 km/h Rear Delta V Sled Performance: Table 2 summarizes key seat and biomechanical responses for the rear crash simulations. The crash pulse was 9.8 g for 100 ms. Maximum responses up to head restraint contact are interesting. The 1990s stiff, strong foreign benchmark developed the highest torso acceleration and head-neck velocity of all seats modeled. It also had the second highest neck and torso displacement. This was better than the 1960-70s yielding seat, but not as good as the 1980-90s yielding seats even though the maximum seatback angle was greater with both earlier yielding seats. However, these maximum angles are severe enough were loss of occupant retention becomes a factor.

The high retention seat produced uniformly better results than any other seat modeled. The most pronounced effect was an earlier head restraint contact with lower loads on the torso. This resulted in the lowest neck responses at and after head restraint contact. While the ABTS seat had the lowest seatback rotation, its biomechanical responses were increased by the early torso acceleration related to its stiffness, even as there was an early head restraint contact and small gap.

Table 2: Seat and Biomechanical Responses in 36 km/h Rear Delta V Simulations.

	1960-70s Yielding Seats	1980-90s Yielding Seats	1990s Stiff, Strong Seats	Yielding, Strong Seats	ABTS Stiff Seat
Seat Properties					
Seat Stifffness (kN/m)	20	20	40	20	40
Frame Stiffness (deg/kN)	3.4	3.4	1.8	1.4	1.0
jk (deg/m)	68	68	72	28	40
Load Limit (kN)	3.0	5.0	7.7	10.0	20.0
H-Point Moment (Nm)	660	1100	1700	2200	4400
36 km/h Rear Impact Responses					
Head Restraint Contact (ms)	92	70	66	48	52
Max up to Head Restraint Contact					
Torso Acceleration (g)	6.3	8.4	13.5	4.7	9.7
Head-Neck Velocity (m/s)	1.82	1.78	2.78	0.72	1.66
Torso Displacement (mm)	100.1	41.2	61.9	9.8	25.5
Neck Displacement (mm)	73.1	35.2	53.6	9.1	23.4
Seatback Angle (deg)	57.0	39.3	36.2	28.1	29.6
Added Gap (mm)	324.1	164.7	131.1	37.6	55.5
Head Restraint Drop (mm)	281.2	103.5	77.6	18.7	28.6
Maximum Values					
Max. Seatback Angle (deg)	72.4	63.2	47.4	37.8	34.6
Torso Load (kN)	3.0	5.0	7.7	8.0	9.6
Neck Displacement (mm)	88.5	72.7	118.8	16.6	54.2

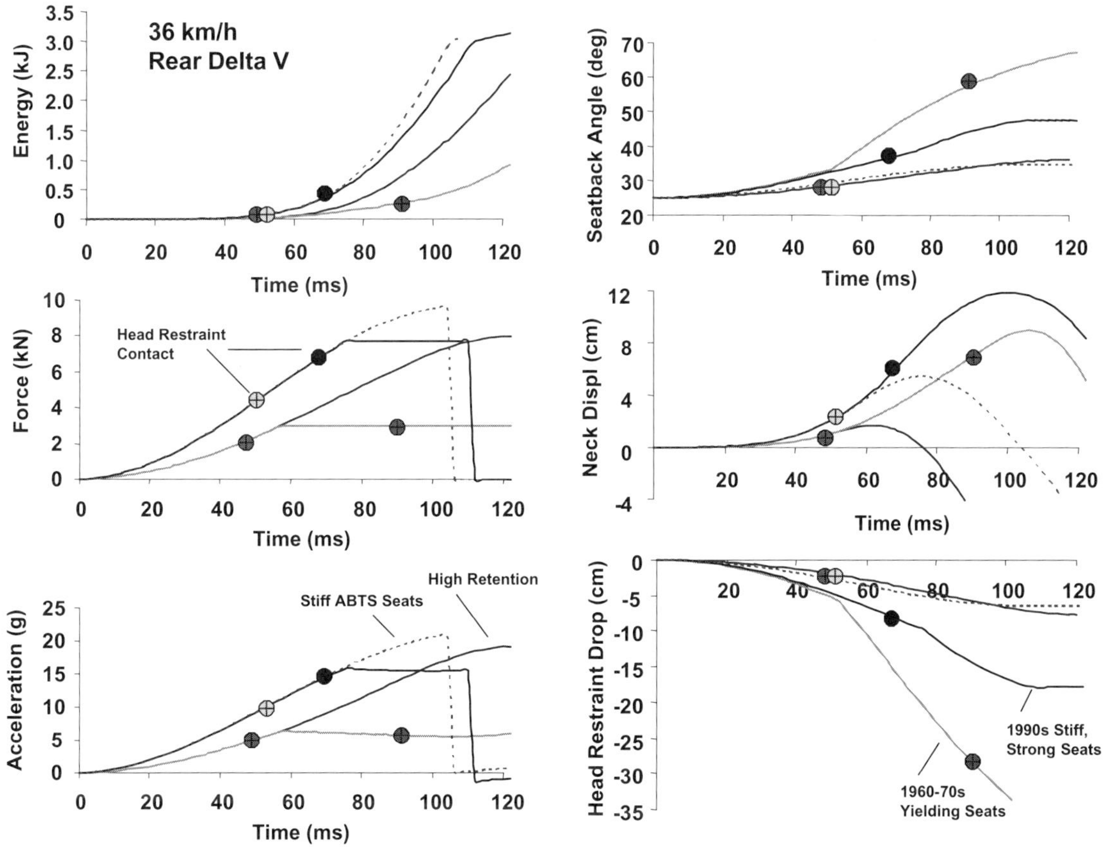

Figure 13: Six graphs of the 36 km/h rear impact simulations.

Table 3: Seat and Biomechanical Responses in 16 km/h Rear Delta V Simulations.

	1960-70s Yielding Seats	1980-90s Yielding Seats	1990s Stiff, Strong Seats	Yielding, Strong Seats	ABTS Stiff Seat
Seat Properties					
Seat Stifffness (kN/m)	20	20	40	20	40
Frame Stiffness (deg/kN)	3.4	3.4	1.8	1.4	1.0
jk (deg/m)	68	68	72	28	40
Load Limit (kN)	3.0	5.0	7.7	10.0	20.0
H-Point Moment (Nm)	660	1100	1700	2200	4400
16 km/h Rear Impact Responses					
Head Restraint Contact (ms)	126	100	92	72	76
Max up to Head Restraint Contact					
Torso Acceleration (g)	6.0	6.3	8.1	4.1	7.0
Head-Neck Velocity (m/s)	1.50	1.45	2.02	0.83	1.60
Torso Displacement (mm)	153.0	68.8	90.0	20.3	45.9
Neck Displacement (mm)	87.0	49.4	66.9	17.2	37.8
Seatback Angle (deg)	42.3	35.7	32.2	27.7	28.4
Added Gap (mm)	195.1	125.7	86.3	33.4	41.6
Head Restraint Drop (mm)	129.7	73.5	47.0	16.6	20.9
Maximum Values					
Max. Seatback Angle (deg)	42.7	37.6	32.8	30.1	29.3
Torso Load (kN)	3.0	3.7	4.3	3.7	4.3
Neck Displacement (mm)	103.8	71.7	96.9	27.2	65.0

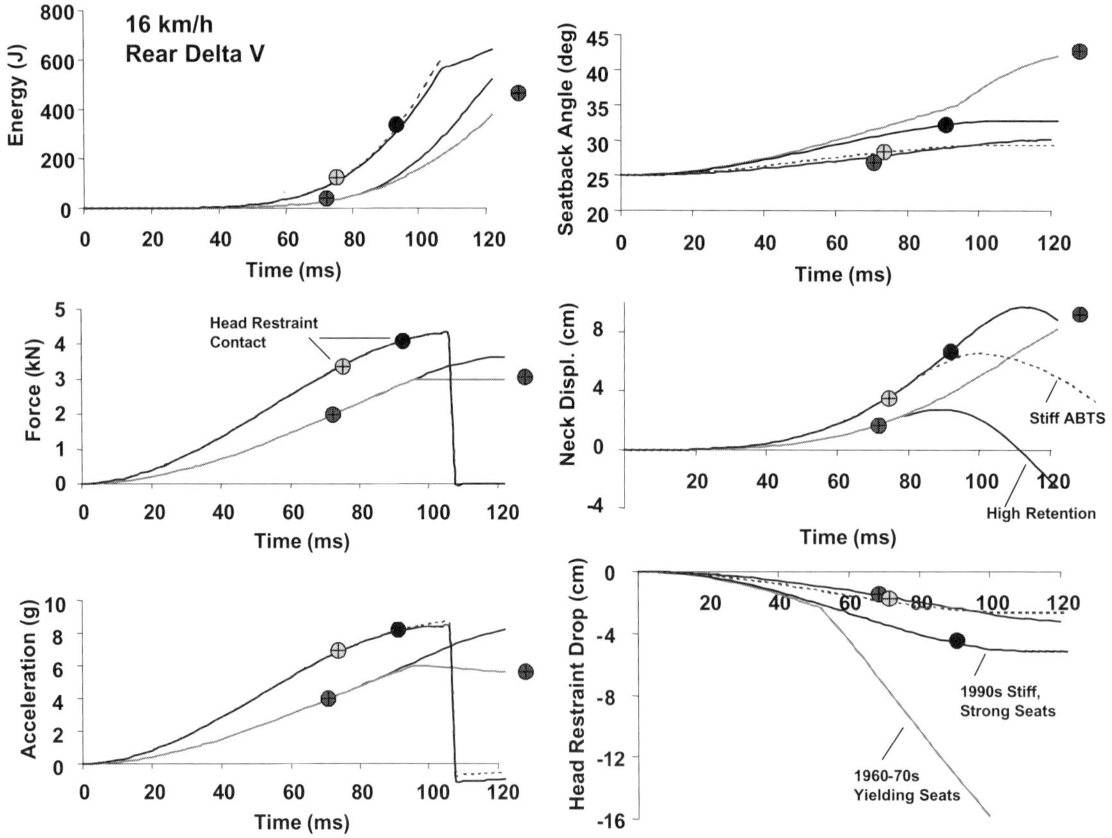

Figure 14: Six graphs of the 16 km/h rear impact simulations.

Table 4: Draft Performance Targets and Criteria for Rear Crash Safety.

	Target	Criteria
Quasistatic Seat Test		
Seat Stiffness		<25 kN/m
Frame Stiffness		<2.0 deg/kN
Load Limit	>10.0 kN	>7.7 kN
H-point Moment	>2200 Nm	>1700 Nm
Upward H-point Displacement	<50 mm	
Twist Angle up to 15 deg rotation	<10 deg	<15 deg
No >2 kN drop with rotation of	>5 deg	>10 deg
Rear Sled Performance		
30-36 km/h Rear Delta V		
Seatback Angle	<45 deg	<60 deg
Twist Angle	<10 deg	<15 deg
Neck Displacement Criteria		
Neck Extension	<35 deg	<45 deg
x-Displacement	<40 mm	<60 mm
z-Displacement		<-30 mm
Neck Moment	<15 Nm	<20 Nm
Upward H-point Displacement	<50 mm	
Rebound	tbd	
10-16 km/h Rear Delta V		
Neck Displacement Criteria		
Neck Extension	<10 deg	<20 deg
x-Displacement	<20 mm	< 30 mm
z-Displacement		<-15 mm
Torso Acceleration	<10 g	<15 g
Neck Moment	<5 Nm	<10 Nm
Upward H-point Displacement	<50 mm	
Rebound	tbd	

Figure 13 shows six responses from the 36 km/h rear impact simulations. The energy transfer plot shows that all seats engage head restraint contact before the majority of energy is transferred to the occupant. However, the highest energy transfer before head restraint contact occurs with the 1990s stiff, strong seat. The force and torso acceleration plots demonstrate the timing of head restraint contact with respect to load on the body. This influences neck responses in terms of relative velocity between torso and head and neck displacement. In the rear crash, the seat is pushed forward loading the torso in proportion to seat stiffness. Stiffer seats develop earlier forces and push the torso forward increasing neck responses because the inertia of the head causes it to remain in place until sufficient force develops in the neck to cause extension and bending loads or head restraint contact occurs. Only the F_L = 3 kN load limit simulation is shown for the yielding seats of the 1960-70s. The head restraint trajectory plot shows the substantial seatback rotation with the 1960-70s yielding seats. Much less occurs with seats of the 1980s but the 1990s stiff, strong seats had greater seatback rotation than high retention seats largely because of higher loads on the torso and jk product.

These responses demonstrate the importance of a low seat and frame stiffness and reasonable load limit. Clearly the high limit of the ABTS is beyond the needs of even the most severe rear crashes if sufficient yielding behavior is introduced with the seat. This type of modeling cannot address issues of early ramping, seatback twist and other effects that are important to the overall seat response in rear crashes. However, they do provide insight into the principle factors.

16 km/h Rear Delta V Sled Performance: Table 3 summarizes the key responses for the 16 km/h rear impacts. A comparison with the 36 km/h response shows some are larger for this crash severity, reflecting the greater time to develop biomechanical responses in lower speed rear crash. Responses can further increase as the severity of rear crash delta V decreases (Viano 2003c). This crash is at the upper end of the exposures associated with whiplash based on insurance and NHTSA analyses of whiplash injuries.

First, the time to head restraint contact is increased from the higher severity crash at 36 km/h. This gives more time for the development of responses. While the maximum torso acceleration and relative velocity between the head and neck are lower than in the 36 km/h crash, the neck displacements and seatback rotation at head restraint contact are uniformly greater at 16 km/h. Nonetheless, the trend in response among seats is similar to that for higher speed rear crash. The lowest biomechanical responses occur with the high retention seat that combines a yielding seat stiffness (low loads on the torso until head contact) and high retention frame (low rotation with low added gap and head restraint drop). These characteristics are essential to the improved performance in low and high severity rear crashes.

A comparison of the 1980-90s yielding seat with the 1990s stiff, strong seat demonstrates the points made by many researchers about the risks of increased whiplash with rigid (stiff) seats. Except for the larger seatback rotation, the yielding seat uniformly outperforms the strong, stiff seat in low speed crashes. Overall, yielding seats of the 1980-90s show an good performance in controlling neck responses and transferring crash energy to the occupant. This is consistent with the many field crash studies demonstrating good performance in a range of rear crashes (Malliaris 1990, Digges et al. 1993), although the model does not reflect the lower head restraint height in use during that period of time. This would increase biomechanical responses as most head restraints are not positioned up and forward to the most favorable position (O'Neill et al. 1972, Viano, Gargan 1996).

Figure 14 summarizes six responses for a lower speed rear impact typical of the upper end of the whiplash spectrum of crashes. Here, the head restraint contact occurs later reflecting the great time for the head and neck to response as the severity of crash decreases (Viano 2003c). The energy transfer plot shows nearly 80% of the crash energy is transferred to the occupant before head restraint contact with the 1990s stiff, strong seat. In contrast, the high retention seat transfers only a small fraction of energy up to head restraint contact. Most of the transfer occurs after support for the head, neck and torso. This reflects a reason for the higher biomechanical responses with stiff, strong seats. Again, only the F_L = 3 kN load limit simulation is shown for the yielding seats of the 1960-70s. The seatback rotation and head restraint trajectory plots show the value of high retention and ABTS seats. They involve the lowest seatback rotation and smallest additional head restraint gap at head contact. This has a direct benefit in occupant protection in rear crashes.

DRAFT PERFORMANCE CRITERIA: ESSENTIAL SEAT PROPERTIES AND BIOMECHANICAL RESPONSES

Overview of Tests, Targets and Criteria: After considering performance criteria for rear crash safety that include static and dynamic testing, the following targets and criteria are proposed to improve seat performance in rear crash safety. They involve provisions in the FMVSS 202 and 207 with reference to 301 (FMVSS 1968). An extensive study of Quasistatic Seat Test (QST) requirements has been already completed, so the inclusion here are stricter targets and minor revisions to criteria after eight years of seat develop to standards established in 1995 (Viano 2002).

A second set of targets and criteria are proposed for two different ranges of rear delta V sled tests. The original work in establishing QST requirements was intended to ensure good overall occupant kinematics in the full range of rear crashes in a 33.7 mph (54.2 km/h) due care RMB test. The procedure is identical to the FMVSS 301 RMB test except at a higher impact speed. The delta Vs for a series of passenger cars showed a range of 17.5-22 mph (28.2-35.4 km/h) with an average of 19.3 mph (31.1 km/h). The 30-36 km/h rear sled test is chosen to reflect the range in severity for the due care test with passenger vehicles in the fleet. While this delta V range is higher than previous development and evaluation testing (Viano 2002), the new generation of high retention seats is intended to perform well through that crash severity range.

The 10-16 km/h test is chosen to match conditions including the whiplash evaluation procedure being developed by IIHS and the international insurance community (Langwieder, Hell 2002). The severity range covers the conditions for whiplash and it gives a reasonable evaluation of occupant interactions from low to moderate severity crashes. As shown by Viano (2003c), many biomechanical responses increase with lower severity rear impacts because of the greater duration until head restraint contact. This fact was considered in the development of final specifications for crash severity. New instrumentation and testing methods are now available to record dynamic responses of seatback rotation and occupant motion, which allow a more detailed evaluation of occupant interactions. Some targets and criteria are specified up to head restraint contact, and others reflect the maximum value experienced throughout the exposure.

The intention of the proposal is to leave some guidelines as targets to challenge the development of new seat systems, while specifying some as minimum performance requirements for new seats. It also leaves open the actual specification of test speeds. While it would be helpful to eliminate one of the three proposed tests, it is not advisable until a fuller understanding of whiplash and occupant interactions with the seat is gained in real world crashes, or when instrumentation allows the measurement of equivalent responses so that a test provides redundant information.

Draft Targets and Criteria: Table 4 shows criteria and targets for seat performance in QST rearward dummy loading into the seatback. The average response between 1-5 kN is used to determine the seat stiffness and frame rotation to calculate k and j, respectively. As shown in Table 3, this force range covers the low-speed rear impact exposures associated with whiplash. Limiting j and k have additional benefits in higher severity crashes. The load limit is set at a minimum of 7.7 kN with a higher target of 10 kN. This ensures enough energy transfer capability of the seat to limit seatback rotation in the more severe rear crashes. The other QST criteria ensure minimal seatback twist and a gradual deformation of seat components. An additional target is included to limit the upward motion of the H-point. This encourages efforts for an open path for the pelvis and a downward trajectory, which provides safety benefits in low- through high-speed rear crashes.

The high severity rear sled test ensures an overall seat system performance. The draft requirements on seatback rotation and twist are set to reduce the risk of loss of retention. The draft criteria on neck biomechanical responses ensure early head support and interaction with the head restraint with responses within established limits (SAE J1460-2 1998, SAE J885, 1986, Viano, Davidsson 2002). The target levels are included to emphasize further safety improvements may be realized with lower seat and neck response levels.

The 10-16 km/h rear delta V sled test addresses whiplash performance. It was initially considered that some criteria could be specified up to head restraint contact. This would show the importance of the early responses, which can increase the speed of the head into the restraint. However, only maximum values are specified at this time for the tests. The draft targets and criteria address the overall seat and biomechanical response in the tests, including the stored energy, which increases rebound.

LIMITING ISSUES AND FURTHER CONSIDERATIONS

While an upgrade in rear crash safety can be realized by meeting the proposed performance requirements, there remain a number of additional issues that are either not addressed by the requirements or confounded in particular crashes. The following are some examples of issues needing further study and design alternatives.

Severe Rear Crashes

- Heavy trucks and tractor-trailers can over-ride rear structures inducing downward pitch of the vehicle rear and intrusion into the rear seating area.
- Intrusion of rear structures which compromises rear space and interacts with the seat.
- Complex multi-crash conditions, which increase the risk of unbelted occupant ejection.
- Out-of-position occupants with inboard or forward lean involving torso load mainly on the seat frame and neck extension.
- Rollover crashes secondary to rear impacts involving unbelted occupant impacts with seat frame structures.
- Out-of-position, unbelted rear occupants interacting with the seat frame structures and involving head and other body region impacts on seat structures and trim.
- The role of occupant restraints, belt pretensioning and other restraint enhancements and designs, which may influence occupant retention and restraint by the seat.

Low-Speed Rear Crashes

- Rear impacts can be followed by front crashes involving rebound impacts on front structures.
- Out-of-position occupants leaning forward or inboard at impact increasing injury risks.
- Head restraint adjustment up to the most favorable position, which is infrequently done by occupants and leads to increased whiplash risks.
- FMVSS 202 head impact performance does not apply below the beltline or to seat frame structures.
- Active head restraints have been introduced, which in some cases include pyrotechnical activation of safety systems. These technologies require special considerations in sled evaluations, where the vehicle structures and triggering systems are not included and performance outside the deployment conditions adds another factor.
- Issues of occupant age, size and gender influencing the alignment with seat components, determination of seat-occupant interaction properties (j, k and F_L) and overall safety evaluation methods.

REFERENCES

Backaitis S, Mertz H. Hybrid III: The First Humanlike Crash Test Dummy. SAE Special Publication PT-44, Society of Automotive Engineers, Warrendale, PA. 1994.

Benson BR, Smith GC, Kent RW, Monson CR. Effect of seat stiffness in out-of-position occupant response in rear-end collisions. SAE 962434, 40th Stapp Car Crash Conference, Society of Automotive Engineers, Inc., Warrendale, PA, USA, 1996.

Blaisdell DM, Levitt AE, Varat MS. Automotive seat design concepts for occupant protection. SAE 930340, Society of Automotive Engineers, Inc., Warrendale, Pennsylvania, USA, 1993.

Digges, KH, Morris JH, Malliaris AC. Safety performance of motor vehicle seats. SAE 930348, Society of Automotive Engineers, Inc., Warrendale, PA, USA, 1993.

Federal Motor Vehicle Safety Standards No. 202, 207, and 301. 49 Code of Federal Regulations Section 571.21, 1968.

James MB, Strother CE, Warner CY, et al. Occupant Protection in Rear Collisions: Safety and Priorities and Seat Belt Effectiveness, 21st Stapp Car Crash Conf, SAE 912913, Society of Automotive Engineers, Warrendale, PA, 369-376, 1991.

Langwieder K, Hell W. Proposal of an International Harmonized Dynamic Test Standard for Seat/Head Restraints. Traffic Injury Prevention 3(3): 150-158, 2002.

Malliaris AC. Current Issues of Occupant Protection in Car Rear Impacts. Malliaris, Inc., NHTSA Docket 89-20-No1-021, February, 1990.

O'Neill B, Haddon W, Kelly AB, et al. Automobile Head Restraints: Frequency of Neck Injury Claims in Relation to the Presence of Head Restraints. Am J Public Health 62, 399-406; 1972.

Prasad P, Kim A, Weerappuli DPV, Robert V, Schneider D. Relationship Between Passenger Car Seat Back Strength and Occupant Injury Severity in Rear End Collisions: Field and Laboratory Studies. SAE 973343, Society of Automotive Engineers, Warrendale, PA. 1997.

Saczalski KJ, Syson SR, Hille RA, Pozzi MC. Field accident evaluations and experimental study of seat back performance relative to rear-impact occupant protection. SAE 930346, Society of Automotive Engineers, Inc., Warrendale, PA, USA, 1993.

SAE J1460-2. Human Mechanical Impact Response Characteristics-Responses of the Human Neck to Inertial Loading by the Head for Automotive Seated Postures. Society of Automotive Engineers, Warrendale, PA, 1998.

SAE J885. Human Tolerance to Impact Conditions as Related to Motor Vehicle Design. Society of Automotive Engineers, Warrendale, PA, 1986.

Severy DM, Brink HM, Baird JD. Preliminary findings of head support designs. 11th Stapp Car Crash Conf, SAE 670921, SAE Warrendale PA, 337-405, 1967a.

Severy DM, Harrison MB, Baird JD. Collision Performance, LM Safety Car. SAE 670458, Society of Automotive Engineers, Warrendale, PA. 1967b.

Severy DM, Brink HM, Baird JD. Backrest and Head Restraint Design for Rear-end Collision Protection. SAE 680079, Society of Automotive Engineers, Warrendale, PA, 1968a.

Severy DM, Brink HM, Baird JD. Vehicle Design for Passenger Protection from High Speed Rear End Collisions. SAE 680774, Society of Automotive Engineers, Warrendale, PA, 1968b.

Stephens GD, Long TJ, Blaisdell DM. Energy Analysis of Automotive Seat Systems, SAE 2000-01-1380, SP-1494, Society of Automotive Engineers, Warrendale, PA, 2000.

Strother CE, James MB. Evaluation of Seat Back Strength and Seat Belt Effectiveness in Rear End Impacts. 31st Stapp Car Crash Conference, SAE 872214, Society of Automotive Engineers, Warrendale, PA, 225-244, 1987.

USPTO. Vehicle Seat with Perimeter Frame and Pelvic Catcher. US Patent 5,509,716, April 23, 1996.

Viano DC, Gargan MF. Seating Position and Head Restraint Location During Normal Driving: Implications to Neck Injury Risks in Rear Crashes. Accid Anal Prev. 28(6):665-674, 1996.

Viano DC. Role of the Seat in Rear Crash Safety. ISBN 0-7680-0847-6, Society of Automotive Engineers, Warrendale, PA, SAE R-317:1-491, 2002.

Viano DC, Davidsson J. Neck Displacements of Volunteers, BioRID P3 and Hybrid III in Rear Impacts: Implications to Whiplash Assessment by a Neck Displacement Criterion (NDC). Traffic Injury Prevention 3(2):105-116, 2002.

Viano DC. High Retention Seat Performance in Quasistatic Seat Tests. SAE 2003-01-0173, Society of Automotive Engineers, Warrendale, PA, 2003a.

Viano DC. Energy Transfer to an Occupant in Rear Crashes: Effect of Stiff and Yielding Seats. SAE 2003-01-0180, Society of Automotive Engineers, Warrendale, PA, 2003b.

Viano DC. Seat Properties Affecting Neck Responses in Rear Impacts: A Possible Reason Why Whiplash Has Increased. In print Traffic Injury Prevention, 2003c.

Viano DC. Seat Influences on Female Neck Responses in Rear Impacts: A Reason Why Women Have Higher Whiplash Rates. In print Traffic Injury Prevention, 2003d.

Viano DC. Influence of Seat Properties on Occupant Dynamics in Severe Rear Crashes. In review Traffic Injury Prevention, 2003e.

Viano DC. Effectiveness of High-Retention Seats in Preventing Fatality: Initial Results and Trends. SAE 2003-01-1351, Society of Automotive Engineers, Warrendale, PA, 2003f.

Viano DC. Analysis of Driver and RF Passenger Fatalities in Rear-Impacted Vehicles with High Retention Seats. SAE 2004-01-0XXX, Society of Automotive Engineers, Warrendale, PA, 2004.

Warner C, Strother C, James MB, Decker RL. Occupant Protection in Rear-End Collisions II: The Role of Seat Back Deformation in Injury Reduction. 35th Stapp Car Crash Conf, SAE 912914, Society of Automotive Engineers, Warrendale, PA., 1991.

CHAPTER 2:

INFLUENCE OF SEAT PROPERTIES ON OCCUPANT DYNAMICS IN SEVERE REAR CRASHES

Chapter 2

Influence of Seat Properties on Occupant Dynamics in Severe Rear Crashes

David C. Viano
ProBiomechanics LLC

ABSTRACT

Seat performance in retaining an occupant, transferring energy and controlling neck responses is often questioned after severe rear crashes when fatal or disabling injury occur. It is argued that a stiffer seat would have improved occupant kinematics. However, there are many factors in occupant interactions with the seat. This study evaluates four different seat types in 26 and 32 mph (42 and 51 km/h) rear crash delta Vs.

Two seats were yielding with k = 20 kN/m occupant load per displacement. One represented a 1970s yielding seat with j = 3.4 ^{0}/kN frame rotation per occupant load and 3 kN maximum load (660 Nm moment), and the other a high retention seat phased into production since 1997 with j = 1.4 ^{0}/kN and 10 kN maximum load (2200 Nm). Two seats were stiff with k = 40 kN/m. One represented a 1990s foreign benchmark with j = 1.8 ^{0}/kN and a 7.7 kN maximum load (1700 Nm), and the other an all belts to seat (ABTS) with j = 1.0 ^{0}/kN and 20 kN maximum load (4400 Nm). The crash was a constant acceleration of 11.8 g or 14.5 g for 100 ms. Occupant interactions with the seat were modeled using a torso mass, flexible neck and head mass.

By analysis of the equations of motion, the initial change in seatback angle ($\Delta\theta$) is proportional to $jk(y-x)$, the product jk and the differential motion between the vehicle (seat cushion) and occupant. The transition from 1970s-80s yielding seats to stronger seats of the 1990s involved an increase in k stiffness; however, the jk property did not change as frame structures became stronger.

The yielding seats of the 1970s had jk = 68 ^{0}/m, while the stiff foreign benchmark seat had jk = 72 ^{0}/m. The foreign benchmark rotated about the same as the 1970s seat up to 50 ms in the severe rear crashes. While it was substantially stronger, it produced higher loads on the occupant, and the higher loads increased seatback rotations and neck responses. The ABTS seat had the lowest rotations but also caused high neck responses because of the greater loads on the torso. Neck displacement (d) is initially proportional to $(k/m_T)\iint y$, seat stiffness times the second integral of vehicle displacement divided by torso mass. As seat stiffness increases, head-torso acceleration, velocity and neck displacement increase.

This study shows that the jk seat property determines the initial seatback rotation in rear crashes. If a stronger seat has a higher stiffness, it rotates at higher loads on the occupant, reducing the overall benefit of the stronger frame while increasing neck responses related to whiplash or neck extension prior to subsequent impacts. The aim of seat designs should be to reduce jk, provide pocketing of the pelvis and give head-neck support for the best protection in severe rear crashes. For low-speed crashes, a low k is important to reduce early neck responses related to whiplash.

INTRODUCTION

Yoganandan et al. (1989) summarized the epidemiology and biomechanics of spinal injury in motor vehicle crashes. They evaluated hospital records and found quadriplegia primarily occurring from flexion-compression at C5-C6, burst injury at C6 and facet locking at C5. With incomplete quadriplegia, cervical injuries occurred over a larger range of vertebrae with extension injury primarily at C3.

Based on an analysis of NASS, the annual incidence of AIS 3+ cervical injury was estimated at 6,200 with injuries predominantly in front, side and rollover crashes. Table 1 shows the estimates with the lowest incidence of AIS 3+ cervical injury in rear crashes. For the 188 estimated cases, quadriplegia may occur in less than 45 individuals based on proportions from the hospital cases. In the overall NASS sample, about half the cases involved head impact. The data also shows a substantial reduction in AIS 3+ cervical injury risk with restraint use (2.1 vs. 0.64 per 1000 exposed); only 20% of the NASS sample was restrained. These data are consistent with earlier studies and the review by McElhaney, Myer (1993).

Table 1: Incidence of Cervical Injury in Motor Vehicle Crashes (Adapted from Yoganandan et al. 1989)

Crash Type	Annual Incidence per 1000 Exposed			Annual Incidence		
	AIS 1	AIS 2	AIS 3+	AIS 1	AIS 2	AIS 3+
Front	62	1.7	1.4	189,765	5,508	2,693
Side	65	1.6	1.8	107,704	2,570	1,691
Rear	270	2.0	0.65	179,507	1,224	188
Rollover	82	8.9	7.3	30,773	2,693	1,628
Total	479	14.2	11.2	507,800	12,000	6,200
Unrestrained	82	2.1	2.1			
Restrained	120	2.6	0.64			

The most common source for quadriplegia appears to be flexion-compression due to impact at the vertex of the head. In severe rear crashes, there is little time to develop neck extension and with yielding seats of the 1970-90s the seatback rotates with the potential for ramping and head impact in the rear area of the vehicle while the head is rather upright.

Figure 1a shows the injury kinematics with impact on the vertex or occiput while the torso continues to move rearward. Anterior dislocation of the mid-cervical vertebrae can lead to unilateral or bilateral facet locking with a high risk of spinal cord injury leading to quadriplegia. This kinematic is similar to diving injuries seen in shallow water with head impact and the body continuing forward causing a flexion-compression injury. However, this injury is rare, and the overall safety performance of 1970-90s yielding seats has been shown to be good in various studies of rear crash injuries (Malliaris 1990, James et al. 1991, Digges, Malliaris 1993). Even the NASS estimates in Table 1 reflect a high degree of safety in rear crashes with an incidence of 27% AIS 1 and 0.065% of AIS 3+ cervical injury.

With minimal seatback rotation and the torso supported, the inertial load from the head can lead to hyper-extension (Figure 1b). In severe rear crashes, secondary head impact or just the head inertia may cause cervical fractures. This is the most vulnerable orientation of the neck with head impact. This kinematic is also one of many factors used to argue against stiffening seatbacks to reduce rotation in rear impacts, and it fostered a four-decade debate on yielding and stiff seats.

Over the years, the debate furthered the notion that there was an underlying design conflict between occupant retention with stiff or rigidized seats in severe but infrequent rear crashes, and the need for a yielding seatback to prevent whiplash in the frequent, minor rear crashes. There were other arguments against stiff seats, including ramping with head impact, out of position exposures with seat impact, and rebound impacts. Petitions for NHTSA rulemaking to increase seatback strength in FMVSS 207 tests were submitted by Saczalski (1989) and Cantor (1989); and, various product liability cases were pursued on seatback strength (Rake and Boehm 1990). These proposals and cases furthered the debate and relative positions of each side. Government sponsored analyses of field crashes showed yielding seats performing well (Malliaris 1990, Digges, Malliaris 1993), and they did not revise the safety standards and pursued additional research on seats in rear crashes.

Stiff or rigidized seats were one aspect of the pioneering work of Severy et al. (1967a,b, 1968a,b) on occupant dynamics in rear crashes. These studies showed the need for controlled deformation of the seatback and support of the head. Other concept studies were conducted by Saczalski et al. (1993) and Blaisdell et al. (1993). However, the stiff seat concepts were not brought to production because of various criticisms on overall performance.

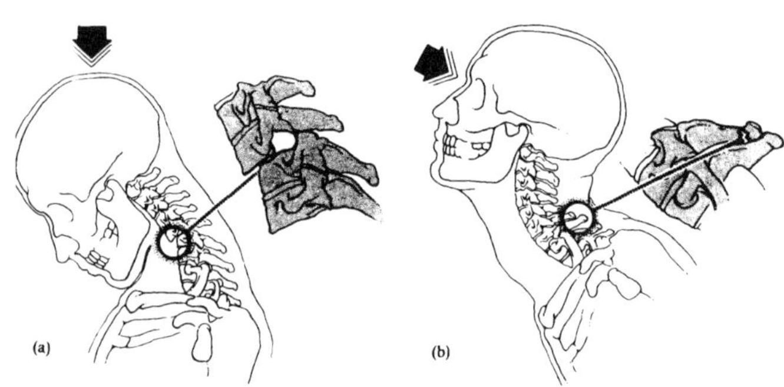

Figure 1: Fracture-dislocation of the cervical spine due to flexion-compression with impact on the vertex of the head (a left) and hyper-extension with forehead impact or inertial loading (b right).

Strother, James (1987), Warner et al. (1991) and Prasad et al. (1997) provided engineering analyses and rationale for the yielding seatback in rear crashes. They countered the stiff seat approach with the yielding seat concept as the best means to provide overall occupant safety in rear crashes by controlled seatback rotation that limited forces on the neck. This involved an energy absorbing restraint as the seatback yielded rearward in a crash and followed the historic trend in load-limiting and energy absorbing deformation of interior components, such as the steering system, padded dashboard and high-penetration resistant windshield. A seatback moment near the FMVSS 207 criterion was argued as necessary to balance the needs for whiplash prevention in low-speed crashes and absorb energy in severe rear crashes.

Viano (2002) studied the role of the seat in rear crashes, and addressed the yielding seat theory. The research showed that there was not a design conflict between a strong seat frame and a yielding seat, if compliance of the seatback between the side frame structures was considered in the load-limiting performance of the seat. The research led to new seat requirements for rear impacts and the development of a new generation of yielding seats phased into production since 1997 (Viano 2002). The seats are called high retention (HR) because of a strong frame structure with j = 1.4 ^{0}/kN, but they yield by deformation of the seat trim. They meet requirements in a Quasistatic Seat Test (QST) and yield rearward at a stiffness close to k = 20 kN/m (Viano 2003a). The field performance of high retention seats has been evaluated and they provide a reduction in fatality risk in severe rear crashes (Viano 2003e).

Viano (2003b,c) recently found additional evidence supporting yielding seats and showed an increased risk of whiplash with stiff seats in low-speed rear crashes. Seat stiffness was found to increase neck responses, particularly in low speed crashes where there is greater time to head contact and proportionately more energy transfer to the occupant before head restraint contact. The stiff seat causes higher torso acceleration leading to higher relative acceleration and velocity between the head and torso, and neck displacement. Stiff seats also increase whiplash risks in women (Viano 2003d).

This study evaluated four seat types with two different stiffnesses. Yielding seats have 20 kN/m stiffness. One represented the typical yielding seat of the 1970s-80s with a load limit of 3 kN giving a 600 Nm moment. The other represented a high retention seat with a load limit of 10 kN and 2000 Nm moment. Stiff seats have 40 kN/m stiffness. One represented a foreign benchmark of European and Asian luxury seats of the 1990s. They have a load limit of 7.7 kN and 1700 Nm moment. The last seat represented an ABTS (all belts to seat) with a load limit of 20 kN and 4000 Nm moment. These exemplar seats cover a range of seat types, which were evaluated in 42 and 51 km/h (26 and 32 mph) rear crashes.

METHODOLOGY

Nomenclature for Data Analysis:

ABTS	All belts to seat
d	Neck displacement, d = x - z
F	Force on Occupant's Torso in kN
G	Force on Neck in kN
i	Time integer
j	Slope of Occupant Load versus Seatback Angle in ^{0}/kN
k	Stiffness of the Seat due to Occupant Loading in kN/m
k_R	Stiffness due to Seatback Rotation in kN/m
k_C	Stiffness due to Suspension Compliance in kN/m
k_{HR}	Stiffness of the Head Restraint in kN/m
k_N	Stiffness of Neck in kN/m
m	Effective Mass of the Occupant Loading the Seat in kg
m_T	Mass of the Torso in kg
m_H	Mass of the Head-Neck in kg
p	Length of the Seatback from the Pivot to the Top of the Head Restraint in m.
r	Gap due to Seatback Rotation at the Center of Pressure of Occupant Loading in m
s	Actual Head Restraint Gap w/r moving Vehicle and Head in m
Δt	Delta time in s
u	Drop in Head Restraint Height Due to Seatback Rotation in m
v	Rearward Displacement of Head Restraint Due to Seatback Rotation in m
w	Head Restraint Gap in m
x	Horizontal Displacement of the Torso in m
$\dot{x}$	Velocity of the Torso in m/s
$\ddot{x}$	Acceleration of the Torso in m/s^2
x_R	Torso Displacement due to Seatback Rotation
x_C	Torso Displacement due to Seat Suspension Compliance
y	Horizontal Displacement of the Vehicle in m
$\dot{y}$	Velocity of the Vehicle in m/s
$\ddot{y}$	Acceleration of the Vehicle in m/s^2
z	Horizontal Head Displacement in m
$\dot{z}$	Head Velocity in m/s
$\ddot{z}$	Head acceleration in m/s^2
θ	Seatback Angle w/r Vertical in deg

Analysis Procedures

Vehicle Dynamics: Acceleration of the vehicle is prescribed in equal time steps, so that a spreadsheet calculation can be made of the vehicle and occupant dynamics. Given the acceleration function for the vehicle, it is straightforward to determine its velocity and displacement, which displaces the seat forward. Some aspects of the equations of motion are repeated here to give context to the results. The full analysis procedure can be found in earlier studies (Viano 2003b-d). The solution is calculated to 200 ms.

Occupant Dynamics: The occupant is displaced forward in a rear crash by the motion of the vehicle and deformation of the seat during the impact. Figure 2 defines parameters for the modeling. Force develops on the torso by the stiffness of the seat and relative torso displacement with respect to the vehicle:

$$F_i = k(y_i - x_i) \qquad (1)$$

Force on the occupant increases with deformation of the seat, which accelerates the occupant in proportion to the effective mass of the torso loaded by the seatback. Based on sled testing with the Hybrid III dummy, Viano (2002) determined that the effective mass (m) loaded by the seatback was 70% of the total dummy mass, or 52.5

kg for the 50th percentile Hybrid III dummy. This agrees with the estimate by Severy (1968b) using an earlier vintage dummy. The mass of the head-neck is 5.5 kg, based on 4.5 kg for the head plus two-thirds of the neck mass or 1.0 kg (Backaitis, Mertz 1994), so torso mass is 47.0 kg. The acceleration of the torso is:

$$\ddot{x}_i = F_i / m_T = F_i / 47.0 \qquad (2)$$

where Equation (2) assumes that the eventual load on the neck has a second order effect on the calculation and can be neglected, and there are no inertial effects of the seatback mass. The occupant's torso acceleration determines the velocity and displacement of the torso.

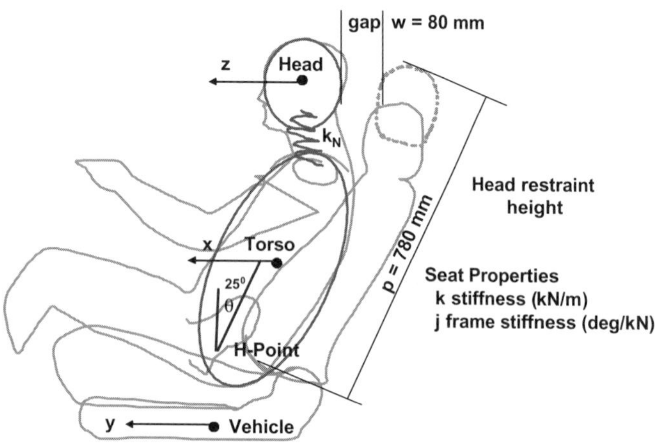

Figure 2: Two mass representation of the occupant with flexible neck loaded by the seatback.

Seat Properties: Seat Stiffness (k), Frame Rotation Stiffness (j) and Load Limit: Seat stiffness (k) and frame rotation stiffness (j) are determined in QSTs between occupant loads of 1 kN to 5 kN (Viano 2002, 2003a). This covers the majority of loading up to head restraint contact in rear crashes. The benchmark average of 14 foreign seats from high-end European and Asian manufacturers was k = 34.2 ± 6.5 kN/m with a range from k = 26-47 kN/m stiffness. The average of the five stiffest seats was k = 40.9 kN/m. The foreign benchmark was exemplified with k = 40 kN/m and j = 1.8 ^{0}/kN. The foreign benchmark seat (called stiff seat) is assumed to have linear stiffness up to a peak force of 7.7 kN and beyond that deformation, the load is held constant. The load limit is an additional seat property. The nominal peak bending moment was 1700 Nm (7.7 kN with a 0.22 m moment arm).

The 1990 C car seat has a seat stiffness that is typical of yielding seats of the 1970s-80s and has an average stiffness of k = 21.1 ± 2.4 kN/m. Yield is essentially due to seatback rotation; however, the average yield load was 1100 Nm. A yielding seat from the 1970s was exemplified with a stiffness of k = 20 kN/m and j = 3.4 ^{0}/kN, but unlike the 1990 C car seat, the maximum seat load was 3 kN to simulate a moment of 660 Nm.

Table 2: Seat Properties Used in the Modeling.

Seat Type	k kN/m	j deg/kN	jk deg/m	Force kN	Moment Nm
1970s Yielding Seat	20	3.4	68	3.0	660
Foreign Benchmark	40	1.8	72	7.7	1700
High Retention Seat	20	1.4	28	10.0	2200
ABTS	40	1.0	40	20.0	4400

The average stiffness of high retention seats, such as the 1997 W (Pontiac Grand Prix) and 1997 P-90 (Chevrolet Malibu), was k = 22.1 ± 5.4 kN/m and achieved a 5 kN load at 32^0, nearly 10^0 less seatback rotation than the 1990 C car seat. The high retention seat was exemplified with k = 20 kN/m and j = 1.4 ^{0}/kN. The maximum load was 10 kN giving a moment of 2200 Nm. High retention (HR) seats use a strong seat frame and recliners for occupant retention in severe rear crashes to limit seatback rotation. They also have an open seatback providing compliance or yield by deformation of the seat trim and an EA pelvic strap to pocket the occupant's pelvis and lower back in a rear crash (Viano 2002, 2003a).

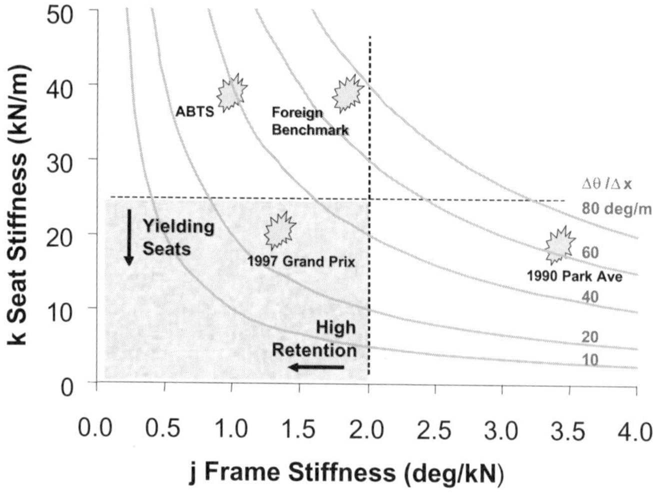

Figure 3: Seat j and k properties for the four exemplar seat types. This also defines a yielding seat with k < 25 kN/m and high retention seat with j < 2.0 ^{0}/kN.

In a recent study of high retention seats, Viano (2003a) found that the average stiffness of more than 20 high retention seats was 25.3 ± 4.9 kN/m, which was statistically similar to the average 24.7 ± 4.9 kN/m for pre-high retention seats. While the seatback stiffness was similar and the effects of the yielding performance were maintained, the high retention seats had a 35% lower seatback rotation at 1700 Nm. The yielding was given by compliance of the open perimeter frame. High retention (HR) seats had an average H-point moment of 2537 ± 703 Nm representing a 2.3-times higher moment than pre-HR designs of the 1980s-90s. A 2.9-times higher energy transfer capability was achieved by the HR seats to 3659 ± 1140 J. The sample included several ABTS seats, which have stronger frames and higher stiffness on average.

The fourth seat evaluated was an ABTS (All Belt to Seat) seat with the stiffness of the foreign benchmark but a lower frame rotation stiffness. The ABTS seat had k = 40 kN/m and j = 1.0 ^{0}/kN. The maximum force was 20 kN and beyond that deformation, the load was held constant at 20 kN giving a moment of 4400 Nm.

The four exemplar seats are defined in Table 2 and shown in Figure 3 following the earlier modeling studies of Viano (2003b-d). Figure 3 shows two stiff seats with k = 40 kN/m. One represents the foreign benchmark with j = 1.8 ^{0}/kN and the other an ABTS seat with j = 1.0 ^{0}/kN. There are two yielding seats with k = 20 kN/m. One represents seats like the 1990 C car (Buick Park Avenue) with j = 3.4 ^{0}/kN, a relatively high j factor. However, in this study, a lower load limit of 3 kN is used to represent a 1970s yielding seat. The other represents the 1997 W car (Pontiac Grand Prix), a high retention seat with j = 1.4 ^{0}/kN with a 10 kN load limit, typical of production high retention seats.

For seat rotations in the constant load limit condition, the change in seatback angle was determined by the change in occupant displacement converted to a seatback angle at the height of the center of pressure. This ensured equilibrium of motion during the load limiting phase.

Head Restraint Trajectory: Torso load causes seatback rotation, which is generally linear up to the load limit. The seatback angle is given by:

$$\theta_i = \theta_0 + jF_i \qquad (3)$$

where $\theta_0 = 25^0$ for typical seats and j is the slope of the line relating occupant load to seatback angle. As the occupant loads the seatback and it rotates, there is a downward and rearward displacement of the head restraint is given by:

$$u_i = p[\cos(\theta_i) - \cos(\theta_0)] \qquad (4)$$
$$v_i = p[\sin(\theta_i) - \sin(\theta_0)] \qquad (5)$$

where p = 780 mm is used to represent the head restraint height up the seatback of modern seats. This value will be used for the analysis of the 1970s yielding seat even though the actual head restraint was lower for the production seat. This allows a direct comparison of the different seat stiffness types. The gap to the head restraint varies with the initial gap behind the head and seatback angle changes. It can be determined from the previous relationships with:

$$w_i = w_0 + v_i \qquad (6)$$

where $w_o = 80$ mm is assumed typical of modern head restraint placements. This relationship does not address the effects of head-neck displacement. The actual gap between the back of occupant head and the head restraint is a function of the head and vehicle displacement, and other geometric relationships defining the head restraint position with seatback rotation. It is given by:

$$s_i = w_i - (y_i - z_i) \qquad (7)$$

For the exemplar seats, the change in seatback angle for occupant displacement into the stiff (foreign benchmark) seat was $\Delta\theta/\Delta x = 72$ ^{0}/m based on QSTs. The 1990 C car and 1970s yielding seats had $\Delta\theta/\Delta x =$ 68 ^{0}/m. The 1997 W car and high retention seats had $\Delta\theta/\Delta x = 28$ ^{0}/m, and the ABTS seat had $\Delta\theta/\Delta x = 40$ ^{0}/m with:

$$\Delta\theta/\Delta x = jk \qquad (8)$$

where $\Delta x = (y - x)$ is the relative displacement between the vehicle (seat) and torso.

Head-Neck Dynamics: Force develops on the neck due to differential motion between the torso and head. It is related to the stiffness of the neck due to a combination of bending, tension and shear:

$$G_i = k_N(x_i - z_i) \qquad (9)$$

Force on the neck increases with the relative displacement between the head and torso, which accelerates the head in proportion to the 5.5 kg effective mass of the head and neck:

$$\ddot{z}_i = G_i/m_H = G_i/5.5 \qquad (10)$$

Head acceleration can be used to determine the velocity and displacement of the head with respect to ground, which adds to the occupant energy transfer. After head restraint contact, there is an additional force on the head in proportion to its penetration into the head restraint. A linear relationship is assumed for current head restraints. For this condition, the force on the head in Equation (9) becomes:

$$G_i = k_N(x_i - z_i) - k_{HR}s_i \qquad (11)$$

Based on subsystem tests of head restraints (Viano 2002, Chapter 13), a 50 mm penetration of the back of the Hybrid III head causes a 2.8 kN load giving a stiffness of 80 kN/m; however, with the head restraint installed in the seat, the stiffness reduces to $k_{HR} = 20$ kN/m. This value was selected after comparison with model validation tests.

Effective Neck Stiffness: The determination of neck stiffness has been described in earlier studies (Viano 2003c-d) and verified by validation tests using the 50% male and 5% female Hybrid III dummies. Neck stiffness of 5 kN/m is used to approximate the 50% male neck extension due to bending, shear and tension.

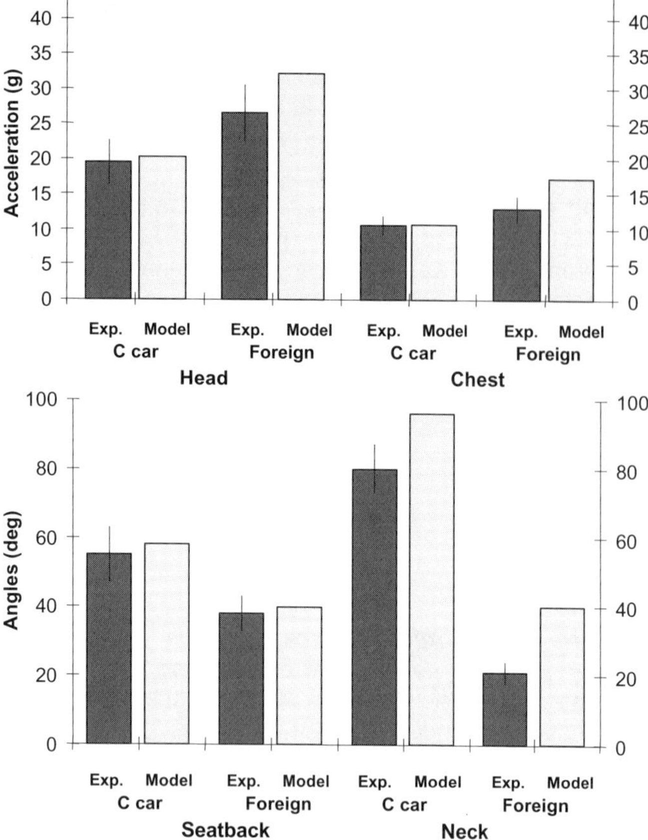

Figure 4: Comparison of rear sled tests and model responses in a 19.5 mph exposure, which includes peak head and chest acceleration, and seatback and neck angle change.

Model Validation: A series of rear sled tests has been conducted at 31.4 km/h (19.5 mph) with various seats using the 50% male Hybrid III dummy (Viano 2002). It included several C car and five foreign benchmark (European and Asian luxury) seats, which were used to validate the mathematical model. For this comparison, the average value in seat stiffness k = 34 kN/m and yield load 8.7 kN was used for the foreign seats, which had an average H-point moment of 1655 Nm. The average C car seat had k = 22 kN/m and a 7.1 kN load limit with an H-point moment of 1140 Nm.

Using these parameters in the model, Figure 4 compares the peak chest and head acceleration, seatback angle change and neck extension angle. The agreement is quite good in amplitude of responses. The lower prediction in change of seatback angle is explained by applying load to maximum seatback angle. After that time, the torso force is reduced to zero after the torso load starts to drop. In actual tests, there is elastic energy stored during the occupant loading that is returned to the occupant as rebound velocity. That involves greater loading and rotation of the seatback in the test, but the effect is not included in the model. Its inclusion would increase the simulated seatback rotation. Also, the head restraint height in the model reflected the lower position of the 1990 C car, which delayed head support.

Differences in Neck Responses with Stiff and Yielding Seats: The initial neck response differences were investigated using the underlying biomechanics of the loading and seat properties. This involved the equations of motion to estimate response ratios with stiff and yielding seats. Several neck responses were considered, including head-torso acceleration and neck displacement. Head-torso velocity was also considered because of its importance to eventual loading of the head restraint.

Vehicle displacement occurs first in the dynamic response and drives all subsequent interactions. Based on Equations (1), (2), (9) and (10), head-torso acceleration is:

$$(\ddot{x}-\ddot{z}) = (k/m_T)y - (k/m_T + k_N/m_H)x + (k_N/m_H)z \quad (12)$$

Early in the seat loading the z displacement can be ignored and the relative acceleration depends primarily on the vehicle displacement, as torso displacement is also a second order influence. Equation (12) can be sequentially approximated as:

$$(\ddot{x}-\ddot{z}) = (k/m_T)y - (k/m_T + k_N/m_H)x \quad (13)$$

$$(\ddot{x}-\ddot{z}) = (k/m_T)y \quad (14)$$

Torso displacement is delayed from vehicle displacement, which influences the early neck responses. The ratio in head-torso acceleration for occupants in vehicles with stiff (subscript s) and yielding (subscript y) seats can be approximated by vehicle displacement as a function of time:

$$(\ddot{x}_S-\ddot{z}_S)/(\ddot{x}_Y-\ddot{z}_Y) = [(k_S/m_{TS})/(k_Y/m_{TY})]y_S/y_Y \quad (15)$$

where head-torso acceleration initially depends on the ratio $(k_S/m_{TS})/(k_Y/m_{TY})$, seat stiffness divided by torso mass for the stiff and yielding seats. Assuming no shift in torso mass for vehicles with stiff and yielding seats, the head-torso acceleration ratio is proportional to:

$$(\ddot{x}_S-\ddot{z}_S)/(\ddot{x}_Y-\ddot{z}_Y) = k_S y_S / k_Y y_Y \quad (16)$$

If the vehicles with stiff and yielding seats have a similar crash pulse, the ratio further reduces to k_S/k_Y. The initial difference in head-torso acceleration depends on the ratio of seat stiffness. The velocity difference between the torso and head, and neck displacement (d) can be determined from equation (14) by single and double integration:

$$\int(\ddot{x}-\ddot{z}) = (k/m_T)\int y \quad (17)$$

$$d = \iint(\ddot{x}-\ddot{z}) = (k/m_T)\iint y \quad (18)$$

Neck displacement initially depends on the second integral of vehicle displacement, seat stiffness and inversely torso mass. Assuming a constant vehicle acceleration, neck displacement depends on the acceleration level and time to the fourth power. The ratio of neck displacement for stiff and yielding seats is:

$$d_S / d_Y = [(k_S / m_{TS})/(k_Y / m_{TY})]\iint y_S / \iint y_Y \quad (19)$$

If the torso mass remains constant for occupants in stiff and yielding seat vehicles, Equation (19) reduces to:

$$d_S / d_Y = [k_S / k_Y]\iint y_S / \iint y_Y \quad (20)$$

where the ratio depends on the second integral of vehicle displacement. Assuming constant vehicle accelerations, Equation (20) depends on the ratio of crash pulse amplitude and seat stiffness. If the crash pulses are similar for vehicles with stiff and yielding seats, the ratio further reduces to:

$$d_S / d_Y = k_S / k_Y \quad (21)$$

Angle Change of the Seatback: An important issue in severe rear crashes is the angle change of the seatback. Seat stiffness is only one factor in the angle change. Using Equations (1), (5) and (6), the seatback angle is related to seat properties:

$$\Delta\theta = jF = jk(y - x) \quad (22)$$

so the ratio of seatback angle change for stiff and yielding seats in the same vehicle and crash severity is:

$$\Delta\theta_S / \Delta\theta_Y = (j_S k_S / j_Y k_Y)(y - x)_S / (y - x)_Y \quad (23)$$

Assuming torso displacement and the seatback load limit give second order effects later, the initial change in seatback angle can be approximated with:

$$\Delta\theta_S / \Delta\theta_Y = j_S k_S / j_Y k_Y \quad (24)$$

so the change in seatback angle is related to the ratio of the product of j and k for the stiff and yielding seats.

Seatback Compliance and Frame Structure Stiffness: Seat deformation includes the suspension between the side frame structures, trim and rotation of the seatback frame. Based on the measured stiffness and seatback rotation, it is possible to determine the individual stiffness components for the suspension and seatback frame, which work in series to support occupant load.

Based on the seatback angle at 5 kN occupant load, the contribution to the horizontal motion of the dummy from seatback rotation is determined using the center of occupant pressure on the seatback. The horizontal displacement due to seatback rotation is:

$$x_R = 0.22[\tan(\theta_{5kN}) - \tan(\theta_0)] \quad (25)$$

where x_R is the displacement due to seatback rotation at 22 cm above the H-point. Stiffness due to frame rotation is:

$$k_R = F / x_R = 5 / x_R \quad (26)$$

$$k_R = \Delta\theta / jx_R = 260 / j \quad (27)$$

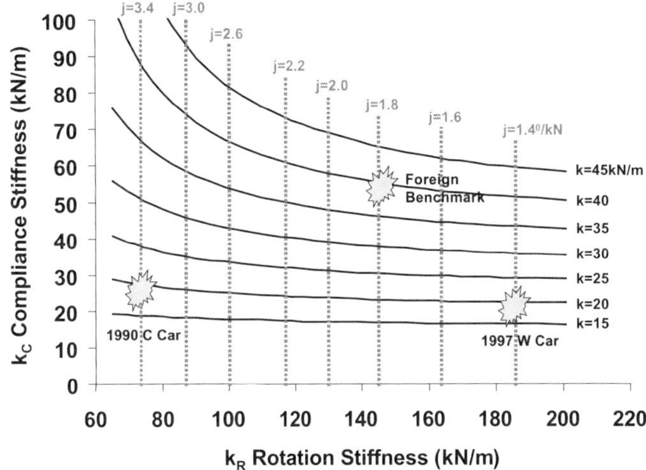

Figure 5: Relationship between seat properties j and k, and the frame stiffness and trim compliance for three of the exemplar seats.

Seat frame stiffness depends solely on the j factor. Seat suspension compliance can be determined from the measured seatback stiffness (k) and the differential displacement due to frame rotation:

$$x_C = (5/k) - x_R \quad (28)$$

so that:

$$k_C = F / x_C = 5 / x_C \quad (29)$$

$$k_C = kk_R / (k_R - k) = 260k / (260 - jk) \quad (30)$$

The trim and suspension stiffness (seat compliance) is determined by both the j and k factors. Seat stiffness (k) can be verified by the relationship:

$$k = k_R k_C / (k_R + k_C) \quad (31)$$

Figure 5 shows the relationship between seat stiffness and the j and k factors. The ABTS results are off the graph since j = 1.0 ^{0}/kN.

RESULTS

42 km/h (26 mph) Rear Crash Severity: Figure 6 shows the force developed on the occupant by the forward motion of the vehicle. The stiff (foreign benchmark) seat reaches the 7.7 kN load limit and plateaus. Head contact is at 62 ms shown as a target on the response curve. The highest force is developed by the ABTS seat at 11.1 kN and head contact is earlier at 48 ms. The stiff and ABTS responses are initially similar due to the 40 kN/m seat stiffness (see Equation 1). The 1970s yielding seat develops only 3 kN load and has a lower slope because of the 20 kN/m stiffness. The high retention seat never reaches its load limit and peaks at 9.2 kN. It has the earliest head contact at 46 ms. The response data are summarized in Table 3 up to head restraint contact and some maxima are also given after head contact.

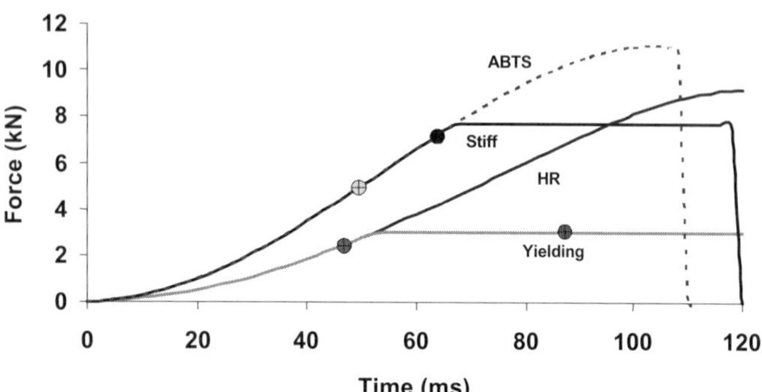

Figure 6: Torso force developed by the four exemplar seats in a 42 km/h (26 mph) rear crash.

Figure 7 shows seatback rotation, which initially follows Equation (22) with the product of seat property jk defining the response. Since the 1970s yielding seat and the stiff seat have similar jk's, they have a similar response until the load limit is reached with the yielding seat. The ABTS and high retention seats also follow a similar response, but since neither seat reaches the load limit, they are similar throughout the occupant loading. The much higher strength of the ABTS seat is not reflected in lower seatback rotation because of its higher stiffness and loads on the occupant.

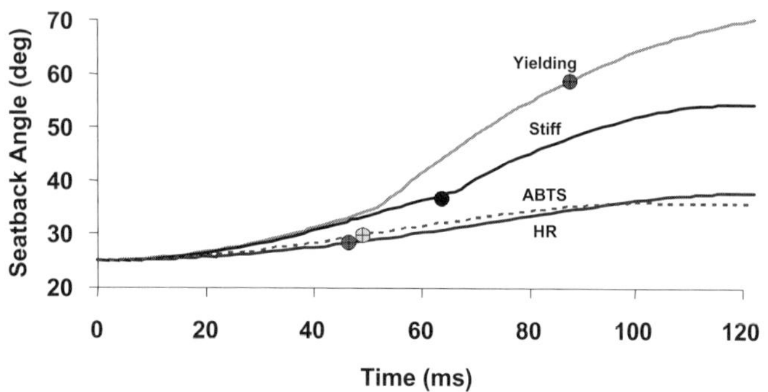

Figure 7: Seatback rotation with the four exemplar seats in a 42 km/h (26 mph) rear crash.

Figure 8 shows torso acceleration, which is higher with the stiffer seats. They peak at 15-25 g. The yielding seats initially have lower acceleration, however later in the response and after head support, the high retention seat has high torso acceleration. Seat stiffness k is the determining factor in the initial torso acceleration. Higher torso acceleration causes higher neck responses.

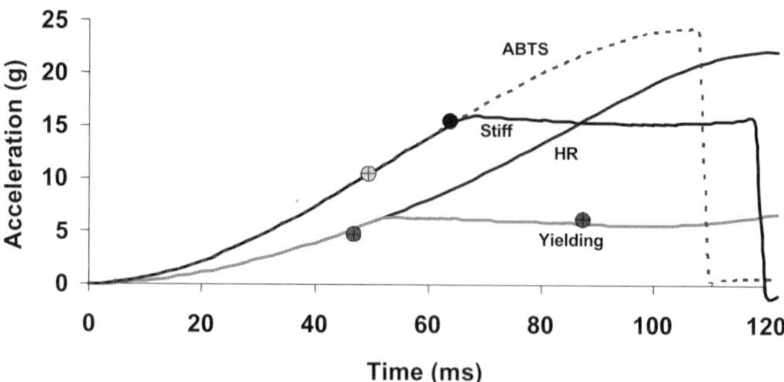

Figure 8: Torso acceleration with the four exemplar seats.

Figure 9 shows the relative acceleration between the torso and head. With the stiff seats, the response is higher than with the yielding seats. The initially higher relative acceleration causes a higher relative velocity between the head and neck before head contact. This response drives higher neck displacements with the stiffer seats (Figure 10).

The higher neck displacements with the stiff seats reflect the higher relative velocity between the head and torso. The high retention and ABTS seats have the earliest head contact with the 1970s yielding seat having the latest due to seatback rotation that pushes the head restraint away increasing the gap.

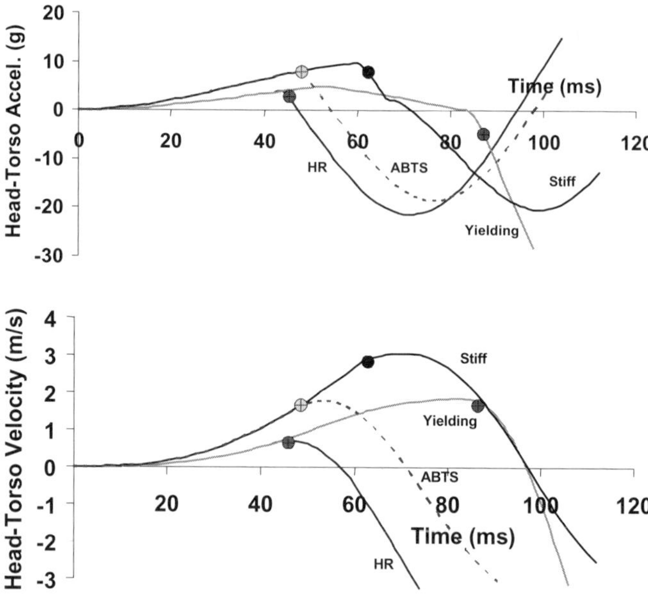

Figure 9: Relative acceleration and velocity between the head and torso for the four exemplar seats.

Figure 11 shows the head restraint drop due to seatback rotation. There is a similar increase in the gap behind the head. At head contact, the drop is only 20 mm with the high retention seat, while it is 83 mm and 294 mm with the stiff and 1970s yielding seats, respectively. This dramatically reduces the effective height of the head restraint. The drop with the foreign benchmark seat reflects the higher loads on the torso, which increase seatback rotation and neck responses up to head restraint contact.

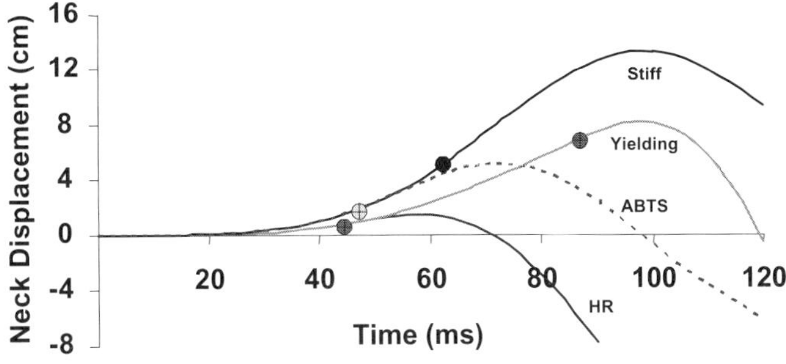

Figure 10: Neck displacement with the four exemplar seats.

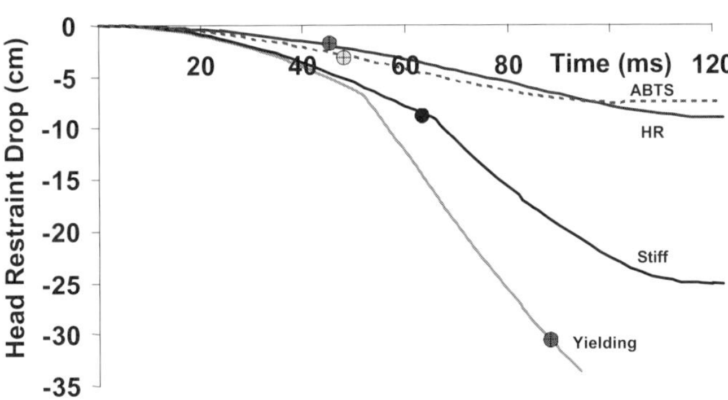

Figure 11: Head restraint drop due to seatback rotation with the four exemplar seats.

Figure 12 shows the peak at head contact or maxima for seatback angle, neck displacement and head velocity with respect to the torso and vehicle. The neck displacement is proportionately greater for the stiff seat due to a combination of much higher head-torso velocity and higher loads on the torso with seatback rotation that pushes the head restraint away and delays head contact. The late head contact and large seatback rotation with the 1970s yielding seat also involve high neck displacement.

The combination of early head contact and low seatback rotation with the high retention seat gives the lowest neck displacement. This also reflects the low head-torso velocity due to low initial loads and acceleration of the torso. The ABTS seat has the highest torso loads, which increase neck displacements but the early head contact and low seatback rotation limits the values. If, however, the head restraint is further rearward or down, the neck displacements can be substantial.

Figure 13 shows the seatback angle change with the addition of a simulation using a yielding seat with a 5 N load limit. This represents a load limit less than the 1990 C car. The response follows the foreign benchmark to a higher seatback angle near 50^0. This comparison shows that seats of the foreign benchmark design have seatback rotations in rear crash that are initially rather similar to the yielding seats of the 1980s-90s. Even though they have greater strength, they are also stiffer, which increased loads on the occupant and increased seatback rotation. This countered the benefits of the stronger seat frame while increasing neck responses.

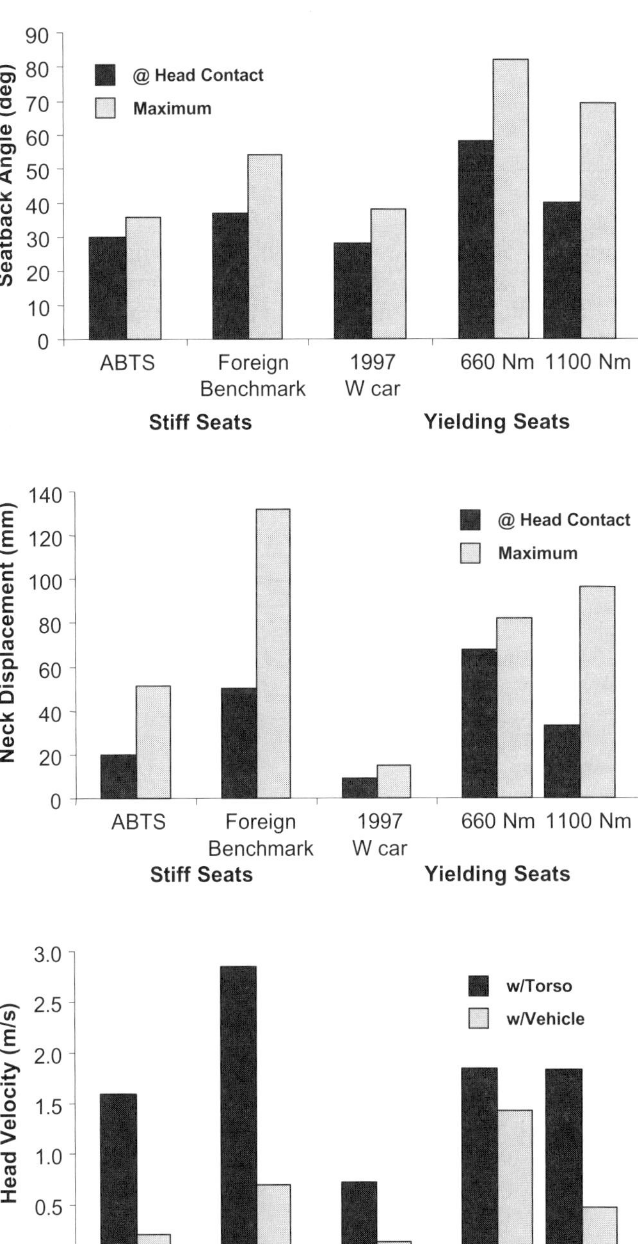

Figure 12: Peak values at head contact or maxima for the seatback angle, neck displacement and head velocity with respect to the torso and vehicle.

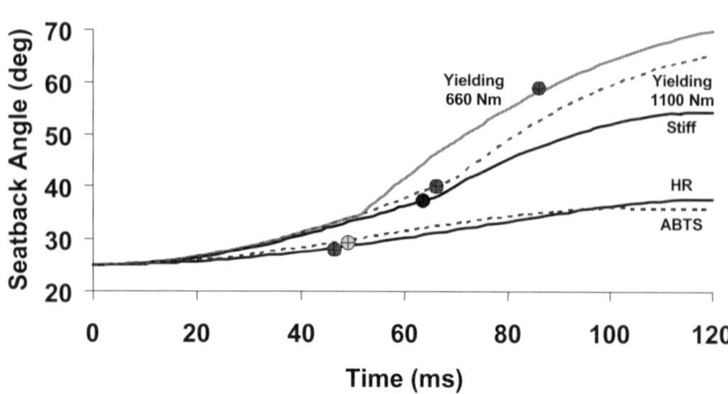

Figure 13: Seatback rotation with the exemplar seats in a 42 km/h (26 mph) rear crash. An additional simulation was run with a load limit of 5 kN (1100 Nm moment).

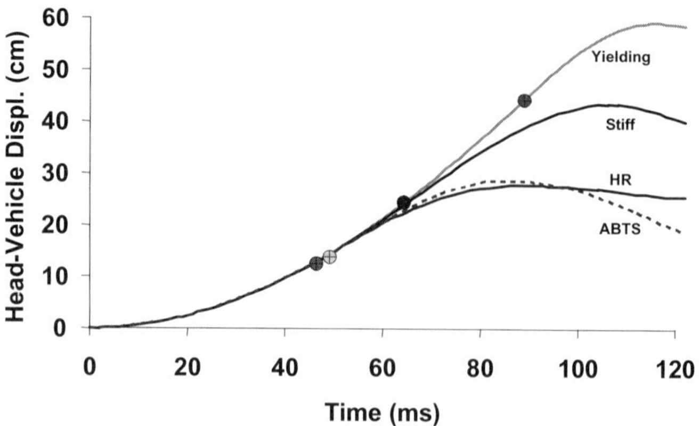

Figure 14: Displacement of the head with respect to the vehicle for the four exemplar seats.

Table 3: Summary of Vehicle and Occupant Dynamics in the Rear Crash Simulations.

	42 km/h (26 mph) Rear Delta V				51 km/h (32 mph) Rear Delta V			
	ABTS	**Stiff Seat**	**High Retention**	**Yielding Seat**	**ABTS**	**Stiff Seat**	**High Retention**	**Yielding Seat**
At or Before Head Contact								
Head Contact (ms)	48	62	46	86	44	56	40	78
Force (kN)	4.7	6.9	2.4	3.0	5.0	7.3	2.3	3.0
Seatback Angle (deg)	29.7	36.9	28.3	58.1	30.0	37.5	28.1	59.4
Head Restraint Drop (mm)	-29	-83	-20	-294	-31	-88	-19	-309
Dynamic Gap (mm)	137	219	120	412	140	225	118	422
Torso Acceleration (g)	10.0	14.5	5.1	5.8	10.6	15.4	4.8	5.9
Energy Transfer (J)	71	281	15	257	66	245	10	215
Neck Force (kN)	102	251	45	338	91	216	33	297
Neck Displacement (mm)	20.3	50.3	9.0	67.6	18.1	43.2	6.5	59.4
Head-Torso Velocity (m/s)	1.61	2.85	0.72	1.84	1.58	2.82	0.65	1.87
Head-Torso Acceleration (g)	7.73	9.69	3.96	4.94	8.41	11.08	3.97	5.18
Maximum								
Force (kN)	11.1	7.7	9.2	3.0	13.5	7.7	10.0	3.0
Seatback Angle (deg)	35.9	54.3	37.8	75.7	38.5	63.3	49.2	79.1
Energy (J) @ Max SB Angle	3904	4327	2893	2701	6025	6092	4709	3636
Neck Displacement (mm)	51.6	132.8	14.9	81.8	47.5	136.7	12.8	72.9

notes: ABTS has 20.0 kN (4500 lb) maximum force giving 4400 Nm or 39,000 inlb moment
Stiff Seat has 7.7 kN (1730 lb) maximum force giving 1700 Nm or 15,000 inlb moment
High Retention has 10.0 kN (2250 lb) maximum force giving 2250 Nm or 20,000 inlb moment
Yielding Seat has 3.0 kN (675 lb) maximum force giving 660 Nm or 5,830 inlb moment

Figure 14 shows head displacement with respect to the vehicle. Seatback rotation with the yielding and stiff seats involves the greatest rearward displacement of the head in the vehicle. The high retention and ABTS seats had about the same displacements.

Figure 15 shows the energy transfer to the occupant in the rear crash with the four exemplar seats. At this crash speed all of the seats transfer most of the crash energy after head contact; however, the lowest values occur with the high retention and ABTS seats. Only the yielding seat does not have the capacity to transfer the crash energy before seatback angle exceed 60^0. This means that ramping will occur before the crash energy is transferred to the occupant. This increases secondary impacts in the rear seating area of the vehicle.

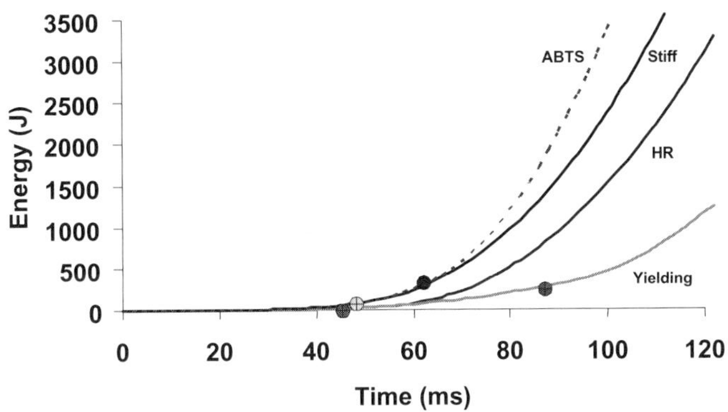

Figure 15: Energy transfer to the occupant with the four exemplar seats.

Figure 16 shows a series of simulations without head restraint contact. This reflects the underlying support of the torso with inertial effects from the head. The highest neck displacement occurs with the ABTS and stiff seat because of the higher torso loads, and the lowest is with the yielding seat. This simulation reflects the situation with lower head restraint placement in the 1970s-90s.

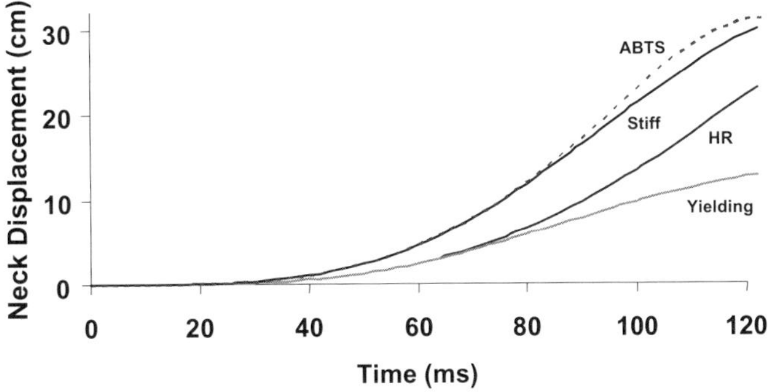

Figure 16: Neck displacement without head restraint contact in the 42 km/h rear crash.

Table 3 gives the peak responses at head contact and some maxima throughout the simulation. Data are given for the 42 and 51 km/h (26 and 32 mph) rear crashes. The 50% higher kinetic energy of the 51 km/h (32 mph) crash increases the occupant response, although the time for head contact is reduced by the higher crash speed.

DISCUSSION

Arguments Against Stiff Seats: This study was motivated by the claims for stiffer seats and the counter arguments. Over the past years there have been many proposals to strengthen seats and increase bending resistance to prevent loss of occupant retention in severe rear crashes. These proposals have been refuted by concerns for increased whiplash in low-speed rear crashes and other adverse kinematics with stiffer seats. This study and others (Viano 2003c,d) support the earlier concerns with stiff seats, but show a way to use a strong seat frame if the seat stiffness is kept low. The research provides a design direction benefiting both improved safety in severe rear crashes and whiplash prevention in low speed impacts, but seat strength (j) and stiffness (k) must be independently controlled.

Stiff seats develop proportionately higher loads on the occupant and therefore require more structure to sustain the forces. This leads to heavier and more costly designs; and, the seats degrade safety performance. While this may seem counter-intuitive, the analysis of the equations of motion and field injury patterns over the past thirty years seem to support the concern with stiff seats; and, it is not clear how stiff seats may be redesigned to address the problems identified.

ABTS designs need to support shoulder belt loads in frontal crashes and require considerable seat frame structure. This adds substantial strength for rear crashes, but also higher weight and cost in comparison to conventional seats. However, the benefits of ABTS may not be realized if it is too stiff, because load builds up early accelerating the occupant and rotating the seatback. While the rotations may only be 5^0-10^0 in severe rear crashes, the higher stiffness increases neck responses and whiplash risks in lower-speed crashes. Also, if the added seat frame structure reduces the space for occupant displacement between the side frames, there can be difficulty in providing low stiffness.

The safety benefits of a stronger seat frame is clear in severe rear crashes if the seat is compliant; but, the benefits are not realized if the seat also gets stiffer. High stiffness merely results in higher loads on the occupant, which rotate the seatback and increase torso and neck responses. Analysis of the equations of motion show that seatback rotation is proportional to jk(y-x), indicating the important parameter to reduce seatback rotation and improve occupant retention is the seat property jk, and j is the seat frame strength property. Also, load on the seat is proportional to the seat stiffness k, so stiffness is a major determinant of occupant force on the seat.

The long-standing debate over stiff and yielding seats prompted this study, and its conclusions explain the key seat properties related to safety performance. The clear design direction is to reduce jk for occupant retention in severe crashes and k for more gradual occupant loading and whiplash prevention. The load limit is also a factor.

This study also confirms that the transition to stiffer seats over the past thirty years may have inadvertently increased whiplash. This fulfills the earlier concerns of Strother, James (1987), Warner et al. (1991) and Prasad et al. (1997) about stiff seats. However, they did not develop the theoretical arguments against stiff seats in the way provided here or show a design solution to improve the safety of 1970s-90s yielding seats.

Evolution of Yielding Seats: This study underscores the importance of the yielding seat approach to occupant safety in rear crashes. Historically, the yielding behavior was achieved by seatback rotation at relatively low loads. The typical H-point moment was 660 Nm in the 1970s increasing to over 1100 Nm in the 1980s-90s in QSTs. These moments result in large seatback rotations in severe rear crashes allowing the occupant to ramp up the seatback. The 1970s-90s seat designs also had relatively stiff structural cross-members in the seatback, which made much of the compliance related to recliner and seat frame rotation, although the foam trim covering the frame provided some deformation distance until the motion was stopped by seatback structures. Often, these structures were just behind the pelvis and upper back, which limited penetration into the seatback and increased the tendency for ramping.

The 1990 C car seat had a stiffness of 21 kN/m, which was somewhat lower than the average seat of that era. It reflected the yielding behavior of seat designs of the 1970s-90s, which showed good field performance in rear crashes (Malliaris 1990, James et al. 1991, Digges, Malliaris 1993). However, it became clear that there were shortcomings by allowing the occupant to rotate the seatback rearward in severe rear crashes, since there could be a loss of occupant retention and the energy transfer capability was limited. The way to maintain the yielding behavior of the seat without significant seatback rotation was conceived in the 1990s after a series of sled tests with the C car seat. That seat had an open perimeter frame, which increased occupant retention to higher delta Vs than designs with substantial cross-frame structures. This was achieved by more pelvic pocketing into the seatback. The C car seat was the genesis of a new generation of yielding seats.

High retention seats are a new generation of yielding seats that has an open perimeter frame with wide side and high upper cross members. This approach allows the occupant to penetrate between the side frame structures. It uses a strong frame and recliners, but allows a yielding performance by letting the occupant deform the seat trim and suspension to lower the stiffness. This was the innovation that allowed a yielding stiffness with lower seatback rotation. High retention seats also have a well positioned head restraint to support the head and neck in rear crashes. This design improved performance over the foreign benchmark seats while maintaining the yield behavior of 1970s-90s seats (Viano 2003a). This study and recent field data (Viano 2003e) show that the high retention design has improved safety in high-severity rear crashes.

Yielding seats allow greater displacement of the occupant with respect to the vehicle in rear crashes than with stiff seats. This gives a more gradual load and acceleration of the occupant forward and lowers neck responses. However, head displacement also depends on seatback rotation. The later acceleration of the occupant delays significant loading until head restraint support. There are two benefits of the delayed loading. One, there is lower force on the upper back early in the crash, which is a primary factor in neck loading before head restraint contact. The lower the upper back load, the lower the force causing neck shear, bending and extension and whiplash. Two, the head receives full head support before the higher forces occur on the occupant with the yielding seat. This allows energy transfer at a time when the head, neck, back and pelvis have uniform support. This reduces the potential for differential motion of the head and torso, which can cause injury.

Optimum Seat Properties for Rear Crash Safety: The practical lower limit for jk has not been investigated; however, there must be some constraints of packaging and design that limit how low a value is feasible in modern vehicles. Similarly how low a value of k is practical has not been determined. Lowering both properties reduces responses related to injury. Furthermore, given a low k, there is no reason not to increase seat strength, thus lowering j is beneficial. From the 1970s-90s seat strength has increased (Viano 2003a). However, an evaluation of 1990s foreign benchmark seats showed a proportional increase in seat stiffness. The trend was to follow a jk of 65-75 ^{0}/m as seat strength increased. The introduction of high retention seats in 1997 gave an entirely new design direction for seats by providing high strength (low j) while lowering the jk property to 25-35 ^{0}/m. This was a unique design direction for seats and provided a new generation of yielding seats.

One possibility in the future is to use a progressively greater k value as occupant load builds up on the seat. In this way, the lowest levels are engaged in the lowest speed rear crashes and stiffness progressively increases to limit occupant displacement in the vehicle. Thus, the lowest k is used where the highest whiplash risks occur. Viano (2003c) showed the peak neck displacement at head contact increased with lower speed rear crashes because of the greater time to head contact and the greater proportion of energy transfer to the occupant. With the stiff foreign benchmark, the peak occurred at 10-12 km/h. It might be possible to flatten the response curve by introducing a progressive stiffness, which would lower responses in the lowest speed rear crashes where whiplash risks are the highest. There may be other design directions for the future, but lowering k and jk seems fundamental.

Implications to FMVSS 207 and 202: FMVSS 207 is a seat frame test that partly addresses the j property of seats. However, this study and earlier investigations of whiplash have emphasized the importance of reducing seat stiffness to manage whiplash risks. For whiplash prevention and FMVSS 202, seat stiffness needs to be evaluate with a realistic occupant load (Viano 2003c); and, it is important that evaluations are made with the 5% female dummy to ensure that seat stiffness is proportionately reduced for her torso mass. It is not clear that the 95% male is a necessary evaluation based on field injuries or analysis of the underlying parameters defining neck responses in rear crashes.

Viano (2003c) showed that neck responses are proportional to ky/m_T, seat stiffness times vehicle displacement divided by torso mass. Neck responses increase with seat stiffness and decrease with heavier occupants. Assuming also a stronger neck in the large male, the relative whiplash risk is even lower. It seems clear that low-speed seat evaluations are needed with the 5% female, more so than any other dummy.

The recent study of Viano (2003d) showed that neck responses were 30% higher in the female than male through most of a rear impact. They are proportional to $(k_F/m_{TF})/(k_M/m_{TM})$, which is the ratio of seat stiffness divided by torso mass for the female and male. The higher level in females implies that seat stiffness is not sufficiently low in proportion to the female mass in comparison to the male. It is also unclear that a single sled severity is adequate to address whiplash risks. Viano (2003c) showed that neck responses increase in lower severity crashes because of the longer time to head contact. They peak in 6-12 km/h rear crashes depending of seat type and stiffness. These crashes are below the crash severity being considered for the evaluation of whiplash risks (Langwieder, Hell 2002).

Role of Yielding Seats in Reducing Whiplash: The compliance of the high retention seat is an important factor in the reduction of whiplash risks and the lower forces on the back are a primary means of lowering loads on the base of the neck (Viano, Olsen 2001). This reduces displacement of the head with respect to the neck before head contact. Displacement of the base of the neck with respect to ground is an important factor in whiplash. The greater the relative occupant displacement with respect to the vehicle, the lower the loads on the torso and neck. There are also benefits from the open perimeter frame seatback, which maintains a more upright posture until the significant loads that rotate the seatback in a rear crash while the occupant's pelvis, back and head have uniform support.

Viano (2003c,d) found that the shift to stiffer seats increased neck displacements by a factor of two based on the exemplar seats used in this analysis. The phase-in of stiffer seats from the 1970s-90s could be partly responsible for the higher rates of whiplash seen over the past two decades and increased the whiplash risks proportionately more in women, since there has been a gradual shift to stiffer seats as seats have gotten stronger during that time frame. The high retention seat defines a new design direction for rear crash safety in low to high severity rear crashes.

Limitations of the Study: Obviously, the model developed and used in this study is not a full occupant dynamics simulation, which would offer the possibility to explore responses in much more detail. The model does provide a convenient means of addressing the fundamental parameters related to seat deformation and torso-head dynamics in rear crashes, and the solution provides quite reasonable responses when compared with sled test data.

As seatback rotation increase, the effects of gravity cause the occupant to fall rearward. These effects are not included in the model. Neither are the effects of ramping which is a factor in occupant interactions with the seat in severe rear crashes. These effects need to be investigated with a more complex model and are beyond the scope of this study.

Foreign benchmark seats have a wide range of characteristics in seat strength and stiffness. The exemplar stiff seat with 40 kN/m and j = 1.8 ^{0}/kN does not represent any specific product in use. It is a composite of the high-end performing seats. The actual designs varied considerably. Nonetheless, this study shows the importance of minimizing jk and k to improve occupant dynamics in rear crashes. These parameters define the essential characteristics of the seat for rear impact loading of the occupant.

ACKNOWLEDGMENTS

This paper was published in Traffic Injury Prevention 4(4) 2003, and is reprinted here with permission from the publisher, Taylor & Francis Group. Their agreement for its use in this book is greatly appreciated.

REFERENCES

Backaitis S, Mertz H. Hybrid III: The First Humanlike Crash Test Dummy. SAE Special Publication PT-44, Society of Automotive Engineers, Warrendale, PA. 1994.

Benson BR, Smith GC, Kent RW, Monson CR. Effect of seat stiffness in out-of-position occupant response in rear-end collisions. SAE 962434, 40th Stapp Car Crash Conference, Society of Automotive Engineers, Inc., Warrendale, PA, USA, 1996.

Blaisdell DM, Levitt AE, Varat MS. Automotive seat design concepts for occupant protection. SAE 930340, Society of Automotive Engineers, Inc., Warrendale, Pennsylvania, USA, 1993.

Cantor A. Petition for Rulemaking to Amend FMVSS 207 to Prohibit Ramping Up the Seat Back of an Occupant During a Collision. National Highway Traffic Safety Administration, December 28, 1989.

Digges, KH, Morris JH, Malliaris AC. Safety performance of motor vehicle seats. SAE 930348, Society of Automotive Engineers, Inc., Warrendale, PA, USA, 1993.

Federal Motor Vehicle Safety Standards No. 202, 207, and 301. 49 Code of Federal Regulations Section 571.21, 1968.

James MB, Strother CE, Warner CY, et al. Occupant Protection in Rear Collisions: Safety and Priorities and Seat Belt Effectiveness, 21st Stapp Car Crash Conf, SAE 912913, Society of Automotive Engineers, Warrendale, PA, 369-376, 1991.

Langwieder K, Hell W. Proposal of an International Harmonized Dynamic Test Standard for Seat/Head Restraints. Traffic Injury Prevention 3(3): 150-158, 2002.

Malliaris AC. Current Issues of Occupant Protection in Car Rear Impacts. Data Link Inc., NHTSA Docket 89-20-No.1-021, February 1990.

McElhaney JH, Myer BS. "Biomechanics of Cervical Trauma," in Accidental Injury: Biomechanics and Prevention, Nahum AM, Melvin JW, Editors, pp 311-361, Springer-Verlag, New York, 1993.

Prasad P, Kim A, Weerappuli DPV, Robert V, Schneider D. Relationship Between Passenger Car Seat Back Strength and Occupant Injury Severity in Rear End Collisions: Field and Laboratory Studies. SAE 973343, Society of Automotive Engineers, Warrendale, PA. 1997.

Rake B, Boehm S. Auto Seat Systems: Dangerous Safety Restraints? Trial April Issue, 80-87, 1990.

Saczalski KJ, Syson SR, Hille RA, Pozzi MC. Field accident evaluations and experimental study of seat back performance relative to rear-impact occupant protection. SAE 930346, Society of Automotive Engineers, Inc., Warrendale, PA, USA, 1993.

Saczalski KJ. Petition to Improve FMVSS 207. National Highway Traffic Safety Administration, April 18, 1989.

SAE J1460-2. Human Mechanical Impact Response Characteristics-Responses of the Human Neck to Inertial Loading by the Head for Automotive Seated Postures. Society of Automotive Engineers, Warrendale, PA, 1998.

SAE J885. Human Tolerance to Impact Conditions as Related to Motor Vehicle Design. Society of Automotive Engineers, Warrendale, PA, 1986.

Severy DM, Brink HM, Baird JD. Backrest and Head Restraint Design for Rear-end Collision Protection. SAE 680079, Society of Automotive Engineers, Warrendale, PA, 1968a.

Severy DM, Brink HM, Baird JD. Preliminary findings of head support designs. 11th Stapp Car Crash Conf, SAE 670921, SAE Warrendale PA, 337-405, 1967a.

Severy DM, Brink HM, Baird JD. Vehicle Design for Passenger Protection from High Speed Rear End Collisions. SAE 680774, Society of Automotive Engineers, Warrendale, PA, 1968b.

Severy DM, Harrison MB, Baird JD. Collision Performance, LM Safety Car. SAE 670458, Society of Automotive Engineers, Warrendale, PA. 1967b.

Stephens GD, Long TJ, Blaisdell DM. Energy Analysis of Automotive Seat Systems, SAE 2000-01-1380, SP-1494, Society of Automotive Engineers, Warrendale, PA, 2000.

Strother CE, James MB, Gordon JJ. Response of out-of-position dummies in rear impact. SAE 941055, Society of Automotive Engineers, Inc., Warrendale, PA, USA, 1994.

Strother CE, James MB. Evaluation of Seat Back Strength and Seat Belt Effectiveness in Rear End Impacts. 31st Stapp Car Crash Conference, SAE 872214, Society of Automotive Engineers, Warrendale, PA, 225-244, 1987.

Viano DC, Olsen S. The Effectiveness of Active Heads Restraint in Preventing Whiplash. Journal of Trauma, 51:959-969, 2001.

Viano DC. Role of the Seat in Rear Crash Safety. ISBN 0-7680-0847-6, Society of Automotive Engineers, Warrendale, PA, SAE R-317:1-491, 2002.

Viano DC. High Retention Seat Performance in Quasistatic Seat Tests. SAE 2003-01-0173, Society of Automotive Engineers, Warrendale, PA, 2003a.

Viano DC. Energy Transfer to an Occupant in Rear Crashes: Effect of Stiff and Yielding Seats. SAE 2003-01-0180, Society of Automotive Engineers, Warrendale, PA, 2003b.

Viano DC. Seat Properties Affecting Neck Responses in Rear Impacts: A Possible Reason Why Whiplash Has Increased. In print Traffic Injury Prevention, 2003c.

Viano DC. Seat Influences on Female Neck Responses in Rear Impacts: A Reason Why Women Have Higher Whiplash Rates. In print Traffic Injury Prevention, 2003d.

Viano DC. Effectiveness of High-Retention Seats in Preventing Fatality: Initial Results and Trends. SAE 2003-01-1351, Society of Automotive Engineers, Warrendale, PA, 2003e.

Warner C, Strother C, James MB, Decker RL. Occupant Protection in Rear-End Collisions II: The Role of Seat Back Deformation in Injury Reduction. 35th Stapp Car Crash Conf, SAE 912914, Society of Automotive Engineers, Warrendale, PA., 1991.

Yoganandan N, Pintar, FA, Haffner M, et al. Epidemiology and Injury Biomechanics of Motor Vehicle Related Trauma to the Human Spine. 33rd Stapp Car Crash Conference, SAE 892438, Society of Automotive Engineers, Warrendale, PA., 1989.

CHAPTER 3:

SEAT PROPERTIES AFFECTING NECK RESPONSES IN REAR IMPACTS: A REASON WHY WHIPLASH HAS INCREASED

Chapter 3

Seat Properties Affecting Neck Responses in Rear Crashes: A Reason Why Whiplash Has Increased

David C. Viano[1,2,3]
[1]ProBiomechanics LLC
[2]Crash Safety Division, Chalmers University of Technology
[3]Safety Integration Center, General Motors Corporation

ABSTRACT

Whiplash has increased over the past two decades. This study compares occupant dynamics with three different seat types (two yielding and one stiff) in rear crashes. Responses up to head restraint contact are used to describe possible reasons for the increase in whiplash as seat stiffness increased in the 1980s-90s.

Three exemplar seats were defined by seat stiffness (k) and frame rotation stiffness (j) under occupant load. The stiff seat had k = 40 kN/m and j = 1.8 ^{0}/kN representing a foreign benchmark. One yielding seat had k = 20 kN/m and j = 1.4 ^{0}/kN simulating a high retention seat. The other had k = 20 kN/m and j = 3.4 ^{0}/kN simulating a typical yielding seat of the 1980s-90s. Constant vehicle acceleration for 100 ms gave delta Vs of 6, 10, 16, 24 and 35 km/h. The 1D model included a torso mass loading the seatback, head motion through a flexible neck, and head restraint drop and rearward displacement with seatback rotation.

Neck displacement was greatest with the stiff seat due to higher loads on the torso. It peaked at 10 km/h rear delta V and was lower in higher severity crashes. It averaged 32% more than neck displacements with the 1980s yielding seat. The high retention seat had 67% lower neck displacements than the stiff seat because of yielding into the seatback, earlier head restraint contact and less seatback rotation, which involved 16 mm drop in head restraint height due to seatback rotation in the 16 km/h rear delta V. This was significantly lower than 47 mm with the foreign benchmark and 73 mm with the 1980s yielding seat. Early in the crash, neck responses are proportional to ky/m_T, seat stiffness times vehicle displacement divided by torso mass, so neck responses increase with seat stiffness.

The trend toward stiffer seats increased neck responses over the yielding seats of the 1980s-90s, which offers one explanation for the increase in whiplash over the past two decades. This is a result of not enough seat suspension compliance as stronger seat frames were introduced. As seat stiffness increased, so has neck displacements and NIC. High retention seats reduce neck biomechanical responses by allowing the occupant to displace into the seatback at relatively low torso loads until head restraint contact and then transferring crash energy. High retention seats resolve the historic debate between stiff (rigid) and yielding seats by providing both a strong frame (low j) for occupant retention and yielding suspension (low k) to reduce whiplash.

INTRODUCTION

Whiplash Rates Have Increased Over the Past Two Decades: Whiplash claims in rear crashes have increased worldwide over the past two decades. In Germany, Richter et al. (2000) found that whiplash disorders among injured drivers had increased from less than 10% in 1985 to more than 30% in 1997. In Sweden, Holm et al. (1999) found that whiplash-related impairment had increased from 16% in 1989 to 28% in 1994. In Japan, Ono and Kanno (1996) found that whiplash injuries have increased from 44% to 51% in car-to-car crashes from 1985 to 1991. In the U.K., Galasko et al. (1993) found that neck injuries have steadily increased in rear crashes. Over the same period of time, seats have become stronger to improve occupant retention in severe rear crashes (Viano 2003a).

The Debate Between Yielding and Stiff (Rigid) Seats: Severy et al. (1955, 1967a,b, 1968a,b) conducted pioneering research on seat and occupant dynamics in rear crashes and showed the need for controlled deformation of the seatback and support of the head. He also worked on seat concepts, including stiff (rigid) seatbacks to enhance occupant retention in severe crashes. However, the concepts were not brought to production because of various criticisms on overall performance. Other rigid seat concepts were studied by Saczalski et al. (1993) and Blaisdell et al. (1993).

Strother, James (1987), Warner et al. (1991) and Prasad et al. (1997) provided engineering analyses and rationale for the yielding seatback in rear crashes. This was argued as the best way to manage occupant energy, reduce whiplash risks and provide overall occupant safety in rear crashes. Controlled seatback rotation was used to limit forces on the neck. Energy absorbing restraint was provided by the rearward yielding of the seatback in a crash and followed the historic trend in load-limiting and energy absorbing deformation of interior components. These researchers countered the rigid seat approach and described issues of ramping, rebound and whiplash. A seatback moment near the FMVSS 207 criterion was argued as necessary to balance the needs for whiplash prevention in low-speed crashes and occupant retention in severe rear crashes. The long-standing debated furthered the notion that there was an underlying design conflict between occupant retention by stiff (rigid) seats in infrequent, severe rear crashes, and the need for a yielding seatback to prevent whiplash in the more frequent, minor rear crashes.

From 1990 through 1995, research was pursued on the role of the seat in rear crashes (Viano 2002). This addressed the yielding seat theory and showed there was no design conflict between a stiff (rigid) and yielding seat, if compliance of the seatback between the side frame structures was considered in the load-limiting performance of the seat. The study led to new seat requirements for rear impact performance and a new generation of yielding seats, called high retention seats, which have been phased into production from 1997 through 2002. These seats meet performance requirements in a Quasistatic Seat Test (QST), yield at a stiffness close to k = 20 kN/m rearward occupant loading, have a strong frame structure and have an improved head restraint placement (Viano 2002, 2003a).

Previous and Current Modeling Studies: In a recent study, Viano (2003b) developed a torso dynamic model of rear crashes, which evaluated the energy transfer to an occupant by stiff (rigid) and yielding seats in 16 and 35 km/h rear crashes. Based on benchmark tests, the stiff seatback was defined as one having k = 40 kN/m stiffness in rearward loading by a Hybrid dummy. The yielding seatback had k = 20 kN/m stiffness, which was comparable to the 1997 W car (Pontiac Grand Prix) seat, the first production high retention seat. The yielding seatback developed 15%-16% lower forces on the occupant and acceleration with a delayed loading as compared to the stiff seat. The time delay allowed energy transfer and force to be applied on the occupant after the head, neck and torso received uniform support by the seatback and head restraint. The yielding seatback provided gradual acceleration and lower loads on the occupant with greater occupant displacement with respect to the vehicle than the stiff seat.

The current study extends the previous model by introducing a head and flexible neck on the occupant's torso. This allowed an evaluation of head restraint trajectory in rear crashes, and head and neck dynamics with seatback rotation. Three seat types were evaluated. One represented a stiff seat comparable to the stiffest foreign benchmark of European and Asian luxury vehicles. They had k = 40 kN/m stiffness. Another seat represented the high retention design of the 1997 W car with a perimeter frame. It had k = 20 kN/m stiffness and a compliant seat suspension that increased the flexibility of the seat trim, foam and suspension hardware in the seatback. The third seat was typical of the 1990 C car (Buick Park Avenue) with k = 20 kN/m stiffness provided by rotation of the seatback frame. This seat is typical of yielding seats of the 1980s-90s. The current study evaluated occupant dynamics in 6-35 km/h rear crashes up to head restraint contact. It addressed a possible reason why whiplash has increased over the past two decades as seat strength and stiffness have increased.

There has been speculation that whiplash can occur before contact with the head restraint (Barnsely et al. 1994 and Yoganandan, Pintar 2000); and, the Neck Injury Criterion (NIC) proposed by the Chalmers University group addresses the early acceleration differences between T1 and the head, which promote the S-shaped response up to head contact (Bostrom et al. 1996). This study addressed torso, head and neck dynamics before head restraint contact, because the initial neck displacements and orientation of head restraint loading are a factor in the eventual neck responses and injury risks. While there is potential for whiplash before head contact that needs more evaluation, it is also important to establish a better understanding of the early phase responses that influence the overall impact biomechanics.

MEHTODOLGY

Nomenclature for Analysis:

- d Neck displacement, d = x - z
- E Energy Transfer to Occupant by the Seat in J
- ΔE Incremental Energy Transfer to the Occupant per Δt in J
- F Force on Occupant's Torso in kN
- G Force on Neck in kN
- HR High Retention seats meeting the 1995 specifications for seat strength in QST and head restraint placement.
- i Time integer
- j Slope of Occupant Load versus Seatback Frame Angle in ^{0}/kN
- k Stiffness of the Seat due to Occupant Loading in kN/m
- k_{HR} Stiffness of the Head Restraint in kN/m
- k_N Stiffness of Neck in kN/m
- m Effective Mass of the Occupant Loading the Seat in kg
- m_T Mass of the Torso in kg
- m_H Mass of the Head-Neck in kg

p Length of the Seatback from the Pivot to the Top of Head Restraint in m.

QST Quasistatic Seat Test

r Gap due to Seatback Rotation at the Center of Pressure of Occupant Loading in m

s Actual Head Restraint Gap w/r moving Vehicle and Head in m

SAHR Self Aligning Head Restraint or Saab Active Head Restraint.

Δt Delta time in s

u Drop in Head Restraint Height Due to Seatback Rotation in m

v Rearward Displacement of Head Restraint Due to Seatback Rotation in m

w Head Restraint Gap in m

x Horizontal Displacement of the Torso in m

$\dot{x}$ Velocity of the Torso in m/s

$\ddot{x}$ Acceleration of the Torso in m/s^2

y Horizontal Displacement of the Vehicle in m

$\dot{y}$ Velocity of the Vehicle in m/s

$\ddot{y}$ Acceleration of the Vehicle in m/s^2

z Horizontal Head Displacement in m

$\dot{z}$ Head Velocity in m/s

$\ddot{z}$ Head acceleration in m/s^2

θ Seatback Angle w/r Vertical in deg

Analysis Procedures

Vehicle Dynamics: Acceleration of the vehicle is assumed to be prescribed in equal time steps, so that a spreadsheet calculation can be made of the vehicle and occupant dynamics. This part of the analysis follows Viano (2003b). The 1D model is shown in Figure 1. Given the acceleration function for the vehicle, it is straightforward to determine the velocity at any time i by the relationship:

$$\dot{y}_i = \dot{y}_{i-1} + \ddot{y}_i \Delta t \quad (1)$$

where Δt is the incremental time step for the calculation, $\dot{y}_0 = \ddot{y}_0 = 0$ and i ranges from 0 to 100 giving a solution for 200 ms with a time step of $\Delta t = 2$ ms. Based on Equation (1), the displacement of the vehicle can be determined from:

$$y_i = y_{i-1} + \dot{y}_i \Delta t \quad (2)$$

Occupant Dynamics: The occupant is displaced forward in a rear crash by the motion of the vehicle and deformation of the seat during the impact. Force develops on the occupant's torso by the stiffness of the seat and relative displacement of the torso with respect to the vehicle:

$$F_i = k(y_i - x_i) \quad (3)$$

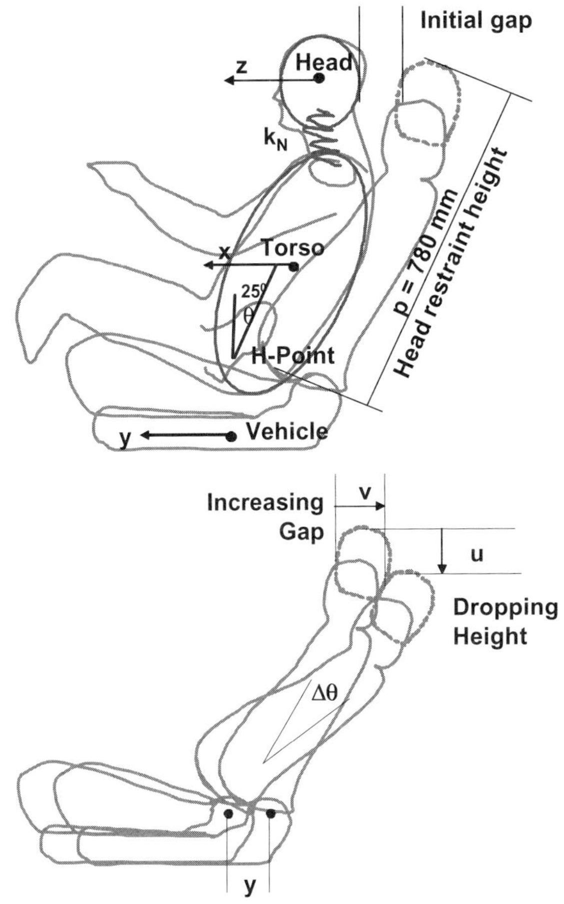

Figure 1: Neck moment and extension angle showing SAE biomechanical corridors for the 50% male, experimental data on the Hybrid III neck and idealized neck stiffness (line through the triangle). The conversion of moment to force and angle to displacement is also shown.

Force on the occupant increases with seat deformation, which accelerates the occupant in proportion to the effective mass of the torso loaded by the seatback. Based on sled testing with the Hybrid III dummy, Viano (2002) determined that the effective mass (m) loaded by the seatback was 70% of the total dummy mass, or 52.5 kg for the 50th percentile Hybrid III dummy. The mass of the head-neck is assumed to be 5.5 kg, so the torso mass is 47.0 kg. The effective head mass is based on 4.5 kg for the head plus two-thirds of the neck mass or 1.0 kg (Backaitis, Mertz 1994). The acceleration of the occupant's torso is:

$$\ddot{x}_i = F_i / m_T = F_i / 47.0 \quad (4)$$

where Equation (4) assumes that the eventual load on the neck has a second order effect on the calculation and there are no inertial effects of the seatback mass. The occupant's torso acceleration can be used to determine the torso velocity and displacement:

$$\dot{x}_i = \dot{x}_{i-1} + \ddot{x}_{i-1} \Delta t \quad (5)$$

$$x_i = x_{i-1} + \dot{x}_{i-1}\Delta t \quad (6)$$

where $\ddot{x}_0 = \dot{x}_0 = x_0 = 0$.

The incremental energy transfer to the occupant can be determined by the average load and incremental displacement of the occupant with respect to ground:

$$\Delta E_i = 0.5(F_i + F_{i-1})(x_i - x_{i-1}) + 0.5(G_i + G_{i-1})(z_i - z_{i-1}) \quad (7)$$

where $\Delta E_0 = 0$, and the second term is the incremental energy transfer from the head-neck, which will be described later. The cumulative energy transfer is a function of time and is given by:

$$E_i = E_{i-1} + \Delta E_i \quad (8)$$

where $E_0 = 0$.

Head Restraint Trajectory: The relationship between applied torso load and seatback rotation is determined in QSTs. It is generally linear for seats up to the yield load, so the seatback angle is given by:

$$\theta_i = \theta_0 + jF_i \quad (9)$$

where $\theta_0 = 25^0$ for typical seats and j is the slope of the line relating occupant load to the frame angle of the seatback. As the occupant loads the seatback and it rotates, there is a downward and rearward displacement of the head restraint. Figure 1 shows this effect, which is given by:

$$u_i = p[\cos(\theta_i) - \cos(\theta_0)] \quad (10)$$
$$v_i = p[\sin(\theta_i) - \sin(\theta_0)] \quad (11)$$

where p = 780 mm is used to represent the head restraint height up the seatback of modern seats. This value will be used for the analyses of the 1990 C car seat even though the actual head restraint was lower for the production seat. This allows a direct comparison of the different seat stiffness types. The gap to the head restraint varies with the initial gap behind the head and seatback angle changes. It can be determined from the previous relationships with:

$$w_i = w_0 + v_i \quad (12)$$

where $w_o = 80$ mm is the initial gap and is assumed typical of modern head restraint placements. This relationship does not address the effects of head-neck displacement in the crash. The actual gap between the back of occupant head and the head restraint is a function of the head and vehicle displacement, and other geometric relationships defining the head restraint position with seatback rotation. It is given by:

$$s_i = w_i - (y_i - z_i) \quad (13)$$

Seat Properties: Seat Stiffness (k) and Frame Rotation Stiffness (j) Under Occupant Load: Seat stiffness (k) and frame rotation stiffness (j) are determined in QSTs (Quasistatic Seat Tests) between occupant loads of 1 kN to 5 kN (Viano 2002, 2003a). This covers the majority of loading up to head restraint contact in rear crashes associated with whiplash. The benchmark average of 14 foreign seats from European and Asian manufacturers was k = 34.2 ± 6.5 kN/m with a range from k = 26-47 kN/m stiffness. The average of the five stiffest seats was k = 40.9 kN/m, so the foreign benchmark was exemplified with k = 40 kN/m. Figure 2 shows the stiff, foreign benchmark seat with k = 40 kN/m and j = 1.8 ^{0}/kN.

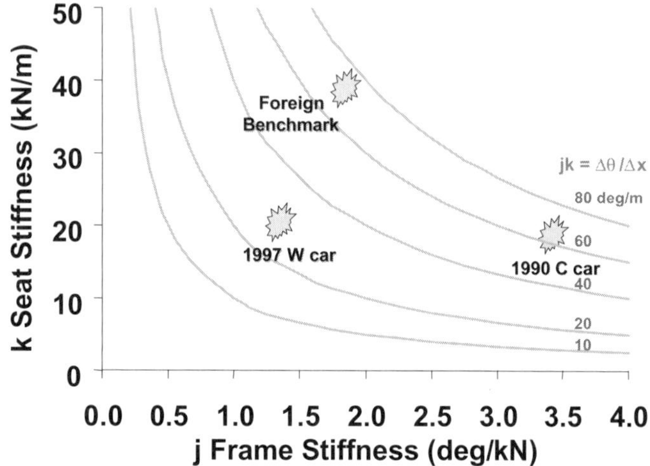

Figure 2: Comparison of seat stiffness (k) and frame rotation stiffness (j) at 5 kN load for three types of production seats, a foreign benchmark of European and Asian luxury seats, 1997 W car (Pontiac Grand Prix) and 1990 C car (Buick Park Avenue). The star shows the general characteristic of these seats, but is not intended to give the range in values.

The 1990 C car (Buick Park Avenue) seat has an average seat stiffness of k = 21.1 ± 2.4 kN/m and is typical of yielding seats of the 1980s-90s. Yield is largely due to seatback rotation. The 1980s-90s yielding seat in this analysis was assumed to have k = 20 kN/m and j = 3.4 ^{0}/kN. This is a relatively high j factor indicating yielding by seatback rotation.

The average stiffness of the 1997 W (Pontiac Grand Prix) and 1997 P-90 (Chevrolet Malibu) was k = 22.1 ± 5.4 kN/m. The 1997 W and P-90 were the first high retention seats that used a strong seat frame and recliners for occupant retention in severe rear crashes. This limited seatback rotation. They also had an open seatback providing compliance or yielding by deformation of the seat trim and an energy absorbing pelvic strap to pocket the occupant's pelvis and lower back in a rear crash (Viano 2002, 2003a). The high retention seat was exemplified with k = 20 kN/m and j = 1.4 ^{0}/kN. Three exemplar seats were used in this analysis and followed an earlier study (Viano 2003b). For the exemplar seats, the initial change in seatback angle for occupant displacement into the seat is given by:

$$\Delta\theta = jk\Delta x \tag{14}$$

where $\Delta\theta$ is the change in seatback frame angle and $\Delta x = y - x$. The foreign benchmark has $\Delta\theta / \Delta x$ = 72 ^{0}/m based on QSTs. The 1990 C car seat had $\Delta\theta / \Delta x$ = 68 ^{0}/m and the 1997 W car seat had $\Delta\theta / \Delta x$ = 28 ^{0}/m. The foreign benchmark and 1997 W car seats are assumed to have linear stiffness up to a peak force of 10 kN and beyond that deformation, the load is held constant at 10 kN. The 1990 C car seat has a linear stiffness up to 5 kN and beyond that deformation, the load is held constant at 5 kN. The nominal peak bending moment of the first two seats is 2200 Nm (10 kN with a 0.22 m moment arm) and the third seat has a 1100 Nm. When load on the torso decreased in the simulation, the seatback angle was held constant at the maximum rotation angle. For seat rotations in the constant load limit condition, the change in seatback angle was determined by the change in occupant displacement converted to a seatback angle at the height of the center of pressure. This ensured equilibrium of motion during the load limiting phase.

Head-Neck Dynamics: Force develops on the neck due to differential motion of the torso and head. It is related to neck stiffness due to a combination of bending, tension and shear:

$$G_i = k_N(x_i - z_i) \tag{15}$$

Force on the neck increases with the relative displacement between the head and torso, which accelerates the head in proportion to the 5.5 kg effective mass of the head and neck:

$$\ddot{z}_i = G_i / m_H = G_i / 5.5 \tag{16}$$

Head acceleration can be used to determine the velocity and displacement of the head with respect to ground:

$$\dot{z}_i = \dot{z}_{i-1} + \ddot{z}_{i-1}\Delta t \tag{17}$$

$$z_i = z_{i-1} + \dot{z}_{i-1}\Delta t \tag{18}$$

where $\ddot{z}_0 = \dot{z}_0 = z_0 = 0$. Displacement of the head under load adds a second term to the energy transfer to the occupant in Equation (7).

After head restraint contact, there is an additional force on the head in proportion to its penetration into the head restraint. A linear relationship is assumed for current head restraints. For this condition, the force on the head in Equation (15) becomes:

$$G_i = k_N(x_i - z_i) - k_{HR} s_i \tag{19}$$

Based on subsystem testing by Viano (2002, Chapter 13), a 50 mm penetration of the back of the Hybrid III head typically causes 2.8 kN load giving a head restraint stiffness of 80 kN/m. However, this test involves rigidly supported head restraint posts so a value of $k_{HR} = 20$ kN/m was used to reflect the stiffness installed on the seatback.

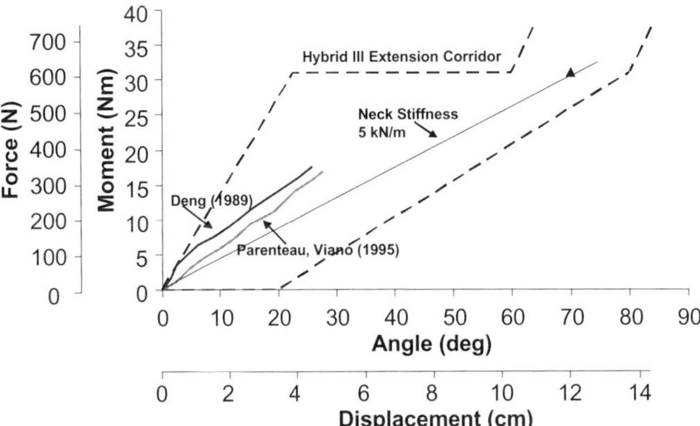

Figure 3: Neck moment and extension angle showing SAE biomechanical corridors for the 50% male, experimental data on the Hybrid III neck and idealized neck stiffness (line through the triangle). The conversion of moment to force and angle to displacement is also shown.

Effective Neck Stiffness: Figure 3 shows the neck extension corridors based on human volunteer responses (Mertz, Patrick 1971, SAE J1460-2, 1998). Moment-angle data from two studies of the Hybrid III neck are also shown for reference (Deng 1989, Parenteau, Viano 1995). In this analysis, the 50% male neck response is approximated by a straight line from zero through 31 Nm and 70^0, which is slightly softer than the Hybrid III neck but fits within the corridor.

The moment-angle was converted to force-displacement using dimensions for the Hybrid III head and neck and published response data. The moment about the occipital condyles was converted to force at the head center of gravity by using the 5 cm distance between the head cg (center of gravity) and the occipital condyles (Mertz et al. 1989, Backaitis, Mertz 1994). This gives a load of 620 N for 31 Nm moment. The extension angle of 70^0 involves an x-displacement of the head cg of approximately 12 cm. This is based on data from Mertz et al. (1973), which showed a 6.4 cm x-displacement of the head cg at a neck extension angle of 40^0 and 15.7 cm at 78^0. A conversion of 5.8^0/cm was used reflecting more of the lower extension behavior. These conversions give a neck stiffness of 5 kN/m in extension due to bending, shear and tension.

Model Validation: A series of rear sled tests has been conducted with the 1990 C car seat using the 50% male Hybrid III dummy (Viano 2002). Figure 4 shows the averages for repeated tests with the C car seat at 13 and 19 km/h (8 and 12 mph) rear delta V. Peak chest acceleration, seatback rotation and neck extension are shown. The average results from the model are also shown for the yielding seat with k = 20 kN/m and j = 3.4 ^{0}/kN. Peak chest acceleration occurred before head restraint contact, and the comparison between the experiment and model is quite good. The maximum seatback rotation and neck extension occurred after head restraint contact. The agreement is also reasonable.

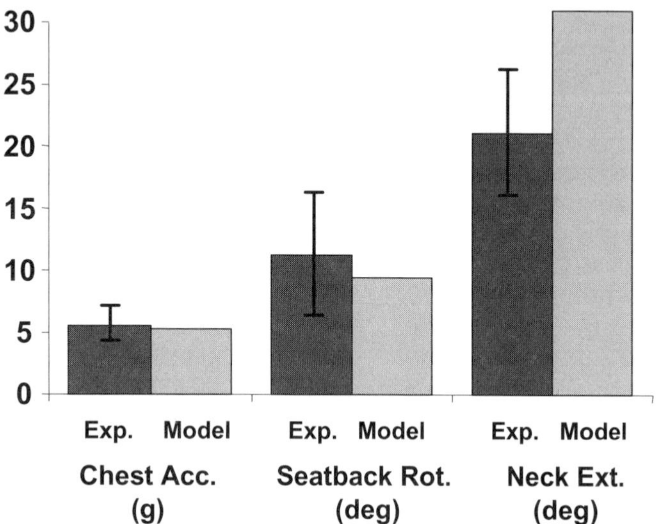

Figure 4: Comparison of peak chest acceleration, seatback rotation and neck extension for rear sled tests with the 50% Hybrid III male dummy in the 1990 C car type seat with model results for the yielding seat with k = 20 kN/m and j = 3.4 0/kN. The results are averages for rear delta Vs of 10-20 km/h and the bars represent one standard deviation in response.

Seat and Occupant Biomechanical Responses Up to Head Contact in 6, 10, 16, 24 and 35 km/h Rear Crashes with Three Different Seat Types: Five different crash pulses were used to determine the head restraint position at head contact and occupant dynamics. This involved a 100 ms crash pulse with constant accelerations of 1.6, 3.0, 4.5, 6.7 and 10 g to reach the increasing delta Vs. The calculations were done in Excel. The solution was determined for increasing time steps until the occupant's head contacted the head restraint. This was determined when Equation (13) passed through zero, indicating that the forward motion of the occupant and head restraint had closed the initial 80 mm head-restraint gap and any subsequent gap caused by seatback rotation. Head contact occurred in all rear crashes, except for two of the 6 km/h conditions, indicated by nc (no contact). After head restraint contact, an additional force was introduced to push the head forward due to head restraint loading.

Differences in Neck Responses with Stiff and Yielding Seats: The determination of injury tolerances for whiplash is formidable because there is no consensus on the underlying mechanism, appropriate criteria or tolerance levels. The initial neck response differences were investigated using the underlying biomechanics of the seat loading and the seat properties. This involved the equilibrium Equations (3-14) to estimate response ratios with different seat types, occupants and vehicles.

RESULTS

Occupant Dynamics in a 6 km/h Rear Crash Delta V: Figure 5 shows the vehicle and occupant dynamics in a 6 km/h rear crash with a 1990 C car seat. In this crash, the head does not contact the head restraint. The results are similar for the 1997 W car; however, head contact occurs at 134 ms. The plots also include the relative acceleration, velocity and displacement between the head and torso, and the relative displacement between the torso and vehicle.

The dynamics of the crash will be described by the three plots. First, the vehicle is accelerated forward, which pushes the seat forward. This introduces a relative displacement between the occupant and the seat (vehicle), and develops force in proportion to seat stiffness. This force accelerates the occupant's torso forward causing a delayed displacement with respect to the vehicle. There is a secondary effect with torso displacement as it stretches the neck and eventually accelerates the head. This involves relative velocity and displacement between the head and neck. The motion sequence for the occupant involves early peak head-torso acceleration at 76 ms, a later peak relative head-torso velocity of 0.58 m/s at 114 ms, and then a peak neck displacement of 45.5 mm at 142 ms. With the 1990 C car seat, the maximum seatback rotation was 4.6^0 with a 28.7 mm drop in head restraint height and a 55 mm additional gap due to seatback rotation. The peak load was 1.34 kN. The 1997 W car seat had a 1.9^0 seatback rotation with a 11.3 mm drop in head restraint height and a 23 mm additional gap with a 1.34 kN occupant load (Table 1). The changes in line slope of some responses are associated with the simplifying assumptions in the model.

16 km/h Rear Crash: Selected time histories are shown in Figure 6 for the 16 km/h severity, since this condition reflects the prevailing consensus for evaluation of whiplash risks in rear crashes (Langwieder, Hell 2002, Hell et al. 2002). First, the 1990 C car seat experiences a 10.6^0 angle change of the seatback at head contact. The seatback rotation increases the initial 80 mm gap by 125 mm, and delays head restraint contact until 100 ms. There is also a 72.8 mm drop in height of the head restraint due to seatback rotation. The yielding behavior of the seat limits load on the occupant, which results in a neck displacement of 48.9 mm at head restraint contact. There is a peak head-torso relative acceleration of 2.27 g and velocity of 1.44 m/s with a head impact velocity of 1.80 m/s. Since the torso load continues to rotate the seatback near head contact, the trajectory of the head restraint is -20.1^0 at head contact.

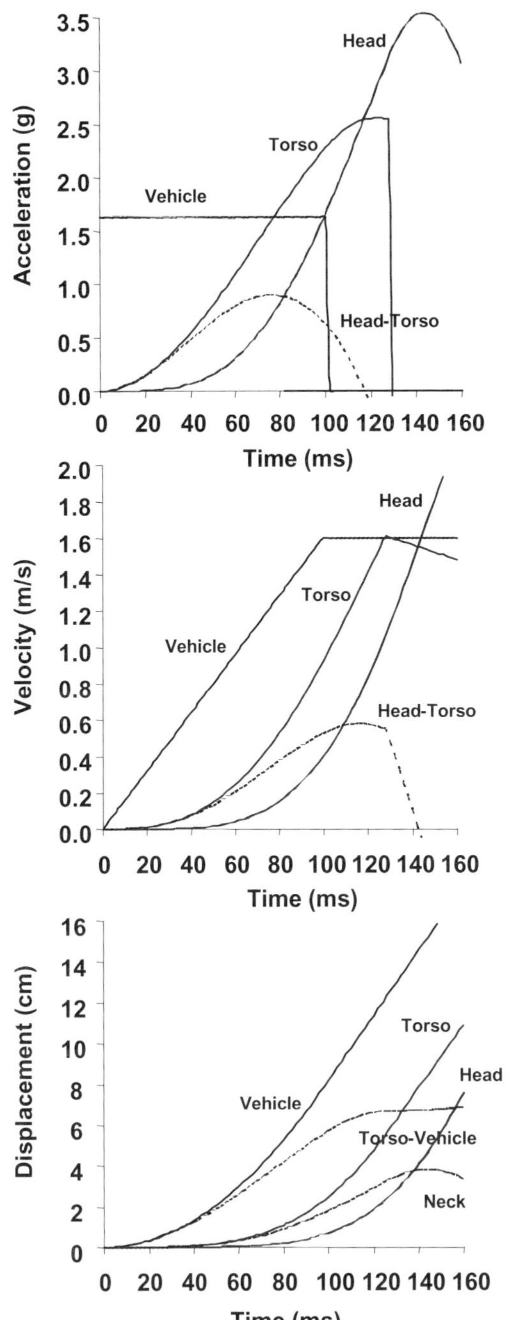

Figure 5: Vehicle and occupant biomechanical responses for a 6 km/h rear crash delta V with the 1990 C car seat.

The foreign benchmark seat has a much stronger frame and stiffness than the 1990 C car, and it builds up higher loads on the occupant. This results in a 4.0 kN torso load and a seatback rotation of 7.3^0 at 92 ms at head restraint contact. It involves a 47.4 mm drop in head restraint height and increased gap of 85.5 mm prior to head contact, although the head restraint trajectory is only -6.9^0. The higher torso load and relatively late contact result in a 66.3 mm neck displacement at head contact. There was also a relative head-torso acceleration of 3.76 g and velocity of 2.02 m/s, all higher than with the 1990 C car.

In contrast to the previous seats, the 1997 W car seat allows the torso to penetrate into the seatback at a relatively lower load of 1.9 kN, which reduces seatback rotation to only 2.7^0 with a 16.4 mm head restraint drop and 33.1 mm additional gap. Most importantly, the time of head restraint contact is lowered to 72 ms as a result of a substantially lower induced gap by seatback rotation and lower torso accelerations. The lower torso loads also reduce neck responses. The neck displacement is only 17.1 mm with a relative head-torso acceleration of 2.42 g and velocity of 0.82 m/s. The head impact velocity is 2.16 m/s; however, the neck loads and accelerations are reduced at head contact.

Figure 6 shows that over half (278 J) of the kinetic energy is transferred to the occupant prior to head contact and support from the head restraint with the foreign benchmark seat. This increases loads and deformation of the neck early in the crash and aggravates potential whiplash injury. In contrast, the 1997 W car transfers only 27 J, 10% of that of the foreign benchmark seat and 5% of the occupant's crash energy prior to head contact. This means that the majority of energy is transferred after the head receives support from the head restraint. The 1990 C car response is in between the other two seats.

The occupant responses with the 1990 C car seat follow those of the 1997 W car to a point in the right hand plots of Figure 6, but the head contact is much later with the C car seat, which extends the responses considerably. The 1990 C car has greater seatback rotation since that was the means of reducing torso loads and neck biomechanical responses. However, the initial change in seatback angle is similar to the foreign benchmark. The lowest seatback rotation is with the 1997 W car seat, which coupled with greater suspension compliance involves lower loads on the upper back and neck. The foreign benchmark seat has much higher neck loads than the 1997 W car, because of higher torso loads. Torso velocity with respect to the vehicle peaks early, at about 50 ms, with the foreign benchmark seat and then decreases until head restraint contact. The opposite occurs with the 1997 W car seat as head restraint contact occurs just after the peak value, however the velocity and associated accelerations are far below human tolerances and are not a factor related to injury.

Biomechanical Responses Up to Head Restraint Contact: Figure 7 shows peak biomechanical responses for the three exemplar seats and five severities of rear crash. Neck displacements are greatest with the foreign benchmark seat and increase with decreasing crash severity to a peak at about 10 km/h rear delta V.

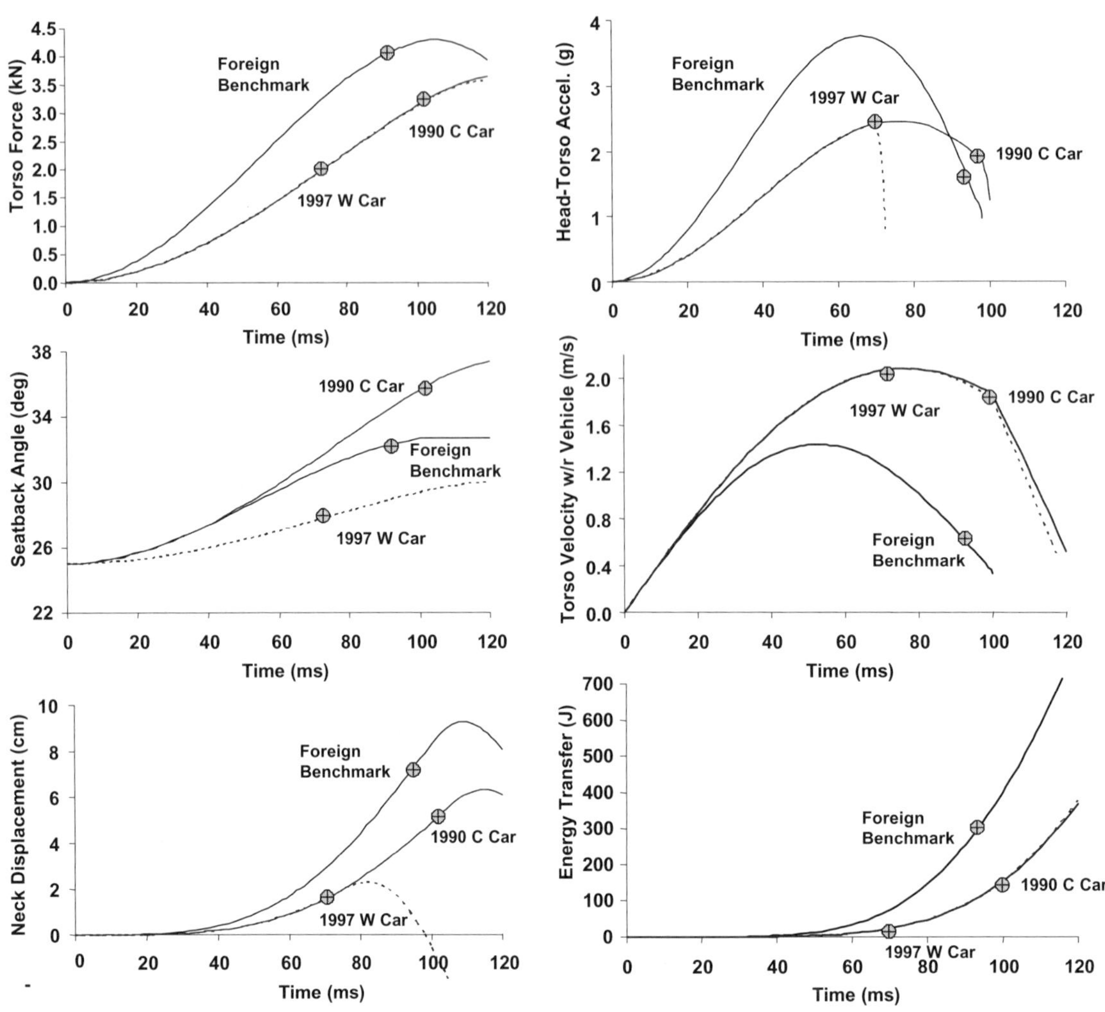

Figure 6: Selection of seat and biomechanical responses in the 16 km/h rear crash with the three exemplar seat types. The circles show the time of head restraint contact for each seat type and response. The dots on each trace represent the time of head restraint contact.

This reflects the greater duration of loading prior to head restraint contact in the lower speed crashes, and the higher loads on the torso than with the other seats. The higher neck displacements are a result of the stiffer seatback pushing the occupant forward. The 1990 C car seat shows the same trend as the foreign benchmark seat, but it has a consistently lower neck displacement, about 26% lower on average due to its yielding behavior. In contrast, the 1997 W car has the lowest neck displacements reflecting the greater seatback compliance and lower loads on the torso prior to head restraint contact. Neck displacements also increase with decreasing crash delta V. The average reduction in neck displacement compared with the foreign benchmark seat is 66%.

The peak relative velocity of the head with respect to the torso prior to or at head restraint contact increases with crash severity reflecting the occupant acceleration forward and reaction loads on the neck. The foreign benchmark seat has the greatest relative head-torso velocity followed by the 1990 C car seat. The 1997 W car has a plateau in velocity between the 10-16 km/h severity range with slightly lower responses at higher and lower crash severities.

The relative acceleration between the head and torso peaks early in the crash. For the 6-24 km/h crashes, each seat increases peak head-torso acceleration with higher rear delta V, however the stiff seat has higher values than the two yielding seats. NIC shows the same trend as the relative head-torso acceleration, since the

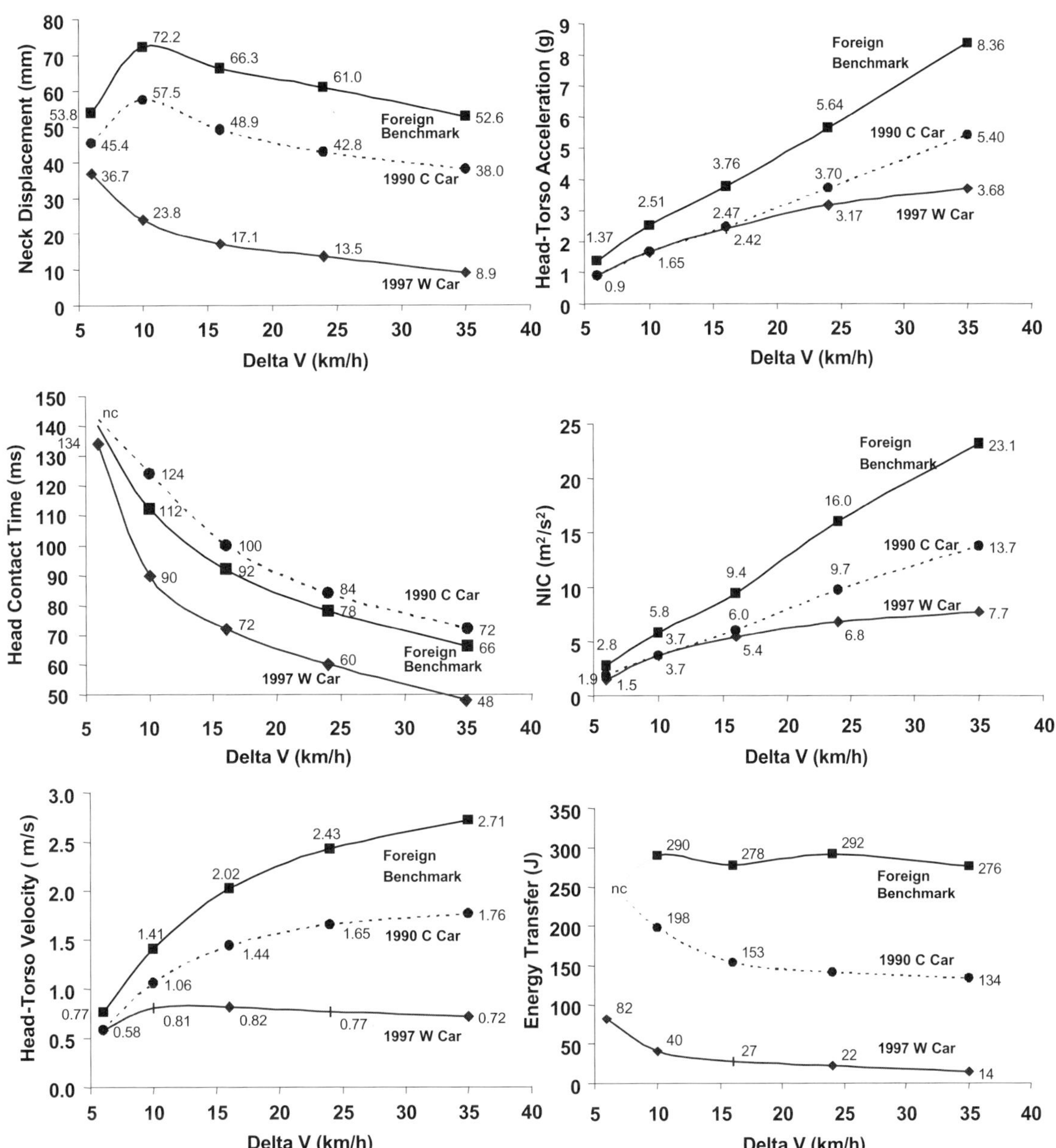

Figure 7: Peak biomechanical responses up to head restraint contact as a function of rear crash delta V for the three exemplar seat types, including neck displacement (top left), head restraint contact time (middle left), head-torso velocity (bottom left), head-torso acceleration (top right), NIC (middle right), and energy transfer (bottom right). No head restraint contact is indicated by nc.

velocity squared term has a small effect on the calculation. Finally, the foreign benchmark seat transfers considerably more energy to the occupant before head restraint contact than the yielding seats. The yielding seats transfer crash energy after the head, neck and torso are supported. Table 1 summarizes other responses up to head contact that are not included in Figure 7.

Head Restraint Trajectory: Figure 8 shows the head restraint trajectory for 10, 16 and 35 km/h rear crash delta Vs. The trajectory of the foreign benchmark and 1990 C car seats are interlaced by crash severity, even though there is a difference in seat strength as seat stiffness is an additional factor. The initial seatback rotation is proportional to jk. The forward displacement is with respect to ground or an off-board view of the crash. This includes the rearward displacement of the head restraint due to seatback rotation and the forward displacement of seat cushion. The foreign benchmark seat involves much higher occupant loads, which increase seatback rotation and reduce the height of the head restraint. This increases neck responses with the stiffer seat.

Table 1: Peak Seat and Occupant Biomechanical Responses
At or Before Head Restraint Contact in a 6, 10, and 16, 24 and 35 km/h Rear Delta V Crashes.

	1997 W car	**Foreign**	**1990 C car**
k Seat Stiffness (kN/m)	20	40	20
j Frame Stiffness (deg/kN)	1.4	1.8	3.4
jk (deg/m)	28	72	68
Yield Load (kN)	10	10	5
1.6 g / 6 km/h			
Head-Vehicle Vel. (m/s)	1.20	1.00	1.20
Head Impact Vel. (m/s)	0.96	0.71	0.62
Torso-Vehicle Displ. (mm)	67	39	67
Seatback Rot. (deg)	1.9	2.8	4.6
Torso Load (kN)	1.3	1.6	1.3
Head Restraint Drop (mm)	-11.3	-16.9	-28.7
HR Trajectory (deg)	0.0	-	0.0
3.0 g / 10 km/h			
Head-Vehicle Vel. (m/s)	2.16	1.83	2.21
Head Impact Vel. (m/s)	1.68	1.45	1.17
Torso-Vehicle Displ. (mm)	91	70	123
Seatback Rot. (deg)	2.5	5.1	8.3
Torso Load (kN)	1.8	2.8	2.5
Head Restraint Drop (mm)	-15.3	-32.1	-55.2
HR Trajectory (deg)	-6.5	0	-2.8
4.5 g / 16 km/h			
Head-Vehicle Vel. (m/s)	2.89	2.75	3.30
Head Impact Vel. (m/s)	2.16	1.98	1.80
Torso-Vehicle Displ. (mm)	97	101	156
Seatback Rot. (deg)	2.7	7.3	10.6
Torso Load (kN)	1.9	4.0	3.1
Head Restraint Drop (mm)	-16.4	-47.4	-72.8
HR Trajectory (deg)	-8.8	-6.9	-20.1
6.7 g / 24 km/h			
Head-Vehicle Vel. (m/s)	3.71	4.08	4.74
Head Impact Vel. (m/s)	2.69	2.51	2.36
Torso-Vehicle Displ. (mm)	108	131	185
Seatback Rot. (deg)	3.0	9.2	12.5
Torso Load (kN)	2.2	5.2	3.7
Head Restraint Drop (mm)	-18.3	-61.5	-88.4
HR Trajectory (deg)	-10.5	-16.0	-29.8
10 g / 35 km/h			
Head-Vehicle Vel. (m/s)	4.61	5.67	6.38
Head Impact Vel. (m/s)	3.27	3.07	2.95
Torso-Vehicle Displ. (mm)	108	159	216
Seatback Rot. (deg)	3.0	11.0	14.7
Torso Load (kN)	2.2	6.4	4.3
Head Restraint Drop (mm)	-18.4	-75.8	-106.6
HR Trajectory (deg)	-11.9	-25.9	-37.8

In contrast, the 1997 W car seat has a much lower drop in head restraint height due to greater seat suspension compliance and lower loads on the occupant prior to head restraint contact. Also, the responses are clustered closer together demonstrating a more consistent response irrespective of crash severity. The foreign benchmark seat has nearly an 80 mm drop in head restraint height at head contact in the 35 km/h rear crash. This substantially lowers the effective height of the head restraint from the initial position. The 1997 W car seat has less than 20 mm drop.

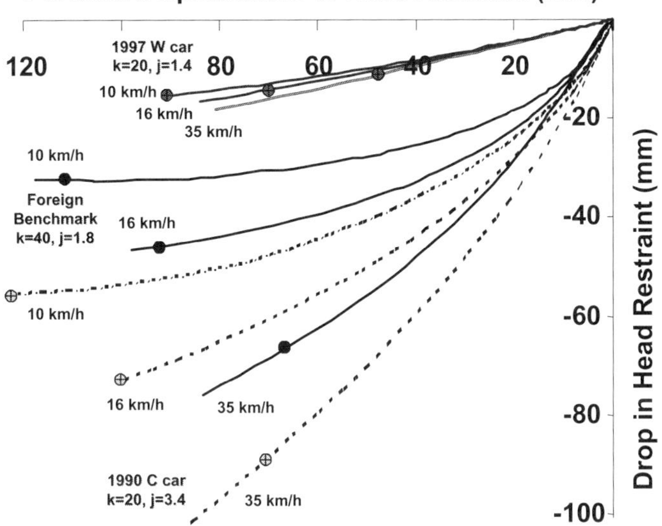

Figure 8: Trajectory of the head restraint for the three exemplar seats and three rear crashes. The data are from an inertial perspective or an off-board view of the rear crash. The dots on each trace represent the time of head restraint contact.

Implications of the Model - Differences in Neck Responses with Stiff and Yielding Seats: Figures 5 and 6 show that vehicle displacement occurs first in the dynamic response, and the neck response is initially greater with the stiff seat than the 1980s yielding seat. This is true for neck displacement, and the relative velocity and acceleration between the head and torso, which derive from the difference in head-torso acceleration due primarily to torso displacement. Based on Equations (3), (4), (15) and (16), head-torso acceleration is:

$$(\ddot{x}-\ddot{z}) = (k/m_T)y - (k/m_T + k_N/m_H)x + (k_N/m_H)z \quad (20)$$

Early in the seat loading the z displacement can be ignored and the relative acceleration depends primarily on the vehicle displacement, as torso displacement is also a second order influence. Equation (20) can be sequentially approximated as:

$$(\ddot{x}-\ddot{z}) = (k/m_T)y - (k/m_T + k_N/m_H)x \quad (21)$$

$$(\ddot{x}-\ddot{z}) = (k/m_T)y \quad (22)$$

Figure 5 shows that torso displacement is delayed from vehicle displacement, which influences the early neck responses. The ratio in head-torso acceleration for occupants in vehicles with stiff (subscript s) and yielding (subscript y) seats can be approximated by:

$$(\ddot{x}_S - \ddot{z}_S)/(\ddot{x}_Y - \ddot{z}_Y) = [(k_S/m_{TS})/(k_Y/m_{TY})]y_S/y_Y \quad (23)$$

where head-torso acceleration depends on the ratio $(k_S/m_{TS})/(k_Y/m_{TY})$, seat stiffness divided by torso mass for the stiff and yielding seats and vehicle displacement is a function of time. Assuming no shift in torso mass for the vehicles with stiff and yielding seats, the head-torso acceleration ratio is proportional to:

$$(\ddot{x}_S - \ddot{z}_S)/(\ddot{x}_Y - \ddot{z}_Y) = k_S y_S / k_Y y_Y \quad (24)$$

If the vehicles with stiff and yielding seats have a similar crash pulse, the ratio further reduces to k_S/k_Y. This indicates that the initial difference in head-torso acceleration depends on the ratio of seat stiffness.

While there is uncertainty about when whiplash occurs in a crash, there are three possibilities. First, injury may occur early in the crash up to head restraint contact when the greatest relative acceleration differences occur between the torso and head. Torso loading of the seat is critical to whiplash up to head restraint contact. The second phase involves head loading of the head restraint. This depends on the velocity of head impact, head restraint properties, neck stiffness and the trajectory of head restraint with seat displacement forward. In the third phase, rebound occurs. This involves the torso rebounding from the seat and the head from the head restraint where phasing is a factor.

NIC is determined by the head-torso acceleration early in the crash plus the square of head-torso velocity. Equation (22) gives the main contributor to NIC, which is proportional to seat stiffness, vehicle displacement and inversely torso mass. Figure 7 shows the NIC response differences are influenced by seat stiffness. Also during the early loading, there are differences in relative velocity and displacement between the head and torso, which are derived by integration these equations. The velocity difference between the torso and head, and neck displacement (d) can be determined from Equation (22) by single and double integration:

$$\int(\ddot{x}-\ddot{z}) = (k/m_T)\int y \quad (25)$$

$$d = \iint(\ddot{x}-\ddot{z}) = (k/m_T)\iint y \quad (26)$$

Neck displacement initially depends on the second integral of vehicle displacement, seat stiffness and inversely torso mass. Assuming a constant vehicle acceleration, neck displacement depends on time to the fourth power. Figure 7 shows delay in head restraint contact for the lower severity rear crashes, which gives more time to develop neck displacement. Neck displacements are lower in the high speed impacts because of the shorter time to head contact.

The ratio of neck displacement for stiff and yielding seats is:

$$d_S / d_Y = [(k_S / m_{TS})/(k_Y / m_{TY})]\iint y_S / \iint y_Y \qquad (27)$$

If the torso mass is similar for occupants in stiff and yielding seat vehicles, Equation (27) reduces to:

$$d_S / d_Y = [k_S / k_Y]\iint y_S / \iint y_Y \qquad (28)$$

where the ratio depends on the second integral of vehicle displacement. Assuming constant vehicle acceleration, Equation (28) depends on the ratio of crash pulse amplitude and seat stiffness. If the crash pulses are similar for vehicles with stiff and yielding seats, the ratio further reduces to:

$$d_S / d_Y = k_S / k_Y \qquad (29)$$

This is the ratio of seat stiffness. The shift to stiffer seats increased neck displacements by a factor of two based on the exemplar seats used in this analysis, so the phase-in of stiffer seats could be partly responsible for the higher rates of whiplash seen over the past two decades. There has been a gradual shift to stiffer seats during that time frame.

DISCUSSION

One Reason Why Whiplash Rates Have Increased Over the Past Decades: Based on the seat testing by Severy, Strother and Warner (summarized in Warner et al. 1991), there has been an increase in seat strength for rearward occupant loading. The average strength of seats from the 1960s was 2.6 ± 0.6 kN based on body block loading up to 45° seatback angle. In similar seat teats from the 1970s and 1980s, the strength increased to 3.2 ± 1.2 kN, a 23% increase over the 1960s seats. The QST tests by Viano (2002) indicated that seats of the 1980s and early 1990s supported even higher loads. Domestic (Ford and Chrysler) seats supported a torso load of 5.8 ± 1.2 kN and foreign seats 7.7 ± 2.0 kN. While the loading interface in the QST is a Hybrid dummy, the results are comparable to the earlier body block tests. The 1980s and early 1990s seats represent a 2.2-times increase in domestic seat strength versus the average 1960s seat, and a 3.0-times increase for the average foreign manufacturer's seats. The luxury foreign benchmark performance is even higher. Similar increases occurred in the peak FMVSS 207 rearward load and moment. Seat stiffness increased with seat strength. The 1970s-90s yielding seats had k = 20 kN/m, while the stronger foreign benchmark seats had k = 40 kN/m.

The stiffer foreign benchmark seat resists seatback rotation by occupant loading. However, it also builds up greater force on the occupant, which causes more seatback rotation. This reduces the overall benefit of the stronger seat frame. There is also a significantly greater torso acceleration and load at the base of the neck with the foreign benchmark seat. The neck displacement responses in Figure 7 and relative velocity between the head and torso clearly demonstrate how whiplash risk may have increased by shifts from the yielding seats of the 1980s to the stiffer (stronger) seat frames represented by the exemplar foreign benchmark seat. These data suggest a 32% increase in the key neck biomechanical responses with the shift from the 1980s-90s yielding seats to the foreign benchmark over a range in rear crash severities. These results seem to validate the concerns raised by Warner et al. (1991) and coworkers that stronger seats may increase whiplash.

Figure 8 shows the head restraint trajectory of the foreign benchmark seat is more like the yielding seats of the 1980s-90s because of similar seatback rotation. This is a consequence of the initial seatback rotation being proportional to the product of jk. The foreign benchmark has jk = 72 °/m and the 1980s yielding seats jk = 68 °/m, essentially similar values. The high retention seat has jk = 28 °/m. Since seat rotation depends on the forward displacement of the vehicle in a rear crash, the foreign benchmark rotates like the 1980s-90s yielding seats and does not benefit from the stronger frame because of a high jk product. However, the extent of similarity is limited by the yield load of the seat frame.

The importance of independently managing seat suspension stiffness and frame rotation stiffness has not been sufficiently appreciated in earlier studies. They have a significant effect on whiplash risks and seem to be the basis for controlling neck biomechanics up to head restraint contact. These two factors were controlled in the 1997 Saab 9-5 seat, which also introduced an active head restraint (SAHR). Extensive modifications were made in the seat frame to allow pocketing of the torso to lower the k factor (Viano 2002, Wiklund et al. 1997). The WHIP seat (Lundell 1998) introduced a low k by allowing compression of the seatback trim and movement of the seatback pivot until head support.

The 1997 W car seems to be one of the first seats to reduce both the j and k. The high-retention seat lowered neck biomechanical responses by gradually accelerating the torso, which reduces the displacement of the base of the neck and lowers energy transfer until head restraint contact. By providing a yielding behavior through deformation of the seat suspension, the occupant experiences head restraint contact earlier and with lower torso loads. As shown in Figure 7, this reduces energy transfer to the occupant before head contact from the stiff seat designs of the foreign benchmark. This allows the major energy transfer to occur after support is provided to the head, neck and torso, which is a fundamentally different approach to seat design than the yielding seats of the 1980s-90s and the stronger foreign luxury seats. High retention seats also have 5.5-times greater strength than those of the 1960s to improve occupant retention in the most severe rear crashes (Viano 2003a).

The peak neck displacements in Figure 7 are intriguing for another reason. There is a peak at 10 km/h rear crash delta V for the foreign benchmark and 1990 C car seats. Neck displacements are lower in higher severity crashes primarily due to an earlier head restraint contact. Langwieder, Hell (2002), Hell et al. (2002) among others have shown that the highest incidence of whiplash is for crashes below 15 km/h, which is consistent with the trend in neck displacements.

Neck Displacements and NIC: Figure 7 shows the peak neck displacements with the three exemplar seats at head restraint contact. The 1997 W car experiences a 67% lower neck displacement than the stiffer foreign benchmark seat for 6-35 km/h rear crashes. There is a clear distinction in the responses based on neck displacement. Viano et al. (2002) built on the hypotheses of Panjabi et al. (1999) and recommended evaluating neck extension and displacement of the occipital condyles with respect to T1 as an essential response related to whiplash. This is consistent with the neck displacements reported here. Viano, Davidsson (2002) proposed a Neck Displacement Criterion (NDC). In contrast to neck displacement, the relative head-torso acceleration and NIC increase with impact severity. The foreign benchmark seat has higher relative head-torso accelerations and NIC than the yielding seats. Most responses increase with rear delta V.

Benefits of Yielding Seatbacks: This analysis underscores the importance of the yielding seat approach to occupant safety in rear crashes. Historically, the yielding behavior was achieved by seatback rotation at relatively low loads. The typical H-point moment would be about 1000 Nm in the QST test of 1980s and early 1990s seats. These seatback designs had relatively strong structural cross-members in the seatback, which made much of the compliance related to recliner and seat frame rotation, although the foam trim covering the frame provided deformation distance until the motion was stopped by seatback structures. Often, these structures were just behind the pelvis, which limited penetration into the seatback and increased the tendency for ramping, and adjacent to the upper back increasing load on the torso.

The 1990 C car seat had a stiffness of 21 kN/m, which was lower than the average of GM (General Motors) seats of that era. These seats reflected the yielding behavior of seat designs of the 1970s to early 1990s. However, it became clear that there were shortcomings by allowing the occupant to rotate the seatback rearward in severe rear crashes, since there could be a loss of occupant retention; and, the energy transfer capability was limited. The way to maintain the yielding behavior of the seat without significant seatback rotation was conceived in the early 1990s after a series of sled tests with the C car seat. That seat had an open perimeter frame, which increased occupant retention to higher delta Vs than designs with substantial cross-frame structures. This was achieved by more pelvic pocketing into the seatback and it promoted pelvic drop with occupant displacement rearward. This is the opposite of ramping and improves head restraint support.

The C car seat was the genesis of a new generation of seats based on the open perimeter frame and strong seat frame and recliners, which allow yielding performance by letting the occupant penetrate between the frame side structures through compliance of the seat suspension. However, the C car seat had a relatively low upper cross member, so load was applied to the mid-back. The high retention seat raised the upper cross member to at or above the shoulders to allow penetration of the upper and lower torso. This allowed earlier head contact with the 1997 W car seat. The head impact velocity was slightly higher than with the other seats but was considered manageable, since the other neck responses are substantially reduced.

FMVSS 207 Versus QST Seat Tests: FMVSS 207 is a seat frame test that does not determine seat stiffness k, which involves the compliance of the seat suspension or trim. It involves a point load on the upper structural cross-member of the frame, so it does not represent an occupant interface. Only the seat structure and recliner are determined by the frame test. The QST and body block test involves a fully trimmed seat and uses a Hybrid dummy for the seat loading in a rearward direction. It determines the deformation of the seat suspension, trim and frame structure under a loading condition representative of that in a rear sled or barrier crash test. It can be used to determine the k and j factors.

Seat strength has steadily increased in FMVSS 207 tests from the 1960s to today. However, this test addresses only the j factor of seat frame strength, but there apparently has been a coincident increase in k stiffness with modern seats (Viano 2002, 2003a). High k stiffness increases torso and neck loads in rear crashes and may aggravate whiplash based on these results. A subsystem test like the QST is needed to assess both k and j under occupant loading, since a reduction in both factors is needed to ensure occupant retention in high severity crashes and lower whiplash risks in low speed crashes. A revision to FMVSS 207 is needed to include an occupant load on the seat and determination of j and k. These characteristics cannot be as easily determined from sled or barrier tests as with a QST-like test.

Role of High Retention (Yielding Seatback) Seats in Reducing Whiplash: Seat stiffness seems to be an important factor in reducing whiplash risks. The lower forces on the back with the yielding seat are a primary means of lowering loads on the base of the neck (Viano,

Olsen 2001). This reduces neck displacements before head restraint contact.

Displacement of the base of the neck with respect to ground appears to be a factor in controlling whiplash. If it is lowered by reducing seat stiffness, the gap is closed with lower neck loading until head restraint support. There are also benefits from the open perimeter frame seatback, which reduces seatback rotation because a low jk product maintains a more upright posture until head and neck support. Older seat designs rotate rearward and yield as energy is transferred to the occupant. New seats are stiffer but load more the torso and transfer much of the crash energy before head contact.

Figure 7 shows the energy transfer to the occupant up to head restraint contact. There is a high energy transfer with stiff seats. This seems to be contrary to what is needed to reduce whiplash risks. The 1997 W car perimeter frame design involves very low energy transfer until after head restraint contact when the head, neck and torso and rotate more in unison as energy is transferred. It also emphasizes pelvic drop as a means of lowering the occupant's head with respect the head restraint during seatback loading (Viano 2002). This is a further benefit in managing neck responses.

The perimeter frame design was also the basis for the introduction of the Saab Active Head Restraint (SAHR), since the early occupant penetration into the seatback was used to raise and bring forward the active head restraint. The effectiveness of SAHR was recently determined in field crashes, which found a 75% reduction in whiplash injury and underscored the importance of the new generation of yielding seats using the high retention and perimeter frame design (Viano, Olsen 2001). SAHR is intended to compensate for the head restraint drop due to seatback rotation and achieve an even earlier contact with a more horizontal trajectory than with high retention seats. The upward motion of SAHR provides a vector of head restraint loading that acts upward behind the head center of gravity causing flexion loads on the head and neck (Viano 2002).

Limitations of the Study: Obviously, the model developed and used in this analysis is not a full occupant dynamics simulation, which would offer the possibility to explore the responses in much more detail. The model provides a convenient means of addressing the fundamental responses related to seatback deformations and torso-head dynamics in rear crashes; and, the solution provides quite reasonable response when compared with test data on volunteers and BioRID in low-speed rear sled tests (Davidsson 2000). However, the solution does not reflect the influence of occupant ramping or pelvic drop with occupant displacement into the seat or the effects of gravity acting on the occupant. This study does not address variations in the initial head restraint height or gap behind the head, or head restraint interactions. Most modern seats have a much more favorable head restraint placement than seat designs of the 1980s. These effects are beyond the scope of this analysis, but could be pursued using more sophisticated occupant dynamic simulations with these results as a basis.

The loading interface on the upper seatback influences the force distribution on the base of the neck. This may require more than a single stiffness value to adequately describe the interaction. Also, the flexible spine of the human is not comprehended as the lordotic curvature straightens in a rear crash. The flexible spine of BioRID also involves different accelerations of T1 and T6, which are not comprehended with a single torso mass. Finally, the idealized constant acceleration rear crash pulse is not representative of actual vehicles, which show a double acceleration pulse (Viano 2002, Langwieder, Hell 2002).

The 14 seats in the foreign benchmark group have a wide range of characteristics in stiffness, compliance and frame strength. The exemplar seat with k = 40 kN/m and j = 1.8 °/kN does not represent any specific product in use. It is a composite of the high-end performing seats, but the actual designs varied considerably. Nonetheless, the study shows the importance of adjusting the seat and frame stiffness (the k and j factors) to improve occupant dynamics in rear crashes. These parameters define the essential characteristics of the seat for rear impact loading of the occupant.

ACKNOWLEDGMENTS

This paper was published in Traffic Injury Prevention 4(3) 2003, and is reprinted here with permission from the publisher, Taylor & Francis Group. Their agreement for its use in this book is greatly appreciated.

REFERENCES

1. Backaitis S, Mertz H. (1994) Hybrid III: The First Humanlike Crash Test Dummy. SAE Special Publication PT-44, Society of Automotive Engineers, Warrendale, PA.
2. Barnsley L, Lord S, Bogduk N. (1994) Whiplash injury. Pain Sep;58(3):283-307.
3. Bostrom O, Svensson M, Aldman B, et al. (1996) A New Neck Injury Criterion Candidate Based on Injury Findings in the Cervical Spine Ganglia after Experimental Sagittal Whiplash. IRCOBI Conf 123-136.
4. Blaisdell DM, Levitt AE, Varat MS. (1993) Automotive seat design concepts for occupant protection. SAE 930340, Society of Automotive Engineers, Inc., Warrendale, Pennsylvania, USA.
5. Davidsson J. (2000) Development of a Mechanical Model for Rear Impacts: Evaluation of Volunteer Responses and Validation of the Model. ISBN 91-7197-924-7, Doctoral Thesis, Crash Safety Division, Chalmers University of Technology, Gothenburg, Sweden.

6. Deng YC. (1989) Anthropometric Dummy Neck Modeling and Injury Considerations. Accident Analysis and Prevention 21(1):85-100.
7. Galasko CSB, Murray PM, Pitcher M, et al. (1993) Neck Sprain After Road Traffic Accidents: a Modern Epidemic. Injury 24 (3), 155-157.
8. Hell W, Schick S, Langwieder K, Zellmer H. (2002) Biomechanics of Cervical Spine Injuries in Rear End Car Impacts: Influence of Car Seats and Possible Evaluation Criteria. Traffic Injury Prevention 3(3): 127-140.
9. Holm L, Cassidy JD, Sjogren Y, Nygren A. (1999) Impairment and Work Disability Due to Whiplash Injury Following Traffic Collisions. An Analysis of Insurance Material from the Swedish Road Traffic Injury Commission. Scand J Public Health. 27(2):116-23.
10. Langwieder K, Hell W. (2002) Proposal of an International Harmonized Dynamic Test Standard for Seat/Head Restraints. Traffic Injury Prevention 3(3): 150-158.
11. Lundell B, Jakobsson L, Alfredsson B, et al. (1998) Guidelines for and the design of a car seat concept for improved protection against neck injuries in rear-end car impacts. SAE 980301, Society of Automotive Engineers, Warrendale, Pennsylvania, USA.
12. Mertz, HJ, Patrick L. (1971) Strength and Response of the Human Neck. SAE 710855, Society of Automotive Engineers, Inc., Warrendale, Pennsylvania, USA.
13. Mertz HJ, Neathery RF, Culver CC. (1973) Performance Requirements and Characteristics of Mechanical Necks. in Human Impact Response: Measurement and Simulation, 263-288, Plenum Press, New York.
14. Mertz HJ, Irwin AL, Melvin JW, Stalnaker RL, Beebe MS. (1989) Size, Weight and Biomechanical Impact Response Requirements for Adult Size Small Female and Large Male Dummies. SAE 890756, Society of Automotive Engineers, Warrendale, PA.
15. Ono K, Kanno M. (1996) Influence of the Physical Parameters on the Risk to Neck Injuries in Low Impact Speed Rear Collisions. Accid Anal Prev 28(4):493-499.
16. Panjabi MM, Wang JL, Delson N. (1999) Neck Injury Criterion Based on Intervertebral Motion and its Evaluation Using an Instrumented Neck Dummy. 1999 IRCOBI Conf, 179-190.
17. Parenteau CS, Viano DC. (1995) A New Method to Determine the Biomechanical Properties of Human and Dummy Joints. 1995 IRCOBI Conference, pp. 183-198, September.
18. Prasad P, Kim A, Weerappuli DPV, Robert V, Schneider D. (1997) Relationship Between Passenger Car Seat Back Strength and Occupant Injury Severity in Rear End Collisions: Field and Laboratory Studies. SAE 973343, Society of Automotive Engineers, Warrendale, PA.
19. Richter M, Otte D, Pohlemann T, Krettek C, Blauth M. (2000) Whiplash-type Neck Distortion in Restrained Car Drivers: Frequency, Causes and Long-term Results. Eur Spine J. 9(2):109-17.
20. Saczalski KJ, Syson SR, Hille RA, Pozzi MC. (1993) Field accident evaluations and experimental study of seat back performance relative to rear-impact occupant protection. SAE 930346, Society of Automotive Engineers, Inc., Warrendale, Pennsylvania, USA.
21. SAE J1460-2. (1998) Human Mechanical Impact Response Characteristics-Responses of the Human Neck to Inertial Loading by the Head for Automotive Seated Postures. Society of Automotive Engineers, Warrendale, PA.
22. Severy DM, Mathewson JH and Bechtol CO. (1955) Controlled Automotive Rear-End Collisions, An Investigation of Related Engineering and Medical Phenomena. Canadian Services Medical Journal, VII, 727-759.
23. Severy DM, Brink HM, Baird JD. (1967a) Preliminary findings of head support designs. 11th Stapp Car Crash Conf, SAE 670921, SAE Warrendale PA, 337-405.
24. Severy DM, Harrison MB, Baird JD. (1967b) Collision Performance, LM Safety Car. SAE 670458, Society of Automotive Engineers, Warrendale, PA.
25. Severy DM, Brink HM, Baird JD. (1968a) Backrest and Head Restraint Design for Rear-end Collision Protection. SAE 680079, Society of Automotive Engineers, Warrendale, PA.
26. Severy DM, Brink HM, Baird JD. (1968b) Vehicle Design for Passenger Protection from High Speed Rear End Collisions. SAE 680774, Society of Automotive Engineers, Warrendale, PA.
27. Strother CE, James MB. (1987) Evaluation of Seat Back Strength and Seat Belt Effectiveness in Rear End Impacts. 31st Stapp Car Crash Conference, SAE 872214, Society of Automotive Engineers, Warrendale, PA, 225-244.
28. Viano DC, Olsen S. (2001) The Effectiveness of Active Head Restraint in Preventing Whiplash. J Trauma 51:959-969, November.
29. Viano DC. (2002) Role of the Seat in Rear Crash Safety. SAE Book, ISBN 0-7680-0847-6, Society of Automotive Engineers, Warrendale, PA, SAE R-317:1-491.
30. Viano DC, Olsen S, Locke GS, Humer M. (2002) Neck Biomechanical Responses with Active Head Restraint Systems: Rear Barrier Tests with BioRID and Sled Tests with Hybrid III. SAE 2002-01-0030, Society of Automotive Engineers, Warrendale, PA.
31. Viano DC, Davidsson J. (2002) Neck Displacements of Volunteers, BioRID P3 and Hybrid III in Rear Impacts: Implications to Whiplash Assessment by a Neck Displacement Criterion (NDC). Traffic Injury Prevention 3(2):105-116.

32. Viano DC. (2003a) High Retention Seat Performance in Quasistatic Seat Tests. SAE 2003-01-0173, Society of Automotive Engineers, Warrendale, PA.
33. Viano DC. (2003b) Energy Transfer to an Occupant in Rear Crashes: Effect of Stiff and Yielding Seats. SAE 2003-01-0180, Society of Automotive Engineers, Warrendale, PA.
34. Warner C, Strother C, James MB, Decker RL. (1991) Occupant Protection in Rear-End Collisions II: The Role of Seat Back Deformation in Injury Reduction. 35th Stapp Car Crash Conf, SAE 912914, Society of Automotive Engineers, Warrendale, PA.
35. Wiklund K, Nilson G, Larsson H, Humer M. (1997) Saab Active Head Restraint (SAHR) - Seat Design to Reduce the Risk of Neck Injuries in Rear Impacts. SAE 980297, Society of Automotive Engineers, Warrendale, PA.
36. Yoganandan N, Pintar FA. editors. (2000) Frontiers in Whiplash Trauma: Clinical and Biomechanical. IOS Press Amsterdam, The Netherlands, Library of Congress No. 00-101366.

CHAPTER 4:

SEAT INFLUENCES ON FEMALE NECK RESPONSES IN REAR CRASHES: A REASON WHY WOMEN HAVE HIGHER WHIPLASH RATES

Chapter 4

Seat Influences on Female Neck Responses in Rear Crashes: A Reason Why Women Have Higher Whiplash Rates

David C. Viano[1,2]

[1]ProBiomechanics LLC
[2]Crash Safety Division, Chalmers University of Technology

ABSTRACT

Since the earliest crash investigations, whiplash has been found to occur more often in women than men. This study addresses seat properties that may explain a reason for the higher rates in women, and changes in whiplash in general over the past two decades.

Three exemplar seats were defined on the basis of seat stiffness (k) and frame rotation stiffness (j) for rearward occupant load. Stiff seats have k = 40 kN/m and j = 1.8 ^{0}/kN representing a foreign benchmark loaded by a male. One yielding seat had k = 20 kN/m and j = 1.4 ^{0}/kN simulating a high retention seat (1997 Grand Prix) and another k = 20 kN/m and j = 3.4 ^{0}/kN simulating a 1980s-90s yielding seat (1990 Buick Park Avenue). Constant vehicle acceleration for 100 ms gave delta Vs of 6, 10, 16 and 24 km/h. The 1D model included a torso mass loading the seatback with flexible neck and head mass. Based on biomechanical data and scaling, neck stiffness was 5 kN/m and 3 kN/m for the male and female, respectively. Based on validation tests, seat stiffness was 25% less with the female. Occupant dynamics were simulated in a step-forward solution based on the differential displacement between the head, torso and seat up to head restraint contact.

Neck responses were 30% higher in the female than male through most of the rear impact and are proportional to $(k_F / m_{TF})/(k_M / m_{TM})$, which is the ratio of seat stiffness divided by torso mass for the female and male. Neck displacements were higher with the stiff seat than the 1990 C car seat for both the female and male. They peaked at 10 km/h and dropped off for higher severity crashes due to the shorter time to head contact. Neck displacements were greater in the female than male for the lowest severity crashes with the stiff and 1990 C car seats, when displacement was scaled for equal tolerance. The female in 1997 W car seat had the lowest neck displacements.

Stiff seats increased neck displacements over the yielding seats of the 1980s in rear crashes. The trend is similar in men and women, but early neck displacements are greater in women because of a higher ratio of seat stiffness to torso mass. This implies that seat stiffness is not sufficiently low in proportion to the female mass in comparison to males. The j and k seat properties influence neck biomechanics and occupant dynamics, but k is important in determining early response differences between males and females.

INTRODUCTION

Women Have Higher Whiplash Rates Than Men: From the earliest crash investigations, women have been found to have higher whiplash rates than men (Kihlberg 1969, States et al. 1969, O'Neill et al. 1972, Huelke, Marsh 1974). Although these studies address crash data from the United States, studies in Europe and other countries since the 1980s have shown similar results (Thomas et al. 1982, Evans 1992, Laberge-Nadeau et al. 1993, Dolinis 1997, Richter et al. 2000). Some studies found only a slightly greater risk in women, while others found 40-50% higher rates, and even others found more than double the rate. Dolinis (1997) conducted a statistical analysis showing 2.08-times the relative risk of whiplash in women as compared to men.

Since whiplash has an uncertain diagnosis and both insurance claims and medical records are variable in defining it, the range in relative risk is somewhat understandable since the type of outcome claim can make a difference in the calculation of rates. For example, Larsen, Holm (2000) found that women 20-59 year had the highest risk of neck pain. This is consistent with findings by Chapline et al. (2000) that there was a two-fold incidence of neck pain in women after rear crashes than men. However, Viano, Olsen (2001) found that neck pain lasting less than a week resolved without long-term consequences. Their conclusion was that neck pain lasting less than a week should not be included as whiplash injury since one week differentiated patients with long-term injury. Nonetheless, women seem to experience more whiplash injury, but how much greater is uncertain. Cassidy et al (2000) found that women had a higher incidence of whiplash claims than men 6 months after a crash, even when insurance claim remuneration was eliminated.

The issue of higher whiplash rates in women than men has received attention over the past years, partly because of the general increase in whiplash. Whiplash claims have increased worldwide over the past two decades. In Germany, Richter et al. (2000) found that drivers with whiplash had increased from less than 10% in 1985 to more than 30% in 1997. In Sweden, Holm et al. (1999) found that whiplash-related medical impairment and work disability had increased from 16% in 1989 to 28% in 1994. In Japan, Ono and Kanno (1996) found that whiplash injuries have increased from 44% to 51% in car-to-car crashes from 1985 to 1991. In the U.K., Galasko et al. (1993) found that neck injuries have steadily increased in rear crashes.

There is a question whether the incidence of whiplash in women may have disproportionately increased with modern seats that have higher stiffness to occupant load (Viano 2003c). The earlier designs of the 1980s had a yielding seatback that rotated under occupant load protecting the neck in low-speed rear crashes. As seats have become stiffer, questions arise whether the higher head restraint placement with stiffer seats has been sufficient to control neck responses and lower injury risks in general, and in women in particular. Clearly, the field data shows it has not and that women remain more at risk than men. Warner et al (1991), Prasad et al (1997) and others raised concerns that a transition from the yielding to stiff seats may increase whiplash rates. Viano (2003c) seems to have confirmed their concerns. Yet, none of these studies addressed the differential risk in women.

While there are hypotheses for the higher whiplash rates in women than men, there have been scant scientific studies giving proof for any of the theories or pointing out seat properties that may be responsible for the differential risk in women. This study attempts to present one reason for the higher incidence of whiplash in women than men and what properties of the seat may be responsible.

Transition to Stiffer Seats During the Past Two Decades: Warner et al. (1991) summarized the seat testing of Severy et al. (1967) and Strother, James (1987) that used a body block loading in a rearward direction. They showed that seats had increased in strength from 2.6 ± 0.6 kN in the 1960s to 3.2 ± 1.2 kN in the 1970s and early 1980s, a more than 20% increase. Viano (2003a) found that Ford and Chrysler seats of the 1980s-1990s supported torso loads of 5.8 ± 1.2 kN and General Motors seats 6.6 ± 1.2 kN, a 2.2-2.5-fold increase, yet most of these seats yielded by frame rotation. A sample of mostly European and Asian luxury seats supported even higher loads of 7.7 ± 2.0 kN, a 3-fold increase. These seats have strong frames and provide greater occupant retention in high-speed rear crashes. However, the strength increase resulted in greater seat stiffness, which is the change in force with rearward occupant displacement. The 1980s-90s yielding seats nominally had k = 20 kN/m, which increased to 40 kN/m with the stiffest foreign benchmark seats (Viano 2003c). The implications of seat strength and stiffness to whiplash have fostered a long-standing debate.

The Debate Over Yielding and Stiff (Rigid) Seats: Severy et al. (1967) conducted pioneering research on seats and occupant dynamics in rear crashes, and showed the need for controlled seatback deformation and head support. He also worked on stiff (rigid) seat concepts to enhance occupant retention in severe crashes. However, the concepts were not brought to production because of various criticisms on overall performance. Strother, James (1987), Warner et al. (1991) and Prasad et al. (1997) argued for yielding seatbacks as the best way to manage occupant energy, reduce whiplash risks and provide overall occupant safety in rear crashes. It limits forces on the neck. This long-standing debate furthered the notion that there was a design conflict between occupant retention by stiff (rigid) seats in infrequent, severe rear crashes, and yielding seatbacks to prevent whiplash in the frequent, minor rear crashes.

High retention seats started appearing in vehicles in the late 1990s (Viano 2002). They have 5.5-times greater strength to resist rotation in severe rear crashes than the 1960s designs, but have a low stiffness due to compliance of the seat trim and suspension across the side frames. The low stiffness is comparable to earlier yielding seats of the 1970s-80s (Viano 2003a). The high retention design demonstrated that there was not a design conflict between a rigid and yielding seat system, if compliance of the seatback between the side frame structures was considered in the load-limiting performance of the seat. These seats have a strong frame and a yielding stiffness under occupant load.

By providing yielding through deformation of the seat suspension, the occupant experiences head restraint contact earlier and with lower loads on the torso than with a stiff seat. This reduced the energy transfer to the occupant before head support and allowed the major energy transfer to occur after head restraint contact (Viano 2003b,c). This is a fundamentally different approach to seat designs of the 1960s to early 1990s, including the foreign benchmark group. However, questions remain on how the high retention design performs in the field with male and female occupants and how the transition to stiffer seats influences whiplash rates in the field. A recent study by Viano (2002c) attempted to address the latter question; and, a study of FARS has shown a 50% reduction in fatality risk with high retention seats in single vehicle crashes and two-vehicle crashes involving a light striking vehicle (Viano 2003d). The current study focuses on the difference between men and women interacting with seats in 6-24 km/h rear crashes.

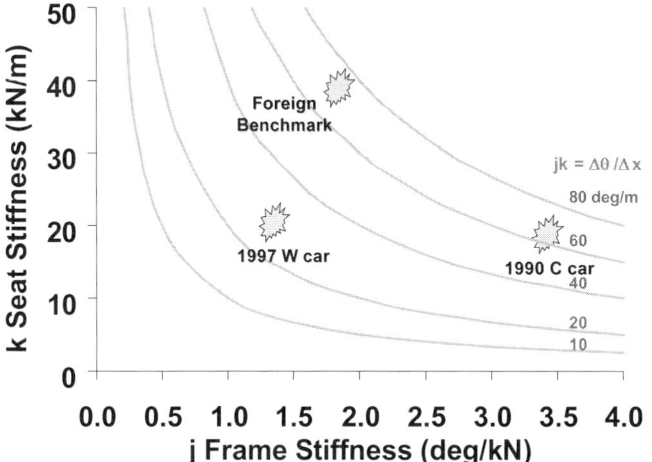

Figure 1: Seat properties related to occupant interactions in rear crash include seat stiffness (k) and frame structure stiffness (j) for three exemplar seat types. Iso-stiffness lines are shown in the background for constant jk = $\Delta\theta / \Delta x$ (from Viano 2003c with permission).

METHODOLOGY

Occupant interactions with the seat are modeled using two seat properties that define the response to load for three exemplar seat types. Figure 1 shows j and k factors for the exemplar seats. Seat stiffness (k) is the load developed by the seat as the occupant displaces rearward. A yielding seat is defined as one with k < 25 kN/m, whereas a stiff seat is one with k > 35 kN/m. The second seat property is frame rotation stiffness (j), which is the amount of seatback rotation occurring with a rearward occupant load. j < 2.0 ^{0}/kN provides high retention of the occupant by resisting rotation in severe rear crashes whereas, j > 3.0 ^{0}/kN is typical of yielding seats of the 1980s-90s, which provide low seat stiffness primarily by seatback rotation.

The 1D mathematical model involves a torso mass, a flexible neck and head mass that interact with the seatback in a rear crash (Viano 2003b,c). This allows the evaluation of head and neck dynamics with seatback deformation and the influence of head restraint trajectory, which is related to head restraint drop and rearward displacement with seatback rotation. The three exemplar seats were evaluated in 6-24 km/h rear crashes up to head restraint contact with a male and female occupant.

Nomenclature for Analysis:

F	Force on the occupant's torso in kN
G	Force on the neck in kN
i	Time integer
j	Seat frame rotation stiffness, slope of occupant load versus seatback angle in ^{0}/kN
k	Seat stiffness due to occupant load in kN/m, subscript F is female and M is male.
k_N	Stiffness of neck in kN/m.
m	Effective mass of the occupant loading the seat ($m = m_T + m_H$) in kg
m_T	Mass of the torso in kg
m_H	Mass of the head-neck in kg
p	Length of the seatback from the pivot to the top of head restraint in m.
QST	Quasistatic Seat Test (Viano 2002)
R	Ratio of female to male neck responses.
s	Actual head restraint gap w/r moving vehicle and head in m
Δt	Delta time in s
u	Drop in head restraint height due to seatback rotation in m
v	Rearward displacement of head restraint due to seatback rotation in m
x	Displacement of the torso in m
x-z	Neck displacement in m
Δx	Occupant displacement in QSTs in m
y	Displacement of the vehicle in m
z	Displacement of the head in m
θ	Seatback angle w/r vertical in deg
$\Delta\theta$	Change in seatback angle in QSTs in deg
ϕ	Head restraint trajectory w/r to ground in deg

Occupant Dynamics and Seat Interaction: The model of Viano (2003b,c) was used to evaluate neck responses for females and males due to seat interactions in rear impacts. Details on the model can be found in the earlier studies and only those equations relevant to interpretation of the results and discussion will be included here. The occupant is displaced forward in a rear crash by the motion of the vehicle and deformation of the seat during impact. Force develops on the occupant's torso in proportion to the seat stiffness and torso displacement with respect to the vehicle:

$$F_i = k(y_i - x_i) \qquad (1)$$

Force accelerates the occupant in proportion to the effective mass of the torso loaded by the seatback.

$$\ddot{x}_i = F_i / m_T \qquad (2)$$

where $\ddot{x}$ is torso acceleration. Viano (2002) determined that the effective mass (m) loaded by the seatback was 70% of the dummy mass from sled testing with the Hybrid III dummy. Since the head and neck are indirectly accelerated by motion of the torso, their mass is not initially loaded by the seat. Using information from Backaitis, Mertz (1994), the 50% male and 5% female dummy torso masses are m_T = 47.0 kg and 27.2 kg, respectively and the head and neck masses are m_H = 5.5 kg and 4.2 kg, giving an upper body mass of m = 52.5 kg and 31.4 kg loading the seat. The effective male head mass was determined using 4.5 kg for the head plus two-thirds of the neck mass or 1.0 kg. The effective female head mass was similarly determined.

Head-Neck Dynamics: Force develops on the neck due to differential motion between the torso and head. It is related to the neck stiffness due to a combination of bending, tension and shear:

$$G_i = k_N(x_i - z_i) \qquad (3)$$

where the quantity in parenthesis in neck displacement and k_N is neck stiffness. Force on the neck increases with the relative displacement between the head and torso (neck displacement), which accelerates the head ($\ddot{z}$) in proportion to its effective mass (m_H):

$$\ddot{z}_i = G_i / m_H \qquad (4)$$

Head Restraint Contact: A relationship exists between torso load and seatback rotation. It is generally linear up to the yield load, so seatback angle is given by:

$$\theta_i = \theta_0 + jF_i \qquad (5)$$
$$\Delta\theta_i = jF_i \qquad (6)$$

where $\theta_0 = 25^0$ for typical seats, and j is the slope of the line relating occupant load to seatback angle. As the seatback rotates, it causes a downward and rearward displacement of the head restraint with respect to the vehicle (occupant):

$$u_i = p[\cos(\theta_i) - \cos(\theta_0)] \qquad (7)$$
$$v_i = p[\sin(\theta_i) - \sin(\theta_0)] \qquad (8)$$

where u and v define the head restraint trajectory and p = 780 mm is used to represent typical modern seats. This p value will also be used for the 1990 C car seat even though the actual head restraint was substantially lower. This allows a direct comparison of the different seat stiffness types irrespective of the initial head restraint position. The gap to the head restraint varies with the initial gap behind the head and seatback angle. It is a function of the head and vehicle displacement, and other geometric relationships defining the head restraint position with seatback rotation:

$$s_i = s_0 + v_i - (y_i - z_i) \qquad (9)$$

where $s_o = 80$ mm will be assumed typical of modern head restraint placements.

Seat Properties: Seat Stiffness (k) and Frame Rotation Stiffness (j) to Occupant Loading: Seat stiffness (k) and frame rotation stiffness (j) are determined between occupant loads of 1-5 kN. This covers the majority of loading up head restraint contact in rear crashes. Three exemplar seats were used, following earlier studies where more details can be found (Viano 2003b,c).

The stiff seat has k = 40 kN/m and j = 1.8 ^{0}/kN representing a foreign benchmark seat from Europe and Asia (Figure 1). One yielding seat had k = 20 kN/m and j = 1.4 ^{0}/kN simulating the 1997 W car seat, which was the first high retention seat (Viano 2002). The other had k = 20 kN/m and j = 3.4 ^{0}/kN simulating the 1990 C car type seat, which yielded by seatback rotation and is typical of volume produced seats in the 1980s-90s.

For the exemplar seats, the change in seatback angle for occupant displacement into the seat was $\Delta\theta / \Delta x$ = 72 ^{0}/m for the foreign benchmark. The 1990 C car seat had $\Delta\theta / \Delta x$ = 68 ^{0}/m and the 1997 W car seat had $\Delta\theta / \Delta x$ = 28 ^{0}/m, where

$$\Delta\theta = jk\Delta x \qquad (10)$$

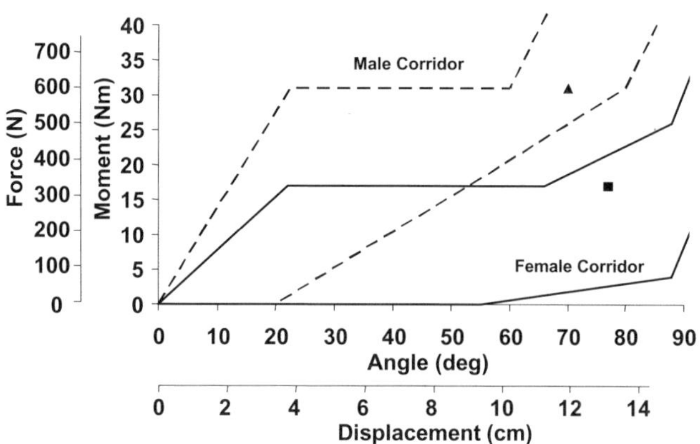

Figure 2: Neck moment and extension angle showing SAE biomechanical corridors for the 50% male and 5% female. The conversion of moment to force and angle to displacement is also shown with the average neck stiffness used in the analysis with the point used for the male (triangular point) and female (square point).

Effective Neck Stiffness: Figure 2 shows the neck extension corridors for the 50% male and the 5% female, which are based on human volunteer responses and scaling (Mertz, Patrick 1971, SAE 1998, 1998, Mertz et al. 1989). These have widespread acceptance. In the earlier analysis (Viano 2002c), the male neck response was approximated by a straight line from zero through 31 Nm and 70^0 (triangular point in the plot).

The moment-angle was converted to force-displacement using dimensions for the Hybrid III head and neck, and published response data. The moment about the occipital condyles was converted to force at the head center of gravity (cg) by using the 5 cm distance between the head cg and the occipital condyles (Mertz et al. 1989, Backaitis, Mertz 1994). This gives a load of 620 N for 31 Nm moment. The extension angle of 70^0 involves an x-displacement of the head cg of approximately 12 cm. This is based on data from Mertz et al. (1973), which showed a 6.4 cm x-displacement of the head cg at a neck extension angle of 40^0 and 15.7 cm at 78^0. A conversion of 5.8^0/cm was used reflecting more of the lower extension behavior. These conversions gave a male neck stiffness of 5 kN/m in extension due to bending, shear and tension. The female neck stiffness was adjusted from the male stiffness using scale factors from Mertz et al. (1989) as used to develop the female corridors (SAE 1998). This adjusted the triangular point to the square point for the female, which is at 17 Nm and 78^0, and reduces the male neck stiffness to 3 kN/m for the female.

In terms of the relative neck response differences, the recent FMVSS 208 sets forth a neck injury criterion Nij and tolerances for the male and female dummies (NHTSA 2002). The peak extension moment involves a female to male ratio of 0.50 (My = 67 Nm/135 Nm). The tension limit has a ratio of 0.63 (Fz = 4287 N/ 6806 N). These levels are consistent with the scaled response corridors in Figure 2. They indicate substantially lower neck loads at somewhat higher neck extension angles for the female. Data on the natural range of motion from a number of published studies is included in an SAE (1998) task force report. Table 1 shows rather similar neck extension and flexion ranges for the female and male. There is an age effect, which reduces the range.

Table 1: Differences in the Natural Range of Motion of Females and Males (Derived from information in SAE 1998)

Source	Age	Extension		Flexion-Extension		Ratio
		Male	Female	Male	Female	Female/
	yrs	deg	deg	deg	deg	Male
Buck et al. (1959)	18-23	73 ± 9	81	139	150	1.09
Granville, Kreezer (1937)	20-40	61 ± 27		121		1.98
Foust et al. (1973)	18-24			139 ± 9	136 ± 18	0.98
	35-44			109 ± 18	123 ± 20	1.13
	67-72			83 ± 22	104 ± 17	1.25
Schneider et al. (1975)	18-24			129	124	0.96
	35-44			103	105	1.02
	62-74			77	84	1.09

Female and Male Interactions with the Seat and Model Validation with Experiments: For female seat loading, a modification is needed in the seat stiffness (k) to reflect the lighter weight and size of the woman. The smaller size of the female dummy means a lower force on the seatback for a given deformation. This reduces the effective stiffness of the seat. For this study, the k = 40 kN/m seat stiffness for the male was reduced to k = 30 kN/m for the female, and the k = 20 kN/m seat is reduced to k = 15 kN/m, a 25% reduction in stiffness. The amount of reduction was determined by comparison of model responses with experimental data from rear sled tests.

A series of rear sled tests was conducted with the 1990 C car seat with a 50% male and 5% female Hybrid III dummy (Viano 2002). Figure 3 shows average peak values for the tests between 10-19 km/h. Comparable results are shown from the model using the previously defined occupant parameters and a yielding seat with k = 20 kN/m and j = 3.4 ^{0}/kN for the female. Peak chest acceleration, seatback rotation and neck extension angle show good correlation in amplitude and response differences between the two dummies. Figure 4 shows the ratio between female and male responses for the experiment and model. It shows similar trends and good agreement. The female chest acceleration is somewhat higher than that of the male, but the seatback rotation and neck extension angles are larger in the male.

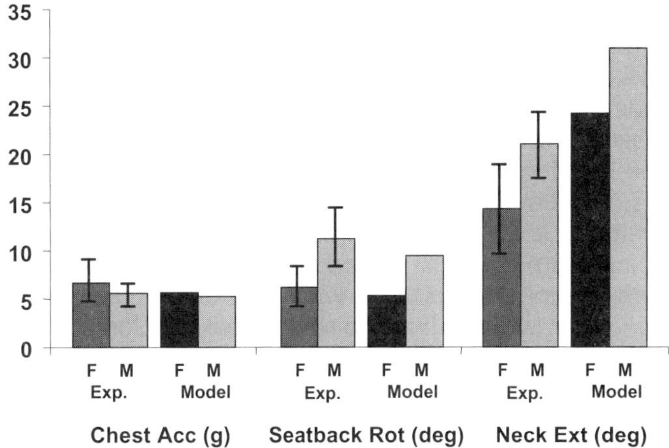

Figure 3: Comparison of average peak values for chest acceleration, seatback rotation and neck extension from sled experiments and model results for 10-19 km/h rear delta V exposures. Both 50% male and 5% female dummy results are shown.

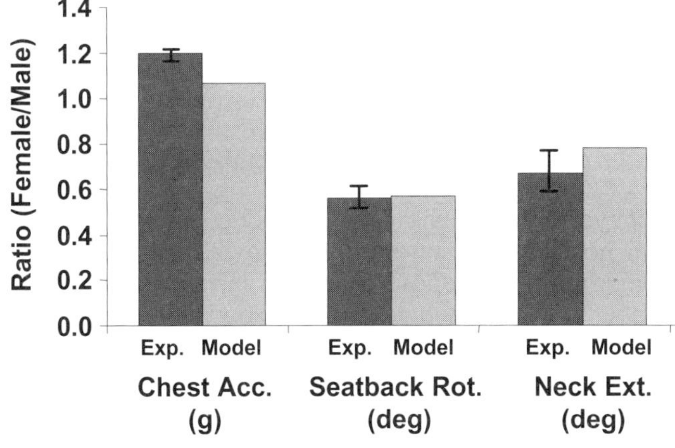

Figure 4: Ratio of female to male peak values from sled experiments and model results for 10-19 km/h rear delta V exposures shown in Figure 3. The bars represent one standard deviation in experimental response.

Determination of Seat and Occupant Biomechanical Responses Up to Head Contact in 6, 10, 16 and 24 km/h Rear Crashes with Three Different Seat Types: Four different crash pulses were used to determine the neck displacements up to head contact and other aspects of the occupant interaction with the seat. Using a constant acceleration over 100 ms of 1.6, 3.0, 4.5 and 6.7 g gives delta Vs of 6, 10, 16 and 24 km/h. Occupant dynamics and interaction differences were demonstrated for the 10 km/h rear crash of the male and female occupant dynamics.

All of the calculations were done in Excel. The solution was determined for increasing time steps until the occupant's head experienced contact with the head restraint. This was determined when the forward motion of the vehicle closed the initial 80 mm gap to the head restraint and any subsequent gap caused by seatback rotation. Head contact occurred in all rear crashes, except for two cases at 6 km/h labeled nc (no contact).

Energy and delta V transfer to the occupant were checked at head contact. Once the seatback angle reached a maximum value, it was held constant as the torso load reduced. After head contact, an additional force was introduced on the head. This simulated the stiffness of head restraint deformation. A value of 20 kN/m was used as well as other characteristics of the exemplar seats that can be found in Viano (2003c).

Differences in Female and Male Neck Responses and Tolerances: The determination of injury tolerances for whiplash in males and females is formidable because there is no consensus on the underlying mechanism, appropriate criteria or tolerance levels. Relative risk was analyzed several ways. First, the initial neck response differences between the female and male were investigated using the underlying biomechanics of the seat loading and the seat properties. This involved the equilibrium equations to estimate the initial response ratios. Second, an attempt was made to estimate the relative tolerance difference between the males and females. This would allow an adjustment of male neck displacements to reach a similar risk as in the female. The adjustment used was 0.82, equivalent to the x-direction scale factor determined by Mertz et al. (1989) for the neck.

RESULTS

Figure 5 shows occupant accelerations for the 5% female and 50% male dummy in a 10 km/h rear crash delta V with the exemplar foreign benchmark seat. The female experiences higher torso acceleration early in the loading. This increases the relative acceleration between the head and torso with a 30% greater value in the small female. This response peaks at 65 ms. Head restraint contact is later in the male.

Figure 6 compares occupant responses for the female and male in the 10 km/h rear crash for the three exemplar seats. The foreign benchmark seat involves the higher initial loads on the torso with larger forces in the male than female. Even though the male forces are higher, the resulting torso accelerations are higher and earlier with the female due to a proportionately lower torso mass. This results in more severe neck responses in the female because of her lower neck stiffness.

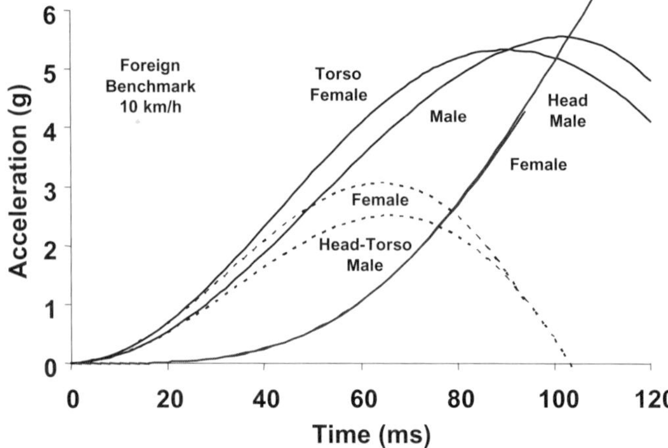

Figure 5: Female and male responses in the 10 km/h rear crash with the foreign benchmark seat with k = 40 kN/m and j = 1.8 ^{0}/kN.

The yielding seats involve lower relative acceleration between the head and torso. The 1997 W car involves the earliest head contact due to torso penetration into the seatback. The head-torso velocity increases with a larger response in the female. The greater relative velocity and acceleration between the head and torso leads to earlier and greater neck displacements in the female. The head-torso acceleration peaks at 65 ms with the foreign benchmark seat, and involves a similar peak in torso acceleration of 5.6 g for both the male and female occupants. For the 10 km/h crash, the female has greater neck displacements than the male in the 1997 W car. The greater compliance of the 1997 W car seat also results in higher torso velocity with respect to the vehicle and head impact velocity on the head restraint with the male having a greater response than the female.

Table 2 gives the peak values at or before head restraint contact for the 6-24 km/h rear crashes involving the male and female (nc denotes no head restraint contact). Since the 16 km/h rear crash delta V reflects the prevailing consensus for evaluation of whiplash risks in rear crashes (Langwieder, Hell 2002, Hell et al. 2002), the following description of the seat interactions with the occupants will focus on that crash severity.

The 1990 C car seat experiences a 10.6^0 and 5.8^0 angle change of the seatback with the male and female in the 16 km/h rear delta V crash at head contact. The seatback rotation increases the initial 80 mm gap to 205 mm and 150 mm respectively, and delays head restraint

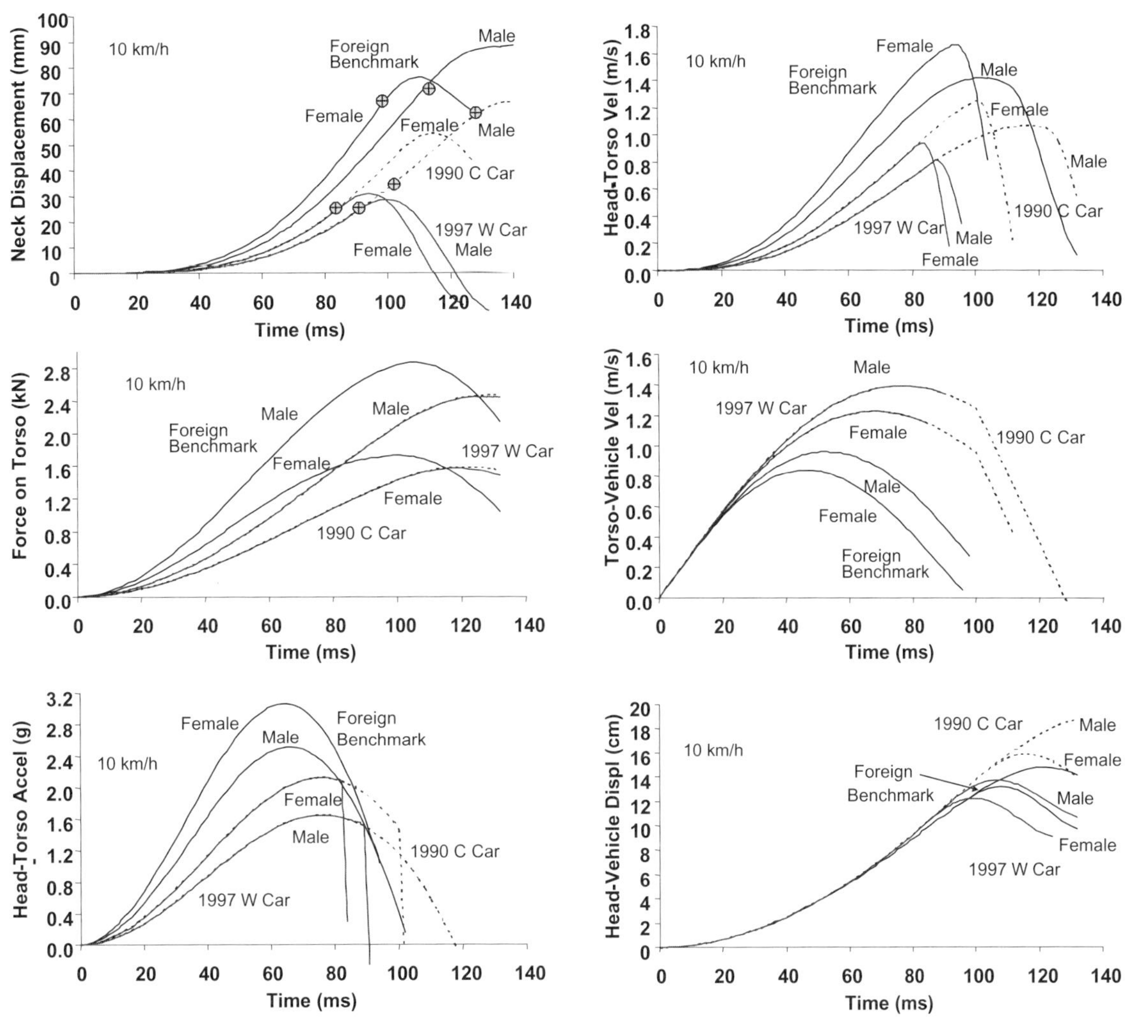

Figure 6: Selection of seat and biomechanical responses in the 10 km/h rear crash comparison of the female and male with the three exemplar seat types. The time for head restraint contact is shown with small targets on the responses. The dots in the neck displacement plot represent head contact time (see Table 2).

contact. There is a 72.8 mm and 37.2 mm drop in height of the head restraint due to seatback rotation. The yielding behavior of the seat limits load on the occupant, which results in a neck displacement of 48.9 mm and 36.9 mm respectively at head restraint contact. The peak head-torso relative acceleration is higher in the female at 3.2 g versus 2.5 g in the male but the relative velocity is similar at 1.4 m/s. Since the torso load continues to rotate the seatback near head contact, the trajectory of the head restraint is -20^0 and is -14^0 respectively at head contact.

The foreign benchmark seat has a much stronger frame than the 1990 C car, and builds up higher loads on the occupant with its stiff suspension. These effects result in a 4.0 kN torso load in the male and 2.3 kN in the female, which cause seatback rotations of 7.3^0 and is 4.1^0 respectively at head restraint contact. This involves a 47.4 mm and 25.3 mm drop in head restraint height and increased gap to 167 mm and 130 mm respectively that needs to be closed for head contact, although the head restraint trajectory is only -6.9^0 and -5.2^0. The higher torso load and relatively late contact result in a 66.3 mm and 56.2 mm neck displacement at head contact. The relative head-torso acceleration is higher in the female at 4.7 g versus 3.8 g for the male, and the relative head-torso velocity is 2.0 m/s versus 2.1 m/s, all higher than with the 1990 C car.

In contrast, the 1997 W car seat allows the torso to penetrate into the seatback at lower loads of 1.9 kN and 1.3 kN, which reduce seatback rotation to only 2.7^0 and 1.8^0 respectively with a 16.4 mm and 10.5 mm head restraint drop and 113 mm and 102 mm gap. Most importantly, the time of head restraint contact is earlier

Table 2: Peak Seat and Occupant Biomechanical Responses At or Before Head Restraint Contact for the Female and Male in 6-24 km/h Rear Delta V Crashes.

	1997 W car		**Foreign Benchmark**		**1990 C car**	
k Seat Stiffness (kN/m)	20	15	40	30	20	15
j Rotation Stiffness (deg/kN)	1.4	1.4	1.8	1.8	3.4	3.4
jk (deg/m)	28	21	72	54	68	51
Yield Load (kN)	10	10	10	10	5	5
1.6 g / 6 km/h	**Male**	**Female**	**Male**	**Female**	**Male**	**Female**
Head Contact (ms)	134	118	nc	156	nc	152
Head-Vehicle Vel. (m/s)	1.26	1.35	1.00	1.12	0.58	1.40
Head Vel. (m/s)	0.96	1.12	0.71	1.12	0.62	0.86
Head-Torso Accel. (g)	0.89	0.85	1.37	1.32	0.90	1.00
Head-Torso Vel. (m/s)	0.58	0.58	0.77	1.95	0.58	0.74
Neck Displacement (mm)	36.7	24.4	53.8	85.5	45.4	51.0
Torso-Vehicle Displ. (mm)	67	75	39	50	67	81
Seatback Rot. (deg)	1.9	1.6	2.8	2.7	4.6	4.2
Torso Load (kN)	1.34	1.13	1.56	1.49	1.34	1.23
Head Restraint Drop (mm)	-11	-9	-17	-16	-29	-26
Head Restraint Gap (mm)	103	99	114	113	136	131
HR Trajectory (deg)	0	-3	nc	0	nc	0
Energy Transfer (J)	82	30	236	177	257	91
3.0 g / 10 km/h						
Head Contact (ms)	90	84	112	96	124	102
Head-Vehicle Vel. (m/s)	2.16	2.10	1.83	1.84	2.21	2.20
Head Vel. (m/s)	1.68	1.78	1.31	1.58	1.17	1.57
Head-Torso Accel. (g)	1.65	2.12	2.51	3.06	1.65	2.12
Head-Torso Vel. (m/s)	0.81	0.93	1.41	1.66	1.06	1.24
Neck Displacement (mm)	23.8	24.7	72.2	61.9	57.5	44.7
Torso-Vehicle Displ. (mm)	91	77	70	57	123	97
Seatback Rot. (deg)	2.5	1.6	5.1	3.1	8.3	4.9
Torso Load (kN)	1.81	1.15	2.81	1.72	2.45	1.45
Head Restraint Drop (mm)	-15	-10	-32	-19	-55	-31
Head Restraint Gap (mm)	111	100	142	117	179	139
HR Trajectory (deg)	-7	-4	0	-1	-3	-8
Energy Transfer (J)	40	24	290	117	198	64
4.5 g / 16 km/h						
Head Contact (ms)	72	68	92	80	100	84
Head-Vehicle Vel. (m/s)	2.89	2.79	2.75	2.73	3.31	3.14
Head Vel. (m/s)	2.16	2.30	2.00	2.19	1.80	2.05
Head-Torso Accel. (g)	2.42	3.08	3.76	4.73	2.47	3.17
Head-Torso Vel. (m/s)	0.82	0.96	2.02	2.11	1.44	1.42
Neck Displacement (mm)	17.1	18.1	66.3	56.2	48.9	36.9
Torso-Vehicle Displ. (mm)	97	84	101	77	156	115
Seatback Rot. (deg)	2.7	1.8	7.3	4.1	10.6	5.8
Torso Load (kN)	1.94	1.26	4.04	2.32	3.12	1.72
Head Restraint Drop (mm)	-16	-11	-47	-25	-73	-37
Head Restraint Gap (mm)	113	102	167	130	205	150
HR Trajectory (deg)	-9	-6	-7	-5	-20	-14
Energy Transfer (J)	27	18	278	121	153	55
6.7 g / 24 km/h						
Head Contact (ms)	60	56	78	66	84	70
Head-Vehicle Vel. (m/s)	3.71	3.52	4.08	3.82	4.74	4.23
Head Vel. (m/s)	2.69	2.82	2.51	2.69	2.36	2.53
Head-Torso Accel. (g)	3.17	3.80	5.64	6.88	3.70	4.63
Head-Torso Vel. (m/s)	0.77	0.87	2.43	2.31	1.65	1.49
Neck Displacement (mm)	13.5	13.6	61.0	45.0	42.8	30.1
Torso-Vehicle Displ. (mm)	108	92	131	98	185	132
Seatback Rot. (deg)	3.0	1.9	9.2	5.1	12.5	6.7
Torso Load (kN)	2.15	1.39	5.24	2.93	3.69	2.0
Head Restraint Drop (mm)	-18	-12	-62	-32	-88	-44
Head Restraint Gap (mm)	117	104	188	142	226	161
HR Trajectory (deg)	-11	-7	-16	-11	-30	-20
Energy Transfer (J)	22	14	292	108	141	47

because of the substantially lower gap induced by seatback rotation and lower torso accelerations. The lower torso loads also reduce neck loads. This results in smaller neck displacements although the female response is larger than the male at 18.1 mm versus 17.1 mm, respectively. The relative head-torso acceleration is higher in the female at 3.1 g versus 2.4 g, as is the relative head-torso velocity at 0.96 versus 0.82 m/s in the male. The head impact velocities are similar; however, the neck loads and accelerations are reduced from those with the foreign benchmark or C car seats.

Figures 5 and 6 show that female neck responses are greater than in the male for the early loading. This is true for neck displacement, and the relative velocity and acceleration between the head and torso. Using Equations (1-4), the relative acceleration between the head and torso is:

$$(\ddot{x} - \ddot{z}) = (k/m_T)y - (k/m_T + k_N/m_H)x + (k_N/m_H)z \quad (11)$$

The ratio between female and male responses can be estimated early in the crash by Equation (11) based solely on the y displacement from the vehicle crash and second order effects from torso and head displacement:

$$\begin{aligned}&(\ddot{x}_F - \ddot{z}_F)/(\ddot{x}_M - \ddot{z}_M) = \\ &[(k_F/m_{TF})/(k_M/m_{TM})](y_F/y_M) - f(x_F, x_M, z_F, z_M)\end{aligned} \quad (12)$$

where $f(x_F, x_M, z_F, z_M)$ is a function of torso and head displacement for the male and female, which depends on the ratios k/m_T and k_N/m_H. The higher responses of the female than male early in the crash are related to the difference in seat stiffness and torso mass, which influence occupant acceleration. For the assumptions used in this analysis, the ratio is 1.30 indicating that neck displacement for the female are 30% greater than that of the male as a function of vehicle displacement (time). The relative acceleration and velocity between the head and neck for the female are also 30% greater in the female. Figures 5 and 6 show this ratio holds up to about 90 ms for the 10 km/h rear crash. After that time, other effects of the torso displacement, neck stiffness and head mass play a role in the neck response.

Figure 7 shows peak neck displacement for the female and male adjusted to a similar tolerance. This involved multiplying the male displacement by 0.82 to reflect a similar risk as the female based on geometric scaling. The result is consistent with higher whiplash risks in women for the foreign benchmark and 1980s-90s yielding seats for crashes below 10 km/h. Responses with the W car seat are rather comparable throughout the crash range, although the trend is for lower responses for women in the lowest severity crashes.

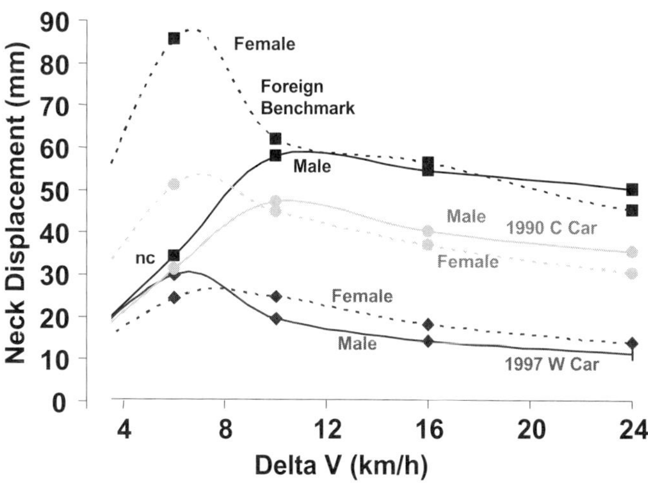

Figure 7: Peak neck displacements for the female (dashed lines) and male (solid lines) up to head restraint contact for 6-24 km/h delta V rear crashes with the three exemplar seats. The male neck displacement has been scaled by 0.82 to reflect a similar tolerance to that of the female. No contact with the head restraint is noted by nc.

DISCUSSION

One Reason Why Women Have Higher Whiplash Rates than Men: There are many factors that influence the difference in whiplash between men and women. This study highlights one. Early in the crash, the female has higher neck responses than the male because of the relative difference in seat stiffness and torso mass, which influence torso acceleration and displacement. For the assumptions here, there are 30% higher relative responses, and likely a higher risk. Interestingly, the first and second order effects in the loading are solely based on seat stiffness and torso mass, and then neck stiffness and head mass ratios between the female and male. The j factor of seat frame rotation stiffness is not a direct factor in the relative response. This indicates that seat frame rotation is not a key to controlling the higher whiplash risks in females except as it indirectly influences the k factor in actual seats. Rather, adjusting down the ratio $(k_F/k_M)(m_{TM}/m_{TF})$ is fundamental. Obviously, more work is needed to study the relative difference in seat stiffness as a function of the size of females and males, but this study points to the importance of lowering k_F as a means of reducing whiplash risks in females.

An effort was made to adjust the peak neck displacements of the male to a comparable tolerance of the female using the 0.82 geometric factor, but the uncertainty in this type scaling is great. When the male neck displacements were scaled, peak neck displacement in the female was higher than the male for crash below 10 km/h with the 1990 C car seat. Since this exemplifies the yielding seats of the 1980s-90s, the ratio is consistent with the field experience. There are several additional observations. If whiplash injury occurs early in the loading, the neck response differences are related to the ratio $(k_F / k_M)(m_{TM} / m_{TF})$, which shows a 30% higher neck response in the female. Also, the geometric scaling may not be sufficient to reflect the underlying risk differences between females and males, a different adjustment may be needed, and the adjustment may be time or rate dependent.

This study shows that the lower neck stiffness of the female is a factor only late in the loading, yet it cannot be discounted as influencing the higher incidence of whiplash in women. In a rear crash, the seat is pushed forward, and the size of the occupant does not influence the seat cushion motion. Since the seatback has a lower stiffness with the female, less force is developed as compared to a male with the same seat deformation. However, the lighter weight of females means that higher accelerations will occur for the same force as in a male. The female experiences higher accelerations than the male, and it occurs earlier with the female. This also gives similar or higher impact velocity into the head restraint, which may be an additional factor related to injury.

Higher torso acceleration in the female is consistent with data from Hyge sled tests of the 1990 C car seat (Figure 3) and human volunteers 6.5 km/h and 9.5 km/h delta V (Hell et al. 2002). Thus, lower torso loads on the female are offset by a lighter torso mass. This results in higher torso accelerations for the female, and earlier and greater displacement of the base of the neck. In this study, the early displacements were 30% greater.

There are other differences between men and women in vulnerability to whiplash. Women tend to drive smaller cars than men, and may be exposed to higher crash severities based on the lighter vehicles they use for travel (Viano, Gargan 1996). Yet the smaller vehicles are more often fit with integrated head restraints, which tend to provide better head restraint height. Women are also lighter than men and sit less deep into the seat cushion. This allows easier ramping up the seatback, which causes the head and neck to rise with respect to the head restraint.

The issue of potential gender differences is perplexing and clearly not well understood. There are uncertainties in comparable risk on the basis of exposure, tolerance, and driving situation. It could be argued that while women drive substantially fewer miles per year, they may do more driving in urban areas where the risks of a rear crash are greater. Women also tend to drive with a more upright posture to better position their eyes for driving visibility, but this tends to increase the gap behind the head because they do not always rotate the seatback more upright. A larger gap increases neck responses in rear sled tests with male and female dummies (Viano 2002).

The shift from yielding seats of the 1980s to stiffer designs has increased neck responses in both women and men. This is a consequence of higher seat stiffness and higher loads on the torso that accelerate the base of the neck, leading to higher neck displacements before head restraint contact. The move to a higher placement of head restraints and closer proximity to the back of the head reduce neck displacements up to head contact. However, evaluating these effects is beyond the scope of this study.

Essential Biomechanical Responses Related to Whiplash: Panjabi et al (1999) recommended evaluating neck extension angles for whiplash assessment and Viano, Davidsson (2002), Viano et al. (2002) built on that proposal by adding neck displacements of the occipital condyles (OC) with respect to T1. These criteria are consistent with the neck displacements reported here. Figure 7 shows differences in neck displacement for the three exemplar seats and the amplitude for each increases with decreasing rear crash delta V. This trend is consistent with field data showing the greatest incidence of whiplash in lower speed crashes (Langwieder, Hell 2002, Hell et al. 2002). The highest risks are found for the foreign benchmark seat in low rear delta V crashes.

The Neck Injury Criterion (NIC) addresses the early acceleration differences between T1 and the head, which promote the S-shaped response before head contact (Bostrom et al. 1996). NIC consistently shows a higher response in females than males. For example, the relative acceleration between the head and torso is 30% higher in the female than male for the three exemplar seats. However, NIC shows an increase with crash delta V, which is not consistent with field experience. The head to torso velocity difference is also 30% higher in the female, reflecting other possible injury mechanisms (Yoganandan, Pintar 2000). What is clear is that the ratio of peak neck displacements for the female and male is not consistent with the field experience with higher injury rates in women.

Role of Yielding Seatbacks and SAHR Active Head Restraints in Reducing Whiplash: This analysis underscores the importance of the yielding seat to whiplash prevention in males and females. Historically, the yielding behavior was achieved by seatback rotation at relatively low loads, which is associated with other types of injury in high-speed rear crashes. These designs had relatively stiff structural cross-members in the seatback, which made most of the compliance due to recliner and seat frame rotation although the foam trim covering the frame provide some deformation distance until the motion was stopped by seatback structures. Often, these structures were just behind the pelvis, which limited penetration into the seatback and increased the tendency for ramping. A way to maintain the earlier yielding behavior without significant seatback rotation involved an open perimeter frame, which used a strong seat frame and recliner, but allowed yielding of the occupant between the frame side structures through compliance of the seat suspension (Viano 2002).

A suspension compliance with low k in the high retention seat is an important factor in the reduction of whiplash risks, and even lower k values within other constraints would be beneficial for whiplash injury prevention. The lower forces on the back with the yielding seatback are a primary means of lowering loads on the base of the neck (Viano, Olsen 2001). This reduces neck displacements before head restraint contact. Displacement of the base of the neck with respect to ground is a key factor in whiplash as shown by the analysis here. More displacement of the occupant with respect to the vehicle lowers the accelerations needed to reach the vehicle delta V. There are also benefits from the open perimeter frame seatback, which maintains a more upright posture until head and neck support. After that point, the seatback rotates as energy is transferred to the occupant.

The Saab Active Head Restraint (SAHR) was designed to move the head restraint toward the head early in a crash. The system involves a seatback design that lowers loads on the upper torso and allows the body to penetrate into the seat. This early displacement activates the SAHR mechanism, which rotates the head restraint upward and forward. The two main principles of SAHR are lowering the stiffness of the upper seatback to reduce forces on the upper back and shoulders, and the early movement of the head restraint to close the gap and change its trajectory. The field data on SAHR performance is noteworthy, since it showed no women with whiplash injury with SAHR, although neck pain lasting less than a week occurred (Viano, Olsen 2000).

Limitations of the Study: Obviously, the model developed and used in this analysis is not a full occupant dynamics simulation, which would offer the possibility to explore responses is more detail. The model does provide sufficient detail. It is a convenient means of addressing the fundamental responses related to seatback deformations and torso-head dynamics in rear crashes, and it showed a reasonably good response comparison to experimental data with the 50% male and 5% female dummy in rear impacts. However, the solution does not reflect the influence of occupant ramping or pelvic drop with occupant displacement into the seat or the effects of gravity acting on the occupant or the inertial effects of the seat. Also, this study does not address variations in the initial head restraint height or gap behind the head, or head restraint interactions, which are factors in real-world injury. Most modern seats have a much more favorable head restraint placement than seat designs of the 1980s. These effects do not markedly affect the analysis and are beyond the scope of this analysis. However, they could be pursued using this model or more sophisticated occupant dynamic simulations with these results as a basis.

The loading interface on the upper seatback influences the force distribution on the base of the neck. This may require more than a single stiffness value to adequately describe the interaction. Also, the flexible spine of the human is not comprehended as the lordotic curvature straightens in a rear crash. The flexible spine of BioRID also involves different accelerations of T1 and T6, which are not comprehended with a single torso mass. The whole topic of scaling anthropometry, impact responses and tolerances between male and female occupants needs more evaluation. Finally, the idealized constant acceleration rear crash pulse is not representative of actual vehicles, which show a double acceleration pulse (Langwieder, Hell 2002, Viano 2002). However a more complex crash pulse would not greatly alter the response comparisons. Viano (2002b) showed that an initially higher acceleration did not change the comparison between stiff and yielding seats in the earlier study.

Obviously, the parameters defining neck stiffness, tolerance and seat stiffness are critical to the analysis of differences between male and female occupants. The most important is the neck displacement tolerance scaling. While accepted scaling techniques were used, more evaluation of the sensitivity to variations and research on underlying differences in needed. For example, the difficulty in clear diagnosis of whiplash complicates scaling based on unknown mechanical principles. Some parameter variation was investigated and did not alter the conclusions of the study.

The 14 seats in the foreign benchmark group have a wide range of characteristics in stiffness, compliance and frame strength. The exemplar seat with k = 40 kN/m and j = 1.8 ^{0}/kN stiffness does not represent any specific product in use. It is a composite of the top stiffness seats, and the actual designs varied considerably. Nonetheless, the study shows the importance of adjusting the seat frame stiffness and seat suspension compliance (the j and k factors) to improve occupant dynamics in rear crashes. These parameters define the essential characteristics of the seat for rear impact loading of the occupant. This study points to the importance of seat stiffness and torso mass in the early neck responses and differences between the male and female.

ACKNOWLEDGMENTS

This paper was published in Traffic Injury Prevention 4(3) 2003, and is reprinted here with permission from the publisher, Taylor & Francis Group. Their agreement for its use in this book is greatly appreciated.

REFERENCES

1. Backaitis S, Mertz H. (1994) Hybrid III: The First Humanlike Crash Test Dummy. SAE Special Publication PT-44, Society of Automotive Engineers, Warrendale, PA..
2. Bostrom O, Svensson M, Aldman B, et al. (1996) A New Neck Injury Criterion Candidate Based on Injury Findings in the Cervical Spine Ganglia after Experimental Sagittal Whiplash. IRCOBI Conf 123-136.

3. Buck CA, Dameron FB, Dow MJ, Skowlund HV. (1959) Study of Normal Range of Motion in the Neck Utilizing a Bubble Goniometer. Archives of Physical Medicine and Rehabilitation, 40 September.
4. Cassidy JD, Carroll LJ, Cote P, Lemstra M, Berglund A, Nygren A. (2000) Effect of eliminating compensation for pain and suffering on the outcome of insurance claims for whiplash injury. N Engl J Med Apr 20;342(16):1179-86.
5. Chapline JF, Ferguson SA, Lillis RP, Lund AK, Williams AF. (2000) Neck pain and head restraint position relative to the driver's head in rear-end collisions. Accid, Anal Prev Mar;32(2):287-97.
6. Dolinis. (1997) Risk Factors for "Whiplash" in Drivers: a Cohort Study of Rear-end Traffic Crashes. Injury. Apr 28(3):173-79.
7. Evans RW. (1992) Some Observations on Whiplash Injuries. Neurol Clin. 10(4):975-97.
8. Foust DR, Chaffin DB, Snyder RG, Baum JK. (1973) Cervical Range of Motion and Dynamic Responses and Strength of Cervical Muscles. SAE 730975, 17th Stapp Car Crash Conference, Society of Automotive Engineers, Inc., Warrendale, PA.
9. Galasko CSB, Murray PM, Pitcher M, et al. (1993) Neck Sprain After Road Traffic Accidents: a Modern Epidemic. Injury 24 (3), 155-157.
10. Granville AD, Kreezer G. (1937) The Maximum Amplitude and Velocity of Joint Movements in Normal Male Human Adults. Human Body, Vol. 9.
11. Hell W, Schick S, Langwieder K, Zellmer H. (2002) Biomechanics of Cervical Spine Injuries in Rear End Car Impacts: Influence of Car Seats and Possible Evaluation Criteria. Traffic Injury Prevention 3(3): 127-140.
12. Holm L, Cassidy JD, Sjogren Y, Nygren A. (1999) Impairment and Work Disability Due to Whiplash Injury Following Traffic Collisions. An Analysis of Insurance Material from the Swedish Road Traffic Injury Commission. Scand J Public Health. 27(2):116-23.
13. Huelke DF, Marsh JC. (1974) Analysis of Rear-End Accident Factors and Injury Patterns. Proc. 18th AAAM, Assoc for the Adv of Automotive Medicine, Des Plaines, IL, 174-199.
14. Kihlberg JK. (1969) Flexion-Torsion Neck Injury in Rear Impacts. 13th Proc. AAAM, Assoc for the Adv of Automotive Medicine, Des Plaines, IL.
15. Laberge-Nadeau, C, Tao XT, Maag U. (1993) Neck Injuries Amongst Motor Vehicle Occupants Involved in a Collision. AAA Foundation for Traffic Safety, Washington DC, April.
16. Langwieder K, Hell W. (2002) Proposal of an International Harmonized Dynamic Test Standard for Seat/Head Restraints. Traffic Injury Prevention 3(3): 150-158.
17. Larsen LB, Holm R. (2000) Prolonged neck pain following automobile accidents. Gender and age related risk calculated on basis of data from an emergency department. Ugeskr Laeger Jan 10;162(2):178-81.
18. Mertz, HJ, Patrick L. (1971) Strength and Response of the Human Neck. SAE 710855, Society of Automotive Engineers, Inc., Warrendale, Pennsylvania, USA.
19. Mertz HJ, Neathery RF, Culver CC. (1973) Performance Requirements and Characteristics of Mechanical Necks. in Human Impact Response: Measurement and Simulation, 263-288, Plenum Press, New York.
20. Mertz HJ, Irwin AL, Melvin JW, Stalnaker RL, Beebe MS. (1989) Size, Weight and Biomechanical Impact Response Requirements for Adult Size Small Female and Large Male Dummies. SAE 890756, Society of Automotive Engineers, Warrendale, PA..
21. NHTSA website at www.nhtsa.dot.gov, 2002.
22. O'Neill B, Haddon W, Kelley AB, Sorrenson WW. (1972) Automobile Head Restraints - Frequency of Neck Injury Claims in Relation to the Presence of Head Restraints. Amer Journal of Public Health 62:399- 405.
23. Ono K, Kanno M. (1996) Influence of the Physical Parameters on the Risk to Neck Injuries in Low Impact Speed Rear Collisions. Accid Anal Prev 28(4):493-499.
24. Panjabi MM, Wang JL, Delson N. (1997) Neck Injury Criterion Based on Intervertebral Motion and its Evaluation Using an Instrumented Neck Dummy. 1999 IRCOBI Conf, 179-190.
25. Prasad P, Kim A, Weerappuli DPV, Robert V, Schneider D. (1997) Relationship Between Passenger Car Seat Back Strength and Occupant Injury Severity in Rear End Collisions: Field and Laboratory Studies. SAE 973343, Society of Automotive Engineers, Warrendale, PA.
26. Richter M, Otte D, Pohlemann T, Krettek C, Blauth M. (2000) Whiplash-type Neck Distortion in Restrained Car Drivers: Frequency, Causes and Long-term Results. Eur Spine J. 9(2):109-17.
27. SAE J1460-2. (1998) Human Mechanical Impact Response Characteristics-Responses of the Human Neck to Inertial Loading by the Head for Automotive Seated Postures. Society of Automotive Engineers, Warrendale, PA.
28. Schneider LW, Foust DR, et al. (1975) Biomechanical Properties of the Human Neck in Lateral Flexion. SAE 751156, 19th Stapp Car Crash Conference, Society of Automotive Engineers, Inc., Warrendale, PA.
29. Severy DM, Brink HM, Baird JD. (1967) Preliminary findings of head support designs. 11th Stapp Car Crash Conf, SAE 670921, SAE Warrendale PA, 337-405.
30. States JD, Korn MW, Masengill JB. (1969) The Enigma of Whiplash Injuries. 13th Proc. AAAM, Assoc for the Adv of Automotive Medicine, Des Plaines, IL, 83-108.

31. Strother CE, James MB. (1987) Evaluation of Seat Back Strength and Seat Belt Effectiveness in Rear End Impacts. 31st Stapp Car Crash Conference, SAE 872214, Society of Automotive Engineers, Warrendale, PA, 225-244.
32. Thomas C, Faverjon G, Hartemann F, Tarriere C, et al. (1982) Protection Against Rear-End Accidents. Proc. IRCOBI 17-29.
33. Viano DC, Gargan MF. (1996) Seating Position and Headrest Location During Normal Driving: Implications to Neck Injury Risks in Rearend Crashes. Accident Analysis & Prevention, 28(6):665-674.
34. Viano DC, Olsen S. (2001) The Effectiveness of Active Head Restraint in Preventing Whiplash. J Trauma 51:959-969, November.
35. Viano DC. (2002) Role of the Seat in Rear Crash Safety. SAE Book, ISBN 0-7680-0847-6, Society of Automotive Engineers, Warrendale, PA, SAE R-317:1-491.
36. Viano DC, Olsen S, Locke GS, Humer M. (2002) Neck Biomechanical Responses with Active Head Restraint Systems: Rear Barrier Tests with BioRID and Sled Tests with Hybrid III. SAE 2002-01-0030, Society of Automotive Engineers, Warrendale, PA.
37. Viano DC, Davidsson J. (2002) Neck Displacements of Volunteers, BioRID P3 and Hybrid III in Rear Impacts: Implications to Whiplash Assessment by a Neck Displacement Criterion (NDC). Traffic Injury Prevention 3(2):105-116.
38. Viano DC. (2003a) High Retention Seat Performance in Quasistatic Seat Tests. SAE 2003-01-0173, Society of Automotive Engineers, Warrendale, PA.
39. Viano DC. (2003b) Energy Transfer to an Occupant in Rear Crashes: Effect of Stiff and Yielding Seats. SAE 2003-01-0180, Society of Automotive Engineers, Warrendale, PA.
40. Viano DC. (2003c) Seat Properties Affecting Neck Responses in Rear Crashes: A Reason Why Whiplash Has Increased. In print, Traffic Injury Prevention 4(3).
41. Viano DC. (2003d) Effectiveness of High-Retention Seats in Preventing Fatality: Initial Results and Trends. SAE 2003-01-1351, Society of Automotive Engineers, Warrendale, PA.
42. Warner C, Strother C, James MB, Decker RL. (1991) Occupant Protection in Rear-End Collisions II: The Role of Seat Back Deformation in Injury Reduction. 35th Stapp Car Crash Conf, SAE 912914, Society of Automotive Engineers, Warrendale, PA.
43. Yoganandan N, Pintar FA. editors. (2000) Frontiers in Whiplash Trauma: Clinical and Biomechanical. IOS Press Amsterdam, The Netherlands, Library of Congress No. 00-101366.

CHAPTER 5:

ENERGY TRANSFER TO AN OCCUPANT IN REAR CRASHES: EFFECT OF STIFF AND YIELDING SEATS

2003-01-0180

Energy Transfer to an Occupant in Rear Crashes: Effect of Stiff and Yielding Seats

David C. Viano[1,2,3]
[1]ProBiomechanics LLC
[2]Crash Safety Division, Chalmers University of Technology
[3]Safety Integration Center, General Motors Corporation

ABSTRACT

For several decades, there has been a debate on the safety merits of yielding and rigidized (stiff) seats. In 1995, GM adopted requirements for high retention seats and introduced a new generation of yielding seatbacks. These seats have the same stiffness as the yielding seats of the 1980s and early 1990s, but have a strong frame structure and recliners to substantially limit seatback rotation in severe rear crashes. The yielding behavior is given by compliance of the seat suspension across the side structures and an open perimeter frame, which allows the occupant to penetrate into the seatback.

The purpose of this study is to compare the energy transfer characteristics and occupant dynamics of yielding and stiff seats in 35 km/h and 16 km/h rear crashes. Based on benchmarking tests, the stiff seatback is defined as one having a 40 kN/m stiffness in rearward loading by a Hybrid dummy. The yielding seatback has a 20 kN/m stiffness, which is comparable to the 1997 W car seat, the first high retention seat introduced by GM. For this study, the vehicle acceleration was a constant 10 g or 4.5 g over 100 ms giving the delta V of 35 km/h or 16 km/h. An Excel spreadsheet analysis was conducted to determine occupant dynamics in a step-forward solution based on the differential displacement between the occupant and vehicle. Force and acceleration of the occupant are proportional to seatback stiffness and the differential displacement.

The yielding seatback developed 15%-16% lower forces on the occupant and acceleration with a delayed loading as compared to the stiff seat. The time delay improved the efficiency of the energy transfer and allowed force to be applied on the occupant after the head, neck and torso received uniform support by the seatback and head restraint. Even with an initial 15 cm gap behind the occupant or a higher initial vehicle acceleration pulse, the yielding seatback provided a more gradual acceleration and lower loads on the occupant with 68% greater occupant displacement with respect to the vehicle.

This study shows the benefits of a yielding seatback over a stiff seat on occupant dynamics in rear crashes. The more gradual ride-up of the crash delta V allows a later occupant loading during the time the head, neck and back have more uniform support from the seat and head restraint. In contrast, the stiff seat has a load of 4 kN when the relative penetration of the occupant is 10 cm, which is before the likely time of head restraint contact. The comparable load with the yielding seatback is only 2 kN indicating much lower forces on the back and base of the neck.

INTRODUCTION

Severy et al. (1955, 1967a,b, 1968a,b) conducted pioneering research on seat and occupant dynamics in rear crashes and showed the need for controlled deformation of the seatback and support of the head. He also worked on concept seats, including rigidized or stiff seatbacks. However, many of the concepts were not brought to production because of various criticisms on overall performance. Other seat concept studies were conducted by Saczalski et al. (1993) and Blaisdell et al. (1993).

Strother, James (1987), Warner et al. (1991) and Prasad et al. (1997) provided engineering analyses and rationale for the yielding seatback in rear crashes. They countered the rigidized seat approach with the yielding seat concept as the best means to provide overall occupant safety in rear crashes by controlled seatback rotation that limited forces on the neck. This involved an energy absorbing restraint as the seatback yielded rearward in a crash and followed the historic trend in load-limiting and energy absorbing deformation of interior components, such as the steering system, padded dashboard and high-penetration resistant windshield (see summary in Viano 1986). A seatback moment near the FMVSS 207 criterion was argued as necessary to balance the needs for whiplash injury prevention in low-speed crashes and occupant retention in severe rear crashes.

The debated furthered the notion that there was an underlying design conflict between occupant retention by stiff or rigidized seats in severe, but infrequent rear crashes, and the need for a yielding seatback to prevent whiplash in the frequent, minor rear crashes. Petitions for NHTSA rulemaking to increase seatback strength in the FMVSS 207 test were submitted by Saczalski (1989) and Cantor (1989); and, various product liability cases were pursued on seatback strength (Rake and Boehm 1990). These proposals and cases furthered the debate and relative positions of each side.

From 1990 through 1995, General Motors pursued a study on the role of the seat in rear crashes, and addressed the yielding seat theory. The research showed that there was not a design conflict between a rigidized and yielding seat system, if compliance of the seatback between the side frame structures was considered in the load-limiting performance of the seat. The research led to new seat requirements for rear impact performance, which were adopted in 1995 for all new seat systems introduced after 1997. The seat requirements led to the development of a new generation of yielding seats that have been phased into production (Viano 2002, 2003).

High Retention Seat Requirements: The new generation of seats was designed to specifications for high retention (HR) in a Quasistatic Seat Test (QST). The QST involves occupant loading of the seat in a rearward direction and targets peak H-point moment to >1700 Nm giving an energy transfer capability of 2000 J (Viano 2002). Based on QST tests of seats from the late 1980s and early 1990s, and a series of Hyge sled tests, performance specifications were established for seats to achieve high retention. The specifications are summarized in Table 1 and provide criteria for front bucket or split bench seats.

Table 1: Seat Technical Specifications for High Retention in the Quasistatic Seat Test

1. >1700 Nm (15,000 in-lb) Moment About H-Point
2. No Force Drops >2 kN (450 lb) in 50 ms Causing >10° Change in Seatback Angle
3. Seatback Twist <15° for Seatback Angles up to 60°
4. Head Restraint Height to B-Plane and Front Surface to Within 20 mm of Back-of-Head Ellipse for the 95th Percentile Occupant

The minimum H-point bending moment of 1700 Nm (15,000 in-lb) provides energy transfer to the occupant without excessive rotation of the seatback and possibility of loss of retention in severe rear crashes. This provides energy management in rear sled tests that are consistent with the exposure in a 33.7 mph (54.2 km/h) FMVSS 301-type rear moving barrier test. This test results in an average delta V of 31.1 km/h (19.3 mph) with a range of 27.4-35.4 km/h (17-22 mph) for a series of passenger cars of different mass. FMVSS 301 involves a 30 mph (48.3 km/h) flat moving barrier of 2273 kg (5000 lb) mass. The average energy transfer at 60^{0} for seats that provide occupant retention in the higher-speed rear test is about 2000 J, which corresponds to a velocity performance of 31.4 km/h (19.5 mph) based on a kinetic energy calculation and a 70% effective occupant mass loading the seatback, or 52.5 kg for the 50th percentile male occupant (Viano 2002).

Second, deformation of seat hardware should provide smooth control of occupant kinematics and uniform resistance. Thus, there should not be a drop in ram load of more than 2 kN in a 50 ms running time-window that causes more than a 10° change in seatback angle before load reverses in the QST. This criterion assures uniform loading during a test and eliminates rapid changes, which may influence occupant kinematics and retention.

Third, this criterion was originally set to <15° twist of the seatback up to 60° average seatback angle and was based on data from pre-HR seats. As seats developed to the new specification, the higher rotation limit became impractical, as HR seats met the moment criterion at much lower rotation angles than pre-HR seats (Viano 2003). The revised criterion limits seatback twist to <15° up to 40° average seatback angle from vertical or a peak H-point moment of 2000 Nm to prevent seatback twist and loss of retention by lateral motion of the occupant.

Fourth, the head restraint needs to be positioned high and close enough to support the head and neck as greater energy transfer is achieved by the seatback. This assures that retention and occupant containment are not associated with an increased risk of whiplash injuries in low-speed through high-speed crashes, even with forward seating positions and large gaps to the head restraint.

A head restraint design criterion was adopted after Hyge sled testing. The height of the top of head restraint should be above the B-plane (a horizontal plane through the head center of gravity of the 95th percentile seated occupant) and closer than 20 mm rearward of the back-of-head ellipse for the 95th percentile seated occupant.

The first three criteria are determined in a QST test. The fourth criterion is specified by the design layout for the seated occupant. Evaluation of performance to the criteria can be done during design and early prototyping of a seat, before final vehicle crash testing.

High Retention Seat Performance: In a recent study of high retention seats, Viano (2003) found that the average stiffness of more than 20 GM high retention seats was 25.3 ± 4.9 kN/m, which was statistically similar to the average 24.7 ± 4.9 kN/m for pre-high retention GM seats. While the seatback stiffness was similar and the effects of the yielding performance were maintained, the high retention seats had a 35% lower seatback rotation at the 1700 Nm criterion. The yielding was given by compliance of the open perimeter frame design that will be described later.

The high retention (HR) seats had an average H-point moment of 2537 ± 703 Nm and represented a 2.3-times higher moment than pre-HR designs of the 1980s and early 1990s. A 2.9-times higher energy transfer capability was achieved by the HR seats to 3659 ± 1140 J providing calculated rotations <60^0 for crashes up to 42.5 km/h (26.4 mph) on average with the 50th percentile male. This was a 69% increase over pre-HR seats (Viano 2003).

Seat Stiffness and Compliance to Occupant Loading: By using the Quasistatic Seat Test (QST), the seat stiffness is determined between occupant loads of 1 kN to 4 kN (see Viano 2002). This defines the compliance of the seat system to occupant loading and is generally reported in kN/m. Figure 1 shows the benchmark average of 14 high-end competitive seats from foreign manufacturers. The competitive seat stiffness averaged 34.2 ± 6.5 kN/m with a range from 26-47 kN/m stiffness at a seatback angle of 34^0 at 5 kN.

The 1990 C car seat used an open seatback frame and allowed the occupant to penetrate between the side structures. This seat had an average stiffness of 21.1 ± 2.4 kN/m and was much more compliant to occupant loading than other seats on the market. However, it rotated to 42^0 under a 5 kN occupant load. This seat was the genesis for a new generation of seats using a perimeter frame design that allows the occupant to penetrate between the side frame structures. The average stiffness of the first GM seats meeting the high retention requirements was 22.1 ± 5.4 kN/m for the 1997 W and P-90 seats and achieved a 5 kN load at 32^0, nearly 10^0 less seatback rotation than the 1990 C car seat.

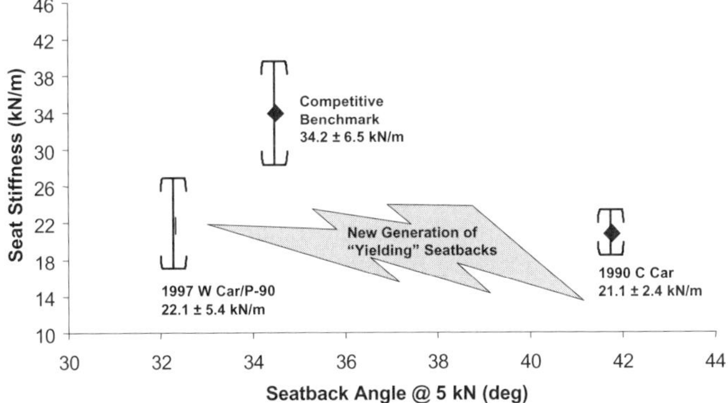

Figure 1: Comparison of seat stiffness of a competitive benchmark of high-end seats versus the 1990 C car seat, which was the genesis for the first high retention seat used by GM in the 1997 W car.

Figure 2 shows an example of a high retention seat using the perimeter frame approach. This design uses a strong seat frame and recliners for occupant retention in severe rear crashes. This limits seatback rotation. However, the open seatback allows a compliant design, which gives a yielding behavior by deformation of the seat trim and EA pelvic strap to pocket the occupant's pelvis and lower back in a rear crash.

For the purpose of a theoretical analysis of the yielding and stiff seatbacks, two levels of stiffness were used. One represented the benchmark stiffness of luxury competitive seats with a 40 kN/m stiffness; and, the other represented the high retention design using a perimeter frame with 20 kN/m stiffness. This study addresses vehicle and occupant dynamics in a 35 km/h and 16 km/h rear crash. Occupant responses and energy transfer were determined for the stiff and yielding seatbacks in the rear crashes. The work of Stephens et al. (2000) is pertinent.

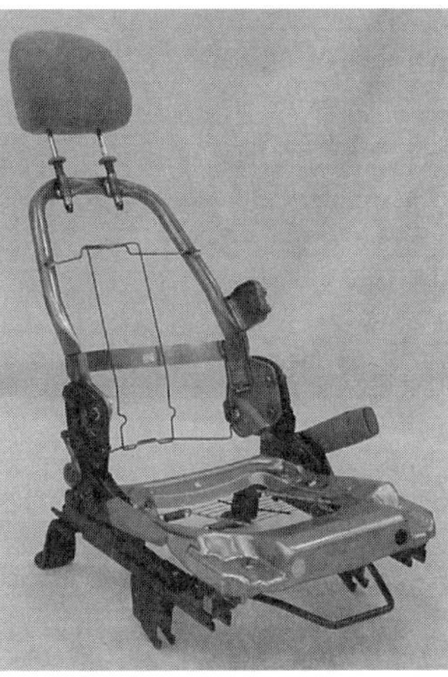

Figure 2: New generation of yielding seatbacks using the perimeter frame seat concept with open and compliant seatback (modified from Viano 2002).

METHODOLOGY

Nomenclature for Data Analysis:

F	Force on Occupant in kN
E	Energy Transfer to Occupant by the Seat in J
ΔE	Incremental Energy Transfer to the Occupant per Δt in J
x	Displacement of the Occupant in m
$\dot{x}$	Velocity of the Occupant in m/s
$\ddot{x}$	Acceleration of the Occupant in m/s^2
y	Displacement of the Vehicle in m
$\dot{y}$	Velocity of the Vehicle in m/s
$\ddot{y}$	Acceleration of the Vehicle in m/s^2

i	Time integer
Δt	Delta time in s
k	Stiffness of the seat in kN/m
m	Effective Mass of the Occupant Loading the Seat in kg
HR	High Retention seats meeting the 1995 specifications for seat strength in QST and head restraint placement.

Analysis Procedures

Vehicle Dynamics: In this analysis, the acceleration of the vehicle is assumed to be prescribed in equal time steps, so that a spreadsheet calculation can be made of the vehicle and occupant dynamics. Given the acceleration function for the vehicle, it is straightforward to determine the velocity at any time i by the relationship:

$$\dot{y}_i = \dot{y}_{i-1} + \ddot{y}_i \Delta t \quad (1)$$

where Δt is the incremental time step for the calculation, $\dot{y}_0 = \ddot{y}_0 = 0$ and i ranges from 0 to 100, giving a solution for 200 ms with a time step of Δt =2 ms. Based on equation (1), the displacement of the vehicle can be determined from:

$$y_i = y_{i-1} + \dot{y}_i \Delta t \quad (2)$$

Occupant Dynamics: The occupant is displaced forward in the rear crash by the motion of the vehicle and seat during the impact. For this analysis, a force develops on the occupant by the stiffness of the seat and relative displacement of the occupant with respect to the vehicle:

$$F_i = k(y_i - x_i) \quad (3)$$

Force on the occupant increases with deformation of the seat, which accelerates the occupant in proportion to the effective mass of the occupant loaded by the seatback. Based on sled testing with the Hybrid III dummy, Viano (2002) determined that the effective mass (m) loaded by the seatback was 70% of the total dummy mass, or 52.5 kg for the 50th percentile Hybrid III dummy. This gives the following acceleration of the occupant:

$$\ddot{x}_i = F_i / m = F_i / 52.5 \quad (4)$$

The occupant acceleration can be used to determine the velocity and displacement of the occupant:

$$\dot{x}_i = \dot{x}_{i-1} + \ddot{x}_{i-1} \Delta t \quad (5)$$

$$x_i = x_{i-1} + \dot{x}_{i-1} \Delta t \quad (6)$$

where $\ddot{x}_0 = \dot{x}_0 = x_0 = 0$.

Now, the incremental energy transfer to the occupant can be determined by the average load and incremental displacement of the occupant with respect to ground:

$$\Delta E_i = 0.5(F_i + F_{i-1})(x_i - x_{i-1}) \quad (7)$$

where $\Delta E_0 = 0$. The cumulative energy transfer is a function of time and is given by:

$$E_i = E_{i-1} + \Delta E_i \quad (8)$$

where $E_0 = 0$.

Determination of Energy Transfer in a 35 km/h and 16 km/h Rear Crash with Two Stiffness Seats: For this analysis, two different seat stiffnesses are used. One reflects the typical stiff seat property of 40 kN/m and the other, a yielding seat stiffness of 20 kN/m. The seats were assumed to have a linear stiffness up to a peak force of 10 kN and beyond that deformation the load was held constant at 10 kN. This reflects the typical elastic-plastic behavior of seats under occupant loading.

The two different crash pulses were used to determine the energy transfer to the occupant. One involved a severe rear crash with constant 10 g acceleration of the vehicle for 100 ms, giving a delta V of 9.8 m/s (35 km/h) and the other a lower severity crash with a 4.5 g constant vehicle acceleration for 100 ms, giving a 4.4 m/s (16 km/h) delta V.

All of the calculations were done in Excel and plots were made of key results. The solution was determined for increasing time steps until the occupant experienced the delta V of the vehicle. The energy transfer to the occupant was also checked to make sure it reached E = 2508 J for the 35 km/h delta V and 508 J for the 16 km/h delta V rear impact exposures. Once the solution reached the prescribed occupant velocity change and energy transfer, the applied load was reduced to zero. With this approach, no rebound effects were determined in the analysis, and only the loading phase was evaluated up to the vehicle delta V.

Out of Position Occupant Exposures: In a second analysis, the same seat stiffnesses and crash delta Vs were used with the occupant leaning forward at the onset of the crash. This addressed issues raised by Strother et al. (1994), Benson et al. (1996) and Viano (2002). A 15 cm gap was assumed between the occupant and seatback. All other aspects of the simulation were the same and the occupant experienced load and acceleration once the initial gap was closed by vehicle (seat) motion forward in the crash. This involved a 17 ms and 25 ms delay in loading for the high and low severity rear crash exposures, respectively.

Influence of an Initially Higher Acceleration Pulse: In a third analysis, the same seat stiffnesses were studied with a vehicle pulse that was initially higher at 9 g for the first 25 ms followed by 75 ms of 3 g acceleration to reach the 16 km/h rear crash delta V. The results of this crash pulse were compared with the constant 4.5 g pulse in the initial studies at 16 km/h.

RESULTS

35 km/h Crash Severity: Figure 3 shows the force developed on the occupant by the forward motion of the vehicle. The higher stiffness seat reaches the 10 kN force limit and plateaus until the occupant delta V is reached. The 20 kN/m stiffness seat develops 15% lower force and is delayed in the loading. It reaches a peak force of 8.5 kN to achieve the necessary energy transfer. The lower force involves a 15% lower acceleration of the occupant with a peak of 16.6 g versus 19.5 g with the stiffer seat. The data are summarized in Table 2.

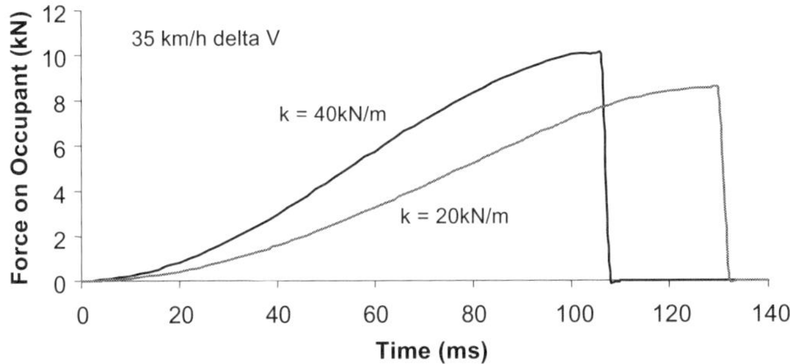

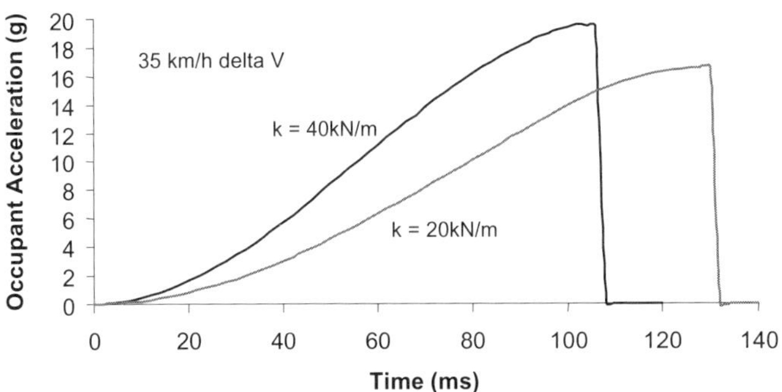

Figure 3: Force and acceleration of the occupant by the seat in a 35 km/h rear crash with two different seat stiffnesses.

Figure 4 shows the incremental energy transfer as a function time and the applied load on the occupant. The yielding seat stiffness involves a longer time for energy transfer and is delayed over the stiffer seat performance. This causes the energy transfer to occur while the occupant has a greater displacement with respect to the ground per unit of time, so the incremental energy transfer is greater at any given load on the occupant. This implies a more efficient transfer to the occupant with lower loads and accelerations. The yielding seatback improves the occupant loading in a rear crash and provides an improved ride-up of the crash delta V with a 42% greater vehicle displacement to reach the occupant delta V than with the stiffer seatback.

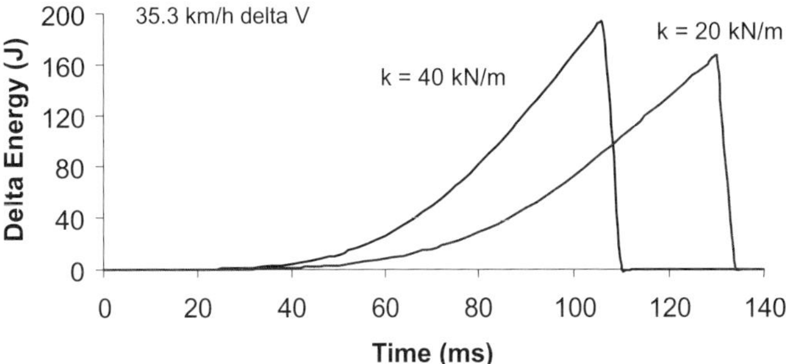

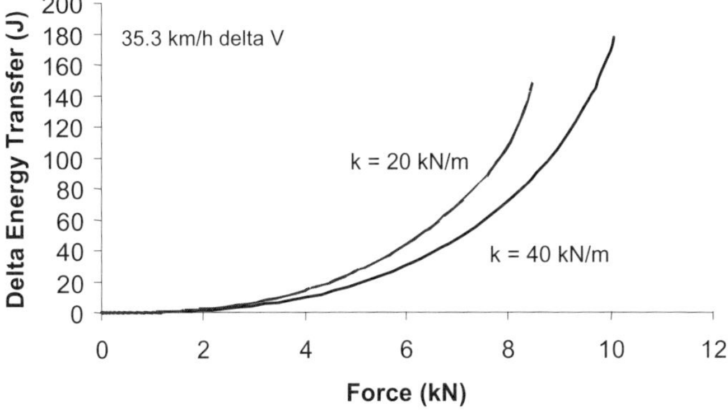

Figure 4: Incremental energy transfer to the occupant with the yielding and stiff seatback stiffness.

16 km/h Crash Severity: Figure 5 shows the acceleration of the occupant for the 16 km/h rear delta V exposure. As in the previous example, the 40 kN/m seat stiffness causes higher and earlier accelerations of the occupant. This gives higher forces during the time the head is not in contact with the head restraint, and is likely a factor in the potential for whiplash occurring during the S-shaped loading of the neck. The more gradual acceleration of the occupant with the yielding seat, results in 16% lower peak accelerations and forces on the occupant to achieve the prescribed delta V. The peak acceleration with the stiffer seat was 8.8 g as compared with 7.4 g with the yielding seatback.

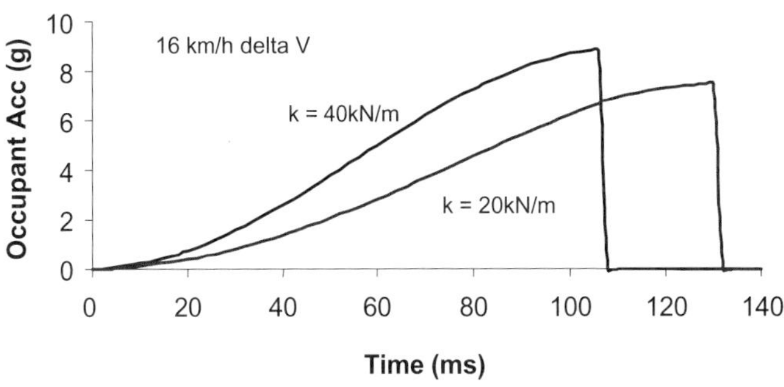

Figure 5: Acceleration of the occupant in the 16 km/h rear delta V exposure with the stiff and yielding seatback.

Table 2: Summary of Peak Vehicle and Occupant Dynamics in the Rear Crash Simulations.

	35.3 km/h Delta V		% diff	16 km/h Delta V		% diff
Vehicle Dynamics						
Acceleration (m/s2)	98	98		44	44	
Duration (ms)	100	100		100	100	
Velocity (m/s)	9.8	9.8		4.4	4.4	
Seat Property						
Stiffness (kN/m)	40	20	-50%	40	20	-50%
Occupant Dynamics						
Normal Seating Position						
Force (kN)	10.0	8.5	-15%	4.5	3.8	-16%
Acceleration (g)	19.5	16.6	-15%	8.8	7.4	-16%
Max Relative Displ. (m)	0.253	0.426	68%	0.114	0.191	68%
Veh. Displ. at Occ. delta V (m)	0.559	0.794	42%	0.251	0.356	42%
Out-of-Position Seating						
Force (kN)	10.0	8.7	-13%	5.0	4.0	-20%
Acceleration (g)	19.4	17.0	-12%	9.7	7.8	-20%
Max Relative Displ. (m)	0.281	0.452	61%	0.14	0.216	54%
Veh. Displ. at Occ. delta V (m)	0.578	0.794	37%	0.26	0.365	40%
Initially Stiff Crash Pulse						
Force (kN)				4.4	3.7	-16%
Acceleration (g)				8.6	7.3	-15%
Max Relative Displ. (m)				0.111	0.187	68%
Veh. Displ. at Occ. delta V (m)				0.241	0.345	43%

Table 3: Summary of Hyge Sled Tests at 31.1 km/h (19.3 mph) with Various Seats (from Viano 2002).

	SEAT Type	**HEAD HIC**	**Res. Acc. (g)**	**Ang. Acc. -(r/s/s)**	**Ang. Vel. -(r/s)**	**NECK Moment My -(Nm)**	**Force Fz (N)**	**Ext. Ang. (deg)**	**CHEST Res. Acc. (g)**	**SEAT Seatback Ang. (deg)**
	1990s C car	66	16	1401	25	20	700	69	11	88
Generic Tests	**1250 Nm/Std**	144	21	1007	33	38	975	80	12	83
Generic Tests	**1250 Nm/High**	101	26	786	30	20	1254	37	13	85
Generic Tests	**1700 Nm/Std**	140	31	2401	31	33	480	68	15	56
Generic Tests	**1700 Nm/High**	78	28	1829	26	15	380	28	15	57
Generic Tests	**2350 Nm/Std**	122	33	990	30	37	1300	60	20	47
Generic Tests	**2350 Nm/High**	76	26	710	19	17	748	30	19	46
	Foreign Benchmark	84	26	1778	27	13	1500	21	13	64
	1997 W Prototype	34	22	995	22	8	920	5	16	53
	diff from Foreign	-60%	-15%	-44%	-19%	-41%	-39%	-76%	23%	-17%

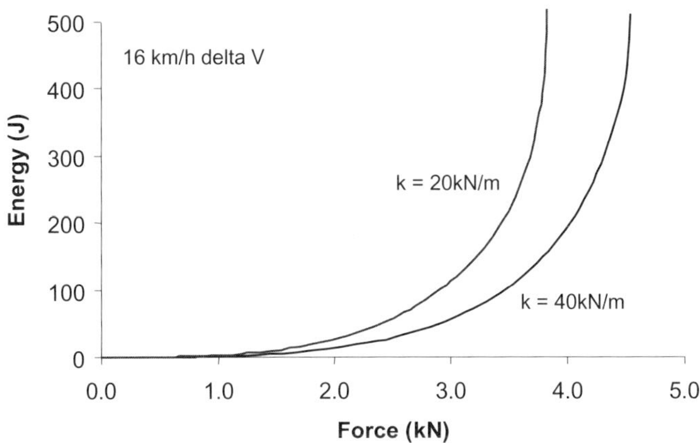

Figure 6: Over all energy transfer to the occupant with the stiff and yielding seatback in the 16 km/h exposure.

Figure 6 shows the energy transfer to the occupant as a function of the applied force. The yielding seatback transfers more energy at lower loads than the stiffer seat system and requires lower force and acceleration to achieve the 16 km/h delta V. Again, there is an improved efficiency with the yielding seatback stiffness.

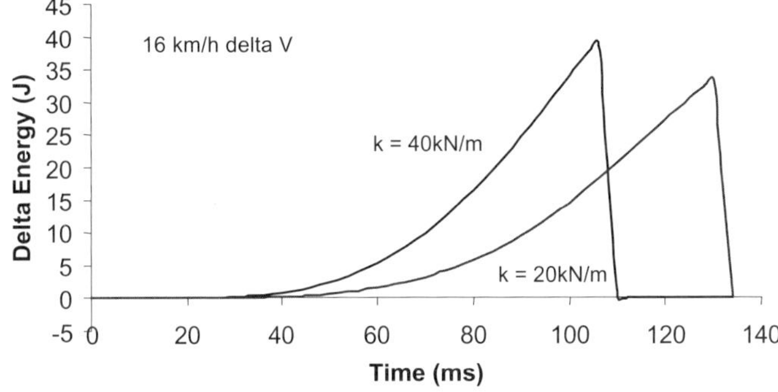

Figure 7: Incremental energy transfer to the occupant as a function of time for the stiff and yielding seatback.

Figure 7 shows the time history of the incremental energy transfer to the occupant at 16 km/h. There is a more gradual transfer with the yielding seatback as it reaches peak transfer at about 130 ms as compared to a peak transfer at 105 ms with the stiffer seat.

Figure 8 shows the incremental energy transfer to the occupant as a function of the applied load. As in the 35 km/h crash, there is an improved efficiency of energy transfer with the delayed loading by the yielding seatback. Seatback compliance increases the energy transfer at a given load.

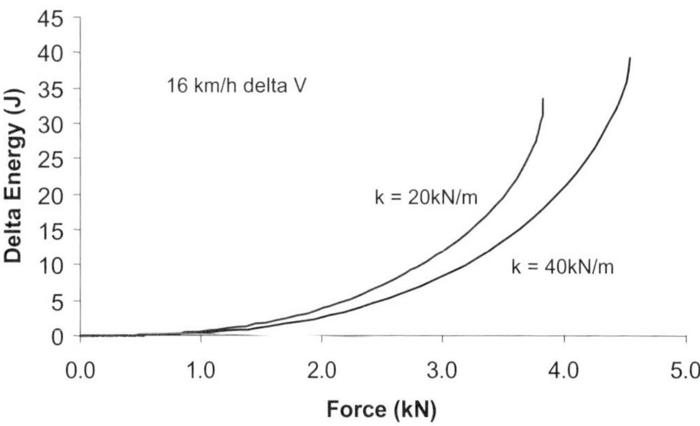

Figure 8: Incremental energy transfer to the occupant as a function of the applied load for the stiff and yielding seatback in the 16 km/h exposure.

Comparison of 16 km/h and 35 km/h Crash Severities: Figure 9 compares the force and acceleration of the occupant in the 16 km/h and 35 km/h rear delta V impacts. The stiff seat in the 16 km/h impact develops force and acceleration much like the yielding seat in the 35 km/h exposure for the first 80 ms. Thereafter they differ because of the much greater energy in the higher speed impact. Nonetheless, the occupant initially experiences higher forces with the stiff seat, which makes the low severity crash response with the stiff seat initially seem like a more severe impact with the yielding seat.

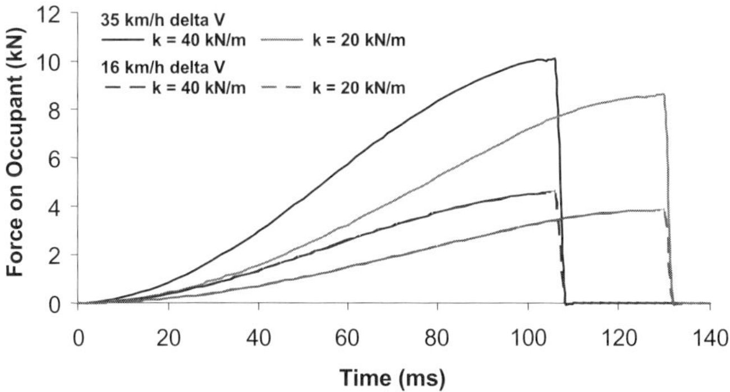

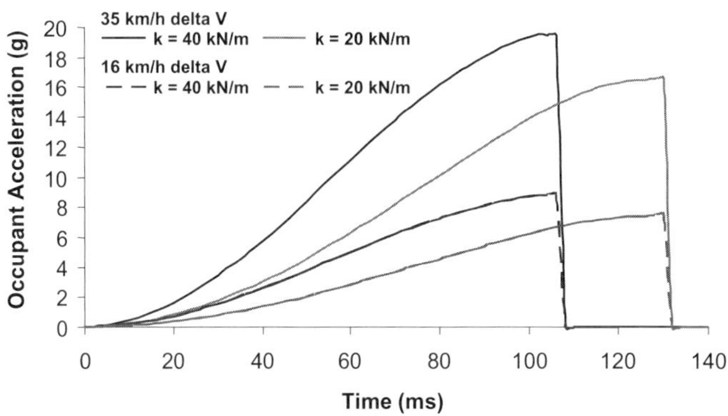

Figure 9: Comparison of the force and acceleration of the occupant in the 35 km/h and 16 km/h rear impacts with the stiff and yielding seatback.

Out-of-Position Performance in the 16 km/h and 35 km/h Crash Severities: Figures 10 and 11 show the occupant responses for the out-of-position seating with an initial gap of 15 cm. The results are essentially similar to the in-position seating situation, although there is a slight improvement of the yielding seat performance over the stiff seat in this condition from that seen with normal seating.

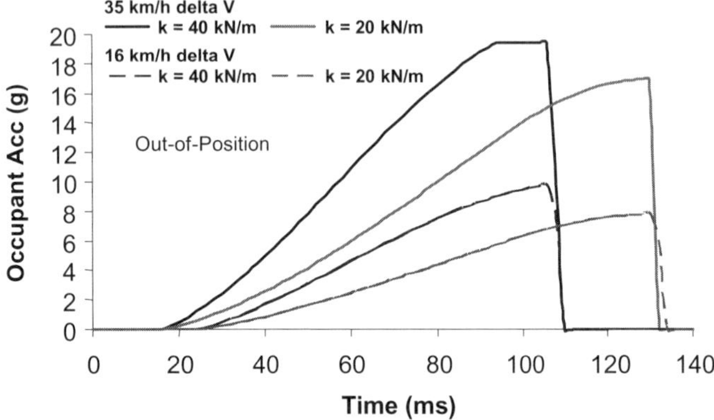

Figure 10: Acceleration of the occupant for the out-of-position simulations with an initial 15 cm gap with two seat stiffnesses and crash severities.

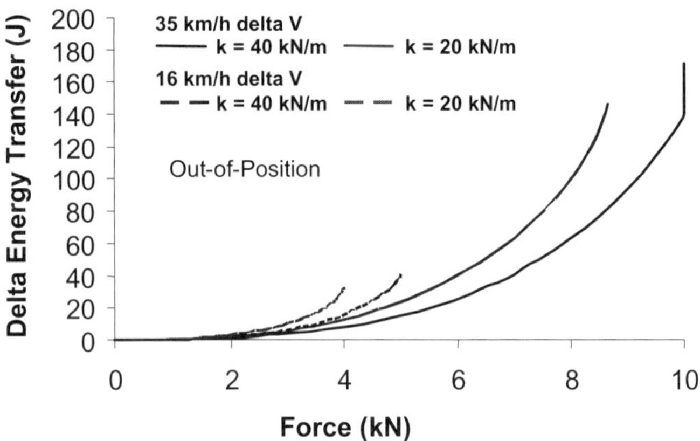

Figure 11: Incremental energy transfer to the occupant in out-of-position simulations with a stiff and yielding seatback in two rear crash severities.

Influence of an Initially Higher Acceleration Pulse: With an initial 9 g pulse for 25 ms, there is an earlier acceleration of the vehicle and forward motion. This engages the occupant earlier in the 16 km/h crash. Figure 12 shows the occupant acceleration with the constant and initially higher vehicle acceleration pulse in the 16 km/h rear crash delta V. While the onset of acceleration is earlier with the initially higher pulse, the peak levels are similar. Other aspects of the comparison are given in Table 2, but show comparable results.

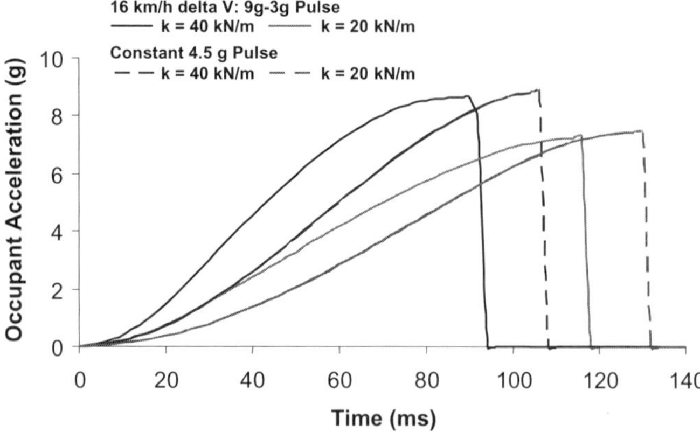

Figure 12: Comparison of 16 km/h rear impacts with a constant and initially high vehicle acceleration pulse.

DISCUSSION

The more compliant, yielding seatback allows 68% greater displacement of the occupant with respect to the vehicle in a rear crash than with the stiff seat. This gives a more gradual load and acceleration of the occupant forward. In this example, there was a 15% lower peak force and occupant acceleration at 35 km/h and 16% in the 16 km/h rear crash with the yielding seatback.

The later acceleration of the occupant delays significant loading until after the head contacts the head restraint and gets support. In low-speed rear sled tests, Viano (2002) found a head contact time of 110-130 ms for a range of seats. There are two benefits of the delayed loading. One, there is lower force on the upper back early in the crash, which is a primary factor in neck loading before head restraint contact. The lower the upper back load, the lower the forces causing neck shear, bending and extension. Two, the head receives full head restraint support before the higher forces occur on the occupant with the yielding seatback design. This allows energy transfer at a time when the head, neck, back and pelvis have uniform seatback and head restraint support. This reduces potential differential motion of the body, which can be responsible for injury.

The occupant reaches the crash delta V with much more forward displacement of the vehicle with the compliant seatback. The vehicle averaged 42% greater displacement with the compliant seat for the two crash severities. This is a result of more compliance and a greater occupant penetration with respect to the seat (vehicle). This gives a more efficient energy transfer as a greater increment of energy is transferred at a lower load with more displacement of the occupant with respect to ground. This also occurs as the vehicle has a higher speed with respect to ground increasing the incremental energy transfer per unit of time.

Occupant dynamics in the out-of-position situations are similarly influenced by the yielding seatback stiffness. There seems to be no disbenefit by having a higher compliance with an initial gap behind the back. In fact, the forces and accelerations of the occupant were proportionately lower with the initial 15 cm gap and the yielding seatback stiffness. The high, early vehicle acceleration pulse at 16 km/h resulted in an earlier loading of the occupant by about 20 ms, however the peak responses were essentially similar to the constant acceleration crash pulse.

Benefits of Yielding Seatbacks: This theoretical analysis underscores the importance of the yielding seat approach to occupant safety in rear crashes. Historically, the yielding behavior was achieved by seatback rotation at relatively low loads. The typical H-point moment would be about 1000 Nm in the QST test of 1980s and early 1990s seats. This moment would result in seatback rotations of 35^0, adding to the initial seatback angle of 25^0 and allowing the occupant to ramp up the seatback. These seatback designs had relatively stiff structural cross-members in the seatback, which made much of the compliance related to recliner and seat frame rotation, although the foam trim covering the frame provided some deformation distance until the motion was stopped by seatback structures. Often, these structures were just behind the pelvis, which limited penetration into the seatback and increased the tendency for ramping.

The 1990 C car seat had a stiffness of 21 kN/m, which was somewhat lower than the average of GM seats of that era. These seats reflected the yielding behavior of seat designs of that time. However, it became clear that there were shortcomings by allowing the occupant to rotate the seatback rearward in severe rear crashes, since there could be a loss of occupant retention; and, the energy transfer capability was limited.

The way to maintain the yielding behavior of the seat without significant seatback rotation was conceived in the 1990s after a series of sled tests with the GM C car seat. That seat had an open perimeter frame, which increased occupant retention to higher delta Vs than designs with substantial cross-frame structures. This was achieved by more pelvic pocketing into the seatback.

The C car seat was the genesis of a new generation of seats based on the open, perimeter frame design as shown in Figure 2. These new designs used a strong seat frame and recliner, but allowed a yielding performance by letting the occupant penetrate between the frame side structures through compliance of the seat suspension. This was the innovation that allowed a yielding stiffness without seatback rotation. Figure 1 shows the relative seatback stiffness of the 1990 C car seat, and the 1997 W car/P-90 designs that were the first high retention seats introduced on the market.

Table 3 shows the comparative performance of the original 1990 C car and the 1997 W car seats in 31.1 km/h (19.3 mph) rear sled tests. The compliance of the 1997 W car seat and high retention structure provided improved performance over a benchmark, competitive group of luxury foreign seats. The 1997 W car seat achieved a neck extension angle of only 5^0 as compared to 69^0 for the 1990 C car seat and 21^0 for the benchmark group. Neck extension moments were also reduced. There was also a reduction in seatback rotation to 53^0 from vertical with the W car versus 88^0 for the C car and 64^0 for the benchmark group. The generic seat tests were used to compare results with various seat strengths and head restraint heights (see Viano 2002).

FMVSS 207 Versus QST Seat Tests: FMVSS 207 is a seat frame test that does not determine the compliance of the seat suspension or trim. It involves a point load on the upper structural cross-member of the frame, so it does not represent an occupant interface. Only the seat structure and recliner are determined by the frame test, and the load to achieve a moment is much higher on the seat frame than the center of pressure from an occupant. This results in much lower horizontal forces on the seat frame in the FMVSS 207 test than occur with an occupant loading to the same H-point moment (Viano 2002).

The QST involves a fully trimmed seat and uses a Hybrid dummy for the loading of the seat in a rearward direction. It determines the deformation of the seat suspension, trim and frame structure under a loading condition representative of that in a rear sled or barrier crash test. Since the dummy is freely suspended in the loading, it can move lateral and up the seatback in a manner that is more relevant to a real life crash situation.

The QST was the principal subsystem test used to develop the new generation of yielding seats. The high retention seats involved a greater degree of compliance between the seat frame structures to give a deformation distance to gradually load the occupant while pocketing the pelvis and lower back into the seat suspension. The QST test also enabled the shifting of frame structure to areas loaded by the occupant, thus helping to achieve goals for light weight and low cost.

The compliant seatbacks with the new generation of yielding seatbacks require lower force to accelerate the occupant through a delta V in a rear crash than with stiffer seats. This increases the efficiency of the energy transfer with the yielding seatback, while increasing also occupant retention by pocketing the pelvis and lower back. This enables the seats to achieve higher delta V performance at lower force on the seat frame and lower H-point moments, while reducing neck loads and head displacements.

Role of Yielding Seatbacks in Reducing Whiplash: The compliance of the high retention seat is an important factor in the reduction of whiplash risks. The lower forces on the back with the yielding seatback are a primary means of lowering loads on the base of the neck (Viano, Olsen 2001). This reduces displacements of the head with respect to the neck before head restraint contact. Displacement of the base of the neck with respect to ground is an additional factor in whiplash. The greater the relative displacement of the occupant with respect to the vehicle, the lower the loads to achieve an occupant delta V. There are also benefits from the open perimeter frame seatback, which maintains a more upright posture until the significant loads that rotate the seatback in a rear crash while the occupant's pelvis, back and head have seat and head restraint support.

REFERENCES

Benson BR, Smith GC, Kent RW, Monson CR. Effect of seat stiffness in out-of-position occupant response in rear-end collisions. SAE 962434, 40th Stapp Car Crash Conference, Society of Automotive Engineers, Inc., Warrendale, Pennsylvania, USA, 1996.

Blaisdell DM, Levitt AE, Varat MS. Automotive seat design concepts for occupant protection. SAE 930340, Society of Automotive Engineers, Inc., Warrendale, Pennsylvania, USA, 1993.

Cantor A. Petition for Rulemaking to Amend FMVSS 207 to Prohibit Ramping Up the Seat Back of an Occupant During a Collision. National Highway Traffic Safety Administration, December 28, 1989.

Digges, KH, Morris JH, Malliaris AC. Safety performance of motor vehicle seats. SAE 930348, Society of Automotive Engineers, Inc., Warrendale, Pennsylvania, USA, 1993.

Federal Motor Vehicle Safety Standards No. 202, 207, and 301. 49 Code of Federal Regulations Section 571.21, 1968.

Prasad P, Kim A, Weerappuli DPV, Robert V, Schneider D. Relationship Between Passenger Car Seat Back Strength and Occupant Injury Severity in Rear End Collisions: Field and Laboratory Studies. SAE 973343, Society of Automotive Engineers, Warrendale, PA. 1997.

Rake B, Boehm S. Auto Seat Systems: Dangerous Safety Restraints? Trial April Issue, 80-87, 1990.

Saczalski KJ. Petition to Improve FMVSS 207. National Highway Traffic Safety Administration, April 18, 1989.

Saczalski KJ, Syson SR, Hille RA, Pozzi MC. Field accident evaluations and experimental study of seat back performance relative to rear-impact occupant protection. SAE 930346, Society of Automotive Engineers, Inc., Warrendale, Pennsylvania, USA, 1993.

Severy DM, Brink HM, Baird JD. Backrest and Head Restraint Design for Rear-end Collision Protection. SAE 680079, Society of Automotive Engineers, Warrendale, PA, 1968a.

Severy DM, Brink HM, Baird JD. Preliminary findings of head support designs. 11th Stapp Car Crash Conf, SAE 670921, SAE Warrendale PA, 337-405, 1967.

Severy DM, Brink HM, Baird JD. Vehicle Design for Passenger Protection from High Speed Rear End Collisions. SAE 680774, Society of Automotive Engineers, Warrendale, PA, 1968b.

Severy DM, Harrison MB, Baird JD. Collision Performance, LM Safety Car. SAE 670458, Society of Automotive Engineers, Warrendale, PA. 1967.

Severy DM, Mathewson JH and Bechtol CO. Controlled Automotive Rear-End Collisions, An Investigation of Related Engineering and Medical Phenomena. Canadian Services Medical Journal, VII, 727-759, 1955.

Stephens GD, Long TJ, Blaisdell DM. Energy Analysis of Automotive Seat Systems, SAE 2000-01-1380, SP-1494, Society of Automotive Engineers, Warrendale, PA, 2000.

Strother CE, James MB. Evaluation of Seat Back Strength and Seat Belt Effectiveness in Rear End Impacts. 31st Stapp Car Crash Conference, SAE 872214, Society of Automotive Engineers, Warrendale, PA, 225-244, 1987.

Strother CE, James MB, Gordon JJ. Response of out-of-position dummies in rear impact. SAE 941055, Society of Automotive Engineers, Inc., Warrendale, Pennsylvania, USA, 1994.

USPTO. Vehicle Seat with Perimeter Frame and Pelvic Catcher. US Patent 5,509,716, April 23,1996.

Viano DC. Cause and Control of Automotive Trauma. Bulletin NY Academy of Medicine, Second Series, 64(5):376-421, June, 1988.

Viano D, Olsen S. The Effectiveness of Active Head Restraint in Preventing Whiplash. J Trauma 51:959-969, November, 2001.

Viano DC. Role of the Seat in Rear Crash Safety. ISBN 0-7680-0847-6, Society of Automotive Engineers, Warrendale, PA, SAE R-317:1-491, 2002.

Viano DC, Olsen S, Locke GS, Humer M. Neck Biomechanical Responses with Active Head Restraint Systems: Rear Barrier Tests with BioRID and Sled Tests with Hybrid III. SAE 2002-01-0030, Society of Automotive Engineers, Warrendale, PA, 2002.

Viano DC. High-Retention Seat Performance in Quasistatic Seat Tests. SAE 2003-01-0173, Society of Automotive Engineers, Warrendale, PA, 2003.

Warner C, Strother C, James MB, Decker RL. Occupant Protection in Rear-End Collisions II: The Role of Seat Back Deformation in Injury Reduction. 35th Stapp Car Crash Conf, SAE 912914, Society of Automotive Engineers, Warrendale, PA., 1991.

SECTION 2:

A NEW GENERATION OF YIELDING SEATS WITH PREIMETER FRAME AND HIGH RETENTION

CHAPTER 6:

HIGH RETENTION SEAT PERFORMANCE IN QUASISTATIC SEAT TESTS

2003-01-0173

High Retention Seat Performance in Quasistatic Seat Tests

David C. Viano[1,2]
[1]ProBiomechanics LLC
[2]Vehicle Structure and Safety Integration, General Motors Corporation

ABSTRACT

A new generation of seats has been designed to specifications for high retention (HR) in a Quasistatic Seat Test (QST). The QST involves occupant loading of the seat in a rearward direction and targets peak H-point moment to >1700 Nm giving an energy transfer capability of 2000 J. QST tests from 1998-2000 were compared to results from pre-HR seat designs of the late 1980s and early 1990s to determine performance improvements.

Twenty-seven QST tests of HR seats were randomly selected from a larger series and were evaluated for strength and seat deformation under occupant loading. They represented 20 different seat types from four suppliers. Averages and standard deviations in QST results were computed. In addition, eight repeat tests were conducted with one seat to determine repeatability of the QST. These data were compared to an earlier repeatability study of the 1994 W pre-HR seat, which was evaluated at two facilities. Finally, 12 QST tests were conducted where variability was introduced in the seat back angle, track position, offset and orientation, and in the seatback angle transducers.

The average H-point moment of the HR seats was 2537 ± 703 Nm and represented a 2.3-times higher moment than pre-HR designs. This was achieved with a 35% reduction in seatback angle change at the 1700 Nm target, but a similar initial stiffness or "yield" behavior of the seats. A 2.9-times higher energy transfer capability was achieved by the seats to 3659 ± 1140 J providing calculated rotations <60^0 for crashes up to 42.5 km/h on average with the 50th percentile male, a 69% increase over pre-HR seats. The repeatability tests demonstrated a coefficient of variation below 3% for the peak force and seatback angle change at the 1700 Nm target moment. The comparable result from the earlier study was within 12%. In conclusion, the new generation of high retention seats provides occupant retention to a significantly higher rear crash delta V than pre-HR designs. These improvements were refined by QST testing and were achieved while reducing the mass and cost of the conventional (non-ABTS) seat designs.

INTRODUCTION

In 1995, General Motors decided that all new seat programs after 1997 would meet performance targets in a Quasistatic Seat Test (QST). Descriptions of the test and the rationale for the specifications can be found in a previous publication (Viano 2002), so only a brief overview will be given here.

THE QUASISTATIC SEAT TEST (QST)

The Quasistatic Seat Test (QST) was developed to assess occupant interactions with the seat in rear impacts. The sequence of seat loading is shown in Figure 1. QST involves a Hybrid dummy placed in the design seating position and subjected to rear displacement by a hydraulic ram loading through the flexible lumbar spine element. This distributes force through the buttocks and upper torso and simulates occupant loading of the seat in a rear impact. The hydraulic ram is supported on sliding rails so the dummy can freely move upward and sideways as the seat deforms. The system is counterbalanced so only the mass of the dummy vertically loads the seat. The kinematic sequence in the test mimics the dynamics of torso loading and ramping in a rear crash. The QST test does not address the head/neck interactions with the upper seatback and head restraint.

More than fifteen channels of response data are collected in a QST. They include the trajectory of the occupant, occupant loads into the seat, forces supporting the front and rear attachments to the vehicle floor, and rotation of the right and left side of the seatback. Other responses are calculated from the force and displacement data, including the H-point moment (M_h) and seatback pivot moment (M_p):

$$M_h = F_x(z_h + z) \quad (1)$$

$$M_p = F_x(z_p + z) \quad (2)$$

where F_x is the horizontal ram load, z_h is the H-point offset, z_p is the z offset from the ram to the seat pivot point and z is the vertical travel of the ram. For the pivot point moment, the positive and negative change in z position of the ram is included in the M_p computation.

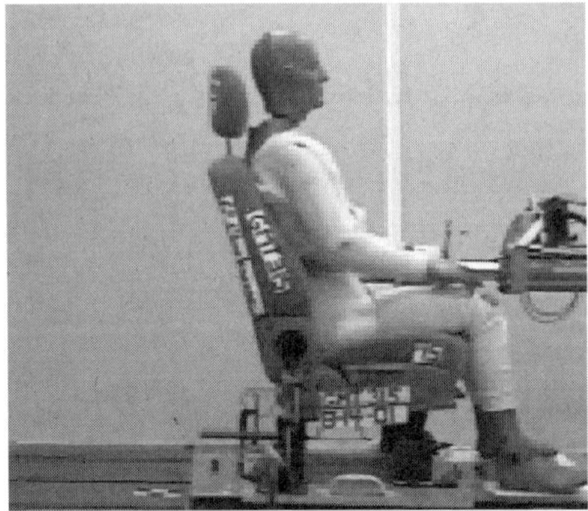
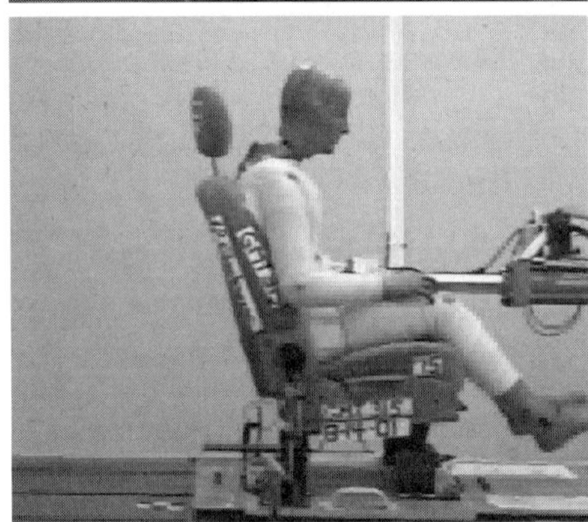
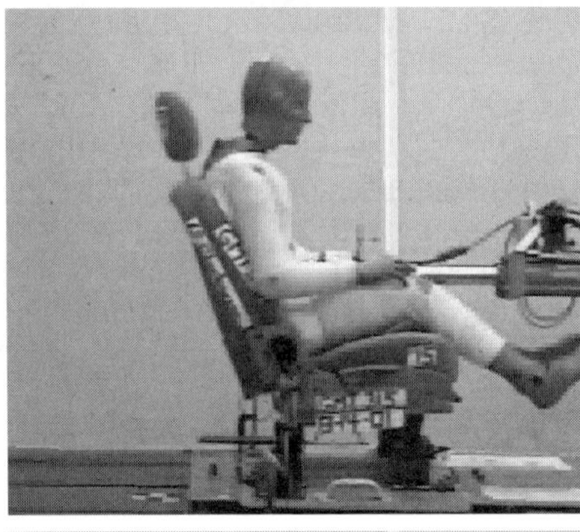
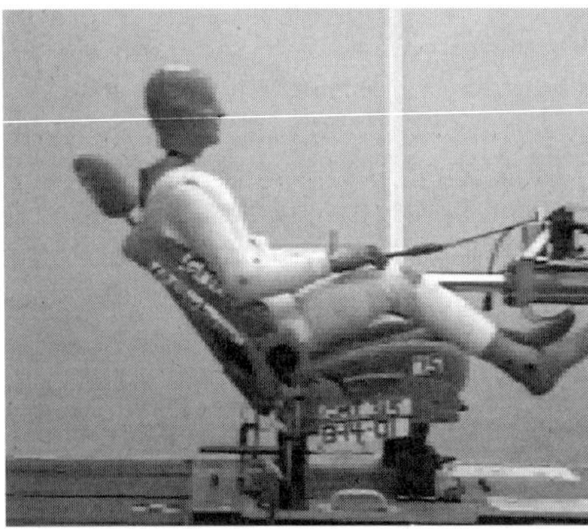

Figure 1: Sequence from a QST test of the GMT 315 seat.

In contrast, the H-point moment calculation involves only the upward component of ram displacement. The z position is held at zero when the H-point drops in position. In practical testing, the sum of the horizontal loads from the seat base measurements are used rather than the ram load as this gives greater accuracy at high loads.

Twist of the seatback is computed as the difference in angle from the right and left sides of the seatback, and the torsional load on the seatback (T) is computed as:

$$T = F_x y \tag{3}$$

where y is the lateral motion of the ram and dummy. The complete analysis of seat responses provides a detailed understanding of the compliance and strength of the seat to rear loading by an occupant. Summary data are compared among seats using either the peak value of responses or levels at 60° average seatback angle from vertical.

Since the QST measures occupant load and displacement, the energy transfer to the seat (E) can be computed by integration:

$$E = \int F_x dx \tag{4}$$

where F_x is the horizontal dummy load and x is ram displacement measured by a string potentiometer. A rear crash involves energy transfer to the occupant by the seatback, cushion and head restraint. The amount of energy is given by the kinetic energy (E_c) of an occupant in a crash:

$$E_c = \frac{1}{2} m V^2 \tag{5}$$

where m is the effective mass of the occupant loading the seatback, V is the change in velocity during the crash or the delta V of the struck vehicle and there is a negligible stored energy in seatback deformation. This is the reduced formulation of kinetic energy with $\lambda = 0$ in the energy formulation given by (Stephens et al. 2000). Since some parts of the body are not initially loaded by the seatback in a rear crash, the effective mass of the occupant is less than the total body weight.

The effective mass loading the seatback in a rear crash is difficult to determine and may vary through a crash; however, the seatback is primarily interacting with the pelvis, torso, upper body, and legs. The feet, lower legs, hands, and the lower arms are not involved in the initial energy transfer. This gives an effective mass of 52.5 kg or 70% of the overall dummy mass.

It is possible to calculate a threshold velocity (V_t) for a seat from Equations (4) and (5). This velocity is the maximum severity of crash delta V that a seat can perform with a certain weight occupant and rearward rotation of the seatback. It is calculated by assuming the occupant energy is based on the effective mass interacting with the seatback in a rear crash and is equal to the energy transfer capability of a seat at 60° seatback angle with respect to vertical:

$$V_t = (2E/m)^{0.5} \quad (6)$$

where V_t is the velocity a seat can transfer energy to an occupant without exceeding the specified seatback angle, in this case 60°. Energy transfer is related to the amount of rear displacement of the occupant or the angle of the seatback from vertical.

It is also possible to relate energy to the H-point bending moment and seatback angle change. Energy transfer to the occupant is given in Equation (4). The displacement can be approximated by the arc length of motion about the H-point and the angle change:

$$dx = (z_h + z)d\theta \quad (7)$$

where the radius is the sum of the H-point offset (z_h) and the z-displacement (z) during a test. Combining (4) and (7) gives the H-point moment in Equation (1) integrated over the change in seatback angle:

$$E = \int F_x(z_h + z)d\theta = \int M_h d\theta \quad (8)$$

QST SPECIFICATIONS FOR HIGH RETENTION

Based on QST tests of seats from the late 1980s and early 1990s and a series of Hyge sled tests described in Viano (2002), performance specifications were established for seats to achieve high retention. The specifications are summarized in Table 1 and provide criteria for front bucket or split bench seats.

Table 1: Seat Technical Specifications for High Retention in the Quasistatic Seat Test

1. >1700 Nm (15,000 in-lb) Moment About H-Point
2. No Force Drops >2 kN (450 lb) in 50 ms Causing $>10^{o}$ Change in Seatback Angle
3. Seatback Twist $<15^{o}$ for Seatback Angles up to 60°
4. Head Restraint Height to B-Plane and Front Surface to Within 20 mm of Back-of-Head Ellipse for the 95th Percentile Occupant

First, the minimum H-point bending moment is 1700 Nm (15,000 in-lb) to provide energy transfer to the occupant without excessive rotation of the seatback and possibility of loss of retention in severe rear crashes. This provides energy management in rear sled tests that are consistent with the exposure in a 33.7 mph (54.2 km/h) FMVSS 301-type rear moving barrier test. This test results in an average delta V of 31.1 (19.3 mph) for passenger cars with a range from 27.4-35.4 km/h (17-22 mph) based on vehicle mass. FMVSS 301 involves a 30 mph (48.3 km/h) flat moving barrier of 2273 kg (5000 lb) mass. The average energy transfer at 60^{0} for seats that provide retention in the high-speed rear test is about 2000 J, which corresponds to a velocity performance of 31.4 km/h (19.5 mph) based on the kinetic energy calculation and a 70% effective occupant mass for the 50th percentile male occupant.

Second, deformation of seat hardware should provide smooth control of occupant kinematics and uniform resistance. Thus, there should not be a drop in ram load of more than 2 kN in 50 ms that causes more than a 10° change in seatback angle before load reverses in the QST. This criterion assures uniform loading during a test and eliminates rapid changes, which may influence occupant kinematics and retention.

Third, a limit of 15° twist of the seatback up to 60° from vertical is specified to prevent excessive twisting of the seatback and loss of retention by lateral motion of the occupant. This criterion is based on the data from pre-HR seats. No particular group of seats performed better than another.

Fourth, the head restraint needs to be positioned high and close enough to support the head and neck as greater energy management is achieved by the seatback. This assures that retention and occupant containment are not associated with an increased risk of whiplash injuries in low-speed through high-speed crashes, even with forward seating positions and large gaps to the head restraint.

A head restraint design criterion was adopted after Hyge sled testing. The height of the top of head restraint should be above the B-plane (a horizontal plane through the head center of gravity of the 95th percentile seated occupant) and closer than 20 mm rearward of the back-of-head ellipse for 95th percentile seated occupant.

The first three criteria are determined in a QST test. The fourth criterion is specified by the design layout for the seated occupant. Evaluation of performance to the criteria can be done during design and early prototyping of a seat, before final vehicle crash testing.

THE CURRENT STUDY

The performance of seats was assessed after 5 years of testing to the QST specifications for high retention. This involved 20 different seat types from four seat suppliers involving tests run from February 1998 through October 2000. These results were compared to the original QST data on seats designed before the 1995 adoption of seat specifications in the QST.

In a second analysis, the original repeatability study of the 1994 W seat conducted in 1995 was compared to a second repeatability study conducted in 2001. The first study involved repeated tests at two laboratories with the QST test, so it addressed repeatability at a test site and between two test facilities. Eight new repeatability tests were conducted with a conventional dual recliner seat that met HR requirements. Finally, a variability study was conducted, where variations in the seat and instrumentation setup were made to assess the sensitivity of responses in the QST to wide variations in the test setup.

METHODOLOGY

HIGH RETENTION SEAT PERFORMANCE IN QST

The QST tests were conducted at the Research Safety, Health and Environment (RSHE) facility at the General Motors Technical Center in Warren, Michigan from February 1998 through October 2000. Twenty-seven tests representing 20 different seat systems were randomly selected from a larger series of approximately 150 QST tests. The seats were from four suppliers: Faurecia, Johnson Control Inc. (JCI), Lear and Magna. The seats included prototype and pre-production seats being evaluated in the QST test. Most of the seats were conventional high retention designs with perimeter frames. There were six ABTS (All Belts to Seat) seats in the group.

Table 2: Setup Differences in the QST Variability Study

1. Baseline w/4^0 added recline (21.5^0)
2. Baseline w/3^0 incline (14.5^0)
3. Baseline w/seat full rear on tracks
4. Same as 3. with goniometers offset of:
   right: 25 mm up and rear
   left: 25 mm down and forward
5. Baseline w/seat full forward
6. Same as 3. with test plate 200 mm forward
7. Baseline w/test plate up 16mm
8. Fixture 6^0 cw
9. Fixture 6^0 ccw
10. Fixture 6^0 ccw, seat full rear
11. Full rear, fixture 200 mm fwd, 4^0 recline (21.5^0)
12. Seat 16^0 moved 90 mm inboard

QST REPEATABILITY STUDIES

2001 GMT 315 Repeatability Tests at RSHE

Eight pre-production seats were tested at RSHE from June through August 2001. The base level seat has dual recliners and a perimeter frame design with low rear profile. The baseline position in the repeatability study had a mid-track location and seatback angle of 17.5^0.

1994 W Seat Studies at RSHE/Delphi

At the time of approval of QST specifications in 1995, a repeatability study was conducted at two test facilities. The 1994 W seat was tested at the RSHE facility and at the Delphi Interior facility in Warren, Michigan. The seat has a single-side recliner and was developed before the HR QST requirements. A total of 10 repeated tests were conducted, 4 at RSHE and 6 at Delphi in February 1995.

QST VARIABILITY STUDY

Twelve identical GMT 315 seats were included in the variability study. The variations in setup were selected to represent the extreme in setup variability and involved various seat positions and instrumentation misalignments. The 12 variations included in the study are described in Table 2. Again, the baseline position was a mid-track location and seatback angle of 17.5^0. The variations included changes in initial seatback angle, position on the seat track, attachment to the base-plate fixture and orientation on the fixture. The difference between full rear and forward on the track was approximately ±100 mm of the mid-track position.

RESULTS

HIGH RETENTION SEAT PERFORMANCE IN QST

Table 3 summarizes the results for 27 QST tests from February 1998 through October 2000. The offset records the vertical distance between the centerline of the ram and the seat pivot point in the recliner. For these seats, the ram load averaged 14.4 ± 3.7 kN and occurred just below 50^0 from vertical. The seatback stiffness represents the average value between ram loads of 1.0 to 4.0 kN. It was 25.3 kN/m. The peak moment about the recliner averaged 3809 ± 1165 Nm and the H-point moment averaged 2537 ± 703 Nm. The seats reached the target specification H-point moment of 1700 Nm at 36^0 ± 13^0. The energy transfer capability of the seats averaged 3859 ± 1140 J up to a seatback angle of 60^0. The average maximum torque about the long axis of the seatback was 549 ± 704 Nm up to 60^0 and resulted in a maximum average twist of 19^0 ± 16^0 up to 60^0 seatback angle.

Seventeen out of 27 seats had force drops greater the 2 kN within a 50 ms time window during a test. The average force drop was 8.6 ± 4.6 kN. The average angle change during the force drop was 4.4^0 ± 2.8^0 so none failed the 10^0 limit until load resumes. Loads under the front and rear attachments of the seats are also included, and represent the loads that must be held by the installation in the vehicle body.

Table 3: QST Responses for High Retention Seats

Test #	Z Offset mm	Ram Load X kN	Ram Load θ deg	Ram Load X <60 kN	Seatback Stiffness kN/m	Peak Force Tang. kN	Peak Force Nor. kN	Peak Mom. Nm	Angle @ Pk Mom deg	Ram X @ Pk Mom mm	Pk Mom H-pt Nm	Angle @ H-pt deg	Angle at 1700 deg
478	285	23.0	37	23.0	24.0	13.9	18.4	6228	37	394	3541	37	24
491	205	10.5	43	10.5	23.2	7.2	7.9	3693	50	736	3217	50	37
493	312	25.9	42	25.9	26.4	17.3	19.3	7472	42	375	3981	42	40
494	203	16.0	68	13.0	9.7	14.8	9.3	3126	68	754	2463	68	24
512	223	12.0	54	12.0	27.8	9.7	7.7	2769	54	341	1943	54	47
546	238	11.8	27	11.8	29.4	5.4	11.1	2783	27	318	1822	27	19
548	229	13.8	31	13.8	30.0	7.1	11.8	3732	31	355	2697	31	23
551	382	12.5	55	12.5	24.8	10.3	7.3	2246	55	382	1931	55	51
552	283	16.1	36	16.1	31.4	10.0	13.0	3459	36	283	2540	36	26
557	241	13.7	73	10.6	20.2	13.1	7.5	3502	73	642	2314	73	62
560	231	13.4	51	13.4	22.5	11.3	8.5	4474	58	621	3453	58	30
561	222	16.3	34	16.3	29.6	11.4	13.6	4601	60	523	3709	60	27
562	257	14.6	37	14.6	24.7	9.2	11.9	3545	37	303	2246	37	29
569	269	13.3	33	13.3	24.4	7.3	11.1	3758	33	340	2232	33	27
570	235	13.5	33	13.5	25.0	7.3	11.4	3370	33	309	2273	33	28
573	236	19.8	91	12.8	20.2	10.9	7.9	3199	52	400	2172	52	41
576	214	17.5	64	14.2	25.7	15.7	7.8	5257	64	649	4209	64	39
581	233	12.7	35	12.7	27.6	8.1	10.4	3037	35	270	2038	35	30
583	233	12.5	58	12.5	22.0	10.6	7.5	2617	50	397	1921	58	50
592	259	16.5	83	8.8	17.3	16.4	6.3	2250	43	325	2545	83	73
595	225	9.4	60	9.4	27.9	8.2	5.3	3724	60	673	3065	59	38
601	253	12.3	48	12.3	24.6	9.6	8.3	2930	48	311	1887	48	43
602	250	11.4	52	11.4	22.1	9.1	7.5	2880	52	354	1811	59	47
607	227	11.5	44	11.5	27.9	8.1	9.6	2811	46	326	1990	46	32
608	260	14.5	46	14.5	29.1	10.5	10.2	3700	46	340	2235	46	36
610	247	14.2	38	14.2	32.0	9.0	11.2	3317	38	321	2184	38	30
611	237	10.8	34	10.8	32.7	6.1	9.0	2976	35	282	2086	35	27
Avg	248	14.4	48.4	13.5	25.3	10.3	10.0	3609	47	419	2537	49	36
sd	36	3.7	16.4	3.6	4.9	3.1	3.3	1165	12	152	703	14	13

Test #	Total Engy J	Total Engy @ 60 J	Peak Torque Nm	Peak Twist deg	Torque < 60 deg Nm	Twist deg	Force Drop kN	Peak θ deg	# Frc Drps	FS Load X kN	FS Load Z kN	RS Load X kN	RS Load Z kN
478	7895	7895	871	36	871	36	8.6	2.0	3	-6.0	-17.2	-17.0	-16.3
491	3789	3789	1822	41	1822	41				-1.2	-9.3	-9.8	-9.1
493	5945	5162	815	28	815	28	19.5	5.9	2	-11.4	-17.6	-14.4	-12.7
494	5912	4793	980	43	268	43				-12.6	-7.5	-3.5	-6.6
512	3026	2394	242	36	58	17				-5.5	-8.6	-6.8	-8.4
546	3213	3204	287	12	287	12				-11.3	-8.8	-1.0	-7.8
548	3797	3797	986	15	986	15	9.7	6.2	1	-10.5	-11.2	-3.2	-10.1
551	3930	2509	268	38	268	8				-14.3	-8.6	2.0	-7.6
552	4700	3581	351	63	351	56	7.2	3.9	1	-18.1	-11.7	2.6	-9.8
557	5009	2858	634	7	75	6	5.1	0.2	1	-6.6	-6.8	-4.0	-6.4
560	5409	4878	2281	75	2281	48	10.6	8.5	2	-3.9	-10.7	-9.7	-10.4
561	5470	4371	140	4	105	4	7.7	5.1	3	-1.5	-12.1	-14.8	-12.1
562	3796	3796	308	14	308	14				-2.1	-11.4	-12.4	-10.9
569	3747	3747	1358	27	1358	27	7.1	4.2	3	12.3	-9.8	-25.6	-10.0
570	3558	3558	1049	23	1049	23	9.4	6.2	1	13.4	-9.9	-26.9	-9.9
573	6465	3255	249	14	64	4	7.3	7.0	4	-5.7	-8.2	-7.1	-7.8
576	5981	4191	3122	47	2537	44	15.5	8.8	2	-3.1	-12.1	-11.1	-11.3
581	2343	2343	304	29	304	29	3.3	3.0	1	-9.5	-9.9	-3.3	-9.1
583	4963	3382	397	14	85	8	4.6	0.2	1	-8.6	-7.3	-3.9	-6.7
592	4467	2457	800	17	71	4	15.4	7.3	1	-6.8	-6.1	-1.9	-6.0
595	3264	3221	319	20	318	20				-1.1	-8.2	-8.7	-7.7
601	3694	2610	140	13	106	6	2.4	1.9	1	-5.6	-7.9	-6.7	-7.5
602	4293	2850	465	26	63	3	5.2	3.2	2	-6.9	-7.8	-4.5	-7.6
607	3567	3567	140	9	140	9				-4.6	-8.6	-7.1	-8.5
608	3922	3922	132	8	132	8	8.4	1.1	1	-5.0	-9.9	-9.6	-9.5
610	4937	3848	67	6	67	3				-4.1	-9.0	-10.1	-9.0
611	2824	2824	58	5	58	5				-2.6	-7.9	-8.2	-7.9
Avg	4441	3659	688	25	550	19	8.6	4.4		-5.3	-9.8	-8.4	-9.1
sd	1271	1140	733	18	704	16	4.6	2.8		6.7	2.7	7.0	2.2

Table 4: Comparison of High Retention Seat Responses to Pre-HR Responses in QST Tests

	GM High Retention	GM Pre-HR	% Diff	Foreign	Domestic
# Tests	27	29		18	8
H-point Moment (Nm)	2537	1092	132%	1384	935
	±703	±189		±393	±147
Angle up to 1700 Nm (deg)	36	56	-35%	56	60
	±13	±12		±10	±7
Energy @ 60 deg (J)	3659	1274	187%	1669	1214
	±1140	±390		±546	±586
Delta V up to 60 deg (km/h)	42.5	25.1	69%	28.7	24.5
Peak Force (kN)	14.4	6.6	118%	7.8	5.1
	±3.7	±1.3		±2	±1.4
Stiffness (N/mm)	25.3	24.7	2%	31.7	23.2
	±4.9	±4.9		±7	±5.6

Table 4 summarizes the key average responses from the recent tests with data from earlier seat testing. The older data were reported by Viano (2002) on GM, domestic and foreign seats of the late 1980s and early 1990s. The peak H-point moment of the high retention seats is 2.32 times higher (132% higher) than the average of the pre-HR GM seats prior to setting the 1700 Nm specification. The seatback angle to reach the moment specification was 35% lower than earlier seat designs and occurred at only 36^0, a 16^0 angle change from the nominal 20^0 initial seatback angle. The energy transfer capability of the high retention seats was 2.87-times greater than pre-HR (high retention) seats.

Figure 2 shows the energy transfer of the seats up to 60^0 seatback angle versus the peak H-point moment in the QST test. The high retention seats consistently perform above the target specification of 1700 Nm H-point moment, and many of the seats perform to a substantial margin. This causes many of the seats to have a high energy transfer capability. The threshold velocity change for the high retention seats averaged 42.5 km/h (26.4 mph) as compared to 25.1 km/h (15.6 mph) for the pre-HR seats. This represents a 69% improvement.

The peak ram load increased 2.18-times (118%) to 14.4 kN reflecting the greater occupant retention capability of the seats. These improvements in strength were achieved without a change in the initial stiffness of the seat or the "yielding" behavior. This reflects a similar compliance, which is important for control of occupant kinematics in the lower severity crashes where whiplash injuries dominate.

Figure 3 shows the peak ram load versus the angle of the seatback at the 1700 Nm target moment specification (or the peak moment if the seat did not meet the specification). The high retention seats reach the target moment at much lower angle from vertical, reflecting a much stronger seatback. However, the initial "yielding" behavior of the seats was similar to pre-HR designs.

QST REPEATABILITY STUDIES

2001 GMT 315 Repeatability Tests at RSHE

Table 5 summarizes the QST responses for the repeatability study with the GMT 315 seat. The average ram force at the 1700 Nm target moment was 11.1 ± 0.2 kN. The coefficient of variation for the ram load is 1.6%. A slightly higher variation was seen in the seatback angle. The average angle was $27.6^0 \pm 0.8^0$ giving a coefficient of variation of 2.9%. When the peak H-point moment was considered, the coefficient of variation was 6%, but the peak moment happened well after the 1700 Nm specification.

Figure 4 shows the individual responses for H-point moment and seatback twist versus the seatback angle in the repeatability QST tests of the GMT 315 seat. Only one test had a higher moment response than the other seven in the tests, and the average response is also shown. There is a remarkable consistency in the results. There is very small twist with the dual recliner seat, which has good compliance in the seatback suspension and a strong seatback frame, which resist rotation under occupant load.

1994 W Seat Studies at RSHE/Delphi

Table 6 summarizes the QST responses for the repeatability study with the 1994 W seat. The peak ram load was 6.7 ± 0.6 kN giving a coefficient of variation of 9% among the seat tests. The average H-point moment was 1107 ± 127 Nm with a 11% coefficient of variation. There were no drops in force that met the >2 kN force drop within a 50 ms time-window, but the twist was high at $35^0 \pm 10^0$ reflecting the single-side recliner.

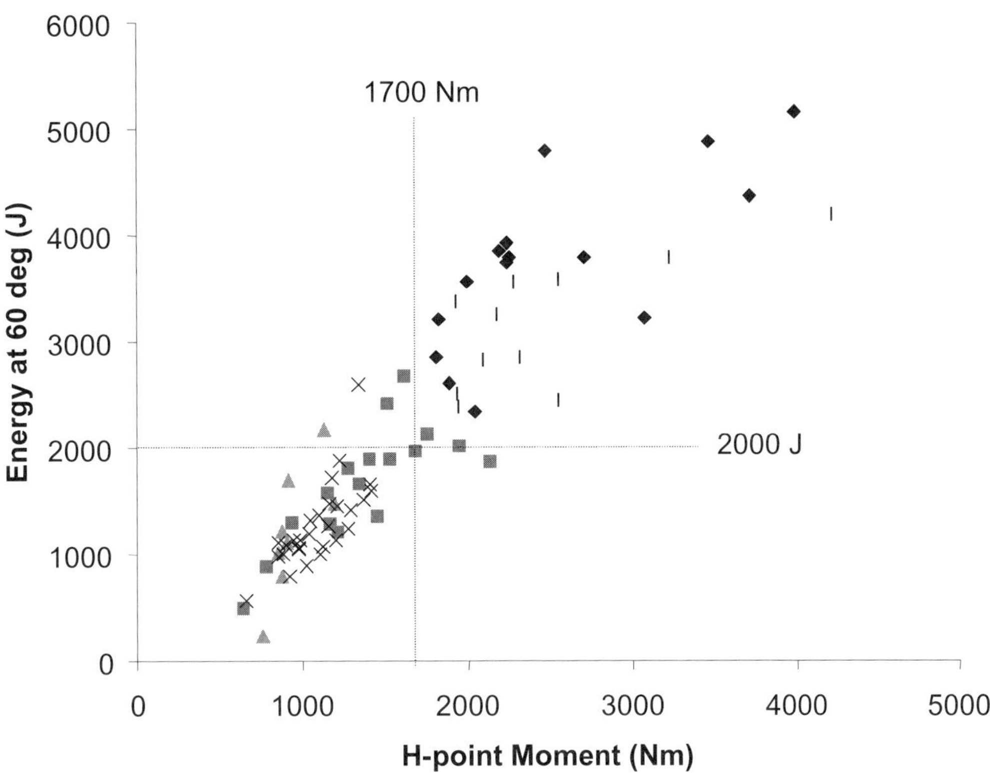

Figure 2: Energy transfer at 60˘ seatback angle versus peak H-point moment for high retention and pre-HR seats.

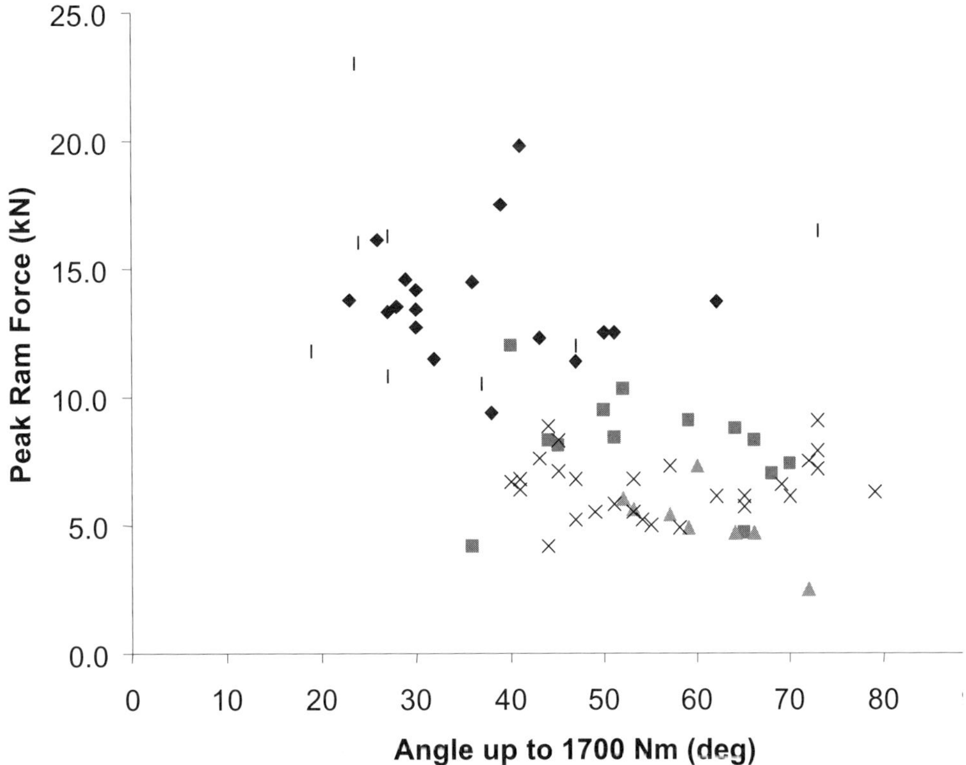

Figure 3: Ram load versus seatback angle up to the 1700 Nm H-point moment for high retention and pre-HR seats.

Table 5: Results of the Repeatability QST Testing of the GMT 315 Seat.

Repeatability Test #	Force kN	Angle deg	z-displ. mm	Twist deg	Peak Moment Nm
	Responses at 1700 Nm				
1	11.18	27.5	-1.37	0.94	2074
2	10.69	26.0	6.04	0.17	2421
3	11.27	27.7	-1.94	0.37	2246
4	11.15	28.8	-0.73	-1.78	2068
5	11.14	27.3	0.16	-0.06	2135
6	11.12	28.1	0.16	0.87	2335
7	11.09	27.8	1.21	0.13	2125
19	11.10	27.8	-0.56	1.59	2103
Avg	11.09	27.6	0.37	0.28	2188
sd	0.17	0.80	2.49	0.99	131
CV	1.6%	2.9%			6.0%

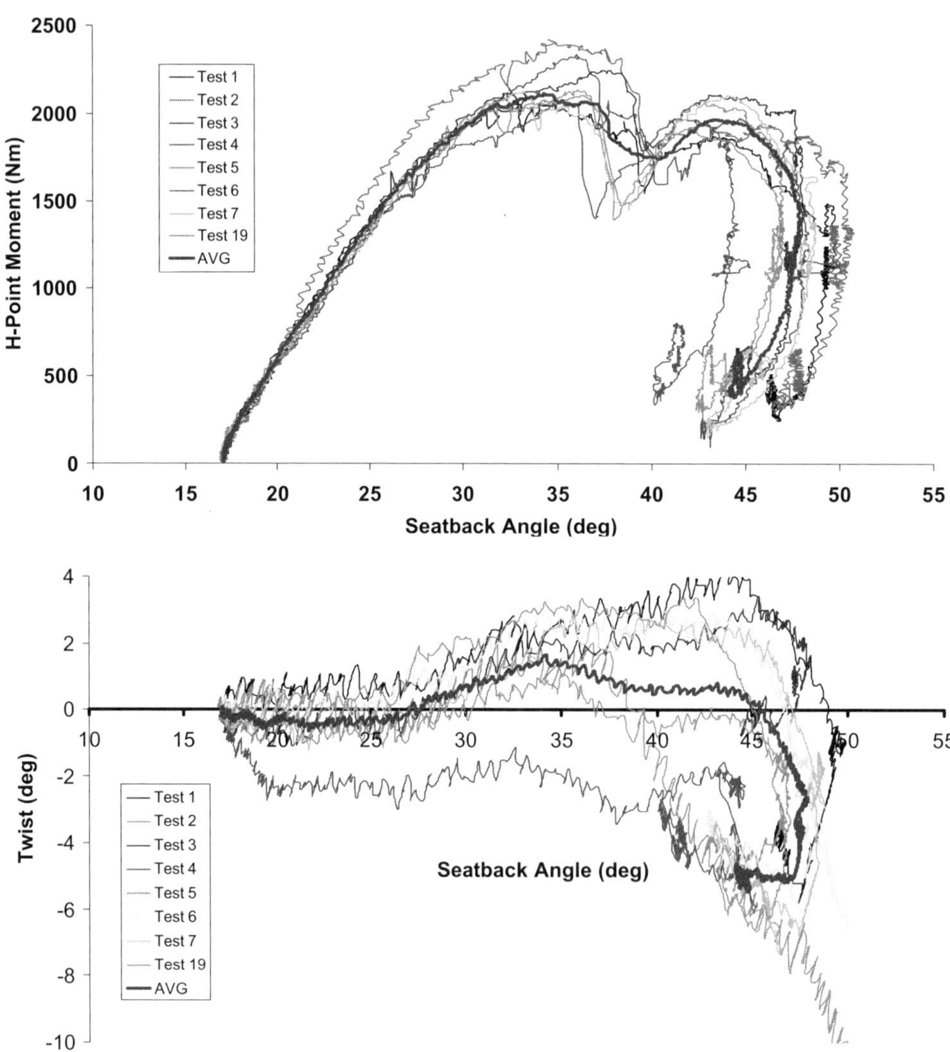

Figure 4: H-pt moment (top) and twist (bottom) versus the seatback angle for the repeatability QST tests.

Table 6: Results of the Repeatability QST Testing of the 1994 W Seat.

Test #	Z Offset mm	Ram Load X kN	Ram Load θ deg	Seatback Stiffness kN/m	Peak Force Tang. kN	Peak Force Nor. kN	Peak Mom. Nm	Angle @ Pk Mom deg	Ram X @ Pk Mom mm	Pk Mom H-pt Nm	Angle @ H-pt deg	Total Engy J	Engy @ 60 deg J	Torque < 60 deg Nm	Twist < 60 deg deg
QST Testing at RSHE Facility															
244	276	5.7	47	27.3	4.2	3.9	1487	47	231	881	47	1859	1859	417	45
246	272	7.1	60	25.2	6.2	4.0	2138	60	440	1303	60	2499	1866	305	35
247	273	5.8	64	26.8	5.2	3.8	1763	64	442	1077	64	2087	1356	240	19
248	281	7.4	58	23.0	6.3	4.0	2256	58	404	1317	58	2460	1916	392	35
Avg	276	6.5	57	25.6	5.5	3.9	1911	57	379	1145	57	2226	1749	339	34
sd	4	0.9	7	1.9	1.0	0.1	352	7	100	207	7	307	263	81	11
CV	1%	13%	13%	8%	18%	2%	18%	13%	26%	18%	13%	14%	15%	24%	31%
QST Testing at the Delphi Facility															
1	280	6.6	46	20.2	4.8	4.6	1882	46	413	1071	52	1895	1895	286	35
2	276	7.2	43	17.3	4.9	5.3	1939	43	412	1108	43	2079	2079	292	31
3	278	6.6	41	35.3	4.3	5.0	1818	41	184	1057	47	2177	2177	271	36
4	276	6.4	41	35.4	4.3	4.8	1745	42	198	1049	48	2016	1928	351	56
5	276	6.3	40	16.7	4.1	4.9	1699	41	398	1060	48	1823	1823	278	28
6	273	7.4	42	16.0	5.0	5.6	1968	42	385	1146	42	2119	2119	275	30
Avg	277	6.8	42	23.5	4.6	5.0	1842	43	332	1082	47	2018	2004	292	36
sd	2	0.4	2	9.3	0.4	0.4	107	2	110	38	4	136	141	30	10
CV	1%	7%	5%	40%	8%	7%	6%	4%	33%	3%	8%	7%	7%	10%	29%
Overall Statistics for Both Facilities															
Avg	276	6.7	48	24.3	4.9	4.6	1870	48	351	1107	51	2101	1902	311	35
sd	3	0.6	9	7.1	0.8	0.6	221	9	103	127	7	231	227	57	10
CV	1%	9%	19%	29%	16%	14%	12%	18%	29%	11%	15%	11%	12%	18%	28%

Table 7: Results of the variability QST Testing of the GMT 315 Seat.

Repeatability Test #	Responses at 1700 Nm Force kN	Angle deg	z-displ. mm	Twist deg	Peak Moment Nm
8	10.9	35	-0.4	-1.1	1675
9	11.2	25	-0.7	-0.1	2186
10	11.9	34	-11.1	-1.0	1797
11	11.3	33	-2.7	3.3	1973
12	12.4	32	-15.9	0.0	1810
13	5.6	50	151.5	2.5	1738
14	11.2	29	-1.1	0.6	2025
15	11.1	28	-0.9	1.5	2114
16	11.2	28	-2.2	0.4	2012
17	11.2	36	-2.0	-1.0	1793
18	11.8	41	-9.8	1.6	1709
20	11.2	27	-1.3	-1.1	2397
Avg	10.9	33	8.6	0.5	1936
sd	1.7	7	45.3	1.5	221
CV	16%	21%			11%
excluding #13					
Avg	11.4	32			
sd	0.5	5			
CV	4%	15%			

The key seat responses of load, moment and energy transfer had a coefficient of variation of 9-12% and were not statistically different between the test facilities. The results at the Delphi facility were more consistent.

QST VARIABILITY STUDY

Table 7 summarizes the QST responses for the variability study with the GMT 315 seat. The average ram force at the 1700 Nm target moment (or to the peak H-point moment) was 10.9 ± 1.7 kN. The coefficient of variation for the ram load is 15.9%. However the large variation was dominated by one test. If test 13 is excluded, the average load was 11.4 ± 0.5 kN and gave a coefficient of variation of 4%. A higher variation was seen in the seatback angle. The average angle was $33.2^0 \pm 7.0^0$ giving a coefficient of variation of 21%. When the peak H-point moment was considered, the CV was 11%.

Figure 5 shows the individual responses for H-point moment and seatback twist versus the seatback angle in the variability QST tests of the GMT 315 seat. The variation in the initial seatback angle is clearly seen in the top plot. Again, there is very small twist with the dual recliner seat.

DISCUSSION

HIGH RETENTION SEAT PERFORMANCE IN QST

Substantial improvements in seat strength have been realized with the introduction in QST requirements. Many of the twenty different seats far exceeded the moment and energy transfer capability set by the QST specifications in 1995. The performance enables the occupant to experience a high-severity rear crash while being contained in the seat without seatback rotations above 60^0 from vertical.

The design strategy for most of the conventional (non-ABTS) designs followed the approach of a perimeter seat frame with open rear profile to allow the occupant's pelvis to displace rearward into the seat frame. This approach allows the initial compliance of the seatback to be similar to earlier seats, thus ensuring a comparable "yielding" performance even though the strength of the high retention seats is almost 3-times greater than earlier seats. By contrast, the pre-HR designs "yield" by seatback rotation rather than seatback compliance within the seatback frame.

Based on the energy transfer capability of the high retention seats up to 60^0 seatback angle, the delta V performance is up to 42.4 km/h (26.4 mph). The studies of Malliaris (1990) showed that this delta V is above the 95th percentile crash severity for vehicles of mass 1150-1600 kg. In lighter cars (<1150 kg), it represents the 90th percentile crash severity, but approaches the 99th percentile for the heavier vehicles (>1600 kg). This performance covers a greater range of real-world crashes than that of earlier seat designs.

The higher ram load and H-point moment of the high retention seats are achieved at lower seatback angles than earlier seat designs. An average 35% reduction in seatback angle at the 1700 Nm target was found. This means that in rear crashes the dynamic seatback rotation is reduced, thus ensuring greater survival space for rear seated occupants and reducing the risk of on-loading of front seat occupants on rear occupants that can be a source of injury in some crashes. Also, by achieving the higher loads and lower seatback angles, the potential for loss of retention for both the belted and unbelted occupant is substantially reduced. With the pocketing of the pelvis in the seatback and the slight reduction in height of the H-point, the overall crash performance of the high retention seats give a wide range of potential field crash improvements.

The higher ram loads do not increase the risk of acceleration injury for the occupant. In fact, the average peak load of 11.4 kN represents a 22 g whole-body acceleration assuming the 52.5 kg effective occupant mass loading the seatback. This acceleration is well below the 60 g tolerance specified in FMVSS 208. It is also known that human tolerance to acceleration is greater in the rearward loading direction, so the occupant can well tolerate the higher loads when they are uniformly distributed over the head, neck, torso and pelvis (Snyder 1970).

By using the perimeter seat frame approach and ensuring equivalent compliance in seatback stiffness, the conventional high-retention seats achieve the safety improvements without an increase in seat mass or cost. The QST is a pivotal subsystem test that helped refine the seat structures. It allows mass shifts to better accommodate the occupant load, rather than merely increasing FMVSS 207 performance, which tends to put structure in areas not optimum for occupant interactions.

Finally, all of the high retention seats include a head restraint that accommodates the 95th percentile seated occupant. This places the head restraint in the up position at the head center of gravity for the 95th percentile occupant and includes a small gap behind the head. The head restraint is about 75 mm (3") higher and 62 mm (2.5") more forward than pre-HR head restraints. This placement provides better head and neck support, as the seats are stronger and maintains the occupant in a more upright seating position than pre-HR designs in high-severity rear crashes.

QST REPEATABILITY STUDIES

2001 GMT 315 Repeatability Tests at RSHE

The GMT 315 seat is a high retention design with dual recliners. There was a remarkable consistency in

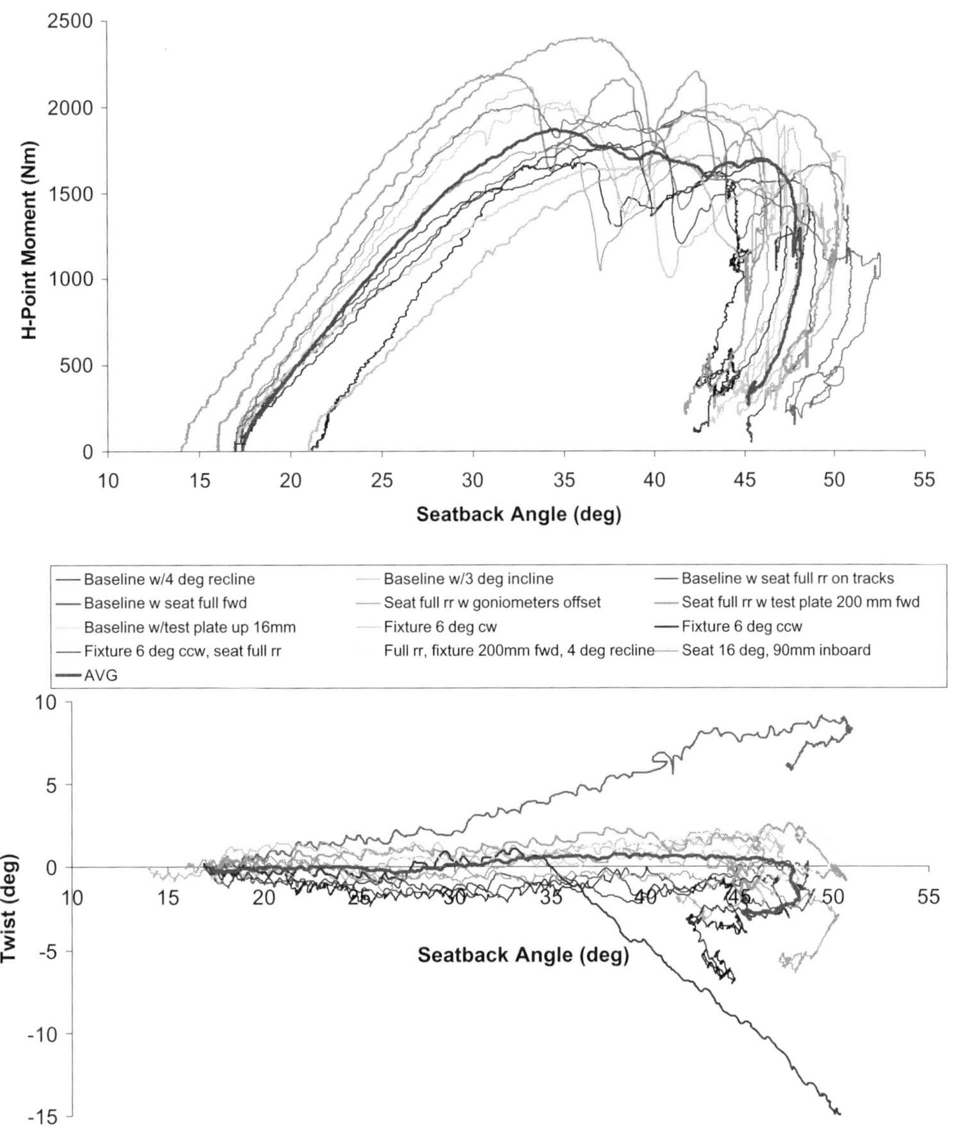

Figure 5: H-pt moment (top) and twist (bottom) versus the seatback angle for the variability QST tests.

performance among the 8 seat tests, particularly at the 1700 Nm target specification for H-point moment. Overall, seat performance in the test is very repeatable. The initial stiffness of the seat is determined by the compliance of the seat suspension and occurs with minimal seatback rotation. This is a stark contrast to early seat designs, which were dominated by seatback rotation and twist up to a 4 kN occupant load.

The deformation pattern of the seats was also quite similar. The perimeter frame design let the occupant penetrate into the seatback. This loaded the frame and deformed the tubular structure above the recliners, which resisted the bending moment on the seatback. Figure 6 shows the deformation pattern of the GMT 315 seat frame. The mass of the fully trimmed seat was 19.8 kg.

1994 W Seat Studies at RSHE/Delphi

The results of the initial repeatability study showed key responses within 9-12% among the seats and two test facilities. This was judged sufficient to adopt the test practice as much of the variation was considered due to seat structure and recliner differences to the occupant loading. For example, the variations in seat stiffness are due to a combination of seatback rotation and twist, which occurred early in the test. Since the stiffness is calculated as the average performance between ram loads of 1 and 4 kN, it was influenced by both modes of seat deformation. The testing led to the development of a test practice and procedures, which reduced the possibility of setup and measurement variations.

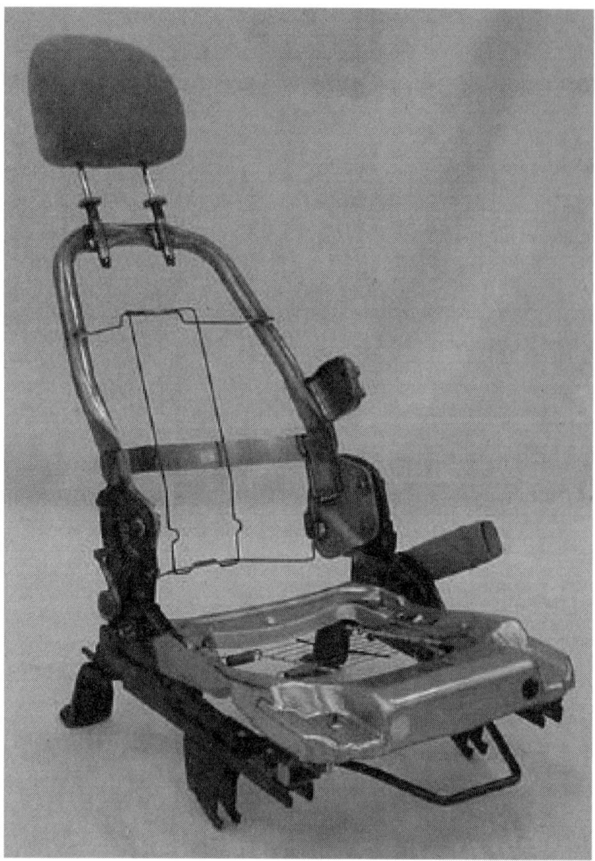

Figure 6: GMT 315 seat after QST testing. The seat is a perimeter frame design with low rear profile to allow occupant penetration into the seatback. This gives the "yielding" behavior of the seat without rearward rotation of the frame. Also, the frame deformation is above the recliners.

QST VARIABILITY STUDY

The twelve tests introduced a wide range of variability into the seat setup and instrumentation. These variations are much larger than would normally occur under normal conditions of setting up a test. Thus, they represent an extreme in misalignment of components. Nonetheless, it is known that seat performance in the QST depends on the position in the seat track and the type of adjusters used in the seat construction. For example, it is expected that different QST responses would be seen with manual and power adjusted products due to the different seat construction of the adjusters.

In normal testing, the seat is positioned in mid-track and all comfort systems are placed in the de-inflated or most relaxed position, so that lumbar bladders are deflated and straps are relaxed in the tests.

TRENDS IN SEAT STRENGTH

Based on the current seat testing and results from previous studies by Severy (1967), Strother et al. (1987) and Warner et al. (1991), Figure 7 shows the change in seat strength over the past four decades. By the 1980s-90s, seats had increased in strength to 6-8 kN peak load from the lows of about 3 kN in the 1960s-70s. High retention seats further increased the strength of seats to over 14 kN, but they did so in a way that the seat stiffness was similar to the yielding seats of the 1980s-90s. This fulfilled the strategy undertaken by the development of high retention seats using a perimeter frame design. It was a means of providing both a strong seat frame and a yielding suspension, thus overcoming the debate between rigid seats for occupant retention or yielding seats for whiplash prevention of the past decades.

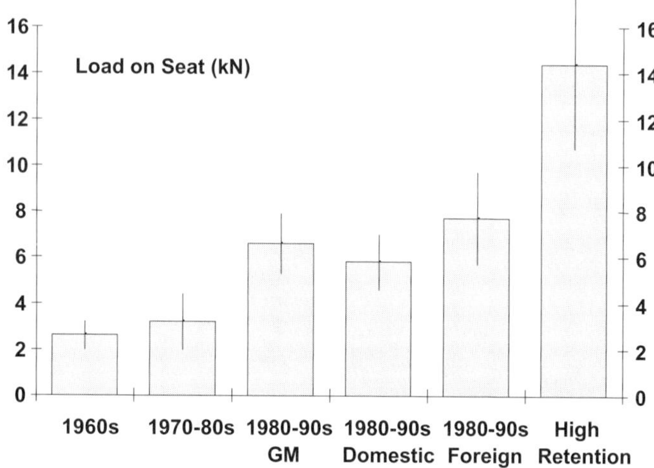

Figure 7: Trends in peak force supported by seats loaded toward the rear.

ACKNOWLEDGMENTS

Many individuals contributed to the testing included in this study. Joe McCleary, Ken Desaele and John Hiben of the GM R&D Center conducted the 1998-2000 QST tests of high retention seats. Mladen Humer made data available from the original repeatability study of the W seat. He was at Delphi seating at the time and is now at Lear Corporation.

Jim Pywell and John Hibens of GM Vehicle Structure and Safety Integration conducted the 2001 repeatability and variability studies. Their efforts are greatly appreciated. In particular, new instrumentation and data analysis methods were introduced in 2000. Jim developed the replacement software to analyze QST responses.

REFERENCES

Malliaris AC. Current Issues of Occupant Protection in Car Rear Impacts. Malliaris, Inc., NHTSA Docket 89-20-No1-021, February, 1990.

Severy DM, Brink HM, Baird JD. Preliminary findings of head support designs. 11th Stapp Car Crash Conf, SAE 670921, SAE Warrendale PA, 337-405, 1967.

Snyder RG. Human Impact Tolerance: American Viewpoint. International Automobile Safety Conference, SAE 700398, P-30 Society of Automotive Engineers, Inc., Warrendale, PA, 1970.

Stephens GD, Long TJ, Blaisdell DM. Energy Analysis of Automotive Seat Systems, SAE 2000-01-1380, SP-1494, Society of Automotive Engineers, Warrendale, PA, 2000.

Strother CE, James MB. Evaluation of Seat Back Strength and Seat Belt Effectiveness in Rear End Impacts. 31st Stapp Car Crash Conference, SAE 872214, Society of Automotive Engineers, Warrendale, PA, 225-244, 1987.

Viano DC. Role of the Seat in Rear Crash Safety. Book, ISBN 0-7680-0847-6, Society of Automotive Engineers, Warrendale, PA, SAE R-317:1-491, 2002.

Warner C, Strother C, James MB, Decker RL. Occupant Protection in Rear-End Collisions II: The Role of Seat Back Deformation in Injury Reduction. 35th Stapp Car Crash Conf, SAE 912914, Society of Automotive Engineers, Warrendale, PA., 1991.

CHAPTER 7:

EFFECTIVENESS OF HIGH-RETENTION SEATS IN PREVENTING FATALITY: INITIAL RESULTS AND TRENDS

Chapter 7

2003-01-1351

Effectiveness of High-Retention Seats in Preventing Fatality: Initial Results and Trends

David C. Viano[1,2]
[1]ProBiomechanics LLC
[2]formerly Vehicle Structure and Safety Integration
General Motors Corporation

ABSTRACT

In 1995, new seat specifications were adopted by GM to provide high retention and improve occupant safety in rear crashes. With more than five years of phase-in of high retention (HR) seats, an analysis of FARS was undertaken to determine the initial field performance of HR seats in preventing fatalities. The 1991-2000 FARS was sorted for fatal rear-impacted vehicles. Using a VIN decoder, GM vehicles with HR front seats were sorted from those with baseline (pre-HR) seats. The fatal rear-impacted vehicle crashes were subdivided into several groups for analysis: 1) single-vehicle rear impacts, 2) two-vehicle rear crashes involving light striking vehicles, and 3) two-vehicle crashes involving heavy trucks and tractor-trailers, and multi-vehicle (3+) rear crashes.

While more field data is needed to increase confidence in the results, FARS analysis shows that high-retention seats reduce the risk of driver fatalities in single-vehicle rear crashes by 50% (-26%, 80%, ±95% CI) and driver/RF deaths by 59% (11%, 81%). For two-vehicle rear-impact crashes, high-retention seats reduce the risk of driver death by 54% (-23%, 83%) and driver/RF deaths by 35% (-38%, 70%) when impacted by light vehicles. There is no difference in risk with heavy truck or tractor-trailer impacts or in multi-vehicle (3+ vehicles involved) crashes. The new generation of seats was developed for high retention based on a perimeter seat frame design that allows the occupant to penetrate between the side frames of the seatback giving a "yielding" performance and pocketing of the pelvis. This increases retention and provides uniform support for the spine. The new generation of seats also includes a higher and more forward placement of the head restraint to reduce neck extension-related injury. That aspect of the seat performance cannot be studied with the FARS data.

The initial trends show that high-retention seats are effective in reducing the risk of fatal injury in single vehicle and light vehicle-to-vehicle rear impacts.

INTRODUCTION

Pioneering safety research on seat designs was conducted by Severy et al. (1955, 1967a,b, 1968a,b) on seatback rotation and support of the head and torso by seatback designs in rear crashes. However, many design approaches were not brought to production because of various criticisms on overall performance. Subsequently, there have been many studies on the epidemiology of rear crash injuries and the required engineering performance of seats and head restraints.

Strother and James (1987) and Warner et al. (1991) provided an engineering analysis and rationale for seatback strength in rear crashes. Their studies also addressed petitions submitted to NHTSA to substantially increase the seatback moment required in FMVSS 207. The authors countered the position in the petitions with the "yielding" seatback concept as the best approach to providing overall occupant safety in rear crashes. This approach provided controlled seatback rotation and energy-absorbing restraint as the seatback yields rearward in a crash. Seatback moments near the FMVSS 207 criterion were argued as necessary to balance the needs for whiplash injury prevention and occupant retention in severe rear crashes.

The Strother and James (1987), Warner et al. (1991) and Prasad et al. (1997) studies furthered the idea that there is an underlying design conflict between occupant retention by a "stiff" or "rigid" seat in severe but relatively infrequent rear crashes, and the need for a "yielding" seatback to prevent whiplash in the more frequent, minor rear impacts. More recent studies have shown the importance of head restraint placement irrespective of seat strength in lowering neck biomechanical responses (Viano 2002). Severe rear crashes were defined to involve vehicle delta Vs of 27.4-35.4 km/h (17 to 22 mph) as occur in FMVSS 301-type rear barrier impacts conducted at 54.2 km/h (33.7 mph). This test has more than 25% higher kinetic energy than the federal standard and the kinetic energy transfer to the occupant in these crashes is well beyond the capability of seats that exceed even twice the moment specified in FMVSS 207 (1968).

Federal Motor Vehicle Safety Standard FMVSS 207 was adopted from recommended practices dating to the mid-1960s. It involves a concentrated load on the upper cross-member of the seatback structure and is run without the seat suspension or trim. The load is centered and oriented perpendicular to an axis through the occupant torso and seating reference point (or H-point). The Standard requires the seat to support a moment of 3,300 in-lb (373 Nm), which is calculated by multiplying the applied load and the distance from the H-point. For years, GM had a performance requirement of 6,600 in-lb (746 Nm) in the test. A different performance applies outside of North America. ECE R17 specifies a minimum moment of 530 Nm (4,691 in-lb) in a test similar to FMVSS 207.

Petitions for NHTSA rulemaking on FMVSS 207 have been submitted by Saczalski (1989) and Cantor (1989); and, various product liability cases have been pursued on seatback strength (Rake and Boehm 1990). These actions have prompted studies on the performance of seats in rear crashes and injury statistics. Malliaris (1990) conducted a thorough analysis of FARS and NASS field accident data, which indicated that nearly 5% (1700 cases) of all fatal passenger crashes are rear impacts involving 3.5% (843) of all fatalities. Yet, when all crashes are considered, rear impacts involve 12% (1,433,000) of exposed occupants and 23% (613,000) of all those injured, so the seat performance in low-speed impacts is a major consideration.

In 1990, GM initiated a study of the role of the seat in rear crash safety. The results of the work have been published in various journal articles (Viano, Olsen 2001, Viano, Davidsson 2002, Viano et al. 2002, Viano, Hardy, King 2002, Linder et al. 2002, Viano 2003) and in a book (Viano 2002) on the development of new seat concepts and requirements. It also led to several innovations in seat design for rear crash safety (USPTO 1994, 1996, 1997, 1998b) and a Quasistatic Seat Test (USPTO 1998a), which applies a rearward load by a dummy on the fully trimmed seat. This is a more realistic alternative to the FMVSS 207 procedure for safety evaluations of seats.

In 1995, GM adopted new seat specifications for rear impact based on the Quasistatic Seat Test (QST). This established requirements for high-retention seat performance and a higher and more forward placement of the head restraint (see Viano 2002). The first production seat to comply with the new specifications was in the 1997 Pontiac Grand Prix and through 2002 essentially all seats now surpass the high retention requirement (Viano 2003). This has resulted in a new generation of seats for rear crash safety.

This study investigates five years of field performance with high-retention (HR) seats. It uses the 1991-2000 FARS and compares the fatality risk of front seat occupants involved in fatal rear-impacted vehicles. Separate analyses were performed on the fatal rear-impact crashes, which were subdivided into three groups for analysis: 1) single-vehicle rear impacts, 2) two-vehicle rear crashes involving light striking vehicles, and 3) two-vehicle crashes involving heavy trucks and tractor-trailers, and multi-vehicle (3+) rear crashes. These analyses compare GM vehicles with high-retention seats that comply with the new seat specifications to GM vehicles equipped with baseline or pre-high retention (pre-HR) seats meeting the 6,600 in-lb (746 Nm) requirement based on FMVSS 207 type testing. The analyses estimate the change in fatality risk with high-retention seats.

METHOD

Data: The fatal crash information consisted of samples from FARS for years 1991-2000. The data was compared for passenger cars and light truck vehicles (LTV), which include vans, sport utility vehicles and small trucks that had high-retention seats with vehicles equipped with non-high-retention or baseline seats introduced before the phase-in of the new seat requirements.

Nomenclature for Data Analysis:

- F — Number of Rear Struck Vehicles in Fatal Crashes
- D — Number of Rear Struck Vehicles with Either a Driver or Driver/RF Passenger Fatality Depending on the Analysis Conducted
- C — Number of Severe Rear Impact Crashes

- R_D — Risk of Driver/RF Passenger Fatality in a Fatal Rear-Impacted Vehicle
- R_R — Risk of Rear Passenger Fatality in a Fatal Rear-Impacted Vehicle
- R_{OD} — Risk of Driver/RF Passenger Fatality in the Striking Vehicle
- R_{OR} — Risk of Rear Passenger Fatality in the Striking Vehicle

$l = R_R/R_D$
$m = R_{OR}/R_{OD}$
$\alpha = D/F$

Subscript /B is baseline vehicle seats
Subscript /H is high retention vehicle seats

$j = R_{OD/H}/R_{OD/B}$
$k = R_{R/H}/R_{R/B}$
$r = l_H/l_B$
$s = m_H/m_B$

Δ = Reduction in risk with high-retention seats

Analysis Procedures

Vehicle-to-Vehicle Fatal Crashes: In the FARS data, there are four sources of fatality in fatal vehicle-to-vehicle rear crashes. The occupants in the struck vehicle experience a rear impact in which there is risk for the driver or right-front passenger to be fatally injured. These two occupants are grouped in the analysis and considered to have a similar risk. There is also a risk for fatal injury to rear seated occupants in the struck vehicle. Again any occupant in the rear seat is included in this risk, which includes both the underlying risk of fatal injury for a rear occupant exposed to a severe rear crash, and the effect of occupancy where a rear occupant is not always present or there may be several in a crash. This gives a cumulative risk based on the underlying risk in a seating position times an occupancy rate. For this analysis an effective risk is used.

In the striking vehicle there is a risk of fatality to the driver and right-front passenger, and similarly as in the struck vehicle, a different risk for rear seated passengers. Again, these risks are effectively the combination of the underlying risk for the seating position times the occupancy rate. Based on these four risks, the number of fatal rear-impacted vehicle crashes is related to the risk of death in the struck and striking vehicle and the seating position in the vehicle times the number C of severe rear crashes. The following relationship applies:

$$F = (R_D + R_R + R_{OD} + R_{OR})C \qquad (1)$$

where F is the number of fatal rear-impacted vehicle crashes. With Equation (1) only fatal crashes of the following four types are considered: all crash fatalities in the front seats of the rear-struck vehicle, all crash fatalities in the rear seats of the rear-struck vehicle, all crash fatalities in the front seats of the striking vehicle, and all crash fatalities in the rear seats of the striking vehicle. This analysis does not include joint probabilities of the death of two occupants. This would include, for example, the joint risk of a driver and rear passenger death in the struck vehicle.

The number of rear struck vehicles with driver and RF passenger deaths is related to the underlying risk times the exposure:

$$D = R_D C \qquad (2)$$

Combining Equation (1) with (2) gives a relation for the risk of a driver/RF passenger fatality in the rear struck vehicle:

$$R_D = (R_D + R_R + R_{OD} + R_{OR})D/F \qquad (3)$$

Manipulation of Equation (3) gives the following:

$$R_D(F - D)/F = (R_R + R_{OD} + R_{OR})D/F \qquad (4)$$

Previous analyses of FARS data have shown a relationship between the risk of death in the front and rear seats. This allows the following relationships to be defined: $R_R = lR_D$ and $R_{OR} = mR_{OD}$, where l and m are parameters relating the two variables. In the struck vehicle, the vehicle specific risk in the rear seating position is less than that for front seated occupants in rear crashes, so l <1, since it involves occupancy and the underlying risk in that seating position. In the striking vehicle, the risk in the rear seat is also less than in the front, so m <1, but both the occupancy and underling risk are lower than for the front seat occupants. The following simplification can be made to the risk of driver and RF passenger death in the struck vehicle, Equation (4):

$$R_D = \alpha(1+m)R_{OD}/[1-\alpha(1+l)] \qquad (5)$$

where $\alpha = D/F$.

Now, considering the two field crash situations of vehicles equipped with baseline seats (B) and those equipped with high-retention seats (H), Equation (5) becomes two relationships based on the available field crash data:

$$R_{D/B} = \alpha_B(1+m_B)R_{OD/B}/[1-\alpha_B(1+l_B)] \qquad (6)$$

$$R_{D/H} = \alpha_H(1+m_H)R_{OD/H}/[1-\alpha_H(1+l_H)] \qquad (7)$$

The change in fatality risk (Δ) for the driver and RF passenger of the struck vehicle is given by the following relationship:

$$\Delta = 1 - (R_{D/H}/R_{D/B}) \qquad (8)$$

It can be assumed that the risk of a driver or RF passenger death in the striking vehicle is similar for the vehicles with baseline seats and high-retention seats; however, the vehicle ages differ on average by 5 years. It is more reasonable to assume that safety improvements have reduced the driver and right front (RF) passenger risks in the striking vehicle, so let $j = R_{OD/H}/R_{OD/B}$. With frontal crash safety improvements, j <1. This reduces Equation (8) to:

$$\Delta = 1 - \frac{j\alpha_H}{\alpha_B}\left(\frac{1+m_H}{1+m_B}\right)\left[\frac{1-\alpha_B(1+l_B)}{1-\alpha_H(1+l_H)}\right] \qquad (9)$$

Some assumptions can be made about the parameters in Equation (9). Based on FARS data, the risk of rear occupant death in rear crashes is 1.5 times greater than for front occupants (Parenteau, Viano 2003). However, the occupancy is lower in the rear seat at 0.24. If these parameters are used, l_B = 0.36. In the striking vehicle, the risk of fatal injury in the rear seat in a front impact is 0.60 times that in the front seat and again assuming a

0.24 occupancy rate, the effective risk gives $m_B = 0.14$. Using these values and the relationships for r and s, reduces Equation (9) to:

$$\Delta = 1 - \frac{j\alpha_H}{\alpha_B}\left(\frac{1+0.14s}{1.14}\right)\left[\frac{1-1.36\alpha_B}{1-(1+0.36r)\alpha_H}\right] \quad (10)$$

If the relative risks are similar in the baseline and high retention vehicles, $r = s = 1.0$ and the safety effectiveness in Equation (10) reduces to:

$$\Delta = 1 - \frac{j\alpha_H}{\alpha_B}\left[\frac{1-1.36\alpha_B}{1-1.36\alpha_H}\right] \quad (11)$$

By assuming r = 1, high retention seats are assumed to reduce risks for the front and rear seated occupant. With lower rotations of the seatback, on-loading of the rear seated occupant is reduced in rear crashes, which lowers risk.

Vehicle-to-Vehicle Fatal Crashes Involving Driver or Driver/RF Passenger Deaths: It is possible to determine effectiveness of the high-retention seat by looking at fatal crashes involving only driver or driver/RF passenger deaths in the struck and striking vehicle. This is a simplification of the previous analysis, and can be determined by setting m and l to zero, which reflects no rear occupants. Equations (6) and (7) reduce to:

$$R_{D/B} = \alpha_B R_{OD/B} / [1-\alpha_B] \quad (12)$$

$$R_{D/H} = \alpha_H R_{OD/H} / [1-\alpha_H] \quad (13)$$

and Equation (9) reduces to:

$$\Delta = 1 - \frac{j\alpha_H}{\alpha_B}\left[\frac{1-\alpha_B}{1-\alpha_H}\right] \quad (14)$$

This relationship is the same as the following based on odds ratios, which also allows the determination of ±95% confidence intervals:

$$\Delta = 1 - j[D_H/(F_H - D_H)]/[D_B/(F_B - D_B)] \quad (15)$$

$$\pm 95\%CI = 1 - j(1-\Delta)\exp(\pm 1.96\lambda^{0.5})$$
$$\lambda = 1/D_H + 1/(F_H - D_H) + 1/D_B + 1/(F_B - D_B) \quad (16)$$

Single-Vehicle Fatal Rear Crashes: In some cases, the rear-impacted vehicles were involved in a single-vehicle crash. This is a special case of the generalized analysis of FARS data where there is no striking vehicle. The rear-impacted vehicle has the possibility of driver/RF fatality and rear-seated passenger deaths. There are two risks when the driver and RF passenger are combined, and all rear occupants are considered. Here, the risk to the driver is considered similar to that for the RF passenger. This effectively combines the underlying risk for the seating position and the occupancy rate. Based on the two risks, the number of fatal rear-impacted vehicle crashes is related to the risk of death in the struck vehicle and the seating position in the vehicle times the number of severe rear crashes C, a modification of Equation (1):

$$F = (R_D + R_R)C \quad (17)$$

Combining Equation (14) with (2) and using other relationships previously defined, gives the following change in risk:

$$\Delta = 1 - \frac{k\alpha_H}{\alpha_B}\left[\frac{1-\alpha_B}{1-\alpha_H}\right] \quad (18)$$

where k is the change in risk for rear-seated passengers from the baseline vehicles to those equipped with high-retention seats. Due to the higher occupant retention of front-seat occupants, the possibility of ramping to the rear is reduced with high-retention seats. This improves the safety of the rear occupants. In this case, k <1.

RESULTS

The bottom row in Table 1 shows that there have been 41 deaths of drivers and RF passengers in 191 fatal rear-impact crashes of GM vehicles equipped with high-retention seats. The comparison group of GM vehicles involved 988 deaths of drivers and RF passengers in baseline seats in 3504 fatal rear-impact crashes. The data was further classified and analyzed by single-vehicle and multi-vehicle crashes. Figure 1 shows that 25% of the fatal crashes were single vehicle crashes, 41% two-vehicle crashes and 34% involving 3+ vehicles. In the Appendix, Tables A1 and A2 give the full data by year of crash, carline and case summary.

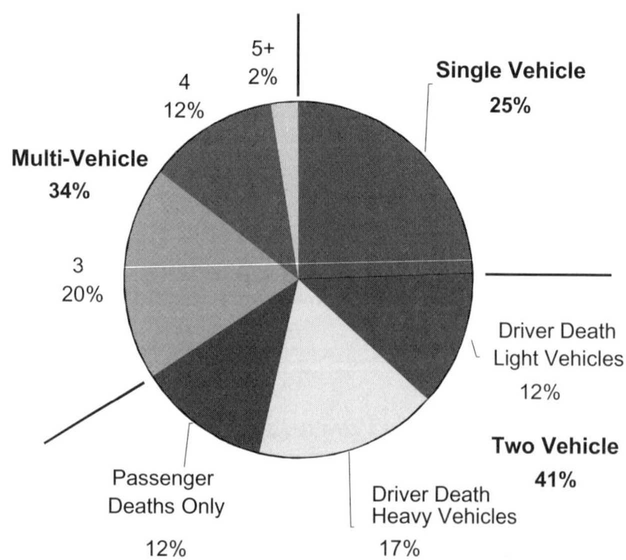

Figure 1: Distribution of fatal rear-impacted GM vehicles with high-retention seats (see also Table 1).

Table 1: Fatal Rear-Impacted Vehicles (FARS 1991-2000).

Rear-Impact Crashes		**High-Retention**	**Baseline Seats**	**Reduction in Risk with High-Retention Seats j,k = 1.0**	**Confidence -95%**	**Confidence +95%**	**Reduction in Risk with High-Retention Seats j,k = 0.9**	**Confidence -95%**	**Confidence +95%**
Single-Vehicle Crashes									
Driver Only	Fatal	8	249						
	Non-Fatal	11	170	50.3%	-26.0%	80.4%	55.3%	-2.1%	84.2%
Driver & RF Passenger	Fatal	10	348						
	Non-Fatal	20	287	58.8%	10.5%	81.0%	62.9%	27.5%	84.6%
Multi-Vehicle Crashes									
Two Light Vehicles									
Driver Only	Fatal	5	227						
	Non-Fatal	22	459	54.0%	-22.9%	82.8%	58.6%	0.4%	86.1%
Driver & RF Passenger	Fatal	9	325						
	Non-Fatal	30	703	35.1%	-38.3%	69.5%	41.6%	-12.0%	75.3%
Heavy Vehicle Rear Crashes									
Driver Only	Fatal	5	77						
	Non-Fatal	15	192	16.9%	-136.6%	70.8%	25.2%	-91.6%	76.3%
Driver & RF Passenger	Fatal	8	105						
	Non-Fatal	20	270	-2.9%	-140.7%	56.1%	7.4%	-95.0%	64.4%
Multi-Vehicle (3+) Crashes									
Driver Only	Fatal	12	149						
	Non-Fatal	55	880	-28.9%	-146.4%	32.6%	-16.0%	-99.6%	45.4%
Driver & RF Passenger	Fatal	14	210						
	Non-Fatal	80	1256	-4.7%	-88.1%	41.8%	5.8%	-52.4%	52.8%
All Multi-Vehicle Crashes									
Driver Only	Fatal	22	453						
	Non-Fatal	92	1531	19.2%	-30.2%	49.8%	27.3%	-5.5%	59.4%
Driver & RF Passenger	Fatal	31	640						
	Non-Fatal	130	2229	16.9%	-24.1%	44.4%	25.3%	-0.5%	55.0%
All Fatal Rear-Impacted Vehicles									
Driver Only	Fatal	30	702						
	Non-Fatal	103	1701	29.4%	-7.0%	53.4%	36.5%	13.3%	62.3%
Driver & RF Passenger	Fatal	41	988						
	Non-Fatal	150	2516	30.4%	0.9%	51.1%	37.4%	19.7%	60.4%

Single-Vehicle Fatal Rear Crashes: Using the FARS data for driver fatalities only in single-vehicle rear crashes, the following rates were calculated for the alpha parameters: $\alpha_H = 0.421$ $\alpha_B = 0.594$. The data for the calculation is shown in Table 1. Using Equation (18) gives the change in fatality risk:

$$\Delta = 1 - 0.496k \tag{19}$$

and

$$\Delta = 50\% - 55\% \tag{20}$$

with k = 1.0-0.9.

Using the FARS data for driver and RF passenger fatalities in single-vehicle rear crashes, the following rates were calculated for the alpha parameters: $\alpha_H = 0.333$ $\alpha_B = 0.548$. Using Equation (15) gives the change in fatality risk:

$$\Delta = 1 - 0.412k \tag{21}$$

and

$$\Delta = 59\% - 63\% \tag{22}$$

with k = 1.0-0.9.

Light Vehicle-to-Vehicle Fatal Crashes: Using the FARS data for driver only fatalities in vehicle-to-vehicle crashes, the following rates were calculated for the alpha parameters when only light vehicles were considered for the striking vehicle: $\alpha_H = 0.185$ $\alpha_B = 0.331$. For this calculation, only passenger cars, SUVs, vans and light duty trucks were considered as the striking vehicle. This eliminated crashes involving heavy trucks and tractor- trailers. Using Equations (14-16) gives the change in fatality risk:

$$\Delta = 1 - 0.459j \tag{23}$$

and

$$\Delta = 54\% - 59\% \tag{24}$$

with j = 1.0-0.9.

Using driver and RF passenger fatalities in either vehicle, the following rates were calculated for the alpha parameters when only light vehicles were considered for the striking vehicle: $\alpha_H = 0.231$ $\alpha_B = 0.316$. Using Equation (13) gives the change in fatality risk:

$$\Delta = 1 - 0.650j \tag{25}$$

and

$$\Delta = 35\% - 42\% \tag{26}$$

with j = 1.0-0.9.

All Vehicle-to-Vehicle Fatal Crashes: In this analysis, all striking vehicles and occupants are considered in fatal rear impacts by vehicles. For the purpose of calculation, $r = s = 1.0$ is assumed. Table 1 shows the results of this analysis with risk reductions of 19%-27% for driver and 17%-25% for driver/RF passengers based on Equation (15). The effectiveness levels are lower when all striking vehicles are considered.

DISCUSSION

In this study, fatal crash data was analyzed to assess the field performance of high-retention seats in rear-impacted vehicles. Both vehicle-to-vehicle and single-vehicle crashes were considered. This gave several estimates for the reduction in risk with high-retention seats based on the classification of data. The results indicate a safety improvement with high-retention seats. Data was obtained from the FARS databases.

High-Retention Seat Design Principles: For impact protection in high-severity rear crashes, several modifications were made in the typical seat frame designs of the late 1980s and early 1990s. These approaches included a perimeter seat frame, open and compliant seatback suspension, low profile rear seat cushion frame and a deformable, pelvic catcher strap, which are approaches to seatback design that allow the occupant to displace between the side frames of the seatback by deformation of the seat suspension and an energy absorbing pelvic strap.

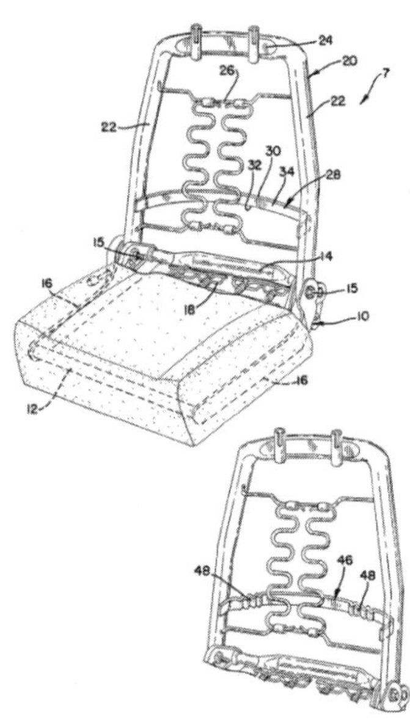

Figure 2: Perimeter frame seat concept with features to allow the occupant to penetrate between the side structures of the seatback (from USPTO 1996).

This approach is shown in Figure 2 and allows a very strong seat frame to minimize seatback rotation and a yielding deformation of the suspension to gradually accelerate the occupant forward in a severe rear crash. It also enhances occupant retention by pocketing the pelvis and lower back for improved containment on the seat even without the use of safety belts. The low rear profile allows the pelvis to drop as it displaced rearward.

This has the secondary benefit of lowering the head with respect to the head restraint providing better head and neck support. These features are included in a US patent granted on the perimeter seat design (USPTO 1996) and allowed the use of a strong seat frame while preserving the "yielding" performance of earlier seat designs. This created a new generation of seats.

The perimeter frame and pelvic catcher approach proved to be one way to address the historic concerns for a rigidized seat raised by Strother and James (1987), Warner et al. (1991), James et al. 1991 and Prasad et al. (1997). These included the potential for higher injury risks with a very strong seatback by rebound, out-of-position loading and neck extension causing whiplash. Sled tests were conducted from low to high severity rear delta V and with normal and out-of-position male and female Hybrid III dummies. This work showed that a strong seatback could be developed with good performance in all conditions (Viano 2002).

The perimeter frame seat was innovative and included design features new to the industry. The introduction in the 1997 Grand Prix saw a seat that was nearly three-times stronger than its predecessor for occupant retention in rear crashes up to 22 mph (35.4 km/h) delta V, and it included a head restraint that was 3" higher and 2.5" more forward to reduce whiplash risks in low to high speed crashes. What was additionally important was that the seat weighed less and cost less than the 1996 Grand Prix seat with comparable features and had similar yielding performance through the pocketing action to gradually accelerate the occupant. Much of this refinement in design and performance was due to developments using the QST test. The perimeter frame design introduced a high volume seat with improved rear crash safety based on laboratory tests that will be discussed in the next sections.

Determining Seat Strength and Energy Management: FMVSS 207 is a subsystem test of the seat, but it does not involve an occupant-loading interface, so many details of occupant interactions in a real-world crash are not comprehended in the test. This is particularly true of the manner in which the occupant's pelvis, lower and upper back and shoulders displace into the seatback, ramp and move on the seat, and load the seat. The occupant exerts a horizontal load and moment low on the seatback, whereas the FMVSS 207 loads high on the upper cross-member in the seatback frame only without suspension or trim, so there can be a similar moment about the recliner but a much lower horizontal load than would be applied by an occupant for the same moment. This is a key distinction, which is relevant to real-world occupant loading of the seat and seat deformation modes in rear crashes.

The Quasistatic Seat Test (QST) was developed to better assess occupant interactions with the seat in rear impacts than FMVSS 207 (Viano 2002). It involves a Hybrid dummy placed in the design seating position and subjected to rear displacement by a hydraulic ram loading through the flexible lumbar spine element. The QST distributes force through the buttocks and upper torso, and simulates occupant loading of the seat in a rear impact. The kinematic sequence in the test mimics the dynamics of torso loading and ramping in a rear crash. It is a destructive test of the seat as the dummy is pushed until failure of the seat system. The test does not address the head/neck interactions with the upper seatback and head restraint.

Seat responses are calculated from the force and displacement of the hydraulic ram loading the dummy into the seat. They include the H-point moment (M_h) and seatback pivot moment (M_p):

$$M_h = F_x(z_h + z) \qquad (27)$$

$$M_p = F_x(z_p + z) \qquad (28)$$

where F_x is the horizontal ram load, z_h is the H-point offset between the centerline of the ram and the dummy H-point, z_p is the z-offset from the ram centerline to the seat pivot point and z is the vertical travel of the ram. The complete analysis of seat responses provides an understanding of the compliance and strength of the seat to rear loading by an occupant. Summary data are compared among seats using either the peak value of responses or levels at 60° average seatback angle from vertical.

Since the QST measures occupant load and displacement, the energy transfer to the seat (E) can be computed by integration:

$$E = \int F_x dx \qquad (29)$$

where x is ram displacement measured by a string potentiometer. A rear crash involves energy transfer to the occupant by the seatback, cushion and head restraint. The amount of energy is given by the kinetic energy (E_c) of an occupant in a crash:

$$E_c = \frac{1}{2} mV^2 \qquad (30)$$

where m is the effective mass of the occupant loading the seatback and V is the change in velocity during the crash or the delta V of the struck vehicle. This is the reduced formulation of kinetic energy with $\lambda = 0$ in the energy Equation given by (Stephens et al. 2000). Since some parts of the body are not initially loaded by the seatback in a rear crash, the effective mass of the occupant is less than the total body weight and varies through a crash. Since the seatback is primarily interacting with the pelvis, torso, upper body, and legs, the effective mass is approximately 52.5 kg, 70% of the dummy mass.

It is possible to calculate a threshold velocity (V_t) for a seat from Equations (26) and (27). This velocity is the maximum severity of crash delta V that a seat can perform with a certain weight occupant and rearward rotation of the seatback. It is calculated by assuming the occupant energy is based on the effective mass interacting with the seatback in a rear crash and is equal to the energy transfer capability of a seat at 60° seatback angle with respect to vertical:

$$V_t = (2E/m)^{0.5} \quad (31)$$

where V_t is the velocity a seat can transfer energy to an occupant without exceeding the specified seatback angle, in this case 60°. Energy transfer is related to the amount of rear displacement of the occupant or the angle of the seatback from vertical.

It is also possible to relate energy to the H-point bending moment and seatback angle change. Energy transfer to the occupant is given in Equation (26). The displacement can be approximated by the arc length of motion about the H-point and the angle change:

$$dx = (z_h + z)d\theta \quad (32)$$

where the radius is the sum of the H-point offset (z_h) and the z-displacement (z) during a test. Combining (26) and (29) gives the H-point moment in Equation (24) integrated over the change in seatback angle:

$$E = \int F_x(z_h + z)d\theta = \int M_h d\theta \quad (33)$$

Thus, the energy transfer capability of a seat is related to the H-point moment integrated over the angle change of the seatback.

New Seat Performance Criteria: Based on QST tests, sled and barrier tests, and geometric relations, Table 2 defines high-retention seat requirements for front bucket or split bench seats adopted by GM in 1995.

Table 2: Seating System Technical Specifications Using the Quasistatic Seat Test

(1) 1700 Nm (15,000 in-lb) Moment About H-Point

(2) No Force Drops >2 kN (450 lb) in 50 ms Moving Time-Window Causing >10° Change in Seatback Angle

(3) Seatback Twist <15° for Seatback Angles <60°

(4) Head Restraint Height to B-Plane and Front Surface to Within 20 mm of Back-of-Head Ellipse for the 95th Percentile Occupant

First, the minimum H-point bending moment is 1700 Nm (15,000 in-lb) to provide energy transfer to the occupant without excessive rotation of the seatback and possibility of loss of retention in severe rear crashes. This provides energy management in rear sled tests that are consistent with the exposure in a 33.7 mph (54.2 km/h) FMVSS 301-type rear moving barrier test. For reference, FMVSS 301 involves a 30 mph (48.3 km/h) flat moving barrier of 2273 kg (5000 lb) mass. Tests on a number of passenger vehicles provided an average 31.1 km/h (19.3 mph) delta V for the struck vehicle in the 33.7 mph rear barrier test. The range in delta V was 27.4 km/h (17 mph) to 35.4 km/h (22 mph).

The H-point moment specification of 1700 Nm provides a nominal 2000 J energy transfer capability for high-retention seats with seatback rotations <60° from vertical. This corresponds to a velocity performance of 31.4 km/h (19.5 mph) based on a kinetic energy calculation and 70% effective occupant mass for the 50th percentile male. In practice, the seats should have an energy transfer capability up to a 35.4 km/h (22 mph) rear delta V crash.

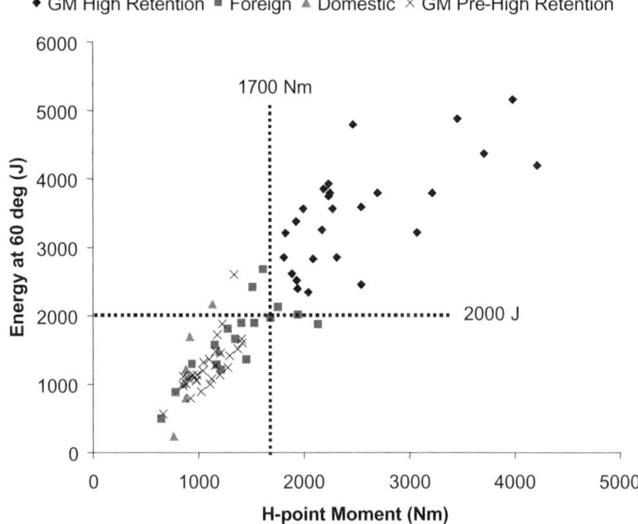

Figure 3: Energy transfer at 60^0 seatback angle versus peak H-point moment for high retention and earlier seat designs.

Second, deformation of seat hardware should provide smooth control of occupant kinematics and uniform resistance. Thus, there should not be a drop in ram load >2 kN in 50 ms moving time-window and more than a 10° change in seatback angle before load reverses in the QST. This assures uniform loading during a test and eliminates rapid changes, which may influence occupant kinematics and retention.

Third, a limit <15° torsion of the seatback up to 60° from vertical prevents excessive twisting of the seatback and potential loss of retention by lateral motion of the occupant. As HR seats have developed, this criterion has been modified to <15^{0} twist for a 20^{0} change in average seatback angle or an H-point moment of 2000 Nm.

Fourth, the head restraint needs to be positioned high and close enough to support the head and neck as greater energy management is achieved by the seatback. This assures that retention and occupant containment are associated with lower risks of whiplash. The height of the top of head restraint should be above the B-plane (a horizontal plane through the head center of gravity of the 95th percentile seated occupant) and closer than 20 mm rearward of the back-of-head ellipse for 95th percentile seated occupant (see Viano 2002).

Performance of High-Retention Seats: The QST performance of GM seats was evaluated after 5 years of testing to the new seat specifications for high retention (Viano 2003). More than twenty different seat types were evaluated from the four primary seat suppliers. Figure 3 shows that seats have substantially increased in energy transfer capability, H-point moment and ability to retain an occupant in severe rear crashes. The average H-point moment of HR seats was 2537 ± 703 Nm and represented a 2.3-times higher moment than pre-HR designs. This was achieved with a 35% reduction in seatback angle change at the 1700 Nm target, and a similar initial stiffness or "yield" behavior of the seats.

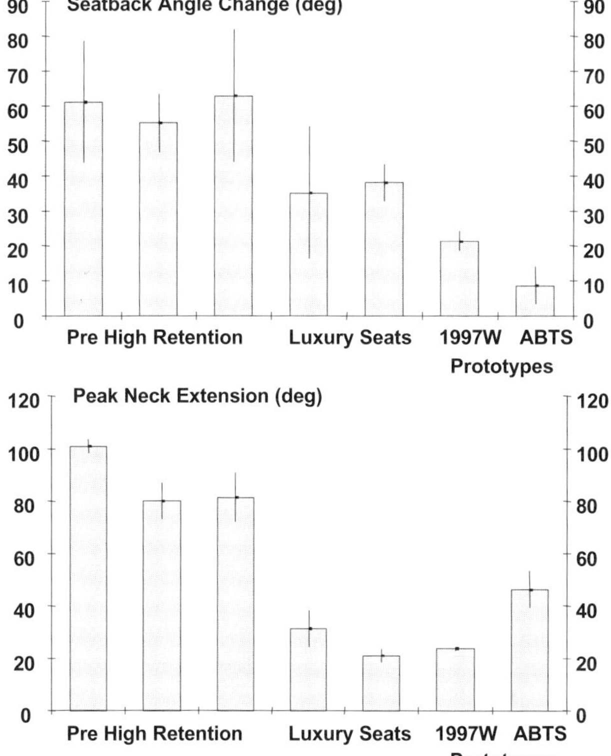

Figure 4: Change in seatback angle and peak neck extension in severe rear sled tests of various seats.

There was a 2.9-times higher energy transfer capability for the seats tested from 1998-2000 as compared to pre-HR designs of the late 1980s and early 1990s. The average energy transfer was 3659 ± 1140 J providing calculated seatback rotations <60^{0} for crashes up to 42.5 km/h on average from Equation (28) with the 50th percentile male occupant. This is a 69% increase in the threshold velocity over pre-HR seat performance. Importantly, the "yielding" behavior of the high-retention seats was similar to the pre-HR designs, because the new generation of seats incorporated a perimeter frame seatback and low rear profile design that provides compliance to pocket the lower torso within the side frames of the seatback.

Hyge Sled Tests of High-Retention Seats: Before the introduction of high-retention seats, a series of rear sled tests was conducted on production seats from various OEMs, GM pre-HR seats and prototypes of the perimeter frame and ABTS (All Belts to Seat) seats meeting high retention requirements. The rear ΔVs in the sled tests varied from 30.6-36.5 km/h (19.0-22.7 mph). A fully instrumented Hybrid III 50th percentile male dummy was place in H-point position on an automotive seat, which was attached to a rigid flat plate on the sled. Seat design positions were used in the setup. An open fixture was used to assess occupant kinematics and seat performance. High-speed movies and video recorded occupant kinematics. In some cases, a lap belt was used to simulate a belt-restrained driver and in other cases, an unrestrained dummy was loosely tethered on the seat to simulate an unbelted occupant, but to prevent ejection from the sled fixture. There were seven foreign luxury seats and a number of domestic seats evaluated.

Figure 4 shows the average and standard deviation in maximum seatback angle change and peak neck extension angle in the rear sled tests. First, the pre-HR seats typically rotated to a horizontal position since the average 60^{0} seatback angle change is added to the initial recline angle of about 25^{0}. This gives an 85^{0} angle from vertical. In these seats the occupant ramped up the seatback engaging the secondary tethers that kept the dummy on the sled. The peak neck extension angles varied from 80^{0}-100^{0}.

Several foreign luxury seats from Europe demonstrated the ability to retain a belted occupant on the seat in a severe rear crash. The second group of luxury seats limited seatback rotation to a 38^{0} +/- 5^{0} angle change. This resulted in an average 63^{0} from vertical, which was close to the value estimated from the energy transfer capability measured in QST tests of the same seats (Viano 2002). Overall, the luxury seat performance was good, except for one seat frame that buckled causing twist, rotation of the seatback, and higher than average dummy responses due to lateral displacement. Neck extension averaged 21^{0} +/- 3^{0} for the foreign luxury seats.

In these tests, the peak head, chest and pelvic accelerations were relatively low in comparison to current human tolerance limits, so restraint of the dummy on the seat involved a low risk of life-threatening injury to the head and torso. However, potential injury associated with loss of retention and subsequent rear interior impact of the head/neck was not assessed in this study and may result in substantial biomechanical responses.

The prototype perimeter frame seat for the 1997 Grand Prix (W car) performed well with low seatback rotation and good head and neck kinematic control with the high positioned head restraint. Changes in seatback angle were just above 20^0 giving an overall angle with respect to vertical below 60^0. A prototype ABTS was tested, which gave the lowest seatback rotations because of the strong seat frame structure to support belt loads in a frontal crash. However, the head restraint used in these tests demonstrated higher in neck responses. In general, seats with low head restraint resulted in the highest neck biomechanics irrespective of seatback rotation. Seats with the lowest neck biomechanics were those with a strong seatback and high head restraint. The combination of low seatback strength and high head restraint fell in between the other two seat groups.

The sled and QST data help expand the understandings first offered by Severy and later addressed by Strother and James (1987), Warner et al. (1991) and Prasad et al. (1997). With a high positioned head restraint, control of neck biomechanics can be achieved irrespective of seatback strength. With this observation, there is no reason not to have the added performance given by a stronger seat in enhancing occupant retention. Thus, there is no fundamental injury-related, design conflict with higher and more forward head restraints and strong, perimeter frame seatback designs. Improved seat performance can be achieved by evaluation to performance requirements in the QST test, higher and more forward head restraint position and seatback designs with EA compliance.

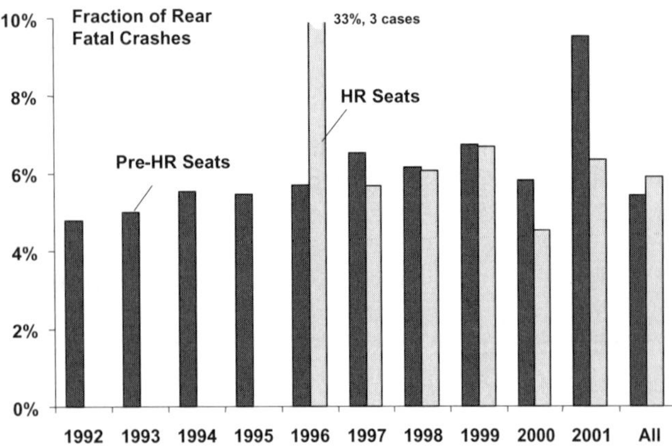

Figure 5: Fraction of the fatal crashes that are rear impacts in FARS.

The QST and sled tests demonstrated that the high-retention seats should improve occupant safety in severe rear crashes. This was achieved by increasing occupant retention on the seat to a higher crash-severity range, maintaining the "yielding" behavior of earlier seats for a gradual acceleration of the occupant, providing more head and neck support with higher and more forward head restraints and providing more uniform support of the spine to reduce local distortions of the vertebrae that may be a source of injury. The laboratory results and field performance of the GM high-retention seats have shown that the new generation of high-retention seats has improved occupant safety in rear crashes.

Combining High-Retention Seats with an Active Head Restraint: The 2000 Buick LeSabre was the first vehicle with a seat that combined the perimeter frame design for high retention with a SAHR active head restraint (Self Aligning Head Restraint or later called a Saab Active Head Restraint). It included an ABTS design, which provides substantially greater strength in the seat frame to support frontal crash restraint loads. The SAHR system provides a means to rotate the head restraint forward and upward by occupant penetration into the seatback for earlier and higher head and neck support. In a recent field effectiveness study, the Saab SAHR active head restraint system has been shown to reduce the risk of moderate to long term whiplash injury by 75 ± 11% (Viano, Olsen 2001).

Limitations of the Study: Since the FARS data is only a tabulation file of the fatal crash, it does not include photographs of the impacted vehicle or details of the fatal injuries to the occupants. It is therefore not possible to determine cause and effect of the seat performance as the primary factor in the fatal injury in the crash. Nonetheless, the effectiveness estimate are likely quite reasonable as this database has been reliably used to determine the effectiveness of safety belts, airbags, occupant age and gender on fatality risk, and other influencing factors, such as alcohol use (see Evans 1991).

The demographics of occupants in the pre-HR and high-retention seat vehicles and those in the striking vehicle may differ. This influences the risk of a crash and fatality in many ways. This is particularly true of the FARS cases for the older vehicles involved in the more recent fatal crashes. The extra years of field use often involve second and third owners, which are known to be younger and less affluent. This influences crash and fatality risks in various ways.

The laboratory tests with dummies do not fully reflect the responses seen in human volunteers or in real world crashes. Most of the differences for rear impacts are in the rigid spine of the dummy and upright posture of the pelvis; however, these aspects are not expected to unduly influence the assessment of occupant retention on the seat. Also, the sled tests are essentially pure rear impacts and do not involve intrusion effects, which are known to affect fatality risks.

Front passengers and rear occupants may be children. There is a recent trend to put children in the rear seat due to concerns with front airbag systems. It is know that the risk of fatality is 50% greater in rear seats than the front in fatal rear impacts since the occupants are closer to the zone of vehicle deformation. Higher rear seat occupancy may increase the number of vehicles involved in fatal rear impacts. The FARS data seems to support this notion.

Figure 5 and Table A1 show the fraction of fatal rear impacted vehicles in FARS crashes. There is a gradual upward trend indicating a relatively larger proportion of fatalities in rear crashes. For the pre-HR vehicles, the five-year fraction from 1992-96 had an average 5.3% rear crashes, whereas the four years from 1997-2000 had 6.3%, a 19% increase. This can be due to many factors, including improved safety in front and side crashes, as well as a higher rear seat occupancy in rear impacts. For these reasons, the driver only and driver/RF passenger statistical studies were conducted. However, it will take more years of data to have sufficient field exposure for a stable estimate of effectiveness of the high-retention seats in this condition.

The fatal rear crashes involving impacts by heavy trucks and tractor-trailers often involve extensive damage to the struck vehicle and intrusion. The FARS data shows that intrusion may be the dominant factor in these fatalities since there is no difference in the fatality rates for the driver and RF passenger between the pre-HR and HR equipped vehicles. Similarly, multi-impacts involving three or more vehicles are more complex than a single rear impact and often involve multiple directions of load on the vehicle. For these crashes there is no difference in risk for the pre-HR and HR equipped vehicles. In fact, the contrary trend is seen at this point, however the data is limited and the seat may not be the critical component involved in the fatal injury.

ACKNOWLEDGMENTS

The helpful comments from Charles Farmer of the Insurance Institute for Highway Safety were appreciated as was the FARS data collection by Joe Lavelle of General Motors.

REFERENCES

Cantor A. Petition for Rulemaking to Amend FMVSS 207 to Prohibit Ramping Up the Seat Back of an Occupant During a Collision. National Highway Traffic Safety Administration, December 28, 1989.

Evans L. Traffic Safety and the Driver. Van Nostrand Reinhold, NY, ISBN 0-442-00163-0, 1991

Federal Motor Vehicle Safety Standards No. 202, 207, and 301. 49 Code of Federal Regulations Section 571.21, 1968.

James MB, Strother CE, Warner CY, et al. Occupant Protection in Rear-end Collisions: Safety and Priorities and Seat Belt Effectiveness, 21 Stapp Car Crash Conf, SAE 912913, Society of Automotive Engineers, Warrendale, PA, 369-376, 1991.

Linder A, Bergman U, Svensson M, Viano DC. Evaluation of the BioRID P3 and the Hybrid III in Pendulum Impacts to the Back – A Comparison to Human Subject Test Data. Proc. Assoc. Adv. Automot. Med. Conf. 44:283-97, 2000 and *Traffic Injury Prevention*, 3(2):158-165, 2002.

Malliaris AC. Current Issues of Occupant Protection in Car Rear Impacts. Data Link, Inc., NHTSA Docket 89-20-No1-021, February, 1990.

Parenteau C, Viano DC. Field Data Analysis of Rear Occupant Injuries Part II: Children, Toddlers and Infants. SAE 2003-03B-145, Society of Automotive Engineers, Warrendale, PA, 2003.

Prasad P, Kim A, Weerappuli DPV, Robert V, Schneider D. Relationship Between Passenger Car Sat Back Strength and Occupant Injury Severity in Rear End Collisions: Field and Laboratory Studies. SAE 973343, Society of Automotive Engineers, Warrendale, PA. 1997.

Rake B, Boehm S. Auto Seat Systems: Dangerous Safety Restraints? Trial April Issue, 80-87, 1990.

Saczalski KJ. Petition to Improve FMVSS 207. National Highway Traffic Safety Administration, April 18, 1989.

Severy DM, Brink HM, Baird JD. Backrest and Head Restraint Design for Rear-end Collision Protection. SAE 680079, Society of Automotive Engineers, Warrendale, PA, 1968a.

Severy DM, Brink HM, Baird JD. Preliminary findings of head support designs. 11th Stapp Car Crash Conf, SAE 670921, SAE Warrendale PA, 337-405, 1967a.

Severy DM, Brink HM, Baird JD. Vehicle Design for Passenger Protection from High Speed Rear End Collisions. SAE 680774, Society of Automotive Engineers, Warrendale, PA, 1968b.

Severy DM, Harrison MB, Baird JD. Collision Performance, LM Safety Car. SAE 670458, Society of Automotive Engineers, Warrendale, PA. 1967b.

Severy DM, Mathewson JH and Bechtol CO. Controlled Automotive Rear-End Collisions, An Investigation of Related Engineering and Medical Phenomena. Canadian Services Medical Journal, VII, 727-759, 1955.

Stephens GD, Long TJ, Blaisdell DM. Energy Analysis of Automotive Seat Systems, SAE 2000-01-1380, SP-1494, Society of Automotive Engineers, Warrendale, PA, 2000.

Strother CE, James MB. Evaluation of Seat Back Strength and Seat Belt Effectiveness in Rear End Impacts. 31st Stapp Car Crash Conference, SAE 872214, Society of Automotive Engineers, Warrendale, PA, 225-244, 1987.

USPTO. High-Retention Seat Back. US Patent 5,295,729, March 22, 1994.

USPTO. Seatback Load Applying Device. US Patent 5,379,646, January 10, 1995a; European Patent EP 0646 782, January 28, 1998a.

USPTO. Vehicle Pivotal Headrest. US Patent 5,378,043, January 3, 1995b; European Patent EP 0627 340, October 15, 1997.

USPTO. Vehicle Seat with Integral, Load Limiting Belt System. US Patent 5,823,627, October 20, 1998. European Patent EP 0870 652 A2, October 14, 1998b.

USPTO. Vehicle Seat with Perimeter Frame and Pelvic Catcher. US Patent 5,509,716, April 23,1996; European Patent EP 0715 985 A2, June 12, 1996.

Viano D, Davidsson J. Neck Displacements of Volunteers, BioRID P3 and Hybrid III in Rear Impacts: Implications to Whiplash Assessment by a Neck Displacement Criterion (NDC). *Traffic Injury Prevention* 3(2):105-116, 2002.

Viano D, Olsen S. The Effectiveness of Active Head Restraint in Preventing Whiplash. *Journal of Trauma* 51:959-969, November, 2001.

Viano DC, Hardy WN, King AI. Human Head, Neck and Torso Response to Pendulum Impacts on the Back. *Crash Prevention and Injury Control* 2(4):289-306, 2001.

Viano DC. Role of the Seat in Rear Crash Safety. SAE Book, ISBN 0-7680-0847-6, Society of Automotive Engineers, Warrendale, PA, SAE R-317:1-491, 2002.

Viano DC. High-Retention Seat Performance in Quasistatic Seat Tests. SAE 2003-01-0173, Society of Automotive Engineers, Warrendale, PA, 2003.

Viano, D.C., Olsen, S., Locke, G.S., Humer, M., "Neck Biomechanical Responses with Active Head Restraint Systems: Rear Barrier Tests with BioRID and Sled Tests with Hybrid III." SAE 2002-01-0030, Society of Automotive Engineers, Warrendale, PA, 2002.

Warner C, Strother C, James MB, Decker RL. Occupant Protection in Rear-End Collisions II: The Role of Seat Back Deformation in Injury Reduction. 35th Stapp Car Crash Conf, SAE 912914, Society of Automotive Engineers, Warrendale, PA., 1991.

Appendix

Table A1: Summary Data from the 1991-2000 FARS.

		1992	1993	1994	1995	1996	1997	1998	1999	2000	2001	All
Vehicles in Fatal Crashes	**B**	13656	12026	11707	11126	7125	5005	3243	1605	516	21	66030
	H	0	0	0	0	6	511	659	1185	773	63	3197
Rear-Impacted Vehicles in Fatal Crashes	**B**	656	600	648	610	407	327	199	108	30	2	3587
	H	0	0	0	0	2	29	40	79	35	4	189
Rear-Impacted Vehicles with a Fatal Driver	**B**	173	151	198	198	109	82	40	33	8	0	992
	H	0	0	0	0	0	9	8	18	5	1	41
Rear-Impacted Vehicles in Fatal Crashes/Vehicles in Fatal Crashes	**B**	4.8%	5.0%	5.5%	5.5%	5.7%	6.5%	6.1%	6.7%	5.8%	9.5%	5.4%
	H	0.0%	0.0%	0.0%	0.0%	33.3%	5.7%	6.1%	6.7%	4.5%	6.3%	5.9%
Rear-Impacted Vehicles with a Fatal Driver/Vehicles in Fatal Crashes	**B**	1.3%	1.3%	1.7%	1.8%	1.5%	1.6%	1.2%	2.1%	1.6%	0.0%	1.5%
	H					0.0%	1.8%	1.2%	1.5%	0.6%	1.6%	1.3%
Rear-Impacted Vehicles with a Fatal Driver/Rear-Impacted Vehicles in Fatal	**B**	26.4%	25.2%	30.6%	32.5%	26.8%	25.1%	20.1%	30.6%	26.7%	0.0%	27.7%
	H					0.0%	31.0%	20.0%	22.8%	14.3%	25.0%	21.7%

Table A2: Summary of Fatalities in High Retention Vehicle Crashes.

			Year State Case	Vehicle Number	Person Number	Model Year	Vehicle	Driver	Right Front	Single Vehicle	Multiple Vehicles	2 Vehicle - At Least One Fatal Driver	2 Vehicle (Both Light) - At Least One Fatal Driver
More than 2 Vehicles			2000_290468	2	1	1998	Chevrolet Malibu	1			1	8 vehicles	
			1997_360486	1	1	1997	Oldsmobile Silhouette	1			1	4 vehicles	
			2000_090225	3	1	2001	Oldsmobile Aurora	1			1	4 vehicles	
			2000_130197	1	1	1999	Cadillac DeVille	1			1	4 vehicles	
			2000_180510	1	4	2000	Buick LeSabre		1		1	4 vehicles	
			1999_061545	1	1	1998	Chevrolet Malibu	1			1	4 vehicles	
			1999_270347	2	1	1999	Cadillac DeVille	1			1	3 vehicles	
			1999_360023	2	1	1998	Buick Park Avenue	1			1	3 vehicles	
			1999_360866	1	1	1999	Buick Century	1			1	3 vehicles	
			2000_040171	2	1	1999	Oldsmobile Silhouette	1			1	3 vehicles	
			2000_122315	2	1	1999	Oldsmobile Alero	1			1	3 vehicles	
			2000_270367	1	1	1997	Chevrolet Venture	1			1	3 vehicles	
			2000_391239	1	1	2000	Pontiac GrandAm	1			1	3 vehicles	
			2000_481782	2	2	1998	Buick Century		1		1	3 vehicles	
2 Vehicles	Neither Driver Killed		1998_483142	1	2	1997	Cadillac DeVille		1		1	Neither Driver Killed	
			2000_062680	2	2	1999	Pontiac GrandAm		1		1	Neither Driver Killed	
			2000_260899	1	2	1999	Buick Park Avenue		1		1	Neither Driver Killed	
			2000_280182	2	2	1999	GMC Sierra-800		1		1	Neither Driver Killed	
			1999_120386	2	4	1999	Pontiac Grand Prix		1		1	Neither Driver Killed	
	At Least One Driver Killed	Other Vehicle Not a Light Vehicle	2000_171185	1	1	1998	Chevrolet Malibu	1			1	1	Unknown Vehicle
			1999_540233	1	1	1999	Buick Century	1			1	1	Heavy Truck
			1999_540233	1	2	1999	Buick Century		1		1	1	Heavy Truck
			2000_200036	1	1	1998	Chevrolet Malibu	1			1	1	Heavy Truck
			2000_200036	1	2	1998	Chevrolet Malibu		1		1	1	Heavy Truck
			2000_280458	1	1	1999	GMC Sierra-800	1			1	1	Heavy Truck
			1999_480408	1	1	1999	Pontiac Bonneville	1			1	1	Bus
		Both Vehicles are Light Vehicles	1998_130733	1	1	1997	Pontiac Grand Prix	1			1	1	1
			1999_170689	1	1	1997	Pontiac Grand Prix	1			1	1	1
			1999_361411	1	1	1999	Buick Park Avenue	1			1	1	1
			2000_270040	2	1	2000	Buick Century	1			1	1	1
			2000_481629	2	1	2000	Chevrolet Silverado-800	1			1	1	1
Single Vehicle			2000_270421	1	1	2000	Pontiac GrandAm	1		1			
			2000_080513	1	2	1997	Pontiac Montana/Transport		1	1			
			1998_470603	1	1	1997	Pontiac Grand Prix	1		1			
			1999_220709	1	2	1999	Chevrolet Silverado-800		1	1			
			1999_420702	1	1	1999	Pontiac GrandAm	1		1			
			2000_061587	1	1	1997	Pontiac Grand Prix	1		1			
			2000_200115	1	1	1999	Pontiac GrandAm	1		1			
			2000_360998	1	1	1997	Pontiac Grand Prix	1		1			
			2000_481036	1	1	1998	Chevrolet Malibu	1		1			
			2000_482056	1	1	1999	Cadillac Eldorado	1		1			
								30	11	10	31	12	5

CHAPTER 8:

CASE STUDY OF DRIVER AND RF PASSENGER FATALITIES IN REAR-IMPACTED VEHICLES WITH HIGH RETENTION SEATS

Case Study of Driver and RF Passenger Fatalities in Rear-Impacted Vehicles with High Retention Seats

David C. Viano
ProBiomechanics LLC

ABSTRACT

High retention seats have been phased into production since 1997 to improve occupant protection in rear crashes. A recent analysis of FARS found a significant reduction in fatality risk with the seats in single-vehicle and two-vehicle rear impacts by light vehicles. This study investigates the 41 fatalities of drivers or right front passengers in high retention seats. The 1997-2000 FARS was sorted for fatal rear-impacted vehicles. Using a VIN decoder, vehicles with high retention seats were determined and 39 fatal rear-impacted vehicles were subdivided into single-vehicle, two-vehicle and multi-vehicle rear crashes. FARS data was evaluated for each crash.

There were 10 fatalities in single-vehicle rear crashes primarily with young male drivers (26 ± 12 yrs). Nine were unbelted, seven were totally ejected and six of the vehicles eventually overturned. In nine cases, the vehicle was going too fast for the road conditions. There were 15 fatalities in two-vehicle rear crashes with older drivers (51 ± 25 yrs). Over half of the fatalities were unbelted but only two were ejected. The cause of these crashes was typically inattention, driving too fast or a rainy weather condition involving the other driver. There were 16 fatalities in multi-vehicle rear crashes, again with older drivers (60 ± 22 yrs). Five of these cases involved an initial front or side impact leading to a primary rear crash in the sequence. Sixteen of the 2+ vehicle crashes involved a tractor-trailer or bus impact where inattention, speeding, or a traffic violation by the other driver or vehicle maintenance was a factor.

The performance of the high retention seat may not be a factor in many of the fatalities in the case study. In single-vehicle fatalities, there was frequent occupant ejection and a subsequent overturn of the vehicle. For the 2+ vehicle rear crashes, extreme crash forces may be a factor with half of the crashes involving a striking tractor-trailer or bus. In 13 of the 2+ vehicle crashes, the victims were over 65 yrs old, which may be an additional factor related to impact tolerance. The practical limit of safety performance with high retention seats in fatal rear crashes is 50%-60% in single- and light-vehicle rear crashes, and no or only small differences when the impact is by a heavy vehicle or is multi-vehicle. The overall effectiveness is 30% based on the initial FARS analysis. The cause and circumstances of fatal rear crashes with high retention seats seem to often involve other factors than the seat, including aggressive vehicle use by young, unbelted drivers and 2+ vehicle crashes involving a striking tractor-trailer.

INTRODUCTION

With more than five years of use of high retention seats, Viano (2003a) evaluated the fatality risk between high retention (HR) and pre-HR seats in rear impacts. Table 1 summarizes the results of the FARS analysis. In single-vehicle rear crashes, high retention seats reduce fatality risk by 50%-59% and in two-vehicle rear crashes they reduce risk by 35%-54% with a light striking vehicle. When the rear impact involves a heavy striking vehicle or there is a multi-vehicle crash, there is no statistical difference between HR and pre-HR retention seats. This is likely the results of extreme crash forces that involve much more than the seat in the occurrence of fatal injuries.

James et al. (1997) investigated severe and fatal injury in rear crashes. Most severe and fatal rear impact injuries were found to involve very high energy crashes where seatback deformation was not a factor, either because the injury was the result of other impacts, or because the seat back was supported by deformed vehicle structure. The majority of rear impact Harm was concentrated in non-severe and non-fatal injuries resulting from much more frequent low- and moderate-speed impacts. Their conclusion was that stronger seatback designs aimed only at reducing severe and fatal injuries may increase exposure to less severe injuries and actually increase the total rear impact Harm. In spite of contrary arguments from Slavik (1997), the conclusions have been supported by recent studies showing that as seats have gotten stronger, they have also gotten stiffer, and the higher stiffness has increased neck responses in low speed crashes (Viano 2003c,f). This could be a factor in the increase in whiplash over the past decades. However, James et al. (1997) conclusions are only apply to the seats of that timeframe, as new seat designs have been introduced that are compliant and use a strong frame. High retention seats have high strength but low stiffness, and reduce injury risks over the entire range of rear crashes and injury severities (Viano 2003a-f). The key to designing low seat stiffness with a strong frame was an open perimeter frame that allows the occupant to displace between the side frame structures (Viano 2002). This gave a new generation of yielding seats.

Table 1: Fatality Reduction with High Retention Seats (modified from Viano 2003a)

Crashes	Reduction in Risk with High-Retention Seats		
	Avg	-95%	+95%
Single-Vehicle			
Driver Only	50.3%	-26.0%	80.4%
Driver & RFP	58.8%	10.5%	81.0%
Multi-Vehicle			
Light Vehicle			
Driver Only	54.0%	-22.9%	82.8%
Driver & RFP	35.1%	-38.3%	69.5%
Heavy Vehicle			
Driver Only	16.9%	-136.6%	70.8%
Driver & RFP	-2.9%	-140.7%	56.1%
3+ Vehicles			
Driver Only	-28.9%	-146.4%	32.6%
Driver & RFP	-4.7%	-88.1%	41.8%
All Multi-Vehicle			
Driver Only	19.2%	-30.2%	49.8%
Driver & RFP	16.9%	-24.1%	44.4%
All Crashes			
Driver Only	29.4%	-7.0%	53.4%
Driver & RFP	30.4%	0.9%	51.1%

Studies of fatal car crashes have focused on various aspects of the causes and circumstances for the impacts. Heavy truck crashes most often involve underride by a passenger car that strikes the rear or side of the truck and plows under the higher structure of the tractor or trailer (Blower, Campbell 2000, Braver et al. 1996, Li, Waller 1981). In a case study of fatal crashes of lap-shoulder belted occupants, heavy truck impacts were a factor in oblique rear crashes where intrusion and extreme crash forces were factors (Viano 1992). Ejection is another factor in fatal crashes, which has received in-depth analysis and review by Malliaris et al. (1996), and post-crash fires are also a consideration (Ragland, Hsia 1998).

In other research, the role of ABS braking systems has been investigated in fatal crashes (Farmer, Lund 1997). The early work showed a reduction in frontal crashes with ABS, but an increase in rear impacts because of more efficient braking. There has also been a concern for more off-road departures and rollovers, but more recent studies have seen those effects diminish (Farmer 2001). Farmer and Lund (2002) have also investigated driver and environmental factors in fatal rollovers.

For single-vehicle crashes, the role of young drivers, high-speed road departures and alcohol have been investigated (Ostrom, Eriksson 1993). There have also been studies on older drivers (Sjogren, et al. 1993), including nighttime driving (Mortimer, Fell 1989). For two-vehicle fatal crashes, the role of the driver has also been evaluated (Perneger, Smith 1991), including older driver problems at intersections (Ridella, Viano 1996) and special issues with female drivers (Viano, Ridella 1996). Kraus et al. (1993) completed an investigation of fatal and severe injury in freeway crashes, showing the influence of occupant age on outcomes in rear crashes. Finally, O'Connor (2002) and Huelke et al. (1981) have summarized serious neck injuries occurring in automotive crashes.

In 1990, a study was initiated on the role of the seat in rear crash safety. The work led to the development of new seats and rear crash requirements, which were adopted in 1995 for new models introduced after 1997. The new seat specifications involved Quasistatic Seat Test (QST) requirements for high-retention performance and a higher and more forward placement of the head restraint (Viano 2002). The first production seats to comply with the new specifications were in the 1997 Pontiac Grand Prix and Chevrolet Malibu, and through 2002, essentially all seats now surpass high retention requirements. This has resulted in a new generation of yielding seats with a strong frame and compliant seatback for rear crash safety (Viano 2003f). This study investigates the fatalities occurring in vehicles with high retention seats to explore the factors involved in the fatality, the role of the seat and any additional seat changes that may need to be considered.

METHODOLOGY

The fatal crashes consisted of cases from FARS for years 1997-2000 involving driver and right-front passenger fatalities in rear impacts with high retention seats. The principal direction of force was 5:00 through 7:00 o'clock. Information about the cause and circumstances for the crash and fatalities was determined for each of the 41 fatalities and 39 vehicle crashes. The data was sorted into single and multi-vehicle crashes, where multi-vehicle crashes were further sorted in 2 and 3+ involved vehicles.

RESULTS

The bottom row in Table 2 shows the 41 deaths with 30 drivers and 11 RF passengers in 191 fatal rear-impact crashes of vehicles equipped with high-retention seats.

These data have been statistically analyzed by Viano (2003f). There were 39 vehicles involved in the 41 deaths. The individual crash and occupant data was further classified and analyzed by single-vehicle, two-vehicle and multi-vehicle crashes. Figure 1 shows that 26% of the fatal crashes were single-vehicle crashes, 38% two-vehicle crashes and 36% involving 3+ vehicles.

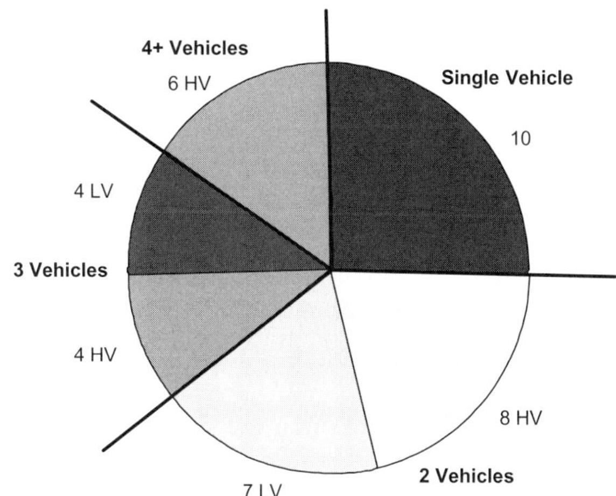

Figure 1: Distribution of 39 fatal rear-impacted vehicles with high-retention seats where HV is a heavy striking vehicle and LV a light striking vehicle (see also Table 2).

Single Vehicle Rear Crashes: Table 3 summarizes the ten single vehicle rear crashes with 8 driver of 2 right front passenger fatalities in a high retention seat. The average age of the drivers is 26 ± 12 years. Seven out ten victims were males, and there were two female drivers. Nine out of ten deaths occurred with an unbelted occupant and seven were totally ejected from the vehicle. While the primary impact was at 5:00-7:00 o'clock, six of the vehicle eventually overturned. In these cases, a tree or pole impact was the most harmful event. In nine of ten cases, the vehicle was going too fast for the road conditions, and there was an impact with a roadside object such as a guardrail, shrubbery or fence leading to a complex crash sequence due to the high speed. Drunk driving was a factor in three cases (30%). With so many cases of total ejection and the frequent vehicle overturn, it is not clear that the performance of the seat was the determining factor in the fatalities.

Two Vehicle Rear Impacts: Table 4 summarizes the 15 two-vehicle rear crashes with a driver of right front passenger fatality in a high retention seat. The average age of these victims was 51 ± 25 years with a higher age and a large spread than for the single vehicle crashes. Four of the deaths were females. Over half of the fatalities were unbelted but there were only two cases of ejection with an unbelted occupant. The primary impact was the rear of the vehicle although there were two cases with an initial side impact followed by the most harmful rear crash. In two crashes, both the driver and RF passenger were killed. The cause of the crashes was typically inattention of the other driver, driving too fast or a rainy weather condition. Four crashes occurred on an Interstate highway. In four cases, the other vehicle was a tractor-trailer or bus.

Three or More Vehicle Rear Impacts: Table 5 summarizes the 16 multi-vehicle rear crashes with a driver of right front passenger fatality in a high retention seat. The average age of these victims was 60 ± 22 years, again a higher age and a large spread than for the single vehicle crashes. 57% (8 of 14) of the victims were females. Five of the cases involved an initial front or side impact leading to a primary rear crashes in the sequence. Ten of the crashes involved a tractor-trailer or bus impact where inattention, speeding, traffic violation or vehicle maintenance issue was a factor in the crash. Six of 14 cases (43%) happened on the Interstate highway.

Table 6: Involvement of Heavy Vehicles in the Fatal Rear Crashes

Striking Vehicle	Multi-Vehicle Rear Impacts				Single Vehicles	Total
	2	3	4+	Total		
	Fatal Crashes					
Light	9	4	0	13	10	23
Heavy	6	4	6	16	0	16
Total	15	8	6	29	10	39
% Heavy	40%	50%	100%	55%		41%
	Driver-RFP Deaths					
Light	9	4	0	13	10	23
Heavy	6	6	6	18	0	18
Total	15	10	6	31	10	41
% Heavy	40%	60%	100%	58%		44%

Table 6 summarizes the proportion of crashes involving heavy striking vehicles. All crashes involving four or more vehicles involved a tractor-trailer, indicating the extreme energy in those heavy vehicles and their ability to engage many other cars in the crash. Half of the two-vehicle crashes involved a tractor-trailer or other heavy vehicle. In the entire sample of multi-vehicle crashes, 55% of crashes and 58% of driver or RF passenger deaths involved a heavy vehicle.

Eight of the 14 victims were over 65 years old and 3 were over 80 years old. The extreme severity of many of these crashes and the decreasing tolerance to impact with age may be an additional factor in the multi-vehicle fatal rear crashes.

Table 2: Fatal Rear Crashes with High Retention Seats (FARS 1997-2000).

Group	Year State Case	Vehicle Number	Person Number	Model Year	Vehicle	Driver	Right Front	# Vehicles	Heavy Vehicle	Drink Driving	Interstate Hwy
More than 2 Vehicles	2000_290468	2	1	1998	Chevrolet Malibu	1		8	Y		Y
	1997_360486	1	1	1997	Oldsmobile Silhouette	1		4	Y		
	2000_090225	3	1	2001	Oldsmobile Aurora	1		4	Y		Y
	2000_130197	1	1	1999	Cadillac DeVille	1		4	Y		Y
	2000_180510	1	4	2000	Buick LeSabre		1	4	Y		
	1999_061545	1	1	1998	Chevrolet Malibu	1		4	Y		Y
	1999_270347	2	1	1999	Cadillac DeVille	1		3			
	1999_360023	2	1	1998	Buick Park Avenue	1		3	Y		
	1999_360866	1	1	1999	Buick Century	1		3	Y		Y
	2000_040171	2	1	1999	Oldsmobile Silhouette	1		3		Y	Y
	2000_122315	2	1	1999	Oldsmobile Alero	1		3			
	2000_270367	1	1	1997	Chevrolet Venture	1		3	Y		
	2000_391239	1	1	2000	Pontiac GrandAm	1		3	Y		
	2000_481782	2	2	1998	Buick Century		1	3			
2 Vehicles – RF Passenger Killed	1998_483142	1	2	1997	Cadillac DeVille		1	2		Y	Y
	2000_062680	2	2	1999	Pontiac GrandAm		1	2		Y	
	2000_260899	1	2	1999	Buick Park Avenue		1	2	Y		Y
	2000_280182	2	2	1999	GMC Sierra-800		1	2			Y
	1999_120386	2	4	1999	Pontiac Grand Prix		1	2			
2 Vehicles – At Least One Driver Killed – Other Vehicle Not a Light Vehicle	2000_171185	1	1	1998	Chevrolet Malibu	1		2	Y	Y	
	1999_540233	1	1	1999	Buick Century	1		2	Y		Y
	1999_540233	1	2	1999	Buick Century		1	2	Y		Y
	2000_200036	1	1	1998	Chevrolet Malibu	1		2	Y	Y	
	2000_200036	1	2	1998	Chevrolet Malibu		1	2	Y		
	2000_280458	1	1	1999	GMC Sierra-800	1		2	Y		
	1999_480408	1	1	1999	Pontiac Bonneville	1		2	Y		
2 Vehicles – At Least One Driver Killed – Both Vehicles are Light Vehicles	1998_130733	1	1	1997	Pontiac Grand Prix	1		2			
	1999_170689	1	1	1997	Pontiac Grand Prix	1		2			
	1999_361411	1	1	1999	Buick Park Avenue	1		2			
	2000_270040	2	1	2000	Buick Century	1		2			
	2000_481629	2	1	2000	Chevrolet Silverado-800	1		2			
Single Vehicle	2000_270421	1	1	2000	Pontiac GrandAm	1		1		Y	
	2000_080513	1	2	1997	Pontiac Montana/Transport		1	1			
	1998_470603	1	1	1997	Pontiac Grand Prix	1		1		Y	
	1999_220709	1	2	1999	Chevrolet Silverado-800		1	1			Y
	1999_420702	1	1	1999	Pontiac GrandAm	1		1			
	2000_061587	1	1	1997	Pontiac Grand Prix	1		1		Y	
	2000_200115	1	1	1999	Pontiac GrandAm	1		1			Y
	2000_360998	1	1	1997	Pontiac Grand Prix	1		1			
	2000_481036	1	1	1998	Chevrolet Malibu	1		1			
	2000_482056	1	1	1999	Cadillac Eldorado	1		1			
						30	11				

Table 3: Single-Vehicle Fatal Rear Crashes with High Retention Seats (FARS 1997-2000).

#	Case ID	Case Vehicle		Occupant					Crash Circumstance	
		Year	Type	Position	Age Sex	Restraint Use	Ejection	Injury Severity	Time	1st Event
1	1998_470603	1997	Pont Grand Prix	Dr	32M	None	Totally	Fatal	21:10	Rollover
2	1999_420702	1999	Pont Grand Am	Dr	20M	None	Totally	Fatal	15:35	Guardrail Impact
3	2000_061587	1997	Pont Grand Prix	Dr	28M	LS	No	Fatal	4:15	Shrubbery Impact
4	2000_200115	1999	Pont Grand Am	Dr	54M	None	Totally	Fatal	17:02	Rollover
5	2000_270421	2000	Pont Grand Am	Dr	19M	None	Totally	Fatal	3:08	Pole Impact
				RF	18F	None	Totally	Incap.		
6	2000_481036	1998	Chevy Malibu	Dr	19M	None	No	Fatal	2:35	Fixed Obj. Impact
7	2000_482056	1999	Cad Excalade	Dr	35F	None	Totally	Fatal	23:08	Guardrail Impact
8	1999_220709	1999	Chevy CK	Dr	19M	None	No	Non Incap.	1:55	Culvert Impact
				RF	19F	None	Totally	Fatal		
9	2000_080513	1997	Pont Trans Sport	Dr	42F	LS	No	Non Incap.	9:15	Fence Impact
				RF	16F	None	Totally	Fatal		
10	2000_360998	1997	Pont Grand Prix	Dr	18M	None	No	Fatal	19:21	Tree Impact
				RF	16F	None	No	Incap.		
				2nd	17F	None	No	Fatal		

#	Case ID	Crash Circumstances			Vehicle Maneuvers/Driving Factors			
		1st Impact Area	Primary Impact Area	Most Harmful Event	Vehicle Actions		Driving Factors	
1	1998_470603	5:00	5:00	Overturn	Negotiating Curve	Run off Road	Driving too Fast	
2	1999_420702	5:00	5:00	Overturn	Negotiating Curve	Run off Road	Driving too Fast	
3	2000_061587	6:00	6:00	Tree	Going Straight	Run off Road	Driving too Fast	Not Licensed
4	2000_200115	3:00	5:00	Overturn	Going Straight	Reckless/Erratic	Driving too Fast	
5	2000_270421	7:00	7:00	Pole	Going Straight	Drinking	Driving too Fast	82 mph
6	2000_481036	6:00	6:00	Fixed Obj	Going Straight	Drinking	Driving too Fast	>97 mph
7	2000_482056	7:00	700	Overturn	Going Straight		Driving too Fast	
8	1999_220709	1:00	5:00	Overturn	Negotiating Curve	Erratic/Reckless	Unsafe/Reckless Citation	55 mph
9	2000_080513	5:00	5:00	Overturn	Going Straight	Over-correction Inattention Citation	Driving too Fast	61 mph
10	2000_360998	6:00	6:00	Tree	Going Straight		Driving too Fast	73 mph

Table 4: Two-Vehicle Fatal Rear Crashes with High Retention Seats (FARS 1997-2000).

#	Case ID	Case Vehicle Year	Case Vehicle Type	Position	Age Sex	Occupant Restraint Use	Ejection	Injury Severity	Time	Crash Circumstance 1st Event	1st Impact Area	Primary Impact Area
11	1998_130733	1997	Pont Grand Prix	Dr	31F	LS	No	Fatal	17:48	Rear Impact	6:00	6:00
12	1998_483142	1997	Cad DeVille	Dr	63F	LS	No	Incap	17:00	Rear Impact	6:00	6:00
				RF	70M	LS	No	Fatal				
13	1999_120386	1999	Pont Grand Am	Dr	53M	LS	No	No	16:57	Rear Impact	6:00	6:00
				RF	73M	No	No	Fatal				
				2nd Left	48F	No	No	Incap.				
				2nd Rt	68F	No	No	Fatal				
14	1999_170689	1997	Pont Grand Prix	Dr	56F	No	No	Fatal	20:00	Rear Impact	6:00	6:00
15	1999_361411	1999	Buick Park Ave	Dr	86M	No	No	Fatal	9:50	Side-Rear Impact	9:00	6:00
16	1999_480408	1999	Pont Bonneville	Dr	39M	No	Totally	Fatal	14:53	Side-Rear Impact	3:00	7:00
17	1999_540233	1999	Buick Century	Dr	25M	LS	No	Fatal	12:24	Rear Impact	6:00	6:00
				RF	30F	LS	No	Fatal		Rollover		
				2nd Rt	9M	LS	No	Fatal				
18	2000_062680	1999	Pontiac	Dr	50F	LS	No	Non-Incap.	18:27	Rear Impact	6:00	6:00
				RF	91M	LS	No	Fatal				
19	2000_171185	1998	Chevy Malibu	Dr	30M	No	No	Fatal	1:30	Side-Rear Impact	3:00	5:00
20	2000_200036	1998	Chevy Malibu	Dr	16M	LS	No	Fatal	21:40	Rear Impact	5:00	5:00
				RF	16M	LS	No	Fatal				
				2nd Left	17M	LS	No	Fatal				
21	2000_260899	1999	Buick Park Ave	Dr	48M	LS	No	Non-Incap.	13:05	Rear Impact	6:00	6:00
				RF	48F	LS	No	Fatal				
22	2000_270040	2000	Buick Century	Dr	89M	No	No	Fatal	13:33	Rear Impact	7:00	7:00
				RF	17F	LS	No	Non-Incap.				
23	2000_280182	1999	GMC CK	Dr	34M	LS	No	Incap.	16:33	Rear Impact	5:00	5:00
				RF	53M	No	No	Fatal				
24	2000_280458	1999	GMC CK	Dr	60M	LS	No	Fatal	14:21	Rear Impact	6:00	6:00
25	2000_481629	2000	Chevy CK	Dr	46M	No	Totally	Fatal	17:10	Rear Impact	5:00	5:00
				RF	39M	No	Totally	Incap.		Rollover		

Table 4 (cont.): Two-Vehicle Fatal Rear Crashes with High Retention Seats (FARS 1997-2000).

#	Case ID	Crash Circumstances		Other Vehicle		Other Occupants			Other Driving Factors
		Most Harmful Event	**Driving Factors**	**Year Type**	**Primary Impact**	**Position**	**Age Sex**	**Injuries**	
11	1998_130733	Impact	Run off Road	1985 Jeep	12:00	Dr	61M	Incap.	Water, Rainy
12	1998_483142	Impact	None	1995 Chevy CK	12:00	Dr	28M	Incap.	Drinking, Hit and Run
13	1999_120386	Impact	Stopped in Lane	1996 Ford E-Van	11:00	Dr	26M	No	Erratic/Reckless Charged with Reckless Driving
14	1999_170689	Impact	Braking	1993 Chevy Baretta	12:00	Dr RF	21F 22F	Incap. Incap.	Raining, Braking
15	1999_361411	Impact	None	1991 Nissan Pathfinder	12:00	Dr	22M	No	Inattention, Driving on Suspended License
16	1999_480408	Impact	Driving too Fast License Suspended Fleeing Police	1991 Grum Bus	6:00	Dr	25M	No	Stopped in Lane
17	1999_540233	Impact	None	1986 White CBE	Rollover	Dr	36M	Lap	Erratic. Reckless Driving too Fast
18	2000_062680	Impact	Going Straight	2000 Lincoln LS	12:00	Dr	42M	No	Improper Lane Change Erratic, Reckless Drving too Fast License Expired Charged with Manslaughter
19	2000_171185	Impact	Going Straight	Unknown	12:00	Dr	39M	Incap.	License Suspended Charges with Intoxication
20	2000_200036	Impact	Going Straight	2000 FRHT CBE	12:00	Dr	65M	No	Going Straight
21	2000_260899	Impact	Run Off Road Driving too Fast	1997 Ford CBE	12:00	Dr	53M	No	Going Straight
22	2000_270040	Impact	U-Turn Failure to Yield Driver Charged	1998 Ford Expedition	11:00	Dr RF	16M 17F	No No	Going Straight Brake & Steer Avoid.
23	2000_280182	Impact	Inattention 75 mph	1999 Ford Taurus	11:00	Dr RF	54M 47F	Incap. Non-Incap.	Going Straight, 70 mph
24	2000_280458	Impact	Passing with Insuff. Distance	1993 Volvo	12:00	Dr	54M	No	None
25	2000_481629	Impact	Going Straight	1988 Mazda RX7	11:00	Dr	21M	LS	Drowsy/Asleep Improper Lane Change Charged with Manslaughter

Table 5: Multi-Vehicle Fatal Rear Crashes with High Retention Seats (FARS 1997-2000).

#	Case ID	Case Vehicle Year	Case Vehicle Type	Occupant Position	Age Sex	Restraint Use	Ejection	Injury Severity	Time	Crash Circumstance 1st Event	1st Impact Area	Primary Impact Area
26	1997_360486	1997	Olds Silhouette	Dr RF 2nd Rt	68F 41M 2F	LS LS Ch-Seat	No No No	Fatal Non-Incap. Fatal	15:34	Rear Impact	7:00	7:00
27	1999_061545	1998	Chevy Malibu	Dr	29M	LS	No	Fatal	8:50	Guardrail Impact Rear Impact	12:00	6:00
28	1999_270347	1999	Cad DeVille	Dr	71F	LS	No	Fatal	16:57	Rear Impact	6:00	6:00
29	1999_360023	1998	Buick Park Ave	Dr	80M	LS	No	Fatal	16:24	Front-Rear Impact	12:00	6:00
30	1999_360866	1999	Buick Century	Dr	52F	LS	No	Fatal	14:15	Rear Impact	6:00	6:00
31	2000_040171	1999	Olds Silhouette	Dr	25F	LS	No	Fatal	22:10	Rear Impact	6:00	6:00
32	2000_090225	2001	Olds Bravada	Dr	59M	No	No	Fatal	17:25	Rear Impact	6:00	6:00
33	2000_122315	1999	Olds Alero	Dr	78F	No	Totally	Fatal	12:10	Front-Rear Impact	1:00	5:00
34	2000_130197	1999	Cad DeVille	Dr RF	82F 88M	No No	Totally No	Fatal Non-Incap.	22:19	Rear Impact	7:00	7:00
35	2000_180510	2000	Buick LeSabre	Dr RF 2nd Rt 2nd Left	45M 82M 81F 50M	LS LS Lap Lap	No No No No	Non-Incap. Fatal Fatal Non-Incap.	11:15	Side-Rear Impact	3:00	5:00
36	2000_270367	1997	Chevy Venture	Dr	76M	LS	No	Fatal	14:29	Rear Impact	6:00	6:00
37	2000_290468	1998	Chevy Malibu	Dr 2nd Row	37M 8M	No No	No No	Fatal Fatal	6:44	Rear Impact	6:00	6:00
38	2000_391239	2000	Pont Grand Am	Dr 2nd Row	27F 73F	LS LS	No No	Fatal Non-Incap.	9:10	Rear Impact	5:00	5:00
39	2000_481782	1998	Buick Century	Dr RF	74M 69F	No No	No No	Non-Incap. Fatal	21:10	Front-Rear Impact	12:00	6:00

Table 5 (cont.): Multi-Vehicle Fatal Rear Crashes with High Retention Seats (FARS 1997-2000).

#	Case ID	Crash Circumstances: Most Harmful Event	Crash Circumstances: Driving Factors	Other Vehicle: Year Type	Other Vehicle: Primary Impact	Other Occupants: Position	Other Occupants: Age Sex	Other Occupants: Injuries	Other Driving Factors
26	1997_360486	Impact	Failure to Yield	1990 Ford Crown Vic	12:00	Dr	47M	No	None
						RF	48F	No	
				1984 PTRB CBE	12:00	Dr	47M	No	Charged with Violation
				1985 Buick Century	12:00	Dr	43M	No	None
27	1999_061545	Impact	Improper Turn	1978 Conv. Bus	12:00	Dr	35M	No	None
			Run Off Road	1993 Dodge Caravan	11:00	Dr	47M	No	None
			Not Licensed			6 Other Occup's		No	
				1995 Honda Civic	12:00	Dr	37M	No	None
						3 Other Occup's		No	
28	1999_270347	Impact	Failure to Yield	1995 Cad DeVille	12:00	Dr	48M	Incap.	Steering & Braking
						2 Other Occup's		Incap.	
				1998 Chevy CK	12:00-7:00	Dr	57M	Incap.	Steering & Braking
						2 Other Occup's		No	
29	1999_360023	Impact	Stopped in Road	1984 Chevy S10	12:00	Dr	43F	Incap.	Making Left Turn
			Ice on Road	1996 KW CBE	9:00	Dr	44M	No	Making Left Turn
30	1999_360866	Impact	Stopped in Road	1987 Chrys. New Yorker	6:00	Dr	41M	No	Stopped in Lane
						4 Other Occup's		No	
				1995 KW CBE	12:00	Dr	55M	LS	Improper Trailing Charged with License Violation
31	2000_040171	Impact	Stopped in Road	1998 Chevy CK	12:00	Dr	29F	Non-Incap.	Driving too Fast Suspended License Charged with Homicide
				1998 Volvo	6:00	Dr	21F	No	
						RF	38M	No	
32	2000_090225	Impact	Changing Lanes	1997 Pontiac Sunfire	6:00	Dr	43F	No	Slowing
			Braking	1999 48401	6:00	Dr	35F	Incap.	Slowing, Braking
						RF	12F	Incap.	
				2000 INTL CBE	12:00	Dr	47M	Non-Incap.	Braking
33	2000_122315	Impact	Left Turn	1998 Dodge Ram	12:00	Dr	22M	Incap.	Going Straight, 15 mph
		Side Window	15 mph			RF	19F	Incap.	Failure to Obey
		Ejection		1999 Dodge Dakota	12:00	Dr	44F	No	None
						RF	52F	No	
34	2000_130197	Impact	Stopped in Lane	1995 Mack CBE	1:00	Dr	46M	No	Going Straight
				1996 KW CBE	4:00	Dr	45M	No	Braking, No License
				1991 Chevy Lumina	1:00	Dr	41F	Incap	None
						2 Other Occup's		No	
35	2000_180510	Impact	Left Turn	1990 Buick Regal	12:00	Dr	19M	Non-Incap.	None
			Failure to Yield	1987 Pontiac 6000	9:00	Dr	31M	No	Stopped in Lane
			Failure to Obey	1994 Mack COE	12:00	Dr	24M	No	Stopped in Lane
36	2000_270367	Impact	Stopped in Lane	1999 INTL CBE	12:00	Dr	28M	No	Inattention, Violation
				1991 INTL CBE	8:00	Dr	41M	No	None
37	2000_290468	Impact	Stopped in Lane	1998 FRHT CBE	12:00	Dr	36M	No	Inattention, Violation Driving too Fast
				1997 Ford F	1:00	Dr	33M	Non-Incap.	None
				1993 Ford F	12:00	Dr	27M	Non-Incap.	
						RF	55M	Non-Incap.	
				1990 Ford Tempo	12:00	Dr	29M	Non-Incap.	
				1998 Ford Taurus	6:00	Dr	16M	Non-Incap.	
				1998 Mercury Tracer	6:00	Dr	29F	Non-Incap.	
				1992 WHIT CBE	8:00	Dr	45M	No	
38	2000_301239	Impact	Going Straight	1993 WHIT CBE	11:00	Dr	39M	Non-Incap.	Impropoer Trailing
				2000 Ford F	12:00	Dr	66M	Non-Incap.	None
39	2000_481782	Impact	Improper U-Turn	1993 Buick Regal	1:00	Dr	38F	No	Suspended License
			Inattention			2 Other Occup's		No	Charged with Vio;ations
			Improper Lights	1996 Dodge Caravan	12:00	Dr	17M	Non-Incap.	None

DISCUSSION

In this study, individual fatal crashes were analyzed to investigate the field performance of high-retention seats in rear-impacted vehicles. Both single-vehicle and vehicle-to-vehicle crashes were considered. This gave insights into the type and circumstances of fatal crashes with the new generation of yield seats.

Young Unbelted Drivers in High-Speed Single-Vehicle Crashes: The single vehicle rear crashes had similar circumstances. They involved young, unbelted drivers traveling too fast for road conditions. The vehicle strikes an object setting up a complex sequence of events leading to a primary rear impact. The crash sequence continues leading to ejection and overturn of the vehicle. For these crashes, the underlying cause of death was ejection and subsequent impacts. Ejection is a sign of a severe, high energy crash and Malliaris et al. (1996) have shown the initial vehicle impacts would be survivable if the occupant were to remain in the vehicle. The lack of seat belt use is a key factor in the ejection and death. Drunk driving was a factor in 30% of the crashes.

Lack of Safety Belt Use: Only one out of ten (10%) of the occupants killed in single-vehicle rear crashes was using the lap-shoulder belt. Belts have an overall crash effectiveness of 42%, but this level increases to 77% in rollover crashes were the prevention of ejection by the lap belt contributes 63% of the effectiveness and is a major reason for the safety performance. Restraint use was 47% in two-vehicle fatalities and 36% in 3+ vehicle crashes. Obviously, encouraging belt use is an important measure to reduce fatal crash injuries even where ejection is not a factor. Lap-shoulder belts are 49% effective in preventing fatalities in rear crashes with 23% is due to the prevention of ejection.

Role of Tractor-Trailers in the Fatal Multi-Vehicle Rear Crashes: 84% of the 3+ vehicle and 62% of 2+ vehicle rear crash fatalities involved a tractor-trailer or other heavy striking vehicle. In most cases, the struck vehicle was stopped in lane, going straight or braking when impacted. Inattention or error by the other driver was the main cause for the crashes. When the impact involved 3+ vehicles, it always involved a heavy vehicle. There can be many factors responsible for the errors of the tractor-trailer operator, and the braking efficiency of large vehicles is not as good as passenger cars, so a sudden emergency braking situation can have dramatic consequences to the traffic stopped or slowing ahead. With the very high mass of the tractor-trailers, impacts with other vehicles only marginally reduce the kinetic energy of the heavy vehicle, so it can continue forward engaging other vehicles in traffic.

Older Occupants in High-Severity Rear Crashes: It is known that human tolerance to impact decreases with age, particularly in the elderly (Bedard et al. 2002). In 8 (57%) of the 3+ vehicle crashes, the victim was over 65 yrs old. Three (21%) were over 80 yrs old. Since these crashes also involved a heavy vehicle impact, the delta V of the struck vehicle is close to that of the striking vehicle. James et al. (1997) found that most fatal crashes involved very high impact speed, so human tolerance of older occupants comes into play in these crashes.

ACKNOWLEDGMENTS

The assistance of Joe Lavelle of General Motors Corporation is appreciated for collecting the FARS cases with high retention fatalities in rear impacts.

REFERENCES

Bedard M, Guyatt GH, Stones MJ, Hirdes JP. The independent contribution of driver, crash, and vehicle characteristics to driver fatalities. Accident Analysis and Prevention 34 (2002) 717–727.

Blower D, Campbell KL. Underride in fatal rear-end truck crashes. SAE 2000-01-3521, SAE International Truck and Bus Meeting and Exposition, Society of Automotive Engineers, Inc., Warrendale, PA, 2000.

Braver ER, Preusser D, Williams AF, Weinstein H. Major types of fatal crashes between large trucks and cars. 1996-12-0011, 40th AAAM Proceedings, Association for the Advancement of Automotive Medicine, Des Plaines, Illinois, USA, 1996.

Farmer CM, Lund AK, Trempel RE, Braver ER. Fatal crashes of passenger vehicles before and after adding antilock braking systems. Accid Anal Prev. 1997 Nov;29(6):745-57.

Farmer CM, Lund AK. Rollover risk of cars and light trucks after accounting for driver and environmental factors. Accid Anal Prev. 2002 Mar;34(2):163-73.

Farmer CM. New evidence concerning fatal crashes of passenger vehicles before and after adding antilock braking systems. Accid Anal Prev. 2001 May;33(3):361-9.

Huelke DF, O'Day J, Mendelsohn RA. Cervical injuries suffered in automobile crashes. J Neurosurg. 1981 Mar;54(3):316-22.

James MB, Decker RL, Warner CY. Severe and fatal injuries in rear impacts. 1997-12-0020, 41st AAAM Proceedings, Association for the Advancement of Automotive Medicine, Des Plaines, Illinois, USA, 1997.

James MB, Strother CE, Warner CY, et al. Occupant Protection in Rear-end Collisions: Safety and Priorities and Seat Belt Effectiveness, 21 Stapp Car Crash Conf, SAE 912913, Society of Automotive Engineers, Warrendale, PA, 369-376, 1991.

Kraus JF, Anderson CL, Arzemanian S, Salatka M, Hemyari P, Sun G. Epidemiological aspects of fatal and severe injury urban freeway crashes. Accid Anal Prev. 1993 Jun;25(3):229-39.

Li LK, Waller PF. Heavy trucks and fatal crashes: an unresolved dilemma. SAE 810518, Society of Automotive Engineers, Inc., Warrendale, PA, USA, 1981.

Malliaris AC, DeBlois JH, Digges KH. Light vehicle occupant ejections--a comprehensive investigation. Accid Anal Prev. 1996 Jan;28(1):1-14. Review.

Malliaris AC. Current Issues of Occupant Protection in Car Rear Impacts. Data Link, Inc., NHTSA Docket 89-20-No1-021, February, 1990.

Mortimer RG, Fell JC. Older drivers: their night fatal crash involvement and risk. Accid Anal Prev. 1989 Jun;21(3):273-82.

O'Connor P. Injury to the spinal cord in motor vehicle traffic crashes. Accid Anal Prev. 2002 Jul;34(4):477-85.

Ostrom M, Eriksson A. Single-vehicle crashes and alcohol: a retrospective study of passenger car fatalities in northern Sweden. Accid Anal Prev. 1993 Apr;25(2):171-6.

Perneger T, Smith GS. The driver's role in fatal two-car crashes: a paired "case-control" study. Am J Epidemiol. 1991 Nov 15;134(10):1138-45.

Prasad P, Kim A, Weerappuli DPV, Robert V, Schneider D. Relationship Between Passenger Car Sat Back Strength and Occupant Injury Severity in Rear End Collisions: Field and Laboratory Studies. SAE 973343, Society of Automotive Engineers, Warrendale, PA. 1997.

Ragland CL, Hsia HS. A case study of 214 fatal crashes involving fire. SAE 986076, 16th International Technical Conference on the Enhanced Safety of Vehicles, National Highway Traffic Safety Administration, Washington, D.C., USA, 1998.

Ridella SA, Viano DC. Significance of Intersection Crashes for Older Drivers. SAE 960457, Society of Automotive Engineers, Warrendale, PA, 1996.

Sjogren H, Bjornstig U, Eriksson A, Sonntag-Ostrom E, Ostrom M. Elderly in the traffic environment: analysis of fatal crashes in northern Sweden. Accid Anal Prev. 1993 Apr;25(2):177-88.

Slavik DH. Discussion of "severe and fatal injuries in rear impacts" 1997-12-0021, 41st AAAM Proceedings, Association for the Advancement of Automotive Medicine, Des Plaines, Illinois, USA, 1997.

Strother CE, James MB. Evaluation of Seat Back Strength and Seat Belt Effectiveness in Rear End Impacts. 31st Stapp Car Crash Conference, SAE 872214, Society of Automotive Engineers, Warrendale, PA, 225-244, 1987.

Thurman DJ, Burnett CL, Beaudoin DE, Jeppson L, Sniezek JE. Risk factors and mechanisms of occurrence in motor vehicle-related spinal cord injuries: Utah. Accid Anal Prev. 1995 Jun;27(3):411-5.

Viano DC, Ridella SA. Crash Causation: A Case Study of Fatal Accident Circumstances and Configurations. SAE 960458, Society of Automotive Engineers, Warrendale, PA, 1996.

Viano DC, Ridella SA. Fatal Crashes of Female Drivers Wearing Safety Belts. SAE 960459, Society of Automotive Engineers, Warrendale, PA, 1996.

Viano DC. Crash Injury Prevention: A Case Study of Fatal Crashes of Lap-Shoulder Belted Occupants. 36th Stapp Car Crash Conference, P-261:179-192, SAE 922523, Society of Automotive Engineers, Warrendale, PA, 1992.

Viano DC. Restraint Effectiveness, Availability and Use in Fatal Crashes: Implications to Injury Control. Journal of Trauma, 38(4): 538-546, 1995.

Viano DC. Role of the Seat in Rear Crash Safety. ISBN 0-7680-0847-6, Society of Automotive Engineers, Warrendale, PA, SAE R-317:1-491, 2002.

Viano DC. High Retention Seat Performance in Quasistatic Seat Tests. SAE 2003-01-0173, Society of Automotive Engineers, Warrendale, PA, 2003a.

Viano DC. Energy Transfer to an Occupant in Rear Crashes: Effect of Stiff and Yielding Seats. SAE 2003-01-0180, Society of Automotive Engineers, Warrendale, PA, 2003b.

Viano DC. Seat Properties Affecting Neck Responses in Rear Impacts: A Possible Reason Why Whiplash Has Increased. for Traffic Injury Prevention, 2003c.

Viano DC. Seat Influences on Female Neck Responses in Rear Impacts: A Reason Why Women Have Higher Whiplash Rates. for Traffic Injury Prevention, 2003d.

Viano DC. Effectiveness of High-Retention Seats in Preventing Fatality: Initial Results and Trends. SAE 2003-01-1351, Society of Automotive Engineers, Warrendale, PA, 2003e.

Viano DC. Influence of Seat Properties on Occupant Dynamics in Severe Rear Crashes. In print, Traffic Injury Prevention 4(4), 2003f.

Warner C, Strother C, James MB, Decker RL. Occupant Protection in Rear-End Collisions II: The Role of Seat Back Deformation in Injury Reduction. 35th Stapp Car Crash Conf, SAE 912914, Society of Automotive Engineers, Warrendale, PA., 1991.

SECTION 3:

OVERVIEW OF SEAT SAFETY PATENTS

CHAPTER 9:

SEAT SAFETY PATENTS

Chapter 9

Seat Safety Patents

David C. Viano
ProBiomechanics LLC

ABSTRACT

This Chapter reviews the history of seat safety patents and includes examples of concepts specific to rear crash protection. It also covers seat concepts for front, side and rollover protection where the technology may have application to rear crashes. The first page from 63 USPTO patents is included to show information on seat concepts and technologies considered for crash safety. Ideas for swinging, tilting and movable seat cushions/seats have a rich history with early patents from the 1930s and many following patents through today. There are variations that include load limiting, energy absorbing and rotating structures to limit and control occupant motion. These approaches were later merged with various seat belt systems to further enhance occupant protection. Finally, a series of patents is included with wrap-around wings, barriers and panels to constrain occupant motion in a variety of crashes; and, other seat approaches with padding and energy absorption, which address rear seat occupant impacts on the seat frame and trim. The patent history and example concepts for seat safety show a rich history of ideas for occupant protection and containment in the seat.

METHODOLOGY

US patents were located by searching primary, secondary and tertiary references for seat safety related topics using the US Patent and Trademark Office website: http://patft.uspto.gov/netahtml/search. Efforts were made to locate the earliest and most interesting examples of patents on particular technologies, but there is every reason to assume that many other concepts and technologies related to seat designs for crash safety were not located.

Patents prior to 1975 are accessible only by a patent number or classification, which made the search depend on secondary and tertiary references in patents that were located. Many of these patents included cited references to earlier patents, which were investigated. However, the earliest US patents did not include cited references to prior art that were reviewed by the patent examiner. This complicated the search for patents before the 1950s. Any patent of significance to the history of seats and crash safety that is not included was an inadvertent oversight. The items located and reported should be considered examples of the history of technology and patents on seat safety. They do show a long-standing history of concepts for occupant safety by seat design and integration of other safety features.

RESULTS

A wide range of seat safety concepts were located for rear crash safety, and even more for frontal crash protection. The frontal safety systems are included since many of the concepts have similar application in rear crashes, and reference to crash protection in many impact types was made in the patents.

There is a rich history on movable seats that rotate, tip or otherwise change position or shape under crash conditions. This turned out to be a considerable area of patent interest even though the designs have not been realized in production. It is interesting to review the various concepts that have been considered over 60 years.

There is another line of patenting with the integration of seat belts, barriers, wings or wrap-around cushions that are intended to control occupant motion and provide restraint in various crashes.

The last area of patenting involves energy absorbing structures and designs for occupant impacts on seat trim and structures. These concepts show an understanding of the need for padded impact surfaces to reduce risks of head and other body region injury including rear seat occupant protection when interacting with front seats.

EXAMPLES OF BREAK-AWAY, ROTATING, MOVING OR TIPPING SEATS FOR REAR CRASH SAFETY

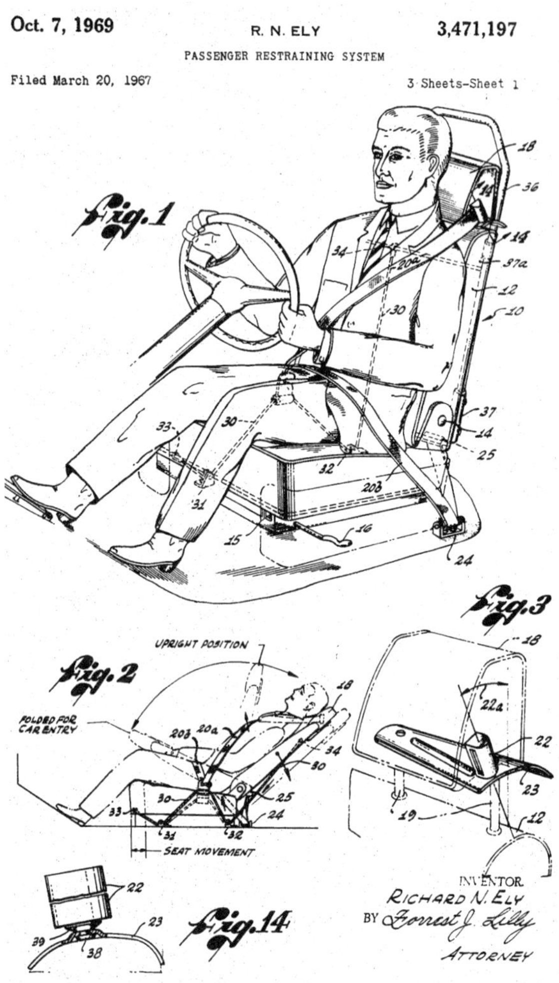

United States Patent [11] 3,578,376

[72] Inventors Tatsuo Hasegawa
Okazaki-shi;
Kohichi Yoshie; Yasuhiro Kamijima,
Toyota-shi, Japan
[21] Appl. No. 848,302
[22] Filed Aug. 7, 1969
[45] Patented May 11, 1971
[73] Assignee Toyota Jidosha Kogyo Kabushiki Kaisha
Toyota-shi, Japan
[32] Priority Sept. 20, 1968
[33] Japan
[31] 43/67598

[54] **SEAT CONSTRUCTION OF A VEHICLE**
6 Claims, 31 Drawing Figs.

[52] U.S. Cl. **296/65,** 297/216
[51] Int. Cl. **B60r 21/10,** B60n 1/02
[50] Field of Search .... 296/65, 65 (.1); 297/216; 244/122, 141

[56] References Cited
UNITED STATES PATENTS

2,682,931	7/1954	Young	297/216(X)
2,735,476	2/1956	Fieber	297/216
2,818,909	1/1958	Burnett	297/216
2,823,730	2/1958	Lawrence	297/216

Primary Examiner—Kenneth H. Betts
Assistant Examiner—Leslie J. Paperner
Attorney—George B. Oujevolk

ABSTRACT: A seat assembly for a vehicle is composed of a floor, a seat, a movable support member supporting the seat to permit the seat to move relative to the floor, a pin or stationary member fixing the seat on the floor to prevent the seat from moving relative to the floor which includes a weak link or releasable section, and an energy absorbing member disposed between the floor and the seat. When an impact in the forward direction is applied to the seat by a rear-end-collision with a following car, if the impact force exceeds a predetermined permissible value, the stationary member is released to permit the seat to move rearwards relative to the floor by the inertia of the seat and passengers, and, simultaneously the energy absorbing member is deformed so as to reduce the impact imposed on the passengers and prevent them from being injured.

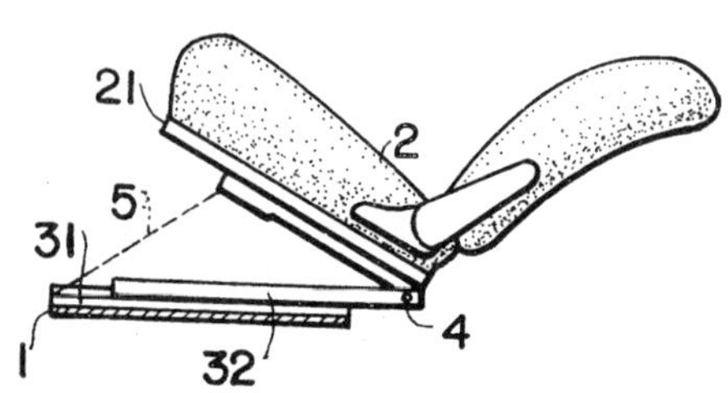

United States Patent [19]
Viano

[11] Patent Number: **5,295,729**
[45] Date of Patent: **Mar. 22, 1994**

[54] **HIGH RETENTION SEAT BACK**
[75] Inventor: **David C. Viano,** Bloomfield Hills, Mich.
[73] Assignee: **General Motors Corporation,** Detroit, Mich.
[21] Appl. No.: **838,184**
[22] Filed: **Feb. 18, 1992**
[51] Int. Cl.[5] **B60N 2/42**
[52] U.S. Cl. **297/216.14;** 297/216.18
[58] Field of Search 297/216, 472, 354

[56] References Cited
U.S. PATENT DOCUMENTS

2,682,931	7/1954	Young	188/1
2,735,476	2/1956	Fieber	155/9
2,823,730	2/1958	Lawrence	155/9
3,269,774	8/1966	Hildebrandt et al.	297/386
3,471,197	10/1969	Ely	297/216
3,552,795	1/1971	Perkins	297/216
3,578,376	5/1971	Hasegawa et al.	296/65
3,734,562	5/1973	Fourrey	297/216
3,806,190	4/1974	Winslow	297/216
3,832,002	8/1974	Eggert, Jr. et al.	297/216
3,853,298	12/1974	Libkie et al.	248/429
3,957,304	5/1976	Koutsky et al.	297/385
4,183,582	1/1980	Taki	297/216
4,325,238	4/1982	Scherbing	70/18
4,349,167	9/1982	Reilly	297/216
4,390,208	6/1983	Widmer et al.	297/216
4,488,754	12/1984	Heesch et al.	297/216
4,775,182	10/1988	von Hoffman	297/45
4,824,171	4/1989	Hollingsworth	297/351

Primary Examiner—Peter M. Cuomo
Assistant Examiner—Darnell M. Boucher
Attorney, Agent, or Firm—William A. Schuetz

[57] **ABSTRACT**

A high retention seat assembly has a seat cushion unit and a back rest unit which moves rearwardly relative to the seat cushion unit when subjected to severe loading by the occupant during a rear collision. A collapsible retention connected with the seat cushion unit and back rest unit is tensioned or pulled taut in response to rearward movement of the back rest unit relative to the seat cushion unit to limit rearward deflection of the back rest unit during such a rear collision.

10 Claims, 5 Drawing Sheets

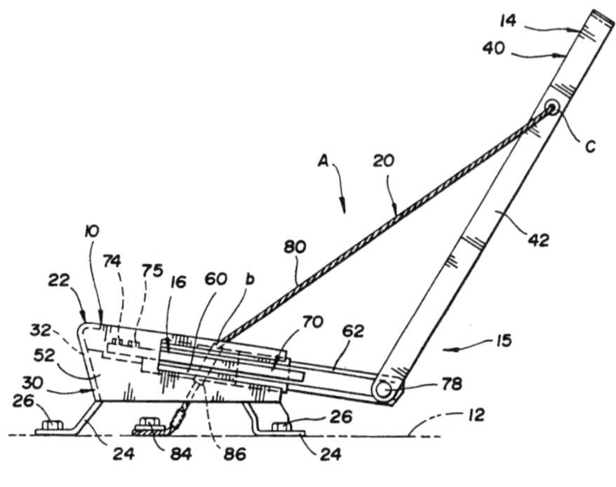

(12) United States Patent
Mueller

(10) Patent No.: **US 6,244,656 B1**
(45) Date of Patent: **Jun. 12, 2001**

(54) **VEHICLE SEAT AND OPERATION METHOD OF SUCH A VEHICLE SEAT**
(75) Inventor: **Olaf Mueller**, Ruesselsheim (DE)
(73) Assignee: **Inova GmbH Technische Entwicklung**, Ruesselsheim (DE)
(*) Notice: Subject to any disclaimer, the term of this patent is extended or adjusted under 35 U.S.C. 154(b) by 0 days.
(21) Appl. No.: **09/197,498**
(22) Filed: **Nov. 23, 1998**
(30) **Foreign Application Priority Data**
Nov. 21, 1997 (DE) .... 297 20 734 U
(51) Int. Cl.[7] **B60N 2/42**
(52) U.S. Cl. **297/216.13**; 297/216.19
(58) Field of Search 297/216.13, 216.14, 297/216.18, 216.19

(56) **References Cited**
U.S. PATENT DOCUMENTS

5,290,089 *	3/1994	Oleszko et al.	297/216.14
5,437,494 *	8/1995	Beauvais	297/216.19
5,449,214 *	9/1995	Totani	297/216.13 X
5,676,421 *	10/1997	Brodsky	297/216.13
5,722,722 *	3/1998	Massara	297/216.13 X
5,810,417 *	9/1998	Jesaanont	297/216.13
5,823,619 *	10/1998	Helig et al.	297/216.13

* cited by examiner

Primary Examiner—Anthony D. Barfield
(74) *Attorney, Agent, or Firm*—Martin Fleit

(57) **ABSTRACT**

A motor vehicle seat has a seat part and a backrest part, as well as protection devices which contain a backrest adjusting device which can be released in the event of a rear impact and which, after its release, controls the backrest part such that the latter counteracts an occupant's sliding-up along the backrest part as the result of the rear impact. In addition, an operating process for a motor vehicle seat with a seat part and a backrest part is provided. In the event of a rear impact, the backrest part is controlled into a protective position so that it counteracts an occupant's sliding-up along the backrest part as the result of the rear impact.

39 Claims, 6 Drawing Sheets

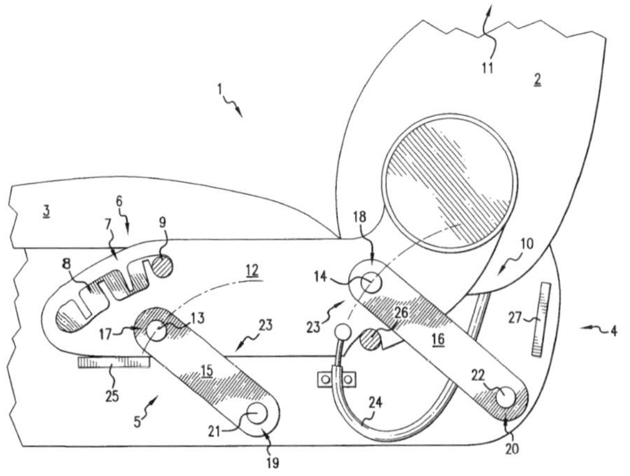

EXAMPLES OF ROTATING, TIPPING, ENERGY ABSORBING OR LOAD-LIMITING SEATS

(12) **United States Patent**
Aufrere et al.

(10) **Patent No.:** **US 6,254,181 B1**
(45) **Date of Patent:** **Jul. 3, 2001**

(54) **VEHICLE SEAT HAVING A MOVABLE SAFETY BAR**

(75) Inventors: **Christophe Aufrere**, Marcoussis; **Gerald Morin**, Paris; **Erik Levernieux**, Etampes, all of (FR)

(73) Assignee: **Bertrand Faure Equipements SA**, Boulogne (FR)

(*) Notice: Subject to any disclaimer, the term of this patent is extended or adjusted under 35 U.S.C. 154(b) by 0 days.

(21) Appl. No.: **09/489,298**

(22) Filed: **Jan. 21, 2000**

(51) **Int. Cl.**[7] **B60N 2/427**
(52) **U.S. Cl.** **297/216.1**; 297/216.13
(58) **Field of Search** 297/216.1, 216.13, 297/216.14, 284.11; 296/68.1

(56) **References Cited**

U.S. PATENT DOCUMENTS

4,623,192		11/1986	Koide et al.	297/216
5,340,185		8/1994	Vollmer	286/68.1
6,050,635	*	4/2000	Pajon et al.	297/216.1
6,113,185	*	9/2000	Yamaguchi et al.	297/216.1

FOREIGN PATENT DOCUMENTS

36 31 881	4/1988	(DE) .
38 41 688	6/1990	(DE) .
42 12 254	4/1993	(DE) .
298 15 521	1/1999	(DE) .
2 641 244	6/1990	(FR) .
2 747 080	10/1997	(FR) .
2 747 081	10/1997	(FR) .

OTHER PUBLICATIONS

First page of WO 00/12350.

* cited by examiner

Primary Examiner—Peter R. Brown
(74) *Attorney, Agent, or Firm*—Marshall, O'Toole, Gerstein, Murray & Borun

(57) **ABSTRACT**

A vehicle seat has a horizontal safety bar which restricts the displacement of the occupant in the event of an accident. This bar is moved towards the occupant by means of an activating device in the event of an accident, following a certain activation trajectory, after which the bar is displaced in the opposite direction applying deceleration on a dissipation trajectory, the initial portion of which follows the final portion of the activation trajectory.

13 Claims, 6 Drawing Sheets

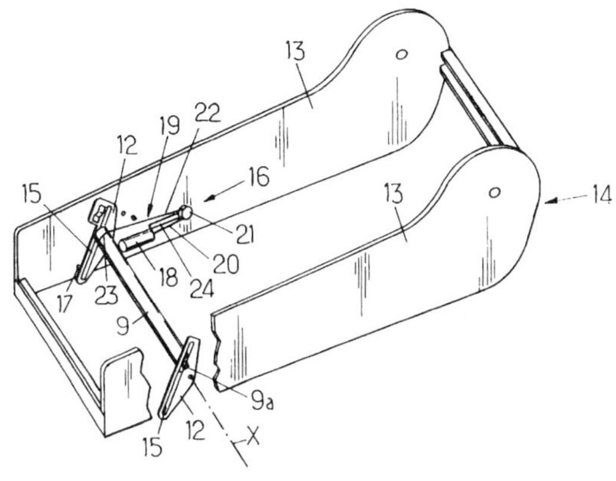

(12) **United States Patent**
Nilsson

(10) **Patent No.:** **US 6,435,591 B1**
(45) **Date of Patent:** **Aug. 20, 2002**

(54) **DEVICE FOR AVOIDING WHIPLASH INJURIES**

(75) Inventor: **Kent Nilsson**, Billdal (SE)

(73) Assignee: **Autoliv Development AB** (SE)

(*) Notice: Subject to any disclaimer, the term of this patent is extended or adjusted under 35 U.S.C. 154(b) by 0 days.

(21) Appl. No.: **09/380,715**

(22) PCT Filed: **Mar. 11, 1998**

(86) PCT No.: **PCT/SE98/00435**

§ 371 (c)(1), (2), (4) Date: **Sep. 3, 1999**

(87) PCT Pub. No.: **WO98/40238**

PCT Pub. Date: **Sep. 17, 1998**

(30) **Foreign Application Priority Data**

Mar. 11, 1997 (SE) 9700907
Jan. 14, 1998 (SE) 9800054

(51) **Int. Cl.**[7] **B60N 2/12**; B60N 2/427; B60N 2/42
(52) **U.S. Cl.** **296/68.1**; 297/216.14; 297/216.15; 297/216.16; 297/216.18; 297/216.19
(58) **Field of Search** 296/68.1; 297/216.1, 297/216.14, 216.15, 216.16, 216.18, 216.19

(56) **References Cited**

U.S. PATENT DOCUMENTS

2,227,717	A	*	1/1941	Jones	296/68.1
3,423,124	A	*	1/1969	Hewitt	297/216.19
3,452,834	A	*	7/1969	Gaut	296/68.1
3,552,795	A	*	1/1971	Perkins et al.	297/216.18
3,578,376	A		5/1971	Hasegawa	
3,582,133	A		6/1971	DeLanenne et al	
3,610,679	A	*	10/1971	Amato	296/146.7
3,731,972	A	*	5/1973	McConnell	297/216.19
3,732,944	A	*	5/1973	Kendall	297/216
3,761,127	A	*	9/1973	Giese et al.	296/68.1

(List continued on next page.)

FOREIGN PATENT DOCUMENTS

JP	405178437	A	*	7/1993	297/216.18
WO	9301950			2/1993	
WO	9422692			10/1994	
WO	9616834			6/1996	
WO	9710117			3/1997	

OTHER PUBLICATIONS

Steffan et al., "Comparison of Head–Neck Kinematics During Rear End Impact Between Standard Hybrid III, RID Neck, Volunteers, and PMTO's" Proc. Jan. 1995 Int. IRCOBI Conf. on the Biomechanics of Impacts, Brunnen, Switzerland.

Primary Examiner—Joseph D. Pape
Assistant Examiner—Hilary Gutman
(74) *Attorney, Agent, or Firm*—Klarquist Sparkman, LLP

(57) **ABSTRACT**

A device for counteracting whiplash injury to a person sitting in a seat (1) which can occur due to a rapid change in velocity, such as a collision, primarily from the rear, or similar. The device is characterized by containing means of allowing controlled displacement of the entire seat (1) against the direction of travel during rapid change of velocity, that the seat (1) is provided with control devices which are arranged to give the seat and its occupant an essentially simultaneous and essentially linear movement against the direction of movement and/or an essentially arcuate movement, and that the seat (1) is arranged to stay in a rear position after the change of velocity.

18 Claims, 6 Drawing Sheets

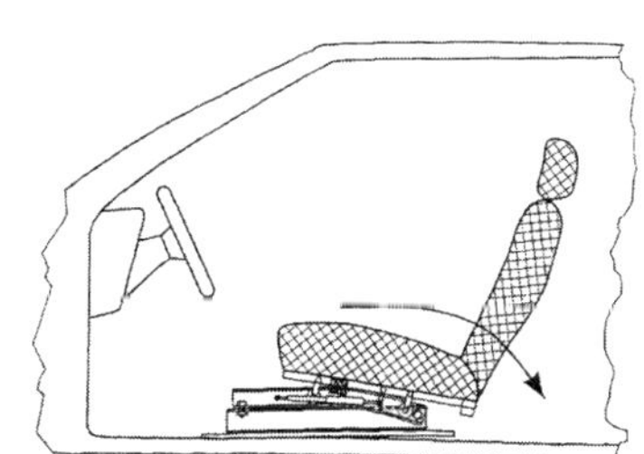

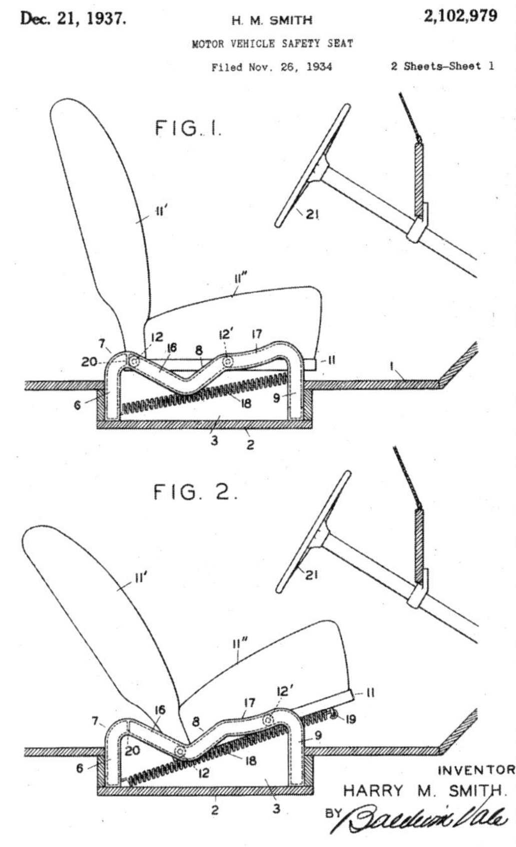

March 1, 1938.   I. KOVACH   2,109,728

SAFETY REAR AUTO SEAT

Filed July 27, 1936

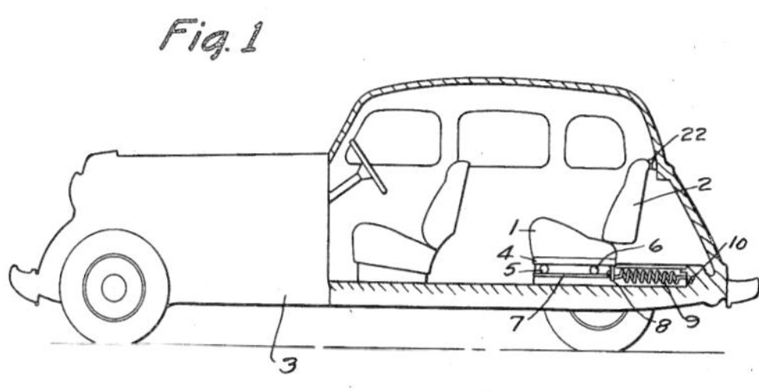

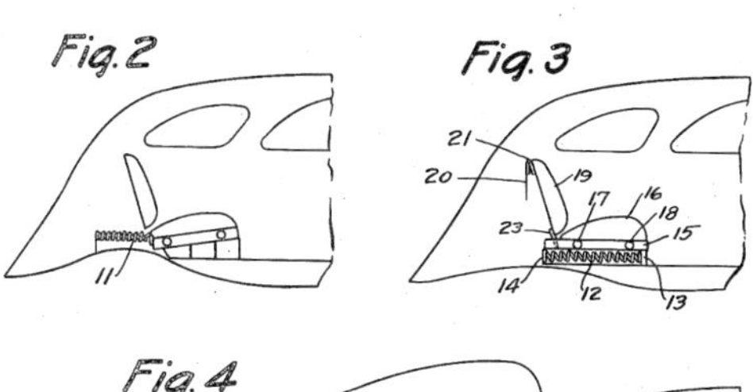

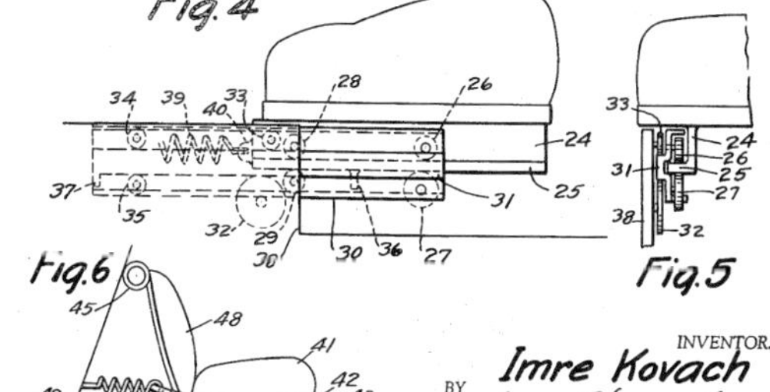

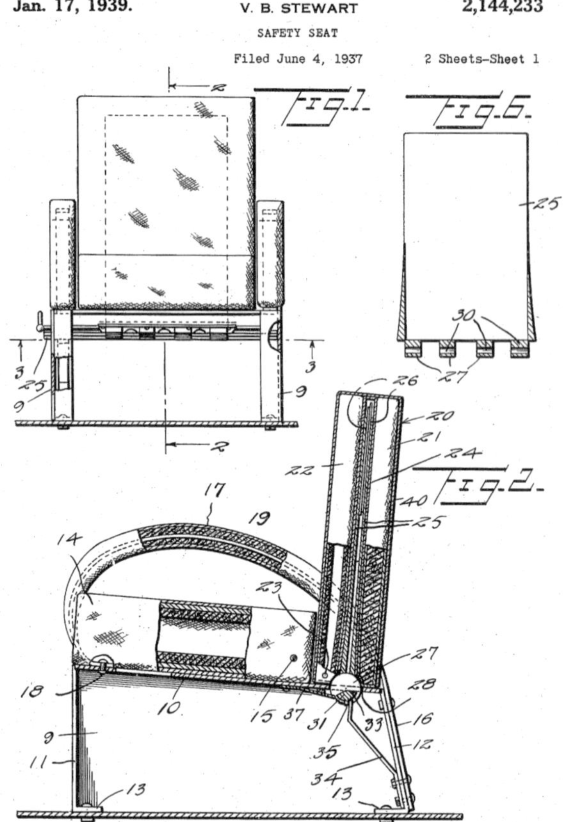
Jan. 17, 1939.
V. B. STEWART
2,144,233
SAFETY SEAT
Filed June 4, 1937
2 Sheets-Sheet 1
Vernon B. Stewart
INVENTOR
BY Victor J. Evans & Co.
ATTORNEY

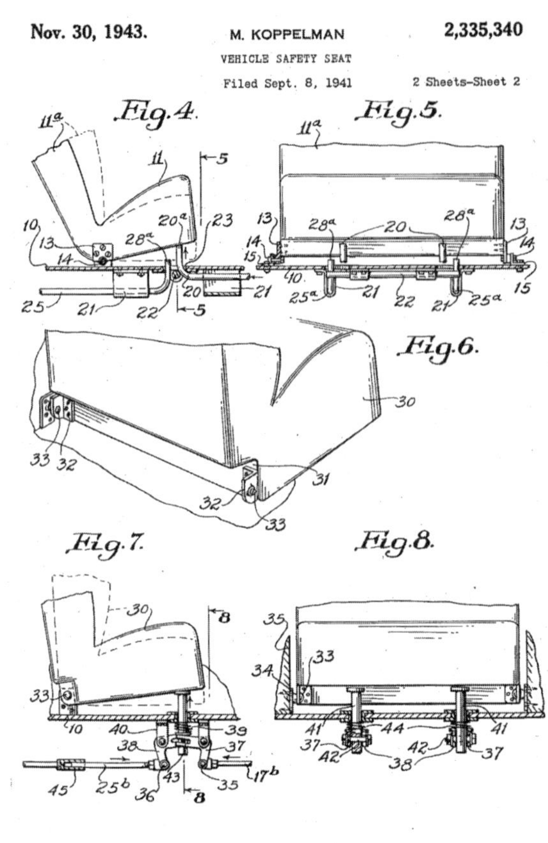
Nov. 30, 1943.
M. KOPPELMAN
2,335,340
VEHICLE SAFETY SEAT
Filed Sept. 8, 1941
2 Sheets-Sheet 2
Fig. 4.
Fig. 5.
Fig. 6.
Fig. 7.
Fig. 8.
INVENTOR.
Morris Koppelman
BY
C. Lauren Maltby
ATTORNEY

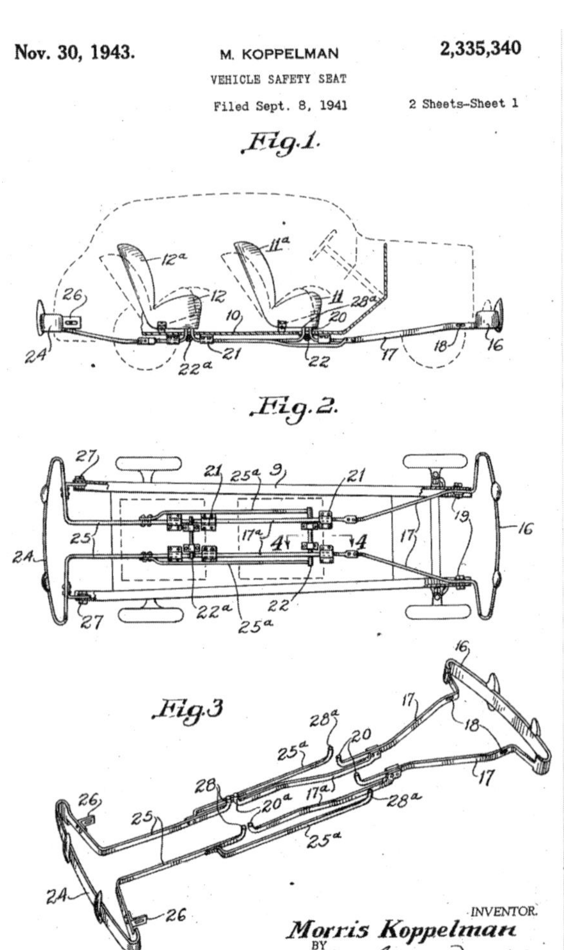
Nov. 30, 1943.
M. KOPPELMAN
2,335,340
VEHICLE SAFETY SEAT
Filed Sept. 8, 1941
2 Sheets-Sheet 1
Fig. 1.
Fig. 2.
Fig. 3
INVENTOR.
Morris Koppelman
BY
C. Lauren Maltby
ATTORNEY.

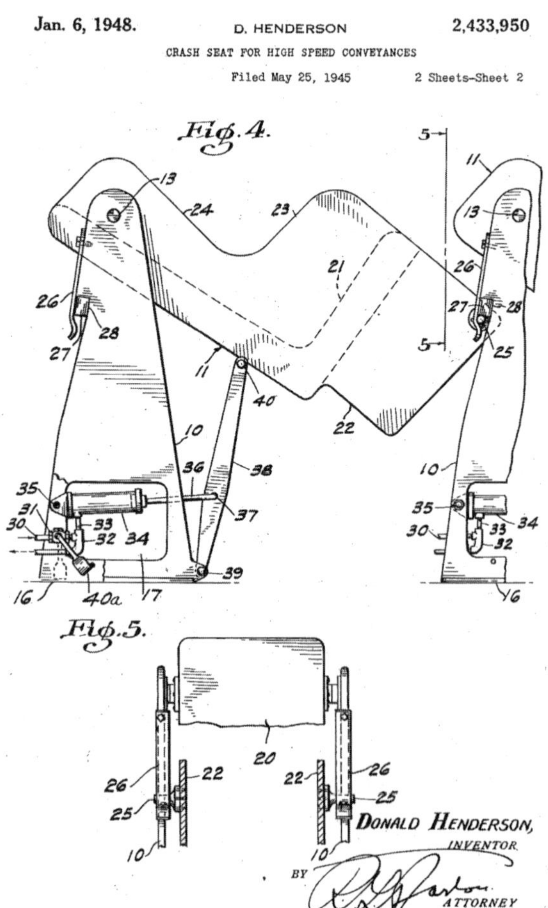
Jan. 6, 1948.
D. HENDERSON
2,433,950
CRASH SEAT FOR HIGH SPEED CONVEYANCES
Filed May 25, 1945
2 Sheets-Sheet 2
Fig. 4.
Fig. 5.
DONALD HENDERSON,
INVENTOR.
BY
ATTORNEY

Nov. 24, 1953 L. A. WOODSWORTH 2,660,222

SAFETY SEAT FOR VEHICLES

Filed May 29, 1952 3 Sheets-Sheet 2

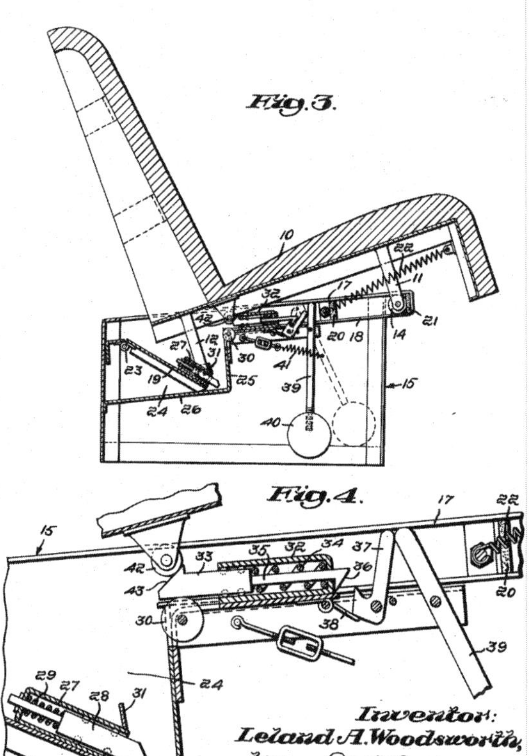

July 6, 1954 V. M. YOUNG 2,682,931

MEANS FOR ABSORBING ENERGY DUE TO SUDDEN IMPACT

Filed March 3, 1950 2 Sheets-Sheet 1

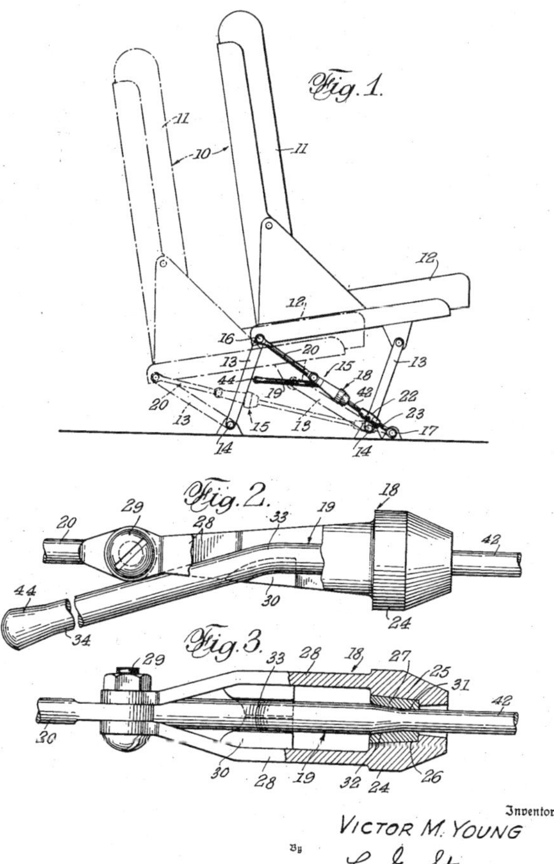

Feb. 21, 1956 H. P. FIEBER 2,735,476

CRASH SEAT

Filed Jan. 4, 1955 2 Sheets-Sheet 1

Fig. 1

Fig. 3

Herman P. Fieber

INVENTOR.

BY

Attorneys

Feb. 28, 1956 M. HARTL 2,736,566

SAFETY SEAT CONSTRUCTION FOR VEHICLES

Filed March 25, 1953 5 Sheets-Sheet 1

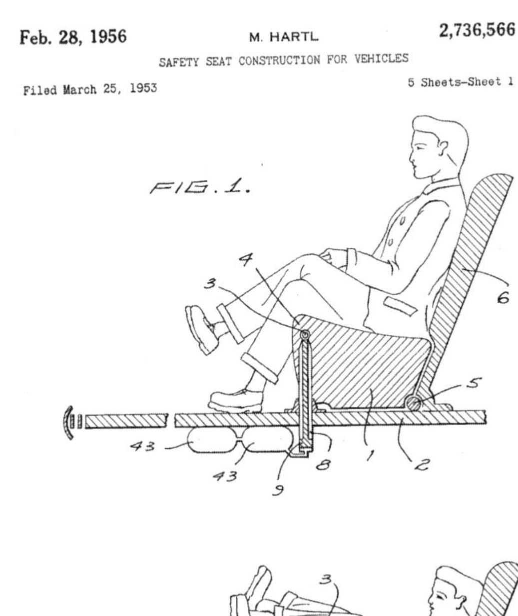

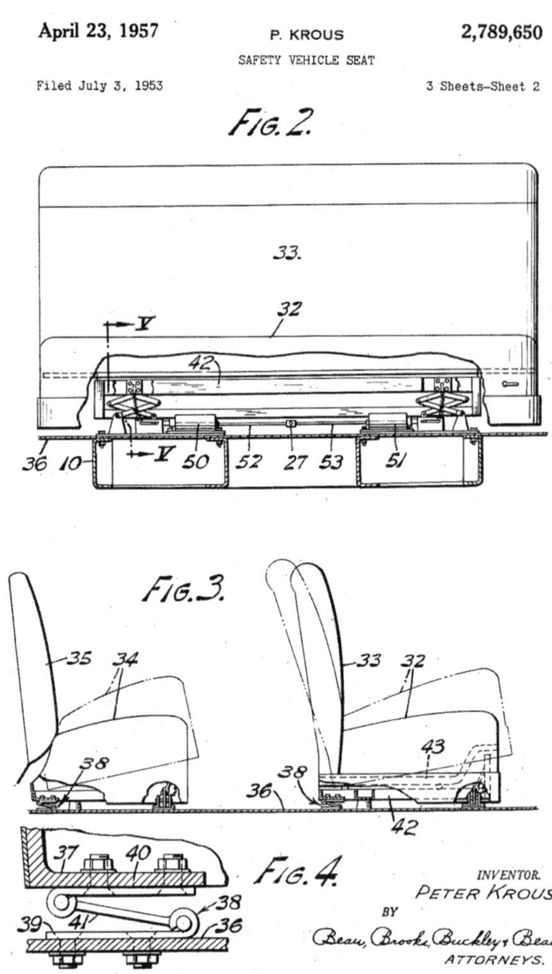

United States Patent [19]

Koutsky et al.

[11] **3,957,304**

[45] **May 18, 1976**

[54] **RESTRAINT FOR A VEHICLE SEAT AND SEAT BELT**

[75] Inventors: **L. John Koutsky,** Milan, Ill.; **John W. Carter,** Davenport, Iowa

[73] Assignee: **Sears Manufacturing Company,** Davenport, Iowa

[22] Filed: **May 30, 1975**

[21] Appl. No.: **582,084**

[52] **U.S. Cl.** **297/385;** 248/393; 248/399; 297/216

[51] **Int. Cl.²** **B60R 21/10;** A62B 35/00

[58] **Field of Search** .... 297/385, 216; 248/399, 248/400, 401, 402, 393; 280/150.5 B

[56] **References Cited**

UNITED STATES PATENTS

3,288,422	11/1966	Krause	297/385 X
3,377,102	4/1968	Hendrickson	297/385
3,572,832	3/1971	Graham	297/385 X
3,692,361	9/1972	Ivarsson	297/385
3,743,230	1/1973	Freedman	248/399
3,785,701	1/1974	Gillmore	297/385
3,799,609	3/1974	Cunningham	248/399 X
3,811,727	5/1974	Rumpel	248/393 X
3,845,987	11/1974	Bashford	297/385

Primary Examiner—James T. McCall
Attorney, Agent, or Firm—Glenn H. Antrim

[57] **ABSTRACT**

Vehicle seat assemblies, including seats and their suspensions that are to be mounted to rigid structure of vehicles, have restraint ties completely contained within the assemblies. The ties restrict the amount of upward and forward movement of the seats when they and their occupants are subject to abnormal forces when the vehicles roll over or are involved in other accidents. Each restraint tie is made from strong, flexible material such as wire rope; one end of a tie is connected to one frame of an assembly at the top of its suspension, and the other end is connected to another frame at the bottom of the suspension. The latter frame may be the movable member of a rugged fore-and-aft adjusting mechanism. In addition, pelvic restraint belts are connected to a frame at the top of the suspension.

4 Claims, 5 Drawing Figures

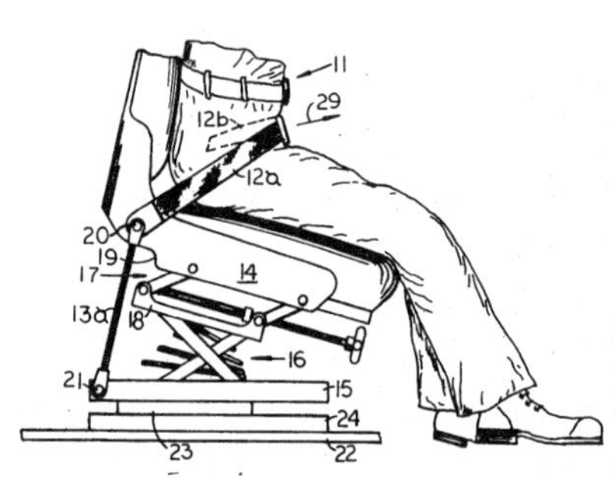

United States Patent [19]

Winslow

[11] **3,806,190**

[45] **Apr. 23, 1974**

[54] **ENERGY ABSORBING SEAT SUPPORT ASSEMBLY**

[75] Inventor: **Gerald R. Winslow,** Royal Oak, Mich.

[73] Assignee: **General Motors Corporation,** Detroit, Mich.

[22] Filed: **July 20, 1972**

[21] Appl. No.: **273,437**

[52] **U.S. Cl.** **297/216**

[51] **Int. Cl.** **B60n 1/08**

[58] **Field of Search** .... 297/216; 188/1 C

[56] **References Cited**

UNITED STATES PATENTS

2,953,189	9/1960	Barash	297/216
2,998,281	8/1961	Stower et al.	297/216 X
3,730,586	5/1973	Eggert	297/216
3,022,976	2/1962	Zia	297/216
3,552,795	1/1971	Perkins et al.	297/216
2,346,895	4/1944	Bergmann	297/216
3,473,775	10/1969	Rice	297/216
3,524,677	8/1970	Louton	297/216
3,059,966	10/1962	Spielman	297/216
3,604,285	9/1971	Olsson	188/1 C
3,696,891	10/1972	Poe	188/1 C
3,507,472	4/1970	Agee et al.	297/216
2,818,909	1/1958	Burnett	297/216
2,682,931	7/1954	Young	297/216

Primary Examiner—Paul R. Gilliam
Attorney, Agent, or Firm—Herbert Furman

[57] **ABSTRACT**

An assembly for supporting a vehicle seat for energy absorbing movement within a vehicle in response to abrupt vehicle acceleration or deceleration. The assembly includes a first track member mounted on the vehicle body floor extending longitudinally of the vehicle and having a pair of plastically deformable flange portions spaced laterally relative to each other and extending along the length of this track member with generally continuous contours interrupted by arcuate depressions. A second track member of the assembly supports the seat and has laterally spaced flange portions which mount plastic guide members that slidably receive the flange portions of the first track member to support the seat for movement longitudinally of the vehicle. A pair of bolts mount the base portion of a die member on the second track member with arcuate deforming portions of the die member respectively received within the arcuate depressions in the flange portions of the first track member so as to plastically deform these flange portions as the seat moves longitudinally of the vehicle in response to abrupt changes in the rate of vehicle movement. Spacers of varying thicknesses are readily interposed between the second track member and the base portion of the die member to allow control of the degree to which the deforming portions are received within the depressions and the consequent degree of plastic deformation and quantity of energy absorbed during a given unit length of seat movement.

4 Claims, 5 Drawing Figures

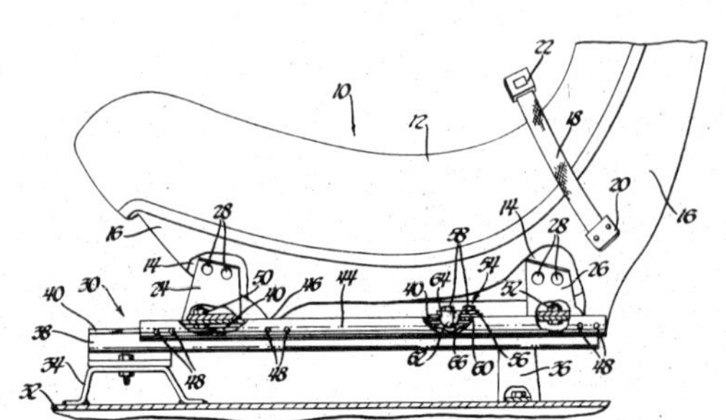

United States Patent [19]

Libkie et al.

[11] **3,853,298**

[45] **Dec. 10, 1974**

[54] **ENERGY ABSORBING SEAT ADJUSTER**

[75] Inventors: **Herbert A. Libkie,** Marlette; **Norbert T. Okoniewski,** Utica, both of Mich.

[73] Assignee: **General Motors Corporation,** Detroit, Mich.

[22] Filed: **June 11, 1973**

[21] Appl. No.: **368,606**

[52] **U.S. Cl.** .... **248/429,** 188/1 C, 296/65 A, 297/386

[51] **Int. Cl.** **B60r 21/10**

[58] **Field of Search** .... 248/429, 430, 420; 297/216, 386; 188/1 C; 296/65 A; 280/150 B

[56] **References Cited**

UNITED STATES PATENTS

1,964,405	6/1934	Nenne	248/429
3,189,313	6/1965	Burns et al.	248/429
3,578,376	5/1971	Hasegawa et al.	296/65 A
3,696,891	10/1972	Poe	188/1 C

FOREIGN PATENTS OR APPLICATIONS

528,799	11/1940	Great Britain	248/429
1,211,230	11/1970	Great Britain	248/429

Primary Examiner—Marion Parsons, Jr.
Attorney, Agent, or Firm—Herbert Furman

[57] **ABSTRACT**

A vehicle seat adjuster includes a first track member attached to the seat and a second track member attached to the vehicle body. The track members are slidably interengaged to permit longitudinal movement therebetween whereby the seat may be adjusted fore and aft. A friction shoe mounted on one of the tracks is vertically movable into frictional engagement with the other. An operating lever is attached to a cam which permits the seat occupant to move the friction shoe to selectively lock and unlock the seat for fore and aft adjusting movement. The force of kinetic impact energy overcomes the frictional engagement of the friction shoe against the other track of the seat adjuster to permit energy absorbing movement of the seat.

1 Claim, 4 Drawing Figures

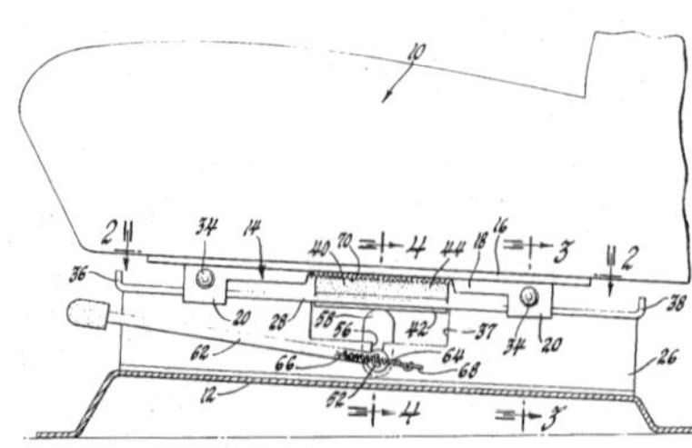

United States Patent [19]

Eggert, Jr.

[11] **3,858,934**

[45] **Jan. 7, 1975**

[54] **VEHICLE SEAT SAFETY SYSTEM**

[75] Inventor: **Walter S. Eggert, Jr.,** Huntington Valley, Pa.

[73] Assignee: **The Budd Company,** Troy, Mich.

[22] Filed: **July 19, 1973**

[21] Appl. No.: **380,912**

[52] **U.S. Cl.** **297/216,** 248/429
[51] **Int. Cl.** **B60r 21/10**
[58] **Field of Search** 297/216, 346; 248/424, 248/429, 430, 420, 422

[56] **References Cited**

UNITED STATES PATENTS

2,240,143	4/1941	Lustig	248/430
2,665,740	1/1954	Rappl	248/430
2,959,207	11/1960	Brewster	297/216
3,059,966	10/1962	Spielman	297/216
3,184,209	5/1965	Colautti	248/429
3,313,512	4/1967	Colautti	248/424
3,437,302	4/1969	Homier	248/422
3,524,678	8/1970	DeLavenne	297/216
3,582,033	6/1971	LaFleche	248/420
3,603,638	9/1971	McGregor	248/429

Primary Examiner—Francis K. Zugel

[57] **ABSTRACT**

Vehicle seat safety system which comprises a yieldable attenuating restraint device, as of the draw type with draw die and elongated drawable element, in which the attenuating device forms a linkage which is pivotally connected to a seat at one end and pivotally connected to a movable seat-positioning slide at the other end, with manual or power means to adjust and hold positions of a seat mounted on retaining tracks.

6 Claims, 7 Drawing Figures

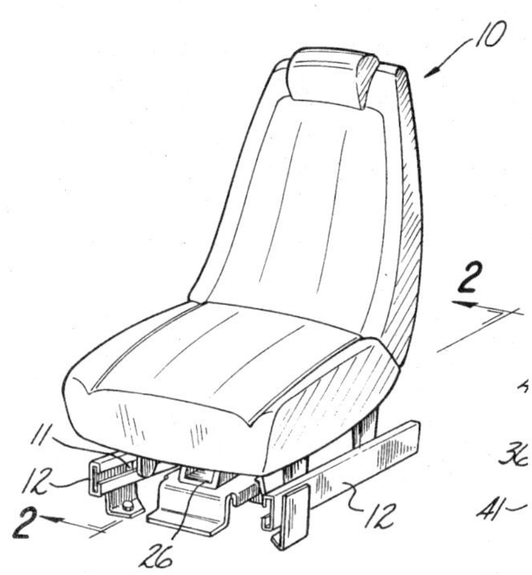

United States Patent [19]

Reilly

[11] **4,349,167**

[45] **Sep. 14, 1982**

[54] **CRASH LOAD ATTENUATING PASSENGER SEAT**

[75] Inventor: **Mason J. Reilly,** Media, Pa.

[73] Assignee: **The Boeing Company,** Seattle, Wash.

[21] Appl. No.: **89,695**

[22] Filed: **Oct. 30, 1979**

[51] **Int. Cl.**[3] **B64D 25/04;** B64D 11/06
[52] **U.S. Cl.** .... **244/122 R;** 297/216; 188/372; 296/65 A
[58] **Field of Search** .... 244/122 R; 188/1 C; 297/314, 216; 248/548, 636; 296/65 A

[56] **References Cited**

U.S. PATENT DOCUMENTS

1,511,264	10/1924	Carter	188/1 C
2,933,127	4/1960	Brewster	244/122 R
3,603,638	9/1971	McGregor	297/216
3,697,128	10/1972	Strier et al.	297/216
3,922,034	11/1975	Eggert	244/122 R
3,968,863	7/1976	Reilly	188/1 C
4,150,805	4/1979	Mazelsky	244/122 R

OTHER PUBLICATIONS

Mason J. Reilly, "Crashworthy Troup Seat Investigation" USAAMROL-TR-74-93, Dec. 1974 317 pp.

Primary Examiner—Galen L. Barefoot
Attorney, Agent, or Firm—Felix J. D'Ambrosio; Jack D. Puffer

[57] **ABSTRACT**

A light-weight aircraft passenger seat, which includes a plurality of integrally mounted wire-bending energy attenuators, and which can be floor mounted to face either forward or backward. Four seat legs, pivotally connected between the seat and floor, extend upward at the same angle from the floor, which is determined by two energy attenuators also pivotally connected between the seat and the floor, and extending upward from the floor in an opposite direction from the seat legs. When facing forward, the seat is stroked forward and downward. When facing backward, the seat is stroked backward and downward. In either a forward or backward facing disposition, the free ends of the energy attenuations extend into the back of the seat. Also the seat is flexibly connected to the seat legs and energy attenuators, which, in turn are connected by swivels to the floor to provide seat articulation as the floor distorts during a crash.

7 Claims, 9 Drawing Figures

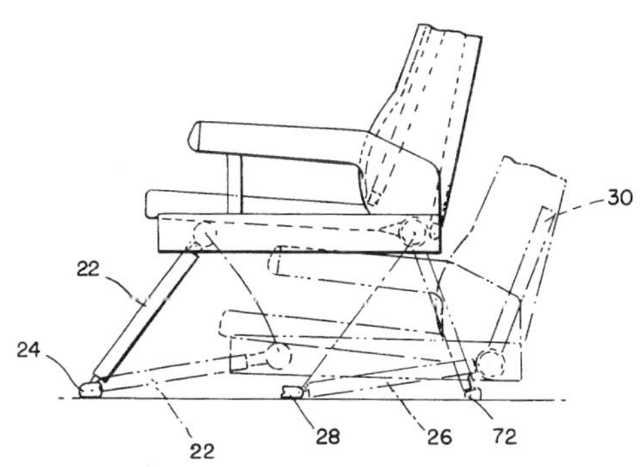

U.S. Patent May 15, 1979 Sheet 2 of 3 **4,154,472**

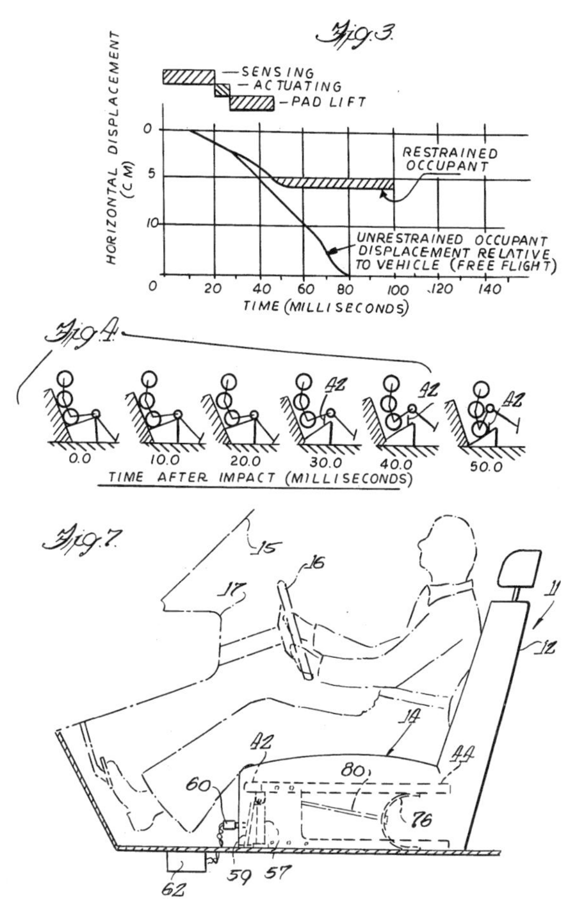

United States Patent [19]

Fox

US005125598A

[11] **Patent Number: 5,125,598**

[45] **Date of Patent: Jun. 30, 1992**

[54] **PIVOTING ENERGY ATTENUATING SEAT**

[75] Inventor: **Roy G. Fox,** Weatherford, Tex.

[73] Assignee: **Bell Helicopter Textron Inc.,** Fort Worth, Tex.

[21] Appl. No.: **443,761**

[22] Filed: **Dec. 7, 1989**

[51] **Int. Cl.**[5] **B64D 25/04**
[52] **U.S. Cl.** .... **244/122 R;** 297/216; 280/806
[58] **Field of Search** .... 244/122 R, 120 AG, 122 R, 244/121; 296/65 R, 65 A, 68; 297/216; 280/806, 805

[56] **References Cited**

U.S. PATENT DOCUMENTS

2,736,566	2/1956	Hartl	297/216
3,420,475	1/1969	Castillo et al.	244/122 R
4,154,472	5/1979	Bryll	296/65
4,257,626	3/1981	Adomeit	297/216
4,408,738	10/1983	Mazelsky	244/122 R
4,423,848	1/1984	Mazelsky	244/122 R
4,474,347	10/1984	Hazelsky	244/122 R
4,523,730	6/1985	Martin	244/122 R
4,525,010	6/1985	Trickey et al.	297/216
4,655,416	4/1987	Carnell et al.	244/121
4,720,139	1/1988	McSmith	297/216

FOREIGN PATENT DOCUMENTS

565702	3/1958	Belgium	296/68.1

Primary Examiner—Galen Barefoot
Attorney, Agent, or Firm—Ross, Howison, Clapp & Korn

[57] **ABSTRACT**

An energy attenuating seat for a vehicle includes a seat member having a seat back and a seat pan. The seat pan includes a front and rear portion. The seat back is rigidly attached to a rigid seat structure or directly to the vehicle. Support structure is provided which is rigidly attached to the vehicle and which is disposed adjacent to the front portion of the seat pan for supporting the seat pan front portion. Structure is provided for hingedly interconnecting the seat pan front portion and the support structure, such that the seat pan rotates about this interconnection. An energy attenuator is interconnected between the seat back and the rear portion of the seat pan, such that vertical load forces experienced by an occupant of the vehicle during a crash causes the seat pan to pivot downwardly whereby the rear portion of the seat pan is disposed closer to the vehicle floor than the seat pan front portion. A vertical height adjustment is provided by allowing the seat to pivot upward and downward.

10 Claims, 1 Drawing Sheet

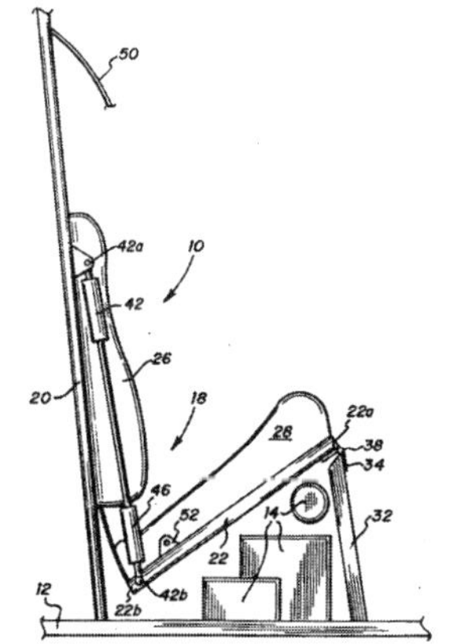

United States Patent [19]

Woolley

[11] **Patent Number: 5,149,165**

[45] **Date of Patent: Sep. 22, 1992**

[54] **STRUCTURES FOR LIFTING OUTBOARD SIDE OF SEAT OF MOTOR VEHICLE DURING SIDE IMPACT COLLISION**

[76] Inventor: **Ronald L. Woolley**, 2510 N. University Ave., Provo, Utah 84604

[21] Appl. No.: **696,103**

[22] Filed: **May 6, 1991**

[51] **Int. Cl.**[5] **B60N 2/42**

[52] **U.S. Cl.** **296/68.1**; 180/274; 297/216; 280/748

[58] **Field of Search** 296/68.1; 180/274; 297/216; 280/748

[56] **References Cited**

U.S. PATENT DOCUMENTS

2,736,566 2/1956 Hartl .... 296/68.1 X
2,970,862 2/1961 Racine 296/68.1
3,743,347 7/1973 Shaw .... 296/68.1 X
3,998,291 12/1976 Davis .... 296/68.1 X
4,396,220 8/1983 Dieckmann et al. 296/68.1
4,966,388 10/1990 Warner et al. .... 280/734 X

FOREIGN PATENT DOCUMENTS

2314990 10/1974 Fed. Rep. of Germany 296/68.1

Primary Examiner—Dennis H. Pedder
Attorney, Agent, or Firm—Terry M. Crellin

[57] **ABSTRACT**

A safety system used in motor vehicles for raising the outboard side of a passenger seat upon side impact of the vehicle by another object. The system comprises a sensor for detecting impact of the side of the vehicle by another object. The sensor activates a trigger mechanism that in turn activates a device for raising the outboard side of the seat. The outboard side of the seat, in addition to moving upwardly may also move inwardly in an arc toward the longitudinal center portion of the interior of the vehicle.

2 Claims, 2 Drawing Sheets

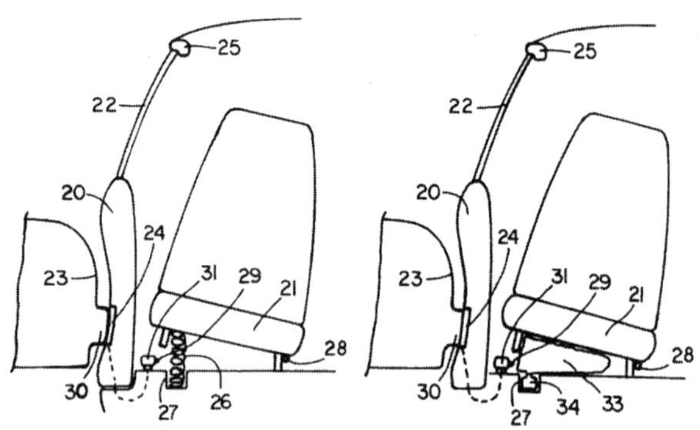

United States Patent [19]

Serber

[11] **Patent Number: 5,460,427**

[45] **Date of Patent: Oct. 24, 1995**

[54] **SEAT ASSEMBLY AND METHOD**

[76] Inventor: **Hector Serber**, 200 Gate 5 Rd., Suite 211, Sausalito, Calif. 94965

[21] Appl. No.: **119,713**

[22] Filed: **Sep. 9, 1993**

Related U.S. Application Data

[63] Continuation of Ser. No. 894,160, Jun. 1, 1992, Pat. No. 5,244,252, which is a continuation of Ser. No. 604,134, Oct. 29, 1990, abandoned.

[51] **Int. Cl.**[6] **A47C 1/02**

[52] **U.S. Cl.** **297/216.19**; 297/216.1; 297/216.15; 297/282; 297/284.7; 297/411.41; 297/316

[58] **Field of Search** 297/276, 282, 297/284.7, 284.4, 320, 411.36, 216.15–216.20

[56] **References Cited**

U.S. PATENT DOCUMENTS

2,024,170 12/1935 Kruse .... 257/411.36 X
2,102,979 12/1937 Smith .
2,736,566 2/1956 Hartl 297/216.15
2,922,461 1/1960 Braun 297/216.15
3,357,736 12/1967 McCarthy 294/216.15
3,610,679 10/1971 Amato .
3,998,291 12/1976 Davis .
4,085,963 4/1978 Bullerdieck .... 297/216.2 X
4,383,714 5/1983 Ishida 297/325
4,650,249 3/1987 Serber .
4,790,599 12/1988 Goldman .... 297/282 X
5,022,707 6/1991 Beauvais et al. .
5,244,252 9/1993 Serber 297/216.19

FOREIGN PATENT DOCUMENTS

537159 4/1955 Belgium .
1430028 3/1969 Germany 297/216.15

Primary Examiner—Peter A. Aschenbrenner
Attorney, Agent, or Firm—Flehr, Hohbach, Test, Albritton & Herbert

[57] **ABSTRACT**

A seat assembly for a vehicle which includes a seat, a seat back, and mounting means mounting the seat for movement in fore and aft directions in the vehicle along an upwardly concaved arcuate path. The arcuate path has a radius of curvature which is sufficiently large to timely move in front of and to contain the person's buttocks, to significantly slow the rate of deceleration of the person seated on said seat, and being sufficiently small to convert a portion of the linear momentum of the person's buttocks to angular momentum thus maintaining frictional contact of the person's buttocks with the seat during stopping at a high rate of deceleration. A seat assembly which includes a lower lumbar support member that is coupled from movement in response to pelvic tilting movement of the seat also is shown. The method includes the step of employing a seat mounted for movement along an upwardly concaved, arcuate path in the fore-aft direction in a vehicle to effect reduction of injury of a person seated on the seat during a frontal collision of the vehicle with an object.

7 Claims, 10 Drawing Sheets

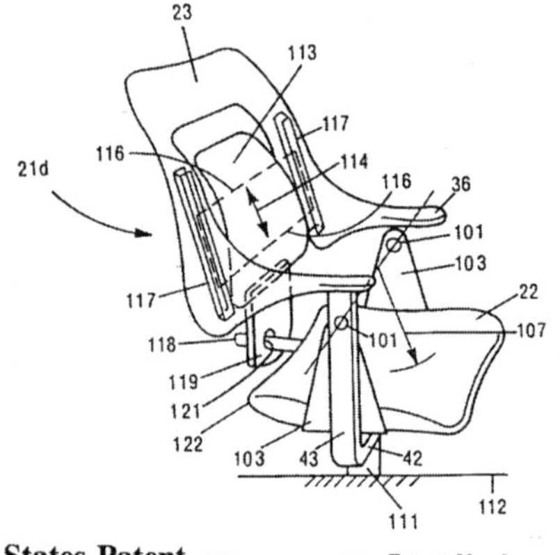

US005244252A

United States Patent [19]

Serber

[11] **Patent Number: 5,244,252**

[45] **Date of Patent: Sep. 14, 1993**

[54] **SEAT ASSEMBLY AND METHOD**

[76] Inventor: **Hector Serber**, 200 Gate 5 Rd., Ste. 211, Sausalito, Calif. 94965

[21] Appl. No.: **894,160**

[22] Filed: **Jun. 1, 1992**

Related U.S. Application Data

[63] Continuation of Ser. No. 604,134, Oct. 29, 1990.

[51] **Int. Cl.**[5] **A47C 1/02**

[52] **U.S. Cl.** **297/216.19**; 297/282; 297/325

[58] **Field of Search** 297/216, 337, 349, 325, 297/326, 282

[56] **References Cited**

U.S. PATENT DOCUMENTS

2,102,979 12/1937 Smith 297/216
2,736,566 2/1956 Hartl 297/216
3,357,736 12/1967 McMarthy .... 297/216 X
3,610,679 10/1971 Amato .... 297/216 X
3,998,291 12/1976 Davis .... 297/216 X
4,085,963 4/1978 Bullerdieck .... 297/216 X
4,383,714 5/1983 Ishida 297/325
4,650,249 3/1987 Serber .
4,790,599 12/1988 Goldman .... 297/282 X

FOREIGN PATENT DOCUMENTS

537159 4/1955 Belgium 297/216

Primary Examiner—Peter A. Aschenbrenner
Attorney, Agent, or Firm—Flehr, Hohbach, Test, Albritton & Albritton

[57] **ABSTRACT**

A seat assembly for a vehicle which includes a seat, a seat back, and mounting means mounting the seat for movement in fore and aft directions in the vehicle along an upwardly concaved arcuate path. The arcuate path has a radius of curvature which is sufficiently large to timely move in front of and to contain the person's buttocks, to significantly slow the rate of deceleration of the person seated on said seat, and being sufficiently small to convert a portion of the linear momentum of the person's buttocks to angular momentum thus maintaining frictional contact of the person's buttocks with the seat during stopping at a high rate of deceleration. A seat assembly which includes a lower lumbar support member that is coupled from movement in response to pelvic tilting movement of the seat also is shown. The method includes the step of employing a seat mounted for movement along an upwardly concaved, arcuate path in the fore-aft direction in a vehicle to effect reduction of injury of a person seated on the seat during a frontal collision of the vehicle with an object.

12 Claims, 10 Drawing Sheets

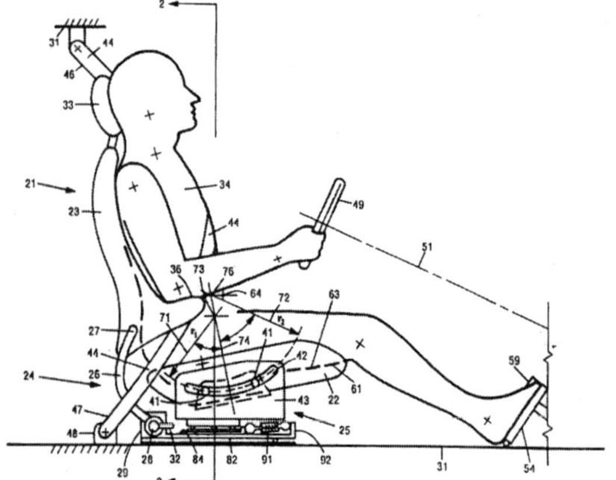

United States Patent [19]

Mikami

[11] **Patent Number: 5,556,160**

[45] **Date of Patent: Sep. 17, 1996**

[54] **SEAT WITH USER PROTECTING MEANS**

[76] Inventor: **Tatuya Mikami**, Crest Suehiro 302, 2-13-20, Suehiro, Ichikawa-shi, Chiba-ken, Japan

[21] Appl. No.: **317,873**

[22] Filed: **Oct. 4, 1994**

[51] **Int. Cl.**[6] **B60N 2/42**

[52] **U.S. Cl.** **297/216.1**; 297/313

[58] **Field of Search** 297/216.1, 330, 297/339, 337, 313, 316–318, 216.15, 216.16, 216.19, 216.2, 325, 329, 344.1; 296/68.1

[56] **References Cited**

U.S. PATENT DOCUMENTS

2,102,979 12/1937 Smith 297/216.19
2,736,566 2/1956 Hartl .... 297/216.1 X
2,942,647 6/1960 Pickles .... 297/330 X
3,173,720 3/1965 Nada .... 297/330 X
3,610,679 10/1971 Amato 297/216.16
3,731,972 5/1973 McConnell 297/216.19
3,858,930 1/1975 Calandra et al. .... 297/216.19 X
3,998,291 12/1976 Davis .... 297/216.19 X
5,244,252 9/1993 Serber 297/216.19

Primary Examiner—Milton Nelson, Jr.
Attorney, Agent, or Firm—Nixon & Vanderhye

[57] **ABSTRACT**

A seat with a user protecting device, includes a leg member, a backrest member attached to the leg member so as to extend upward therefrom, a seat member attached to the leg member so as to be between a seating position in which the seat member is arranged substantially horizontally and an inclined position in which at least a rear end portion of the seat member located on a back rest side of the leg member is arranged below the seating position or a front end portion of the seat member located opposite to the rear end portion thereof is arranged above the seating position, and a seat member rotation mechanism for holding the seat member in the seating position and allowing the seat member to rotate selectively from the seating position to the inclined position.

4 Claims, 1 Drawing Sheet

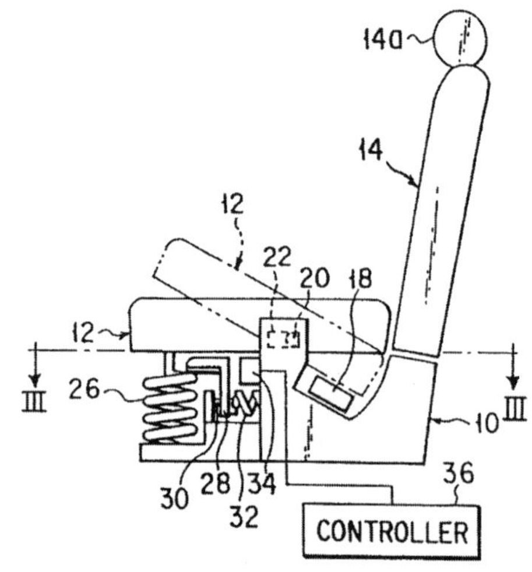

United States Patent [19]

McCarthy

[11] **Patent Number: 5,567,006**

[45] **Date of Patent: Oct. 22, 1996**

[54] **VEHICLE SEAT WITH ARTICULATED SECTIONS**

[76] Inventor: **Joseph McCarthy**, 1725 North Talbot, R.R. #1, Windsor, Ontario, Canada, N9A 6J3

[21] Appl. No.: **130,579**

[22] Filed: **Oct. 1, 1993**

[51] **Int. Cl.**[6] **B60N 2/42**

[52] **U.S. Cl.** **297/216.15**; 297/216.1; 297/216.16

[58] **Field of Search** 297/216.1, 216.15, 297/284.1, 284.3, 284.11, 317, 318, 322, 464, 473, 469, 480, 216.19, 216.18; 296/68.1, 65.1

[56] **References Cited**

U.S. PATENT DOCUMENTS

2,725,921	12/1955	Markin	155/9
2,736,566	2/1956	Hartl	280/29
2,796,112	6/1957	Barsky	155/9
2,943,866	7/1960	Witter	280/150
3,427,070	2/1969	Wallach	297/216
3,556,584	1/1971	Simon	296/65
3,591,232	7/1971	Simon	297/216
3,697,128	10/1972	Strien et al.	297/216
3,732,944	5/1973	Kendall	180/103
3,802,737	4/1974	Mertens	297/216
3,832,002	8/1974	Eggert, Jr. et al.	297/216
3,845,987	11/1974	Bashford	297/216.1 X
3,998,291	12/1976	Davis	296/68.1 X
4,154,472	5/1979	Bryll	296/65 A
4,257,626	3/1981	Adomeit	280/806
4,301,983	11/1981	Horan	244/122 R
4,335,918	6/1982	Cunningham	297/216
4,349,167	9/1982	Reilly	244/122 R
4,363,377	12/1982	Van Gerpen	180/282
4,408,738	10/1983	Mazelsky	244/122 R
4,738,485	4/1988	Rumpf	297/216
5,125,472	6/1992	Hara	180/271
5,149,165	9/1992	Woolley	296/68.1
5,152,578	10/1992	Kiguchi	297/216
5,167,421	12/1992	Yunzhao	297/216
5,294,175	3/1994	Elton	297/216.1
5,340,185	8/1994	Vollmer	297/284.11 X
5,366,269	11/1994	Beauvais	297/216.19

FOREIGN PATENT DOCUMENTS

358053525	3/1983	Japan	297/216.1
9301950	2/1993	WIPO	297/216.19

Primary Examiner—Milton Nelson, Jr.
Attorney, Agent, or Firm—Kenyon & Kenyon

[57] **ABSTRACT**

A seat for supporting a passenger in a vehicle includes a seat base having first and second sections. The first and second sections are coupled to the vehicle such that, upon a sudden acceleration of the vehicle, at least a portion of the acceleration being directed from forward to rearward relative to the orientation of the passenger, the first section of the seat base moves away from a rest position and the second section moves away from a rest position along a second predetermined path, so that as the first and second sections move away from their respective rest positions, the distance between a point on the first section and a point on the second section changes and at least a portion of the second section is raised relative to a portion of the first section.

15 Claims, 22 Drawing Sheets

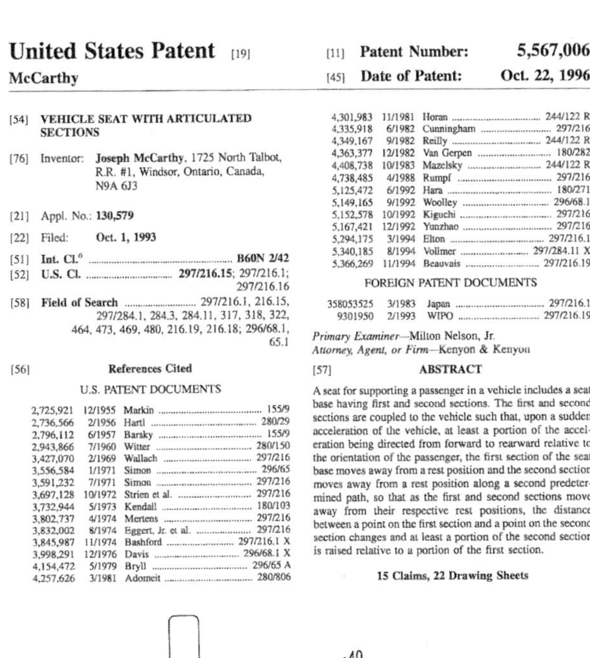

United States Patent [19]

Singer et al.

[11] **Patent Number: 5,636,424**

[45] **Date of Patent: Jun. 10, 1997**

[54] **SAFETY SEAT**

[75] Inventors: **Neil C. Singer**, New York, N.Y.; **Steven J. Gordon**, Jamaica Plain, Mass.; **Christopher T. Zirps**, Milton, Mass.; **Massimo A. Russo**, Brookline, Mass.

[73] Assignee: **Massachusetts Institute of Technology**, Cambridge, Mass.

[21] Appl. No.: **182,511**

[22] Filed: **Jan. 13, 1994**

Related U.S. Application Data

[63] Continuation of Ser. No. 732,860, Jul. 19, 1991, abandoned.

[51] **Int. Cl.**[6] **B23P 11/00**; B60N 2/42

[52] **U.S. Cl.** **29/407.01**; 29/434; 297/216.19; 364/424.055

[58] **Field of Search** 29/404, 407, 407.01, 29/434, 407.08; 364/424.05; 297/216.19

[56] **References Cited**

U.S. PATENT DOCUMENTS

2,102,979	12/1937	Smith	297/216.19
2,335,340	11/1943	Koppelman .	
2,736,566	2/1956	Hartl .	
2,777,531	1/1957	Erickson .	
2,818,909	1/1958	Burnett .	
2,823,730	2/1958	Lawrence .	
2,863,496	12/1958	Pinkel .	
2,922,461	1/1960	Braun .	
2,978,273	4/1961	Racine .	
2,993,732	7/1961	Walker .	
3,112,955	12/1963	Stolz .	
3,357,236	12/1967	McCarthy	297/216 X
3,423,124	1/1969	Hewitt	297/216 X
3,427,070	2/1969	Wallach .	
3,452,834	7/1969	Gaut .	
3,556,584	1/1971	Simon .	
3,591,232	7/1971	Simon .	
3,610,679	10/1971	Amato	297/216 X
3,697,128	10/1972	Strien et al. .	
3,731,972	5/1973	McConnell .	
3,802,737	4/1974	Mertens .	
3,858,930	1/1975	Calandra et al. .	
3,981,520	9/1976	Pulling .	
3,998,291	12/1976	Davis	297/216.19 X
4,085,963	4/1978	Bullerdieck .	
4,249,769	2/1981	Barecki .	
4,349,167	9/1982	Reilly .	
4,634,169	1/1987	Hasstedt .	
4,738,485	4/1988	Rumpf	297/216
5,398,185	3/1995	Omura	364/424.05

FOREIGN PATENT DOCUMENTS

823912	1/1938	France .	
1163168	2/1964	Germany .	
2456028	8/1976	Germany	297/216
7903166	10/1980	Netherlands	297/216
1161419A	6/1985	U.S.S.R. .	

Primary Examiner—Joseph M. Gorski

[57] **ABSTRACT**

The vehicle safety seat supports an occupant and includes structure interconnecting the seat and the vehicle. The interconnecting structure is adapted to constrain the seat, upon vehicle deceleration, to follow a trajectory with respect to the vehicle which substantially minimizes a cost function associated with occupant injury. In a preferred embodiment, the structure constrains the mass center and seat angle to follow trajectories which substantially minimize primarily forward motion of the occupant in the vehicle frame of reference.

6 Claims, 3 Drawing Sheets

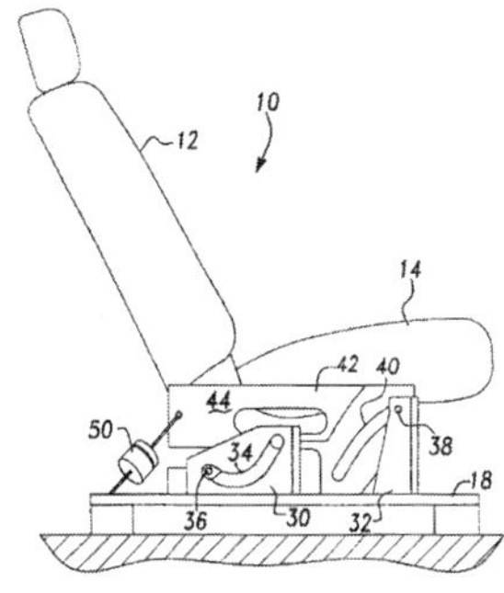

United States Patent [19]

Al-Abdullateef

[11] **Patent Number: 5,605,372**

[45] **Date of Patent: Feb. 25, 1997**

[54] **SAFETY SEAT SYSTEM**

[76] Inventor: **Abdulghafour Al-Abdullateef**, P.O. Box 88897, Los Angeles, Calif. 90045

[21] Appl. No.: **324,245**

[22] Filed: **Oct. 14, 1994**

[51] **Int. Cl.**[6] **B60N 2/42**

[52] **U.S. Cl.** **297/216.16**; 297/216.1

[58] **Field of Search** 297/216.1, 216.15, 297/216.16, 216.19, 216.2, 317, 318, 344.1, 325, 329, 344.14, 261, 264–267, 261.1, 261.2, 261.3, 264.1, 265.1, 266.1, 267.1; 296/68.1

[56] **References Cited**

U.S. PATENT DOCUMENTS

144,603	11/1873	Enger	297/266 X
1,774,555	9/1930	Horsley et al.	297/216.17 X
2,102,979	12/1937	Smith .	
2,227,717	1/1941	Jones	296/68.1 X
2,401,748	6/1946	Dillon .	
2,736,566	2/1956	Hartl .	
3,357,736	12/1967	McCarthy .	
3,531,153	9/1970	Mohs	296/68.1
3,610,679	10/1971	Amato .	
3,998,291	12/1976	Davis .	
4,085,963	4/1978	Bullerdieck .	
4,249,769	2/1981	Barecki .	
4,650,249	3/1987	Serber .	
4,790,599	12/1988	Goldman .	
4,842,232	6/1989	Pipon et al. .	
5,022,707	6/1991	Beauvais et al.	297/216.19 X
5,110,182	5/1992	Beauvais .	
5,167,421	12/1992	Yunzhao .	
5,244,252	9/1993	Serber .	
5,366,269	11/1994	Beauvais	297/216.19

FOREIGN PATENT DOCUMENTS

537159	3/1959	Belgium	296/68.1
2596338	10/1987	France	297/216.19

Primary Examiner—Milton Nelson, Jr.
Attorney, Agent, or Firm—Fulwider Patton Lee & Utecht, LLP

[57] **ABSTRACT**

An automotive safety seat is provided that reduces g-loads imparted to an occupant during a collision by absorbing some of the energy of impact. Kinetic energy is converted to potential energy with the elevation of the seat and its occupant and the compression of springs.

8 Claims, 2 Drawing Sheets

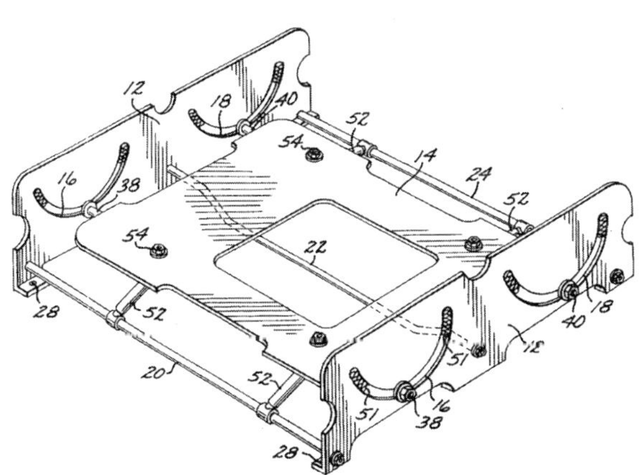

United States Patent [19]

Hubbard

[11] **Patent Number: 5,947,543**

[45] **Date of Patent: Sep. 7, 1999**

[54] **VEHICLE SAFETY SYSTEM**

[76] Inventor: **Leo James Hubbard**, 29 Bridgetown Rd., Hilton Head Island, S.C. 29928

[21] Appl. No.: **08/833,040**

[22] Filed: **Apr. 3, 1997**

[51] **Int. Cl.**[6] **B60N 2/42**

[52] **U.S. Cl.** **296/68.1**; 296/216.18; 296/216.19

[58] **Field of Search** 296/68.1; 297/216.16, 297/216.18, 216.19, 216.1

[56] **References Cited**

U.S. PATENT DOCUMENTS

2,736,566	2/1956	Hartl	296/68.1
3,832,000	8/1974	McDonnell .	
3,858,930	1/1975	Calandra et al.	296/68.1
3,992,046	11/1976	Braess	296/68.1
5,125,472	6/1992	Hara	180/271
5,167,421	12/1992	Yunzhao	296/68.1 X
5,344,204	9/1994	Yunzhao	296/68.1
5,398,185	3/1995	Omura	180/268
5,413,378	5/1995	Steffens, Jr. et al.	280/735
5,439,249	8/1995	Steffens, Jr. et al.	280/735
5,573,269	11/1996	Gentry et al.	280/735

FOREIGN PATENT DOCUMENTS

94022692	10/1994	WIPO	297/216.18

Primary Examiner—Dennis H. Pedder
Attorney, Agent, or Firm—M. LuKacher; R. C. Brown

[57] **ABSTRACT**

An improved system for passenger safety in a front-end vehicle collision, whereby the kinetic energy of a passenger may be reduced before the passenger is exposed to the crash deceleration. An electrical sensor mounted at the extreme front end of the vehicle senses the onset of a front-end collision and sends a crash signal to the passenger compartment. Preferably, the sensor is a normally-closed momentum-activatable switch. A passenger seat is mounted on translating means permitting rearward motion and has a harness to secure a passenger to the seat. The seat is held in place by a shearable pin. Beneath the seat, attached between the floor of the vehicle and the frame of the seat, is a linear actuator powered by an explosive charge which is ignitable electrically in response to the crash signal. Upon firing of the actuator, a guillotine edge on the seat frame shears the pin, and the seat and passenger are thrust rapidly rearward so that the passenger is travelling rearward with respect to the vehicle and more slowly forward with respect to the ground. Thus, the kinetic energy of the passenger is reduced and the intensity of his deceleration from the crash pulse is minimized. The seat back and cushion may be hinged for rotation backwards on the seat frame. A second explosive actuator causes rotation about the hinge coincident with the rearward translation of the frame. Rotation of the seat can orient the passenger more favorably to withstand the remaining shock of the crash pulse.

21 Claims, 6 Drawing Sheets

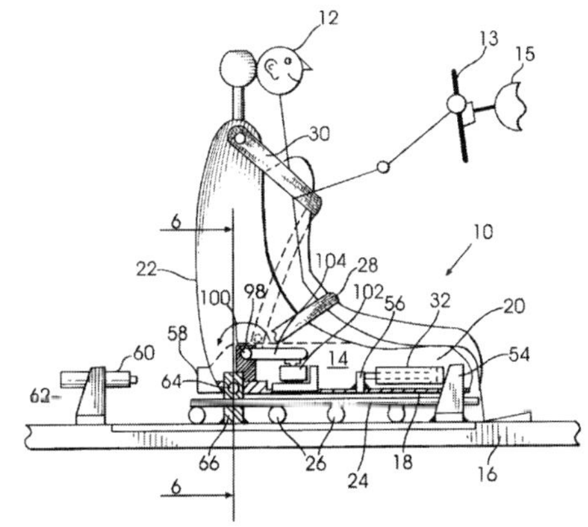

United States Patent [19]
Yoshida et al.

[11] **Patent Number:** **5,967,604**
[45] **Date of Patent:** **Oct. 19, 1999**

[54] **VEHICLE SEAT APPARATUS**

[75] Inventors: **Tadasu Yoshida**, Kariya; **Yukifumi Yamada**, Toyota, both of Japan

[73] Assignee: **Aisin Seiki Kabushiki Kaisha**, Kariya, Japan

[21] Appl. No.: **08/991,191**

[22] Filed: **Dec. 16, 1997**

[30] **Foreign Application Priority Data**

Dec. 17, 1996 [JP] Japan 8-337201

[51] **Int. Cl.**[6] **B60N 2/42**; B60R 21/00
[52] **U.S. Cl.** **297/216.19**; 297/344.1; 297/216.2; 297/216.1; 297/216.16; 296/68.1; 248/429
[58] **Field of Search** 297/216.19, 216.1, 297/216.16, 216.2, 344.1; 296/68.1, 65.13, 65.14; 248/548, 900, 429

[56] **References Cited**

U.S. PATENT DOCUMENTS

2,735,476	2/1956	Fieber	297/216.19
2,736,566	2/1956	Hartl	296/68.1 X
2,789,650	4/1957	Krous	297/216.19 X
2,818,909	1/1958	Burnett	296/68.1 X
2,823,730	2/1958	Lawrence	296/68.1 X
3,451,719	6/1969	Lorean	297/216.1 X
3,452,834	7/1969	Gaut	296/68.1 X
3,669,397	6/1972	Le Mire	297/216.19 X
3,724,603	4/1973	Shiomi et al.	297/216.19 X
3,802,737	4/1974	Mertens	297/216.2
3,933,331	1/1976	Blom	297/344.1 X
4,597,552	7/1986	Nishino	297/344.1 X
4,615,551	10/1986	Kinaga et al.	248/429 X
4,634,169	1/1987	Hasstedt	296/68.1
4,676,556	6/1987	Yamanoi et al.	297/216.1 X
4,993,776	2/1991	Acuto et al.	297/216.1
5,207,480	5/1993	Johnson et al.	297/344.1
5,292,178	3/1994	Loose et al.	297/344.1
5,575,449	11/1996	Shinbori et al.	297/344.1 X
5,662,376	9/1997	Breuer et al.	297/216.16 X

FOREIGN PATENT DOCUMENTS

5-338484 12/1993 Japan .

Primary Examiner—Peter M. Cuomo
Assistant Examiner—Rodney B. White
Attorney, Agent, or Firm—Burns, Doane, Swecker & Mathis, L.L.P.

[57] **ABSTRACT**

A seat apparatus for a vehicle includes a seat cushion for being secured to a vehicle floor, a seat back arranged on a rear portion of the seat cushion, a sliding unit mounted on the seat cushion to permit the vehicle seat to slide relative to the vehicle floor, a pair of front brackets adapted to be fixed to the vehicle floor and connected to the sliding unit at the front end of the seat cushion and a pair of rear brackets adapted to be fixed to the vehicle floor and connected to the sliding unit at the rear end of the seat cushion. The front brackets are adapted to be deformed during a collision to permit the seat cushion to move upward. The front brackets are configured to include a first attaching portion and a second attaching portion that are spaced apart from one another.

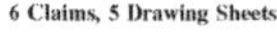
6 Claims, 5 Drawing Sheets

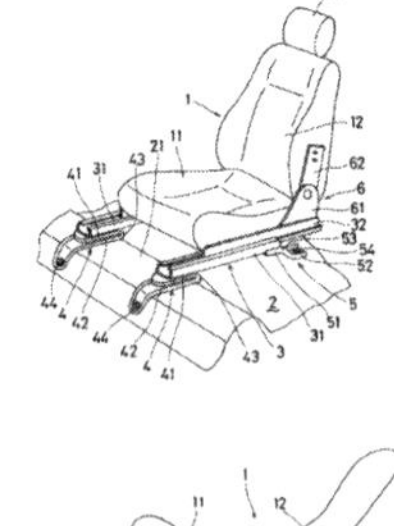

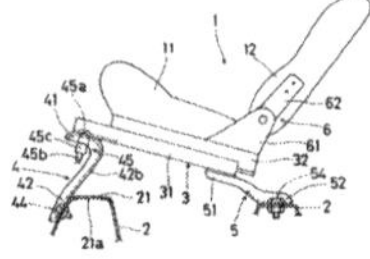

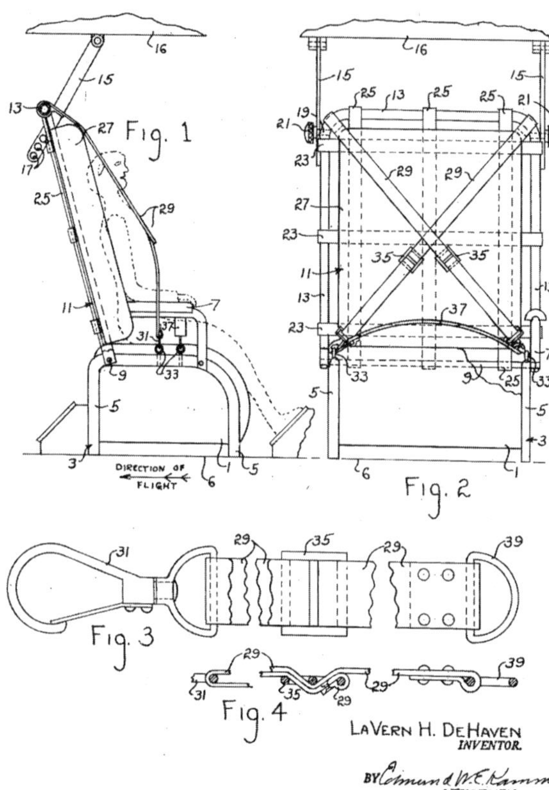

EXAMPLES OF BELT RESTRAINTS, WINGS, BARRIERS AND WRAP-AROUND DESIGNS FOR CRASH SAFETY

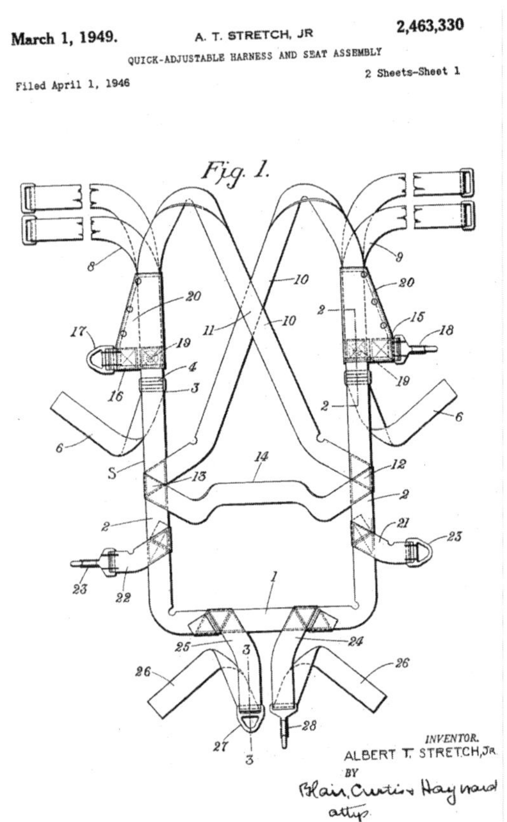

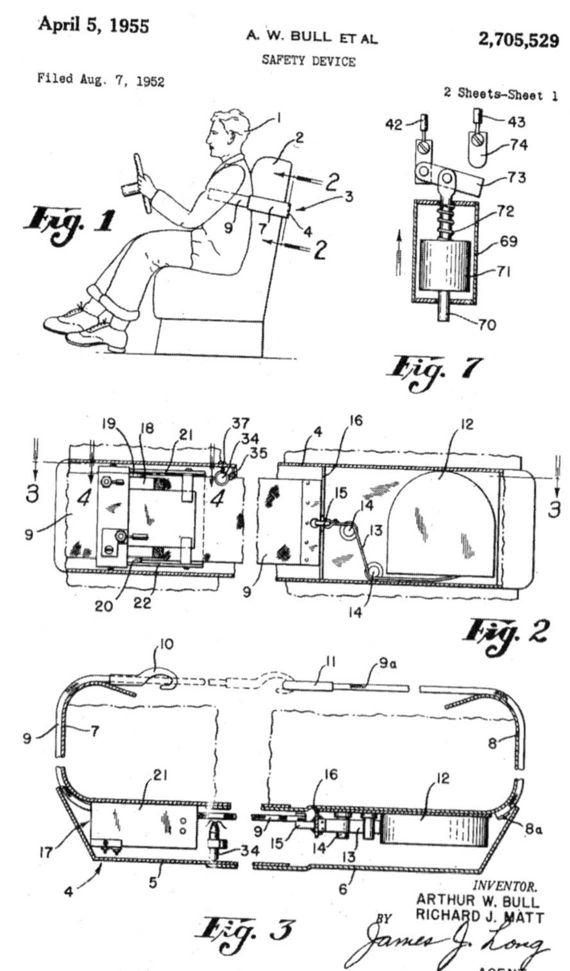

Dec. 31, 1957 N. E. MANOS **2,818,274**

VEHICLE SAFETY BAR FOR PREVENTING PIVOTED SEAT BACK AGAINST FOLDING

Filed Nov. 2, 1955

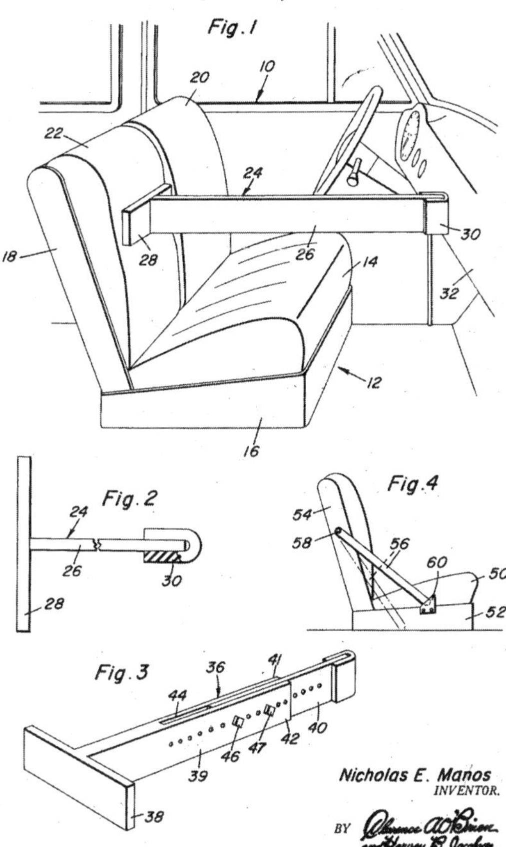

Sept. 13, 1960 I. I. PINKEL **2,952,304**

BACK STRUCTURE FOR CRASH RESISTANT SEAT

Original Filed April 25, 1955 5 Sheets-Sheet 1

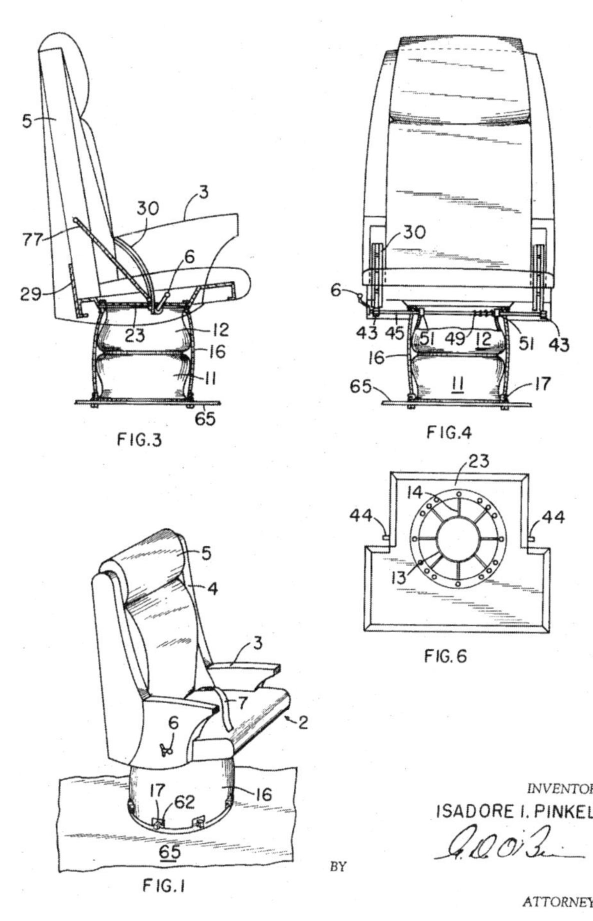

Feb. 10, 1959 L. PÉRAS **2,873,122**

SAFETY SEAT FOR MOTOR VEHICLES

Filed Oct. 9, 1956

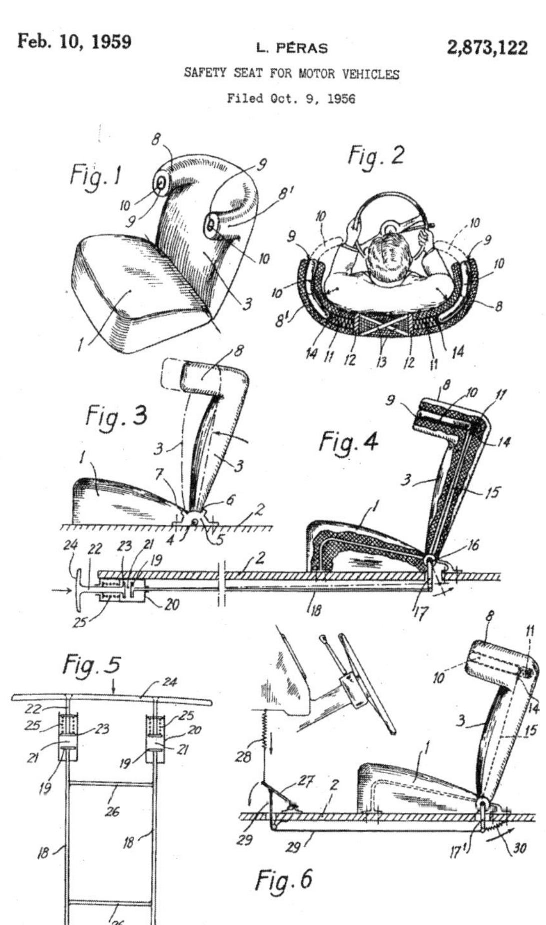

Aug. 30, 1966 F. HILDEBRANDT ETAL **3,269,774**

MOVABLE SEAT INCLUDING SHOCK-ABSORBING MEANS AND SAFETY BELT

Filed March 30, 1964 2 Sheets-Sheet 1

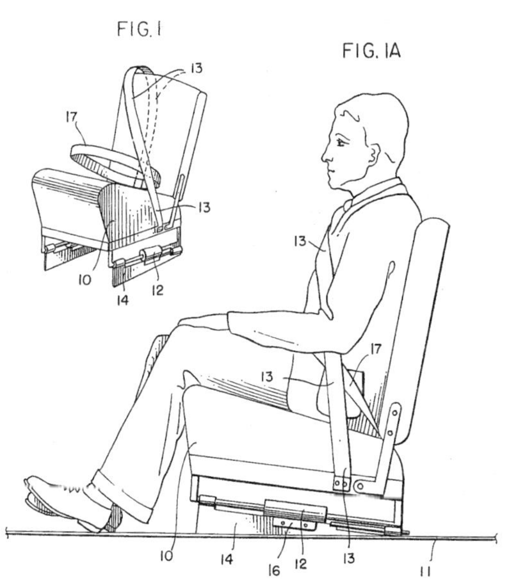

United States Patent [19]

Miller

[11] **3,713,694**

[45] **Jan. 30, 1973**

[54] **BODY RESTRAINING DEVICE FOR VEHICLE**

[76] Inventor: **Ralph A. Miller**, 3832 Burton St., Toledo, Ohio 43612

[22] Filed: **April 20, 1971**

[21] Appl. No.: **135,576**

[52] **U.S. Cl.** ....**297/390,** 280/150 SB, 297/216, 297/389

[51] **Int. Cl.** **A62b 35/60**

[58] **Field of Search**......297/390, 384, 417, 418, 216; 280/150 SB

[56] **References Cited**

UNITED STATES PATENTS

3,466,091	9/1969	De Grusso	297/390
3,262,716	7/1966	Graham	297/390 X
3,376,064	4/1968	Jackson	297/410
2,873,122	2/1959	Peras	297/384 X
3,165,357	1/1965	Rudemann, Jr.	297/384
3,474,471	10/1969	Matibag	297/417 X

FOREIGN PATENTS OR APPLICATIONS

1,008,155	2/1952	France	297/390

Primary Examiner—James T. McCall
Attorney—Paul F. Stutz

[57] **ABSTRACT**

An improved seat construction for vehicle occupants which comprises, in addition to the usual horizontal platform and upstanding back support, a hand-operable lever arm and a pair of shoulder restraining members, the latter being connected via a connector rod to the former whereby shifting of the lever arm causes the shoulder restraining arms to shift between an operative occupant restraining position and an inoperative position, in which the occupant is not restrained, for convenience of ingress and egress of the occupant; said device including, in a preferred embodiment, a shiftable armrest arrangement.

8 Claims, 2 Drawing Figures

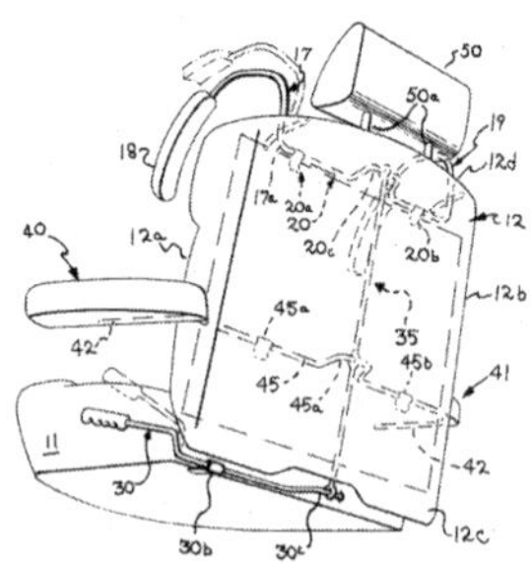

Jan. 15, 1957 N. R. ERICKSON 2,777,531

EMERGENCY CRASH SEAT

Filed Oct. 22, 1954 2 Sheets-Sheet 1

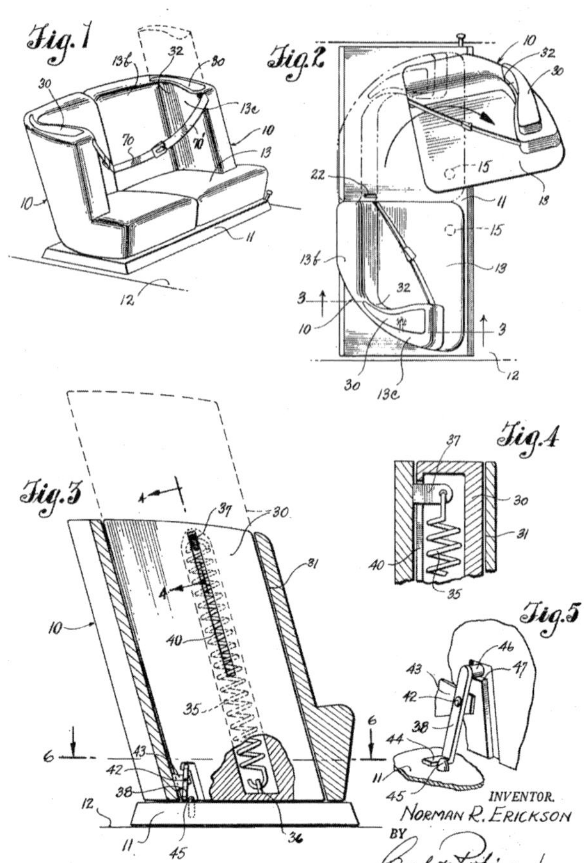

United States Patent [19]

Krejci, III

[11] **3,740,091**

[45] **June 19, 1973**

[54] **ENERGY ABSORBING SEAT ASSEMBLY**

[76] Inventor: **Joseph A. Krejci, III,** 1967 N. 18th Avenue, Apt. 5, Melrose Park, Ill. 60160

[22] Filed: **June 3, 1971**

[21] Appl. No.: **149,567**

[52] **U.S. Cl.** **296/68,** 296/146, 297/216

[51] **Int. Cl.** **B60n 1/04**

[58] **Field of Search**................. 296/68, 65 R, 65 A, 296/146; 297/216, 349, 240

[56] **References Cited**

UNITED STATES PATENTS

3,322,458	5/1967	Bachmann	296/65 R
2,952,304	9/1960	Pinkel	297/216
2,874,993	2/1959	Probst	297/349 X
3,165,355	1/1965	Hitchcock	297/216

FOREIGN PATENTS OR APPLICATIONS

306,860	2/1917	Germany	296/68

Primary Examiner—Leo Friaglia
Assistant Examiner—John A. Pekar
Attorney—Richard R. Mybeck

[57] **ABSTRACT**

An energy absorbing seat assembly for use in mobile vehicles to protect the occupant from injury in the event of a collision. The assembly comprises an energy absorbing seat mounted on a frame which is selectively pivotal relative to the vehicle.

7 Claims, 4 Drawing Figures

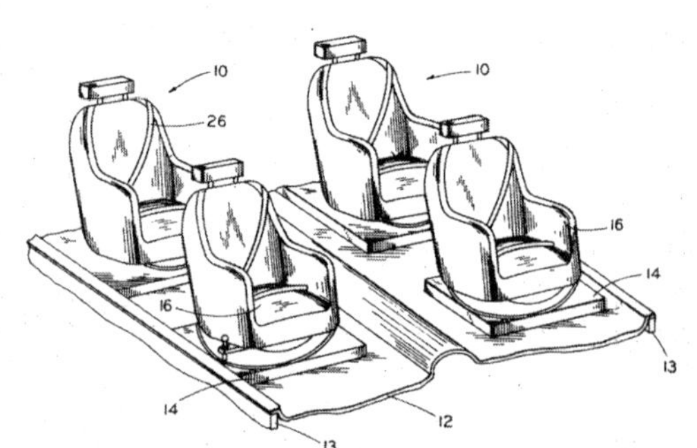

United States Patent [19]

Fourrey

[11] **3,734,562**

[45] **May 22, 1973**

[54] **AUTOMOTIVE SEATS WITH SAFETY HARNESSES**

[75] Inventor: **Francois Fourrey,** Billancourt, France

[73] Assignees: **Regie Nationale des Usines Renault,** Billancourt; **Automobiles Peugeot,** Paris, France

[22] Filed: **June 28, 1971**

[21] Appl. No.: **157,462**

[30] **Foreign Application Priority Data**

Aug. 11, 1970 France..............................7029492

[52] **U.S. Cl.** **297/216,** 297/385, 297/386, 297/389

[51] **Int. Cl.** **A62b 35/60**

[58] **Field of Search**......................297/216, 386, 389, 297/385, 378; 280/150 SB; 188/1 C

[56] **References Cited**

UNITED STATES PATENTS

3,582,133	6/1971	DeLavenne	297/216
3,309,137	3/1967	Wiebe	297/302
3,186,760	6/1965	Lohr et al.	297/216
2,795,266	6/1957	Walther	297/216
3,603,638	9/1971	McGregor	297/216
3,382,527	5/1968	Strien et al.	297/378 X
2,953,189	9/1960	Barash	297/216
3,269,774	8/1966	Hilderbrandt	297/386
3,586,131	6/1971	Le Mire	297/386

Primary Examiner—James T. McCall
Attorney—Stevens et al.

[57] **ABSTRACT**

A seat provided with a safety harness, notably for automotive vehicles and of the type comprising a squab, a back provided with at least one anchor point for the shoulder-belt of said safety harness, said anchor point being located on the rear face of said back, an abdominal belt connected to said shoulder-belt, at least one kinetic energy absorption device interposed between said squab and a seat element adapted, in case of crash, to move in relation to said squab, said device being adapted, during this relative movement, to absorb one fraction of the kinetic energy and wherein the anchor point located on said back is connected through said shoulder-belt to the abdominal belt of the safety harness, characterized in that at least one end of said abdominal belt is connected to the seat element movable in relation to the squab during a crash and that said seat element is mechanically connected to said squab in a manner known per se.

8 Claims, 13 Drawing Figures

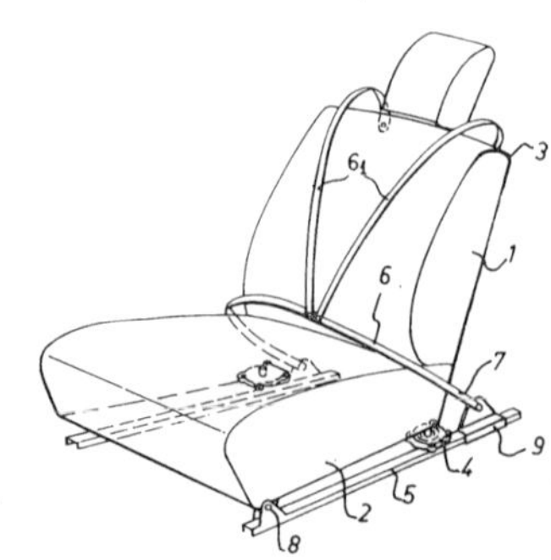

United States Patent [19]

Eggert, Jr. et al.

[11] **3,832,002**

[45] **Aug. 27, 1974**

[54] **AUTOMOTIVE RESTRAINT SYSTEM**

[75] Inventors: **Walter S. Eggert, Jr.**, Huntington Valley; **Michael J. Pavlik**, Norristown, both of Pa.

[73] Assignee: **The Budd Company**, Troy, Mont.

[22] Filed: **Mar. 21, 1973**

[21] Appl. No.: **343,321**

[52] **U.S. Cl.** **297/216,** 297/385, 297/389
[51] **Int. Cl.** **B60r 21/10**
[58] **Field of Search** .... 297/216, 385, 386, 389; 280/150 SB

[56] **References Cited**

UNITED STATES PATENTS

2,899,146	8/1959	Barecki	297/389 X
3,089,144	5/1963	Cherup	2/3 R
3,414,322	12/1968	Linderoth	297/385 X
3,424,495	1/1969	Cherup	297/385 X
3,501,200	3/1970	Ohta	297/216 X
3,552,795	1/1971	Perkins et al.	297/216
3,730,586	5/1973	Eggert	297/216

Primary Examiner—James C. Mitchell

[57] **ABSTRACT**

Vehicle seat safety system in which the occupant or passenger is closely restrained in the seat; the seat back is yieldably restrained in attenuation at a first and lower impact range for forward tilting movement with the occupant; and the seat is yieldably restrained in attenuation at a second and higher range for forward movement of the body and seat as a whole—to first allow the upper portion of the body or torso to move forward under lighter impact loadings, and to allow the body as a whole to move forward under greater impact loadings and thus to minimize, as far as possible, the impact loads on the body and to reduce forward and reverse or snap-back loads on the body torso and neck.

12 Claims, 10 Drawing Figures

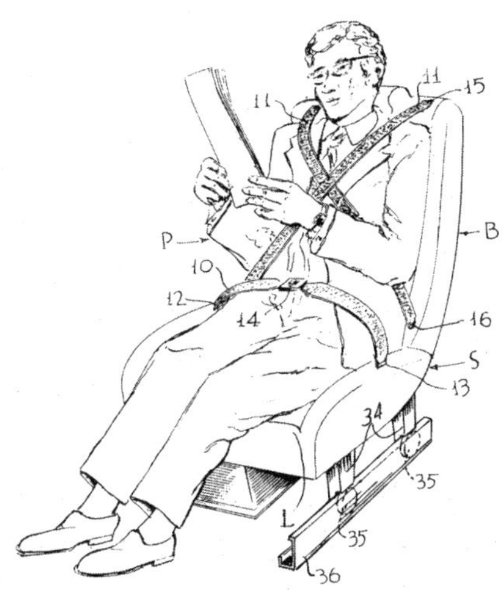

United States Patent [19]

Stedman

BEST AVAILABLE COPY

[11] **3,922,030**

[45] **Nov. 25, 1975**

[54] **PROTECTIVE SEAT**

[75] Inventor: **Robert N. Stedman,** Chillicothe, Ill.

[73] Assignee: **Caterpillar Tractor Co.,** Peoria, Ill.

[22] Filed: **June 24, 1974**

[21] Appl. No.: **482,510**

[52] **U.S. Cl.** .... **296/65 A;** 267/131; 297/216
[51] **Int. Cl.²** **B60N 1/02**
[58] **Field of Search** .... 297/216, 349, 353, 385, 297/285, 452; 248/399, 400; 267/131; 296/65 R, 65 A, 63, 35 B

[56] **References Cited**

UNITED STATES PATENTS

3,191,400	6/1965	Swenson	296/65 R
3,236,556	2/1966	Lathers	296/65 R
3,291,527	12/1966	Hall et al.	297/353
3,730,588	5/1973	Braun	297/452
3,792,896	2/1974	Eggert, Jr.	248/399

Primary Examiner—Robert J. Spar
Assistant Examiner—Lawrence J. Oresky
Attorney, Agent, or Firm—Phillips, Moore, Weissenberger, Lempio & Strabala

[57] **ABSTRACT**

The invention provides a vehicle safety seat particularly applicable to the relative exposed operator cabs of earthmoving vehicles. The seat includes angularly disposed resilient members for improved absorption of shock from both vertical and horizontal loading, and an improved adjustable shoulder support assembly having variably resilient wing portions. The safety seat further includes structure for permitting a desired degree of freedom of movement of the operator or other occupant, which means includes operator-controlled structure for rotating the seat about a central vertical axis.

16 Claims, 2 Drawing Figures

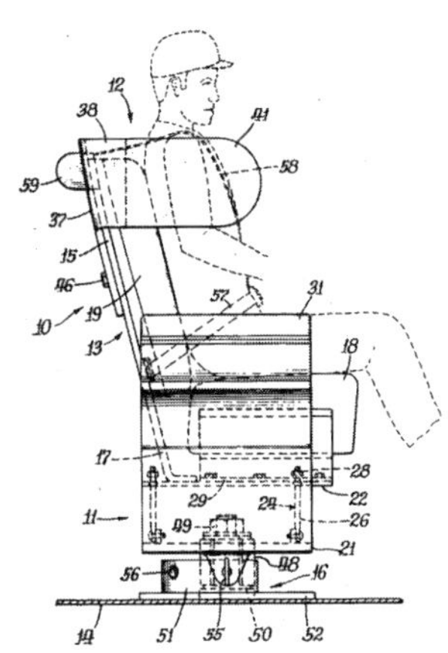

United States Patent [19]

Nonaka et al.

[11] **3,836,168**

[45] **Sept. 17, 1974**

[54] **PERSONAL SAFETY DEVICE FOR USE IN VEHICLES**

[75] Inventors: **Kohei Nonaka; Masaho Tanaka,** both of Tokyo; **Takeshi Maki,** Kawasaki; **Ikuo Harada,** Tokyo; **Masafumi Hamasaki,** Nobeoka, all of Japan

[73] Assignee: **Asaki Kasei Kogyo Kabushiki Kaisha,** Osaka, Germany

[22] Filed: **Nov. 23, 1971**

[21] Appl. No.: **201,322**

[30] **Foreign Application Priority Data**

Nov. 30, 1970 Japan 45-104913

[52] **U.S. Cl.** ..... **280/150 AB,** 180/82 C, 280/150 B, 297/216, 297/390
[51] **Int. Cl.** **B60r 21/08**
[58] **Field of Search** .... 280/150 AB, 150 B, 150 SB; 297/216, 384, 390; 180/82 C

[56] **References Cited**

UNITED STATES PATENTS

2,025,822	12/1935	Pryor	280/150 B
3,022,089	2/1962	Botar	280/150 SB
3,411,602	11/1968	Royce	280/150 SB
3,588,142	6/1971	Gorman	280/150 AB
3,623,768	11/1971	Capener	280/150 SB
3,633,936	1/1972	Huber	280/150 B
3,643,971	2/1972	Kushnick	280/150 AB
3,650,542	3/1972	Shimano et al.	280/150 B
3,664,682	5/1972	Wycech	280/150 AB
3,675,942	7/1972	Huber	280/150 AB
3,682,498	8/1972	Rutzki	280/150 SB
3,695,629	10/1972	Schlanger	280/150 B
3,715,130	2/1973	Harada et al.	280/150 AB

Primary Examiner—David Schonberg
Assistant Examiner—John P. Silverstrim
Attorney, Agent, or Firm—Stevens, Davis, Miller & Mosher

[57] **ABSTRACT**

The invention relates to a personal safety device for protecting an occupant of a vehicle in the event of an accident. The device includes confining means, e.g., a net or cloth for retaining the occupant in close proximity to his seat, the confining means being normally in a gathered or folded condition adjacent to the occupant's seat. Gas generating means, e.g., a cylinder of compressed gas or a combustible chemical, is provided for projecting the confining means in front of the occupant and the device embodies means for drawing and tightening the confining means around the occupant's body. Operation of the device takes place automatically on detection by sensing means in the vehicle of an acceleration exceeding a predetermined magnitude.

2 Claims, 11 Drawing Figures

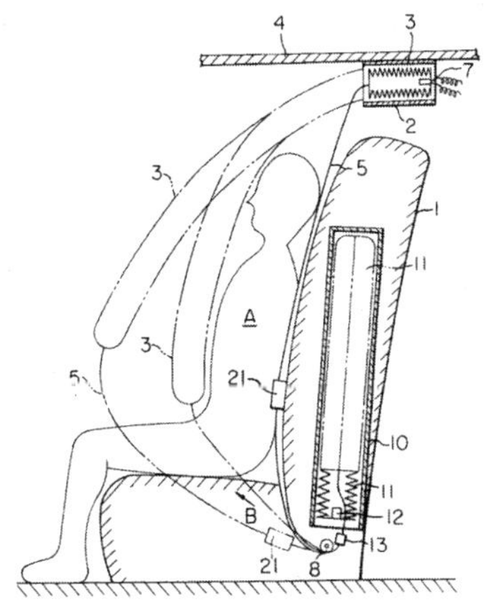

United States Patent [19]

Non

[11] **3,940,164**

[45] **Feb. 24, 1976**

[54] **PASSIVE RESTRAINT SAFETY BELT SYSTEM**

[76] Inventor: **Tse Quong Non,** 2315 S. Wentworth Ave., Chicago, Ill. 60616

[22] Filed: **Oct. 16, 1974**

[21] Appl. No.: **515,304**

[52] **U.S. Cl.** .... **280/150 SB;** 180/82 C; 297/216; 297/390
[51] **Int. Cl.²** **B60R 21/02**
[58] **Field of Search** .... 280/150 SB, 150 B; 180/82 C; 297/385, 388, 390, 216

[56] **References Cited**

UNITED STATES PATENTS

2,873,122	2/1959	Péras	297/385
3,022,089	2/1962	Botar	280/150 SB
3,545,789	12/1970	Graham	280/150 B
3,700,281	10/1972	Servadio	280/150 SB X
3,822,896	7/1974	Hallberg	280/150 SB X

Primary Examiner—Leo Friaglia
Assistant Examiner—Michael J. Forman
Attorney, Agent, or Firm—Fitch, Even, Tabin & Luedeka

[57] **ABSTRACT**

A passive restraint safety belt system for a vehicle comprises at least one belt movable from a hidden stowed position to a restraining position about the occupant of the vehicle at the time of an accident. Means are provided for sensing an impact or sudden deceleration or acceleration of the vehicle and are operable to shift the belt from the stowage position to the restraining position. Means are also provided for returning automatically the belt from the restraining position to the stowage position so that the belt may be readily reused. Preferably, the safety belt is substantially hidden from view and does not interfere with ingress or egress from the vehicle.

11 Claims, 5 Drawing Figures

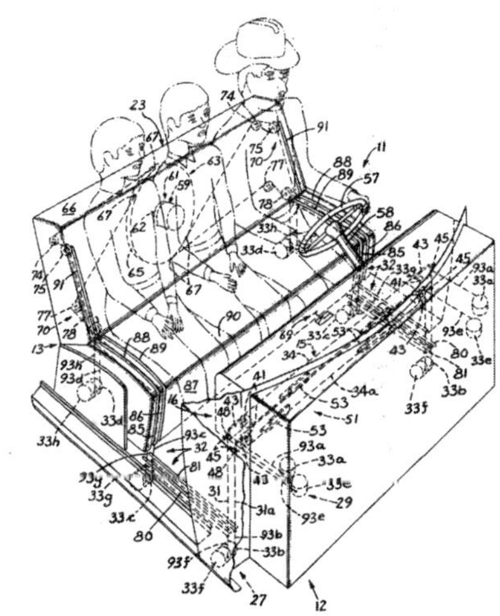

United States Patent [19]

Pulling

[11] **3,981,520**

[45] **Sept. 21, 1976**

[54] **PROTECTIVE ENCLOSURE FOR PASSENGERS OF TRANSPORT DEVICES**

[75] Inventor: **Nathaniel H. Pulling,** Framingham, Mass.

[73] Assignee: **Liberty Mutual Insurance Company,** Boston, Mass.

[22] Filed: **May 5, 1975**

[21] Appl. No.: **574,286**

[52] **U.S. Cl.** .... **280/748;** 180/103 A; 280/730; 296/65 R; 297/390; 297/216

[51] **Int. Cl.**[2] **B60R 21/10**

[58] **Field of Search** .... 280/150 B, 150 AB, 748; 180/82 R, 103 R, 103 A; 296/65 A, 65 R; 297/390, 216

[56] **References Cited**

UNITED STATES PATENTS

2,736,566	2/1956	Hartl	180/82 R
2,818,909	1/1958	Burnett	296/65 A X
3,129,017	4/1964	Graham	296/65 A
3,423,124	1/1969	Hewitt	296/65 A
3,591,232	7/1971	Simon	296/65 A X
3,610,679	10/1971	Amato	296/65 A
3,623,768	11/1971	Capener	280/150 AB X
3,832,000	8/1974	McDonnell	296/65 A
3,837,422	9/1974	Schlanger	280/150 AB X
3,853,298	12/1974	Libkie	296/65 A

FOREIGN PATENTS OR APPLICATIONS

537,159	4/1955	Belgium	296/65 A
1,005,756	4/1952	France	296/65 A

Primary Examiner—M. H. Wood, Jr.
Assistant Examiner—John P. Silverstrim
Attorney, Agent, or Firm—John E. Toupal

[57] **ABSTRACT**

A safety system for preventing injury to passengers of transport vehicles. In response to a collision each seat of the vehicle is automatically moved into a position wherein its occupant is protected within a structurally stable protective enclosure. Forming the enclosure are both movable portions secured to the seats and fixed portions secured to the vehicle. In a preferred embodiment of the invention, the movable seat is provided with a guidance system that normally facilitates selective adjustment of the seat into positions comfortable for its occupant but additionally produces automatic movement of the seat into the protective enclosure in response to a collision.

9 Claims, 10 Drawing Figures

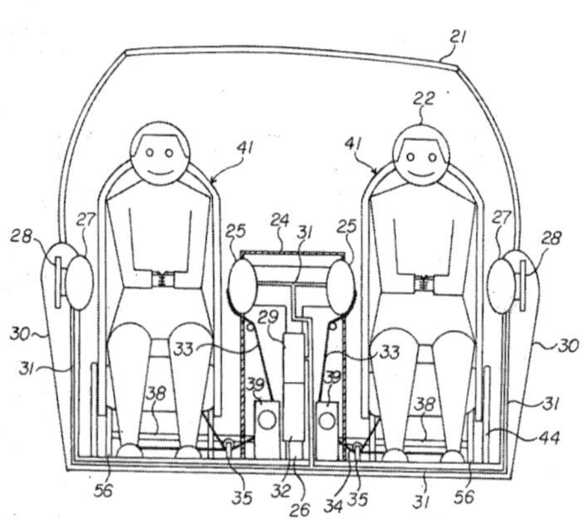

United States Patent [19]

Mason

[11] **4,402,548**

[45] **Sep. 6, 1983**

[54] **SAFETY SEATS FOR VEHICLES**

[75] Inventor: **Stuart V. Mason,** Biggleswade, England

[73] Assignee: **Britax-Excelsior Limited,** England

[21] Appl. No.: **235,649**

[22] Filed: **Feb. 18, 1981**

[30] **Foreign Application Priority Data**

Feb. 27, 1980 [GB] United Kingdom 8006516

[51] **Int. Cl.**[3] **A47C 31/00;** B60R 21/00

[52] **U.S. Cl.** **297/464;** 297/216; 297/250; 297/484

[58] **Field of Search** 297/467, 464, 484, 254, 297/255, 256, 250, 216

[56] **References Cited**

U.S. PATENT DOCUMENTS

3,596,986	8/1971	Ragsdale	297/467 X
3,934,934	1/1976	Farrell, Jr. et al.	297/467
4,047,755	9/1977	McDonald et al.	297/464 X
4,186,961	2/1980	Farrell, Jr. et al.	297/467 X
4,190,287	2/1980	Lemisch et al.	297/467
4,234,228	11/1980	Flamm	297/464

Primary Examiner—James T. McCall
Attorney, Agent, or Firm—Hayes, Davis & Soloway

[57] **ABSTRACT**

A safety seat for a vehicle of the type comprising a shell **10** forming a seat portion **12** a back portion **14** and two side portions **16** and **18**, which are covered by a fabric seat cover **50**, is made adjustable in width by providing a plurality of filler strips **54, 56** which fit between the cover **50** and the side portions **16** and **18** of the shell so as to adjust the interior width of the seat. Lateral head restraint pads **58** and **60** may also be provided. The seat is particularly suitable for use by a growing physically handicapped child who requires lateral support when seated in a motor vehicle at an age when his contemporaries are capable of sitting in a vehicle without such support.

4 Claims, 2 Drawing Figures

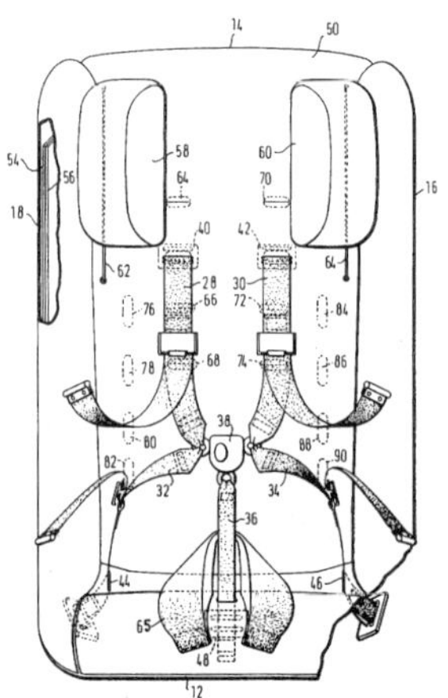

United States Patent [19]

Putsch et al.

[11] **Patent Number: 4,679,854**

[45] **Date of Patent: Jul. 14, 1987**

[54] **VEHICLE SEAT**

[75] Inventors: **Peter-Ulrich Putsch,** Rockenhausen; **Heinz P. Cremer,** Kaiserslautern, both of Fed. Rep. of Germany

[73] Assignee: **Keiper Recaro GmbH & Co.,** Fed. Rep. of Germany

[21] Appl. No.: **845,629**

[22] Filed: **Mar. 28, 1986**

[30] **Foreign Application Priority Data**

Mar. 28, 1985 [DE] Fed. Rep. of Germany 3511216

[51] **Int. Cl.**[4] **B60R 21/00**

[52] **U.S. Cl.** **297/486;** 297/216; 297/464

[58] **Field of Search** .... 297/464, 486, 468, 296, 297/473, 216, 445, 391, 411; 296/65 A; 248/429

[56] **References Cited**

U.S. PATENT DOCUMENTS

2,136,852	11/1938	Knauth	297/486
2,873,122	2/1959	Peras	297/486 X
3,466,091	9/1969	De Grusso	297/486
3,627,379	12/1971	Faust	297/284
3,713,694	1/1973	Miller	397/486
3,901,550	8/1975	Hamy	297/486
3,922,030	11/1975	Stedman	297/468
4,145,083	3/1979	Urban	297/486

Primary Examiner—James T. McCall
Attorney, Agent, or Firm—Wigman & Cohen

[57] **ABSTRACT**

A vehicle seat having an upholstery support that supports the upholstery of the back rest has at least one arm provided on the upholstery support which extends over the shoulder of a seat user and restrains the user in the upward and outward directions during vehicle collision. A side restraint member is connected with the arm and configured so as to protect the user in a lateral direction without interfering with the freedom of movement of the user.

12 Claims, 4 Drawing Figures

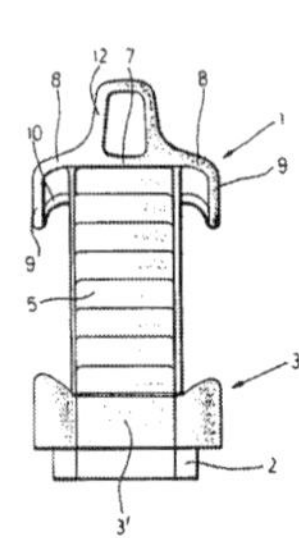

US005423598A

United States Patent [19]

Lane, Jr. et al.

[11] **Patent Number: 5,423,598**

[45] **Date of Patent: Jun. 13, 1995**

[54] **SAFETY APPARATUS**

[75] Inventors: **Wendell C. Lane, Jr.,** Romeo; **Robert M. Varga,** Rochester Hills, both of Mich.

[73] Assignee: **TRW Vehicle Safety Systems Inc.,** Lyndhurst, Ohio

[21] Appl. No.: **75,686**

[22] Filed: **Jun. 11, 1993**

[51] **Int. Cl.**[6] **A47C 31/00**

[52] **U.S. Cl.** **297/479;** 297/474; 297/480; 280/806

[58] **Field of Search** 297/477, 468, 469, 474, 297/479, 480, 486; 280/806

[56] **References Cited**

U.S. PATENT DOCUMENTS

3,439,932	4/1969	Lewis et al. .	
3,442,529	5/1969	Lewis et al.	297/479
3,535,001	10/1970	Lewis et al.	297/479 X
3,547,490	12/1970	Stoffel	297/479
3,667,806	6/1972	Sprecher	297/479
3,695,697	10/1972	Stoffel	297/479
4,159,848	7/1979	Manz et al. .	
5,029,896	7/1991	Ernst	297/479 X

Primary Examiner—Laurie K. Cranmer
Attorney, Agent, or Firm—Tarolli, Sundheim & Covell

[57] **ABSTRACT**

An improved safety apparatus **(10)** is used in association with a belt **(16)** which restrains movement of an occupant of a vehicle. The belt **(16)** is taken up by and paid out from a retractor assembly **(22)**. Upon the occurrence of vehicle deceleration greater than a predetermined deceleration, a solenoid assembly **(132)** operates a clamp assembly **(28)** to grip the belt **(16)** and prevent paying out of the belt. In one embodiment of the invention, the retractor assembly **(22)** includes a sensor assembly **(54)** which operates a retainer assembly **(56)** to prevent withdrawal of the belt from the retractor assembly when the deceleration of the vehicle is in excess of a predetermined deceleration. The sensor assembly **(54)** is mounted on a movable carriage **(32)** in the retractor assembly **(22)**. Although the clamp assembly **(28)** and the retractor assembly **(22)** may be used in many different locations in a vehicle, they are advantageously located on the back **(24)** of a vehicle seat **(12)**.

23 Claims, 4 Drawing Sheets

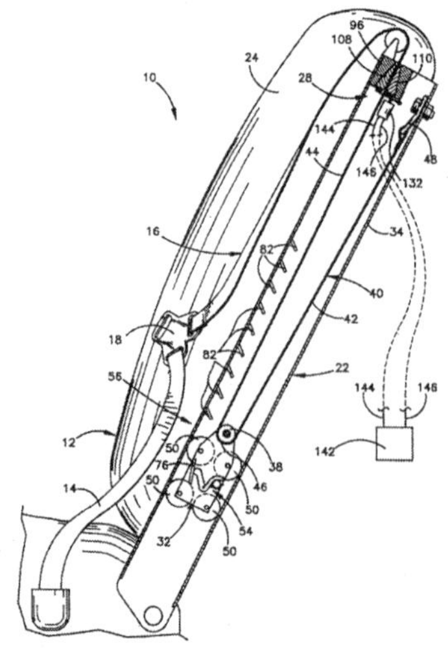

United States Patent [19]

Pokhis

[11] **Patent Number: 5,806,891**

[45] **Date of Patent: Sep. 15, 1998**

[54] **SAFETY BELTS FOR MOTOR VEHICLES**

[76] Inventor: **Naum Pokhis**, 1132 S. Doheny Dr., #303, Los Angles, Calif. 90035

[21] Appl. No.: **862,411**

[22] Filed: **May 23, 1997**

[51] **Int. Cl.**[6] **B60R 22/24**

[52] **U.S. Cl.** **280/801.1**; 297/216.13; 297/483

[58] **Field of Search** 280/801.1, 808, 280/801.2, 804; 297/468, 483, 484, 216.1, 216.13

[56] **References Cited**

U.S. PATENT DOCUMENTS

2,606,727	8/1952	De Haven	297/216.13
2,740,642	4/1956	Atwood	297/216.13
2,873,122	2/1959	Peras	297/216.13
3,762,505	10/1973	Morse	297/216.1
3,995,885	12/1976	Plesniarski	297/483
4,474,347	10/1984	Hazelsky	297/216.1
4,974,876	12/1990	Svensson et al.	280/801.1
5,123,673	6/1992	Tame	297/468

Primary Examiner—Paul N. Dickson

Attorney, Agent, or Firm—Ilya Zborovsky

[57] **ABSTRACT**

A safety system operable during vehicle accidents comprises a seat located inside a vehicle and having a back, a safety belt assembly operative for holding a user to the seat, and means for connecting the back of the seat to a vehicle roof.

5 Claims, 1 Drawing Sheet

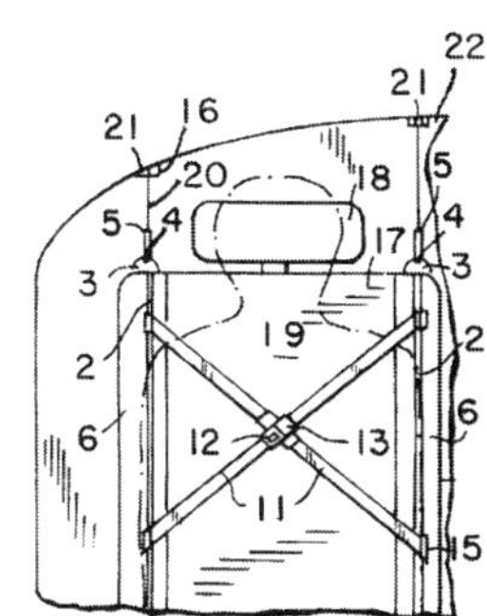

United States Patent [19]

Elqadah et al.

[11] **Patent Number: 6,033,017**

[45] **Date of Patent: *Mar. 7, 2000**

[54] **VEHICLE OCCUPANT PROTECTION APPARATUS**

[75] Inventors: **Wael S. Elqadah**; **Xingyuan Sun**, both of Gilbert; **Jess A. Cuevas**, Scottsdale, all of Ariz.

[73] Assignee: **TRW Inc.**, Lyndhurst, Ohio

[*] Notice: This patent issued on a continued prosecution application filed under 37 CFR 1.53(d), and is subject to the twenty year patent term provisions of 35 U.S.C. 154(a)(2).

[21] Appl. No.: **09/057,815**

[22] Filed: **Apr. 9, 1998**

[51] **Int. Cl.**[7] **B60N 2/42**

[52] **U.S. Cl.** **297/216.1**; 297/216.13; 297/408; 297/486

[58] **Field of Search** 297/216.1, 216.12, 297/216.13, 216.15, 474, 484, 486, 391, 406, 408, 409

[56] **References Cited**

U.S. PATENT DOCUMENTS

2,873,122	2/1959	Peras .	
3,397,911	8/1968	Brosius, Sr. .	
3,420,572	1/1969	Bisland .	
3,623,768	11/1971	Capener et al. .	
3,713,694	1/1973	Miller .	
3,832,003	8/1974	Horvat	297/216.13
3,899,042	8/1975	Bonar .	
3,901,550	8/1975	Hamy .	
5,052,754	10/1991	Chinomi .	
5,378,043	1/1995	Viano et al. .	
5,484,189	1/1996	Patterson .	
5,669,661	9/1997	Pajon .	
5,836,648	11/1998	Karschin et al.	297/216.13

Primary Examiner—Laurie K. Cranmer

Attorney, Agent, or Firm—Tarolli, Sundheim, Covell, Tummino & Szabo L.L.P.

[57] **ABSTRACT**

An apparatus includes a vehicle seat **(10)** and an actuating assembly **(80)**. The seat **(10)** has a wing **(34)** which is movable from a retracted position to a deployed position. When the wing **(34)** is in the deployed position, it projects adjacent to an occupant the seat **(10)** to restrain movement of the occupant off the seat **(10)**. The actuating assembly **(80)** moves the wing **(34)** from the retracted position to the deployed position upon the occurrence of a vehicle crash.

19 Claims, 4 Drawing Sheets

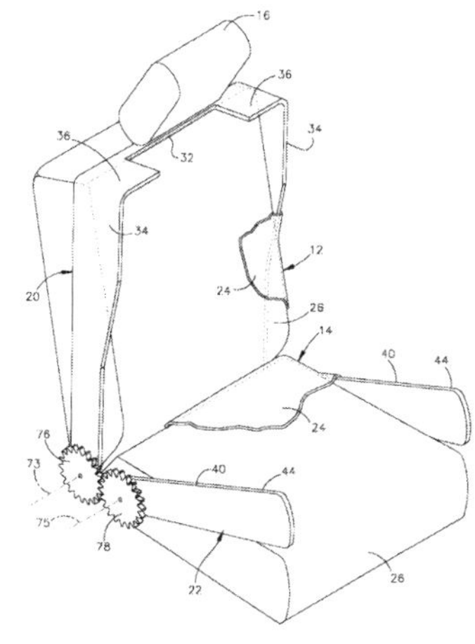

United States Patent [19]

US005806923A

Tschäschke et al.

[11] **Patent Number: 5,806,923**

[45] **Date of Patent: Sep. 15, 1998**

[54] **IMPACT PROTECTION DEVICE IN A MOTOR VEHICLE HAVING A VEHICLE SEAT**

[75] Inventors: **Ulrich Tschäschke**, Ehningen; **Frank Zerrweck**, Stuttgart, both of Germany

[73] Assignee: **Daimler-Benz Aktiengesellschaft**, Stuttgart, Germany

[21] Appl. No.: **891,876**

[22] Filed: **Jul. 9, 1997**

[30] **Foreign Application Priority Data**

Jul. 12, 1996 [DE] Germany 196 28 108.3

[51] **Int. Cl.**[6] **B60N 2/42**

[52] **U.S. Cl.** **297/216.13**; 297/464

[58] **Field of Search** .... 280/730.1, 730.2, 280/748, 749; 297/216.1, 216.13, DIG. 3, 452.41, 468, 487, 464

[56] **References Cited**

U.S. PATENT DOCUMENTS

2,025,822	12/1935	Pryor .
2,606,727	8/1952	De Haven .
3,096,957	7/1963	Peterson et al. .
3,957,304	5/1976	Koutsky et al. .
4,592,523	6/1986	Herndon .
5,295,729	3/1994	Viano .
5,366,268	11/1994	Miller et al. .
5,468,045	11/1995	Weber .
5,474,329	12/1995	Wade et al. .
5,651,582	7/1997	Nakano .

FOREIGN PATENT DOCUMENTS

0565501A1	10/1993	Germany .
4307421A1	10/1993	Germany .
0688702A2	12/1995	Germany .
4436139C1	3/1996	Germany .

Primary Examiner—Milton Nelson, Jr.

Attorney, Agent, or Firm—Evenson, McKeown, Edwards & Lenahan, P.L.L.C.

[57] **ABSTRACT**

An impact protection device in a motor vehicle having a vehicle seat has a taut tightening strap which reaches from the upper area of the backrest into the forward area of the seat cushion, so that the backrest is prevented from folding backwards in a vehicle crash. A protective element is positioned laterally on the vehicle seat, so that the impact of a sitting person moving toward the side during a vehicle crash can be damped.

27 Claims, 2 Drawing Sheets

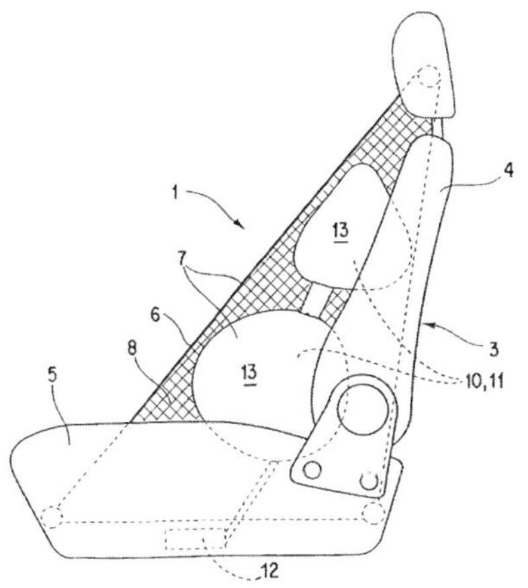

United States Patent [19]

Cantor et al.

[11] **Patent Number: 6,155,601**

[45] **Date of Patent: Dec. 5, 2000**

[54] **SEAT-MOUNTED OCCUPANT CRASH PROTECTION SYSTEM**

[75] Inventors: **Alan Everett Cantor**, Ivyland; **Louis Anthony D'Aulerio**, Horsham, both of Pa.; **Leon P. Domzalski**, Mount Laurel, N.J.; **Bruce Peter Holmberg**, Rockville, Md.; **Michael Leonid Markushewski**, Huntingdon Valley, Pa.; **Larry Andrew Sicher**, Quakertown, Pa.; **Gary Robert Whitman**, Jamison, Pa.; **John Richard Yannaccone**, Perkasie, Pa.

[73] Assignee: **ARCCA Incorporated**, Penns Park, Pa.

[21] Appl. No.: **09/303,303**

[22] Filed: **Apr. 30, 1999**

[51] **Int. Cl.**[7] **B60R 22/00**

[52] **U.S. Cl.** **280/806**

[58] **Field of Search** 280/806

[56] **References Cited**

U.S. PATENT DOCUMENTS

5,423,598	6/1995	Lane, Jr. et al.	280/806
5,492,368	2/1996	Pywell et al.	280/806
5,553,924	9/1996	Cantor et al.	297/452.27
5,571,253	11/1996	Blackburn et al.	280/806
5,634,664	6/1997	Seki et al.	280/806
5,676,398	10/1997	Nurtsch	280/806

Primary Examiner—Kenneth R. Rice

Attorney, Agent, or Firm—Gifford, Krass, Groh, Sprinkle, Anderson & Citkowski, P.C.

[57] **ABSTRACT**

A vehicle seat assembly for protecting an occupant of a vehicle from injury includes a seat assembly having a seat back and a seat bottom, a pair of side bolsters disposed on each side of the seat back, a combined seat/lap belt and shoulder belt restraint which is affixed directly to the seat assembly, and a secondary shoulder restraint affixed to the seat assembly. The seat bottom includes a front portion and a ramp upwardly sloped towards the front portion of the seat bottom. A cushion constructed of a rate sensitive compression material having a compressive response to a slow application of force and a rigid response to a rapid application of force is disposed adjacent to the seat ramp. Pretensioning devices are operatively connected to both the combined seat/lap belt and shoulder belt restraint and to the supplemental shoulder belt restraint and are adapted to activate in response to a signal from a crash sensor to remove slack from the combined seat/lap belt and shoulder belt restraint and/or the supplemental shoulder belt restraint.

14 Claims, 1 Drawing Sheet

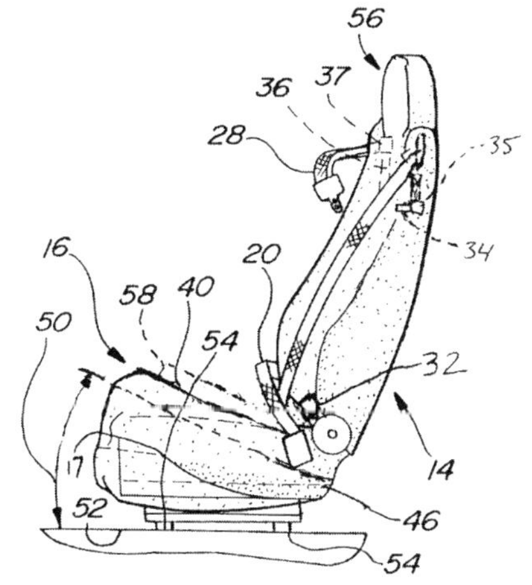

EXAMPLES OF MOVING, ENERGY ABSORBING AND INFLATABLE SYSTEMS FOR SEATS, INCLUDING REAR OCCUPANT IMPACT SAFETY

Aug. 18, 1959 W. J. BLAKE **2,900,036**

SAFETY SEAT FOR AUTOMOBILES

Filed June 25, 1957

Fig.1

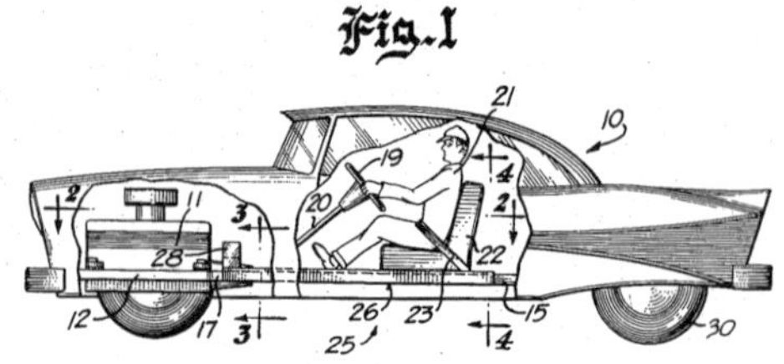

Fig.2

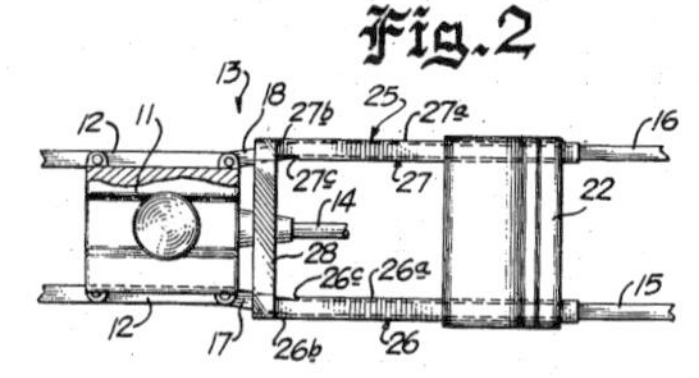

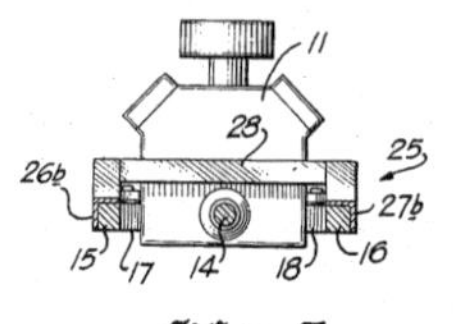

Fig.3

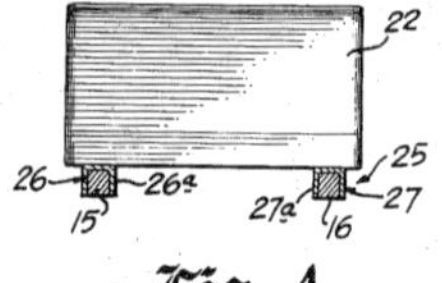

Fig.4

INVENTOR
WILLIAM J. BLAKE
by
Mason, Kolehmainen, Rathburn and Wyss
ATTORNEYS.

United States Patent [19] [11] **3,832,003**
Horvat [45] **Aug. 27, 1974**

[54] **ENERGY ABSORBING SEAT ASSEMBLY**

[75] Inventor: **Rudolph M. Horvat,** Allen Park, Mich.

[73] Assignee: **The Ford Motor Company,** Dearborn, Mich.

[22] Filed: **Aug. 6, 1973**

[21] Appl. No.: **383,945**

[52] **U.S. Cl.** .... **297/216,** 297/379, 297/386, 296/65 A
[51] **Int. Cl.** **B60r 21/10**
[58] **Field of Search** .... 297/454, 216, 378, 379, 297/380, 381, 386; 296/65 A

[56] **References Cited**

UNITED STATES PATENTS

2,401,748	6/1946	Dillon	297/216
2,636,552	4/1953	Long	297/216
3,321,242	5/1967	Ferrara	297/379 X
3,501,200	3/1970	Ohta	297/454
3,761,127	9/1973	Giese et al.	297/386

FOREIGN PATENTS OR APPLICATIONS

1,630,891	3/1971	Germany	296/65 A

Primary Examiner—James T. McCall
Attorney, Agent, or Firm—Keith L. Zerschling; John J. Roethel

[57] **ABSTRACT**

A vehicle seat assembly includes a generally horizontal seat structure and an upstanding backrest structure pivotally mounted on the seat structure for tiltable movement forwardly over the latter. An energy absorbing device mounted on one of the structures is releasably coupled through a latch device means to the other of the structures. The energy absorbing device is operative, when coupled through the latch device means to the other of the structures, to control forward tilting movement of the backrest structure in response to high external forces acting on the latter.

4 Claims, 2 Drawing Figures

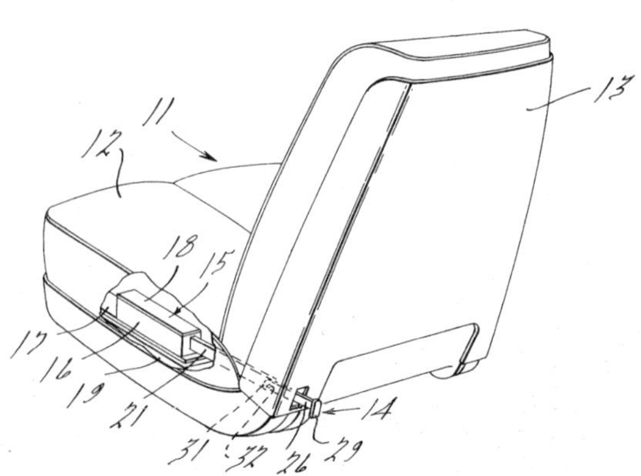

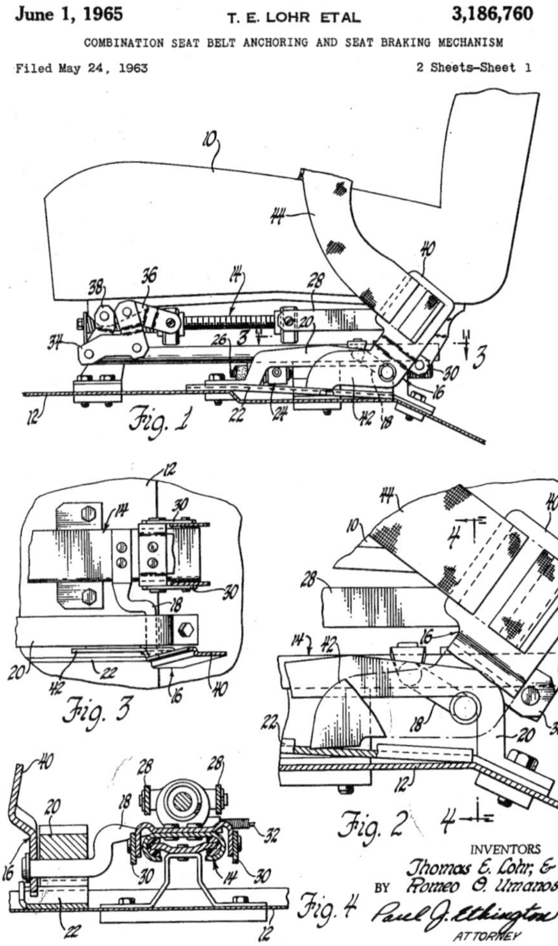

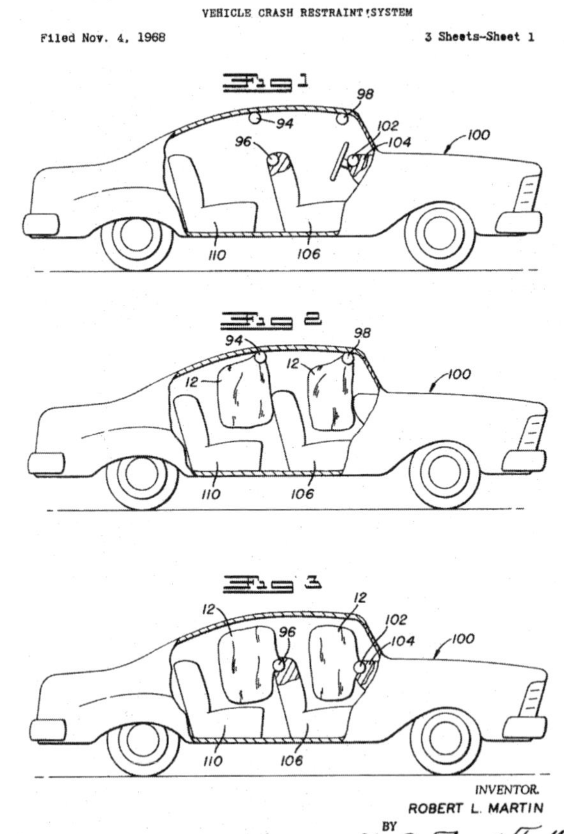

United States Patent [19]

Sakurai et al.

[11] **3,877,749**

[45] **Apr. 15, 1975**

[54] **REAR SEAT PASSENGER-PROTECTING SEAT BACK STRUCTURE FOR AUTOMOTIVE VEHICLE**

[75] Inventors: **Katsuo Sakurai; Takeshi Yamawaki,** both of Toyota, Japan

[73] Assignee: **Toyota Jidosha Kogyo Kabushiki Kaisha,** Toyota-shi, Aichi-ken, Japan

[22] Filed: **Apr. 23, 1973**

[21] Appl. No.: **353,702**

[30] **Foreign Application Priority Data**

July 20, 1972 Japan 47-72697

[52] **U.S. Cl.** **297/216;** 297/390
[51] **Int. Cl.** **B60r 21/02**
[58] **Field of Search** 297/216, 390, 384, 452, 297/460

[56] **References Cited**

UNITED STATES PATENTS

2,098,456	11/1937	Leader	297/452
2,833,339	5/1958	Liljengren	297/216
3,464,751	9/1969	Barecki	297/216
3,537,751	11/1970	Inoue et al.	297/452
3,669,498	6/1972	Meyers	297/452
3,695,707	10/1972	Barecki	297/460
3,802,737	4/1974	Mertens	297/216

FOREIGN PATENTS OR APPLICATIONS

555,246	4/1958	Canada	297/216

Primary Examiner—Francis K. Zugel
Attorney, Agent, or Firm—Stevens, Davis, Miller & Mosher

[57] **ABSTRACT**

A seat back structure in an automobile or the like for protecting a passenger on a rear seat, comprising a main frame board of a front seat back, an upper buffer body secured integrally to the upper part of the main frame board and projecting towards the rear seat, and a lower buffer board secured integrally to the lower part of the main frame board, said lower part having an opening. When, in case of a collision of the automobile or the like, the knees of the passenger violently collides against that part of the rear of the front seat back which corresponds to the lower buffer board, said lower buffer board undergoes a plastic deformation towards the opening, thereby to absorb the impact load of the knees. When the breast and head of the passenger violently collide against that part of the front seat back which corresponds to the upper buffer body, said upper buffer body undergoes a plastic deformation, and subsequently the main frame board undergoes a frontward plastic deformation, thereby to absorb the impact load of the breast and head.

8 Claims, 8 Drawing Figures

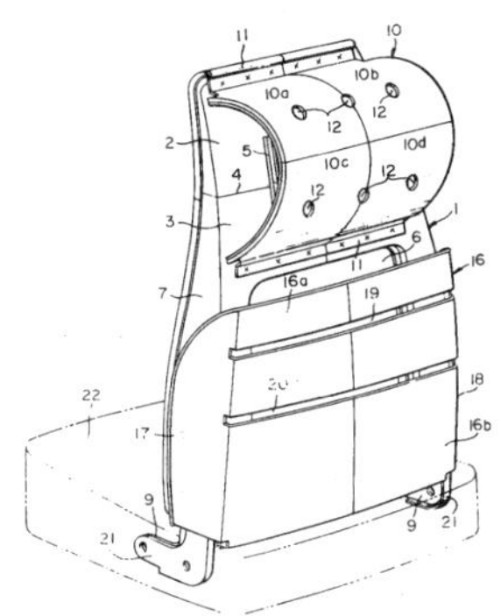

United States Patent [19]

Glance

[11] **3,761,125**

[45] **Sept. 25, 1973**

[54] **ENERGY ABSORBING SEAT BACK**

[75] Inventor: **Patrick M. Glance,** Plymouth, Mich.

[73] Assignee: **Ford Motor Company,** Dearborn, Mich.

[22] Filed: **May 12, 1972**

[21] Appl. No.: **252,805**

[52] **U.S. Cl.** **297/216,** 297/390
[51] **Int. Cl.** **B60r 21/02**
[58] **Field of Search** 297/216, 396, 284, 297/379, 373, 374, 375, 408

[56] **References Cited**

UNITED STATES PATENTS

2,636,552	4/1953	Long	297/216
2,813,577	11/1957	Weekly	297/396
3,544,164	12/1970	Ohta	297/216

Primary Examiner—James T. McCall
Assistant Examiner—Darrell Marquette
Attorney—Keith L. Zerschling et al.

[57] **ABSTRACT**

A vehicle seat in which the side frame members are articulated intermediate their ends and are pivotally mounted on the seat cushion structure. The pivot joints are friction joints that are preloaded to predetermined torsional resistances. The area of the backrest above the upper or articulating pivots provides an impact area adapted under vehicle collision conditions to receive an impact load from a rear seat occupant's head. The area of the backrest between the backrest to cushion pivots and the articulating pivots functions similarly with respect to back seat occupant knee impacts. The friction joints permit rotation of the upper area of the backrest or rotation of the entire backrest relative to the seat cushion, the rotation occurring at a controlled rate as predetermined by the preloading on the pivot joints, thereby controlling the deceleration of the occupant impacting thereagainst.

5 Claims, 3 Drawing Figures

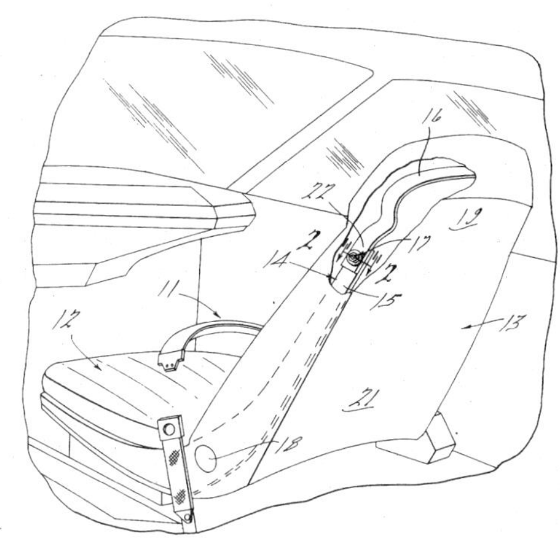

Nov. 3, 1970 MASAHIKO INOUE ET AL 3,537,751

SEAT CONSTRUCTION

Filed Aug. 2, 1968

2 Sheets-Sheet 1

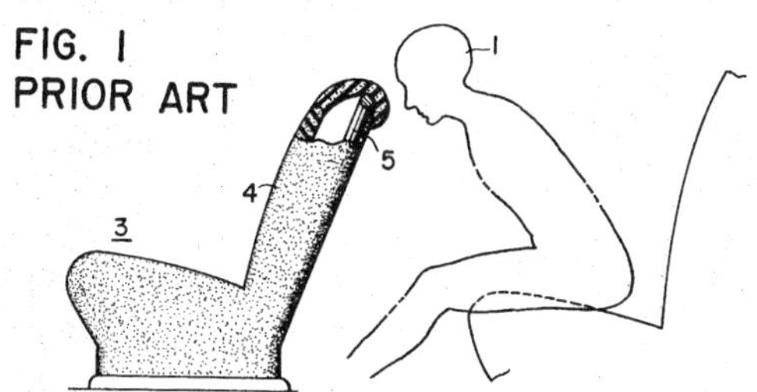

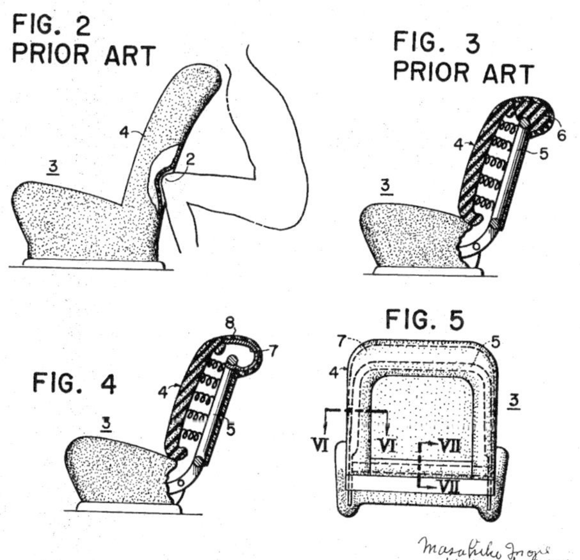

CHAPTER 10:

YIELDING AND PERIMETER FRAME SEAT PATENTS

Chapter 10

Yielding and Perimeter Frame Seat Patents

David C. Viano
ProBiomechanics LLC

ABSTRACT

This Chapter reviews the patent history of perimeter frame and yielding seatbacks and includes examples of concepts leading up to the 1996 perimeter frame and pelvic catcher seat patent. It also covers seat concepts for frame structures with integrated shoulder belts and lumbar supports. The first page from 31 USPTO patents is included to show seat concepts and technologies considered for automotive applications. Seatback frame and suspension concepts have their roots earlier than the 1920s with refinements for furniture leading to automotive seats through the 1950s. Bucket seat concepts were showcased in the 1960s with integrated shoulder belts through the 1970s. The use of resilient and movable lumbar supports was also demonstrated in seatbacks leading to the perimeter frame and pelvic catcher seat patent. This included a low rear profile to allow the occupant to displace rearward without ramping, a perimeter frame to give yielding distance with a strong seat frame and an open seatback for low initial stiffness under occupant load.

METHODOLOGY

US patents were located by searching primary, secondary and tertiary references for perimeter frame, energy absorbing and yielding seat related safety topics using the US Patent and Trademark Office website: http://patft.uspto.gov/netahtml/search. Efforts were made to locate the earliest and most interesting examples of patents on particular technologies, but there is every reason to assume that many other concepts and technologies related to seat designs for crash safety were not located.

Patents prior to 1975 are accessible only by a patent number or classification, which made the search depend on secondary and tertiary references in patents that were located. Many of these patents included cited references to earlier patents, which were investigated. However, the earliest US patents did not include cited references to prior art that were reviewed by the patent examiner. This complicated the search for patents before the 1950s. Any patent of significance to the history of seats and crash safety that is not included was an inadvertent oversight. The items located and reported should be considered examples of the history of technology and patents on seat safety. They do show a long-standing history of concepts for occupant safety by seat design and integration of other safety features.

RESULTS

Early automotive seats included a frame structure and resilient suspension of various designs. Many of the concepts emphasized a strong seat frame with flexible cross-members for comfort. By the 1960s, this included a steel frame structure and flexible metal spring supporting a mat with covering trim. While a variety of other concepts were patented, this basic design continues today as the typical seat frame.

In the 1970s, integrated shoulder belts appeared for increased safety with growing interest in seat belt restraints. This required the addition of cross-frame structures to support the greater crash loads with the shoulder belt attached to the seatback. A number of alternative designs were considered to connect the upper seatback to the vehicle structures. This would transfer shoulder belt load to the vehicle pillars and rails.

By the 1990s, additional frame structures appeared to increase lateral support during vehicle maneuvering and side impacts. These approaches have continued through the 2000s.

Sept. 11, 1928. S. J. LEACH ET AL **1,684,062**

BACK FOR AUTOMOBILE SEATS AND THE LIKE

Filed Dec. 22, 1924

INVENTORS
Sagilo J. Leach and
BY Alfred C. Eberle.
Fay, Oberlin & Fay
ATTORNEYS.

Nov. 21, 1950 J. TURNER **2,530,924**

RECLINING CHAIR

Filed Feb. 27, 1945

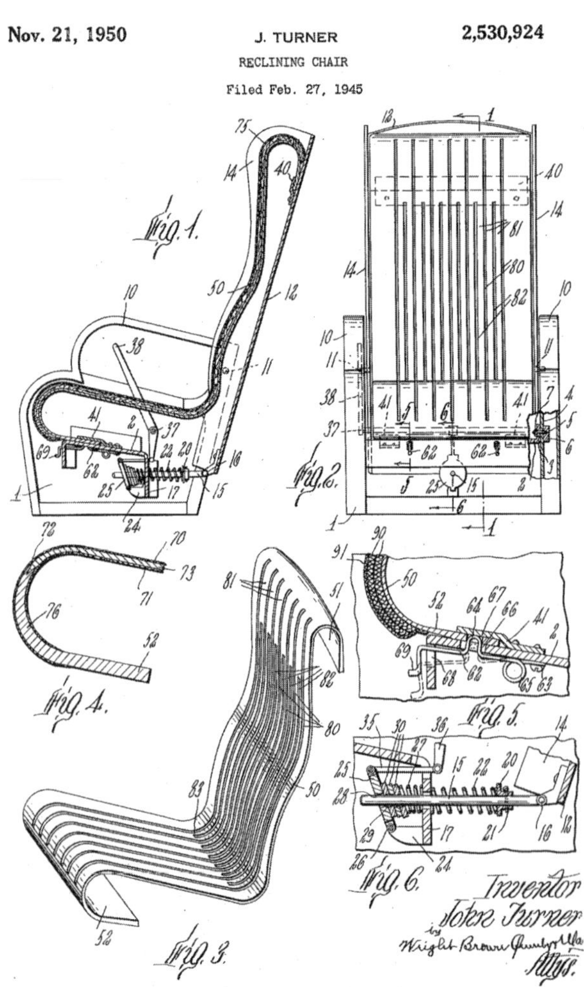

July 15, 1958 A. E. A. BARVAEUS **2,843,195**

SELF-ADJUSTING BACK SUPPORT

Filed Jan. 25, 1956

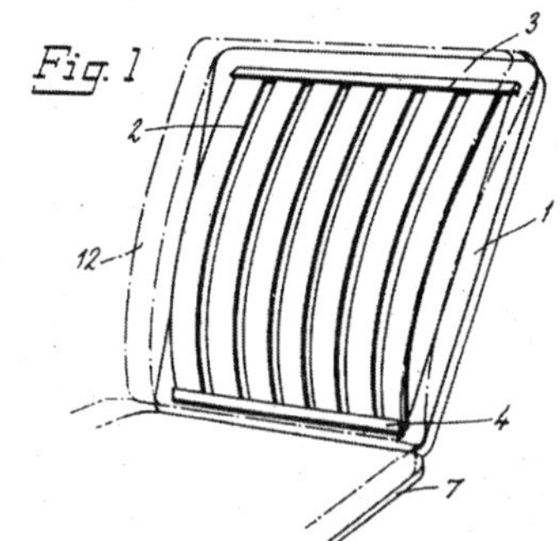

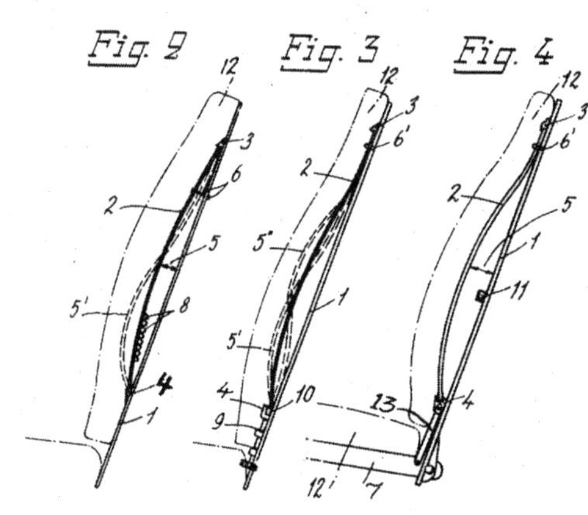

Inventor
Alvar Barvaeus

United States Patent [11] 3,545,808

[72] Inventors **Ernst Gescheidle**
Gross-Gerau, Germany;
[21] Appl. No. **763,188**
[22] Filed **Sept. 27, 1968**
[45] Patented **Dec. 8, 1970**
[73] Assignee **General Motors Corporation**
Detroit, Mich.
a corporation of Delaware
[32] Priority **Nov. 8, 1967**
[33] **Germany**
[31] **No. P1,630,891.1**

[54] **MOTOR VEHICLE SEAT**
7 Claims, 5 Drawing Figs.

[52] **U.S. Cl.** **297/216,** 297/454, 297/460
[51] **Int. Cl.** **A47c 7/02,** B60v 21/00; B60n 1/02
[50] **Field of Search** 297/216, 445, 451, 452, 455, 459, 460

[56] **References Cited**

UNITED STATES PATENTS

2,959,207	11/1960	Brewster	297/216
3,206,250	9/1965	Komenda	297/456
3,437,367	4/1969	Blank	297/216
3,446,469	5/1969	Whitten	297/216X
3,460,791	9/1969	Judd	297/216X

Primary Examiner—Casmir A. Nunberg
Attorneys— E. J. Biskup and J. L. Carpenter

ABSTRACT: A vehicle seat having a backrest provided with laterally spaced uprights which extend substantially up to the top edge of the backrest and possess a resistance moment of a certain magnitude matching the pressure exerted by the back of the seat occupant upon the backrest. The seat is characterized in that the uprights are provided with energy absorbing means which, during impact on the uprights in the direction of their longitudinal axis, cause them to become shortened.

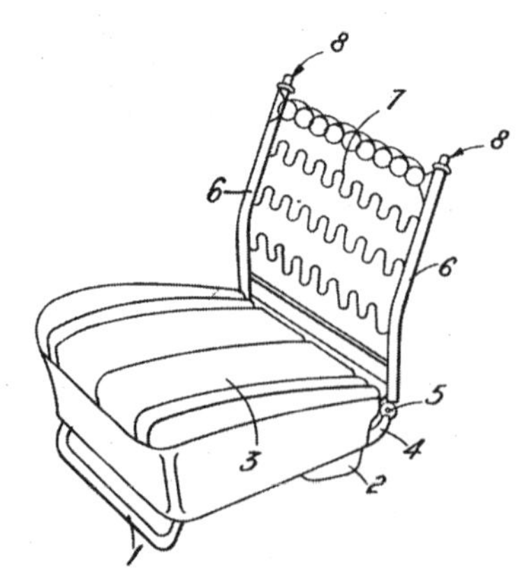

March 31, 1964 G. L. STINE **3,127,220**

SEAT CUSHION AND SPRING CONSTRUCTION

Filed Aug. 9, 1961

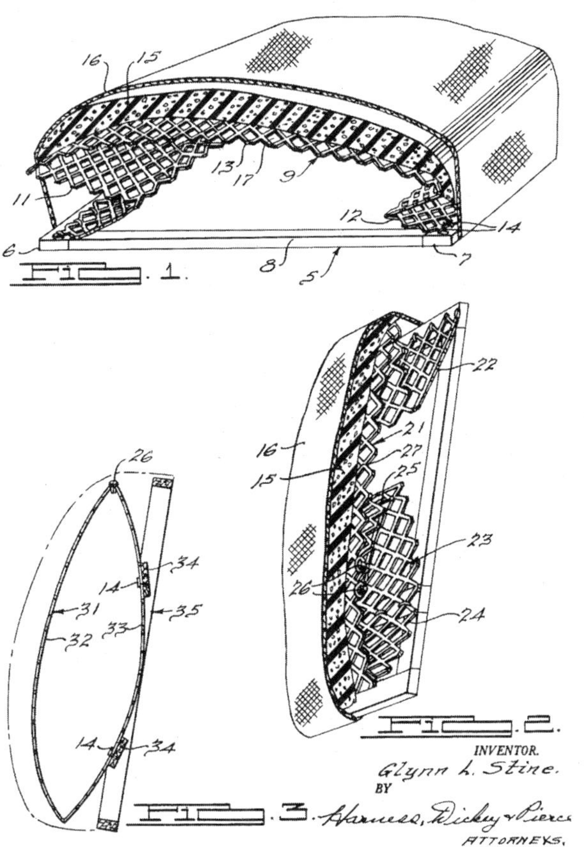

United States Patent [11] **3,586,376**

[72] Inventor **Noel Le Mire**
Billancourt, France
[21] Appl. No. **813,103**
[22] Filed **Apr. 3, 1969**
[45] Patented **June 22, 1971**
[73] Assignees **Regie Nationale Des Usines Renault**
Billancourt, France;
Automobiles Peugeot
Paris (Seine), France
[32] Priority **Apr. 4, 1968**
[33] **France**
[31] **147018**

[54] **SAFETY SEAT OF VEHICLES**
1 Claim, 3 Drawing Figs.
[52] **U.S. Cl.** **297/452,** 297/216
[51] **Int. Cl.** **A47c 7/02,** B60r 21/00, B60n 1/02
[50] **Field of Search** .... 297/216, 302, 452, 454—460
[56] **References Cited**
UNITED STATES PATENTS

3,437,367	4/1969	Blank	297/216 X
3,501,200	3/1970	Ohta	297/216 X

Primary Examiner—Casmir A. Nunberg
Attorney—Stevens, Davis, Miller & Mosher

ABSTRACT: A vehicle seat having an impact-absorbing back formed by upright members interconnected by a transverse bar. The upright members are formed in sections assembled on sleeves and held by retaining means in a normal position. The back will deform upon application of a force thereto by the sections sliding into the sleeves when the force overcomes the retaining means.

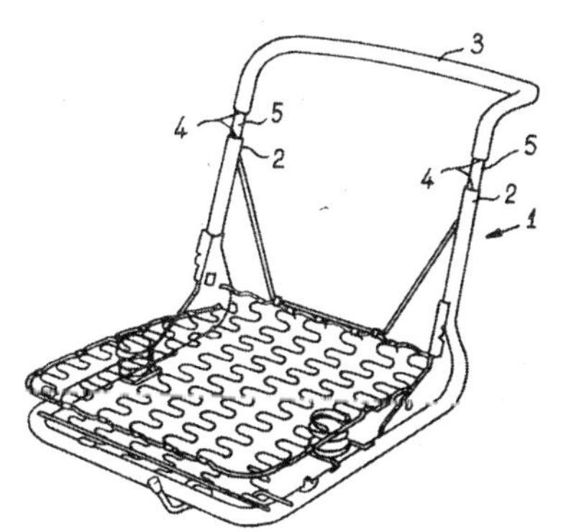

United States Patent [11] **3,627,379**

[72] Inventor **Eberhard Faust**
Bernhausen, Germany
[21] Appl. No. **805,644**
[22] Filed **Mar. 10, 1969**
[45] Patented **Dec. 14, 1971**
[73] Assignee **Recaro Aktiengesellschaft**
Glarus, Switzerland
[32] Priority **Mar. 13, 1968**
[33] **Germany**
[31] **P 17 53 009.5**

[54] **BACKREST FOR A SEAT ESPECIALLY OF A MOTOR VEHICLE**
1 Claim, 5 Drawing Figs.
[52] **U.S. Cl.** **297/284,** 297/391
[51] **Int. Cl.** **A47c 3/00**
[50] **Field of Search** 297/284, 460, 455, 410, 391, 396, 353, 458, 230, 231
[56] **References Cited**
UNITED STATES PATENTS

1,444,536	2/1923	Boyce	297/230 X
1,669,567	5/1928	Linder	297/230 X
2,831,533	4/1958	Pasquarelli	297/231
3,086,817	4/1963	Wilfert	297/284
3,259,435	7/1966	Tordon	297/455

FOREIGN PATENTS

533,992	2/1941	Great Britain	297/231

Primary Examiner—James C. Mitchell
Attorney—Ernest G. Montague

ABSTRACT: A padded backrest for a seat which has a frame with an upper and a lower crossbar and lateral arms secured to the crossbars, the upper crossbar extending within the upper part of the padding and the lateral arms being bent forwardly and downwardly from the upper crossbar and carrying lateral padding elements for supporting at least a part of the upper body of an occupant of the seat in lateral directions, and a shell-like padding support secured to the upper and lower crossbars and consisting of a stiff but deformable material. At least the central part of the padding which is adapted to engage with the back of the occupant is slidable upwardly and downwardly along the padding support relative to the frame. At least this central part is provided with transverse pockets into which, if desired, inserts may be placed to change the cross-sectional shape of the backrest.

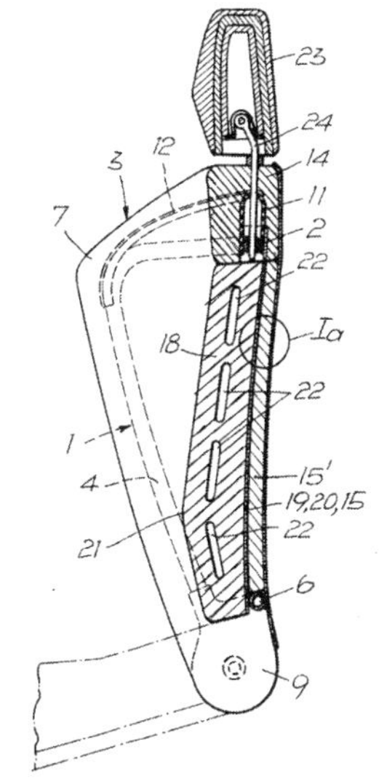

United States Patent [11] **3,604,752**

[72] Inventor **Albert J. Macknick**
North Chicago, Ill.
[21] Appl. No. **858,423**
[22] Filed **Sept. 16, 1969**
[45] Patented **Sept. 14, 1971**
[73] Assignee **United States Steel Corporation**

[54] **SUPPORT MEMBER FOR A VEHICLE SEAT**
13 Claims, 12 Drawing Figs.
[52] **U.S. Cl.** **297/455,** 287/113
[51] **Int. Cl.** **A47c 7/02,** F16b 7/00
[50] **Field of Search** .... 287/49, 53, 113; 211/182; 297/445, 455; 24/125, 81 CC
[56] **References Cited**
UNITED STATES PATENTS

2,106,724	2/1938	Cope	24/125 UX
3,238,584	3/1966	Lassen et al.	24/125
3,259,435	7/1966	Jordan, Jr.	297/455

Primary Examiner—Casmir A. Nunberg
Attorney—Robert J. Leek, Jr.

ABSTRACT: A support member for a vehicle seat having a stationary member is disclosed. The support member has a support portion provided with a support torsion leg, opposed support cantilever legs projecting from the support torsion leg, and a second support cantilever leg connecting the support cantilever legs. One means of a first clamp means and a pivot means secures the support torsion leg to the stationary member. A spring portion of the support member has a spring torsion leg disposed adjacent one leg of the support torsion leg and the second support cantilever leg. Opposed spring cantilever legs project from the spring torsion leg and a spring connecting leg connects the spring cantilever legs. A second clamp means connects the spring torsion leg to the one leg.

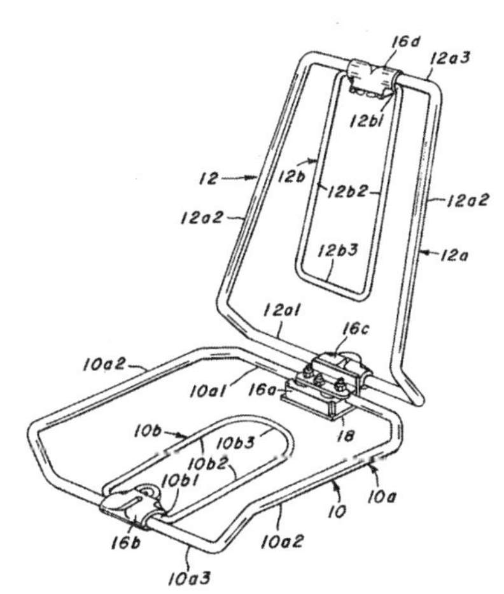

United States Patent [19]

Eiselt et al.

[11] **3,967,852**

[45] **July 6, 1976**

[54] **UPHOLSTERING BODY HAVING A SUPPORTED CORE IN THE FORM OF A YIELDABLE PLATE OF SYNTHETIC MATERIAL AND A RESILIENT LAYER ON THE CORE**

[75] Inventors: **Günter Eiselt; Rudolf Hossbach,** both of Unkel, Germany

[73] Assignee: **Günter Eiselt,** Unkel, Germany

[22] Filed: **Feb. 26, 1975**

[21] Appl. No.: **553,258**

Related U.S. Application Data

[62] Division of Ser. No. 377,735, July 9, 1973, abandoned.

[30] **Foreign Application Priority Data**

July 7, 1972 Germany 2233389
Oct. 13, 1972 Germany 2250249

[52] **U.S. Cl.** **297/452;** 297/455; 297/DIG. 1
[51] **Int. Cl.**[2] **A47C 7/02**
[58] **Field of Search** .... 297/DIG. 1, DIG 2, 452, 297/455, 456, 458–460

[56] **References Cited**

UNITED STATES PATENTS

1,461,497	7/1923	Robbins	267/41 X
2,530,924	11/1950	Turner	297/458 X
3,127,220	3/1964	Stine	297/456 X
3,140,086	7/1964	Lawson	297/DIG. 1
3,270,393	9/1966	Levenson	297/DIG. 2

Primary Examiner—James C. Mitchell
Attorney, Agent, or Firm—Walter Becker

[57] **ABSTRACT**

An upholstering body for use as the seat, or back, or as a combination seat and back in an article of furniture in which the body has an outer layer of resilient material such as a foamed elastomeric material and a supporting core member in the form of a sheet, or plate, of synthetic material, polyester resin, for example. The sheet is preferably reinforced with glass fibers, which may be in the form of glass cloth. The sheet is adapted for anchoring to a furniture frame in such a manner that the sheet is yieldable in the direction in which a load is normally applied thereto while being generally unyieldable in other directions.

19 Claims, 29 Drawing Figures

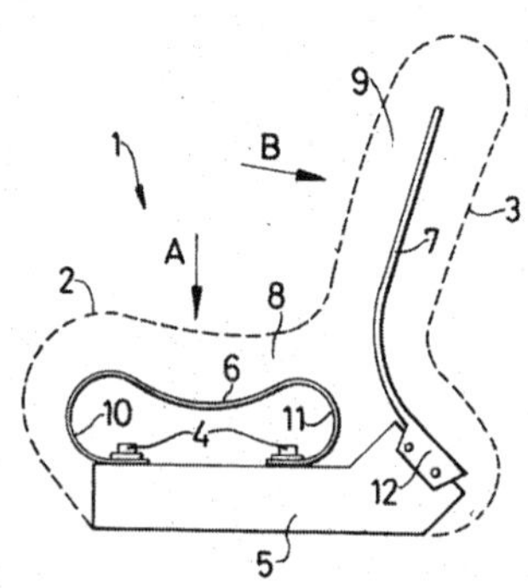

United States Patent [19]

Barecki et al.

[11] **4,109,959**

[45] **Aug. 29, 1978**

[54] **TRANSPORTATION SEAT WITH ENERGY ABSORPTION**

[75] Inventors: **Chester J. Barecki; Donald R. Lewis,** both of Grand Rapids; **Terry L. Camp,** Hudsonville, all of Mich.

[73] Assignee: **American Seating Company,** Grand Rapids, Mich.

[21] Appl. No.: **787,449**

[22] Filed: **Apr. 14, 1977**

[51] **Int. Cl.**[2] **B60R 21/10**
[52] **U.S. Cl.** **297/216;** 297/344; 297/396; 297/445; 297/455
[58] **Field of Search** 248/424, 429, 430; 297/216, 344, 391, 396, 399, 445, 452, 454–456

[56] **References Cited**

U.S. PATENT DOCUMENTS

1,237,956	8/1917	Pue	297/344 X
1,987,331	1/1935	Floraday	248/429
2,281,341	4/1942	Turner	297/445 X
2,346,895	4/1944	Bergman	297/216
3,243,234	3/1966	Fehlner	297/445 UX
3,460,791	8/1969	Judd	297/216 X
3,501,200	3/1970	Ohta	297/216 X
3,523,710	8/1970	Barecki et al.	297/445 X
3,729,226	4/1973	Barecki	297/445 X
3,827,752	8/1974	Bissinger	297/216 X

Primary Examiner—James C. Mitchell
Attorney, Agent, or Firm—Price, Heneveld, Huizenga & Cooper

[57] **ABSTRACT**

A transportation seat for a vehicle and which is capable of absorbing impact energy during a crash or rapid maneuver to thereby reduce or eliminate the severity of and occurrence of injury to a passenger includes a support frame having a leg structure and fixedly secured to the floor of the vehicle. A seat frame adapted to support the seat cushions is slidably connected to the support frame. A plurality of springs positioned between the support frame and the slidable seat frame permit controlled movement of the seat in a forward direction when impacted from the rear to thereby absorb energy of the impact. Further, the head impact area of the seat back is padded and the knee impact area of the seat back is covered by a plastic back panel which encloses a closed cell foam. The upper frame is capable of absorbing energy by deaccelerating the passenger in a crash situation in a controlled manner.

18 Claims, 21 Drawing Figures

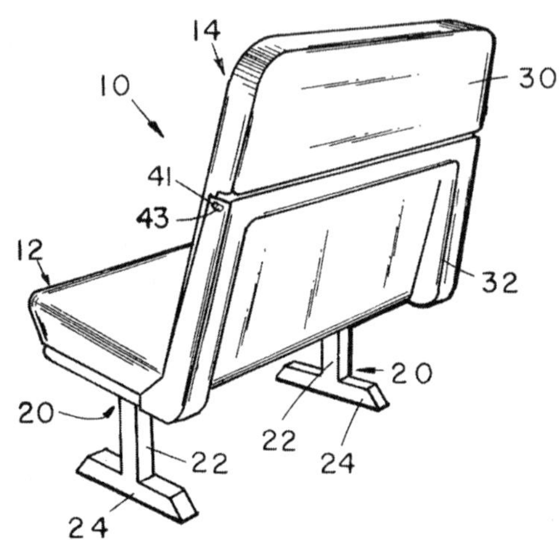

United States Patent [19]

Satzinger

[11] **4,076,306**

[45] **Feb. 28, 1978**

[54] **VEHICLE SEAT REST**

[76] Inventor: **Roland Satzinger,** Hammelburger Str. 21a, D-8731 Euerdorf, Germany

[21] Appl. No.: **721,722**

[22] Filed: **Sep. 9, 1976**

[30] **Foreign Application Priority Data**

Oct. 15, 1975 Germany 2546157

[51] **Int. Cl.**[2] **B60R 21/10**
[52] **U.S. Cl.** **297/216;** 297/389
[58] **Field of Search** .... 297/216, 450, 389, 385, 297/384; 296/35 R; 280/744, 746, 747

[56] **References Cited**

U.S. PATENT DOCUMENTS

3,145,051	8/1964	Rausch	297/216
3,501,200	3/1970	Ohta	297/216 X
3,544,164	12/1970	Ohta	297/216
3,545,808	12/1970	Gescheidle	297/216
3,619,006	11/1971	Barecki	297/450
3,877,748	4/1975	Eggert	297/216

Primary Examiner—James T. McCall
Attorney, Agent, or Firm—Burgess, Dinklage & Sprung

[57] **ABSTRACT**

An improvement in a vehicle seat rest having a shoulder belt, one end of which is fastened to said rest, toward the top thereof, said belt running obliquely across said rest to a belt lock on the opposed side thereof, said vehicle seat rest improved in that the vehicle seat rest has a rectangular frame and a rigid cross-stay running from the region of the point to which said belt is attached at the top of said rest obliquely in the same direction as said belt to the region of said belt lock. There is also disclosed the use of a cross-stay which is incompressible under maximum loading. Also disclosed is a rectangular frame having a deformable section therein.

4 Claims, 3 Drawing Figures

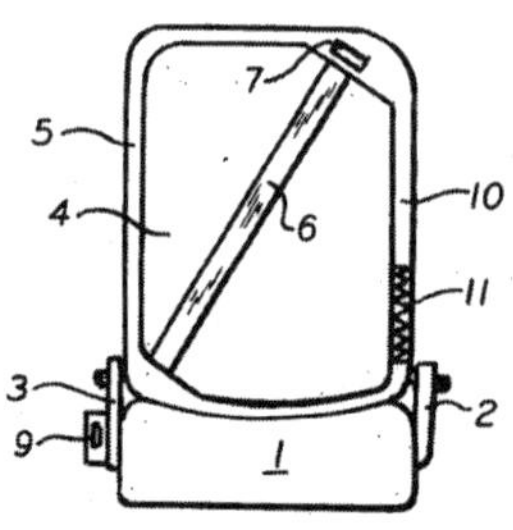

United States Patent [19]

Higuchi et al.

[11] **4,192,545**

[45] **Mar. 11, 1980**

[54] **SEAT-BACK FRAME FOR AN AUTOMOTIVE VEHICLE**

[75] Inventors: **Kazuo Higuchi,** Tokyo; **Katsuhiko Matsumoto,** Kunitachi, both of Japan

[73] Assignee: **Honda Giken Kogyo Kabushiki Kaisha,** Tokyo, Japan

[21] Appl. No.: **944,395**

[22] Filed: **Sep. 21, 1978**

[30] **Foreign Application Priority Data**

Sep. 29, 1977 [JP] Japan 52-129988

[51] **Int. Cl.**[2] **B60R 21/10**
[52] **U.S. Cl.** **297/216;** 297/483
[58] **Field of Search** 297/216, 389, 452

[56] **References Cited**

U.S. PATENT DOCUMENTS

4,076,306 Sa tzinger 297/216

Primary Examiner—James C. Mitchell
Attorney, Agent, or Firm—Polster, Polster and Lucchesi

[57] **ABSTRACT**

A seat-back frame for automotive vehicles comprising a rectangular frame of a rigid material, the frame having two upper and two lower corners, a seat belt connected to one upper corner and a tensile member obliquely extending between an upper and a diagonally lower corner of the frame to reinforce it against rhombic deformation due to external force applied in a lateral direction of the frame due to load applied to the frame, the tensile member being flexed or bowed to some extent in the rearward direction of the seat-back frame to avoid any interference with a cushioning material and cushioning springs when they are pushed backward by an occupant leaning on the seat-back and being flexible enough to straighten under tensile load.

7 Claims, 7 Drawing Figures

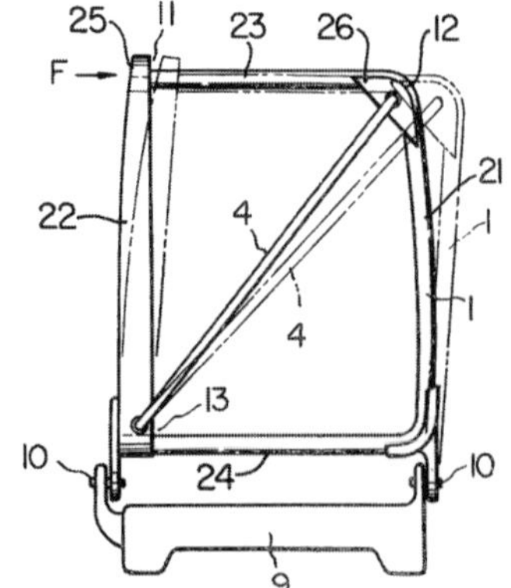

United States Patent [19]

Barecki

[11] **4,249,769**

[45] **Feb. 10, 1981**

[54] **PASSIVE RESTRAINT FOR A VEHICLE**

[75] Inventor: **Chester J. Barecki**, Grand Rapids, Mich.

[73] Assignee: **American Seating Company**, Grand Rapids, Mich.

[21] Appl. No.: **3,668**

[22] Filed: **Jan. 15, 1979**

[51] **Int. Cl.**³ **B60N 1/02**
[52] **U.S. Cl.** .... **296/65 A**; 297/216
[58] **Field of Search** .... 296/63, 65 R, 65 A; 108/108, 152; 297/450, 216

[56] **References Cited**

U.S. PATENT DOCUMENTS

3,310,342	3/1967	Drelichowski	297/216
3,762,764	10/1973	McJunkin	296/63
3,802,738	4/1974	Tantlinger	297/450
4,109,959	8/1978	Barecki et al.	297/216
4,145,081	3/1979	Withers	297/216

Primary Examiner—John J. Love
Assistant Examiner—Ross Weaver
Attorney, Agent, or Firm—Price, Heneveld, Huizenga & Cooper

[57] **ABSTRACT**

An upper suspended cantilevered public transportation vehicle seat is provided including a passive restraint for absorbing the impact energy of a passenger during an accident or rapid maneuver. The vehicle seat includes a bench and an upstanding back portion, both the seat bench and back portions being supported by a sidewall of the vehicle. The seat bench is securely mounted to the sidewall of the vehicle. The upstanding seat back includes top and bottom portions, the bottom portion of the back depending from the seat bench. The top portion of the back is slidably, or pivotably and slidably mounted to the sidewall of the vehicle. This mounting arrangement allows the back of the cantilevered seat to deflect about the bench to absorb the impact energy of a passenger and prevent or reduce serious injury to the passenger during an accident or rapid maneuver.

17 Claims, 11 Drawing Figures

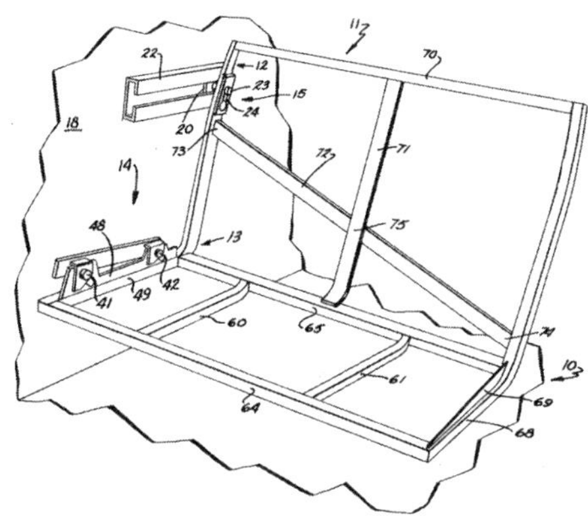

United States Patent [19]

Terada et al.

[11] Patent Number: **4,519,650**

[45] Date of Patent: **May 28, 1985**

[54] SEAT BACK FRAME ASSEMBLY FOR AUTOMOBILE SEATS

[75] Inventors: **Takami Terada; Hiroshi Nawa; Reiki Kawamura**, all of Toyota, Japan

[73] Assignees: **Aisin Seiki Kabushiki Kaisha; Toyota Jidosha Kabushiki Kaisha**, both of Aichi, Japan

[21] Appl. No.: **473,678**

[22] Filed: **Mar. 8, 1983**

[30] **Foreign Application Priority Data**

Mar. 9, 1982 [JP] Japan .... 57-32699[U]

[51] **Int. Cl.**³ **A47C 7/02**
[52] **U.S. Cl.** **297/452**; 297/410
[58] **Field of Search** 297/452, 451, 450, 410, 297/391

[56] **References Cited**

U.S. PATENT DOCUMENTS

2,572,591	10/1951	Booth	297/450 X
3,173,723	3/1965	Hoven et al.	297/452 X
3,427,073	2/1969	Downs et al.	297/410
3,512,833	5/1970	Sugiura	297/410
3,572,831	3/1971	Barecki et al.	297/410
3,630,566	12/1971	Barecki	297/450
3,874,731	4/1975	Jordan	297/452
4,305,617	12/1981	Benoit	297/452
4,351,563	9/1982	Hattori	297/391

FOREIGN PATENT DOCUMENTS

2426728	12/1975	Fed. Rep. of Germany	297/410
1156528	6/1969	United Kingdom .	

Primary Examiner—Francis K. Zugel
Attorney, Agent, or Firm—Finnegan, Henderson, Farabow, Garrett and Dunner

[57] **ABSTRACT**

A seat back frame assembly, comprising a substantially U-shaped pipe member having a horizontal base and a pair of vertical legs extending upwardly from the horizontal base and terminating at a pair of upper ends, a transverse member having a first horizontal portion welded at the opposite ends to the upper ends of the pipe member, a first vertical portion depending from the first horizontal portion to extend along one side of the pipe member and having a lower end continuing to a second horizontal portion lying beneath the first horizontal portion between the vetical legs of the pipe member, and a second vertical portion depending from the second horizontal portion along the other side of the pipe member, each of the first and second vertical portions having opposite edge portions welded to the legs of the pipe member, the first horizontal portion having a pair of transversely spaced holes, the second horizontal portion having a pair of holes which are respectively aligned in vertical direction with the holes in the first horizontal portion for receiving headrest receptacles.

2 Claims, 5 Drawing Figures

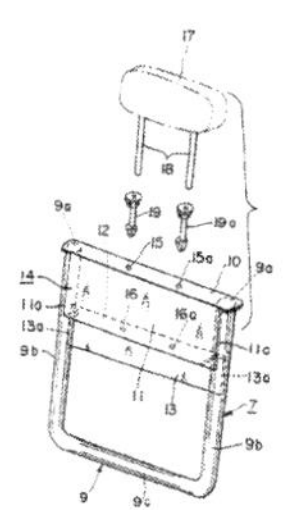

United States Patent [19]

Maeda et al.

[11] Patent Number: **4,512,604**

[45] Date of Patent: **Apr. 23, 1985**

[54] **VEHICULAR SEAT ARRANGEMENT**

[75] Inventors: **Kouzo Maeda; Harutoshi Tsujimura**, both of Yokosuka, Japan

[73] Assignee: **Nissan Motor Company, Limited**, Yokohama, Japan

[21] Appl. No.: **366,218**

[22] Filed: **Apr. 7, 1982**

[30] **Foreign Application Priority Data**

Apr. 20, 1981 [JP] Japan .... 56-57349[U]

[51] **Int. Cl.**³ **B60N 1/02**
[52] **U.S. Cl.** .... **296/65 A**; 297/216; 280/748
[58] **Field of Search** .... 296/63, 65 R, 65 A, 296/188, 189, 30; 280/748; 297/216, 320, 343; 244/122 R

[56] **References Cited**

U.S. PATENT DOCUMENTS

2,872,241	2/1959	Sheldon	296/146
3,357,736	12/1967	McCarthy	296/65 A
3,885,810	5/1975	Chika	296/65 A
4,231,607	11/1980	Bohlin	296/63

Primary Examiner—Robert B. Reeves
Assistant Examiner—Dennis H. Pedder
Attorney, Agent, or Firm—Lane, Aitken & Kananen

[57] **ABSTRACT**

A vehicular seat arrangement for use in an automotive vehicle, comprising at least two seat assemblies positioned on the floor panel in parallel with each other laterally of the vehicle body, each seat assembly having a seat cushion portion and a seat back portiqn upstanding from the seat cushion portion and rockable with respect to the seat cushion portion about a pivot axis in a lateral direction of the seat assembly, a seat reinforcing device including an elongated reinforcement member extending in a lateral direction of the vehicle body between the side wall portions of the vehicle body and passed in part through the seat back portion of one of the seat assemblies and in part through the seat back portion of the other of the seat assemblies, and tilt regulating means permitting the seat back portion of one of the seat assemblies to tilt with respect to the seat cushion portion of the seat assembly about the pivot axis independently of the seat back portion of the other of the seat assemblies.

18 Claims, 19 Drawing Figures

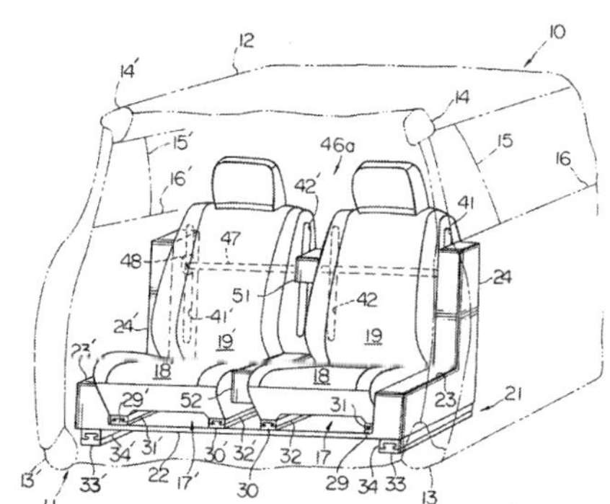

United States Patent [19]

Muraishi

[11] Patent Number: **4,695,097**

[45] Date of Patent: **Sep. 22, 1987**

[54] **SEAT FOR USE IN MOTOR VEHICLE**

[75] Inventor: **Masakazu Muraishi**, Isehara, Japan

[73] Assignee: **Nissan Motor Company, Limited**, Japan

[21] Appl. No.: **909,120**

[22] Filed: **Sep. 17, 1986**

[30] **Foreign Application Priority Data**

Sep. 18, 1985 [JP] Japan 60-207484

[51] **Int. Cl.**⁴ **A47C 7/02**
[52] **U.S. Cl.** **297/452**; 248/429; 297/344
[58] **Field of Search** 297/344, 330, 452; 248/429, 430, 393

[56] **References Cited**

U.S. PATENT DOCUMENTS

3,924,892	12/1975	Geier	297/452
4,509,796	4/1985	Takagi	297/452
4,602,817	7/1986	Raftery	297/440
4,606,532	8/1986	Kazaoka et al.	297/452

FOREIGN PATENT DOCUMENTS

59-14275	4/1985	Japan .
1088119	10/1967	United Kingdom .

Primary Examiner—James T. McCall
Attorney, Agent, or Firm—Lowe, Price, LeBlanc, Becker & Shur

[57] **ABSTRACT**

The movable rail of a seat sliding mechanism is detachably connected through bolts to a side frame of a rectangular seat cushion frame. Thus, fixing of a trimmed edge of an outer skin member of the seat cushion to the proper position of the side frame can be easily achieved with the movable rail unfastened from the side frame.

11 Claims, 6 Drawing Figures

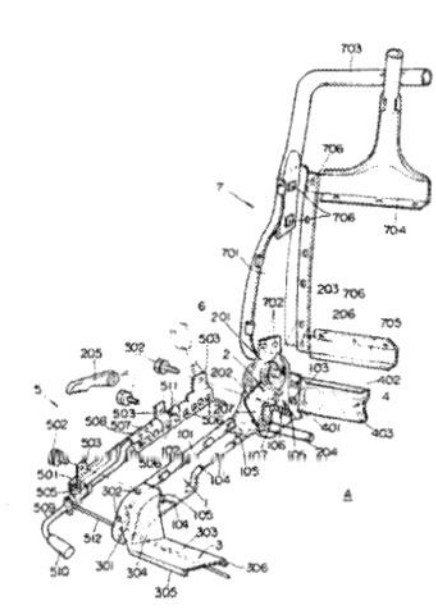

United States Patent [19]
Saito

[11] **Patent Number: 4,796,954**
[45] **Date of Patent: Jan. 10, 1989**

[54] **SEAT BACK REAR STRUCTURE**

[75] Inventor: **Tetsuo Saito**, Tokyo, Japan

[73] Assignee: **Tachi-S Ltd.**, Akishima, Japan

[21] Appl. No.: **14,816**

[22] Filed: **Feb. 13, 1987**

[51] **Int. Cl.**[4] **A47C 7/18**
[52] **U.S. Cl.** **297/452**; 297/460; 297/DIG. 1
[58] **Field of Search** .... 297/DIG. 1, DIG. 2, 297/452, 459, 460, 218, 219

[56] **References Cited**

U.S. PATENT DOCUMENTS

3,874,731	4/1975	Jordan	297/452
4,365,840	12/1982	Kehl et al.	297/460 X
4,615,561	10/1986	Nomura	297/460 X

FOREIGN PATENT DOCUMENTS

68211	1/1983	European Pat. Off.	297/460
1914725	10/1970	Fed. Rep. of Germany	297/452
2262595	7/1974	Fed. Rep. of Germany	297/460
2944054	5/1981	Fed. Rep. of Germany	297/452
1132647	3/1957	France	297/460
102628	6/1984	Japan	297/219

Primary Examiner—Kenneth J. Dorner
Assistant Examiner—James R. Brittan
Attorney, Agent, or Firm—Oldham & Oldham

[57] **ABSTRACT**

A rear structure of a seat back for an automotive seat is disclosed. In order to increase the depth dimension of a seat cushion, a rearwardly curved reinforcing rod is welded to the lower portion of an inverted U-shaped frame forming the seat, and the shape holding portions of a cushion member thicker than the turned-back portions thereof covering the upper portion of the frame are applied against the frame with the rod welded thereto. The outer surfaces of the shape holding portions are arranged to form the same plane with the outer surfaces of the turned-back portions, so that the right and left corners of the rear portion of the seat back can be formed square.

4 Claims, 3 Drawing Sheets

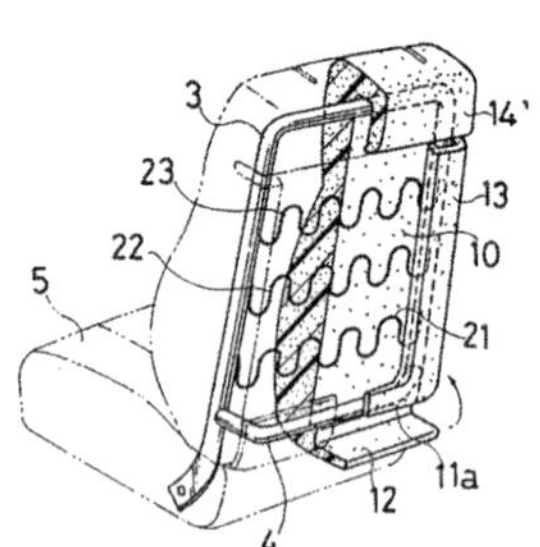

United States Patent [19]
Yokota

[11] **Patent Number: 5,044,693**
[45] **Date of Patent: Sep. 3, 1991**

[54] **SEAT BACK STRUCTURE OF AN AUTOMOTIVE SEAT**

[75] Inventor: **Masaaki Yokota**, Akishima, Japan

[73] Assignee: **Tachi-S Co., Ltd.**, Tokyo, Japan

[21] Appl. No.: **429,422**

[22] Filed: **Oct. 31, 1989**

[51] **Int. Cl.**[5] **A47C 7/02**
[52] **U.S. Cl.** **297/452**; 297/284
[58] **Field of Search** 297/452, 234, 285, 460

[56] **References Cited**

U.S. PATENT DOCUMENTS

3,762,769	10/1973	Poschl	297/460 X
4,231,615	11/1980	Griffiths	297/452
4,565,406	1/1986	Suzuki	297/284
4,588,172	5/1986	Fourrey et al.	297/289 X
4,682,763	7/1987	Kazaoka et al.	297/452 X
4,725,095	2/1988	Benson et al.	297/460 X

FOREIGN PATENT DOCUMENTS

1914154	6/1971	Fed. Rep. of Germany	297/284
2724725	12/1977	Fed. Rep. of Germany	297/284
3223815	1/1983	Fed. Rep. of Germany	297/460
0018428	1/1985	Japan	297/284
654734	3/1986	Switzerland	297/452

Primary Examiner—Jose V. Chen
Attorney, Agent, or Firm—Browdy and Neimark

[57] **ABSTRACT**

A seat back structure of an automotive seat comprising a back support plate, plural springs, and torsion bar. The back support plate is supported in a seat back frame by the springs and torsion bar in a more positively resilient way. The provision of the torsion bar avoids the sudden backward movement of the back support plate which has been found in a conventional seat back structure wherein the support plate is only supported by springs.

1 Claim, 2 Drawing Sheets

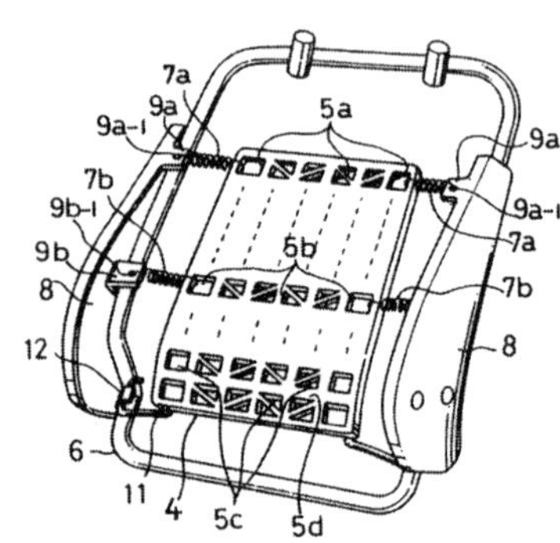

United States Patent [19]
Schmale et al.

[11] **Patent Number: 4,938,527**
[45] **Date of Patent: Jul. 3, 1990**

[54] **PLASTIC BODY FOR VEHICLE SEATS**

[75] Inventors: **Gerhard Schmale**, Huckeswagen; **Siegfried Peters**, Remscheid, both of Fed. Rep. of Germany

[73] Assignee: **Keiper Recaro GmbH & Co.**, Fed. Rep. of Germany

[21] Appl. No.: **319,690**

[22] Filed: **Mar. 7, 1989**

[30] **Foreign Application Priority Data**

Mar. 12, 1988 [DE] Fed. Rep. of Germany 3808316

[51] **Int. Cl.**[5] **A47C 7/02**
[52] **U.S. Cl.** **297/216**; 428/462; 428/521
[58] **Field of Search** 297/216, 471, 472; 428/462, 521

[56] **References Cited**

U.S. PATENT DOCUMENTS

3,420,572	1/1969	Bisland	297/216
3,562,089	2/1971	Warnaka et al.	428/462 X
3,578,376	5/1971	Hasegawa	297/216
3,627,379	12/1971	Faust	297/216
3,802,737	4/1974	Mertens	297/216
4,119,344	10/1978	Kondo	297/471
4,128,217	12/1978	Mazelsky	297/216
4,192,545	3/1980	Higuchi et al.	297/216
4,505,971	3/1985	Martin et al.	428/462 X
4,505,984	3/1985	Stelzer et al.	428/462 X
4,824,705	4/1989	Persson et al.	428/521 X

Primary Examiner—Peter A. Aschenbrenner
Attorney, Agent, or Firm—Wigman & Cohen

[57] **ABSTRACT**

A plastic body for use in vehicle seats, having at least one element joined therewith to form a single structural unit, which element is made from a material having different physical characteristics than those of the plastic. The element is formed as a dampening element made from a plastically deformable metal securely connected with the plastic body and/or as an insert in the plastic body. Between the dampening element and/or insert and the plastic body an intermediate layer is provided, the intermediate layer being made from a material that permits relative movements between the plastic body and the dampening element and/or insert.

36 Claims, 4 Drawing Sheets

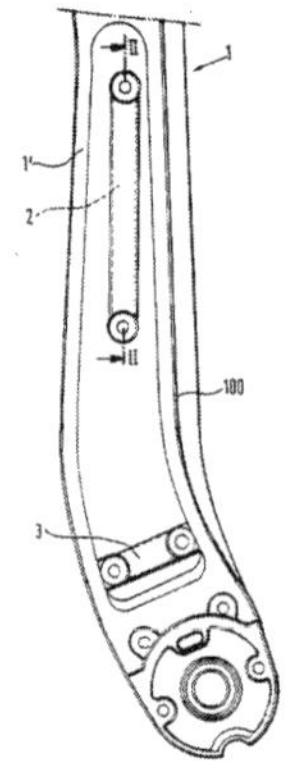

United States Patent [19]
Granzow et al.

[11] **Patent Number: 5,123,706**
[45] **Date of Patent: Jun. 23, 1992**

[54] **AUTOMOBILE SEAT WITH A BACKREST FRAME**

[75] Inventors: **Manfred Granzow**, Dörentrup; **Hans-Peter Mischer**, Horn-Bad Meinberg; **Christian Süss**, Bad Salzuflen, all of Fed. Rep. of Germany

[73] Assignee: **Gebr. Isringhausen**, Lemgo, Fed. Rep. of Germany

[21] Appl. No.: **460,944**

[22] PCT Filed: **Jun. 20, 1989**

[86] PCT No.: **PCT/EP89/00692**

§ 371 Date: **Apr. 23, 1990**

§ 102(e) Date: **Apr. 23, 1990**

[87] PCT Pub. No.: **WO89/12563**

PCT Pub. Date: **Dec. 28, 1989**

[30] **Foreign Application Priority Data**

Jun. 22, 1988 [DE] Fed. Rep. of Germany 3821554

[51] **Int. Cl.**[5] **A47C 7/02**
[52] **U.S. Cl.** **297/452**; 297/355; 297/460
[58] **Field of Search** .... 297/452, 284, 355, 460, 297/377, 361, 284 R; 248/429, 420

[56] **References Cited**

U.S. PATENT DOCUMENTS

2,998,281	8/1961	Stoner et al.	297/374
3,079,118	2/1963	Pickles	248/420
4,585,273	4/1986	Higgs et al.	297/452

FOREIGN PATENT DOCUMENTS

2126476	3/1984	United Kingdom	297/361

Primary Examiner—Kenneth J. Dorner
Assistant Examiner—Milton Nelson, Jr.
Attorney, Agent, or Firm—Salter, Michaelson & Benson

[57] **ABSTRACT**

An automobile seat includes seat frame and backrest frame portions, each of which includes a pair of opposite side braces of generally U-shaped configuration which face away from the seating surfaces of their respective side frame and back rest frame portions. The side braces of the side frame and back frame portions define open channel-like areas for accommodating adjusting units of the seat.

3 Claims, 2 Drawing Sheets

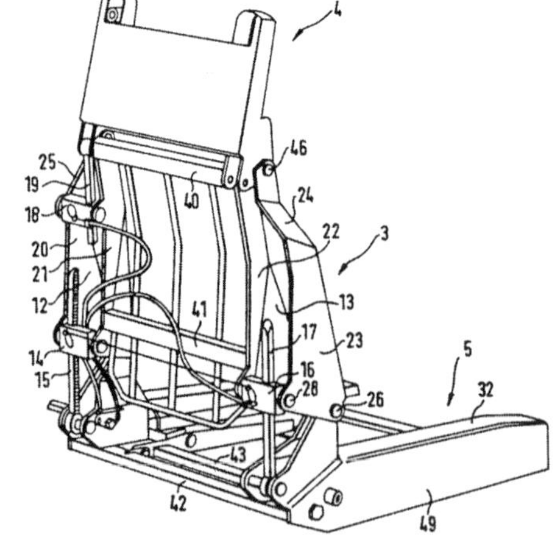

United States Patent [19]

Yamauchi

[11] **Patent Number: 5,129,707**

[45] **Date of Patent: Jul. 14, 1992**

[54] **SEATBACK FRAME HAVING RETRORSE CONNECTION OF A CONCAVE RESILIENT LUMBAR SUPPORT MEMBER**

[75] Inventor: **Yoshihiko Yamauchi**, Ayase, Japan

[73] Assignee: **Ikeda Bussan Company Ltd.**, Ayase, Japan

[21] Appl. No.: **716,020**

[22] Filed: **Jun. 17, 1991**

[30] **Foreign Application Priority Data**

Jun. 28, 1990 [JP] Japan 2-68807

[51] **Int. Cl.**[5] **A47C 7/02**
[52] **U.S. Cl.** **297/460**; 297/216
[58] **Field of Search** 297/460, 216, 452

[56] **References Cited**

U.S. PATENT DOCUMENTS

3,877,749	4/1975	Sakurai et al.	297/216
3,967,852	7/1976	Eiselt et al.	297/452
4,883,320	11/1989	Izumida et al.	297/284 R X
4,928,334	5/1990	Kita	297/452 X
5,040,848	8/1991	Irie et al.	297/460 X
5,054,845	10/1991	Vogel	297/216

Primary Examiner—Kenneth J. Dorner
Assistant Examiner—James M. Gardner
Attorney, Agent, or Firm—Brooks & Kushman

[57] **ABSTRACT**

A seatback frame **(21)** is disclosed as having an inverted U-shaped frame member **(22)** including an elongated and generally horizontal upper member **(23)** and a pair of elongated side members **(24)** and **(25)** respectively extending generally downwardly from the extremities of the upper member. A pair of side panels **(28 and 29)** are respectively secured to and extend generally forwardly from each side member and have inwardly facing surfaces that oppose each other. A resilient lumbar supporting panel **(35)** having lateral extremities is forwardly spaced from the side members proximate the side panels and is generally concave therebetween. It has a central portion that extends generally upwardly and is secured to a central portion of the upper member. The lateral extremities form retrorse portions **(41)** that extend generally rearwardly along the opposed surfaces of the side panels, and the rearmost margins **(43)** of the retrorse portions are attached to the side panels. The resilience of the lumbar supporting panel allows the retrorse portions to flex generally about their attached margins to damp any acceleration of a passenger toward the seatback frame of vice versa.

15 Claims, 2 Drawing Sheets

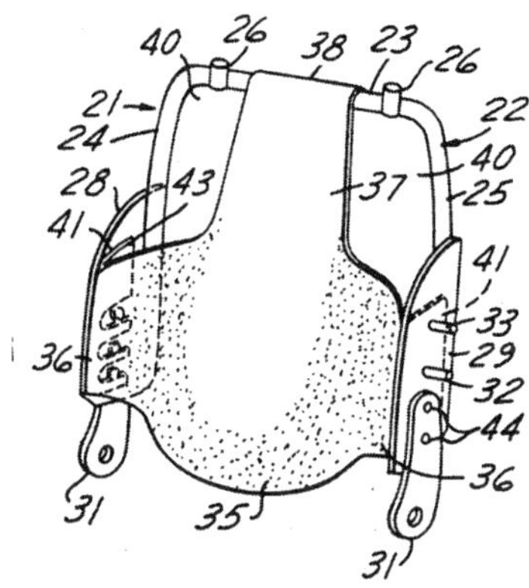

United States Patent [19]

Fujimori et al.

[11] **Patent Number: 5,310,247**

[45] **Date of Patent: May 10, 1994**

[54] **VEHICLE SEATS**

[75] Inventors: **Takashi Fujimori**, Toyota; **Kazuhisa Tatematsu**, Nagoya; **Kunio Nishiyama**, Susono; **Takakazu Mori**, Toyota, all of Japan

[73] Assignees: **Araco K.K.; Toyota Jidosha K.K.**, both of Aichi, Japan

[21] Appl. No.: **35,578**

[22] Filed: **Mar. 23, 1993**

[30] **Foreign Application Priority Data**

Mar. 23, 1992 [JP] Japan .... 4-015185[U]
Jun. 5, 1992 [JP] Japan 4-145492
Jun. 5, 1992 [JP] Japan 4-145499

[51] **Int. Cl.**[5] **B60N 2/22**
[52] **U.S. Cl.** **297/378.12**; 297/216.14; 297/452.18
[58] **Field of Search** 297/216.14, 378.12, 297/452.18, 452.20

[56] **References Cited**

U.S. PATENT DOCUMENTS

2,712,344	7/1955	Weber	297/378.12 X
3,761,127	9/1973	Giese et al.	297/389
5,246,271	9/1993	Boisset	297/378.12 X

FOREIGN PATENT DOCUMENTS

60-21710	2/1985	Japan .
1-77533	7/1988	Japan .

Primary Examiner—Peter R. Brown
Attorney, Agent, or Firm—Fisher & Associates

[57] **ABSTRACT**

A vehicle seat having a backrest hinged to a seat squab frame by means of a pair of reclining mechanisms to be tilted forwardly or backwardly with respect to the seat squab frame, the reclining mechanisms each including a ratchet member secured to each side of the lower end portion of the backrest frame to be locked to a support member secured to each side of the rear end portion of the seat squab frame, which vehicle seat includes a single strut member rotatably connected at its upper end to an upper end of the backrest at one side thereof to be moved forwardly or backwardly, and a resilient member mounted on the upper end of the backrest frame for biasing the strut member forwardly, wherein the ratchet member located at the same side as the one side of the backrest is formed with a semi-circular elongated hole of which the radius center is located at a rotation fulcrum of the strut member, and wherein the lower end of the strut member is slidably engaged with the semi-circular elongated hole and is normally positioned at a forward end of the elongated hole under load of said resilient member.

3 Claims, 7 Drawing Sheets

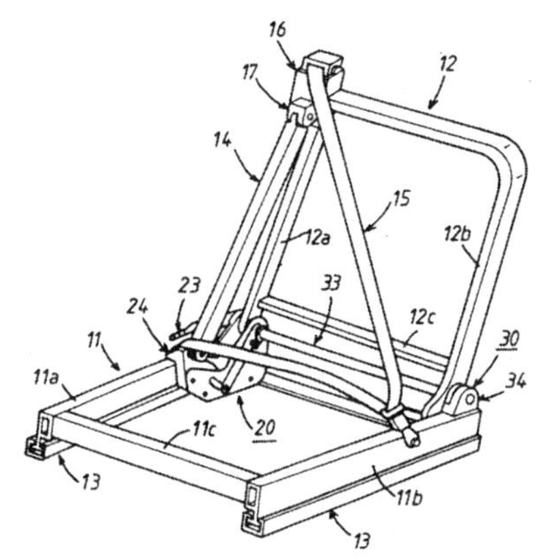

United States Patent [19]

Oleszko et al.

[11] **Patent Number: 5,290,089**

[45] **Date of Patent: Mar. 1, 1994**

[54] **SEAT BELLOWS ENERGY ABSORBER**

[75] Inventors: **Mark A. Oleszko**, Warren; **Kenneth A. Gassman**, Waterford; **Mladen Humer**, East Detroit, all of Mich.

[73] Assignee: **General Motors Corporation**, Detroit, Mich.

[21] Appl. No.: **997,696**

[22] Filed: **Dec. 28, 1992**

[51] **Int. Cl.**[5] **B60N 2/42**
[52] **U.S. Cl.** **297/216.14**; 297/216.1
[58] **Field of Search** 297/216, 472, 355, 354, 297/361, 470, 216.1, 216.13, 216.14, 216.15, 216.16, 354.12, 354.1, 361.1

[56] **References Cited**

U.S. PATENT DOCUMENTS

3,582,133	6/1971	DeLavenne	297/216 X
3,724,603	4/1973	Shiomi et al.	297/216 X
3,838,870	10/1974	Hag	297/472
4,733,911	3/1988	Fulcheri	297/345 X

FOREIGN PATENT DOCUMENTS

0275230	11/1989	Japan	297/216
1039131	8/1966	United Kingdom	297/216

Primary Examiner—Kenneth J. Dorner
Assistant Examiner—Milton Nelson, Jr.
Attorney, Agent, or Firm—Ernest E. Helms

[57] **ABSTRACT**

A recliner arrangement for a vehicle seat is provided which, in a preferred embodiment, includes a seat cushion unit, a seat backrest unit pivotally connected to the seat cushion unit, a linear recliner for selectively adjusting the pivotal location of the backrest unit, and an energy absorber interposed between the linear recliner and the seat backrest unit, the energy absorber being operative upon a predetermined magnitude of force exerted thereon.

7 Claims, 2 Drawing Sheets

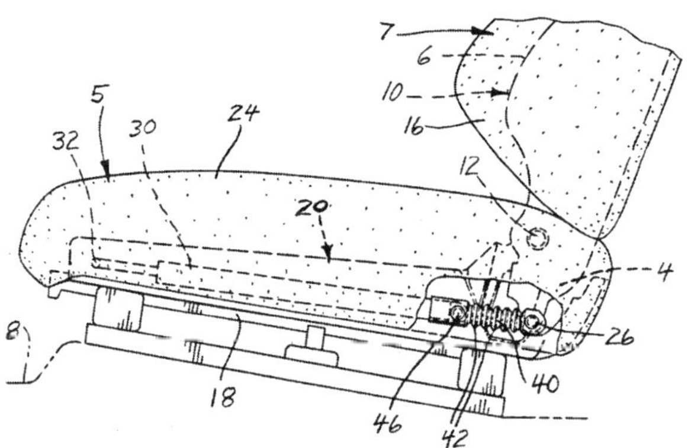

United States Patent [19]

Griswold et al.

[11] **Patent Number: 5,318,341**

[45] **Date of Patent: Jun. 7, 1994**

[54] **VEHICLE SEAT ASSEMBLY WITH STRUCTURAL SEAT BACK TO ACCOMMODATE SEAT BELT LOADS APPLIED TO SEAT BACK**

[75] Inventors: **Les Griswold**, Ann Arbor; **Marc D. Hewko**, Canton; **Robert D. Elton; Paul M. Grippo**, both of Ann Arbor, all of Mich.; **John Krieger**, Dublin, Ohio

[73] Assignee: **Hoover Universal, Inc.**, Plymouth, Mich.

[21] Appl. No.: **826,858**

[22] Filed: **Jan. 28, 1991**

Related U.S. Application Data

[63] Continuation-in-part of Ser. No. 659,500, Feb. 22, 1991.

[51] **Int. Cl.**[5] **B60N 2/22**
[52] **U.S. Cl.** **297/362.11**; 297/216.13; 297/284.4; 297/452.18; 297/483
[58] **Field of Search** .... 297/216, 285 C, 348, 297/345, 346, 338, 130, 483, 475, 452, 460; 297/284.4, 379, 361, 362

[56] **References Cited**

U.S. PATENT DOCUMENTS

1,788,088	1/1931	Fabio .	
3,583,764	6/1971	Lohr .	
3,663,057	5/1972	Lohr et al. .	
3,761,127	9/1973	Giese et al. .	
3,917,211	11/1975	Daunderer et al. .	
4,192,545	3/1980	Higuchi et al.	297/216
4,218,091	8/1980	Webster	297/361
4,515,339	5/1985	Kluting et al. .	
4,579,386	4/1986	Rupp et al.	297/361
4,725,095	2/1988	Benson et al. .	
4,726,617	2/1988	Nishimura	297/473
4,781,354	11/1988	Nihei et al.	297/379
4,886,316	12/1989	Suzuyome et al.	297/284.4
4,887,864	12/1989	Ashton .	
4,889,389	12/1989	White .	
4,940,285	7/1990	Suzuki et al.	297/473
5,014,958	5/1991	Harney	297/346 X

FOREIGN PATENT DOCUMENTS

143144	7/1985	Japan	297/284.4

Primary Examiner—Peter R. Brown
Attorney, Agent, or Firm—Harness, Dickey & Pierce

[57] **ABSTRACT**

A vehicle seat assembly with a structural seat back to accommodate the seat belt loads applied to the seat back by a belt system which is carried by the seat assembly. The seat back includes a reinforced beam along one side to accommodate these loads with a recliner mechanism along the same side of the seat assembly coupled to the reinforced member to resist forward rotation of the seat back caused by seat belt loads applied thereto. The seat assembly is configured to be as similar to conventional seat assemblies as possible to minimize the need to redesign the non-structural components of the seat assembly. The seat assembly includes many features found in current production seat assemblies such as a fore and aft adjuster, an adjustable lumbar support, a seat cushion lift mechanism and a seat back recliner. The seat belt loads have been efficiently managed to avoiding adding unnecessary structure and weight.

17 Claims, 8 Drawing Sheets

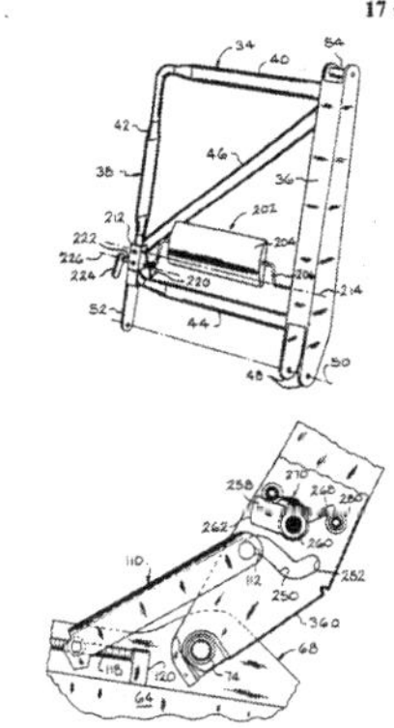

United States Patent [19]

Nishiyama

[11] **Patent Number: 5,328,248**

[45] **Date of Patent: Jul. 12, 1994**

[54] **SEAT FRAME FOR A VEHICLE**

[75] Inventor: **Kunio Nishiyama**, Sosono, Japan

[73] Assignee: **Toyota Jidosha Kabushiki Kaisha**, Toyota, Japan

[21] Appl. No.: **915,613**

[22] Filed: **Jul. 21, 1992**

[30] **Foreign Application Priority Data**

Jul. 23, 1991 [JP] Japan 3-182435
Jul. 23, 1991 [JP] Japan 3-182436

[51] **Int. Cl.**[5] **A47C 7/02**
[52] **U.S. Cl.** **297/452.56**; 297/452.18
[58] **Field of Search** 297/452, 441, 443, 218, 297/216

[56] **References Cited**

U.S. PATENT DOCUMENTS

3,289,220 12/1966 Grimshaw 297/452
3,329,466 7/1967 Getz 297/452
4,037,829 7/1977 Crosby 297/452
4,147,336 4/1979 Yamawaki 297/452
4,509,796 4/1985 Takagi 297/452
4,695,097 9/1987 Muraishi 297/452
4,834,458 5/1989 Izumida 297/452
4,861,104 8/1989 Malak 297/218
4,863,219 9/1989 Ochiai 297/443
5,013,089 5/1991 Abu-Isa 297/452

FOREIGN PATENT DOCUMENTS

62-111227 7/1987 Japan .
62-111228 7/1987 Japan .
179474 5/1962 Sweden 297/218

Primary Examiner—Flemming Saether
Attorney, Agent, or Firm—Cushman, Darby & Cushman

[57] **ABSTRACT**

A front-lower arm portion is disposed in the forward direction of a vehicle, while a rear-lower arm portion is disposed in the rearward direction of the vehicle. The rear-lower arm portion is formed of a material having greater tensile strength and compressive strength than that of the front-lower arm portion. A rear end portion of the front-lower arm portion is overlaid with a front end portion of the rear-lower arm portion, thereby securing both end portions together by means of a rivet.

21 Claims, 13 Drawing Sheets

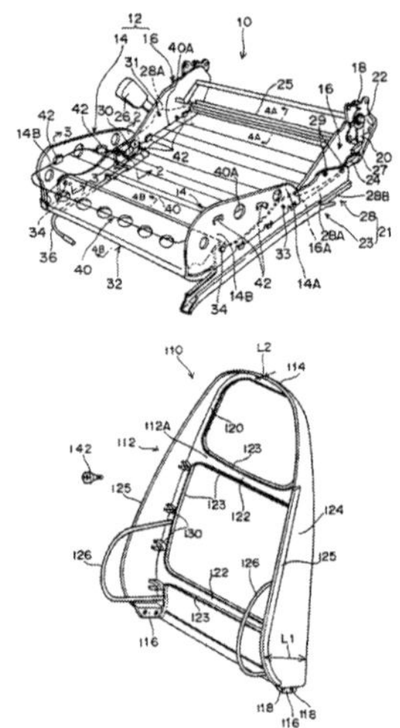

United States Patent [19]

Tadokoro

[11] **Patent Number: 5,810,446**

[45] **Date of Patent: Sep. 22, 1998**

[54] **FRAME STRUCTURE OF SEATBACK**

[75] Inventor: **Takumi Tadokoro**, Machida, Japan

[73] Assignee: **Ikeda Bussan Co., Ltd.**, Ayase, Japan

[21] Appl. No.: **677,938**

[22] Filed: **Jul. 10, 1996**

[30] **Foreign Application Priority Data**

Jul. 14, 1995 [JP] Japan 7-201334

[51] **Int. Cl.**[6] **A47C 7/02**
[52] **U.S. Cl.** **297/452.18**; 297/452.2; 297/452.36
[58] **Field of Search** 297/452.18, 452.2, 297/452.31, 452.3, 452.34, 452.36; 138/121; 29/91, 91.1

[56] **References Cited**

U.S. PATENT DOCUMENTS

976,060 11/1910 Fulton 138/121
2,695,038 11/1954 Parce et al. 138/121
3,604,752 9/1971 Macknick 297/452.2
4,695,097 9/1987 Muraishi 297/452.2
5,129,707 7/1992 Yamauchi 297/452.18
5,131,721 7/1992 Okamoto 297/452.18
5,499,863 3/1996 Nakane et al. 297/452.18
5,509,716 4/1996 Kolena et al. 297/452.18

FOREIGN PATENT DOCUMENTS

7031526 A1 2/1995 Japan .

Primary Examiner—Peter M. Cuomo
Assistant Examiner—Anthony D. Barfield
Attorney, Agent, or Firm—Foley & Lardner

[57] **ABSTRACT**

A reversed U-shaped base frame for a seatback frame structure is of a monoblock structure, which includes an upper horizontal tubular portion and two side vertical portions. The two side vertical portions extend downward from axially opposed ends of the upper horizontal tubular portion. The base frame is produced from a shaped metal sheet by pressing and curling the same in such a manner that a given portion of the shaped metal sheet, which is shaped to produce the upper horizontal tubular portion, is curled to have a substantially circular cross section. The feature of the invention is that the curled given portion has circumferentially opposed edges which are overlapped each other.

5 Claims, 3 Drawing Sheets

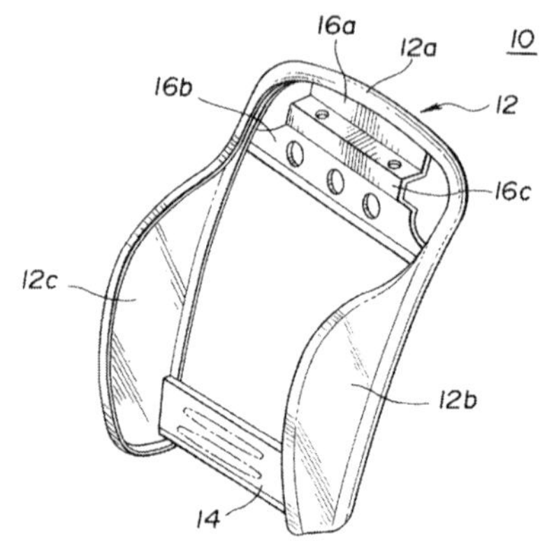

United States Patent [19]

Kolena et al.

[11] **Patent Number: 5,509,716**

[45] **Date of Patent: Apr. 23, 1996**

[54] **VEHICLE SEAT WITH PERIMETER FRAME AND PELVIC CATCHER**

[75] Inventors: **David P. Kolena**, Bloomfield Hills; **Paul A. Glinski**, Chesterfield; **Robert S. Crane**, Waterford; **Mladen Humer**, Eastpointe; **David C. Viano**, Bloomfield Hills; **Richard J. Neely**, Casco, all of Mich.

[73] Assignee: **General Motors Corporation**, Detroit, Mich.

[21] Appl. No.: **335,591**

[22] Filed: **Nov. 8, 1994**

[51] **Int. Cl.**[6] **B60R 21/00**
[52] **U.S. Cl.** **297/216.13**; 297/452.18; 297/216.1
[58] **Field of Search** 297/216.1, 216.13, 297/216.14, 452.18, 452.2, 354.12, 452.53, 452.52

[56] **References Cited**

U.S. PATENT DOCUMENTS

1,598,468 8/1926 Whall .
1,684,062 9/1928 Leach et al. .
2,833,339 5/1958 Liljengren 155/179
3,545,808 12/1970 Gescheidle 297/216
3,586,376 6/1971 Le Mire 297/452.2
4,076,306 2/1978 Satzinger 297/216
4,192,545 3/1980 Higuchi et al. 297/216
4,249,769 2/1981 Barecki .... 296/65 A
4,512,604 4/1985 Maeda et al. .... 296/65 A
4,519,650 5/1985 Terada et al. 297/452
4,526,423 7/1985 Meinershagen et al. 297/440
4,695,097 9/1987 Muraishi 297/452
4,796,954 1/1989 Saito 297/452.2
4,938,527 7/1990 Schmale et al. 297/216
5,044,693 9/1991 Yokota 297/452
5,054,845 10/1991 Vogel 297/216
5,123,706 6/1992 Granzow et al. 297/452
5,129,707 7/1992 Yamauchi 297/460
5,290,089 3/1994 Oleszko et al. 297/216.14
5,310,247 5/1994 Fujimori et al. 297/378.12
5,318,341 6/1994 Griswold et al. 297/362.11
5,328,248 7/1994 Nishiyama 297/452.56

FOREIGN PATENT DOCUMENTS

1130717 5/1962 Germany .
3010662 9/1981 Germany 297/216.14
1011411A 4/1983 U.S.S.R. .

Primary Examiner—Milton Nelson, Jr.
Attorney, Agent, or Firm—Ernest E. Helms

[57] **ABSTRACT**

A vehicle seat is providing including a seat bottom frame; a generally U-shaped seat back frame with an upper cross member with legs pivotally mounted with respect to the seat bottom frame, the seat back cross member being generally at least approximately 470 millimeters along a line generally parallel to the torso of a seated occupant from an H point of the seated occupant, and the cross member being concavely bowed if under approximately 525 millimeters from the H point of the seated occupant; and a deformable lower cross member having ends fixably connected to the seat back legs being concavely bowed, the cross member having a major dimension oriented generally parallel to the torso of the seated occupant, the lower cross member having an upper and a lower end, the lower cross member upper end being vertically above the H point of a seated occupant when the seat back frame is positioned in a normal seating position and where in a rear crash situation, the lower cross member deforms to pivot its lower end further away from the seat back frame legs than its top end to capture the seated occupant's pelvic region between the lower cross member and the bottom frame.

6 Claims, 2 Drawing Sheets

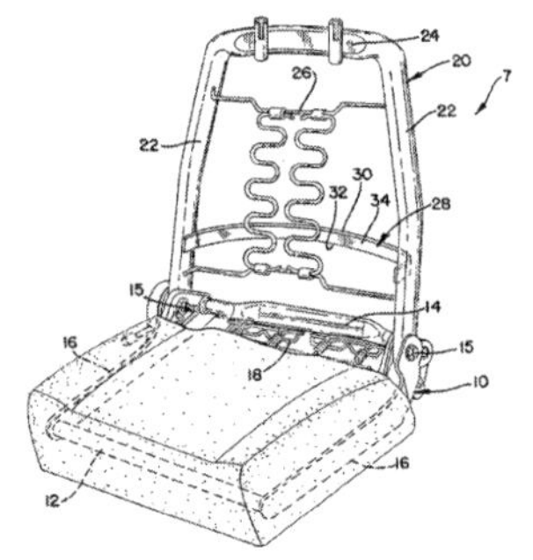

United States Patent [19]

Aufrere et al.

[11] **Patent Number: 5,988,756**

[45] **Date of Patent: Nov. 23, 1999**

[54] **METHOD OF MAKING A VEHICLE SEAT, AND A VEHICLE SEAT MADE BY THE METHOD**

[75] Inventors: **Christophe Aufrere**, Marcoussis; **Mark Moller**, Massy; **François Sensby**, Chatenay Malabry; **Joël Canteleux**, Armenonville, all of France

[73] Assignee: **Bertrand Faure Equipements SA**, Boulogne, France

[21] Appl. No.: **08/941,411**

[22] Filed: **Sep. 30, 1997**

[30] **Foreign Application Priority Data**

Oct. 2, 1996 [FR] France 96 12005

[51] **Int. Cl.**[6] **A47C 7/02**
[52] **U.S. Cl.** **297/452.18**; 297/452.2; 297/391; 72/58
[58] **Field of Search** 297/452.18, 452.2, 297/216.1, 216.13, 216.14, 216.15, 216.16, 216.19, 391; 72/58, 59, 57, 56; 29/421.1

[56] **References Cited**

U.S. PATENT DOCUMENTS

4,192,545 3/1980 Higuchi et al. 297/216.1
4,373,235 2/1983 Volpe 29/416
5,050,932 9/1991 Pipon et al. .... 297/452.18 X
5,070,717 12/1991 Boyd et al. 72/55
5,107,693 4/1992 Olszewski et al. 72/58
5,233,854 8/1993 Bowman et al. 72/58
5,239,852 8/1993 Roper 72/58
5,246,271 9/1993 Boisset .... 297/452.2 X
5,310,247 5/1994 Fujimori et al. .... 297/452.18 X
5,318,341 6/1994 Griswold et al. .... 297/452.18 X
5,333,775 8/1994 Bruggemann et al. 228/157
5,339,667 8/1994 Shah et al. 72/58
5,362,132 11/1994 Griswold et al. .... 297/452.2 X
5,447,360 9/1995 Hewko et al. 297/452.18
5,452,941 9/1995 Halse et al. .... 297/452.2 X
5,468,053 11/1995 Thompson et al. .... 297/452.18 X
5,501,509 3/1996 Urrutia 297/452.18
5,509,716 4/1996 Kolena et al. .... 297/452.18 X
5,547,214 8/1996 Zimmerman, II et al. .... 280/730.2 X
5,547,259 8/1996 Frederick 297/452.18
5,561,902 10/1996 Jacobs et al. .... 72/61 X
5,564,785 10/1996 Schultz et al. 297/452.2
5,567,017 10/1996 Bourgeois et al. .... 297/452.18 X
5,626,396 5/1997 Kuragano et al. .... 297/391 X
5,660,443 8/1997 Pedronne 297/452.2
5,685,614 11/1997 Chabanni 297/452.2
5,697,670 12/1997 Husted et al. .... 297/452.18 X
5,711,577 1/1998 Whalen .... 297/452.18 X

FOREIGN PATENT DOCUMENTS

2 677 935 6/1991 France .
44 00 419 A1 7/1995 Germany .

Primary Examiner—Peter M. Cuomo
Assistant Examiner—Rodney B. White
Attorney, Agent, or Firm—Marshall, O'Toole, Gerstein, Murray & Borun

[57] **ABSTRACT**

A method of making a vehicle seat in which the strength member of the back has a first upright connected to the strength member of the seat proper by a hinge mechanism, a second upright that is not as strong as the first and that is hinged to the strength member of the seat proper, and two horizontal cross-members, the two cross-members and the top portion of the second upright being constituted by a tube folded into a general U-shape and of section that varies along its length. The method includes a step which consists in imparting the final shape to the tube by hydroforming, starting from an initial tube of substantially constant section.

7 Claims, 2 Drawing Sheets

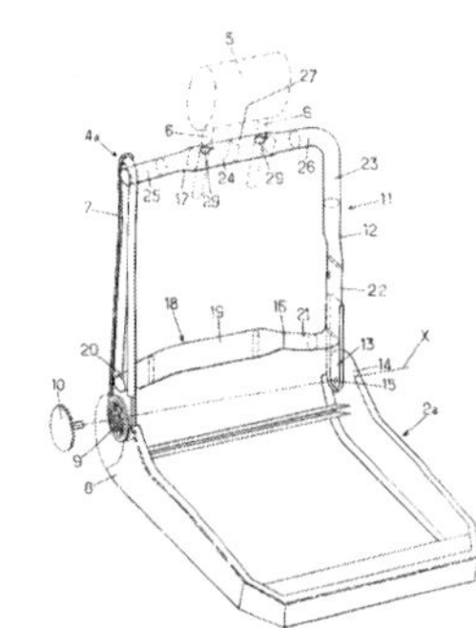

United States Patent [19]

Aumont et al.

[11] **Patent Number: 6,082,823**

[45] **Date of Patent: Jul. 4, 2000**

[54] **BACKREST FRAMEWORK OF AN AUTOMOBILE VEHICLE SEAT**

[75] Inventors: **Jean-Claude Aumont**, Etrechy; **Patrick Daniel**, Paris; **Christophe Aufrere**, Marcoussis, all of France

[73] Assignee: **Betrand Faure Equipments S.A.**, Boulogne Cedex, France

[21] Appl. No.: **09/286,657**

[22] Filed: **Apr. 6, 1999**

[30] **Foreign Application Priority Data**

Apr. 17, 1998 [FR] France 98 05058

[51] **Int. Cl.**[7] **A47C 7/02**; B60N 2/42

[52] **U.S. Cl.** **297/452.2**; 297/452.18; 297/216.13; 297/483; 297/284.1

[58] **Field of Search** 297/452.2, 216.13, 297/483, 452.18, 284.1, 484

[56] **References Cited**

U.S. PATENT DOCUMENTS

4,040,660	8/1977	Barecki	297/216.13
4,626,028	12/1986	Hatsuita et al.	297/284.1 X
4,804,226	2/1989	Schmale	297/216.13
4,889,389	12/1989	White	297/483 X
5,058,953	10/1991	Takagi et al.	297/284.1 X
5,123,706	6/1992	Granzow et al.	297/452.18
5,246,271	9/1993	Boisset	297/483 X
5,253,924	10/1993	Glance	297/216.13 X
5,310,247	5/1994	Fujimori et al.	297/483 X
5,390,982	2/1995	Johnson et al.	297/483 X
5,447,360	9/1995	Hewko et al.	297/216.13 X
5,509,716	4/1996	Kolena et al.	297/216.13
5,516,195	5/1996	Canteleux	297/284.1
5,599,070	2/1997	Pham et al.	297/483
5,641,198	6/1997	Steffens, Jr.	297/483 X
5,645,316	7/1997	Aufrere et al.	297/216.13
5,658,048	8/1997	Nemoto	297/410
5,658,051	8/1997	Vega et al.	297/483
5,681,081	10/1997	Lindner et al.	297/216.13
5,697,670	12/1997	Husted et al.	297/216.13
5,772,280	6/1998	Massara	297/216.13 X
5,823,619	10/1998	Heilig et al.	297/216.13 X
5,823,627	10/1998	Viano et al.	297/216.13 X
5,836,648	11/1998	Karschin et al.	297/216.13 X
5,851,055	12/1998	Lewis	297/216.13 X
5,927,804	7/1999	Cuevas	297/216.13 X

FOREIGN PATENT DOCUMENTS

0 511 100	10/1992	European Pat. Off. .
0 661 190	7/1995	European Pat. Off. .
29 52 064	6/1981	Germany .
36 13 830	10/1987	Germany .
195 01 087	7/1996	Germany .
196 52 939	1/1998	Germany .
WO 97/30865	8/1997	WIPO .

Primary Examiner—Jose V. Chen
Assistant Examiner—Rodney B. White
Attorney, Agent, or Firm—Pollock, Vande Sande & Amernick

[57] **ABSTRACT**

A vehicle seat frame assembly includes two lateral upright side members and a brace connected between the upright side members to form a lower backrest frame. At least one of the upright side members has an upper end and it is rigid from a bottom of the lower backrest frame to the upper end, the upper end further including a point for attaching or passing a seat belt. An upper backrest frame is located above the lower backrest frame and is positioned below the upper end of the at least one upright side member. A hinge is provided for mounting the upper backrest frame at a lower edge thereof to the upright side members for allowing pivotal movement of the upper backrest frame relative to the lower backrest frame.

7 Claims, 3 Drawing Sheets

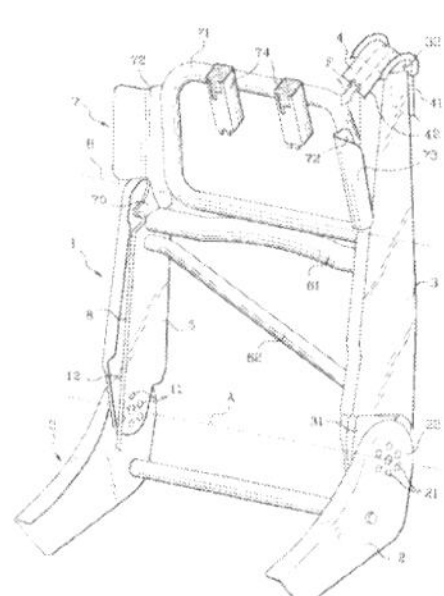

(12) United States Patent

Yoshimura

(10) **Patent No.: US 6,286,902 B1**

(45) **Date of Patent: Sep. 11, 2001**

(54) **SEAT BACK FRAMEWORK OF SEAT**

(75) Inventor: **Masakazu Yoshimura**, Akishima (JP)

(73) Assignee: **Tachi-S Co. Ltd.**, Tokyo (JP)

(*) Notice: Subject to any disclaimer, the term of this patent is extended or adjusted under 35 U.S.C. 154(b) by 0 days.

(21) Appl. No.: **09/458,985**

(22) Filed: **Dec. 10, 1999**

(51) **Int. Cl.**[7] **A47C 7/02**

(52) **U.S. Cl.** **297/452.18**; 297/452.2

(58) **Field of Search** 297/452.18, 452.2, 297/449.1, 452.52

(56) **References Cited**

U.S. PATENT DOCUMENTS

3,767,261	*	10/1973	Rowland	297/452.52 X
3,995,893	*	12/1976	De La Taille et al.	397/452.2 X
4,796,954	*	1/1989	Saito	297/452.2
4,969,688	*	11/1990	Chinomi et al.	297/452.2 X
5,490,718	*	2/1996	Akizuki et al.	297/452.18 X
5,509,716	*	1/1989	Kolena et al.	297/452.18 X

FOREIGN PATENT DOCUMENTS

227800	7/1990	(JP) .
4125749	11/1992	(JP) .

* cited by examiner

Primary Examiner—Peter M. Cuomo
Assistant Examiner—Stephen Vu
(74) *Attorney, Agent, or Firm*—Browdy and Neimark

(57) **ABSTRACT**

A seat back framework of a seat which includes a main frame having two lateral frame sections, a side bracket of a channel cross-section attached to each of the two lateral frame sections, with one end thereof projecting therefrom, and a support wire member extended between the two lateral frame sections, wherein a recession or difference in level is created between that one end of side bracket and one end of the support wire member. In this seat back framework, such one end of the support wire member is welded on and along each of the two lateral frame sections in close proximity to the end of side brackets, thereby reducing or eliminating the foregoing recession or difference in level.

11 Claims, 4 Drawing Sheets

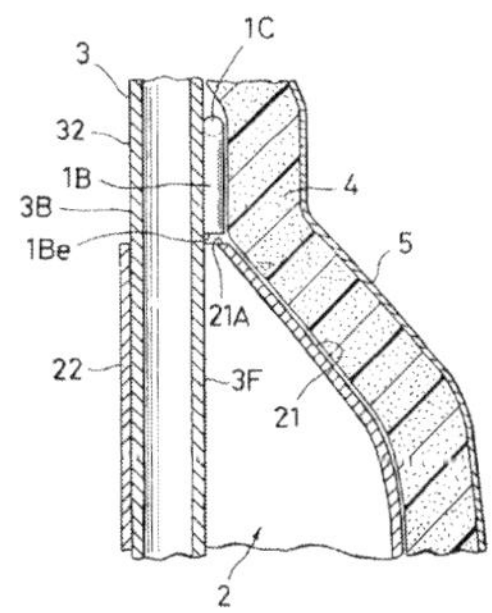

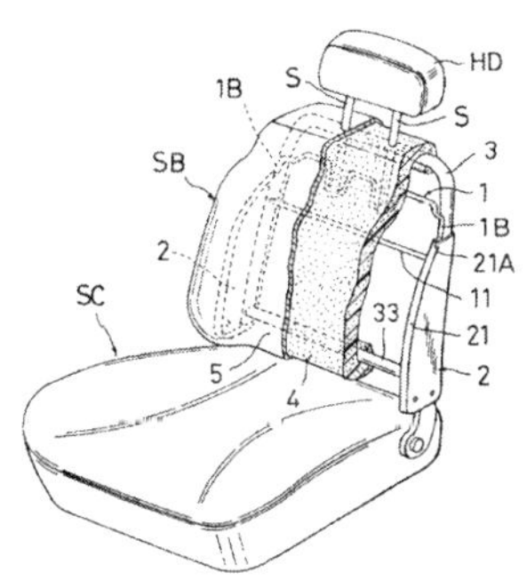

(12) United States Patent

Tsukada

(10) **Patent No.: US 6,290,292 B1**

(45) **Date of Patent: Sep. 18, 2001**

(54) **SEAT BACK STRUCTURE OF HINGED VEHICLE SEAT**

(75) Inventor: **Mitsuru Tsukada**, Akishima (JP)

(73) Assignee: **Tachi-S Co., Ltd.**, Tokyo (JP)

(*) Notice: Subject to any disclaimer, the term of this patent is extended or adjusted under 35 U.S.C. 154(b) by 0 days.

(21) Appl. No.: **09/546,212**

(22) Filed: **Apr. 10, 2000**

(51) **Int. Cl.**[7] **B60N 2/433**

(52) **U.S. Cl.** **297/216.14**; 297/378.12

(58) **Field of Search** 297/216.1, 216.13, 297/216.14, 367, 378.12

(56) **References Cited**

U.S. PATENT DOCUMENTS

5,370,440	*	12/1994	Regala	297/216.14
5,509,716	*	4/1996	Kolena et al.	297/216.13
5,673,971	*	10/1997	Wieclawshi	297/216.14 X
5,749,624	*	5/1998	Yoshida	297/216.13 X
5,884,972	*	3/1999	Deptolla	297/216.14 X

FOREIGN PATENT DOCUMENTS

1-128639	9/1989	(JP) .
3-100530	10/1991	(JP) .

* cited by examiner

Primary Examiner—Peter R. Brown
(74) *Attorney, Agent, or Firm*—Browdy and Neimark

(57) **ABSTRACT**

A structure of seat back frame in a seat back of a hinged vehicle seat, in which a connecting rod is extended therein for actuating a locking mechanism in one of locking and unlocking direction. In the structure, a stopper element is provided vertically of the seat back frame at a point adjacent to a crank-like portion of the connecting rod, with an engagement element disposed between the stopper element and crank-like portion. Upon an external load being applied to a rear side of the seat back, the stopper element is thereby deformed in a direction to the crank-like portion of connecting rod and brought to engagement therewith via the engagement element, so that the connecting rod is prevented against undesired rotation in the unlocking direction.

12 Claims, 2 Drawing Sheets

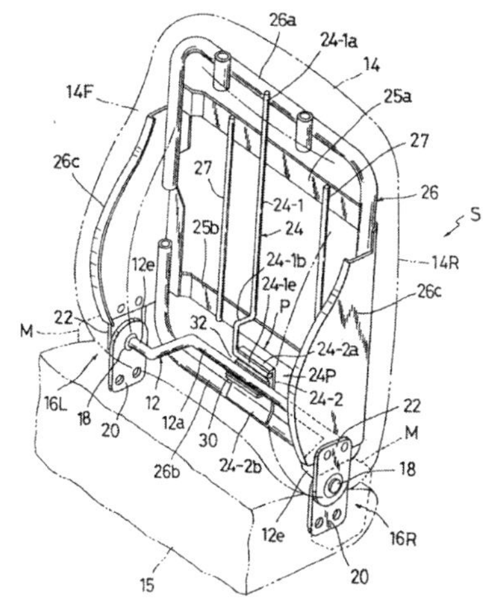

(12) United States Patent

Aufrere et al.

(10) **Patent No.: US 6,361,115 B1**

(45) **Date of Patent: Mar. 26, 2002**

(54) **VEHICLE SEAT WITH ANCHORING WIRE TO SECURE AN ELEMENT ONTO THIS SEAT**

(75) Inventors: **Christophe Aufrere**, Marcoussis; **Joël Canteleux**, Gallardon, both of (FR)

(73) Assignee: **Bertrand Faure Equipements SA**, Boulogne (FR)

(*) Notice: Subject to any disclaimer, the term of this patent is extended or adjusted under 35 U.S.C. 154(b) by 0 days.

(21) Appl. No.: **09/490,510**

(22) Filed: **Jan. 25, 2000**

(30) **Foreign Application Priority Data**

Feb. 3, 1999 (FR) 99 01228

(51) **Int. Cl.**[7] **A47C 7/02**

(52) **U.S. Cl.** **297/452.18**; 297/216.11; 297/256.16

(58) **Field of Search** 297/452.18, 250.1, 297/130, 256.16, 216.11

(56) **References Cited**

U.S. PATENT DOCUMENTS

4,850,644	A	*	7/1989	Kazaoka et al.	297/452.18 X
5,383,708	A		1/1995	Kaisha	
5,509,716	A	*	4/1996	Kolena et al.	297/452.18 X
5,626,395	A	*	5/1997	Aufrere	297/452.18
5,918,934	A	*	3/1999	Siegrist	297/216.11 X
5,941,601	A	*	8/1999	Scott et al.	297/250.1 X
6,030,046	A	*	2/2000	Dorow	297/253 X
6,183,044	B1	*	2/2001	Koyanagi et al.	297/253 X

FOREIGN PATENT DOCUMENTS

DE	196 50087 c	12/1997
DE	198 067 838	1/1999
EP	0 560 184 A	9/1993
EP	0 694 436 A	1/1996
EP	0 858 928 A	8/1998

* cited by examiner

Primary Examiner—Peter M. Cuomo
Assistant Examiner—Stephen Vu
(74) *Attorney, Agent, or Firm*—Marshall, Gerstein & Borun

(57) **ABSTRACT**

A seat for a motor vehicle has a seat with two side flanges mutually joined by a transverse tube. An anchoring wire for securing an element designed to be placed on the seat is joined at two ends to a flap extending in the vicinity of the transverse tube. Extending between the two ends is an open loop and a fixing section. The anchoring wire is deformable under the action of a strong force exerted on the fixing section, causing the loop to contract around the tube and transmit the force to the tube.

12 Claims, 4 Drawing Sheets

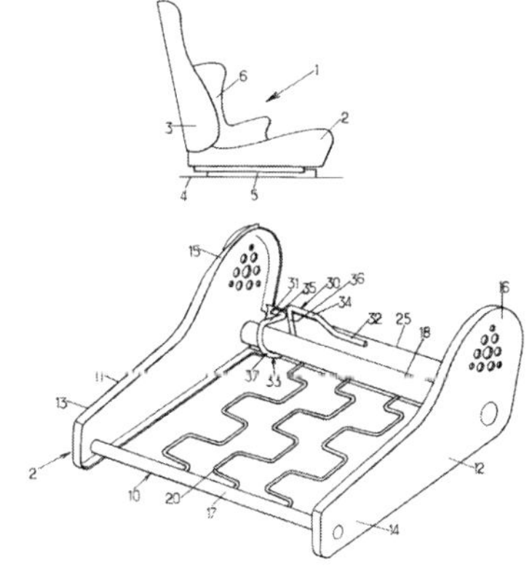

(12) **United States Patent**
Iwamoto et al.

(10) **Patent No.: US 6,371,561 B1**
(45) **Date of Patent: Apr. 16, 2002**

(54) **FRAME STRUCTURE FOR A SEATBACK OF A SEAT**

(75) Inventors: **Yoshiyuki Iwamoto; Shuji Kumano; Kazuhiro Matsuhashi**, all of Hiroshima (JP)

(73) Assignee: **Mazda Motor Corporation**, Hiroshima (JP)

(*) Notice: Subject to any disclaimer, the term of this patent is extended or adjusted under 35 U.S.C. 154(b) by 0 days.

(21) Appl. No.: **09/590,134**

(22) Filed: **Jun. 9, 2000**

(30) **Foreign Application Priority Data**

Jun. 11, 1999 (JP) 11-165047

(51) **Int. Cl.**[7] **A47C 7/02**
(52) **U.S. Cl.** **297/452.18**
(58) **Field of Search** 297/284.1, 284.4, 297/452.18

(56) **References Cited**

U.S. PATENT DOCUMENTS

5,101,811 A * 4/1992 Brunswick
5,509,716 A * 4/1996 Kolena et al.
5,697,670 A * 12/1997 Husted et al.
6,186,594 B1 * 2/2001 Valiquette et al.

FOREIGN PATENT DOCUMENTS

JP 10-85079 4/1998

* cited by examiner

Primary Examiner—Milton Nelson, Jr.
(74) *Attorney, Agent, or Firm*—Nixon Peabody LLP; Donald R. Studebaker

(57) **ABSTRACT**

A seatback frame structure (F) for a seatback (**2**) of a seat (S) comprises a generally rectangular seatback frame (**4**) and a lower or primary wire frame (**9**) which has base frame sections (**9***a*) extending from opposite vertical side frame sections (**4***a*) of the rectangular seatback frame (**4**) at a height of lumbar vertebras, respectively, an upward extending support frame section (**91**) which extends upward from the base frame sections (**9***a*) and comprises opposite vertical side frame sections (**91***b*) extending from the base sections (**4***a*), respectively, and a horizontal frame section (**91***a*) extending between the opposite vertical side frame sections (**91***b*) at a height of shoulder blades, and an upper or subsidiary wire frame (**8**) extending between the opposite vertical side frame sections (**4***a*) behind the upward extending support frame section (**91**). The upward extending support frame section (**91**) at its upper part is bent backward so as to extend along the backbone of a person P sitting on the seat S.

20 Claims, 15 Drawing Sheets

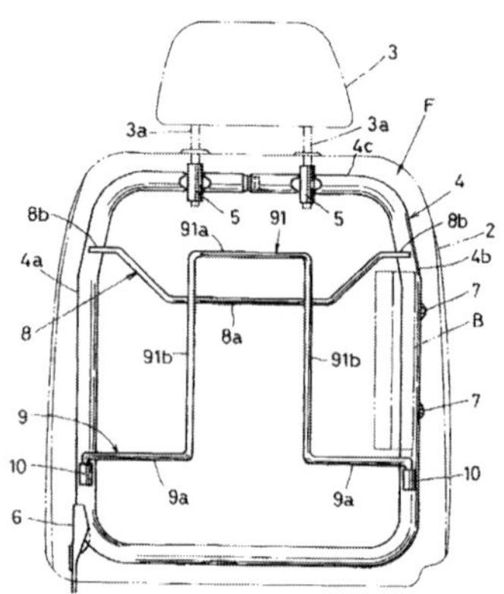

CHAPTER 11:

HEAD RESTRAINT PATENTS

Chapter 11

Head Restraint Patents

David C. Viano
ProBiomechanics LLC

ABSTRACT

This Chapter reviews the history of automotive seat head restraint patents and includes examples of various concepts and designs. It includes examples of designs for adjustable and integrated head restraints in the seatback. The typical two post approach to adjusting the head restraint on the seatback remains in use today. By the 1970s a range of power adjustments were added to the seat and head restraint to vary the height and fore-aft position. By the 1980s, a range of inflatable airbags and bellows were added to enhance control of the occupant's head and neck in rear, side and rollover crashes. The first page from 31 USPTO patents is included to show head restraint concepts and technologies considered for crash safety and comfort. The patent history and example concepts for head restraints show a rich history of ideas for head support and occupant protection.

METHODOLOGY

US patents were located by searching primary, secondary and tertiary references for head restraints safety topics using the US Patent and Trademark Office website: http://patft.uspto.gov/netahtml/search. Efforts were made to locate the earliest and most interesting examples of patents on particular technologies, but there is every reason to assume that many other concepts and technologies related to seat designs for crash safety were not located.

Patents prior to 1975 are accessible only by a patent number or classification, which made the search depend on secondary and tertiary references in patents that were located. Many of these patents included cited references to earlier patents, which were investigated. However, the earliest US patents did not include cited references to prior art that were reviewed by the patent examiner. This complicated the search for patents before the 1950s. Any patent of significance to the history of seats and crash safety that is not included was an inadvertent oversight. The items located and reported should be considered examples of the history of technology and patents on seat head restraint safety. They do show a long-standing history of concepts for occupant safety by seat design and integration of other safety features.

RESULTS

A wide range of head restraint patents have been granted for integrated and adjustable designs. By the 1970s, power adjustable concepts emerged. They were eventually designed to automatically adjust the height and fore-aft position behind the occupant's head. In the 1980s a range of inflatable bellows and airbag designs were added to the head restraint for protection in rear, side and rollover crashes.

The designs of the 1990s-2000s further refine automatic adjustability, and integration of safety functions with the head restraint to provide a systems approach to safety. This includes automatic adjustment of height and gap, and dynamic adjustments triggered by crash sensors coupled with activation of inflatable airbags for head and neck support.

April 1, 1958 C. J. BARECKI ET AL 2,828,810

ADJUSTABLE HEADREST

Filed April 9, 1956 3 Sheets-Sheet 1

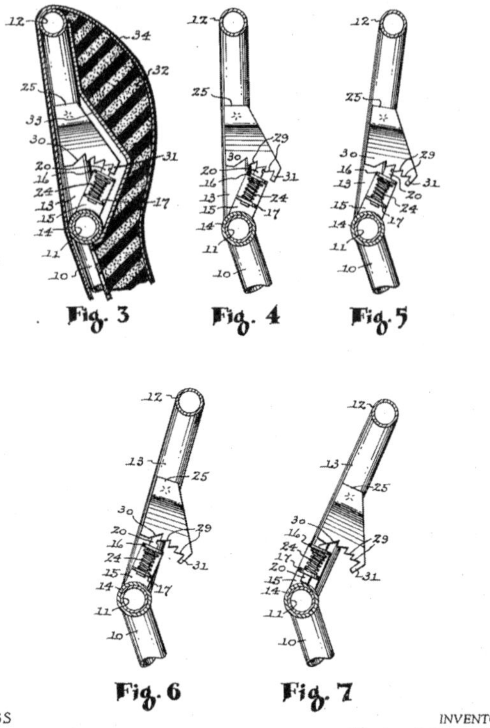

Fig. 3
Fig. 4
Fig. 5
Fig. 6
Fig. 7

WITNESS

Harvey J. Rowly

INVENTOR

Chester J. Barecki
Oscar J. Nelson and

BY John S. Braddock

ATTORNEY

Jan. 7, 1969 T. BISLAND 3,420,572

AUTOMATIC HEADREST AND NECK PROTECTOR

Filed Jan. 20, 1964 Sheet 1 of 3

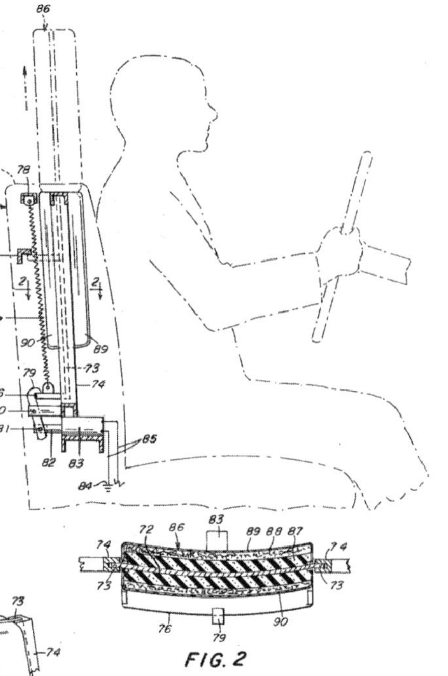

FIG. 1
FIG. 2

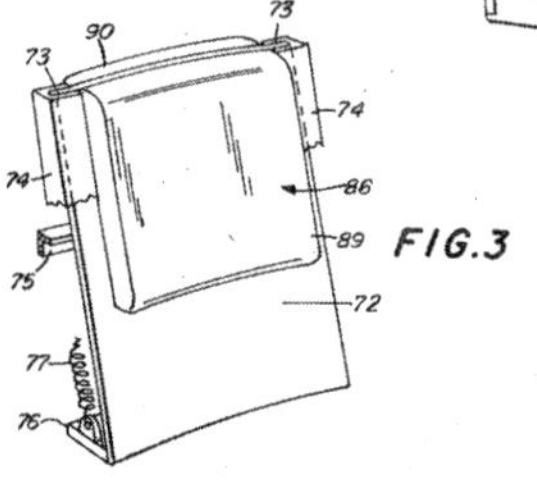

FIG. 3

INVENTOR

THEODORE BISLAND

BY Fisher, Christen, Sabol & Caldwell

ATTORNEYS

June 17, 1958 J. J. BRANDON 2,839,125

ADJUSTABLE HEAD ROLL FOR SEAT STRUCTURE

Filed May 19, 1954

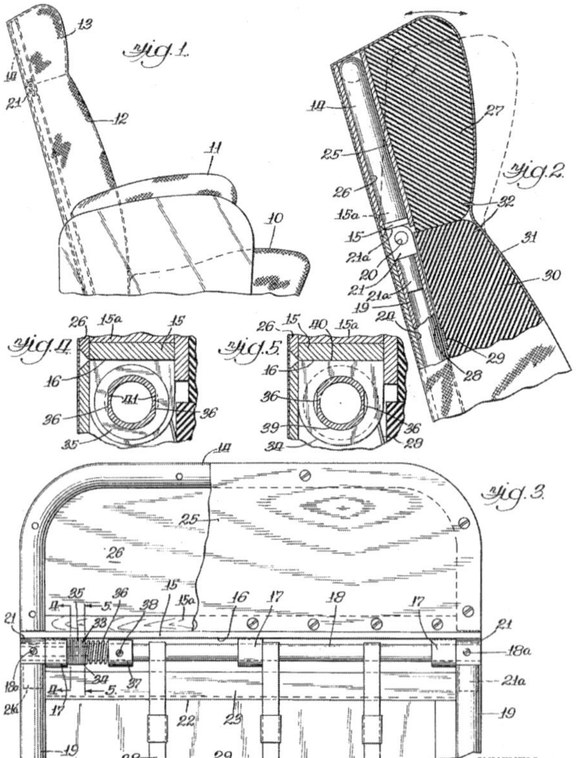

Fig. 1.
Fig. 2.
Fig. 4.
Fig. 5.
Fig. 3.

INVENTOR.

Joseph J. Brandon

BY

Harvey M. Gillespie

Atty.

Feb. 11, 1969 W. T. DOWNS ET AL 3,427,073

SEAT ASSEMBLY

Filed Oct. 22, 1965 Sheet 1 of 7

Fig. 1
Fig. 2

INVENTORS

William T. Downs

BY & Robert I. Homier

Barnard, McGlynn & Reising

ATTORNEYS

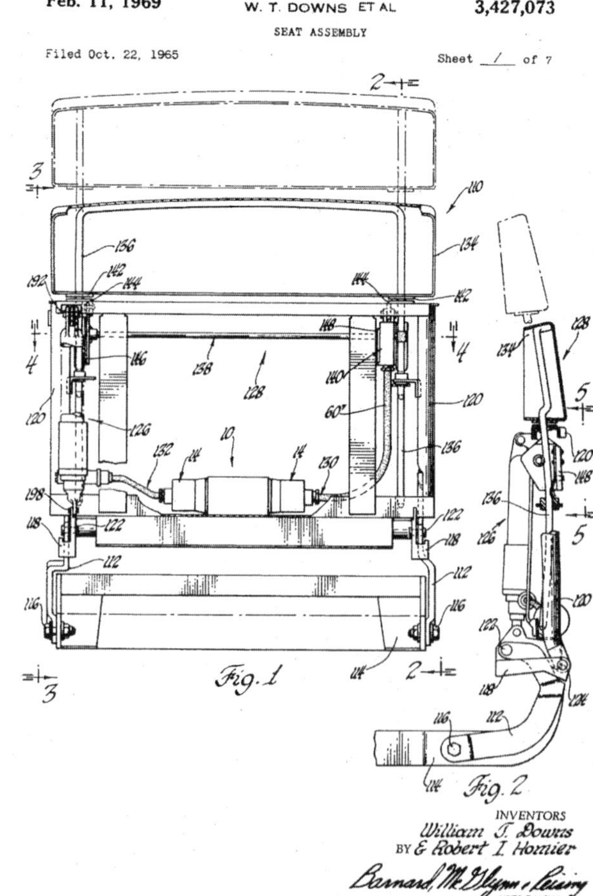

March 17, 1970 H. A. BURGERT 3,501,137

UPHOLSTERING SPRING, ESPECIALLY FOR THE BACK OF MOTOR VEHICLE SEATS

Filed Oct. 13, 1967 2 Sheets-Sheet 1

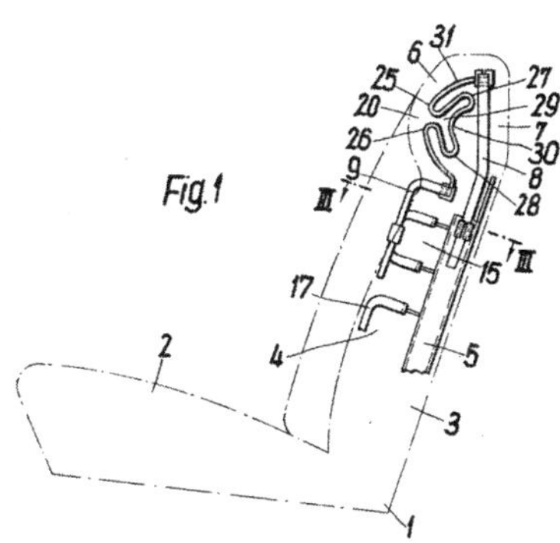

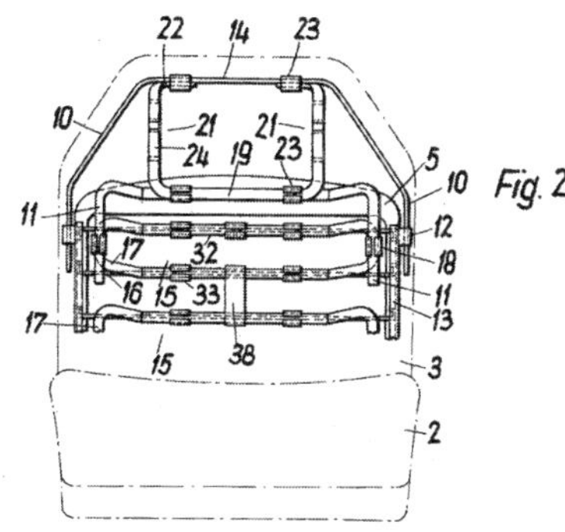

Inventor:
Herbert A. Burgert
By [signature]

United States Patent

Schiesterl et al.

[11] 3,703,313

[45] Nov. 21, 1972

[54] SEAT, ESPECIALLY FOR MOTOR VEHICLES

[72] Inventors: **Gerhard Schiesterl**, Stuttgart; **Helmut Wulf**, Nellingen, both of Germany

[73] Assignee: **Daimler-Benz Aktiengesellschaft**, Stuttgart, Germany

[22] Filed: **July 29, 1971**

[21] Appl. No.: **167,217**

[30] **Foreign Application Priority Data**

July 29, 1970 Germany..........P 20 37 565.9

[52] U.S. Cl. ....**297/391**, 280/150 AB, 297/384, 297/395

[51] Int. Cl. **A47c 7/36**

[58] Field of Search......297/384, 388, 397, 390, 385, 297/395, 404; 280/150 SB, 150 AB, 150 B

[56] **References Cited**

UNITED STATES PATENTS

3,397,911 8/1968 Brosius......................297/216
3,510,150 5/1970 Wilfert......................297/391
3,525,535 8/1970 Kobori........................297/395

FOREIGN PATENTS OR APPLICATIONS

6,615,335 5/1967 Netherlands........280/150 AB

Primary Examiner—Bobby R. Gay
Assistant Examiner—Darrell Marquette
Attorney—Craig, Antonelli & Hill

[57] **ABSTRACT**

A seat, especially for motor vehicles, in which a gas cushion acting as headrest and automatically inflatable in case of an accident is arranged at the backrest of the seat; at least one belt is arranged behind the gas cushion for absorbing the rebound forces of the head of the passenger impinging against the gas cushion whereby the belt is arranged, on the one hand, within the area of the vehicle roof, and on the other, at the seat.

5 Claims, 2 Drawing Figures

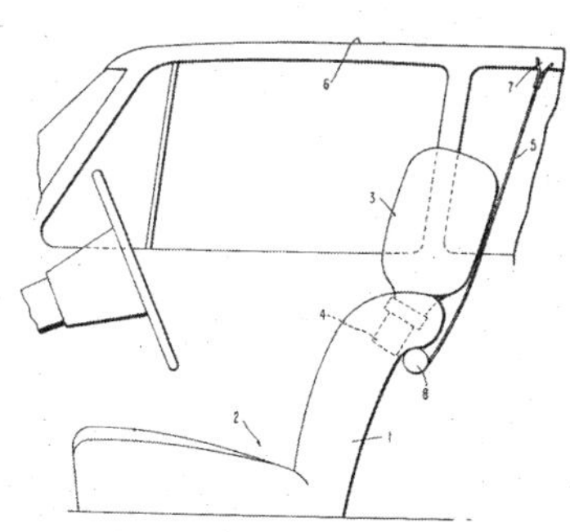

United States Patent

[11] 3,592,508

[72] Inventor **Frederick Druseikis** **Kettering, Ohio**
[21] Appl. No. **2,862**
[22] Filed **Jan. 14, 1970**
[45] Patented **July 13, 1971**
[73] Assignee **General Motors Corporation** **Detroit, Mich.**

[54] **POWER ACTUATED HEADREST ASSEMBLY**
4 Claims, 6 Drawing Figs.

[52] **U.S. Cl.**.. 297/410
[51] **Int. Cl.**.. **A47c 7/36**
[50] **Field of Search**.. 297/410, 397, 353, 330, 348; 182/40, 41, 208

[56] **References Cited**

UNITED STATES PATENTS

2,961,060 11/1960 Taylor.......................... 182/40
2,985,229 5/1961 Shamblin...................... 297/410
3,017,780 1/1962 Kienhofer...................... 297/410
3,311,413 3/1967 Martens........................ 297/410
3,385,397 5/1968 Robinsky...................... 182/41
3,427,073 2/1969 Downs.......................... 297/410
3,462,193 8/1969 Tamura......................... 297/410

Primary Examiner—Francis K. Zugel
Attorneys—J. L. Carpenter, E. J. Biskup and Peter D. Sachtjen

ABSTRACT: A power actuated headrest assembly for a motor vehicle includes a headrest having a downwardly extending support member that slidingly engages a guide member carried interior of a seat back. A plastic tape extends along the support member and has a lower end connected to the guide member and a free upper end that extends into the headrest. An electric motor carried by the headrest includes gear means that drivingly engage the tape for selectively vertically moving the headrest relative to the seat back with the tape serving as a columnar load bearing member.

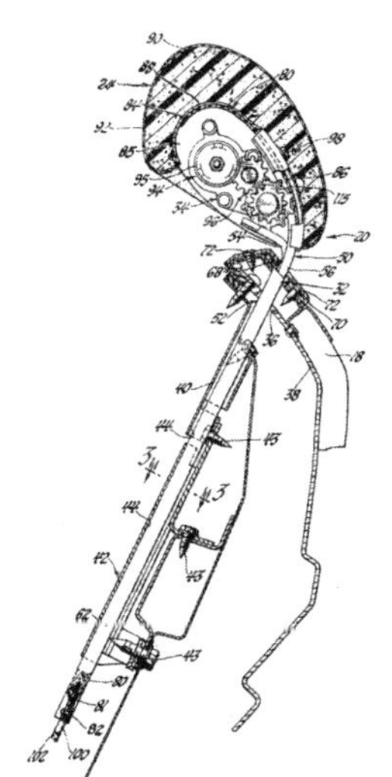

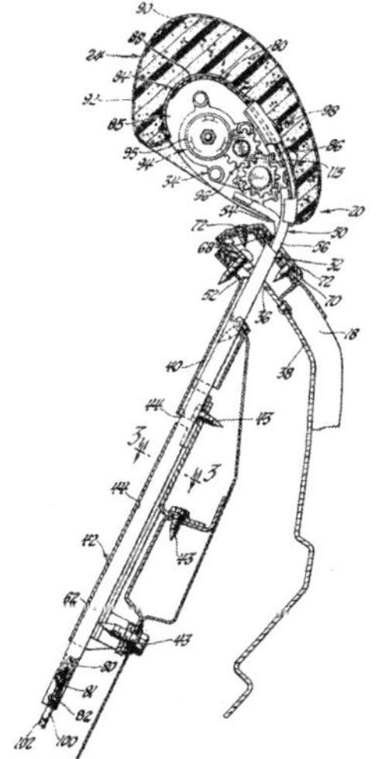

United States Patent [19]

Inoue et al.

[11] 3,729,228

[45] Apr. 24, 1973

[54] VEHICLE SEAT PROVIDED WITH A HEAD REST

[75] Inventors: **Masahiko Inoue, Takashi Saito**, both of Toyota, Japan

[73] Assignee: **Toyota Jidosha Kogyo Kabushiki Kaisha**, Toyota, Japan

[22] Filed: **Ap. 1, 1971**

[21] Appl. No.: **130,214**

[30] **Foreign Application Priority Data**

Apr. 10, 1970 Japan..................................45/30592

[52] **U.S. Cl.**...**297/396**
[51] **Int. Cl.**...**A47b 11/00**
[58] **Field of Search**.....................291/216, 397, 396, 291/398, 399, 452

[56] **References Cited**

UNITED STATES PATENTS

3,264,382 8/1966 Angell et al.........................297/396
3,537,751 11/1970 Inoue et al..........................297/452
3,515,434 6/1970 Sugiura et al........................297/396
3,498,672 3/1970 Leichtl...............................297/397
3,429,615 2/1969 Belk...................................297/397

Primary Examiner—Bernard A. Gelak
Assistant Examiner—Garry Moore
Attorney—Toren & McGeady

[57] **ABSTRACT**

A seat having a head rest wherein the head rest is formed integrally with a back of the seat. Said seat comprises a reinforcement for the head rest provided on the top of a back frame, a shock absorber for the head rest formed as a unit in the shape of a bag for covering said reinforcement for the head rest and an upper portion of said back frame, and means for firmly securing said shock absorber for the head rest to the back.

1 Claim, 6 Drawing Figures

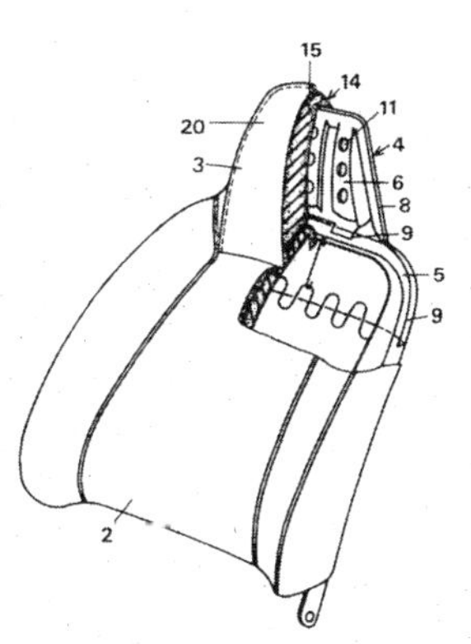

United States Patent [19]

Caldemeyer

[11] **3,738,706**

[45] **June 12, 1973**

[54] **MOTOR DRIVEN HEADREST AND BACK FOR RECLINER CHAIR**

[76] Inventor: **Daniel F. Caldemeyer,** 4300 Jennings Lane, Evansville, Ind. 47712

[22] Filed: **Aug. 30, 1971**

[21] Appl. No.: **175,997**

[52] **U.S. Cl.** **297/410,** 297/354
[51] **Int. Cl.** **A47c 7/36**
[58] **Field of Search** 297/216, 330, 353, 297/363, 410; 312/306

[56] **References Cited**

UNITED STATES PATENTS

3,343,875 9/1967 Ferrara 297/410
3,286,971 11/1966 Walter et al. 297/363

FOREIGN PATENTS OR APPLICATIONS

1,218,541 1/1971 Great Britain 297/410

Primary Examiner—Casmir A. Nunberg
Attorney—Woodard, Weikart, Emhardt & Naughton

[57] **ABSTRACT**

A chairback has adapter plates pivotally attached to each side thereof and readily securable to a chair base or seat mount structure. Sector gears on the chairback engage pinions in gearheads mounted to the adapter plates for motor driven tilting of the chairback. A headrest linearly movable vertically in the chairback has a gear rack post extending into the chairback centrally thereof and engaged by a pinion in a gearhead secured to the lower back frame to drive the headrest up or down. Widely spaced slides on the headrest serve as guides and supports therefor. One embodiment includes a chairback cushion vertically movable in response to headrest extension.

26 Claims, 17 Drawing Figures

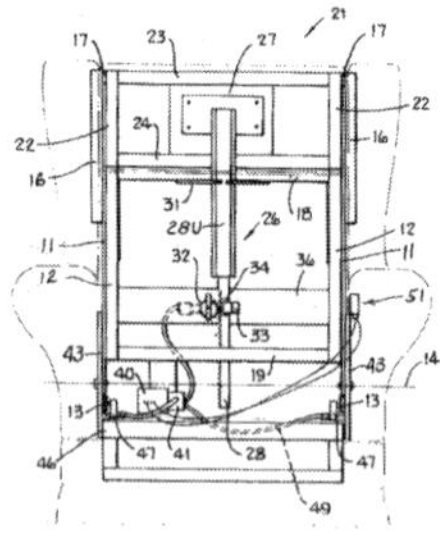

United States Patent [19]

Nishino

[11] **Patent Number: 4,549,766**

[45] **Date of Patent: Oct. 29, 1985**

[54] **VEHICLE SEAT HEADREST**

[75] Inventor: **Takaichi Nishino,** Tokyo, Japan

[73] Assignee: **Tachikawa Spring Co., Ltd.,** Tokyo, Japan

[21] Appl. No.: **413,503**

[22] Filed: **Aug. 31, 1982**

[51] **Int. Cl.⁴** **A47C 7/36**
[52] **U.S. Cl.** **297/396;** 297/391; 403/208
[58] **Field of Search** 297/396, 216, 452, 397, 297/399, 391; 403/208; 248/214

[56] **References Cited**

U.S. PATENT DOCUMENTS

1,775,158 9/1930 Campbell 403/208
3,498,670 3/1970 Finch et al. 297/396
3,729,228 4/1973 Inoue et al. 297/396

Primary Examiner—William E. Lyddane
Assistant Examiner—Mark W. Binder
Attorney, Agent, or Firm—Sandler & Greenblum

[57] **ABSTRACT**

A seat structure containing a frame for a back seat, and a panel for a headrest, which panel is welded in a surrounding manner about said frame.

4 Claims, 5 Drawing Figures

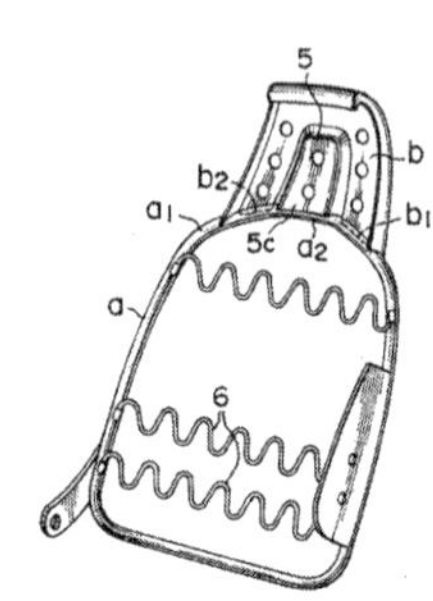

United States Patent [19]

Kapanka

[11] **4,113,310**

[45] **Sep. 12, 1978**

[54] **HEADREST FOR VEHICLES**

[75] Inventor: **Harley L. Kapanka,** Utica, Mich.

[73] Assignee: **General Motors Corporation,** Detroit, Mich.

[21] Appl. No.: **862,954**

[22] Filed: **Dec. 21, 1977**

[51] **Int. Cl.²** **A47C 1/10**
[52] **U.S. Cl.** **297/408**
[58] **Field of Search** 297/391, 408, 112, 114, 297/409, 410

[56] **References Cited**

U.S. PATENT DOCUMENTS

2,886,097 5/1959 Katz .... 297/410 X
3,948,562 4/1976 Graebner et al. 297/408

Primary Examiner—James G. Mitchell
Attorney, Agent, or Firm—John P. Moran

[57] **ABSTRACT**

The drawings illustrate an angularly adjustable head restraint assembly for use with the backrest of a vehicle seat. The assembly includes a post carried by the backrest, a headrest pivotally connected to the post and adapted to being manually pivotally moved from a horizontally oriented reclined attitude to a vertically oriented head support attitude, locking means for locking the headrest in the vertically oriented head support attitude, a cam with cooperating overcenter spring means for causing the cam to snap overcenter once it is provided forwardly from the head support attitude by the headrest to urge the locking means out of its locked relationship with the headrest and permit the headrest to be lowered toward the reclined attitude, and a projection formed on the locking means for contacting the cam and causing it to snap overcenter in the other direction just prior to the headrest attaining the reclined attitude into position for the next sequence.

2 Claims, 5 Drawing Figures

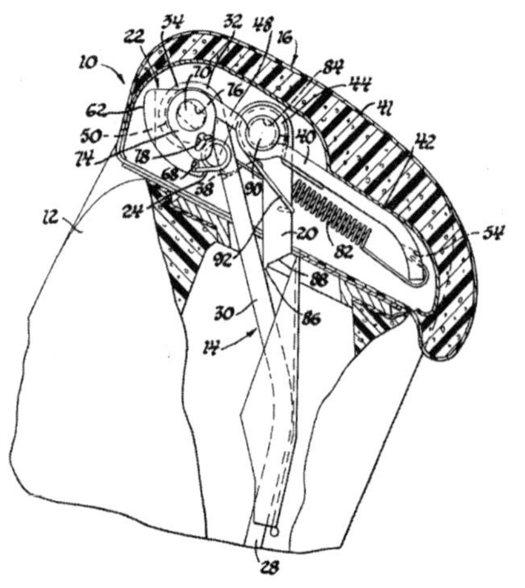

United States Patent [19]

Russo et al.

[11] **Patent Number: 4,693,515**

[45] **Date of Patent: Sep. 15, 1987**

[54] **HEADREST FOR AN AUTOMOTIVE VEHICLE SEAT**

[75] Inventors: **Vincent Russo,** Cary, N.C.; **Max O. Heesch,** Brooklyn, Mich.

[73] Assignee: **ITT Corporation,** New York, N.Y.

[21] Appl. No.: **923,335**

[22] Filed: **Oct. 27, 1986**

[51] **Int. Cl.⁴** **A47C 1/10**
[52] **U.S. Cl.** **297/391;** 297/284; 297/408; 297/409; 297/410
[58] **Field of Search** 297/408, 409, 391, 410, 297/284

[56] **References Cited**

U.S. PATENT DOCUMENTS

3,427,073 2/1969 Downs et al. .... 297/410 X
3,592,508 7/1971 Druseikis 297/410
3,738,706 6/1973 Caldemeyer 297/410
4,514,010 4/1985 Gonzales 297/284

FOREIGN PATENT DOCUMENTS

2932345 2/1981 Fed. Rep. of Germany 297/410

Primary Examiner—James T. McCall
Attorney, Agent, or Firm—James B. Raden; Donald J. Breh

[57] **ABSTRACT**

A headrest for an automotive vehicle seat is disclosed including a headrest frame including a lower portion extending along an upper exterior portion of the seatback and mounted to a mounting frame for generally pivotable forward and rearward movement. A first track provides for the forward and rearward movement and a bladder mounted between the mounted frame and headrest frame selectively expands and contracts to move the headrest frame along the track. The mounting frame is provided with a elongated portion extending into the interior of the seatback and includes a linear gear operated by a reversible motor driven gear for selectively vertically extending and retracting the headrest.

14 Claims, 8 Drawing Figures

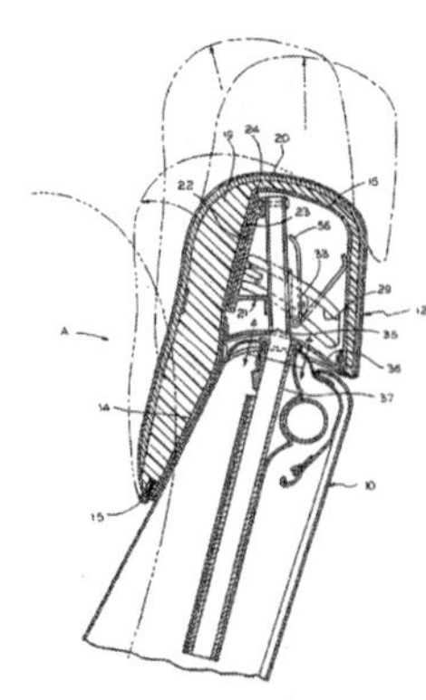

United States Patent [19]

Collier et al.

[11] **Patent Number: 4,779,928**

[45] **Date of Patent: Oct. 25, 1988**

[54] **AUTOMOTIVE HEAD RESTRAINT**

[75] Inventors: **John Collier**, Brantford; **Leonard Neal**, Windsor, both of Canada

[73] Assignee: **Tamco Limited**, Windsor, Canada

[21] Appl. No.: **92,405**

[22] Filed: **Sep. 1, 1987**

[30] **Foreign Application Priority Data**

Sep. 9, 1986 [CA] Canada 517782

[51] **Int. Cl.**[4] **A47C 1/10**
[52] **U.S. Cl.** **297/391**; 297/408; 297/216
[58] **Field of Search** 297/391, 410, 216, 408

[56] **References Cited**

U.S. PATENT DOCUMENTS

3,190,686	6/1965	Smiler	297/391
3,205,005	9/1965	Brown	297/397
3,223,447	12/1965	Terracini	297/397
3,794,382	2/1974	Bloomfield et al.	297/216 X
4,527,834	7/1985	Zyngier	297/410
4,549,766	10/1985	Nishino	297/396
4,598,950	7/1986	Fourrey	297/216 X

FOREIGN PATENT DOCUMENTS

826463	11/1969	Canada .	
1039639	10/1978	Canada .	
2204627	8/1973	Fed. Rep. of Germany	297/391
2839015	3/1980	Fed. Rep. of Germany	297/391
1009546	11/1965	United Kingdom	297/216

Primary Examiner—James T. McCall
Attorney, Agent, or Firm—Browdy and Neimark

[57] **ABSTRACT**

A head restraint for a vehicle seat is disclosed. The restraint comprises a support member for insertion into the vehicle seat and a shock absorbing assembly secured to the upper end of the support member and which comprises a wire lattice form having vertical and horizontal members defining the height, width and depth of said assembly and being welded to the support member at points of contact therewith. An elesticized foam padding is positioned over the lattice form.

10 Claims, 11 Drawing Sheets

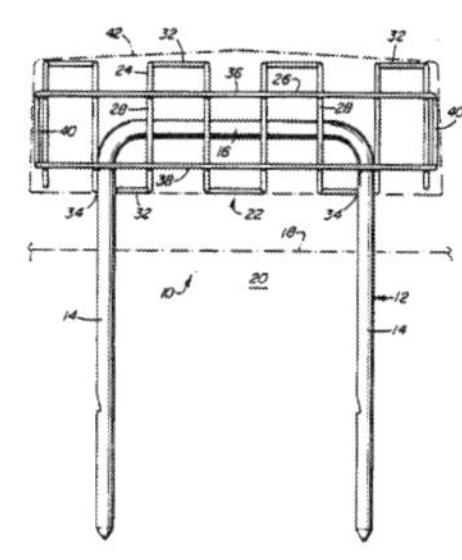

United States Patent [19]

Nemoto

[11] **Patent Number: 4,865,388**

[45] **Date of Patent: Sep. 12, 1989**

[54] **HEADREST FOR AUTOMOTIVE SEAT**

[75] Inventor: **Akira Nemoto**, Akishima, Japan

[73] Assignee: **Tachi-S Co., Ltd.**, Tokyo, Japan

[21] Appl. No.: **332,137**

[22] Filed: **Apr. 3, 1989**

[51] **Int. Cl.**[4] **A47C 7/36**
[52] **U.S. Cl.** **297/403**; 297/408; 297/DIG. 3
[58] **Field of Search** .... 297/403, 408, 391, DIG. 3

[56] **References Cited**

U.S. PATENT DOCUMENTS

4,017,118	4/1977	Cawley	297/391
4,123,104	10/1978	Andres et al.	297/391
4,415,203	11/1983	Cawley	297/DIG. 3
4,612,677	9/1986	Crossett	297/DIG. 3
4,720,146	1/1988	Mawbey et al.	297/408
4,761,011	8/1988	Sereboff	297/DIG. 3
4,778,218	10/1988	Suman	297/391

Primary Examiner—Francis K. Zugel
Attorney, Agent, or Firm—Browdy and Neimark

[57] **ABSTRACT**

A headrest for an automotive seat which comprises a head contact portion, a stretchable rear portion which includes an elastic webbing, and an air bag disposed within those two portions, all of them being fixed upon the top of seat back of the seat. The expansion or shrivelling of the air bag, by using an air pump, causes the headrest to raise for turning upright the head contact portion, or permits the air bag to be compressed by the contracting force of the webbing, thereby lowering the headrest.

6 Claims, 1 Drawing Sheet

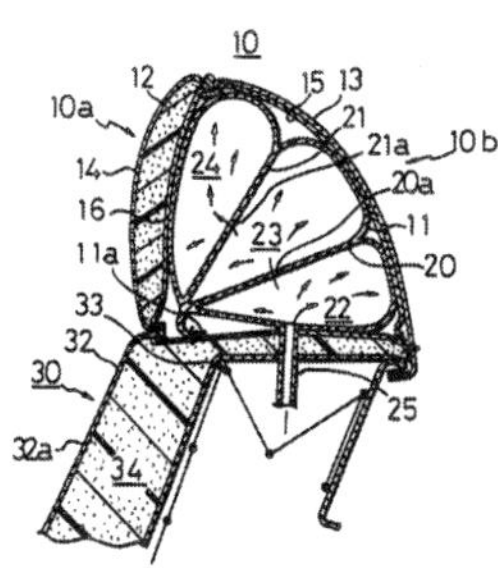

United States Patent [19]

O'Sullivan et al.

[11] **Patent Number: 4,856,848**

[45] **Date of Patent: Aug. 15, 1989**

[54] **MANUAL HEADREST**

[75] Inventors: **Terence J. O'Sullivan**, Sterling Height; **George W. Mould**, Utica; **David E. Armstrong**, Rochester; **Graham S. Foulkes**, Sterling Height; **Michael B. Davis**, Bloomfield Hills, all of Mich.

[73] Assignee: **General Motors Corporation**, Detroit, Mich.

[21] Appl. No.: **213,668**

[22] Filed: **Jun. 30, 1988**

[51] **Int. Cl.**[4] **A47C 7/36**
[52] **U.S. Cl.** **297/391**; 297/409
[58] **Field of Search** 297/409, 408, 410, 356, 297/391

[56] **References Cited**

U.S. PATENT DOCUMENTS

652,001	6/1900	Keil	297/356 X
1,471,168	10/1923	Katz	297/409 X
1,746,091	2/1930	Skidmore	297/409 X
2,828,810	4/1958	Barecki et al.	297/408 X
4,191,422	3/1980	Inasawa et al.	297/409 X
4,265,482	5/1981	Nishimura et al.	297/409 X
4,640,549	2/1987	Yokota	297/408 X
4,674,797	6/1987	Tateyama	297/408
4,762,367	8/1988	Denton	297/409

Primary Examiner—Francis K. Zugel
Attorney, Agent, or Firm—Ernest E. Helms

[57] **ABSTRACT**

The present invention in a preferred embodiment provides an apparatus and method of utilization there of of a fore and aft adjustable vehicle seat headrest. The headrest includes a base providing means of attachment of the headrest to said vehicle seat. A frame is connected with the base. A cushion support with an attached shaft is slidably mounted with the frame for fore and aft movement. The shaft has a series of unidirectional cam slots. A spring biases the cushion support to a rearward position with respect to the frame. Levers pivotally connected with the frame engage with the shaft slots to set the position of the headrest. Release means are provided to remove the levers and retain the levers from reengagement with the slots upon extreme forward movement of the cushion support thereby allowing the spring to return the cushion support to an extreme rearward position. Additionally, means are provided to reengage the levers with the slots whereby the cushion support is reset in the extreme rearward position.

8 Claims, 7 Drawing Sheets

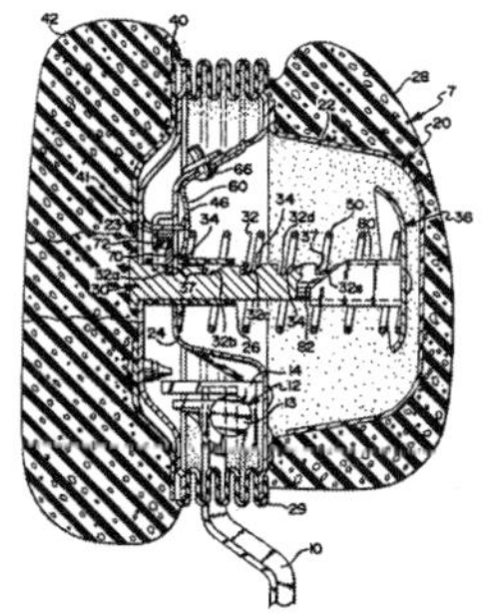

United States Patent [19]

Sekido et al.

[11] **Patent Number: 5,082,326**

[45] **Date of Patent: Jan. 21, 1992**

[54] **VEHICLE SEAT WITH AUTOMATIC ADJUSTMENT MECHANISMS UTILIZING INFLATABLE AIR BAGS**

[75] Inventors: **Hiroshi Sekido**, Chiba; **Tadashi Sakuma**, Kanagawa; **Toshimichi Hioki**, Gifu, all of Japan

[73] Assignee: **Okamoto Industries, Inc.**, Tokyo, Japan

[21] Appl. No.: **514,000**

[22] Filed: **Apr. 25, 1990**

[30] **Foreign Application Priority Data**

Apr. 28, 1989	[JP]	Japan	1-50325
Apr. 28, 1989	[JP]	Japan	1-50326
Apr. 28, 1989	[JP]	Japan	1-50327
Apr. 28, 1989	[JP]	Japan	1-50328
Apr. 28, 1989	[JP]	Japan	1-50329
Apr. 28, 1989	[JP]	Japan	1-50330
Apr. 28, 1989	[JP]	Japan	1-50331

[51] **Int. Cl.**[5] **A47C 3/00**
[52] **U.S. Cl.** .... **297/284 R**; 297/345; 297/430; 297/435; 297/408; 297/DIG. 3
[58] **Field of Search** .... 297/284, 338, 460, DIG. 8, 297/DIG. 3, 408, 430, 434, 435, 431, 344, 396

[56] **References Cited**

U.S. PATENT DOCUMENTS

4,538,854	9/1985	Wilson	297/DIG. 3
4,629,248	12/1986	Mawbey	297/284
4,629,253	12/1986	Williams	297/DIG. 3
4,720,146	1/1988	Mawbey et al.	297/284
4,865,388	9/1989	Nemoto	297/DIG. 3
4,965,899	10/1990	Sekido et al.	297/284

FOREIGN PATENT DOCUMENTS

0068211	1/1983	European Pat. Off. .
0097722	1/1984	European Pat. Off. .
1458099	12/1976	United Kingdom .
2168893	7/1986	United Kingdom .

Primary Examiner—Kenneth J. Dorner
Assistant Examiner—Milton Nelson, Jr.
Attorney, Agent, or Firm—Darby & Darby

[57] **ABSTRACT**

An automatic vehicle seat capable of separately controlling the position or gradient of each portion of a vehicle seat using air bags. Automatic adjustment mechanisms utilizing inflatable air bags are included in the vehicle seat and are capable of adjusting and controlling the height of the seat level, the gradient of a headrest positioned on the seat back, the thigh support of the seat and the mobility of the seat body in a longitudinal direction.

13 Claims, 16 Drawing Sheets

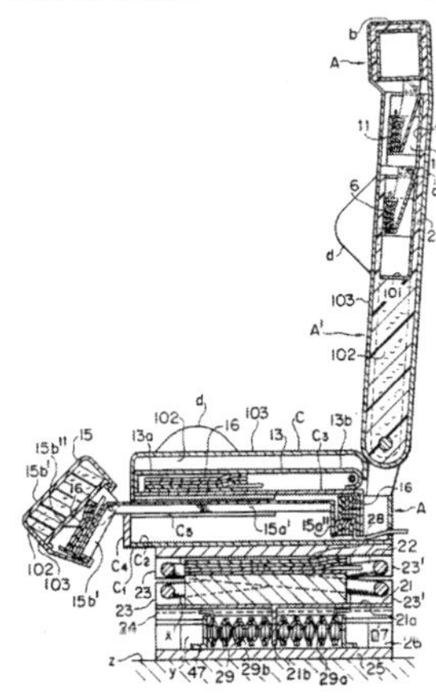

United States Patent [19]

Dellanno et al.

[11] **Patent Number: 5,290,091**

[45] **Date of Patent: Mar. 1, 1994**

[54] **APPARATUS FOR PREVENTING WHIPLASH**

[76] Inventors: **Ronald P. Dellanno**, 40 Fox Run, North Caldwell, N.J. 07006; **Quentin E. Gualtier**, 16 Winding Way, North Caldwell, N.J. 07006

[21] Appl. No.: **8,209**

[22] Filed: **Jan. 25, 1993**

Related U.S. Application Data

[63] Continuation-in-part of Ser. No. 771,827, Oct. 7, 1991, Pat. No. 5,181,763, which is a continuation of Ser. No. 585,392, Sep. 20, 1990, abandoned.

[51] **Int. Cl.**[5] **A47C 7/36**
[52] **U.S. Cl.** **297/391**; 297/404; 297/408; 297/216.12
[58] **Field of Search** 297/391, 396, 408, 216, 297/404, 410; 5/636, 632

[56] **References Cited**

U.S. PATENT DOCUMENTS

D. 276,938	12/1984	Pedersen	D24/36
2,973,029	2/1961	Schlosstein	297/216
2,990,008	6/1961	Bien	297/397
3,071,412	1/1963	Meade	297/391
3,706,472	12/1972	Mertens	297/397
4,256,341	3/1981	Goldner et al.	297/410
4,285,081	8/1981	Price	297/391 X
4,424,599	1/1984	Hannouche	5/632
4,466,662	8/1984	McDonald et al.	297/391 X
4,693,515	9/1987	Russo et al.	397/391
4,754,513	7/1988	Rinz	5/636 X
4,829,614	5/1989	Harper	5/632
4,832,007	5/1989	Davis, Jr. et al.	5/632 X
4,865,388	9/1989	Nemoto	297/403
4,944,554	7/1990	Gross et al.	297/391 X
5,181,763	1/1993	Dellanno et al.	297/391

Primary Examiner—James R. Brittain
Assistant Examiner—Milton Nelson, Jr.
Attorney, Agent, or Firm—Klauber & Jackson

[57] **ABSTRACT**

Apparatus for preventing whiplash-related injuries to a passenger in a vehicle. The apparatus includes a frame supported on a seat of the vehicle and located behind the cranium and cervical spine of a passenger on the seat. A layer of resilient material is supported on the frame, and defines a supporting means located behind the cranium and cervical spine of the passenger which defines one or more contours that interfit with the posterior contour of the passenger's cranium and cervical spine. The apparatus thereby contacts and supports the posterior portion of the passenger's cervical spine substantially simultaneously with contact and support of the passenger's cranium, thereby substantially simultaneously decelerating the cranium and cervical spine during a vehicle collision to prevent whiplash-related injuries to the passenger.

11 Claims, 6 Drawing Sheets

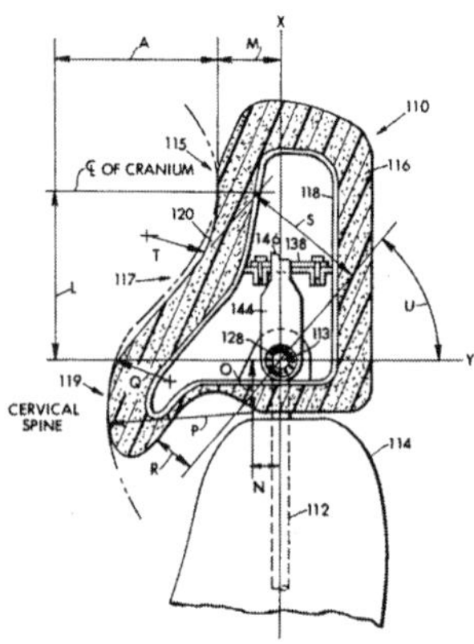

United States Patent [19]

Sakamoto et al.

[11] **Patent Number: 5,642,918**

[45] **Date of Patent: Jul. 1, 1997**

[54] **MOVEABLE HEADREST**

[75] Inventors: **Manabu Sakamoto**, Yokohama; **Koichi Iwasaki**, Ayase, both of Japan

[73] Assignee: **Ikeda Bussan Co., Ltd.**, Ayase, Japan

[21] Appl. No.: **604,649**

[22] Filed: **Feb. 21, 1996**

[30] **Foreign Application Priority Data**

Feb. 21, 1995 [JP] Japan 7-56746

[51] **Int. Cl.**[6] **A47C 7/36**; A47C 1/10
[52] **U.S. Cl.** **297/408**; 297/391
[58] **Field of Search** 297/408, 400, 297/391, 366, 367

[56] **References Cited**

U.S. PATENT DOCUMENTS

2,828,810	4/1958	Barecki et al.	297/408
4,674,792	6/1987	Tamura et al.	297/408
4,674,797	6/1987	Tateyama	297/408
4,678,232	7/1987	Ishida et al.	297/408
4,830,434	5/1989	Ishida et al.	297/408
5,150,632	9/1992	Hein	297/366

FOREIGN PATENT DOCUMENTS

3-39078	8/1991	Japan .
2 176 098	12/1986	United Kingdom .

OTHER PUBLICATIONS

U.K. Search Report, GB 9603578.7, Apr. 30, 1996.

Primary Examiner—Peter M. Cuomo
Assistant Examiner—Anthony D. Barfield
Attorney, Agent, or Firm—Foley & Lardner

[57] **ABSTRACT**

A moveable headrest is equipped with a lower headrest frame shaft fixedly connected to a headrest stay detachably connected to the upper end of the seat back, and a pair of ratchet bases rotatably supported on respective shaft ends of the lower frame shaft. The ratchet bases are coupled with each other at their upper ends through a moveable upper headrest frame. A ratchet-and-pawl mechanism is attached to at least one of the base members for selectively engaging the upper frame with the lower frame shaft and for producing a forward-and-backward angular adjustment of the headrest. The shaft end of the lower frame shaft is formed with an annular stepped portion extending continuously in a peripheral direction of the lower frame shaft nearby the top end, so that the annular stepped portion abuts the inside wall surface of the ratchet base. A pawl bracket is fixedly connected to the shaft end of the lower frame shaft by all-around welding, while sandwiching the ratchet base between the stepped portion and the pawl bracket.

4 Claims, 6 Drawing Sheets

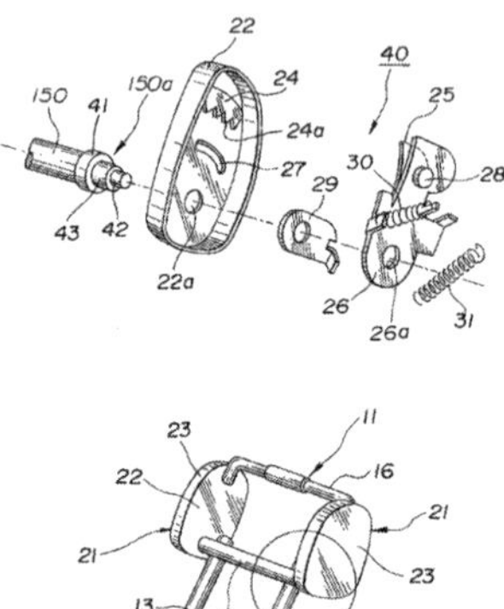

United States Patent [19]

Kuranami

[11] **Patent Number: 5,364,164**

[45] **Date of Patent: Nov. 15, 1994**

[54] **HEADREST FOR SEATS**

[75] Inventor: **Shunji Kuranami**, Kanagawa, Japan

[73] Assignee: **Koito Industries, Ltd.**, Kanagawa, Japan

[21] Appl. No.: **920,435**

[22] PCT Filed: **Nov. 11, 1991**

[86] PCT No.: **PCT/JP91/01541**
§ 371 Date: **Aug. 25, 1992**
§ 102(e) Date: **Aug. 25, 1992**

[87] PCT Pub. No.: **WO92/10963**
PCT Pub. Date: **Jul. 9, 1992**

[30] **Foreign Application Priority Data**

Dec. 25, 1990 [JP] Japan 2-412896

[51] **Int. Cl.**[5] **A47C 7/36**
[52] **U.S. Cl.** **297/408**; 297/391; 16/334
[58] **Field of Search** 297/391, 408, 220; 403/93, 95, 96; 16/324, 334

[56] **References Cited**

U.S. PATENT DOCUMENTS

2,869,622	1/1959	Petersen et al.	297/408
3,484,831	12/1969	Higuchi	297/356
4,370,898	2/1983	Maruyama	403/93
4,674,797	6/1987	Tateyama	297/408
5,145,233	9/1992	Nagashima	297/408

FOREIGN PATENT DOCUMENTS

0240996	6/1960	Australia	16/334
3131597	2/1983	Germany	297/408
0186746	10/1984	Japan	297/408
64-35949	3/1989	Japan .	
4999912	7/1965	United Kingdom	16/334

Primary Examiner—Peter M. Cuomo
Assistant Examiner—Darnell M. Boucher
Attorney, Agent, or Firm—Edwin E. Greigg; Ronald E. Greigg

[57] **ABSTRACT**

A headrest for seats which is recessed in a central recess of the seat and is capable of tilting forward so as to be used as a headrest, and automatically returning this headrest to its original position by tilting it forward more than a predetermined angle with respect to the seat back. The headrest is formed on an upper portion of a seat back and is swingably supported by means of a hinge mechanism equipped with a ratchet mechanism. The hinge mechanism 8 includes a movable hinge portion fixed on a lower portion of a supporting frame of the headrest, and a hinge main body fixed on a seat back frame so as to be hingedly supported on a rotational shaft together with the movable hinge portion. Further, there is provided a return spring together with the supporting frame of the headrest that returns the headrest to its recessed position. Moreover, the ratchet mechanism is constituted in such a manner than, when the headrest is tilted forward beyond a predetermined angle, the ratchet gear is disengaged from the ratchet claw member so that the headrest can be returned to its original position by the pulling force of the return spring.

3 Claims, 5 Drawing Sheets

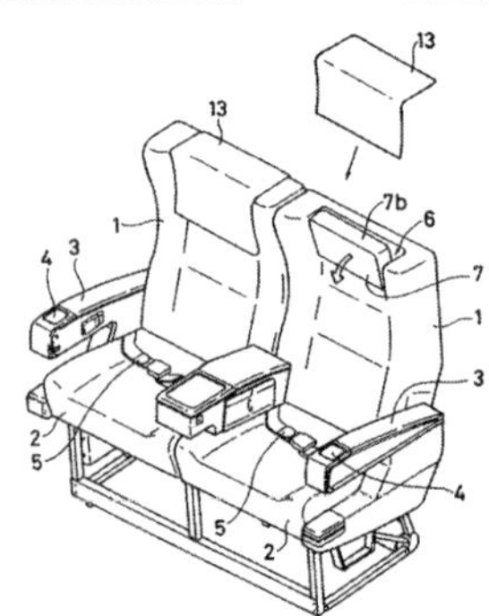

United States Patent [19]

Lee

[11] **Patent Number: 5,669,666**

[45] **Date of Patent: Sep. 23, 1997**

[54] **DEVICE AND METHOD FOR AUTOMATICALLY CONTROLLING A HEADREST**

[75] Inventor: **Hyung-Ho Lee**, Kyungnam-Do, Rep. of Korea

[73] Assignee: **Hyundai Motor Company, Ltd.**, Seoul, Rep. of Korea

[21] Appl. No.: **628,340**

[22] Filed: **Apr. 5, 1996**

[30] **Foreign Application Priority Data**

Apr. 6, 1995 [KR] Rep. of Korea 95-7932

[51] **Int. Cl.**[6] **A47C 1/10**; A47C 7/36
[52] **U.S. Cl.** **297/408**; 297/410
[58] **Field of Search** 297/408, 403, 297/410

[56] **References Cited**

U.S. PATENT DOCUMENTS

4,576,413	3/1986	Hatta	297/408
4,606,578	8/1986	Yasui	297/408
4,668,014	5/1987	Boisset	297/410 X
4,765,683	8/1988	Hattori	297/408 X
4,830,434	5/1989	Ishida et al.	297/408
4,861,107	8/1989	Vidwans et al.	297/408
5,011,225	4/1991	Nemoto	297/408
5,052,754	10/1991	Chinomi	297/408
5,054,856	10/1991	Wang	297/408
5,145,233	9/1992	Nagashima	297/408
5,222,784	6/1993	Hamelin	297/408

Primary Examiner—Peter M. Cuomo
Assistant Examiner—Rodney B. White

[57] **ABSTRACT**

A device for automatically controlling a headrest, includes a motor, a rotatable gear connected to the motor so as to provide rotating movement, a rack gear movably engaging with the rotatable gear to provide lineal movement, a rotator rotatably connected to the rack gear, a first connection mechanism for connecting the rack gear with the rotator, a second connection mechanism for connecting the rotator with the headrest, and a switch circuit for controlling rotation of the motor, whereby tilting of the headrest is automatically controlled by the switch circuit.

20 Claims, 1 Drawing Sheet

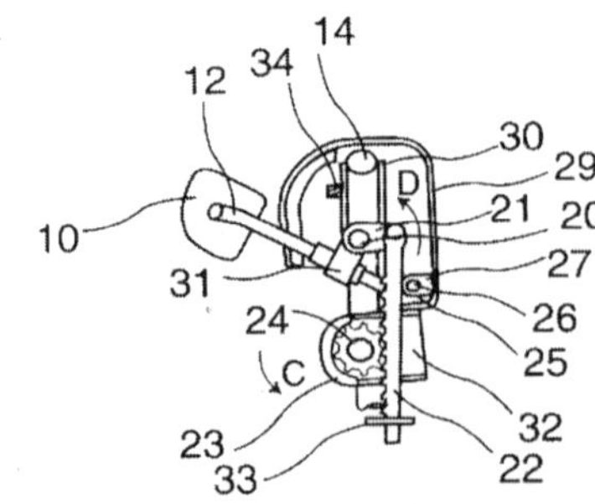

US005738407A

United States Patent [19]

Locke

[11] **Patent Number: 5,738,407**

[45] **Date of Patent: Apr. 14, 1998**

[54] **HEAD RESTRAINT AIRBAG ASSEMBLY**

[75] Inventor: **Gerald S. Locke**, Lake Orion, Mich.

[73] Assignee: **Lear Corporation**, Southfield, Mich.

[21] Appl. No.: **783,223**

[22] Filed: **Jan. 14, 1997**

[51] **Int. Cl.**[6] **B60R 21/22**

[52] **U.S. Cl.** **297/216.12**; 297/216.13; 280/730.1

[58] **Field of Search** 297/216.1, 216.12, 297/216.13; 280/730.1

[56] **References Cited**

U.S. PATENT DOCUMENTS

3,703,313	11/1972	Schiested et al.	280/730.1 X
3,779,577	12/1973	Wilfert	280/730.1
5,536,043	7/1996	Lang et al.	280/730.1 X
5,556,129	9/1996	Coman et al.	297/216.12 X

FOREIGN PATENT DOCUMENTS

2152202	4/1973	Germany	297/216.12
2841729	4/1980	Germany	280/730 R

Primary Examiner—Peter R. Brown
Attorney, Agent, or Firm—Brooks & Kushman P.C.

[57] **ABSTRACT**

A vehicle seat assembly includes a seat back frame and a headrest movable with respect to the back frame. An airbag is positioned on the back frame and deployable vertically with respect to the back frame such that the airbag is positioned forward-in-vehicle from the headrest when fully deployed.

9 Claims, 1 Drawing Sheet

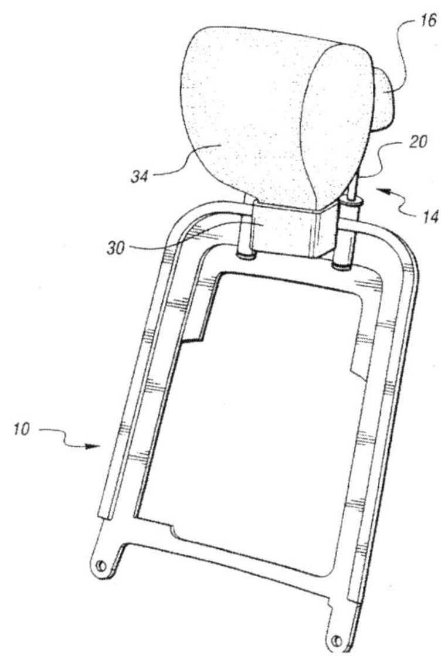

US005738412A

United States Patent [19]

Aufrere et al.

[11] **Patent Number: 5,738,412**

[45] **Date of Patent: Apr. 14, 1998**

[54] **HEADREST FOR AN AUTOMOTIVE VEHICLE**

[75] Inventors: **Christophe Aufrere**, Marcoussis; **Patrick Daniel**, Paris; **Adolfo Castro**, Fresnes, all of France

[73] Assignee: **Bertrand Faure Equipements S.A.**, Boulogne Cedex, France

[21] Appl. No.: **815,106**

[22] Filed: **Mar. 11, 1997**

[30] **Foreign Application Priority Data**

Mar. 18, 1996 [FR] France 96 03484

[51] **Int. Cl.**[6] **A47C 7/36**

[52] **U.S. Cl.** **297/408**; 297/410; 297/216.12

[58] **Field of Search** 297/408, 409, 297/391, 410, 216.1, 216.12

[56] **References Cited**

U.S. PATENT DOCUMENTS

4,256,341	3/1981	Goldner et al.	297/408 X
4,861,107	8/1989	Vidwans et al.	297/391 X
5,199,765	4/1993	Garmendia et al.	297/391
5,290,091	3/1994	Dellanno et al.	297/408 X

FOREIGN PATENT DOCUMENTS

0 254 808	2/1988	European Pat. Off. .
0 564 962	10/1993	European Pat. Off. .
0 577 517	7/1994	European Pat. Off. .
2686383	7/1993	France .
2721267	12/1995	France .
26 44 485	4/1978	Germany .
27 08 461	8/1978	Germany .

Primary Examiner—Milton Nelson, Jr.
Attorney, Agent, or Firm—Pollock, Vande Sande & Priddy

[57] **ABSTRACT**

The headrest (**1**) comprises a translation guidance assembly (**10**) guided along a spindle (**2**) attached to the seat back and an articulation assembly (**20**) that is mounted swivelling on the guidance assembly, around an axis (**7**).

A locking member (**30**) swivels on the guidance assembly (**10**) around an axis (**8**) that is parallel to the axis (**7**), between a locked and an unlocked position. The locking member comprises a pad (**31**) shaped to press on the spindle (**2**) in the locked position, and a head (**32**) that is shaped to lock on a curved section (**23**), attached to the articulation assembly. In this locked position, that the locking member takes under the effect of inertia after a collision, the headrest can neither pivot backwards nor slide downwards.

Application of the headrest is directed to automotive vehicle seats.

10 Claims, 5 Drawing Sheets

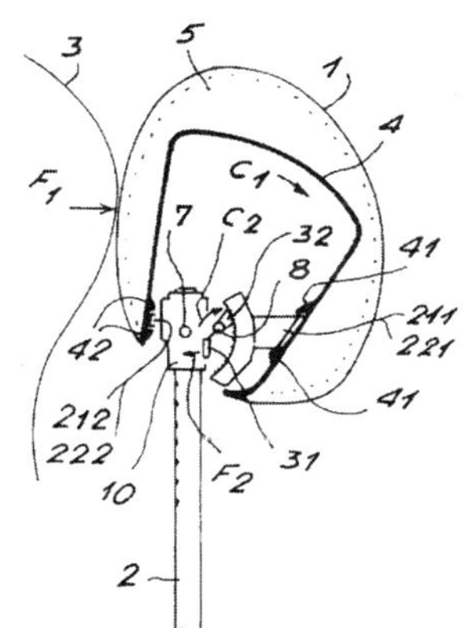

United States Patent [19]

Knoll et al.

[11] **Patent Number: 5,842,738**

[45] **Date of Patent: Dec. 1, 1998**

[54] **HEADREST FOR A VEHICLE SEAT**

[75] Inventors: **Heinz Knoll**, Stuttgart; **Rolf Mitschelen**, Kirchheim/Teck; **Martin Steiner**, Grafenau, all of Germany

[73] Assignee: **Mercedes-Benz AG**, Stuttgart, Germany

[21] Appl. No.: **764,084**

[22] Filed: **Dec. 6, 1996**

[30] **Foreign Application Priority Data**

Dec. 22, 1995 [DE] Germany 195 48 339.1

[51] **Int. Cl.**[6] **B60N 2/48**

[52] **U.S. Cl.** **297/216.12**; 297/408

[58] **Field of Search** 297/216.2, 408; 280/730.1

[56] **References Cited**

U.S. PATENT DOCUMENTS

3,779,577	12/1973	Wilfert	280/730.1
3,838,870	10/1974	Hug	297/216.12 X
4,720,146	1/1988	Mawbey et al.	297/408 X
4,865,388	9/1989	Nemoto	297/408 X
5,082,326	1/1992	Sehido et al.	297/408 X

FOREIGN PATENT DOCUMENTS

2152202	4/1973	Germany	297/216.12
2232726	1/1974	Germany	297/216.12
3131633	2/1983	Germany	297/216.12
3900495A1	7/1990	Germany .	
29504287.7	6/1995	Germany .	
91167/84	6/1984	Japan .	
88559/86	6/1986	Japan .	
28246/93	4/1993	Japan .	

Primary Examiner—Peter R. Brown
Attorney, Agent, or Firm—Evenson, McKeown, Edwards & Lenahan P.L.L.C.

[57] **ABSTRACT**

A headrest for a vehicle seat is provided with a supporting loop and a head cushion mounted thereon and pivotable in a direction of a seat bottom. An adjusting device activated in the event of a collision is provided for displacing the head cushion relative to the supporting loop in such fashion that the head cushion directly abuts the nape of the neck and the back of the seat occupant's head. To improve head support comfort by provision for individual adjustment of the head support contour to the needs of the seat occupant when asleep, the adjusting device is additionally equipped with a manual operating member that permits changing the setting of the head cushion without adversely affecting the protective function of the head cushion in the event of a collision.

22 Claims, 2 Drawing Sheets

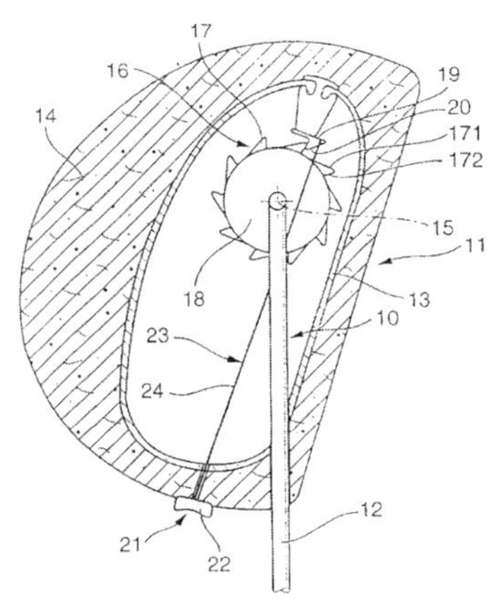

United States Patent [19]

Wu

[11] **Patent Number: 5,904,405**

[45] **Date of Patent: May 18, 1999**

[54] **PILLOW STRUCTURE FOR CHAIRS**

[76] Inventor: **Frank Wu**, 6F-2, No. 6, Lane 712 Chun-Hsiao Rd., Nantze Dist. Kaohsiung, Taiwan

[21] Appl. No.: **09/131,193**

[22] Filed: **Aug. 7, 1998**

[51] **Int. Cl.**[6] **B60N 2/42**; B60N 2/48

[52] **U.S. Cl.** **297/391**; 297/216.12; 297/464; 297/468

[58] **Field of Search** 297/391, 216.12, 297/184.12, 184.13, 393, 408, 464, 468, 475, 487

[56] **References Cited**

U.S. PATENT DOCUMENTS

2,873,122	2/1959	Peras	297/216.12 X
3,098,128	7/1963	Audin	297/391 X
3,372,491	3/1968	Morrison	297/391 X
3,376,064	4/1968	Jackson	297/391
4,565,405	1/1986	Mayer	297/391 X
4,971,393	11/1990	Maisenhalder	297/391
5,411,468	5/1995	Chen	297/391 X
5,806,933	9/1998	Tsui et al.	297/216.12 X

FOREIGN PATENT DOCUMENTS

3829470	10/1989	Germany	297/391

Primary Examiner—Peter R. Brown
Attorney, Agent, or Firm—Bacon & Thomas, PLLC

[57] **ABSTRACT**

A pillow structure includes a main pillow adjustably mounted to a chair and two side pillows respectively mounted two sides of the main pillow and extending in a direction transverse to the main pillow. Each side pillow includes an extension telescopically received therein and slidable along the extending direction of an associated side pillow.

10 Claims, 2 Drawing Sheets

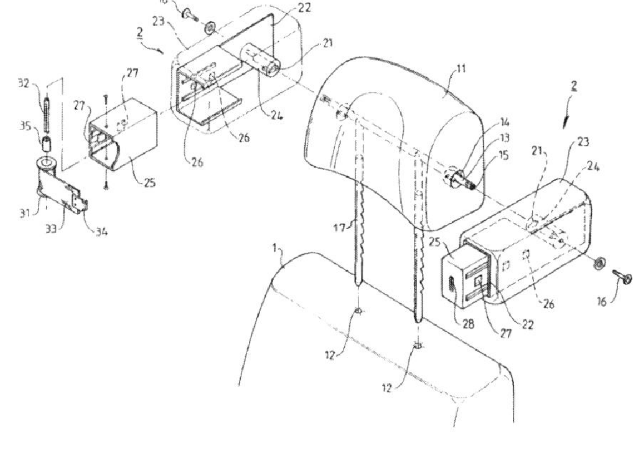

United States Patent [19]

Rückert et al.

[11] **Patent Number: 6,019,424**

[45] **Date of Patent: Feb. 1, 2000**

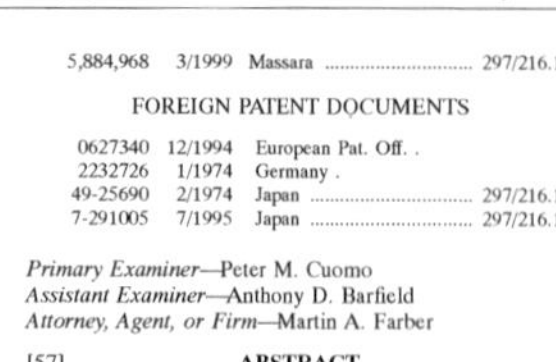

[54] **HEADREST, IN PARTICULAR IN MOTOR VEHICLES**

[75] Inventors: **Edvard Rückert**, Velbert; **Norbert Krüger**, Essen, both of Germany

[73] Assignee: **Ewald Witte GmbH & Co. KG**, Velbert, Germany

[21] Appl. No.: **09/204,098**

[22] Filed: **Dec. 1, 1998**

[30] **Foreign Application Priority Data**

Dec. 3, 1997 [DE] Germany 197 53 540

[51] **Int. Cl.**[7] **B60N 2/42**

[52] **U.S. Cl.** **297/216.12**; 297/216.13; 297/216.14; 297/408

[58] **Field of Search** 297/216.12, 216.13, 297/216.14, 408

[56] **References Cited**

U.S. PATENT DOCUMENTS

3,838,870 10/1974 Hug .
5,378,043 1/1995 Viano et al. 297/216.12
5,884,968 3/1999 Massara 297/216.12

FOREIGN PATENT DOCUMENTS

0627340 12/1994 European Pat. Off. .
2232726 1/1974 Germany .
49-25690 2/1974 Japan 297/216.14
7-291005 7/1995 Japan 297/216.14

Primary Examiner—Peter M. Cuomo
Assistant Examiner—Anthony D. Barfield
Attorney, Agent, or Firm—Martin A. Farber

[57] **ABSTRACT**

A headrest (23) which pivots forward, by lever action, as a result of the mass of the upper body of a vehicle occupant acting on a pressure-bearing surface in the event of an impact from the rear. In order that in case of both a rear-end crash and a head-on crash of the vehicle, the headrest (23) follows the head of an individual sitting on the vehicle seat, the headrest (23) is positioned on a headrest carrier (17) which can be pivoted out of the backrest (3) in the forward direction and can be pivoted outward about a pin (19), this pin being located at a lower level than that of the pivot pin (21) of the headrest (23).

16 Claims, 9 Drawing Sheets

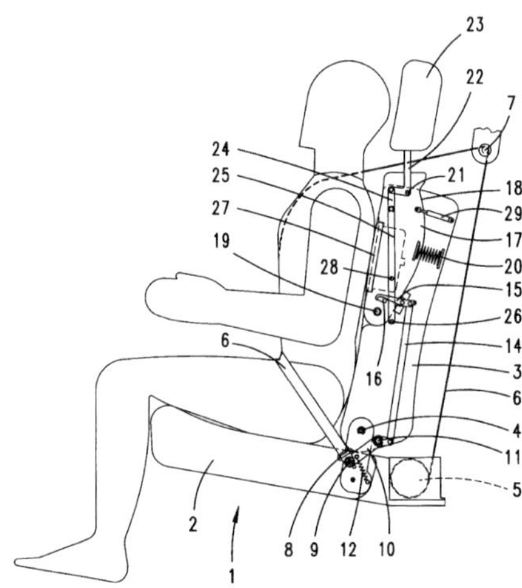

United States Patent [19]

Takeda

[11] **Patent Number: 6,074,010**

[45] **Date of Patent: Jun. 13, 2000**

[54] **HEADREST APPARATUS FOR VEHICLE SEAT**

[75] Inventor: **Nobuhiko Takeda**, Aichi pref., Japan

[73] Assignee: **Aisin Seiki Kabushiki Kaisha**, Kariya, Japan

[21] Appl. No.: **09/027,739**

[22] Filed: **Feb. 23, 1998**

[30] **Foreign Application Priority Data**

Feb. 25, 1997 [JP] Japan 9-041141

[51] **Int. Cl.**[7] **A47C 7/36**

[52] **U.S. Cl.** **297/391**; 297/408; 297/410

[58] **Field of Search** 297/410, 404, 297/408, 391, 403

[56] **References Cited**

U.S. PATENT DOCUMENTS

4,113,310 9/1978 Kapanka 297/408
4,678,232 7/1987 Ishida et al. 297/408
4,711,494 12/1987 Duvenkamp .... 297/408 X
5,011,225 4/1991 Nemoto .... 297/410 X
5,145,233 9/1992 Nagashima 297/408
5,669,668 9/1997 Leuchtmann 297/408
5,895,094 4/1999 Mori et al. .... 297/391 X
5,906,414 5/1999 Rus 297/408

FOREIGN PATENT DOCUMENTS

2-46755 12/1990 Japan .

Primary Examiner—Peter M. Cuomo
Assistant Examiner—Stephen Vu
Attorney, Agent, or Firm—Burns, Doane, Swecker & Mathis, LLP

[57] **ABSTRACT**

A headrest apparatus for a vehicle seat including a guide member for being mounted on a seat-back of the seat, a rotational member rotatably supported to the guide member, a stay slidably supported to the guide member and the rotational member with the capability to be detached from the guide member and the rotational member, a headrest body attached to the stay and positioned to the normal position and a stored position by a rotation of the rotational member relative to the guide member, a spring member connected between the guide member and the rotational member and rotates the rotational member so that the headrest body is positioned to the stored position and an contacting wall formed to the guide member with the capability of making contact with the end of the stay when the headrest body is positioned to the stored position.

12 Claims, 9 Drawing Sheets

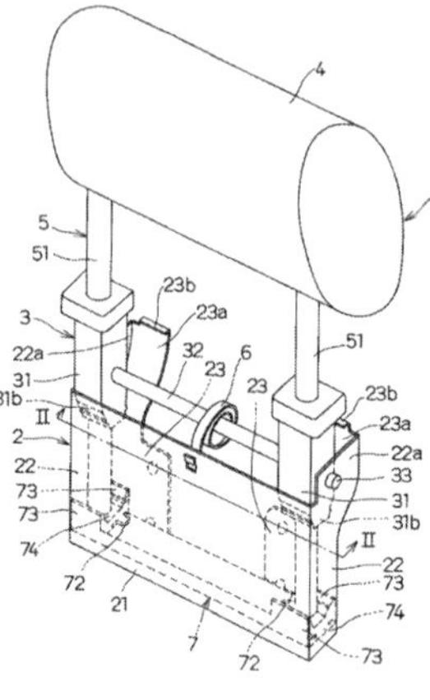

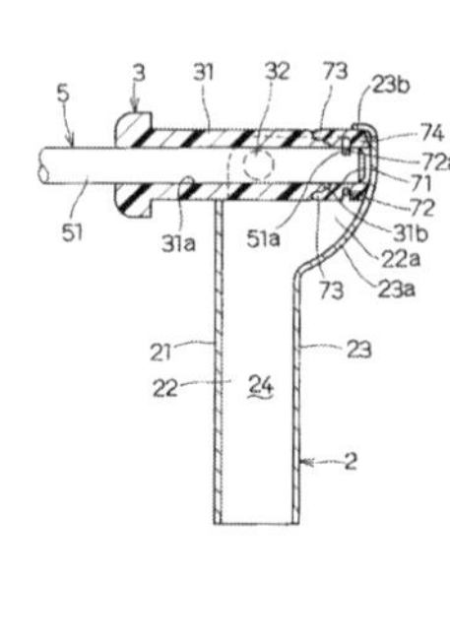

United States Patent [19]

Ikeda et al.

[11] **Patent Number: 6,045,181**

[45] **Date of Patent: Apr. 4, 2000**

[54] **ADJUSTABLE HEADREST**

[75] Inventors: **Noriyuki Ikeda; Moriyuki Eguchi**, both of Kanagawa-ken, Japan

[73] Assignee: **Ikeda Bussan Co., Ltd.**, Ayase, Japan

[21] Appl. No.: **09/270,741**

[22] Filed: **Mar. 16, 1999**

[30] **Foreign Application Priority Data**

Mar. 18, 1998 [JP] Japan 10-089394

[51] **Int. Cl.**[7] **B60N 2/48**

[52] **U.S. Cl.** **297/216.12**; 297/408

[58] **Field of Search** 297/391, 408, 297/403, 410, 216.12

[56] **References Cited**

U.S. PATENT DOCUMENTS

4,674,792 6/1987 Tamura et al. 297/408
4,674,797 6/1987 Tafeyama 297/408
5,642,918 7/1997 Sakamoto et al. 297/408
5,738,412 4/1998 Aufrere et al. 297/408

FOREIGN PATENT DOCUMENTS

2264904 8/1975 Germany 297/216.12
3131633 2/1983 Germany 297/216.12
2316862 11/1998 United Kingdom .

Primary Examiner—Peter R. Brown
Attorney, Agent, or Firm—Nath & Associates; Gary M. Nath; Harold L. Novick

[57] **ABSTRACT**

An adjustable headrest is provided having a main body for supporting the head of the user, a ratchet mechanism within the main body, and a frame rotatably supported on a stay member. The ratchet mechanism includes a base plate fixed to the frame, a ratchet plate with a pawl portion, a support pin connecting the reinforcing plate and the ratchet plate, and a lock member fixed to the frame. The base plate has a receiving hole into which the support pin is inserted. A slit having a width smaller than the outside diameter of the support pin is formed in the base plate and extends backward from the receiving hole such that, when the headrest receives an impact in the backward direction, the support pin moves in the slit while widening the width of the slit under pressure. The resulting backward movement of the pin and the deformation of the slit absorbs the impact.

7 Claims, 7 Drawing Sheets

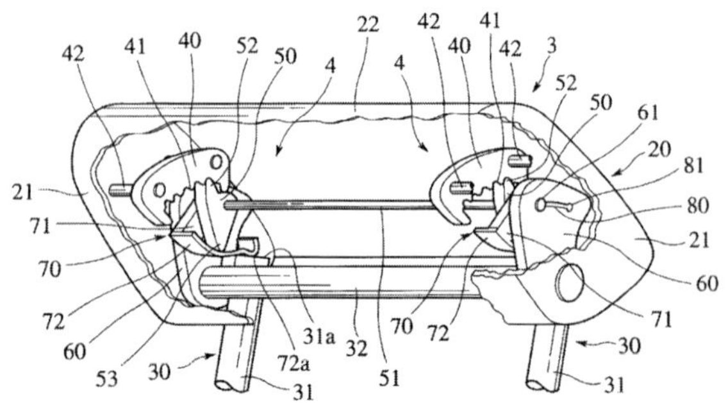

United States Patent [19]

Ptak et al.

[11] **Patent Number: 6,074,011**

[45] **Date of Patent: Jun. 13, 2000**

[54] **AUTOMATIC RETRACTABLE HEAD RESTRAINT**

[75] Inventors: **Kenneth R. Ptak**, Livonia; **Anthony J. DiSalvo**, Allen Park, both of Mich.

[73] Assignee: **Johnson Controls Technology Company**, Plymouth, Mich.

[21] Appl. No.: **09/039,822**

[22] Filed: **Mar. 16, 1998**

[51] **Int. Cl.**[7] **A47C 7/36**; B60N 2/48

[52] **U.S. Cl.** **297/408**; 297/410

[58] **Field of Search** 297/408, 403, 297/410

[56] **References Cited**

U.S. PATENT DOCUMENTS

1,717,568 6/1929 Koenigkramer .... 297/403 X
3,157,434 11/1964 Gianvecchio 297/410
3,506,306 4/1970 Herzer et al. 297/408
3,547,486 12/1970 Herzer et al. .
3,586,366 6/1971 Patrick .... 297/410 X
3,784,253 1/1974 Kohler et al. 297/410
3,813,151 5/1974 Cadiou .... 297/410 X
3,888,540 6/1975 Protze et al. .
4,113,310 9/1978 Kapanka .
4,576,413 3/1986 Hatta .
4,596,403 6/1986 Dieckmann et al. .
4,600,240 7/1986 Suman et al. .
4,623,166 11/1986 Andres et al. .
4,678,232 7/1987 Ishida et al. 297/408
4,711,494 12/1987 Duvenkamp .... 297/408 X
4,798,415 1/1989 Tanino et al. .... 297/408 X
4,807,934 2/1989 Sakakibara et al. .... 297/408 X
4,822,102 4/1989 Duvenkamp .
4,834,456 5/1989 Barros et al. .
4,865,388 9/1989 Nemoto .
5,011,226 4/1991 Ikeda et al. .
5,222,784 6/1993 Hamelin 297/408
5,290,091 3/1994 Dellano et al. .... 297/408 X
5,484,189 1/1996 Patterson .... 297/408 X
5,590,933 1/1997 Andersson .
5,738,411 4/1998 Sutton et al. .... 297/408 X
5,738,412 4/1998 Aufrere et al. 297/408
6,000,760 12/1999 Chung 297/408

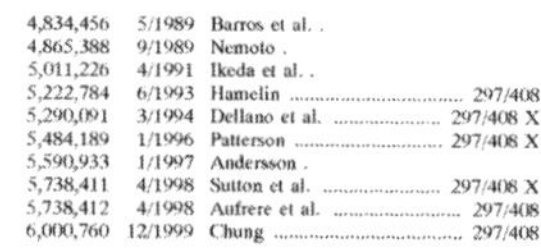

FOREIGN PATENT DOCUMENTS

175959 4/1986 European Pat. Off. 297/408
1066430 10/1959 Germany 297/408
2727987 1/1979 Germany 297/408
3136648 3/1983 Germany .
3129063 4/1983 Germany .
3325927 7/1983 Germany .
4026090 2/1992 Germany .
432747 5/1966 Switzerland 297/410
1200627 7/1970 United Kingdom 297/408
2034803 6/1980 United Kingdom 297/408

Primary Examiner—Peter M. Cuomo
Assistant Examiner—David E. Allred
Attorney, Agent, or Firm—Harness, Dickey & Pierce, P.L.C.

[57] **ABSTRACT**

A vehicle seat assembly with a retractable head rest that rotates from an upright use position to a lowered, retracted, stowed position to improve visibility for a vehicle driver. The motion mechanism for the head rest is contained within the body of the head rest itself thus eliminating modification of the seat back to which the head rest is mounted. The head rest is further vertically adjustable relative to the seat back and is rotatably adjustable within a range of upright use positions.

9 Claims, 2 Drawing Sheets

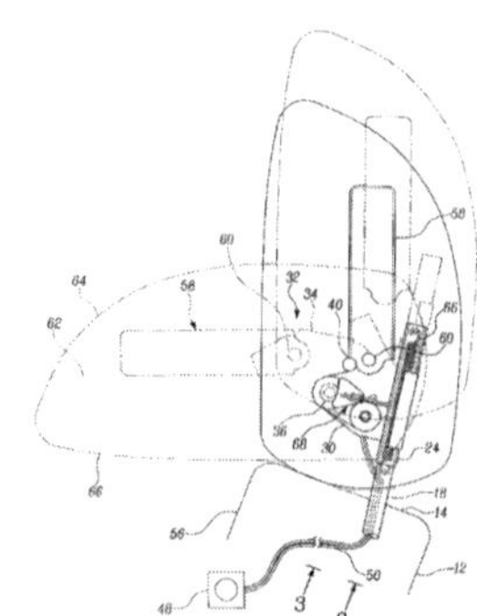

United States Patent [19]

Breitner et al.

[11] **Patent Number: 6,079,776**

[45] **Date of Patent: Jun. 27, 2000**

[54] **HEADREST FOR A VEHICLE SEAT**

[75] Inventors: **Roland Breitner**, Herrenberg; **Thomas Heckmann**, Aidlingen, both of Germany

[73] Assignee: **DaimlerChrysler AG**, Stuttgart, Germany

[21] Appl. No.: **09/148,983**

[22] Filed: **Sep. 8, 1998**

[30] **Foreign Application Priority Data**

Sep. 6, 1997 [DE] Germany 197 39 131

[51] **Int. Cl.**[7] **B60R 21/055**

[52] **U.S. Cl.** **297/216.12**; 297/408

[58] **Field of Search** 297/216.12, 408, 297/391

[56] **References Cited**

U.S. PATENT DOCUMENTS

3,680,912	8/1972	Matsuura	297/216.12 X
4,720,146	1/1988	Mawbey et al.	297/408 X
4,844,544	7/1989	Ochiai	297/408
4,861,107	8/1989	Vidwans et al.	297/408
5,468,045	11/1995	Weber .	
5,683,141	11/1997	Wakamatsu et al.	297/408
5,700,057	12/1997	De Filippo	297/408

FOREIGN PATENT DOCUMENTS

2152202	4/1973	Germany .	
2220267	11/1973	Germany .	
2152202	11/1980	Germany	297/216.12
3900495A1	7/1990	Germany .	
55-26927	2/1980	Japan .	
2-31713	2/1990	Japan .	
5-11858	2/1993	Japan .	
9-149837	6/1997	Japan .	
9-182644	7/1997	Japan .	

OTHER PUBLICATIONS

Search Report, Europe, Dec. 21, 1998.

Primary Examiner—Peter R. Brown
Attorney, Agent, or Firm—Evenson, McKeown, Edwards & Lenahan, P.L.L.C.

[57] **ABSTRACT**

A vehicle seat has a supporting bow which can be fixed on the seat side and has a head cushion which is arranged on the end of the supporting bow away from the seat and which can be adjusted in its inclination as desired by the seat user. For securing the inclination adjustment of the head cushion in the event of a crash, an air-filled, volume-variable hollow body is supported between the supporting bow and a cushion support held on the supporting bow by way of a swivel bearing. The hollow body is provided with a throttle opening which counters a shock-type air discharge with an extremely high flow resistance and an air passage of a low flow rate with only a small flow resistance.

17 Claims, 2 Drawing Sheets

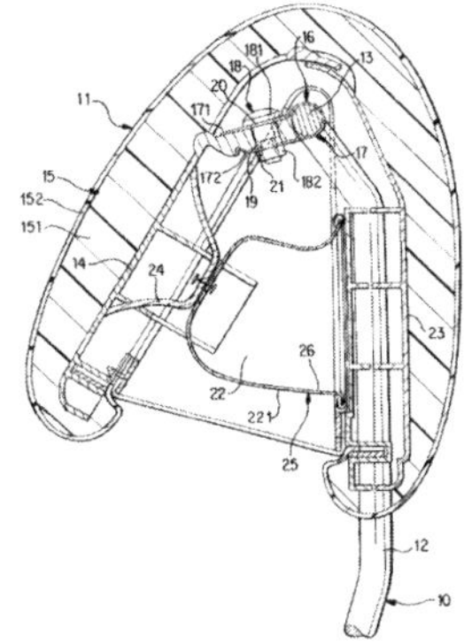

(12) United States Patent

Mayer et al.

(10) **Patent No.: US 6,390,549 B1**

(45) **Date of Patent: May 21, 2002**

(54) **BACKREST FOR A VEHICLE SEAT**

(75) Inventors: **Christian Mayer**, Ditzinger; **Vasilios Orizaris**, Renningen; **Albert Reitinger**, Berglen; **Oana Schüszler**, Wimsheim, all of (DE)

(73) Assignee: **DaimlerChrysler AG**, Stuttgart (DE)

(*) Notice: Subject to any disclaimer, the term of this patent is extended or adjusted under 35 U.S.C. 154(b) by 49 days.

(21) Appl. No.: **09/595,939**

(22) Filed: **Jun. 16, 2000**

(30) **Foreign Application Priority Data**

Jun. 16, 1999 (DE) 199 27 403

(51) **Int. Cl.**[7] **B60N 2/42**

(52) **U.S. Cl.** **297/216.14**; 297/216.12; 297/216.13; 297/216.15; 297/216.16

(58) **Field of Search** 297/216.12, 216.13, 297/216.14, 216.15, 216.16

(56) **References Cited**

U.S. PATENT DOCUMENTS

3,680,912	A	*	8/1972	Matsuura	297/216.12
5,580,124	A	*	12/1996	Dellanno	297/216.12
5,769,489	A	*	6/1998	Dellanno	297/216.14
5,772,280	A	*	6/1998	Massara	297/216.12
5,782,529	A	*	7/1998	Miller, III et al.	297/216.13
5,788,271	A	*	8/1998	Sotelo	297/112
5,833,312	A	*	11/1998	Lenz	297/216.13
5,902,010	A	*	5/1999	Cuevas	297/216.13
6,030,036	A	*	2/2000	Fohl	297/216.14
6,273,511	B1	*	8/2001	Wieclawski	297/463.1

FOREIGN PATENT DOCUMENTS

DE	198 82 039	12/1999
WO	PCT/SE94/01002	10/1994

* cited by examiner

Primary Examiner—Carl D. Friedman
Assistant Examiner—Dennis L. Dorsey
(74) *Attorney, Agent, or Firm*—Crowell & Moring LLP

(57) **ABSTRACT**

A backrest for a vehicle seat, having a head restraint and backrest padding, and having a supporting element which is intended for the seat occupant's back and, is compliant, at least in the upper backrest region, in such a manner that the seat occupant's back is able to shift rearwards relative to the head restraint in the event of a rear-end collision of the vehicle. The supporting element is formed by a shaped cushion that is integrated in the backrest padding in the upper backrest region, and filled with a gaseous medium, in particular air. In the event of a collision this supporting element can suddenly be deflated.

16 Claims, 2 Drawing Sheets

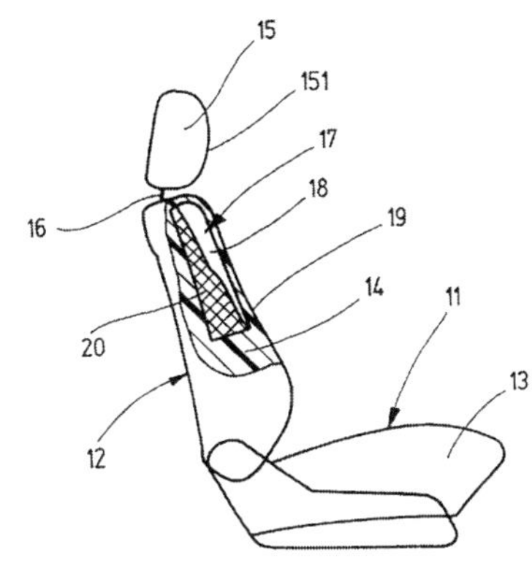

(12) United States Patent

Kienzle et al.

(10) **Patent No.: US 6,375,264 B1**

(45) **Date of Patent: Apr. 23, 2002**

(54) **BACKREST FOR A VEHICLE**

(75) Inventors: **Armin Kienzle**, Rottenburg; **Christian Wobst**, Horb; **Ulrich Ebbeskotte**, Althengstatt, all of (DE)

(73) Assignee: **DaimlerChrysler AG**, Stuttgart (DE)

(*) Notice: Subject to any disclaimer, the term of this patent is extended or adjusted under 35 U.S.C. 154(b) by 0 days.

(21) Appl. No.: **09/617,875**

(22) Filed: **Jul. 17, 2000**

(30) **Foreign Application Priority Data**

Jul. 16, 1999 (DE) 199 33 430

(51) **Int. Cl.**[7] **A47C 1/10**

(52) **U.S. Cl.** **297/403**

(58) **Field of Search** 297/403, 391, 297/408; 280/751; 296/63

(56) **References Cited**

U.S. PATENT DOCUMENTS

4,576,411	A	*	3/1986	Kitamura	297/403 X
4,623,166	A	*	11/1986	Andres et al.	297/403 X
4,711,494	A	*	12/1987	Duvenkamp	297/403
4,807,934	A	*	2/1989	Sakakibara et al.	297/403
5,669,668	A	*	9/1997	Leuchtmann	297/403 X
5,681,079	A	*	10/1997	Robinson	297/403 X

FOREIGN PATENT DOCUMENTS

DE	3545142		6/1987	
DE	3600411	*	7/1987	297/403
DE	4030949		4/1992	
DE	4243192	*	5/1993	297/403
DE	4220222		8/1993	
DE	19705867		4/1998	

* cited by examiner

Primary Examiner—Jerry Redman
(74) *Attorney, Agent, or Firm*—Crowell & Moring LLP

(57) **ABSTRACT**

A backrest for a vehicle seat, in particular a rear seat, has a head restraint which can be pivoted away from an approximately vertical operative position into an approximately horizontal inoperative position on the rear side of the backrest. The head restraint has a head cushion and at least one supporting rod which holds the head cushion and passes through a slot cutout in the backrest and is fixed to a pivoting shaft arranged in the backrest. In order to avoid openings on the upper side of the backrest for the passage of the supporting rod, the slot cutout is arranged in a rear wall of the backrest, and the supporting rod is designed in such a manner that it emerges out of the head cushion towards the rear side of the backrest with a curved rod section running concentrically to the pivoting shaft, and extends as far as the pivoting shaft with a straight rod section which is angled off at the end of the curved rod section.

13 Claims, 1 Drawing Sheet

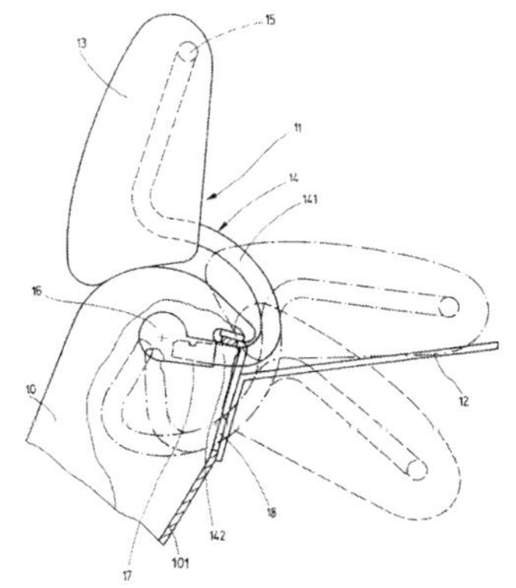

(12) United States Patent

Wooten

(10) **Patent No.: US 6,428,043 B1**

(45) **Date of Patent: Aug. 6, 2002**

(54) **VEHICLE HEAD RESTRAINT DEVICE**

(76) Inventor: **Arnold Gray Wooten**, P.O. Box 12252, Winston Salem, NC (US) 27117-2252

(*) Notice: Subject to any disclaimer, the term of this patent is extended or adjusted under 35 U.S.C. 154(b) by 0 days.

(21) Appl. No.: **09/790,934**

(22) Filed: **Feb. 22, 2001**

(51) **Int. Cl.**[7] **B60R 21/02**

(52) **U.S. Cl.** **280/748**; 2/410; 2/411; 2/416; 180/271

(58) **Field of Search** .... 280/748; 180/271, 180/274, 282; 2/425, 424, 414, 411, 410, 416; 244/121, 122 R, 122 A, 122 AG; 297/464, 216.1, 216.12

(56) **References Cited**

U.S. PATENT DOCUMENTS

4,477,041	A	*	10/1984	Dunne	244/122
4,664,341	A	*	5/1987	Cummings	244/122 AG
4,909,459	A	*	3/1990	Patterson	244/122 AG
4,923,147	A	*	5/1990	Adams et al.	244/122 AG

FOREIGN PATENT DOCUMENTS

GB	2252895	*	8/1992	244/122 AG
WO	WO88/01968	*	3/1988	244/122 AG

* cited by examiner

Primary Examiner—Eric Culbreth

(57) **ABSTRACT**

A head restraint device is provided for vehicle drivers and the like. The device allows the head to turn from side to side but restricts the forward and rearward movement as may occur during an abrupt stop while traveling at a high velocity. Cables are slidably affixed to the passenger compartment frame and resilient members dampen sudden forward or rearward head movement during vehicle impact.

13 Claims, 4 Drawing Sheets

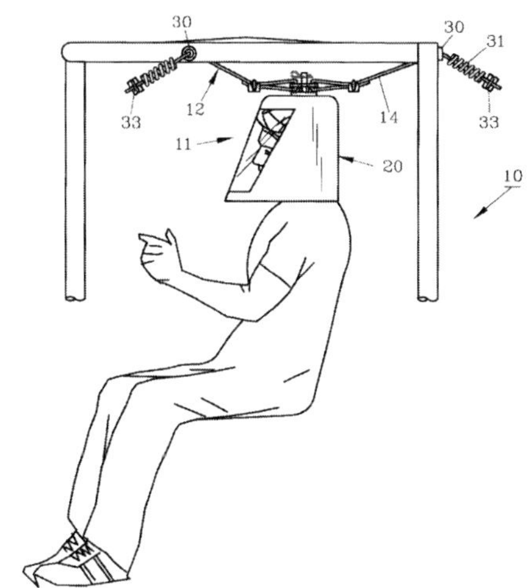

(12) **United States Patent**
Dinkel et al.

(10) **Patent No.: US 6,511,130 B2**
(45) **Date of Patent: Jan. 28, 2003**

(54) **HEAD RESTRAINT FOR A VEHICLE SEAT**

(75) Inventors: **Emil Dinkel**, Renningen (DE); **Vasilios Orizaris**, Renningen (DE); **Juergen Schrader**, Weil im Schoenbuch (DE)

(73) Assignee: **DaimlerChrysler AG**, Stuttgart (DE)

(*) Notice: Subject to any disclaimer, the term of this patent is extended or adjusted under 35 U.S.C. 154(b) by 44 days.

(21) Appl. No.: **09/809,434**

(22) Filed: **Mar. 16, 2001**

(65) **Prior Publication Data**

US 2002/0043860 A1 Apr. 18, 2002

(30) **Foreign Application Priority Data**

Mar. 16, 2000 (DE) 100 12 973

(51) **Int. Cl.**[7] **A47C 7/36**
(52) **U.S. Cl.** **297/410**; 297/408; 297/409
(58) **Field of Search** 297/410, 408, 297/409

(56) **References Cited**

U.S. PATENT DOCUMENTS

4,693,515 A * 9/1987 Russo et al. .... 297/408 X
4,856,848 A * 8/1989 O'Suillivan .... 297/409 X
5,181,763 A * 1/1993 Dellanno et al. .... 297/408 X
5,290,091 A * 3/1994 Dellanno et al. .... 297/408 X
5,364,164 A * 11/1994 Kuranami 297/408
5,642,918 A * 7/1997 Sakamoto et al. 297/408
5,669,666 A * 9/1997 Lee 297/408
5,738,412 A * 4/1998 Aufrere et al. 297/408
5,829,838 A * 11/1998 Offenbacher 297/408
5,842,738 A * 12/1998 Knoll et al. .... 297/408 X
6,033,018 A * 3/2000 Fohl .... 297/408 X
6,045,181 A * 4/2000 Ikeda et al. 297/408
6,074,010 A * 6/2000 Takeda .... 297/408 X
6,074,011 A * 6/2000 Ptak et al. 297/408
6,079,776 A * 6/2000 Breitner et al. .... 297/408 X
6,199,947 B1 * 3/2001 Wiklund .... 297/408 X

FOREIGN PATENT DOCUMENTS

DE 3404379 9/1985
DE 3900495 7/1990
DE 4305909 9/1994
DE 29504287.7 6/1995
DE 19520893 12/1996
DE 19722515 9/1998

* cited by examiner

Primary Examiner—Rodney B. White
(74) *Attorney, Agent, or Firm*—Crowell & Moring LLP

(57) **ABSTRACT**

A head restraint for a vehicle seat has a holding element which can be inserted into a seat backrest in a manner allowing it to be displaced vertically. A padding support is arranged on the holding element which holds a padded headrest intended for supporting the head. A pivoting device is arranged between the holding element and padding support and is intended for setting the inclination of the padded headrest about a rotational axis orientated transversely to the seat depth. In order to provide a head restraint which is anatomically shaped and can be set in height and inclination in such a manner that in every position it forms a continuation of the contour of the seat occupant's back, the padded headrest has a padded tongue which is pulled out downwards over the padding support for the purpose of supporting the back of the neck. The contour of the front surface of the padded headrest and padded tongue reproduces the profile of the back of the head and back of the neck. The pivoting device is designed so that the rotational axis of the padded headrest lies outside the head-restraint structure and, in all possible vertical settings of the head restraint, always lies on the level backrest surface.

49 Claims, 7 Drawing Sheets

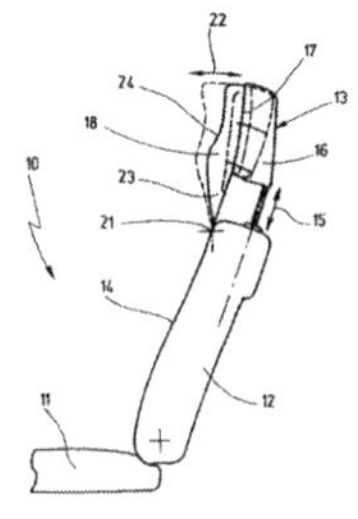

(12) **United States Patent**
Marinelli

(10) **Patent No.: US 6,533,341 B2**
(45) **Date of Patent: Mar. 18, 2003**

(54) **ADJUSTABLE PIVOTING MEANS FOR A VEHICLE SEAT BACKREST**

(75) Inventor: **Francesco Marinelli**, San Secondo di Pinerolo (IT)

(73) Assignee: **Lear Corporation**, Southfield, MI (US)

(*) Notice: Subject to any disclaimer, the term of this patent is extended or adjusted under 35 U.S.C. 154(b) by 0 days.

(21) Appl. No.: **09/892,827**

(22) Filed: **Jun. 27, 2001**

(65) **Prior Publication Data**

US 2002/0070578 A1 Jun. 13, 2002

(30) **Foreign Application Priority Data**

Jun. 30, 2000 (IT) TO00A0649

(51) **Int. Cl.**[7] **B60N 2/22**
(52) **U.S. Cl.** **296/65.16**; 296/69; 297/354.1; 297/383
(58) **Field of Search** 296/63, 65.01, 296/65.05, 65.16, 69; 297/383, 354.1, 342

(56) **References Cited**

U.S. PATENT DOCUMENTS

1,685,770 A * 10/1928 Bowen 297/326
1,966,343 A * 7/1934 Hallowell et al. 297/327
4,763,952 A * 8/1988 Gaudreau, Jr. 297/217.2
6,257,664 B1 * 7/2001 Chew et al. 297/284.9

* cited by examiner

Primary Examiner—D. Glenn Dayoan
Assistant Examiner—Patricia L. Engle
(74) *Attorney, Agent, or Firm*—Bill C. Panagos

(57) **ABSTRACT**

The backrest of a rear seat of a motor vehicle presents a frame, which lower fastening means pivoting on the body of the motor vehicle on a transverse axis and which upper fastening means to the body can be released to rotate the backrest forwards for extending the motor vehicle luggage compartment. The lower pivots are shaped so to pivot the backrest on the body of the motor vehicle with the lower part of the backrest arranged either in a first most advanced operative position or in a second most retracted operative position.

16 Claims, 3 Drawing Sheets

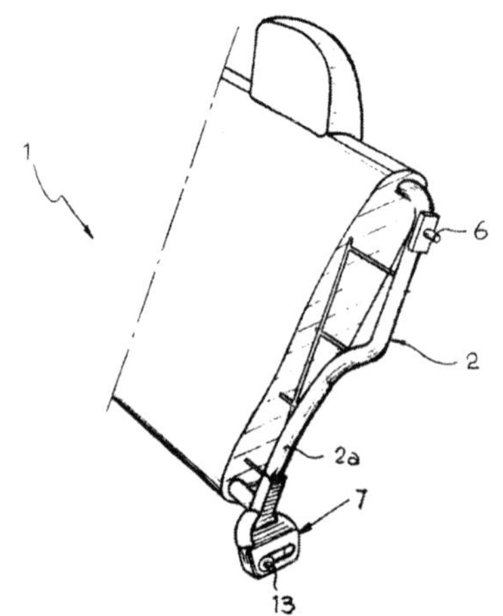

CHAPTER 12:

ACTIVE HEAD RESTRAINT PATENTS

Chapter 12

Active Head Restraint Patents

David C. Viano
ProBiomechanics LLC

ABSTRACT

This Chapter reviews active head restraints from the first patent in 1995 for a pivotal headrest. This concept moved the head restraint up and forward in a rear crash to prevent whiplash. 36 USPTO patents have been granted on various refinements in the approach to activate head restraint motion in rear crashes. The majority of patents deal with variations in the mechanism and linkage used to raise and move the head restraint forward. This motion provides earlier head support in a rear crash. These concepts include integrated designs and telescoping arms to adjust the head restraint position. Other safety functions have also been added to the pivoting head restraint including rotating side seatback structure, contouring for yield distance and combinations with inflatable airbags. With the determination that the Saab SAHR active head restraint reduces whiplash in real world crashes, there has been considerable thought given to active head restraints and integration in the seatback.

METHODOLOGY

US patents were located by searching primary and secondary references for active head restraints using the US Patent and Trademark Office website: http://patft.uspto.gov/netahtml/search. Efforts were made to locate patent examples following the first pivotal head restraint patent granted in 1995 (USPTO 5,378,043). Patents were identified that cited the 1995 concept. Then, secondary references were evaluated to provide broader coverage of the topic. Finally, key words were searched on active, inflatable and safety head restraints to identify other concepts.

Any patent of significance to the history of active head restraints for rear crash safety that is not included was an inadvertent oversight. Nonetheless, the items located and reported should be considered examples of the history of technology and patents on active head restraints.

RESULTS

In 1995, the first patent for a pivotal head restraint was grant for a concept that raised and moved forward the head restraint to provide earlier head support in a rear crash (USPTO 5,378,043). That system went into production in 1997 and has been found to reduce whiplash by 75% in real-world rear crashes (Viano, Olsen 2001). Over the last 7 years, there have been 36 USPTO patents that have followed the first active head restraint design. They include refinements in the design and integration of other safety functions.

A majority of the patents address variations in the linkage and mechanism that integrates the head restraint with the seatback and the means to activate the system in rear crashes. The original design used the mass of the occupant and energy of the crash to raise and move the head restraint forward. Many of the patents demonstrate innovative mechanisms and means to facilitate head restraint movement while maintaining an open seatback. Some involve pyrotechnic activation and the associated rear crash triggering.

Later patents have integrated inflatable airbags for wrap around protection and greater neck support, as well as moving side structures to support the torso in side and rollover crashes. These refinements have intended goal to reduce whiplash and control occupant kinematics in rear crashes.

REFERENCE

Viano, D.C., and Olsen, S., "The Effectiveness of Active Head Restraint in Preventing Whiplash," J. Trauma 51(5):959–969, November, 2001.

United States Patent [19]

Viano et al.

[11] **Patent Number: 5,378,043**

[45] **Date of Patent: Jan. 3, 1995**

[54] **VEHICLE PIVOTAL HEADREST**

[75] Inventors: **David C. Viano**, Bloomfield Hill; **Richard J. Neely**, Casco; **Mladen Humer**, East Detroit, all of Mich.

[73] Assignee: **General Motors Corporation**, Detroit, Mich.

[21] Appl. No.: **69,317**

[22] Filed: **Jun. 1, 1993**

[51] **Int. Cl.**[6] **B60R 21/00**

[52] **U.S. Cl.** **297/408**; 297/216.12

[58] **Field of Search** 297/391, 399, 404, 405, 297/408–410, 216.1, 216.12

[56] **References Cited**

U.S. PATENT DOCUMENTS

2,636,552	4/1953	Long	297/216.12 X
2,973,029	2/1961	Schlosstein	155/173
3,065,029	11/1962	Spound et al.	297/391
3,310,342	3/1967	Drelichowski	297/395
3,449,012	6/1969	Caron	297/403
3,488,090	1/1970	Douglas	297/389
3,586,366	6/1971	Patrick	297/391
3,655,241	4/1972	Herzer et al.	297/396 X
3,866,723	2/1975	Smith	297/216.1 X
3,929,374	12/1975	Hogan et al.	297/61
3,964,788	6/1976	Kmetyko	297/395
4,082,354	4/1978	Renner et al.	297/410
4,099,779	7/1978	Goldner	297/408
4,278,291	7/1981	Asai	297/391
4,285,545	8/1981	Protze	297/483
4,312,538	1/1982	Kennedy et al.	297/408
4,511,180	4/1985	Klaus	297/408
4,623,166	11/1986	Andres et al.	297/403 X
4,645,233	2/1987	Bruse et al.	280/808
4,720,146	1/1988	Mawbey et al.	297/409
4,762,367	8/1988	Denton	297/409
4,822,102	4/1989	Duvenkamp	297/403
4,861,107	8/1989	Vidwaus et al.	297/391 X
4,977,973	12/1990	Takizawa	180/271
5,181,763	1/1993	Dellanno et al.	297/391
5,205,585	4/1993	Reuber et al.	280/753

FOREIGN PATENT DOCUMENTS

2206329	9/1972	Germany .	
2232726	1/1974	Germany	297/216.12
2430572	1/1976	Germany	297/391
2644485	4/1978	Germany	297/391
3131633	2/1983	Germany	297/403

Primary Examiner—Peter R. Brown
Assistant Examiner—Milton Nelson, Jr.
Attorney, Agent, or Firm—Ernest E. Helms

[57] **ABSTRACT**

A vehicle seat and headrest arrangement is provided including a seat bun frame having fore and aft ends, a seatback frame joined to the bun frame adjacent the aft end of the bun frame, and a headrest pivotally attached with the seatback frame along a pivotal axis generally perpendicular to the fore and aft direction whereby, upon a rear vehicle impact, the headrest moves in a forward direction toward the head of a vehicle seat occupant.

12 Claims, 3 Drawing Sheets

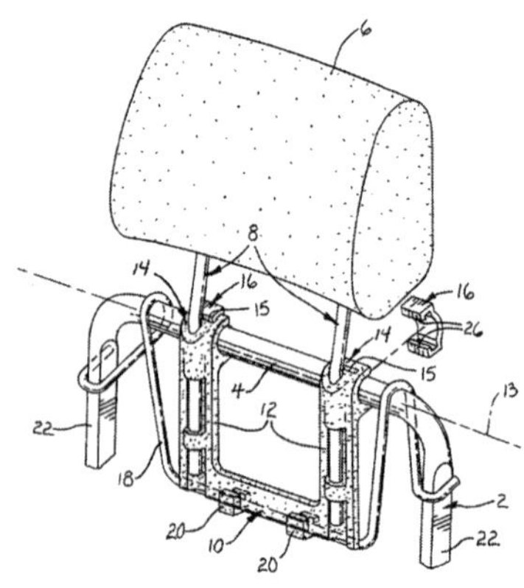

United States Patent [19]

Heilig et al.

[11] **Patent Number: 5,820,211**

[45] **Date of Patent: Oct. 13, 1998**

[54] **HEADREST FOR A VEHICLE SEAT**

[75] Inventors: **Alexander Heilig**, Wissgoldingen; **Helmut Maiwald**, Schechingen, both of Germany

[73] Assignee: **TRW Occupant Restraint Systems GmbH**, Alfdorf, Germany

[21] Appl. No.: **901,376**

[22] Filed: **Jul. 28, 1997**

[30] **Foreign Application Priority Data**

Aug. 16, 1996 [DE] Germany .... 296 14 238 U

[51] **Int. Cl.**[6] **B60N 2/42**

[52] **U.S. Cl.** **297/216.12**; 297/404; 297/408

[58] **Field of Search** 297/216.1, 216.12, 297/216.14, 408, 404, 391

[56] **References Cited**

U.S. PATENT DOCUMENTS

2,636,552	4/1953	Long	297/216.12 X
2,973,029	2/1961	Schlosstein	297/216.12
3,186,763	6/1965	Ferrara	297/216.12 X
5,181,763	1/1993	Dellanno et al.	297/404 X
5,290,091	3/1994	Dellanno et al.	297/216.12 X
5,378,043	1/1995	Viano et al.	297/216.12 X

FOREIGN PATENT DOCUMENTS

2152202	11/1980	Germany	297/216.12

Primary Examiner—Milton Nelson, Jr.
Attorney, Agent, or Firm—Tarolli, Sundheim, Covell, Tummino & Szabo

[57] **ABSTRACT**

A headrest for a vehicle seat having a backrest comprises a frame securable to the backrest of the vehicle seat, a pad part swivably secured to the frame and a device engaging the pad part which swivels the pad part in the direction of the head of a vehicle occupant in a rear end collision. The device comprise at least one tensioned spring for storing energy. The spring is integrated in the headrest, is relaxed in the case of a rear end collision and results in a swivel movement of the pad part due to the stored energy being liberated.

10 Claims, 1 Drawing Sheet

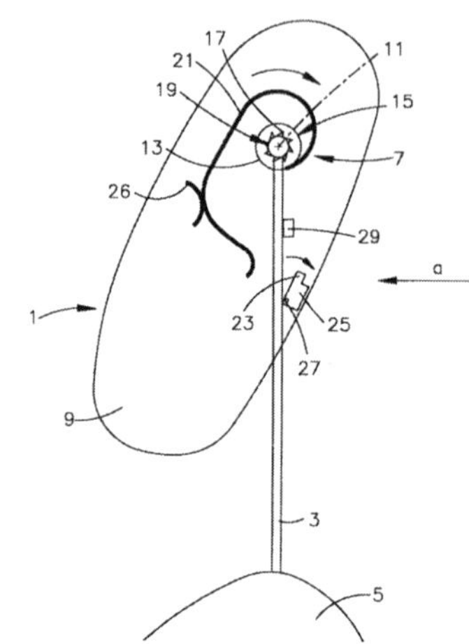

United States Patent [19]

Wieclawski

US005795019A

[11] **Patent Number: 5,795,019**

[45] **Date of Patent: Aug. 18, 1998**

[54] **VEHICLE SEAT AND HEADREST ARRANGEMENT**

[75] Inventor: **Stanislaw Andrzej Wieclawski**, Gross-Gerau, Germany

[73] Assignee: **Delphi Automotive Ssystems Deutschland GmbH**, Wuppertal, Germany

[21] Appl. No.: **827,812**

[22] Filed: **Apr. 11, 1997**

[30] **Foreign Application Priority Data**

Oct. 9, 1996 [GB] United Kingdom 9621082

[51] **Int. Cl.**[6] **B60N 2/42**

[52] **U.S. Cl.** **297/216.12**; 297/216.14

[58] **Field of Search** 297/391, 404, 297/408, 216.12, 216.13, 216.14, 216.1, 284.1

[56] **References Cited**

U.S. PATENT DOCUMENTS

2,636,552	4/1953	Long	297/216.12 X
3,838,870	10/1974	Hug	280/150
5,145,233	9/1992	Nagashima	297/408
5,378,043	1/1995	Viano et al.	297/408

FOREIGN PATENT DOCUMENTS

0627340	12/1994	European Pat. Off. .
WO87/03256	6/1987	WIPO .
WO96/06752	3/1996	WIPO .

Primary Examiner—Milton Nelson, Jr.
Attorney, Agent, or Firm—Patrick M. Griffin

[57] **ABSTRACT**

A vehicle seat and headrest arrangement **(7)** comprising a seat back frame **(2)** having an upper part **(24)** and a lower part **(26)**, the upper part including a cross-frame member **(4)**, and the lower part including at least a lower portion **(28)** of a pair of spaced side members **(22)**; a headrest **(6)** mounted on the cross-frame member; an impact plate **(42)** positioned between the side members of the lower part and secured to the upper part; pivot hinges **(32)** connecting the upper part with the lower part and defining a pivot axis **(38)** about which the upper part and the headrest can pivot in a forward direction, and the impact plate can pivot in a rearward direction, from a normal position; and a spring device **(40)** biasing the upper part, the headrest and the impact plate to their normal position. The upper part of the seat back and the headrest both pivot towards a vehicle occupant during a rear impact on the vehicle.

1 Claim, 5 Drawing Sheets

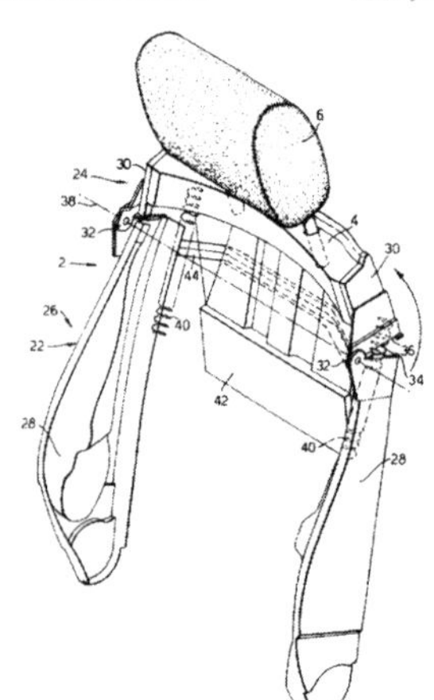

United States Patent [19]

Heilig et al.

[11] **Patent Number: 5,823,619**

[45] **Date of Patent: Oct. 20, 1998**

[54] **VEHICLE SEAT**

[75] Inventors: **Alexander Heilig**, Wissgoldingen; **Helmut Maiwald**, Schechingen, both of Germany

[73] Assignee: **TRW Occupant Restraint Systems GmbH**, Alfdorf, Germany

[21] Appl. No.: **806,734**

[22] Filed: **Feb. 27, 1997**

[30] **Foreign Application Priority Data**

Mar. 4, 1996 [DE] Germany 296 03 991.8
Jun. 3, 1996 [DE] Germany 296 09 786.1

[51] **Int. Cl.**[6] **B60N 2/42**

[52] **U.S. Cl.** **297/216.12**; 297/216.13; 297/61

[58] **Field of Search** 297/61, 216.12, 297/216.13, 408

[56] **References Cited**

U.S. PATENT DOCUMENTS

3,397,911	8/1968	Brosius, Sr.	297/216.12
3,713,694	1/1973	Miller	297/216.13 X
5,378,043	1/1995	Viano et al.	297/216.12 X
5,484,189	1/1996	Patterson	297/216.12 X
5,669,661	9/1997	Pajon	297/216.13

Primary Examiner—Peter R. Brown
Attorney, Agent, or Firm—Tarolli, Sundheim, Covell, Tummino & Szabo

[57] **ABSTRACT**

A vehicle seat comprises a back rest with a headrest. A flexible traction element extends through the interior of the back rest and is adapted to respond to the movement of an inertial mass shifted in a rear end impact and to cause a displacement of an attachment means of the traction element on the headrest. The displacement is converted in a mechanical setting mechanism, coupled with the traction element, into a movement of the headrest toward the head of the occupant to reduce the distance between the headrest and the head of the occupant.

18 Claims, 4 Drawing Sheets

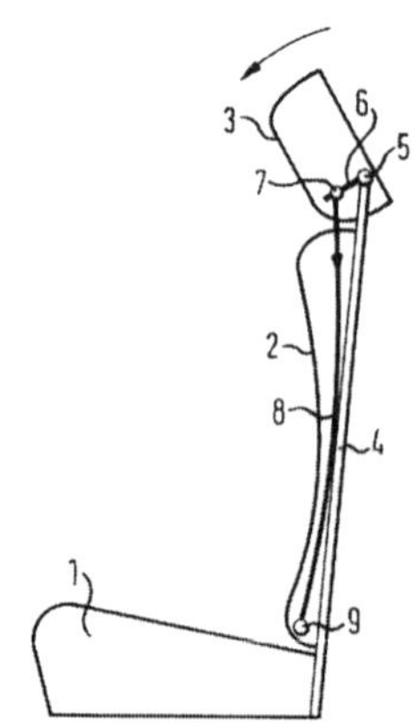

United States Patent [19]

Offenbacher

[11] **Patent Number: 5,829,838**

[45] **Date of Patent: Nov. 3, 1998**

[54] **INTEGRAL AUTOMOTIVE SEATBACK AND HEADREST CUSHION**

[75] Inventor: **Lon A. Offenbacher**, Rochester, Mich.

[73] Assignee: **General Motors Corporation**, Detroit, Mich.

[21] Appl. No.: **907,977**

[22] Filed: **Aug. 11, 1997**

[51] **Int. Cl.**[6] **A47C 7/36**
[52] **U.S. Cl.** **297/408**; 297/396
[58] **Field of Search** 297/408, 410, 297/391, 396

[56] **References Cited**

U.S. PATENT DOCUMENTS

2,828,810	4/1958	Barecki et al.	297/408 X
2,839,125	6/1958	Brandon	297/408 X
3,224,808	12/1965	Spielman	297/396 X
3,501,137	3/1970	Burgert	297/410 X
3,729,228	4/1973	Inoue et al.	297/396
4,549,766	10/1985	Nishino	297/403
4,856,848	8/1989	O'Sillivan et al.	297/391
4,865,388	9/1989	Nemoto	297/408 X
5,082,326	1/1992	Sekido et al.	297/408 X
5,378,043	1/1995	Viano et al.	297/408
5,464,269	11/1995	Mizelle	397/408
5,520,435	5/1996	Fujimoto et al.	297/408 X

Primary Examiner—Milton Nelson, Jr.
Attorney, Agent, or Firm—Patrick M. Griffin

[57] **ABSTRACT**

An integral seatback frame and headrest frame cushion and cover that accommodates angular and/or upward motion of the headrest relative to the seatback. A lower, primary cushion and integral, upper front flap cover the front of the seatback frame and headrest frame respectively, concealing the gap between the two. A rear flap covers most of the back surface of the headrest frame, and the two flaps are not secured tightly to the surfaces of the headrest frame, but have some internal clearance therefrom. When the headrest frame rocks forward, the two covering flaps travel with it without binding or stretching the cushion material.

3 Claims, 2 Drawing Sheets

United States Patent [19]

Cuevas

[11] **Patent Number: 5,927,804**

[45] **Date of Patent: Jul. 27, 1999**

[54] **VEHICLE OCCUPANT PROTECTION APPARATUS**

[75] Inventor: **Jess A. Cuevas**, Scottsdale, Ariz.

[73] Assignee: **TRW Inc.**, Lyndhurst, Ohio

[21] Appl. No.: **09/022,095**

[22] Filed: **Feb. 11, 1998**

[51] **Int. Cl.**[6] **B60N 2/42**
[52] **U.S. Cl.** **297/216.12**; 297/61; 297/408; 297/216.13
[58] **Field of Search** 297/61, 216.12, 297/216.13, 216.14, 406, 408

[56] **References Cited**

U.S. PATENT DOCUMENTS

2,973,029	2/1961	Schlosstein	297/216.12
3,339,911	9/1967	Brosius, Sr. .	
3,420,572	1/1969	Bisland .	
3,713,694	1/1973	Miller .	
5,052,754	10/1991	Chinomi .	
5,181,763	1/1993	Dellanno et al.	297/408 X
5,378,043	1/1995	Viano et al.	297/216.12 X
5,484,189	1/1996	Patterson .	
5,669,661	9/1997	Pajon .	
5,694,320	12/1997	Breed .	
5,716,102	2/1998	Ray et al.	297/216.13 X
5,823,619	10/1998	Heilig et al.	297/216.12
5,833,312	11/1998	Lenz	297/216.12 X
5,836,648	11/1998	Karschin et al.	297/216.13

FOREIGN PATENT DOCUMENTS

2232726	1/1974	Germany	297/216.12

Primary Examiner—Laurie K. Cranmer
Attorney, Agent, or Firm—Tarolli, Sundheim, Covell, Tummino & Szabo

[57] **ABSTRACT**

An apparatus (**10**) comprises a vehicle seat (**12**) including a frame (**20**) and a headrest (**14**). A linkage (**16**) is connected between the frame (**20**) and the headrest (**14**) to move the headrest (**14**) relative to the frame (**20**) in a forward direction. The linkage (**16**) is actuatable under the influence of a rear-end vehicle crash force applied to the linkage (**16**) by an occupant of the seat (**12**). The apparatus (**10**) further comprises a motor (**70**) which is connected to the linkage (**16**) to actuate the linkage (**16**). The motor (**70**) is actuated in response to a frontal vehicle crash condition.

9 Claims, 3 Drawing Sheets

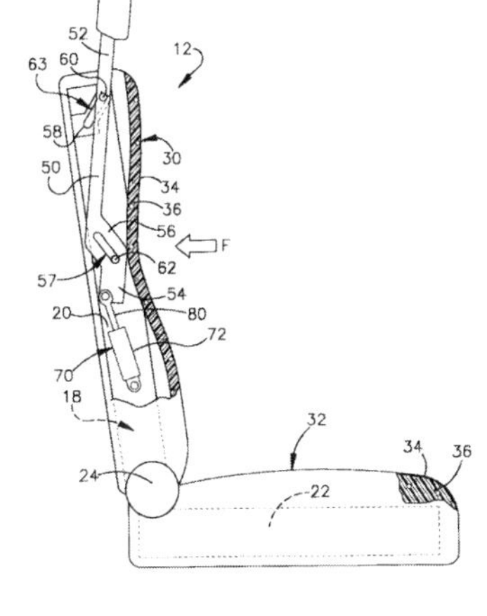

United States Patent [19]

Karschin et al.

[11] **Patent Number: 5,836,648**

[45] **Date of Patent: Nov. 17, 1998**

[54] **VEHICLE SEAT WITH MULTIFUNCTION BACKREST**

[75] Inventors: **Kurt Karschin**, Kirchheim/Teck; **Rene Heller**, Mengen-Rulfingen, both of Germany

[73] Assignee: **Recaro GmbH & Co.**, Kirchheim/Teck, Germany

[21] Appl. No.: **962,063**

[22] Filed: **Oct. 31, 1997**

[30] **Foreign Application Priority Data**

Oct. 31, 1996 [DE] Germany 196 43 977.9

[51] **Int. Cl.**[6] **B60N 2/42**
[52] **U.S. Cl.** **297/216.14**; 297/216.12; 297/216.13; 297/354.12; 297/408
[58] **Field of Search** 297/216.12, 216.13, 297/216.14, 353, 354.1, 354.11, 354.12, 374, 408

[56] **References Cited**

U.S. PATENT DOCUMENTS

3,186,763	6/1965	Ferrara	297/216.12 X
3,652,128	3/1972	Schwarz	297/216.12 X
3,761,125	9/1973	Glance	297/216.12 X
4,640,547	2/1987	Fromme	297/353 X
4,787,676	11/1988	Neve de Mevergnies	297/353
5,290,091	3/1994	Dellanno et al.	297/216.12 X
5,378,043	1/1995	Viano et al.	297/216.12 X

FOREIGN PATENT DOCUMENTS

4120608	1/1993	Germany .
9215255	7/1993	Germany .

Primary Examiner—Peter R. Brown
Attorney, Agent, or Firm—Roylance,Abrams,Berdo & Goodman, L.L.P.

[57] **ABSTRACT**

A vehicle seat has a seat part and a backrest adjustable in inclination relative to the seat part. The backrest is configured in two parts, a top backrest part and a bottom backrest part. The top backrest part is arranged pivotally relative to the bottom backrest part, when the holding force of a clamping device is overcome in the case of a crash under the effect of an actuating force coming from the person in the seat. The top backrest part executes an adjustment movement intercepting the person in the seat. The vehicle seat improves the safety and security of the seat occupant upon a frontal crash or rear crash.

11 Claims, 3 Drawing Sheets

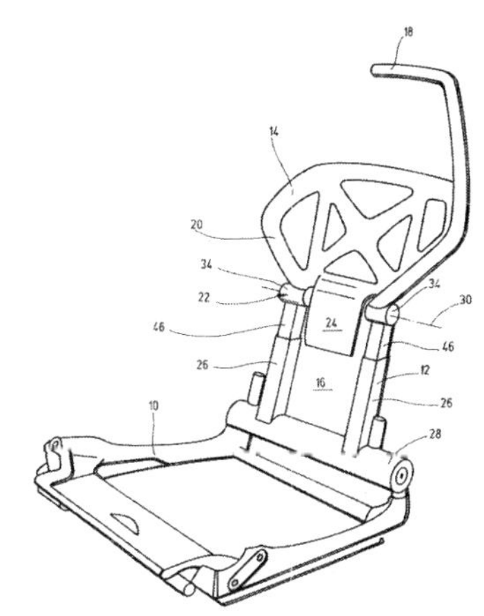

United States Patent [19]

Föhl

[11] **Patent Number: 5,934,750**

[45] **Date of Patent: Aug. 10, 1999**

[54] **VEHICLE SEAT**

[75] Inventor: **Artur Föhl**, Schorndorf, Germany

[73] Assignee: **TRW Occupant Restraint Systems GmbH**, Alfdorf, Germany

[21] Appl. No.: **08/800,023**

[22] Filed: **Feb. 13, 1997**

[30] **Foreign Application Priority Data**

Feb. 26, 1996 [DE] Germany 296 03 467

[51] **Int. Cl.**[6] **B60N 2/42**
[52] **U.S. Cl.** **297/216.12**; 297/408
[58] **Field of Search** 297/216.1, 216.12, 297/408

[56] **References Cited**

U.S. PATENT DOCUMENTS

3,838,870	10/1974	Hug	297/216.12 X
4,236,750	12/1980	Moritz	297/216.12 X
4,674,792	6/1987	Tamura et al.	297/408
5,290,091	3/1994	Dellanno et al.	297/408 X
5,378,043	1/1995	Viano et al.	297/216.12 X

FOREIGN PATENT DOCUMENTS

0627340	12/1994	European Pat. Off. .
2301906	12/1996	United Kingdom .
9606752	3/1996	WIPO .

Primary Examiner—Peter R. Brown
Attorney, Agent, or Firm—Tarolli, Sundheim, Covell, Tummino & Szabo

[57] **ABSTRACT**

In the event of a rear end collision the headrest of a vehicle seat is moved from a normal initial position to a position approximate to the vehicle occupant's head under the effect of inertial forces occurring at a mass coupled to the headrest, thereby reducing the distance between the vehicle occupant's head and the headrest.

20 Claims, 3 Drawing Sheets

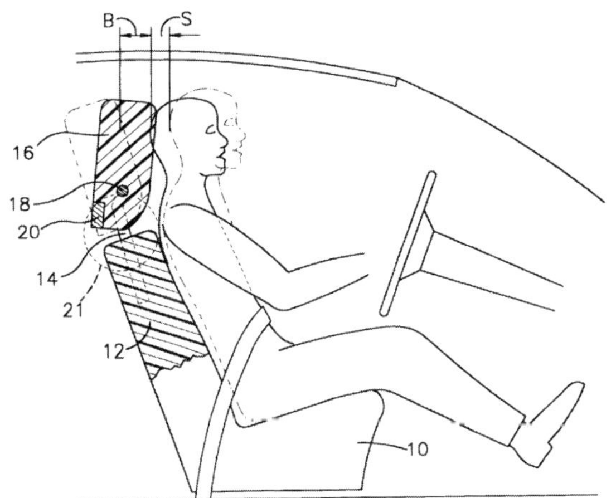

United States Patent [19]

Schubring et al.

[11] **Patent Number: 5,938,279**

[45] **Date of Patent: Aug. 17, 1999**

[54] **DYNAMIC VEHICLE HEAD RESTRAINT ASSEMBLY**

[75] Inventors: **James Douglas Schubring**, Swartz Creek; **James Bolsworth**; **Todd Hughes Smith**, both of Sterling Heights; **Radu Gabriel Munteanu**, Warren, all of Mich.

[73] Assignee: **Lear Corporation**, Southfield, Mich.

[21] Appl. No.: **09/070,992**

[22] Filed: **May 1, 1998**

[51] **Int. Cl.**[6] **B60N 2/42**

[52] **U.S. Cl.** **297/216.12**; 297/408; 297/410

[58] **Field of Search** 297/410, 408, 297/216.12

[56] **References Cited**

U.S. PATENT DOCUMENTS

5,378,043 1/1995 Viano et al. .

Primary Examiner—Peter M. Cuomo
Assistant Examiner—David E. Allred
Attorney, Agent, or Firm—Brooks & Kushman PC

[57] **ABSTRACT**

A dynamic head restraint mount (**18**) includes an low swinging impact plate (**22**) which swings back below a seat back frame top tube (**10**) in response to a sudden vehicle deceleration. As it swings back, the plate (**22**) causes the upper ends of a pair of upstanding head restraint support post guide tubes (**24**) to rock forwardly of the top tube (**10**), tilting the head restraint **14** forward. Concurrently, a linkage activated by the same impact plate (**22**) motion independently causes the support posts (**12**) of the head restraint (**14**) to slide up to a consistent highest position above the top tube (**10**), regardless of the adjusted position within the guide tubes (**24**) from which the support posts (**12**) began.

3 Claims, 6 Drawing Sheets

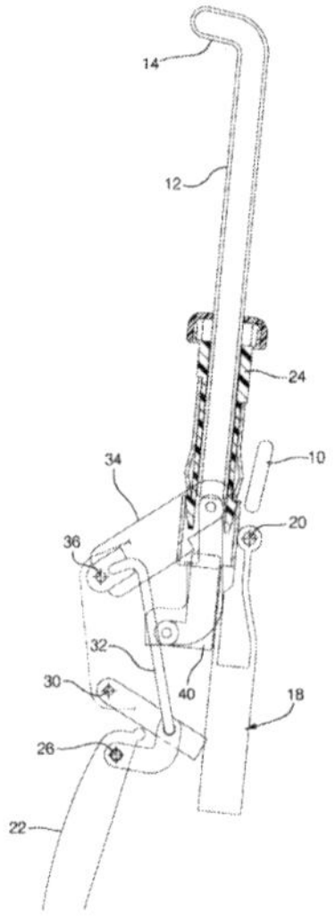

United States Patent [19]

Charras et al.

[11] **Patent Number: 6,024,406**

[45] **Date of Patent: Feb. 15, 2000**

[54] **VEHICLE SEAT PROVIDED WITH A DEVICE FOR PROTECTING THE NECK IN THE EVENT OF IMPACT FROM BEHIND**

[75] Inventors: **Fabrice Charras**, Paris; **Bernard Courtois**, Morigny; **Yacine Ziar**, Juvisy sur Orge, all of France

[73] Assignee: **Bertrand Faure Equipements S.A.**, Boulogne, France

[21] Appl. No.: **09/108,280**

[22] Filed: **Jul. 1, 1998**

[30] **Foreign Application Priority Data**

Jul. 3, 1997 [FR] France 97 08435

[51] **Int. Cl.**[7] **B60N 2/42**; B60R 21/00

[52] **U.S. Cl.** **297/216.14**; 297/216.1; 297/216.13; 297/472

[58] **Field of Search** 297/216.14, 216.13, 297/216.1, 472

[56] **References Cited**

U.S. PATENT DOCUMENTS

1,529,248 3/1925 Greene et al. .
3,578,376 5/1971 Hasegawa et al. 297/216
3,802,737 4/1974 Mertens 297/216.1
3,998,291 12/1976 Davis 297/216.1
5,219,202 6/1993 Rink et al. 297/216.13
5,286,058 2/1994 Wier .... 297/472 X
5,366,269 11/1994 Beauvais 297/216.19
5,378,043 1/1995 Viano et al. 297/216.12
5,437,494 8/1995 Beauvais 297/216.19
5,462,332 10/1995 Payne et al. 297/216.1
5,509,716 4/1996 Kolena et al. 297/216.13
5,593,210 1/1997 Schwarzbich 297/361.1
5,645,316 7/1997 Aufrere et al. 297/216.13
5,676,421 10/1997 Brodsky 297/216.13
5,738,407 4/1998 Locke .... 297/216.13 X
5,772,280 6/1998 Massara .... 297/216.13 X
5,795,019 8/1998 Wieclawski 297/216.12
5,927,804 7/1999 Cuevas .... 297/216.13 X

FOREIGN PATENT DOCUMENTS

843668 7/1939 France 5/3
2 590 529 5/1987 France B60N 1/02
2 602 133 2/1988 France A47C 7/46
2 754 221 4/1998 France B60N 2/48
37 34 363 A1 4/1989 Germany B60N 1/08
WO 87/03256 6/1987 WIPO B60N 1/06
WO 97/10117 3/1997 WIPO B60N 2/42

Primary Examiner—Peter M. Cuomo
Assistant Examiner—Rodney B. White
Attorney, Agent, or Firm—Marshall, O'Toole, Gerstein, Murray & Borun

[57] **ABSTRACT**

A motor vehicle seat has a strength-member for its back that includes an upper portion pivotally mounted on a lower portion and carrying a headrest, the two portions of the strength-member are connected together by a locking device which holds them together and which is dimensioned to give way when the back of a passenger bears against a thrust member of the upper portion of the strength-member in the event of an impact from behind, such that the upper portion of the seat-back strength-member and the upper region of the front face of the seat-back then pivot forwards together with the headrest, with this movement being braked by an energy absorber device.

16 Claims, 8 Drawing Sheets

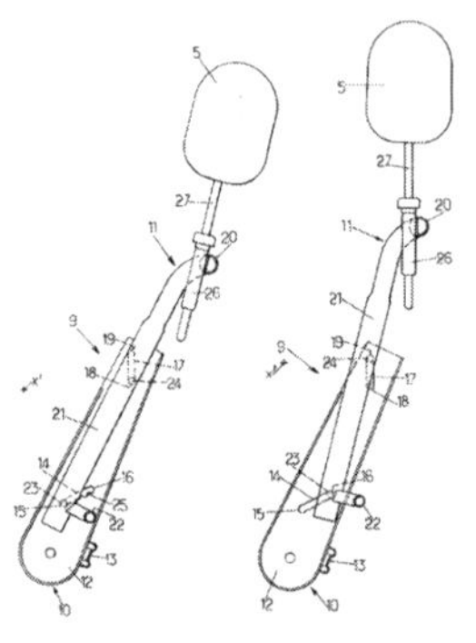

United States Patent [19]

Rückert et al.

[11] **Patent Number: 6,019,424**

[45] **Date of Patent: Feb. 1, 2000**

[54] **HEADREST, IN PARTICULAR IN MOTOR VEHICLES**

[75] Inventors: **Edvard Rückert**, Velbert; **Norbert Krüger**, Essen, both of Germany

[73] Assignee: **Ewald Witte GmbH & Co. KG**, Velbert, Germany

[21] Appl. No.: **09/204,098**

[22] Filed: **Dec. 1, 1998**

[30] **Foreign Application Priority Data**

Dec. 3, 1997 [DE] Germany 197 53 540

[51] **Int. Cl.**[7] **B60N 2/42**

[52] **U.S. Cl.** **297/216.12**; 297/216.13; 297/216.14; 297/408

[58] **Field of Search** 297/216.12, 216.13, 297/216.14, 408

[56] **References Cited**

U.S. PATENT DOCUMENTS

3,838,870 10/1974 Hug .
5,378,043 1/1995 Viano et al. 297/216.12
5,884,968 3/1999 Massara 297/216.12

FOREIGN PATENT DOCUMENTS

0627340 12/1994 European Pat. Off. .
2232726 1/1974 Germany .
49-25690 2/1974 Japan 297/216.14
7-291005 7/1995 Japan 297/216.14

Primary Examiner—Peter M. Cuomo
Assistant Examiner—Anthony D. Barfield
Attorney, Agent, or Firm—Martin A. Farber

[57] **ABSTRACT**

A headrest (**23**) which pivots forward, by lever action, as a result of the mass of the upper body of a vehicle occupant acting on a pressure-bearing surface in the event of an impact from the rear. In order that in case of both a rear-end crash and a head-on crash of the vehicle, the headrest (**23**) follows the head of an individual sitting on the vehicle seat, the headrest (**23**) is positioned on a headrest carrier (**17**) which can be pivoted out of the backrest (**3**) in the forward direction and can be pivoted outward about a pin (**19**), this pin being located at a lower level than that of the pivot pin (**21**) of the headrest (**23**).

16 Claims, 9 Drawing Sheets

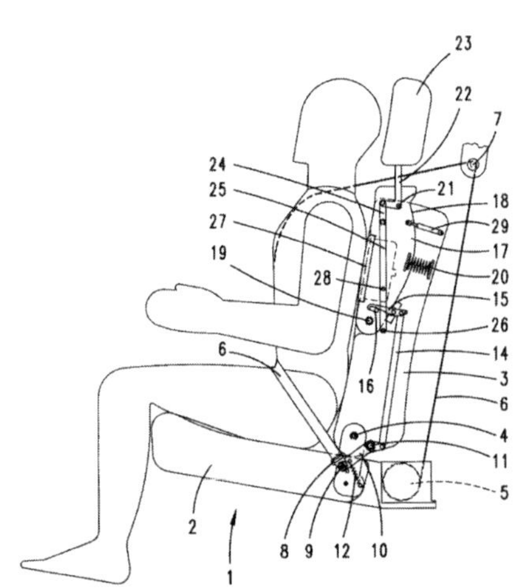

United States Patent [19]

Elqadah et al.

[11] **Patent Number: 6,033,017**

[45] **Date of Patent: *Mar. 7, 2000**

[54] **VEHICLE OCCUPANT PROTECTION APPARATUS**

[75] Inventors: **Wael S. Elqadah**; **Xingyuan Sun**, both of Gilbert; **Jess A. Cuevas**, Scottsdale, all of Ariz.

[73] Assignee: **TRW Inc.**, Lyndhurst, Ohio

[*] Notice: This patent issued on a continued prosecution application filed under 37 CFR 1.53(d), and is subject to the twenty year patent term provisions of 35 U.S.C. 154(a)(2).

[21] Appl. No.: **09/057,815**

[22] Filed: **Apr. 9, 1998**

[51] **Int. Cl.**[7] **B60N 2/42**

[52] **U.S. Cl.** **297/216.1**; 297/216.13; 297/408; 297/486

[58] **Field of Search** 297/216.1, 216.12, 297/216.13, 216.15, 474, 484, 486, 391, 406, 408, 409

[56] **References Cited**

U.S. PATENT DOCUMENTS

2,873,122 2/1959 Peras .
3,397,911 8/1968 Brosius, Sr. .
3,420,572 1/1969 Bisland .
3,623,768 11/1971 Capener et al. .
3,713,694 1/1973 Miller .
3,832,003 8/1974 Horvat 297/216.13
3,899,042 8/1975 Bonar .
3,901,550 8/1975 Hamy .
5,052,754 10/1991 Chinomi .
5,378,043 1/1995 Viano et al. .
5,484,189 1/1996 Patterson .
5,669,661 9/1997 Pajon .
5,836,648 11/1998 Karschin et al. 297/216.13

Primary Examiner—Laurie K. Cranmer
Attorney, Agent, or Firm—Tarolli, Sundheim, Covell, Tummino & Szabo L.L.P.

[57] **ABSTRACT**

An apparatus includes a vehicle seat (**10**) and an actuating assembly (**80**). The seat (**10**) has a wing (**34**) which is movable from a retracted position to a deployed position. When the wing (**34**) is in the deployed position, it projects adjacent to an occupant the seat (**10**) to restrain movement of the occupant off the seat (**10**). The actuating assembly (**80**) moves the wing (**34**) from the retracted position to the deployed position upon the occurrence of a vehicle crash.

19 Claims, 4 Drawing Sheets

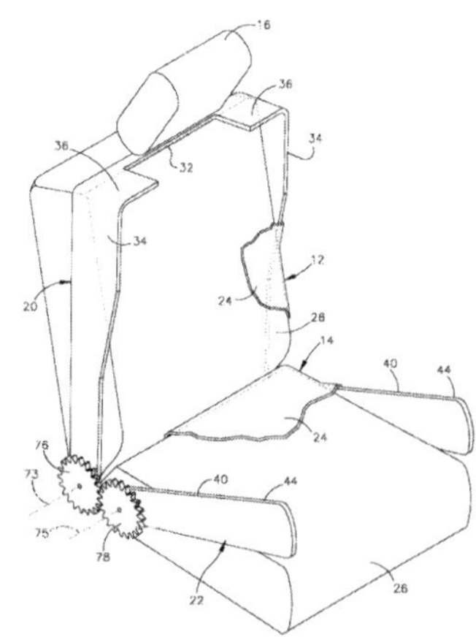

U.S. Patent Mar. 7, 2000 Sheet 1 of 4 6,033,018

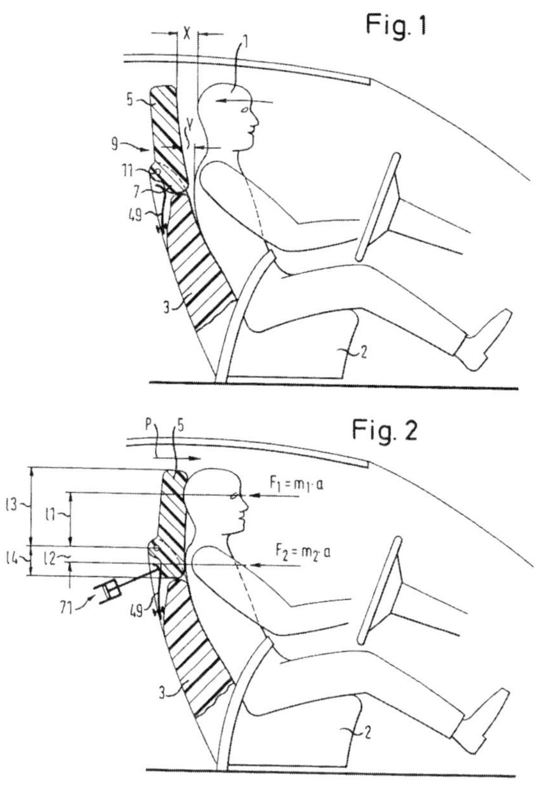

U.S. Patent May 16, 2000 Sheet 2 of 3 6,062,642

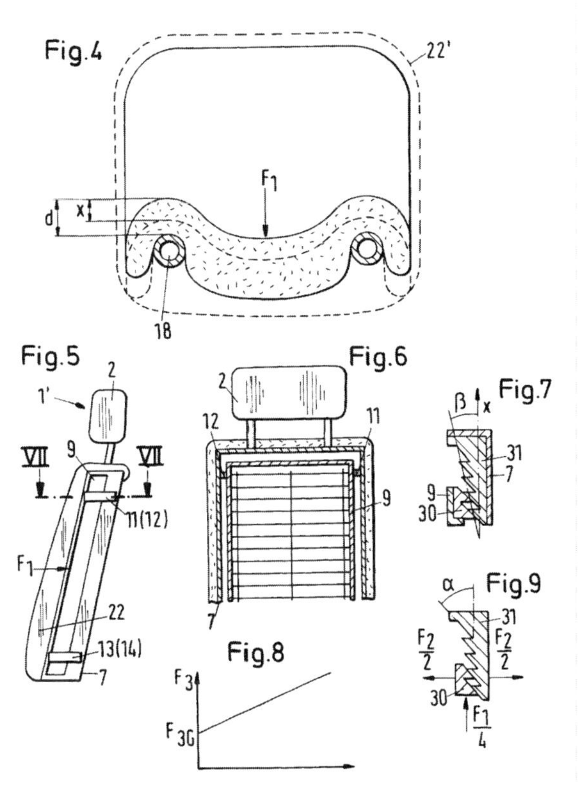

United States Patent [19]

Sinnhuber et al.

[11] **Patent Number: 6,062,642**

[45] **Date of Patent: May 16, 2000**

[54] **VEHICLE SEAT**

[75] Inventors: **Ruprecht Sinnhuber**, Gifhorn; **Tim-Bosse Vogel**, Ulm; **Thomas Wohllebe**, Braunschweig, all of Germany

[73] Assignee: **Volkswagen AG**, Wolfsburg, Germany

[21] Appl. No.: **09/054,023**

[22] Filed: **Apr. 2, 1998**

[51] **Int. Cl.**[7] **B60N 2/42**

[52] **U.S. Cl.** **297/216.13**; 297/216.1; 297/216.14

[58] **Field of Search** 297/216.12, 216.13, 297/452.5, 452.52, 216.1

[56] **References Cited**

U.S. PATENT DOCUMENTS

2,608,243	8/1952	Kostrowski	297/452.5 X
2,836,226	5/1958	Fridolph	297/452.5
4,368,917	1/1983	Urai	297/452.54
5,253,924	10/1993	Glance	297/216.13 X
5,310,247	5/1994	Fujimori et al.	297/378
5,378,043	1/1995	Viano et al.	297/216.12 X
5,769,489	6/1998	Dellanno	297/216.12 X

FOREIGN PATENT DOCUMENTS

0754590	1/1997	European Pat. Off. .	
1470423	2/1967	France	297/452.5
2263931	7/1973	Germany	297/452.5
91036119	9/1991	Germany .	
4031285	4/1992	Germany .	
92152546	8/1993	Germany .	
4238549	5/1994	Germany .	
4337019	5/1995	Germany .	
4421946	6/1995	Germany .	
29610078	11/1996	Germany .	
9511818	5/1995	WIPO .	
9748570	12/1997	WIPO .	

Primary Examiner—Peter R. Brown
Attorney, Agent, or Firm—Baker Botts, L.L.P.

[57] **ABSTRACT**

In the embodiment described in the specification, a vehicle seat has a framework arrangement with both backrests and seat parts on which a cushion is mounted. The framework arrangement includes a first framework and a second framework which can be displaced relative to the first framework. To reduce the risk of injury to a vehicle seat occupant in the event of rear-impact accidents and/or accidents involving a vehicle turning over, the second framework is supported from the first framework by an energy-absorption structure.

27 Claims, 3 Drawing Sheets

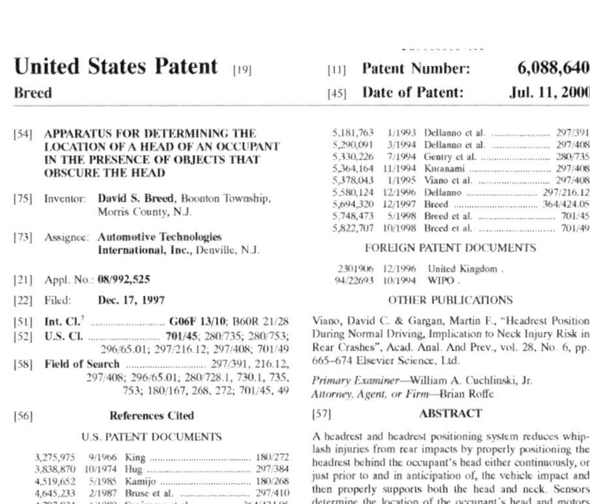

United States Patent [19]

Breed

[11] **Patent Number: 6,088,640**

[45] **Date of Patent: Jul. 11, 2000**

[54] **APPARATUS FOR DETERMINING THE LOCATION OF A HEAD OF AN OCCUPANT IN THE PRESENCE OF OBJECTS THAT OBSCURE THE HEAD**

[75] Inventor: **David S. Breed**, Boonton Township, Morris County, N.J.

[73] Assignee: **Automotive Technologies International, Inc.**, Denville, N.J.

[21] Appl. No.: **08/992,525**

[22] Filed: **Dec. 17, 1997**

[51] **Int. Cl.**[7] **G06F 13/10**; B60R 21/28

[52] **U.S. Cl.** **701/45**; 280/735; 280/753; 296/65.01; 297/216.12; 297/408; 701/49

[58] **Field of Search** 297/391, 216.12, 297/408; 296/65.01; 280/728.1, 730.1, 735, 753; 180/167, 268, 272; 701/45, 49

[56] **References Cited**

U.S. PATENT DOCUMENTS

3,275,975	9/1966	King	180/272
3,838,870	10/1974	Hug	297/384
4,519,652	5/1985	Kamijo	180/268
4,645,233	2/1987	Bruse et al.	297/410
4,797,824	1/1989	Sugiyama et al.	364/424.05
4,811,226	3/1989	Shinohara	364/424.05
4,853,687	8/1989	Isomura et al.	364/424.05
4,935,680	6/1990	Sugiyama	364/424.05
4,995,639	2/1991	Breed	280/735
5,003,240	3/1991	Ikeda	364/424.05
5,006,771	4/1991	Ogasawara	318/568.1
5,008,946	4/1991	Ando	180/167
5,071,160	12/1991	White et al.	280/735
5,074,583	12/1991	Fujita et al.	280/735
5,095,257	3/1992	Ikeda et al.	364/424.05
5,151,944	9/1992	Yamamura et al.	381/151
5,161,820	11/1992	Vollmer	280/730.1
5,181,763	1/1993	Dellanno et al.	297/391
5,290,091	3/1994	Dellanno et al.	297/408
5,330,226	7/1994	Gentry et al.	280/735
5,364,164	11/1994	Kuranami	297/408
5,378,043	1/1995	Viano et al.	297/408
5,580,124	12/1996	Dellanno	297/216.12
5,694,320	12/1997	Breed	364/424.05
5,748,473	5/1998	Breed et al.	701/45
5,822,707	10/1998	Breed et al.	701/49

FOREIGN PATENT DOCUMENTS

2301906	12/1996	United Kingdom .
94/22693	10/1994	WIPO .

OTHER PUBLICATIONS

Viano, David C. & Gargan, Martin F., "Headrest Position During Normal Driving, Implication to Neck Injury Risk in Rear Crashes", Acad. Anal. And Prev., vol. 28, No. 6, pp. 665–674 Elsevier Science, Ltd.

Primary Examiner—William A. Cuchlinski, Jr.
Attorney, Agent, or Firm—Brian Roffe

[57] **ABSTRACT**

A headrest and headrest positioning system reduces whiplash injuries from rear impacts by properly positioning the headrest behind the occupant's head either continuously, or just prior to and in anticipation of, the vehicle impact and then properly supports both the head and neck. Sensors determine the location of the occupant's head and motors move the headrest both up and down and forward and back as needed. In one implementation, the headrest is continuously adjusted to maintain a proper orientation of the headrest to the rear of the occupant's head. In another implementation, an anticipatory crash, is used to predict that a rear impact is about to occur, in which event, the headrest is moved proximate to the occupant. A pre-inflated airbag within the headrest automatically distributes the pressure to evenly support both the head and neck.

19 Claims, 10 Drawing Sheets

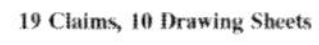

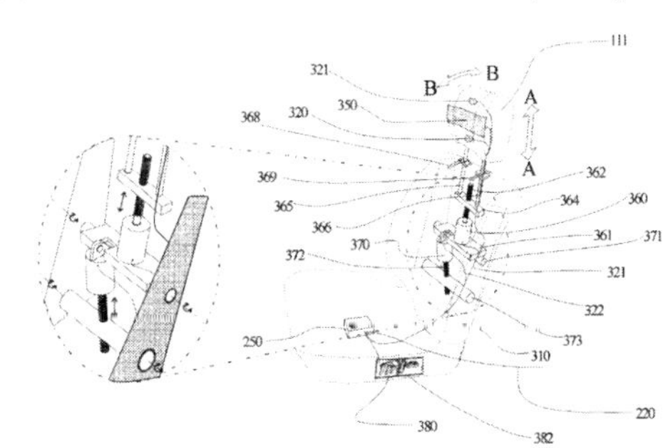

United States Patent [19]

Wu et al.

[11] **Patent Number: 6,109,690**

[45] **Date of Patent: Aug. 29, 2000**

[54] **PIVOTING SEAT BACK**

[75] Inventors: **Wei-Pin Wu**, Canton; **John J. Flannery**, Fenton; **Robert H. Dietze, Sr.**, Brighton; **Joseph P. Vitale**, South Lyon; **Mark A. Pattok**, Canton; **Jeffrey A. Lindberg**, Southgate; **Robert M. Healy**, Garden City, all of Mich.

[73] Assignee: **Johnson Controls Technology Company**, Plymouth, Mich.

[21] Appl. No.: **09/216,658**

[22] Filed: **Dec. 18, 1998**

[51] **Int. Cl.**[7] **B60N 2/42**

[52] **U.S. Cl.** **297/216.13**; 297/378.11; 297/216.14

[58] **Field of Search** 297/216.14, 216.13, 297/378.11, 374, 354.1, 354.11, 468, 216.12, 216.19

[56] **References Cited**

U.S. PATENT DOCUMENTS

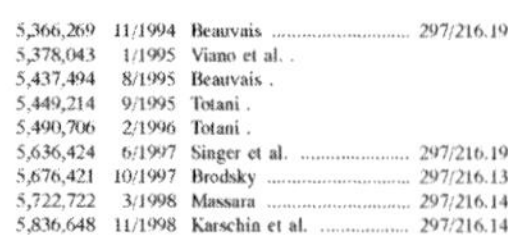

3,357,736	12/1967	McCarthy	297/216.19
5,366,269	11/1994	Beauvais	297/216.19
5,378,043	1/1995	Viano et al. .	
5,437,494	8/1995	Beauvais .	
5,449,214	9/1995	Totani .	
5,490,706	2/1996	Totani .	
5,636,424	6/1997	Singer et al.	297/216.19
5,676,421	10/1997	Brodsky	297/216.13
5,722,722	3/1998	Massara	297/216.14
5,836,648	11/1998	Karschin et al.	297/216.14

Primary Examiner—Anthony D. Barfield
Attorney, Agent, or Firm—Harness, Dickey & Pierce, P.L.C.

[57] **ABSTRACT**

A vehicle seat assembly having a seat back which pivots during a rear impact vehicle collision in Such a manner that the upper end of the seat back rotates foreword. The pivoting seat back includes a releasable latch mechanism to hold the seat back frame fixed in a predetermined position about a pivot axis. The latch mechanism is releasable by inertia forces in response to a rear vehicle collision, to release the latch member and free the seat back frame for rotation about the pivot axis.

46 Claims, 6 Drawing Sheets

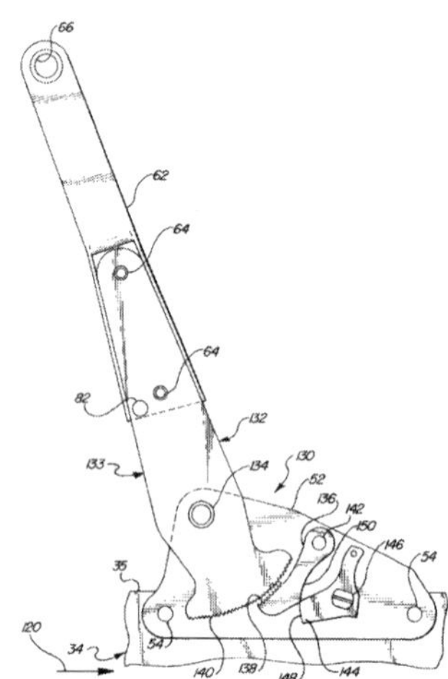

(12) United States Patent

Nakano et al.

(10) **Patent No.: US 6,250,714 B1**

(45) **Date of Patent: *Jun. 26, 2001**

(54) **SEATBACK FOR AUTOMOBILE**

(75) Inventors: **Nobuyuki Nakano**; **Shigeki Hirabayashi**, both of Kanagawa-ken (JP)

(73) Assignee: **Ikeda Bussan Co., Ltd.**, Kanagawa-ken (JP)

(*) Notice: This patent issued on a continued prosecution application filed under 37 CFR 1.53(d), and is subject to the twenty year patent term provisions of 35 U.S.C. 154(a)(2).

Subject to any disclaimer, the term of this patent is extended or adjusted under 35 U.S.C. 154(b) by 0 days.

(21) Appl. No.: **09/384,232**

(22) Filed: **Aug. 27, 1999**

(30) **Foreign Application Priority Data**

Aug. 28, 1998 (JP) P10-243751
Aug. 28, 1998 (JP) P10-243752

(51) **Int. Cl.**[7] **B60N 2/42**

(52) **U.S. Cl.** **297/216.12**; 297/216.13

(58) **Field of Search** 297/216.13, 216.12, 297/216.1, 284.4, 408

(56) **References Cited**

U.S. PATENT DOCUMENTS

5,378,043	*	1/1995	Viano et al.	297/216.12 X
5,823,619	*	10/1998	Heilig et al.	297/216.12
5,884,968	*	3/1999	Massara	297/216.12
5,927,804	*	7/1999	Cuevas	297/216.12
5,938,279	*	8/1999	Schubring et al.	297/216.12
6,019,424	*	2/2000	Rückert et al.	297/216.12
6,024,406	*	2/2000	Charras et al.	297/216.138

FOREIGN PATENT DOCUMENTS

0 627 340 A1	12/1994	(EP) .
2 318 045	4/1998	(GB) .
98 09838	3/1998	(WO) .

* cited by examiner

Primary Examiner—Peter R. Brown
(74) *Attorney, Agent, or Firm*—Nath & Associates PLLC; Gary M. Nath; Marvin C. Berkowitz

(57) **ABSTRACT**

There are provided a headrest; a seatback frame; a supporting member which was kept by the seatback frame, supports the headrest to be swung, and is arranged at least inside a shoulder point of AFO 5% tile mannequin from a front view thereof; and a pressure receiving member which is formed of a rigid body, is supported by the supporting member, and is to receive a predetermined pressure to swing the supporting member. Preferably, the pressure receiving member has a cross-sectional configuration formed in an almost U-shape, and is spaced from a back face of a pad arranged in front of the pressure receiving member.

7 Claims, 8 Drawing Sheets

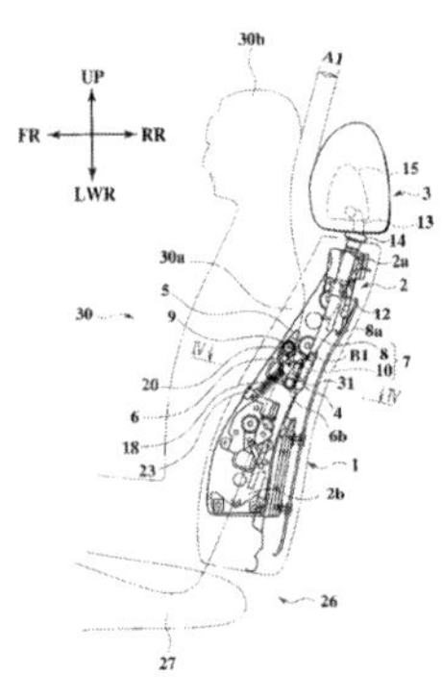

(12) United States Patent

Andersson

(10) **Patent No.: US 6,179,379 B1**

(45) **Date of Patent: Jan. 30, 2001**

(54) **SAFETY SEAT**

(75) Inventor: **Sture Andersson**, Nacks (SE)

(73) Assignee: **Autoliv Development AB**, Vårgårda (SE)

(*) Notice: Under 35 U.S.C. 154(b), the term of this patent shall be extended for 0 days.

(21) Appl. No.: **09/254,397**

(22) PCT Filed: **Sep. 4, 1997**

(86) PCT No.: **PCT/SE97/01473**

§ 371 Date: **Oct. 22, 1999**

§ 102(e) Date: **Oct. 22, 1999**

(87) PCT Pub. No.: **WO98/09836**

PCT Pub. Date: **Mar. 12, 1998**

(30) **Foreign Application Priority Data**

Sep. 6, 1996 (GB) 9618671

(51) **Int. Cl.**[7] **B60N 2/42**; B60R 21/00

(52) **U.S. Cl.** **297/216.13**; 297/216.12

(58) **Field of Search** 297/216.13, 216.12, 297/216.14, 216.1

(56) **References Cited**

U.S. PATENT DOCUMENTS

3,929,374		12/1975	Hogan et al. .	
4,040,661		8/1977	Hogan et al. .	
5,290,091	*	3/1994	Dellanno et al.	297/216.12 X
5,378,043	*	1/1995	Viano et al.	297/216.12 X
5,676,421	*	10/1997	Brodsky	297/216.13
5,746,467	*	5/1998	Jesadanont	297/216.13 X
5,772,280	*	6/1998	Massara	297/216.13 X
5,823,619	*	10/1998	Heilig et al.	297/216.13 X
5,884,968	*	3/1999	Massara	297/216.12
5,927,804	*	7/1999	Cuevas	297/216.13 X
5,934,750	*	8/1999	Fohl	297/216.12
6,019,424	*	2/2000	Ruckert et al.	297/216.13 X
6,022,074	*	2/2000	Swedenclef	297/216.13 X
6,024,406	*	2/2000	Charras et al.	297/216.13 X
6,033,017	*	3/2000	Elqedah et al.	297/216.13 X
6,033,018	*	3/2000	Fohl	297/216.13

FOREIGN PATENT DOCUMENTS

WO 95/11818 5/1995 (WO) .

* cited by examiner

Primary Examiner—Peter M. Cuomo
Assistant Examiner—Rodney B. White
(74) *Attorney, Agent, or Firm*—Venable; Robert Kinberg

(57) **ABSTRACT**

A safety seat for use in a motor vehicle has a squab (**101**) and a back (**104**) and a head-rest (**105**). There is a pivotal connection between the back (**104**) and the squab (**101**). The head-rest (**105**) is mounted on a series of levers (**112**). If the vehicle is involved in a rear impact, the back (**104**) of the seat may move rearwardly. Simultaneously, because of the levers (**112**) the head-rest (**105**) will move forwardly relative to the back of the seat. Thus, the head-rest will come into contact with the head of the occupant of the seat, thus minimizing the risk of "whiplash" injuries.

13 Claims, 6 Drawing Sheets

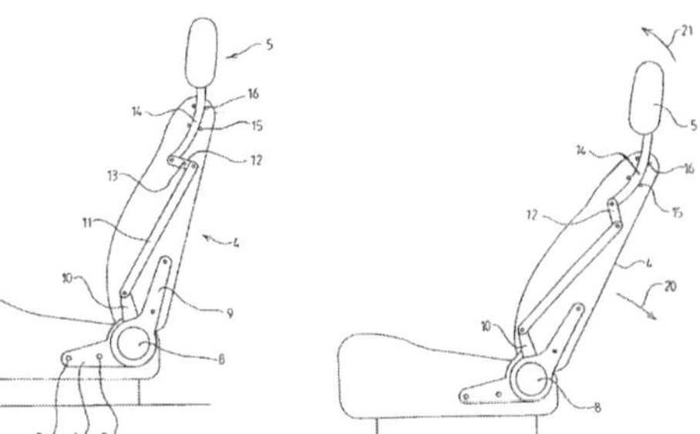

FIG.1

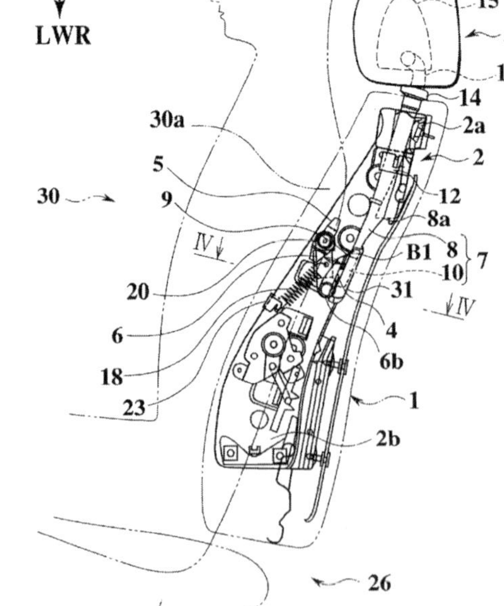

(12) **United States Patent**
Breed

(10) **Patent No.: US 6,331,014 B1**
(45) **Date of Patent: Dec. 18, 2001**

(54) **VEHICULAR SEATS INCLUDING OCCUPANT PROTECTION APPARATUS**

(75) Inventor: **David S. Breed**, Boonton Township, Morris County, NJ (US)

(73) Assignee: **Automotive Technologies International Inc.**, Denville, NJ (US)

(*) Notice: Subject to any disclaimer, the term of this patent is extended or adjusted under 35 U.S.C. 154(b) by 0 days.

(21) Appl. No.: **09/613,924**

(22) Filed: **Jul. 11, 2000**

(51) **Int. Cl.**[7] **B60R 21/22**
(52) **U.S. Cl.** **280/730.1**; 297/216.12; 297/216.13
(58) **Field of Search** 280/730.1, 730.2; 297/216.12, 216.13, 410, 408

(56) **References Cited**

U.S. PATENT DOCUMENTS

3,275,975	9/1966	King	180/272
3,838,870	10/1974	Hug	297/384
4,519,652	5/1985	Kamijo	180/268
4,645,233	2/1987	Bruse et al.	297/410
4,797,824	1/1989	Sugiyama et al.	364/424.05
4,811,226	3/1989	Shinohara	364/424.05
4,853,687	8/1989	Isomura et al.	364/424.05
4,865,388 *	9/1989	Nemoto	280/408
4,935,680	6/1990	Sugiyama	364/424.05
4,995,639	2/1991	Breed	280/735
5,003,240	3/1991	Ikeda	364/424.05
5,006,771	4/1991	Ogasawara	318/568.1
5,008,946	4/1991	Ando	180/167
5,071,160	12/1991	White et al.	280/735
5,074,583	12/1991	Fujita et al.	280/735
5,095,257	3/1992	Ikeda et al.	364/424.05
5,151,944	9/1992	Yamamura et al.	381/151
5,161,820	11/1992	Vollmer	280/730.1
5,181,763	1/1993	Dellanno et al.	297/391
5,290,091	3/1994	Dellanno et al.	297/408
5,330,226	7/1994	Gentry et al.	280/735
5,364,164	11/1994	Kuranami	297/408
5,378,043	1/1995	Viano et al.	297/408
5,580,124	12/1996	Dellanno	297/216.12
5,694,320	12/1997	Breed	364/424.05
5,748,473	5/1998	Breed et al.	701/45
5,769,489	6/1998	Dellanno	297/216.14

(List continued on next page.)

FOREIGN PATENT DOCUMENTS

2301906	12/1996	(GB) .
94/22693	10/1994	(WO) .

OTHER PUBLICATIONS

Viano, David C. & Gargan, Martin F., "Headrest Position During Normal Driving, Implication to Neck Injury Risk in Rear Crashes", Acad. Anal. And Prev., vol. 28, No. 6, pp. 665–674 Elsevier Science, Ltd.

Primary Examiner—Eric Culbreth
(74) *Attorney, Agent, or Firm*—Brian Roffe

(57) **ABSTRACT**

A seat for a vehicle for protecting an occupant of the seat in a crash including a headrest portion, an expandable bladder arranged at least partially in the headrest portion and arranged to conform to the shape of a neck and head of the occupant upon expansion, and an igniter for causing expansion of the bladder upon receiving a signal that protection for the occupant is desired. A fluid-containing chamber is coupled to the igniter and in flow communication with the bladder whereby the igniter causes fluid in the chamber to expand and flow into the bladder to expand the bladder. A control valve associated with the bladder enables the release of fluid from the bladder. A crash sensor system determines that a crash requiring protection for the occupant, e.g.,. a rear impact, is desired and generates and directs the signal to the igniter. Instead of the expandable bladder and igniter, the seat may include an airbag arranged at least partially in the headrest portion whereby the headrest portion is pivotable with respect to a backrest portion of the seat toward the occupant.

32 Claims, 14 Drawing Sheets

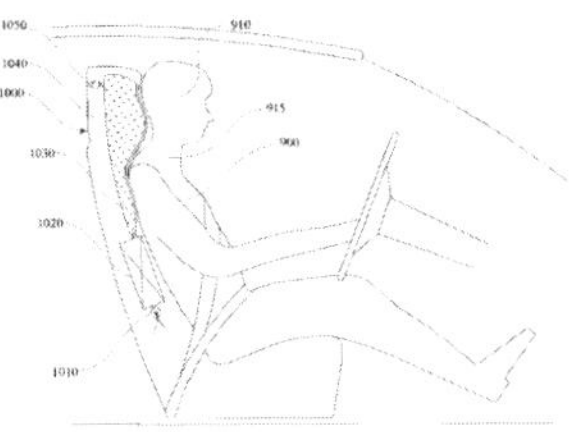

(12) **United States Patent**
Bigi et al.

(10) **Patent No.: US 6,402,238 B1**
(45) **Date of Patent: Jun. 11, 2002**

(54) **CAR SEAT WITH A HEAD REST**

(75) Inventors: **Dante Bigi**, Torino (IT); **Alexander Heilig**, Wissgoldingen (DE)

(73) Assignee: **TRW Occupant Restraint Systems GmbH & Co. KG**, Alfdorf (DE)

(*) Notice: Subject to any disclaimer, the term of this patent is extended or adjusted under 35 U.S.C. 154(b) by 0 days.

(21) Appl. No.: **09/446,041**

(22) PCT Filed: **Jun. 16, 1998**

(86) PCT No.: **PCT/EP98/03625**

§ 371 (c)(1), (2), (4) Date: **Dec. 15, 1999**

(87) PCT Pub. No.: **WO98/57818**

PCT Pub. Date: **Dec. 23, 1998**

(30) **Foreign Application Priority Data**

Jun. 16, 1997 (DE) .... 297 10 511 U

(51) **Int. Cl.**[7] **B60R 21/22**
(52) **U.S. Cl.** **297/216.12**; 280/730.1
(58) **Field of Search** 297/216.12, 391, 297/410; 280/730.1

(56) **References Cited**

U.S. PATENT DOCUMENTS

3,420,572 A		1/1969	Bisland	
3,680,912 A		8/1972	Matsuura	
5,056,816 A		10/1991	Lütze et al.	
5,110,185 A		5/1992	Schmutz et al.	
5,378,043 A		1/1995	Viano et al.	
5,694,320 A	*	12/1997	Breed	297/216.12 X
5,833,312 A	*	11/1998	Lenz	297/216.12 X
6,158,812 A	*	12/2000	Bonke	297/216.12 X
6,199,900 B1	*	3/2001	Zeigler	297/216.12 X

FOREIGN PATENT DOCUMENTS

DE	3900495	7/1990
DE	29603991	8/1996
EP	0593845	4/1994
EP	0627340	12/1994

* cited by examiner

Primary Examiner—Peter R. Brown
(74) *Attorney, Agent, or Firm*—Tarolli, Sundheim, Covell, Tummino & Szabo L.L.P.

(57) **ABSTRACT**

A vehicle seat (**10**) with a back rest (**12**) and with an adjustable head rest (**15**) having a supporting surface (**16**) which the head of the vehicle occupant can touch in the case of restraint is characterized in that the head rest (**16**) has an expansion device which increases the size of the supporting surface (**16**) of the head rest (**15**) in the case of restraint. Furthermore, the head rest (**15**) is vertically adjusted in relation to the back rest (**12**) at least when the head rest (**15**) is arranged so close to the back rest (**12**) that the back rest (**12**) would prevent a complete increase in size of the supporting surface (**16**).

17 Claims, 4 Drawing Sheets

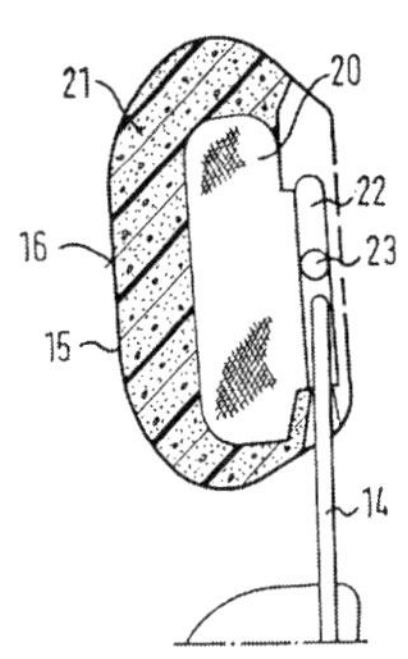

(12) **United States Patent**
Kore

(10) **Patent No.: US 6,385,517 B1**
(45) **Date of Patent: May 7, 2002**

(54) **PASSENGER PROTECTING APPARATUS FOR USE IN A VEHICLE**

(75) Inventor: **Haruhisa Kore**, Fuchu-cho (JP)

(73) Assignee: **Mazda Motor Corporation** (JP)

(*) Notice: Subject to any disclaimer, the term of this patent is extended or adjusted under 35 U.S.C. 154(b) by 0 days.

(21) Appl. No.: **09/504,331**

(22) Filed: **Feb. 14, 2000**

(30) **Foreign Application Priority Data**

Feb. 16, 1999 (JP) 11-037784

(51) **Int. Cl.**[7] **B60R 21/01**
(52) **U.S. Cl.** **701/45**; 180/271; 180/282
(58) **Field of Search** 280/806; 180/268, 180/271, 273, 282; 701/45

(56) **References Cited**

U.S. PATENT DOCUMENTS

3,420,572 A		1/1969	Bisland	
3,838,870 A		10/1974	Hug	
4,645,233 A	*	2/1987	Bruse et al.	280/808
5,186,494 A	*	2/1993	Shimose	
5,378,043 A	*	1/1995	Viano et al.	297/408
5,552,986 A		9/1996	Omura et al.	
5,626,359 A	*	5/1997	Steffens, Jr. et al.	280/735
5,670,853 A	*	9/1997	Bauer	318/286
5,694,320 A		12/1997	Breed	
5,748,473 A	*	5/1998	Breed et al.	364/424.055
5,785,347 A	*	7/1998	Adolph et al.	280/735
5,788,281 A		8/1998	Yanagi et al.	
5,822,707 A	*	10/1998	Breed et al.	701/49
5,823,619 A		10/1998	Heilig et al.	
5,890,084 A	*	3/1999	Halasz et al.	701/45
5,927,804 A	*	7/1999	Cuevas	297/216.12
6,088,640 A	*	7/2000	Breed	701/45
6,142,524 A	*	11/2000	Brown et al.	280/806
6,155,601 A	*	12/2000	Cantor et al.	280/806
6,209,909 B1	*	4/2001	Breed	280/735
6,213,511 B1	*	4/2001	Downie et al.	280/806
6,213,512 B1	*	4/2001	Swann et al.	280/806
6,278,360 B1	*	8/2001	Yanagi	340/436

FOREIGN PATENT DOCUMENTS

DE	197 54 311 A1	12/1998
EP	0 627 340 A1	12/1994
JP	10000973	1/1998
JP	10006832	1/1998
WO	WO 98/09838	3/1998

* cited by examiner

Primary Examiner—J. J. Swann
Assistant Examiner—J. Allen Shriver
(74) *Attorney, Agent, or Firm*—Brooks & Kushman P.C.

(57) **ABSTRACT**

A passenger protecting apparatus is provided with a head rest driving mechanism for driving a head rest forward according to a pressing contact force generated when a passenger's back is pressed against a seat back during a rear crash of the vehicle, a rear crash predictor for predicting a rear crash of the vehicle, and a passenger moving mechanism for moving the passenger toward the seat back when the rear crash of the vehicle is predicted by the rear crash predictor. The passenger's head can be effectively supported by the head rest during the rear crash of the vehicle.

13 Claims, 10 Drawing Sheets

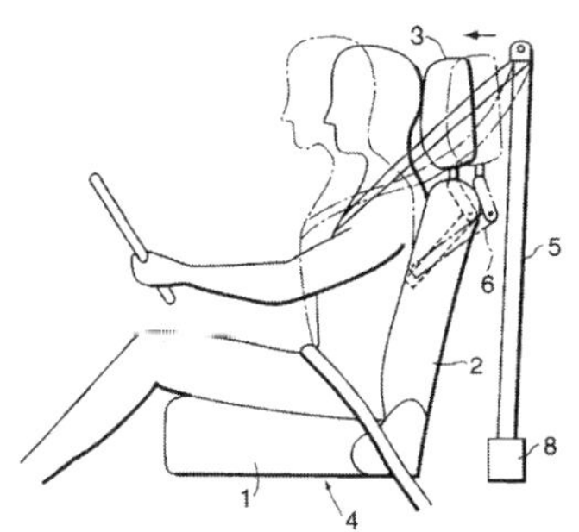

(12) **United States Patent**
Shah et al.

(10) **Patent No.: US 6,416,125 B1**
(45) **Date of Patent: Jul. 9, 2002**

(54) **VEHICLE SEAT CONNECTION ASSEMBLY**

(75) Inventors: **Suresh D. Shah**, Troy; **Mladen Humer**, Eastpointe, both of MI (US)

(73) Assignee: **Lear Corporation**, Southfield, MI (US)

(*) Notice: Subject to any disclaimer, the term of this patent is extended or adjusted under 35 U.S.C. 154(b) by 0 days.

(21) Appl. No.: **09/577,940**

(22) Filed: **May 23, 2000**

(51) **Int. Cl.**[7] **B60N 2/42**
(52) **U.S. Cl.** **297/216.12**; 297/408; 297/463.1; 297/463.2; 403/164
(58) **Field of Search** .... 297/408, 391, 297/216.12, 216.13, 463.1, 463.2; 403/164, 165; 16/225, DIG. 13

(56) **References Cited**

U.S. PATENT DOCUMENTS

4,685,737 A		8/1987	Deley et al.	
4,900,060 A	*	2/1990	Yamamoto et al.	297/469
5,145,233 A		9/1992	Nagashima	
5,378,043 A		1/1995	Viano et al.	
5,669,667 A	*	9/1997	Schmidt	297/408 X
5,711,579 A		1/1998	Albrecht	
5,733,009 A		3/1998	De Filippo	
5,795,019 A		8/1998	Wieclawski	
5,836,648 A		11/1998	Karschin et al.	
5,927,804 A		7/1999	Cuevas	
6,019,424 A		2/2000	Rückert et al.	
6,024,406 A		2/2000	Charras et al.	
6,199,947 B1	*	3/2001	Wilklund	297/216.12

* cited by examiner

Primary Examiner—Anthony D. Barfield
(74) *Attorney, Agent, or Firm*—Brooks & Kushman P.C.

(57) **ABSTRACT**

A vehicle seat connection (**18**) has particular utility for mounting a headrest support assembly (**14**) on a seat back frame (**12**). The connection assembly (**18**) includes first and second plastic seat components embodied by first and second connectors (**50, 52**) that are injection molded in situ over enlarged flanges (**56, 58**) of first and second ends (**44, 46**) of a connector link (**42**) which has an intermediate portion (**48**) extending between its ends. The connector link is movable with respect to both the first and second connectors (**52, 52**) to move the headrest upwardly and forwardly upon rearward movement of a vehicle seat occupant.

16 Claims, 2 Drawing Sheets

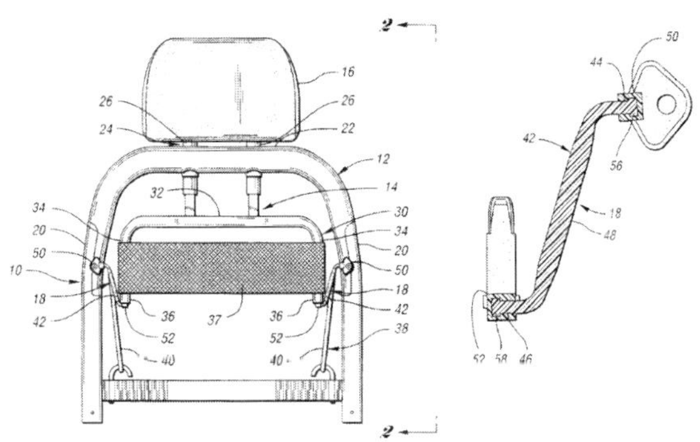

U.S. Patent Jul. 9, 2002 Sheet 1 of 2 US 6,416,125 B1

Fig. 1

Fig. 2

United States Patent [15] 3,703,313

Schiesterl et al. [45] Nov. 21, 1972

[54] **SEAT, ESPECIALLY FOR MOTOR VEHICLES**

[72] Inventors: **Gerhard Schiesterl**, Stuttgart; **Helmut Wulf**, Nellingen, both of Germany

[73] Assignee: **Daimler-Benz Aktiengesellschaft**, Stuttgart, Germany

[22] Filed: **July 29, 1971**

[21] Appl. No.: **167,217**

[30] **Foreign Application Priority Data**

July 29, 1970 Germany..........P 20 37 565.9

[52] **U.S. Cl.** ....**297/391,** 280/150 AB, 297/384, 297/395

[51] **Int. Cl.** **A47c 7/36**

[58] **Field of Search**......297/384, 388, 397, 390, 385, 297/395, 404; 280/150 SB, 150 AB, 150 B

[56] **References Cited**

UNITED STATES PATENTS

3,397,911 8/1968 Brosius......................297/216

3,510,150 5/1970 Wilfert......................297/391

3,525,535 8/1970 Kobori.......................297/395

FOREIGN PATENTS OR APPLICATIONS

6,615,335 5/1967 Netherlands........280/150 AB

Primary Examiner—Bobby R. Gay
Assistant Examiner—Darrell Marquette
Attorney—Craig, Antonelli & Hill

[57] **ABSTRACT**

A seat, especially for motor vehicles, in which a gas cushion acting as headrest and automatically inflatable in case of an accident is arranged at the backrest of the seat; at least one belt is arranged behind the gas cushion for absorbing the rebound forces of the head of the passenger impinging against the gas cushion whereby the belt is arranged, on the one hand, within the area of the vehicle roof, and on the other, at the seat.

5 Claims, 2 Drawing Figures

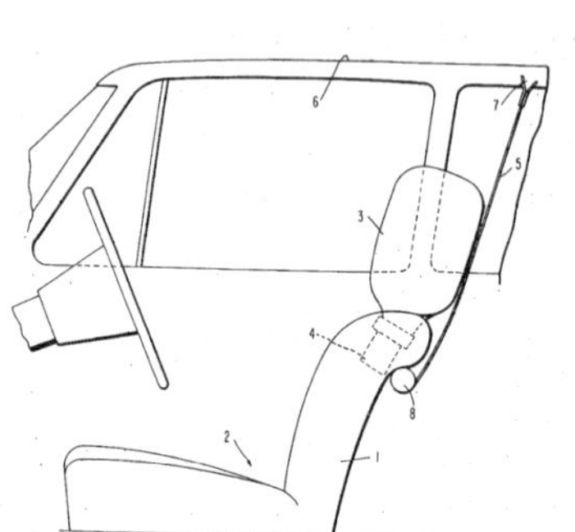

(12) United States Patent

Wiklund

(10) **Patent No.:** **US 6,199,947 B1**
(45) **Date of Patent:** **Mar. 13, 2001**

(54) **VEHICLE SEAT PROVIDED WITH A HEADREST**

(75) Inventor: **Kristina Wiklund**, Trollhattan (SE)

(73) Assignee: **SAAB Automobile AB** (SE)

(*) Notice: Subject to any disclaimer, the term of this patent is extended or adjusted under 35 U.S.C. 154(b) by 0 days.

(21) Appl. No.: **09/560,251**

(22) Filed: **Apr. 28, 2000**

Related U.S. Application Data

(63) Continuation of application No. 09/254,341, filed as application No. PCT/SE97/01487 on Sep. 5, 1997, now abandoned.

(30) **Foreign Application Priority Data**

Sep. 6, 1996 (SE) 9603238

(51) **Int. Cl.**[7] **B60R 21/055**; B60N 2/48

(52) **U.S. Cl.** **297/216.12**; 297/408

(58) **Field of Search** 297/216.17, 391, 297/408

(56) **References Cited**

U.S. PATENT DOCUMENTS

3,802,737 * 4/1974 Mertens 297/216.12

5,378,013 * 1/1995 Viano et al. 297/408

5,823,619 * 10/1998 Herlig et al. 297/216.12

* cited by examiner

Primary Examiner—Peter R. Brown
(74) *Attorney, Agent, or Firm*—Ostrolenk, Faber, Gerb & Soffen, LLP

(57) **ABSTRACT**

A vehicle chair has a neck support (**4**), which by means of a support mechanism (**5**) is mounted movable relative to the back (**3**) of the chair, so that during rear end collisions with the help of the back of the traveller and a maneuvering means (**10**) in the chair to be forced forwards in order to meet and intercept the head of the traveller. This maneuvering means (**10**) is connected to the frame (**9**) in the back of the chair by means of a link mechanism (**11**), wherein the connection (**13**) of the link mechanism (**11**) to the frame (**9**) is situated higher than the connection (**14**) of the link mechanism (**11**) to the maneuvering means (**10**).

13 Claims, 2 Drawing Sheets

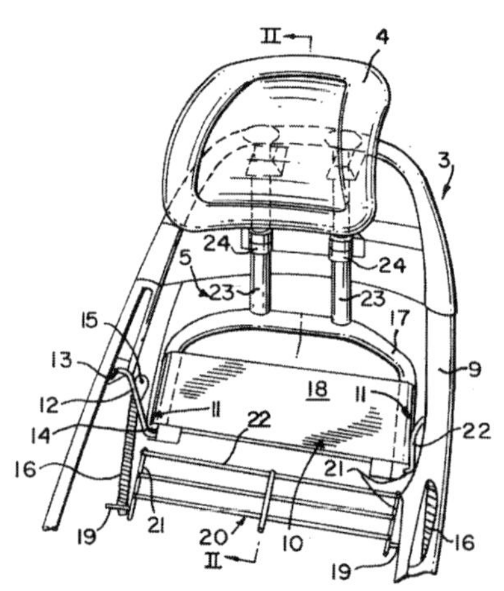

United States Patent [19]

Wilfert

[11] **3,779,577**
[45] **Dec. 18, 1973**

[54] **SAFETY INSTALLATION FOR THE PASSENGERS OF VEHICLES, ESPECIALLY PASSENGER MOTOR VEHICLES**

[75] Inventor: **Karl Wilfert**, Gerlingen-Waldstadt, Germany

[73] Assignee: **Daimler-Benz Aktiengesellschaft**, Stuttgart-Unterturkheim, Germany

[22] Filed: **Nov. 23, 1971**

[21] Appl. No.: **201,407**

[30] **Foreign Application Priority Data**

Nov. 28, 1970 Germany.................. P 20 58 608.7

[52] **U.S. Cl.** .... **280/150 AB**

[51] **Int. Cl.** **B60r 21/10**

[58] **Field of Search**.................. 280/150 AB, 150 B

[56] **References Cited**

UNITED STATES PATENTS

3,642,303 2/1972 Irish et al. .... 280/150 AB

3,603,535 9/1971 DePolo.................. 280/150 AB

3,510,150 5/1970 Wilfert.................. 280/150 AB

3,663,035 1/1970 Norton.................. 280/150 AB

2,418,798 4/1947 Whitmer.................. 280/150 AB X

3,476,402 11/1969 Wilfert.................. 280/150 AB

Primary Examiner—Kenneth H. Betts
Attorney—Craig, Antonelli & Hill

[57] **ABSTRACT**

A safety installation for the passengers of motor vehicles, especially of passenger motor vehicles in which the backrests of the front seats are equipped with a headrest and accommodate within the area of each backrest underneath the headrest two inflatable gas cushions which are so arranged that upon inflation one gas cushion unfolds in front and one gas cushion unfolds behind the headrest.

6 Claims, 2 Drawing Figures

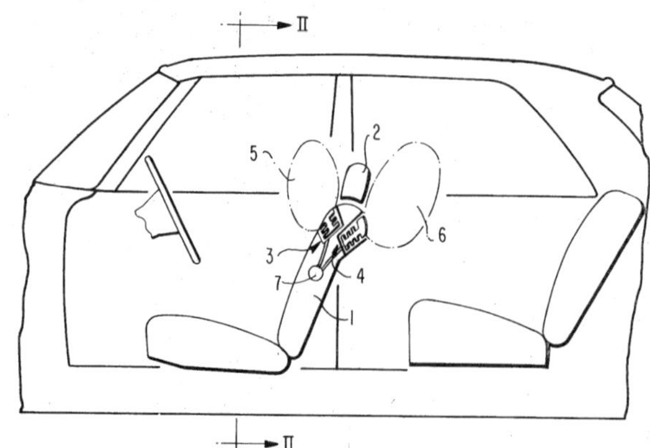

United States Patent [19]

Kashiwamura et al.

[11] **Patent Number: 4,655,505**

[45] **Date of Patent: Apr. 7, 1987**

[54] **PNEUMATICALLY CONTROLLED SEAT FOR VEHICLE**

[75] Inventors: **Takayoshi Kashiwamura; Ryoichi Iwasaki,** both of Kawasaki, Japan

[73] Assignee: **NHK Spring Co., Ltd.,** Japan

[21] Appl. No.: **798,896**

[22] Filed: **Nov. 18, 1985**

[30] **Foreign Application Priority Data**

Dec. 13, 1984	[JP]	Japan	59-264319
Feb. 7, 1985	[JP]	Japan	60-022497
Apr. 15, 1985	[JP]	Japan	60-079499
Jun. 20, 1985	[JP]	Japan	60-134838

[51] **Int. Cl.**[4] **A47C 7/46; B60N 1/06**

[52] **U.S. Cl.** **297/284;** 297/330; 297/DIG. 3

[58] **Field of Search** .... 297/284, 330, DIG. 3

[56] **References Cited**

U.S. PATENT DOCUMENTS

3,867,732	2/1975	Morrell	297/DIG. 3 X
4,190,286	2/1980	Bentley	297/DIG. 3 X
4,467,252	8/1984	Takeda et al.	297/330 X
4,552,402	11/1985	Huber et al.	297/DIG. 3 X

FOREIGN PATENT DOCUMENTS

2746630	4/1979	Fed. Rep. of Germany	297/330
2912755	10/1980	Fed. Rep. of Germany	297/330
2536975	6/1984	France	297/DIG. 3
138024	10/1981	Japan	297/330
206426	12/1983	Japan	297/330

Primary Examiner—Kenneth J. Dorner
Assistant Examiner—Peter R. Brown
Attorney, Agent, or Firm—Jeffers, Irish & Hoffman

[57] **ABSTRACT**

Disclosed is a seat for vehicle which is pneumatically controlled for a desired body pressure distribution of a person seated in the seat, comprising: a plurality of air bags embedded in a seat; an air pressure source; a single first conduit connected the air pressure source at its one end and branched off into a plurality of second conduits leading to the air bags at its other end; a plurality of on-off valves each provided in the corresponding one of the second conduits leading to the air bags; an exhaust valve connected to the first conduit at its one end and to the atmosphere at its other end; a single pressure sensor provided in the first conduit; and control means connected to the air pressure source, the on-off valves, the exhaust valve and the pressure sensor; the control means comprising means for storing predetermined air pressure values for the air bags, means for selectively opening and closing the on-off valves and the exhaust valve, and means for preventing the valve opening and closing means for opening the valves according to the result of comparison between the output from the pressure sensor and predetermined air pressure values. Thus, the structure for air pressure control is simplified and the body pressure distribution of the passenger may be adapted to the acceleration of the vehicle and other conditions.

12 Claims, 9 Drawing Figures

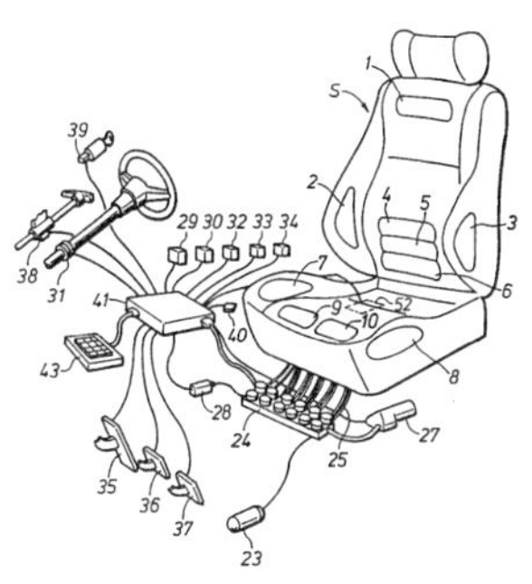

United States Patent [19]

Coman et al.

[11] **Patent Number: 5,556,129**

[45] **Date of Patent: Sep. 17, 1996**

[54] **VEHICLE SEAT BACK WITH A HEAD SIDE IMPACT AIR BAG**

[75] Inventors: **Sorin Coman,** Madison Hts; **Frank Wu,** Novi; **Robert L. Demick,** Eastpointe, all of Mich.

[73] Assignee: **Hoover Universal,** Plymouth, Mich.

[21] Appl. No.: **441,813**

[22] Filed: **May 16, 1995**

[51] **Int. Cl.**[6] **B60R 21/22**

[52] **U.S. Cl.** **280/730.2;** 297/216.12

[58] **Field of Search** 280/728.1, 730.1, 280/730.2; 297/483, 216.12, 216.1

[56] **References Cited**

U.S. PATENT DOCUMENTS

4,946,191	8/1990	Putsch .	
4,989,896	2/1991	DiSalvo et al.	280/728.3
5,112,079	5/1992	Haland et al. .	
5,251,931	10/1993	Semchena et al. .	
5,322,322	6/1994	Bark et al. .	
5,348,342	9/1994	Haland et al. .	
5,362,132	11/1994	Griswold et al. .	
5,390,950	2/1995	Barnes et al.	280/728.2
5,458,396	10/1995	Rost	297/216.12

FOREIGN PATENT DOCUMENTS

4218252	12/1992	Germany	280/730
4231050	3/1994	Germany	280/807
4-166452	6/1992	Japan	280/730.2

Primary Examiner—Karin Tyson
Attorney, Agent, or Firm—Harness, Dickey & Pierce, P.L.C.

[57] **ABSTRACT**

A seat back for a motor vehicle having a tower extending upwardly from the seat back upper end along one lateral side of the seat back. The tower contains an inflatable air bag which is deployable to provide head side impact protection for a seat occupant. By mounting the air bag within a raised tower, the air bag is deployed from a higher location in the seat back to reduce the possibility of interference between the air bag and a shoulder belt. Furthermore, by not mounting the air bag to the seat back head restraint, it is possible to provide a fully adjustable head restraint attached to the seat back and extending upwardly from the seat back upper end. A removable side bolster is also provided which covers the air bag canisters mounted to the seat back frame. Following deployment of an air bag, the bolster and air bag canister are removed and replaced with new components. This enables the seat assembly to be repaired without replacement of the entire seat assembly or seat back.

6 Claims, 9 Drawing Sheets

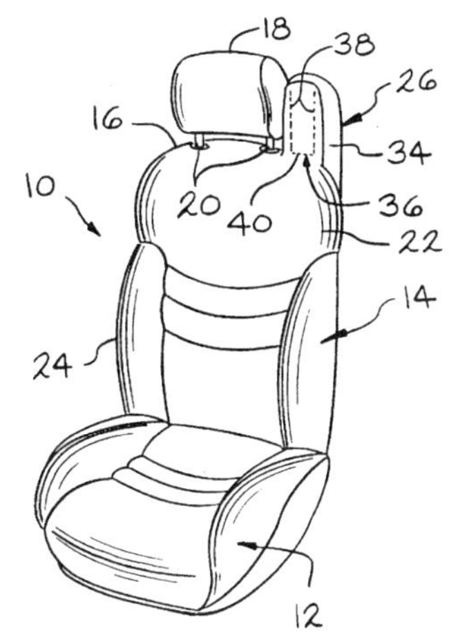

United States Patent [19]

Stawicki

[11] **Patent Number: 5,330,255**

[45] **Date of Patent: Jul. 19, 1994**

[54] **SEAT INTEGRATED INFLATABLE NECK SUPPORT**

[75] Inventor: **Edwin V. Stawicki,** Highland, Mich.

[73] Assignee: **Davidson TEXTRON INC.,** Dover, N.H.

[21] Appl. No.: **974,705**

[22] Filed: **Nov. 12, 1992**

[51] **Int. Cl.**[5] **A47C 1/10**

[52] **U.S. Cl.** **297/391;** 297/397

[58] **Field of Search** .... 297/391, 393, 397, DIG. 3; 5/636, 644

[56] **References Cited**

U.S. PATENT DOCUMENTS

3,017,221	1/1962	Emery	297/397
3,185,497	5/1965	Lagace	297/397 X
3,510,150	5/1970	Wilfert	297/391 X
3,680,912	8/1972	Matsuura	297/391
4,097,086	6/1978	Hudson	297/391 X
4,114,948	9/1978	Perkey	297/391 X
4,123,104	10/1978	Andres et al.	297/391
4,178,038	12/1979	Yada et al.	297/391
4,206,945	6/1980	Kifferstein	297/397 X
4,285,081	8/1981	Price	297/393 X
4,738,488	4/1988	Camelio	297/393 X
4,744,601	5/1988	Nakanishi	297/DIG. 3
4,838,611	6/1989	Talaugon	297/391
4,865,388	9/1989	Nemoto	297/DIG. 3
4,946,191	8/1990	Putsch	297/DIG. 3
5,015,036	5/1991	Fergie	297/391 X

FOREIGN PATENT DOCUMENTS

39-0050	9/1988	Fed. Rep. of Germany .
61-1835	8/1986	Japan .
63-31248	4/1988	Japan .
63-6384	8/1988	Japan .

Primary Examiner—Peter M. Cuomo
Assistant Examiner—Jerry Redman
Attorney, Agent, or Firm—Reising, Ethington, Barnard, Perry & Milton

[57] **ABSTRACT**

An inflatable neck support is a U-shaped collar comprising an inflatable bladder and a decorative cover that provides a wide range of styling choices. The inflatable neck support is attached to a seat back or a head restraint in a variety of ways so that it is easily adjusted by the seat occupant to meet the seat occupant's comfort needs. When not in use, the inflatable neck support is deflated and stored in a number of convenient ways.

18 Claims, 3 Drawing Sheets

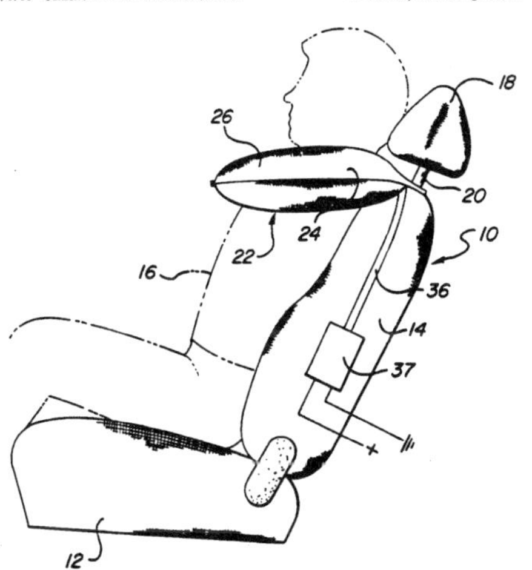

United States Patent [19]

Breed

[11] **Patent Number: 5,694,320**

[45] **Date of Patent: Dec. 2, 1997**

[54] **REAR IMPACT OCCUPANT PROTECTION APPARATUS**

[75] Inventor: **David S. Breed,** Boonton Township, N.J.

[73] Assignee: **Automotive Technologies Intl, Inc.,** Morris Township, N.J.

[21] Appl. No.: **476,882**

[22] Filed: **Jun. 7, 1995**

[51] **Int. Cl.**[6] **B60R 21/055**

[52] **U.S. Cl.** **364/424.055;** 297/216.12; 297/408; 318/568.1

[58] **Field of Search** 364/424.055, 460, 364/461; 297/403, 408, 410, 216.12; 318/560, 568.1, 466, 467, 468, 469

[56] **References Cited**

U.S. PATENT DOCUMENTS

3,838,870	10/1974	Hug	297/384
4,645,233	2/1987	Bruse et al.	297/410
4,797,824	1/1989	Sugiyama et al.	364/424.05
4,811,226	3/1989	Shinohara	364/424.05
4,853,687	8/1989	Isomura et al.	364/424.05
4,935,680	6/1990	Sugiyama	364/424.05
5,003,240	3/1991	Ikeda	364/424.05
5,006,771	4/1991	Ogasawara	318/568.1
5,095,257	3/1992	Ikeda et al.	364/424.05
5,151,944	9/1992	Yamamura et al.	381/151
5,290,091	3/1994	Dellanno et al.	297/408
5,330,226	7/1994	Gentry et al.	280/735
5,364,164	11/1994	Kuranami	297/408
5,378,043	1/1995	Viano et al.	297/408
5,580,124	12/1996	Dellanno	297/216.12

FOREIGN PATENT DOCUMENTS

WO 94/22693 A1 10/1994 WIPO .

OTHER PUBLICATIONS

Viano, David C. & Gargan , Martin F., "Headrest Position During Normal Driving, Implication to Neck Injury Risk in Rear Crashes" in Acad. Anal. and Prev., vol. 28 No. 6 pp. 665–674, 1996 Elsevier Science, Ltd.

Primary Examiner—Gary Chin
Attorney, Agent, or Firm—Samuel Shipkovitz

[57] **ABSTRACT**

A headrest and headrest positioning system reduces whiplash injuries from rear impacts by properly positioning the headrest behind the occupant's head either continuously, or just prior to and in anticipation of, the vehicle impact and then properly supports both the head and neck. Sensors determine the location of the occupant's head and motors move the headrest both up and down and forward and back as needed. In one implementation, the headrest is continuously adjusted to maintain a proper orientation of the headrest to the rear of the occupant's head. In another implementation, an anticipatory crash, is used to predict that a rear impact is about to occur, in which event, the headrest is moved proximate to the occupant. A pre-inflated airbag within the headrest automatically distributes the pressure to evenly support both the head and neck.

20 Claims, 11 Drawing Sheets

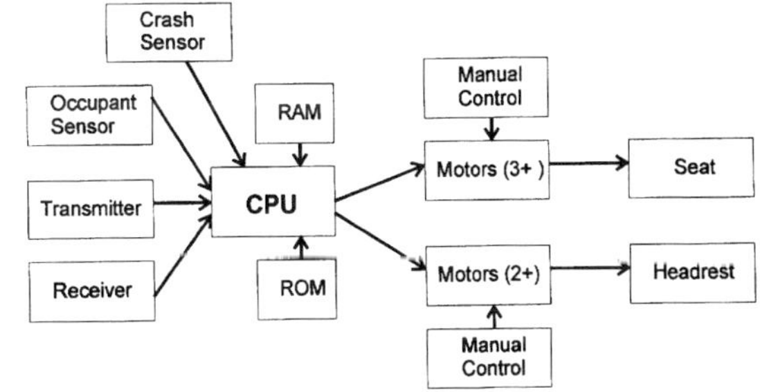

United States Patent [19]

Locke

[11] **Patent Number: 5,738,407**

[45] **Date of Patent: Apr. 14, 1998**

[54] **HEAD RESTRAINT AIRBAG ASSEMBLY**

[75] Inventor: **Gerald S. Locke**, Lake Orion, Mich.

[73] Assignee: **Lear Corporation**, Southfield, Mich.

[21] Appl. No.: **783,223**

[22] Filed: **Jan. 14, 1997**

[51] **Int. Cl.**^{6} **B60R 21/22**

[52] **U.S. Cl.** **297/216.12**; 297/216.13; 280/730.1

[58] **Field of Search** 297/216.1, 216.12, 297/216.13; 280/730.1

[56] **References Cited**

U.S. PATENT DOCUMENTS

3,703,313 11/1972 Schiesterl et al. .... 280/730.1 X
3,779,577 12/1973 Wilfert 280/730.1
5,536,043 7/1996 Lang et al. .... 280/730.1 X
5,556,129 9/1996 Coman et al. .... 297/216.12 X

FOREIGN PATENT DOCUMENTS

2152202 4/1973 Germany 297/216.12
2841729 4/1980 Germany .... 280/730 R

Primary Examiner—Peter R. Brown
Attorney, Agent, or Firm—Brooks & Kushman P.C.

[57] **ABSTRACT**

A vehicle seat assembly includes a seat back frame and a headrest movable with respect to the back frame. An airbag is positioned on the back frame and deployable vertically with respect to the back frame such that the airbag is positioned forward-in-vehicle from the headrest when fully deployed.

9 Claims, 1 Drawing Sheet

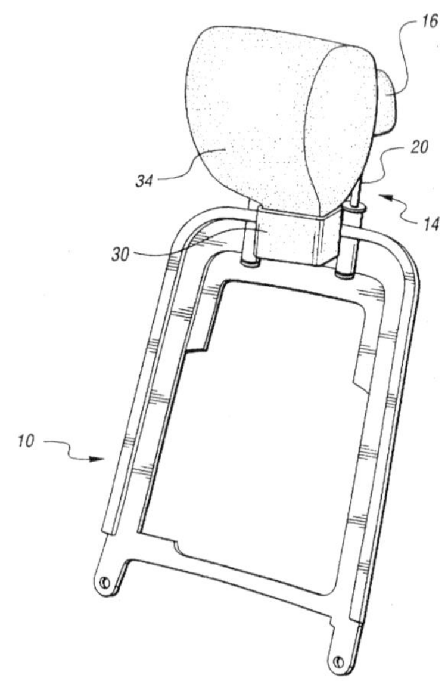

United States Patent [19]

Miller, III et al.

[11] **Patent Number: 5,782,529**

[45] **Date of Patent: Jul. 21, 1998**

[54] **INFLATABLE SEAT BACK**

[75] Inventors: **H. John Miller, III**, Hillside; **Gerald Keller**, Blue Bell, both of Mich.

[73] Assignee: **AlliedSignal Inc.**, Morristown, N.J.

[21] Appl. No.: **803,563**

[22] Filed: **Feb. 20, 1997**

[51] **Int. Cl.**^{6} **B60R 21/22**

[52] **U.S. Cl.** **297/216.13**; 297/216.12; 280/730.1

[58] **Field of Search** 297/216.1, 216.12, 297/216.13, 216.14, 284.6, 452.41; 280/730.1, 730.2

[56] **References Cited**

U.S. PATENT DOCUMENTS

1,976,320 10/1934 Austin .... 297/452.41 X
3,698,670 10/1972 Ewing .... 297/216.13 X
3,779,577 12/1973 Wilfert 280/730.1
4,514,010 4/1985 Gonzalez .... 297/452.41 X
5,562,324 10/1996 Massara et al. .... 297/452.41 X

FOREIGN PATENT DOCUMENTS

2841729 4/1980 Germany .... 280/730 R

Primary Examiner—Peter R. Brown
Attorney, Agent, or Firm—Markell Seitzman

[57] **ABSTRACT**

An occupant protection system for protecting a seated occupant comprising: seat (**20**) having a seat cushion (**22**) and a seat back (**24**), frame means (**26**, **30**) to support the seat back, a headrest extending above the seat back, cushion material (**32**) for providing resiliency to the seat and a protective cover material (**34**) covering the cushion material; an air bag (**50**,**50'**), having a first portion extending generally along the length of the seat back and a second portion which when inflated extends above the seat back to be positioned generally in front of the headrest, whereby upon inflation of the first portion of the air bag a portion of the seat back is gently urged against the spine of the occupant and the second portion expands upwardly to provide a protective layer between the headrest and the occupant's head.

15 Claims, 6 Drawing Sheets

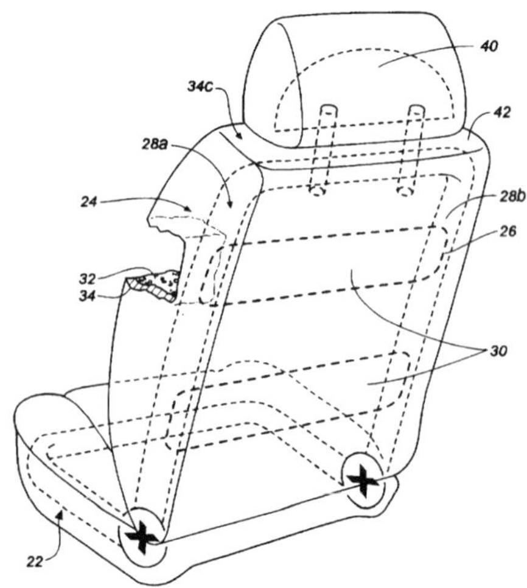

United States Patent [19]

Geuss et al.

[11] **Patent Number: 5,975,637**

[45] **Date of Patent: Nov. 2, 1999**

[54] **ADJUSTABLE VEHICLE SEAT**

[75] Inventors: **Hartwich Geuss**, Stuttgart; **Werner Reichelt**, Esslingen; **Helge Schmidt-Spalding**, Stuttgart, all of Germany

[73] Assignee: **DaimlerChrysler AG**, Stuttgart, Germany

[21] Appl. No.: **09/213,237**

[22] Filed: **Dec. 17, 1998**

[30] **Foreign Application Priority Data**

Dec. 19, 1997 [DE] Germany 197 56 700

[51] **Int. Cl.**^{6} **A47C 7/36**

[52] **U.S. Cl.** **297/391**; 297/410; 297/344.1; 297/354.12; 297/216.12; 297/DIG. 3; 297/284.6

[58] **Field of Search** .... 297/216.13, 216.1, 297/216.12, 126.14, 344.1, 344.12, 325, DIG. 3, 452.41, 354.12, 391, 410, 284.6, 284.4, 284.1; 248/429

[56] **References Cited**

U.S. PATENT DOCUMENTS

3,680,912 8/1972 Matsuura .... 297/216.12 X
4,222,608 9/1980 Maeda 297/410
4,655,505 4/1987 Kashiwamura et al. 297/284.6
4,720,146 1/1988 Mawbey et al. .
4,923,250 5/1990 Hattori 297/410
5,330,255 7/1994 Stawicki 297/391
5,439,271 8/1995 Ryan .... 297/344.1 X
5,458,396 10/1995 Rost .... 297/216.13 X
5,782,529 7/1998 Miller, III et al. 297/216.13

FOREIGN PATENT DOCUMENTS

2152202 11/1980 Germany 297/216.12
3141515 4/1983 Germany 297/410
37 18 126 A1 12/1988 Germany .
44 09 046 C2 9/1995 Germany .

Primary Examiner—Milton Nelson, Jr.
Attorney, Agent, or Firm—Evenson, McKeown Edwards & Lenahan P.L.L.C.

[57] **ABSTRACT**

A vehicle seat includes a seat part displaceable at least in the seat longitudinal direction; a back rest that is held on the seat part; and an adjustable head restraint held on the back rest and having a head cushion. For automatically adapting the head restraint correctly in terms of safety to seat adjustments carried out by a seat user, a medium reservoir of variable volume, communicating with a filling and discharge device, is integrated in the head cushion so that the cushion front face is shifted forwards during an increase in volume and is drawn back again as a result of a reduction in volume. The filling quantity introduced into the medium reservoir or drawn off from the medium reservoir is controlled depending on the seat setting by means of a control unit. The medium reservoir also may include a pressure limiter that allows a controlled outflow of the reservoir when a predetermined pressure value, caused by the impact of the head, is reached.

22 Claims, 3 Drawing Sheets

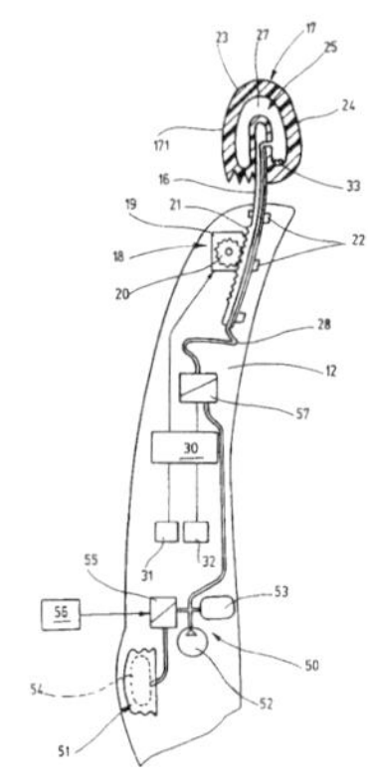

United States Patent [19]

Bonke

[11] **Patent Number: 6,158,812**

[45] **Date of Patent: Dec. 12, 2000**

[54] **INDIVIDUALLY ADAPTABLE HEADREST FOR SEATS WITH BACKS**

[76] Inventor: **Christoph Bonke**, Wendelsteinstrasse 15, D-83126 Flintsbach, Germany

[21] Appl. No.: **08/913,640**

[22] PCT Filed: **Feb. 29, 1996**

[86] PCT No.: **PCT/DE96/00367**

§ 371 Date: **Feb. 11, 1998**

§ 102(e) Date: **Feb. 11, 1998**

[87] PCT Pub. No.: **WO96/28318**

PCT Pub. Date: **Sep. 19, 1996**

[30] **Foreign Application Priority Data**

Mar. 13, 1995 [DE] Germany .... 295 04 287 U

[51] **Int. Cl.**^{7} **A47C 7/36**

[52] **U.S. Cl.** **297/391**; 297/216.12; 297/397

[58] **Field of Search** 297/216.12, 216.13, 297/216.14, 397, 391; 280/730.1, 730.2

[56] **References Cited**

U.S. PATENT DOCUMENTS

3,779,577 12/1973 Wilfert 280/730.1
4,440,443 4/1984 Nordskog 297/397
4,744,601 5/1988 Nakanishi .... 297/216.12 X
5,133,084 7/1992 Martin .... 280/730.1 X
5,290,091 3/1994 Dellanno et al. 297/216.12
5,499,840 3/1996 Nakano 280/730.1
5,580,124 12/1996 Dellanno 297/216.12
5,630,651 5/1997 Fishbane .... 297/391 X
5,694,320 12/1997 Breed .... 297/216.12 X
5,833,312 11/1998 Lenz .... 297/216.14 X

FOREIGN PATENT DOCUMENTS

0113645 7/1984 European Pat. Off. .
2395729 1/1979 France .
3042802 6/1982 Germany .
2246292 1/1992 United Kingdom .
2246292A 1/1992 United Kingdom .

Primary Examiner—Peter M. Cuomo
Assistant Examiner—Stephen Vu
Attorney, Agent, or Firm—Domingue & Waddell, PLC

[57] **ABSTRACT**

A head restraint for seats with a seat back with an upper support body (**1**) which is provided in the extension of the seat back (**3**) above the upper edge (**2**) of the seat back (**3**) at head height, and with the longitudinal axis (**6**) aligned horizontally and parallel to the longitudinal axis (**7**) of the seat back (**3**), whereby the underside of the upper support body (**1**) is in contact with, or spaced apart from, the upper edge (**2**) of the seat back (**3**). At least two horizontally spaced apart structures (**5**) for lateral support of the head are provided at the front (**9**) of the upper support body (**1**) and/or integrated into it. At the height of the cervical lordosis of a seated person a lower, essentially cylindrical support body (**4**) for supporting the cervical spine is attached; when viewed from the side, the cross-section of the said support body (**4**) is essentially round, oval, elongated or semicircular in shape, and the longitudinal axis (**13**) of the said support body (**4**) is aligned horizontally and parallel to the longitudinal axis (**6**) of the upper support body (**1**).

17 Claims, 8 Drawing Sheets

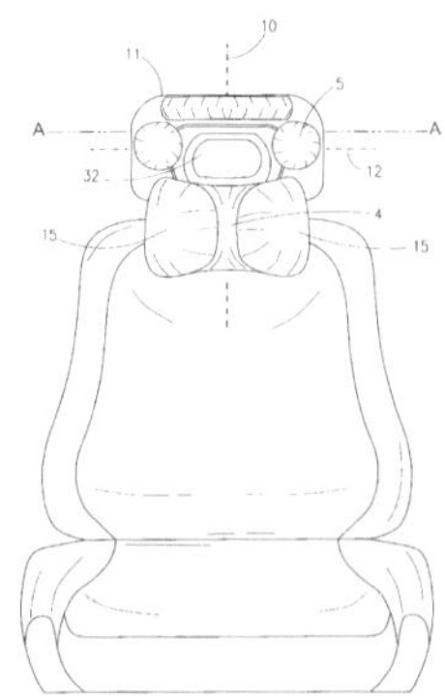

SECTION 4:

REPRINTS OF SAE PAPERS AND AAAM ARTICLES ON SEATS

2000-01-1380

Energy Analysis of Automotive Seat Systems

Gregory D. Stephens, Timothy J. Long and David M. Blaisdell
Collision Research & Analysis, Inc.

ABSTRACT

Collision performance of automotive seat systems has been a subject of inquiry since crash research was in its infancy. However, when federal standards were initiated in 1968 regarding seat system performance, they became the baseline for automotive design, and later became the topic of numerous debates in terms of occupant crash force and energy management. This subject of energy management as it relates to seat design has been extended and expanded in the current time period.

This paper will discuss current design trends in automotive seat design collision performance in terms of new data recently becoming available. Also, due to recent proposals and discussion regarding modification of FMVSS 207, a review of seatback performance data in a dynamic environment will be presented. Any proposals regarding the modification of FMVSS 207 require careful evaluation and quantification of seat system goals. Identification and establishment of beneficial characteristics of seat structures and how these characteristics can be evaluated in both static and dynamic environments is crucial before any discussion of seat system performance goals can be established.

Since approximately 1993, the Federal Government has conducted a substantial number of rear-end crash tests that provide new data that can be utilized to evaluate and better understand the static and dynamic performance of automotive seat systems. Procedures for evaluation and analysis of this data will be discussed, along with a review of current energy management capabilities of seat systems. Recent research on the energy analysis of rear impact collision exposures will also be included and discussed.

INTRODUCTION

Research on the collision performance of seat systems has spanned nearly five decades and has analyzed various aspects of the crash environment as it relates to occupant-seat interaction. The federal government began addressing the issue in the mid-1960's. FMVSS 207, which specified a minimum seat strength and performance requirement, was introduced along with dozens of other standards relating to various safety aspects of automotive design. The initial seat standard followed a GSA requirement for seats in vehicles sold to the government and was based on Society of Automotive Engineers recommended practice J879.

Although research on various aspects of seat design had been conducted since the mid-1950's [1]*, the requirements of FMVSS 207 and FMVSS 202 introduced a number of changes into seat designs, including improved head support and increased strength requirements for seats. This government interest, as well as an increase in awareness of the extent to which motorists were incurring injuries in rear-end collisions, generated additional interest in the occupant kinematics or motions of the occupant in the seat system during these types of crashes. Full scale crash testing conducted in the 1960's [2,3,4] evaluated a number of seat-related issues, including an evaluation of various seat design concepts. This early research identified the head motions that applied forces to the neck and caused hyperextension or "whiplash." The evaluation of dummy motion under various crash conditions with various seat restraint systems gave researchers the foundation necessary to better understand occupant protection requirements for rear-end collision exposures. Some recommendations from these earlier studies however, proved to be impractical and introduced potential new problems that many felt could increase, rather than reduce, the potential for serious injury during certain crash exposures.

Early crash test research evaluated a wide range of severity levels for various seatback heights and strengths, as well as the effect of various occupant anthropometry and positions within the occupant compartment. One observation from this work indicated that an occupant who is effectively "tied" to the automotive structure by an integrated belt system that was attached to a rigid seat structure could ride out a collision, depending almost entirely on the vehicle structure for energy management [5]. Unfortunately, if the occupant was separated or offset in some manner from this rigid seat system, the absence of a yielding seatback that could moderate the crash forces, substantially increased

* Numbers in brackets designate references at end of paper

acceleration levels and resulting injury potential [6]. These observations reinforced the conclusion that yielding seatback structures were required to protect occupants over a wide range of collision exposures.

STATISTICS

Since 1968 there has been an enormous amount of data collected that can assist in analyzing the rear-end collision with respect to its associated injury exposure. Available statistical data is segmented into various categories including overall frequency, severity levels, serious injuries (AIS 3+), and HARM.

REAR IMPACT FREQUENCY – As a category of collision type, the rear-end impact is the most common type of collision. Although it accounts for nearly a third of all collisions, it accounts for only 3 to 5 percent of all serious or fatal injuries. Figure 1 depicts the overall frequency of rear-end collision exposure to risk of AIS 3+ injuries [7,8]. The data depicted is from NASS files, which represents collisions where at least one vehicle is towed away. It is observed in this data, that almost 70 percent of all rear-end collisions have occurred before serious injury risk becomes an issue. Further analysis of these same data indicates that 85 percent of the rear end collision exposures are at a delta-V or crash severity of approximately 32 kph (20 mph) or less. Again, it is noted that this data is taken from collisions where the severity level was sufficiently high to have had at least one vehicle towed away from the accident scene. Because of that omission, the frequency data is biased toward higher severity levels than would be the case if the large number of non-tow-away rear-end collisions were included in this data. This high number of low-speed collisions results in a large number of lower severity injuries (AIS-1).

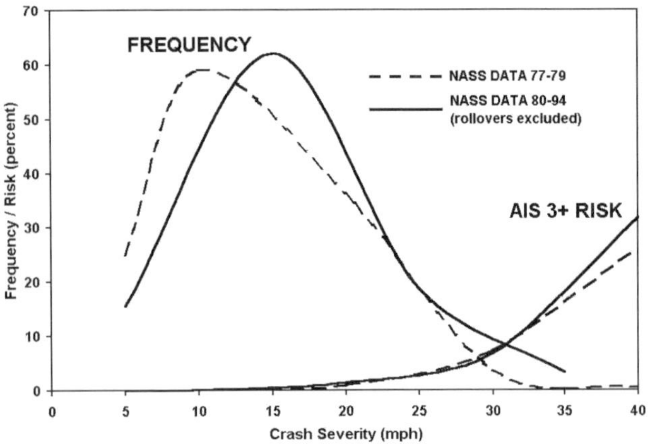

Figure 1. Crash Severity vs Frequency / Risk

Published research by Malliaris and others [9,10,11] presented the technique of HARM to evaluate injury exposures in various impact modes. This method employed the frequency and societal cost factors associated with injuries as a function of various severity levels. Part of the research included an evaluation of the rear impact mode and its associated HARM. It was observed that an overwhelming majority of the injuries was occurring from "non-contact" and from frontal structures. As seen in Figure 2, the "non-contact" HARM was seen to be dominating the influence of injury exposure. This finding is important due to the lower severity implication of the "non-contact" and "front" structures. Meaning if one assumes these exposures are due to "whiplash" and rebound effects, then the stiffness of the seat plays a significant role in those injury exposures. Other researchers have also reported this observation [10,12].

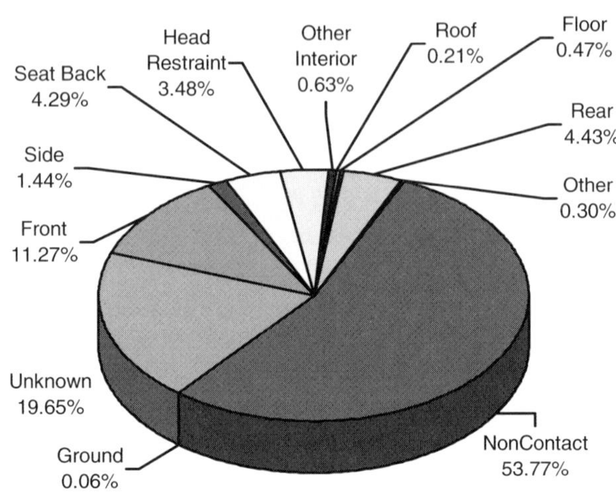

Figure 2. Distribution of HARM (NASS 1976-86) to Front Seat Unrestrained Occupants by Source of Most Severe Injury (Single Rear-Impact Event)

Low Severity Injuries in Crashes – Stiffer or "rigidified" seats have been shown by a number of research studies to have a greater potential for "whiplash" associated injuries in rear-end collisions [7,13,14,15]. While the exact mechanisms concerning these injuries are still somewhat unclear, studies back in 1967, as well as today, have shown that the general kinematics and injury potential associated with the "whiplash" phenomenon appear more frequently in stiffer and stronger seat designs versus the more yielding seat designs. This tendency for increased neck injury becomes an important consideration when one evaluates the effect of strengthening current seat designs. If the force levels required to plastically deform a seat are increased, then the forces currently responsible for many of the lower severity injury and HARM are increased as well. This could have substantial implications on the frequent, lower severity collisions where current seat designs are performing reasonably well.

FEDERAL STANDARDS

There are three Federal Standards directly addressing rear impact collision performance. These are Federal Motor Vehicle Safety Standard number 207 (FMVSS 207), which generally addresses seat performance, FMVSS 202 that deals with head restraints and FMVSS 301, which generally addresses fuel system performance but provides a body of data that is useful in seat collision performance analysis.

FMVSS 207 – FMVSS 207 was initiated in 1968 to require automotive seats to withstand certain forces related to the rear-impact collision environment [16]. The recommended practice and federal standard generally specifies the following:

1. Seat anchorages and seatbacks must be able to withstand 20 times their weight in the forward and rearward directions.
2. The seatback must be able to withstand a moment about a seating reference point of 373 newton-meters (3300 inch-pounds) in a rearward direction.
3. If belt restraints are directly attached to the seat, the seat system must also be able to meet belt attachment requirements.

The prescribed method of compliance testing generally requires attachment of a force application device to an upper cross-member of the seat frame structure. The procedure dictates that the force required to deform the seat rearward then be utilized to compute the moment about the H-Point described in the standard.

Other research has evaluated various methods of producing more realistic seat testing procedures. In 1967, Severy [2] described a testing method that has been utilized over the years, and in 1987, Strother [9] described a slightly modified version of this test. These testing procedures evaluate the seat in a fully trimmed condition, and apply the load near the center of gravity of the occupant torso in a manner that more closely simulates crash conditions. Additionally, by utilizing a "body block", representative of an occupant torso, this evaluation technique provides some basis for evaluating seat trim resistance and pocketing effects.

Currently, there are ongoing inquiries and research to revisit Federal standards and the applicable Society of Automotive Engineers' recommended practices regarding seat crash performance to determine if updates are appropriate. Most seats currently exceed the existing standards by large margins [17,18]. Recent proposals and discussions have centered on the development of a static or dynamic testing procedure to potentially require a certain minimum level of energy absorption by the seat. Other suggestions have been made to limit rearward motion. A later section will discuss occupant energy dissipation by seating systems during automotive crashes. Also, the potential effect of limiting rearward motion is addressed by evaluating dummy occupant performance during rear-end crash tests.

FMVSS 202 – FMVSS 202 is the federal standard that specifies requirements for head restraints to reduce the frequency and severity of neck injury in various collisions [19]. This standard essentially addresses low to moderate severity rear-end collision exposures. The federal standard specifies that for individual seat systems the headrest portion (in a fully extended condition) be no less than 27½ inches from the seating reference point and 6¾ inches wide. Additionally, under prescribed conditions of backrest deflection, the headrest must be able to withstand a 200 pound force over no more than 4 inches of headform deflection.

This standard has general seat geometric and static load performance requirements. With all of the new data becoming available with respect to "whiplash" kinematics and the general focus on addressing lower severity injuries, the standard is currently under review for revision purposes. Part of the review and consideration is to make it more compatible with FMVSS 207, such that one performance standard would not reduce the benefits of the other one. In light of substantial amounts of new biomechanical data on lower speed head and neck kinematics, this effort is proving to be an extremely difficult task.

FMVSS 301 – FMVSS 301 is the safety standard related to the general performance of the vehicle fuel systems [20]. It was developed on the premise of limiting fuel spillage during and following a crash. The federal standard generally specifies several tests including a rear moving barrier crash at 48 kph (~30 mph). These tests, especially as currently structured, provide substantial seat performance data.

Seat and Occupant Performance – In the rear moving barrier crash, the vehicle is impacted by a 4000 pound rigid barrier moving at approximately 48 kph (~30 mph). Normally, the struck vehicle undergoes a 24 to 32 kph (15 – 20 mph) delta-V (ΔV). Over the years, some of the tests have utilized instrumented dummies. First, Hybrid II anthropometric test devices (ATDs) were utilized and then Hybrid III ATDs became the standard. When instrumentation has been present, it generally has shown accelerations and neck motions that did not produce injurious values. In many of these tests, the severity level imparted on the vehicle and occupant typically is high enough to fully flex the seat system. This occurrence has been termed by some as "seat back collapse" and has been considered to be a seat failure, doing nothing towards absorbing the energy of the occupant. Contrary to these opinions, data from the rear moving barrier crash tests shows that actual energy absorption by the combination of vehicle structural crush and seat deformation has served to absorb occupant energy and to limit collision forces that otherwise would be transferred directly to the occupant.

GOVERNMENT RESEARCH REGARDING SEAT CRASH PERFORMANCE – The federal government (NHTSA) has revisited the issue of "seat back failure" on various occasions. In the mid 1970s, the issues relating to seat yield as a function of injury potential were visited. After considerable discussion and analysis, it was determined that seats as currently designed were targeting the severity levels that had the greatest potential for injury, and therefore had good strength characteristics for mitigating those injuries. In 1989 and continuing to the present time, NHTSA has been evaluating seat crash performance. In 1992 they issued a statement reaffirming the advantages of yielding seat designs. They also recommended additional research to further evaluate the question of potential improvements in seat design to additionally reduce injury exposures [21].

Notwithstanding the good crash performance of seats in most collision exposures, a number of seat designs are showing a trend toward incorporation of restraint systems and additional inflatable restraints for side protection. Other designs attempt to better control occupant motion with the intended purpose of reducing neck hyperextension. Most of these new features require additional and more complex structures built into seating systems; some of these additions result in seat designs becoming stiffer and more bulky. As a result of these changes, the rear energy management capabilities and associated problems are changing and becoming more complex.

Modified FMVSS 301 – Since about 1993, NHTSA has modified the test procedure of the rear moving barrier crash test to obtain further information regarding occupant–seat forces and motion. This new procedure essentially requires instrumentation of the driver Hybrid III, the seat and the seatbelt in order to gather information regarding performance of the seating system (see Appendix 1).

A small number of these tests have been evaluated by the authors to provide a better understanding of the occupant kinematics and associated timing. Additional analysis of a greater number of tests will be required to more completely develop the trends and determine more detailed relationships. However, the evaluation of the data thus far generally supports the older research as it relates to injury exposure. The injury levels as predicted by the ATD instrumentation and associated calculations of injury criteria generally show values well within human tolerance limits. Although this finding is not new, it is felt that with the acceleration and seat rotation data now being obtained, it will be possible to better evaluate the force and energy levels as a function of seatback rotation and deformation during the collision.

Generally, the seat structure absorbs energy and resists dummy inertial force through a substantial portion of the rotation. Some testing shows that this phenomenon is evident through a backrest rotation in excess of 60 degrees as measured from vertical [22]. Earlier research has reported seat effectiveness through only 45-50 degrees in terms of retention abilities in a rear impact mode [5].

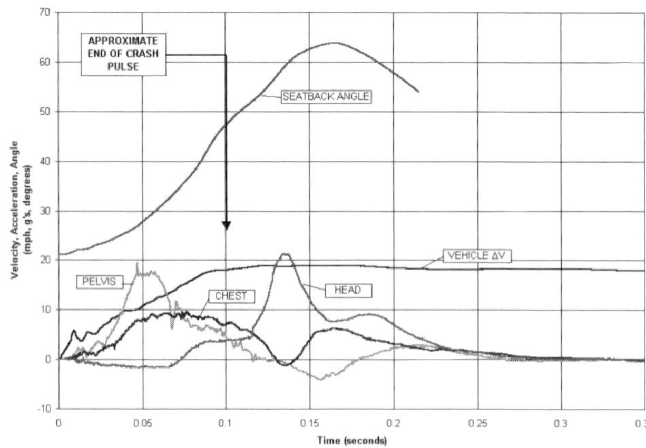

Figure 3. Graph of Seatback Angle and Crash Accelerations as a Function of Time

Figure 3 depicts selected data obtained from an instrumented FMVSS 301 rear crash test and plotted on the same graph to show various relationships. One is able see the relationship of the dummy acceleration levels to the seat rotation and vehicle velocity data. This data shows that the seat resists force over a longer time period than the vehicle crash pulse, thereby absorbing a significant amount of energy. The concept of how this energy absorption occurs has been over-simplified in previous research [23].

Static Seat Testing – In November 1998, Molino of NHTSA [24] published a report concerning a project that included testing of current production automotive seats. In this project, 48 seats representing 25 different configurations were tested. The average seatback moment was approximately 1273 newton-meters (11,266 inch-pounds). A typical test example is depicted in Figure 4. Both the driver and passenger seats were tested and are represented on the same graph. One of the two seats was deformed through its ultimate strength and beyond, usually to an angle of 60 to 70 degrees with respect to vertical, if the seat structure permitted. The sister seat was then tested through 75 percent of the rotation of the first seat, and was then unloaded to determine its restitution properties. It is important to note that the restitution of the seat depends on its relationship to the ultimate strength and not necessarily on its complete rotation [25]. A graph depicting Molino's restitution analysis is shown in Figure 5. This information was determined through rotation-dependant data as opposed to more appropriate strength-based data. Furthermore, the restitution characteristics of a given seat will change depending on the applied load.

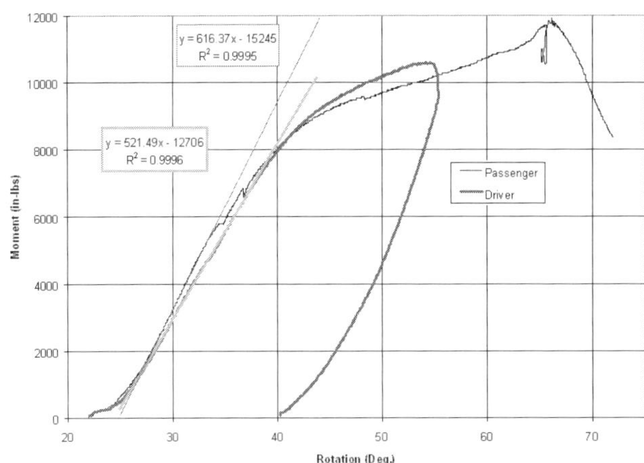

Figure 4. Static Seat Test Data Shows Elastic Properties of Seat

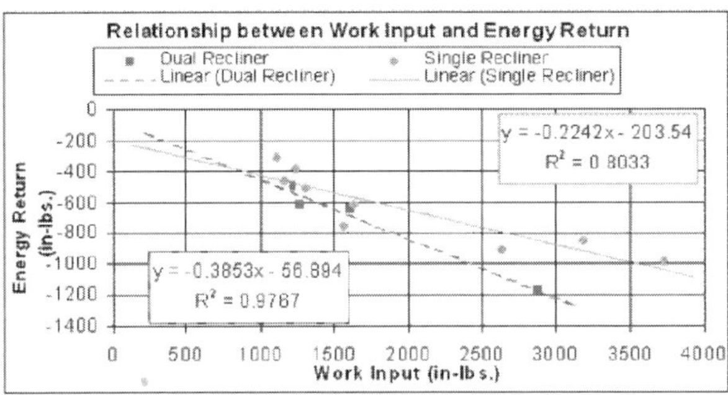

Figure 5. Molino Restitution Data

In the report, Molino defines the yield strength as being the force at which the seat structure has moved into the plastic regime, causing the seat-loading curve to shift by approximately 5 percent away from the elastic slope. As seen in the load-deflection curve of Figure 4, loading curves typically have a linear portion in the first part of the curve that is typical of the elastic range of a material. It then translates into the non-linear range of its force-deflection curve. It was reported that on average, the yield strength of the seats tested was in the range of approximately 60 percent of the ultimate load. This finding is an important factor when considering the crash severity level at which seats will elastically spring back to their upright and undeformed position. Since the forces applied to an occupant will increase as the severity level of a crash increases, if the seat does not reach permanent yield at an appropriate severity level, potentially more severe injuries due to rebound effects could occur. Considering the elastic range of recent seat designs and if the trend increases toward stiffer seats, there is a valid concern regarding injuries increasing in frequency and in severity level (i.e. AIS 1 increased to AIS 2, etc.).

EVALUATION OF ENERGY CHARACTERISTICS IN SEATING SYSTEMS

The evaluation of energy management in seating systems requires the identification and evaluation of multiple characteristics. These items not only include the static moment strength, but also the dynamic moment capabilities of the seat system, seatback restitution, seatback stiffness and overall energy capacity. Relationships between static and dynamic characteristics have to be identified before any evaluation can be completed. Direct application of static seating performance as a predictor of dynamic seat performance cannot be completed without careful analysis and verification of applied assumptions.

Static seat testing can be performed to determine the ultimate moment capabilities of a seat. General practice has been the application of a horizontal load located at the upper cross member, or using a body block that applies a horizontal load approximately 9"-14" above the H-point. This moment arm is typically 10"-17" above seat pivot points. This type of evaluation also allows the measurement of the stored and returned energy. The ratio of stored energy to energy returned is an important characteristic in the evaluation of seat performance. This characteristic cannot be determined for the entire deformation range of the seat through a single static seat test since it has been found that restored energy is a function of the applied moment. In order to adequately determine these characteristics, the seatback must be loaded and unloaded from various applied forces, measuring the hysteresis at each level.

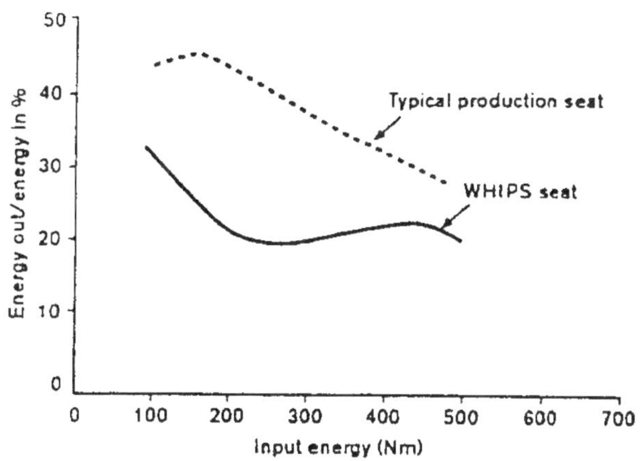

Figure 6. Lundell Restitution Data

Research (WHIPS) completed by Lundell in 1998 [25], wherein a typical production seat was statically tested through a series of increasing moment applications measuring the hysteresis at each level, demonstrated the relationship between the applied load and restitution. As illustrated in Figure 6, restitution is a function of the load applied, peaking at lower loads while dropping off linearly as seat back angle increases. Excessive elastic restitution is an undesirable characteristic that has to be identified and attenuated, if possible. Evaluation of the energy capacity of a seat through static seat tests requires the evaluation of certain variables and clear identification of important cause and effect relationships between static and dynamic loads.

The assumption of a single moment arm representing occupant loading has to be examined and re-evaluated. A general assumption made in the past has been to apply 60 to 70 percent of the occupant weight with a moment arm of approximately 10"-14". Figure 7 illustrates the moment arms about the H-point of a Hybrid III at three center of gravity (CG) locations; the head, the chest and the pelvis (lower torso). For this analysis it will be assumed that the lower legs and feet contribute little to the effective moment applied to the seatback.

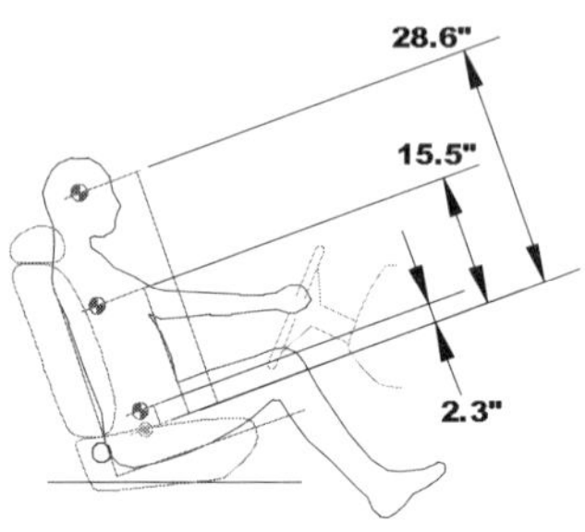

Figure 7 – Diagram of Hybrid III CG Reference Points from H-Point

The weights at each CG location used in this evaluation of the dynamic applied moment are 11 pounds, 56.2 pounds, and 76.3 pounds for the head, chest and pelvis, respectively. It is important to note that while the total weight equals 143.5 pounds, it cannot be assumed that each CG inertial force is applied during the same time interval. Utilizing dynamic rear moving barrier crash tests, and specifically the acceleration data at each location, actual moments can be calculated as a function of time. The weight of the seatback was neglected in these calculations.

The energy capacity of a seat derived from static seat tests has to be carefully evaluated and analyzed. The relationship between the dissipation of the kinetic energy of an occupant and the static energy capacity of a seat has been addressed incorrectly in past research [23]. The assumption that the kinetic energy of an occupant is solely dissipated into the seat structure, as postulated by some researchers, proves to be an incomplete assumption.

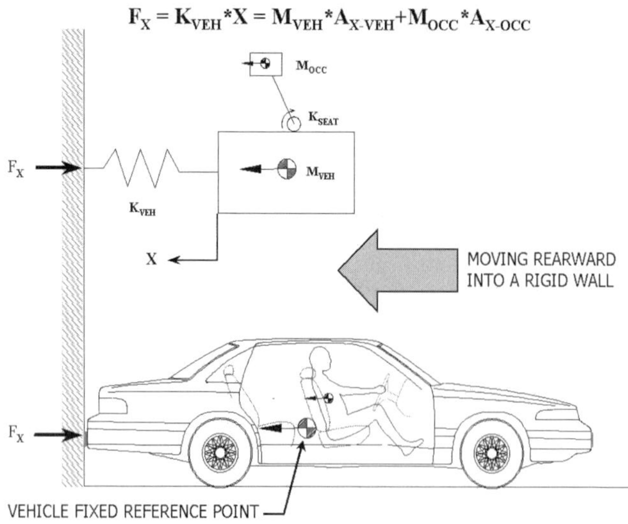

Figure 8. Example of Model Representing Fixed Rear Barrier Impact

As an example, in Figure 8 a vehicle is shown traveling rearward into a fixed barrier. The vehicle, seat and occupant can be represented by a spring-mass system within a spring-mass system (diagramed above the vehicle in Figure 8). The vehicle and the occupant are traveling with a velocity V. The kinetic energy of the system can be found to be as follows:

$$KE_{sys} = 1/2\sum mv^2 = 1/2\left(M_{veh} \cdot V_{veh}{}^2 + M_{occ} \cdot V_{veh}{}^2\right)$$

As can be seen in Figure 8, the occupant's deceleration contributes to the force acting on the rear structure (K_{veh}) and therefore the work done. If the seat design within the vehicle is perfectly rigid, the kinetic energy of the occupant is transferred through the rigid seat system and consequently is absorbed by the rear structure of the vehicle. However, if the seat system absorbs energy by some form of yielding and deformation, then there is energy dissipated by the seat that does not go into the rear structure energy absorption. This concept is modeled in the equation below and seen in the barrier test example (Figure 3), where the seat system continues absorbing energy well after the crash pulse has ended.

$$E_{veh} = 1/2\, K_{veh} X^2 = 1/2\left(M_{veh} \cdot V_{veh}{}^2 + \lambda \cdot \left(M_{occ} \cdot V_{veh}{}^2\right)\right)$$

Where:

E_{veh} = Energy absorbed by the vehicle structure

λ = Percent of occupant kinetic energy absorbed by the vehicle structure (rigid seat: λ =1)

Hence, the energy absorbed by the seat can be described by the following:

$$E_{seat} = 1/2\left(M_{occ} \cdot V_{veh}^{\ 2}\right) \cdot (1-\lambda) = M_o \cdot \theta$$

Where:

E_{seat} = Energy absorbed by the seat system

M_o = Moment about seat pivot point

θ = Angular displacement of seatback

Because of this relationship between occupant, seat and vehicle, the energy capacity of a seating system cannot be completely evaluated without looking at the collision environment. The postulation that the energy absorbed in a static test can be used directly to determine maximum delta-V, or threshold velocity where occupant retention is lost, does not take into account the vehicle factors for evaluation of energy management as shown by the mathematical model.

ENERGY MANAGEMENT CONSIDERATIONS

Although crash analysis research [26] as well as existing statistics point to the effectiveness of current seat designs, considerable discussion and study continues to focus on the question of injury reduction with respect to AIS-1 injuries. Additionally, the less frequent but potentially more severe question of spinal and head injuries at higher crash exposure levels, as well as the potential adverse effects of current yielding seatback interaction with other occupants within the passenger compartment remains to be solved.

Some have suggested that both the low-speed AIS-1 and the higher-speed AIS-3+ injury problem can be solved by merely increasing the strength of seatbacks to some substantially higher level and then somehow limit the total amount of deflection that can occur, no matter how severe the collision [23]. This "solution" would be similar to a restraint design for frontal collisions that would sense the severity of the collision and would instantly adapt itself by changing the elongation characteristics of the belt webbing and the deployment characteristics of the airbag. Current technology does not allow this type of crash sensing and corresponding component modification that would be necessary for such a solution to be properly applied to seating systems.

In order to gain insight into the relative advantages of various design approaches for seat systems, it is necessary to not only consider the various types of injury exposures, but also to evaluate the crash environment in more detail. For example, if we consider the overall environment for a front-seat occupant in a rear-end collision, we see that several factors may come into play, depending on the seat system and the crash exposure considered.

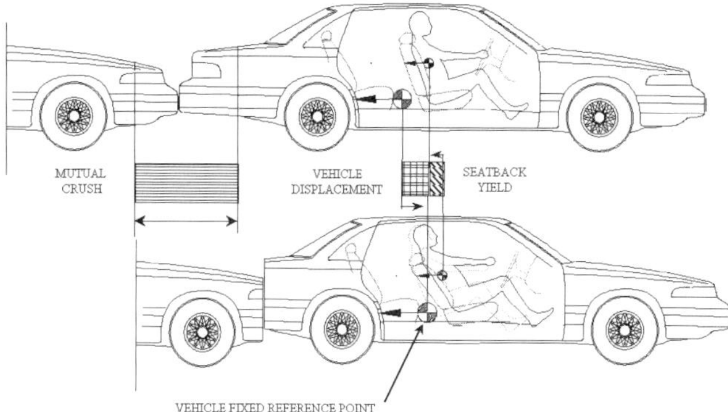

Figure 9. Diagram of Different Crush and Displacements in a Vehicle-to-Vehicle Crash

The following sections represent four conceptual examples of different seat-vehicle combinations. Figure 9 depicts the crush and displacements referenced in the examples that follow. The mutual crush represents the combination of striking vehicle and struck vehicle dynamic crush damage. During the crash, the struck vehicle and the occupant are displaced forward certain distances indicated in each of the following conceptual examples.

RIGID SEAT IN A CONVENTIONAL PASSENGER CAR – First, consider a system where the seat is a theoretically rigid attachment to the car and does not provide any structural energy absorption during the crash.

Figure 10 shows a simplified crash profile for such a condition. In this example, the mutual vehicle crush provides the primary energy absorption mechanism for both vehicle and occupant. During the crush phase, the struck vehicle is accelerated and is displaced forward a distance of approximately 1½ feet. In this hypothetical crash, the occupant is inertially displaced a small amount, (a distance of three inches) into the seat upholstery but does not deform the backrest structure and therefore the seat structure does not absorb any energy. In this hypothetical crash example with a delta-V of 20 MPH, the acceleration rate for the occupant is approximately the same as that of the struck car which is 20 g's. The occupant's effective delta-V is also only slightly higher than that of the vehicle C.G. since there is little elastic rebound from the seat upholstery.

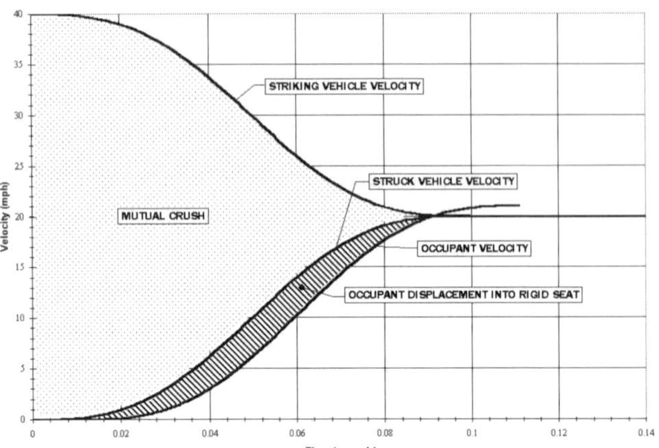

Figure 10. Crash Velocity Profile of an Occupant in Close Proximity to a Rigid Seat at Impact

YIELDING SEAT IN A CONVENTIONAL PASSENGER CAR – The next condition to be evaluated, represented in Figure 11, is the same crash pulse as the previous example, except that the seat structure is allowed to deform during occupant inertial loading. In this example, the seatback deforms such that the occupant's upper torso displaces rearward a distance of approximately one foot while the car is accelerated forward approximately 2½ feet. Note that by allowing the seat back to yield, the resulting longer acceleration time and distance for the occupant acceleration phase lowers his acceleration to 10 g's, or half the rate of that with a non-yielding seat.

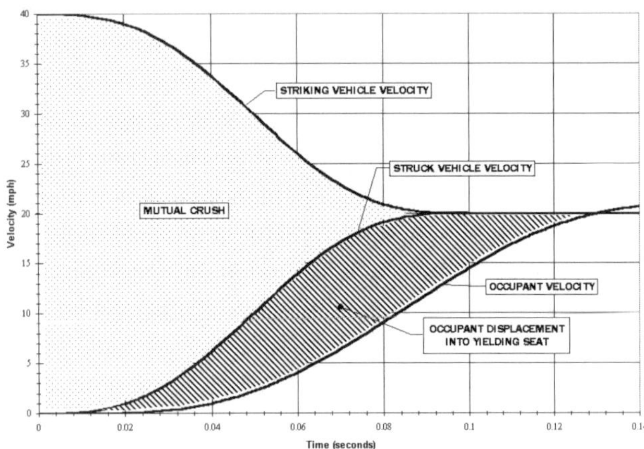

Figure 11. Crash Velocity Profile of an Occupant in Close Proximity to a Yielding Seat at Impact

RIGID SEAT IN A CONVENTIONAL PASSENGER CAR WITH FORWARD DISPLACED OCCUPANT – Both of the previous examples assume proper seating position for the occupant. In this example, a starting position with the torso six inches ahead of the seatback is assumed. As depicted in Figure 12, the occupant acceleration is delayed by its 6 inch offset and then has a similar 3 inch cushion/body compliance displacement into the rigid seat structure utilized in the first example. However, now the occupant has to "catch up" to the vehicle velocity in approximately a quarter of the time of the first example. The resulting accelerations for the occupant increase dramatically to more than double the rate in the first example where the occupant started its impact in contact with the rigid seat. This increase occurs because the occupant is not attached to the car during the critical first portion of the crash phase.

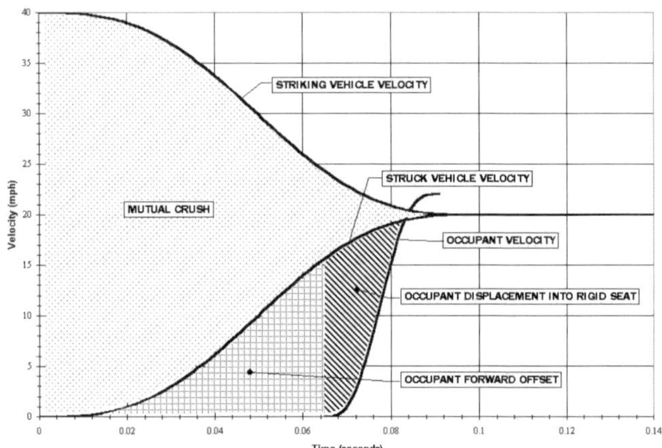

Figure 12. Crash Velocity Profile of an Occupant in a Rigid Seat, Slightly Out of Position at Impact

YIELDING SEAT IN A CONVENTIONAL PASSENGER CAR WITH FORWARD DISPLACED OCCUPANT – Lastly, consider the same starting position of the previous example except the seat structure is allowed to deform. As depicted in Figure 13, the effect of a similar six inch offset on the occupant in a yielding seat is relatively small. The seatback yield properties serve to limit the rate of acceleration that can be applied to the occupant, even when he is not able to utilize all of the vehicle crash pulse. It should be noted that under this "out of position" condition for the yielding seat occupant, the actual rearward displacement has increased, due to the delayed interaction with the seat. Therefore, issues of ramping and increased neck motion may exist as a result of this increased displacement within the passenger compartment. However, when the occupant becomes "out of position" the pelvis will interact with the lower seat structure before the upper body, thereby decreasing the relative energy left to be absorbed by the upper torso and head. As the angle between the torso and seatback increases, so does the effective force component of the upper torso and head acting through the pelvis.

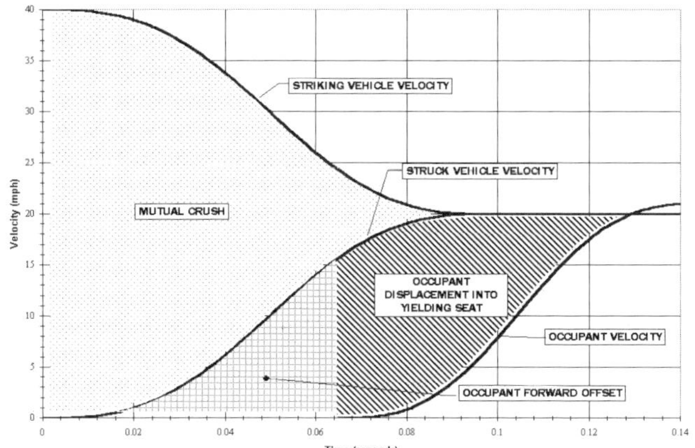

Figure 13. Crash Velocity Profile of an Occupant in a Yielding Seat, Slightly Out of Position at Impact

EVALUATION OF ACTUAL CRASH PULSES INVOLVING MOVING BARRIER TESTS – It can be seen from the preceding analyses that vehicle structural characteristics play a significant part in determining seat performance during a crash. The following analysis of actual crash data demonstrates various factors that must be considered when evaluating seat design modifications. Data files have been obtained from the referenced crash test reports and overlaid in graphical form to depict the relationship of the accelerations normal to the seatback, velocities, and seatback angle, all as a function of time [27,28,29].

Any comparison of moving barrier tests with actual car-to-car crashes must consider the fact that the barrier by design has no energy absorbing capability. For this reason, the extent to which the striking vehicle in a car-to-car crash absorbs energy and thereby extends the crash pulse must also be considered.

Conventional Seat – Conventional Vehicle – The following example demonstrates crash charact-eristics of a conventional 4-door passenger car. The vehicle is of current design and normal construction for such a vehicle. As depicted in Figure 14, the delta-V of the vehicle is achieved in a time typical for this type of vehicle when struck by a moving barrier. Note the peak values for the three acceleration curves. The pelvis peaks at approximately 50 milliseconds, the chest peaks at approximately 80 milliseconds, and the head peaks at approximately 130 milliseconds.

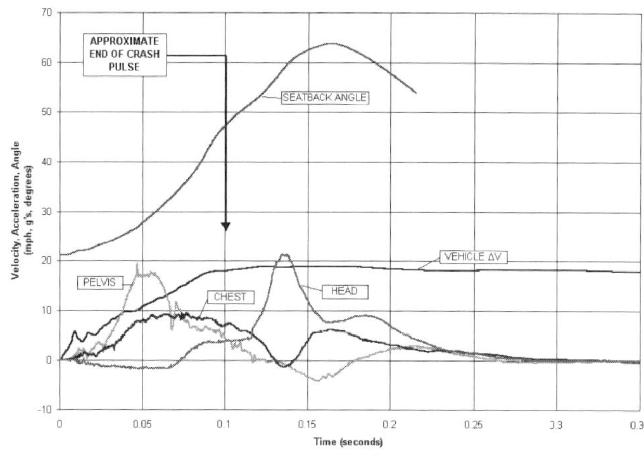

Figure 14. Graph of Seatback Angle and Crash Accelerations as a Function of Time – Conventional Vehicle with a Yielding Seat.

The phase lag shown in the curves demonstrates how extending the load normal to the seatback over a longer period of time contributes to the effectiveness of the strength and performance characteristics of the seat system. Examination of the curve representing the seatback angle over the crash duration demonstrates this performance. In Figure 15, the dynamic seatback moment about the H-point as a function of time as well as the energy absorbed as a function of seatback angle are plotted.

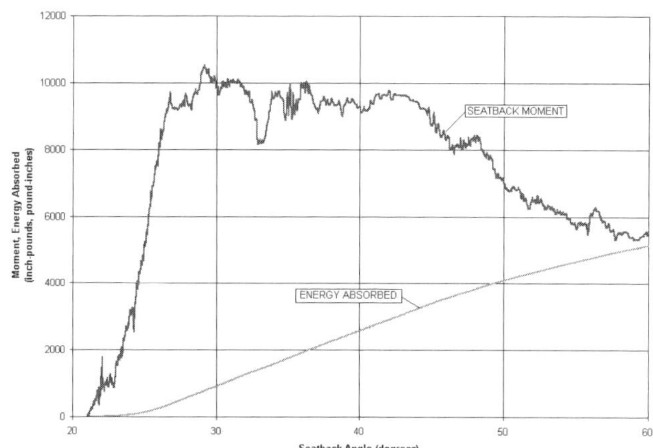

Figure 15. Seatback Moment and Absorbed Energy as Seat Yields Rearward

This representation is a preliminary analysis of the seatback performance based on measured values of acceleration along with analyzed values of geometric position. It is observed that due to early pelvis interaction with the lower seat geometry, initial loading into the seat system occurs quickly. However, due to the lag and follow-up of the chest and head, the load is sustained over a longer period of time adding to the effective energy absorption.

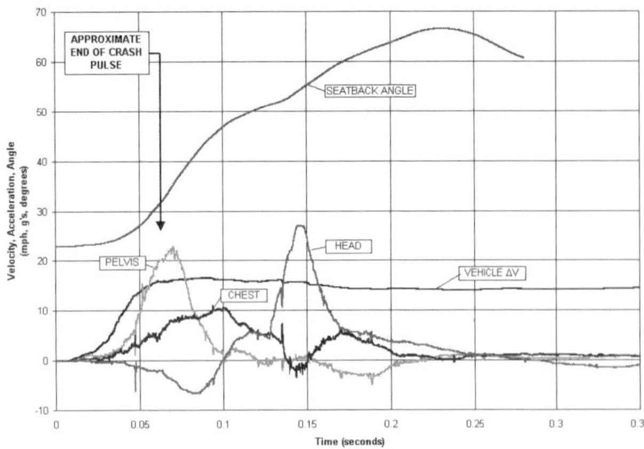

Figure 16. Graph of Seatback Angle and Crash Accelerations as a Function of Time – Stiff Vehicle with a Yielding Seat

Conventional Seat – Stiff Vehicle – This example depicts a stiffer SUV-type vehicle. This vehicle represents the kind of structure and associated response that one could expect out of SUVs, trucks, minivans and vans when struck by a moving barrier. As depicted in Figure 16, the delta-V for the struck vehicle is achieved in a little more than 50 milliseconds. It should be noted that the barrier is not absorbing any energy, hence the response in the vehicle structure and associated occupant is more severe than in a normal vehicle-to-vehicle crash. However, again it is interesting to note that three discrete peaks in acceleration exist for the pelvis (~70 milliseconds), chest (~100 milliseconds), and the head (~150 milliseconds).

As depicted in the graph, the occupant responses are slightly higher than in the previous example. The seatback angle graph depicts a general change in slope at approximately 100 milliseconds. It can be hypothesized that the steeper portion of the seatback angle curve is due to the early loading and interaction with the mass of the pelvis. Furthermore, a second point of inflection occurs just before 150 milliseconds. This appears to correlate well with the mass of the head providing additional loading.

During the crash, the occupant is retained inside the seat system and reaches the vehicle velocity at the end of the seat deformation. The moment in this dynamic environment is calculated using the previously described weights and moment arm distances illustrated in Figure 7. As was also indicated earlier, the moments were calculated using the "x" axis accelerations, since it is assumed that these accelerations remain approximately normal to the seatback. The energy absorbed by the seat in this dynamic environment is obtained by integrating the moment versus angular displacement. As indicated in Figure 17, the seat system resists force and absorbs energy through a substantial amount of rotation. In the static seat test seen in Figure 18, total energy absorbed was evaluated up to an angle of 60 degrees and was found to be 3662 in-lbs. Equating this energy to the kinetic energy of the occupant (λ =0) the maximum delta-V or threshold velocity, where seat energy absorption would be exhausted under a simplified energy analysis, would be 7.4 mph. If 70% of the weight of the occupant were used, as proposed by some researchers, the maximum delta-V would be 10.6 mph. Utilizing the delta-V of the vehicle from the dynamic test of 16.5 mph, these numbers indicate that after complete seat yield the occupant would have a residual velocity of 5.9 to 9.1 mph in a rearward direction relative to the vehicle. However as previously noted, examination of the film from the onboard cameras indicate very little, if any, residual velocity of the occupant. This observation is confirmed by examination of the neck data that indicate no increased loading at the point of maximum rearward excursion.

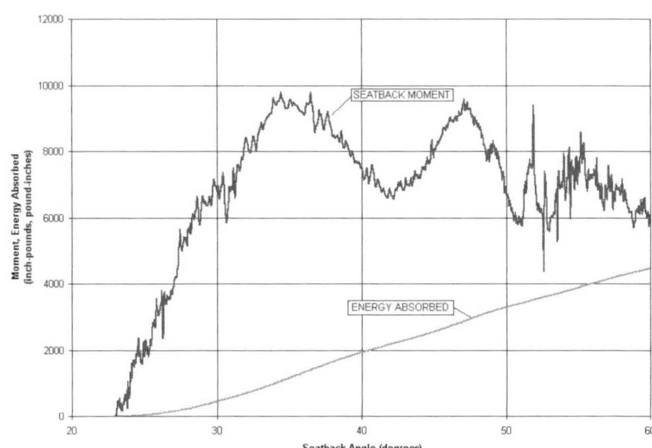

Figure 17. Seatback Moment and Absorbed Energy as Seat Yields Rearward

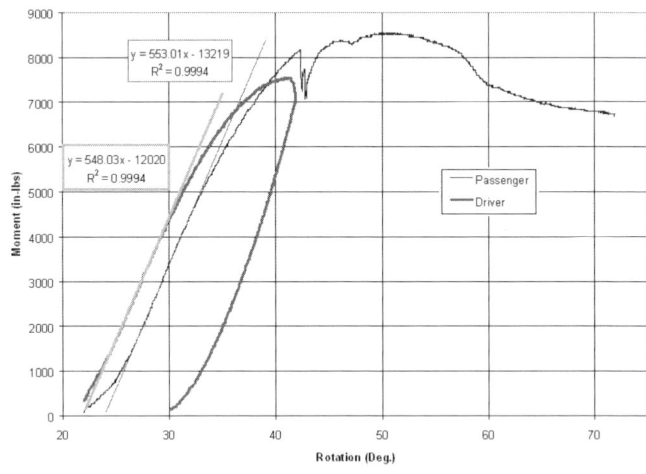

Figure 18. Graph of Static Seat Test Data for SUV Vehicle

From analysis of these tests it can been seen that the basis for the previously referenced over-simplified calculations is incorrect and it under-estimates the actual energy absorption of the seat/vehicle system as it pertains to occupant retention. Therefore, while energy absorption measurement in a static test indicates energy absorption characteristics of the seat structure itself, it cannot be applied directly as an overall evaluation of seat collision performance.

Deflection-Limited Seat – Stiff Vehicle – The next example is of a two-seat vehicle with a rear bulkhead located a short distance behind the two seats. As in the conceptual examples, this crash test demonstrates the responses of a compact stiff vehicle with seatback rotation limitation. Figure 19 shows a pre-test photograph alongside a post-test photograph of the driver dummy. As shown in the photograph the occupant head has an initial separation from the seat headrest of approximately 4 inches. The graphical data depictions of the various acceleration values are shown in Figure 20.

Figure 19. Pre and Post Test Photographs of the Driver Dummy position

It is observed that the characteristic phase lags of the three discrete acceleration peaks are substantially reduced. The peaks of the pelvis, chest and head have occurred within 10 milliseconds of each other, as opposed to the 30-50 millisecond lag found previously. It is reported in the crash test report and observed in the onboard camera film that the dummy head sustains its high accelerations through contact with the seat headrest.

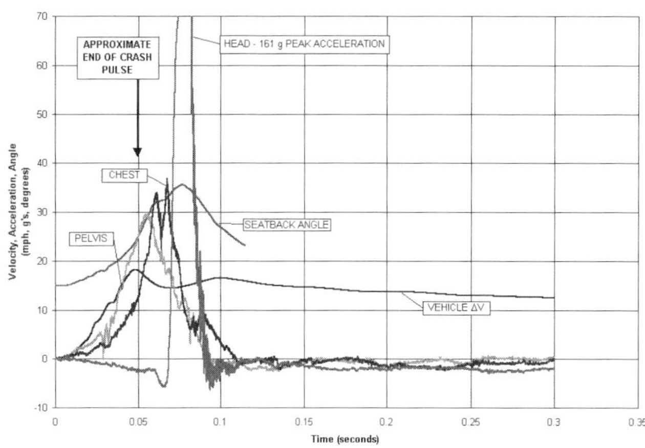

Figure 20. Graph of Seatback Angle and Crash Accelerations as a Function of Time – Stiff Vehicle with Limited-Yield Seat.

Depicted in Figure 21 is the calculation of the seatback moment as a function of rotation angle. The 30,000+ inch-pound resistance indicated in Figure 21 is provided by interaction with the rear bulkhead, which limits motion in a rearward direction. The indicated energy absorption of the occupant is no more than that of a conventional passenger vehicle configuration. However, this energy absorption occurs in less than half of the time and distance than a conventional seatback, thereby producing high acceleration values on the dummy occupant.

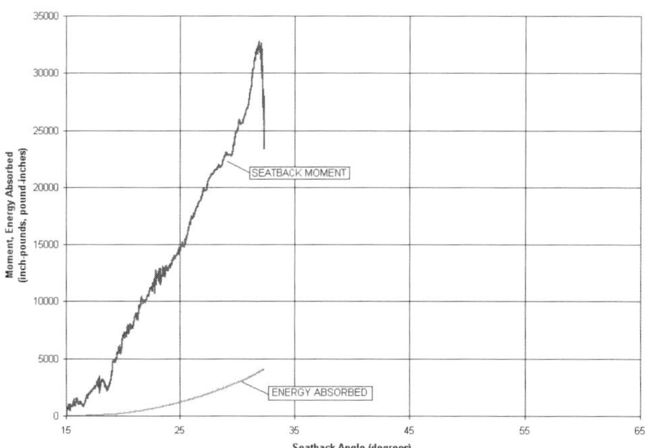

Figure 21. Seatback Moment and Absorbed Energy as Seat Rearward yield is limited by rear Bulkhead.

Figure 22 depicts a sequence of interior video stills of the driver dummy taken from the onboard camera. The dummy head at the start of the crash is shown to be forward of the headrest. The seat rearward deflection is limited to approximately 18 degrees of seatback rotation. Early in the crash sequence the occupant deforms the backrest in a normal manner. The seat rotation is then stopped and the occupant "bottoms-out" against the structure. As a result of the limited amount of available deformation, or yielding, the accelerations experienced by the occupant are significantly increased. The peak acceleration level recorded by the head is 161 g's. This high acceleration value leads to a HIC which exceeds recommended limits at a crash severity where head accelerations are generally not severe.

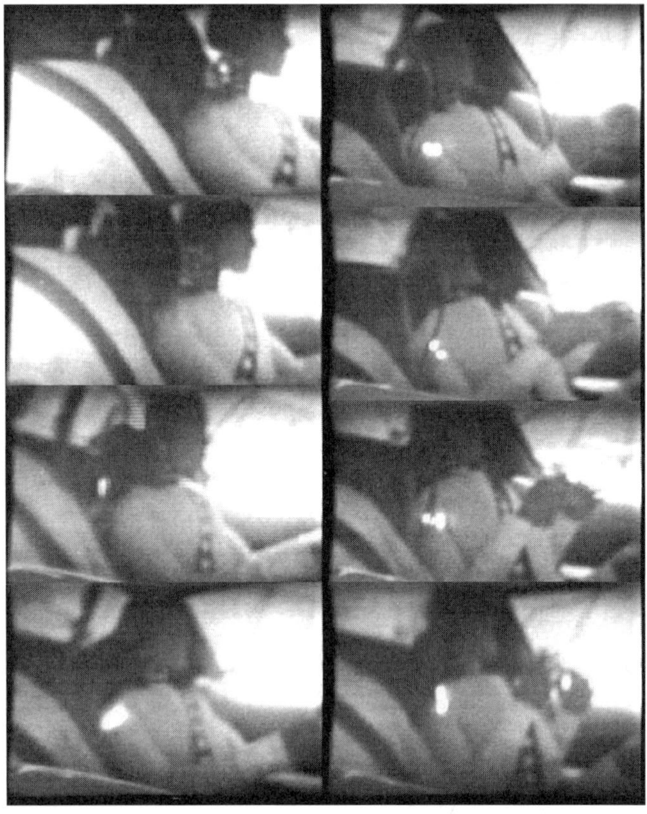

Figure 22. Occupant interacts with deflection-limited seat and rebounds rapidly.

CONCLUSIONS

- FMVSS 207 and FMVSS 202 are the Federal Standards directly addressing the performance of seatbacks and head restraints. These standards are currently under review to be updated. Any change in the requirements should take into account the energy management considerations of current seat designs when considering potential benefits arising from stronger seats.
- New data from recent government static seat testing shows a moderate trend toward stronger seat designs. Any effect of this trend on injury rates remains to be determined.
- Restitution analysis conducted in the government seat research does not take into account the relationship of restitution to the applied load. However, the elastic to plastic transition in the load-deflection curves occurred at an average of 60 percent of the ultimate load of the tested seats which has important implications of the elasticity and resulting rebound effects of stronger seat designs.
- Early research focused on identification and introduction of basic concepts for controlling occupant kinematics. More recent research has identified additional details concerning occupant kinematics and injury potential adding to the ability to better understand rear impact kinematics.
- Evaluation of the frequency of rear-impact collisions shows this impact mode to be a very common collision; however, the majority of rear collisions take place at severity levels where risk of serious injury and fatality are extremely low.
- Lower-severity types of injuries are heavily represented and the effect of strengthening seats has important implications on this injury level range.
- As new seat designs become more complex and stronger to incorporate new features and safety devices, the potential adverse consequences not only to the occupant within the seat, but also on those behind and adjacent to the seat need to be evaluated.
- A mathematical model was presented depicting the various factors involved with an energy analysis of seat crash performance. Energy management analysis and thus seatback collision performance cannot be completely evaluated without careful analysis of both the seat and the vehicle collision environments.
- Velocity-Time graphical analysis is an important tool in evaluation of seat performance and occupant force levels. A conceptual analysis approach was presented depicting the results of different seat-occupant combinations. It was shown that as an occupant becomes "out of position" for the rigid or "rigidified" seat systems, the force and resulting injury levels increase dramatically.
- An evaluation of new data surrounding the recently modified FMVSS 301 standard was presented. It was shown that current seat designs are performing well by absorbing significant energy at this severity exposure. Detailed examination of the data revealed the timing of load application from dummy accelerometer data. This analysis shows substantial energy absorption characteristics due to delayed interaction and duration of loading.
- An example was shown that depicted a deflection-limited seat with a slightly "out of position" occupant. This example directly addresses the problem of arbitrarily limiting rearward motion given the kinematics of potential head to seat interactions.

- Additional analysis must be done to further identify the relationships between static and dynamic data. Further analysis will also be required to evaluate the neck injury properties and associated relationships to the presented data.
- A new concept in energy analysis for automotive seats has been presented. This analysis took a basic understanding of energy dissipation factors present in rear collisions and constructed an overview of energy management analysis. It was shown that a simple application of the energy equation for an assumed mass of the occupant to obtain a "threshold" delta V does not take into account how the crush of the vehicle affects the energy of the occupant and how the stiffness of the seat, in terms of rear deflection affects occupant motion and injury potential.

ACKNOWLEDGMENTS

Grateful appreciation is extended to Richard Kent of the University of Virginia's Automotive Safety Lab and to the SAE reviewers who provided valuable assistance with this paper.

REFERENCES

1. D.M. Severy, J. Mathewson, O. Bechtol, "Controlled Automobile Rear-End Collisions and Investigation of Related Engineering and Medical Phenomena", Canadian Services Medical Journal, XI: 727 – 759, November, 1955.
2. D.M. Severy, H.M. Brink, J.D. Baird, "Preliminary Findings of Head Support Designs", Proceedings, 11th Stapp Car Crash Conference, SAE #670921, October 10-11, 1967
3. D.M. Severy, H.M. Brink, J.D. Baird, "Backrest and Head Restraint Design for Rear-End Collision Protection", Automotive Engineering Congress, Detroit, SAE #680080, January 8-12, 1968.
4. D.M. Severy, H.M. Brink, J.D. Baird, "Vehicle Design for Passenger Protection from High-Speed Rear-End Collisions", Proceedings, 12th Stapp Car Crash Conference, SAE #680774, October 22-23, 1968.
5. D.M. Severy, H.M. Brink, J.D. Baird, D.M. Blaisdell, "Safer Seat Designs", Proceedings 13th Stapp Car Crash Conference, 1969,SAE #690812
6. C. Strother, M. James, "Response of Out-of-Position Dummies in Rear Impact", SAE #941055, 1995
7. P. Prasad, A. Kim, D. Weerappulli, V. Roberts, D. Schneider, "Relationships Between Passenger Car Seat Back Strength and Occupant Injury Severity in Rear End Collisions: Field and Laboratory Studies", 1997 , SAE #973343
8. L.L. Ricci, "NCSS Statistics: Passenger Cars", UMTRI, NHTSA Report # DOT-HS 805 531, June 1980
9. A.C. Malliaris, et al, "Harm Causation and Ranking in Car Crashes", International Congress and Exposition, Detroit, SAE #850090, February 25-March 1, 1985.
10. M. James, C. Strother, M. James, R. Decker, "Occupant Protection in Rear-End Collisions: I. Safety Priorities and Seat Belt Effectiveness", SAE #912913, 1991
11. R. Kent, M. James, R. Norhagen, "Seat Design and Occupant Protection in Rear Impacted Vehicles", Paper #98SAF033, Proceeedings of the 31st International Symposium on Automotive Technology and Automation, June 1998
12. C. Strother, M. James, "Evaluation of Seat Back Strength and Seat Belt Effectiveness in Rear End Impacts", SAE #872214, 1987
13. M. Svensson, "Neck-Injuries in Rear-End Car Collisions", Chalmer Institute, Goteborg, Sweden, 1993
14. S. Parkin, M. Mackay, A. Hassan, R. Graham "Rear End Collisions and Seat Performance- To Yield or not to Yield", 39th Annual Proceedings AAAM, October 16-18, 1995, Chicago, Illinois
15. O. Bostrom, M. Krafft, B. Aldman, A. Eichberger, R. Fredriksson, Y. Haland, P. Lovsund, H. Steffan, M. Svensson, C. Tingvall, "Prediction of Neck Injuries in Rear Impacts Based on Accident Data and Simulations", IRCOBI Conference, September 1997
16. Current version of Federal Motor Vehicle Safety Standard 207, 49 CFR. 571.207.
17. C. Warner, C. Strother, M. James, R. Decker, "Occupant Protection in Rear-End Collisions: II. The Role of Seat Back Deformation in Injury Reduction", SAE #912914, 1991.
18. L. Molino, "Determination of Moment-Deflection Characteristics of Automobile Seatbacks", NHTSA-1998-406
19. Current version of Federal Motor Vehicle Safety Standard 202, 49 CFR. 571.202.
20. Current version of Federal Motor Vehicle Safety Standard 301, 49 CFR. 571.301.
21. 39-FR—10268 1974, proposal to amend FMVSS 202 and 207 established in Docket 74-13, March 19,1974
22. D. Viano, "Influence of Seatback Angle on Occupant Dynamics in Simulated Rear-End Impacts", SAE #922521, 1992
23. K. Saczalski, S. Syson, R. Hille, M. Pozzi, "Field Accident Evaluations and Experimental Study of Seat Back Performance Relative to Rear-Impact Occupant Protection", 1993 , SAE #930346
24. L. Molino, "Determination of Moment-Deflection Characteristics of Automobile Seatbacks", NHTSA-1998-4064
25. B. Lundell, L. Jakobson, B. Alfredsson, C. Jernstrom, I. Hellman, "Guidelines for and the Design of a Car Seat Concept for Improved Protection against Neck Injuries in Rear End Car Impacts", International Congress and Exposition, Detroit, SAE #980301, 1998

26. D.M. Blaisdell, A.E. Levitt, M.S. Varat, "Automotive Seat Design Concepts for Occupant Protection", SAE #930340, 1993

27. Final Report of FMVSS 301 Compliance Testing of a 1995 Mazda Protégé 4-Door Sedan NHTSA No. CS5405, April 1995

28. Final Report of FMVSS 301 Compliance Testing of a 1995 Honda Passport MPV NHTSA No. CS5304, May 1995

29. Safety Compliance Test: Fuel System Integrity: 1993 Honda Civic Del Sol 2 Door Coupe NHTSA No. CP5310, July 1993

CONTACT

Questions and comments are welcome. The authors are all Research Engineers at:

Collision Research & Analysis, Inc.
www.collisionresearch.com

APPENDIX 1

APPENDIX 2 OF FEDERAL MOTOR VEHICLE SAFETY STANDARD 301

MODIFIED FMVSS 301 TEST, REAR IMPACT ONLY

For the purpose of acquiring information for research and development, fmvss 301, fuel system integrity test - rear impact, is modified with the following additional test and data acquisition requirements:

A2.1 TEST EQUIPMENT DESCRIPTION – The following is a list of the minimum additional test equipment needed to perform the modified FMVSS 301 test.

A. A anthropomorphic test dummy, Part 572E (Hybrid III).
B. Head Center of Gravity (C.G.) Triaxial Accelerometers for the test dummy specified above.
C. Chest C.G. Triaxial Accelerometers for the test dummy specified above.
D. Pelvis C.G. Triaxial Accelerometers for the test dummy specified above.
E. A H-Point machine (H-Point Template shown below.

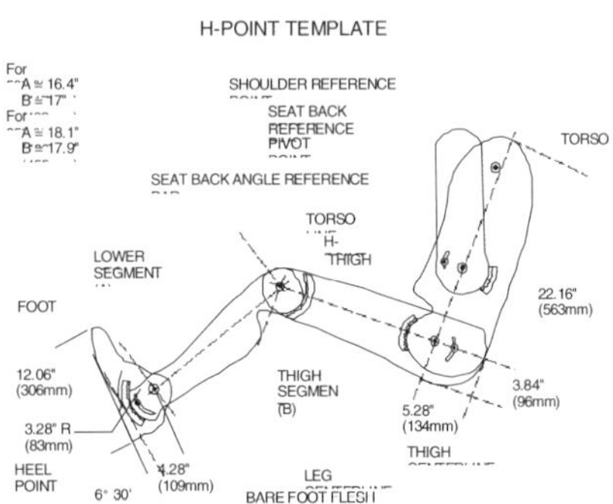

F. Sensors to measure vehicle acceleration, driver seat back angular displacement and acceleration, driver seat belt motion (playout), and driver seat belt load.
G. Data recording equipment having sufficient channels to record the necessary time history. Each data channel shall consist of a sensor, signal conditioner, data acquisition device, and all interconnecting cables and must conform to the requirements of SAE Recommended Practice J211.

A2.2 ADDITIONAL INSTRUMENTATION – Instrument the Hybrid III dummy as follows:

A. Head C.G. Triaxial Accelerometers
B. Chest C.G. Triaxial Accelerometers
C. Pelvis C.G. Triaxial Accelerometers
D. Upper Neck
   1. 3-axis Neck Force Transducers (GFE*)
   2. 3-axis Neck Moment Transducers (GFE*)
      • Government Furnished Equipment

Dummy calibration, according to Part 572E Dummy Performance Calibration Procedure, to be performed prior to the start of testing, after every fifth test, and at the completion of testing. Instrument the test vehicle to record the following:

A. Primary and redundant accelerometers to record occupant compartment acceleration
B. Driver seat back angular displacement
C. Driver seat back angular acceleration
D. Driver seat belt motion (playout)
E. Driver seat belt load

NOTE: Location of transducers will be determined in coordination with the Office of Vehicle Safety Compliance (OVSC) and the Office of Crashworthiness Research (OCR).

Onboard Camera Requirement

One high speed camera (1,000 frames/second) placed in the vehicle to acquire improved film coverage of seat/dummy/belt dynamics in a rear crash. Actual camera location will be chosen in coordination with OVSC and OCR.

Final Report and Film Requirements

A. Plots to be included in test report —

    Head X,Y,Z resultant acceleration vs. time

    Chest X,Y,Z resultant acceleration vs. time

    Pelvis X,Y,Z resultant acceleration vs. time

    Upper neck X,Y,Z force vs. time

    Upper neck X,Y,Z moment vs. time

    Vehicle acceleration vs. time

    Seat back angular displacement vs. time

    Seat back angular acceleration vs. time

  Seat belt force vs. time

  Seat belt playout vs. time

B. Two copies of high speed film delivered with Final Test Report

A2.3 DUMMY POSITIONING AND MEASUREMENTS – Anthropomorphic Test Dummy (not furnished by the Government)(Amending Section 10.8 Part 572B Test Dummies) - Seat a Part 572E (Hybrid III) test dummy in the driver location and measure its seating according to Appendix C and Pages 75-83 of Test Procedure TP-208-09 dated March 15, 1993. An H-point machine not be furnished by the Government. An un-instrumented Part 572B Test Dummy shall be placed in the front passenger location as ballast.

A2.4 DATA TAPES FOR ALL STANDARDS – NHTSA Tape Reference Guide dated August 1993.

973343

Relationships Between Passenger Car Seat Back Strength and Occupant Injury Severity in Rear End Collisions: Field and Laboratory Studies

P. Prasad, A. Kim, and D.P.V. Weerappuli
Ford Motor Company

V. Roberts
Institute for Product Safety

D. Schneider
Biokinetic Engineering, Inc.

ABSTRACT

Recent design characteristic changes in a small segment of production passenger car front seats have focused attention on the influence of seat back strength on occupant kinematics and potentially injurious loads placed on occupants during rear end collisions.

The National Accident Sampling Study database from the years of 1980 to 1993 was interrogated to determine the relationship between vehicle change in velocity, and the nature and severity of injuries sustained by passengers occupying those seats in rear end collisions.

The results of the NASS data analysis show that the yielding seats in most current automobiles perform well as a passive restraint system. When the yielding passenger car seats are compared to the stiffer seat/cab, the passenger car seats offered improved protection. Additionally, the data indicate that the three point restraint system provides protection and restraint for front seat occupants in rear impact.

To determine what effect seat back stiffness and other seat characteristics have on occupant responses, several seats were evaluated in dynamic Hyge rear impact sled simulations. Two types of production seats, with yielding seat backs, were modified to obtain higher seat back stiffness characteristics. Other production seats were chosen for various design features (i.e. head restraint and seat geometry, recliner systems, integrated seat belts, etc.), as well as seat back stiffness. All the seats were tested in a rigid environment with a restrained Hybrid III dummy. Seats A, A-M, and A-R, were tested at ΔVs of 9, 16, and 24 kph; all the other seats were tested at these speeds and also at a 40 kph ΔV.

The results of the sled tests indicated that stiffer seats do not have any consistent advantages over yielding seats for the complete range of speeds tested. When the production seats were modified to obtain higher seat back stiffness, higher responses were seen at the neck, thoracic spine, and lumbar spine. The other, stiffer production seats of different design also showed no consistent improvement in occupant protection across the speed range. Even a seat with a head restraint located closer to the center of gravity of the occupant's head did not perform better than the baseline seat. Throughout the testing, the most sensitive response to seat design and crash severity was the lower neck extension moment.

INTRODUCTION

Collisions that involve forces directed from the rear of a vehicle have received a revival of research interest in recent years. The recent research involves the measurement of the dynamic response of the head/neck complex, the mechanism of soft tissue neck injuries and injury mitigation concepts. The majority of the new research is aimed at the identification and mitigation of low level, AIS 1, neck injuries that are generally associated with low speed rear impacts. However, rear impacts also occur at higher speeds in which serious neck injuries can occur. Analysis of accident data in the US and UK [33] show that the incidence of AIS 3+ neck injuries in rear impact is low, 1%. In an attempt to further reduce the incidence of these serious neck injuries, stiffer seat backs have been suggested [2],but the effect of increased seat back stiffness on the predominant soft tissue neck injuries has not been studied.

The goal of this paper is to present the results of an experimental study aimed at determining the effect of seat back stiffness on the dynamic response of front seat vehicle occupants in a range of impact velocities covering low speed to high speed rear impact collisions. The first part of this paper reports on a literature survey conducted to identify past studies relative to the role of seat back stiffness in rear impact. The second part of this paper examines NASS data to extract overall effect of seat back stiffness on front seat occupant injuries, and the third part outlines and discusses the experimental study performed by the authors of this paper.

LITERATURE SURVEY

The impact response of an occupant exposed to a rear-end collision in the struck vehicle has been a topic of discussion in the scientific literature for over 30 years. Investigations have focused principally on two classes of occupant response and associated injury risk. At lower impact severities, the so called "whiplash" injury syndrome can be produced consisting of soft tissue injuries to the head and neck. At significantly higher levels of impact severity, often associated with seat back deformation, occupants can be exposed to more severe cervical spine injuries. The real world incidence of these severe injuries in rear impact is low, approximately 1% [33]. These two injury syndromes have been studied predominantly in independent research investigations, although they are closely linked by the impact response of the seat structure and the safety belt restraint system which dictates the nature of the crash forces imparted to the body in either crash severity. This review of the work done to date is limited in scope and will be divided into two sections. The first topic deals with whiplash injuries. This is followed by a discussion of the research describing the more severe rear end collision characteristics.

WHIPLASH EXPOSURE - The first class of injury commonly reported by early epidemiological investigators was the so called "whiplash" injury to the neck produced by relative motion between the head and the upper torso during a rear impact [3,4,5,6,7]. These injuries include muscle and ligamentous strain producing transient cervical pain, symptoms of headache and concussion, and injury to the intervertebral disk infrequently requiring cervical fusion. Many automotive seat designs during this period lacked a means of support for the head and neck during the collision, allowing extension/flexion movement of the neck which was assumed to cause the whiplash injury pattern. Following the introduction of head support in seating systems, further field studies continued to document the production of inertial injuries to the head and neck in low speed collisions in spite of the design change.

In a study of over 339,000 accidents, Nygren *et al* [8] found that 10% of those complaining of pain after a rear-end collision had remaining problems five years after the accident, while fitted headrests had little effect in preventing whiplash injuries. An in-depth study of 33 occupants exposed to a rear end collision was conducted by Olsson *et al* [9]. The duration of neck symptoms was found to be related to the estimated horizontal gap between the head and headrest. The sensitivity of this type of injury to the relationship of occupant and seat structure was recently reported by Viano and Gargan [10]. In a study of over 1900 vehicles he found that 90% of occupants had a less than optimum relationship to the headrest with respect to vertical elevation and horizontal space between the head and headrest.

Quantitative investigations of the whiplash phenomena have also been conducted on a variety of test surrogates and human volunteers. These studies have provided data on both the injury mechanisms and occupant kinematics and kinetics associated with rear end collisions. Ommaya *et al* [11] exposed primates to "whiplash" inertial loading of the head and neck and produced a stunned or dazed condition of the test subjects simulating cerebral concussion. Similar work by Martinez [12] utilizing Belgian hares related the production of cranial and cervical pathology to a whiplash inertial acceleration. These and similar studies allowed the physiologic mechanism of these injuries to be investigated using a variety of species. The relationship of whiplash type head and neck injuries to the human in an automotive environment requires quantification of the impact conditions present in a rear end collision, including the mechanical response of the occupant's head and neck. To this end, Mertz and Patrick [13] conducted a series of crash simulator tests at speeds of 16 and 37 kph using anthropomorphic dummies (ATDs), human cadavers, and a volunteer. A means of analytically evaluating the neck reaction forces was developed. Test results indicated that the dynamically equivalent torque, due to muscle forces and cervical spine distortions calculated at the occipital condyles of the test subjects, is the major factor in producing neck injury. The same authors [14] expanded this investigation to the human injury threshold and observed that dynamic torques at the occipital condyles in extension and flexion of 47 and 88 Nm, respectively, could be tolerated. Torque versus angular position of the head response envelopes were proposed for ATD mechanical necks when in dynamic extension and flexion.

A complete understanding of the various injury mechanisms that may be related to a rear impact whiplash exposure is not known. Consistent with this lack of knowledge regarding the injuries themselves is a similar void in completely understanding the interaction of the occupant and the seat in such exposures. Numerous human volunteer studies have been undertaken to relate the occupant kinematics to the supporting seat structures and to define the tolerances in rear impact collisions. McConnell *et al* [15] exposed four male volunteers to rear impact ΔVs of 3 to 8 kph in four different production vehicles. Results of these low speed test series suggest a compression-tension neck injury mechanism and concluded that this mechanism was distinct from the classic hyperextension phenomena. A postulated relationship between lower jaw movement during whiplash and the production of temporomandibular joint injury was suggested by Schneider *et al* [16]. Szabo *et al* [17] did not find any substantiation for this claim in an experimental study of 16 kph car-to-car rear impacts with six different test subjects. These individuals, up to 58 years of age, suffered no ill effects in this series employing subcompact vehicles as the struck car. X-ray analysis of human head/neck motion was performed by Matsushita *et al* [18]. Results of this study suggest that the most significant factor in head/neck kinetic response is the initial curvature of the cervical and thoracic spine. Further evidence of the relationship between the occupant and seat was confirmed by Viano and Gargan [10] who demonstrated in Hyge sled tests using a Hybrid III ATD that non-optimum head to headrest geometry produced higher values of neck extension when compared to the proper positioning. McConnell *et al* [19] conducted a series of human volunteer tests. Fourteen car-to-car tests involving six male test subjects were performed. Although hyperextension limits were not reached, subjects exposed multiple times did exhibit typical whiplash symptoms suggesting that the injuries may have occurred before the headrest was contacted.

For the last decade, there has been an increase in European interest in the low speed collision phenomenon. In an annual sample of Danish cervical spine injured patients, Juhl and Seerup [20] found that 46% of spinal lesions arose from traffic collisions. Thomas *et al* [21] examined characteristics of rear-end impacts occurring in France and observed that 70% of all collisions produce a struck vehicle ΔV less than 25 kph, and that cervical pain risk is high for females when the seat remains intact at a low ΔV. A simulation of low speed rear collisions was carried out by Svensson *et al* [1] using a new rear impact dummy neck (RID) mounted on a Hybrid III ATD. Test results showed that head-neck motion is influenced by the stiffness and elasticity of the backrest as well as the properties of the head restraint. An aggravation of the whiplash motion produced by the elastic rebound of the seat was also observed. Svensson *et al* [2] later used the same RID neck to evaluate modifications to the seat back and head restraint, virtually eliminating neck extension with the proper choice of seat properties.

HIGH SPEED REAR END COLLISION - The second class of occupant response and attendant injury exposure is encountered in the high speed rear end collision where the resulting ΔV of the struck vehicle is sufficient to produce gross deformation of the occupant's seat back and significant rotation of the head, neck, and torso. These collisions expose the occupants to an entirely different field of accelerations, accompanying forces, and potential injuries. Again, the supporting seat structure, including the head restraint, plays an important role in mitigating injury.

Severy *et al* [22] conducted car-to-car collisions at 32 and 48 kph with ATDs and concluded that head and chest accelerations were higher if the rear seated occupant was postured leaning forward, thus separating the ATD from the relatively rigid seat back. In this same series, front seated occupants had reduced head and chest accelerations resulting from the yielding front seat back even if initially separated from the seat back structure. The value of a headrest in preventing head and torso differential motion was also quantified. This work was continued [23] in an evaluation of whiplash protective devices. Some degradation of head support performance was observed as the relative impact speed increased for all designs considered.

Berton [24] presented the results of a study relating seat back rotation, head restraint position, and rear end collision speed on crash dummy response. Velocity changes from 16 to 48 kph were employed in crash simulator tests and up to 80 kph closing speeds in car-to-car crashes. Berton observed an increase in head vertical accelerations associated with a rigid seat back and a reduction in head response if the seat back rotated rearward during the collision. He recommended that seat back structure deformation should be plastic or damped to minimize the forward acceleration of the occupant's torso and therefore reduce the amount of cervical spine extension. Employing a lumped parameter model of head/neck motion in a rear end collision, Martinez [25] recognized the reduction in occupant acceleration resulting from a yielding seat back structure.

Human cadaver surrogates were used by Clemens and Burow [26] and Hu *et al* [27] to study the nature of neck injuries sustained in 19-25 kph velocity changes in rear-end collisions. Disc disruption, ligamentous tears, and fracture dislocations were observed by Clemens in 53 exposures where the torso was constrained from moving. Hu also produced similar AIS 3 injury results in a smaller sample of six subjects supported by rigid and yielding seat backs. These were followed by new research which explored the relationship between the vehicle and seat structures, and the occupant's response in a rear end collision. Williams and McKenzie [28] analytically determined that increased seat back stiffness produced greater loads on the cervical spine. Mechanical testing of seats by Strother and James [29] documented the characteristics of production seating systems, while additional analysis of accident field data with a largely unbelted population showed that existing seat designs were providing a high level of occupant protection at moderate to severe levels of impact severity. Later, Warner *et al* [30] contrasted the consequences of a rigid versus yielding seat back design on occupant safety. Data were presented showing the stiffness of production vehicle seat backs had not changed significantly over a three decade period and elastic rebound associated with stiffer seat designs can produce subsequent injuries. The out-of-position occupant was identified as at risk when combined with a seat back of elevated stiffness. Strother *et al* [31] performed 30 rear impact tests with a Hybrid III ATD comparing yielding and rigid seat back designs at speeds from 8 to 32 kph. Yielding seat designs produced lower injury measures for all out-of-position subjects whether belted or unbelted. James *et al* [32] interrogated the NASS and FARS databases and examined the societal harm of various injury mechanisms associated with current seat back and restraint system designs. Harm related to occupant ejection or secondary vehicle contact was found to be minimal, while restraint usage had a substantial injury reducing effect. Parkin *et al* [33] analyzed 323 accidents in the United Kingdom and observed that plastic yielding of the seat back is beneficial in a rear end collision, producing a lower incident of AIS 1 injuries when compared with rigid seat backs. In this sample, less than 1% of occupants had injuries of AIS 3+. Severe head and neck injuries, AIS 3+, also comprised less than 1% of exposed occupants in the Thomas *et al* study [21].

Severe injuries of the cervical spine can occur in rear collisions with a high ΔV for the struck vehicle. ΔV of the struck vehicle has long been recognized as an important variable in describing the severity of impact for the struck vehicle [34]. Inertial loading of the neck occurs when the impact conditions allow the torso to be supported and the head free to move rearward. The neck loads increase with additional increments of ΔV. The existence of various types of inertially created serious cervical injuries has been known and reported extensively by Portnoy *et al* [35], Moffatt *et al* [36], and Huelke *et al* [37]. Other types of head and cervical trauma can be produced through contact loading of the head/neck/torso system. In rear collisions these impact circumstances can arise in severe crashes where there is movement of an unrestrained occupant from the seated position, permitting contact with deforming vehicle structures, or a significant lateral component of velocity moves the occupant into a vehicle side structure resulting in an injury producing contact. Aspects of these crash kinematics and injuries were discussed by Saczalski *et al* [38] in a presentation of 23 anecdotal case studies in support of increasing the overall stiffness of seat back design. No data were presented to evaluate the consequences of such a design change on injury production at the more prevalent low speed impact condition. In a report by Marriner [39], several analyses of real world accident data have concluded that the incidence of serious or greater injury to the head/neck complex has been, and continues to be an infrequent consequence of rear end collision.

Strother and James [29] interrogated the 1970-1985 FARS and NCSS databases. Fatalities resulting from rear impacts were significantly under represented when compared to frontal and side modes of collision. Injury rates were also found to be significantly lower when compared to the other collision types, leading the authors to conclude that suggested changes in seat design would probably be of limited benefit for occupant protection. In 1991, James *et al* [32] followed up on this study and applied the societal harm concept [40] to rear impact injuries. Injuries resulting from contact with rear structures represented less than three percent of total harm in rear collisions, and a similar quantitative representation was found for unrestrained occupants who were totally ejected. Digges *et al* [41], in a review of 1979-1986 NASS data, concluded that for rear impacts, non-contact injuries produced the largest portion of injury harm. Contact injuries in this study were produced predominantly by the seat itself and frontal components in the car. The 1988-1990 NASS data were examined for specific crash characteristics associated with seat performance for rear impacts with ΔVs between 48 and 62 kph. 92% of the observed injuries had an AIS less than or equal to 2, and the only injury with an AIS greater than 3 was a burn. A second subset of 72 occupants exposed to ΔVs greater than 48 kph also indicated that 7 of 16 restrained occupants in yielding seats received injuries from frontal components. Three of seventeen occupants in non-yielding seats sustained cervical fractures, all at ΔVs in excess of 53 kph. No cervical fractures were observed in seats that yielded.

The reader is referred to the publications of Strother and James [29], James *et al* [32], and Warner *et al* [30], as well the Comments to Docket 89 20-N01-014 [42] for additional bibliographic information and an expanded discussion of seat back design considerations.

NASS DATA REVIEW

To define the nature and extent of the injuries associated with rear impact collisions, a subset of the NASS data, the Vehicles Crashworthiness Database, was employed [3,4]. This database includes only those accidents contained within the National Accident Sampling System from 1980-1994 wherein a valid ΔV was obtained by the investigator and injury data for the vehicle occupants were available. Vehicles that were involved in rollovers as initial or subsequent events were excluded.

When rear impacts were filtered from the database, 4,424 of the vehicles containing 7,518 occupants were available for analysis. These constituted 11% of the available vehicles and 11.7% of the occupants in the planar dataset.

Figure 1 shows the cumulative distribution by ΔV of all of the vehicles in the database where the Principal Direction of Force (PDOF) was 5,6,7 o' clock. The distribution of the accidents indicates that 50% of all rear impact collisions involve a ΔV of 21 kph or less and that 94% of all such collisions occur at ΔV s of 40 kph or less. The data also show that 86 % of all rear end collisions occur at velocity changes less than 26-32 kph, the range generally produced in the struck vehicle in the crash tests performed for FMVSS 301.

When we examine the distribution of injuries occurring with the occupants of the front, outboard seating positions of those passenger cars contained in the database, we find the distribution of injuries to be that shown in Figure 2. Figure 2 illustrates the distribution, as a function of ΔV, for occupants of the front outboard seating positions of automobiles whose respective maximum AIS = 0, 1 or 2, and 3 or greater. The data show that, for those occupants who received maximum injuries of AIS 3 or more, less than 1% of all occupants in rear impacts received injuries with changes in velocity of the struck vehicle in excess of 40 kph.

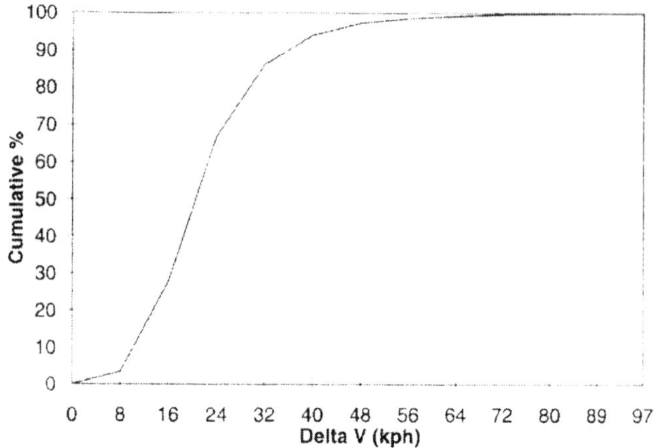

Figure 1. - Distribution of vehicles vs. ΔV for rear impacts (n=4,424).

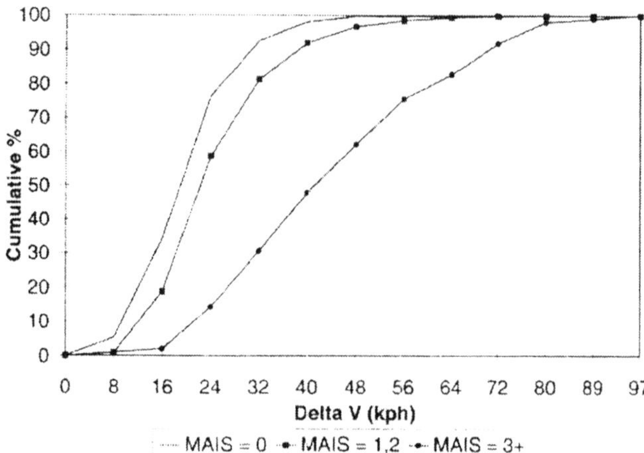

Figure 2. - Distribution of injured occupants vs. ΔV for rear impacts: MAIS = 0 (n=2,161), MAIS = 1 or 2 (n=3,124), and MAIS = 3+ (n=98).

It has been suggested that the study of injuries to the occupants of pickup trucks in rear impacts would show the efficacy of stiff seats since the close proximity of the back of the cab to the seat backs should prevent seat back deflection and produce a seat that is relatively more rigid than passenger car seats. Figure 3 examines this hypothesis. Figure 3 compares the injuries to the occupants of passenger cars and pickup trucks in rear impact collisions. The potential for minor to moderate injury when seated in a pickup truck is greater, for any change in velocity, than when seated in the front seats of passenger cars. The same relationship holds true whether the occupants are restrained or unrestrained.

Automobile occupants generally have a lower risk of injury when exposed to a rear end collision of the same severity as one which produces minor to moderate injuries in occupants of pickup trucks. This may indicate that the movement toward stiffer seat backs could very well increase rather than decrease the injuries that they are trying to avoid.

Similarly, it has also been suggested that the deflection of the seat back will allow the occupants of those seats to be ejected from the vehicle through the rear window. A review of all cases of ejection of front seated occupants in rear impacts who were utilizing three point restraints found that only one, out of 2,224 occupants, was ejected through the rear window. That occupant was originally seated in the front seat of a pickup truck.

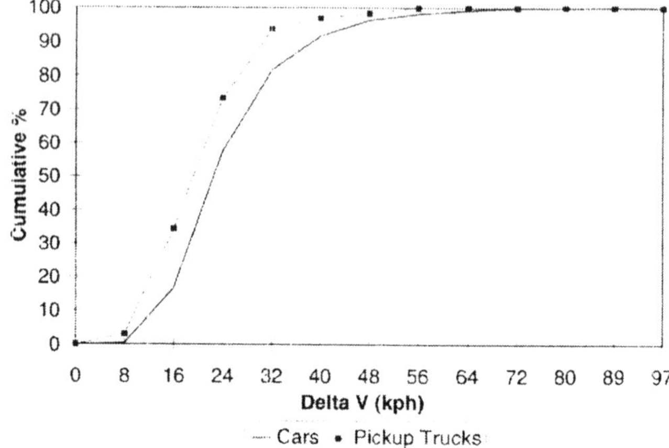

Figure 3. - Distribution of occupants with AIS = 0, 1, or 2, vs. ΔV for rear impacts: automobiles (n=1,318) and pickup trucks (n=64).

Finally it is worth examining the performance of current automotive seat backs in rear impacts with the performance of three-point active restraints in frontal collisions. Figure 4 compares the two impact environments. It illustrates the distribution of drivers and right front seat passengers in frontal and rear impacts with the same maximum AIS, three or greater. The difference in the two groups, other than the direction of impact, is the use or non-use of the three point restraint system. The occupants in the frontal collisions were using the three-point active restraints while those in the rear impacts were unrestrained. The distribution of the two groups is essentially the same, showing yielding seat backs provide a similar amount of protection to the unrestrained occupant in rear impacts that a three-point belt system provides in frontal impacts.

DYNAMIC HYGE SLED TESTING

A total of three series of Hyge sled tests were conducted to determine the effect that seat back stiffness, and other factors, have on occupant kinematics and responses in rear impacts. In the first two test series, the only difference between the tested seat configurations was seat back stiffness; all other factors were held constant. In the third series, production front seats of various designs and stiffnesses were tested to see if any of them provided the occupant with a

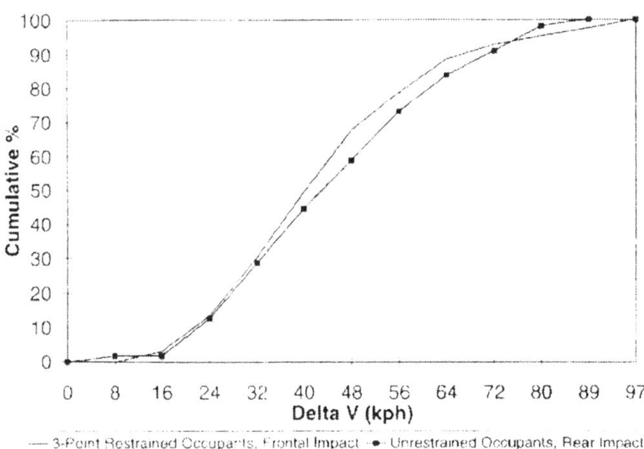

Figure 4. - Distribution of occupants with MAIS = 3+ vs. ΔV: three-point belt restrained occupants in frontal impacts (n=552) and unrestrained occupants in rear impacts (n=56).

significant advantage in rear impacts. The same test protocol was followed in all three series.

TEST METHOD - For all three series, the seats were installed in either a rigidized car buck or on a rigid platform buck, eliminating any performance differences due to structural characteristics or floor deformation. The height of the platform buck's floor, with respect to the seat, was adjusted to approximate the real world geometry for each seat type. All the seats were placed in mid-seat track position, except where noted. If the seat cushion was adjustable, it was placed in the full down position and the seat back was set at an angle between 20° and 21°. Adjustable head restraints were placed in their upper most vertical setting, and if they rotated, in their most rearward position. Rear seats were not installed to allow for maximum front seat back deflection. If a seat showed any signs of permanent deformation or damage, it was replaced before the next test. The Hyge sled acceleration pulses used were derived from a 80.5 kph car-to-car rear impact and scaled to give the required ΔVs (Fig.5).

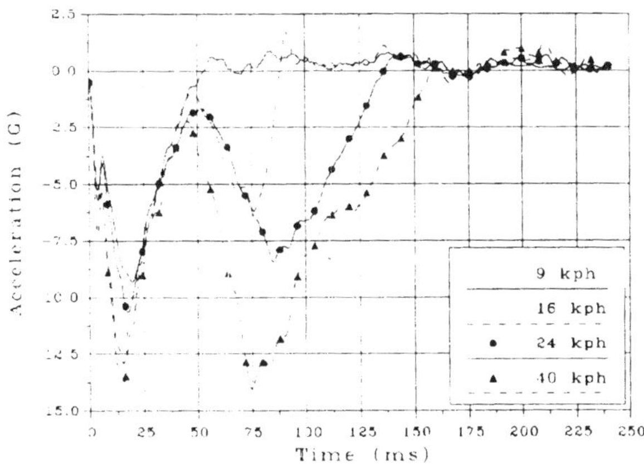

Figure 5. - Hyge sled acceleration pulses.

Every seat was tested with a Part 572(b) Hybrid III dummy whose suitability for rear impact testing has been established by Prasad *et al* [46]. The Hybrid III dummies were instrumented with head, chest, and pelvis accelerometers; upper and lower neck load cells; and lumbar spine load cells. If available, thoracic spine load cells were also used. All the instrumentation was oriented according to SAE J211 convention. The Hybrid III dummies were consistently placed in the normal seating position for each type of seat. All the ATDs were restrained with the corresponding seat belt system except where noted.

RESULTS:

Series 1 - In order to see the effect seat back stiffness has in rear impacts without the complication of confounding factors, three versions of a high volume, American production front bucket seat were tested at ΔVs of 9,16, and 24 kph. For these configurations, all the factors, such as head restraint and seat geometry, cushion stiffness, etc., were identical; only the seat back stiffness was changed. Seat A, the baseline, is the standard, unmodified production seat with a yielding, deformable seat back and a single, outboard recliner (Fig. 6). Seat A-M is a version of the baseline seat which has been modified to produce a higher seat back stiffness. Figure 7 shows the relative seat back stiffness functions of seats A and A-M as measured in an upper-bar test. The functions were obtained by applying a rearward force to the seat back frame through a rigid bar approximately 0.6 m above the seating reference point and measuring the bar's displacement. The increase in seat back rigidity was obtained by welding a steel bracket to the seat back and seat cushion frames on the inboard side of the baseline seat. Seat A-R is the rigid configuration of the base seat. For A-R, a metal bar, which contacted the seat frame below shoulder level, was placed behind the seat and rigidly attached to the sled buck, preventing any rearward deflection or movement of the seat back (Fig. 8). For all three seat versions, the 50th% Hybrid III was restrained with the corresponding three-point seat belt system.

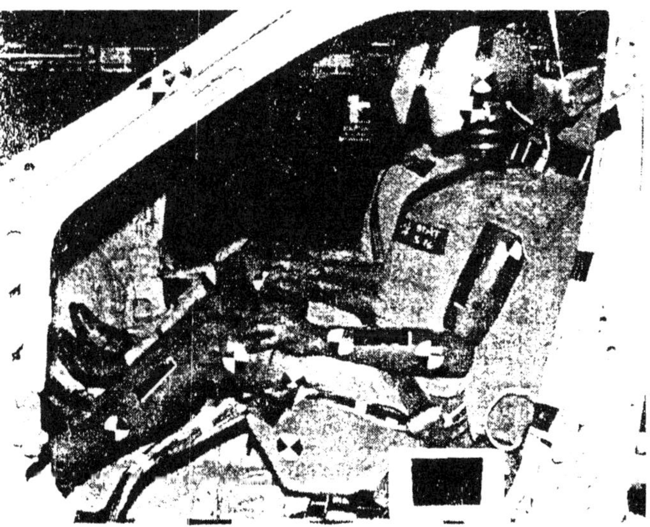

Figure 6. - Hybrid III in seat A.

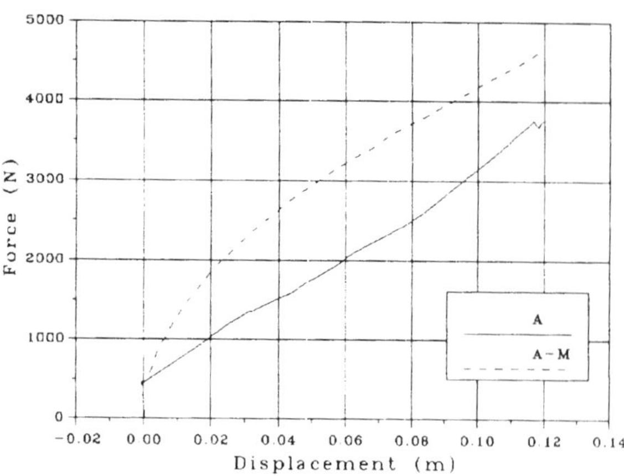

Figure 7. - Seat back force-deflection curves of seats A and A-M.

Figure 8. - Hybrid III in seat A-R, rigid seat modification.

Figure 9 shows the normalized peak magnitudes of the pertinent neck responses for seats A, A-M, and A-R. Each response was normalized with respect to its Injury Assessment Reference Value (IARV), listed in Table 1, therefore, any value greater than one indicates that there is a risk of severe neck injury to a small portion of the population. The upper neck IARVs were previously established by Mertz [45], and for the shear and tensile responses of the lower neck, the corresponding upper neck IARVs were used. For the lower neck extension moment, an IARV of 154 Nm was chosen [46].

Table 1. - Injury Assessment Reference Values (IARV)

Response	IARV
Upper Neck Fx	+/- 3100 N
Upper Neck Fz	+3300 / - 4003 N
-Corrected Upper Neck My	-57 Nm
Lower Neck Fx	+/- 3100 N
Lower Neck Fz	+3300 / - 4003 N
-Lower Neck My	-154 Nm

As Figure 9 shows, at a ΔV of 9 kph, none of the seats have any neck responses which approach a possibly injurious level. Because the ΔV is so low, most of the occupant's energy is absorbed by the seat cushions, not by seat back deflection, so there is not much differentiation between the responses with these three seat configurations. Compared to the baseline, A, A-M has slightly lower tensile peak magnitudes at the upper and lower neck and A-R has slightly lower peaks for the lower neck tension force and corrected upper neck extension moment (moment at the occipital condyles), and a slightly higher lower neck shear force peak. The peak magnitudes of the ATDs' responses for the 9 kph ΔV are listed in Appendix A - Table 1. The lower neck extension moments, which are not significantly different for the three seat variations at this speed, also correspond to the moment obtained with a seat with an average Neck Injury Factor score (NIF), defined by Eichberger *et al* [47], of one, at a similar speed as discussed in [46].

At the higher ΔV of 16 kph, the differences in seat back stiffness between the three configurations affect the Hybrid III's responses, as can be seen in Figure 10. For the neck responses of concern, the baseline seat A, which has the lowest seat back stiffness as indicated by dynamic deflection, has the best overall performance. Seat A-M has three responses with higher peaks (the corrected upper neck extension moment, the lower neck shear force, and the lower neck extension moment) than seat A, and only one response, the lower neck tensile force, with a lower peak. The completely rigid seat, A-R, has four higher responses (the upper and lower neck tensile forces, the lower neck shear force, and the lower neck extension moment) and only one lower response, the corrected upper neck extension moment. The lower neck extension moment reflects the trend of increasing seat back stiffness. This, in turn, indicates that as the seat back stiffness of the base seat is increased, the NIF rating of the seat will also increase [46]. Even though none of these channels indicate that there is a risk of severe neck injury, the higher neck forces and moments seen with seats A-M and A-R also indicate that as seat back stiffness increases, the risk of low severity injuries, such as whiplash, also increases. The ATD's responses are listed in Appendix A - Table 2. It is interesting to note that increasing the seat back stiffness also resulted in some significantly increased thoracic and lumbar spine loads and torques.

Figure 11 clearly illustrates that, for a 24 kph ΔV, the overall neck performance obtained when the Hybrid III was tested with the seat with the lowest seat back stiffness, A, is better than those obtained with the stiffer seats. Seats A-M and A-R both have significantly higher lower neck shear forces and noticeably higher lower neck extension moments than seat A. In addition, the peak corrected upper neck extension moment measured with seat A-R is significantly higher than that measured with seat A although both peak values are fairly low. Again, none of these neck responses indicate that there is a risk of severe neck injuries, but seats A-M and A-R may

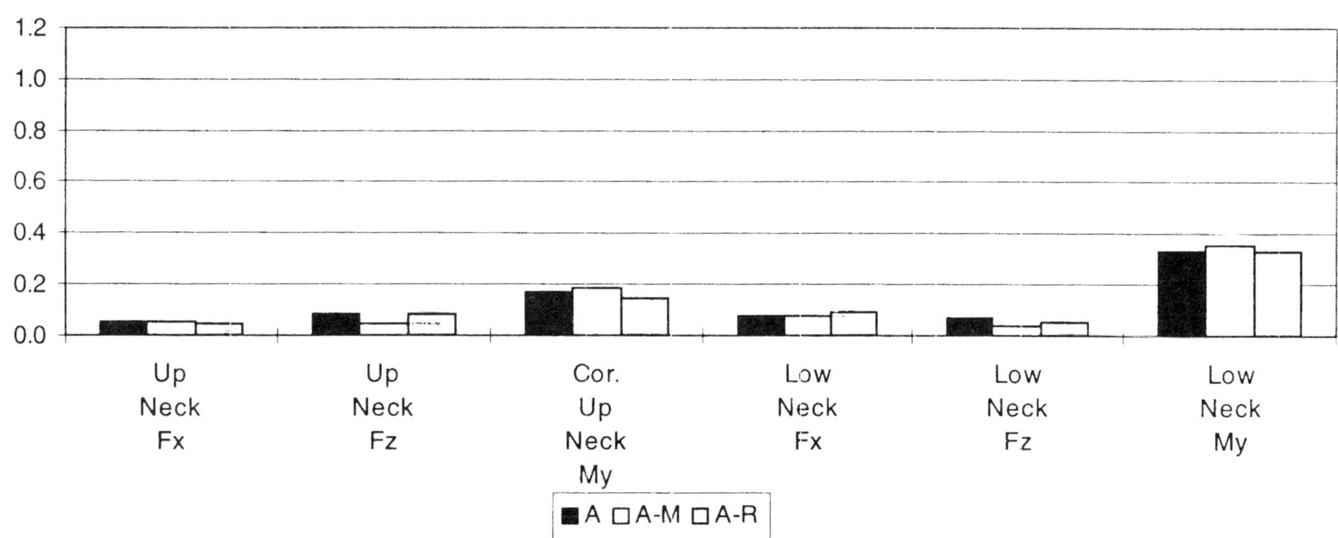

Figure 9. - Neck responses normalized with respect to IARV's: ΔV= 9 kph, seats A, A-M, and A-R.

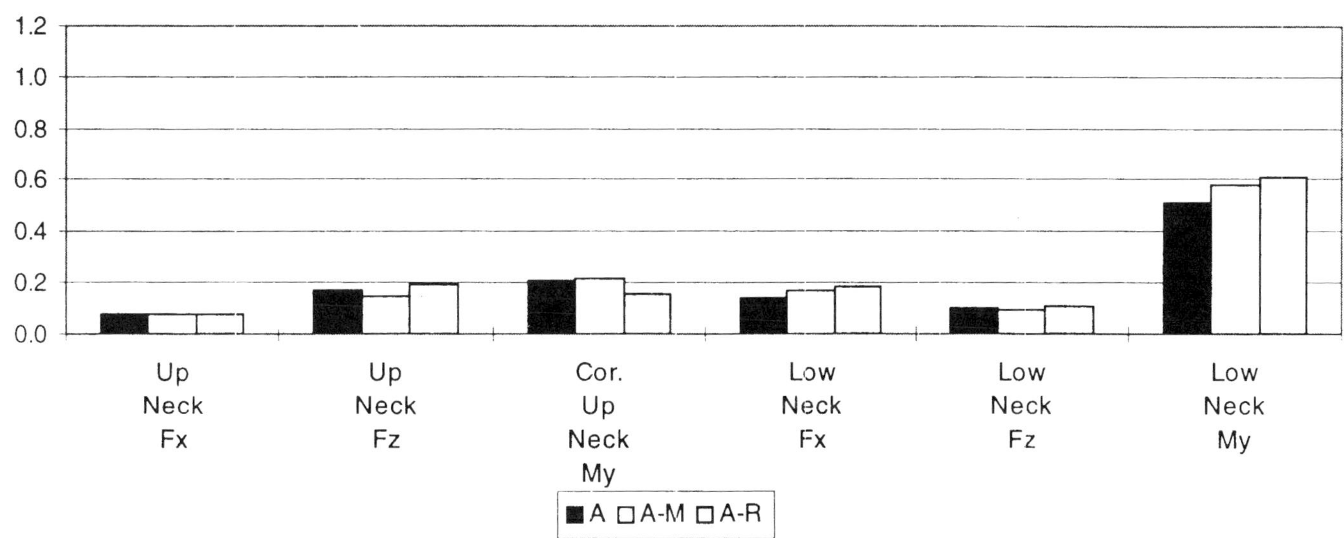

Figure 10. - Neck responses normalized with respect to IARV's: ΔV= 16 kph, seats A, A-M, and A-R.

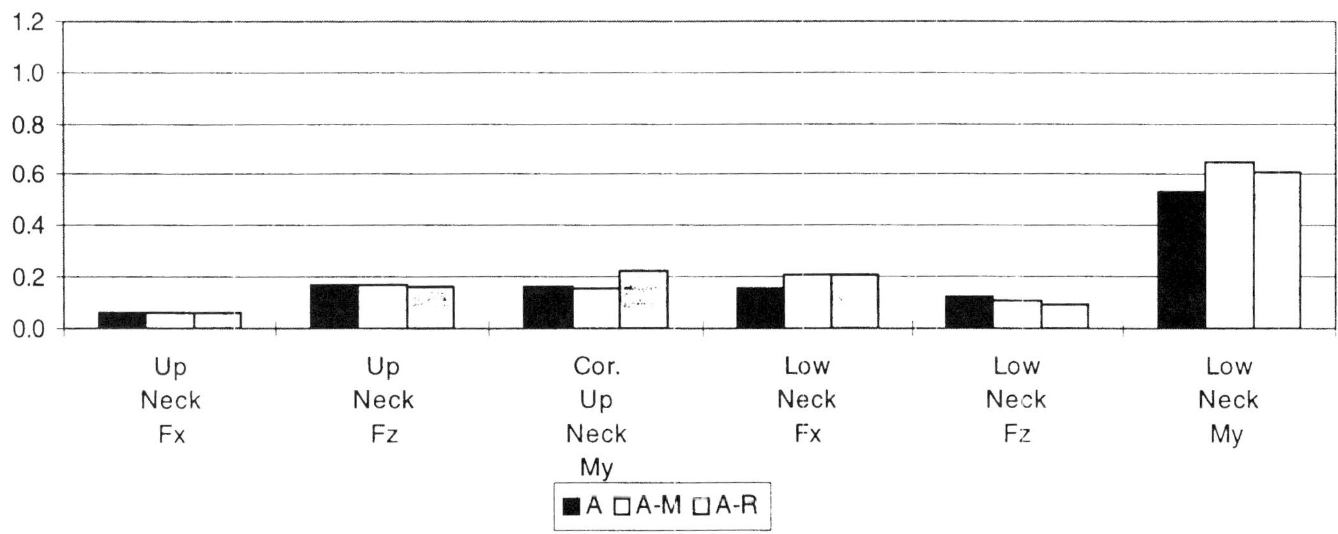

Figure 11. - Neck responses normalized with respect to IARV's: ΔV= 24 kph, seats A, A-M, and A-R.

present a higher risk of low severity neck injuries (whiplash), to the occupant than seat A. The ATD responses for this test speed, listed in Appendix A - Table 3, also show a significant increase in the thoracic and lumbar shear spine forces with the two stiffer seat configurations.

Series 1 clearly shows that for a range of speeds up to a 24 kph ΔV, which covers 67% of all rear impacts reported in NASS (Fig. 1), that as seat back rigidity increases, the occupant's neck loads and moments also increase. Even though none of these forces or moments indicate a risk of severe neck injury, the higher magnitudes indicate a higher risk of low severity, soft tissue neck injuries such as whiplash.

Series 2 - A second, high volume, American production front bucket seat with a single, outboard recliner was also studied to determine the effect of seat back stiffness on occupants under more severe conditions (Fig. 12). This second seat was tested not only at the lower speeds (ΔVs of 9,16, and 24 kph), but also at a higher ΔV of 40 kph, covering 94% of rear impact accidents. Only two variations of this seat were tested: seat B, the standard, production seat, and seat B-R, a rigid version of the seat. The same external modification which was used to make seat A-R rigid was also used for seat B-R. All other factors were held constant. The corresponding belt system for these seats consisted of a passive shoulder and active lap belt combination.

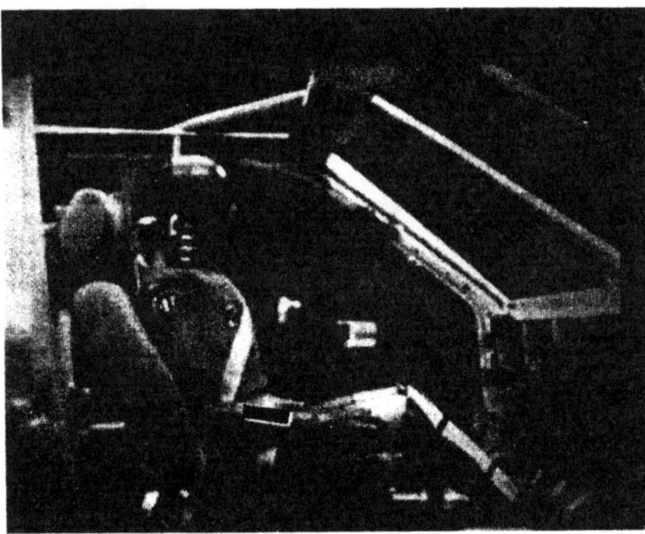

Figure 12. - Hybrid III in seat B.

A comparison of the pertinent neck responses obtained for the two different seat configurations are shown in Figures 14-16 and 18 for the respective ΔVs of 9,16, 24, and 40 kph. At the 9 kph ΔV, the similarity of the peak magnitudes of the responses indicates that compression of the head restraint and seat cushions is still the primary mechanism of energy absorption. The rigid seat, B-R, did have a higher corrected upper neck extension moment, but none of the scores indicate that there is a risk of severe neck injuries. Some of the Hybrid III lumbar and thoracic spine responses increased significantly with seat B-R in comparison to seat B (Appendix A - Table 4).

More differentiation is seen between the two seat configurations at the 16 kph ΔV (Fig. 15). Three of the Hybrid III's neck responses (the corrected upper neck extensions moment, lower neck tensile force, and lower neck extension moment), have slightly higher peaks with seat B-R than with seat B. There is no indication of a risk of severe neck injuries with either seat at this speed. The peak magnitudes of the negative thoracic shear and compressive lumbar forces are dramatically higher with the rigid seat, as shown in Appendix A -Table 5.

Figure 16 clearly shows that the yielding, deformable seat, B, mitigates neck loads and moments better than the rigid seat, B-R, at a 24 kph ΔV. The higher responses with B-R also indicate that the rigid seat presents the occupant with a greater risk of low severity neck injuries, such as whiplash, than the production seat. There were also significant increases in some of the thoracic and lumbar spine channels (Appendix A -Table 6).

At the highest ΔV, 40 kph, the lower neck extension moment measured with seat B-R exceeds the IARV of 154 Nm, indicating that there is a risk of severe neck injury to some portion of the population with this rigid seat (Fig. 18). The production seat, B, whose seat back has yielded completely, has no responses which exceed their respective IARVs and, in general, had lower responses than B-R. Plastic deformation of the seat back limited the occupant's neck loads and moments; and despite the large deflection, the restrained Hybrid III dummy was retained in the seat (Fig. 13). Significantly higher shear and compressive lumbar spine forces were also noted with the stiffer seat (Appendix - Table 7).

Figure 13. - Hybrid III in seat B: ΔV = 40 kph, @ 185 ms.

The results of series 2 clearly indicate that by deforming and absorbing energy, the production seat protects the occupant better than the rigid seat in rear impacts. Throughout the range of speeds tested, the rigid seat produced slightly higher neck responses than the production seat and at the highest speed, one of the responses exceeded its IARV. The production seat

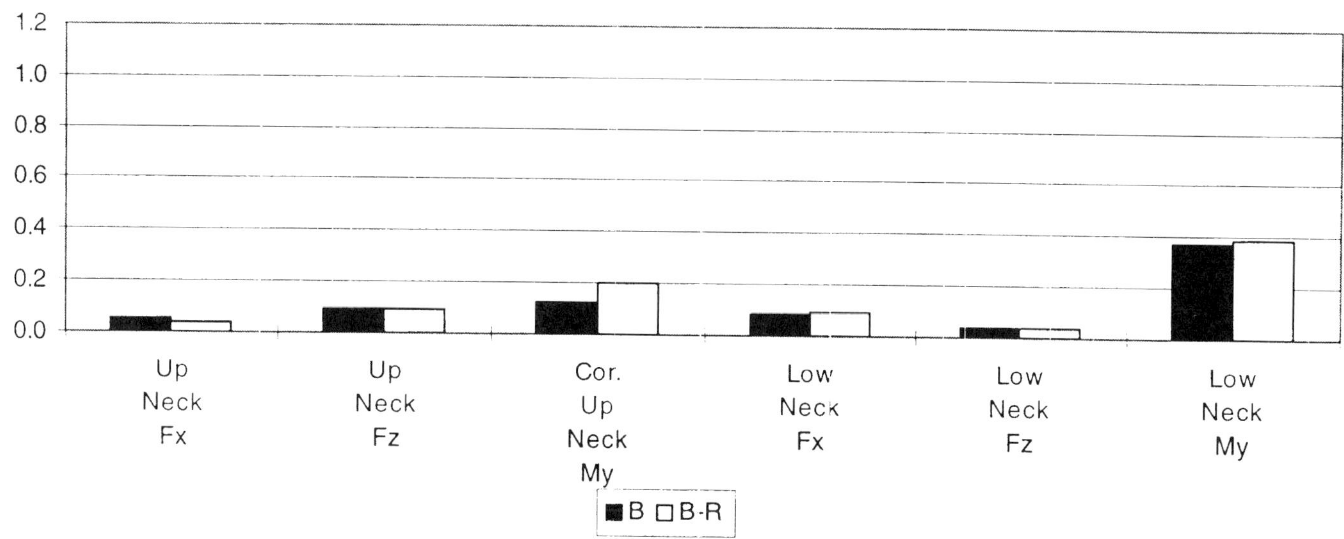

Figure 14. - Neck responses normalized with respect to IARV's: ΔV= 9 kph, seats B and B-R.

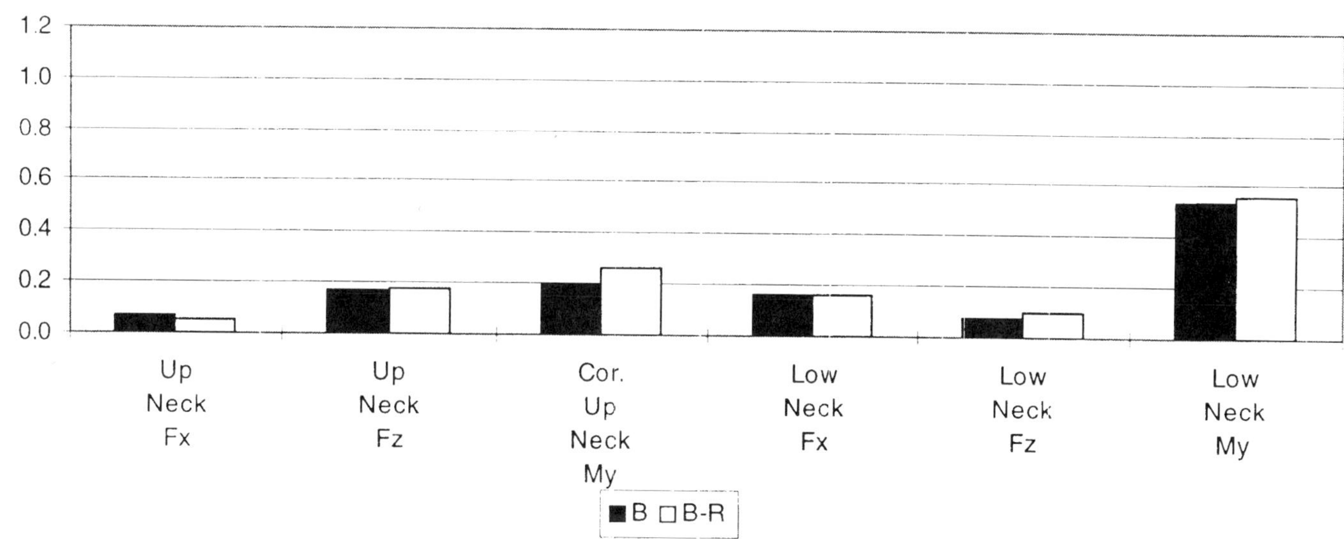

Figure 15. - Neck responses normalized with respect to IARV's: ΔV= 16 kph, seats B and B-R.

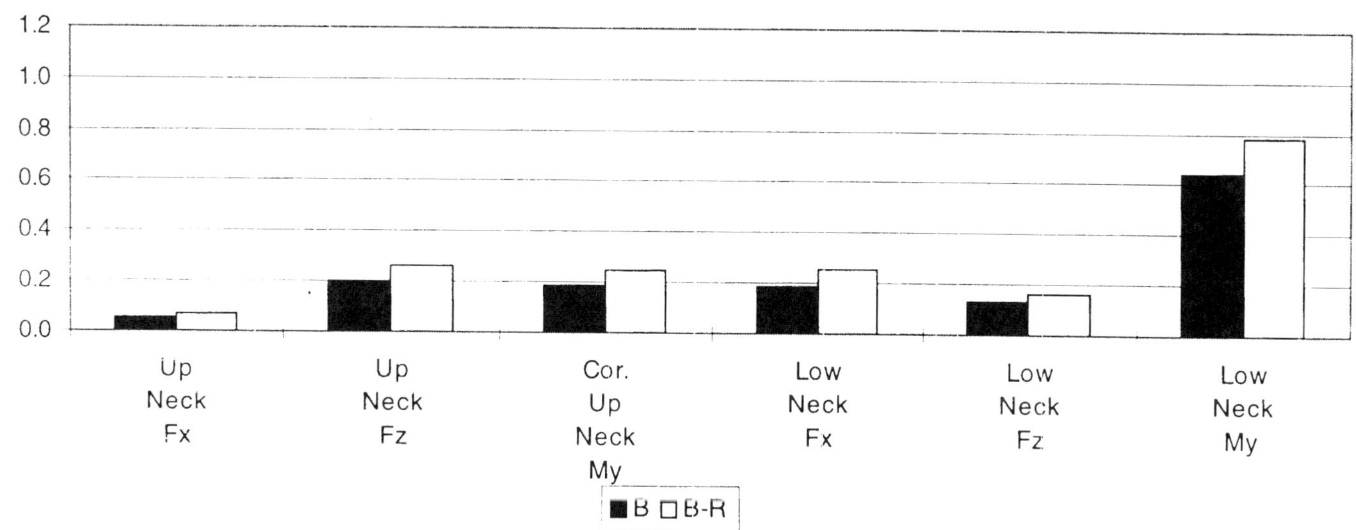

Figure 16. - Neck responses normalized with respect to IARV's: ΔV= 24 kph, seats B and B-R.

presents the occupants with lower risks of both low and high severity neck injuries.

Series 3 - Four different types of production front seats were subjected to the same simulated Hyge sled rear impacts for the ΔVs of 9, 16, 24, and 40 kph. These seats were chosen for various characteristics in their designs, such as head restraint geometry, cushion stiffness and geometry, seat back stiffness, and corresponding restraint systems, which differed from those of the production seats used in series 1 and 2. The Hybrid III responses obtained with these seats are compared to those obtained with seat B as a baseline (see Appendix A - Tables 4-7).

Seat C, shown in Figure 17, is a European, production front bucket seat whose fixed head restraint was given a "good" rating by the IIHS in comparison to the "poor" rating seat B received [48]. The seat back has a dual recliner system and a higher stiffness characteristic, as determined by dynamic deflection, than seat B. A three-point belt system from another vehicle was used to restrain the ATDs because the original three-point belt system for seat C was unavailable.

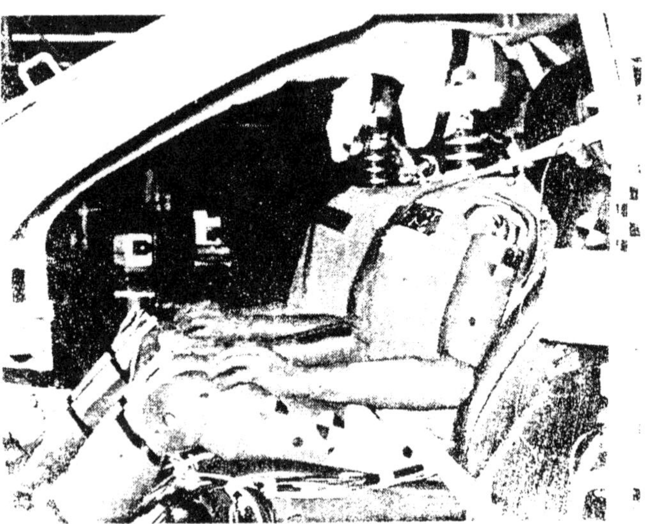

Figure 17. Hybrid III in seat.

At the lowest test speed, 9 kph ΔV, a comparison of the neck responses do not point out any major differences between the performances of seats B and C (Fig. 19). The significantly lower corrected upper neck extension moment obtained with seat C is balanced by the significantly higher lower neck shear force. The other four responses are very similar to those of seat B. The thoracic and lumbar spine forces and moments for both seats were also comparable.

A more noticeable difference between the two seats is seen at the 16 kph ΔV, Fig. 20. Seat C, which deflects about half as much as seat B dynamically, has four higher neck responses despite having a better ranked head restraint. The lower neck tensile force and extension moment are significantly higher than those in the baseline seat. Neither seat produces any neck response which indicates a risk of severe neck injury, but the increased magnitudes seen with seat C indicate that it may present the occupant with a higher risk of low severity neck injuries, i.e. whiplash.

At the higher ΔV of 24 kph, five of the six neck responses with seat C have higher peak magnitudes than with seat B (Fig. 21). None of the neck responses indicate that there is a risk of high severity neck injuries, but the stiffer seat, C, again has a higher risk of whiplash type injuries despite the difference in seat design. The large increases in some of the thoracic and lumbar spine responses seen with increasing seat rigidity in series 1 and 2 are not evident here.

At a ΔV of 40 kph, substantial yielding occurred in both seats B and C, but the deformation modes were different. For seat B, the seat back itself deformed about the seat pivot whereas with seat C, the front, outboard seat track bracket separated from its attachment to the sled. The neck responses are shown in Figure 22. The lower neck extension moment of seat C signifies a risk of serious neck injury to a portion of the population since the IARV is exceeded and the higher peaks of the other responses indicate that C presents a higher risk of low severity neck injuries than seat B.

Seat D, a captain's seat with a pedestal base and no seat track adjustment, is available for installation in customized American vans (Fig. 24). It should be noted that its vertically adjustable head restraint, in its upper most position, is much higher with respect to the Hybrid III's head than that of seat B. Seat D was chosen for this study not only because its different design, but also because of its reinforced seat back. A seat belt, which is connected to the seat frame, runs diagonally up the seat back from the inboard to the outboard side and attaches to the vehicle's structure via the D-ring. The purpose of this added restraint is to limit the seat back deflection in an impact situation.

With seat D at a ΔV of 9 kph, the three upper neck responses and the lower neck extension moment have higher ratios to their IARVs than with seat B (Fig. 23). At this low speed, this is the most noticeable deviation in performance from the baseline seat that is seen and a major reason for this was probably the increased cushion depth of seat D. There are also significant increases in some of the thoracic and lumbar spine responses.

At the next speed, a 16 kph ΔV, the peak upper and lower neck tensile forces increased significantly with seat D but, at the same time, the corrected upper neck extension moment and the lower neck shear force decreased compared to the baseline responses (Fig. 26). Several of the thoracic and lumbar spine responses also increased significantly while only one decreased. Neither of these seats present the occupant with a risk of serious neck injury at this speed.

In comparison to seat B, seat D has three lower neck responses (the lower neck shear force and the corrected upper neck and lower neck extension moments), and two higher responses (the upper and lower neck tensile forces), and may offer the occupant slightly more protection at a ΔV of 24 kph (Fig. 27). This cannot be attributed to the increased seat back rigidity

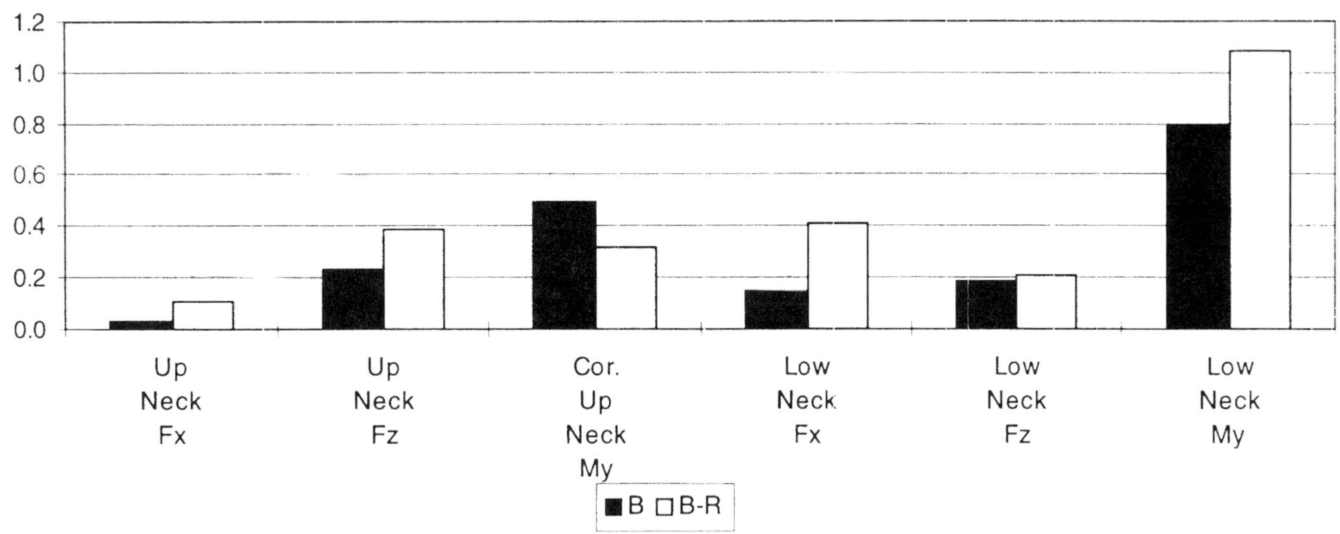

Figure 18. - Neck responses normalized with respect to IARV's: ΔV= 40 kph, seats B and B-R.

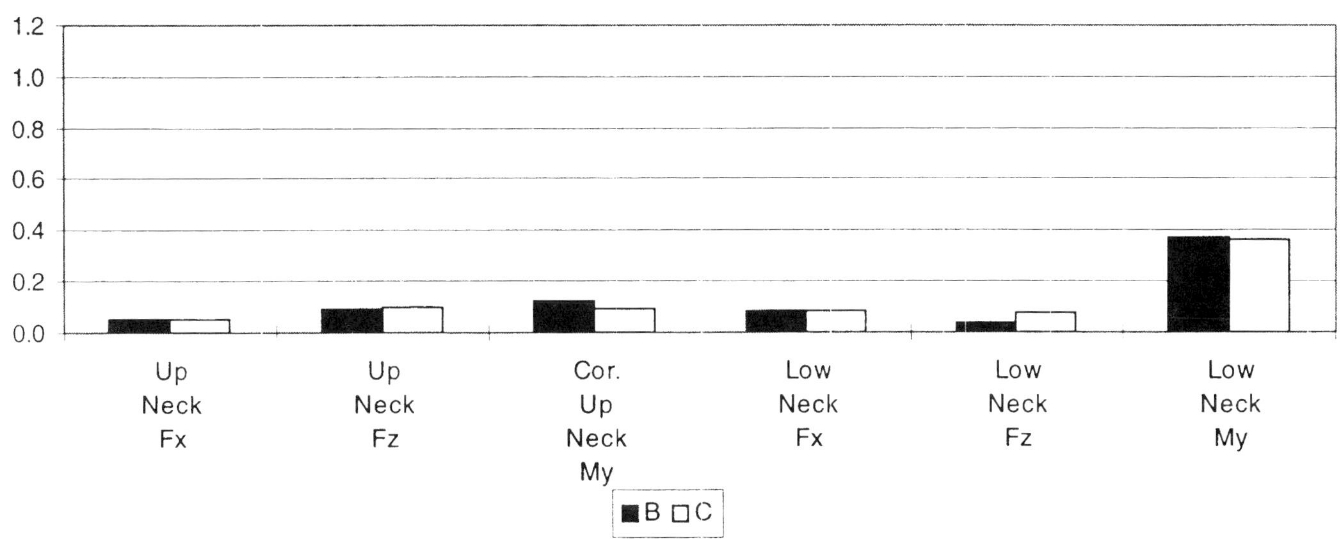

Figure 19. - Neck responses normalized with respect to IARV's: ΔV= 9 kph, seats B and C.

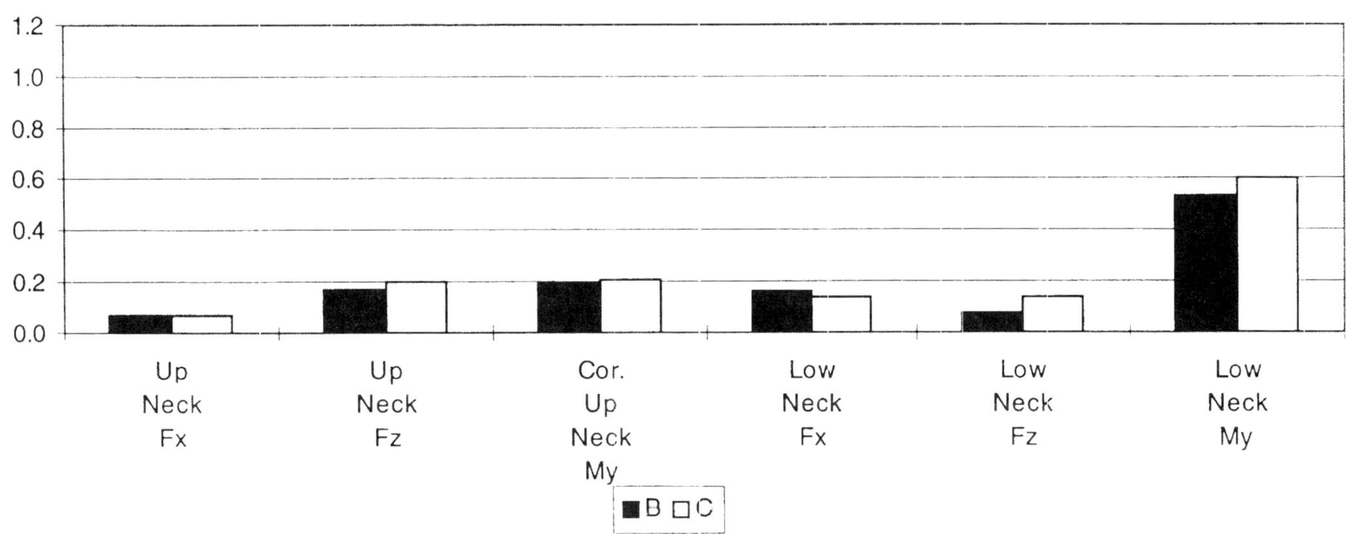

Figure 20. - Neck responses normalized with respect to IARV's: ΔV= 16 kph, seats B and C.

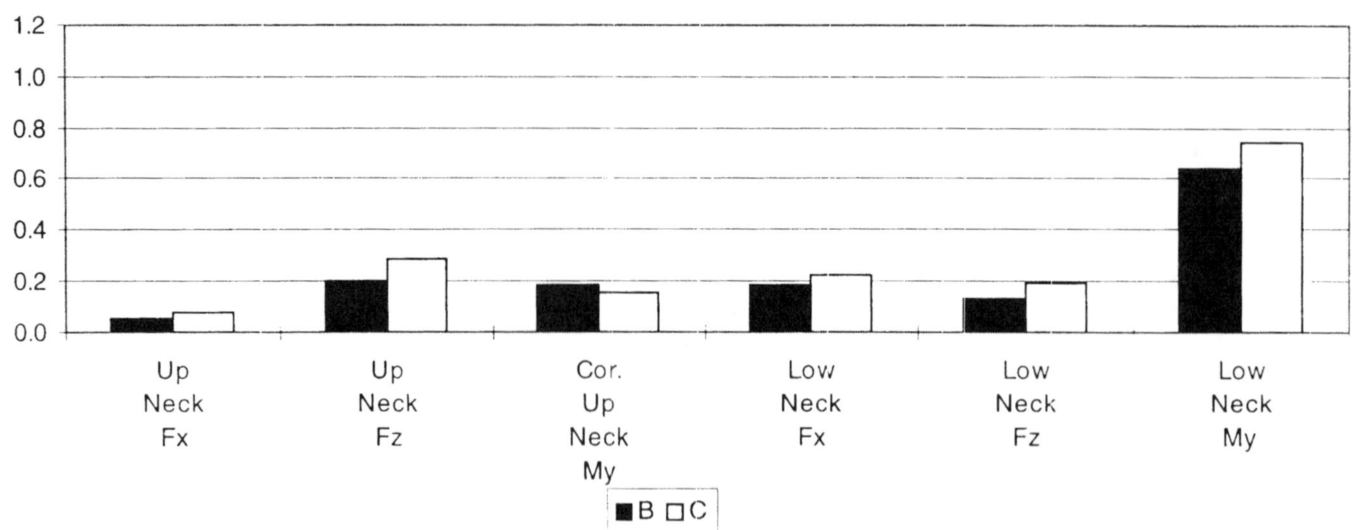

Figure 21. - Neck responses normalized with respect to IARV's: ΔV= 24 kph, seats B and C.

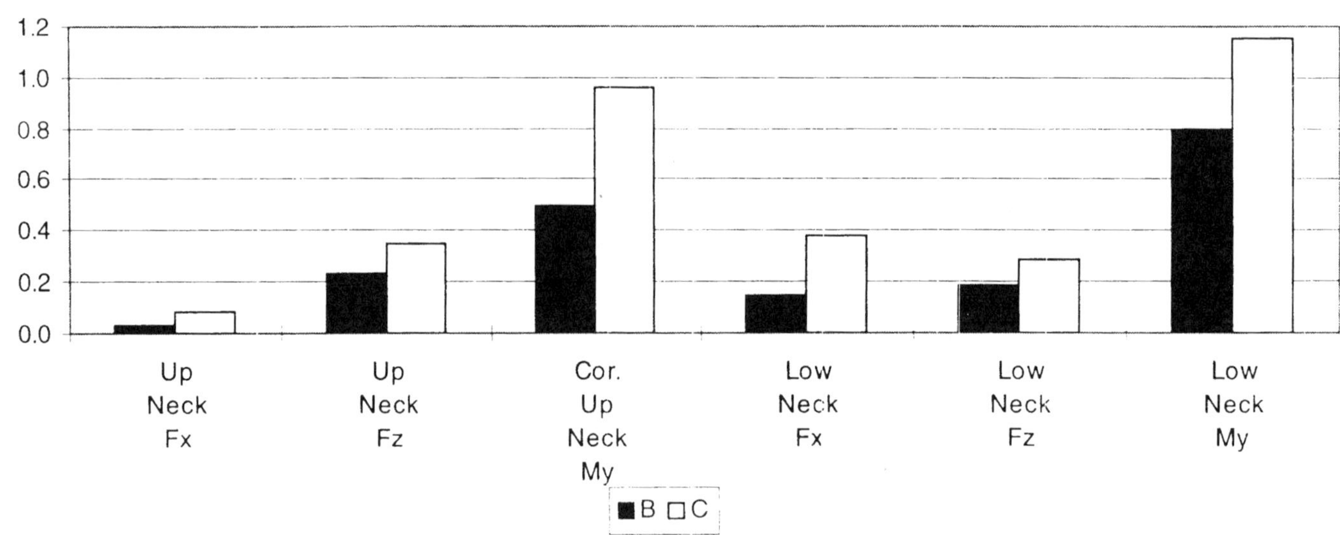

Figure 22. - Neck responses normalized with respect to IARV's: ΔV= 40 kph, seats B and C.

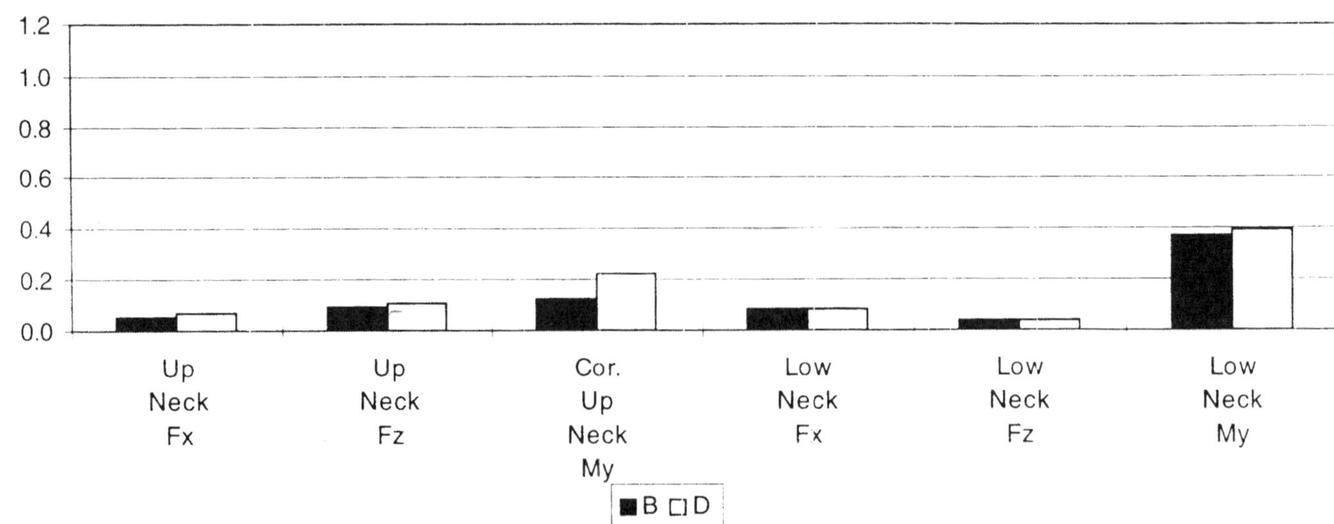

Figure 23. - Neck responses normalized with respect to IARV's: ΔV= 9 kph, seats B and D.

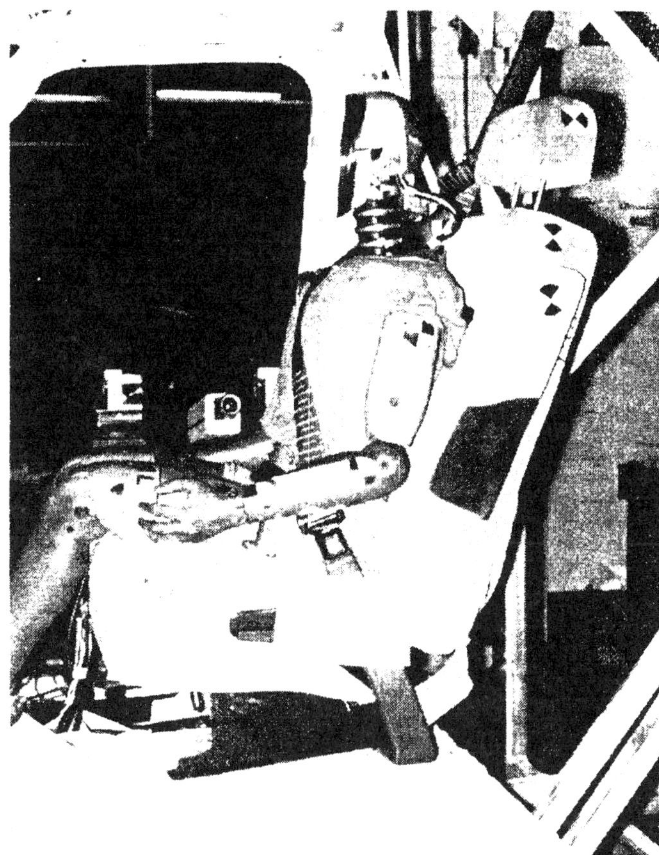

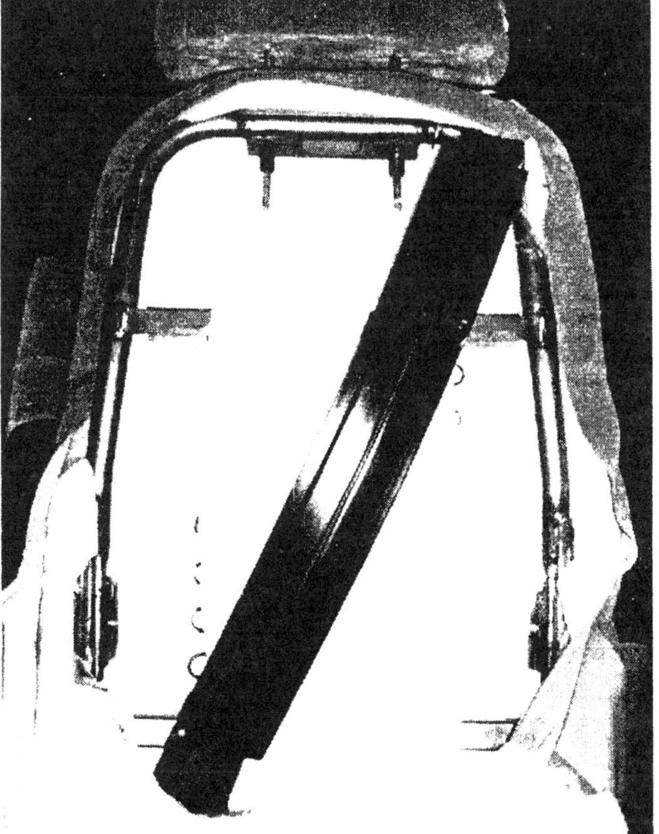

Figure 24. - Hybrid III in seat D.

since comparable dynamic deflections were measured for both seats C and D and yet, their performances, compared to the baseline, were drastically different. The slight improvement at this speed may be the result of the increased head restraint height, some other design characteristic, or the fact that the head restraint support deformed slightly, absorbing some energy (Fig. 25). With D, there are also large increases in the peak tensile forces of the thoracic and lumbar spine compared to those measured with seat B.

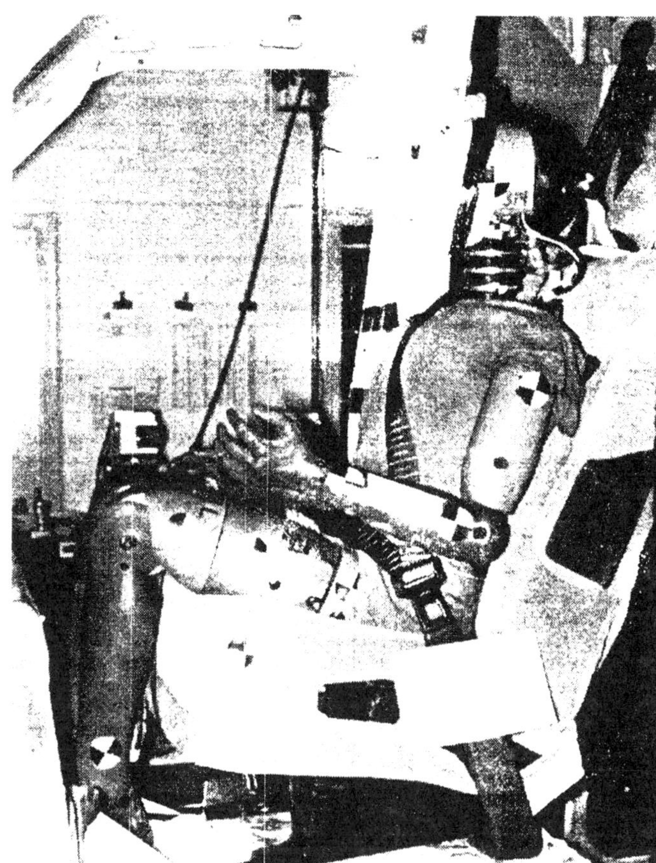

Figure 25. - Hybrid III in seat D: $\Delta V = 24$ kph.

At the highest ΔV of 40 kph, seat B clearly offers more protection to the occupant than D (Fig. 28). With seat D, five of the neck responses have higher ratios to their IARVs than with seat B. One, the lower neck extension moment, even exceeded its reference value signifying a risk of severe neck injury to some percentage of the population. The extreme deformation of the head restraint support, seen in Figure 29, illustrates the high energy level of this impact. At this speed, restricting the deflection of the seat back, via the internal seat belt connected to the seat frame, did not benefit the occupant.

Seat E is a production, European front bucket seat with a dual recliner system and integrated seat belts (Fig. 30). The head restraint, which is mounted to the seat by one outboard support arm versus two center rods, can be rotated manually to adjust for tilt. The vertical position of the head restraint automatically adjusts for the occupant as the seat cushion is positioned. From the dynamic deflections measured for this seat, it is clear that it has a much higher seat back stiffness characteristic than seat B.

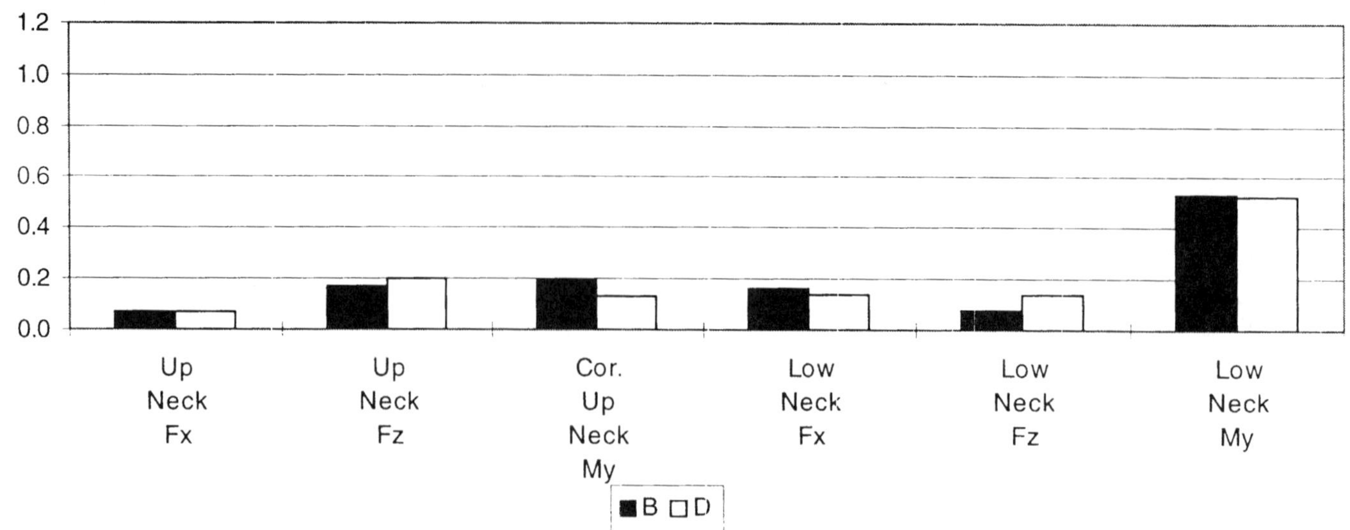

Figure 26. - Neck responses normalized with respect to IARV's: ΔV= 16 kph, seats B and D.

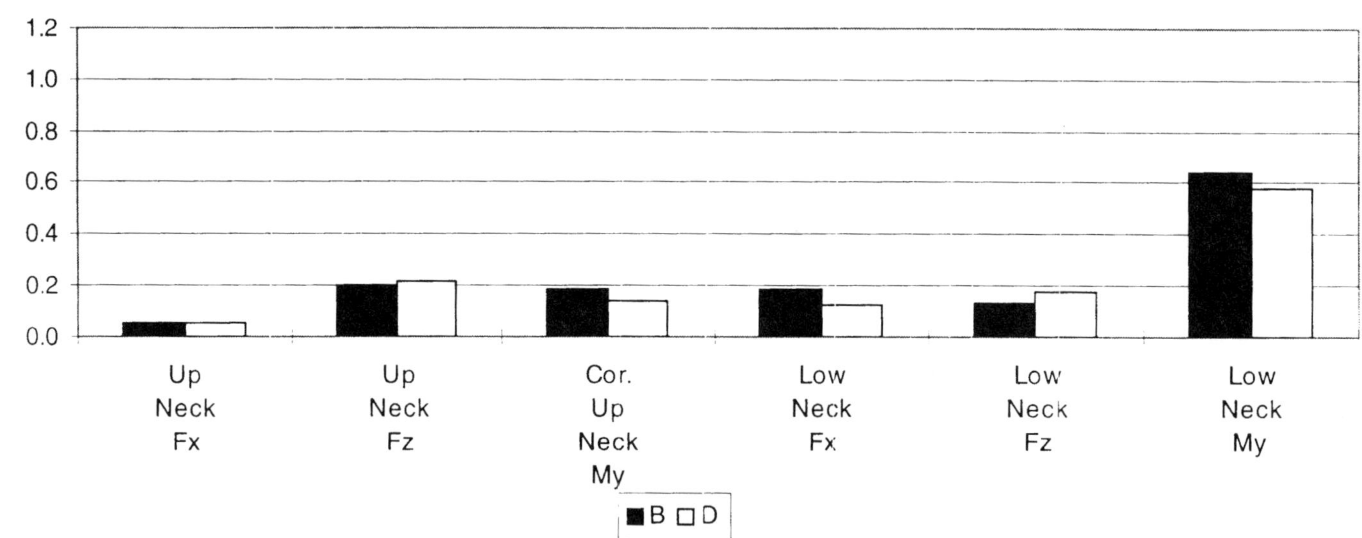

Figure 27. - Neck responses normalized with respect to IARV's: ΔV= 24 kph, seats B and D.

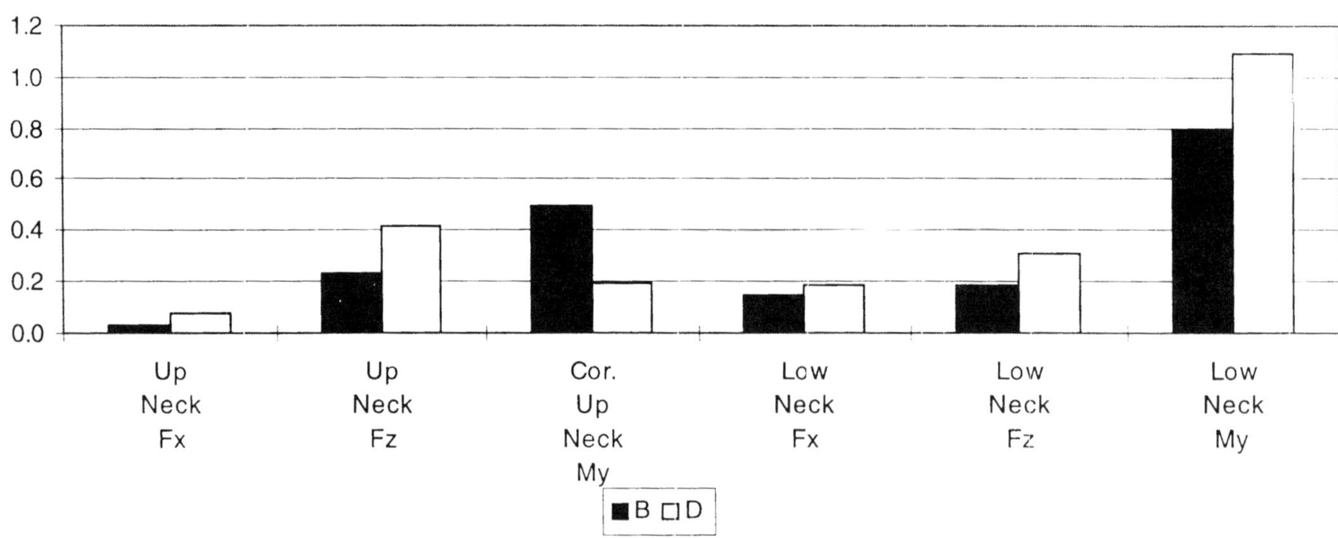

Figure 28. - Neck responses normalized with respect to IARV's: ΔV= 40 kph, seats B and D.

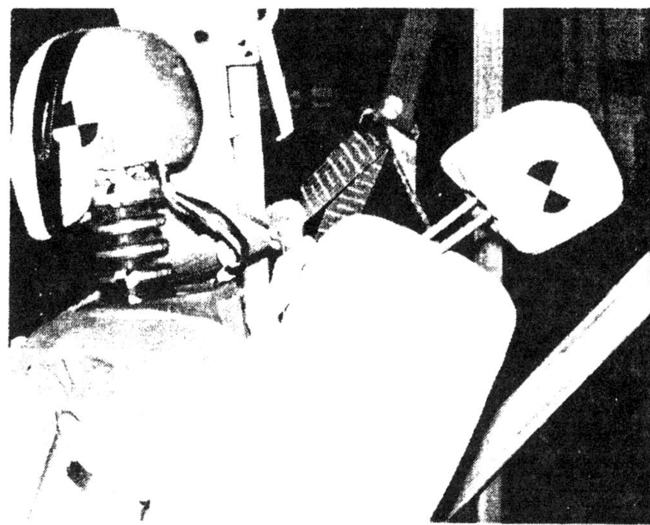

Figure 29. Hybrid III in seat D: ΔV = 40 kph.

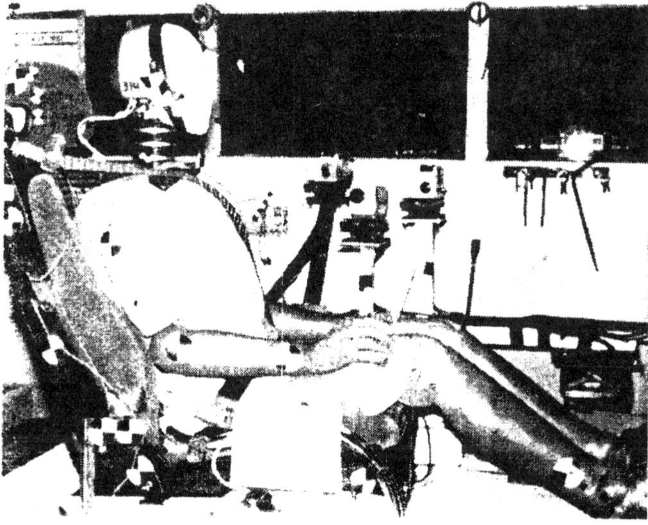

Figure 30. - Hybrid III in seat E.

Despite the vast differences between the two seat designs, at the 9 kph ΔV, no drastic differences are seen in the neck responses (Fig. 31). However, three of the neck responses (the lower neck tensile force, and corrected upper neck and lower neck extension moments) are higher with seat E than B. These responses indicate that the increase in seat back rigidity, or some other design characteristic, may be having a slight effect on the neck responses even at this low speed. There are significant increases in the peak magnitudes of some thoracic and lumbar spine responses as well.

At the 16 kph ΔV, the upper neck shear force, and the lower neck tensile force and extension moment measured with seat E are slightly higher than those of seat B (Fig. 32). This increase is offset by the significantly lower corrected upper neck extension moment. Both of these seats do not present a risk of severe neck injuries to the occupant, and as indicated by the similar forces and moments, would present a similar risk of low severity injuries such as whiplash at this speed. Some of the thoracic and lumbar spine responses have significantly higher peak magnitudes with seat E than with seat B, but none of the values are injurious.

Even with the increased seat back rigidity, at a ΔV of 24 kph, Figure 33 indicates that seat E provides the occupant with a slight advantage over seat B. The lower peak magnitudes of the upper and lower neck tensile forces and lower neck extension moment seem to outweigh the higher peaks of the upper neck shear force and corrected extension moment. Although there is no risk of severe neck injuries with either seat, the risk of low severity injuries such as whiplash, is probably slightly lower with seat B at this speed. Some of its characteristics, such as increased head restraint height, and the fact that its design allows the head restraint to move rearwards with the occupant's head, thereby absorbing energy, could account for this. With regard to the lower torso, the amount of significantly higher and lower responses seen with seat E is about even and only the negative lumbar shear force has increased dramatically compared to the baseline.

At the highest speed, 40 kph ΔV, the results are mixed (Fig. 34). With seat E, the upper and lower neck shear forces, the upper neck tensile force, and lower neck extension moment are all higher than with seat B. However, with seat B, the corrected upper neck moment has a significantly higher peak and the lower neck tensile force peak is also higher than with seat E. Without knowing the exact mechanism that causes whiplash or other neck injuries, it is hard to determine which one of these seats presents the occupant with a higher risk for these types of injuries. Since none of the IARVS was exceeded, there is no risk of severe neck injuries with either seat at this speed.

Seat F, which is also a production, European front bucket seat, has a design very similar to seat E (Fig. 37). It has a dual recliner system, a head restraint which is mounted from a single, outboard arm and adjusts automatically along its vertical axis as the seat cushion is positioned, and an integrated belt system. The head restraint can also be rotated manually to adjust for tilt, like that of seat E. However, by comparing the dynamic deflection with seat E, it can be determined that seat F has a seat back stiffness characteristic which is much stiffer than seat B's or E's.

Overall, as seen in Figure 35, the performance of seat F is slightly worse than seat B's at the 9 kph ΔV. Although only the lower neck tensile force has a significantly larger peak, a total of five of the six neck responses have higher peaks than the baseline. This may indicate a slight increase in the risk of whiplash type injuries with seat F even at this low speed. Also, the peak negative shear and compressive forces of the lumbar spine are approximately three times as large as those measured with seat B.

At a ΔV of 16 kph, four of the six responses (the upper neck shear force, the upper and lower neck tensile forces, and the lower neck extension moment) have higher ratios to their IARVs with the stiffer seat F, than with seat B (Fig. 36). In total, accounting for both the positive and negative directions, 11 of the 14 neck

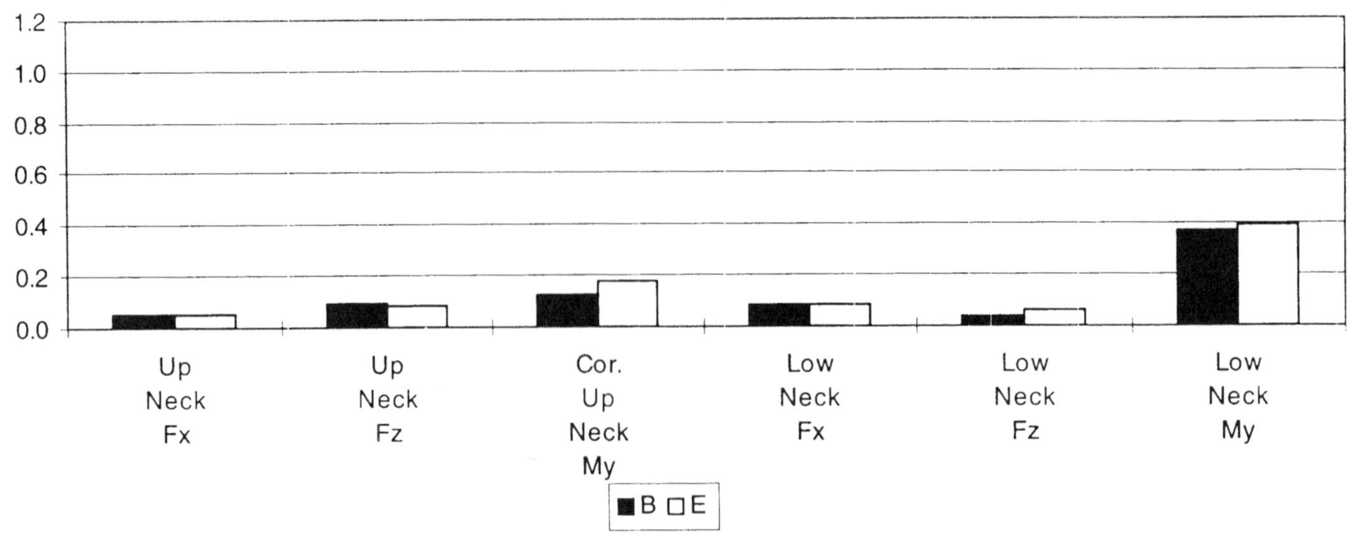

Figure 31. - Neck responses normalized with respect to IARV's: ΔV= 9 kph, seats B and E.

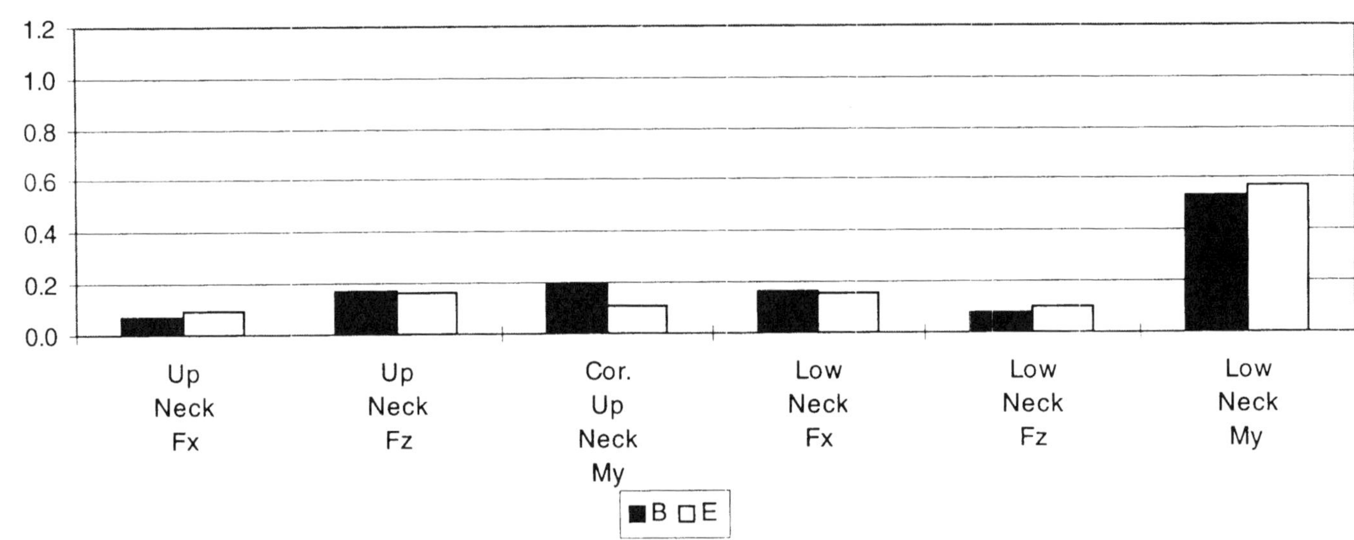

Figure 32. - Neck responses normalized with respect to IARV's: ΔV= 16 kph, seats B and E.

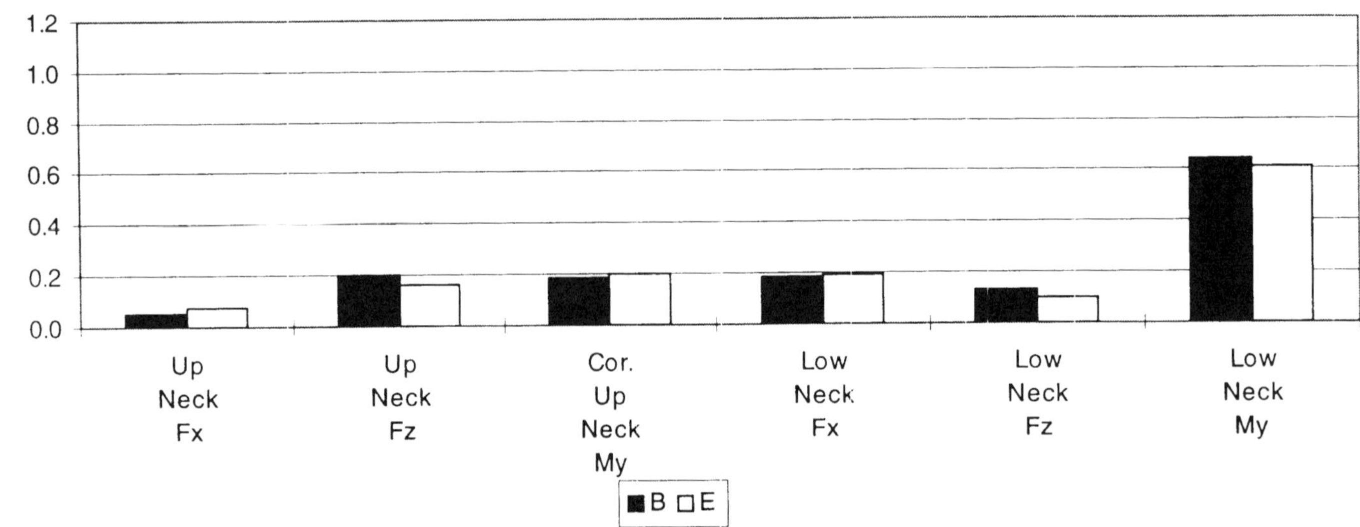

Figure 33. - Neck responses normalized with respect to IARV's: ΔV= 24 kph, seats B and E.

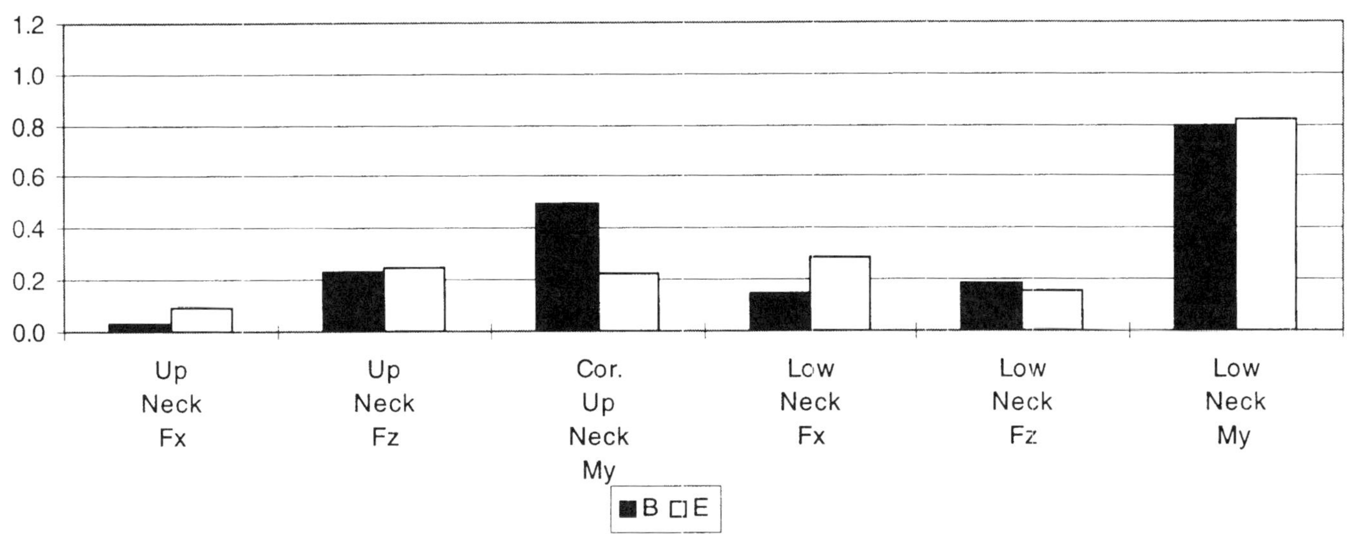

Figure 34. - Neck responses normalized with respect to IARV's: ΔV= 40 kph, seats B and E.

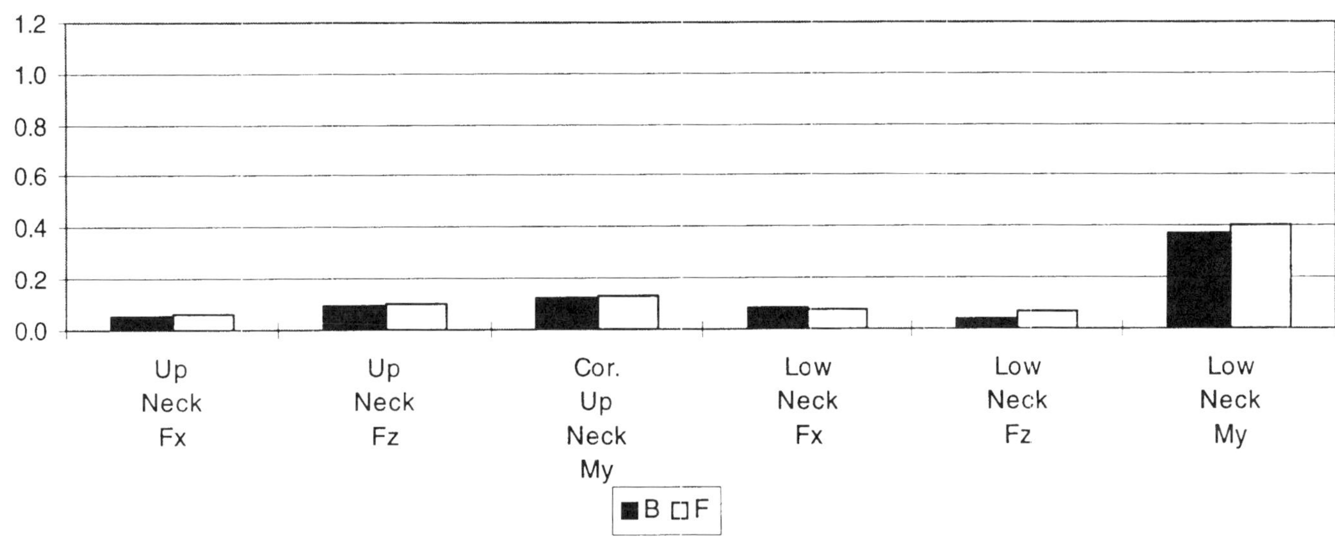

Figure 35. - Neck responses normalized with respect to IARV's: ΔV= 9 kph, seats B and F.

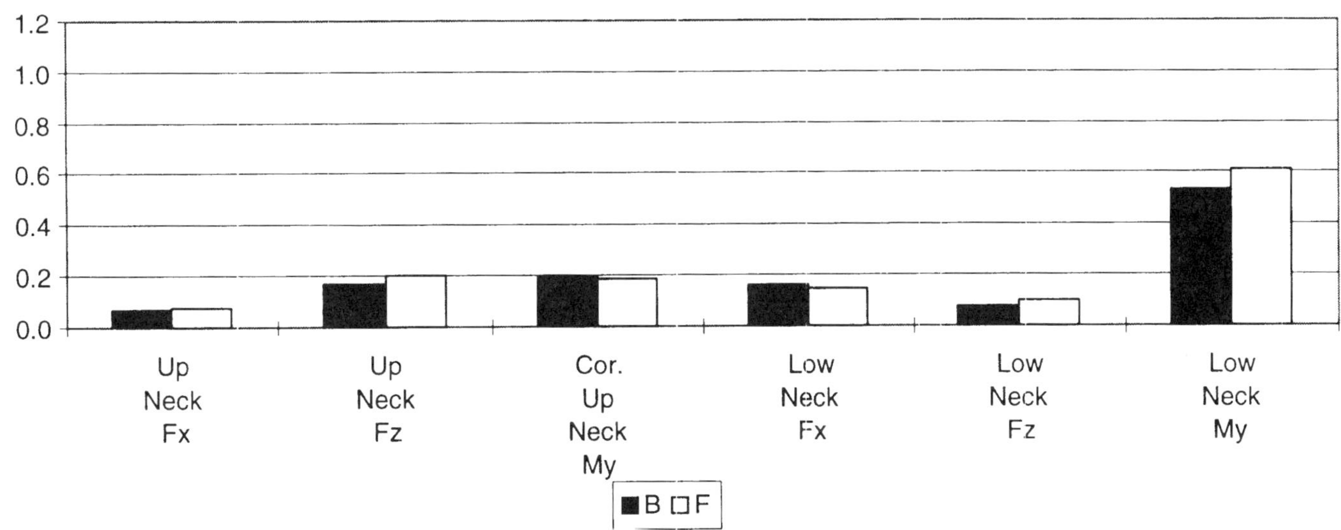

Figure 36. - Neck responses normalized with respect to IARV's: ΔV= 16 kph, seats B and F.

channels have significantly higher peaks. Although all the values are low, the increases suggest that the occupant may have a higher risk for whiplash type injuries with the more rigid seat, F, than with B, the yielding seat, at this speed. Also, with seat F, the peak compressive lumbar spine force is approximately 4 times that of the baseline's.

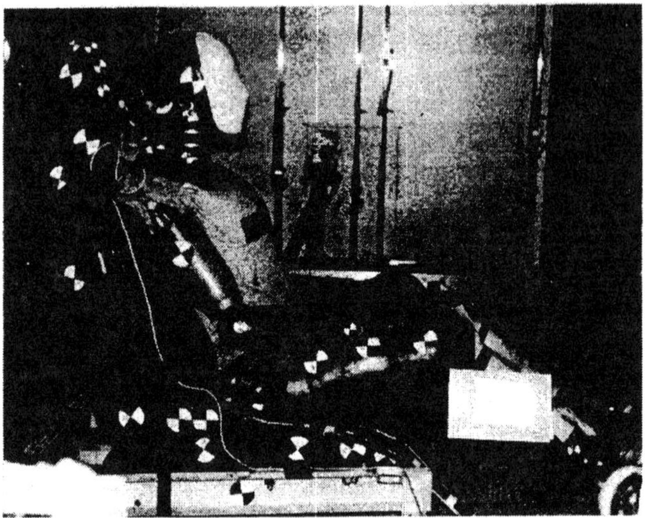

Figure 37. - Hybrid III in seat F.

In contrast to the 16 kph ΔV response, seat F provides better protection to the occupant's neck than seat B at a ΔV of 24 kph (Fig. 38). Four of the neck channels (the upper and lower neck tensile forces, the lower neck shear force, and the lower neck extension moment) have lower ratios with seat F than with seat B, although none of the peak magnitudes are significantly lower. At this speed, the occupant has no risk of severe neck injuries with either seat but may be more at risk for low severity injuries, such as whiplash, with the baseline seat. The negative shear and compressive lumbar spine forces increased dramatically with seat F.

At the highest ΔV of 40 kph, the yielding, deformable seat B provides the occupant with better protection than the more rigid seat F (Fig. 39). Four of the responses (the upper neck and lower neck shear forces, the upper neck tensile force, and the lower neck extension moment) are higher with seat F. The fact that the lower neck extension moment is slightly below its IARV with seat F alone indicates that seat B presents a smaller risk of neck injury to the occupant.

Summary - Seat C, which has a more conventional design than E or F and a higher seat back stiffness than B, did not provide as much protection to the occupant as seat B, despite the improved static position of its head restraint. This comparison shows that good static head restraint position does not necessarily guarantee good dynamic performance, even for low speed rear impacts.

Seat D, which has much thicker seat cushions than any of the other seats and a seat back restraint which limits rearward movement, seems to confirm the fact that at the lowest speed, a 9 kph ΔV, the primary mechanism of controlling neck loads and moments is deflection of the seat cushions, rather than the seat back itself. Its neck responses at the 16 and 24 kph ΔVs indicate that despite the increased seat back stiffness (which is comparable to that of seat C), equivalent performances to that of seat B were obtained. At the highest ΔV, however, a significant decrease in performance was observed, possibly indicating that increased seat back rigidity is more detrimental in severe rear impacts.

Seats E and F, although very similar in design to each other, were designed very differently than the other tested seats. They also had much higher seat back stiffnesses, especially seat F, which was almost rigid. At the two lowest speeds, 9 and 16 kph ΔV, both seats produced results which were either equivalent or slightly worse than that of seat B. These are similar to the results of series 1 and 2 which indicated that a higher seat back stiffness may increase the occupant's risk of whiplash type injuries at low speeds. However, unlike series 1 and 2, at a ΔV of 24 kph, both seats E and F provided the occupant with protection equal or slightly better than that offered by seat B. The obvious conclusion is that other seat characteristics can be designed to either compensate for, or work with increased seat back rigidity, to provide the occupant with an acceptable level of protection. At the highest ΔV of 40 kph, seat E provided the occupant with protection similar to that of seat B, but seat F, the more rigid seat, did not.

Overall, the results of series 3 seem to indicate that no one seat design characteristic determines occupant performance in rear impacts for the range of ΔVs tested. Seat back stiffness, however, seems to be a very influential factor. For the lower severity rear impacts, other seat characteristics (such as head restraint geometry, cushion stiffness, seat back geometry, etc.), can be designed to provide the occupant with protection similar to that offered by a seat with a yielding, deformable seat back. However, at the more severe impacts, controlled deformation of the seat back seems to absorb some of the occupant's energy, thereby, mitigating some of the neck loads and moments while stiffer seat backs may aggravate the condition. In the three cases where the lower neck extension IARV was exceeded at a 40 kph ΔV, two of the seats (B-R and D) were stiffer than the baseline, B. The third seat, C, which did yield at the 40 kph ΔV, was also stiffer than the baseline seat at the lower velocities tested.

DISCUSSION

Dummy response data from three series of tests with various seats have been reported in this paper. Based on the forces and moments developed at the occipital condyles of the Hybrid III dummy, it can be said that in the majority of the tests across the speed range, between ΔVs of 9 and 40 kph, no loads were noticed that exceeded known IARVs at the head/neck junction. Only in one case, seat C, at the 40 kph ΔV, the upper neck

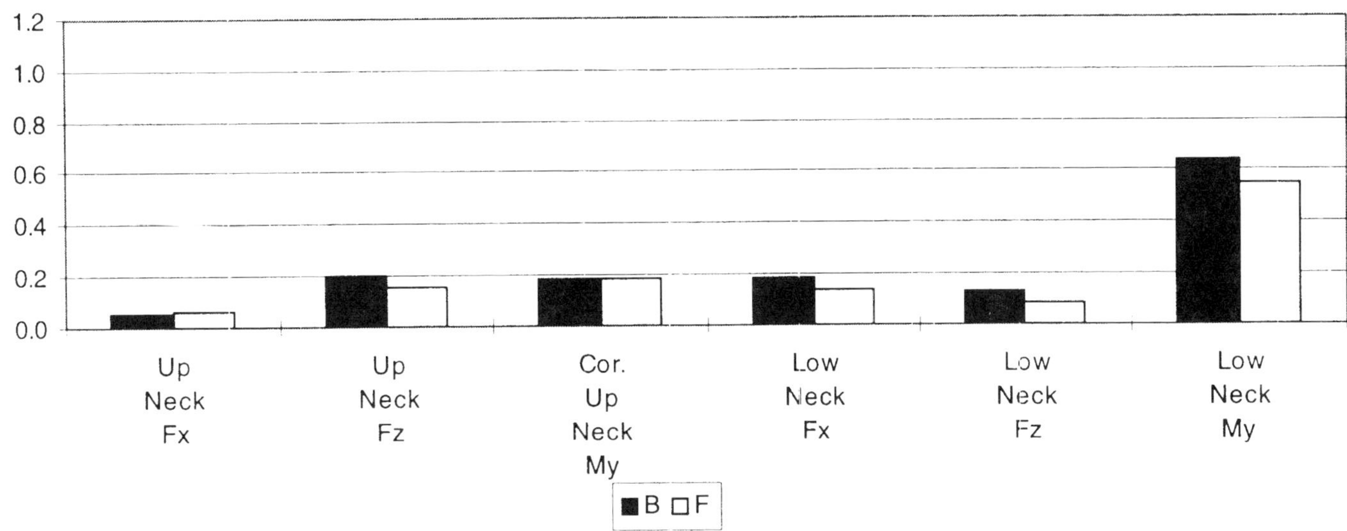

Figure 38. - Neck responses normalized with respect to IARV's: ΔV= 24 kph, seats B and F.

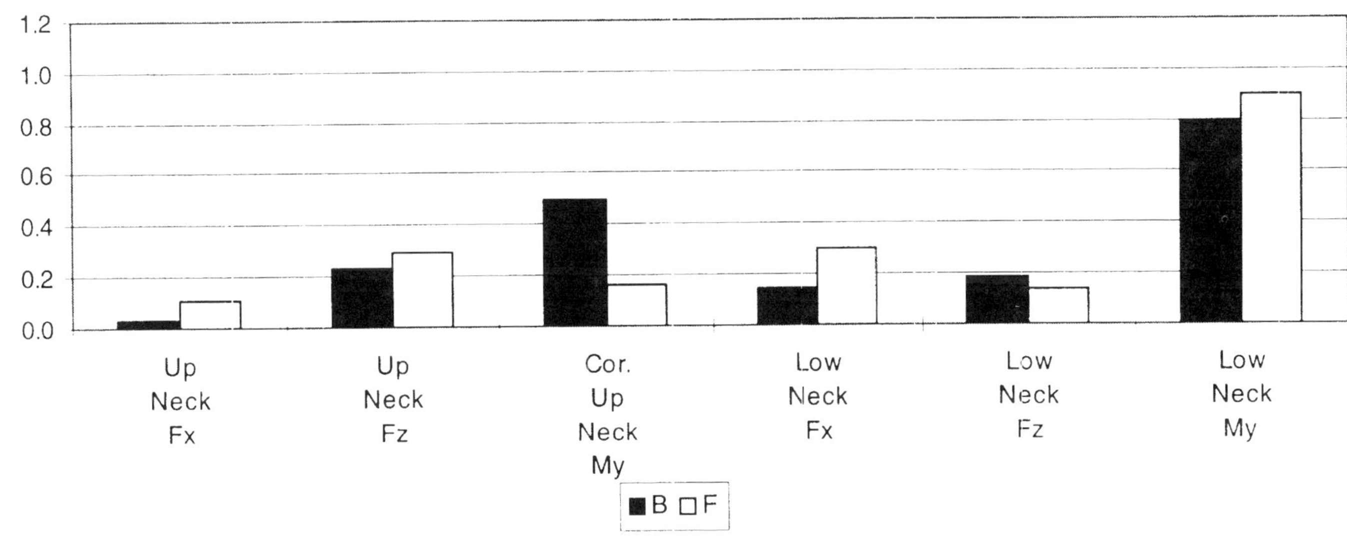

Figure 39. - Neck responses normalized with respect to IARV's: ΔV= 40 kph, seats B and F.

extension moment come close to, but did not exceed, 57 Nm. In most cases, the forces and moments at the occipital condyles were substantially below their known IARVs. It is believed that these responses were low, regardless of seat back stiffness, due to the head restraints being positioned at their highest position in the tests and because the dummies were belted and were retained in the seat even at the highest ΔV where two of the seat backs had deformed to a nearly horizontal position.

The neck response most sensitive to impact severity and seat back stiffness appears to be the extension moment measured at the location of the lower neck load cell. Most of the other neck responses measured at both the upper and the lower neck load cell locations were not as sensitive to impact speed or seat back stiffness. The extension moment measured at the lower neck load cell, for various test severities and seat back stiffnesses, is shown in Figure 40. Also shown in this figure are the extension moments at the occipital condyles. In Figure 40 seat B is a production seat that yields substantially as impact severity increases. Seat B-R is a rigid version of seat B, and seat F is the stiffest production seat tested in this study. The relative insensitivity of the occipital condyles extension moments to impact severity and seat stiffness is obvious from this figure. However, it can be seen that the extension moment at the lower neck load cell location is sensitive to both impact severity and seat back stiffness. It can be seen in the figure that for the baseline seat, the extension moment increases with impact severity - from 57 Nm at 9 kph ΔV to 122 Nm at 40 kph ΔV. This moment increases linearly with impact severity with the rigid seat, B-R, to the extent of exceeding known tolerable moments [46] at the 40 kph ΔV. Seat F, the stiffest production seat, has higher lower neck extension moments at the 9, 16, and 40 kph ΔVs than those of the baseline seat. At the 24 kph ΔV, this moment was the lowest of the three seats shown in this figure. No obvious reasons have been identified for this phenomenon at the 24 kph ΔV. The sensitivity of the lower neck extension response to seat design changes has also been identified in our companion paper [46]. A potential cause of this sensitivity may be the phenomenon of "differential rebound" identified by modeling studies reported by Prasad *et al* [50] in which the torso begins to rebound from the seat back and the head lags behind the torso.

The stiffer seats in general show higher shear and compressive forces in the neck, Appendix A - Table 7, than the baseline production seat. Although the levels of these forces are judged to be low enough to not cause serious neck injuries, these increased force levels may result in increased potential for nerve damage in the articular facets - a hypothesis advanced by Yang and Begeman [51].

Another dummy response that appears to be sensitive to both crash severity and seat characteristics is the peak shear force in the lumbar spine of the dummy. The shear forces developed in four of the seats tested across the entire speed range are shown in Figure 41. From this figure, it can be seen that the baseline seat has the lowest lumbar spine shear force that is maximum at 24 kph ΔV and slightly lower at the 40 kph ΔV. In sharp contrast, seat B-R, the rigid version of seat B, and seat E show substantially higher shear forces at the 9 kph ΔV and increase linearly as the impact severity increases. Seat F is even higher than seats B-R or E at the 9 kph ΔV, and approximately the same as seats B-R or E at the 40 kph ΔV. The significance of these highly elevated shear forces in the stiffer seats need further investigation. Tests reported by Begeman *et al.* [52] on spinal segments from the human cadaveric lumbar region show failures due to shear at loads as low as 1,200 N statically and approximately 1,300 N dynamically. Although some differences between the shear stiffness of the Hybrid III lumbar spine and those of the cadaveric spinal segments are noted by Begeman *et al* [52], such large differences in shear loads in the seats tested deserve further attention. At least, the data indicate that if stiffer seats become more dominant, a new injury mechanism to the lumbar spine may be introduced in the field that is not present with the currently, more popular, yielding seats.

For the baseline seats, A and B, the dummy neck responses at the head/neck junction and at the location of the lower neck load cell, are substantially below known IARVs for velocities up to a ΔV of 40 kph, the 94th%ile of all rear impacts in NASS. Since the baseline seats tested represent the vast majority of seats in the United States, the test data indicate a very low probability of severe neck injuries for belted occupants, even in high severity accidents. The raw data used in Figure 2 show that only 47 out of 5,041 occupants involved in passenger car rear impacts had an AIS 3+ injury for ΔVs up to 40 kph, a less than 1% rate for a AIS 3+ injury. However, all these occupants are not belted and are not in the seating position used in these tests. Therefore, the AIS 3+ injury frequency for belted occupants in the real world would be even less than the one reported above, showing that the test data reflect real world experience. Earlier discussion of the test data of stiffer and rigid seats have shown that stiffer seats lead to higher neck and lumbar spine loadings. This in turn may lead to a higher incidence of low severity and higher severity injuries to in-position belted occupants. If an occupant is not seated normally, but is out-of-position, he will gain more energy before striking the seat back because of the increased travel distance. Because of this, a seat with a stiffer seat back, even one that performed well at the highest speed tested in this study, may present the occupant with a higher risk of severe neck injuries than a plastically, yielding seat which will absorb some of the occupant's energy as it deforms [31,49]. Due to the fact that some of the occupants involved in rear impact accidents are out-of-position when they are struck, this effect must be considered further.

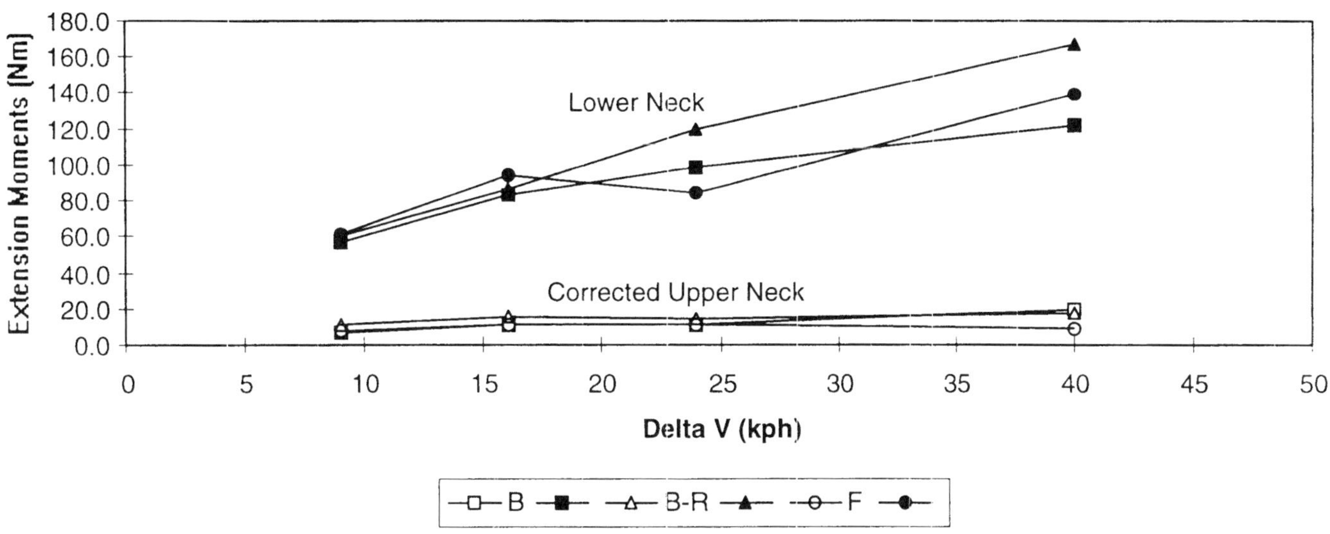

Figure 40. - Corrected upper neck and lower neck extension moments: seats B, B-R, and F.

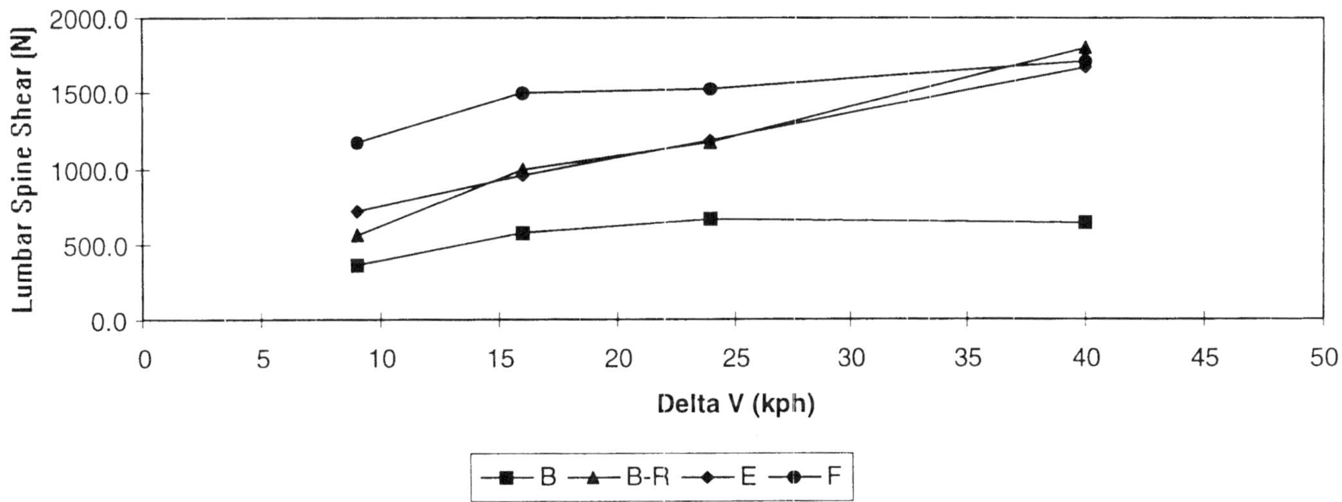

Figure 41. - Lumbar spine negative shear force, -Fx: seats B, B-R, E, and F.

CONCLUSION

Literature survey and analysis of NASS data indicate that high volume production seats, mostly yielding seats in rear impacts, provide excellent passive protection across a large range of impact velocities. Comparison between passenger car seat performance and light truck seat performance indicate that stiffer seats can increase the incidence of neck injuries in the real world. A series of sled tests utilizing yielding and limited yield seats also validated the relative world performance of yielding seats. Based on the results of seats tested and reported in this paper, the following conclusions can be drawn:

1. Stiffening of conventional seats without any other modifications (seats A and B), can result in an overall increase of "whiplash" type of injuries in velocities ranging between 8 to 24 kph ΔVs. At higher velocity impacts, 40 kph ΔV, the potential for serious neck injuries is also increased, since known tolerable loads at the neck bracket location have been exceeded.
2. Stiffer seats also tend to increase lumbar and thoracic spine loads.
3. Static positioning of the head rest - higher and closer to the head - does not necessarily translate into lower neck forces and moments in dynamic tests. This is demonstrated by comparing the dynamic responses recorded in seats B and C.
4. Comparison of the dynamic responses of the Hybrid III dummy in baseline and stiffer, more recently designed seats, e.g. seats D, E, and F, show no distinct advantage of the stiffer seats over the baseline. In fact, seat D could lead to a higher incidence of serious neck injuries at 40 kph ΔV than the baseline seat. Although seats E and F, the stiffest production seats tested, show neck forces and moments that do not exceed known tolerance criteria for the neck, they show higher neck forces and moments in the lower neck at the 40 kph ΔV than those in the baseline seat. Strother *et al.* [31] and Benson *et al.* [49] have shown that neck forces and moments are higher in stiff seats than in yielding seats for out-of-position occupants. Therefore, real world concerns regarding neck injuries for out-of-position occupants exist with the stiffer seats.
5. The dummy response measure most sensitive to seat design changes and crash severity appears to be the extension moment measured at the location of the lower neck load cell. This response also appears to correlate with the Neck Injury Factor, as shown by Prasad *et al.* [46] in a companion paper. Further research is needed to statistically establish injury criteria for this measurement in the dummy for whiplash and serious neck injury.

ACKNOWLEDGMENTS

The authors would like to thank the GTO/Hyge sled group for all the help they gave. We would also like to thank everyone who helped with the production of this paper.

REFERENCES

1. Svensson, M.Y.; Lovsund, P.; Haland, Y.; Larsson, S., "Rear-End Collisions--A Study of the Influence of Backrest Properties on Head-Neck Motion Using a New Dummy Neck," SAE International Congress and Exposition; Society of Automotive Engineers, Inc., Warrendale, PA, USA, SAE Trans., Vol. 102, Section 6, March 1993 930343.

2. Svensson, M. Y.; Lovsund, P.; Haland, Y.; Larsson, S., "The Influence of Seat-Back and Head-Restraint Properties on the Head-Neck Motion During Rear-Impact," International Conference on the Biomechanics of Impacts; International Research Council on Biokinetics of Impacts, Bron, France, September 1993 1993-13-0028 .

3. Newman, M. K.; Biluk, F.; Kyropoulos, P., "Clinical Study of Pain; Traumatic and Control neck Injury Patients; Electromyographic Observations," American Association for Automotive Medicine, annual meeting; Association for the Advancement of Automotive Medicine, Des Plaines, IL, USA, October 1964 1964-12-0001.

4. Janes, J. M.; Hooshmand, H., "Severe Extension-Flexion Injuries of the Cervical Spine," American Association for Automotive Medicine, annual meeting; Association for the Advancement of Automotive Medicine, Des Plaines, IL, USA, October 1964 1964-12-0006.

5. MacNab, I., "Whiplash Injuries of the Neck," American Association for Automotive Medicine; Association for the Advancement of Automotive Medicine, Des Plaines, IL, USA, October 1965 1965-12-0002.

6. States, J.D.; Balcerak, J.C.; Williams, J.S.; Morris, A.T.; Babcock, W.; Polvino, R.; Riger, P.; Dawley, R.E., "Injury Frequency and Head Restraint Effectiveness in Rear-End Impact Accidents," 16th Stapp Car Crash Conference; Society of Automotive Engineers, Inc., Warrendale, PA, USA, SAE Trans., Vol. 81, 1972 720967.

7. Huelke, D.F.; Marsh, J.C., "Analysis of Rear-End Accident Factors and Injury Patterns," American Association for Automotive Medicine, 18th conference; Association for the Advancement of Automotive Medicine, Des Plaines, IL, USA, September 1974 1974-12-0014.

8. Nygren, A.; Gustafsson, H.; Tingvall, C., "Effects of Different Types of Headrests in Rear-End Collisions," Tenth International Technical Conference on Experimental Safety Vehicles (1985), Oxford, England; National Highway Traffic Safety Administration, Washington, DC, USA, 1985 856023.

9. Olsson, I.; Bunketorp, O.; Carlsson, G.; Gustafsson, C.; Planath, I.; Norin, H.; Ysander, L., "An In-Depth Study of Neck Injuries in Rear End Collisions," International IRCOBI Conference on the Biomechanics of Impacts; International Research Council on Biokinetics of Impacts, Bron, France, September 1990 1990-13-0019.

10. Viano, D.C.; Gargan, M.F., "Headrest Position During Normal Driving: Implications to Neck Injury Risks in Rear Crashes," 39th Annual Proceedings of the Association for the Advancement of Automotive Medicine; Association for the Advancement of Automotive Medicine, Des Plaines, IL, USA, October 1995 1995-12-0015.

11. Ommaya, A.K.; Hirsch, A.E.; Martinez, J.L., "Role of "Whiplash" in Cerebral Concussion," 10th Stapp Car Crash Conference; Society of Automotive Engineers, Inc., Warrendale, PA, USA, 1966 660804.

12. Martinez, J.L.; Wickerstrom, J.K.; Barcelo, B.T., "The Whiplash Injury--a Study of Head-Neck Action and Injuries in Animals," ASME Paper No. 65-WA/HUF-6. Presented at Human Factors Division, AMSE Meeting, Chicago, November 1965.

13. Mertz, H.J.; Patrick, L.M., "Investigation of the Kinematics and Kinetics of Whiplash During Vehicle Rear-End Collisions," 11th Stapp Car Crash Conference; Society of Automotive Engineers, Inc., Warrendale, PA, USA, SAE Trans., Vol. 76, 1967 670919.

14. Mertz, H.J.; Patrick,L.M., "Strength and Response of the Human Neck," 15th Stapp Car Crash Conference; Society of Automotive Engineers, Inc., Warrendale, PA, USA, SAE Trans., Vol. 80, 1971 710855.

15. McConnell, W.E.; Howard, R.P.; Guzman, H.M.; Bomar, J.B.; Raddin, J.H.; Benedict, J. V.; Smith, H.L.; Hatsell, C.P., "Analysis of Human Test Subject Kinematic Responses to Low Velocity Rear End Impacts," SAE International Congress and Exposition; Society of Automotive Engineers, Inc., Warrendale, PA, USA, March 1993 930889.

16. Schneider, K.; Zernicke, R.F.; Clark, J.G., "Modelling of the Jaw-Head-Neck Dynamics During Whiplash," Journal of Dental Residency, Vol. 68, 1989.

17. Szabo, T.J.; Welcher, J.B.; Anderson, R.D.; Rice, M.M.; Ward, J.A.; Paulo, L.R.; Carpenter, N.J., "Human Occupant Kinematic Response to Low Speed Rear-End Impacts," SAE International Congress and Exposition; Society of Automotive Engineers, Inc., Warrendale, PA, USA, SAE Trans., Vol. 103, Section 6, February 1994 940532.

18. Matsushita, I.; Sato, T.B.; Hirabayashi, K., "Improvement of Seat Characteristics for Safety to Rear-End Collisions," JSAE Autumn Convention; Society of Automotive Engineers of Japan, Inc., Tokyo, Japan, October 1994 948332.

19. McConnell, W.E.; Howard, R.P.; Van Poppel, J.; Krause, R.; Guzman, He.M.; Bomar, J.B.; Raddin, J.H.; Benedict, J.V.; Hatsell, C.P., "Human Head & Neck Kinematics After Low Velocity Rear-End Impacts--Understanding "Whiplash"," 39th Stapp Car Crash Conference; Society of Automotive Engineers, Inc., Warrendale, PA, USA, November 1995 952724.

20. Juhl,M.; Seerup, K.K., "Cervical Spine Injuries--Epidemiological Investigation, Medical and Social Consequences," International IRCOBI Conference on the Biomechanics of Impacts, 6th; International Reasearch Council on Biokinetics of Impacts, Bron, France, September 1981 1981-13-0005.

21. Thomas, C.; Faverjon, G.; Hartemann, F.; Tarriere, C.; Patel, A.; Got, C., "Protection Against Rear-End Accidents," International IRCOBI Conference on the Biomechanics of Impacts, 7th; International Research Council on Biokinetics of Impacts, Bron, France, September 1982 1982-13-0002.

22. Severy, D.M.; Bring, H.M.; Baird, J.D., "Study of Seat and Head-Support Performance During Full-Scale Rear-End Collisions," 11th Stapp Car Crash Conference; Society of Automotive Engineers, Inc., Warrendale, PA, USA, 1967 670921.

23. Severy, D.M.; Brink, H.M.; Baird, J.D., "Backrest and Head Restraint Design for Rear-End Collision Protection," Automotive Engineering Congress and Exposition; Society of Automotive Engineers, Inc., Warrendale, PA, USA, 1968 680079.

24. Berton, R.J., "Effect of Seatback Rotation, Head Restraint Position, and Collision Speed on Whiplash Injuries," Automotive Engineering Congress and Exposition; Society of Automotive Engineers, Inc., Warrendale, PA, USA, 1968 680080.

25. Martinez, J.L., "Headrest and Seat Back Proposals Designed to Eliminate Head and Neck Injuries," 12th Stapp Car Crash Conference; Society of Automotive Engineers, Inc., Warrendale, PA, USA, 1968 680775.

26. Clemens, H.J.; Burow, K., "Experimental Investigation on Injury Mechanisms of Cervical Spine at Frontal and Rear-Front Vehicle Impacts," 16th Stapp Car Crash Conference; Society of Automotive Engineers, Inc., Warrendale, PA, USA, SAE Trans., Vol. 81, 1972 720960.

27. Hu, A.S.; Bean, S.P.; Zimmerman, R.M., "Response of Belted Dummy and Cadaver to Rear Impact," 21st Stapp Car Crash Conference; Society of Automotive Engineers, Inc., Warrendale, PA, USA, 1977 770929.

28. Williams, J.F.; McKenzie, J.A., "The Effect of Seatback Stiffness and Collision Severity on the Dynamic Behavior of the Head During "Whiplash"," International Conference on the Biokinetics of Impacts; International Research Council on Biokinetics of Impacts, Bron, France, June 1973 1973-13-0026.

29. Strother, C.E.; James, M.B., "Evaluation of Seat Back Strength and Seat Belt Effectiveness in Rear End Impacts," 31st Stapp Car Crash Conference; Society of Automotive Engineers, Inc., Warrendale, PA, USA, 1987 872214.

30. Warner, C.Y.; Strother, C.E.; James, M.B.; Decker, R.L., "Occupant Protection in Rear-End Collisions: II. the Role of Seat Back Deformation in Injury Reduction," 35th Stapp Car Crash Conference; Society of Automotive Engineers, Inc., Warrendale, PA, USA, SAE Trans., Vol. 100, Section 6, 1991 912914.

31. Strother, C.E.; James, M.B.; Gordon, J.J., "Response of Out-of-Position Dummies in Rear Impact," SAE International Congress and Exposition; Society of Automotive Engineers, Inc., Warrendale, PA, USA, SAE Trans., Vol. 103, Section 6, February 1994 941055.

32. James, M.B.; Strother, C.E.; Warner, C.Y.; Decker, R.L.; Perl, T.R., "Occupant Protection in Rear-End Collisions: I. Safety Priorities and Seat Belt Effectiveness," 35th Stapp Car Crash Conference; Society of Automotive Engineers, Inc., Warrendale, PA, USA, SAE Trans., Vol. 100 Section 6, 1991 912913.

33. Parkin, S.; Mackay, G. M.; Hassan, A. M.; Graham, R., "Rear-End Collisions and Seat Performance--To Yield or Not to Yield," 39th Annual Proceedings of the Association for the Advancement of Automotive Medicine; Association for the Advancement of Automotive Medicine, Des Plaines, IL, USA, October 1995 1995-12-0016.

34. Marquardt, J.F., "Collision Severity - Measured by ΔV," American Association for Automotive Medicine, 21st conference; Association for the Advancement of Automotive Medicine, Des Plaines, IL, USA, September 1977 1977-12-0033.

35. Portnoy,H.D.; McElhaney, J.H.; Melvin, J.W.; Croissant, P.D., "Mechanism of Cervical Spine Injury in Auto Accidents," American Association for Automotive Medicine, 15th annual conference; Association for the Advancement of Automotive Medicine, Des Plaines, IL, USA, October 1971 1971-12-0005.

36. Moffatt, E.A.; Siegel, A.W.; Huelke, D.F.; Nahum, A.M., "The Biomechanics of Automotive Cervical Fractures," American Association for Automotive Medicine, 22nd conference; Associoation for the Advancement of Automotive Medicine, Des Plaines, IL, USA, July 1978 1978-12-0014.

37. Huelke, D.F.; Mendelsohn, R.A.; States, J.D.; Melvin, J.W.; "Cervical Fractures and Fracture-Dislocations Sustained Without Head Impact," Journal of Trauma, Vol. 18 Number 7, July, 1978.

38. Saczalski, K.J.; Syson, S.R.; Hille, R.A.; Pozzi, M.C.; "Field Accident Evaluations and Experimental Study of Seat Back Performance Relative to Rear-Impact Occupant Protection," SAE International Congress and Exposition, Society of Automotive Engineers, Inc., Warrendale, PA, USA, SAE Trans., Vol. 102, Section 6, March 1993 930346.

39. Marriner, P.C.; Submission or Report entitled: "Evaluation of Seat Back Failures," Prepared by TES Limited for Transport Canada, Submitted to U.S. DOT Docket 89-20-N01(018), January 1990.

40. Malliaris, A.C.; Hitchcock, R.; Hansen, M., "Harm Causation and Ranking in Car Crashes," International Congress and Exposition; Society of Automotive Engineers, Inc., Warrendale, PA, USA, SAE Trans., Vol. 94, Section 1, 1985-850090.

41. Digges, K.H.; Morris, J.H.; Malliaris, A.C., "Safety Performance of Motor Vehicle Seats," SAE International Congress and Exposition; Society of Automotive Engineers, Inc., Warrendale, PA, USA, March 1993 930348.

42. Comments to Docket 89-20-N01-014.

43. Roberts, V.L.; Compton, C.P., "The Relationship Between ΔV and Injury," 37th Stapp Car Crash Conference; Society of Automotive Engineers, Inc., Warrendale, PA, USA, SAE Trans., Vol. 102, Section 6, November 1993 933111.

44. Roberts, V.L.; Passmore, J.R., "The Development of a PC Vehicle Crashworthiness Database," SAE International Congress and Exposition; Society of Automotive Engineers, Inc., Warrendale, PA, USA, February, 1995 950889.

45. Mertz, H.J., "Anthropomorphic Test Devices," in "Accidental Injury, Biomechanics and Prevention," ed. by Nahum, A.M.; Melvin, J.W., Springer-Verlag, New York, 1993.

46. Prasad, P.; Kim, A.; Weerappuli, D.P.V., "Biofidelity of Anthropomorphic Test Devices for Rear Impact," Proceedings of the 41st STAPP Car Crash Conference, SAE, Inc., Warrendale, PA, USA.

47. Eichberger, A.; Geigl, B.C.; Moser, A.; Fachbach, B.; Steffan, H.; Hell, W.; Langwieder,K, "Comparison of Different Car Seats Regarding Head-Neck Kinematics of Volunteers During Rear End Impact, " Proceedings of

1996 International IRCOBI Conf. On the Biomechanics of Impacts, Dublin, Ireland, 1996.

48. "Whiplash Injuries, " Insurance Institute for Highway Safety, Arlington, VA, Status Report, Vol. 30:8, 1995.

49. Benson, B.R.; Smith, G.C.; Kent, R.W.; Monson, C.R., "Effect of Seat Stiffness in Out-of-Positioin Occupant Response in Rear-End Collisions," Proceeding of the 40th Stapp Car Crash Conference, SAE, Inc., Warrendale, PA, USA, ISBN 1-56091-901-9, paper 962434.

50. Prasad, P.; Mital, N.; King, A..I.; Patrick, L.M., "Dynamic Response of the Spine During $+G_x$ Acceleration, " Proceedings of the 19th STAPP Car Crash Conference, Society of Automotive Engineers, Inc., San Diego, CA, November 1975, 751172.

51. Yang, K.H.; Begeman, P.C.; Muser, M.; Niederer, P.; Walz, F., "On the Role of Cervical Facet Joints in Rear End Impact Neck Injury Mechanisms, " SAE International Congress and Exposition; Society of Automotive Engineers, Inc., Warrendale, PA, USA, ISBN 1-56091-938-8, paper 970497.

52. Begeman, P.C.; Visarius, H.; Nolte, L.P.; Prasad, P., "Viscoelastic Shear Responses of the Cadaver and Hybrid III Lumbar Spine," Proceeding of the 38th Stapp Car Crash Conference, SAE, Inc., Warrendale, PA, USA, ISBN 1-56091-543-9, paper 942205.

APPENDIX A - DATA TABLES

All the tables of dynamic Hyge sled test results include an estimate of the dynamic seat back deflection, obtained from film analysis, and the measured post-test seat back angles to indicate the relative difference in seat back stiffness between the types of seats tested. No deflection measurements were taken for rigid seat configurations. In general, the seats are ordered from lowest to highest seat back stiffness as determined by the dynamic deflection, but in Tables 4-7, B-R (the rigid configuration of seat B) which may be the stiffest seat tested, is listed second. The average peak magnitudes of the 50th% Hybrid III responses and their population standard deviations, if available, are also listed. If the ATD responses obtained with one seat were significantly higher or lower than those obtained with the baseline seat, they are highlighted with an up arrow, ↑, or a down arrow, ↓. The responses were judged to be significantly different if their ranges did not overlap. If the population standard deviation was only available for one seat, it was doubled to determine the range; if no standard deviation was given, a 25% difference from the baseline was judged to be significant.

Table 1. - Dynamic rear impact Hyge sled tests: Series 1, ΔV = 9 kph.

ΔV = 9 kph	Seat A (n = 2)	Seat A-M (n = 1)	Seat A-R (n = 2)
Dynamic Deflection (Deg)	9.0 ± 2.1	7.0	NA
Post-test Seat Back Angle (Deg)	0.5 ± 0.7	0.0	NA
Up Neck Fx (N)	175.63 ± 16.8 -136.09 ± 51.5	166.76 -158.35	155.06 ±23.7 -101.88 ± 19.1
Up Neck Fz (N)	269.13 ± 14.9 -58.22 ± 4.1	159.86 ↓ -89.98 ↑	273.69 ± 25.0 -102.79 ± 25.6 ↑
Corrected Upper Neck My (Nm)	5.83 ± 3.7 -9.78 ± 0.5	12.00 -10.40	7.59 ± 1.7 -8.28 ± 0.8 ↓
Lower Neck Fx (N)	241.50 ± 38.2 -107.64 ± 27.2	250.24 -74.24	296.66 ± 36.3 -100.86 ± 6.5
Lower Neck Fz (N)	217.22 ± 10.0 -94.48 ± 30.3	135.44 ↓ -140.42	185.53 ± 13.0 ↓ -125.03 ± 0.1
Lower Neck My (Nm)	25.61 ± 12.1 -51.79 ± 3.2	29.76 -54.81	20.81 ± 4.3 -51.81 ± 5.1
Corrected Lower Neck My (Nm)	12.46 ± 11.4 -56.32 ± 4.2	29.34 -49.77	23.11 ±7.1 -43.56 ± 5.7 ↓
Thoracic Spine Fx (N)	124.77 ± 94.7 -344.32 ± 90.8	35.23 -604.04 ↑	55.11 ± 10.3 -653.63 ± 139.3 ↑
Thoracic Spine Fz (N)	385.73 ± 96.8 -192.40 ± 175.2	384.80 -352.68	399.21 ± 35.0 -167.31 ± 30.7
Thoracic Spine My (Nm)	32.09 ± 3.0 -43.47 ± 5.7	46.93 ↑ -41.64	27.05 ± 7.3 -72.32 ±1.1 ↑
Lumbar Spine Fx (N)	67.05 ± 21.5 -429.45 ± 30.5	108.66 -368.16	52.47 ± 17.7 -436.59 ± 32.4
Lumbar Spine Fz (N)	363.74 ± 103.3 -272.06 ± 258.8	275.95 -612.49	347.10 ± 22.0 -475.60 ± 98.0
Lumbar Spine My (Nm)	62.32 ± 4.3 -21.34 ± 3.7	54.74 -18.69	44.57 ± 1.3 ↓ -30.83 ± 1.3 ↑

Table 2. - Dynamic rear impact Hyge sled tests: Series 1, ΔV = 16 kph.

ΔV = 16 kph	Seat A (n = 2)	Seat A-M (n = 1)	Seat A-R (n = 2)
Dynamic Deflection (Deg)	15.8 ± 5.3	6.5	NA
Post-test Seat Back Angle (Deg)	6.0 ± 1.4	2.0	NA
Up Neck Fx (N)	244.612 ±1.5 -68.14 ± 6.8	246.06 -214.66 ↑	233.36 ± 43.6 -145.74 ± 6.9 ↑
Up Neck Fz (N)	561.34 ± 2.5 -53.98 ± 7.4	496.84 ↓ -100.35 ↑	634.06 ± 157.0 -151.17 ± 14.9 ↑
Corrected Upper Neck My (Nm)	7.69 ± 1.2 -11.82 ± 0.01	11.46 ↑ -12.45 ↑	9.55 ± 0.8 -8.90 ± 2.3 ↓
Lower Neck Fx (N)	436.64 ± 22.2 -76.68 ± 6.5	538.65 ↑ -82.96	564.23 ± 154.4 -130.06 ± 10.9 ↑
Lower Neck Fz (N)	321.06 ± 22.4 -109.71 ± 35.3	298.33 -39.68	365.23 ± 68.9 -191.73 ± 36.0 ↑
Lower Neck My (Nm)	2.89 ± 1.4 -79.28 ± 1.2	32.46 ↑ -89.10 ↑	27.79 ± 3.7 ↑ -94.45 ± 14.7
Corrected Lower Neck My (Nm)	21.97 ± 0.5 -65.63 ± 4.3	43.31 ↑ -69.14	32.68 ± 2.4 ↑ -59.43 ± 6.3
Thoracic Spine Fx (N)	225.91 ± 220.0 -352.30 ± 83.4	42.34 -904.72 ↑	133.20 ± 107.3 -1697.36 ± 180.5 ↑
Thoracic Spine Fz (N)	536.21 ± 93.4 -193.58 ± 222.6	463.93 -379.46	425.41 ± 82.8 -538.21 ± 40.3 ↑
Thoracic Spine My (Nm)	16.01 ± 0.8 -56.29 ± 25.4	77.58 ↑ -80.58	55.30 ± 0.2 ↑ -179.02 ± 38.9 ↑
Lumbar Spine Fx (N)	127.44 ± 44.4 -559.56 ± 101.3	40.10 -652.97	95.19 ± 6.3 -942.62 ± 199.9 ↑
Lumbar Spine Fz (N)	454.59 ± 3.2 -245.71 ± 204.9	370.70 ↓ -699.23 ↑	354.73 ± 68.1 ↓ -1221.42 ± 118.3 ↑
Lumbar Spine My (Nm)	53.50 ± 0.9 -27.82 ± 2.2	96.99 ↑ -12.61 ↓	98.84 ± 7.8 ↑ -17.85 ± 0.1 ↓

Table 3. - Dynamic rear impact Hyge sled tests: Series 1, ΔV = 24 kph.

ΔV = 24 kph	**Seat A (n = 1)**	**Seat A-M (n = 1)**	**Seat A-R (n = 1)**
Dynamic Deflection (Deg)	26.0	21.0	NA
Post-test Seat Back Angle (Deg)	10.0	4.0	NA
Up Neck Fx (N)	198.51 -24.99	203.41 -145.05 ↑	188.37 -155.15 ↑
Up Neck Fz (N)	574.68 -60.80	566.68 -85.49 ↑	548.44 -111.60 ↑
Corrected Upper Neck My (Nm)	3.80 -9.09	15.86 ↑ -8.63	14.07 ↑ -12.97 ↑
Lower Neck Fx (N)	483.94 -60.00	639.18 ↑ -125.74 ↑	646.29 ↑ -180.37 ↑
Lower Neck Fz (N)	411.26 -103.10	351.44 -73.57 ↓	303.22 ↓ -146.16 ↑
Lower Neck My (Nm)	0.21 -81.87	39.78 ↑ -99.57	38.05 ↑ -93.74
Corrected Lower Neck My (Nm)	9.52 -58.20	37.37 ↑ -68.20	41.47 ↑ -69.22
Thoracic Spine Fx (N)	495.06 -286.67	43.66 ↓ -1075.53 ↑	40.95 ↓ -1309.05 ↑
Thoracic Spine Fz (N)	654.30 -380.13	280.67 ↓ -367.32	281.38 ↓ -598.26 ↑
Thoracic Spine My (Nm)	6.84 -62.87	42.02 ↑ -122.58 ↑	66.01 ↑ -162.80 ↑
Lumbar Spine Fx (N)	131.53 -484.39	36.27 ↓ -819.77 ↑	28.40 ↓ -1009.70 ↑
Lumbar Spine Fz (N)	717.46 -413.40	328.57 ↓ -718.35 ↑	243.53 ↓ -1344.63 ↑
Lumbar Spine My (Nm)	39.77 -57.86	66.39 ↑ -12.65 ↓	101.51 ↑ -16.51 ↓

Table 4. - Dynamic rear impact Hyge sled test results: Series 2 and 3, ΔV = 9 kph.

ΔV = 9 kph	Seat B (n = 3)	Seat B-R (n = 2)	Seat C (n = 2)	Seat D (n = 1)	Seat E (n = 1)	Seat F (n = 1)
Dynamic Deflection (Deg)	10.5 ± 4.8	NA	5.8 ± 1.8	5.0	2.5	4.0
Post-test Seat Back Angle (Deg)	1.8 ± 1.8	NA	1.8 ± 1.8	1.5	1.0	1.0
Up Neck Fx (N)	151.89 ± 10.5 -161.98 ± 77.1	127.75 ± 2.2 ↓ -100.95 ± 41.1	175.36 ± 7.7 ↑ -152.48 ± 13.2	214.26 ↑ -122.59	174.85 ↑ -179.92	145.67 -201.49
Up Neck Fz (N)	305.21 ± 36.8 -33.41± 17.6	317.85 ± 47.2 -78.51 ± 26.7 ↑	326.71 ± 7.7 -38.14 ± 21.4	368.03 -33.74	288.54 -147.72 ↑	328.53 -146.34 ↑
Corrected Upper Neck My (Nm)	9.53 ± 4.5 -6.94 ± 1.1	6.61 ± 2.1 -11.45 ± 1.3 ↑	9.66 ± 0.4 -5.20 ± 0.3 ↓	7.23 -12.95 ↑	13.85 -10.21 ↑	11.15 -7.38
Lower Neck Fx (N)	252.94 ± 24.4 -126.35 ± 11.2	278.04 ± 44.8 -108.09 ± 39.5	253.38 ± 7.8 -48.00 ± 23.6 ↓	254.20 -100.04 ↓	258.47 -123.92	250.87 -105.42
Lower Neck Fz (N)	126.34 ± 49.4 -81.78 ± 52.1	119.52 ± 43.5 -93.43 ± 7.2	245.97 ± 2.5 ↑ -21.93 ± 7.6 ↓	130.90 -158.26	99.90 -237.30 ↑	233.34 ↑ -140.33
Lower Neck My (Nm)	35.63 ± 16.4 -57.07 ± 3.7	23.08 ± 7.9 -59.62 ± 3.9	30.76 ± 1.3 -55.96 ± 1.2	23.08 -60.57	40.57 -61.05	37.51 -62.00
Corrected Lower Neck My (Nm)	35.13 ± 18.2 -46.95 ± 4.0	18.63 ± 11.9 -50.20 ± 6.5	33.06 ± 2.4 -39.26 ± 1.1 ↓	21.76 -53.05	34.86 -56.16 ↑	43.84 -46.12
Thoracic Spine Fx (N)	32.47 ± 20.0* -330.46 ± 80.7*	47.46* -893.60* ↑	48.75 ± 2.8 -450.78 ± 118.6	309.63 ↑ -256.69	227.60 ↑ -381.24	19.69 -434.75
Thoracic Spine Fz (N)	397.18 ± 1.4* -180.94 ± 43.3*	385.11* ↓ -196.87*	377.68 ± 28.6 -256.83 ± 152.0	706.34 ↑ -220.75	620.05 ↑ -495.95 ↑	198.47 ↓ -146.61
Thoracic Spine My (Nm)	41.00 ± 15.9* -51.68 ± 5.1*	49.06* -99.44* ↑	50.38 ± 8.4 -66.81 ± 12.7	34.56 -64.26 ↑	62.91 -60.43	39.93 -55.42
Lumbar Spine Fx (N)	36.90 ± 13.0 -372.13 ± 34.4	106.64 ± 63.0 -559.78 ± 50.6 ↑	109.67 ± 57.8 ↑ -440.91 ± 87.3	23.94 -513.74 ↑	335.11 ↑ -718.35 ↑	18.82 -1173.38 ↑
Lumbar Spine Fz (N)	326.88 ± 36.3 -374.89 ± 89.4	308.82 ± 73.8 -666.53 ± 112.3 ↑	286.38 ± 7.8 -403.88 ± 92.5	472.38 ↑ -289.21	436.70 ↑ -517.75	98.17 ↓ -1302.37 ↑
Lumbar Spine My (Nm)	75.74 ± 16.1 -19.10 ± 10.6	66.58 ±35.9 -21.65 ± 7.9	64.25 ± 7.4 -25.25 ± 6.1	69.55 -90.00 ↑	70.40 -55.69 ↑	95.89 -2.33

* For these responses, the actual number of responses equals n -1.

Table 5. - Dynamic rear impact Hyge sled test results: Series 2 and 3, ΔV = 16 kph.

ΔV = 16 kph	**Seat B (n = 2)**	**Seat B-R (n = 2)**	**Seat C (n = 2)**	**Seat D (n = 1)**	**Seat E (n = 1)**	**Seat F (n = 1)**
Dynamic Deflection (Deg)	15.3 ± 0.4	NA	8.3± 0.4	9.0	6.5	3.0
Post-test Seat Back Angle (Deg)	4.0 ± 1.4	NA	3.0 ± 0.7	4.0	1.0	-1.5
Up Neck Fx (N)	211.88 ± 10.0 -96.77 ± 3.1	176.47 ± 0.2 ↓ -135.40 ± 15.1 ↑	206.39 ± 3.7 -208.86 ± 17.6 ↑	223.87 -126.15 ↑	284.54 ↑ -153.01 ↑	215.99 -235.70 ↑
Up Neck Fz (N)	572.24 ± 35.5 -39.05 ± 11.3	589.58 ± 27.4 -108.91 ± 14.3 ↑	660.53 ± 64.8 -54.03 ± 16.8	655.19 ↑ -60.63	529.31 -113.69 ↑	651.63 ↑ -178.50 ↑
Corrected Upper Neck My (Nm)	7.05 ± 0.7 -10.84 ± 1.8	9.82 ± 0.1 ↑ -14.90 ± 0.3 ↑	9.42 ± 2.2 -11.81 ± 1.8	9.15 ↑ -7.53	11.94 ↑ -6.16 ↓	13.56 ↑ -10.63
Lower Neck Fx (N)	499.07 ± 24.5 -122.36 ± 21.8	499.29 ± 2.2 -113.65 ± 8.7	429.97 ± 22.9 ↓ -80.80 ± 3.0 ↓	425.94 ↓ -96.70	478.60 -115.56	454.14 -173.29 ↑
Lower Neck Fz (N)	251.93 ± 14.7 -57.29 ± 40.0	320.68 ± 19.0 ↑ -178.28 ± 5.9 ↑	457.41 ± 87.9 ↑ -64.14 ± 2.3	471.93 ↑ -121.92	324.97 ↑ -230.41 ↑	332.75 ↑ -227.92 ↑
Lower Neck My (Nm)	19.42 ± 1.8 -82.77 ± 3.1	26.13 ± 3.5 ↑ -86.39 ± 1.0	30.32 ± 12.0 -93.28 ± 0.2 ↑	11.23 ↓ -80.70	30.66 ↑ -88.01	50.10 ↑ -94.36 ↑
Corrected Lower Neck My (Nm)	20.72 ± 5.2 -59.39 ± 3.2	24.82 ± 4.8 -63.53 ± 1.9	37.65 ± 6.1 ↑ -59.48 ± 3.2	19.80 -59.45	30.09 -70.04 ↑	52.40 ↑ -68.66 ↑
Thoracic Spine Fx (N)	32.01 ± 5.9 -587.36 ± 22.3	60.20 ± 29.0 -1689.57 ± 27.4 ↑	73.26 ± 21.5 ↑ -678.32 ± 196.3	415.00 ↑ -270.30 ↓	46.66 ↑ -432.35 ↓	22.04 -476.38 ↓
Thoracic Spine Fz (N)	636.73 ± 14.8 -240.86 ± 13.3	371.47 ± 13.7 ↓ -480.32 ± 57.3 ↑	581.35 ± 135.9 -341.85 ± 18.0 ↑	1028.38 ↑ -352.42 ↑	773.95 ↑ -648.07 ↑	507.96 ↓ -187.79 ↓
Thoracic Spine My (Nm)	28.27 ± 3.2 -74.05 ± 7.5	39.29 ± 4.8 ↑ -201.73 ± 1.4 ↑	68.91 ± 6.7 ↑ -92.69 ± 5.8 ↑	41.43 ↑ -85.06	63.90 ↑ -79.22	31.28 -73.93
Lumbar Spine Fx (N)	49.99 ± 28.3 -570.90 ± 14.8	57.33 ± 14.3 -994.57 ± 8.2 ↑	94.41 ± 82.6 -753.71 ± 10.4 ↑	6.93 -1232.54 ↑	395.03 ↑ -952.32 ↑	148.43 ↑ -1499.87 ↑
Lumbar Spine Fz (N)	547.10 ± 20.8 -441.49 ± 38.7	296.84 ± 23.9 ↓ -1560.36 ± 90.0 ↑	344.74 ± 19.6 ↓ -575.46 ± 192.6	777.51 ↑ -594.25 ↑	365.98 ↓ -807.76 ↑	206.39 ↓ -1724.04 ↑
Lumbar Spine My (Nm)	80.57 ± 3.1 -23.42 ± 2.8	87.77 ± 8.8 -11.43 ± 5.0 ↓	88.27 ± 60.9 -17.06 ± 11.5	78.18 -116.26 ↑	80.80 -28.22	93.97 ↑ -3.03 ↓

Table 6. - Dynamic rear impact Hyge sled test results: Series 2 and 3, ΔV = 24 kph.

ΔV = 24 kph	Seat B (n = 2)	Seat B-R (n = 2)	Seat C (n = 1)	Seat D (n = 1)	Seat E (n = 2)	Seat F (n = 1)
Dynamic Deflection (Deg)	30.3 ± 5.3	NA	17.5	15.0	6.5 ± 1.4	1.5
Post-test Seat Back Angle (Deg)	14.5 ± 4.2	NA	5.0	5.0	1.0 ± 0.0	1.0
Up Neck Fx (N)	161.11 ± 73.4 -35.44 ± 18.9	206.65 ± 41.1 -172.38 ± 12.1 ↑	240.64 -190.91 ↑	167.42 -86.91 ↑	241.42 ± 36.0 -193.20 ± 12.1 ↑	199.98 -175.61 ↑
Up Neck Fz (N)	662.53 ± 108.5 -29.18 ± 13.6	874.48 ± 106.3 -60.18 ± 16.9 ↑	940.75 ↑ -86.02 ↑	724.13 -40.04	541.10 ± 35.5 -122.81 ± 25.7 ↑	505.29 -138.29 ↑
Corrected Upper Neck My (Nm)	4.81 ± 0.01 -10.62 ± 1.3	14.43 ± 0.01 ↑ -14.02 ± 1.0 ↑	13.94 ↑ -8.66	13.25 ↑ -7.75 ↓	19.71 ± 1.4 ↑ -11.29 ± 3.6	12.22 ↑ -10.74
Lower Neck Fx (N)	580.46 ± 84.9 -54.03 ± 32.4	785.96 ± 3.2 ↑ -145.09 ± 2.2 ↑	698.78 -114.22	382.71 ↓ -136.33 ↑	597.74 ± 31.4 -112.69 ± 27.9	433.68 -100.52
Lower Neck Fz (N)	436.79 ± 133.4 -84.25 ± 1.8	545.10 ± 6.0 -117.14 ± 9.3 ↑	633.40 -244.68 ↑	598.70 -100.97 ↑	336.89 ± 27.9 -194.02 ± 16.5 ↑	282.58 -156.26 ↑
Lower Neck My (Nm)	2.13 ± 1.3 -98.81 ± 12.1	40.97 ± 2.8 ↑ -120.04 ± 4.2 ↑	24.98 ↑ -114.34	19.15 ↑ -89.38	45.71 ± 2.4 ↑ -93.88 ± 3.2	33.09 ↑ -84.63
Corrected Lower Neck My (Nm)	4.68 ± 1.5 -67.8 ± 1.4	39.86 ± 1.6 ↑ -79.34 ± 9.1 ↑	37.45 ↑ -75.00 ↑	21.86 ↑ -55.07 ↓	43.14 ± 0.1 ↑ -69.91 ± 4.3	37.93 ↑ -65.00
Thoracic Spine Fx (N)	63.12* -817.99*	33.89* ↓ -2160.39* ↑	37.04 ↓ -841.12	507.07 ↑ -222.40 ↓	45.40 ± 31.4 -420.31 ± 3.2 ↓	21.01 ↓ -498.62 ↓
Thoracic Spine Fz (N)	851.35* -77.57*	348.23* ↓ -463.04* ↑	592.03 ↓ -566.23 ↑	1319.28 ↑ -165.15 ↑	742.15 ± 73.9 -365.23 ± 132.0 ↑	393.16 ↓ -197.71 ↑
Thoracic Spine My (Nm)	8.26* -98.03*	70.47* ↑ -226.52* ↑	51.47 ↑ -89.58	15.47 ↑ -88.21	47.98 ± 15.6 ↑ -90.90 ± 12.5	32.54 ↑ -77.00
Lumbar Spine Fx (N)	79.33 ± 32.7 -660.7 ± 60.7	65.65 ± 28.5 -1176.50 ± 71.1 ↑	56.04 -687.22	10.03 ↓ -994.13 ↑	65.98 ± 55.4 -1190.95 ± 48.1 ↑	21.64 -1532.78 ↑
Lumbar Spine Fz (N)	733.03 ± 90.6 -397.92 ± 133.6	264.30 ± 57.8 ↓ -1294.37 ± 125.8 ↑	455.48 ↓ -837.56 ↑	1241.88 ↑ -263.14	295.88 ± 59.0 ↓ -937.19 ± 98.1 ↑	123.34 ↓ -1821.90 ↑
Lumbar Spine My (Nm)	46.90 ± 20.4 -48.0 ± 7.1	117.15 ± 27.7 ↑ -7.44 ± 2.7 ↓	79.11 -26.00 ↓	70.92 -151.05 ↑	93.45 ± 10.6 ↑ -18.07 ± 16.1 ↓	103.40 ↑ -4.22 ↓

* For these responses, the actual number of responses equals n -1.

Table 7. - Dynamic rear impact Hyge sled test results: ΔV = 40 kph.

ΔV = 40 kph	Seat B (n = 1)	Seat B-R (n = 2)	Seat C (n = 1)	Seat D (n = 1)	Seat E (n = 1)	Seat F (n = 1)
Dynamic Deflection (Deg)	Total	NA	Total	22.0	8.0	1.5
Post-test Seat Back Angle (Deg)	Total	NA	Total	6.5	3.0	0.5
Up Neck Fx (N)	99.99 -68.94	324.99 ± 11.6 ↑ -129.33 ± 5.3 ↑	254.74 ↑ -65.87	238.46 ↑ -123.97 ↑	291.57 ↑ -217.55 ↑	328.17 ↑ -197.40 ↑
Up Neck Fz (N)	768.61 -20.19	1284.58 ± 152.9 ↑ -78.62 ± 3.2 ↑	1160.93 ↑ -124.05 ↑	1388.67 ↑ -45.90 ↑	819.32 -114.27 ↑	958.99 -188.51 ↑
Corrected Upper Neck My (Nm)	4.34 -28.42	16.03 ± 0.04 ↑ -17.96 ± 2.4 ↓	15.63 ↑ -54.91 ↑	15.66 ↑ -11.10 ↓	21.74 ↑ -12.89 ↓	17.03 ↑ -9.12 ↓
Lower Neck Fx (N)	451.92 -35.46	1273.46 ± 18.2 ↑ -132.66 ± 50.3	1172.05 ↑ -470.15 ↑	568.90 ↑ -222.80 ↑	887.38 ↑ -124.46 ↑	947.87 ↑ -127.44 ↑
Lower Neck Fz (N)	611.16 -105.33	682.55 ± 391.0 -106.33 ± 49.7	957.65 ↑ -942.98 ↑	1030.16 ↑ -397.47 ↑	505.29 -234.01 ↑	452.36 ↓ -233.70 ↑
Lower Neck My (Nm)	0.13 -122.36	32.72 ± 4.0 ↑ -167.15 ± 10.3 ↑	2.09 ↑ -177.27 ↑	22.63 ↑ -167.89 ↑	50.64 ↑ -126.20	43.70 ↑ -138.85
Corrected Lower Neck My (Nm)	5.83 -106.30	31.88 ± 5.0 ↑ -108.2 ± 6.1	19.24 ↑ -207.1 ↑	28.93 ↑ -145.90 ↑	50.87 ↑ -88.47	43.62 ↑ -91.44
Thoracic Spine Fx (N)	NA* NA*	-1.02* -3147.85*	448.36 -880.26	645.40 -24.59	NA NA	2.88 -693.44
Thoracic Spine Fz (N)	NA* NA*	416.87* -573.35*	1573.70 -469.71	1679.56 -217.60	1067.96 -317.14	600.48 -279.69
Thoracic Spine My (Nm)	NA* NA*	43.58* -369.56*	11.65 -316.57	20.86 -124.05	62.78 -153.54	36.20 -88.61
Lumbar Spine Fx (N)	105.20 -636.06	42.68 ± 66.3 -1807.44 ± 19.2 ↑	158.26 ↑ -1096.43 ↑	167.87 ↑ -424.92 ↓	18.77 ↓ -1667.56 ↑	4.91 ↓ -1718.71 ↑
Lumbar Spine Fz (N)	1071.08 -521.31	390.67 ± 1.7 ↓ -1963.12 ± 268.9 ↑	1812.56 ↑ -717.91 ↑	1265.01 -58.09 ↓	370.61 ↓ -1111.11 ↑	145.85 ↓ -1946.44 ↑
Lumbar Spine My (Nm)	0.73 -187.21	163.60 ± 14.2 ↑ -10.16 ± 8.1 ↓	0.32 ↓ -238.50 ↑	0.03 ↓ -184.38	124.50 ↑ -3.10 ↓	123.71 ↑ -1.84 ↓

* For these responses, the actual number of responses equals n -1.

SEVERE AND FATAL INJURIES IN REAR IMPACTS

Michael B. James, Robin L. Decker and Charles Y. Warner
Collision Safety Engineering, Inc. --- Orem, Utah, USA

ABSTRACT

Although rear impacts are numerous, most are of low severity and therefore are not frequently sampled in detailed field accident data bases. Only in recent years has there been sufficient data collected to develop a clear picture of injury contact sources and harm distributions for the occupants involved in rear impacts.

Much attention has been focused on reducing severe and fatal injuries that occur in rear impacts by seat design alternatives, primary among which have been proposals to reduce seat back deformation. In this study, field accident data have been evaluated to determine the types of severe and fatal injuries that occur in rear impacts, and their likely injury mechanisms.

Severe and fatal injuries in rear impacts are relatively infrequent. Most severe and fatal rear impact injuries occur in very high energy accidents where seat back deformation is not a factor, either because the injury is the result of other impacts, or because the seat back was supported by deformed vehicle structure. The majority of rear impact Harm is concentrated in non-severe and non-fatal injuries resulting from much more frequent low- and moderate-speed impacts. These data support the conclusion that stronger seat back designs aimed only at reducing severe and fatal injuries may increase exposure to less severe injuries and actually increase the total rear impact Harm.

FIELD ACCIDENT DATA: DEFINITION OF REAR IMPACTS

CLASSIFICATIONS - Comparing field accident data studies can be difficult because there are few standardized definitions and classifications used in various studies of automobile accidents. Descriptions of rear impacts can be particularly confusing. This is partially because there are several different ways to define rear impacts. Some accident studies classify all accidents by "type" such as head-on, side impact, rear impact, rollover, etc. A rear impact, when defined by this method, is usually the front of one vehicle contacting the rear of another vehicle. The occupants and associated injuries for both the frontally-damaged vehicle and the rear-impacted vehicle may be grouped together in some accident data files. Single vehicle accidents are usually placed in a separate category when using this method, regardless of the location of the vehicle damage.

Another approach is to classify vehicles not by accident "type", but by the nature of the impact experienced by each individual vehicle. There are two general ways to define the impact: one is by area of damage, the other by the Principal Direction of Force (PDOF). Both of these parameters are contained in the Collision Deformation Classification (CDC), which provides a detailed uniform method to code the type, location, and extent of damage sustained by a vehicle (SAE 1980). Using this categorization, rear impacts would include vehicles with either back damage (letter #1 = B), or a PDOF between 5 and 7 o'clock. These variables focus on the damage sustained by each vehicle, regardless of accident "type"; therefore no distinction is made between single-vehicle and multi-vehicle accidents.

Appropriate classification of rear impact data depends on the application of the data. Data categorized by accident "type" are useful in analyzing accident causation and accident avoidance issues such as highway design, driver behavior, or early warning systems. Two such areas of recent interest concerning rear impacts have been the effectiveness of center high-mounted brake lights and early warning crash avoidance systems (Knipling et al. 1993; Rausch et al. 1982; Somers and Hansen 1984).

Data categorized by area of damage and PDOF are useful in analyzing crashworthiness issues such as how occupants are injured, the effectiveness of restraint systems, and the distribution of injury exposure. It is also appropriate to use this categorization when estimating the safety effectiveness of different seat designs.

SAMPLING CRITERIA - A complicating factor in descriptions of rear impact field accident data is that the percentage of all impacts which are considered "rear impacts" changes dramatically, depending on the accident sampling base. Rear impacts have been referred to as the most common type of accident, yet there are relatively few data available for study in most field accident data bases. This is because there are many low-speed rear impacts which are too minor to be reported and sampled for collection in established data bases.

There are three broad categories currently in use to define field accident data bases: 1) police-reported accidents, 2) tow-away accidents, and 3) fatal injury accidents. When the sampling criterion requires higher accident severity for inclusion, a higher concentration of the more severe exposures will result. In general, rear impacts tend to occur at relatively low accident severities and rarely involve serious injury. Because of this, as the criteria become more stringent, the percentage of all impacts categorized as rear impacts decreases. Table 1 shows the percentage of rear-impact accidents and rear-impacted vehicles for these three sampling criteria (NHTSA: 18 refs).

Table 1 - Percentages of Rear Impact Accidents and Vehicles: NHTSA Reports: NASS, GES, CDS, FARS, Accident Facts

Percentage of Total Accidents /Vehicles	Police Reported	Tow-aways	Fatals
Rear Impact Accidents	30.7 %	21.0 %	4.6 %
Rear Damaged Vehicles	18.2 %	7.0 %	5.1 %

CRASHWORTHINESS AND FIELD ACCIDENT DATA

Vehicle crashworthiness is typically evaluated through analytical engineering, crash testing, and field accident data analysis. There are strengths and weaknesses to each unique method, and each one is important to a more complete understanding of crashworthiness. Field accident data are particularly useful for determining safety priorities, relating exposure levels to injury mechanisms and evaluating different vehicle design alternatives.

With regard to seat back strength for example, there have historically been two different design philosophies (James et al. 1991; Warner et al. 1991; Strother et al. 1994, NHTSA, 1989; Saczalski et al. 1993; Muzzy et al. 1992). One approach advocates seat backs that can yield and absorb energy during a rear impact, to provide occupant "ride-down." The other approach advocates seat backs that are, in most cases, "rigid" in an attempt to prevent occupant contacts with rear interior structures and ejection through rear portals during severe rear impacts. The concept of ridigized seat backs is based upon the whole body human tolerance data developed for military applications. These data come from tests where the occupant's torso, head and neck are held firmly against the seat back and the occupant is anticipating the impact. This testing established a human tolerance level in rear collision tests of 40 G's. In automobile accidents the occupants will not necessarily be positioned in this fashion. The injury mechanisms will be different, and the injury thresholds will be substantially lower.

Field accident data are useful in evaluating the desirability of yielding versus rigid seat back designs. Many such studies use a concept called Harm. Harm calculation is a method developed by the NHTSA for weighting injuries according to both their frequency and severity to provide a more complete picture of the societal harm that occurs with an injury. (Data Link 1990; Malliaris et al. 1985) The 1990 study by Data Link, Inc. used the 1979 - 1986 NASS data to examine occupant protection issues in rear impacts. This was the first study with sufficient data to present an in-depth evaluation of injury sources in rear impacts. The Harm associated with various areas of occupant contact were tabulated in the Data Link study. Table 2 summarizes this data for front seat occupants in rear impacts. It shows that the injury mechanisms which rigid seat backs are designed to minimize (i.e. occupant contact with structures in the rear of the vehicle and contact with the ground) do not represent a significant portion of the total Harm. The dominant injury sources are non-contact (which usually represents "whiplash"- type neck trauma), contact with frontal interior surfaces (usually from elastic rebounding off the seat back), and contact with the seat structure (seat back or head restraint).

Table 2 - Percent Harm Distribution by Injury Source, Front Seat Occupants in Rear Impacted Vehicles: NASS 1979 - 1986 (from Datalink ,1990) Sample size 5456 unrestrained and 1943 restrained occupants, before inflation for national estimates.

Harm Impact Site	All	Unrestrained	Restrained
Front Interior	14.7	14.3	15.9
Side Interior	6.2	3.9	12.8
Seat Back	5.1	5.1	5.1
Head Restraint	7.1	3.6	16.8
Other Interior	1.3	1.3	1.3
Roof	1.6	0.9	3.6
Floor	0.8	1	0.3
Rear	2.1	2.8	0.1
Other	0.6	0.7	0.1
Non-Contact	37.6	42.2	24.6
Ground	0.3	0.4	0
Unknown Source	22.5	23.6	19.4
Total	100	100	100

An additional study of field accident data also found that rear impacts contribute a very small portion of the total number of serious and fatal injuries associated with automobile accidents (James et al. 1991). Fatalities in rear impacted passenger cars were less than 4% of the total. Only 8% of all AIS 3+

injuries were associated with rear impacts. That study concluded: 1) rear impacts do not account for a significant portion of automobile injuries, 2) current production yielding seat backs provide a high level of protection in rear impacts, and 3) injury mechanisms which might be addressed by rigid seats make up a very small proportion of rear impact injuries.

RECENT FIELD ACCIDENT DATA: NASS, CDS, GES - Additional field accident data have become available since the 1990 Data Link and the 1991 James studies. The number of documented cases has increased as each accident year has been added, and there have been changes in the government-sponsored data collection programs as well. The NASS program began in 1979 to sample motor vehicle accidents. In 1988, the NASS system underwent a reorganization which divided it into two separate programs: the General Estimates System (GES) and the Crashworthiness Data System (CDS). The GES is a probability sample of police reported tow-away accidents. Approximately 48,500 accidents (120,000 occupants) are sampled annually, but no independent NHTSA sponsored accident investigations are performed. All data are taken from the police report of the accident. Injury data are limited to the police reported "KABCO" scale which categorizes each occupant's condition as fatal (K), incapacitating injury (A), non-incapacitating injury (B), possible injury (C), or no injury (O).

The CDS is also a probability sample of police-reported tow-away accidents, but these selected cases receive a full, in-depth accident investigation conducted by specially trained NHTSA investigation teams. Approximately 5,000 accidents (11,300 occupants) are sampled annually in CDS. The investigations are similar to those performed for the 1979-1987 NASS cases, but are more detailed. The CDS sampling protocol biases the selection toward late model vehicles and more seriously injured occupants. The 1988 - 1994 CDS field data contain 3,548 rear impacted passenger cars, vans, light trucks, and utility vehicles (compared to 5,621 rear impacted vehicles contained in the earlier NASS years).

The accumulating data continue to show that current production vehicles are performing quite well in rear impacts. Figure 1 shows the distribution of Harm by the area of vehicle damage using the combined NASS and CDS. Rear impacts account for only 3% of the total Harm. Table 3 shows the distribution of injuries from the 1988-1993 GES for all accident modes. Rear impacted vehicles are over-represented in the "possible injury" category, but are under-represented in the actual injury categories. These injury data are from police reports which contain fewer details, and are less reliable as the NASS and CDS injury data. Figure 2 shows the combined NASS and CDS distribution of severe Maximum Abbreviated Injury Scale (MAIS) 4+ injuries by area of vehicle damage. Rear impacted vehicles account for only 1% of these injuries.

Table 4 shows the distribution of injury sources for front seat occupants in rear impacts using the 1988-1994 NASS CDS data. The distribution from the 1979-1986 NASS presented earlier in Table 2 is included in Table 4 for comparison. The CDS data show a higher percentage of Harm for contacts with structures in front of the occupant, and a lower percentage of Harm in the non-contact category compared to the earlier NASS data. The reason for the different distributions is not known, but it may be because the CDS data represents more severe accidents and later model vehicles. However, both the earlier NASS and more recent CDS data agree that Harm from contacts in the rear of the vehicle and from ejection do not constitute a relatively significant safety concern. In fact, those two categories of Harm together represent only about 3% of the rear impact Harm; or about one tenth of one percent of the total automobile accident Harm.

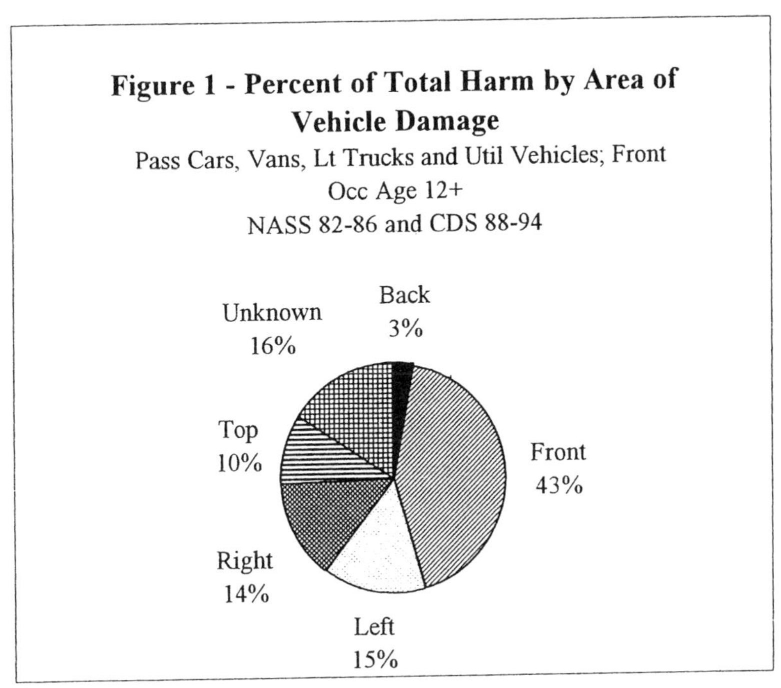

Figure 1 - Percent of Total Harm by Area of Vehicle Damage
Pass Cars, Vans, Lt Trucks and Util Vehicles; Front Occ Age 12+
NASS 82-86 and CDS 88-94

The distribution of accident severity is also important in evaluating crashworthiness. Figure 3 shows the distribution of injury in rear impacts by the change in velocity (Delta-V). Note that the line representing all occupants exposed to rear impacts essentially overlays the line representing occupants with AIS 0-3 injuries. This indicates that almost all injured occupants received injuries of AIS 3 or lower. Figure 3 also shows that approximately 85% of all occupants exposed to a rear impact were in impacts of 32 km/h (20 mph) or lower Delta-V. Similarly, 85% of all occupants with AIS 0-3 injuries were

Table 3 - Percent Injury KABCO Scale vs Area of Vehicle Damage: Front Seated, Outboard Occupants Ages 12+ in Light Vehicles (1988-1993 GES) Sample Size 466,807 occupants before inflation for national estimates.

Injury Reported	No injury	Poss. Injury	Non-Incapac. Injury	Incap-acitating Injury	Fatal	Un-known
KABCO Code	(O)	(C)	(B)	(A)	(K)	-----
Non-Collision	74.9	9.7	9.5	4.7	0.42	0.82
Front	78.9	10.2	6.7	3.2	0.19	0.95
Right Side	83.4	9.0	4.5	2.2	0.19	0.71
Left Side	81.6	10.1	4.8	2.5	0.20	0.79
Back	76.9	18.2	2.7	1.3	0.04	0.97
Top	76.1	7.9	11.3	3.33	0.74	0.66
Under	83.5	7.4	4.9	3.3	0.16	0.70

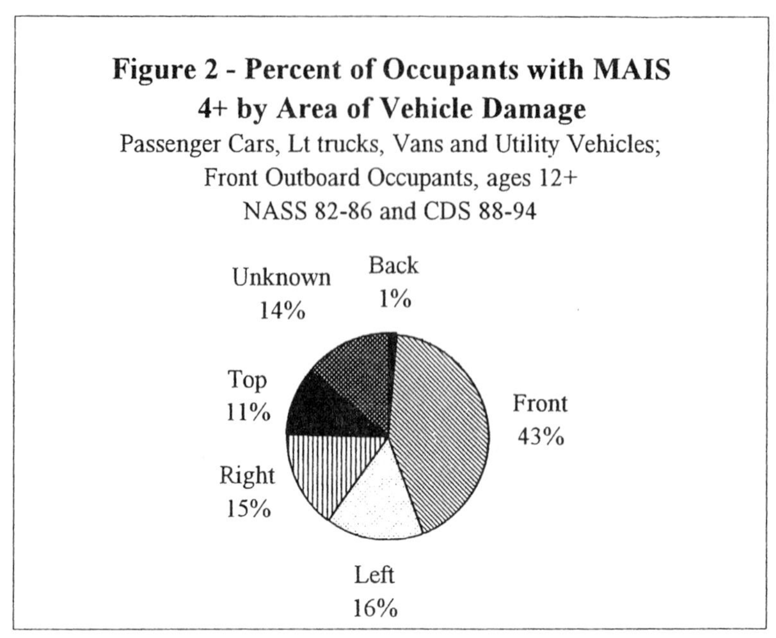

exposed to impacts of 32 km/h (20 mph) or lower Delta-V. Severely injured occupants (AIS 4-6) were involved in more severe rear impacts, 50% of which occurred in impacts of almost 48 km/h (30 mph) or higher Delta-V. These distributions are based upon tow-away accidents only. If non-tow-away accidents could be sampled, the number of severely injured occupants would not be expected to increase significantly, but it is not clear what effect there would be with regard to the number of occupants with minor injuries because there is reason to believe AIS 1 whiplash injuries maybe relatively frequent events in such non-tow-away accidents.

INDIVIDUAL CASE STUDIES - NASS AND CDS

BACK DAMAGED VEHICLES - While severe and fatal injuries in rear impacts are relatively infrequent, mitigating these injuries is the subject of current research. Determining the mechanisms of these injuries, and their relationship to current production yielding seat backs is one way of evaluating the potential benefits of new, alternative seat designs. An extensive search of the 1979-87 NASS and the 1988-94 CDS was made; including all passenger cars, utility vehicles, light trucks and vans with "back damage." Occupants were limited to those seated in the front outboard positions, who were 12 years of age or older. Table 5 indicates that in these 16 years of accident data (which includes 174,545 vehicles), there were only 82 fatally injured occupants and 28 non-fatal occupants with an MAIS of 4+.

It should be noted that NASS cases are selected using a complex sampling protocol, in which accidents involving severe and fatal injuries are sampled more frequently than accidents involving minor injuries. Each case, therefore, is assigned a weighting factor which is the inverse of the probability of selecting that accident. Weighted data can be used to do so, but when there are few data (such as in this situation), it is inappropriate to use weighted data to make national estimates. It must be noted that unweighted data include an unnaturally large number of fatalities and serious injuries, since severe accidents are sampled at a higher rate. Nonetheless, unweighted data are very useful to help understand significant injury patterns.

Table 4 - Percent Harm Distribution by Injury Source, Front Seat Occupants in Rear Impacted Vehicles: NASS CDS 1988-1994 compared to NASS 1979-1986. Sample Size 1988-1994 1074 unrestrained, 2242 restrained occupants before inflation for national estimates

Injury	All		Belted		Unbelted	
	79-86	88-94	79-86	88-94	79-86	88-94
Front	14.7	23.9	15.9	21.9	14.3	26.0
Side	6.2	5.6	12.8	3.1	3.9	8.3
Seat Back	5.1	10.3	5.1	11.3	5.1	9.2
Head Restraint	7.1	6.6	16.8	7.2	3.6	5.9
Other Int.	1.3	4.6	1.3	7.4	1.3	1.7
Roof	1.6	-----	3.6	-----	0.9	-----
Floor	0.8	1.6	0.3	1.3	1.0	1.9
Rear	2.1	2.6	0.1	3.6	2.8	1.6
Other	0.6	2.5	0.1	1.9	0.7	3.1
Non-Contact	37.6	20.3	24.6	20.5	42.2	20.0
Ground	0.3	1.2	0.0	0.0	0.4	2.6
Fire	-----	3.3	-----	4.5	-----	2.0
Unknown	22.5	17.5	19.4	17.3	23.6	17.7
Total	100	100	100	100	100	100

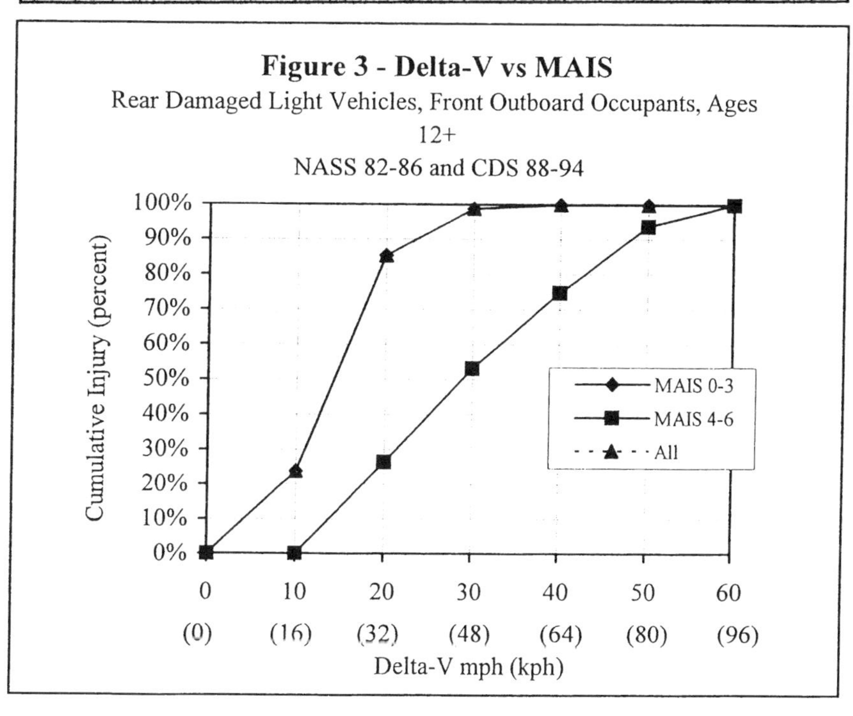

Table 5 - Severe & Fatally Injured Occupants: NASS 1979 - 1987 and CDS 1988 - 1994 [Rear Impacted Pass Cars, Light Trucks, Vans and Utility Vehicles; Front Seated Outboard Occupants, Ages 12+]

Accident Year	Fatal	Non-Fatal MAIS 4+	Accident Year	Fatal	Non-Fatal MAIS 4+
1979	3	1	1988	18	3
1980	1	0	1989	7	3
1981	3	0	1990	7	4
1982	2	0	1991	11	1
1983	0	2	1992	4	3
1984	2	3	1993	7	3
1985	3	1	1994	5	4
1986	7	0	Σ (16 yrs)	82	28
1987	2	0	Ave	5.13	1.75

A subset in which the seat back performance can be reasonably linked to occupant injury potential was selected from the original 110 unweighted cases. Accident vehicles involving a subsequent rollover, fire, or ejection through portals forward of the B-pillar were eliminated, removing 35 of the fatal occupants and six of the MAIS 4+ injured occupants, leaving 47 fatal occupants and 22 MAIS 4+ injured occupants, for a total of 69 cases.

CRASH SEVERITY - Only 23 (33%) of these 69 cases reported a calculated Delta-V. In Table 6 these cases are grouped into three accident severity categories. 70% of these cases with a known Delta-V have a Delta-V greater than 35 km/h (22 mph) and over half have a Delta-V over 64 km/h (40 mph). It is unlikely that seat back stiffness would have any effect on the injury outcome in cases where the Delta-V exceeds 64 km/h.

Table 6 - Delta-V of Serious & Fatally Injured Occupants: NASS 1979 - 1987 and CDS 1988 - 1994 [Rear Damaged Passenger Cars, Light Trucks, Vans and Utility Vehicles; Front Seated Outboard Occupants, Ages 12+; Rollover, Fire and Ejection through Frontal Portals Not Included]

Velocity Change	Fatality and MAIS 4+ Injury	Percent
0-33.6 kph (0-21 mph)	7	30%
35-62 kph (22-39 mph)	4	17%
64 + kph (40 + mph)	12	53%
Total	23	100%

Injury occurrence can also be examined based on another measure of accident severity; the amount of deformation experienced by the rear impacted vehicle. The CDC damage extent code can be used as a measure of deformation, and is reported for all 69 cases. This damage extent code divides the rear half of the vehicle into eight sections from the rear bumper to the B-pillar of the vehicle (SAE,1980). Deformation into the area forward of the B-pillar would be considered Extent Zone 9. The distribution of the CDC Damage Extent code for the 69 cases (including the 23 where the Delta-V was known) is shown in Table 7.

Table 7 - CDC Deformation Extent Zone Distribution: NASS 1979 - 1987 and CDS 1988 - 1994 [Rear Damaged Passenger Cars, Lt Trucks, Vans and Utility Vehicles; Front Seated Outboard Occupants, Ages 12+; Rollover, Fire and Ejection through Frontal Portals Not Included]

CDC Damage Extent Zone	Number of Cases	Percent
0 - 5	18	26.1
6	10	14.5
7-9	41	59.4
Total	69	100.0

Stiffening the seat back would have no effect in cases where the seat back is already supported by vehicle structure. In vehicles where the crush reaches Damage Extent Zone 6 or greater, it is reasonable to assume that most front seat backs would be supported by the deformed rear vehicle structure.

There are 26 injuries to 17 occupants among the 69 cases in Table 7 where the CDC Damage Extent is less than 6 (see Table 8). In 12 of these cases, the vehicle experiences at least one additional impact. Many of the severe injuries are not attributable to the rear impact, but appear to be the consequence of other impacts. Interestingly, <u>none</u> of the AIS 4+ injuries are the result of occupant contact with rear interior surfaces of the vehicle, nor are they related to occupant ejection out of the rear of the vehicle. Only one neck injury is tabulated. It is not clear from these data that any of these injuries would have been mitigated by non-yielding seat backs.

When cases of all impact severities are included (all 69 cases) there are a total of 110 AIS 4+ injuries. Only 10% of these injuries are reported as having resulted from occupant contact with some rear interior component of the vehicle or an object external to the vehicle (see Table 9).

HARD COPY NASS CDS FILES - Cases where the CDC extent is less than 6, or where the known Delta-V is less than 64 km/h (40 mph) would be the most likely candidates for circumstances where stiffer or rigid seats could possibly mitigate severe and fatal injuries. Therefore, the hard copy files for the 19 such cases included among the 69 cases of Table 7 were obtained and subjected to further analysis. The CDS investigator notes and vehicle photographs from the 19 cases were used to evaluate the effect of seat back yield on injury causation.

It was not possible to determine the exact mechanism of injury for every case from the information in the file. Any injury that could potentially be attributable to the rear impact was grouped together with the injuries that clearly were associated with the rear impact. This review led to two significant observations. First, for more than half of these severely injured occupants, the severe injury was not associated with the rear impact. Second, most of the seat backs for these severely injured occupants had either not yielded rearward at all, or if they had yielded rearward it was to an angle less than 45 degrees beyond the vertical. Some of the seat backs were unable to yield rearward because of vehicle structure, such as the bulkhead in a two-seat sports car or the cab in a pickup truck. Other seat backs were prevented from yielding rearward by vehicle crush. Of these 19 cases, injury was associated with the rear impact in nine cases. In three of those cases there was no visible seat back

Table 8. AIS 4+ Injuries in Rear-Impacted Vehicles with CDC Damage Extent 0 to 5: NASS 1979 - 1987 and CDS 1988 - 1994 [Rear Damaged Passenger Cars, Lt Trucks, Vans and Utility Vehicles; Front Seated Outboard Occupants, Ages 12+; Rollover, Fire and Ejection through Frontal Portals Not Included]

Case Number	Body Region	Contact Source	Secondary PDOF
3	Head	Roof	--
10 (4 Injuries)	Chest	Rt. Side Interior	3- O'clock
10	Head	Seat Back Sup't.	3- O'clock
12	Chest	Steering Wheel	1- O'clock
46 (2 Injuries)	Head	Lt. Side Interior	Unknown
52 (2 Injuries)	Abdomen	Unknown	12- O'clock
60	Chest	Rt. Side Interior	3- O'clock
60	Head	Seat Back Sup't.	3- O'clock
85 (2 Injuries)	Head	Lt. B-Pillar	3- O'clock
86	Neck	Head Restraint	1- O'clock
88	Head	Unknown	8- O'clock
92	Chest	Lt. Side Interior	9- O'clock
93	Chest	Lt. Side Hdwe	10- O'clock
97	Chest	Steering Wheel	12- O'clock
98	Chest	Other Occupant	12- O'clock
100 (2 Injuries)	Head	Head Restraint	--
104	Chest	Steering wheel	--
105	Head	B-Pillar	3- O'clock
107	Head	Unknown	--

Table 9. AIS 4+ Injuries in Rear-Impacted Vehicles: NASS 1979 - 1987 and CDS 1988 - 1994 [Rear Damaged Passenger Cars, Lt Trucks, Vans and Utility Vehicles; Front Seated Outboard Occupants, Ages 12+; Rollover, Fire and Ejection through Frontal Portals Not Included]

Injury Source	Number of Injuries	Percent
Steering Assembly	25	23 %
Seat Back / Head Restraint	18	16 %
B-Pillar / Side Structure	13	12 %
Rear Header / External Objects	11	10 %
Roof	9	8 %
Other	3	3 %
Unknown Source	31	28 %

yield, in an additional five of the nine cases, there was seat back yield to an angle less than 45 degrees, and in <u>only</u> <u>one</u> of the nine cases had the seat back yielded to an angle beyond 45 degrees.

These data support three findings:

1) Many severe and fatal injuries in rear impacted vehicles are actually the result of an additional impact from another direction.
2) Most severe and fatal injuries in rear impacts are not caused by rearward yielding of the seat back. They occur in seat backs that have not yielded.
3) Stiffened or rigid seat backs per se will not substantially mitigate severe and fatal injuries in rear impacts.

ACCIDENT DATA AND PROPOSED DESIGN CHANGES

The above findings highlight two important factors with regard to accident severity. First, since most Harm occurs at low accident severity levels, vehicle safety systems designed to protect occupants from high severity impacts should be configured so that they do not tend to increase occupant injury exposure in low and moderate severity impacts. Second, there are practical limits to protecting occupants in high severity impacts.

The laws of physics and physiology are such that a safety system usually cannot give optimum performance at every accident severity and occupant condition. A safety system designed to reduce occupant injury at severe exposures often tends to increase occupant injuries received at less severe exposures. Most design choices are a tradeoff in this regard, and accident severity distributions must be taken into account when making reasoned choices. Horsch explained why a design alternative which lowers occupant injury response at a specific impact severity is not necessarily the best design alternative at other severities (1987). He presents an example of a hypothetical side impact design change that reduces 50% of the injuries occurring at Delta-V's above 43 km/h (27 mph), and adds only 1% to the injury probability for Delta-V's below 21 km/h (13 mph). The overall result would predict a 12% increase in overall side impact injury. This is due to the large number of injuries below 21 km/h (13 mph) and the small number of injuries above 43 km/h (27 mph). Norin et al. presents a similar study using different seat belt systems in frontal impacts (Norin et al. 1991). These researchers have shown that the risk of increasing injury over the entire range of accident severities increases as the design test speed increases.

Accidents involving severe and fatal injuries receive more public attention than those involving minor and moderate injuries. However, severe and fatal injury accidents must be seen in the context of the entire population of accidents. For example, high-energy air bag systems were developed in the mid-seventies with a stated design goal of providing frontal impact protection at Delta-V's of 80 km/h (50 mph). Current redesigned "30 mph" airbag systems have already generated concerns about deployment-induced injuries and have let to "de-powering" of current airbags. One can imagine the injuries that un-moderated "50 mph" high-energy deployments might have created at more-frequently-encountered 24-48 km/h (15-30 mph) crash speeds. This reality is particularly significant in rear impacts, where the majority of total Harm comes from low severity injuries and where the severe and fatal injuries occur in very severe accidents.

Figure 4 shows the distribution of rear impact Harm over the range of accident severities using 16 km/h (10 mph) increments of Delta-V with Harm divided into two categories, that associated with less serious injuries, MAIS 1-3, and that associated with severe injuries, MAIS 4-6. All paralyzing neck and spine injuries, for example, would be contained in the latter category. The

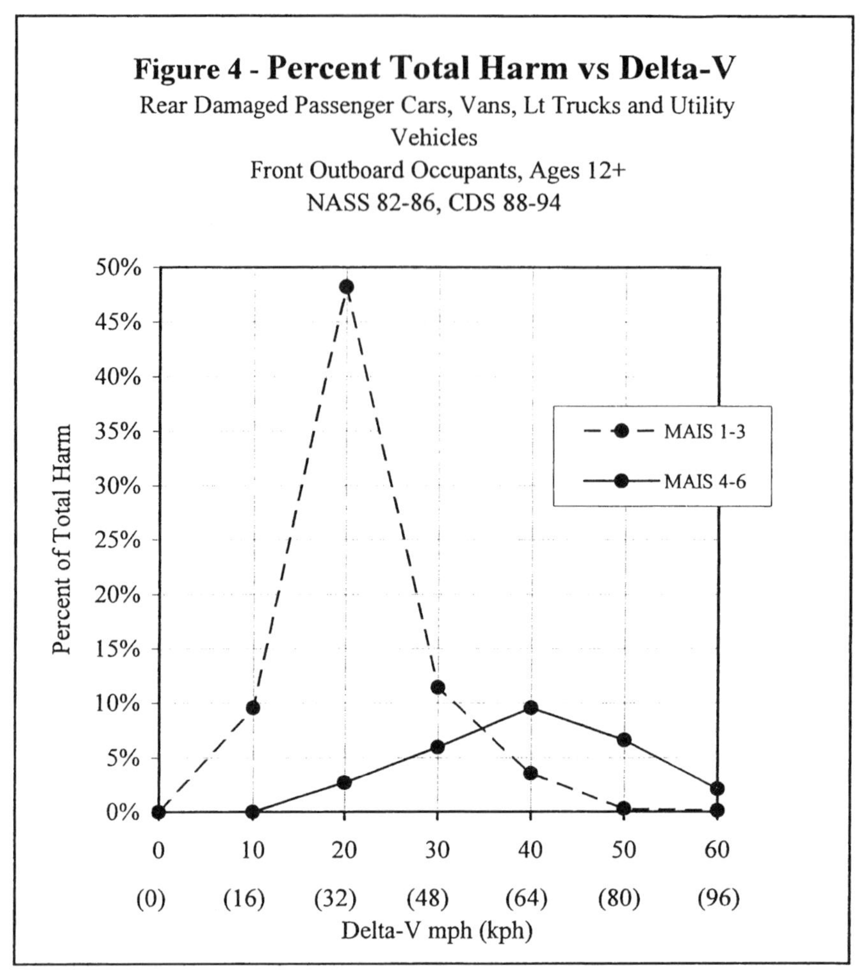

figure indicates that the Harm associated with severe injuries is only a small fraction of the total rear impact Harm. Not surprisingly, the Harm associated with the more severe injuries occurs in relatively high energy impacts, the mean Delta-V being about 64 km/h (40 mph). A much larger portion of the total Harm is associated with MAIS 1-3 injuries. The mean Delta-V associated with these injuries is about 27 km/h (17 mph), an impact severity involving only about 1/5th of the energy of a 64 km/h (40 mph) Delta-V collision.

This distribution of rear impact Harm must be appreciated when addressing potential safety benefits of proposed design alterations to current production seat designs. An hypothetical safety system which could eliminate all MAIS 1- 3 Harm in rear impacts with a Delta-V of 24 km/h or less would reduce the total rear impact Harm by about 30 percent. It would be impossible to achieve this same Harm reduction by concentrating on the relatively rare MAIS 4 to 6 injuries, even if one could eliminate every one of them, and assuming that it could be done without any increase in the MAIS 1 to 3 Harm.

NEW SEAT DESIGNS - Recently, patents, popular press articles, and trade show presentations have described innovative concepts designed to reduce occupant injury exposure in rear impacts. One such concept is a self-aligning head restraint developed by Viano et. al. (Raynal 1996; Birch 1996). As the occupant's shoulders load the seat back, a simple lever mechanism displaces the head restraint forward and upward. It reportedly improves dummy kinematics and reduces Hybrid III neck forces by supporting the head early in the crash sequence. Since stiffer seat backs will require more effective head restraints to avoid increasing whiplash motions, such an articulating head restraint might allow seat backs to be incrementally strengthened while improving whiplash protection.

Another new concept is called the "Catcher's Mitt" seat, also developed by Viano, et al. (Adler 1996; Birch 1996). This concept incorporates a relatively stiff seat back perimeter frame with a yielding seat back cushion suspension, into which the occupant can "pocket". It is suggested that this concept improves "occupant retention" in impact severities up to about 32 km/h (20 mph) Delta-V without increasing head/neck forces.

In contrast to extremely stiff or rigid seat back designs, these concepts have a potential to improve occupant response in the region of severities where rear impact Harm is concentrated. The true effectiveness will have to be established using actual human field performance data, since the mechanisms of "whiplash" injury are still not understood well enough to allow an objective dummy measurement. Recent studies suggest that the sociological factors associated with whiplash may be as significant as the physiological factors (Schrader 1996). However, the field accident data suggest that it is highly unlikely that these new seat concepts will affect the majority of the more serious and fatal rear impact injuries, which most frequently occur at extreme levels of rear impact severity.

CONCLUSIONS

1. Fatalities and MAIS 4+ injuries are relatively rare events in rear impact collisions and tend to occur at very high impact severities.

2. Many severe and fatal injuries in rear impacts are the result of additional impacts from directions other than the rear.

3. Most severe and fatal injuries in rear impacts are not associated with rearward yielding of seat backs; they occur with seat backs that do not yield. Therefore, it is not expected that stiffened or rigid seat backs will substantially mitigate severe and fatal injuries in rear impacts.

4. If alternative designs incorporate a stiffened or rigid seat, care must be taken that overall harm is not increased by exacerbating injuries now seen at low and moderate crash severities.

5. Self-aligning head restraints and other innovative concepts intended to improve seat performance have a greater potential to reduce harm in less severe rear impacts, occurring at lower speeds.

BIBLIOGRAPHY

Adler, A.; Safety Seat. Detroit Free Press. February 27, 1996.

Benson, B.R.; Smith, G.C.; Kent, R.W.; Monsen, C.E. and Warner, C.Y., "Effect of Seat Stiffness in Out-of-Position Occupant Response in Rear-end Collisions", Proc. 40th Stapp Conference, Albuquerque, NM, [SAE paper # 962434], November 1996.

Birch, S.; Self-Aligning Head Restraint - Delphi and Saab get Ahead on Seat Technology., Automotive Engineeringpg. 107. May 1996.

Data Link Inc; Current Issues of Occupant Protection in Car Rear Impacts. Prepared for the Office of Crashworthiness, Rulemaking, NHTSA, Washington, DC. Docket 89-20-NO1-021. February 1990.

Horsch, J.D.; Evaluation of Occupant Protection from Responses Measured in Laboratory Tests. SAE # 870222. SAE International Congress and Exposition. February 1987.

James, M.B.; Strother, C.E.; Warner, C.Y.; Decker, R.L.; Perl, T.R. ; Occupant Protection in Rear-end Collisions: I. Safety Priorities and Seat Belt Effectiveness. SAE #912913. 35th Stapp Conference. November 18, 1991.

Knipling, R.R.; Wang, J.S.; Yin, H.S.; Rear-End Crashes: Problem Size Assessment and Statistical Description. NHTSA. Office of Crash Avoidance Research. Washington, DC, May 1993.

Malliaris, A.C.; Hitchcock, R.,; Hansen, M.; Harm Causation and Ranking in Car Crashes. SAE# 850090. SAE International Congress and Exposition. February 25, 1985.

Muzzy, W.H.; Canton, A.E.; Eisentraut, D.K.; D'Aulerio, L.A.; Seat Back Yielding and Collapse: A Danger to Occupants During Real-World Impacts. 20Th Annual International Workshop: Human Subjects for Biomechanical Research. Seattle, Washington. November 1, 1992

NHTSA; 54 FR 40897. Federal Motor Vehicle Safety Standards; Seating Systems; Occupant Crash Protection; Seat Belt Assemblies. Request for Comments. October 4, 1989

NHTSA; Planning for Safety Priorities 1983 Safety Priorities Plan. DOT HS 806399. Washington, DC, April 1983.

NHTSA; Fatal Accident Reporting System 1988-1991 - A Review of Information on Fatal Traffic Crashes in the United States in 1988 DOT HS 807 507,693,794,&954, Washington, DC, December 1989-March 1993.

NHTSA; General Estimates System 1988-1991 - A Review of Information on Police-Reported Traffic Crashes in the United States DOT HS 807 607,665,&781 Washington, DC, August 1990-November 1991.

NHTSA; National Accident Sampling System 1983 - A Report of Traffic Accidents and Injuries in the U.S. DOT HS 806699, 867, 074, & 296, Washington, DC, January 1985 - July 1988.

NHTSA; National Accident Sampling System Crashworthiness Data System Washington, DC, 1988-1993. DOT HS 808 197 & 298, December 1994 - August 1995.

NHTSA; Traffic Safety Facts 1992-94 - A Compilation of Motor Vehicle Crash Data from the Fatal Accident Reporting System and the General Estimates System (Revised 1992 Data) DOT HS 808 022, 169, & 292, Washington, DC, March 1994-August 1995.

Norin, H.; Jernstrom, C.; Koch, M.; Ryberg, S.; Svensson, S.; Avoiding Sub-Optimized Occupant Safety by Multiple Speed Impact Testing. Proc. 13th International Conference on Experimental Vehicle Safety. Paper No. S9-O-05. Paris, France. November 4-7, 1991.

Rausch, A.; Wong, J.; Kirkpatrick, M.; A Field Test of Two Single Center, High Mounted Brake Light Systems. Accident Analysis and Prevention Vol. 14, No. 4, pp. 287-291. June 1984.

Saczalski, K.; Syson, S.R.; Hille, R.A.; Pozzi, M.C.; Field Accident Evaluations and Experimental Study of Seat Back Performance Relative to Rear-Impact Occupant Protection; SAE# 930346; Seating System Comfort and Safety. March 1 1993.

Raynal, W.; GM Head-Restraint System May Cut Whiplash by 40%. The Detroit News. February 27, 1996.

SAE; Collision Deformation Classification. SAE Recommended Practice J224.. March 1980.

Schrader, H.; Obelieniene, D.; Bovim, G.; Surkiene, D.; Mickeviciene, D.; Miseviciene, I.; Sand, T.; Natural Evolution of Late Whiplash Syndrome Outside the Medicolegal Context. The Lancet Vol 247, No. 9010, pp. 1207-1211. May 4, 1996.

Somers, R.L.; Hansen, A.; The Cost of Rear-End Collisions in Denmark and the Potential Savings From a High Center-Mounted Auxiliary Brake Light. Accident Analysis and Prevention Vol. 16, No. 5/6, pp. 421-432. June 1984.

Strother, C.E.; James, M.B.; Gordon, J.J.; Response of Out-of- Position Dummies in Rear Impact. SAE 941055. SAE International Congress and Exposition. February 1994.

Warner, C.Y; Strother, C.E.; James, M.B.; Decker, R.L.; Occupant Protection in Rear-End Collisions: II. The Role of Seat Back Deformation in Injury Reduction. SAE paper # 912913, 35th Stapp Car Crash Conference. November 18, 1991.

962434

Effect of Seat Stiffness in Out-of-Position Occupant Response in Rear-End Collisions

Brent R. Benson, Gregory C. Smith, and Richard W. Kent
Collision Safety Engineering

Charles R. Monson
GMH Engineering

ABSTRACT

Accident data suggest that a significant percentage of rear impacts involve occupants seated in other than a "Normal Seated Position". Pre-impact acceleration due to steering, braking or a prior frontal impact may cause the driver to move away from the seat back prior to impact. Nevertheless, virtually all crash testing is conducted with dummies in the optimum "Normal Dummy Seated Position". A series of 7 rear impact sled tests, having a nominal AV of 21 mph, with Hybrid III dummies positioned in the "Normal Dummy Seated Position", "Out of Position" and slightly "Out of Position" is presented. Tests were performed on yielding production Toyota and Mercedes Benz seats as well as on a much stiffer modified Ford Aerostar seat. Available Hybrid III upper and lower neck as well as torso instrumentation was used to analyze and compare injury potential for each set of test parameters. In all cases, neck forces and moments were found to increase when the dummy's torso was leaned forward at impact. For the out-of-position tests, the results showed that the upper neck loads were clearly related to seat stiffness; stiffer seats produced greater neck loading. The lower neck loads, however, showed no definite correlation with seat back stiffness.

INTRODUCTION AND BACKGROUND

Fatalities and serious injuries in rear-end collisions are relatively infrequent; only about four percent of all fatalities result from very severe collisions in this accident mode [James, 199 1]. About seventy percent of the societal harm from rear-end collisions is concentrated in much larger numbers of low-level injuries caused in crashes with ΔV's of less than 20 mph [James, 1996]. Despite the relatively low priority of rear-impact protection compared to other areas of automobile crash protection, and despite the relatively small percentage of harm associated with rear-impact accidents of severity greater than 20 mph, considerable research into improving crash protection in high-severity rear-end impacts is being conducted within the automotive safety community. Proposals that seats should be stiffened to help retain occupants in these high-severity rear-end crashes are countered by concerns that stiffening seat backs will swell the already much larger amassment of harm from the greater number of minor and moderate injuries that occur in lower-severity rear-end impacts [Warner, 199 1].

Historically, crash testing of occupant protection concepts has been performed using anthropomorphic test devices (crash dummies) which are placed and postured in what is designated the Normal Dummy Seated Position (NDSP). In this NDSP, the dummy sits in an erect posture with its buttocks squarely on the seat and its shoulder blades flush against the seat back. For a single-occupant seat, the head and chest are laterally centered on the headrest and seat back, respectively. Any positioning other than the NDSP can be thought of as Out-of-Position (OOP). Except for a 1994 study at Collision Safety Engineering and a few unintentional exceptions [Strother, 1994], virtually all rear-impact experiments reported in the literature have been conducted with the dummies in the NDSP. This is consistent with the need for standardized test protocols to facilitate performance measurement, maximize test repeatability, and improve comparison between tests. Exclusive NDSP testing may be inconsistent, however, with the need to gain a comprehensive understanding of the most effective methods of protecting occupants in rear-end impacts within the real-life environment. The Normal Seated Position (NSP), with the occupant's head in close proximity to the head restraint, is coincidentally the optimal position for occupant injury protection in rear-end impact, an intuitive fact demonstrated by the measured increase in injury levels for the few OOP tests that have been performed [Strother, 1994].

Data from the National Accident Sampling System (NASS) suggest that pre-impact vehicle motions in a significant percentage of rear-impacts may cause many accident victims to not be in the NSP when the impact occurs, even if they happened to be in such a position at the onset of the accident. Many occupants would be involuntarily moved from their initial positions by accelerations caused by abrupt pre-impact braking and/or steering. Rear impacts often happen because the vehicle in front did something unexpected, such as slamming on brakes, hitting the next car in front, or cutting in from another lane. Such actions logically lead to driver responses that tend to move occupants from their initial seated positions.

Other studies have focused on voluntary causes for occupants in rear-impacted automobiles to not be in the NSP. Mackay, et al found that people don't sit in cars the way dummies do, even when driving straight down the road, looking straight ahead [Mackay, 1995]. They found that drivers generally sit further forward in their seats than do crash dummies in the NDSP. This difference was particularly pronounced for small females, who were found to sit with their heads about three-and-a-half inches further forward than the corresponding crash test dummies. Of the population of drivers studied, nearly 50% sat with their heads six inches or more forward of their head restraints while driving straight down the road.

Personal observation and experience suggest several additional reasons for occupants of automobiles to be out-of-position, aside from the simple fact that they may not be comfortable sitting erect in the NDSP, especially for long periods of time. They may be taking advantage of being stopped at an intersection to stretch and readjust their position, or to retrieve something from a purse or bag or from the glove box. They may be reaching for controls, such as the tape player or air conditioner. They may be shifting gears with a manual stick shift. They may be interacting with children in the rear or passenger-side seat. They may have their head turned looking in the rearview mirror or talking with other occupants in the car or someone at the intersection, They may be adjusting their clothing, A study made by the authors at partially obstructed intersections found it very common for people to lean forward several inches to see around the obstructions before proceeding into traffic. There are obviously numerous causes for occupants to not be in the NSP.

That occupants may not be in the optimum restraint position is a fact that was recognized in the mid-1970's during research into passive frontal impact protection. Early air cushion restraint systems that ignored the out-of-position problem were later found to need modification in order to not increase injury levels to out-of-position occupants in near-threshold-deployment levels of frontal impact. Indeed, even some front passenger air bag systems built today may be overly aggressive when deployed to protect an occupant pitched forward by pre-impact braking, or when deployed into a child sitting on the edge of the seat. Hopefully, the lessons learned during these development years will not be ignored by those proposing dramatic seat redesigns.

Neglect of the OOP problem when evaluating seat designs for occupant protection in rear impacts may result in designs that have the potential to increase the overall harm caused by this accident mode. As is shown in the few OOP tests that have been performed, there can be significant differences between the NDSP and the OOP dummy responses and injury measurements [Strother, 1994].

PURPOSE OF STUDY

The purpose of this study was to examine and quantify occupant response data, as a function of seat back stiffness, experienced by a 50th percentile Hybrid III male anthropomorphic test dummy in a rear end collision. Upper and lower neck forces and moments will be examined for two different seated positions: the out-of-position (OOP) where the dummy was pitched forward and his head was 20 inches from the head restraint and also the normal dummy seated position (NDSP).

Automobile seats with markedly different seat back stiffnesses were used in this examination - two production automobile seats and one modified seat that has been proposed as an improvement in rear-impact protection [Saczalski, 1993].

TEST METHODOLOGY

Tests were carried out with the seats mounted to a sled which was impacted into a fixed barrier at a nominal speed of 19 mph, resulting in a ΔV of approximately 21 mph. As depicted in Figure 1, 86% of AIS O-3 injuries occur at speeds less than 20 mph, while only about 27% of AIS 4-6 (severe or greater) injuries occur in the O-20 mph speed range [James, 1996].

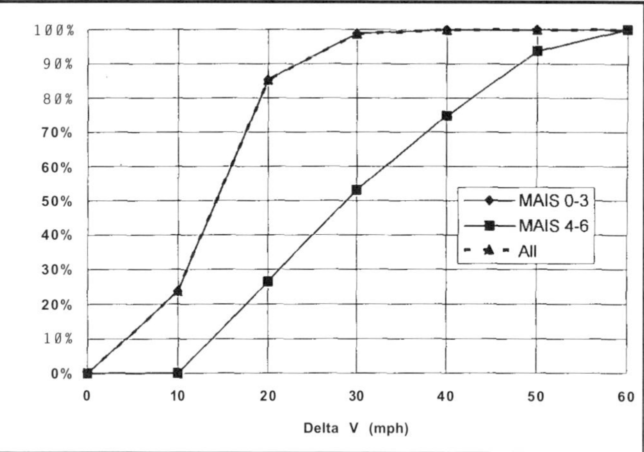

Figure 1: Delta V vs MAIS, NASS 82-86, CDS 88-94; Rear Damaged Light Vehicles; Front Outboard Occupants, Ages 12+

The sled tests were performed at the Collision Safety Engineering crash test facility according *to* the overall methodology outlined in a paper by Strother in 1994. [Strother, 1994]. A Hybrid III 50th percentile male anthropomorphic test device (ATD) was placed in each seat with the following instrumentation:

Upper neck tension-compression force, F_Z ;
Upper neck shear force, F_X ;
Upper neck extension/flexion moment, M_Y ;

Lower neck tension-compression force, F_Z ;
Lower neck shear force, F_X ;
Lower neck extension/flexion moment, M_Y ;

Lumbar tension-compression force, F_Z ;
Lumbar shear force, F_X ;
Lumbar extension/flexion moment, M_Y ;

Head Acceleration, G_X, G_Y, and G_Z ;
Thorax Acceleration in the mid-sagittal plane, G_X and G_Z .

For the OOP tests the Hybrid III dummy was pitched forward with his buttocks still correctly positioned in the seat cushion, A light weight string tether around the neck helped maintain a distance of 20 inches from the back of the head to the head restraint in all out-of-position tests.

The two production seats selected were driver's seats from a 1989 Toyota Corolla and a 1985 Mercedes 190E. The "limited-yield" seat was a 1988 Ford Aerostar driver's seat modified to render it very stiff as compared with the production seats. The seat in each of the tests was placed in the mid-position and the seat back angle was adjusted to 15 degrees. The inboard as well as the outboard aft edges were measured pre and post impact in order to document residual deformation. The static seat pull characteristics of these seats are presented in Table 1. As can be seen, the modified seat represents a significant increase in seat back stiffness, strength and energy absorption capability over the production seats.

	Stiffness (lbf/in)	Maximum Strength (lbf)	Total Absorbed Energy (ft-lbf)	% of Total Energy Restored
1988 Ford Aerostar (modified)	694.3	3632	1249.4	12.0
1985 Mercedes 190E	128.2	1144	299.7	50.4
1989 Toyota Corolla (4-dr)	100.0	784	400.0	28.2

Table 1: Static Seat Pull Characteristics

TEST RESULTS

A total of seven sled tests were performed, three with the modified Aerostar seat, two with the production Mercedes seat, and two with the production Toyota seat. Typical plots of sled acceleration and velocity during impact are shown in Figure 2. A summary of maximum (peak) values of the instrumented dummy measurements for the seven tests, along with the corresponding maximum 50th percentile male injury tolerance limits, is presented in Table 2.

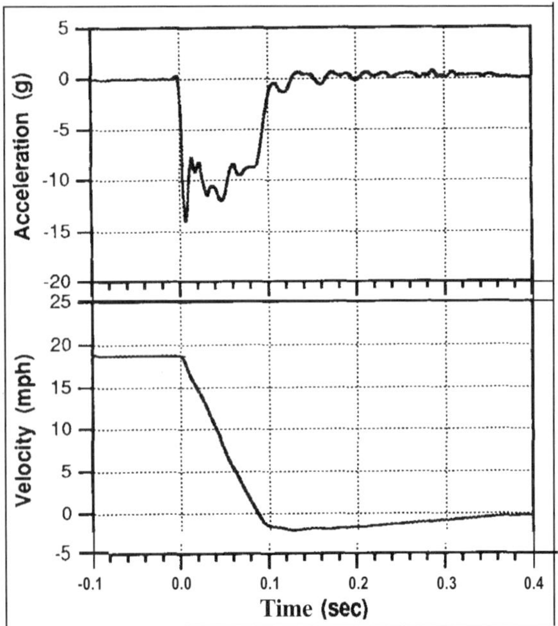

Figure 2: Typical Sled Motion

TESTS WITH MODIFIED AEROSTAR SEAT (S3, S2, S1)

Three modified Aerostar seats were tested, each with the Hybrid-III dummy in a different seated position: (S3) Normal Dummy Seated Position (NDSP), (S2) Torso pitched forward slightly (Slightly OOP), and (S1) Torso pitched forward (OOP).

TEST S3 - Normal Dummv Seated Position: This test was conducted with the back of the dummy against the seat back at impact. In this position, the back of the head was about 4.25 inches forward of the leading surface of the head restraint. Figure 3a through 3f are photographs taken from the high speed film showing the motion of the ATD and seat during the test.

Figure 3a is the photograph taken at time zero, the instant that the sled carriage contacts the fixed barrier. Figure 3b was taken at 46 ms into the event, corresponding to initial rearward seat back deformation, The chest acceleration has reached about 10 g at this point, and the lower neck extension torque is just starting to increase. Note that the head has yet to contact the head restraint. Figure 3c is at about 60 ms, at which time seat back deformation temporarily ceases. The lower neck extension torque is continuing to rise at the same steady rate started at about 46 ms. There is still no head contact with the head restraint at this time.

Figure 3d is taken at about 70 ms into the event. At this time, the dummy's head contacts the head restraint and the seat back deformation resumes. Figure 3e is at 76 ms, at which time the extension torque in the lower neck reaches a peak of about 1226 in-lbf (very nearly reaching the 1240 in-lbf injury limit). Note that at this time the head is in good alignment with the torso.

The last photograph in this series *is* Figure 3f, taken at 100 ms. The seat back dynamic deformation has reached its maximum value of about 7 deg; the seat back would be reclined a maximum of about 22 deg. From this position the seat back rebounds, pitching the dummy forward into the shoulder belt.

TESTS2 - Torso Pitched Forward Slightly: This test was conducted under the same conditions as Test S3, except that the torso of the dummy was leaned forward in the seat slightly, so that the displacement between the back of the head and the leading edge of the head restraint was about 6 inches.

At about 46 ms into the event, the dummy's shoulders are into the seat back and the first phase of seat back deformation begins. The torso deceleration has reached about 11 or 12 g at this time. At about 60 ms, the seat back deformation is temporarily halted. It is about this time that the dummy's head first contacts the head restraint. At about 80 ms into the crash,

Peak Measurement	units	S1		S2		S3		M1		M2		T1		T2		Stat Injury Criteria	
		Lo	Hi	Lo	Hi	Lo	Hi	Lo	Hi	Lo	Hi	Lo	Hi	Lo	Hi	Lo	Hi
Upper Neck Fx	(lbf)	-233	112	-114	52	-50	23	-56	67	-33	24	-108	31	-18	23	-697	697
Upper Neck Fz	(lbf)	-60	**937**	-32	450	-33	389	-46	**809**	-49	332	-106	737	-37	416	-899	742
Upper Neck My	(in-lbf)	**-682**	596	-293	241	-65	113	-261	451	-172	93	-329	571	-89	92	-505	1684
Lower Neck Fx	(lbf)	-117	196	-51	208	-30	147	-65	483	-35	163	-87	219	-15	197	-697	697
Lower Neck Fz	(lbf)	-99	690	-31	387	-43	353	-81	523	-81	245	-80	408	-61	365	-899	742
Lower Neck My	(in-lbf)	**-2838**	913	**-1720**	617	-1226	410	**-2984**	385	**-1269**	320	**-2959**	963	**-1474**	134	-1240	2200
Lumbar Fx	(lbf)	-129	487	-89	95	-44	88	-211	302	-76	96	-114	389	-63	114	-2401	2401
Lumbar Fz	(lbf)	-156	578	-167	529	-196	399	-322	649	-304	704	-247	450	-126	546	-1571	2850
Lumbar My	(in-lbf)	-2241	701	-559	416	-712	332	-1561	508	-994	502	-1905	101	-1254	230	-3269	10912
Peak Head Accel	(g)	*		50		38		57		26		52		30			
Peak Thorax Accel	(g)	29		27		22		21		17		26		15		60	
HIC (36ms)		*		319		149		451		59		437		61		1000	

Table 2: Summary of Test Results (*Head Z-axis accelerometer failed)

A:t=Omsec
Initial contact
Seat back angle 15°

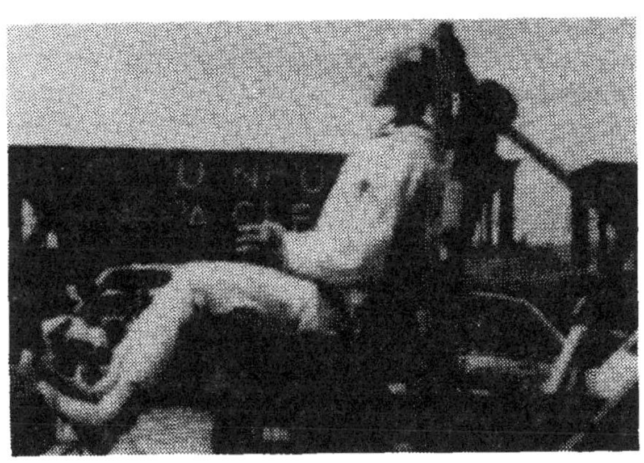

B: t = 46 msec
Seat back begins rearward deformation
Chest acceleration at 10 g's
Lower neck extension starting to increase

C: t = 60 msec
Seat back deformation temporarily ceases
Lower neck extension torque continues to rise

D: t = 90 msec
Dummy's head contacts restraint
Seat back deformation resumes

E: t = 76 msec
Extension torque in lower neck peaks at 1226 in-lbf

F: t = 100 msec
Dynamic seat back deflection ~ 7 °

Figure 3: (TEST S3 - Modified Aerostar Seat) - Normal Dummy Seated Position

the extension moment in the lower neck has reached a maximum of about 1720 in-lbf (exceeding the 1240 in-lbf injury criterion). Again at this time, the head and torso are in reasonably good alignment, with only a slight extension angle evidenced. At 92 ms into the event, the dummy has attained its maximum rearward excursion with little or no seat deformation over that achieved during the 46-60 ms time frame. The seat back recline is estimated from the high speed film at about 22-23 degrees (estimated dynamic seat back deflection of about 7-8 degrees). Finally, at 180 ms, the seat back has sprung forward, recovering a significant portion of its dynamic deformation and propelling the upper body of the dummy forward and slightly upward to re-encounter the shoulder belt. The measured residual seat back deformation was about 4 degrees.

TEST S1 - Torso Pitched Forward: This sled test was conducted under the conditions of Tests S3 and S2, except that the torso of the dummy was pitched forward in a more pronounced attitude. In this position, the back of the head was about 20 inches forward of the head restraint. Figures 4a through 4f document the overall response of the dummy and seat in this test.

Figure 4a is a photograph taken at the onset of the impact. Comparing this photo with Figure 3a, the difference in dummy positioning is clearly seen. Figure 4b is taken at about 62 ms, at which time the seat back yielding begins. The onset of seat deformation occurs later in this test compared with Tests S2 and S3 because of the greater movement required under this condition to place the shoulders of the dummy against the seat back. Figure 4c is at about 72 ms into the crash, at which time the seat back deformation stops. The seat back recline at this instant is estimated from the high speed film at about 23 deg, which would correspond to about 8 degrees of dynamic seat back deflection. As can be seen in this photo, the head is still some distance forward of the head restraint when seat back yield ceases. Figure 4d is at about 100 ms. It is around this time that the tension loads and the neck extension torque in the upper neck, as well as the extension torque in the lower neck begin their climb to values exceeding the injury criteria. Figure 4e is a photograph taken at about 114 ms into the impact - this is the instant where the upper neck tension and the upper and lower neck extension torques reach their above-criteria peak values. Again, the head is in only slight extension at this time. Maximum head extension, in fact, occurs at about 130 ms (Figure 4f) and appears to be less than 45 degrees. From this position, the dummy is thrown forward and slightly upward as energy stored in the seat back is released. About 3 degrees of residual seat back deformation was measured in this test.

SUMMARY - SLED TESTS WITH THE MODIFIED AEROSTAR SEATS

Three sled tests were conducted using the modified Aerostar seats. The dummy position was varied in these tests from the Normal Dummy Seated Position (Test S3) to a position slightly pitched forward (Test S2), and then to a position where the torso is pitched forward to where the head was about 15-16 inches forward of its NDSP position (Test S 1).

In the NDSP test, the dummy lower neck extension moment approached the limiting value for likelihood of injury. Only a slightly different head position in Test S2 relative to that of the NDSP test (Test S3) resulted in about a 50 percent increase in the neck torques, thereby exceeding the injury criterion. With the dummy leaned about 15-16 inches forward of its NDSP at impact (Test S1), the lower neck moment measurement increased markedly, well beyond the injury threshold. In this case, the tension force in the upper neck (937 lbf) also exceeded its respective injury criterion (742 lbf).

TESTS WITH THE MERCEDES SEAT

Two Mercedes seats were tested, one in the Normal Dummy Seated Position (NDSP), and one with the dummy pitched forward (OOP) at impact.

TEST M2 - Normal Dummy Seated Position: Test M2 was conducted with the Mercedes seat with the anthropomorphic dummy in the Normal Dummy Seated Position (NDSP). In this position, the dummy's shoulders are firmly against the seat back, and the back of the dummy's head is about 2 inches forward and upward from the front surface of the head restraint. Figures 5a through 5f document the response of the dummy and the seat during this test.

Figure 5a is at the onset of the deceleration, i.e. at time zero. The seat back is positioned in the erect position so that the initial seat back recline is about 15 deg. In Figure 5b, at time 36 ms, the seat back is just beginning to yield rearward. The chest A-P acceleration is climbing up to a plateau of about 7.5 g, which will be reached at about 48 ms into the event. In Figure 5c, at 124 ms, the head just contacts the head restraint as seat back yield continues in the upper portion of the seat back frame. In Figure 5d, at 132 ms, the lower neck extension moment enters a period of relatively rapid increase, reaching a peak of 1269 in-lbf at about 142 rns, Figure 5e. This torque level is just above the criterion value of 1240 in-lbf.

Finally, at about 180 ms, the yielding of the upper portion of the seat back ceases and the dummy begins its rebound phase of motion (Figure 5f). The maximum head extension angle occurs at about this time. As measured from the high-speed film, the seat back attains a maximum angle of about 50 degrees (outboard side) to 57 degrees (inboard side) for a maximum dynamic deflection of about 35-42 degrees. A residual seat back angle of about 50.6 degrees was measured on the inboard side of the seat post-impact indicating a recovery of approximately 7 degrees.

TEST Ml - Torso Pitched Forward: Test Ml was conducted under the same conditions as Test S 1 except that the Mercedes seat was employed rather than the modified Aerostar seat. That is, the dummy torso was pitched forward so that, at impact, the back of the head was about 20 inches forward of the head restraint and about 18 inches

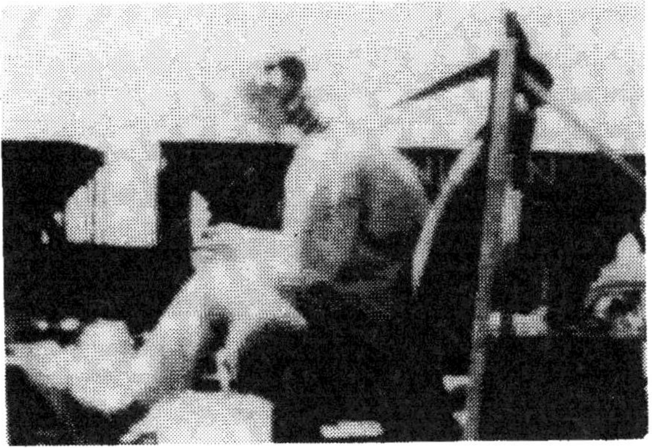

A: t = 0 msec
Initial contact
Seat back angle 15"

D: t = 100 msec
Upper neck tension and extension begin climbing

B: t = 62 msec
Seat back begins to yield

E: t = 114 msec
Injury criteria values are exceeded in the upper neck tension and upper and lower neck extension torques

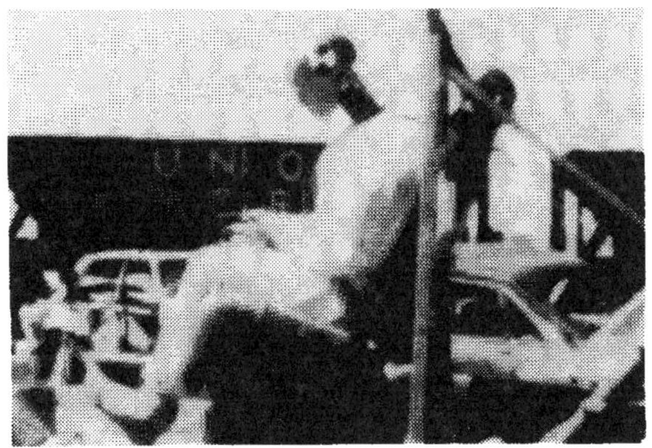

C: t = 72 msec
Seat back deformation stops
Seat back angle at ~ 23 °

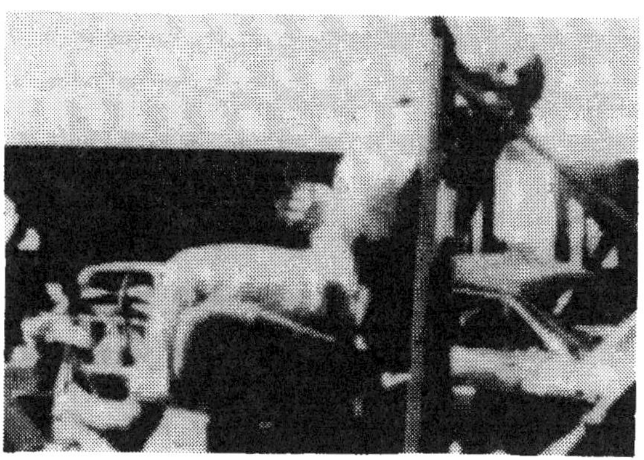

F: t = 130 msec
Maximum head extension

Figure 4: (TEST S1 - Modified Aerostar Seat) - Out-of-Position

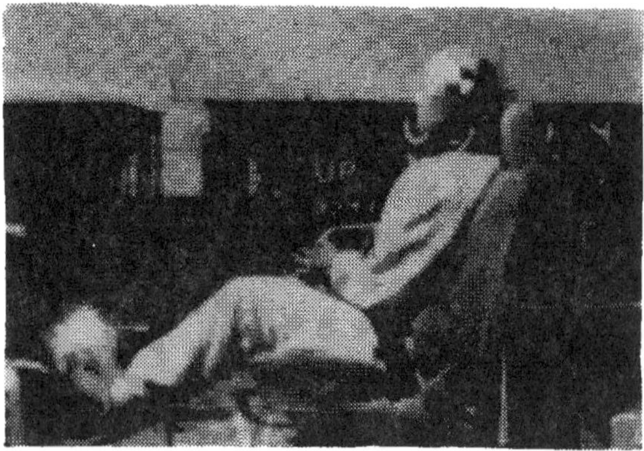

A: t=Omsec
Initial contact
Seat back angle 15°

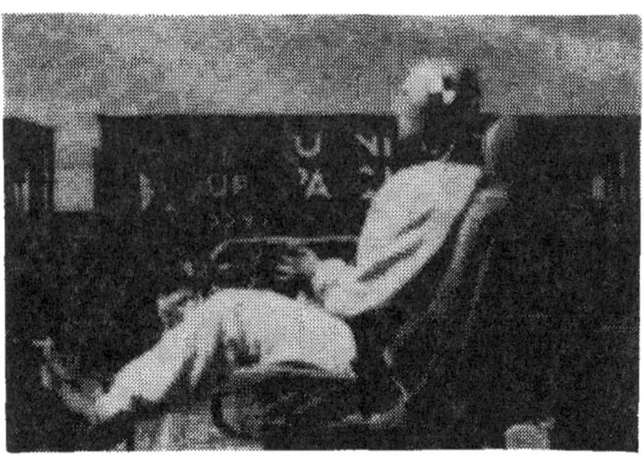

B: t = 36 msec
Seat back begins to yield
Chest acceleration 7.5 g's

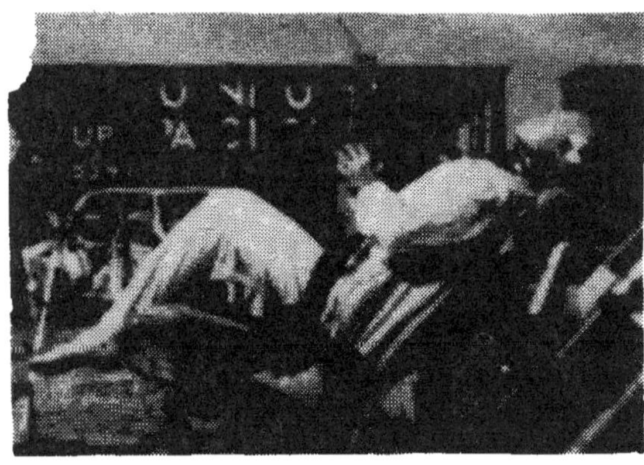

C: t = 124 msec
Head contacts head restraint
Seat back yield continues

D: t = 132 msec
Lower neck extension values increase rapidly

E: t = 142 msec
Lower neck extension peaks at 1269 in-lbf

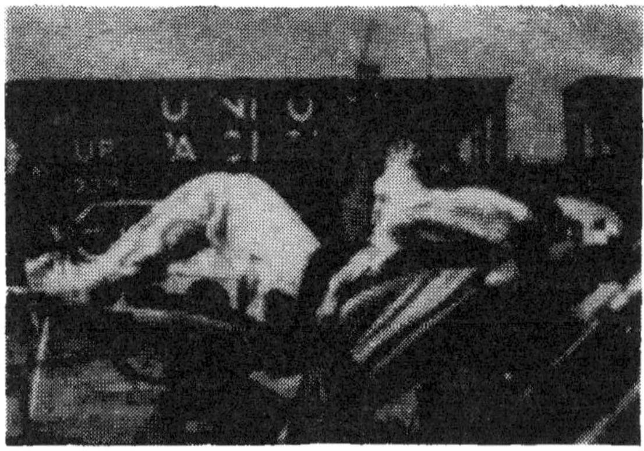

F: t = 180 msec
Seat yielding ceases, dummy begins rebound
Maximum head extension
Dynamic seat back deflection to ~ 35-42°

Figure 5: (TEST M2 - Mercedes Seat) - Normal Dummy Seated Position

A:t=Omsec
Initial contact
Seat back angle 15°

B: t = 70 msec
Shoulders have reached seat back
Seat back begins to yield

C:t= 100 msec
Seat back yield ceases

D: t = 108 msec
Dummy's head rotated into contact with head restraint
Second phase of seat back yield begins

E:t= 150msec
Seat back yielding ceases
Maximum dynamic seat back deformation at -45 °

F: t = 400 msec
Rebound phase

Figure 6: (TEST M 1 - Mercedes Seat) - Out-of-Position

forward of its NDSP. Figures 6a through 6f illustrate the dummy kinematics and seat response during this test.

Figure 6a is taken at the initiation of the impact. Comparing this photo with Figure 5a, one can see the similarity in dummy pre-impact positioning. Figure 6b is at about 70 ms into the event, at which time the shoulder area of the dummy torso has reached the seat back and the seat back starts its initial phase of rearward yield. During this first phase of seat back yield, the upper neck tension force and the lower neck extension moment start increasing (i.e. at about 75-80 ms). Figure 6c is taken about 100 ms into the impact, the time at which this initial phase of seat back yield is over. At about 108 ms (Figure 6d), the dummy's head has rotated into contact with the head restraint and the second phase of seat back yield starts. This second phase of seat back deformation lasts until about 150 ms (Figure 6e), at which time the seat attains its maximum dynamic seat back angle of about 45 degrees (for a maximum dynamic seat back yield of about 30 deg). During this time period, i.e. around 125 ms, the upper neck tension force and the lower neck extension moment reach peaks that exceed their respective injury criteria. Figure 6f is taken at about 400 ms into the event, at which time the dummy is reacting to the release of energy stored in the seat back by rebounding upward and forward. About 17 degrees of permanent seat back deformation was measured after this test.

SUMMARY OF TESTS WITH THE MERCEDES SEATS

Two tests were conducted with the Mercedes seats; one test with the ATD in the Normal Dummy Seated Position (Test M2), and the other test with the dummy torso pitched forward at impact. In both tests, the lower neck extension moment exceeded the injury criterion. In addition, during the OOP test, the upper neck tension reached a peak of just over 800 lbf, exceeding the injury criterion of 742 lbf.

A comparison of the neck forces and torques achieved with the Mercedes seat and those achieved by the modified Aerostar seat, indicates that the forces and torques are comparable, despite the fact that the permanent deformations of the Mercedes seats were significantly greater than those of the modified Aerostar seats. Interestingly, the residual seat deformations were greater in the tests with the NDSP dummies as compared with those experienced in the OOP tests.

TESTS WITH THE TOYOTA SEAT

Two Toyota seats were tested, one in the Normal Dummy Seated Position (NDSP), and one with the dummy pitched forward (OOP) at impact.

TEST T2 - Normal Dummy Seated Position: Test T2 was conducted using the Toyota seat, with the anthropomorphic dummy in the Normal Dummy Seated Position. With the shoulder area of the dummy pressed against the seat back, this puts the rear surface of the head about 1 ½ inches forward and slightly up from the leading surface of the head restraint.

Figures 7a through 7f document the response of the dummy and seat in this test. Figure 7a is at the initiation of the impact. Figure 7b is taken at 40 ms into the event, at which time the seat back begins yielding. The A-P chest deceleration has increased to about 10 g at this point in time, and the onset of seat back deformation is accompanied by a temporary drop in chest deceleration, which returns to the 10 g level at about 50 ms, climbing slightly thereafter. Figure 7c is at 110 ms, at which time the head contacts the head restraint. It is at this time that the lower neck extension moment begins to rise more rapidly to its peak of 1474 in-lbf at about 130 ms (Figure 7d). At about 150 ms the seat back yield ceases as the seat back attains its peak deflection. From film analysis, it is estimated that the maximum seat back angle attained was approximately 60 deg (Figure 7e). With an initial recline of 15 deg, this would correspond to about 45 degrees of dynamic seat back deformation. Post impact seat back recline angle was measured at 60 deg, indicating little seat back recovery.

TEST T1 - Torso Pitched Forward: Test Tl was conducted using the Toyota seat under the same conditions as Tests S 1 and M1, with the dummy torso pitched forward at impact. Figures 8a through 8f document the dummy kinematics and seat response for this test.

Figure 8a is at time zero and Figure 8b is at about 60 ms when the seal back deformation begins. In Figure 8c, at 130 ms, the head first contacts the head rest. During the period between Figures 8c and 8d the lower neck extension torque begins to increase. In Figure 8d, at 140 ms, this extension torque reaches a maximum of about 2959 in-lbf, well above the 1240 in-lbf criterion and at a level comparable to that in the other OOP tests with the modified Aerostar seat and the Mercedes seat. At 170 ms into the event, Figure 8e, the seat back attains its maximum deformation, with the backrest tilted rearward at about a 45 degree angle, corresponding to about 30 degrees of dynamic seat back deformation. As can be seen in Figures 8d and 8e, the head and neck extension is not excessive in this test. Finally, in Figure 8f, the dummy is undergoing its mild rebound motion during seat back unloading. The residual seat back recline measured after the test was about 40 deg, indicating about 5 degrees of seat back angle recovery.

SUMMARY OF TESTS CONDUCTED WITH THE TOYOTA SEATS

Two tests were conducted with the Toyota seats, one with the dummy in the Normal Dummy Seated Position and the other with the dummy torso pitched forward at impact. Looking first at the test in the NDSP, the resulting neck forces and torques in this test (T2) are comparable to those obtained in the earlier tests with the modified Aerostar seat and the Mercedes seat. Examining the tests conducted with the dummy torso pitched forward at impact, the neck loads sustained by the dummy in the Toyota seat are lower than those sustained by the dummy seated in the other two seats. As was the case with the modified Aerostar seat and the Mercedes seat, the test with the NDSP occupant produced

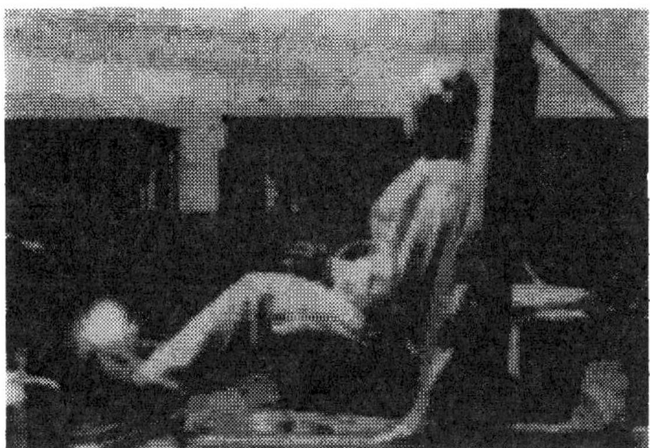

A:t=Omsec
Initial contact
Seat back angle 15°

D: t = 130 msec
Lower neck moment peaks

B: t = 40 msec
Seat back yield begins
Chest acceleration 10 g's

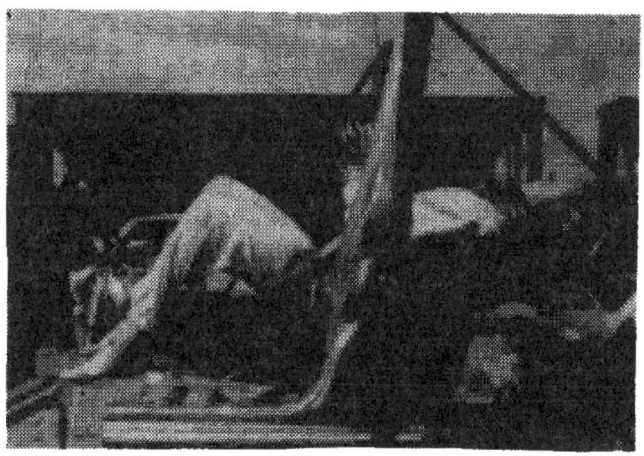

E: t = 150 msec
Seat back yield ceases
Maximum dynamic deflection at -60"

C: t = 110 msec
Head contacts the head rest
Lower neck extension moment begins to rise

Figure 7: (TEST T2 - Toyota Seat) - Normal Dummy Seated Position

A: t = 0 msec
Initial contact
Seat back angle 15°

B: t = 60 msec
Seat back deformation begins

C: t = 130 msec
Head contacts head restraint

D:t= 140msec
Extension torque reaches a maximum

E: t = 170 msec
Dynamic seat back deflection maximum at -45 °

F: t = 210 msec
Dummy is rebounding

Figure 8: (TEST T1 - Toyota Seat) - Out-of-Position

more seat back deformation than the test with the dummy torso pitched forward at impact.

DISCUSSION OF TEST RESULTS

On the basis of the seven tests conducted during this program it appears that, if one ignores the slightly excessive lower neck extension torques (relative to the accepted injury criteria [Mertz, 1968]), these three seat designs are capable of providing reasonable protection to a belted 50th percentile male occupant in a nominally 21 mph AV rear impact where that occupant is in the Normal Dummy Seated Position. When the occupant's upper torso is pitched forward at the instant of impact, however, neck forces and moments increase. This increase is relatively more dramatic as seat stiffness increases, with greater loads produced by stiffer seats. In fact, when the dummy head was simply moved forward less than 2 inches relative to the NDSP, the dummy in the modified Aerostar seat sustained an increase in the lower neck extension torque of almost 50 percent (from 1226 in-lbf to 1720 in-lbf).

Figure 9 shows the measured upper-neck compression and tension loads as functions of the time duration for which the measurements exceeded the plotted values. The time duration-dependant injury criteria for tension and compression of the 50th percentile male upper neck are also plotted for comparison. Tests S 1 (Modified Aerostar Seat), Ml (Mercedes Seat) and T1 (Toyota Seat) have been plotted such that occupant response as a function of the seat stiffness could be examined. A similar plot is presented in Figure 10 which shows the measured upper-neck fore-aft shear loads and corresponding 50th percentile male injury

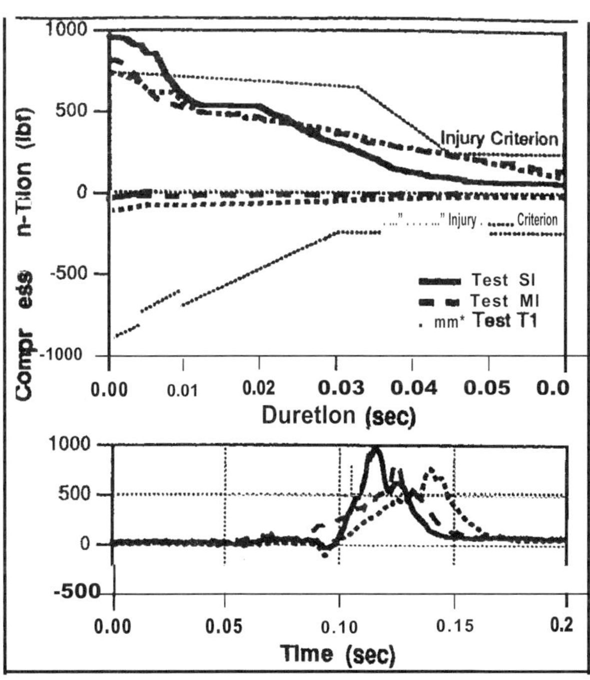

Figure 9: Upper Neck Axial Loads for Tests S1, Ml, & Tl

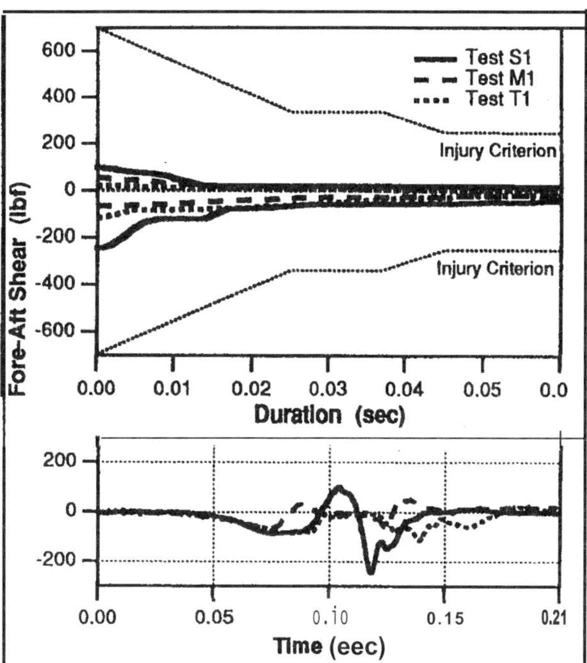

Figure 10: Upper Neck Shear Loads for Tests S 1, M 1 & T 1

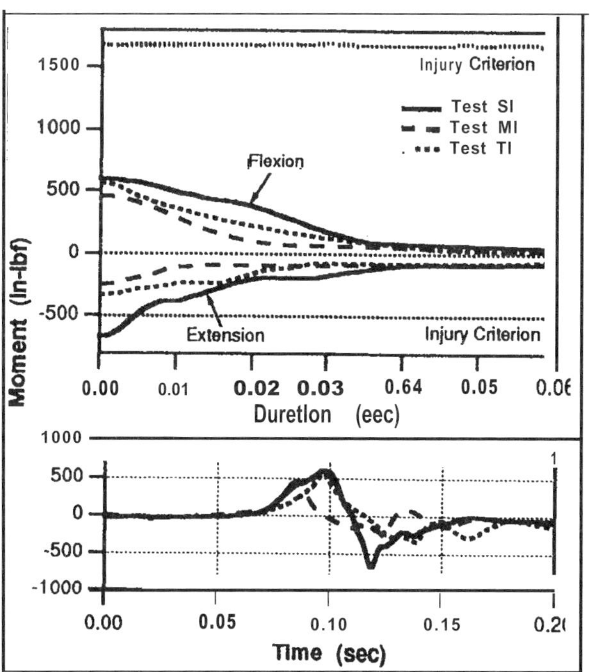

Figure 11: Upper Neck Moment Loads for Tests S 1, M 1 & T 1

criteria for the same three out-of-position tests [Melvin, 1985]. Finally, Figure 11 plots the upper neck moment loads and corresponding injury criteria. It is seen that the stiff modified Aerostar seat produced the greatest upper neck loads for each of the measured modes. This difference is especially remarkable in the neck extension measurements, where the modified Aerostar values are approximately twice those of the two production seats.

Figures 12, 13 and 14 are data observed in the lower neck for the axial loads, shear loads, and moment loads

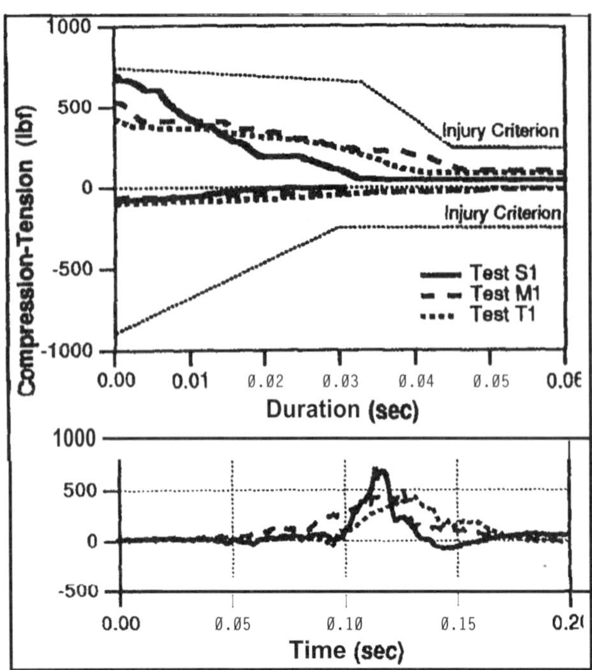

Figure 12: Lower Neck Axial Loads for Tests S 1, M 1 & T 1

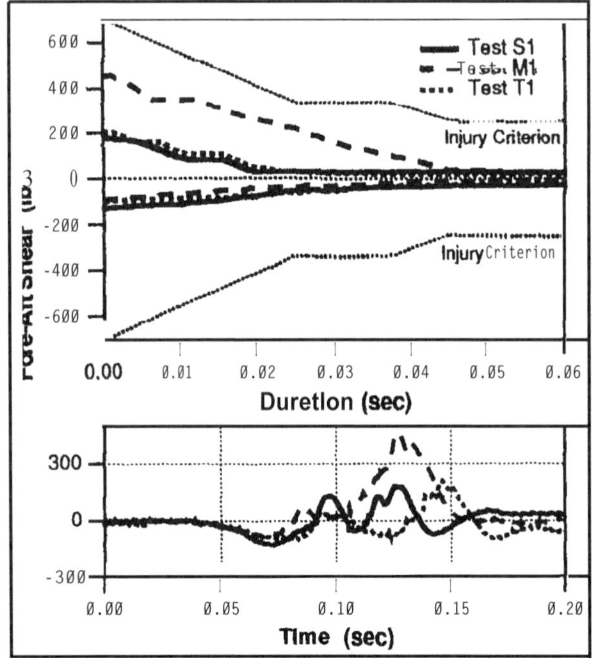

Figure 13: Lower Neck Shear Loads for Tests S 1, M 1 & T1

respectively. In contrast to the upper neck loads, the lower neck loads show no clear correlation with seat stiffness.

As can be seen in Figure 14, in all the OOP tests the lower neck extension torques exceeded the injury criterion by a substantial margin, including those in the test with the relatively less stiff Toyota seat. Since this criterion was exceeded by such a wide margin, either these values suggest very real injury potential, or the injury criterion is too conservative. Moreover, given the fact that these seats have significantly different stiffness properties, it seems that the magnitude of these lower neck torques are somewhat

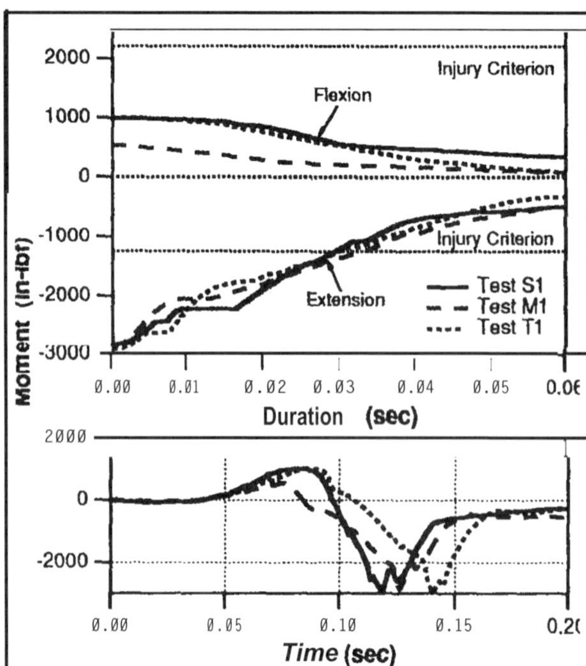

Figure 14: Lower Neck Moment Loads for Tests S 1, M 1 & T1

independent of seat properties; they are certainly not a function of the magnitude of seat deformation.

Interestingly, none of the test dummies underwent *any* extreme hyperextension in any of these tests and the timing of the peak of the lower neck extension moment did not correspond to either the instant of maximum neck extension or the instant of maximum dummy and seat displacement. This demonstrates the inadequacy of evaluating injury potential by simple visual examination of crash-test films without adequate proper dummy neck instrumentation.

To the extent that research efforts are focused on the relatively low-priority area of rear impact protection, further research may be warranted to determine a more appropriate lower neck torque injury criterion and to determine whether or not occupants pitched forward at the instant of rear impact are exposed to any significant lower neck trauma. If it is determined that they are, then perhaps a novel seat design concept like the "catcher's mitt" seat developed by General Motors will show some advantage in this regard.

For all of the seats tested, there was more seat deformation in the tests conducted with the dummy in the Normal Dummy Seated Position as compared to that sustained in tests with the dummy torso pitched forward at impact. This seems counter-intuitive, since relative velocity between the upper body and seat back is allowed to build up in the OOP tests before contact, seemly requiring additional seat back energy absorption. What appears to happen, however, is that during the period of time wherein the torso is rotating into contact with the seat back, ride down energy is being absorbed by the dummy without causing seat back deformation, since the loads into the seat back are occurring relatively close to the seat back anchorage system.

REFERENCES

[James, 1996]
James, Michael B., et. al; "Severe and Fatal Injuries in Rear-End Impacts"; to be published.

[James, 1991]
James, Michael B., et. al; "Occupant Protection in Rear-End Collisions: I. Safety Priorities and Seat Belt Effectiveness"; SAE #9 129 13; 35th Stapp Conference; pg. 369; November 199 1.

[Mackay, 1995]
Mackay, G.M., et. al; "Rear End Collisions and Seat Performance - To Yield or Not To Yield." 39th Annual AAAM Proceedings, p. 23 1. Chicago, IL; October 16-18, 1995.

[Melvin, 1985]
Melvin, John W.; "Advanced Anthropomorphic Test Device (AATD) Development Program, Phase 1 Reports: Concept Definition, UMTRI for US Department of Transportation, NHTSA, p. 101; 1985.

[Mertz, 1968]
Mertz, Harold J.; "Car Occupant Response to Rear-End Collision - A Mathematical Model," 1st International Conference on Vehicle Mechanics, p. 588, Detroit, MI, July 16-18 1968.

[Saczalski, 1993]
Saczalski, Kenneth J., et. al; "Field Accident Evaluations and Experimental Study of Seat Back Performance Relative to Rear-Impact Occupant Protection"; SAE #930346; Seat System Comfort and Safety; pg. 151; March 1993.

[Strother, 1994]
Strother, Charles E., et. al; "Response of Out-of-Position Dummies in Rear Impact", SAE #94 1055; In-Depth Accident Investigations: Trauma Team Findings in Late Model Vehicle Collisions, pg. 65; February 28, 1994.

[Warner, 1991]
Warner, Charles Y., et. al; "Occupant Protection in Rear-End Collisions: II. The Role of Seat Back Deformation in Injury Reduction", SAE #912914; 35th Stapp Conference; pg. 379; March 199 1.

REAR END COLLISIONS AND SEAT PERFORMANCE - TO YIELD OR NOT TO YIELD

S. Parkin, G.M. Mackay, A.M. Hassan and R. Graham
Birmingham Accident Research Centre
University of Birmingham, England

ABSTRACT

THIS PAPER PRESENTS DATA ON the relative frequencies of rear end collisions in comparison to other types of crashes. The severity of rear end impacts is shown to be low, with a mean ETS of about 30 km/hr and 90% of AIS1 injuries occurring below 50 km/hr. The frequency of AIS 3+ injuries in this data set is extremely low at 1%. The results show that plastic yielding of front seat backs is beneficial in that the incidence of AIS1 neck injuries is lower in comparison to seats which do not yield. Trends in seat design are discussed and it is suggested that the very large number of minor injury, low speed, rear end collisions should not be forgotten in design terms if seat back strength is increased.

Debate over the optimisation of seat back strength has not lead to any consensus view, in part because of the absence of real world accident data to illustrate crash severities, frequencies and injury outcomes. Useful reviews by Strother and James (1987) and Blaisdell (1993) illustrate the difficulties. We have therefore analysed our database in Britain to address some of the issues, in particular because our in-depth studies are unusual in that we have representative samples of crashes (weighted by injury severity) with detail on seat back deformation.

Mackay and Ashton (1973) reported that in their United Kingdom on-the-spot accident studies of 759 cars involved in police reported injury accidents, 9.4% sustained the predominant damage to the rear of the vehicle, and of those the Equivalent Test Speed (ETS) was less than 20 km/hr for almost 80%. The Fatal Accident Reporting System (FARS) 1991 data shows that 6.1% of cars involved in fatal crashes were in collision with the rear end of a motor vehicle in transport, and the corresponding figure for light trucks was 17.6%. The General Estimates System 1991, using FARS data, shows that passenger cars' rear ends were the initial point of contact for 18% of the cars where no injury occurred, 24% of cars where there was a minor or moderate injury, and 10% of the cars where there was a severe or fatal injury. The National Accident Sampling System (NASS) data for 1979-1980 reveals that in multi-vehicle accidents, rear end collisions represent 27.1% of the tow-away accidents, and 32.5% of the non-tow-away. Data Link in 1989 showed that accidents from NASS and FARS gave the following results:

Table 1: Rear impacts 1981-1986 NASS/FARS (Data link 1989)

Percentage of the total associated with rear impacts	
All car crashes	11.1%
Fatal car crashes	4.9%
All crash involved occupants	12.2%
All injured occupants	23.3%
Seriously injured occupants (AIS 3+)	7.6%
Occupant fatalities	3.5%

Although 23.3% of all injured occupants were involved in rear impacts, only 7.6% of all AIS 3+ injured occupants are involved in rear impacts, hence the majority of the 23.3% injured occupants suffered minor injury. This was substantiated in the Data link work which produced the following table:

Table 2: Distribution of injuries to occupants in towed rear impacted vehicles

Maximum AIS	Number of occupants	Percentage
Unknown	198	4.0
0	2508	50.7
1	2044	41.3
2	150	3.0
3	35	0.7
4	4	0.1
5	9	0.2
6	2	0.04

These data indicate an extremely small proportion of people (1%) exposed to injury in rear end impacts actually receive an AIS 3 or greater injury with seats as currently designed.

METHODOLOGY

An ongoing study into vehicle crash performance and occupant injury, the Co-operative Crash Injury Study (CCIS), has been underway in Great Britain since November 1983 (Mackay et al, 1985). Data for this paper comprises information on crashes in the East and West Midlands regions of England. The former is predominantly rural and the latter urban. Cars are examined at garages and vehicle dismantlers within a few days of the accident occurring. Only tow-away cars less than six years old at the time of the accident are considered. Injury details are obtained from medical notes, and questionnaires sent to car occupants. All fatal accidents are investigated with some 50% of serious accidents and a further 12% of slight accidents according to the National Road Accident system of classification. The resulting sample represents all levels of injury outcome while being biased towards the more serious and fatal cases. This should be borne in mind when interpreting results presented in this paper. Vehicles with rear end damage were recorded according to the Society of Automotive Engineers' (SAE) Collision Deformation Classification (CDC). Where possible the severity of the impact was estimated using the CRASH3 computer programme to give the speed change, Delta V, and/or the Equivalent Test Speed (ETS). It should be noted, of course, that the CRASH3 algorithm has repeatedly been shown to incorrectly calculate Delta V for impact severities less than or greater than the government compliance test speeds that the stiffness data is generated from. Whilst a large scale statistical study such as this can benefit from the repeatability inherent in a single method reconstruction analysis, the presented Delta V's and ETS's should be thought of as indicative of a trend and not absolutely accurate speeds.

Case selection - This paper presents the results from 2 separate analyses, both of which are concerned with rear impacts. In each case the same database was analysed. The impact had to be the most severe impact in cases where the vehicle had sustained more than one. In addition to the above criteria, seat back damage in vehicles were sampled for study when an occupied front seat had sustained damage of some sort, be it a small degree of seat back deformation, or complete failure of the seat mountings. Within this classification other seats where no damage had occurred were included as they were occupied front seats within a vehicle that had the other front seat damaged.

Seat back deformation - This was classified by the amount that the seat back was bent rearwards compared to the undamaged state (as opposed to the vertical). If the seat back had residual bending of up to 15 degrees it was classified as Low deformation. Between 15 and 30 degrees was Medium deformation, and more than 30 degrees was High deformation, relative to the initial position. If either of the hinges had broken, then the seat back was recorded as failing, in the engineering sense, regardless of the bending that had occurred prior to failure.

Seat damage - was classified by the following: Symmetrical bending, Asymmetrical bending, Complete collapse and Anchorage damage. Anchorage damage was deemed to be damage to the seat runners or mountings whereby the seat base could either lozenge (I.e. one side of the seat base move rearwards whilst the other side remained in place) due to failure of one side of the runner, or pivot rearwards due to failure of the front mounting points. This condition was therefore considered to be an intermediate condition between a yielding seat and a collapsed seat.

RESULTS

Results will be presented in two sections. The first will look at the frequencies and severities of rear end impacts, and comparable results from other studies will be presented alongside. The second section will present results on seat back strength and related injuries versus yielding and non-yielding seats.

Section 1. Frequencies and severities - Table 3 shows the most severe impact types that form the database for this study. The most severe impact was to the rear of the vehicle in only 6.0% of the vehicles selected for the CCIS study.

Table 3: Frequency of most severe impact types

Impact site	Frequency	Percent
Back	323	6.0
Front	3201	59.7
Left side	672	12.5
Right (driver) side	880	16.4
Top	241	4.5
Underneath	13	0.2
Unclassifiable	31	0.6
Total	5361	100.0

The direction of force of the impacts to the rear was analysed and the results are shown in the following table:

Table 4. Direction of force for the rear impacts

Direction of force	Frequency	Percent
Non horizontal	1	0.3
4 O'clock	1	0.3
5 O'clock	33	10.2
6 O'clock	266	82.4
7 O'clock	20	6.2
8 O'clock	2	0.6
Total	323	100.0

It is clear that the most frequent impact direction is purely fore-aft, and that impacts that are non horizontal or outside of 5-7 O'clock are a very rare occurrence, accounting for only 1.2% of impacts to the rear.

Figure 1 shows the frequency of distribution of initial direct contact of the sample.

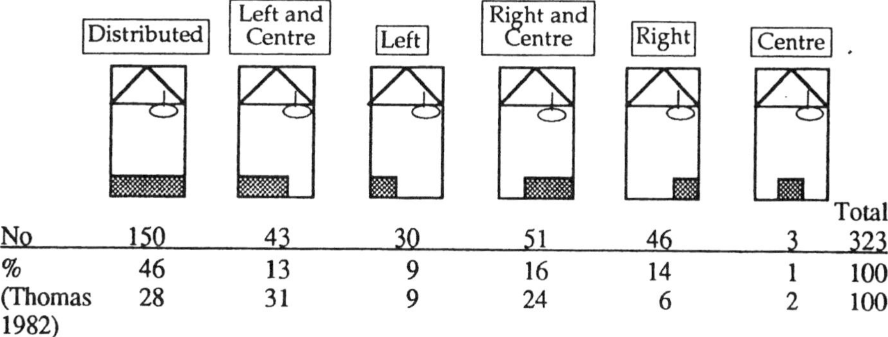

	Distributed	Left and Centre	Left	Right and Centre	Right	Centre	Total
No	150	43	30	51	46	3	323
%	46	13	9	16	14	1	100
(Thomas 1982)	28	31	9	24	6	2	100

Fig. 1 - Distribution of direct contact

Results obtained in a French study by C. Thomas et al (1982) are shown by way of comparison. In France, of course, the rule of the road is

reversed, therefore comparisons with the position of the steering wheel means left and right should be reversed for the French data.
This study found substantial differences in the distribution of the damage to the rear of the vehicles when compared to the French study. This study found that approaching half of the vehicles had fully distributed direct contact, compared with less than a third in the French study. Direct contact with the outer two thirds of the vehicle was a significant impact type in the French study, but there was no appreciable difference in this study between the outer two thirds and the outer third, with all these impact types ranging between 9 and 16% of the total. Both studies found that direct contact with the centre third of the vehicle was not a significant impact type.

Severities - These are shown in the following graphs. Figure 2 displays the Delta V distribution for this study, and the following graph, Figure 3 is the comparable results from the French study. It was only possible to calculate the Delta V for 51 of the Birmingham vehicles. The mean Delta V was 27 km/hr. The mean in the French study was 23 km/hr, and the difference is probably due to the sampling system in the current study, with the bias towards more severe impacts. Figure 4 presents the ETS for 230 of the Birmingham vehicles, and the mean was just less than 30 km/hr. It is clear from the three graphs that rear impacts are generally at a speed change of less than 30 km/hr, and averaging at between 20 and 25 km/hr when sampling biases are removed.

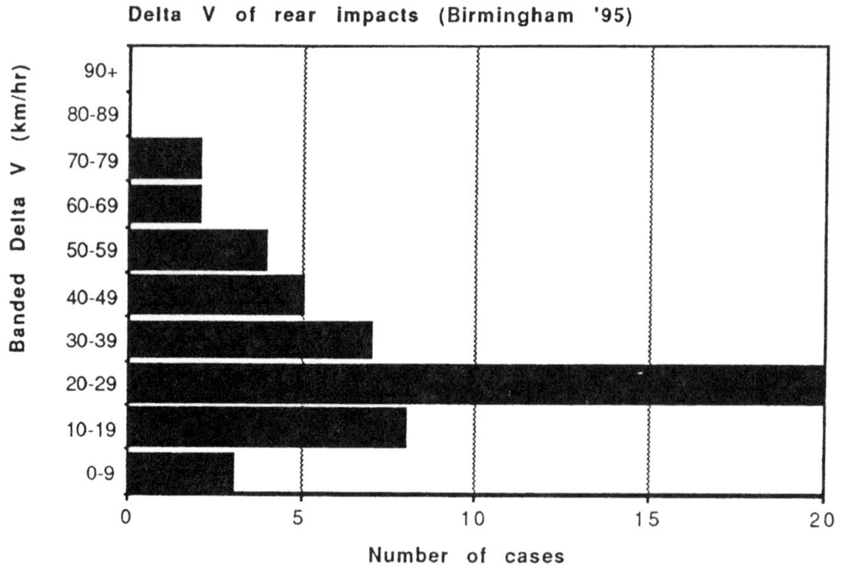

Figure 2: Delta V of 51 Birmingham vehicles

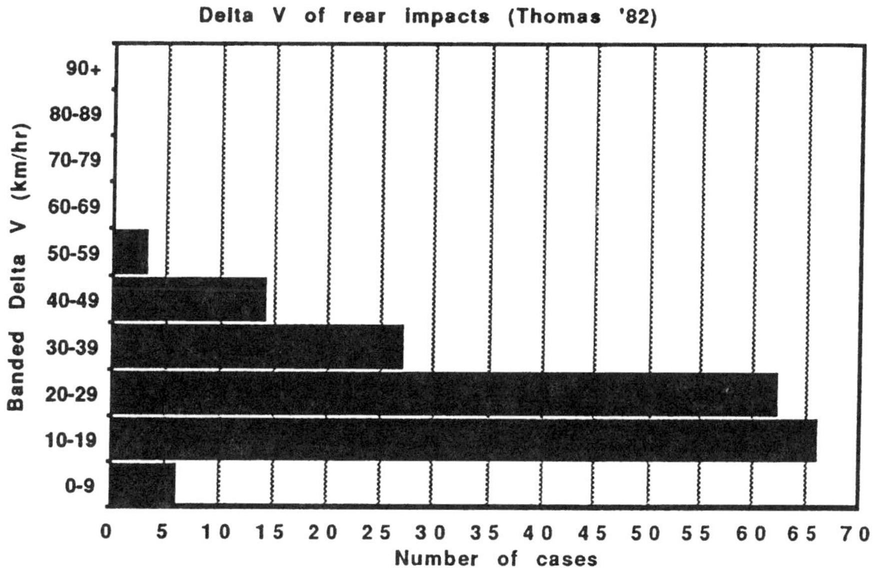

Figure 3: Delta V of 172 French vehicles

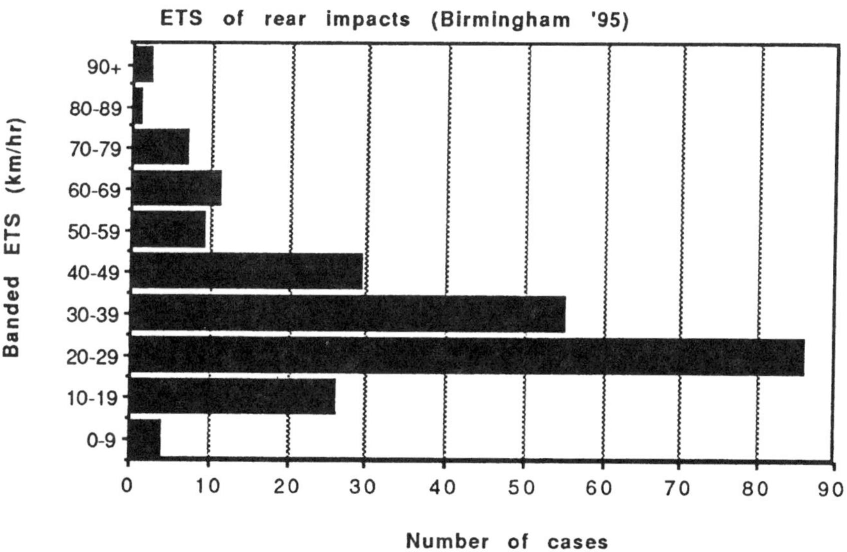

Figure 4: ETS of 230 Birmingham vehicles

Injury severity of the occupants - Table 5 shows the frequency of seat occupancy of the sample of rear end collision vehicles.

Table 5: Seat occupancy in rear impacts

Seat position	Frequency	Percent
Driver (right)	315	62.5
Front left	110	21.8
Rear right	25	5.0
Rear left	31	6.2
Rear centre	8	1.6
Across rear seat	1	0.2
Rear not known	14	2.8
Rear total	79	15.7
Total	504	100.0

Figure 5 shows the injury levels by ETS for the front seat occupants when both the injury level and ETS were known. 29% of the occupants were uninjured. 59% received a maximum injury of AIS 1. 9% suffered a maximum injury of AIS 2, and 3% suffered a MAIS of 3 or more.

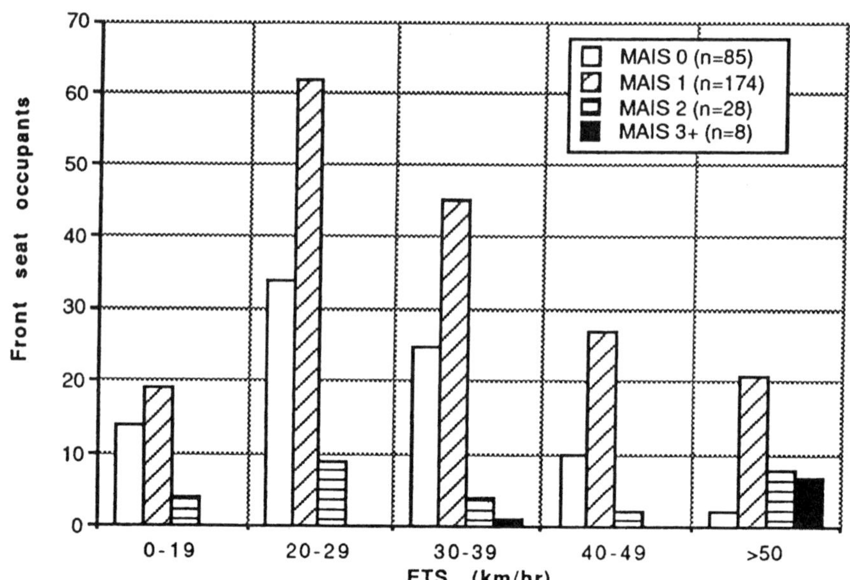

Figure 5: Maximum injury level of front seat occupants by ETS

Section 2. Seat performance and injuries - 356 front seats were sampled for study. Table 6 shows the degree of damage sustained by each of the seats.

Table 6: Damage to front seats

Seat damage	Frequency	Percent
Asymmetric bending	39	11.0
Symmetric bending	95	26.4
Complete collapse	4	1.4
Anchorage damage	37	10.4
Asymmetric + Anchorage	20	5.6
Symmetric + Anchorage	34	9.6
No damage	127	35.7
Total	356	100.0

Figure 6 shows the relative frequency of each type of damage by impact severity (Asymmetric and symmetric bending have been combined.)

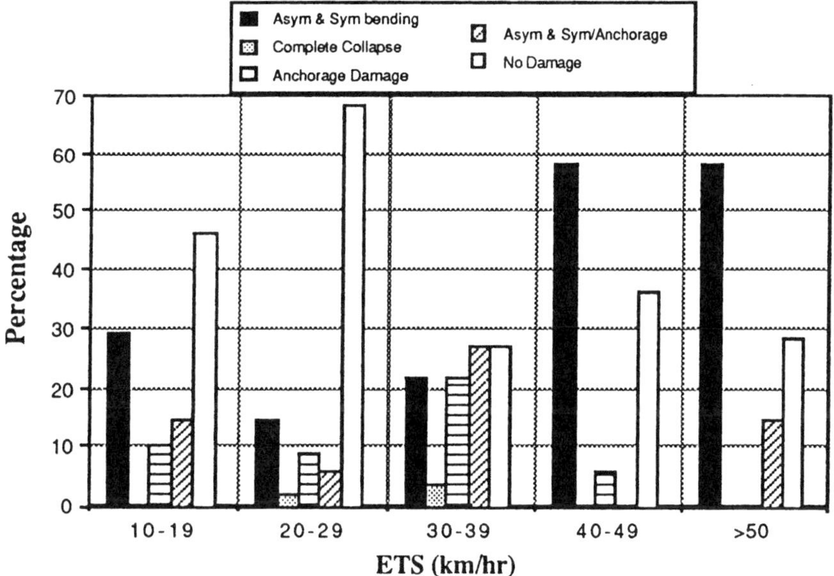

Figure 6: Seat damage by ETS

As is to be expected, seats suffering no damage become generally less frequent as the severity of the collision increases. Complete collapse of the seat is a rare event, even in severe collisions. A substantial proportion of the seats suffered some degree of bending, and the relative proportion of bending increases with crash severity.

Seat damage versus neck injury - The following two figures illustrate the effect of seat damage on AIS1 neck injury. Each type of seat damage and level of deformation is plotted with the frequency of receiving an AIS1 neck injury or not, regardless or any other type of injury. The results indicate that with the presence of yield, the occupant is less likely to suffer an AIS1 neck injury (complete collapse and failure occurred in very few cases, hence the results for those modes are misleading). Asymmetric bending leads to a lower probability of AIS1 neck injury. If the seat bends, and then some form of anchorage fails then AIS1 neck injury is even less likely. A seat which exhibits no damage is considerably more likely to result in AIS1 neck injury. There appears to be little difference in the frequency of AIS1 neck injury versus the amount by which a seat deforms, but AIS1 neck injuries are approximately twice as frequent in an undamaged seat than in a yielding seat.

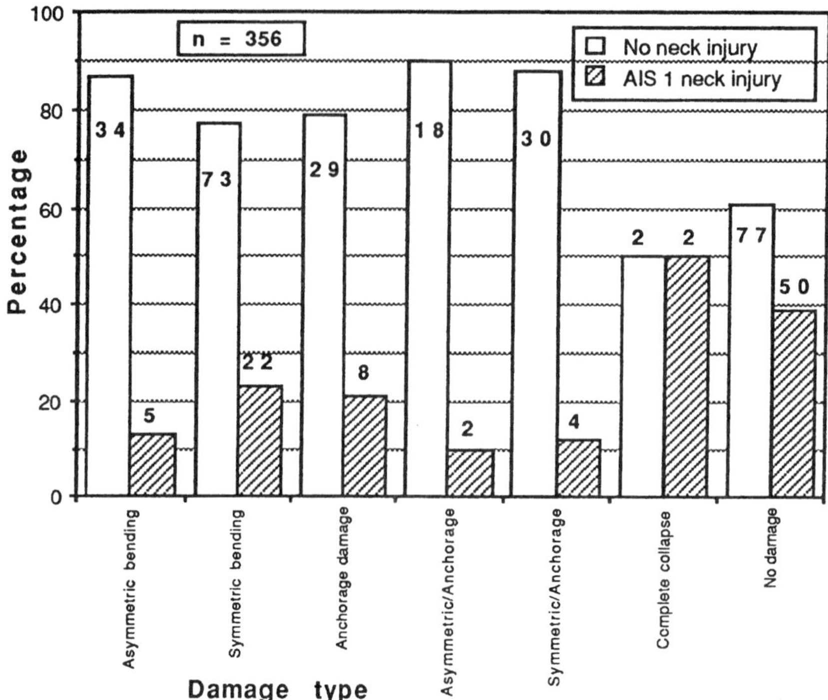

Figure 7: The frequency of AIS1 neck injury for different seat damage modes (actual numbers are also shown)

We performed a significance test to see if an occupant was significantly more likely to receive an AIS1 neck injury if the seat was undamaged than if the seat was damaged in some way, I.e. all the other groups combined. The results of that test are shown in Table 7 (expected values are shown in parentheses).

Table 7. Chi squared significance test for AIS1 neck injury versus seat damage

Variable	No neck injury	AIS1 neck injury	Total
Seat damage	186 (169.2)	43 (59.8)	229
No damage	77 (93.8)	50 (33.2)	127
Total	263	93	356

χ 2 = 17.95, d.f. = 1, p , 0.001

The result is therefore highly significant.

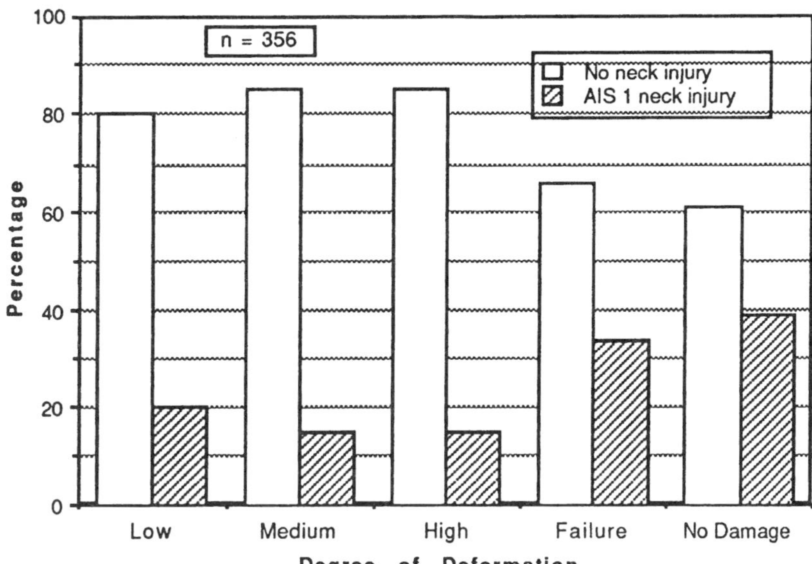

Figure 8: The frequency of AIS1 neck injury for different degrees of deformation

The frequency of AIS 1 neck injury was not seen to be related to impact severity, as shown in figure 9, although there does appear to be a trend towards a lower relative frequency of AIS1 neck injury with increasing crash severity which is also consistent with a higher incidence of seat back yield.

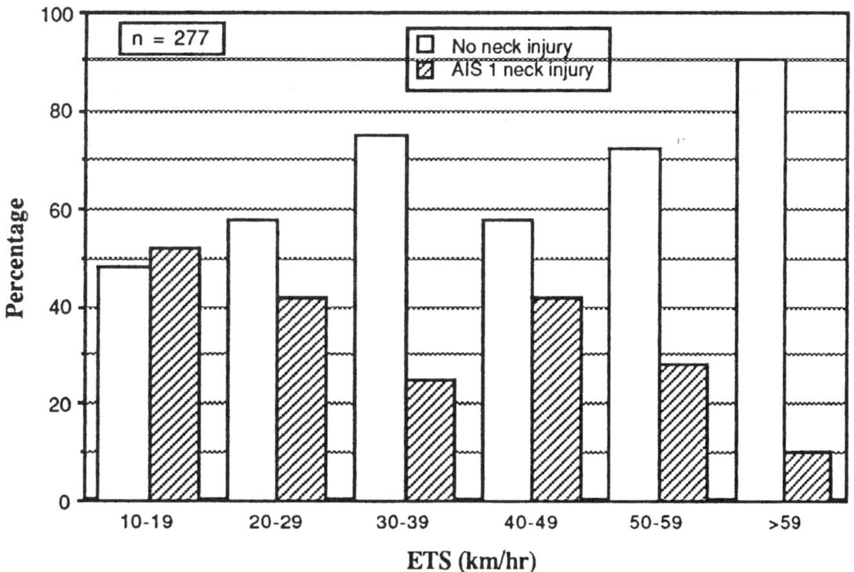

Figure 9: Frequency of AIS1 neck injury by banded ETS

We also conducted an analysis of front and rear seat occupants in the overall injury categories of no injury, AIS1, and AIS2+, according to the crash severity assessed by ETS. The results are shown in the cumulative frequency curves of Figure 10.

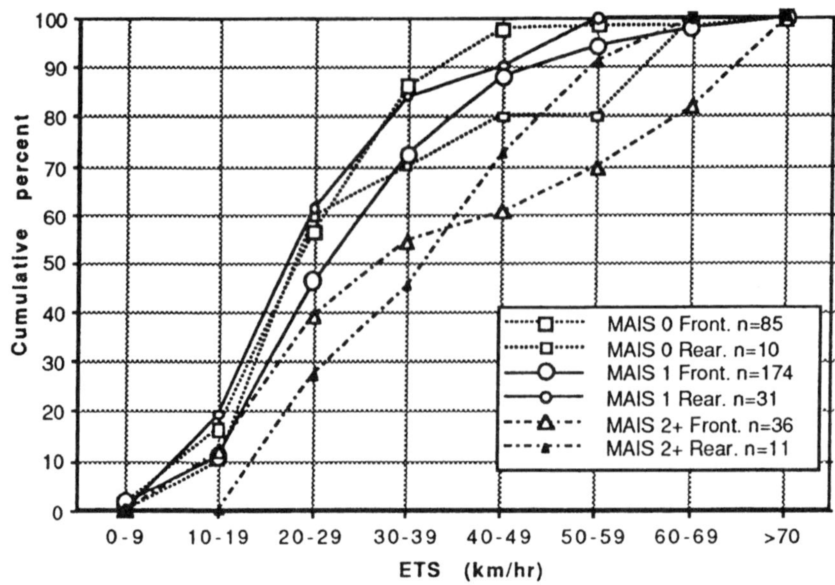

Figure 10. Cumulative frequency of injury severity for front and rear occupants involved in rear impacts

It appears that there are few discernible differences between front and rear seat outcomes from these data. They reflect, of course, that the overwhelming fact that the great majority of rear end collisions are at low speeds, below 40 km/hr. Front and rear seat occupants have very different characteristics; for example about one third of rear seat occupants are children with a sitting height of less than 70 cms (27.5 inches). The numbers of cases with AIS2 and greater injury is very small for both front and rear seats and thus further disaggregation by age, sex, sitting height and crash severity does not yield meaningful cell numbers. These results are still of interest if only to show the overall exposures and outcomes for occupants in rear end collisions.

DISCUSSION

Virtually all of the cars in this study had seats with head restraints of either the fixed or adjustable type, conforming to the UN/ECE Regulation 25 and European Directive 74/408/EEC as amended, which are substantially the same as the Federal Motor Vehicle Safety Standards (FMVSS) 202 and 207. The work of Svenson (1993) and others has demonstrated however that hyperextension of the neck can occur even with head restraints of adequate height. This arises because the horizontal separation between the head restraint and the initial position of the head is so great that excessive relative rearward motion of the head can still occur (Parkin et al, 1993). Headrest position and the implications for neck injury in rear crashes is discussed by Viano (1995) elsewhere in the proceedings. An additional consideration is the load-deflection characteristics of the seat back. It is entirely feasible for the torso to have undergone its complete acceleration from seat back loading and to be rebounding before the head gets any support from the head restraint. This condition is particularly acute when the characteristics of the seat back under load are totally within the elastic range. For that condition the torso can undergo a velocity change approaching twice the velocity change of the car. Such a condition will generate shear forces and bending moments of much greater magnitude in a neck which is being hyperextended, in comparison to a yielding seat.

If it is accepted that the AIS1 soft tissue neck sprains reported in this study are largely due to hyperextension, which seems likely, then the hypothesis that a non-yielding, elastic seat back leads to a greater injury risk than a yielding seat seems likely. A second consideration is the changing pattern to the forces acting on the neck as a result of seat back yield. In an upright position the head mass produces maximum shear and bending moments and little tension in the neck (in a gross sense). As the torso becomes more reclined the forces parallel to the neck (Z upwards relative to the neck) increase but the shear and bending moments decrease, down to near zero when the torso becomes horizontal. These two factors lead to the proposition that a plastically yielding seat diminishes neck injury risk. That is what the data in Figure 7 suggests. The data also suggest AIS1 neck injuries are independent of speed.

Real world crash data always raises more questions than answers. It would be of interest, for example, to explore weight, age and gender differences. Females restrained in frontal crashes have the same injury risk in a crash population 10 km/hr less severe (delta V) than males (Mackay et al, 1994). Many studies have shown that neck sprain is twice as frequent in females than males (Larder et al, 1985). In this rear end collision data set however, such disaggregation quickly results in unsatisfactory cell sizes.

At the present time there is considerable divergence in seat back strength with an increasing trend towards stronger seat backs. This study broadly reflects seat backs of cars of the mid 1980's, and it demonstrates that plastic yield is beneficial. Given the fact that the great majority of rear end collisions are at relatively modest speeds then increasing seat back strength may well lead to an increase in minor AIS1 neck sprains. It may be that greatly increased seat back strength could have some marginal benefits on those very infrequent cases of AIS3+ injury (only 1% Of our sample, which is itself biased towards severe injury). However, this paper illustrates that minor neck injuries in rear end collisions are important as a design consideration.

Given an increase in seat back strength with the ultimate design of integrating the upper anchorage of the seat belt into the seat, then the relative position of the head support in both horizontal and vertical directions becomes increasingly important. Our work on actual head positions (Parkin et al, 1993) suggests that horizontal as well as vertical adjustment may well be appropriate. Beyond that consideration should be given to the design of seats having different compliance and strength in forward and rearward directions. For example, a seat back with sufficient forward strength to support belt loads in frontal impacts, while maintaining beneficial plastic yielding in rearward impacts.

Clearly, this study has not been able to look at the injury consequences of the dynamic force-deflection characteristics of a seat, this being the arena or other groups, and which is covered by the literature. If the trend is to continue towards stronger seats, then consideration must be given to the dynamic force-deflection and the rebound characteristics that a seat has, as there will clearly be significant consequences for injury, particularly to the neck.

ACKNOWLEDGEMENTS

The Co-operative Crash Injury Study is managed by the Transport Research Laboratory (TRL) on behalf of the Department of Transport (Vehicle Standards and Engineering Division) who fund the project with Ford Motor Company Limited, Toyota Motor Company Limited and Rover Group Limited. The data was collected by the Accident Research Units of Loughborough and Birmingham Universities and from the Vehicle Inspectorate.

REFERENCES

Blaisdell D.M., Levitt A.E., Varat M.S. *Automotive seat design concepts for occupant protection.* SAE Technical Paper Series number 930340, Proc. international Congress and Exposition - *Seat System Comfort and Safety.* SP-963, pp109-120. Society of Automotive Engineers, 1993.

COMSIS Corporation. *Report on traffic accidents and injuries for 1979-1980 - The National Accident Sampling System.*

Data Link. *Car crash outcomes in rear impacts.* Appendix A to *Current Issues of occupant protection in car rear impacts.* Data Link inc., Washington D.C., 1989.

Larder D.R., Twiss M.K., Mackay G.M. *Neck injuries to car occupants using seat belts.* Proc. AAAM Conf., Des Plaines Ill., pp153-165, 1985.

Mackay G.M., Ashton S.J. *Injuries in collisions involving small cars in Europe.* Proc International Automotive Engineering Congress. SAE Technical Paper Series number 730284, 1973.

Mackay G.M., Galer M.D., Ashton S.J., Thomas P. *The methodology of in-depth studies of car crashes in Britain.* SAE Technical Paper Series number 850556, Society of Automotive Engineers, 1985.

Mackay G.M., Parkin S., Scott A.W. *Intelligent restraint systems - What characteristics should they have?* Proc. Joint AAAM/IRCOBI Conf., Advances In Occupant Restraint Technologies, Lyon, France, 1994.

Parkin S., Mackay G.M., Cooper A. *How drivers sit in cars.* Proc. AAAM Conf., San Antonio, Texas, pp375-387, 1993.

Strother C.E., James M.B. *Evaluation of seat back strength and seat belt effectiveness in rear end impacts.* SAE Technical Paper Series number 872214, Proceedings of the 31st Stapp Car Crash Conference pp225-244.

Svenson M.Y., Lovsund P., Haland Y., Larsson S. *The influence of seat back and head restraint properties on the head-neck motion during rear impact.* Proc. International IRCOBI Conf., pp395-406, 1993.

Thomas C., Faverjin G., Hartemann F., Tarriere C., Patel A., Got C. *Protection against rear-end accidents.* Proc. International IRCOBI Conf., pp17-29, 1982.

U.S. Department of Transportation, National Highway Traffic Safety Administration Washington D.C. *General Estimates System 1991.*

U.S. Department of Transportation, National Highway Traffic Safety Administration Washington D.C. *Fatal Accident Reporting System 1991.*

U.S. Department of Transportation, National Highway Traffic Safety Administration. *Traffic Safety Facts 1992 (revised 1992 Data).* DOT HS 808 022.

Viano D.C., Gargan M.F. *Headrest position during normal driving: implications to neck injury risks in rear crashes.* Proc. AAAM Conf. Chicago, Ill, 1995.

941055

Response of Out-of-Position Dummies in Rear Impact

Charles E. Strother and Michael B. James
Collision Safety Engineering, Inc.

John Jay Gordon
GMH Engineering, Inc.

ABSTRACT

Field accident data suggest that a significant number of occupants involved in rear impacts may be positioned at impact other than in the "Normal Seated Position" - the optimum restraint configuration that has been used almost exclusively in published seat testing. Pre-impact vehicle acceleration from braking, swerving, or a prior frontal impact could cause an occupant to be leaning forward at the instant of the collision, creating a situation where the vehicle "ride-up" potential would be limited. No rear impact tests involving yielding, production-type seats with forward-leaning dummies are found in the literature. Thirty rear-impact sled tests with a forward-leaning, "Out-of-Position" Hybrid III dummy are presented. Tests were performed with a calibrated seat set in either the rigidified or yielding configuration and with the dummy either unbelted or restrained by a production three-point belt system. Test speeds ranged from 5 to 20 mph. The available Hybrid III neck and lumbar instrumentation were used for comparative purposes. Lower injury measures were recorded with the seat in the yielding configuration at all test speeds and for both belted and unbelted dummies.

INTRODUCTION

In 1987, Collision Safety Engineering (CSE) presented a paper summarizing experimental and analytical research efforts in rear impact protection [Strother/James, 1987]. Specifically, technical papers dealing with the effect of seat back strength on occupant safety were reviewed in an effort to compare the merits of yielding and rigidified seat back design philosophies. Among the conclusions reached in conjunction with that literature review were:

(1) yielding seats generally produce lower dummy injury measures as compared to rigid seats tested under the same or similar conditions, and

(2) the response of occupants in yielding seats seems to be less sensitive to variations in occupant pre-impact seating position when compared to the response of occupants in rigidified seats.

In 1991, another paper discussed the role of seat back deformation on injury reduction in rear impacts [Warner et al, 1991]. It was noted that rigid seats may be appropriate for belted occupants who are sitting in the "Normal Seating Position" (NSP), i.e., with their shoulder blades pressed firmly into the seat back and with their heads close to the head restraint. However, rigid seats may increase the injury potential for occupants in other seating postures. Accident data from the National Accident Sampling System (NASS) files suggested that a significant percentage of rear impact accident victims are not in the NSP at the instant of impact. These data indicated that many occupants who normally sit in the NSP might experience pre-impact decelerations which would alter their position. For example, pre-impact braking could cause the occupant to lean forward and abrupt pre-impact steering could cause the occupant to be displaced laterally.

Almost all rear impact experiments reported in the literature have been conducted with dummies in the NSP. Several "inadvertent" or unintentional Out-of-Position (OOP) rear impact experiments were conducted during a series of sled tests using anthropomorphic dummies seated in the rigidified "Integrated Safety Seat" built by researchers at the University of Michigan [Melvin, 1971] [Hilyard,1973]. Comparison of the response of these OOP dummies, having their upper body pitched forward, with the response of the NSP dummies revealed several important differences:

(1) Although the sled buck was confined to simple one-dimensional motion (i.e. no pitching motion was simulated), the OOP dummies encountered the seat backs relatively high when compared with their NSP counterparts. Consequently, the head of the OOP dummies overrode the head restraint.

(2) Probably as a result of overriding the head restraint, the OOP dummies experienced significantly more ramping than did the NSP dummies.

(3) As an additional consequence of having missed the head restraint, the OOP dummies experienced significant hyperextension. The severity of the hyperextension was aggravated by the fact that there was minimal energy absorption by the rigidified seat back employed in these tests.

(4) A significant portion of the energy stored in the ISS seat back was returned to the dummy occupant in the form of a forward and upward rebound velocity. These rebound velocities were significantly greater than those experienced in the corresponding NSP tests.

While these "inadvertent" OOP tests are important because they demonstrate significant differences between OOP and NSP responses, they are limited to a narrow range of test speeds and seat designs. Further, the tests were conducted at extreme impact severities: the delta-V's were in the range of 30 to 40 miles per hour. The data from the National Crash Severity Study (NCSS), by contrast, indicates that such impacts represent 98-99th percentile tow-away rear collisions, see Figure 1 [Ricci, 1979]. In addition, no tests of any production-like seats are reported in the literature. OOP tests with typical car seats are virtually non-existent.

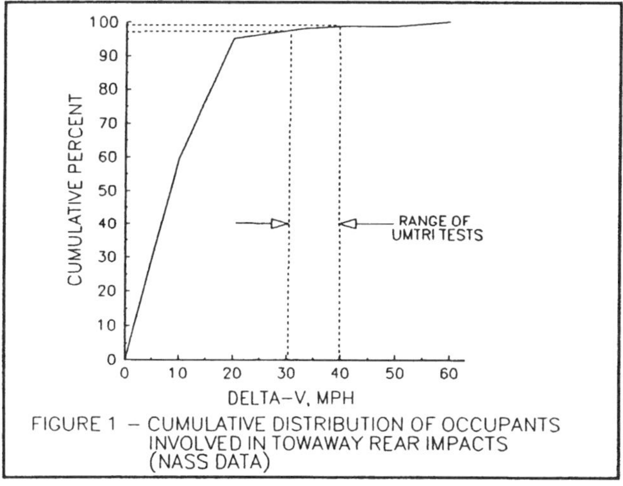

FIGURE 1 – CUMULATIVE DISTRIBUTION OF OCCUPANTS INVOLVED IN TOWAWAY REAR IMPACTS (NASS DATA)

A sled test program was developed to investigate the differences between OOP and NSP dummy responses in rear impact at more reasonable tests speeds, and with production-like seats. It was decided to conduct tests in the accident severity range of most interest, i.e. in that range where a significant portion of the societal losses due to rear-impact injuries is observed.

Data from the NASS accident files, years 1980 to 1986 were used to establish a reasonable impact severity range. All injuries to front seat occupants of vehicles undergoing a Principal Direction of Force of 5 to 7 o'clock were examined. Injuries were first categorized using the Abbreviated Injury Scale (AIS) for each 5 mph accident severity (delta-V) range. The injuries were then combined into a single number using the concept of "Harm" [Malliaris, 1985]. The weighting factors used for each AIS level were those given in the appendix, Subtable B. in the 1989 Data Link Inc. study of rear impact injuries [Data Link, 1989]. These weighting factors are presented in the Appendix of this paper.

The results of this analysis are shown in Figure 2, which is a graph of the cumulative distribution of rear-impact Harm with accident severity. As can be seen in this Figure, nearly 70 percent of the total rear impact Harm is associated with delta-V's of 20 mph or less, and less than 8 percent of the total rear impact Harm occurs at delta-V's of 5 mph or less. It was decided that the subject experimental program would concentrate on rear impacts in the 5-20 mph delta-V range.

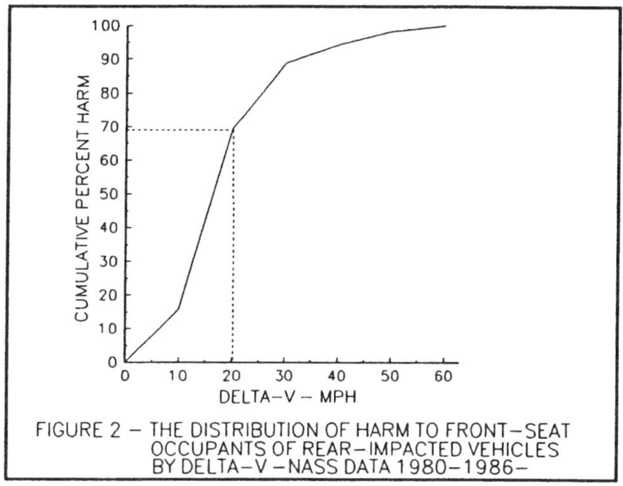

FIGURE 2 – THE DISTRIBUTION OF HARM TO FRONT-SEAT OCCUPANTS OF REAR-IMPACTED VEHICLES BY DELTA-V –NASS DATA 1980–1986–

TEST METHODOLOGY

The Sled Buck

The sled tests were conducted at the CSE crash test facility. A moving barrier, with the modifications described in this section, was used as the sled buck. The "sled" is thus an acceleration-deceleration type, i.e., the buck is relatively slowly accelerated up to the impact speed and is then decelerated by impacting the fixed barrier. The deceleration of the buck was controlled by crushing a set of metal drums attached to the face of the moving barrier, see Figure 3.

The configuration of metal drums was designed to provide a deceleration profile typical of a passenger car undergoing a rear impact involving the entire rear end of the vehicle. Figure 4 is a graph of the buck deceleration profile from the highest (21.7 mph) impact speed test. The buck decelertion levels were approximately 10 g's. The geometry of the array of barrels relative to the center of gravity of the moving barrier produced little pitching motion during impact. Vents were designed in the barrel end panels to eliminate buildup of pneumatic pressure, minimizing rebound and making the barrier deceleration relatively insensitive to impact velocity. The actual buck deceleration profile is not a critical factor in this type of testing since most of the vehicle deceleration has occurred before the OOP occupant encounters the seat back. For tests involving the "yielding" calibrated seat, the buck configuration included a rear seat back as shown in Figure 5a. For tests involving the "rigidified" calibrated seat, the rear seat back was removed (since no interaction with the rear seat

(a) PRE-IMPACT

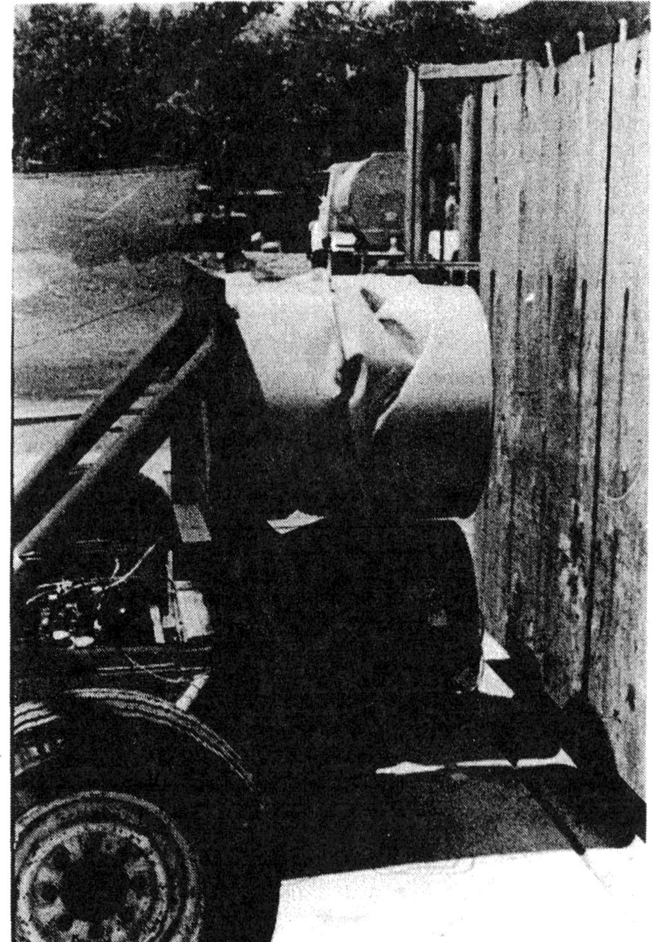

(b) POST-IMPACT

FIGURE 3 - THE SLED BUCK DECELERATION BARRELS

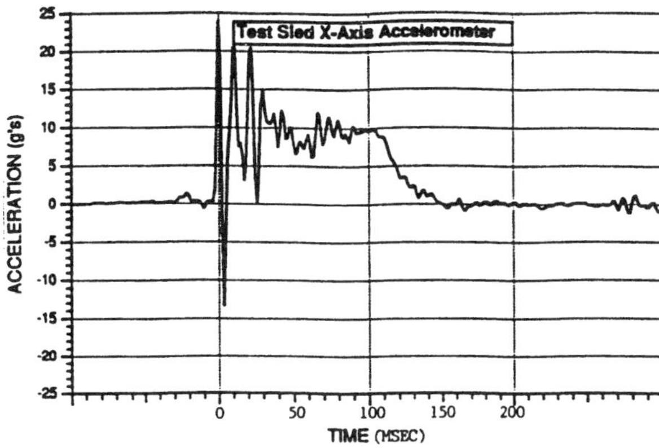

Figure 4 - Sled Buck Deceleration Profile

area was anticipated) to make room for the rigidifying elements. Figure 5b shows the three-point seat belt assembly adapted from a 1985 Chevrolet Cavalier 2-door Hatchback. The Cavalier belt is a continuous loop design incorporating a vehicle-sensitive locking retractor mechanism.

Test Instrumentation and Documentation

A speed trap was used to measure the sled impact velocity. A single axis accelerometer was also mounted to the moving barrier to document the crash pulse. Each test was documented photographically via video and high-speed cameras as depicted schematically in Figure 6. After each test, the amount of barrel crush and the final seat back angles were recorded. Dummy instrumentation is described in a subsequent subsection.

The Calibrated Test Seat

A specially-designed "calibrated" seat was developed for use in this sled test series to minimize the time and expense between the individual sled tests. The calibrated seat was designed to be capable of producing a predetermined torque about the seated reference point by deforming replaceable elements. This allowed the seat to be quickly refurbished between tests, while assuring repeatable deformation characteristics.

Two configurations of the calibrated seat were used in this test series: The "yielding" configuration and the "rigid" configuration. Figure 7 shows the calibrated seat in the "yielding" configuration. Figure 7(a) is an overall view of the seat and Figure 7(b) is a close-up of the energy absorbing elements (A) that link the seat cushion frame to the seat back structure.

The seat back torque in this configuration is solely determined by the geometry of the two yielding elements (A). These elements were designed to produce an ultimate seat back strength typical of a production seat. Figure 8 shows the static force-versus-angle-change characteristic of the "yielding" calibrated seat.

The calibrated seat was designed using parts from a production seat. Modifications that would increase the mass of the seat back were kept to an absolute minimum because an unrealistically high seat back mass could affect the dummy

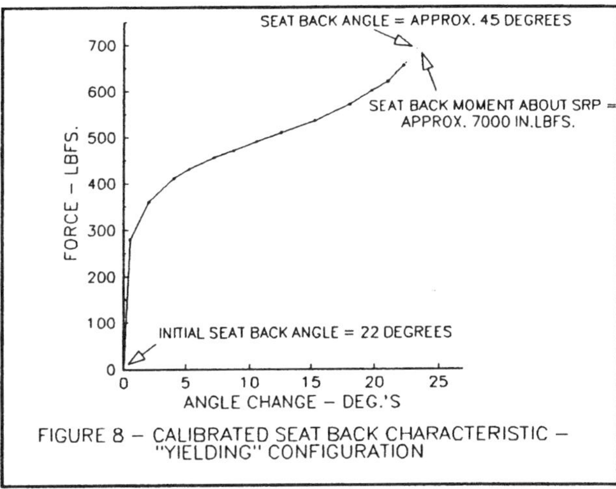

FIGURE 8 – CALIBRATED SEAT BACK CHARACTERISTIC – "YIELDING" CONFIGURATION

response. As configured, the seat back weight for the yielding seat was approximately 18 pounds, a weight not atypical of production seat backs.

The calibrated seat has a 30 1/4 inch high integral head restraint. The early testing indicated the OOP dummy would sometimes deform the head restraint when it encountered the seat back structure. In tests 1 through 7, the head restraint was merely straightened between experiments. This yielding and subsequent straightening logically resulted in a head restraint that was softer as the testing progressed. To ensure that test results were not being affected by the changing compliance of the head restraint portion of the seat back frame, the upper portion of the seat back frame was subsequently modified to allow the replacement of the head restraint after each test. This modification was made after Test 7.

Figure 9 shows the calibrated seat in the "rigid" configuration. Figure 9a is an overall view and Figure 9b is a close-up of the elements (B) which were used to give the seat back the necessary additional strength. The maximum torque capability for this seat configuration was set at 56,000 in-lbfs. In addition, the seat back was given a high initial stiffness to ensure that it would be essentially rigid under the tested conditions. In order to simplify the construction of the rigid seat, the added strength of the seat back was provided via a pair of force-limiting elements B attached to the seat back at about the level of the upper body center of gravity. The piston element, which is pinned to the seat back frame deforms a metal strap of prescribed width and thickness which is attached to the sled buck portion of the apparatus. The static force-versus-angle-change characteristic for the "rigidified" calibrated seat is shown in Figure 10.

The effective weight of the seat back in the "rigid" configuration (i.e. the weight of the seat back plus the weight of the moving portion of the rigidifying elements) was 33 pounds. This is greater than that of the "yielding" calibated seat, but probably not inappropriate. The weight of the "rigid"

(a) THE "UNRESTRAINED OOP" (YIELDING SEAT CONFIGURATION)

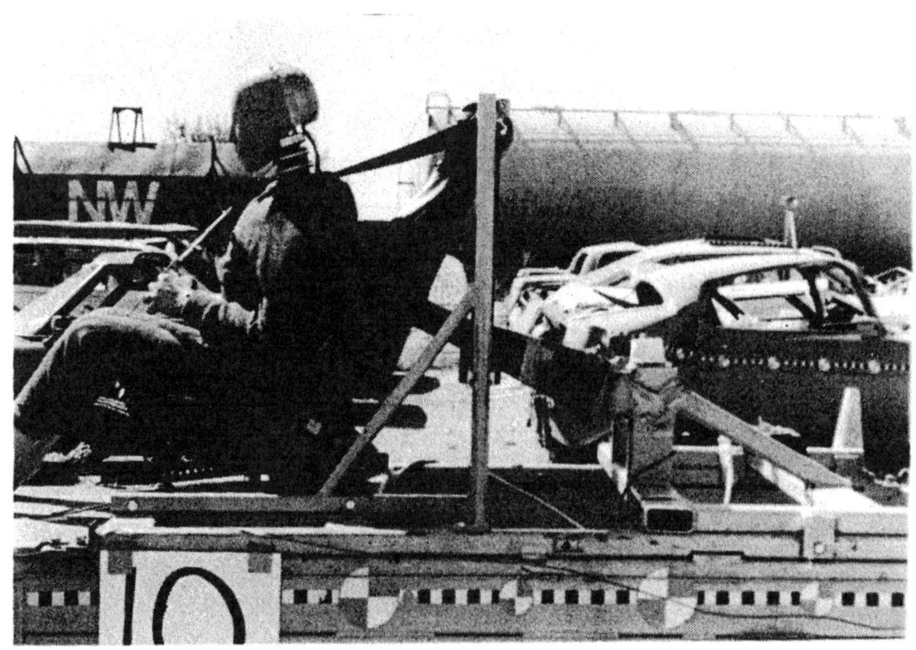

(b) THE "RESTRAINED OOP" (RIGID SEAT CONFIGURATION)

FIGURE 5 - ATD PRE-IMPACT POSITIONS

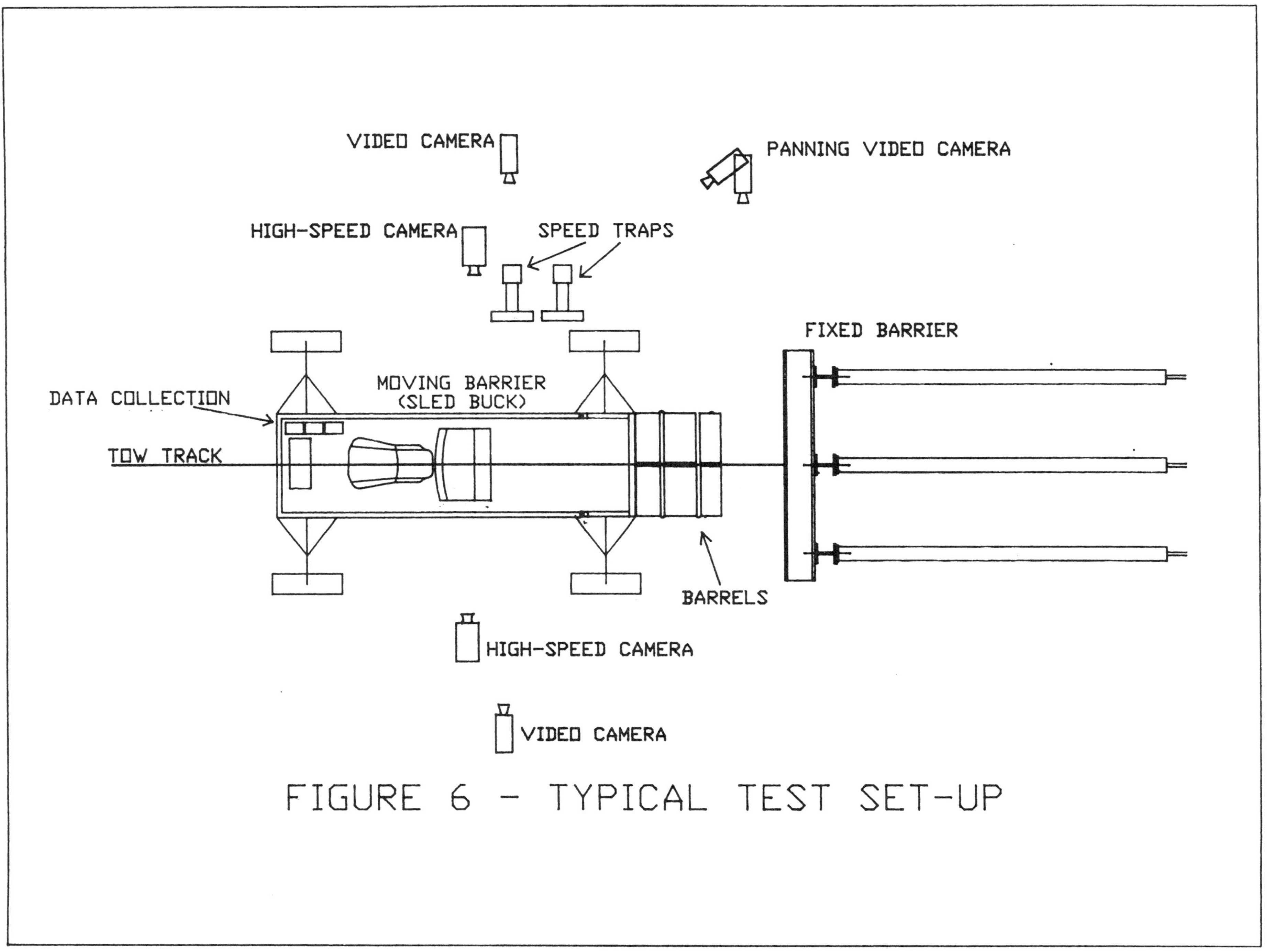

FIGURE 6 - TYPICAL TEST SET-UP

Reinforced, production seat back with replaceable head restraint (init. angle = 22°)

Production seat cushion

(a) OVERALL VIEW

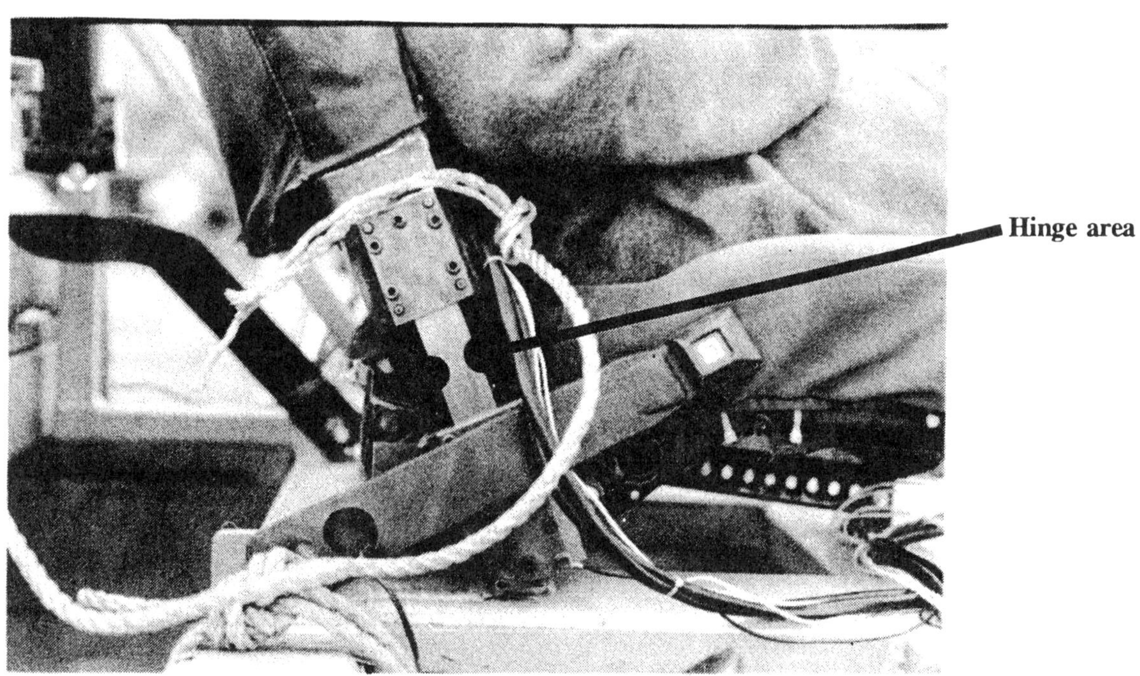

(b) CLOSE-UP OF EA UNITS (A)

FIGURE 7 - CALIBRATED SEAT IN THE "YIELDING" CONFIGURATION

(a) OVERALL VIEW

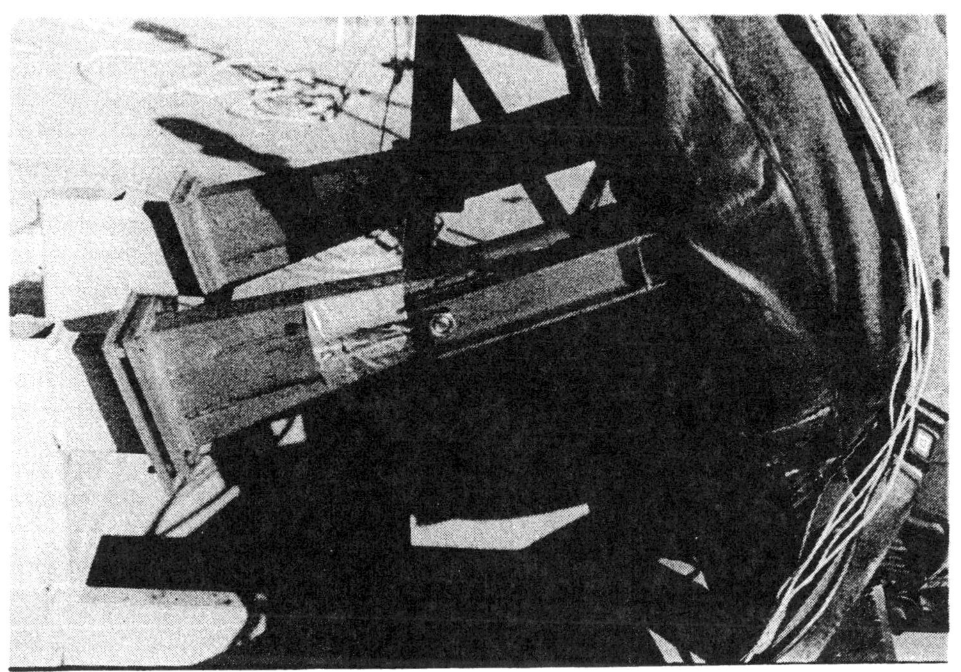

(b) CLOSE-UP OF EA UNITS (B)

FIGURE 9 - CALIBRATED SEAT IN THE "RIGIDIFIED" CONFIGURATION seat back is not as critical a factor as it is in the "yielding" configuration, since, seat back movement in the tests involving the rigidified seat was nil.

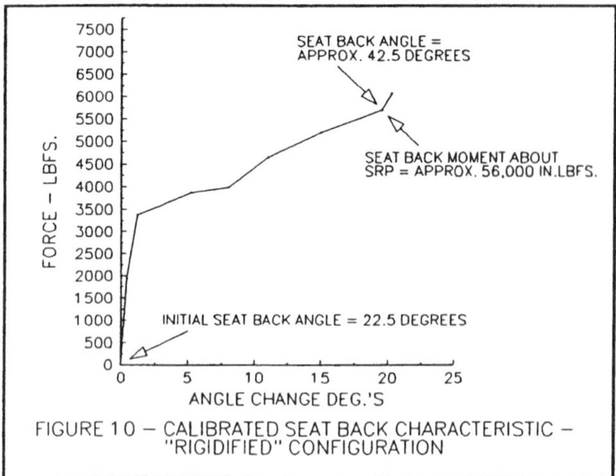

FIGURE 10 – CALIBRATED SEAT BACK CHARACTERISTIC – "RIGIDIFIED" CONFIGURATION

The objective of these sled tests was to compare the response of an OOP occupant with and without seat back yield. Therefore, the "rigidified" calibrated seat was designed so that it possessed both a high initial stiffness and a high moment capacity, typical of a seat back which is supported at a high elevation relative to the seat pivot. The authors recognize that there may be other seat designs proposed that would not fit the two categories of seats tested in this program. Additional OOP testing of these proposed seat designs may be appropriate in the future.

<u>Test Dummy, Associated Instrumentation and Injury Criteria</u>
The anthropomorphic test device (ATD) used was the Hybrid III 50th percentile male dummy. This ATD was selected primarily because of its neck and lumbar instrumentation. While the response of the Hybrid III neck is not ideal, it is generally agreed that it has the greatest biofidelity of any test dummy presently available. Evaluating injury potential from the Hybrid III neck instrumentation has not been well developed; however, a set of Hybrid III neck injury criteria have been issued by the International Standards Organization (ISO)[Mertz,1984]. These criteria are summarized below:

1. The torque at the juncture of the head and neck
   In extension: 504 in-lbfs. (57 Newton-meters)
   In flexion: 1681 in-lbfs.(190 Newton-meters)

2. Neck force at the juncture of the head and neck
   In shear: 247-692 lbfs.(1100 to 3100 Newtons), depending on duration, see Figure 11.
   In compression: 247-900 lbfs.,(1100 to 4000 Newtons), depending on duration, see Figure 12.
   In tension: 247-742.5 lbfs.(1100 to 3300 Newtons), depending on duration, see Figure 13.

The above criteria were used in this program to provide some guidelines regarding the magnitudes of the neck forces and

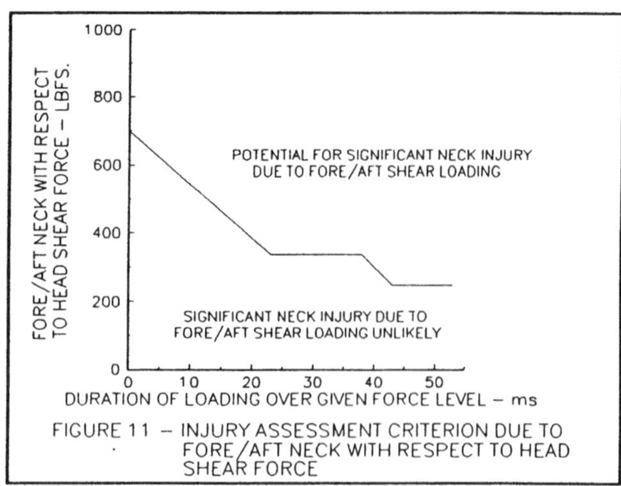

FIGURE 11 – INJURY ASSESSMENT CRITERION DUE TO FORE/AFT NECK WITH RESPECT TO HEAD SHEAR FORCE

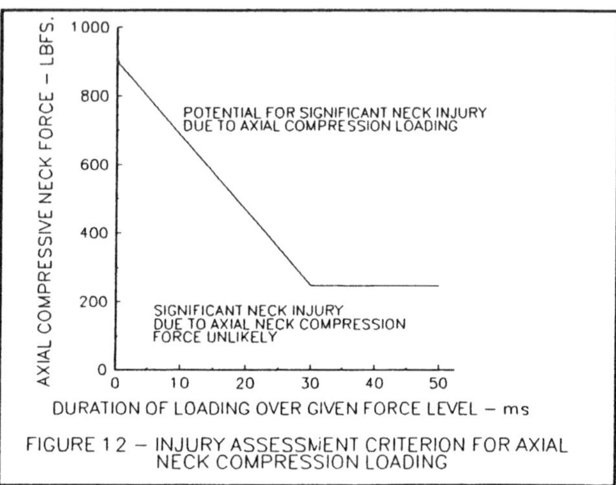

FIGURE 12 – INJURY ASSESSMENT CRITERION FOR AXIAL NECK COMPRESSION LOADING

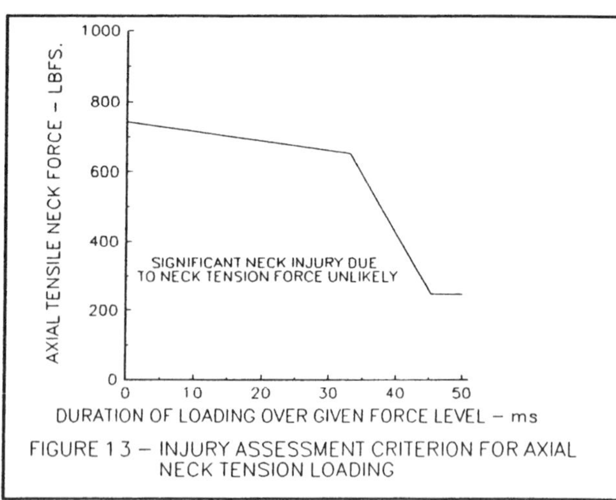

FIGURE 13 – INJURY ASSESSMENT CRITERION FOR AXIAL NECK TENSION LOADING

moments recorded during testing. In addition to determining whether the neck loads are higher or lower for the two seat configurations, it is also useful to determine whether the neck loads represent any potential for injury. While the ISO criteria may not be well enough established to make predictions of specific injury, they do represent the best data available and can be used to suggest the potential for injury.

The ATD instrumentation package is summarized in Figure 14. Since the response of the ATD in these tests was essentially confined to motion in the mid-sagittal or vertical longitudinal plane, a number of channels of dummy data that would otherwise have been necessary were eliminated, simplifying the data collection process.

Dummy Pre-impact Positioning

Films of human subjects involved in emergency braking demonstrations were reviewed for insights into what would constitute reasonable "pitched forward" dummy pre-impact positions. These films came from two sources. The first was a set of demonstrations done by the Ford Motor Company in the mid-70's during their investigation into likely positions of unbelted passengers during air cushion deployments. These tests involved human volunteers [Ford, 197x]. The second set of data was generated in a limited series of experiments conducted by the authors, in which an unbelted Hybrid III dummy was placed in the passenger seat of a late model sedan in the NSP, and motion of the dummy during emergency stops was videotaped and analyzed.

The dummy positions depicted in Figure 5 were selected for use in this program. The position depicted in Figure 5a represents the impact position of an unrestrained passenger either caught unawares by pre-impact emergency braking or thrown forward by a prior, minor frontal impact. Figure 5b represents the same situation for a belt-restrained occupant. It is recognized that these positions are somewhat arbitrary - there are an infinite number of positions that occupants can assume at the moment of impact. For example, sideways occupant motion would be expected in a vehicle which experiences lateral deceleration, such as during a spinout or pre-impact swerving. There was no attempt to simulate these motions in the present test series.

Templates were constructed so that the dummy positions could be reproduced with suitable precision during testing. The construction of the Hybrid III is such that he "wants" to sit upright. Also,the sled buck undergoes a low-level acceleration prior to impact. Therefore, a cable and break-away strap, shown in Figure 15 were used as a tether for the ATD. The tether separated with minimal force; it did not alter the occupant motion.

TEST RESULTS

A total of 32 tests were conducted, however instrumentation problems on tests 2 and 4 reduced the number of useful tests to 30. Twenty of the tests used the "rigidified" calibrated seat, the remaining ten employed the "yielding" calibrated seat. The tests with the "rigidified" seat can be further subdivided into 12 tests with the "unrestrained OOP" dummy and 8 tests with the "restrained OOP" dummy. Similarly, 6 of the tests with the "yielding" seat were with the unrestrained OOP dummy and 4 tests were with the "restrained OOP" dummy. Tables 1-4 summarize the injury measures for these four test groupings.

Unrestrained OOP dummy with the "rigidified" seat - Figure 16 is a series of photographs taken from the high speed camera coverage of Test 14, conducted at an impact speed of 12 mph. This figure illustrates dummy kinematics in a typical test of this configuration. Similar figures will be presented in the three subsequent subsections to illustrate dummy kinematics in tests conducted in the other three test configurations, also at an impact speed of about 12 mph.

Figure 16a is at the moment of impact. The tethering system has kept the ATD in the intended OOP configuration. Figure 16b was taken 86 milliseconds into the event. The shoulder area of the dummy is coming into contact with the rigidified seat back. During these 86 milliseconds the vehicle (and seat back) have decelerated so that a significant relative velocity between the ATD upper body and the seat back has developed prior to the dummy contact. This is in contrast to the NSP situation where the initial close proximity between the ATD upper body and the seat back allows very little relative velocity to build up. This is why the minimal energy absorption capability of a rigid seat might be adequate for NSP occupants. In the OOP case, however, the minimal energy absorption of the rigidified seat combined with the larger relative velocity results in increased ATD accelerations and internal forces. Also, note that during the period of dummy torso rotation into the seat back, the head orientation with respect to the inertial reference system has not changed appreciably. Therefore, when the dummy strikes the rigidized seat back with the shoulders, the head is in a flexed position with respect to the torso.

Figure 16c was taken 128 milliseconds into the event, when the impact between the ATD upper body and the seat back is at its maximum. The head has rotated from its flexed position in Figure 16b to a position of slight extension. Also note that the torso has ramped several inches up the seat back (compare the elevation of the torso in 16b with that in 16c using the background for reference). At this speed, the ramping did not result in the Hybrid III head getting over the head restraint.

Figure 16d shows the dummy rebounding out of the rigidified seat, attaining a maximum head elevation at about 300 milliseconds into the event. It is important to note that the method of rigidifying the seat (by attaching the force pods midway up the seat back) minimizes the amount of stored energy relative to a more customary, cantilevered seat structure. In a cantilevered design, the seat back structure would elastically deform to develop the required moment capacity and store significant energy which would be imparted to the dummy during rebound.

The sled buck does not have a roof structure, however, it appears that, had a roof structure been present, the ATD would have rebounded "forehead first" into it. Although the velocity of this rebound is not great (estimated from the film to be about 3 mph), the potential for significant neck loading due to body augmentation cannot be ignored.

Figures 17 and 18 are graphs of the Hybrid III neck and lumbar response maxima for the unrestrained, rigid seat tests.

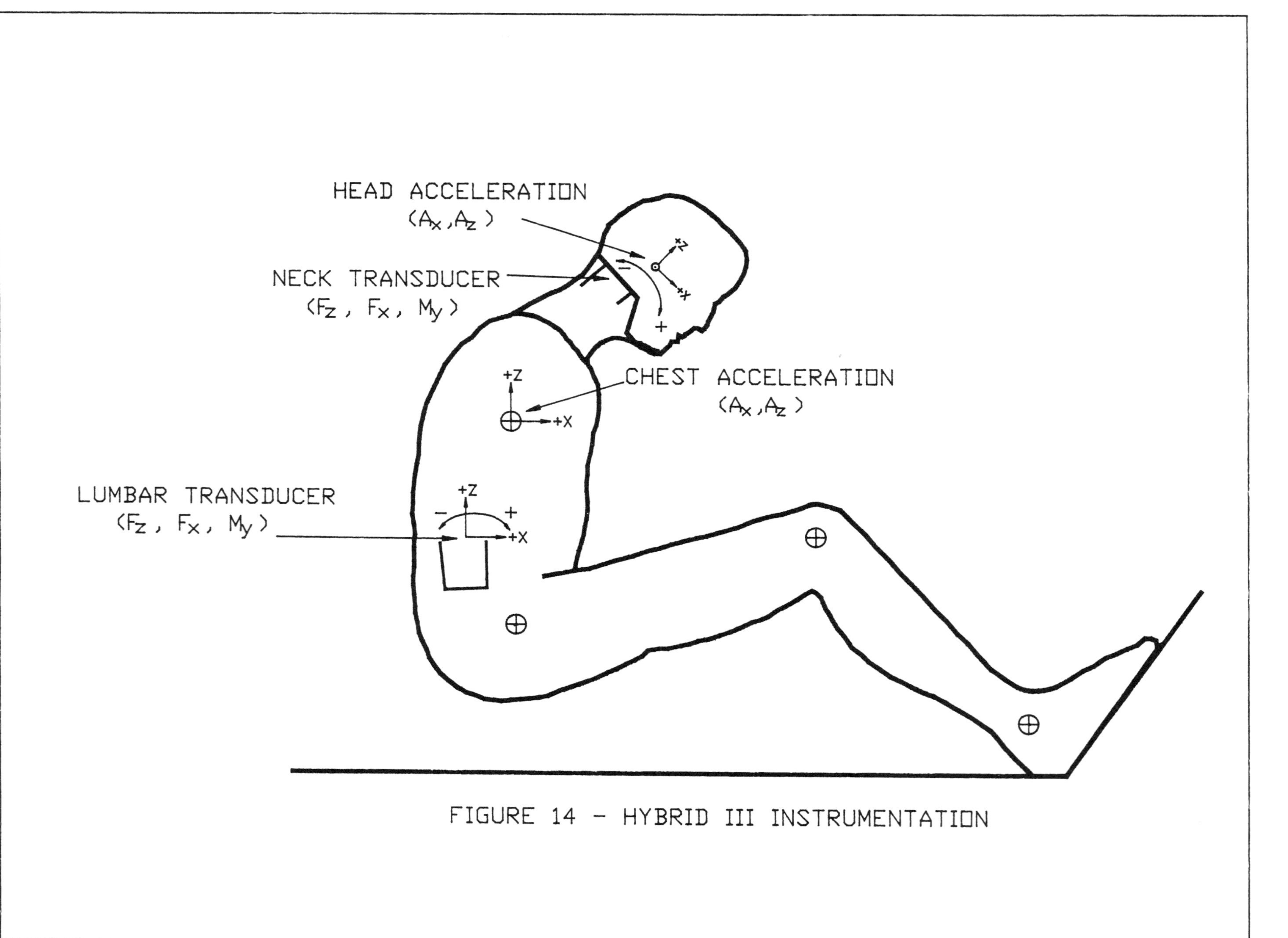

FIGURE 14 - HYBRID III INSTRUMENTATION

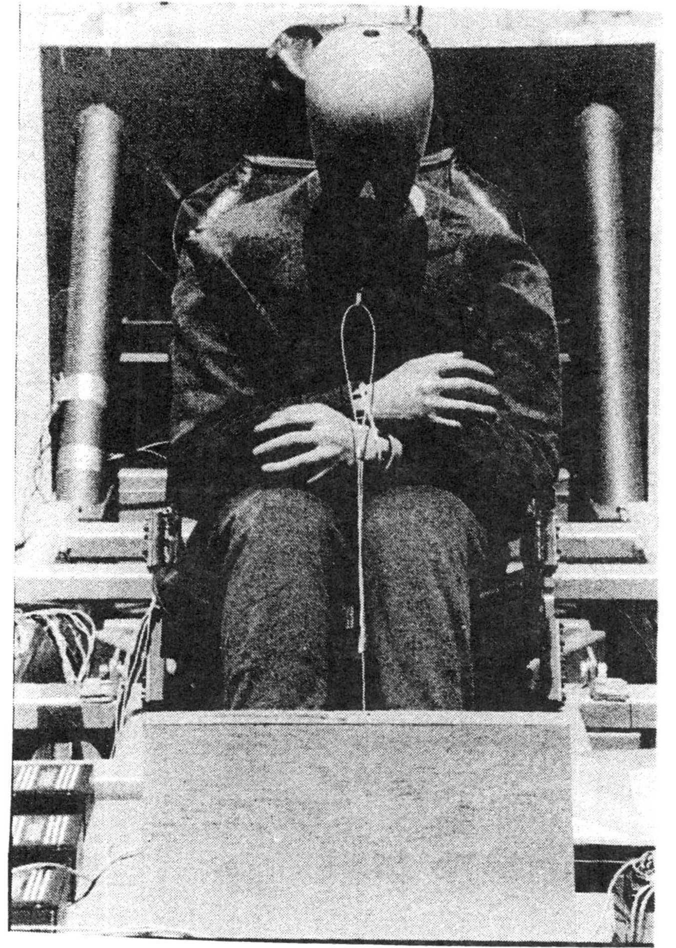

(a)

(b)

FIGURE 15 - ATD TETHERING SYSTEM

TABLE 1 - UNRESTRAINED OOP DUMMY TESTS WITH 56,000 IN.LBF. SEAT									
	TEST	DUMMY RESPONSE							
TEST	VELOCITY	MAX. NECK FORCES AND MOMENTS				LUMBAR			
NO.	(MPH)	Shear Force (x)	Vert. Force (z)	Moment (in.lbfs.)		Shear Force (x)	Vert. Force (z)	Moment (in.lbfs.)	
		lbfs. *	lbfs. **	Ext.	Flex.	lbfs. *	lbfs. **	Ext.	Flex.
1	17.1	+232/-190	+755/-290	830	1180	+285/-405	+530/-40	1950	810
3	20.7	+258/-150	+915/-270	795	945	+140/-585	+725/-102	2790	70
5	19.8	+209/-98	+778/-115	705	490	+295/-575	+695/-235	2790	460
6	17.9	+115/-40	+550/-60	460	530	+165/-600	+498/-450	2990	2
7	16.0	+120/-50	+550/-50	500	495	+130/-590	+410/-405	2875	0
8	18.0	+108/-105	+1340/-200	275	850	+245/-485	+790/-60	2400	490
9	15.2	+249/-140	+1000/-110	498	570	+198/-555	+790/-200	2700	100
14	12.0	+267/-88	+1243/-188	581	559	+218/-553	+718/-153	2525	460
15	14.1	+265/-85	+1306/-200	516	649	+248/-637	+840/-213	2950	470
16	8.9	+113/-46	+829/-59	354	265	+70/-399	+516/-164	2082	0
17	4.9	+57/-32	+47/-47	96	235	+23/-329	+77/-177	1567	0
23	5.8	+57/-33	+141/-29	183	235	+35/-318	+159/-165	1612	0
*	+ denotes forward shear loading								
**	+ denotes tensile loading								

TABLE 2 - UNRESTRAINED OOP DUMMY TESTS WITH 9,000 IN.LBF. SEAT									
	TEST	DUMMY RESPONSE							
TEST	VELOCITY	MAX. NECK FORCES AND MOMENTS				LUMBAR			
NO.	(MPH)	Shear Force (x)	Vert. Force (z)	Moment (in.lbfs.)		Shear Force (x)	Vert. Force (z)	Moment (in.lbfs.)	
		lbfs. *	lbfs. **	Ext.	Flex.	lbfs. *	lbfs. **	Ext.	Flex.
21	5.1	+50/-32	+136/-24	206	199	+47/-301	+153/-124	1388	47
22	9.2	+152/-55	+637/-18	355	325	+94/-435	+500/-71	1941	59
25	11.3	+72/-64	+306/-65	243	245	+53/-329	+248/-88	1671	0
27	13.8	+76/-39	+389/-41	318	296	0/-459	+271/-218	2118	47
29	17.6	+81/-34	+535/-24	332	265	+71/-418	+460/-47	1924	0
31	20.5	+67/-71	+431/-343	273	251	+135/-341	+494/-94	1624	0
*	+ denotes forward shear loading								
**	+ denotes tensile loading								

TABLE 3 - RESTRAINED OOP DUMMY TESTS WITH 56,000 IN.LBF. SEAT									
	TEST	DUMMY RESPONSE							
TEST	VELOCITY	MAX. NECK FORCES AND MOMENTS				LUMBAR			
NO.	(MPH)	Shear Force (x)	Vert. Force (z)	Moment (in.lbfs.)		Shear Force (x)	Vert. Force (z)	Moment (in.lbfs.)	
		lbfs. *	lbfs. **	Ext.	Flex.	lbfs. *	lbfs. **	Ext.	Flex.
10	21.8	+50/-60	+800/-100	285	195	+205/-135	+1000/-560	610	700
11	20.0	+170/-110	+1320/-160	380	660	+298/-450	+620/-110	1950	805
12	16.2	+262/-141	+1324/-288	573	904	+270/-482	+1000/-30	2290	753
13	17.8	+194/-77	+1308/-130	397	670	+224/-453	+850/-47	2100	684
18	12.2	+92/-55	+781/-47	206	413	+212/-318	+459/-247	1435	706
19	9.8	+154/-50	+944/-189	350	510	+194/-376	+610/0	1800	490
20	5.9	+67/-39	+437/-35	200	206	+59/-318	+312/-159	1459	271
24	6.1	+86/-42	+259/-35	361	229	+59/-324	+212/-183	1467	95
*	+ denotes forward shear loading								
**	+ denotes tensile loading								

TABLE 4 - RESTRAINED OOP DUMMY TESTS WITH 9,000 IN.LBF. SEAT									
	TEST	DUMMY RESPONSE							
TEST	VELOCITY	MAX. NECK FORCES AND MOMENTS				LUMBAR			
NO.	(MPH)	Shear Force (x)	Vert. Force (z)	Moment (in.lbfs.)		Shear Force (x)	Vert. Force (z)	Moment (in.lbfs.)	
		lbfs. *	lbfs. **	Ext.	Flex.	lbfs. *	lbfs. **	Ext.	Flex.
26	12.7	+74/-58	+324/-118	301	331	+71/-396	+441/-211	1770	59
28	15.5	+67/-34	+349/-59	340	244	+59/-341	+306/-76	1676	0
30	17.3	+65/-25	+371/-59	258	243	+82/-329	+435/0	1548	0
32	21.2	+92/-30	+572/-454	688	251	+147/-329	+510/-82	1369	0
*	+ denotes forward shear loading								
**	+ denotes tensile loading								

(a) Impact begins (t=0)

(b) Dummy's shoulders contact seat back with head in flexion (t=86ms)

(c) Head rotates into head restraint while torso arrested. Rebound begins (t=128ms)

(d) Dummy attains maximum vertical excursion during rebound (t=300ms)

FIGURE 16 - UNRESTRAINED OOP DUMMY KINEMATICS (RIGIDIFIED SEAT) DURING 12 MPH REAR IMPACT (TEST 14)

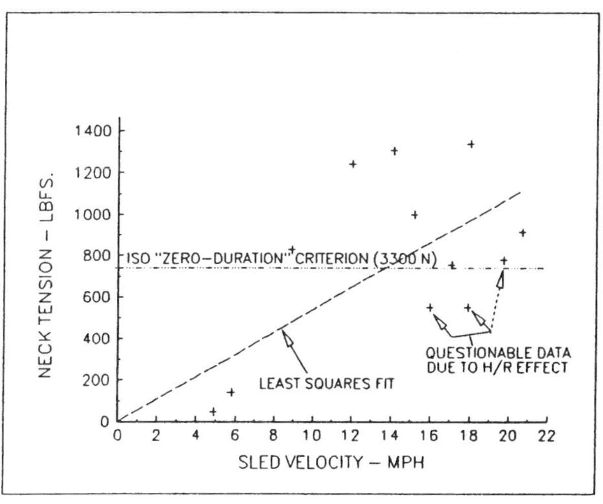

(a) Neck Tension

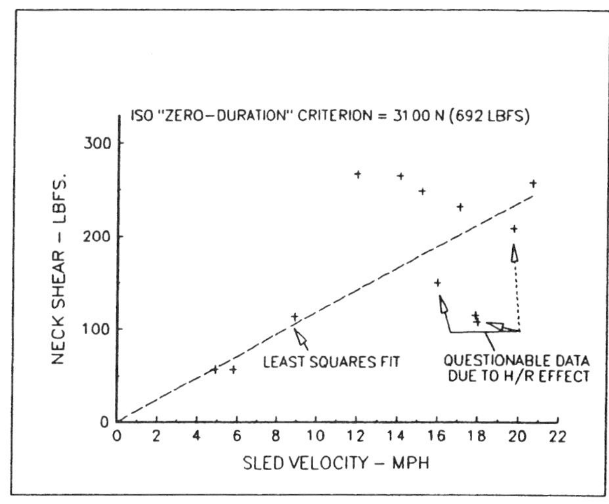

(b) Neck Shear

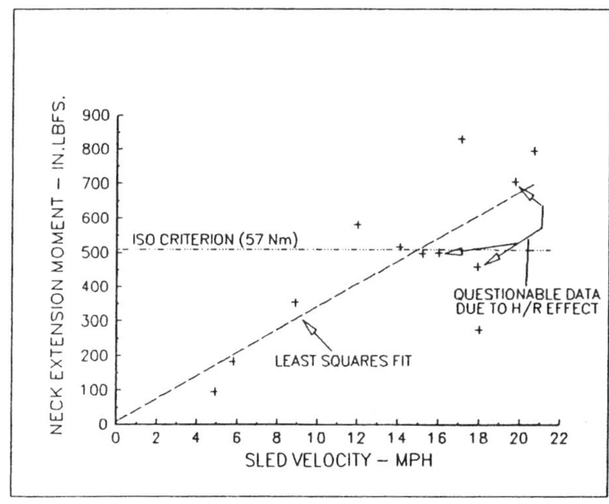

(c) Neck Extension Moment

(d) Neck Flexion Moment

FIGURE 17 - MAXIMUM HYBRID III NECK FORCES & MOMENTS, TEST IN UNRESTRAINED OOP, RIGIDIFIED SEAT

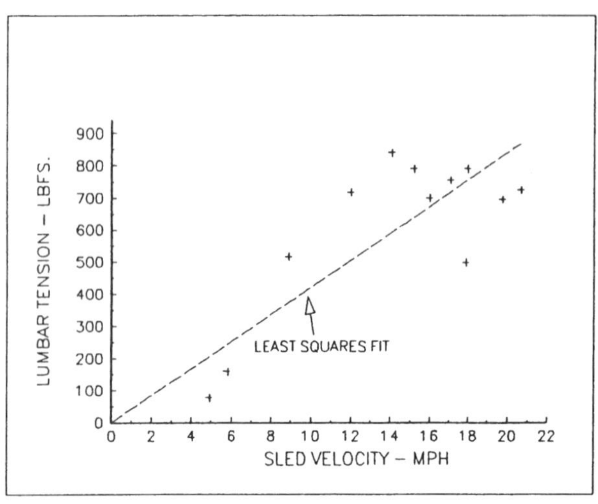

(a) Lumbar Tension

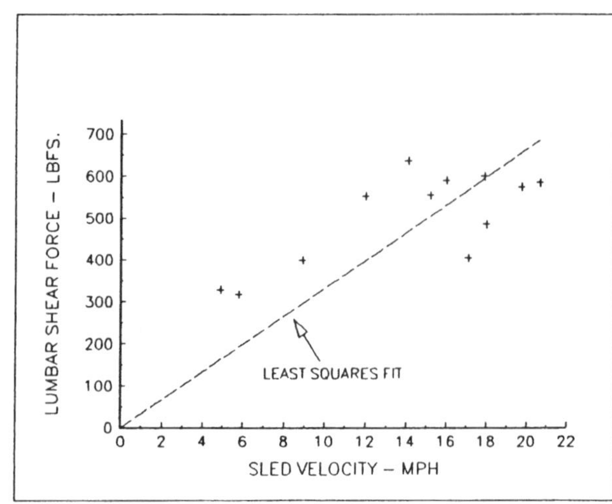

(b) Lumbar Shear

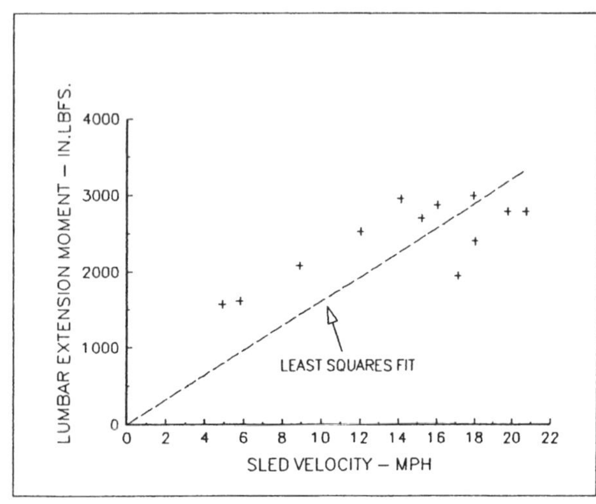

(c) Lumbar Extension Moment

(d) Lumbar Flexion Moment

FIGURE 18 - MAXIMUM HYBRID III LUMBAR FORCES & MOMENTS, TESTS UNRESTRAINED OOP, RIGIDIFIED SEAT

Each plot contains a linear dashed line going through the origin which represents a least squares fit of the data, simply to assist in comparing results from the four test series. Where appropriate, the neck response graphs also contain information concerning the ISO injury criteria. In the case of neck forces, the criteria are time dependent, so only the highest "tolerable" force, namely that at a zero time duration, is shown on the graph. The notations on Figures 17 concerning questionable data refer to issue of the yielding and straightening of the head restraint referred to earlier in the paper.

Unrestrained OOP tests with the Yielding Seat - Figure 19 is a series of photographs taken from the high speed camera coverage of Test 25, a nominally 12 mph impact (actual speed 11.3 mph) involving an unrestrained OOP Hybrid III in the "yielding" calibrated seat. Comparison of these photographs with those of Figure 16 indicate several interesting differences which arise when the seat yields and absorbs energy.

Figure 19a shows the moment of impact. Again, the tethering system has successfully retained the ATD in the desired OOP during the ride up. Figure 19b is at about 64 milliseconds into the impact. The seat back has deflected from its pre-impact position even though the shoulder area of the dummy has not yet made contact with it. There was sufficient torque applied to the lower region of the seat back by the thoracic/lumbar area of the dummy's back to have initiated seat back yield 42 milliseconds into the event. By the time depicted in Figure 19b, this initial period of seat back deflection is over.

Figure 19c shows the event at about 92 milliseconds. The shoulders are just beginning to get involved with the seat back. Again notice that during the first 90 milliseconds, the head orientation has stayed at its pre-impact attitude relative to the ground so that at dummy shoulder contact with the seat back, the head is in flexion relative to the torso.

In Figure 19d, about 112 milliseconds into the event, the head has rotated into a basically neutral position. At this time, the second phase of seat back yielding is initiated. In Figure 19e (t = 144 msecs), this second yielding period comes to an end as the rearward motion of the dummy is arrested by the seat energy absorbing process. Rebound is about to begin.

Figure 19f shows the dummy in the process of rebounding forward. This photo shows the point where the vertical excursion of the head has reached a maximum (t = 280 msecs). Comparison of Figures 16d and 19f reveal that rebound from the yielding seat is at a much lower velocity, and it does not result in the vertical elevation seen in the rigidified seat test. The ATD in this test came to a final rest position in the seat while the dummy in the rigidified seat test rebounded out of the seat.

Figures 20 and 21 summarize the dummy response data obtained during the 6 tests conducted in this series. The corresponding plots from the rigidified seat (Figures 17 and 18) can be compared to these figures to discern the potential of seat back yield in lowering dummy injury measures in the unrestrained OOP situation. The linear regression equations, indicate the following reduction factors for the different dummy response maxima:

Neck Forces
- Tension - 46 percent reduction
- Shear - 53 " "

Neck Moments
- Extension - 41 percent reduction
- Flexion - 56 " "

Lumbar Forces
- Tension - 35 percent reduction
- Shear - 22 " "

Lumbar Moments
- Extension - 25 percent reduction
- Flexion - 93 " "

Restrained OOP tests with the Rigidified Seat - Figure 22 is a series of photographs depicting the dummy kinematics during Test 18, a nominally 12 mph impact (actual speed 12.2 mph) involving a restrained OOP Hybrid III in the rigidified calibrated seat.

The kinematics depicted in Figure 22 are generally similar to those of the unrestrained occupant in the rigidized seat, with the following differences. The restrained dummy is initially closer to the seat back, and hence the impact of the ATD upper body with the seat back occurs earlier in the event and with lesser velocity. Also, the seat belt system is effective in limiting the rebound excursions both vertically and forward.

Figures 23 and 24 contain plots of the dummy response maxima as a function of sled velocity for the tests in this third series. Comparisons between these results and the earlier results from the unrestrained dummy indicate generally lower dummy injury measures when the dummy is restrained. This reduction appears to be a consequence of the closer proximity of the ATD to the rigidified seat back at the moment of impact.

Restrained OOP tests with the Yielding Seat - Figure 25 is the series of photographs from the high speed camera coverage of Test 26, a nominally 12 mph impact (actual speed 12.7 mph) involving a restrained OOP dummy sitting in the yielding calibrated seat. Comparison of these photographs with those of Figure 22 (restrained dummy in the rigid seat) yield conclusions similar to those found when comparing rigid vs. yielding seat kinematics with the unrestrained occupant. (i.e., both the dummy reaction force and the rebound velocity are lower in the yielding seat tests.) Similar to the unrestrained yielding seat tests, seat yield was found to occur in two phases - an initial phase associated with the loading of the seat back by the lower torso, followed (after a short period of no seat yield) by a second period of seat deformation that accompanied the involvement of the upper torso and head with the seat back. The seat belt was effective controlling rebound motion, even though the rebound, as expected, was significantly less than that seen in the rigid seat tests.

(a) Impact begins (t=0)

(b) Seat back yield temporarily stops as dummy approached seat back (t=64ms)

(c) Dummy shoulder contacts seat back (t=92ms) arrested. Rebound begins (t=116ms)

(d) Head contacts head restraint as seat back yield resumes (t=112ms)

(e) Seat yield over, dummy begins mild rebound forward (t=144ms)

(f) Dummy attains maximum vertical excursion during rebound (t=280ms)

FIGURE 19 - UNRESTRAINED OOP DUMMY KINEMATICS (YIELDING SEAT) DURING A NOMINALLY 12 MPH REAR IMPACT (TEST 25)

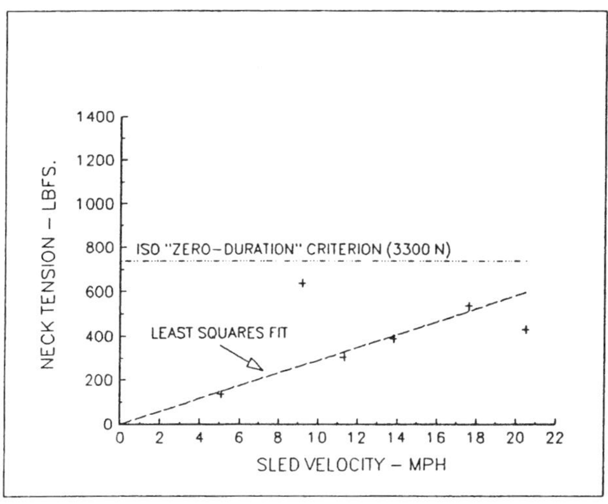

(a) Neck Tension

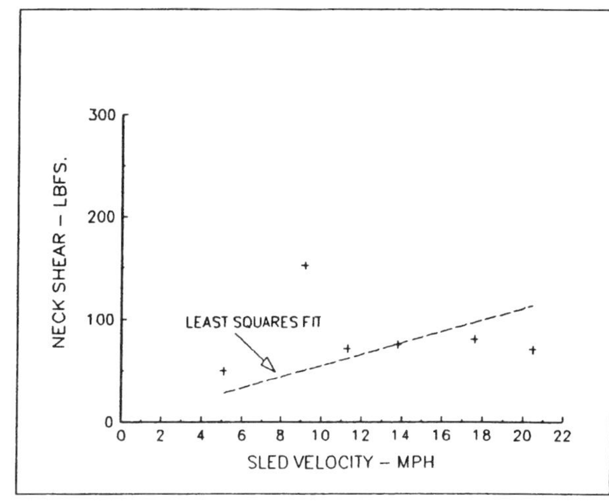

(b) Neck Shear

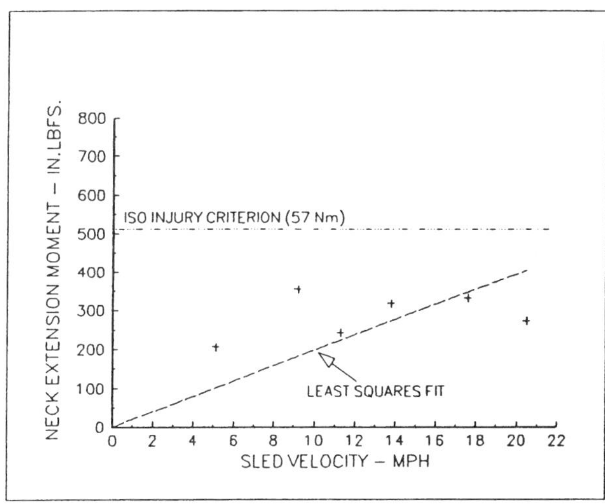

(c) Neck Extension Moment

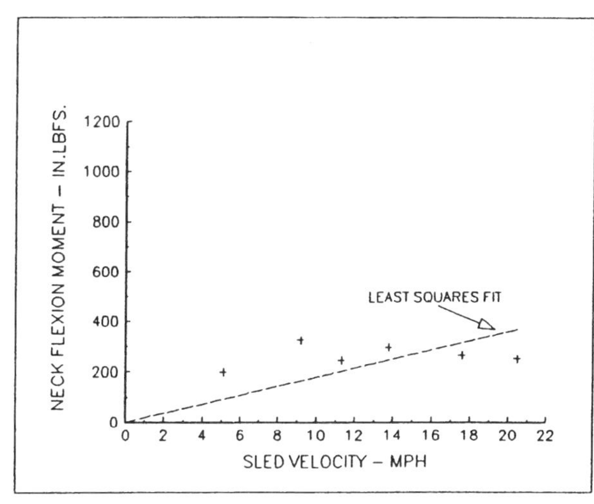

(d) Neck Flexion Moment

FIGURE 20 - MAXIMUM HYBRID III NECK FORCES & MOMENTS, TESTS IN UNRESTRAINED OOP, YIELDING SEAT

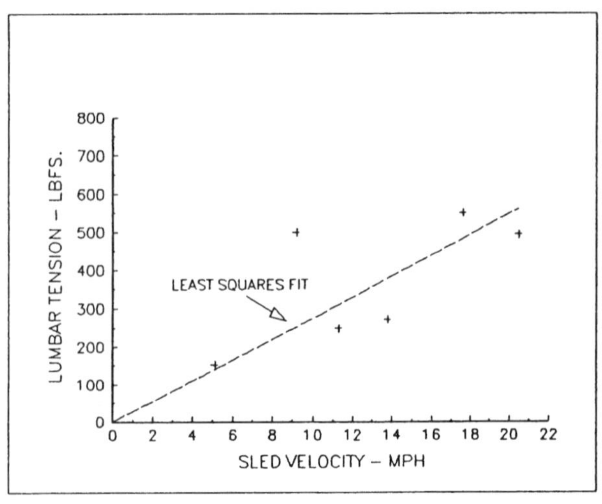

(a) Lumbar Tension

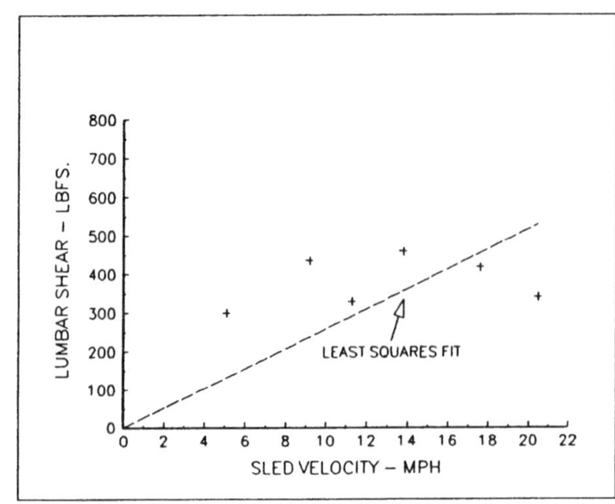

(b) Lumbar Shear

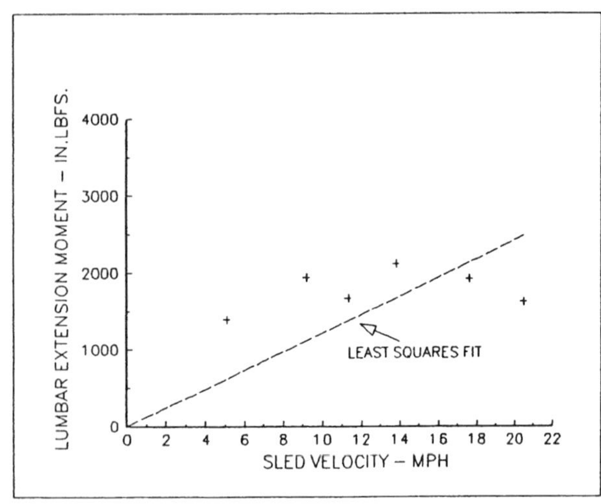

(c) Lumbar Extension Moment

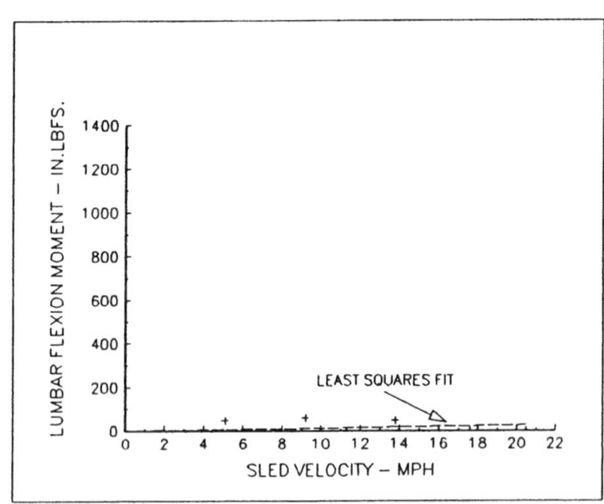

(d) Lumbar Flexion Moment

FIGURE 21 - MAXIMUM HYBRID III LUMBAR FORCES & MOMENTS, TESTS IN UNRESTRAINED OOP, YIELDING SEAT

(a) Impact begins (t=0)

(b) Dummy's shoulders contact seat back with head at original angle

(c) Head rotates into head restraint while torso arrested. Rebound begins (t=116ms)

(d) Dummy attains maximum vertical excursion during rebound (t=206ms)

FIGURE 22 - RESTRAINED OOP DUMMY KINEMATICS (RIGIDIFIED SEAT) DURING 12 MPH REAR IMPACT (TEST 18)

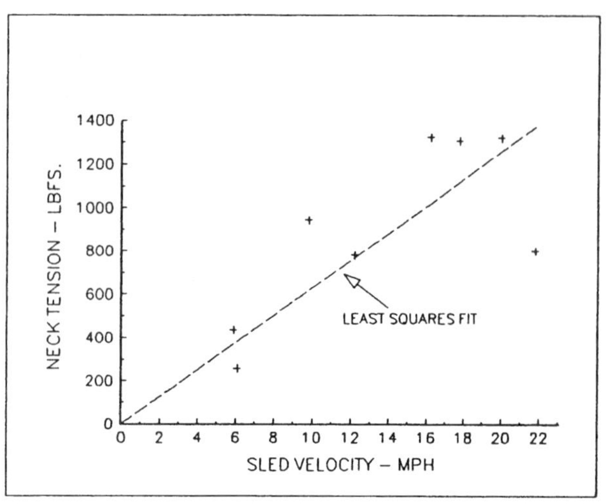

(a) Neck Tension

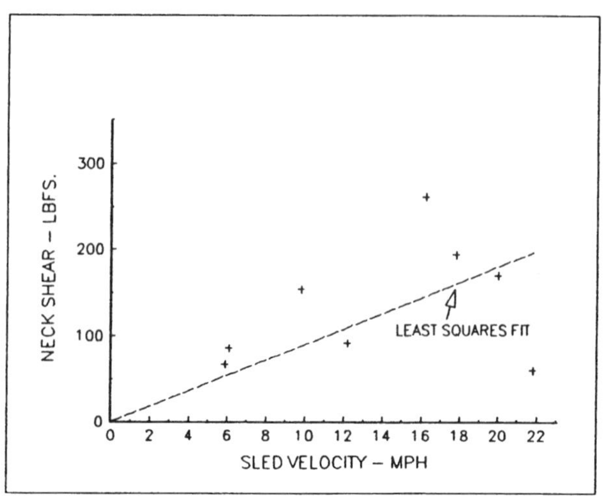

(b) Neck Shear

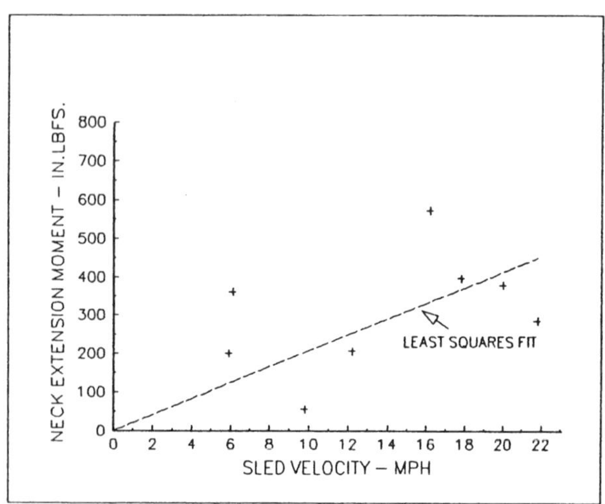

(c) Neck Extension Moment

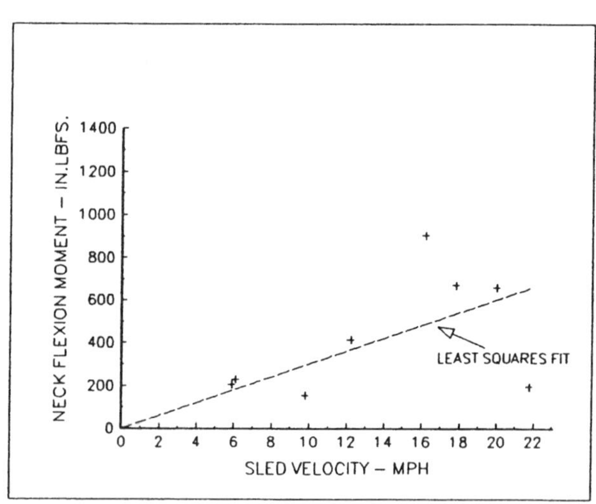

(d) Neck Flexion Moment

FIGURE 23 - MAXIMUM HYBRID III NECK FORCES & MOMENTS, TESTS IN RESTRAINED OOP, RIGIDIFIED SEAT

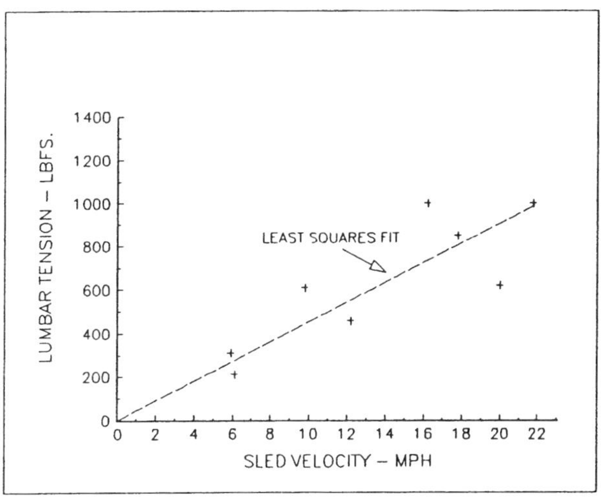

(a) Lumbar Tension

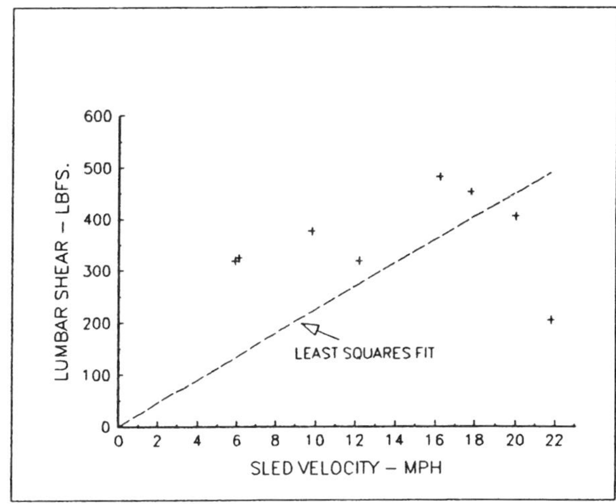

(b) Lumbar Flexion

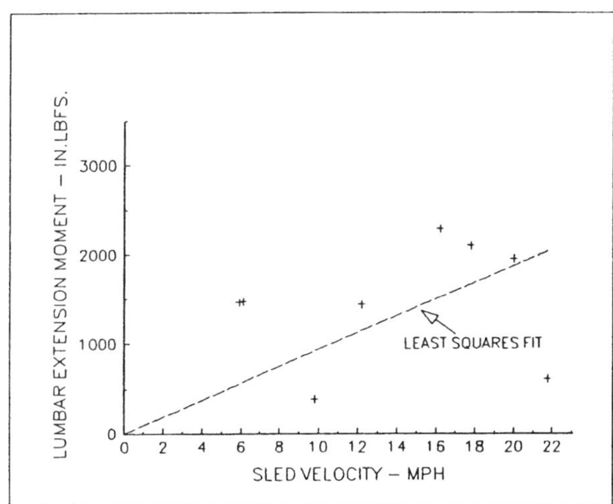

(c) Lumbar Extension Moment

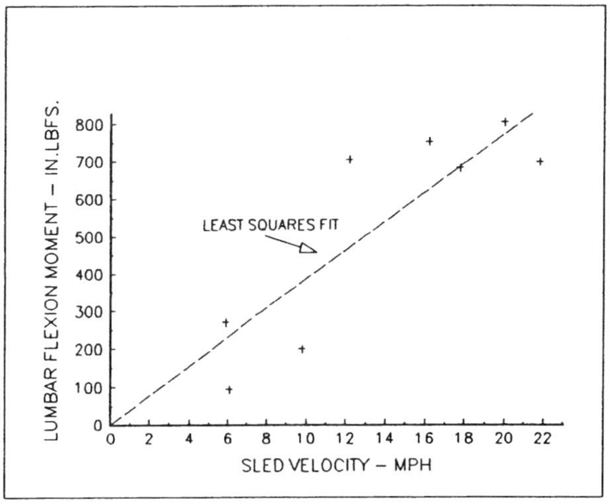

(d) Lumbar Flexion Moment

FIGURE 24 - MAXIMUM HYBRID III LUMBAR FORCES & MOMENTS, TESTS IN UNRESTRAINED OOP, RIGIDIFIED SEAT

(a) Impact begins (t=0)

(b) Shoulders contact yielding seat back (t=86ms)

(c) Seat back yield temporarily ceases as head rotates toward head restraint (t=100ms)

(d) Head contacts head restraint, seat back yield resumes (t=118ms)

(e) Seat back yield ceases as dummy head arrested. Rebound begins (t=160ms)

(f) Dummy attains maximum vertical excursion during rebound (t=300ms)

FIGURE 25 - RESTRAINED OOP DUMMY KINEMATICS (YIELDING SEAT) DURING 12 MPH REAR IMPACT (TEST 26)

Figures 26 and 27 contain the plots of the restrained dummy response maxima which can be compared to the corresponding Figures 23 and 24 to investigate the mitigating potential of seat back yield for the restrained dummy. Again using the linear regression fits to determine the overall reduction in injury parameters, the following percentage reductions are found:

Neck Forces
- Tension - 61 percent reduction
- Shear - 51 " "

Neck Moments
- Extension - 18 percent increase
- Flexion - 49 percent reduction

Lumbar Forces
- Tension - 46 percent reduction
- Shear - 80 " "

Lumbar Moments
- Extension - 3 percent reduction
- Flexion -100 " " (elimination)

SUMMARY

Four series of rear impact sled tests, totalling 30 impacts are presented. All of these tests were conducted with the dummy Out-of-Position (leaning forward). A calibrated seat was set into either the "rigid" or the "yielding" configuration. The rigid seat was designed to have a high initial stiffness and moment capacity, the yielding seat had a peak moment capacity typical of a production bucket seat. Tests were performed with both restrained or unrestrained dummies.

The highest overall dummy neck and lumbar injury measures were observed during tests in the unrestrained mode with the rigid seat. Corresponding tests using the yielding seat back showed reductions in neck forces of on the order of 50 percent and reductions of lumbar forces of approximately 25 percent or more.

Tests conducted in the restrained mode resulted in higher forces with the rigid seat compared to the yielding seat and showed the benefit of seat belt use in controlling both the occupant pre-impact position and post-impact excursions. Injury measures observed in the restrained mode were uniformly lower than those evidenced in the unrestrained mode due to the reduction in the severity of the impact with the seat back that accompanied the more proximate position of the dummy. Comparison between tests with the yielding and rigid seat backs revealed that the neck and lumbar forces were, with the exception of neck and lumbar extension moments, reduced by a factor of about 50 percent or more. The lumbar extension moments were increased by only about 3 percent which, in the absence of an injury criterion, may not be of any real significance.

Seat back yield in the restrained OOP tests resulted in neck extension moments about 18 percent higher than those in the rigid seat tests. At the highest impact severities tested, this increase results in the moments just exceeding the ISO criteria. In view of the fact that the rigid seat dramatically increased neck tension forces under these same conditions (to the point where the ISO criterion is exceeded by about one hundred percent), the authors feel that the yielding seat back probably results in a net safety advantage. However until the biomechanics community has a better understanding of the mechanism of neck injury and human tolerance in rear impacts, an assessment of the percent reduction in the probability of neck injury accompanying seat back yield would not be warranted.

LIST OF REFERENCES

[DataLink, 1989]
"Car Crash Outcomes in Rear Impacts. Appendix A to Current Issues of Occupant Protection in Car Rear Impacts." Data Link, Inc., Washington, D.C. 1989.

[Ford, 197x]
Submission to the NHTSA passive restraint docket.

[Hilyard, 1973]
Hilyard, Joseph F.; John W. Melvin; James H. McElhaney: "Deployable Head Restraints." Highway Safety Research Institute, University of Michigan. DOT-HS-800-802. Jan. 31, 1973.

[Malliaris, 1985]
Malliaris, A.C.; Ralph Hitchcock; and Marie Hansen: "Harm Causation and Ranking in Car Crashes." SAE 850090, SAE International Congress and Exposition. Feb. 25, 1985.

[Melvin, 1971]
Melvin, J.W.; J.H. McElhaney: "Deployable Head Restraints." Final Report, Highway Safety Research Institute. University of Michigan. FH-11-7612. June 30, 1971.

[Mertz, 1984]
Mertz, H.J.: "Injury Assessment Values Used to Evaluate Hybrid III Response Measurements." ISO/TC22/SC12/WG5 N123. May, 1984.

[Ricci, 1979]
Ricci, L.L.: "NCSS Statistics: Passenger Cars." Special Report, Highway Safety Research Institute. University of Michigan. Report UM-HSRI-80-36 Prepared under NHTSA Contract DOT-HS-8-01944.

[Strother,James, 1987]
Strother, C.E.; and Michael B. James: "Evaluation of Seat Back Strength and Seat Belt Effectiveness in Rear End Impacts." SAE 872214, Proceedings of the 31st Stapp Car Crash Conference, New Orleans, Louisiana. October 9-11, 1987.

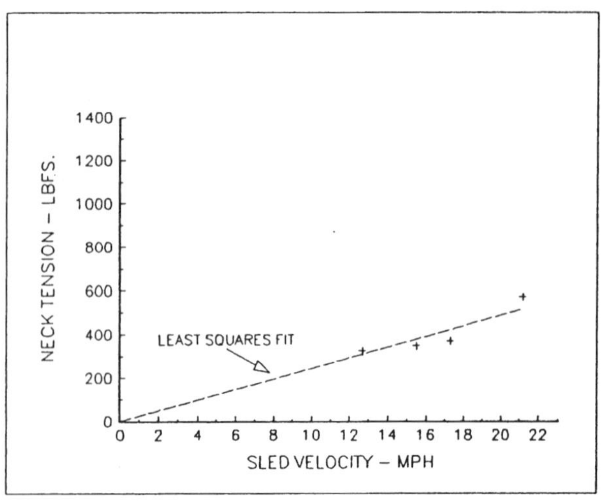

(a) Neck Tension

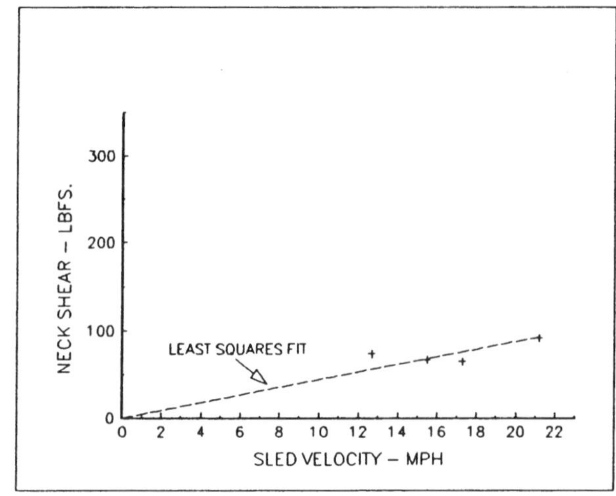

(b) Neck Shear

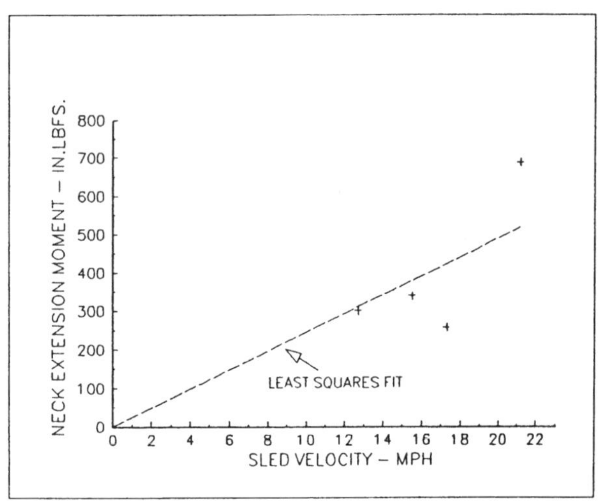

(c) Neck Extension Moment

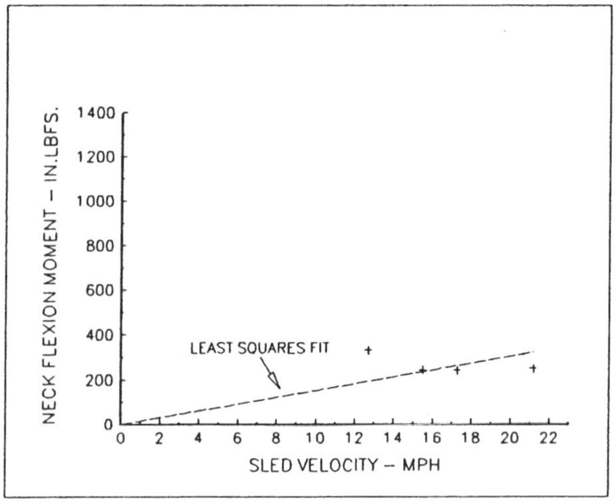

(d) Neck Flexion Moment

FIGURE 26 - MAXIMUM HYBRID III NECK FORCES & MOMENTS, TESTS IN RESTRAINED OOP, YIELDING SEAT

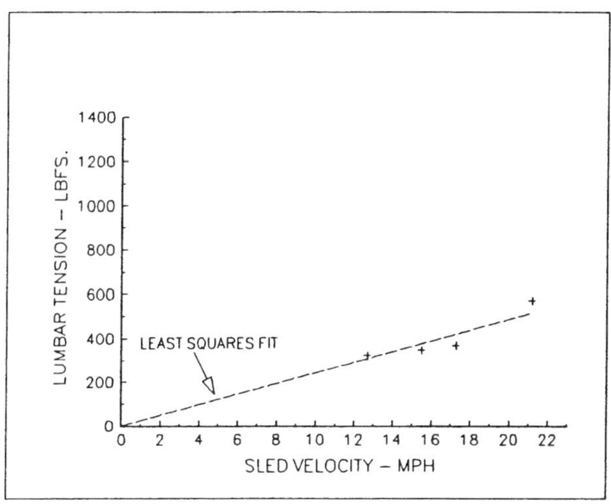

(a) Lumbar Tension

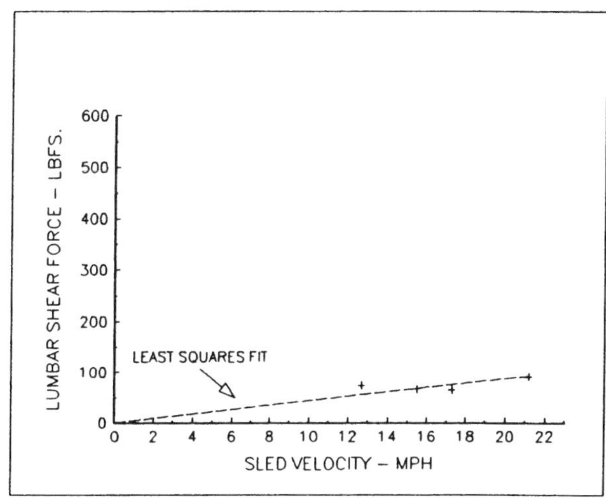

(b) Lumbar Shear

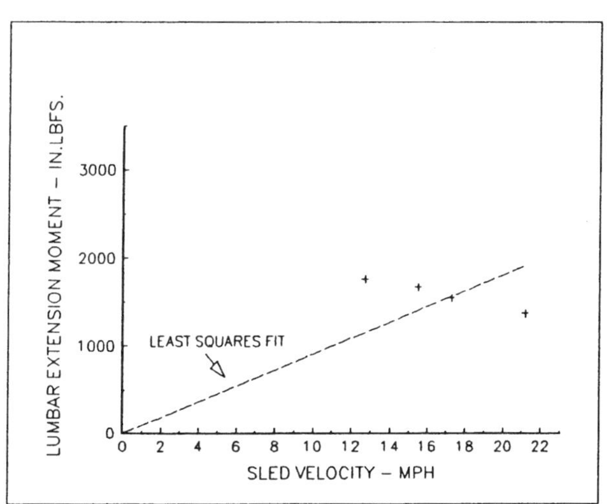

(c) Lumbar Extension Moment

FIGURE 27 - MAXIMUM HYBRID III LUMBAR FORCES & MOMENTS, TESTS IN RESTRAINED OOP, YIELDING SEAT

[Warner et al, 1991]
Warner, C.Y.; C.E. Strother; M.B. James; R.L. Decker: "Occupant Protection in Rear-end Collisions: II. The Role of Seat Back Deformation in Injury Reduction" SAE 912914, Proceedings of the 35th Stapp Car Crash Conference, San Diego, CA. November 1991

APPENDIX

The following weighting factors, taken from the Data Link Study of rear impacts [Data Link, 1989], were used to combine injuries at various AIS levels into a single Harm number:

AIS	Weighting Factor
1	0.4
2	2.7
3	7.1
4	38.8
5+	186.8

930348

Safety Performance of Motor Vehicle Seats

K. H. Digges, J. H. Morris, and A. C. Malliaris
Kennerly Digges & Associates

ABSTRACT

Comfort and safety are major considerations in the design of occupant seats in motor vehicles. In rear impacts, the seat is the principal component of the occupant restraint system. However, it also contributes to occupant restraint in frontal and side impacts, and rollovers.

To determine how well automobile seats protect occupants from injuries in rear impacts, crash data from the national accident files in the United States were analyzed. The distributions and causes of injuries in rear impacts were categorized according to injured body region and source of injury. In addition, forty-nine selected accident cases were reviewed, to determined how the seat performed in crashes. Finally, rear impact crash tests of passenger cars were analyzed to determine dummy motion and seat performance.

Non-contact injuries were associated with the largest portion of injury harm. The seat and frontal components were the two largest sources of contact injury harm. For restrained occupants, the harm distribution was: Non-contact - 25%; Seat - 22%; Front - 16%. For the 49 cases reviewed, nearly half of the occupants in severe rear impacts received injuries from frontal components inside the car. Three occupants in severe crashes received AIS 2 neck fractures. All three were in seats with backs which did not deform during the crash.

Opportunities for safety improvements include improved head restraints, seats designed to absorb energy, and belt/seat systems which act together to protect the occupant in both rear and frontal impacts.

SAFETY RESEARCH IN SEATING SYSTEMS

Some of the earliest automotive safety research sponsored by the federal government was directed to improve seating systems. An early program at UCLA, sponsored by the Public Health Service was summarized in an SAE paper (Severy 1968). The authors concluded that elastic rebound from seat backs increases the chance of multiple impact injuries. Also, they concluded that increasing the seat rigidity reduces occupant rebound.

Severy applied his experience from the UCLA program to design an integrated seat for the Liberty Mutual Safety Car. The resulting design was constructed and crash tested. The design was a capsule seat with integrated double shoulder belt and lap belt restraints, and side wings for side impact protection. The seat was mounted on a pedestal base, which was designed to flex at the floor pan, thereby mitigating impact energy and reducing occupant rebound. The design is shown in an SAE paper (Severy 67).

Later studies, funded by the Federal Highway Administration were awarded to Cornell Aeronautical Laboratory, and HSRI. The final reports, "Integrated Seat and Occupant Restraint System Performance" were published in 1967. These studies examined modeling, injury criteria, and cost benefits analysis for integrated seat concepts. Recommendations were made for elaborate follow-on research. In follow-on research, HSRI designed an integrated safety seat, which employed two shoulder belts and a lap belt which formed an "A" configuration. In this design, the single upper anchorage was attached to the roof. The HSRI design is described by Melvin (71,72) and is shown in Figure 1.

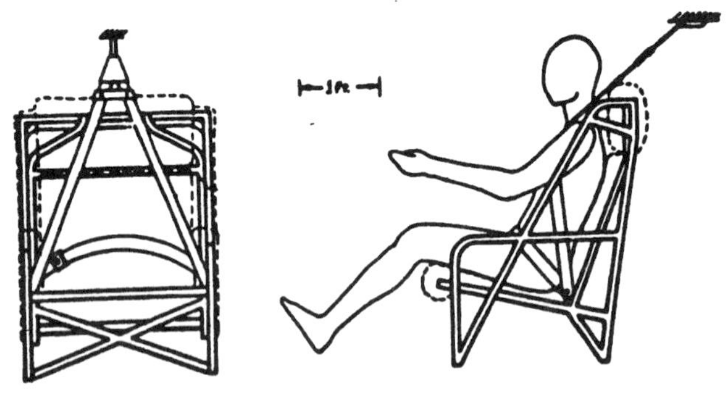

Figure 1 The HSRI Safety Seat

The HSRI seat was further developed and tested in a NHTSA research program to develop a deployable head restraint. The results of this research are reported by Melvin (71) and Hilyard (73). The HSRI team demonstrated effective protection in severe rear-end collisions at vehicle-to-vehicle closing speeds of 80 mph with 40G peak crush structures. Both deployable and non-deployable head restraints were designed and tested. The authors stressed the importance of matching head restraints and seat back structure. They also found benefits in minimizing elastic energy storage in the seat back by utilizing a basically rigid load carrying seat structure.

Rear impact tests with dummies and cadavers were reported by Hu (78 and 80). In these NHTSA sponsored programs, sixteen dummy and nine cadaver tests were documented. The cadaver tests were at a delta-v of 16 to 17 mph. Five tests incorporated deflecting seatbacks, and six had rigid seatbacks. Neck fractures of AIS 3 level were observed on all but two specimens. The remaining were uninjured. The difference in specimens and test variables for the cadaver tests were such that insufficient data was developed to draw conclusions on the relative risk of neck injury for deflecting vs rigid seats.

Several recent papers from Collision Safety Engineering, Inc. provide useful data, references, and technical assessments of safety performance. Strother (87) provides a summary of seat performance in rear impact tests. Warner (91), presents data on the static tests of seats for cars ranging from model year 1964 through 1988. James (91) presents an analysis of field accident studies, and sled tests to show that injuries addressed by stiffening the seatback are a minimal portion of the total injuries. These three papers argue that increasing seatback stiffness would increase injury exposure due to ramping, rebound, out of position occupants, flailing of lower extremities, and non-contact injuries.

A recent trend in seat design is the incorporation of the safety belt anchorages on the seat structure. In many new cars, the anchorage for the lap belt buckle is on the seat structure. Some cars have both lower belt anchorages attached to the seat. Seat manufacturers are now developing structurally efficient designs in which all three anchorages are supported by the seat. These changes in seat and restraint system design should have a very positive effect on comfort, belt use rate and safety.

Daimler Benz, and BMW both offer integrated seats in production cars. The integrated seat solves unfavorable belt geometry problems which exist in convertibles, and cars without B-pillars.

The BMW seat is described in a paper by Haberl (89). Test results produced dramatic reductions ranging from 25% to 57% in HIC, Chest G's and shoulder belt loads, when compared to baseline 50 kph tests. Test results are reported for a 50% and a 95% dummy. An interesting feature of the seat design is the force limiting feature which is accomplished by deformation of the seat back.

At the 1991 ESV Conference Renault displayed an experimental vehicle with an integrated seat. A summary of their research on a seat with an integral belt restraint was presented at the conference (Foret-Bruno, 1991). In this paper, the authors examined how to strengthen the seat for an integrated restraint without increasing the risk of cervical injury in rear impacts. A previous Renault paper (Thomas 1982) had postulated that seat breakage was more effective than a head restraint in preventing neck injuries. In the 1982 paper, the head restraint effectiveness was found to be less than 10%. In the 1991 paper, the effectiveness was found to be 33%. The head restraint effectiveness was found to increase as seat backs became stronger, and broke an higher forces. Based on testing with dummies and one cadaver, the authors recommend head restraint specifications. The authors state that better overall protection should result from the integrated seat.

Air bag protection for rear seat occupants can be expected in the next decade, according to Lee Iacocca (92). This feature will influence the design of the front seat, probably requiring stronger seat backs

ACCIDENT ANALYSIS

An analysis of FARS, and NASS accident data, and Polk exposure data provides insight into opportunities for reducing injuries from rear impacts.

The car exposure and casualties for rear impacts, as a percentage of all events for the combined years 1981-1986 are shown in Figure 2. The Figure shows that rear impacts constitute approximately 11% of all NASS reported car crashes, but involve only 5% of all fatal crashes. Among occupants involved in crashes, 12% are in rear crashes, but 23% of those injured are in rear crashes. However, occupants with serious injuries in rear crashes constitute only 7.6% and 3.5% of all serious and fatal injuries, respectively.

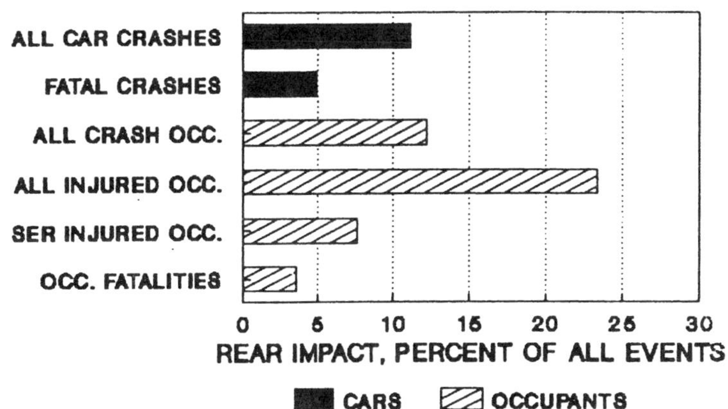

Figure 2

It is evident from these figures that rear impacts cause a disproportionate number of minor injuries. However, only a very small fraction of serious and fatal injuries occur in rear impacts.

In spite of the small numbers, it is desirable to further reduce rear impact casualties where practicable. In order to examine the sources of rear impact injuries, the concept of Harm (Malliaris 82) is useful. The total Harm caused by injuries is computed by estimating the monetary cost of all injuries suffered by the population under consideration. The distribution of Harm according to the injury source for restrained occupants in rear impacts is shown in Figure 3.

Around 20% of the Harm is from unknown sources, illustrating the difficulty in assigning causes of injuries. Non-contact injuries are the cause of another 25% of the Harm. Among contact injuries, the head restraint is the largest source of harm (17%), closely followed by frontal parts of the car interior (16%).

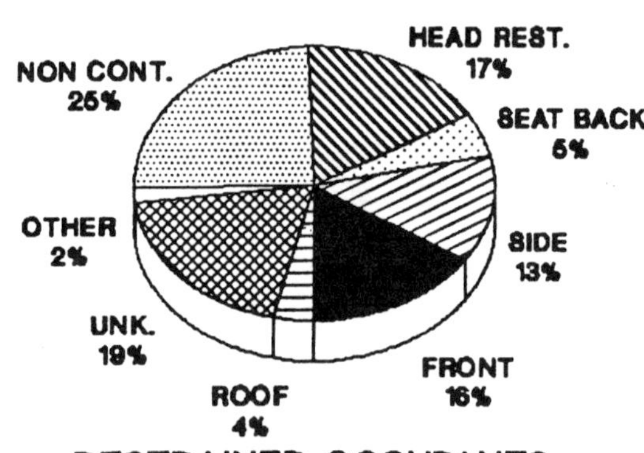

Figure 3

The 16% of the Harm caused by frontal elements, such as steering wheels, dashboards, and windshields may be partially explained by multiple impacts, including those producing frontal deceleration. For the car rear impacts in the NASS 1979-86, only half of the cases involved a car being impacted in the rear by another car, with no subsequent impacts. The other half involved a variety of events, including frequent cases where the impacted car was directed into a third vehicle, resulting in a rear impact followed by a frontal impact.

In order to examine the relative severity of injuries from frontal components, an additional analysis was conducted. The 1st and 2nd most severe injuries to all occupants involved in all single event rear impact crashes were evaluated. The results show that 13% of the most severe injury, and 30% of the 2nd most severe injury were from frontal components in the car. The cause of these injuries cannot be determined from the data available. Potential causes of injuries from frontal components could be: (1) subsequent vehicle decelerations (a rear impact accelerates a vehicle forward, and a resulting rearward acceleration is inevitable); (2) Energy released from elastic deformation of the seat; (3) Flailing of the upper and lower extremities as the occupant moves rearward; and (4) vertical acceleration components which may throw the occupant upward into the steering system. Further research will be required to quantify the extent to which these different components contribute to the "rebound" phenomena.

The side interior is the third largest source of contact Harm to restrained occupants in rear crashes. The side components cause 13% of the Harm. Improvements in the restraint offered by the seat is a possible countermeasure. Improved interior padding will also reduce harm in this category.

The seat back is assigned 5% of the Harm. This appears to be a relatively small amount in view of virtually assured contact with the seat back in a rear collision. However, the contact injuries with the seat back do not necessarily reflect the performance of the seat as an occupant restraint/protection system. Deflection of the seat back could permit the occupant to ramp up the back, exposing him to impact with the roof or hard structure in the rear of the car.

Among other sources of injury, the roof was largest at 3.6%. Rear components were assigned only 0.1% of the Harm. Based on these distributions, impacts with rear components cause negligible Harm in 1979-86 NASS cases for restrained

occupants. Contacts with the roof produce a relatively small amount of harm, when compared with front and side components in the vehicle.
The harm caused by non-contact injuries is shown in Figure 4.

HARM BY BODY REGION
NONCONTACT INJURIES; REAR IMPACTS

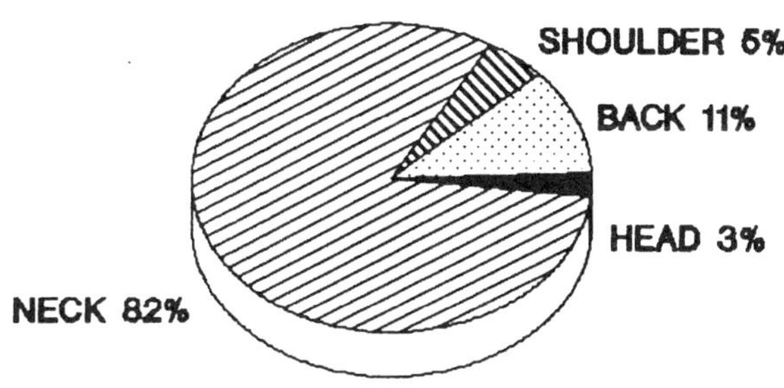

RESTRAINED OCCUPANTS

NASS 1979-86

Figure 4

The data shows that 82% of the non-contact harm was to the neck. This suggests that at least 20% of all rear impact harm to restrained occupants is to the neck. Earlier published data (Malliaris 85) shows that for all occupants and all crash directions, about 9% of harm is to the neck. The high incidence of neck injuries in rear impacts suggests the need for continued improvements in neck injury protection by the seating system. The presence of head restraints as the largest source of contact Harm and the large fraction of neck non-contact harm, suggest opportunities in head restraint design for additional head/neck protection.

INVESTIGATION OF ACTUAL CRASHES

In order to further evaluate how injuries occur in rear crashes, individual cases from the NASS 1988-90 files were reviewed. The first group examined contained all cases of rear impacts in 1988-90 NASS with a delta-v between 30 and 39 mph. The second group was selected based on injury severity.

EIGHTEEN CRASH SEVERITY BASED CASES

This review encompasses all cases in 1988-90 NASS with rear delta V 30 - 39 MPH. Although a small number of cases are present, all the events in the selected severity range are included. However, the unweighted NASS data are only anecdotal evidence. The nationally-weighted data would be needed for any generalizations to national experience. An analysis of the distribution of these cases is presented separate from the other cases investigated. The selection process for the other cases was based on outcome injury severity rather than crash severity.

Of the eighteen cars involved, three had a front bench seat. Two cars burned following the crash event - one occupant died in the fire. Nine of the eighteen cases involved multiple vehicle impacts. Twenty-nine occupants were involved in the eighteen cases of which nineteen were restrained. However, two belt systems released during the crash event - one broke and the other was not properly latched. Out of 25 occupied front seats - 22 yielded. Sixteen seats were deformed by the occupant, and three were deformed by vehicle intrusion. Five seat back locks deformed; two seat adjustors deformed, and 1 track/anchor deformed.

The most frequently injured body part was the head and face, but the highest severity injuries were inflicted to the head and chest. The most frequently recorded probable source of injury was the steering wheel (9), followed by the seat back (8), headrest (5), and flying glass (4). The number of non-contact injuries was 9 - equal to the number of injuries caused by the steering wheel.

In this small sample, the injury rate for AIS 2 and greater injuries for restrained occupants is 26%, while the injury rate for unrestrained occupants is 50%.

The distribution of injuries based on the maximum AIS are shown in Table 1.

Rebound may have contributed to

injuries in twelve of the twenty five cases. In five cases, rebound was definitely involved. In most of these cases, a frontal impact followed the rear impact, thereby adding to the rebound velocity. In two cases, only rear impacts were involved.

TABLE 1 - INJURED OCCUPANTS BY INJURY SEVERITY

	Front Occupants	Rear Occupants
AIS 0	3	1
AIS 1	12	3
AIS 2	6	
AIS 3	2	
AIS 4-5	0	
AIS 6	1 (BURN)	
UNKN.	1	

The earlier reported finding that the steering wheel is the most frequent cause of injury is consistent with the rebound phenomena observed. Most rebound injuries were minor. However, one AIS 3 chest injury may have been aggravated by rebound. This case also had multiple vehicle impacts.

In this limited number of severe rear impact cases, ramping could not be positively identified as contributing to any injuries. It may have contributed to injury in one case. Seat deformation performance may have contributed to one AIS 1 and one AIS 2 injury. However, rebound related injuries were possible in five cases, and probable in seven additional cases. Among these twelve injuries, 8 were AIS 1, 2 were AIS 2, 1 was AIS 3, and 1 was an injury of unknown severity.

THIRTY-ONE SELECTED REAR IMPACT CASES

An additional thirty-one cases were selected for hard copy review. These cases had two selection criteria. First, cases with delta-v greater than 40 mph were included. This produced two fatal cases. A second group included restrained occupants with injuries of AIS 2+ at any delta-v.

For the purposes of the analysis to follow, the eighteen cases previously discussed were included, to make a total of 49 cases. These 49 cases involve 72 front seat occupants and 26 unoccupied seats.

Table 2 is a summary of the seat damage data of occupied front seats. The summary contains the number of damage events from each damage category. Some seats had more than one kind of damage.

TABLE 2 - SEAT DAMAGE

No Seat Damage	25
Adjustors Damaged	3
Folding Locks Damaged	13
Tracks/Anchors Damaged	2
Damage by Occupant	34
Damage by Intrusion	3
Recline Mechanism Damaged	1
Total	81

Nine seats received two types of damage. Therefore, there are 81 instances of damage, or non-damage, for the 72 occupied seats. Bucket seats with folding back lock were the most predominate seats in the cases reviewed (55). There were 26 unoccupied front seats in the cases reviewed. Four of these seats received damage; three from intrusion, and one coded "seat back folding locks failed".

The predominate mode of damage was seat deformation by the occupant, followed by seat back folding lock damage. Three cases of seat adjuster damage were noted. All were associated with seats deformed by occupant impact. Two of the occupants involved had AIS 1 injuries. In the other case, the principal injury was an AIS 3 chest injury of unknown cause. Improved seat restraint might have mitigated this injury, but the evidence is uncertain. No AIS 2+ injuries from seat adjustor failure could be positively identified from the rear impact cases examined.

Fourteen cases of folding lock

damage were reported. Among these cases, five occupants may have had injuries at the AIS 2 level which were exacerbated by the reported lock damage. In four of five cases, the source of the AIS 2 injury is unknown.

Two cases of seat anchorage damage were reported. The maximum injury for the occupants in both of these cases was AIS 1. Seat anchorage damage were not associated with any serious rear impact injuries.

The analysis separates the data into two groups: (1) - no seat yielding, and (2) - cases with permanent seat back yielding.

In cases with no permanent seat back yielding, all but one of the 17 occupants was restrained. No injuries were attributed to ramping or seat related deformation. Three AIS 2 neck fractures were observed. All occurred in rear crashes with delta-v greater than 33 mph. All were in seats without permanent deformation. Frontal injuries were present in 8 of the cases. In most cases, the frontal contact was the result of a subsequent frontal collision.

In cases with permanent seat deformation, 16 of 18 occupants were restrained. Seat deformation may have contributed to 3 of the injuries, and ramping may have contributed to one. Seven of 16 occupants had injuries from frontal components of the vehicle. Four of these cases were single event rear impacts.

For both sets of data, 15 of 35 occupants had injuries from frontal components. These results suggest an opportunity for improving the seat/restraint design to reduce injuries from frontal vehicle components during rear impact collisions. These frontal injuries were more frequent than those associated with seat deformation. This observation is consistent with the earlier finding, based on NASS data analysis, which showed that 16% of Harm in rear crashes is from frontal components of the vehicle. This compares with 10% for the seat back, top and rear components of the vehicle.

FILM ANALYSIS

In order to further investigate the performance of seats in rear impacts, preliminary film analyses of rear impact tests conducted under FMVSS 301 were undertaken. These tests require a 4,000 lb. moving barrier to impact the rear of the car at 30 mph.

Five test films were reviewed. In four of the tests, the seat was permanently deformed. In one test no permanent seat deformation was noted. The delta-v for this test was approximately 17 mph. This test was subjected to film analysis to determine the rebound velocity of the occupant. It was found that the seat underwent extensive elastic deformation. The dummy head nearly disappeared below the rear window sill during the rear impact. The elastic energy of the seat was then transferred to the dummy as rebound velocity. Film analysis indicates that the dummy head reached a rebound velocity of about 11 mph.

Additional film analysis and case studies are required to draw definite conclusions on the rebound performance of existing seats and restraint systems.

CONCLUSIONS

This preliminary analysis suggests that improvements in seat performance is more complex than simply increasing the strength of the seat back. Legitimate concerns exist over the potential increase in neck injuries and other injuries which might accompany strengthened seats. These concerns may be addressed by including dummy test requirements in the specification of safety performance of both seating and restraint systems.

The data analysis from this study did not permit the quantification of neck injury risks for deformed vs non-deformed seats. In the accident cases analyzed, three non-contact neck fractures occurred in seats which did not deform. No non-contact neck fractures were observed in seats which deformed.

Harm analysis provides insights of injury frequency and severity in rear impacts. This analysis shows that non-contact neck injuries constitute more than 20% of the harm to restrained occupants. The head restraint is the largest source of contact harm (17%). Improvements in head restraints to reduce these two frequent types of injuries offers a challenging opportunity for harm reduction.

NASS analysis shows that 16% of the harm in rear impact NASS cases was from frontal contacts. In many of the cases, the rear impact is followed by a frontal impact, either in a line of stopped traffic, or by being accelerated into a fixed object. In these cases, and in cases where no subsequent impact occurs, injuries from impact with frontal components of the vehicle are frequently observed. For the data set of AIS 2+ injuries in selected rear impacts, injuries from frontal components were believed to be present in 15 of 35 occupants. It is not possible to determine how many of these injuries were related to the elastic response of the seat, or to other phenomena. However, some seat induced rebound velocity can be observed by dummy motion analysis of FMVSS 301 rear impact test films. The degree to which existing belt systems mitigate the rebound energy could not be determined from the test films examined. It is evident that the rebound phenomena needs to be researched in conjunction with future seat improvements.

The literature of safety research in seating design shows that very creative ideas have been considered. It now appears that several safety features are emerging in production vehicles. First, the integration of the lower inside anchorage point with the seat is becoming a standard practice. Second, the lower outside anchorage point is also being located on the seat in newer cars. Third, the upper anchorage point is being integrated into the seat and/or head rest on some specialty cars by Mercedes and BMW. Renault reports research to follow this lead. Fourth, rear seat air bags are currently being planned. These bags would most likely deploy from the back of the front seat. Finally, suppliers of automotive seats are actively developing efficient integrated seats as new products.

The last three items will require stronger seat backs. Data from Warner (91) shows that seat strength in US cars has not changed significantly in the past 20 years. The concerns of neck loading, ramping, out-of position occupants, limb flailing and rebound will all have to be addressed for stronger seats. The neck problems will demand adequate head restraints.

These problems need to be researched and resolved. The integrated seat offers promising solutions to many of these challenges. Many innovation of the integrated seat and belt system such as force limiting, automatic size adjustment, reduced slack, and improved fit (all reported by BMW) are already available. Some of these innovations would improve belt performance in all crash modes, including rollover.

A number of exciting technical possibilities are currently under development for improving occupant safety in all crash modes through seat/restraint design.

ACKNOWLEDGEMENT

This research was partially sponsored by the National Highway Traffic Administration, and the Transportation Systems Center, U. S. Department of Transportation. The views presented are those of the authors and not those of the above mentioned sponsors. The authors would like to thank Mr. J. Kossar, Mr. D. Cohen, Dr. W. Liu, and Ms. S. Partyka for their technical assistance during the project. Documentation of the project results may be found in references Digges 1992, and Malliaris 1990.

REFERENCES

Cornell, 1967; Cornell Aeronautical Laboratories, Inc., "Integrated Seat and Occupant Restraint Performance", Technical Report, U.S. Dept. of Transportation, Vol. I, 1967.

Digges 1992; K. Digges, J. Morris, "Upgrade Seating- Patent, Literature Search, and Accident Analysis", NHTSA Docket 89-20, November, 1992.

Foret-Bruno, 1990; J. Foret-Bruno, C. Tarrier, J. LeCoz, C. Got, F. Guillon, "Risk of Cervical Lesions in Real-World and Simulated Collisions," 34th AAAM Conference, p.373, October, 1990.

Foret-Bruno, 1991; J. Forret-Bruno, F. Dauvilliers, C. Tarrier, "Influence of the Seat and Head Rest Stiffness on the Risk of Cervical Injuries in Rear Impacts", 13th ESV Conference, Nov., 1991.

Haberl, 1989; J. Haberl, F. Ritzel, S. Eichinger, "The Effect of Fully Seat-Integrated Front Seat Belt Systems on Vehicles", 12th ESV Conference, May, 1989.

HSRI, 1967; Highway Safety Research Institute, "Integrated Seat and Occupant Restraint Performance", 1967.

Hilyard, 1973; J. Hilyard, J. Melvin, J. McElhaney, "Deployable Head Restraints" Highway Safety research Institute, U.of Mich., DOT-HS 800 802, Jan., 1973.

Hu, 1978; A. Hu, S. Bean, R. Zimmerman, "Response of Belted Dummy and Cadaver to Rear Impact", DOT-HS 803 028, Jan., 1978.

Hu, 1980; A. Hu, S. Bean, "Response of Belted Dummy and Cadaver to Rear Impact", DOT-HS 805 792, Dec.,1980.

Iaccoa, 1992; L. Iacocca, Interview by Larry King, CNN, Dec. 11, 1992.

James, 1991; M. James, C. Strother, C. Warner, R, Decker, T. Pearl, "Occupant Protection in Rear-End Collisions: I. Safety Priorities and Seat Belt Effectiveness", 35th Stapp Conference, Nov., 1991.

Malliaris, 1982; A. Malliaris, R. Hitchcock, and J. Hedlund, "A Search for Priorities in Crash Protection", SAE 820242, 1982.

Malliaris, 1985; A. Malliaris, R. Hitchcock, M. Hansen, "Harm Causation and Ranking in Car Crashes," SAE 850090, 1985.

Malliaris, 1990; A. Malliaris, "Current Issues of Occupant Protection in Car Rear Impacts", Data Link, Inc., NHTSA Docket 89-20-NO1-021, Feb., 1990.

Melvin, 1971; J. Melvin, J. McElhaney, "Deployable Head Restraints", HSRI-71-103 (DOT-HS 800 515), June, 1971.

Melvin, 1972; J. Melvin, J. McElhaney, "Occupant Protection in Rear-end Collisions", SAE 720033, 1972.

Severy, 1967; D. Severy, H. Brink, J. Biard, "Collision Performance of the LM Safety Car", SAE 670912, 1967.

Severy, 1968; D. Severy, H. Brink, J. Biard, "Backrest and Head Restraint Design for Rear-end Collision Protection", SAE 680079, 1968.

Strother, 1987; C. Strother, M. James, "Evaluation of Seat Back strength and Seat Belt Effectiveness in Rear Impacts", 31st Stapp, SAE 872214, Oct., 1987.

Thomas, 1982; C. Thomas, G. Faverjon, F. Hartemann, C. Tarrier, A. Patel, C. Got, "Protection Against Rear-End Accidents", IRCOBI Conference, 1982.

Warner, 1991; C. Warner, C. Strother, M. James, R. Decker, "Occupant Protection in Rear-end Collisions II, The Role of Seat Back Deformation in Injury Reduction", 35th Stapp, SAE 912914.

930346

Field Accident Evaluations and Experimental Study of Seat Back Performance Relative to Rear-Impact Occupant Protection

Kenneth J. Saczalski
Environmental Research & Safety Technologists, Inc.

Stephen R. Syson and Richard A. Hille
Syson-Hille & Associates

Mark C. Pozzi
Biodyne Research, Inc.

ABSTRACT

This study examines in some detail 23 actual rear-impact cases dealing with front seat collapse and compares the findings with similar results from 23 Canadian cases. In addition, seat tests and car-to-car crash tests are utilized to examine the potential hazards and/or benefits of collapsing versus non-collapsing seat systems. Evaluation of the above 46 cases indicates that an extremely high rate of rearward ejection occurs to restrained front-seat occupants subjected to rear impact. The majority of those ejected experienced serious to fatal injuries, either from contact in the rear or outside of the vehicle, when seated in collapsing seats. These results are contrary to some earlier published data and, as such, recommendations are made which could help improve data collection methods so as to better evaluate the issues associated with rear-impact seat strength.

INTRODUCTION

Interest in determining the adequacy of automotive seat back strength for providing occupant protection during rear-end collisions has been renewed in recent years [Saczalski, 1989]. A sparsity of objective field accident data complicates assessment of key issues which, for the most part, seem to partition into two groups: one advocating stronger and less yielding front seat systems, and the other suggesting that the current designs of collapsing front seats are able to absorb sufficient energy so as to minimize harmful loads applied to the occupants during rear impacts.

In this paper, several actual accident cases are reviewed in some detail with major focus on occupant kinematics, kinetics, resulting injuries, locations of impact, and, in some cases, the amounts of energy absorbed during the collision by collapsing seat structures. Experiments on exemplar production seat systems were conducted to determine amounts of possible resistance or energy attenuation provided by the seat structure. In addition, rear-impact, car-to-car experiments, utilizing instrumented anthropometric test devices, were reviewed and compared with the results of a car-to-car crash test which contained both a non-yielding seat and a standard collapsing seat structure, each occupied by an instrumented fiftieth percentile male surrogate. One of the primary objectives of the latter test was to gain insight into the potential hazards of both seat design approaches (i.e., collapsing seats versus stronger, less yielding seat structures).

With respect to the actual accident cases reviewed and discussed in this paper, 23 cases deal with real-world accidents evaluated by the authors, and another 23 cases represent the results of a recent Canadian study on real-world accidents involving seat back failures, examined on behalf of Transport Canada [Marriner, 1990]. Evaluation of the field accident data from the above-cited 46 real-world cases, which focused primarily on the effects of automotive seat structure in rear impacts, indicates that an extremely high rate of partial or total ejection (approximately 50%) occurs to restrained front-seat occupants subjected to rear impact. Furthermore, the majority of the ejected occupants experienced serious to fatal injuries, either from contact in the rear or outside of the vehicle after rearward ejection, and in virtually all cases the occupants were seated in standard collapsing seat systems. These results are contrary to some earlier published data [Data Link, 1990] and analysis [James, 1991; Warner, 1991] which relied upon the National Accident Sampling System (NASS) data base and concluded that, among other things, restraint usage, combined with collapsing seats in rear impacts, has a substantial injury-reducing effect. These conclusions are similar to those presented by Strother in an earlier paper on this subject [Strother, 1987]. For whatever the reason, even though it was publicly available at about the same time as the Data Link study, the Canadian data has not been reviewed or cited in some of the earlier mentioned technical articles dealing with issues of automotive seat strength in rear impacts. As such, one of the objectives of the current paper is to present the Canadian data, along with the previously unpublished data relating to the 23 real-world cases examined by the authors, so as to provide a more complete data base for other automotive engineers and researchers interested in better understanding the hazards and/or benefits associated with each of the two basic schools of thought relating to seat design for rear-impact occupant protection.

There are several potential hazards suggested by the opponents of each of the two basic approaches suggested for rear-impact occupant protection. With respect to collapsing seat structures (which seems to be the "norm" for most current automotive systems), opponents suggest at least six classes of potentially hazardous situations which can occur when a seat back collapses in an uncontrolled manner and these include: (1) loss of vehicle control by a driver when the seat back collapses rearward in an uncontrolled manner during a rear impact; (2) reduced effectiveness of the restraint system when the collapsed seat back allows a front-seat occupant to rotate and slide rearward from under the lap belt during a rear impact, thus enabling potential injurious contact with rear seat objects and passengers; (3) ejection of occupants who have slid out from beneath their lap and shoulder harness system when the seat back easily collapses rearward in an uncontrolled manner from rear impact and allows the passenger to be tossed around unconstrained during the subsequent motions of the vehicle which could experience other impacts, and/or roll-over after the initial impact, since the driver has usually lost control of the vehicle due to the rearward collapse of his or her body; (4) injury to rear-seat passengers who are likely to be struck by the violent rearward motion of the front seat occupant collapsing into the rear seat passenger area where the rear seats do not collapse rearward in a rear impact; (5) reduction or loss of egress capabilities of rear-seat passengers whose bodies are likely to be trapped under the plastically deformed and collapsed front seat backs of occupied seats during rear impact (This situation is especially dangerous in the event of post-crash fires); and (6) injury to fully restrained front-seat passengers during a frontal impact when the seat back easily collapses from the rear loading of a lap belted or unrestrained rear-seat passenger (or heavy object) which can greatly enhance the loading on the thorax, spine, and abdomen of the restrained but compressed front-seat passenger.

With regard to non-collapsing seat structures, opponents suggest at least four classes of potentially hazardous situations and these include: (1) severe hyperextension neck injury to out-of-position occupants (i.e., someone leaning around the edge of the stronger seat back with their head and neck away from the headrest); (2) ramping of unbelted occupants up the seat back and into the roof structure with the potential for injurious compressive neck loads; (3) rebound and possible ejection of unbelted occupants who elastically load-up the stronger seat back structure and then may rebound forward into the windshield and dashboard area; and (4) injury to unrestrained or "lap belted" only rear-seat passengers who, in the event of a frontal impact, can move forward violently into the rear of the stronger non-collapsing seat structure in front.

To be sure, all of the above potentially hazardous situations can possibly occur and, as such, they must be considered in the final analysis as to which seat design approach is best for providing optimum protection to occupants subjected to rear impact, as well as frontal. Thus, another objective of this paper is to address the above-identified potential hazards through the use of the more recently available real-world data, and car-to-car tests with instrumented surrogates, so as to provide additional insight into the weakness or strength of each approach. Historically, much data has already been presented on the merits of stronger seat systems by researchers such as Severy [Severy, 1958; Severy, 1967] and more recently on the merits of collapsing-seat structures by researchers at Collision Safety Engineering [Strother, 1987]. In the following sections, a brief overview of some historically relevant aspects of the seat strength issue is presented along with some data on human tolerance to longitudinal loadings.

SEATS, RESTRAINTS AND WHOLE BODY HUMAN TOLERANCE

Investigations reported by Colonial John P. Stapp [Stapp, 1955] indicate that for spineward (chest-to-back direction) loadings, such as those which may occur in frontal impact, the whole body human tolerance limit is approximately 45 G's for a time duration of about 100 milliseconds, with the body fully restrained in a 7.6 centimeter-wide shoulder harness, lapbelt with thigh straps, and a chest belt. These tolerance limits decrease with longer time durations and may be increased with time durations much shorter than 100 milliseconds.

With respect to rear impact or sternumward (back to sternum direction) loading, it can be expected that the full length overall body support of a non-yielding seat back and headrest should offer an even greater degree of restraint than that of the localized harness system used in the frontal impact by Stapp. In fact, the maximum recorded human tolerance limit for sternumward loading, without permanent injury, in a non-yielding full-length seat, was measured as 83 G's with the time duration of 0.04 seconds and a load on-set rate of 3800 G's per second [Beeding and Mosely, 1960]. Although no permanent injury was suffered by the volunteer in the above test, the subject did experience shock and required on-the-scene medical treatment, thus suggesting that the human tolerance limit for sternumward loading is somewhere between the 83 G and 45 G level. The research by Carr [Carr, 1975] presents an excellent data summary of rear-impact human tolerance and lists data source references.

For the most part, the above G-levels are well above the vehicle deceleration levels experienced in most frontal and rear barrier automotive collisions and suggests that if the head, neck and upper torso are fully supported by a strong seat back and headrest, the occupant should be able to survive moderate rear impacts without experiencing serious injury. The early research by Severy and others, dating back into the 1950 time frame, seemed to confirm the above through investigation of the effects of production seats versus stronger seats and headrests as a potential means for reducing harmful loadings and motions to rear-impacted vehicle occupants. A review of some of this early work was presented in the 1987 paper of Strother and James [Strother, 1987]. A more complete review and update of the research dealing with the issues of seat strength is provided in the Shaw critique of the Strother article [Shaw, 1990].

Briefly summarized, in 1967 as a result of the research conducted with car-to-car crash tests, and the testing performed on the Liberty Mutual Safety car, Severy recommended that seats be designed to withstand 30 G loadings when loaded by an occupant during rear impact [Severy, 1967]. In 1968, as a result of his research, Severy recommended torque resisting levels of 1808 N-m (16,000 inch-pounds) to 3728 N-m (33,000 inch-pounds) about the "H" point with a seat back height of 72 cm. (28 inches) [Severy, 1968]. In 1969, Severy recommended that the seat back should also sustain a torque of 11,298 N-m (100,000 inch-pounds) in the rearward direction without exceeding a 10° deflection in that direction [Severy, 1969]. In contrast to the above, the Federal Motor Vehicle Safety Standard (FMVSS) 207 for seat back strength and torque requirements, adopted in 1971, is only 373 N-m (3,300 inch-pounds).

Numerous other researchers made recommendations for stronger seats, less yielding seats, and dynamic testing during the years following the adoption of the FMVSS 207 requirement but as yet the requirement has not changed from the 373 N-m level.

In 1974, Nash filed a petition with the National Highway Traffic Safety Administration (NHTSA) to include passive occupant crash protection in impacts from the rear of the vehicle [Nash, 1974]. This petition recommended amendments to FMVSS 208 (dynamic frontal impact) and 207 that would have mandated dynamic rear-impact tests compatible with the test requirements of FMVSS 301-75 (vehicle fuel systems dynamic test). These recommended tests were to be conducted with anthropomorphic test devices at each designated seating position. Recommended performance criteria included: ". . . Rearward flexion of the neck shall not rotate rearward through an angle of more than 45 degrees . . . . No part of the seats or head restraints shall become disengaged from their mountings or attachments nor shall any part of the seat or head restraint be distorted so that the anthropomorphic test devices would leave their designated seating positions or have contact with other than laterally adjacent test devices."

During this same time period of the early to mid-1970's, several manufacturers applied for and received patents dealing with methods for making seats stronger and less yielding. For instance, the Ford Motor Company received a patent in 1973 for a "Safety Seat with a Safety Belt, in Particular for Motor Vehicles" [Giese, 1973]. Also, in 1974, Muncharu Urai of Japan filed for and received a patent for a "Seat Mounting Device for Vehicle" which attached a portion of the seat back to the roof of the vehicle for the purpose of "increasing the seat's holding force, to alleviate the shock imparted to the human body in the event of a collision accident" [Urai, 1975]. Figure 1, shown below, illustrates the patent drawing for the Ford Safety Seat with a Safety Belt. Ford also applied for and received a patent in 1973 for an "Energy Absorbing Seat Back" [Glance, 1973]. In this patent, the upper portion of the seat back was designed to have a <u>predetermined</u> torsional resistance, with a controlled rate, to thereby control the deceleration of a rear occupant impacting the front seat during a frontal collision.

In the 1970 time frame, there was also considerable on-going American, European and Japanese automobile industry research which demonstrated through the U.S. Government Research Safety Vehicle (RSV) program and the Experimental Safety Vehicle (ESV) program that stronger, less yielding seats with energy absorbing foam pads could provide much improved occupant protection in all modes of impact, including the rear-impact direction (currently, several automotive racing organizations require non-collapsing seat systems).

Also, in 1974, the NHTSA published a proposed rulemaking to strengthen the seat strength standard (FMVSS 207) and the headrest standard (FMVSS 202) by combining the two requirements into a single rule, and then imposing dynamic crash test requirements on the overall "rear impact restraint system" that would incorporate these components [Federal Register, 1974]. These requirements would have been similar to those in FMVSS 301, which sets test criteria to determine fuel system integrity in rear-end crashes. For various reasons, however, in April 1979, NHTSA terminated the 1974 rulemaking and replaced it with a regulatory plan for overall "significant upgrading" of occupant protection in all directions, including rear, side, front and rollover, and including "new comprehensive standards ... developed in terms of injury levels that occur" in dynamic crashes in all four modes [NHTSA Five Year Plan, 1979]. Ultimately, the 1979 regulatory plan, which would have upgraded seat "restraint" performance in "crash exposures representative of the real world," was abandoned by the Reagan Administration in 1981.

In spite of the research of the 1960's and 1970's which suggested improved safety to occupants in stronger seat systems, there has been, with a few exceptions, rather sparse activity on the part of NHTSA and the industry as to research and development of safer seat systems. Two exceptions are the BMW 850 rigidized seat with integrated restraints [Habrel, 1989] and the Mercedes Benz 500 SL sports car with a rigidized seat and integrated restraint [Mercedes Benz, 1989]. The theoretical and experimental results presented by Habrel regarding the BMW 850 rigidized seat design demonstrates that the stronger seat systems allow for attachment of the belt shoulder harness restraints to the seat and seat back system with concomitant lower HIC, chest G loads, and shoulder belt loads than those likely to be incurred with current belt restraint configurations mounted to the vehicle structure.

The above paper by Habrel also points out that the stiffer seat frame with the seat-integrated belt system (SBS) offers "demonstrably lower loads on the occupants in the event of side collisions and increases the occupant protection effect in the event of a rear-end collision for those occupying the front seats, as well as those in the rear of the car." Habrel also notes that, "Since the SBS seat back deforms only slightly, driver and front passenger are held reliably in their seats, and there is no risk for the rear-seat passengers being trapped behind the front-seat backs; and the good belt wraparound effect of the SBS system -- improves protection in roll-over accidents." He also notes that, "The occupant is held in his seat more effectively, with the risk of contact injuries, particularly as a result of head contact with parts of the roof frame, being further reduced." The results presented in this paper suggest that this BMW seat system can withstand torque values as high as about 5,650 N-m (50,000 inch-pounds). Figure 2 illustrates the BMW 850 rigidized seat [IIHS Status Report, 1989] along side of the Cox Safety Seat tested by Severy in the late 1960 time period [Severy, 1967].

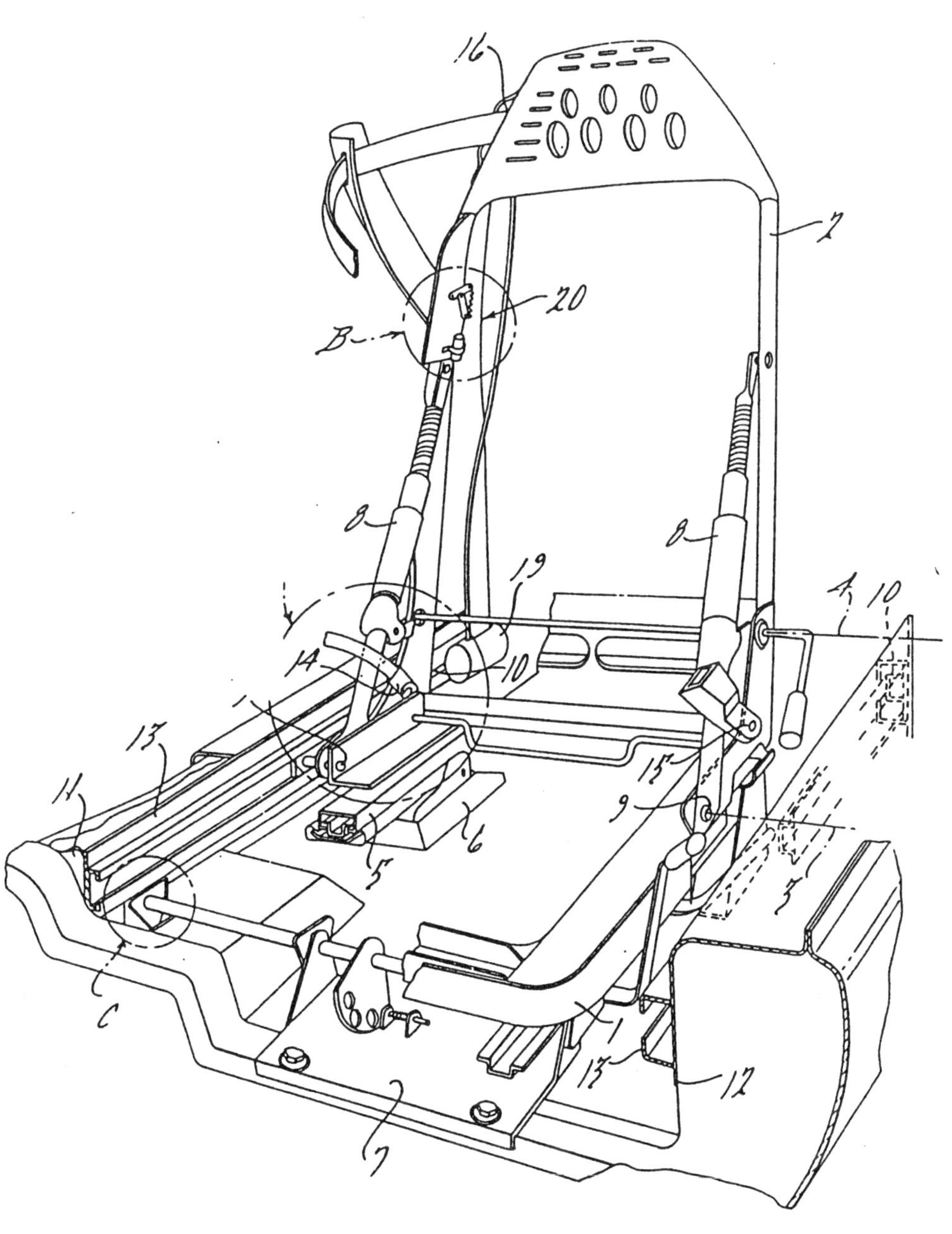

Figure 1 - Ford Safety Seat Patent Drawing [Giese, 1973]

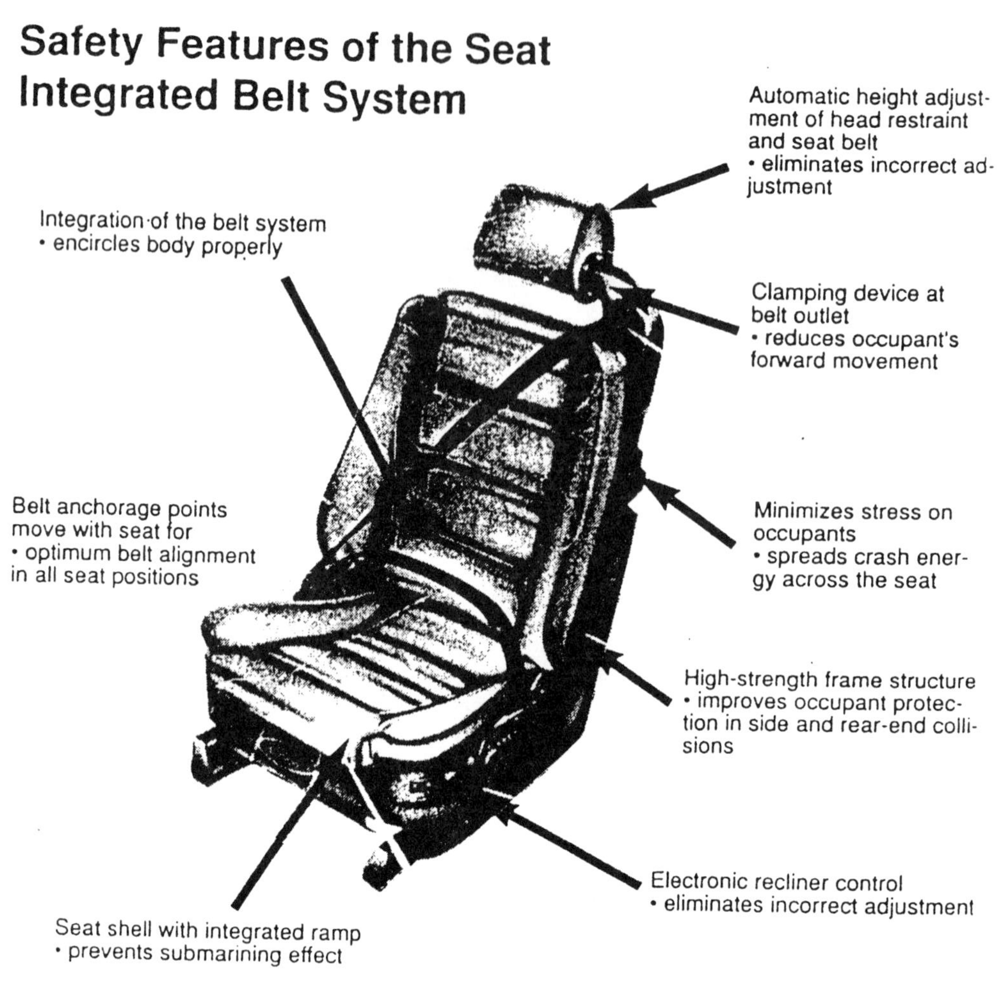

Figure 2 - BMW Seat with Seat Integrated Belt System [IIHS, 1989], [Strothers, 1987].

All in all, however, as reported by Severy and others, there has been little change in the load carrying capability of automotive seat systems between the 1940's and the late 1980's [Severy, 1976; Strother, 1987]. The above factors (i.e., early research indicating improved protection to occupants in stronger seats but relatively little change in seat strength since the 1940's, etc.) coupled with the authors' experiences gathered during investigation of injuries sustained by restrained occupants in real-world rear impacts, and the conflicting data resulting from studies relying on the NASS data base to establish effectiveness of restraints with collapsing seats in rear impacts, prompted the collaboration which led to the field accident analysis set forth in the following sections of this paper.

BELT EFFECTIVENESS BASED ON FIELD DATA

As noted previously, a Canadian field accident study of 23 accidents was conducted to look at accidents involving seat back collapse. Briefly summarized, the case studies represent a wide range of incidents which include failures of seat backs during normal driving or everyday use to failures during crash impact situations. A summary of each case describing the circumstances is contained in Appendix A. Table 1 summarizes key data from the Canadian study which identifies occupant positions, injury severity, occurrence of seat back failure, occurrence of ejection, restraint use, and contribution of seat back failure to the resultant injuries. It is interesting to note that the seat back failures take place in a wide variety of ways which include failures such as floorpan and seat structure distortion, failure of recliner mechanisms, seat track separation and seat frame distortion.

Forty-one occupants were involved in the 23 Canadian case studies and front seat back collapse occurred in at least one of the front occupant positions in each case. In 3 of the cases seat back collapse occurred to drivers during normal occupant loading (i.e., driver reaches for wallet while stopped in a line at a drive-through restaurant, etc.). Fortunately, in all 3 cases there was only one minor injury and all 3 drivers were able to regain control of their vehicles.

Of the remaining 20 cases, there were 28 front seat occupants who experienced some form of seat back collapse or failure, and of these, 20 occupants were identified as being positively restrained. However, nearly 50 percent of those 20 restrained occupants in collapsing seats (9 out of 20) were either partially or totally ejected through the rear area of the vehicle. In addition, over half of those ejected received fatal or major injuries. With respect to the restrained front seat occupants not ejected (11 out of 20) from collapsing front seats, only 18 percent (2 out of 11) received fatal injuries.

Seven of the 28 front seat occupants in collapsing seat systems were unrestrained, and in one other instance it was unknown if the restraint was used or not used. With respect to this group, 86 percent (6 out of 7) were partially or totally ejected with fatal injuries being received by 83% (5 of 6) of those ejected.

With respect to the general effect of seat back failure on passenger ejection, the Canadian summary notes that eleven of the 23 cases examined resulted in one or more of the passengers being ejected from the vehicle (4 occupants were partially ejected), and in 14 cases of the 23 "the occupant was forced out of the seat in a rearward direction because the seat was unable to withstand the force exerted on it by the occupant." The Canadian study concludes that, "In all cases the ejection would probably have been avoided if the seat back remained upright when loaded by the occupant during the ejection."

As a result of these rearward direction loadings of front seat occupants, there were also 3 cases (cases 9, 19 and 23) where the rear seat passenger seated behind the occupant in the collapsing front seat received injuries due to the failure of the seat back in front of them. In 2 of these 3 cases, the rear seat occupants were fatally injured. One of the fatalities occurred to a seven week old infant (case 9) where the headrest detached from the collapsing seat in front and struck the infant. This case appears to illustrate one of the hazards anticipated by Nash in his 1974 NHTSA petition which requested that "no part of the seats or headrests shall become disengaged from their mountings or attachments..." [Nash, 1974]. All total, 60 percent of the rear seat passengers (3 out of 5) were injured when seated behind a front seat passenger with a collapsing seat. All of these Canadian results suggest a much higher percentage of injuries due to rear contact than the 2.8 percent reported in the Data Link study for only unrestrained cases [Data Link, 1989].

Similar results are found from the 23 author researched field accidents summarized in Table 2 and Appendix B. Fifty-three occupants were involved in these 23 case studies and front seat back collapse occurred in at least one of the front occupant positions in each case. There were 36 occupants in collapsing front seats and of these, 19 were identified as being positively restrained. In this case, 58 percent of the 19 restrained occupants in collapsing seats (11 out of 19) were either partially or totally ejected from their restrained front seat positions, through or into the rear of the vehicle. Of those ejected 73 percent (8 out of 11) received moderate to fatal injuries. With respect to the restrained front seat occupants not ejected in this phase of the study (8 out of 19), only 12 percent (1 out of 8) received major injury.

As in the Canadian study, there were seven unrestrained occupants seated in collapsing front seat systems and 10 instances where it was unknown if the restraint was used or not used. With respect to this group of 7 unrestrained front seat occupants, 57 percent (4 out of 7) were partially ejected and, of these, 75 percent (3 out of 4) received major injury. In virtually all of the above cases of front seat occupants occupying collapsing seats it was determined that the collapse of the seat contributed to the injuries received. In addition, it was also found that in at least 3 of the author investigated cases (cases 7, 8, and 14) there was a loss

Table 1 - Summary of Canadian Field Accident Study

Case Vehicle	Occupant Position	Injury Severity	Seat Back Failure	Ejected	Restraint Used	Contribution to Injury
1	Driver	MINOR	YES	YES	YES	YES
	RF	MINOR	YES	NO	YES	YES
	LR	MINOR	YES	YES	NO	YES
2	Driver	MINOR	NO	NO	YES	N/A
	RF	MINOR	YES	PARTLY	YES	Unknown
3	Driver	NONE	NO	NO	YES	N/A
	RF	FATAL	YES	YES	NO	YES
4	Driver	MAJOR	YES	YES	YES	YES
	RF	MINOR	NO	NO	YES	N/A
5	Driver	MINOR	YES	NO	YES	Unknown
6	Driver	MINOR	YES	NO	YES	YES
	RR	Unknown	NO	YES	YES	N/A
7	Driver	MINOR	YES	NO	YES	YES
	RF	MINOR	YES	NO	YES	YES
8	Driver	FATAL	YES	YES	YES	YES
9	Driver	MINOR	YES	YES	NO	YES
	RF	MINOR	YES	NO	NO	Unknown
	LR	FATAL	NO	NO	YES	YES*
10	Driver	FATAL	YES	YES	NO	YES
11	Driver	MINOR	YES	YES	Unknown	YES
12	Driver	MAJOR	YES	YES	YES	YES
	RF	FATAL	YES	YES	NO	Unknown
	RF	MINOR	YES	YES	NO	Unknown
13	Driver	FATAL	YES	YES	YES	YES
	RF	MINOR	NO	NO	YES	N/A
14**	Driver	NONE	YES	NO	Unknown	N/A
	LR	NONE	NO	NO	Unknown	N/A
15	Driver	MINOR	YES	NO	YES	YES
16	Driver	MINOR	YES	PARTLY	YES	YES
17	Driver	MINOR	YES	NO	YES	YES
18**	Driver	Unknown	YES	NO	Unknown	Unknown
19	Driver	MINOR	YES	NO	YES	YES
	RF	MINOR	YES	NO	YES	YES
	RR	MINOR	NO	NO	YES	YES*
20**	Driver	MINOR	YES	NO	Unknown	YES
21	Driver	FATAL	YES	YES	NO	YES
22	Driver	FATAL	PARTLY	YES	YES	YES
	RF	FATAL	YES	NO	YES	Unknown
23	Driver	FATAL	PARTLY	YES	YES	YES
	RF	FATAL	YES	NO	YES	Unknown
	RR	FATAL	NO	NO	NO	YES*

* The passenger was injured due to the failure of a seatback in front of them.

** Vehicle stationary or moving when failure of recliner mechanism occurred without acceleration loads.

Table 2 - Summary of Author Researched Field Accident Study

Case Vehicle	Occupant Position	Injury Severity	Seat Back Failure	Ejected	Restraint Used	Contribution to Injury	Loss of Control
1	DR (1) RF (3)	MINOR MAJOR	NO YES	NO PARTIAL	UNKNOWN UNKNOWN	NO YES	NO
2	DR (1) RF (3) REAR (?) REAR (?)	MAJOR MAJOR MAJOR MAJOR	YES YES UNKNOWN UNKNOWN	NO NO NO NO	NO NO NO NO	POSSIBLE POSSIBLE POSSIBLE POSSIBLE	NO
3	DR (1) RF (3) LR (4) RR (6)	MINOR MINOR MAJOR MINOR	YES YES NO NO	PARTIAL PARTIAL NO NO	YES YES YES YES	YES YES YES YES	NO
4	DR (1)	MODERATE	YES	YES	YES	POSSIBLE	POSSIBLE
5	DR (1) RF (3) RR (6)	MINOR MINOR MAJOR	YES YES NO	NO NO NO	YES YES YES	NO NO NO	NO
6	DR (1) RF (3) RR (6)	MAJOR MINOR MINOR	PARTIAL PARTIAL NO	NO NO NO	YES YES YES	POSSIBLE POSSIBLE NO	NO
7	DR (1)	MAJOR	YES	PARTIAL	YES	YES	YES
8	DR (1) RF (3)	MAJOR MINOR	YES NO	PARTIAL NO	UNKNOWN UNKNOWN	YES NO	YES
9	DR (1) LR (4)	MINOR FATAL	YES NO	NO NO	UNKNOWN UNKNOWN	UNKNOWN YES	NO
10	DR (1)	MAJOR	YES	NO	UNKNOWN	YES	NO
11	DR (1) RF (3)	FATAL MINOR	YES YES	YES PARTIAL	YES YES	YES POSSIBLY	NO
12	DR (1) RF (3)	MAJOR MINOR	YES YES	NO NO	YES YES	YES UNKNOWN	NO
13	DR (1) RF (3) LR (4) REAR (7)	MAJOR FATAL MINOR NONE	YES YES YES NO	PARTIAL PARTIAL NO NO	YES YES YES UNKNOWN	YES YES YES N/A	NO
14	DR (1) RF (3) LR (4) CR (5) RR (6)	FATAL FATAL FATAL FATAL FATAL	YES YES NO NO NO	NO NO NO NO NO	Unknown Unknown Unknown Unknown Unknown	YES YES YES YES YES	YES
15	RF (3) DR (1)	MAJOR MINOR	YES NO	PARTIAL NO	YES YES	YES YES	NO
16	DR (1)	MAJOR	YES	NO	NO	YES	NO
17	DR (1) RF (3)	MAJOR MINOR	YES YES	PARTIAL PARTIAL	NO NO	YES YES	NO
18	DR (1)	MAJOR	YES	PARTIAL	NO	YES	NO
19	DR (1) RF (3) LR (4) RR (6)	MAJOR FATAL MODERATE MINOR	YES YES NO NO	NO NO NO NO	Unknown Unknown Unknown Unknown	YES YES YES YES	NO
20	DR (1) RF (3)	MINOR MAJOR	PARTIAL YES	NO YES	YES YES	Unknown YES	NO
21	DR (1) RF (3)	MINOR MAJOR	PARTIAL YES	NO YES	YES YES	Unknown YES	NO
22	DR (1)	MAJOR	YES	PARTIAL	NO	YES	NO
23	DR (1) RF (3)	MINOR MAJOR	YES YES	NO YES	Unknown Unknown	YES YES	NO

of vehicle control with subsequent injury resulting from the collapse of the driver's seat. Tables 3 and 4 summarize the combined results comparing ejection data for restrained and unrestrained occupants seated in collapsing front seat systems.

Overall, the above results relating to ejection show that there is a lower incidence of ejection if the occupants of a collapsing seat system are restrained. Thus, as was indicated by the Data Link study [Data Link, 1989], there is a certain amount of effectiveness or benefit from wearing a restraint during seat back collapse. However, as also shown by the 5 cases of front seat occupants in non-collapsing seats of this study (cases 3 and 13 of Table 1, and cases 1, 8 and 19 of Table 2) there was a 100 percent retention rate of the occupants - independent of restraint use. This data indicates that, although restraints may have some effect in the case of a collapsing seat system, the non-collapsing seat system clearly provides a much more effective restraint than that of the belts with collapsing seats. The above conclusion seems to be corroborated in the Data Link study when one observes the results for harm versus occupant belt use rates comparing car rear end impact to all car impacts, as shown in Table A-27 of the Data Link study [Data Link, 1989]. This data is summarized below in Table 5.

Table 3 - Ejection Data for Belted Front Seat Occupants *

	Canadian Data	Author Data	Total
Ejected	9	11	20
Not Ejected	11	8	19
Total	20	19	39

*Occupants in collapsing front seats

Table 4 - Ejection Data for Unbelted Front Seat Occupants *

	Canadian Data	Author Data	Total
Ejected	6	4	10
Not Ejected	1	3	4
Total	7	7	14

*Occupants in collapsing front seats

Table 5 - Harm Versus Belt Use [Data Link, 1989]

	Car Rear End Impacts			All Car Impacts		
Occ. Belt Use	Distribution of Harm	Distribution of Occupants	Harm per Occupant	Distribution of Harm	Distribution of Occupants	Harm per Occupant
No	73.4	70.1	1.0	88.6	71.6	1.2
Yes	22.6	29.9	0.9	11.4	28.4	0.4
All	100.0	100.0	1.0	100.0	100.0	1.0

The above data shows that there is only about a 10 percent reduction in harm for rear end impacts when belts are used versus almost 70 percent reduction in harm for all car impacts in crash modes such as front, side and rear.

Further evidence of the safety benefits associated with non-collapsing seats versus collapsing seats is shown by the four cases of this study in which there were two front seat occupants in a vehicle where one seat back failed and the other did not fail (cases 3, 4 and 13 of Table 1, and case 15 of Table 2). In all 4 cases, the occupant in the failed seat suffered major or fatal injuries, even though 3 of the 4 seriously injured persons in the collapsing seats were belted, while those occupants in the non-collapsing seats received only minor or no injuries at all. While the evidence seems to clearly support the benefits of non-collapsing seats, there was one case (case 4 of Table 2) in which a collapsing seat appeared to save the driver's life when the intrusion of a tractor trailer vehicle would most likely have caused fatal injuries if the driver had remained upright in a non-collapsing seat. Instead, the driver received survivable injuries.

As in the Canadian study, the author investigated cases also show a much higher incidence of injuries associated with contact in the rear of the vehicle than the 2.8 percent figure reported in Table 13 of the Data Link study [Data Link, 1989].

One possible reason for some of the differences in the findings of the Data Link study and the current studies reported in this paper may be related to the data collection methods used to obtain NASS data which served as the basis for the Data Link analysis. In particular, the NASS data does not specifically seek out, or attempt to evaluate quantitatively, the often subtle damage in the rear of vehicles, which may be masked beneath the soft foam of seat back cushions and provide evidence of injuries due to contact of front seat occupants deposited in the rear of the vehicle by the collapsing front seats.

An example of how injury contact regions may be masked and need quantitative evaluation, often times through testing of exemplar structures, is given by the first case study in the following section.

SELECTED CASE STUDIES OF REAR IMPACT

Two rear impact accident cases are reviewed in this section of the paper, with focus on occupant kinematics, kinetics, resulting injuries, locations of impact and the amount of energy absorbed by the collapsing seats during the collision.

Case Study I - A comparison of the effects between collapsing and non-collapsing seats in rear impact is illustrated in this first case study. In addition, this case also demonstrates the value of quantitative testing to establish if certain subtle structural damage patterns, often hidden beneath soft foam coverings of rear cushions, are likely to be caused by body contact with forces sufficient to cause the injuries received.

For this case, a 1985 Ford Tempo 4 door sedan was slowing when it was struck in the rear by a tractor trailer vehicle. The target vehicle was being driven by a small adult female (approximately 5th percentile) and the right front passenger seat was occupied by an average adult male (approximately 50 percentile). Both subjects of the target vehicle were belted (3 point lap and shoulder belt) and were slowing their vehicle because a vehicle in front was in the process of making a left-hand turn. Impact between the tractor trailer vehicle and the target vehicle was essentially collinear. After the initial rear end impact, the target vehicle was shoved forward into the left turning vehicle which produced a small amount of damage to each vehicle.

Analysis of the damage to the vehicles indicated a change in velocity of about 30 km/hr (18 mph) to the target vehicle during the rear impact. The driver seat exhibited little damage but the right front passenger seat was broken loose from the seat tracks and deformed rearward. The driver received no serous injuries. The right front passenger; however, was found in the collapsed seat with a compressive neck load.

A survey of NHTSA crash test data revealed a FMVSS 301 rearend crash test on an essentially identical vehicle which by chance also had a 5th percentile female surrogate and 50 percentile male surrogate seated in the front seats, although in opposite positions to those of the actual occupants involved in the crash [Garn, 1983].

The rear end crush of both the accident target vehicle and the NHTSA test vehicle, as illustrated by figure 3, was almost identical. The final rest positions of the surrogates in the test vehicle were essentially the same as the rest positions of the actual occupants, with the heavier subject laying back in the collapsed seat and the lighter subject seated almost upright. Only the smaller (lighter) surrogate was apparently instrumented in the NHTSA test and the recorded loads indicated low non-injury producing acceleration levels and head injury criteria (HIC) as was the case with the small driver of the actual case vehicle. The test report also noted, however, that even though the larger (heavier) surrogate produced significantly more residual deformation in the driver seat of this test, there was no apparent contact of the dummy with the vehicle interior based upon post-test visual inspection.

Similarly, at first glance, the interior of the accident vehicle also showed no apparent contact damage. In light of the compressive load to the neck of the heavier occupant in the collapsed seat, however, it was clear that some action or reaction source must have provided a contact surface with sufficient resistance to allow the neck to be loaded compressively by the torso as the seat collapsed rearward. Removal of the rear seat back cushion revealed what appeared to be a small embossment below the forward edge of the package tray sheet metal structure just behind the head of the seriously injured right front passenger.

Figure 3 - Accident and NHTSA Test Vehicle Damage

In order to determine if the embossment found was caused by a force of the magnitude and direction necessary to be consistent with loads generated by someone sliding rearward from beneath the restraint when the seat back collapses, an exemplar rear clip was purchased and a simulated head push test was conducted. Figure 4 illustrates the structural test set-up to duplicate subtle damage and establish load levels on an exemplar clip. The results of the test confirmed that the load levels necessary to duplicate the damage would also have been large enough to cause the serious neck injury received by the heavier front passenger.

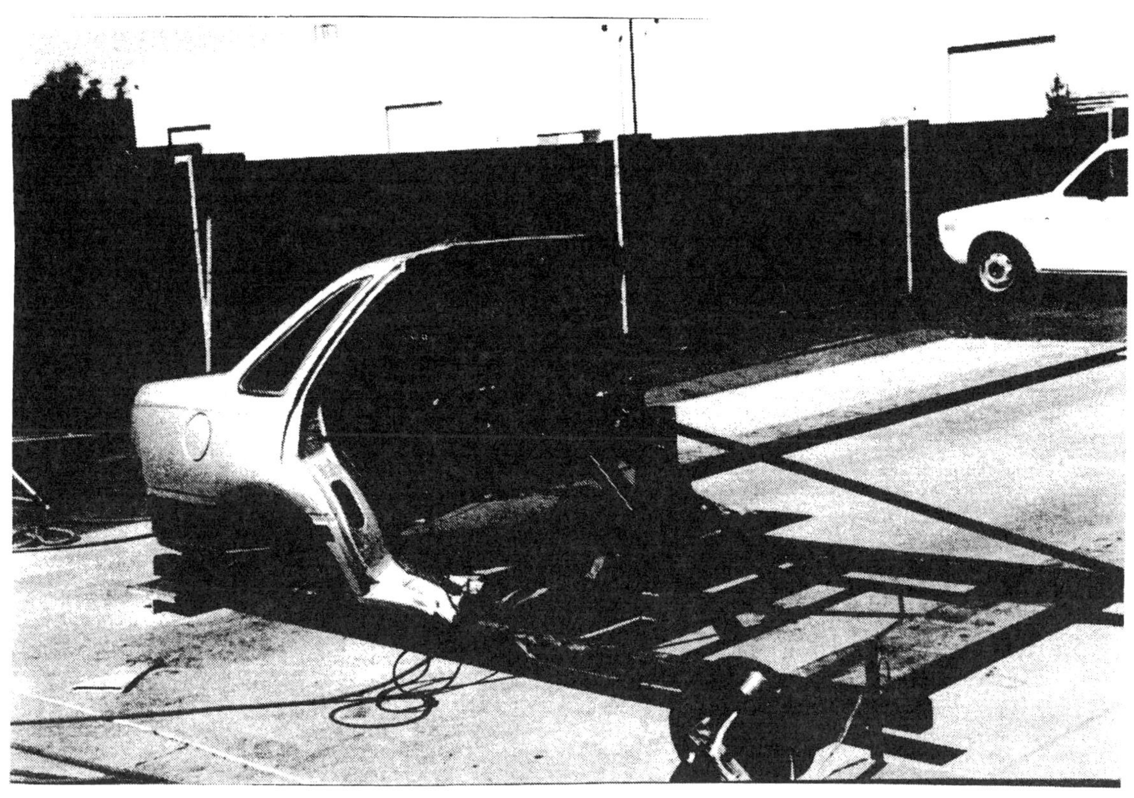

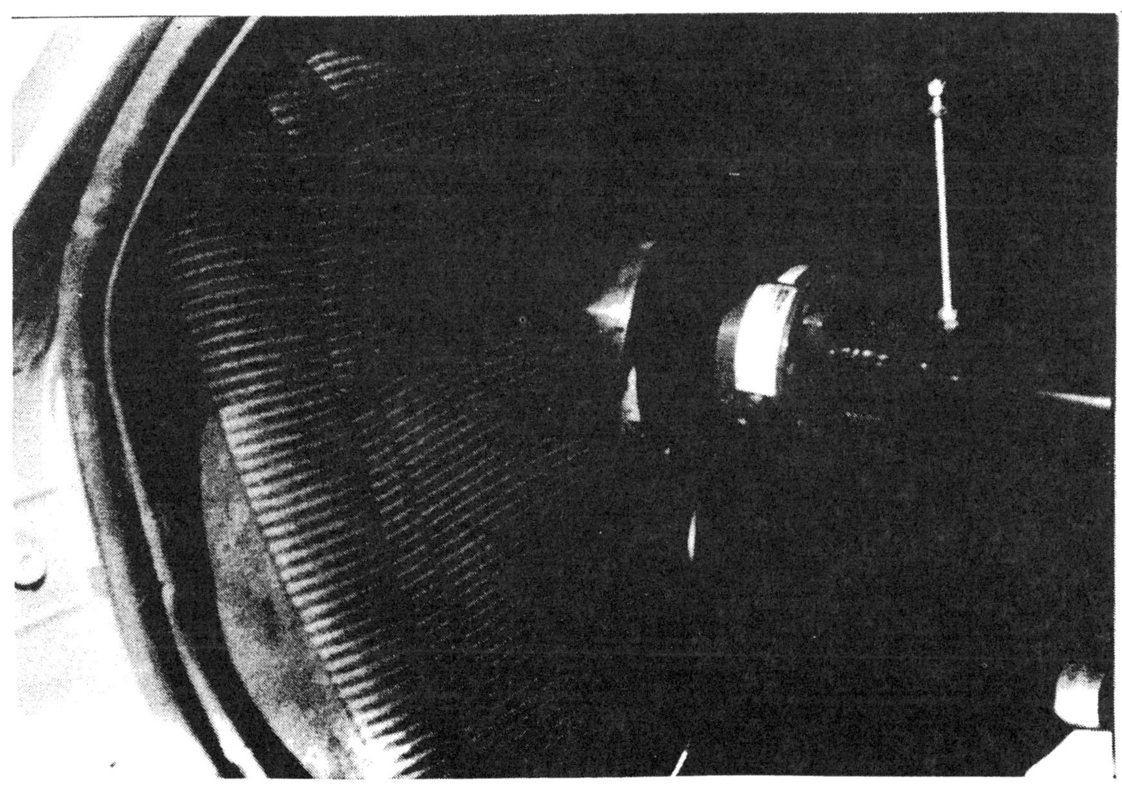

Figure 4 - Head Push Test on Exemplar Rear Clip

With respect to locations of the occupants after the two collisions (i.e., rear impact and then front), it should be recalled that the rear impacted vehicle was shoved forward into the left-turning vehicle in front of the target vehicle, and, this frontal impact, while of a much lower severity than the rear impact, was of sufficient deceleration as to allow the reclined occupant (right front passenger) to slide forward into a position that suggested no contact with the rear of the vehicle. In fact, the driver (lighter occupant) did comment about being thrown forward during this second impact. In contrast to the occupant of the actual accident vehicle, the dummies of the NHTSA test did not apparently move as far rearward which may be due in part to the molded seated pelvic area that is obviously not as flexible as an actual human subject.

Thus, the kinematics of the heavier passenger in the collapsing seat were such that as the seat reclined the passenger slid rearward, about 50 centimeters, struck his head below the top edge of the package tray, received compressive loads to the neck as the lower torso inertia continued to load into the head and neck, and rebounded back forward during the less severe second (frontal) impact.

With regard to the energy absorbing capability of the collapsing seats, tests were run on exemplar seats to duplicate the damage found on the actual seat and determine the approximate amount of energy absorbing benefit received by the restrained passenger with the compressive neck injury. Figure 5 shown below illustrates the torque versus seat back angle curve generated for an exemplar seat which had essentially identical damage to the seat occupied by the seriously injured occupant. The test indicates that the seat absorbed approximately 300 N-m of energy, while the energy generated by moving the 50 percentile occupant's upper body at a velocity of about 30 km/hr would be approximately 1420 N-m. Thus, the collapsing seat would absorb some energy but not enough to prevent the body from continuing rearward with enough energy to forcefully impact into the rear structure of the vehicle.

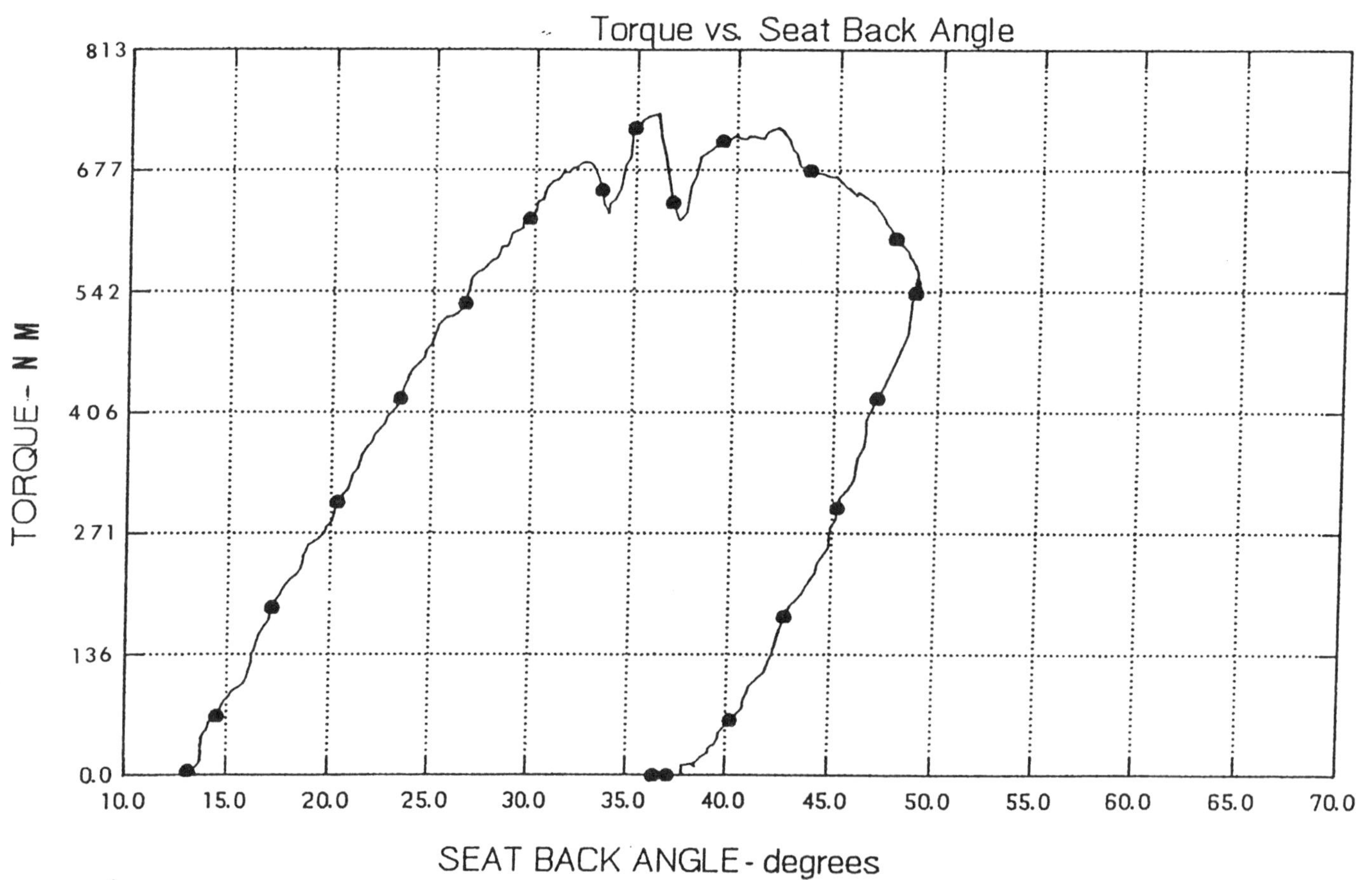

Figure 5 - Torque Versus Deformation Angle for Exemplar Seat of Case I

Case Study II - This case involves a rear impact to a 1977 Toyota Corolla, 2-door sedan struck in the left rear by a 1979 full size van. The target vehicle was occupied by 4 adult male passengers, two in the front and two in the rear. It was unknown if the passengers were restrained or not. The impact caused severe crushing of the left rear area and the calculated change in velocity of the target vehicle (Toyota) was about 44 Km/hr (26.4 mph). Figure 6 illustrates the severe crush on the left rear side of the vehicle.

The seat backs of the target vehicle front bucket seats were mounted asymmetrically to the seat bottom structure by means of a single outboard recliner mechanism that transmitted loads between the seat back and the seat bottom. The inboard connection was a non-torque transmitting hinge pin.

Figure 6 - Damage to Case II Vehicle

Due to the asymmetric load coupling of the two seat components, a greater amount of inboard twisting and bending takes place on the free hinge side of the seat back as compared to the stronger outboard side with the torque load carrying mechanism. This inboard twisting makes it easier for an occupant to deflect rearward and towards the center of the vehicle.

In this particular case the principal direction of force was at about a 15 degree angle from the left rear toward the right front and as a result the right front passenger was deflected not only rearward due to the collapsing seat but also somewhat toward the rear center of the vehicle between the two rear passengers. The 27-year-old right front passenger received massive head injury from impact with a hard surface and died of cardiac arrest secondary to the massive head injury. Figure 7 shows the area of protruding or inward buckle on the rear package tray which lines up with the predicted occupant path of the head of the right front passenger.

The driver of the vehicle moved rearward and inboard toward the load carrying recliner side of his seat back and as such there was not as much deformation noted in his seat. The driver sustained a left temporal head injury and simple fracture of some ribs. The passenger behind the driver was sandwiched between the vehicle crush and the collapsing seat of the driver. This passenger received moderate injuries which included simple fracture of the left clavicle and lower spine injury. The right rear passenger had internal injuries to the abdomen and also had lower spine injury. It is interesting to note that although there were no headrests on the rear seat, neither rear passenger received any neck injury as a result of the accident. The injuries to the rear passengers were attributed to the combined crush of the front passengers in collapsing seats and the rear crush of the vehicle. The injuries to both front passengers were attributed to the deformation of seat backs.

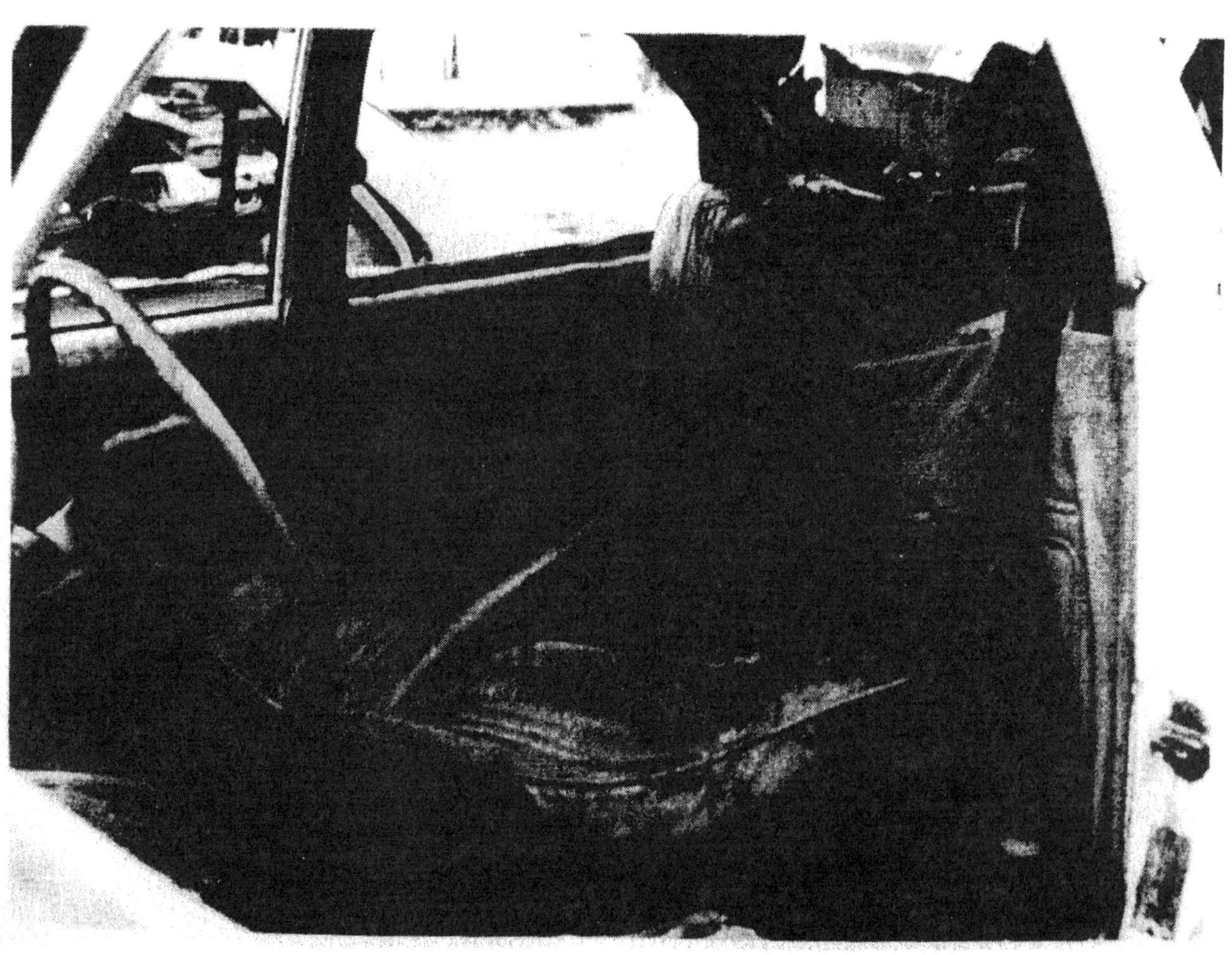

Figure 7 - Interior Impact Region of Case II Vehicle

As in the previous case study, tests on exemplar seats were run to establish the approximate amounts of energy absorption afforded to the front seat passengers. Figure 8 illustrates the torque versus seat back angle data for the passenger seat test. The test data indicates that the passenger seat absorbed about 340 N-m of energy; however, as in the previous case this level was much lower than the kinetic energy associated with the motion of the right front passenger.

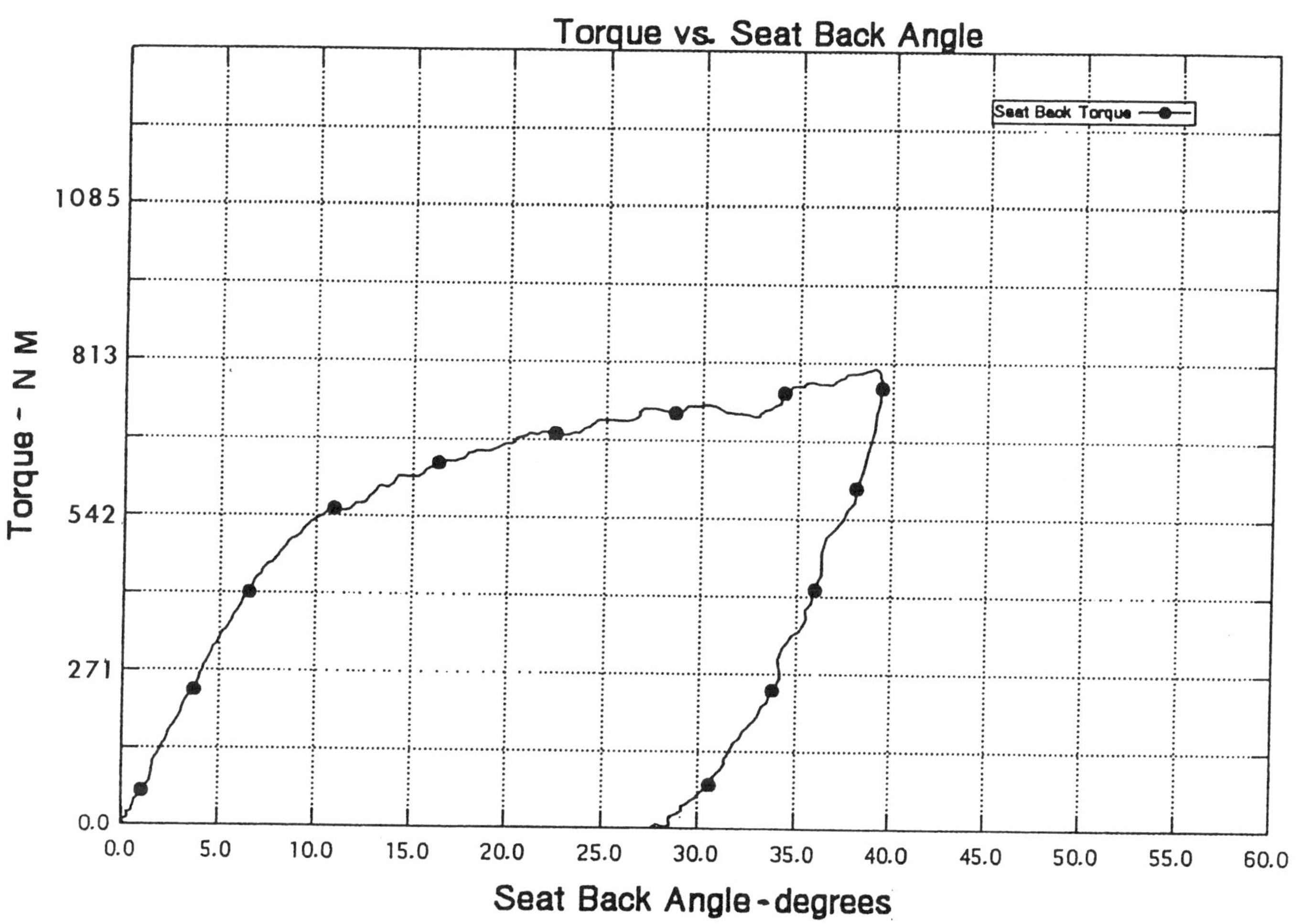

Figure 8 - Torque Versus Deformation Angle For Exemplar Seat of Case II

REAR IMPACT CAR TO CAR CRASH TEST DATA

Additional information on occupant response relative to seat back failure is contained in some NHTSA New Car Assessment Program (NCAP) rear impact tests. The 1979 NCAP tests contained instrumented 50 percentile male dummies. 1980 and later, NCAP tests included non-instrumented 50 percentile male dummies in the front seats, but the dummies were painted to identify head strike zones within the vehicle. Color photographs were taken of these tests to document the head strike zones. The head strike zones indicated that "ramping" of occupants did not occur until the seat backs had reclined fully. No clear documented head strikes on inner roof panels were noted in any tests conducted at Dynamic Science, Inc. during the 1979 or 1980 NCAP program.

As part of the Fuel System Integrity Defect Investigation of Ford Pintos and Chevrolet Vegas, NHTSA also commissioned detailed tests of seat and hatchback crashworthiness in rear barrier and rear vehicle-to-vehicle impact tests [Pirtle, 1978].

At least 19 crash tests in this program were conducted in 1978 and 1979, with instrumented dummies, seats, and vehicle structures. Ground-based and on-board high speed cameras were used to document dummy kinematics. In every test of Pintos and Vegas, permanent deflection of seat structures occurred. In most tests the dummies moved rearward and struck the rear deck or the instrumentation anchorages mounted aft of the driver seat. The only test where seat deflection did not occur was in a 60 Km/hr front-to-rear impact between two 1971 Chevrolet Impalas. Despite having a 50 percentile male driver and 50 percentile male passenger dummy in the front bench seat, no seat failure occurred. This test also resulted in the lowest head and chest injury levels recorded in the test program. The results of this test appear to refute the notion that non-deforming seat backs will result in more injuries.

Similar results were also obtained in a more recently conducted car to car crash test [Slattery, 1990] where the right front passenger dummy, an unrestrained 50 percentile adult male, was seated in a rigidly braced conventional seat, while the driver dummy, also an unrestrained 50 percentile adult male, was seated in the standard collapsing seat. During this test the stationary target vehicle, a 1259 Kg (2770 pound) sports car, was impacted in the right rear (behind the right front passenger seat location) by a 2456 Kg (5404 pound) large sedan traveling at approximately 90 Km/hr (53.5 mph). This impact resulted in severe damage to the vehicle in the area behind the right front passenger seat (non-collapsing) and a vehicle change in velocity of about 62 Km/hr (37 mph).

Film coverage of the event clearly showed that the unbelted passenger dummy in the non-collapsing seat did not ramp up the seat back nor did it violently rebound forward into the dash or windshield area, as is sometimes suggested by advocates of the collapsing seat theory. In addition, the recorded chest loads on the right front passenger in the non-collapsing seat were at about 30 G's and well below the injury thresholds established by human volunteers as discussed earlier in this paper. By contrast the driver dummy flipped violently rearward and buried its head in the shelf of the rear package tray with a peak force of about 225 G's and a pulse width sufficient to indicate potential head injury. As in the car to car crash test results reported by Pirtle [Pirtle, 1978], the Slattery test also refutes the notion that non-deforming or stronger, seat backs will result in more injuries.

The study reported by Pirtle also provided interesting information on the effects of seat deformities resulting from non-symmetric loading on the seats due to the use of a non-load carrying connection or hinge, on the inboard side of many seats.

Seats with no inboard locking seat back latch showed a consistent tendency to fail in a twisting manner, with driver seats twisting clockwise and passenger seats twisting counterclockwise when viewed from behind. This twisting failure mode allows the test dummies to move toward the midline of the vehicle, even during straight in-line rear impacts. This failure mode increases the likelihood of the front seat occupant's torso to move directly away from the shoulder harness. This movement also increases the likelihood of ejection or slippage from the restraint belt system, especially when a single ELR/sliding latch plate system is used. Once tension is relieved on the shoulder portion of this restraint belt, slack can be taken up by the lap portion of the belt. This excess belt slack can and does allow virtually unrestrained rearward motion within the vehicle.

The above study [Pirtle, 1978] also showed that excessive restraint belt slack was induced in all tests where seat back failure occurred. The only tests without significant dummy movement into the rear areas of the vehicle were the two 35 Km/hr rear impact tests that included the Impala to Impala Test (with the 50 and 95 percentile dummies in a bench seat), and the left rear 15 degree angled impact test. This last test shows the large difference in seat deflection when the occupant is moving toward the area of greatest seat back rigidity (i.e. the recliner on the outboard side) compared to the straight rearward or inboard vectors that twist easily due to the lack of torque load carrying capability of the hinge pin connection between the seat back and seat pan structures.

Finally, there are numerous FMVSS 301 rear impact tests and NHTSA defect studies [Pozzi, 1980], that also show the consistent tendency of front seat occupants to strike the rear seat back or rear deck area, even while restrained, in moderate rear collisions. The real world injuries that occur in similar accidents is consistent with these test results.

SUMMARY AND CONCLUSIONS

Among the many results obtained from this study of real world accidents, and associated experiments, it was found that a majority of restrained front seat occupants were either partially or totally ejected from the seat systems during rear impact, even at changes of velocity as low as 30Km/hr (18 mph) or less. These findings are consistent with those observed in the "real world" accident study recently conducted for Transport Canada.

It was also found that many front seat restrained occupants received serious paralyzing injuries in the rear area of the vehicle even though they were found in the front seats post-accident. This suggests that possible improvements in the NASS reporting and evaluation methods might be of value since recent studies using the above data sources show a much lower rate for rear area impact than that observed in this study.

In addition to the above, the results indicate that while collapsing seat structures absorb some amount of energy during impact they also appear to contribute to hazardous situations such as: loss of driver control; entrapment and delay of egress of rear seat passengers; injury to rear seat passengers located in non-collapsing seats; and, as mentioned previously partial and/or total ejection and injury of restrained front seat occupants.

Furthermore, instrumented test dummies in strengthened and collapsing seats suggests that some of the alleged hazards of stronger seats, such as ramping and rebound of unrestrained passengers, may not be a significant problem, even at relatively high velocity changes. As a result of the findings in this study, it is concluded that stronger, less-yielding and non-collapsing seats are more likely to provide improved safety benefits over seat systems which collapse at relatively low energy levels. It is not the intent of the authors to suggest, however, that energy absorption is undesirable in a motor vehicle seat system. Quite the contrary, energy absorption is an important component of crash safety design; but, as shown by the case studies of this paper, many current seat systems only provide a small fraction of the energy absorbing level needed to safely decelerate an occupant in a controlled fashion, even during the moderate impacts such as those discussed in the Case I study. Even in that case, the seat system would have required at least five times more energy absorbing capability to prevent the injurious rear compartment contact of the occupant. What this means is that the seat torque for that specific case would have to be increased from about 730 N-m (6,454 in.-pounds) to at least 3,650 N-m (32,205 in.-pounds) with limited rearward deflection. Also, additional energy absorbing passive occupant protection materials could be mounted beneath the soft comfort foam pads if the seat support structures were designed with increased load and torque capabilities.

Finally, it is noted that the field accident studies investigated by the authors of this paper were based upon forensic analysis and testing during the course of litigation related matters. In such matters there are two sides which spend considerable time, effort and resources to gather valuable quantitative data, such as structural tests, etc., which are ultimately used to determine the most likely cause of a particular system failure and/or injury. In many cases, this data has been made a matter of public record or could be obtained with proper confidentiality agreements. If at all possible, such data should be made available for objective and independent review by data base collection teams that could add the data to a central data base such as NASS, for use by all safety engineers.

The above recommendation would increase the current data base and provide more detailed quantitative data than that usually available to NASS. The initiative for establishing such a program should be taken by groups such as the National Transportation Safety Board (NTSB), NHTSA, and/or the Consumer Product Safety Commission (CPSC).

REFERENCES

[Beeding, 1960]
Beeding, E.L., Jr. and Mosely, J. D.: "Human Deceleration Test," AFMDC-TN-60-2, Holoman Air Force Base, New Mexico, 1960.

[Carr, 1975]
Carr, R. W.: "An Occupant Simulation Handbook," Contract No. N00014-71-C-0318, Dept. of the Navy, Dynamic Science Report No. 6910-75-123, 1975.

[Data Link, 1990]
"Current Issues of Occupant Protection in Car Rear Impacts," Prepared for the Office of Crashworthiness, Rulemaking, NHTSA, Data Link, Inc. Report, Wash. D.C., Feb. 1990, Docket 89-20-N01-021.

[Federal Register, 1974]
Federal Register, Tuesday, March 19, 1974, Volume 39, No. 54, p. 10268.

[Garn, 1983]
Garn, R.: "FMVSS 301 Standards Enforcement Testing of Fuel System Integrity - 1984 Ford Tempo, 4-Door Sedan" Dynamic Science Report No. 301-DYS-83-013, September, 1983.

[Giese, 1973]
Giese, R.; et al; Inventors of Patent No. 3,761,127, Granted to Ford Motor Company for "Safety Seat with a Safety Belt, In Particular for Motor Vehicles," September 25, 1973.

[Glance, 1973]
Glance, Patrick M.; Inventor of Patent No. 3,761,125, Granted to Ford Motor Company for "Energy Absorbing Seat Back," September 25, 1973.

[Habrel, 1989]
Habrel, J.; Ritzl, F.; and Eichinger, S.: "The Effects of Fully Seat Integrated Front Seat Belt System On Vehicle Occupants in Frontal Crashes," ESV Conference, Göteborg, May 1989.

[IIHS Status Report, 1989]
IIHS Status Report, Vol. 24, No. 8, August 26, 1989.

[James, 1991]
James, Michael B.; Strother, Charles E.; Warner, Charles Y.; Decker, Robin L.; and Perl, Thomas R.: "Occupant Protection in Rear-End Collisions: I. Safety Priorities and Seat Belt Effectiveness," SAE 912913, Proceedings of the 35th Stapp Conference, San Diego, CA., November 18-20, 1991.

[Marriner, 1990]
Marriner, Paul C.; Submission of Report entitled: "Evaluation of Seat Back Failures," Prepared by TES Limited for Transport Canada, Submitted to U.S. DOT Docket 89-20-NO1(018), January 1990.

[Mercedes Benz, 1989]
Mercedes Benz advertisement, 1989.

[Nash, 1974]
Nash, Carl: Petition to National Highway Traffic Safety Administration, Docket 74-13-2, February 1974, 208-PRM-000018.

[NHTSA Five Year Plan, 1979]
NHTSA Five Year Plan for Motor Vehicle Safety and Fuel Economy Rulemaking, April 20, 1979.

[Pirtle, 1978]
Pirtle, R.: "Test and Evaluation of Head Restraints, Seat Backs, and Anchorages in Vehicles Subject to Rear Impact Collision," Dynamic Science Rpt. 3005-78-84, DOT-HS-8-01886, Mod. 1, June 1978.

[Pirtle, 1978]
Pirtle, R.: "Test and Evaluation of Seat Belt System and Hatchback Door Lock and Hinge System Integrity in Rear Impact Collision," Phase 1, Dynamic Science Rpt. 8317-78-103, DOT-HS-8-01886, Mod. 2, June 1978.

[Pozzi, 1980]
Pozzi, M.: "Test and Evaluation of Volkswagen Type I Front Seat Integrity in Rear Impact Collisions," Dynamic Science Report No. 8340-80-192, DOT-HS-8-01886 (Mod-12), November 1980.

[Saczalski, 1989]
Saczalski, Kenneth: Petition to National Highway Traffic Safety Administration, Docket 89-20-N01-001, April 18, 1989.

[Severy, 1958]
Severy, D.M.; Mathewson, J.H.: "Automobile Barrier and Rear-End Collision Performance," presented at the SAE Summer Meeting, Atlantic City, N.J., June 8-13, 1958.

[Severy, 1967]
Severy, D.M.; Bruik, H.M; and Baird, J.D.: "Collision Performance, LM Safety Car," SAE No. 670458, presented at SAE Mid-Year Meeting, Chicago, IL., May 9267.

[Severy, 1968]
Severy, D.M.; Bruik H.M.; and Baird, J.D.: "Backrest and Head Restraint Design for Rear-End Collision Protection," SAE Paper No. 680079, presented at the SAE Congress, Detroit, Michigan, January 1968.

[Severy, 1969]
Severy, D.M.; Bruik, H.M.; Baird, J.D.; and Blaisdell, D.M.: "Safer Seat Design," SAE Paper No. 690812, Proceedings of Thirteenth Stapp Car Crash Conference, Harvard School of Public Health, December 1969.

[Severy, 1976]
Severy, D.M.; Blaisdell, D.M.; and Kerkhoff, J.F.: "Automotive Seat Design and Collision Performance," SAE paper No. 760810, Proceedings of the 20th Stapp Car Crash Conference, 1976.

[Shaw, 1990]
Shaw, L.M.; "Critique of SAE Paper No. 872214" submitted to National Highway Traffic Safety Administration," Docket 89-20-No1, October 31, 1990.

[Slattery, 1990]
Private communication and correspondence regarding Failure Analysis Associates car to car crash test No. AZO-3984 conducted 9/12/90 for Attorney Robert Slattery, Milwaukee, Wisconsin.

[Stapp, 1955]
Stapp, J.P.: "Effects of Mechanical Force on Living Tissue; I-Abrupt Deceleration and Wind Blast," Journal of Aviation Medicine, Vol. 26, No. 4, August 1955, pp. 268-288.

[Strother, 1987]
Strother, Charles E.; and James, Michael B.: "Evaluation of Seat Back Strength and Seat Belt Effectiveness in Rear End Impacts," SAE Paper No. 872214, Proceedings of the 31st Stapp Car Crash Conference, New Orleans, Louisiana, October 9-11, 1987.

[Urai, 1975]
Urai, Muncharu, Inventor of Patent No. 3,922,029 Assigned to Takeji Saito for "Seat Mounting Device for Vehicle," November 25, 1975.

[Warner, 1991]
Warner, Charles Y.; Strother, Charles E.; James, Michael B.; and Decker, Robin L.: "Occupant Protection in Rear End Collisions: II, The Role of Seat Back Deformation in Injury Reduction," SAE Paper No. 912914, Proceedings of the 35th Stapp Car Crash Conference, San Diego, Ca., November 18-20, 1991.

930340

Automotive Seat Design Concepts for Occupant Protection

David M. Blaisdell, Andrew E. Levitt, and Michael S. Varat
Collision Research and Analysis Inc.

ABSTRACT

The concept of increasing strength requirements for automotive seats has been proposed as a means of reducing occupant injuries, particularly in the rear-end impact environment. This paper will evaluate various safety trade-offs and practical requirements of seat design brought about by modifications that include the rigidification of seat structures. Rigidified and yielding seat design concepts are evaluated, utilizing analytical procedures as well as data from static and dynamic tests. The effect of seat rigidification is examined in terms of occupant interaction with the surrounding structure and with the restraint system. Potential effects of these modifications on occupant kinematics and resulting injury exposures are also examined. The elastic properties of conventionally rigidified seat structures are compared to rigid seat structures in terms of their effect on occupant motion during collision. Various crash types are evaluated in terms of occupant protection potential relating to seat collision performance; however, the design emphasis is on the rear-end collision exposure.

INTRODUCTION

Considerable attention has been given to the subject of automotive seat design as it relates to occupant protection during rear-end collisions. Much of this attention has been focused on the issue of seatback strength requirements and on the general acceptance by the automotive design community of occupant energy management by means of yielding seat structures. Since seats are the primary structure in direct contact with the motorist, considerable research has been done over the past three decades to evaluate various concepts for safety seating systems that could be developed to provide an additional layer of protection to occupants. The concept of creating a high-G "astronaut seat" to which the motorist would be fastened and which would serve to partially surround him or her with non-penetrable, energy-absorbing structure has been explored by a number of researchers. (1),(2),(3),(4),(5)*

Results of controlled laboratory tests on such structures have generally been positive and have resulted in recommendations for additional research that could lead to implementation of such systems in production automobiles. In the less controlled environment of the production automobile, significant obstacles to such implementation have been noted. The uncontrolled nature of pre-impact motorist positioning and restraint conditions, the multitude of complex interactions between motorist and vehicle interior as well as the sometimes contradictory requirements of crash energy management, has made it difficult to design a seat structure that can increase the overall safety of vehicle occupants for all collision exposures. Additionally, the statistical evidence of good occupant protection from

* Numbers in parenthesis () refer to references at the end of the paper.

current seat designs and the governmental emphasis on development of passive restraints has directed research away from such integrated safety seat design concepts.

SIGNIFICANT RESEARCH INTO AUTOMOTIVE SEAT DESIGN as it relates to occupant crash protection began in the mid-fifties with a series of car to car crash tests conducted at UCLA. That study evaluated the effect on motorists of relatively low-speed rear-end collisions (7 to 20 mph closing speed).(6) It had long been noted that many lower speed rear-end collisions, wherein the vehicle sustained only minor damage, often resulted in disproportionately severe occupant injuries. Although these neck and back injuries were not always immediately evident following a collision, they often resulted in weeks or months of painful recovery. These UCLA crash tests, which utilized human volunteers in addition to anthropometric dummies, disclosed that the primary cause of these injuries was the excessive relative movement that took place between the head and torso during collision. The low-back seats of that era did not support the head, allowing substantial rotation and extension to occur. This hyperextension resulted in stressed and torn tissue and, in severe cases, disc and cord damage. In this early series of tests, the benefit of seatback deflection and yielding during the higher speed collisions was observed for the first time. It was noted that better head and neck alignment with the torso was maintained during the more severe collisions when the backrest yielded rearward, thus resulting in less injury-producing motion than occurred at lower impact speeds.

Additional research and recommendations for better head support to reduce this head excursion resulted in the implementation of FMVSS 202 (7) which required, starting Jan., 1, 1969, that head restraints be installed in passenger cars. This standard required that the top of the head restraint be no less than 27.5 in above the seating reference point when fully extended. At this elevation the head restraint was considered to be sufficiently high to provide protection to a large percentage of the motoring public. This conclusion was based on the determination that most rear-end collisions occur at relatively low speeds. Notwithstanding the presence of a head restraint, the preponderance of data continued to indicate the benefits of allowing seatbacks to yield and absorb energy in addition to limiting excessive head excursions that would otherwise occur whenever the head was not completely supported.

Various proposals have been made to increase seatback strength without adequately evaluating the potential effects of such a change on occupant safety. These recommendations appear to overlook the results of the earlier safety-seat research projects, wherein it was found that merely increasing seatback strength without provision for occupant position control, including but not limited to wrap-around structure and mandatory active restraint of all occupants, could be counterproductive in terms of overall occupant safety. The severe hyperextension exposures reported by Severy(8) and the potentially severe ramping and out-of-position injury potential described by Strother and James(9,10) could increase rather than mitigate overall injuries if such a recommendation were followed without considerable additional research and evaluation. The abandonment of the conventional cantilever seat in favor of the semi-hoop, acceleration "hammock" approach,(11) although solving several of the practical structural problems, still would require occupant position control in conjunction with mandatory restraint usage.

COLLISION PERFORMANCE OF PARTIALLY RIGIDIFIED SEATS

A series of thirteen rear-end collision experiments conducted at UCLA in 1967 evaluated many of the significant variables relating to occupant protection as a function of seat design.(12) In addition, since the experiments involved car-to-car collisions, both frontal and rear-end collision exposures were evaluated. Three of the variables under study were seatback strength, seatback height and collision severity.

Table 1. Selected Data From UCLA Crash-test Series.

Seat Strength, In-Lb.	Delta-V MPH	Dynamic seat angle, Degrees	Perm. Seat Deflect., Degrees	Perm. Deflect. %	Elastic Deflect. Deg.	Elastic Deform-ation, %	Head Offset, In.	Remarks
9100 in-lb	5.5	29.5	1.5	16	8	84	3	Driver X-97
9100 in-lb	11	33.2	8	61	5.2	39	3	Driver X-94
9100 in-lb	16.5	44.5	19	78	5.5	22	3	Driver X-93
16100 in-lb	5.5	27.4	0	0	7.4	100	0	Passenger X-97
16100 in-lb	11	29.5	3	32	6.5	68	0	Passenger X-94
16100 in-lb	16.5	34.9	7	47	7.9	53	0	Passenger X-93
16100 in-lb	16.5	34.5	7	48	7.5	52	3	Driver X-96
16100 in-lb	16.5	35.3	8	52	7.3	48	3	Passenger X-96
16100 in-lb	22	54.9	21	60	13.9	40	3	Passenger X-98
33000 in-lb	16.5	27	0	0	7	100	0	Driver X-99
33000 in-lb	16.5	27.4	0	0	7.4	100	0	Passenger X-99
33000 in-lb	30	55.4	6	17	29.4	83	0	Driver X-103
33000 in-lb	30	57.9	15	40	22.9	60	3	Passenger X-103
33000 in-lb	30	51	8	26	23	74	0	Driver X-104
33000 in-lb	30	48.6	10	35	18.6	65	3	Passenger X-104

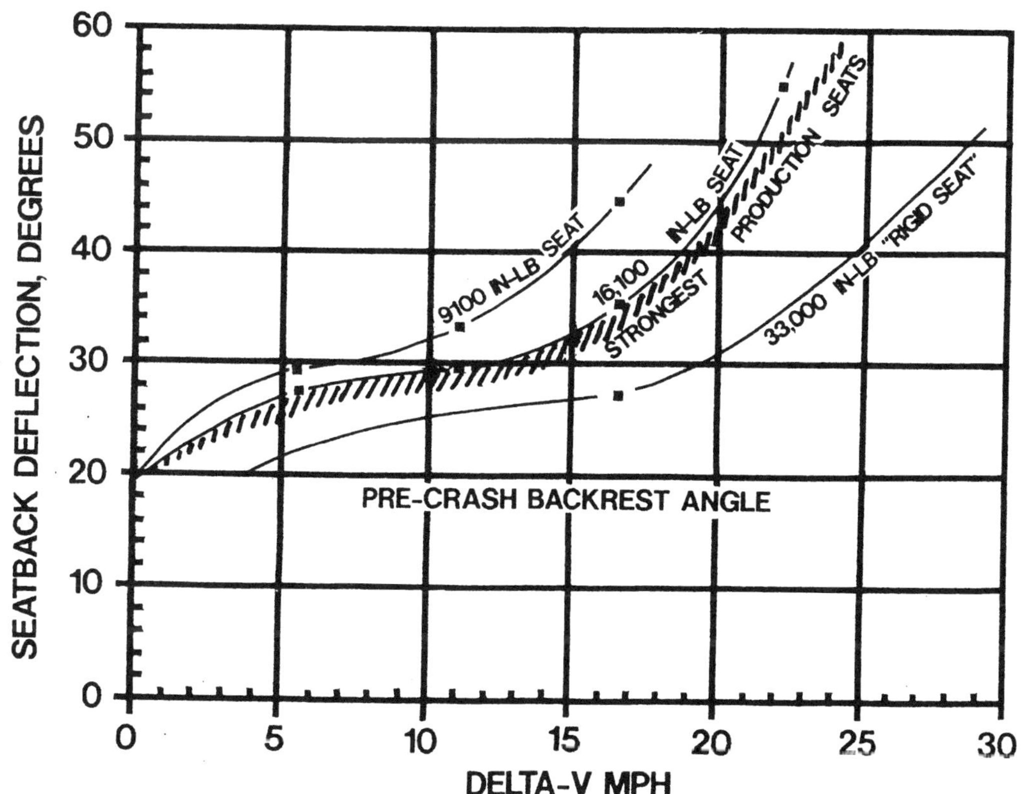

Figure 1. Seatback Deflection for Selected Test Seats

In order to determine a baseline for the strength evaluation, production seats were tested statically to determine then-current production seatback strengths. The procedure for these static tests involved pulling rearward on the seatback from a location 35.6 cm. (14 in.) above the seat cushion. This height approximated torso CG height for a 50th percentile male occupant. Based on these tests a value of 9100 in-lb. was chosen for the baseline seat strength. In order to assure controlled seatback yielding, all adjustment mechanisms were bypassed and yield links replaced the normal backrest supports. This procedure allowed accurate control of seatback strength and yield characteristics. Backrest yield resistance was increased in measured increments to allow adequate evaluation of head-neck hyper-extension characteristics as a function of seatback height and impact severity. Table 1 shows the results of selected tests from the series.

Conventional yielding backrests of the type tested during the UCLA crash test series (9100 in-lb. to 16,100 in-lb.) use most of their energy absorption capability in the 5 to 20 MPH Delta-V range, Figure 1. Throughout this range conventional seats minimize head-to-torso misalignment and protect against the majority of serious injury exposures. As the collision severity exceeds these levels however, the front seat occupants move rearward sufficiently to contact rear seat areas. As collision severity levels increase, and as the energy absorbing capabilities of the passenger compartment structures are exceeded, serious injuries may result. Although the threshold for exceeding vehicle interior energy absorption (E-A) capabilities may be raised by increasing the strength of seatbacks, several safety trade-offs must be considered. These include increased ramping, danger of severe hyperextension

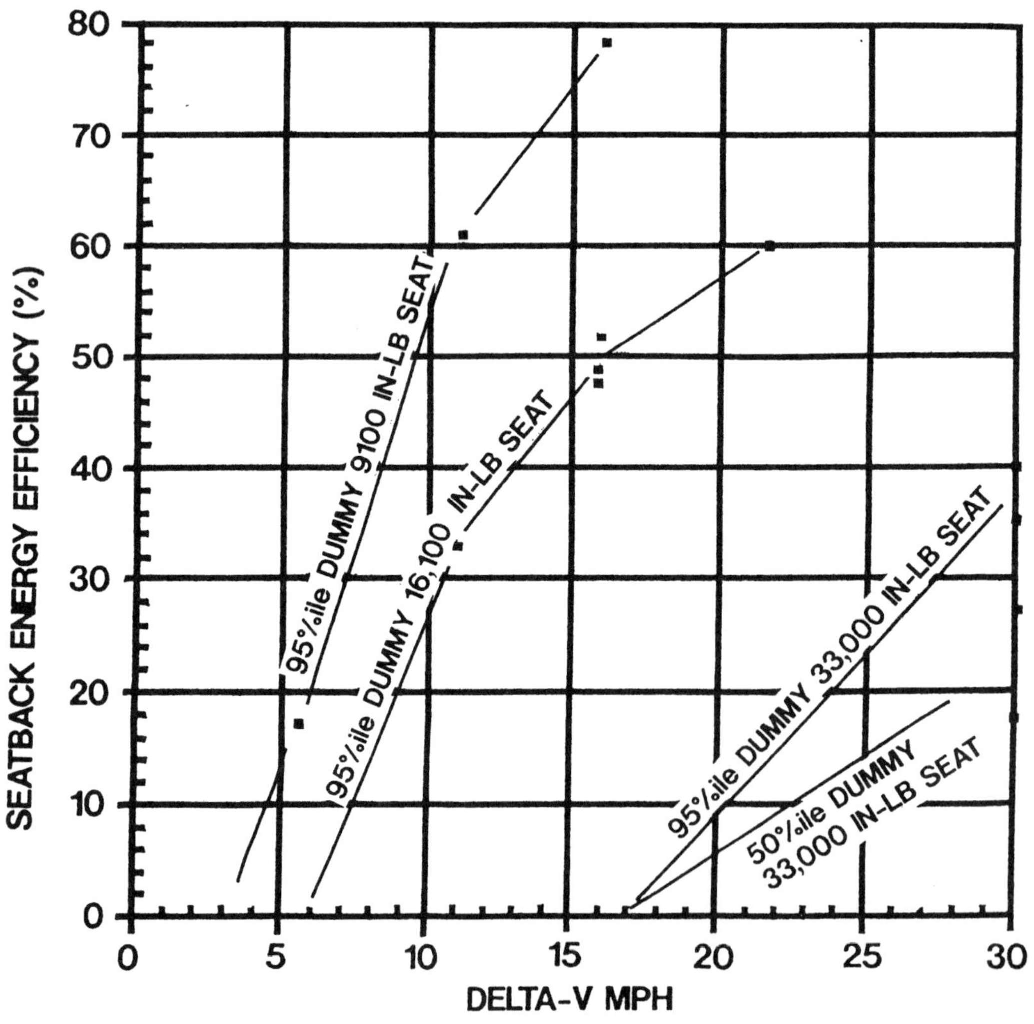

Figure 2. Energy-absorbing Properties of UCLA Test Seat Structures

injuries along with increased rebound potential. As conventional seatback structures are made less yielding, their potential for elastic rebound increases, thus making the structure less energy efficient. This rebound effect, in conjunction with the ramping tendency acts to increase the effective occupant change of velocity during collision (Delta-V) and thereby increases potential for injury. The more easily yielding seats tended to perform more efficiently in terms of energy absorption so that a substantial percentage of their E-A capability was utilized, Figure 2.

EVALUATION TECHNIQUES

Collision severity is generally quantified by determination of Delta-V and vehicle peak deceleration rates. When evaluating occupant response to the collision, the delayed onset of acceleration and force application to the

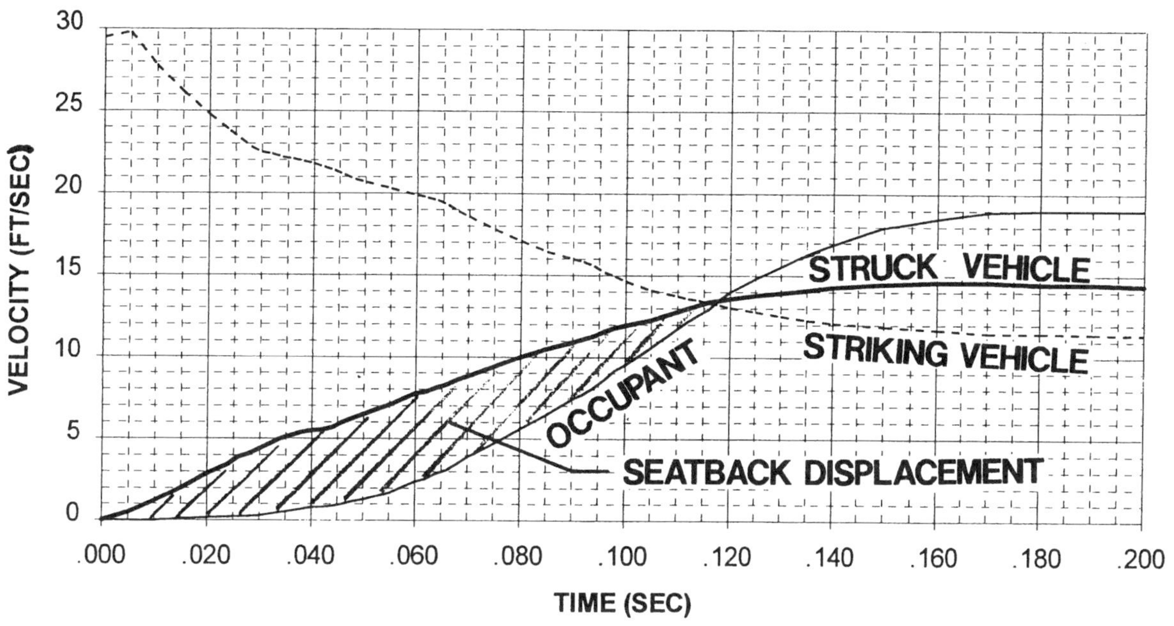

Figure 3A. UCLA Experiment 94.

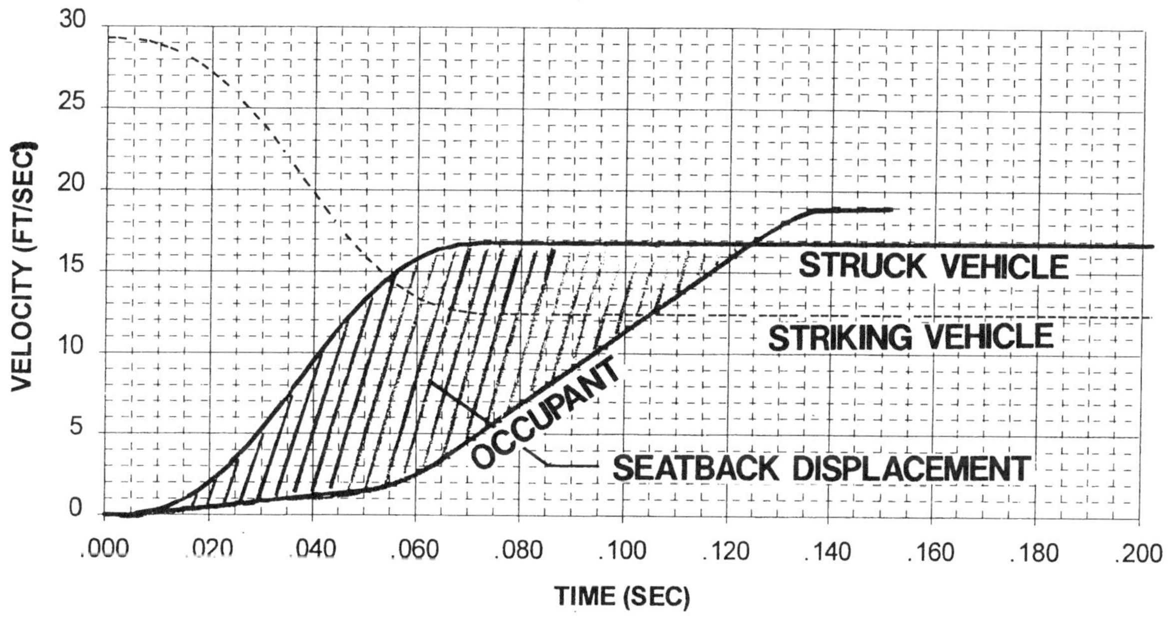

Figure 3B. Modern Compact Vehicle at X-94 Impact Velocity.

occupant as a result of the relative movement between motorist and vehicle must be of prime concern. For the rear-end collision, seat comfort padding, body compliance and structural characteristics are primary mechanisms for this delayed response. As with the evaluation of restraint systems for frontal and side collisions, it is useful to evaluate seat performance by integrating the acceleration pulse, to render the velocity vs. time (V-T) plot for the given collision. Since the area under the velocity-time curve represents displacement, both the effect of crushing structure and the delayed seat loading caused by occupant displacement into the padding and seat structure can be evaluated.

This graphical V-T technique was used to evaluate seat performance for two selected collision severity levels, Figure 3(a-d). Seat collision performance is a function of collision severity, vehicle structural crush and seat yield characteristics. For any given seat design,

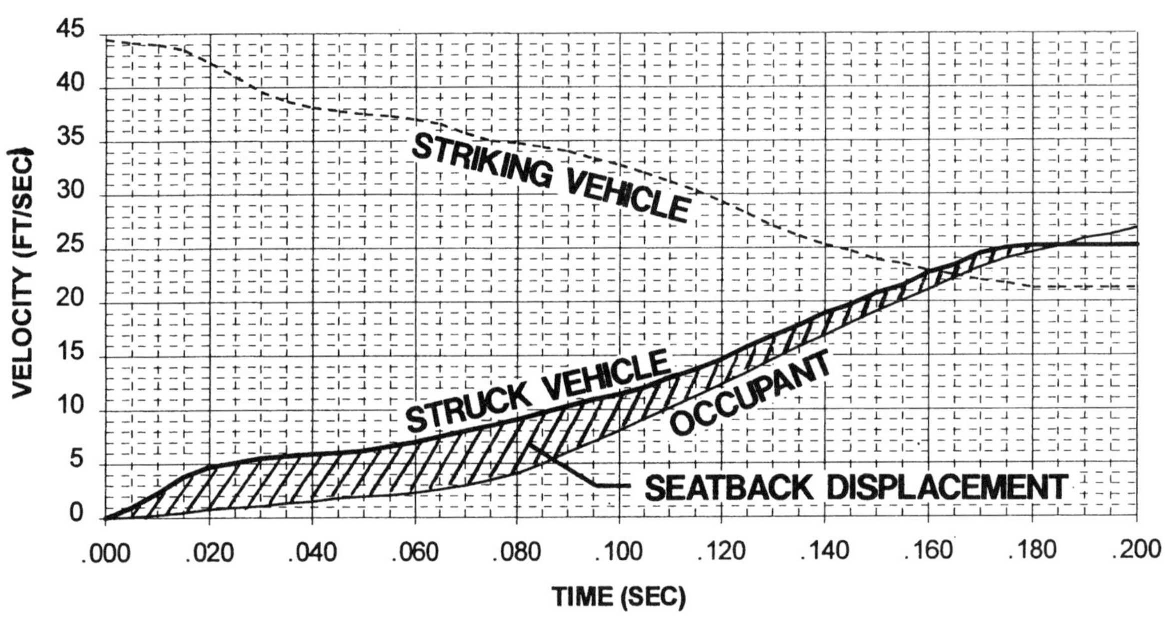

Figure 3C. UCLA Experiment 93.

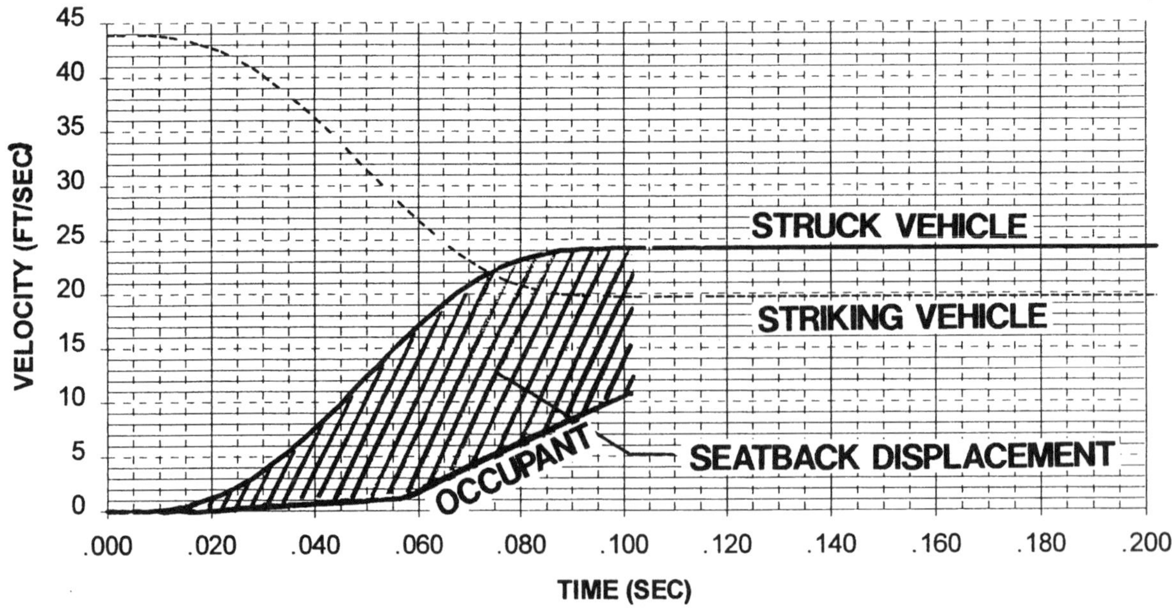

Figure 3D. Modern Compact Vehicle at X-93 Impact Velocity.

seatback yield characteristics determine the manner in which it manages occupant crash energy. The large and readily crushable rear-end passenger car structures of the type tested in UCLA Experiments 93 and 94 have in great part been replaced by smaller, more efficient structures having less available crush space and often with stiffer structures. When comparing collision performance of seats in the larger vehicles, Figs 3a and 3c, with seats in modern cars, Figs 3b and 3d, it can be seen that for a given seatback strength, the shorter crash pulse in the modern vehicle forces the seatback to yield an additional amount to compensate for the shorter distance traveled by the vehicle before it attains its full delta-V from the collision. Higher levels of vehicle structural stiffness provides increased protection from direct passenger compartment intrusion, but from a seat design standpoint, complicates the available design choices.

OCCUPANT RAMPING CONSIDERATIONS

Even when a seatback is sufficiently high to directly oppose rearward head motion at the start of a collision, its effectiveness in limiting head excursion may be compromised by occupant forced movement or **"ramping"** up the plane of the backrest. This condition may exist even in the presence of a substantially rigidified seatback. Under these ramping conditions, a rigidified seatback will provide additional restraint to the shoulders and upper torso while the unsupported head may be exposed to increased levels of angular accelerations and rotation. Without the maintenance of axial alignment of the cervical vertebrae, the seated occupant can be exposed to potentially life threatening cervical extension injuries. The ramping mechanism can be introduced by one or a combination of the following factors:

Ramping as a result of Seatback Geometry

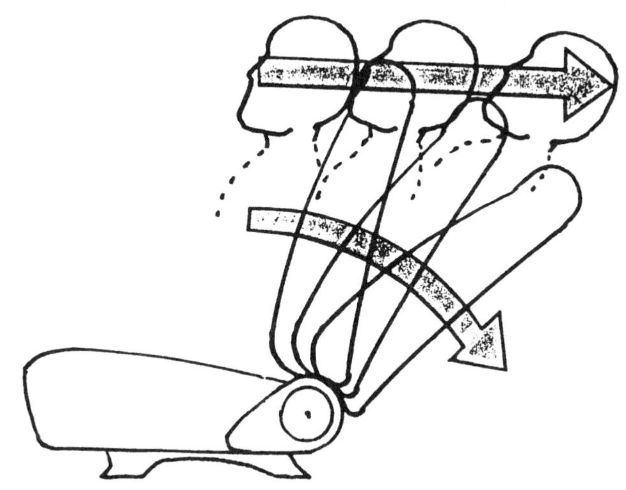

Figure 4. Seatback geometry affects Ramping Potential.

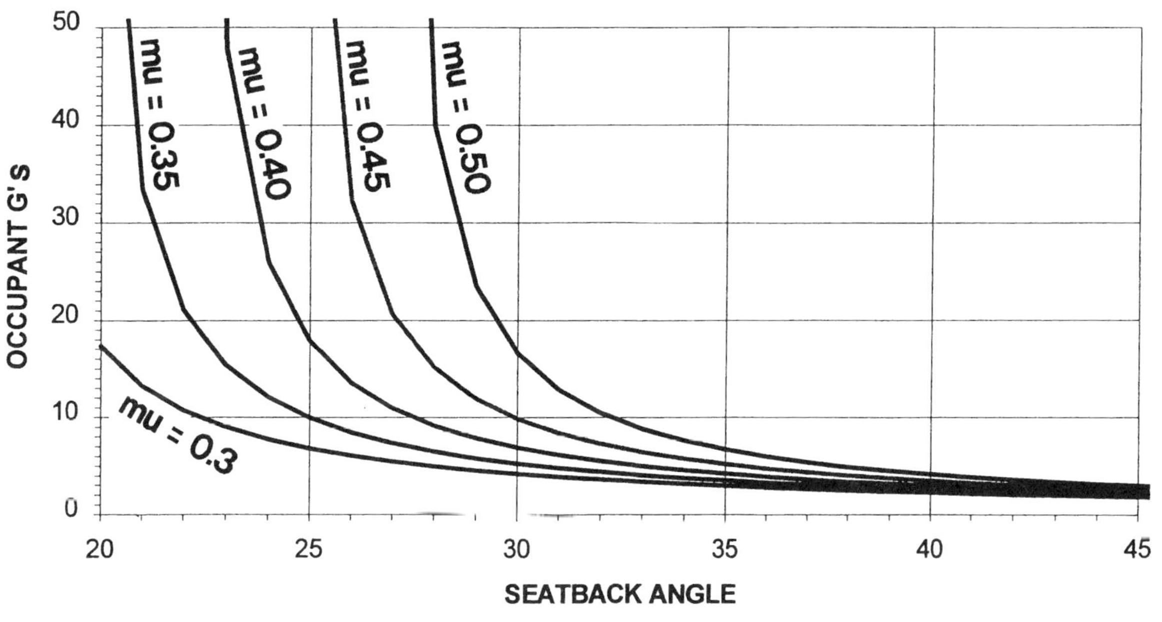

Figure 5. Ramping Forces as a function of Seatback Angle.

When a conventionally cantilevered seatback in a rear-ended car is deflected rearward by occupant and seat inertial loading, it pivots about a point near its base, causing the top of the seat to lower in elevation, Figure 4. Even if restrained by a seat belt, the occupant is caused to move more directly rearward, resulting in the effect of increased body elevation relative to the seatback and a tendency for the head to become unsupported.

Ramping From Non-Horizontal Collision Forces

Collisions between vehicles of mismatched geometry and off-road collisions with roadside objects often involve non planar forces that can displace occupants from their seats. Any downward collision force component that serves to depress the rear of the vehicle will tend to cause elevation or ramping of the occupant relative to the interior. This upward motion can take place irrespective of seatback strength characteristics or crash induced deflection angle.

Collision Forced Ramping on Inclined Plane of Seatback

During a rear-end collision, even if all collision forces are horizontal, components of the forces act to cause ramping. Gravity continues to operate, acting to hold the occupant in the seat; however, all collision forces have a geometric component that acts to cause forced movement of the occupant upward along the plane of the seatback., Figure 5.

Factors that determine the extent to which collision-forced ramping may occur include seatback angle, cushion compression characteristics, seat contour characteristics, seatback-to-body coefficient of friction and collision force magnitude.

OUT OF POSITION CONSIDERATIONS

The most frequent rear-end collision occurs when a momentarily distracted motorist does not correctly evaluate traffic conditions ahead until it is too late to avoid collision. Although brakes are often applied, the vehicle does not stop before contact occurs. In addition to the effect of sudden crash force application on the body of the struck car occupant, this condition presents several potential hazards. For example, driver, when stopped in traffic may take the opportunity to reach for something to the side or behind him. He may also react to the screeching of brakes by turning to look behind. In either case his body is not in an optimum position to resist the onset of collision forces. Such an out of position condition, although not necessarily especially dangerous for current yielding seatbacks could become significantly more serious if a rigidified seat structure provides a non-yielding fulcrum around which the motorist may become forcibly engaged, thus allowing a "karate chop" to the neck or exposure to extreme hyper-extension. An additional complicating factor occurs during multiple collisions. When the first impact is closely followed by being forced into the car ahead, or when the striking car is subsequently rear-ended, the opportunity for significant mis-positioning exists. Unless occupant freedom of movement within the compartment can be severely limited, rigidification of seats may, in many of these cases, increase rather than mitigate injury and may cause additional serious injuries in the low to moderate range of collision severity where few such injuries now occur.

OCCUPANT RESTRAINT CONSIDERATIONS

With current and future inflatable restraint implementation, with current seat belt use laws and with improved public awareness of active restraint system effectiveness, concerns related to uncontrolled occupant interaction with rigidified seat structures have been somewhat diminished. Unfortunately there are still a considerable number of unrestrained motorists.[13] Belt use percentage is even less favorable for rear-seated adult passengers.

Evaluation of rigidified seats must also consider the effect of unrestrained rear seat occupants. Rigidified front seats can act as a barrier to the forward movement of these occupants and may cause them to pivot upward to a potentially

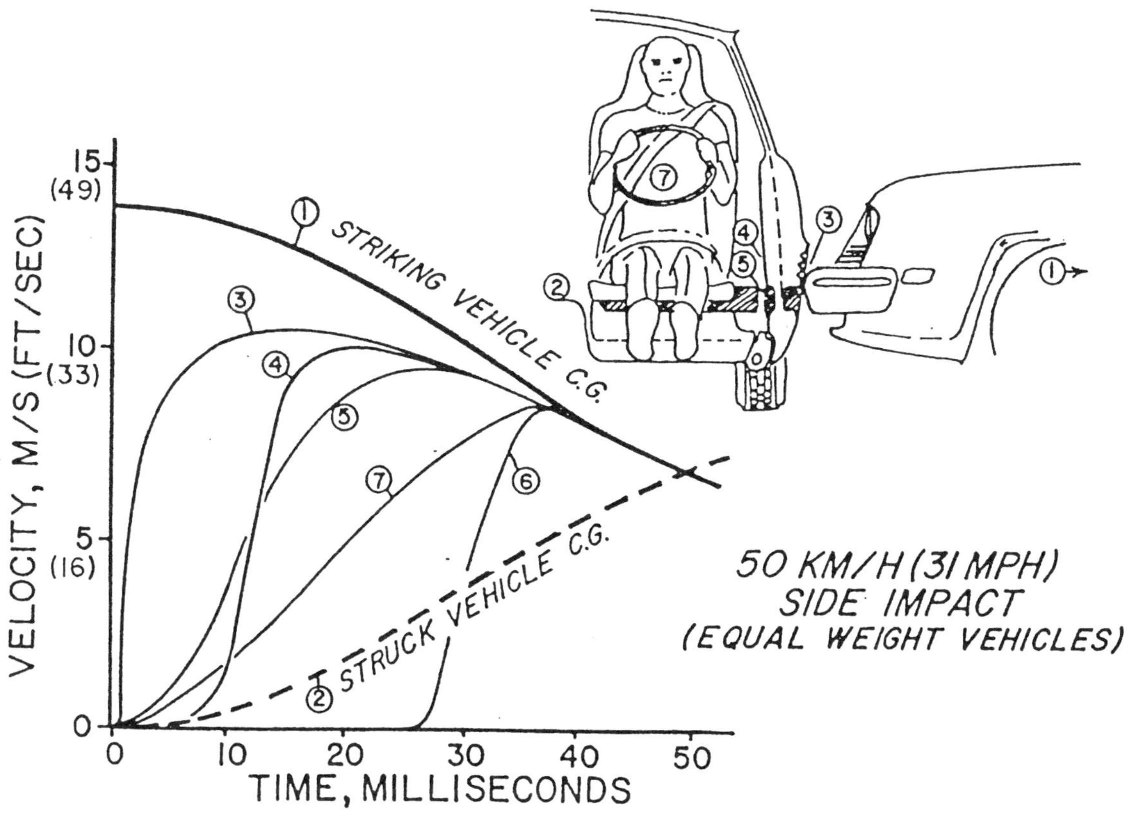

Figure 6. VT Graph for Side Impact With Utilization of Integrated Safety Seat Structure.

serious impact into roof structures. Past recommendations for rigidified seats have included provisions for pocketing the knees of these rear seat occupants in a specially designed front seat backrest structure so that they will not be forcibly ramped into the roof during a serious frontal collision.[14,15] Since the time these suggestions were made, however, concerns have been raised regarding the effect of knees forcibly interacting with restrained front seat occupants through this deforming structure. It has been suggested that such impingement into the backs of front seat occupants could result in serious spinal injury. Additional research would be required to evaluate this concern.

Conventional seat belt geometry allows substantial movement as the various slack-inducing agents operate during a collision. Notwithstanding this restraint compliance, it has been shown that seat belt use limits the rearward travel of front seat occupants in severe rear-end collisions and reduces the likelihood of roof and rear window header impact. The body is thus targeted into the more energy-absorbing areas of the rear seatback. Proposed restraint improvements, including direct shoulder belt attachment to the top of the backrest in conjunction with substantially rigidified seatbacks may additionally limit rearward occupant movement but will not entirely eliminate the ramping problem with its associated injury potential.

GENERAL SEAT DESIGN CONSIDERATIONS

Automotive design evolution has a history of building on proven concepts and procedures that have met the test of time in terms of practicality, durability, efficiency and buyer acceptance. Critics of automotive seat design have generally not been concerned with these factors in their laboratory attempts to utilize seat structures for new crash protection concepts. In other cases they have suggested drastic changes in seat structure to markedly increase seat strength without any significant research project accompanying the suggestions. Any attempt to implement these substantial seat redesigns, whether to create a high-G, integral safety seat or to merely increase backrest strength to some arbitrary higher level, must be preceded by careful analysis and by substantial additional research and testing, both from an engineering and a biomechanical standpoint.

Proponents of the current yielding seatback designs point out that these designs have proven effective over the years both from a utility and from a safety standpoint. When injury moderating effectiveness of current seat designs are studied by looking at crash experience of the past 25 years, it becomes evident that current and past seat designs have served well in protecting motorists from severe and fatal injuries.

Rear-end collisions account for approximately one third of the six million plus automotive collisions that occur annually in the US. In studies of field collision data [16,17], however, it was noted that less than five percent of all motor vehicle collision fatalities are from rear-end collisions and that both fatalities and severe injuries are underrepresented in the statistics. In a separate study [18] it was found that of 21,440 fatally injured front-outboard seated occupants of passenger cars in 1990, cars having rear impact damage accounted for only 3.2% of the total. The statistic given for seriously injured occupants of rear-ended cars (AIS-3 or above) was 3.7%. Since these statistics also include many collisions where seat design is not a factor, it is apparent that current seat designs perform well in protecting motorists.

Notwithstanding the developing evidence that current yielding seats provide good protection over a wide range of collision severities and that the arbitrary rigidification of seats may introduce more severe injury mechanisms than they mitigate, seat redesign should still be considered in the larger picture of occupant protection from all types of collision exposures. If the problems of out of position and unrestrained occupants can be overcome, the integrated safety seat concept could still provide many energy management advantages. The close proximity of the seat structure to the occupant allows improved restraint geometry if the seat can become the anchorage point for all portions of an active restraint. Additionally, energy management in side-impact collision exposures may be substantially improved if seats can contribute more effectively to the rigidification of side structures and can assist in moderating the rate of occupant "ride-up" during a collision, Figure 6. [19]

For the direct side-impact condition, it can be seen that a potential exists for accelerating the occupant of a wrap-around integrated safety seat in a less injury producing manner if he is accelerated earlier and thereby can take advantage of the available crush distance between the opposing vehicle and his torso. The potential for head-torso misalignment also exists for this crash configuration unless a means can be developed to accelerate the head with the torso. Unless it is made crash deployable, upper wrap-around seat structures that could maintain head-body alignment would tend to limit visibility in an unacceptable manner.

Additional research can provide a basis for determination as to the relative benefit of these types of seat redesigns. A determination must be made whether these potential benefits outweigh potential safety hazards as well as the practical disadvantages of increased cost, weight penalty and motorist discomfort before such changes are implemented into production vehicles by industry or are mandated by government.

REFERENCES

1. Hilton, Bernard C., "Design of Low Cost Seating for Effective Packaging of Vehicle Occupants", S.A.E. Paper No. 660797, Society of Automotive Engineers, Warrendale, Pa, 1966.

2. Severy, D.M., Brink, H.M., Baird, J.D., "Collision Performance, L.M. Safety Car", S.A.E. Paper No. 670458, Society of Automotive Engineers, Warrendale, Pa, 1967.

3. Severy, D.M., Brink, H.M., Baird, J.D.,

Blaisdell, D.M., "Safer Seat Designs", S.A.E. Paper No. 690812, Society of Automotive Engineers, Warrendale, Pa, 1969.

4. Melvin, J.W., "Occupant Protection in Rear End Collisions", S.A.E. Paper No. 720033, Society of Automotive Engineers, Warrendale, Pa, 1972.

5. Severy, D.M., Blaisdell, D.M., Kerkhoff, J.F., "Automotive Seat Design and Collision Performance", S.A.E. Paper No. 760810, Society of Automotive Engineers, Warrendale, Pa, 1976.

6. Severy, D.M., Mathewson, J.W. , Bechtol, C.O., "Controlled Automobile Rear-End Collisions, An Investigation of Related Engineering and Medical Phenomena", Canadian Services Medical Journal, 1955.

7. 49 CFR 571.202.

8. Severy, D.M., Brink, H.M., Baird, J.D., "Vehicle Design for Passenger Protection from High-Speed Rear-End Collisions", S.A.E. Paper No. 680774, Society of Automotive Engineers, Warrendale, Pa, 1968.

9. Strother, C.E., James, M.B., "Evaluation of Seat Back Strength and Seat Belt Effectiveness in Rear End Impacts", S.A.E. Paper No. 872214, Society of Automotive Engineers, Warrendale, Pa, 1987.

10. Warner, C.Y., Strother, C.E., James, M.B., Decker, R.L., "Occupant Protection in Rear End Collisions: II. The Role of Seat Back Deformation in Injury Reduction.", S.A.E. Paper No. 912914, Society of Automotive Engineers, Warrendale, Pa, 1991.

11. Refer to Reference 5.

12. Severy, D.M., Brink, H.M., Baird, J.D., "Backrest and Head Restraint Design for Rear-End Collision Protection", S.A.E. Paper No. 680079, Society of Automotive Engineers, Warrendale, Pa, 1968.

13. "Evaluation of the Effectiveness of Occupant Protection. FMVSS 208. Interim Report - June 1992", U.S. D.O.T. NHTSA, Washington, D.C., June 1992

14. Severy, D.M., Brink, H.M., Baird, J.D., "Passenger Protection from Front-End Impacts", S.A.E. Paper No. 690068, Society of Automotive Engineers, Warrendale, Pa, 1969.

15. Refer to Reference 5.

16. Strother, C.E., James, M.B., "Evaluation of Seat Back Strength and Seat Belt Effectiveness in Rear End Impacts", S.A.E. Paper No. 872214, Society of Automotive Engineers, Warrendale, Pa, 1987.

17. James, M.B., Strother, C.E., Warner, C.Y., Decker, R.L., Perl, T.R., "Occupant Protection in Rear-end Collisions: I. Safety Priorities and Seat Belt Effectiveness", S.A.E. Paper No. 912913, Society of Automotive Engineers, Warrendale, Pa, 1991.

18. National Highway Traffic Safety Administration Docket No. 89-20; Notice 3 Federal Motor Vehicle Safety Standards; Seating Systems; Head Restraints, 1992

19. Refer to Reference 5.

922521

Influence of Seatback Angle on Occupant Dynamics in Simulated Rear-End Impacts

David C. Viano
General Motors Research and Environmental Staff

ABSTRACT

In the early 1980's a series of tests was conducted simulating rear-end crashes. The tests demonstrated that a conventional automotive bucket seat adequately retains an unbelted dummy on the seat for rear-end impacts up to 6.4 m/s and 9.5 g severity. For this severity of impact the total rearward rotation of the seatback is less than 60° from the vertical and is associated with a normal acceleration of the dummy's chest into the seatback of up to 10 g. The tangential acceleration of the dummy, which may induce riding up the seat, was generally less than the normal component so that the occupant was prevented from sliding up the deflected seatback. The bucket seat provided adequate containment and control of occupant displacements for each of the initial seatback angles of 9°, 22°, and 35°. Thus, the performance of the seat was not significantly effected by the initial seatback angle when the impact was less than 6.4 m/s (9.5 g) and the total deflection of the seatback was less than 60°.

The bucket seat did not retain the dummy on the seat for rear-end impact simulations of 9.6 m/s and 15.5 g, irrespective of the initial seatback angle of 9°, 22°, or 35°. In these tests there was a total seatback deflection of greater than 60°. The tangential acceleration of the chest was comparable to the normal component of acceleration inducing deformation of the seatback. In this case there was sufficient tangential acceleration to allow the dummy to slide up the seatback and induce extension of the neck greater than 70° as the head rides above the integrated head restraint.

In a limited series of tests, the ability of the seat to adequately retain an occupant surrogate in rear-end impacts was directly associated with the severity of deceleration. Although the seatback may dynamically deform 20°-30° during occupant loading, the tangential acceleration of the occupant is not sufficient to overcome the normal compression of the occupant into the seat at a seatback angle less than 60°. However, in those simulations where the seatback angle was greater than 60°, there was enough tangential acceleration of the occupant to permit riding up the seatback with the potential for secondary impacts of the occupant with the automotive interior.

We also observed that extension of the neck occurs as the anthropomorphic dummy's head rides above the integrated headrest of the seatback. This occurs late in the impact simulation (100-120 ms) near the end of the whole body displacement of the occupant into and deflecting the seatback. More importantly, as the anthropomorphic dummy leaves the seatback, the head may be extended as much as 70° when the surrogate moves toward rearward components of the car interior.

INTRODUCTION

From the pioneering work of the 1950's and 1960's it became clear that the shape and stiffness of the seatback plays a significant role in protecting occupants during rear-end impacts. After extensive accident investigation and review of occupant dynamics in rear-end impact simulations, Severy [1] made the recommendation that the height of the seatback should be a minimum of 70 cm above the H point. This significantly extended the height of the seatback and was a motivating factor for the development of both headrests on the seatback and integrated headrests. Of course the overlying concern for improved design of seatback was motivated by whiplash injury during rear-end impacts. Subsequent to the introduction of headrests there were early claims of no apparent improvement in minimizing whiplash injury with head restraint [2]. However, the criticism was contradicted by numerous accident investigation studies showing a 10-20% reduction in both the incidence of whiplash injury [3,4] and the frequency of claimed neck injuries [5]. Additional match comparisons also showed that the integrated headrest was a superior performing structure during rear-end impact [6,7].

Basic studies of occupant kinematics during rear-end collisions have identified an interaction with the deforming seatback and the role of the yielding seatback [8]. Controlled deformation of the seatback was effective in improving occupant protection [6,9] and a yielding seatback was beneficial in high severity impacts [10]. However, more recent tests have shown that there is no obvious difference in occupant dynamics with the yielding seatback structure, since a matched comparison of rigid and deformable seatbacks showed that the incidence of neck injury was similar in cadaver specimen [11]. Even though it has been difficult to prove in laboratory studies that the yielding seatback improves occupant protection, early accident investigations showed that bending of the seatback or deformation of the seating track was associated with a lower frequency of neck injury in matched accidents [7]. These studies spawned the hope for improved occupant protection with the yielding seatback in the late 1960's and early 1970's [12]. However, the actual role of the yielding seatback in rear-end impacts is still debated.

Review of numerous surveys on the subject of mechanism of neck injury, occupant interaction with the deforming seatback, and factors associated with whiplash injury [13-15] indicates conflicting observation and opinion based on experimental and accident investigation data. It has also been difficult to assess occupant interaction with the seatback in terms of injury potential, even though it has been observed that injury of the neck is minimized by preventing the hyperextension angle to exceed 80° [16].

The most complete study of factors associated with injury in rear-end impacts [17] identified that the following factors were important to reducing potential injury in rear-end impacts. Melvin indicated that attempts should be made to minimize: 1) head displacement, rotation and acceleration, 2) differential motion of the head and torso into the deflected seatback, 3) occupant ramping up the seatback, and 4) occupant rebound. Based on these factors of performance he developed a deployable head restraint which was effective in rear-end impact simulations of 20 m/s and 40 g severity. Although Melvin's study showed the extreme to which occupant protection could be achieved in rear-end impact, the current study sought to provide more information on the potential kinematics of the car occupant's body, neck, and head in rear-end collision using laboratory sled conditions and an anthropomorphic surrogate.

Current accident statistics compiled by Data Link [18, 19] indicate that while 3.5% of fatalities and 7.6% of serious injuries occur in rear-end impacts, 23.3% of all injuries are related to rear-end crashes. The neck is involved in three-quarters of the cases, and the injury is typically related to a non-contact source. The most severe injuries are usually associated with relatively large changes in velocity of the struck vehicle, seatback deflection, and head impact. The possible consequence of cervical injury causing quadriplegia or paraplegia is of concern in the effort to prevent injury and improve the protection of passenger car occupants. However, this study does not specifically address whiplash injury [20, 21].

Sled tests were conducted in the early 1980's with a standard Part 572 dummy to investigate the automotive seat performance in rear-end impacts of increasing severity and to gain a basic insight into occupant kinematics. While the Part 572 dummy is known today to not have the biofidelity of the Hybrid III dummy, the study may provide a useful addition to the technical literature because it primarily focuses on occupant retention and not neck biomechanical responses. In general, the effects of the initial seatback angle of 9°, 22° and 35° from the vertical were investigated, particularly how it can influence the occupant dynamics and interaction with the deforming seatback and whether the seatback adequately retains the occupant and prevents surrogate rideup in a rear-end collision.

METHODOLOGY

In the early 1980's crash simulator tests were conducted on the General Motors Research Laboratories Hyge Sled using a range of impact severities from 4.2-9.6 m/s change in velocity simulating rear-end impact. An automotive bucket seat was rigidly attached to the sled fixture and the seatback modified for an initial angle of 9°, 22° or 35° from the vertical (Figure 1). The force-deflection response of the seat is given in Figure 8 of a companion paper [22] and is for a quasi-static loading of the seatback. A test matrix was developed (Table 1) to include similar rear-end impacts at each initial seatback angle. In all cases a loosely attached lap-belt was used to prevent the dummy from entirely leaving the seat structure should rideup of the seatback occur. Belt loads were not measured as they occurred late in crash sequence.

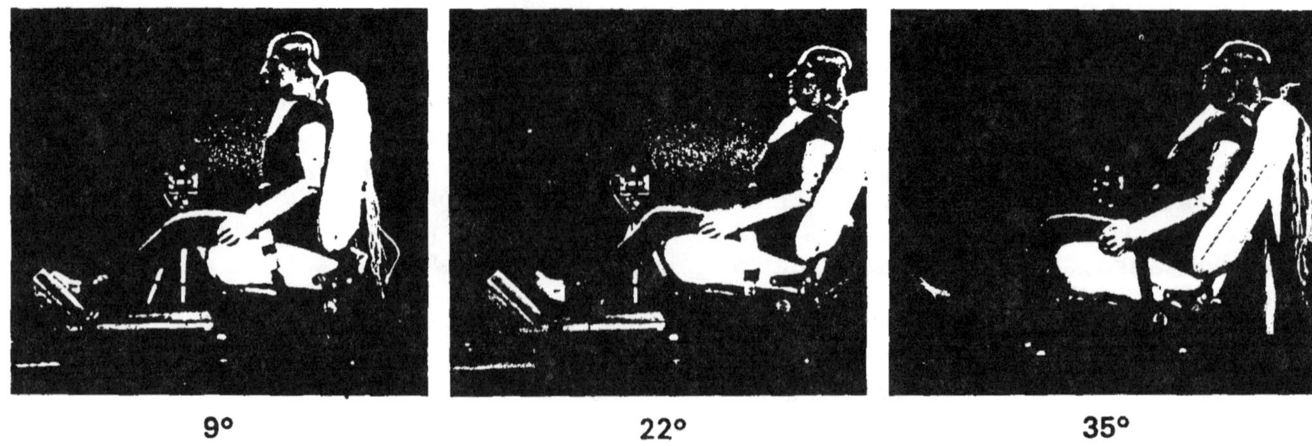

Figure 1: Sled test set-up with initial seatback angle of 9°, 22°, and 35° from vertical.

Table 1

Velocity Change (m/s)	4.2	6.4	9.6
Mean Acceleration (g)	5.5	9.5	15.5
Pulse Duration (ms)	135	100	85

Equivalent stopping distance 60-70 cm.

The Part 572 dummy was positioned forward facing on the bucket seat and was instrumented for biaxial acceleration of the head and chest. The joints were set to 1 g stiffness, the feet positioned on the inclined toe board, and arms and hands at the sides of the dummy thighs. Two high speed cameras were used to provide lateral coverage of occupant kinematics and interaction with the deflecting seatback. An onboard Photosonics 1B was used with a framing rate of 1,000 fps. A Red Lakes LoCam covered the gross occupant kinematics at a framing rate of 500 fps. Analysis of the high speed photographs were made to recover the time history of the deformation of the seatback, extension of the head/neck, and whole body displacement of the head and chest.

RESULTS

Peak values of responses for tests conducted at a 9°, 22°, or 35° initial seatback angle are shown in Table 2 as a function of increasing severity of deceleration. Test data are grouped according to the initial angle of the seatback. At each initial seatback angle an increase in severity of sled acceleration produces an increase in the maximum net induced rotation or deformation of the seatback. This is also associated with an increase in the total rotation of the chest, i.e., the posture of the dummy may be slightly greater in rotation than the total seatback angle because of nonuniform deformation of the seatback, compression of the dummy's upper torso into the elastic seat cushion, and upward movement of the hip relative to the torso.

Component acceleration of the chest and head are measured with reference to a moving center fixed to the dummy. In these tests the X acceleration component is normal and forward facing from the chest so that this acceleration component represents the effect of inertial loading of the chest normal to the seatback during initial loading. It is the major forcing function deforming the seatback. The other component of force inducing deflection of the seatback is the inertial component associated with the mass of the seatback itself. The Z component of chest acceleration is generally tangential to the orientation of the seatback and represents the forcing function causing the occupant to rideup the seatback. In most cases for low severity impacts (< 6.4 m/s and 9.5 g) the rideup acceleration is less than the normal component of acceleration causing seatback deformation. In this case the compression of the occupant into the seatback and the relatively high normal force results in restraint of the occupant in the seat as the total deflection of the seatback reaches up to 60°. However, for higher severity impacts (9.6 m/s and 15.5 g) the tangential rideup acceleration of the dummy becomes comparable and even exceeds the normal component compressing the dummy into the deforming seatback. In these situations the normal component of inertial load on the seatback caused more than 60° seatback deformation. The higher tangential accelerations and upward movement of the hips cause occupant rideup on the seatback. In these cases the occupant was not retained on the seat by the deformed seatback. Since the occupant would ride off of the seatback for the highest severity impact with the initial seatback angle of 35°, this test was not performed.

The resultant acceleration and displacement of the head were quite similar for each initial seatback angle for a particular impact severity. However, the components of displacement in both the X (horizontal) and Z (vertical) show a quite different trajectory of the head depending on the initial configuration of the seatback. In all cases of adequate retention, the total extension angle of the neck (rotation of the head with respect to the axis of the chest) was less than 60° (ranging from 24° to 52°). However, when the seat did not retain the occupant, rideup allowed the head to go above the top edge of the integrated head restraint and allowed greater extension of the neck. In these cases the total extension angle of greater than 70° (ranging from 70° to 86°) was observed. This means that the orientation of the head with respect to the torso, when ride-off of the seat occurs, approaches hyperextension -- a mode of greater vulnerability to neck injury if subsequent head impact occurs. However, the results are limited by the standard Part 572 neck, which does not have the biofidelity of the Hybrid III.

The overall trajectory of the head with respect to the horizontal and vertical planes of the test fixture are shown for increasing severity of impact for the three initial angles of seatback in Figure 2. As an example of the kinematics of the head during rear impact simulations of increasing severity note the 22° initial seatback angle responses. As the severity of impact increases there is more whole body displacement and rotation into the deforming seatback prior to rotational deformation of the seat i.e. there may be a 20 cm increase in rearward displacement of the head prior to substantial downward rotation of the seatback as the impact severity increases. In this case the total rearward displacement of the head increases from 33 to 82 cm with increasing severity of impact. There is also a greater downward rotation of the head from 12 to 50 cm.

Dynamics of the occupant interaction with the deflecting seatback during rear-end simulations is demonstrated by the time history of interaction responses in Figure 3. It is clear that up to about 100 or 120 ms the deformation of the seatback and chest are similar, and there is little or no extension of the neck. However, after this time the seatback reaches its maximum rotation as the chest of the occupant continues to rotate backwards slightly. This increased rotation of the chest is accompanied by extension of the neck i.e. rearward rotation of the head with respect to the axis of the chest. Because of the significant elasticity of the dummy neck extension returns back to 0 by about 220 ms. It is also quite clear that the pattern of occupant response is similar irrespective of the initial seatback angle in our test range of 9° to 35°.

Table 2

Sled Parameters		Seatback			Chest			
Velocity (m/s)	Mean Acceleration (g)	Initial Angle (°)	Max. Net Rotation (°)	Max. Gross Angle (°)	Maximum Rotation (°)	Acceleration X (g)	Z (g)	R (g)
4.2	5.5	9	26	35	32	7.0	2.5	7.4
6.4	9.5	9	34	43	46	10.0	5.5	11.4
9.6	15.5	9	52	61	100	15.0	15.0	21.2
4.2	5.5	22	13	35	30	7.0	5.0	8.6
6.4	9.5	22	24	46	54	10.0	7.5	12.5
9.6	15.5	22	46	68	88	16.0	17.0	23.4
4.2	5.5	35	20	55	37	3.0	4.0	5.0
6.4	9.5	35	24	59	42	10.0	7.0	12.2

Sled Parameters		Head							
Velocity (m/s)	Mean Acceleration (g)	Acceleration X (g)	Z (g)	R (g)	Displacement X (cm)	Z (cm)	R (cm)	Neck Extension (°)	Restrained by Seat
4.2	5.5	15.0	3.0	15.5	37.5	5.2	37.9	24	+
6.4	9.5	28.0	15.0	31.0	60.0	13.6	61.6	45	+
9.6	15.5	33.0	39.0	45.0	83.2	50.6	92.4	70	-
4.2	5.5	15.5	4.0	17.0	33.0	12.0	35.1	27	+
6.4	9.5	15.0	12.5	17.0	55.1	23.7	60.6	44	+
9.6	15.5	27.5	38.0	40.0	81.4	49.4	91.2	86	-
4.2	5.5	10.5	5.0	12.0	33.3	15.3	36.7	20	+
6.4	9.5	17.0	12.5	18.0	49.0	27.1	56.0	52	+

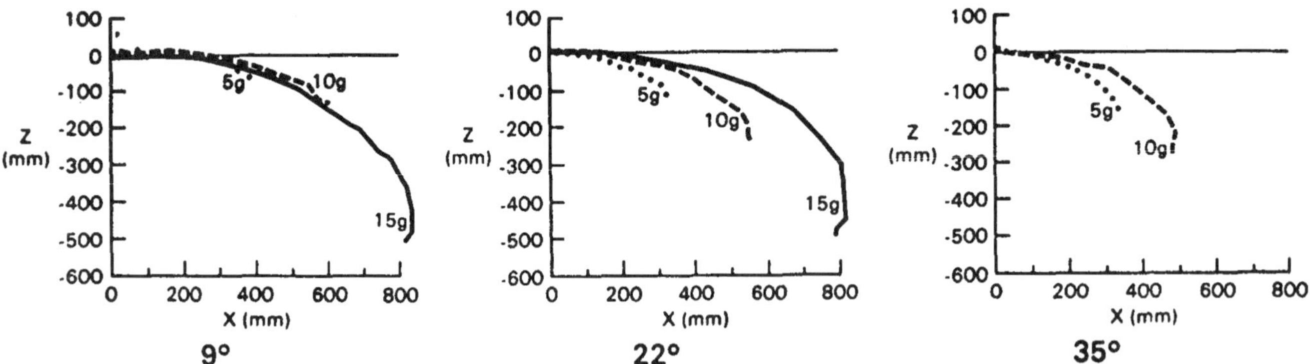

Figure 2: Displacement of the head in the midsagittal plane during rearend impact.

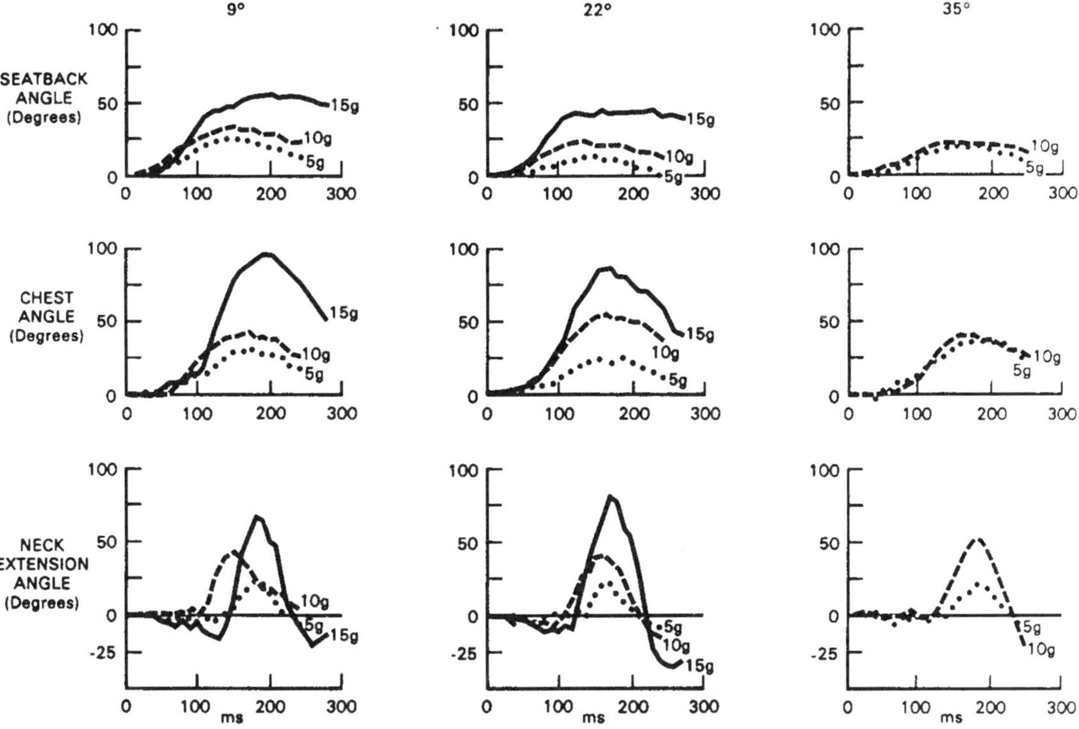

Figure 3: Occupant interaction with the seatback during rearend impact simulations is shown by time histories of increasing seatback angle, chest rotation, and neck extension.

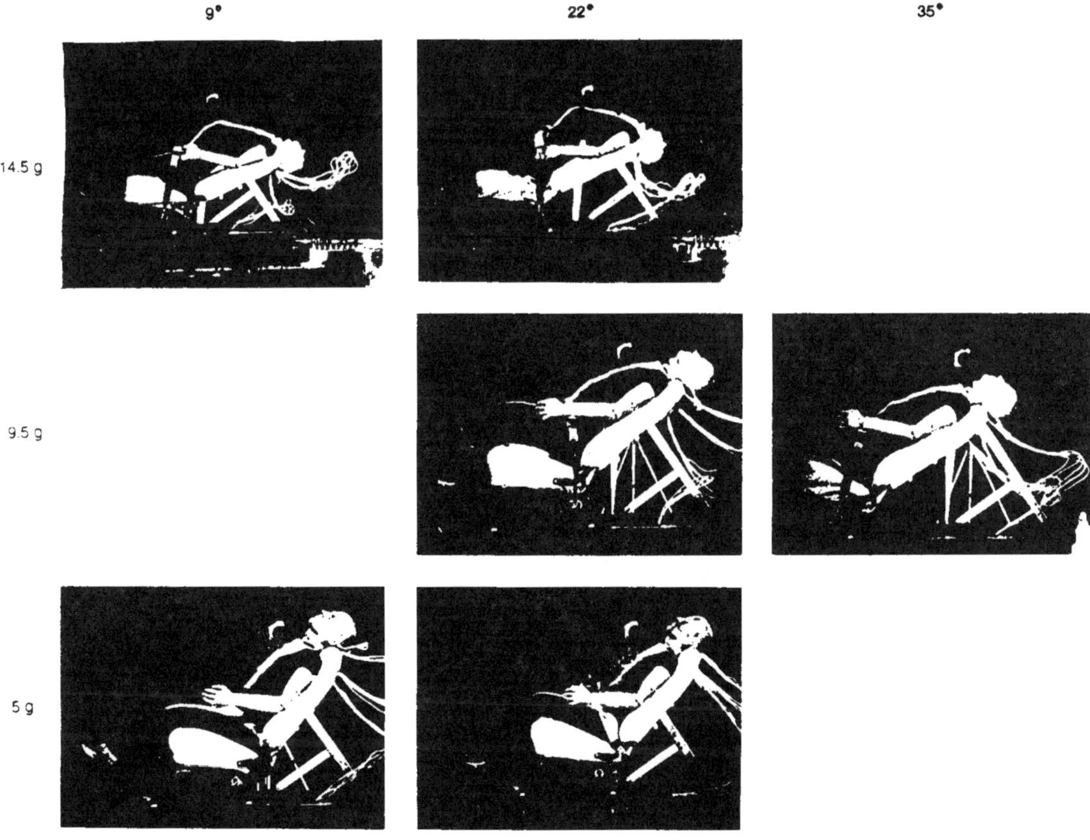

Figure 4: Occupant kinematics at time of maximum neck extension.

Selected photographs (Figure 4) at the time of maximum neck extension demonstrate that the neck extends over the deflected top edge of the integrated head restraint of the bucket seat. At this point and time the dummy can slide up the deflected seatback if there is significant total excursion of the seatback and the component of tangential acceleration of the chest is equal to or greater than the component normal to the seatback. This holds true unless the pelvis lifts too greatly off the seat causing a misalignment of the torso axis with the seatback. There was a modest rebound of the dummy, except when the occupant was not retained on the seat and there was subsequent restraint by tightening of the loose lap belt.

DISCUSSION

Occupant retention by the seat during rear-end impacts depends primarily on the severity of Delta V for a particular seat. In the low to moderate severity accelerations the occupant translates rearward and inertially loads the seatback. This loading coupled with the inertial response of the seat mass deforms the seatback rearward. In these exposures however the total rearward rotation of the seatback was less than 60° and the normal component of occupant acceleration into the seatback was greater than the tangential component. The latter acceleration was not sufficient to overcome friction and cause the occupant to ramp up the seatback.

In the moderate to high acceleration exposures the dummy translated and rotated rearward as the seatback deformed. In the present tests, the total deformation of the seatback was greater than 60° and was associated with a tangential component of occupant acceleration that was greater than the dynamic friction and the occupant rode up the seatback. With sufficient rideup the head was free to extend over the integrated headrest. The dummy kinematics resulted in greater than 70° of Part 572 dummy neck extension as the surrogate displaced rearward with respect to the deformed seat structure. Since the neck structure of the Part 572 dummy is stiffer than the Hybrid III, these extension angles are a conservative estimate of the neck response. When the anthropomorphic dummy is not retained on the seat as the seatback deforms, the occupant translates rearward with potential impact against interior components of the car.

As the head extended over the top of the integrated seatback and the dummy surrogate rode up and out of the seat, there was a possibility that the anthropomorphic dummy head could strike interior components while the head was in an unnatural configuration with respect to the chest. It was possible to assume that if similar kinematics occurred in real life situation, the head impact resulting in compression of the neck, while the neck is in extension, could result in neck injury at lower force levels than when in a more natural position.

Obviously the total kinematics of the occupant surrogate must be included in a detailed analysis of the occupant dynamics, and the unrealistic conditions of the rather primitive simulation of the anthropomorphic dummy head and neck structures must be considered before these observations can be projected into the real life situation. In particular, it would be important to recognize the timing of the rebound of the head and neck from extension as the head rides over the integrated headrest. There may be enough time for the head to rebound to a more normal posture prior to head contact with interior components. It would also appear that improved occupant protection may be afforded by reducing neck extension if the occupant rides off the seatback during rear-end impact.

REFERENCES

1. Severy, D.M., Brink, H.M. and Baird, J.D., "Rigid Seats with 28-in. Seatback Effectively Reduce Injuries in 30+ mph Rear-End Impacts." SAE Journal, SAE 69-532, Society of Automotive Engineers, Warrendale, PA, April 1969.
2. Fell, J.C., "Data Relevant to the Performance of Head Restraints in Collisions." DOT Report 82HO2590, Department of Transportation, Washington, DC, 1972.
3. States, J.D., and Balcerak, J.C., "The Effectiveness of Head Restraints in Rear-End Impacts. Final Report." DOT-HS-167-2-261, Department of Transportation, Washington, DC, 1973.
4. States, J.D., Balcerak, J.C., and Williams, J.S., et al, "Injury Frequency and Head Restraint Effectiveness in Rear-End Impact Accidents." In Proceedings of the 16th Stapp Car Crash Conference, SAE Technical Paper #720967, Society of Automotive Engineers, Warrendale, PA, 1972.
5. Haddon, W. Jr., O'Neill, B., Kelley, A.B., and Sorenson, W.W., "Automobile Head Restraints -- Frequency of Neck Injury Claims in Relation to the Presence of Head Restraints." *Amer J Pub Health*, 62:399-406, 1972.
6. Martinez, J.L., "Yielding Seatback is Effective in Protecting Car Occupant from Whiplash if Properly Damped." SAE Journal, Society of Automotive Engineers, Warrendale, PA, December, 1969.
7. Lange, W., "Mechanical and Physiological Response of the Human Cervical Vertebral Column to Severe Impacts Applied to the Torso." Max-Planck-Inst. Fur Arbeitsphysiologie, Dortmund, 1970.
8. Furusho, H., Yokoya, K., Nishino, S., and Fujiki, S., "Analysis of Occupants Movements in Rear-end Collision." *Bull JSAE*, n. 3, Society of Automotive Engineers, Warrendale, PA, 1971.

9. Martinez, J.L., and Garcia, D.J., "A Model for Whiplash." *J Biomech*, 1:23-82, January 1968.
10. States, J.D., Korn, M.W., and Masengill, J.B., "The Enigma of Whiplash Injuries." In Proceedings of the 13th Annual Conference of the American Association for Automotive Medicine, Association for the Advancement of Automotive Medicine, Des Plaines, IL, 1969.
11. Hu, A.S., Bean, S.P., and Zimmerman, R.M., "Response of Belted Dummy and Cadaver to Rear Impact." Report No. PSL-PR00848, Contract DOT-HS-5-01201, Monitor DOT-HS-803-028, Department of Transportation, Washington, DC, 1976.
12. "The Effects of Seatback Stiffness and Collision Severity on the Dynamic Behavior of the Head During 'Whiplash'." In Proceedings of the International Conference on the Biokinetics of Impacts, IRCOBI, Lyon, France, 1973.
13. McPherson, D. and Simpson, J., "Mechanisms of Spinal Cord Injury (SCI) in Vehicle Occupants." In Proceedings of the 21st Conference of the American Association for Automotive Medicine, Association for the Advancement of Automotive Medicine, Des Plaines, IL, 1977.
14. States, J.D., "Soft Tissue Injuries of the Neck." SAE-79-135, Society of Automotive Engineers, Warrendale, PA, 1979.
15. Degravelles, W.D. Jr., Kelley, J.H., Kenkel, W.F., and Burchinal, L.G., Whiplash Injuries. Charles C. Thomas, 1969.
16. Mertz, H.J. and Patrick, L.M., "Investigation of the Kinematics and Kinetics of Whiplash." SAE Transactions, Vol. 76, SAE Technical Paper #670919, Society of Automotive Engineers, Warrendale, PA, 1967.
17. Melvin, J.W. and McElhaney, J.H., "Occupant Protection in Rear-End Collisions." SAE Technical Paper #720033, Society of Automotive Engineers, Warrendale, PA, 1972.
18. Data Link Inc., "Car Crash Outcomes in Rear Impacts. Appendix A to Current Issues of Occupant Protection in Car Rear Impacts." Data Link Inc., Washington, DC, 1989.
19. Data Link Inc., "Current Issues of Occupant Protection in Car Rear Impacts." Prepared for the Office of Crashworthiness, Rulemaking, NHTSA. Docket 89-20-NO1-021, Data Link Inc., Washington, DC, 1990.
20. Hohl, M., "Soft Tissue Injuries of the Neck in Automobile Accidents." *J Bone Joint Surg*, 56A:1675-1686, 1974.
21. Viano, D.C., "Restraint of a Belted or Unbelted Occupant by the Seat In Rear-End Impacts." In Proceedings of the 36th Stapp Car Crash Conference, SAE Technical Paper #922522, Society of Automotive Engineers, Warrendale, PA, 1992.

922522

Restraint of a Belted or Unbelted Occupant by the Seat in Rear-End Impacts

David C. Viano
General Motors Research and Environmental Staff

ABSTRACT

This sled test series involved occupant loading of the seat in rear crashes of 4.3-8.3 m/s (9.6-18.5 mph). The tests were conducted in the early 1980s and involved an unbelted or lap-shoulder belted Part 572 dummy in rear and oblique rear impacts. The research is reported today to provide comparative data for the record and serves as a control benchmark for more current technologies and safety research methodologies on seat performance in rear crashes. Safety belts improved occupant retention on the seat primarily by the lap belt reducing the upward and rearward movement of the pelvis. Tests were also conducted on the mechanisms for energy absorption by seatback deflection.

INTRODUCTION

Sled tests conducted in the early 1980s showed that an unbelted dummy rode up and off the seatback in rear-end crashes when the total rotation of the seatback exceeded 60° [1]. The tests also demonstrated that the severity of sled velocity was the primary factor related to the possibility that the unbelted dummy would or would not be retained on the seat. In those tests the initial seatback angle was varied from 9°-35° and was not a significant factor in occupant retention.

Since the reported tests [1] were conducted at only three sled velocities and utilized an early version bucket seat, another series of tests was conducted in 1981. These tests were written up recently, and involved a wider range of sled velocities, a newer version bucket seat, and both unbelted and lap-shoulder belted occupants. In addition, the belted passenger tests were conducted simulating a pure rearward impact and an orientation 15° oblique of pure rear. The study emphasized the potential role of seatback deflection in retaining the occupant on the seat in rear-end crashes and in partial energy-absorption during a crash, and explored the potential benefits of safety belt use enhancing occupant retention on the seat.

While the tests were conducted over ten years ago with the Part 572 dummy, the data remain unduplicated in the literature today. The information is being reported at this time to provide comparative data for the record and serve as a control benchmark for more current technologies and safety research methodologies on rear-end crash safety research conducted at General Motors Research Laboratories. This research, although conducted years ago, may help pave the way for more current studies on the role of the seat system in belted and unbelted occupant protection. This paper will not attempt to provide a comprehensive review of contemporaneous research on occupant responses and seat performance in rear-end crashes. The interested reader is referred to several recent papers [3-5] to obtain a broader understanding of the subject.

Another stimulus for the current reporting of the research occurred in a recent study of fatal crashes of belt restrained occupants [2]. That study explored the potential for further gains in belted occupant safety, and the nature and circumstances of the fatal crashes. One observation from the study was the uniqueness and violence of many fatal crashes frequently involving high speed. There were several cases in which the belted driver was killed in 4, 5, or 6 o'clock oblique rear-end crashes involving a heavy truck. In these cases, the crash severity and oblique direction of occupant loading caused both bending and twisting of the seatback, and excursion of the driver's head toward the zone of crash deformation. The twisting response of the seatback was reminiscent of the data obtained in this study with oblique loading by the occupant into the seat. These factors stimulated the current report.

METHODOLOGY

Hyge sled tests were conducted in 1981 on the General Motors Research Laboratories facility. An open fixture was utilized to maximize photographic coverage (Figure 1). Rear-end crashes were simulated over a range in sled velocities from 4.2-8.3 m/s (9.3-18.5 mph). An automotive bucket seat, identified as B-2, was used in these tests in contrast to earlier tests with another type of bucket seat, identified as B-1 [1]. The bucket seat was rigidly attached to a flat plate and the seatback positioned in the normal rearward orientation. A new seat was used for each test.

The first series involved a wide range of sled severities. The mean acceleration was approximately 12 g's. The desired velocity was obtained by adjusting the piston stroke. This modified the pulse duration. For these tests, an instrumented Part 572 dummy was positioned on the seat and a loose lap belt was draped over the legs to prevent the dummy from entirely leaving the seat and test fixture in the more severe tests with significant seatback ride-up or ramping.

The second series involved a lap-shoulder belted dummy simulating the right-front passenger. The restraint system was rigidly attached to the sled fixture, and loads were measured in the left and right segments of the lap belt. Tests were conducted using a range in sled velocity from 4.3-8.3 m/s (9.6-18.5 mph). These speeds were achieved using a mean acceleration of 9 or 13 g's. The fixture was oriented to simulate a pure rearward or 15° from rear oblique direction. The pure rear impact represents a 6 o'clock direction of rear loading, and the oblique direction was representative of a 6 - 7 o'clock loading on the left-rear panel of a vehicle.

Biaxial accelerations were recorded from the head and chest of the Part 572 dummy. The test series did not address neck responses which represent an important consideration in the overall interaction of an occupant with the seat. This study specifically focused on occupant retention on the seat. The research was also conducted before the wider availability of the Hybrid III dummy, which has more human-like responses.

Two high-speed cameras were used to obtain photographic coverage of lateral occupant kinematics. One camera was attached to the sled and had a framing rate of 1000 fps. An off-board camera provided broad coverage at a framing rate of 500 fps. Film analysis was conducted from the on-board view to determine rotations of the head, neck, chest and seatback as well as displacements of the head and chest.

RESULTS AND DISCUSSION

Table 1 summarizes the test results from the two series of sled tests on the rear-end impact response of a seated dummy. The first series involved an unbelted dummy exposed to impact velocities ranging from 4.6-8.3 m/s (10.2-18.5 mph) in severity. In all but the highest severity test, the unbelted dummy was retained by the seat. The second series of tests involved a lap-shoulder belted passenger-side dummy. The direction of impact was rearward or 15° toward the driver side simulating a slight oblique orientation. Response data were obtained from analysis of high-speed movies or transducers in the dummy.

BELTED VERSUS UNBELTED RESPONSES

Figure 2 demonstrates the kinematic differences for the unbelted and lap-shoulder belted dummy in three severities of impact. While in all cases the dummy was retained on the seat, there is an obvious increase in the height of the pelvis without the lap belt. Figure 3 provides further explanation for the difference. The maximum seatback angle is similar for both conditions of restraint, but the angle of the chest is significantly greater without safety belt use. The chest has over a 30° greater angle--is more horizontal--in the highest severity test. Since the overall displacement of the head is similar, the hip has risen substantially higher without belt use. The photographs in Figure 2 confirm the difference.

The angle of the seatback and chest increase with the severity of the sled velocity. There seems to be a reasonable relationship between increasing crash energy, seatback deflection, and occupant ride-up of the seatback. Analysis of the dummy kinematic and biomechanical responses indicates that sled velocity was much more important than average sled acceleration. Therefore, the 9 g and 13 g mean acceleration test data were merged for purposes of analysis.

Lap belt loads are given in Table 1 and demonstrate the restraint action provided through belt use in rear-end crashes. There is an increase in belt tension with sled velocity and the greater severity of the exposure. As indicated in the kinematic analysis, one benefit of the lap belt is to hold the occupant down as the angle of the seatback tends to promote ride-up under crash conditions.

FURTHER ANALYSIS OF UNBELTED RESPONSES

Figure 4 summarizes the results of film analysis of the movement of the dummy and seatback in rear-end crashes of increasing severity. The unbelted dummy experiences differential rotation of the head, chest and neck. Data on seatback rotation is also included for comparison with occupant motions.

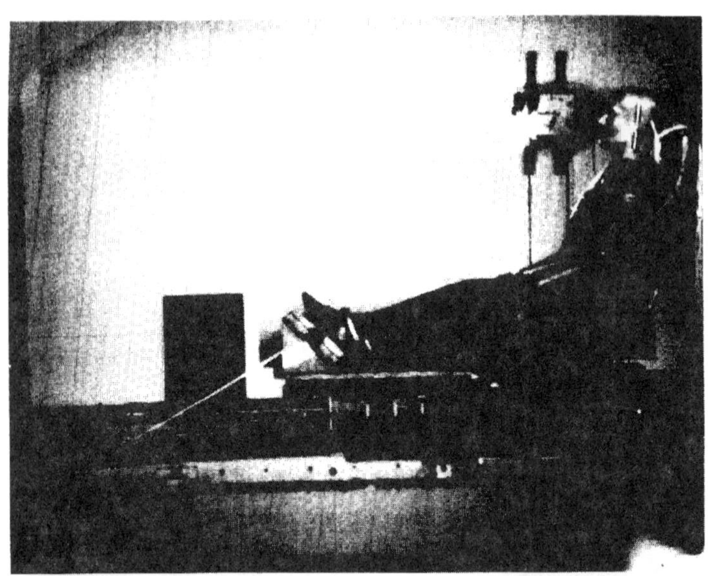

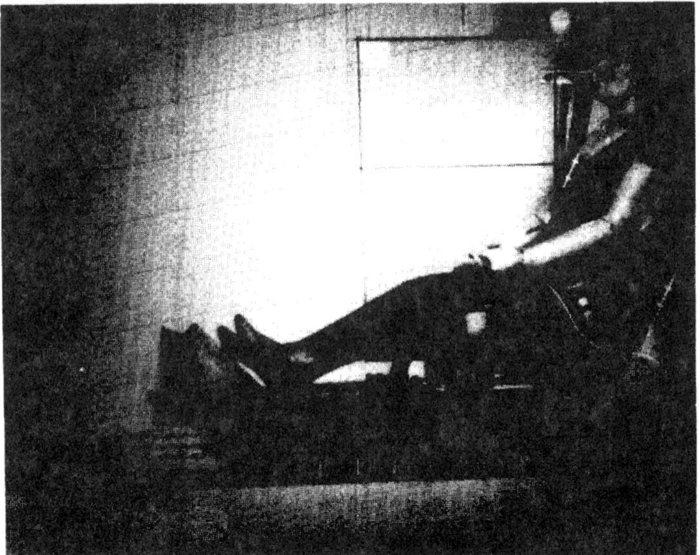

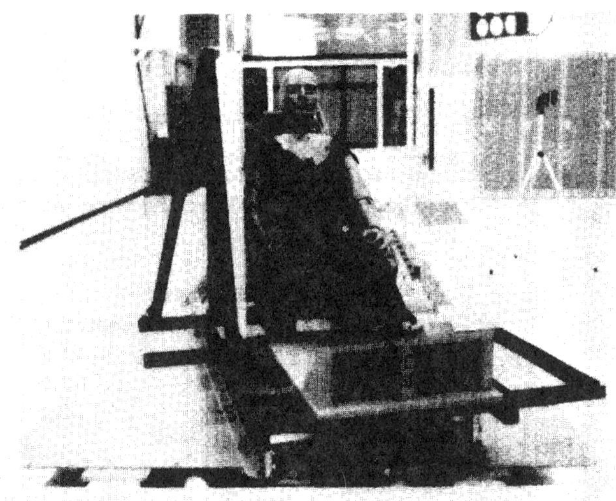

Figure 1: Set-up photographs of the Hyge sled tests involving rear impacts. Top left: Rear impact with an unbelted Part 572 dummy, Bottom left: Rear impact with a lap-shoulder belted dummy, Right: Oblique impact at 15° from rear with a lap-shoulder belted dummy.

Table 1
Rear Impact Sled Tests

Test #	Sled Velocity m/s	Head Accel x-g	z-g	Res-g	Head Displ x-cm	z-cm	Res-cm	Neck Ext Deg
	UNBELTED DUMMY							
	Pure Rearward Impact at 12 g's							
248	4.60	14.0	7.0	16.0	46.3	13.3	48.2	29
253	5.35	11.5	8.5	12.5	51.4	21.4	55.7	27
250	5.45	11.0	9.0	12.5	51.6	21.5	55.9	32
251	6.25	9.0	11.0	11.0	59.4	29.2	66.2	30
252	6.38	8.0	11.0	12.0	60.9	29.5	67.7	30
255	6.52	8.0	11.0	12.0	63.6	34.7	72.5	27
249	7.31	6.5	14.0	14.0	70.1	45.5	83.6	22
254	7.32	7.0	12.0	12.5	70.1	46.0	83.9	25
256	8.33	7.0	12.5	13.5	89.5	50.1	102.6	23

Note: These tests were conducted on a Hyge sled with an acceleration level of 12g. The initial seatback angle varied from 22-26 degrees. These tests involved an unbelted dummy.

Test #	Sled Velocity m/s	Head Accel x-g	z-g	Res-g	Head Displ x-cm	z-cm	Res-cm	Neck Ext Deg
	BELTED DUMMY							
	Pure Rearward Impact at 9 g's							
343	4.3	12.5	7.2	12.5	54.7	29.2	62.0	28
342	5.9	10.0	10.0	10.0	67.8	44.2	81.0	23
341	7.7	7.0	12.5	12.5	79.5	54.4	96.3	23
	Pure Rearward Impact at 13 g's							
331	4.8	17.5	7.5	17.5	52.3	26.7	58.7	36
332	4.8	14.5	7.5	14.8	58.2	29.9	65.4	28
333	6.2	14.0	12.0	14.5	67.5	41.9	79.4	23
334	8.3	12.5	17.5	17.5	81.5	56.5	99.1	21
	15 Degrees from Rearward Impact at 9 g's							
338	4.8	12.0	6.3	12.5	65.4	32.8	73.2	27
339	6.3	9.5	9.5	10.0	76.7	46.7	89.8	24
340	8.3	7.0	12.0	12.0	88.6	65.6	110.2	18
	15 Degrees from Rearward Impact at 13 g's							
337	4.8	12.5	10.0	13.0	62.5	27.5	68.3	25
336	6.3	12.5	10.0	12.5	76.0	44.8	88.2	21
335	8.3	10.0	12.5	12.5	94.0	55.9	109.4	20

Note: These tests were conducted with a lap-shoulder belted dummy on a bucket seat. The tets were run in a pure rearward and 15 degrees from rear orientation. Loads were measured in the lap belt of the three-point restraint system.

Table 1 (continued)
Rear Impact Sled Tests

Test #	Chest Accel x-g	z-g	Res-g	Chest Angle deg	Seatback Angle deg	Occupant Retention	Belt Loads Lt lb	Rt lb
	UNBELTED DUMMY							
	Pure rearward Impact at 12 g's							
248	8.0	5.0	9.4	39	45	Yes	--	--
253	8.0	5.0	9.4	53	52	Yes	--	--
250	7.5	5.0	9.0	55	53	Yes	--	--
251	7.5	5.0	9.0	68	59	Yes	--	--
252	8.0	5.0	9.4	69	60	Yes	--	--
255	7.0	5.0	8.6	73	66	Yes	--	--
249	7.5	5.0	9.0	87	72	Yes	--	--
254	7.5	6.0	9.6	87	71	Yes	--	--
256	6.0	5.0	7.8	96	66	No	--	--
	BELTED DUMMY							
	Pure Rearward Impact at 9 g's							
343	5.0	5.1	7.1	39	56	Yes	50	50
342	5.0	6.2	8.0	57	66	Yes	80	55
341	4.0	9.5	10.3	64	82	Yes	375	350
	Pure Rearward Impact at 13 g's							
331	7.5	6.3	9.8	38	40	Yes	125	100
332	7.5	7.0	10.3	38	43	Yes		
333	7.5	7.5	10.6	49	52	Yes	175	150
334	7.5	7.5	10.6	63	72	Yes	360	370
	15 Degrees from Rearward Impact at 9 g's							
338	6.0	2.8	6.6	34	54	Yes	100	150
339	5.0	6.0	7.8	52	56	Yes	100	180
340	4.0	7.5	8.5	64	69	Yes	325	400
	15 Degrees from Rearward Impact at 13 g's							
337	7.0	5.5	8.9	35	42	Yes	100	100
336	6.3	7.0	9.4	54	59	Yes	150	150
335	6.0	7.5	9.6	64	63	Yes	375	475

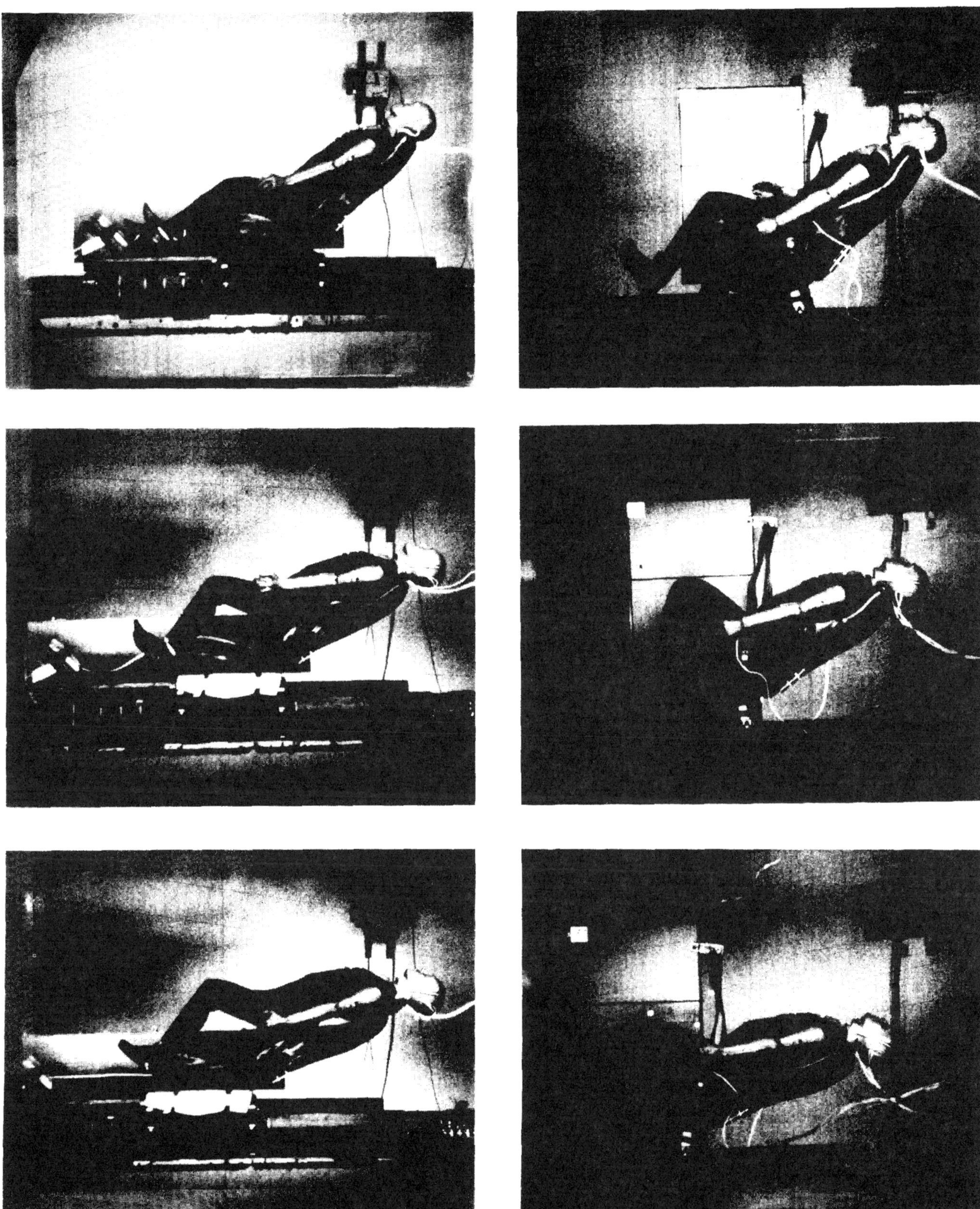

Figure 2: Photographs from high-speed movies of the rear impact kinematics of the Part 572 dummy. Left side involves unbelted responses: Top Test 248 at 4.6 m/s, Middle Test 255 at 6.5 m/s, and Bottom Test 256 at 8.3 m/s. Right side involves belted responses: Top Test 331 at 4.8 m/s, Middle Test 333 at 6.2 m/s and Bottom Test 334 at 8.3 m/s.

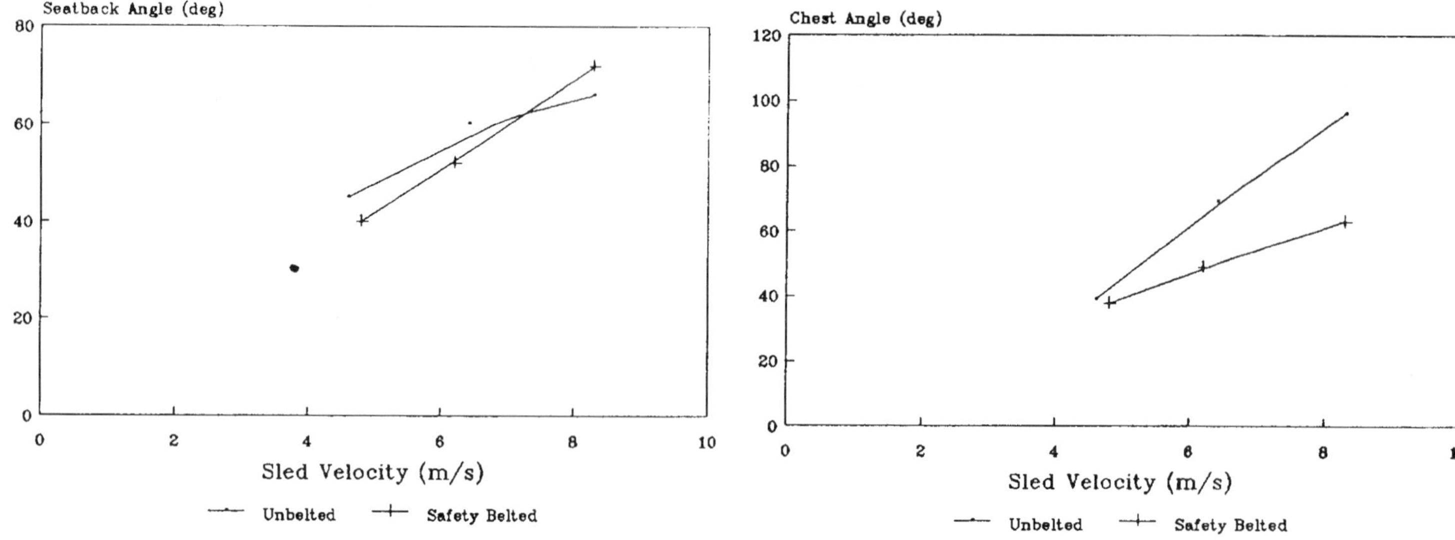

Figure 3: Peak angle changes in the rear impact tests for the Part 572 dummy chest and seatback as a function of sled velocity.

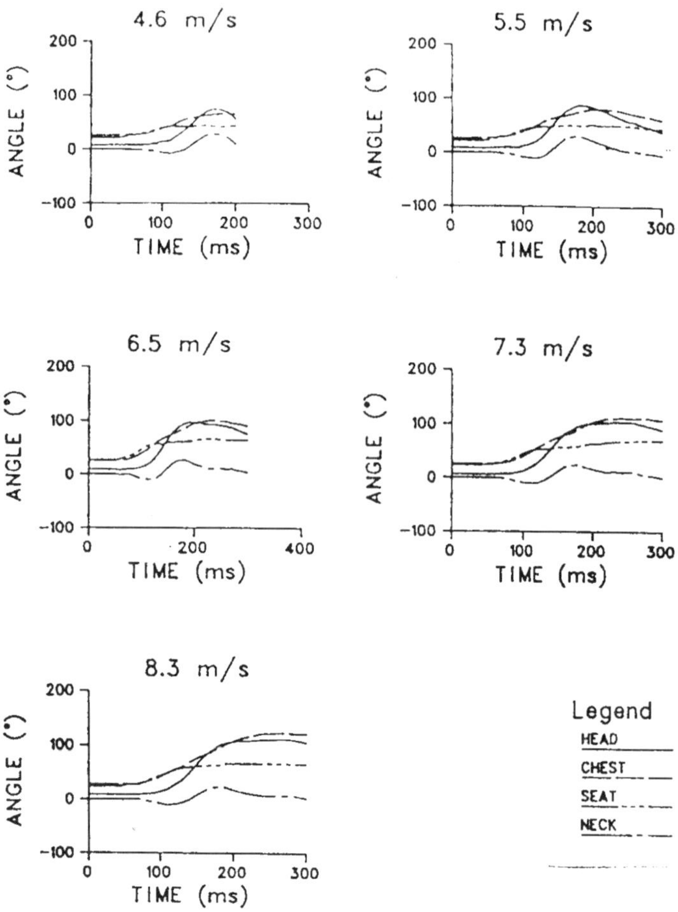

Figure 4: Angle changes for increasing severity of sled velocity for an unbelted Part 572 dummy in rear impacts.

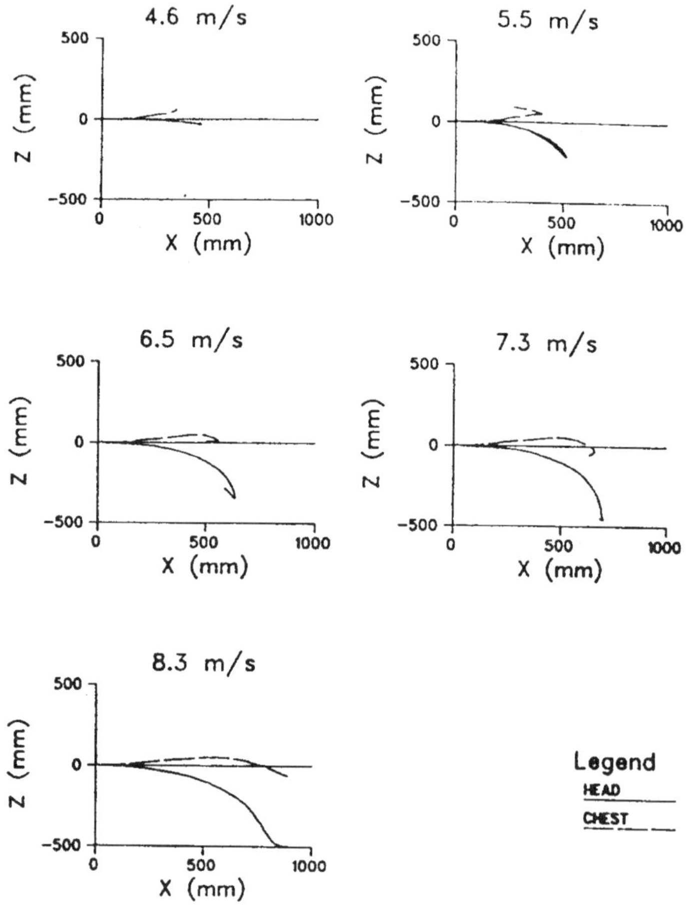

Figure 5: Trajectory information for the head and chest of an unbelted Part 572 dummy in rear impacts.

Figure 5 provides information on the horizontal and vertical displacement of the head and chest. These plots confirm a greater motion with increasing severity of the crash.

Figure 6 plots the velocity of the head as a function of time for the increasing severity of sled velocity. The speed increases from 2.8-4.4 m/s (6.2-9.8 mph) at the maximum. Since the dummy was not restrained by the seatback in the highest severity test, the head velocity represents an estimate of the speed that the occupant would move into the rear compartment of the vehicle, if secondary tethers were not employed in the test or in an analogous vehicle test.

COMPARISON OF CURRENT AND PREVIOUS UNBELTED DUMMY TESTS

A greater degree of occupant restraint was provided by the B-2 bucket seats used in the current study. Previous tests [1] were conducted in the 1980's with an earlier model bucket seat identified as B-1. These tests showed that while restraint was provided in the lower severity tests, loss of occupant retention was observed when the horizontal accelerations of the dummy exceeded the restraint provide by the dynamic friction between the occupant and the seatback. The increased interaction in the current study may be the result of a combination of effects, including greater compliance of the seatback in pocketing the dummy during rearward loading and greater resistance to rearward bending of the seatback. The latter would increase the energy absorbed by seat deflection for the same rotation of the seatback.

In these tests the dummy interacted for a longer period of time with the seatback during the impact, thus increasing the dynamic effects of friction, which act to restrain the occupant. This is a benefit and enhances occupant retention in the seat. However, there was no difference in the two seats when occupant retention is plotted as a function of the severity of the sled velocity (Figure 7).

MECHANISM FOR ENERGY ABSORPTION BY SEATBACK DEFLECTION

The construction of seat B-1 includes a seatback with side supports that pivot about a bolt attachment on both sides of the seat frame. In the normal position, the seatback is supported against two metal posts which protrude from the seat frame. Upon rearward loading, seatback rotation deforms the posts allowing controlled deflection of the seatback. With increasing force, the posts deform more until the seatback disengages from contact with the bent posts. In some tests, there was deformation of the connection supporting the seatback on the seat pivot mechanism.

In contrast, seat B-2 had a different construction and included a single side recliner mechanism. Energy absorption by seatback deflection was provided through tearing of a metal bracket in the recliner mechanism and bending of the sheetmetal frame. However, these EA deformations provided greater levels of resistance to rearward bending than observed in the B-1 seats.

STATIC REAR PULL TESTS WITH A SEATED DUMMY

Figure 8 summarizes the static force-deflection response of seats B-1 and B-2. Load was applied through a seatbelt wrapped around the dummy. While the two bucket seats provide similar resistance to rearward pull during the initial deflection of the seatback, the B-1 seat has substantially high load resistance for deflections greater than 35°. The ultimate load was 2900 N versus 2450 N (15.5% lower) with the B-2 seat. However, the energy absorbed by the B-2 seat was actually greater than with the B-1 type because of a greater ultimate deflection. The final unloaded angle of the B-1 seat was 40° compared with 52° for the B-2 seat.

PURE REARWARD VERSUS 15° OBLIQUE REARWARD IMPACTS

Figure 9 demonstrates belted occupant kinematics in the pure rear and the 15° oblique rear impacts at three severities of sled velocity. In all cases of oblique loading, the belted dummy is retained on the bucket seat. However, some twisting of the upper body and twisting of the seatback about its long axis are apparent. The B-2 bucket seat includes a modest wing near the upper arm, which may help retain the dummy in the oblique impacts.

Figure 10 shows similar rotations of the seatback and chest angle for the pure rearward and 15° oblique tests at three sled severities. Deflections increase in both conditions with greater impact energies. This demonstrates similar occupant interaction with the seatback for the two orientations of impact.

Figure 11 demonstrates differences in maximum displacement of the head for each severity of sled velocity. The oblique tests resulted in an offset which averaged 9.2 ± 3.4 cm for comparable tests. This indicates greater excursion of the head and is consistent with the observed twisting of the seatback due to torque related to an unsymmetric loading as the dummy rides over the edge of the seat. Dummy head and chest accelerations seemed to be reasonably similar for the two loading orientations, at least within the variability of the responses recorded.

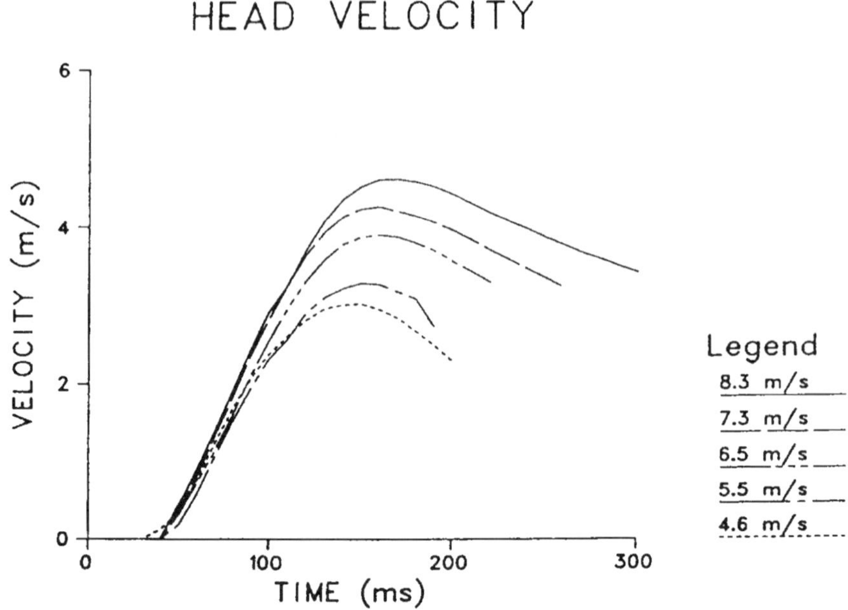

Figure 6: Head velocity from the unbelted Part 572 dummy for increasing severity of sled velocity.

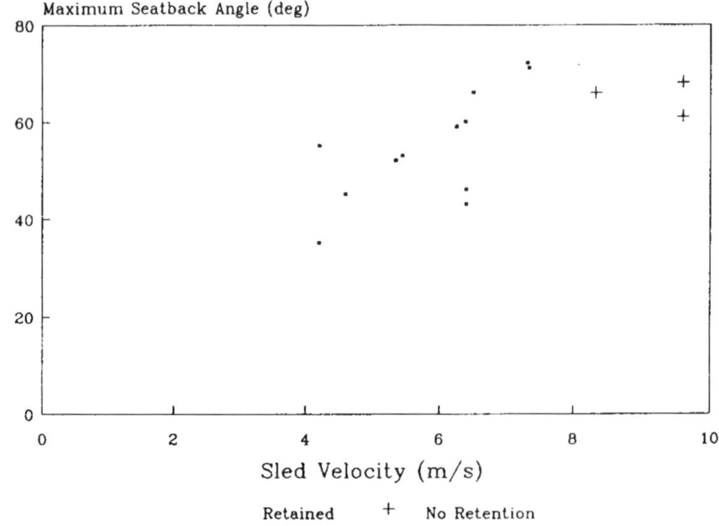

Figure 7: Maximum seatback angle for increasing severity of sled velocity in relation to retention of the Part 572 dummy in rear impacts.

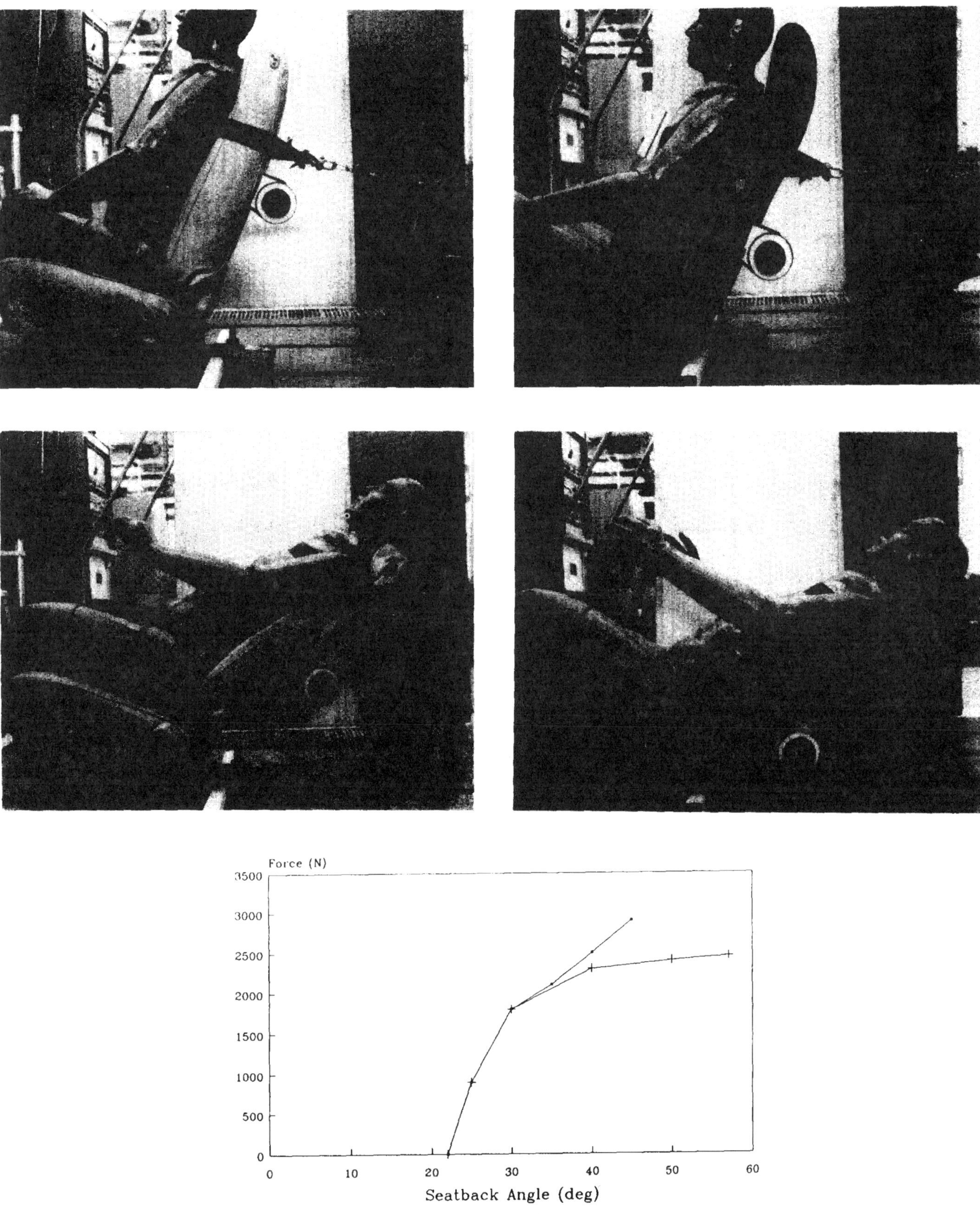

Figure 8: Rear pull tests on the B-1 (Left side) and B-2 (Right side) bucket seats, and Bottom: Force-rotation response related to rearward loading of the seatback.

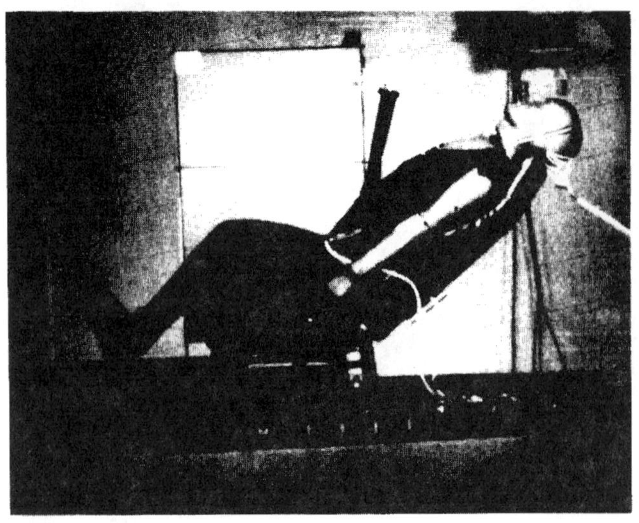

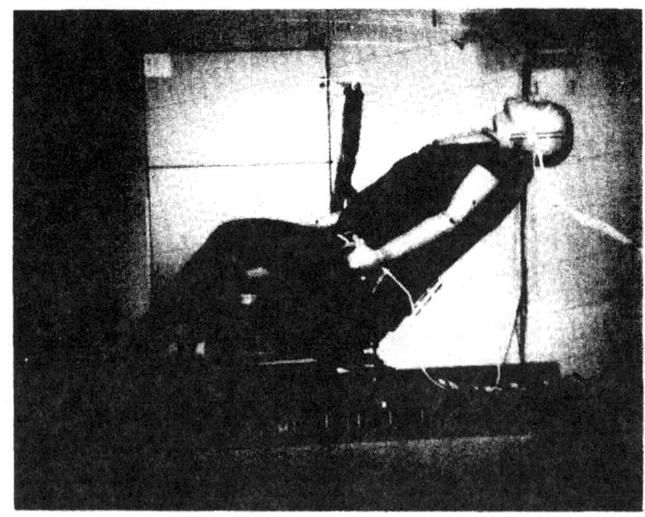

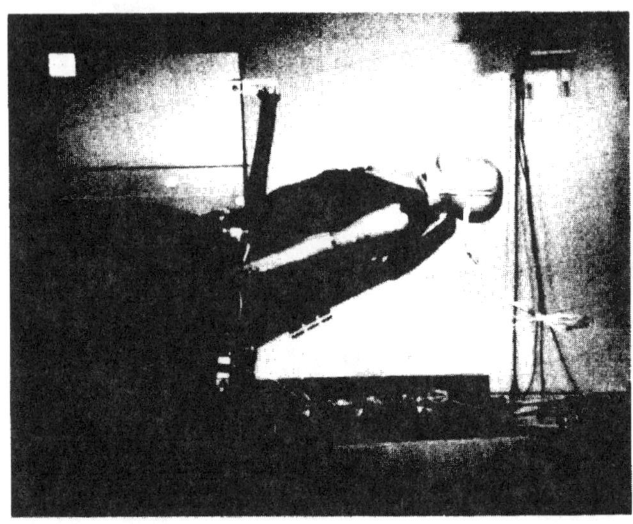

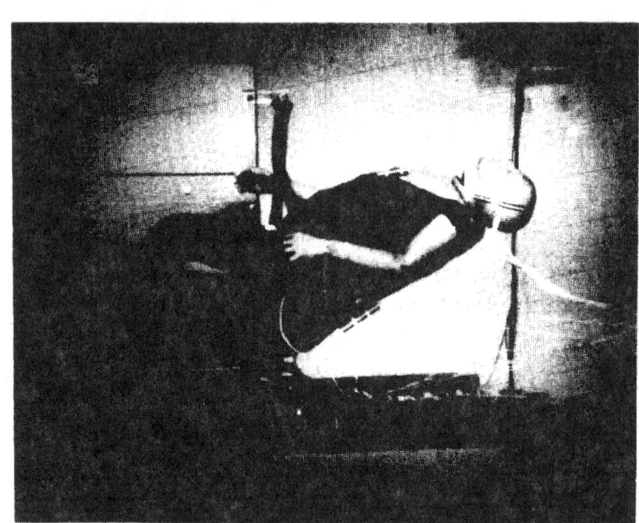

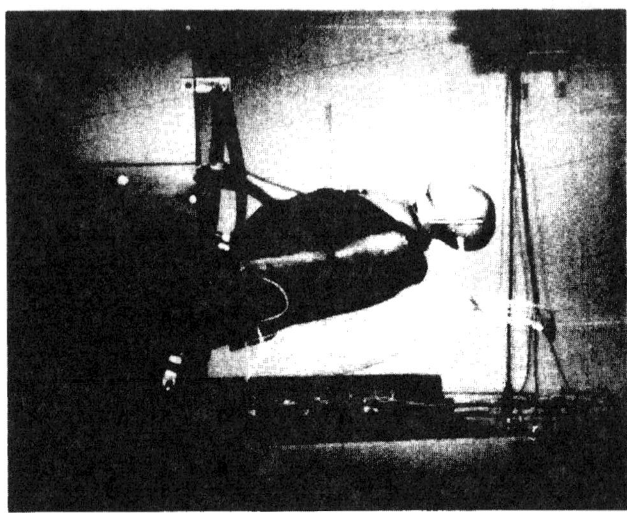

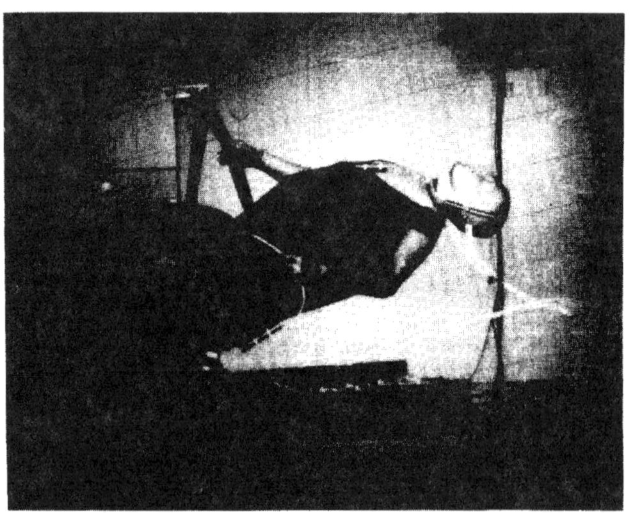

Figure 9: Photographs from high-speed movies of the rear impact kinematics of the Part 572 dummy. Left side involves belted responses for rear impacts: Top Test 343 at 4.3 m/s, Middle Test 342 at 5.9 m/s, and Bottom Test 341 at 7.7 m/s. Right side involves belted responses at 15° oblique rear impact: Top Test 337 at 4.8 m/s, Middle Test 336 at 6.3 m/s, and Bottom Test 335 at 8.3 m/s.

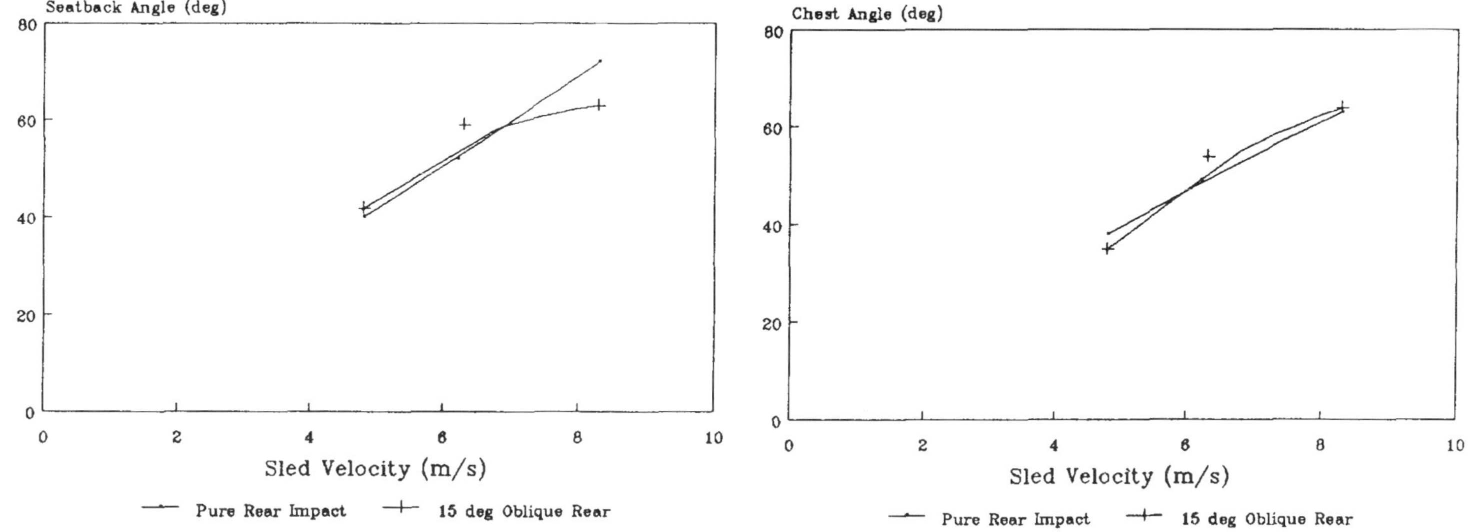

Figure 10: Comparison of maximum seatback and chest angle changes for the rear and 15° oblique rear impact tests with the belted Part 572 dummy.

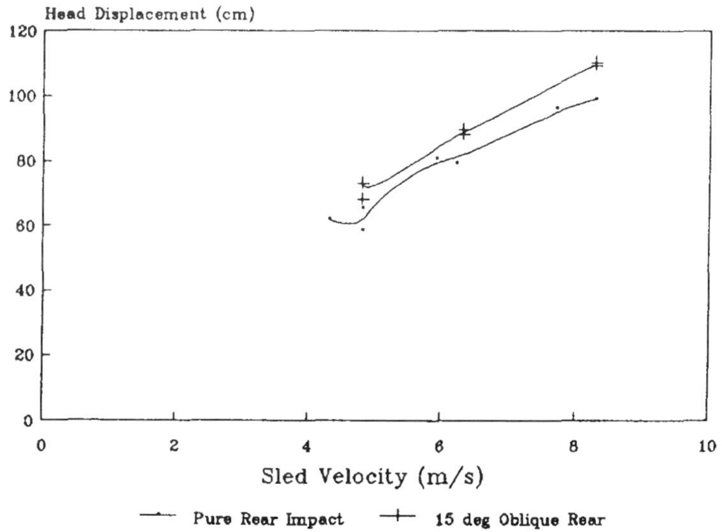

Figure 11: Maximum head displacement for the rear and 15° oblique rear impact tests with the Part 572 dummy.

CONCLUSIONS

In 1981, two series of Hyge sled tests were conducted simulating rear-end crashes. The first series involved an unbelted dummy on a bucket seat exposed to 4.6-8.3 m/s (10.2-18.5 mph) sled velocities. The two test series utilized a newer bucket seat which provided greater retention of the dummy than observed in the first experiments [1]. In all cases, retention occurred for dynamic seatback deflections greater than 60° rearward rotation. Loss of occupant retention occurred only in the highest severity test. As in the previous series, there was a gradual and controlled deformation of the seatback. With this seat, the EA was provided by tearing of metal in the recliner mechanism and bending of sheetmetal in the seat frame, as opposed to bending of the support posts of the seatback in the earlier series. The mechanisms of energy absorption were studied in additional static tests.

In this series of tests, the determining factor for the unbelted occupant to ramp up and off the seatback seems to be sled velocity. This relates to the severity of the exposure and the energy that must be managed by deformation of the seatback. Different mean accelerations of the sled did not have a significant effect on dummy responses but were a factor in seatback rotation for the same sled velocity.

Safety belt use improved occupant retention and seatback interactions for sled velocities of 4.3-8.3 m/s (9.6-18.5 mph). While the maximum displacement of the head was similar for belted and unbelted occupants in comparable tests, chest rotation was substantially lower with belt use. This correlates to the lap belt preventing upward displacement of the pelvis. Chest and head accelerations were slightly greater with safety belt use for similar sled velocities, indicating greater occupant restraint. There was better kinematic control with lap belt loads on the pelvis.

Tests at a 15° oblique rear impact direction demonstrated that safety belts provide important pelvic restraint enhancing occupant interactions with the seatback. In these tests the occupant not only induces bending loads on the seatback but also twisting. The oblique loading slightly increased head excursions but lowered rearward rotation of the seatback. However, adequate occupant restraint was provided in all tests. Lap belt loads increased with the severity of the sled velocity. The two test series demonstrated substantial restraint and energy absorption by controlled seatback deformation in rear-end crashes. The precise mechanism of EA varied with the seat type and deformations. The interactions provided restraining forces until sufficient rotation and twisting of the seatback led to loss of occupant retention in the simulated rear-end collisions.

ACKNOWLEDGEMENT

The tests reported in this study were conducted in the early 1980s by Richard Madeira, now retired from the GM Research Laboratories, Ed Jedrzejczak and others in the Biomedical Science Department. Their help in conducting the tests and analyzing the film and test data is greatly appreciated.

REFERENCES

1. Viano, D.C., "Influence of Seatback Angle on Occupant Dynamics in Simulated Rear-End Impacts." In Proceedings of the 36th Stapp Car Crash Conference, SAE Technical Paper #922521, Society of Automotive Engineers, Warrendale, PA, 1992.
2. Viano, D.C., "Crash Injury Prevention: A Case Study of Fatal Crashes of Lap-Shoulder Belted Occupants." In Proceedings of the 36th Stapp Car Crash Conference, SAE Technical Paper #922523, Society of Automotive Engineers, Warrendale, PA, 1992.
3. Strother, C.E. and James, M.B., "Evaluation of Seat Back Strength and Seat Belt Effectiveness in Rear End Impacts." In Proceedings of the 31st Stapp Car Crash Conference, pp. 225-244, SAE Technical Paper #872214, Society of Automotive Engineers, Warrendale, PA, 1987.
4. James, M.B., Strother, C.E., Warner, C.Y., Decker, R.L., and Peri, T.R., "Occupant Protection in Rear-End Collisions: I. Safety Priorities and Seat Belt Effectiveness." In Proceedings of the 35th Stapp Car Crash Conference, pp. 369-377, SAE Technical Paper #912913, Society of Automotive Engineers, Warrendale, PA, 1991.
5. Warner, C.Y., Strother, C.E., James, M.B., and Decker, R.L., "Occupant Protection in Rear-End Collisions: II. The Role of Seat Back Deformation in Injury Reduction." In Proceedings of the 35th Stapp Car Crash Conference, pp. 379-389, SAE Technical Paper #912914, Society of Automotive Engineers, Warrendale, PA, 1991.

912913

Occupant Protection in Rear-end Collisions: I. Safety Priorities and Seat Belt Effectiveness

Michael B. James, Charles E. Strother, Charles Y. Warner, Robin L. Decker, and Thomas R. Perl
Collision Safety Engrg., Inc.

ABSTRACT

Recent detailed field accident data are examined with regard to injuries associated with rear impacts. The distribution of "Societal Harm" associated with various injury mechanisms is presented, and used to evaluate the performance of current seat back and restraint system designs. Deformation associated with seat back yield is shown to be beneficial in reducing overall Societal Harm in rear impacts. The Societal Harm associated with ejection and contact with the vehicle rear interior (the two injury mechanisms addressed by a rigid seat approach), is shown to be minimal. The field accident data also confirm that restraint usage in rear impacts has a substantial injury-reducing effect. Laboratory tests and computer simulations were run to investigate the mechanism by which seat belts protect occupants in rear impacts.

INTRODUCTION

In 1987, Collision Safety Engineering presented a paper concerning the performance of production front seats and active seat belt systems in rear impacts [Strother, 1987]. Accident statistics from the National Crash Severity Study (NCSS) and the National Accident Sampling System (NASS) were analyzed to understand the relative priority of rear impacts and evaluate the performance of production seats in the rear end crash mode. The effectiveness of seat belts in rear impacts was investigated by reviewing conclusions reached by investigators who had analyzed both restrained and unrestrained subjects in dynamic tests. At that time, the only accident field study available addressing the issue of seat belt effectiveness in rear impacts indicated that belts reduced the incidence of any injury by 12 percent and decreased the incidence of serious injury by over 57 percent.

Since 1987, several studies examining field accident data on rear impacts have been published. These studies provide more detailed insights into the relative importance of this accident mode, as well as information on how occupants are injured in such collisions. They also provide further additional understanding of the role of energy absorbing seat backs in injury mitigation and the potential consequences of rigidized seat structures.

The beneficial effects of seat belt use in rear impacts is also consistently demonstrated by field accident studies conducted during recent years. These studies are in general agreement with each other concerning not only the benefit of seat belt use in rear collisions, but also in the magnitude of the effectiveness of belts in this impact situation.

The mechanism by which seat belts restrain occupants in rear collisions is shown to be due to the forces acting on an occupant's lower body during such an impact. The results of a series of rear impact dynamic tests using cadaver subjects are examined and used as a basis for mathematical occupant simulations based on the Motor Vehicle Manufacturer's Association (MVMA) 2-Dimensional Crash Victim Simulator Program. These simulations demonstrate the effect of a seating position more representative of production automobiles than that employed in the original cadaver tests. In both the cadaver tests and the simulations, the lower body of a belted occupant in a rear impact is observed to articulate such that the lap belt is retained in position on the hips during the collision, limiting the rearward and upward trajectory motion of the torso and head.

REAR IMPACT FIELD ACCIDENT STUDIES

Early field accident data from Fatal Accident Reporting System (FARS) and NCSS indicated that rear impacts deserve a low safety priority by comparison to other accident modes. The 1970-1985 FARS data showed that less than 4% of occupant fatalities occur in impacts with a Principal Direction of Force (PDOF) of 5-7 o'clock. The NCSS data suggested that part of the explanation for the relatively low percentage of serious injuries and fatalities in rear impacts was the generally lower distribution of accident severities in this mode. However, NCSS data also indicated that for a given impact severity the probability of serious injury was significantly less for rear impacts than for side or frontal impacts. This finding

*Information in brackets refers to references listed at end of paper.

suggested that the vehicle protective system, including the vehicle seats, was providing a high level of protection in rear impacts [Strother, 1987].

Recently, investigators have made more detailed analysis of rear impact field accident data. In December of 1987, the National Highway Traffic Safety Administration (NHTSA) analyzed the 1982-1986 FARS data [Partyka, 1987]. The focus of the study was to compare fatality rates in various sizes of vehicles struck in the rear. Not surprisingly, it was concluded that the fatality rate for small cars (i.e. cars weighing less than 3000 pounds) was higher than the fatality rate for larger cars, but overall fatalities for rear-impacted passenger cars were found to be less than 3% of the total (see Figure 1).

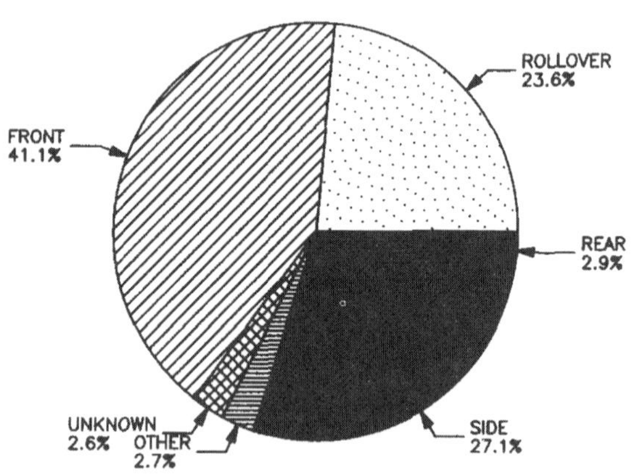

FIGURE 1: FATALITIES BY COLLISION TYPE, FARS 82–86 PASSENGER CAR OCCUPANT FATALITIES [Partyka, 1987]

Additional conclusions may be drawn from the data presented in the 1987 Partyka study. While FARS does not have a coded measure of impact severity (such as the impact velocity change (delta-V)), it is possible to approximate severity levels from other coded data. For example, the majority of rear impact fatalities occurred in a struck vehicle which was stopped, where relatively large closing velocities were more likely. Over 90 percent of the rear-impacted vehicles with fatalities were coded as having "disabling/severe" damage. The rear impact fatalities generally occurred in accidents where a larger vehicle impacted the subject vehicle. Thirty percent of the fatalities were in cars struck by large trucks, and over forty-one percent were in cars struck by either pickup trucks, vans, or large cars.

FARS does have a variable for "travel speed," although there is no basis for estimating its accuracy because there is no accident reconstruction procedure in FARS, as there is in NCSS and NASS. Further, this variable was coded "unknown" for over half of the vehicles. Recognizing these limitations, the FARS data indicated that over 83 percent of the rear impact fatalities occurred in accidents where the striking vehicle was travelling over 40 miles-per-hour. It is also interesting to note that in over 25 percent of the rear impact fatalities in two-vehicle accidents, there was some indication of alcohol involvement on the part of the striking vehicle driver.

Another rear impact study was conducted by Data Link, Inc. in July of 1989 [Data Link, 1989]. The accident data in both FARS and NASS were evaluated. The FARS data indicated that rear end impacts account for 4.4 percent of all fatal crashes involving passenger cars and 3.5 percent of all passenger car occupant fatalities. Like the earlier Partyka study, the Data Link analysis of the FARS data indicated that smaller cars have higher casualty rates than larger cars in rear impacts. This trend is generally true of all accident modes, but in rear impacts the sensitivity to car size was found to be even more pronounced. Further, the number of car fatalities for all impact modes was seen to be decreasing from 1978 to 1988, the absolute rates for fatalities in rear impacts were observed to remain relatively constant.

Table 1, taken from the Data Link study, combines the NASS data with the FARS data. The proportion of injured occupants involved in rear impacts is relatively high. While only 12.2 percent of crash-involved occupants were in a rear impact, 23.3 percent of the injured occupants were in such collisions. However, these injuries are predominately minor. Only 7.6 percent of all

TABLE 1: REAR IMPACTS
1981–1986 NASS/FARS
[From Table A-9, Data Link 1989]

PERCENTAGE OF THE TOTAL ASSOCIATED WITH REAR IMPACTS	
All Car Crashes	11.1%
Fatal Car Crashes	4.9%
All Crash Involved Occupants	12.2%
All Injured Occupants	23.3%
Seriously Injured (AIS 3–6)	7.6%
Occupant Fatalities	3.5%

AIS 3 or greater injuries are associated with rear impacts. Data Link found that the NASS data were in agreement with the FARS data with regard to the over-representation of small cars in rear-impact casualties (Table 2). This over-representation is attributable to the fact that most rear impacts are car-to-car impacts, where conservation of momentum puts the smaller car at a disadvantage in terms of accident severity. Figure 2 shows that the accident severity seen by occupants of small cars in rear impacts is greater than the accident severity level seen by occupants of larger cars. Table 3 shows the distribution of injury severities (AIS) to occupants in towed, rear-impacted vehicles. Note that these NASS and FARS data indicate that only 1 percent of the injuries sustained by occupants in towed rear-impacted vehicles was an AIS 3 or greater injury.

A third analysis of rear impacts, also by Data Link Inc., was conducted in February of 1990 [Data Link, 1990]. This study used 1979-1986 NASS data to investigate a number of specific issues raised in a petition to NHTSA which resulted in the initiation of a public docket on seat issues [54 FR 40897, 1989]. A review of the docketed rulemaking regarding seats is given in a companion paper [Warner, 1991].

The selected NASS files contained 4273 rear-impacted vehicles and 7399 occupants. One of the specific issues analyzed was the yielding or non-yielding of seat backs (the yielding of the seat back was termed "seat back support failure" by Data Link). Since NASS does not specifically code for seat back yield, these data had to be extracted from individual accident investigation files. The post-impact seat back condition was determined by reviewing the investigator notes and by examining the post-crash photographs. To reduce the number of cases

TABLE 2: REAR IMPACTS
1981-1986 NASS/FARS
[From Table A-9, Data Link 1989]

CAR SIZE	Small (<2500lbs)		Medium (2500-3500lbs)		Large (>3500lbs)	
Percentage of:	All(%)	Rear(%)	All(%)	Rear(%)	All(%)	Rear(%)
All Car Crashes	25.1	26.0	40.4	38.9	34.5	35.1
Fatal Car Crashes	29.9	29.9	38.4	38.4	31.7	31.7
All Crash Involved Occupants	24.5	22.3	39.6	39.7	36.0	37.9
All Injured Occupants	30.3	27.2	40.9	41.4	28.8	31.4
Seriously Injured Occupants	34.5	50.0	41.5	29.0	24.0	21.0
Occupant Fatalities	32.4	37.2	41.1	39.9	26.5	22.9
Integrated Casualties (Harm)	32.5	48.5	42.3	33.4	25.2	18.1

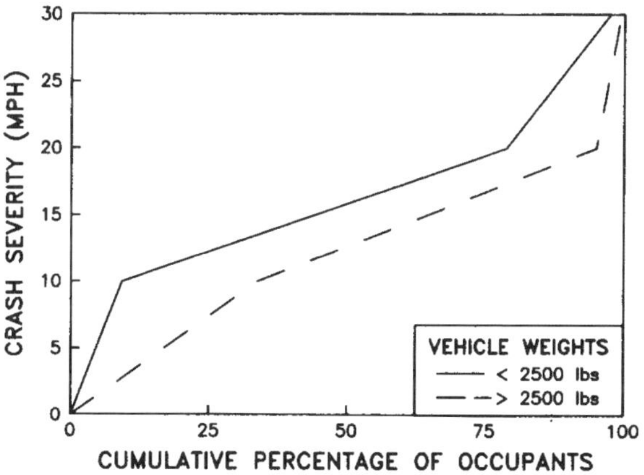

FIGURE 2: DISTRIBUTION OF OCCUPANTS IN REAR IMPACTS NASS 1981-1986 [From Table A-19, Data Link 1989]

TABLE 3: DISTRIBUTION OF INURIES TO OCCUPANTS IN TOWED REAR IMPACTED VEHICLES 1981-1986 NASS/FARS [From Table A-9, Data Link 1989]

MAXIMUM AIS	NUMBER OF OCCUPANTS	PERCENTAGE
UNKNOWN	198	4.0
0	2508	50.7
1	2044	41.3
2	150	3.0
3	35	0.7*
4	4	0.1*
5	9	0.2*
6	2	0.04*
ALL	4950	100.0

*AIS 3+ = 1.0%

which had to be individually reviewed, selection criteria were established to choose the more severe accidents: a delta-V above 15 miles-per-hour or a Collision Deformation Code (CDC) extent above 3, and at least one occupant with an AIS 2 or greater injury, hospitalization, or a fatality. Of the total 116 cases selected for review, about 60 percent indicated some evidence of seat back yield or "failure", about 15 percent had insufficient data to make a determination and the remaining 25 percent indicated no detectable yield. Recognizing the limitations in this analysis, it was nevertheless concluded that seat back yield was quite frequent for the crash severities investigated (i.e. for delta-V's above 15 miles-per-hour). This finding is consistent with earlier findings regarding the relationship between accident severity and seat back yield based on laboratory crash tests [Strother, 1987].

It is interesting that the 1990 Data Link study noted no readily evident relation between the magnitude of seat back deformation and crash severity. It was also noted that the majority of seat backs deformed less than completely. This indicates that the energy absorbing characteristics of production seats were rarely exceeded, even though this sampling of accidents focused on more severe impacts.

In evaluating the consequences associated with seat back yield, the concept of "Harm" was used [Malliaris, 1985]. This concept consists of assigning weighting values to the various levels of injury. The amount of "Harm" associated with specific injury sources can then be calculated by adding the weighted values for all injuries associated with those sources. This allows comparison of high frequency, lower severity injuries to low frequency, high severity injuries.

In evaluating the potential danger of yielding seat backs, of particular interest is the Harm associated with contacts in the rear compartment by front seat occupants, and the Harm associated with front seat occupant ejections. Table 4 shows the Harm associated with front seat, unrestrained occupants, broken down by the area of injury source. In interpreting the data in Table 4, it should be pointed out that Type 1 rear impacts are vehicle-vehicle rear impacts where no other impact takes place. Type 2 rear impacts involve three vehicles in a train-like configuration wherein the subject car is the middle vehicle

TABLE 4: DISTRIBUTION OF HARM TO FRONT SEAT UNRESTRAINED OCCUPANTS BY SOURCE OF MOST SEVERE INJURY, FOR VARIOUS CAR IMPACT TYPES

NASS 1976–1986

[From Table 13, Data Link 1990]

INJURY SOURCE	REAR TYPE 1	REAR TYPE 2	REAR TYPE 3	ALL REAR	ALL OTHER
FRONT	11.27%	30.27%	16.14%	14.3%	33.3%
SIDE	1.44%	1.82%	10.38%	3.9%	10.0%
SEAT BACK	4.29%	7.03%	6.53%	5.1%	0.3%
HEAD RESTR	3.48%	7.93%	2.55%	3.6%	0.1%
OTHER INTERIOR	0.63%	0.28%	3.21%	1.3%	0.7%
ROOF	0.21%	7.49%	0.43%	0.9%	4.1%
FLOOR	0.47%	2.66%	1.71%	1.0%	1.3%
REAR	4.43%	0.00%	0.00%	2.8%	0.1%
OTHER	0.30%	2.27%	1.32%	0.7%	6.2%
NONCONTACT	53.77%	28.61%	19.30%	42.2%	4.4%
GROUND	0.06%	0.00%	1.51%	0.4%	2.0%
UNKNOWN	19.65%	11.64%	36.91%	23.6%	37.6%
ALL	100.00%	100.00%	100.00%	100.0%	100.0%

and sustains a rear impact followed by a frontal impact. Type 3 rear impacts are rear impacts that do not fall into either of the other two categories (single vehicle impacts into fixed objects, etc.).

The Harm associated with occupant contacts with the rear compartment is not a significant portion of the total Harm to unrestrained, front-seat occupants in rear impacts. The NASS category "rear" (referring to contact with rear interior vehicle structures) makes up less than 3 percent of the total Harm in all rear impacts. It is overwhelmed in importance by "non-contact" Harm and is less than one-fourth of the Harm attributed to the frontal structures of the occupant compartment. Non-contact injury is assumed to be predominately "whiplash" type neck trauma. Considering the total Harm attributed to the seat back structure (namely, the "seat back," "head restraint," and "non-contact" categories), it is apparent that these sources combined have about 15 times more significance than occupant impacts with rear structures. Figure 3 (based on the Data Link study) summarizes the Harm associated with ejection for rear impacts. It can be seen that ejections are not a particularly significant problem in rear impacts.

In October of 1990, investigators at Peugot/Renault, working together with two medical doctors, presented data on injuries sustained by restrained motor vehicle occupants in France [Foret-Bruno, 1990]. They found that rear impacts accounted for about 10 percent of all injuries, but only 4 percent of severe injuries and fatalities. Looking at 153 occupants involved in single-impact rear

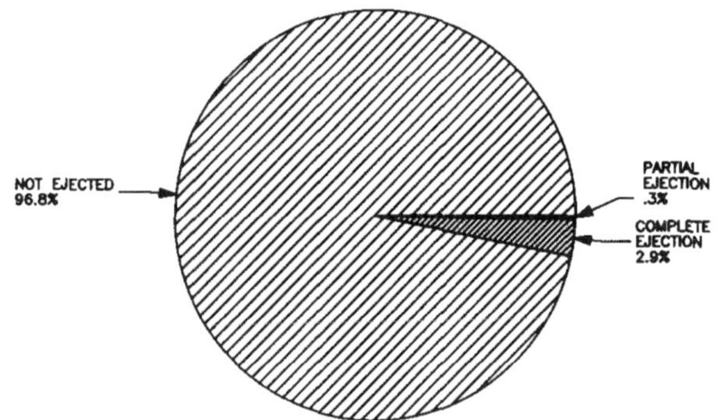

FIGURE 3: DISTRIBUTION OF HARM TO UNRESTRAINED, FRONT SEAT OCCUPANTS IN REAR IMPACTS AS A FUNCTION OF EJECTION [From Table 12, Data Link 1990]

collisions, the most dominant injury category was injury to the cervical spine area (40 cases). All but 2 of these injuries were minor (AIS 1). In fact, looking at 3,781 restrained front-seat occupants in all types of collisions, only 3 cases of very severe cervical lesion (paraplegia) were found, prompting the authors to conclude that these incidents are "extremely rare".

These French investigators recognized that many factors are involved in rear impact injury potential, including the presence of a head restraint, the resistance of the seat back, the impact severity, the gender of the occupant, and the position of the head relative to the head restraint at impact. They concluded that "seat deformation seems more efficient than the head restraint in reducing neck pain," even though it may allow minor head impacts.

The primary focus in evaluating rear impact field accident data has been on front seat occupants. The potential for serious injury to rear seat occupants in rear impacts should also be considered in evaluating the performance of yielding seat backs, because of possible contact as the seat back rotates rearward. NASS data from 1982-1987 contain 876 rear seat occupants in rear-impacted vehicles (about 15 percent of the occupants involved in rear impacts). Table 5 indicates that there were 25 serious injuries (AIS 3 or greater) to rear occupants, but only 2 (8%) can be attributed to contact with the front seat back or head restraint. Both of these

TABLE 5: DISTRIBUTION OF SEVERE INJURIES TO 876 REAR SEAT OCCUPANTS IN REAR IMPACTS, 1982–1987 NASS

AIS	INJURY SOURCE	
	ALL	SEATBACK OR HEAD RESTRAINT
3	15	2
4	4	0
5	1	0
6	5	0
TOTAL	25	2

injuries were AIS 3 femur fractures. The risk of a serious injury to a rear seat occupant in a rear impact from the yielding seat back or head restraint is thus estimated to be very small.

One significant measure designed to reduce the frequency of rear impacts is the center high-mounted stop lamp (CHMSL), which is required on all passenger cars after model year 1986. The early studies evaluating CHMSL demonstrated a reduction as high as 50 percent in the number of rear impacts [Malone, 1978]. While CHMSL will probably be more effective in reducing vehicle damage type rear collisions, it should also have a significant effect of reducing the injury exposure in many car-to-car rear impacts.

SEAT BELT EFFECTIVENESS IN REAR IMPACT

The effectiveness of seat belts in providing front seat occupant protection in rear impacts was addressed in 1987 [Strother, 1987]. The following four potential benefits were attributed to seat belts in a rear impact situation: (1) the belt may control or eliminate the phenomenon of ramping up the seat back, (2) the belt may reduce the velocity of the occupant relative to the vehicle interior and thus reduce injuries resulting from occupant contacts, (3) the belt may minimize the potential for occupants to be out of position at impact, and (4) the belt may be effective in controlling forward rebound of the occupant. The only field accident study available estimated that seat belts reduced the incidence of any injury in rear impacts by 12 percent and the incidence of serious injury by over 57 percent [Levine, 1972]. It was also noted that rear-impact crash tests, done for FMVSS 301 compliance and New Car Assessment Program (NCAP) evaluations, with belted and unbelted dummy occupants showed that the lap belt stayed on the pelvis of dummy occupants and had the effect of limiting their excursion off the seat back.

Claims have been made that seat belts do not provide protection in rear-end crashes because they tend to loosen as the seat deforms rearward, allowing the occupant to slide out from under the belt [Rake, 1990; Saczalski, 1989]. While this is undoubtedly possible in rare cases (i.e., multiple impacts, improper belt use etc.,) field accident data and laboratory tests show that the belt does offer occupant restraint in rear impacts. Rear-end crash tests with restrained dummies characteristically demonstrate the restraining effect of lap belts, notwithstanding seat back yield. Two instances which allowed direct comparison of belt effectiveness in 35 mph Moving Rigid Barrier (MRB) NCAP crashes of Honda and Volvo automobiles, were described in our earlier paper [Strother, 1987]. These two tests involved both a belted and an unbelted front seat occupant and, in both instances, the lap belt was observed to affect the dummy trajectory. It has been asserted that dummies give unrealistic results with respect to lap belt effectiveness, because the Part 572 dummy pelvis structure includes a rubber "flesh" molded into the seated position which causes the dummy's legs to unrealistically rotate up and catch on the lap belt as the dummy moves rearward. While it is clear that the dummy's legs do rotate up, this motion is expected based on the geometry of the seated occupant, irrespective of the dummy hip structure. In most passenger cars, the center of mass of the upper leg is normally above the H-point (Figure 4). In a rear impact,

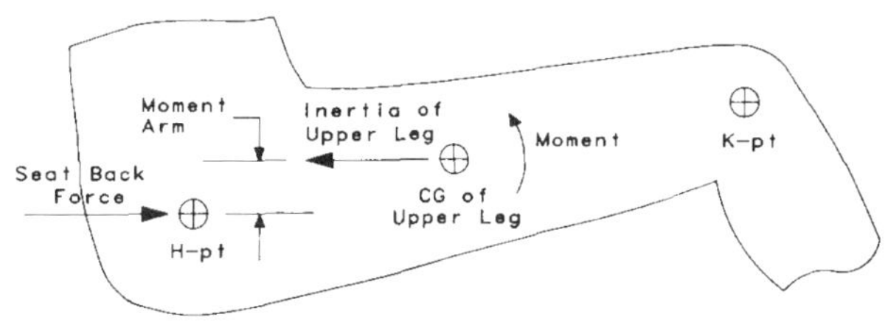

FIGURE 4: UPPER LEG FORCE DIAGRAM

therefore, the leg will tend to rotate upward due to impact forces, irrespective of the hip joint torque. This motion discourages the lap belt from slipping down the legs of the occupant, producing downward components on the hip.

This phenomenon has been observed with cadaver subjects in rear impact sled test simulations (16 mile-per-hour delta-V) [Hu, 1977]. Six tests were available for analysis. Targets at the various occupant joints were used to document occupant motion (see Table 6). In general, the test results indicate an increase in upper leg angulation and a decrease in the angle between the torso and the femur as the impact progresses. It should be noted that the occupants' seated posture in these tests were not particularly realistic for passenger cars. Generally the seated position of the cadaver subjects was significantly more erect than would be expected in today's low profile passenger vehicles. For example, in Test C3R, the cadaver subject's upper legs were actually at a negative angle (i.e. the knee joints were below the H-point) at

TABLE 6: CADAVER SLED TEST RESULTS
[Hu 1977]

TEST NO[1]	AGE/SEX	HT (In)	WT (lbs)	INITIAL TORSO/LEG ANGLE (DEG)	POST-IMPACT TORSO/LEG ANGLE[2] (DEG)	INITIAL LEG ANGLE[3] (DEG)	POST-IMPACT LEG ANGLE[2,3] (DEG)
C1D (6738)	64M	70	126	94.0	91.0	+13.8	+22.0
C2D (6740)	60M	69	134	89.0	105.0	+4.0	+9.5
C3R (6766)	52F	64	132	115.0	99.0	−1.0	−6.0
C4R (6769)	64M	69	159	96.0	83.0	+10.5	+19.5
C5D (6772)	57M	72	195	90.0	120.0	+13.3	+8.0
C6R (6773)	61M	71	165	NO DATA	NO DATA	NO DATA	NO DATA

NOTES

1. Code for test I.D.
   1st Alphabet C=Cadaver
   2nd Numeral = subject number
   3rd Alphabet D=deflecting seatback, R=rigid seatback
   Number in parenthesis = Daisy Track sled run number
2. Data given for time at which head reaches maximum rotational displacement
3. Relative to horizontal, positive indicates knee higher than hip

impact initiation. This was the only test where the upper leg was observed at a negative angle after impact. Also, the tibia was nearly vertical in a number of the tests, which had the effect of allowing the ankle area to slide into contact with the seat structure during the impact phase. This ankle/seat contact would discourage any increase in upper leg angle. Even with all of these factors, however, there was no observed tendency for the lap belt to slip down the legs of the test subjects.

The MVMA 2-Dimensional Crash Victim Simulator program [Bowman, 1985] was employed to investigate occupant kinematics during rear impact of a belted occupant with a more reasonable seated position. The kinematics of Test 6769 [Hu, 1977] were first duplicated using the model. This test was chosen because the subject approximated the dimensions and weight of a fiftieth percentile male. The occupant parameter data set for a standard Hybrid III 50th percentile male, was used as established by University of Michigan Transportation Research Institute (UMTRI). The seat geometry and pre-impact occupant positioning were taken from the high speed film. The seat back angle was approximately 7 degrees back from vertical. Figure 5 shows the upper leg rotation for both the cadaver test and the MVMA 2D simulation and Figure 6 shows the hip joint displacement. The agreement between the sled test and the simulation was judged to be quite good, leading to two findings: (1)

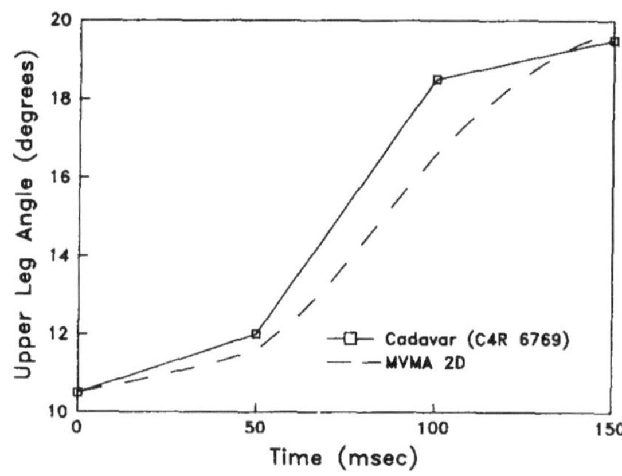

FIGURE 5: UPPER LEG ROTATION

the impact forces do cause the leg to rotate up and catch the lap belt and (2) the joint stiffness for the Hybrid III dummy is an adequate representation of human cadavers for this purpose.

Additional simulations were made with a geometry more representative of typical passenger car production seats. Figure 7 shows a comparison of sled test geometry with a more typical seated geometry. The initial position of the upper leg varied between approximately 10 and 20 degrees from horizontal, and the seat back angle was changed to both 20 and 25 degrees. Figure 8 presents the simulated upper leg motions for the two different seat

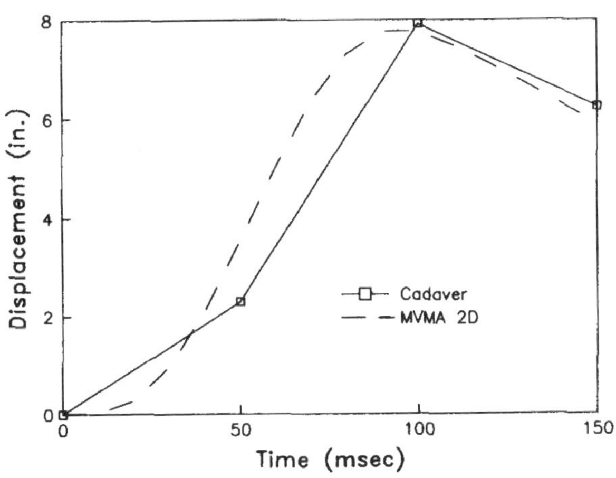

FIGURE 6: HIP JOINT DISPLACEMENT

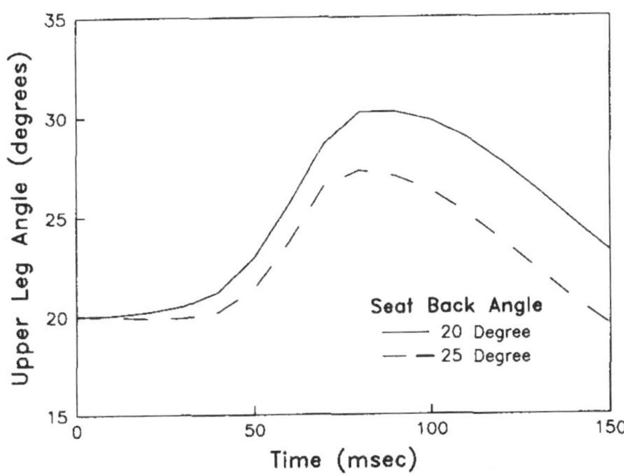

FIGURE 8: UPPER LEG ROTATION MVMA 2D SIMULATION

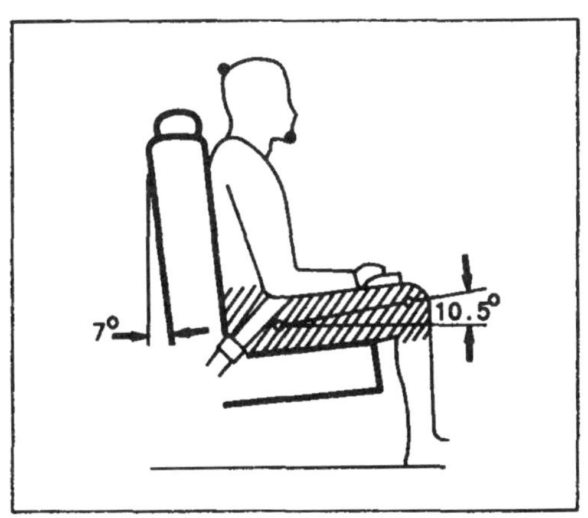

a. Cadaver in Sled Experiments [Hu,1977]

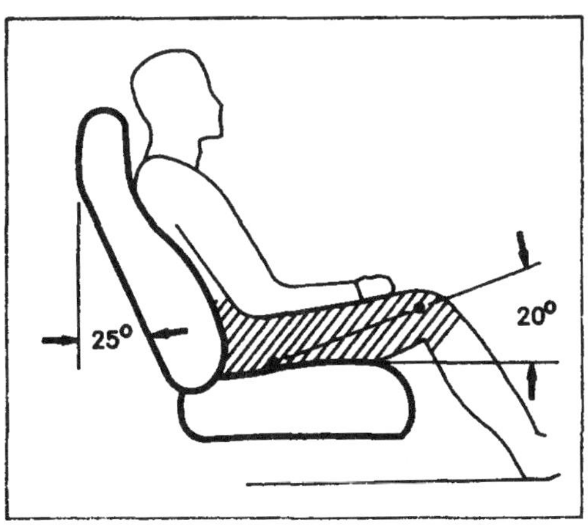

b. Occupant in Typical Passenger Car

FIGURE 7. SEATED GEOMETRY COMPARISONS

back angles. In summary, the beneficial effect of lap belts, observed earlier in crash tests with anthropomorphic dummies, has been shown to be mainly the result of crash dynamics rather than an artifact of dummy construction.

FIELD ACCIDENT DATA ON EFFECTIVENESS OF SEAT BELTS IN REAR IMPACTS

Several recent field accident studies have been conducted which address the issue of seat belt effectiveness in rear impacts since 1987. Partyka evaluated the effectiveness of front seat 3-point belt systems for various impact modes using a matched-pair analysis of the 1987 FARS data [Partyka, 1988]. The effectiveness in terms of fatality prevention in rear impacted passenger cars was found to be 32%. The Data Link study discussed previously [Data Link, 1990], evaluated the effectiveness of seat belts in rear impacts in terms of Harm. They found an overall seat belt effectiveness in rear impacts of 37 percent (see Table 7). In chain-type rear impacts, where the subject car experiences a subsequent frontal impact into another vehicle, the Harm reduction seat belt effectiveness is 62 percent, suggesting that the shoulder harness is also effective in reducing injuries caused by subsequent frontal impacts.

Evans evaluated the 1975-1983 FARS data to determine the effectiveness of belts in various directions of impact [Evans, 1991]. He found a belt effectiveness in rear impacts of about 47%.

Partyka reviewed the NASS data to compare the incidence of injury to the chest, head and face experienced by belted drivers to that experienced by unbelted drivers in rear impacts [Partyka, 1990]. She found that the belted drivers had 41% fewer such injuries than the unbelted drivers.

In summary, this review of accident statistical studies indicates that seat belt effectiveness in rear impacts is in the range of 37-47 percent.

TABLE 7: SAFETY BELT EFFECTIVENESS IN REAR IMPACTS FOR FRONT SEATING POSITIONS

[From Table 11, Data Link 1990]

TYPE OF ACCIDENT	UNRESTRAINED			RESTRAINED			SEAT BELT EFFECTIVENESS
	% OCCUP	% HARM	HARM/OCCUP	% OCCUP	% HARM	HARM/OCCUP	
SINGLE CAR-TO-CAR (1)	39.0	33.6	.27	14.0	6.5	.14	48 %
CHAIN COLLISION (2)	6.0	4.9	.26	2.5	0.8	.10	62 %
ALL OTHER (3)	29.5	44.0	.47	9.0	10.2	.36	23 %
ALL	74.4	82.5	.35	25.6	17.8	.22	37 %

CONCLUSIONS

A review of recently-published field accident studies on rear impact safety issues reveals a general agreement with those conclusions published earlier with regard to the relatively low safety priority of the rear impact mode, and the relatively lower distribution of injuries to occupants of vehicles struck in the rear. These newer studies provide additional insights into the relative importance of different injury mechanisms for occupants who are injured in rear collisions. Data on the distribution of Harm in rear impacts show the relative insignificance of occupant contacts in the rear compartment area and occupant ejection, compared to the more important "non-contact" and seat-induced Harm. Such data indicate that stiffening front seat backs in an effort to reduce rear-compartment contact and ejection injuries would provide very little, if any safety benefits, and may cause additional non-contact neck injury.

The field accident data show: (1) rear impacts do not account for a significant portion of automobile injuries, (2) current production seat backs, provide a high level of protection in rear impacts, and (3) the injury mechanisms which may be addressed by rigid seats make up a minuscule proportion of rear impact injuries. These field accident studies also provide additional information on the effectiveness of seat belts in rear impacts. Reductions in Societal Harm or injury production of 37-47 percent are indicated. The mechanism by which lap belts benefit occupants of vehicles struck in the rear indicates that the lap belt does remain in position on the pelvis, limiting upward and backward excursion.

ACKNOWLEDGEMENTS

Grateful appreciation is expressed to Ron Nordhagen and Nan Strother for assistance in preparing figures and tables, and to Elaine Murray for the typewritten drafts.

REFERENCES

[Bowman, 1985]
Bowman, Bruce M.; Robert O. Bennett, D. Hurley Robbins: "MVMA Two-Dimensional Crash Victim Simulation, Version 5" Volumes 1, 2, 3, University of Michigan Transportation Research Institute. UMTRI-85-24-1,2,3. June 18, 1985

[Data Link, 1989]
"Car Crash Outcomes in Rear Impacts." Appendix A to Current Issues of Occupant Protection in Car Rear Impacts." Data Link, Inc., Washington, D.C. 1989.

[Data Link, 1990]
"Current Issues of Occupant Protection in Car Rear Impacts," prepared for Office of Crashworthiness, Rulemaking, NHTSA. Data Link, Inc., Washington, D.C. Feb., 1990. Docket 89-20-NO1-021.

[Evans, 1991]
Evans, Leonard: Traffic Safety and the Driver, Van Nostrand Reinhold, NY. ISBN 0-442-00163-0. 1991.

[Foret-Bruno, 1990]
Foret-Bruno, J. Y.; C. Tarriere, J. Y. LeCoz, C. Got; F. Guillon: "Risk of Cervical Lesions in Real-World and Simulated Collisions." 34th AAAM Conference Proceedings, Scottsdale, Arizona, p. 373. October, 1990.

[Hu, 1977]
Hu, A; S. Bean; and R. Zimmerman: "Response of Belted Dummy and Cadaver to Rear Impact." SAE 770929, Proceedings of the 21st Stapp Car Crash Conference. 1977.

[Levine, 1972]
Levine, D.N.; Campbell, B.J.: "Effectiveness of Lap Seat Belts and the Energy-Absorbing Steering System

in the Reduction of Injuries," Journal of Safety Research, Vol. 4, No. 3. September 1972.

[Malliaris, 1985]
Malliaris, A. C.; Ralph Hitchcock; and Marie Hansen: "Harm Causation and Ranking in Car Crashes." SAE 850090, SAE International Congress and Exposition. Feb. 25, 1985.

[Malone, 1978]
Malone, Thomas B.; Mark Kirkpatrick; Jeffrey S. Kohl; Clifford Baker: "Field Test Evaluation of Rear Lighting Systems." Essex Corp. DOT HS 803 467. 1978.

[54 FR 40897, 1989]
NHTSA Request for Comments, opened Docket 89-20-NO1. October 4, 1989.

[Partyka, 1987]
Partyka, Susan C.: "Fatalities in Rear-Impact Small Cars From 1982 Through 1987," Papers on Vehicle Size -- Cars and Trucks, p. 125. Dec., 1987.

[Partyka, 1988]
Partyka, Susan C.: "Belt Effectiveness in Pick-up Trucks and Passenger Cars by Crash Direction and Accident Data." Papers on Adult Seat Belts - Effectiveness & Use. National Center for Statistics and Analysis, HS 807 285. June, 1988.

[Partyka, 1990]
Partyka, Susan C.: "Comparisons of Belt Effectiveness in Preventing Chest, Head, and Face Injuries in Front and Rear Impacts." August, 1990.

[Rake, 1990]
Rake, Buddy Jr.; Scott E. Boehm: "Auto Seat Systems - Dangerous Safety Restraints?" Trial, pp. 80-87. April, 1990.

[Saczalski, 1989]
Saczalski, Kenneth: Petition to National Highway Traffic Safety Administration. Docket 89-20-NO1-001. April 18, 1989.

[Strother, 1987]
Strother, Charles E.; and Michael B. James: "Evaluation of Seat Back Strength and Seat Belt Effectiveness in Rear End Impacts." SAE 872214, Proceedings of the 31st Stapp Car Crash Conference, New Orleans, Louisiana. October 9-11, 1987.

[Warner, 1991]
Warner, Charles Y.; Charles E. Strother, Michael B. James, Robin L. Decker: "Occupant Protection in Rear-End Collisions: II. The Role of Seat Back Deformation in Injury Reduction" Proceedings of the 35th Stapp Car Crash Conference, San Diego, CA, November 18-20, 1991.

912914

Occupant Protection in Rear-end Collisions: II. The Role of Seat Back Deformation in Injury Reduction

Charles Y. Warner, Charles E. Stother, Michael B. James, and Robin L. Decker
Collision Safety Engrg., Inc.

ABSTRACT

The National Highway Traffic Safety Administration (NHTSA) has recently opened a rulemaking docket seeking comments on the design of automobile seats and their performance in rear impacts. There are two philosophies of seat design: one advocates rigid seats, the other advocates seats which yield in a controlled manner. A review of the legislative history of seat back design standards indicates that yielding seats have historically been considered a better approach for passenger cars. The design characteristics of current production automobile seats are evaluated and show no significant changes over the past three decades. Concerns about the performance of rigid seat backs in real world rear impacts are discussed, specifically increased injury exposure due to ramping, rebound and out-of-position occupants.

INTRODUCTION

Research relating to the role of seat back deformation in rear impact collisions has been previously summarized [Strother, 1987]. It was found that there are two design philosophies with regard to seat back strength. One prevalent philosophy suggests that seat backs ought to yield in a controlled manner and absorb energy in order to provide the occupant with more "ride-down." The other philosophy suggests that in severe rear impacts, yielding seat backs expose occupants to contact with structures in the rear of the vehicle and to ejection through the rear window, and asserts that rigid, high seat backs would provide an effective restraint for occupants in rear impacts because the occupant would remain in the seat even during severe exposures. The consensus of available testing and modeling efforts comparing rigid and yielding seat backs in 1987 suggested that yielding seat backs reduce injury exposure in rear impacts. Rear crash tests and static pull tests of production seats indicated that all vehicle production seats were designed to yield and provide ride-down. In fact, published efforts to design rigid seats capable of mass production had been largely unsuccessful. It was also noted that most prior research on rigid seat designs did not address the performance of such seat structures in protecting out-of-position occupants, nor did it investigate the implications of such designs on injury exposure to rear seat occupants in frontal impacts.

In severe rear impacts, rigid seat backs may provide protection to belted occupants in the optimal, normal seated position (NSP). However, yielding seat backs of the type found in current production vehicles offer performance advantages in the vast majority of real world exposures. Yielding seat backs are less sensitive to occupant pre-impact position, and they reduce rebounding of the occupant in the forward direction. They also provide a more forgiving structure for unrestrained and lap-belted rear seat occupants in frontal impacts. Field accident data confirm the beneficial effect of yielding seat backs on injury exposure [James, 1991].

The National Highway Traffic Safety Administration (NHTSA) has recently opened a rulemaking docket seeking comments on the design of automobile seats and their performance in rear impacts [54 FR 40897, 1989]. The docket is based on a petition which would require seat backs to be essentially rigid. Before addressing specific seat back design issues, it is instructive to follow the history of rulemaking on this topic.

RULEMAKING HISTORY OF SEAT BACKS AND HEAD RESTRAINTS

In November of 1963, the Society of Automotive Engineers (SAE) adopted Recommended Practice J879, which defined two test procedures for evaluating passenger vehicle seat strength [SAE J879, 1963]. The first procedure was a 20G static pull test for seat anchor strength. The second procedure was the application of a 4250 inch-pound rearward static moment to the seat back.

Two years later, in 1965, the General Services Administration (GSA) of the Federal Government published Federal Standard 515-6 that was to apply to all 1967 model automobiles purchased by the Government [GSA 515-6, 1965]. This standard was essentially the same as SAE Recommended Practice J879.

In 1968 Federal Motor Vehicle Safety Standard (FMVSS) 207 went into effect, covering all 1969 model

*Information in brackets refers to references listed at end of paper.

year passenger cars [32 FR 2415, 1967]. FMVSS 207 was essentially the GSA standard, but with two minor changes. First, forward folding seats were required to withstand a forward load of 20 times the weight of the seat back. Second, the rearward static moment requirement was changed from 4250 inch-pounds to 3300 inch-pounds, to be applied by a force at the uppermost cross member of the seat back frame. The moment for FMVSS 207 was calculated about the Seated Reference Point (SRP), while the moment for the GSA standard had been calculated about a point approximately 8 inches below the SRP. Therefore, this change was basically a redefinition rather than a reduction in requirements.

In 1968, the NHTSA also issued FMVSS 202, requiring head restraints on cars manufactured after January 1, 1969 [33 FR 2945, 1968]. The standard specified that head restraints be capable of sustaining a 200 pound rearward load (or a load resulting in seat back failure) while deflecting less than 4 inches.

In 1974, the NHTSA issued a Proposal to Amend that would substantially modify FMVSS 202 and 207 [39 FR 10268, 1974]. The proposal included a dynamic rear impact test which was to be conducted similarly to the FMVSS 301 rear moving barrier test for fuel system integrity [40 FR 48353, 1975]. The proposal was somewhat unique in that the preamble did not refer to any safety need associated with the proposed changes. Further, there was no discussion of the types of injuries the proposed standard was addressing, the number of such injuries occurring under the existing standard, or how any of those injuries might be mitigated by the proposed standard. These rather substantial rulemaking deficiencies in the proposal were consistently cited in Docket #74-13 responses [39 FR 10268, 1974]. The amended standard would have required seats to be an order of magnitude stiffer than those built to the existing standard, even though numerous prior studies cited in the docket responses indicated that stiffer seat backs might actually increase injury. No study was cited by any respondent which indicated stiffer seats would reduce injury exposure. The few respondents who favored the proposal suggested occupant ejection through the rear window and contact with interior components in the rear seat area were the major injury mechanisms intended to be prevented by stiffer seat backs. No data were presented to verify that seat backs conforming to the proposed standard would reduce exposure to these injury mechanisms, nor were any data presented to establish the frequency of such exposures in real world accidents.

References to field accident data in the docket responses indicated that rear impacts were the least dangerous accident mode, rear impacts contributed only 4.5 percent of total accident injuries and most of the injuries that do occur are minor to moderate in severity [MVMA, 1974], [Ford, 1974]. Respondents suggested that even with an effective countermeasure, the exposure to injury in rear impacts is so low it would be difficult to achieve a measurable safety benefit. To evaluate the effectiveness of the proposed countermeasure (i.e. stiffer seat backs), Chrysler analyzed the Highway Safety Research Institute (HSRI) data base [Chrysler, 1974] They found 207 rear-impacted vehicles with sufficient data to determine the condition of the seat after the accident. Seat track separation occurred in only 11 of those cases (5%), and in each case of track separation the vehicle had experienced over 20 inches of residual rear-end crush, indicating a severe crash exposure. Of the 16 occupants in these 11 vehicles, only 4 had injuries more severe than "minor" (2 had whiplash, 1 bumped his head on the roof rail and 1 hit his chin on the steering wheel). Their analysis indicated stronger seats or seat tracks would not have eliminated any of the injuries.

Chrysler was also able to evaluate the effect of seat back deformation on injury exposure. In 42 percent of the rear impact cases, the seat backs were reported to have had at least some deflection. The cases with reported seat back deflection were divided into two groups: those with less than 15 degrees of residual seat back deflection (less than approximately 40 degrees behind vertical) and those with greater deflection. The data showed no noticeable difference in injury frequency between the two groups. Since greater seat back deflection should logically be correlated with higher accident severity, this was a remarkable finding. The single rear impact fatality in the HSRI data involved a 1971 Oldsmobile Cutlass impacted by a 1968 Lincoln Continental at an estimated closing speed of 70 miles per hour. Interestingly, the seat back deflection in that case was less than 15 degrees, probably due to support by intruding structures. Chrysler hypothesized that the fatal head/neck injury was caused by contact with intruding sheet metal.

The proposed amendments to FMVSS 207 were originally to become effective on September 1, 1976. On July 9, 1976, the NHTSA indicated that no action was to be taken to finalize the proposed amendments "because of questions which arose as a result of comments to the docket for this rulemaking action. We are now conducting research on rear impacts, utilizing seats with cadaver test subjects in order to resolve some of these problems" [Carter, 1976]. The tests were those of Hu et al. [Hu, 1977].

About two years later, the NHTSA published a "Request for Comments" on its proposed 5-Year Plan [43 FR 11100, 1978]. The intent of the plan was to prioritize the then-existing rulemaking efforts according to the following five factors:

1. Number of deaths and injuries,
2. Availability of practical solutions,
3. Research and engineering data on the problem,
4. Anticipated costs to consumers and industry,
5. Lifesaving potential as a safety standard.

The 5-Year Plan included a variety of then-existing rulemaking actions which the NHTSA was considering for termination, including the proposal to require stiffer seats. The appendix to the 5-Year Plan contained a new approach to evaluating rear impact protection, specifically:

" 1. Determine, through accident studies, the extent of the rear end collision problem that may exist in future vehicle population.
2. Through in-depth accident studies, determine the types of trauma encountered in rear end collisions and obtain a more precise description of the nature of the collision.
3. Identify kinematic occupant parameters which best correlate with injury and identify seat design features which can mitigate injury to the head/neck complex.

4. Develop injury criteria and an appropriate dummy head/neck complex necessary to simulate occupant response in rear impacts.
5. Develop and demonstrate seat designs which provide significantly increased rear impact protection.
6. Develop and demonstrate rear end structures in small cars with improved crash energy management, compartment integrity and immunity from the generation of fires.
7. Develop and demonstrate other in-vehicle or component countermeasures, if necessary, to optimize occupant protection."

It is interesting to note that the 1974 proposal assumed that there was a safety problem associated with rear impacts and assumed to know the solution to the problem, i.e. stiffer seat backs. Four years later, after accumulating public comments in that docket and performing testing, the NHTSA proposed to evaluate rear impacts to determine if there was a problem and, if a problem was found to exist, what might be done about it.

When the 5-Year Plan was finally issued, on April 20, 1979, it indicated that there were no public comments regarding the termination of the 1974 proposal to require stiffer seat backs and that it was therefore terminated [NHTSA, 1979].

There were no further published rulemaking activities associated with seat back strength for over a decade. In 1989, the NHTSA published a Request For Comments related to two petitions for rulemaking associated with rear impacts [54 FR 40897, 1989], opening Public Docket 89-20-NO1. The preamble to the Request makes no mention of the 1974 proposal or the proposed evaluation presented in the 5-Year Plan. The first petition proposed to modify FMVSS 207 to require that seat backs withstand 20 times the combined weight of the seat back and an occupant and/or be able to withstand a moment about the seated reference point about 17 times higher than the existing standard (i.e., 56,000 inch-pounds) [Saczalski, 1989]. The purpose of the petition was to reduce the exposure of occupants in rear impacts to sliding out of their lap belts as the seat back yields and either contacting vehicle interior components or being ejected. The second petition proposed to amend FMVSS 208 (Occupant Protection) and FMVSS 209 (Seat Belt Assemblies) to require that all seat belt retractors be webbing sensitive [Horkey, 1988]. Horkey's proposed modification was based on the theoretical possibility that the pendulum in a vehicle-sensitive retractor could somehow be in the neutral position when the occupant rebounds off the seat back in a rear collision, thus allowing the belt to spool out, exposing the occupant to contact with the steering wheel or other frontal vehicle components.

NHTSA's preamble to the 1989 Request for Comments indicates that there are "two competing schools of thought related to the proper performance of seats in rear impacts." One view is that stiffer seats are preferable because they are less likely to deform rearward in rear impacts and therefore may reduce the exposure of occupants in severe impacts to contact with the vehicle's rear interior structures and reduce the incidence of ejection through the rear window. The petition for stiffer seat backs was based on four such cases resulting in serious/fatal injuries which the petitioner had observed during a two year period. "Other researchers," the preamble continues, "advocate seats that absorb energy by yielding in a controlled manner, increasing the occupant's ride down." This school of thought, according to the NHTSA, contends that stiffer seats exacerbate injuries associated with rebound and also result in more serious whiplash and neck injuries.

The responses to the Request for Comments indicate that current production seats are performing well in rear impacts. The comments generally recognize that energy absorption by the seat back mitigates injury exposure and that stiffer seats would increase both rebound effects and whiplash injury potential. A decided minority of respondents argued that energy absorption by seat back yield is not a good design approach.

The most significant docket responses are those which contain field accident data, since the first question to be addressed in any contemplated rulemaking is whether a safety-related problem exists. General Motors analyzed 1488 rear-impacted vehicles (model years 1975-1985) in its Motors Insurance Company (MIC) accident files [General Motors, 1989]. Of these vehicles, 629 (42%) had permanent seat back deformation. Only 1 of these 629 cases resulted in injury from contact with the rear area of the occupant compartment (an Abbreviated Injury Scale (AIS) 1 level injury from contact with the back window or rear header) and only 2 occupants (both unrestrained) were ejected through portals behind the B-pillar. Mercedes Benz indicated in their response that rear-impacted vehicles account for only 4 percent of the serious-or-greater (AIS 3+) injuries to Mercedes occupants [Mercedes Benz, 1989]. It was also noted that AIS level 1 neck injuries are the most common injuries found in rear impacts. Investigation of 54 rear impact accidents involving Mercedes vehicles indicated that permanent seat back deformation occurs only in "higher speed" impacts.

More detailed evaluations of FARS and NASS field accident data are also referenced in the docket [Data Link, 1989], [Data Link, 1990], [Partyka, 1987], [Partyka, 1988], [Partyka, 1990]. These studies (discussed in detail in a companion paper [James, 1991]), indicate: 1) rear impacts do not represent a large percentage of the overall injury exposure, 2) seat backs do yield in rear collisions, 3) seat belts are effective in reducing injury exposure in rear impacts, 4) injuries from occupant contact against rear compartment structures are infrequent, and 5) ejection is not a significant societal injury problem in rear impacts. The last two findings are important because they show that the injury mechanisms which rigid seats are designed to minimize do not represent important injury exposures in real world accidents.

On October 17, 1990, the NHTSA terminated its action on the petition to require vehicle sensitive seat belt retractors based on a review of FARS and NASS data files. These data indicated that present seat belt retractor designs are effective in providing protection in rear impacts [55 FR 42031, 1990], [Partyka, 1987], [Partyka, 1988], [Partyka, 1990]. At this writing, the NHTSA has yet to rule on the petition relating to seat back strength.

STRENGTH OF PRODUCTION PASSENGER VEHICLE SEATS

The average strength of production vehicle seats has not varied dramatically over the past two decades. The strength and energy absorbing properties of 48 production seats based on static pull tests have been published [Strother, 1987]. These tested seats, ranging from model year 1964 to model year 1977, indicated that the average production seat back was capable of sustaining rearward loads about 2 - 3 times that required by FMVSS 207. To determine whether later model passenger car seats exhibited similar properties, thirteen additional static pull tests were performed on 1980's vintage seats.

Table 1 contains a summary of static seat pull test data for the original 48 seats and the additional 13 seats tested more recently. A detailed description of the test protocol is contained in the earlier paper [Strother, 1987]. Briefly, a body block representing the upper torso is positioned in the seat so that a rearward horizontal force can be applied at a level about 14 inches above the seated reference point. Data from a representative test is shown in Figure 1. New averages combining the 48 tests

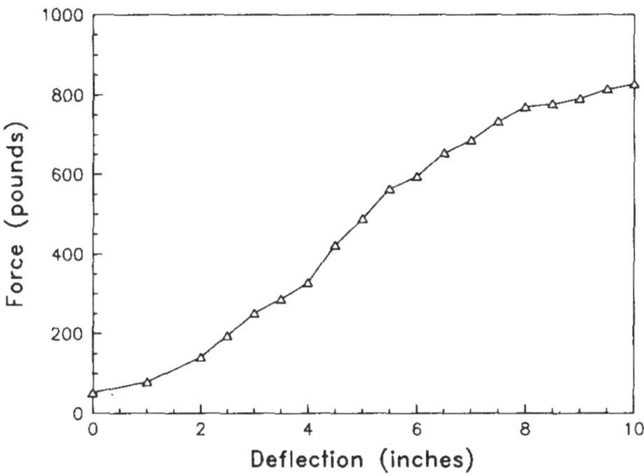

FIGURE 1: TYPICAL SEAT PULL TEST

and the additional 13 tests have been computed for stiffness, ultimate strength, and energy absorption. These new averages are not significantly different from the former average values, and the highest and lowest values in each category remain unchanged.

RIGID SEAT BACK CONCERNS IN REAL WORLD REAR IMPACTS

The concept of a rigid seat back is compelling in theory, at least for severe impacts with well-positioned and restrained occupants. A rigid seat back could distribute forces over a large area of the body, an area much larger than that used by belt systems in frontal collisions. The whole-body tolerance for humans has been shown to be over 35 g's in the sternum-ward direction [Eiband, 1959]. However, the field accident data indicate that injury mechanisms which rigid seats are designed to avoid are statistically not a serious problem. In fact, rigid seats could likely increase the total harm associated with rear impacts because of increased injury exposure in more frequent but less severe rear impacts. There are serious kinematic concerns associated with the performance of rigid seat backs in real world situations. The three most important concerns are: 1) ramping, 2) rebound, and 3) out of-position occupants.

"RAMPING" AND REAL ACCIDENTS - The term "ramping" has been used to describe motion of the occupant parallel to the seat back during a rear impact. This phenomenon has been demonstrated in tests of both yielding and rigid seat backs, particularly in high severity impacts. An occupant who ramps is potentially exposed to head and neck injury by contact with vehicle structures, but such injuries are not statistically significant. Ramping in rigid seats generally leads to contact with the roof structure, while ramping in yielding seats generally leads to contact with the rear seat cushion, back window, or rear header. Lap belts are effective in reducing the potential for ramping by trapping the thighs of the occupant [James, 1991].

Some proponents of rigid seat backs maintain that ramping is not a problem with rigid seat backs, even for unbelted occupants. Others suggest that limiting seat back angle to 40-45 degrees from vertical will prevent ramping of unbelted occupants. However, there is potential for ramping in rigid seats, in seats with limited angulation, and in yielding seats for unbelted occupants.

Factors which can control the onset of ramping include: seat back angle, occupant position, impact severity, seat back/occupant friction, and vertical motion of the vehicle during the impact. The analysis of the ramping forces acting on a seated occupant in a rear impact cannot be simply described, but the effects of specific factors can be evaluated. A simple mathematical model can be used to predict the onset of ramping in an idealized situation based on the seat back angle, seat back/occupant friction, and vehicle acceleration (see Appendix A).

To establish a range of appropriate friction coefficients, static tests were conducted on various clothing-seat upholstery combinations with a range of volunteer subjects. The tests were conducted by positioning the subject in a knees-to-chest configuration on an upholstered flat ramp. The ramp angle was then increased until the subject began to slide (see Figure 2). The angle of incipient motion was then used to determine the static friction coefficient. Results of these experiments indicate that the static friction can vary between 0.3 and 0.48 (Figure 3). Figure 4 graphically presents the seat back angle for initiation of ramping as a function of seat frame (or vehicle compartment) acceleration, based on this simple analysis. However, ramping in real world collisions will occur at even more vertical seat back orientations than those shown in Figure 4 for several reasons. First, friction coefficients between fabrics decrease dramatically with increasing pressure, i.e., as the occupant loads the seat back, the frictional coefficients decrease [Carr, 1988], [Posey, 1984], [Wilson, 1963]. Also, the dynamic, or sliding, friction values of fabrics against each other are less than the static friction values. The vertical and lateral vehicle compartment accelerations associated with real world collisions can readily initiate motion between the occupant and the seat back such that the lower dynamic friction values will prevail. Accident studies show that most severe rear-end collisions occur with larger vehicles striking smaller vehicles in the rear [James, 1991], [Partyka, 1987]. This frequently results in an override situation where significant vertical accelerations and changes in vehicle pitch attitude can initiate occupant motion relative to the seat back. When these effects are present, unrestrained occupants will begin to ramp at seat back angles as low as 25 degrees behind the vertical. Thus, seat backs do not have to be 40 or 45 degrees from vertical before ramping occurs.

TABLE 1 - SUMMARY OF STATIC SEAT PULL TESTS

SEAT DESCRIPTION			SEAT PERFORMANCE				
MODEL YR.	MANUFACTURER	MODEL	STIFFNESS (lbf/in.)[1]	STRENGTH (lbf.)[2]	ABSORBED ENERGY (ft. lbs)[3]	REMARKS	TEST CONDUCTED BY:
'64	MG	–	117.6	430.0	118.6		CSE
'64	AMC	RAMBLER	85.1	560.0	93.6		CSE
'64	VW	TYPE I	137.9	620.0	263.7		CSE
'64	TRIUMPH	–	266.7	600.0	174.3		CSE
'64	PONTIAC	LEMANS	265.0	550.0	89.7		CSE
'65	RENAULT	DAUPHINE	73.2	455.0	100.1		SEVERY
'66	CHEVROLET	CHEVELLE	103.4	505.0	170.5		SEVERY
'66	DATSUN	411	169.3	325.0	174.2		SEVERY
'66	DODGE	DART	85.1	660.0	323.2		SEVERY
'66	FORD	CORTINA	105.8	415.0	114.1		SEVERY
'66	FORD	FALCON RANCHERO	106.0	615.0	244.6	One side of split bench seat	SEVERY
'66	PLYMOUTH	BARRACUDA	121.2	850.0	233.7	One side of split bench seat	SEVERY
'66	SAAB	95	144.9	425.0	175.4		SEVERY
'66	VW	TYPE III	79.4	640.0	204.9		SEVERY
'66	VOLVO	544	160.0	730.0	418.5	Friction clutch at s/back pivot adjusted for maximum torque	SEVERY
'67	FIAT	850	137.9	475.0	163.6		SEVERY
'67	PLYMOUTH	BARRACUDA	105.3	680.0	374.2	Split bench seat, one side	SEVERY
'67	RENAULT	R10	66.7	490.0	191.2		SEVERY
'67	VW	TYPE I	115.6	610.0	322.5		SEVERY
'68	DATSUN	510	102.6	360.0	87.7[5]		SEVERY
'68	FORD	MUSTANG	160.0	787.0	278.4		SEVERY
'68	OPEL	KADETT	88.9	530.0	241.9		SEVERY
'68	SAAB	96	137.9	475.0	277.1		SEVERY
'68	VOLVO	144	103.6	700.0	509.8		SEVERY
'69	VW	TYPE I	78.4	681.0	N/A*		CSE
'69	MERCURY	COUGAR	40.4[5]	715.0	N/A		CSE
'70	OPEL	1900	88.9	570.0	N/A		SEVERY
'70	RENAULT	R10	52.6	525.0	N/A		SEVERY
'70	TOYOTA	COROLLA	40.8	310.0	N/A		SEVERY
'70	SUBARU	–	55.5	300.0[5]	N/A		SEVERY
'70	DATSUN	510	142.9	365.0	N/A		SEVERY
'70	FIAT	850	95.2	873.0	N/A		SEVERY
'70	AUSTIN	AMERICA	222.2	830.0	N/A		SEVERY
'70	AMC	GREMLIN	285.7	1400.0[4]	N/A		SEVERY
'70	CHEVROLET	MONTE CARLO	139.1	772.0	480.2		CSE
'70	VW	TYPE I	210.0	790.0	243.3		CSE
'71	PLYMOUTH	DUSTER	182.4	830.0	535.9[4]		CSE
'71	VOLVO	144	222.2	630.0	373.4	Friction clutch at s/back pivot adjusted for maximum torque	CSE
'72	CHEVROLET	VEGA	160.0	720.0	N/A		CSE
'73	FORD	PINTO	117.6	950.0	370.2		SEVERY
'73	VW	S. BEETLE	606.1	1055.0	376.5		CSE
'74	CHEVROLET	MONTE CARLO	285.7	830.0	243.3		CSE
'74	FORD	MAVERICK	57.1	346.6	145.0		CSE
'76	CHEVROLET	MONTE CARLO	62.9	508.0	132.5	Swivel bucket seat	CSE
'76	CHEVROLET	CHEVETTE	144.9	671.0	207.6		CSE
'76	DATSUN	B210	169.5	650.0	280.6		CSE
'76	VOLVO	242DL	178.6	1113.0	294.2	Seat broke just prior to reaching 45 degrees	CSE
'76	ROLLS ROYCE	S. SHADOW	45.9	500.0	194.9		CSE
'77	VW	RABBIT	444.4[4]	1180.0	N/A		CSE
'77	TOYOTA	COROLLA	100.0	623.0	290.0		CSE
'82	CHEVROLET	CAMARO	120.7	538.0	197.8		CSE
'82	TOYOTA	TERCEL	145.0	413.0	164.3		CSE
'83	VOLVO	700 SERIES	75.3	1254.0	532.8		CSE
'84	FORD	ESCORT	60.7	700.0	269.4		CSE
'86	CHEVROLET	SPECTRUM	64.0	538.0	226.9		CSE
'86	NISSAN	STANZA	77.2	581.0	258.7		CSE
'86	PLYMOUTH	HORIZON	55.7	598.0	192.4		CSE
'86	SUBARU	STATION WAGON	66.3	827.0	399.3		CSE
'86	VW	JETTA	78.4	866.0	336.9		CSE
'86	BMW	733I	82.6	591.0	194.6		CSE
'88	MERCEDES	300	128.2	1144.0	299.7		CSE
AVERAGE VALUES			134.8	660.2	256.9		
[4]HIGHEST VALUE			606.1	1400.0	535.9		
[5]LOWEST VALUE			40.4	300.0	87.7		

*NOT AVAILABLE
[1]AVERAGE OVER FIRST 200 LBFS
[2]MAXIMUM FORCE DURING DEFLECTION, STOPPED AT 45 DEGREES
[3]ENERGY NOT CALCULATED BEYOND 45 DEGREES

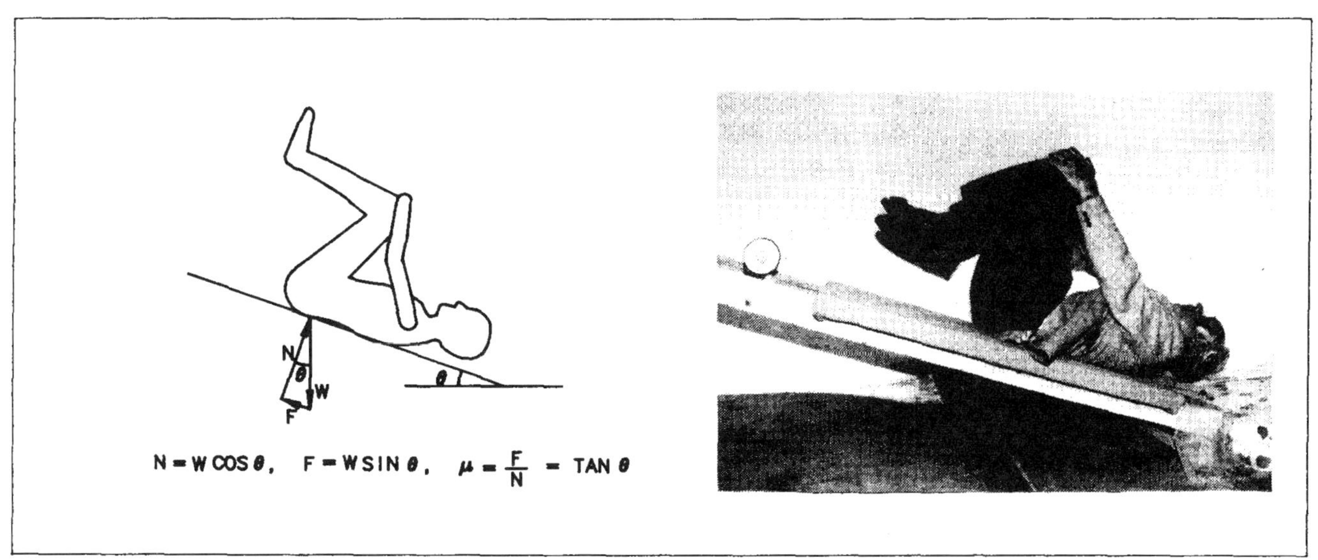

FIGURE 2: OCCUPANT SEATBACK FRICTION EXPERIMENTS

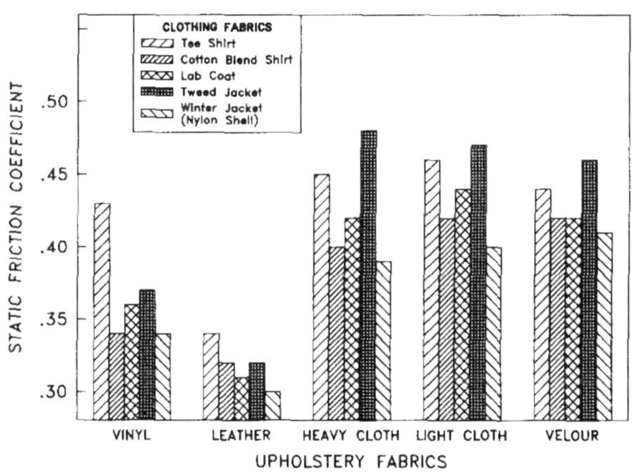

FIGURE 3: STATIC FRICTION COEFFICIENTS BETWEEN UPHOLSTERY AND CLOTHING FABRICS

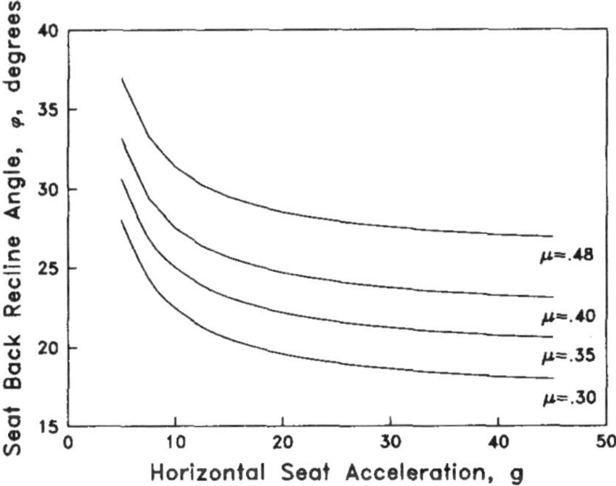

FIGURE 4: SEAT BACK RECLINE ANGLE FOR INCIPIENT RAMPING AS A FUNCTION OF HORIZONTAL SEAT ACCELERATION

Rigid seat backs would have to be uncomfortably vertical in order to prevent initiation of ramping of unrestrained occupants.

REBOUND - A second concern with rigid seat backs is the injury potential associated with rebound. No practical seat can be completely rigid; some elastic deformation will occur during occupant loading. The energy associated with that elastic deformation is stored and returned to the occupant in the forward direction at the end of the impact. Injury from contact with frontal vehicle components is a concern for unbelted occupants. Yielding seat backs absorb energy as they deform, but still have the potential to store some elastic energy. Field accident data show with current, yielding seat backs, the injuries associated with this rebound phenomenon represent a greater risk than the injuries associated with the occupant contacting rear vehicle structures and ejection [Data Link, 1990]. Rigidized seat structures could substantially increase these rebound-related injuries.

Rebound is also an important factor in low severity whiplash injuries. Whiplash injuries are not well understood, and persist in spite of mandatory head restraints. In fact, some researchers feel that whiplash injuries occur before the head ever hits the head restraint [Bogduk, 1986]. A recent evaluation of field accident data in France found that seat back deformation "seems more efficient than the head restraint in reducing neck pain" [Foret-Bruno, 1990]. Even though the exact mechanism for whiplash is not yet known, it is generally agreed that stiffer seat backs will make the whiplash problem worse [Berton, 1968], [Foret-Bruno, 1990], [Horkey, 1989], [Kihlberg, 1969], [Mertz, 1967], [Mertz, 1968], [Thomas, 1982].

In assessing the potential for rebound with rigid seats, it is important to note the method of making a seat rigid can greatly influence the amount of elastic energy stored by the system. Production seats must be longitudinally adjustable. Recline adjustability has also become an expected feature in most vehicles. A rigid seat with these adjustment features will store more elastic energy than a non-adjustable rigid seat. The difficulty of building an adjustable rigid seat has been discussed elsewhere [Strother, 1987]. Thus, the testing of artificially rigidized, non-adjustable seats is of little benefit in assessing the injury potential associated with rebound in practicable production seats.

OUT-OF-POSITION OCCUPANTS - A third, and perhaps the most serious, concern with rigid seat backs is the increased injury potential for out-of-position occupants. Automobile design and testing uses the Normal Seated Position (NSP) for occupants in order to have a benchmark for design and repeatability in testing. Testing done to establish whole body tolerance levels were performed using young healthy males anticipating the acceleration, and perfectly positioned in elaborate restraint systems. Little testing has been done on such tolerance over the range of actual seated positions of occupants in collisions. However, full scale testing and mathematical models of rear impact collisions have shown that only small variations in occupant position can result in large increases of impact forces [Berton, 1968], [Foret-Bruno, 1990], [Hu, 1977], [Olsson, 1990], [Romilly, 1989], [Severy, 1967]. The early air bag research recognized the need to address the out-of-position problem with regard to air bag deployment in frontal impacts. It was recognized that not only do some occupants normally sit in an out-of-position posture, but that pre-impact braking or swerving can cause a normally seated occupant to be out-of-position at impact. One analysis of Japanese traffic accident data found that in 70 percent of the cases studied, severe braking immediately preceded the collision [Takeda, 1980]. These researchers investigated the effect of such braking on occupant position by observing the motion of an unrestrained child dummy and also an unrestrained adult dummy "holding" a child dummy in panic braking situations. In both cases, the child and adult dummies were thrown forward such that their heads and chests contacted the instrument panel. Stalnaker also demonstrated the effect of pre-impact braking on occupant position using anesthetized baboons and child dummies [Stalnaker, 1982].

The 1982-1984 NASS data files were examined to determine how often pre-impact vehicle momentum changes precede rear impacts. These data files listed three actions of the subject vehicle immediately prior to the collision, including braking, swerving, and spinning/yawing. Figure 5 shows that about two-thirds of rear impacted vehicles experienced such actions before impact.

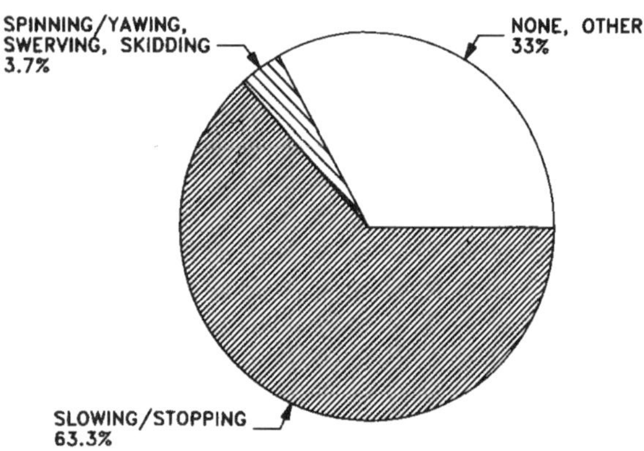

FIGURE 5: ACTION PRIOR TO REAR IMPACT
NASS 1982-1984

Virtually all rear-impact testing with dummy and cadaver subjects has been conducted with "normally seated" occupants, i.e., erect with their backs firmly against the seat back. As an interesting contrast, the response of an out-of-position occupant in a rigid seat was inadvertently tested by researchers at the University of Michigan during their work with the Integrated Safety Seat [Melvin, 1971], [Hilyard, 1973]. An acceleration-deceleration sled was being used to develop and evaluate deployable head restraints. In several of the tests the dummy occupant was not properly tethered into the NSP prior to the test. During the ride-up to impact velocity, the dummy moved forward in the seat. This motion is similar to that which would occur from either braking prior to a rear impact or from a relatively minor frontal impact preceding a rear collision.

Table 2 identifies these out-of-position tests, giving the sled velocity (delta-V), the dummy size, and comments of the investigators concerning the test results. These tests differ markedly from tests run in the same series with dummies in the NSP. First, the out-of-position dummy tends to contact the seat back relatively high, with its head extending over the head restraint. Second, there is more ramping of the out-of-position dummy up the seat back. Third, the out-of-position dummy experiences more severe hyperextension of the head/neck. Fourth, there is an increased upward and forward rebound of the out-of-position dummy out of the seat. Figure 6 presents graphics derived from the high speed film for one of these out-of-position tests.

These sled tests are not representative of real car-to-car rear impacts because the sled buck was constrained to have only linear horizontal motion (i.e., no vertical impulse) and the sled buck was not equipped with a roof structure. The films indicate a high risk of head contact if there had been a roof structure, even without any vertical motion of the vehicle. The lack of neck load transducers, and the poor biofidelity of the head/neck system also makes it difficult to accurately assess the injury potential in these tests. However, it is clear that problems exist for out-of-position occupants during rear impacts with rigidized seats.

No tests were found in the literature involving out-of-position testing with production or other yielding seat backs. Impacts with a yielding seat back structure by an out-of-position occupant would be expected to be of reduced severity because of the energy absorbing properties of the yielding seat, which would attenuate impact effects of the relative velocity built up between the occupant and the vehicle compartment.

CONCLUSIONS

The rulemaking history of seat design standards indicates that the concept of yielding, energy absorbing seat backs has historically been considered the appropriate design approach for passenger cars. Research findings associated with the rulemaking efforts confirm that rear impacts do not represent a substantial contribution to overall automobile accident injuries, and that yielding seat backs do provide occupant protection.

Static pull test data of 61 production seats indicate that the strength of production passenger car seats has not substantially changed over the past three decades.

TABLE 2

"Inadvertent" Out-of-Position Occupant Tests from Deployable Head Restraint Sled Testing

TEST NO	ΔV (mph)	DUMMY SIZE	COMMENTS OF INVESTIGATORS CONCERNING TEST
A-334	30	95M	"Dummy was leaning slightly forward out of position at the start of the crash pulse. Moderate head-neck extension occurred along with ramping. There was significant upward rebound along with forward rebound."[1]
A-350	30	95M	"Dummy was poorly positioned - leaning forward. Upon impact the dummy rotated backward striking the head restraints high with its shoulders. The head missed the [head] restraint resulting in severe head-neck extension followed by violent upward and forward rebound."[1]
A-361	30	95M	"The dummy was leaning forward out of position at the start of the pulse and rotated backward severely, resulting in the head going over the top of the head restraint with severe head-neck hyperextension. There was severe upward and forward rebound."[1]
Z-139	40	50M	"Due to inadequate installation of cables that hold the dummy in the seat during sled acceleration, the dummy was badly out of position at impact. The results of the test are <u>invalid</u> [underlining added]."[2]

1. [Melvin, 1971]
2. [Hilyard, 1973]

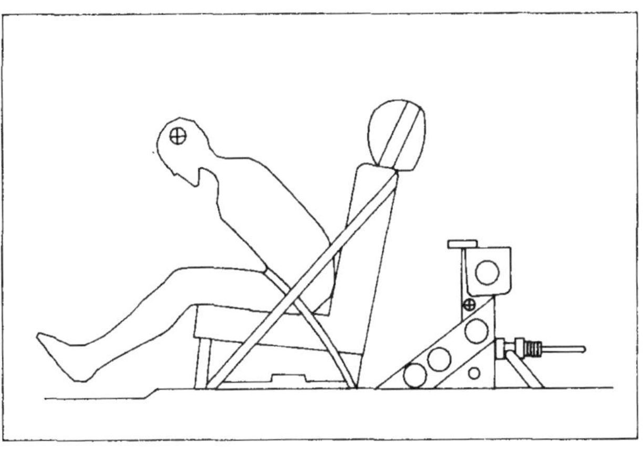

t=0 ms

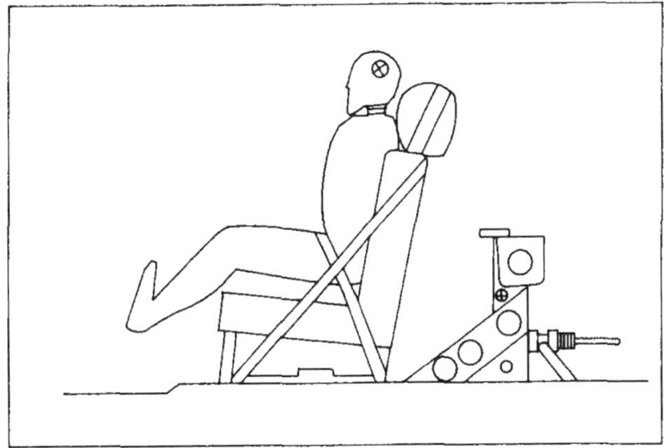

t=53 ms

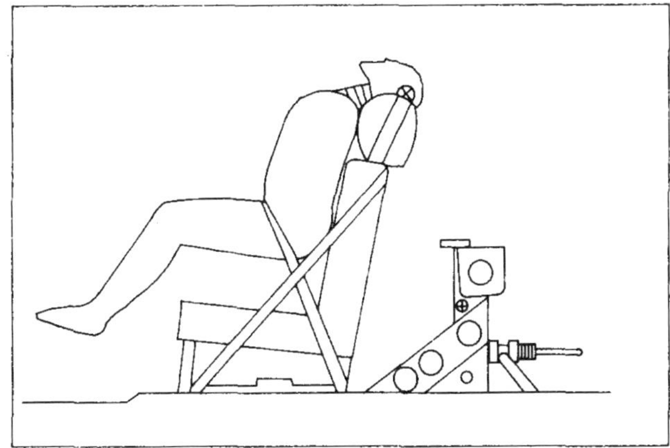

t=78 ms

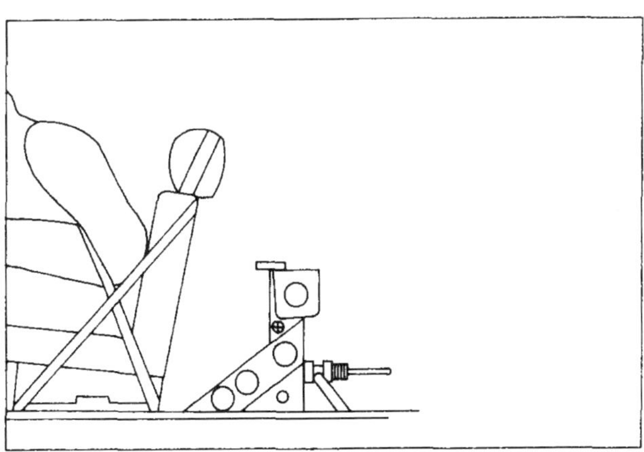

t=160 ms

FIGURE 6: Out-Of-Position Occupant Kinematics With Rigid Seat (based on film of HSRI sled test Z-139) [Hilyard, 1973]

Rigid seat backs have the potential to increase injury exposure in real world impacts due to three major concerns:

1) Rigid seats have the potential of exposing unrestrained occupants to impacts with the roof structure in severe rear impacts. Ramping can be expected with seat back angles as low as 25 degrees behind the vertical.
2) Non-yielding seats, particularly those with adjustable seat backs, can increase occupant rebound because all of the seat deflection will represent elastic energy which will be returned to occupants in the form of rebound velocity. Injuries associated with rebound include whiplash and contacts with frontal vehicle components.
3) Rigid seats are potentially dangerous to occupants not in the Normal Seated Position, especially unrestrained occupants. A majority of rear-impacted vehicles experience pre-impact changes in momentum which could cause occupants to be out-of-position at impact.

ACKNOWLEDGEMENTS

Grateful appreciation is expressed to Doug Allsop, Gerald Carter, and Karl Wenger for assistance with the friction experiments; to Ron Nordhagen and Nan Strother for preparing the figures and tables; and to Elaine Murray for the typewritten drafts.

REFERENCES

[Berton, 1968]
R.J. Berton: "Whiplash: Tests of the Influential Variables." SAE 680080, Automotive Engineering Congress, Detroit, MI. January 8-12, 1968.

[Bogduk, 1986]
Bogduk, Nikolai: "Anatomy and Pathophysiology of Whiplash," Clin Biomech, Vol. 1, No. 2, pp. 92-101, May, 1986.

[Carr, 1988]
Carr, W. W.; J. E. Posey, W. C. Tincher: "Frictional Characteristics of Apparel Fabrics," Textile Research Journal, 0040-5175/88/58003, p. 129, March, 1988.

[Carter, 1976]
Carter, Robert L., Associate, Administrator, NHTSA Motor Vehicle Programs. Letter to M.H. Lanz. Docket 74-13-NO1-044. July 9, 1976.

[Chrysler, 1974]
"Detailed Comments by Chrysler Corporation on Docket 74-13; Notice 1, Notice of Proposed Amendment MVSS 207 - Seating Systems." Docket No. 74-13-NO1-07. June 11, 1974.

[Data Link, 1989]
"Car Crash Outcomes in Rear Impacts. Appendix A to Current Issues of Occupant Protection in Car Rear Impacts." Data Link, Inc., Washington, D.C. 1989.

[Data Link, 1990]
"Current Issues of Occupant Protection in Car Rear Impacts," prepared for Office of Crashworthiness, Rulemaking, NHTSA. Data Link, Inc., Washington, D.C. Feb., 1990. Docket 89-20-NO1-021.

[Eiband, 1959]
Eiband, A. Martin: "Human Tolerance to Rapidly Applied Accelerations: A Summary of the Literature," NASA Memo 5-19-59E. Washington, D.C. June 1959.

[32 FR 20865, 1967]
FMVSS 202 Notice of Proposed Rulemaking, December 28, 1967.

[33 FR 2945, 1968]
FMVSS 202 adopted. February 14, 1968.

[33 FR 5793, 1968]
FMVSS 202 amended. April 16, 1968.

[33 FR 15065, 1968]
FMVSS 202 amended. October 9, 1968.

[32 FR 2415, 1967]
FMVSS 207 original notice. February 3, 1967.

[32 FR 5498, 1967]
FMVSS 207 amended. April 4, 1967.

[32 FR 14281, 1967]
FMVSS 207 ANPR. Request for comments. October 14, 1967.

[33 FR 19723, 1968]
FMVSS 207 amended. December 25, 1968.

[39 FR 10268, 1974]
Proposal to amend FMVSS 207 & 202, establish Docket 74-13. March 19, 1974.

[40 FR 48353, 1975]
FMVSS 301 Original Notice. October 15, 1975.

[43 FR 11100, 1978]
Request for comments. Established Docket 78-07-NO1. NHTSA 5 yr. plan. March 16, 1978.

[54 FR 40897, 1989]
Request for comments. Opened Docket 89-20-NO1. October 4, 1989.

[55 FR 42031, 1990]
Termination of Rulemaking. October 17, 1990.

[Ford, 1974]
"Comments of Ford Motor Company," Docket 74-13-NO1-15. June 17, 1974.

[Foret-Bruno, 1990]
Foret-Bruno, J. Y.; C. Tarriere, J. Y. LeCoz, C. Got; F. Guillon: "Risk of Cervical Lesions in Real-World and Simulated Collisions." 34th AAAM Conference Proceedings, Scottsdale, Arizona, p. 373, October, 1990.

[General Motors, 1989]
"Seatback Performance in Rear Impacts," USG 2751 Attachment. Robert A. Rogers, General Motors Director, Automotive Safety Engineering, Dec. 4, 1989. Docket 89-20-NO1-009.

[GSA 515-6, 1965]
General Services Administration Standard 515-6, "Anchorage of Seats for Automotive Vehicles." Federal Standard, U.S. Government Printing Office. June 30, 1965.

[Hilyard, 1973]
Hilyard, Joseph F.; John W. Melvin; James H. McElhaney: "Deployable Head Restraints." Highway Safety Research Institute, University of Michigan. DOT-HS-800-802. Jan. 31, 1973.

[Horkey, 1988]
Horkey, Edward: Petition to National Highway Traffic Safety Administration. Docket 89-20-NO1-001. March 25, 1988.

[Horkey, 1989]
Horkey, Edward: Letter to Docket 89-20. November 27, 1989. Docket 89-20-NO1-007.

[Hu, 1977]
Hu, A; S. Bean; and R. Zimmerman: "Response of Belted Dummy and Cadaver to Rear Impact." SAE 770929, Proceedings of the 21st Stapp Car Crash Conference. 1977.

[James, 1991]
James, Michael B.; Charles E. Strother, Charles Y. Warner, Robin L. Decker, and Thomas R. Perl: "Occupant Protection in Rear-end Collisions: I. Safety Priorities and Seat Belt Effectiveness." Proceedings of the 35th Stapp Car Crash Conference, San Diego, CA. November 1991.

[Kihlberg, 1969]
Kihlberg, J.K.: "Flexion-Torsion Neck Injury in Rear Impacts," Cornell Aeronautical Laboratory, April 1969.

[MVMA, 1974]
"Seating Systems," Docket 74-13-NO1-016, Motor Vehicle Manufacturers Association of the United States, Inc. June 17, 1974.

[Mercedes Benz, 1989]
"Comments to Docket 89-20, Notice 1 Concerning Standards 207, 208 and 209," Docket 89-20-NO1-014, Mercedes-Benz of North America, Inc. December 7, 1989.

[Melvin, 1971]
Melvin, J.W.; J.H. McElhaney: "Deployable Head Restraints." Final Report, Highway Safety Research Institute. University of Michigan. FH-11-7612. June 30, 1971.

[Mertz, 1967]
Mertz, H.J.; L.M. Patrick: "Investigation of the Kinematics and Kinetics of Whiplash," Proceedings, 11th Stapp Car Crash Conference (SAE 670919), October 1967.

[Mertz, 1968]
Mertz, H.J.: "Car Occupant Response to Rear-end Collisions - A Mathematical Model," 1st International Conference on Vehicle Mechanics, p. 588, July 1968.

[NHTSA, 1979]
"Five Year Plan for Motor Vehicle Safety and Fuel Economy Rulemaking Calendar Years 1980-1984." U.S. Dept. of Transportation. NHTSA, Washington, D.C. April 20, 1979.

[Olsson, 1990]
Olsson, I.; O. Bunketorp: "An In-Depth Study of Neck Injuries in Rear End Collisions," IRCOBI, p. 269, September 12, 1990.

[Partyka, 1987]
Partyka, Susan C.: "Fatalities in Rear-Impact Small Cars From 1982 Through 1987," Papers on Vehicle Size -- Cars and Trucks, p. 125, December, 1987.

[Partyka, 1988]
Partyka, Susan C.: "Belt Effectiveness in Pick-up Trucks and Passenger Cars by Crash Direction and Accident Data." National Center for Statistics and Analysis, DOT HS 807 285. June, 1988.

[Partyka, 1990]
Partyka, Susan C.: "Comparisons of Belt Effectiveness in Preventing Chest, Head, and Face Injuries in Front and Rear Impacts." August, 1990.

[Posey, 1984]
Posey, Jamie E.: "Measurement of Frictional Characteristics of Fabrics." Thesis from Georgia Institute of Technology. March 1984.

[Prasad, 1975]
Prasad, P.; N. Mital, A.I. King, L.M. Patrick: "Dynamic Response of the Spine During Gx Acceleration," (SAE 751172), Proceedings, 19th Stapp Car Crash Conference, 1975.

[Romilly, 1989]
Romilly, D. P.; R. W. Thompson; F. P. D. Navin; M. J. Macnabb: "Low Speed Rear Impacts and the Elastic Properties of Automobiles." Proceedings of the 12th ESV Conference in Gothenburg, Sweden. May, 1989.

[Strother, 1987]
Strother, Charles E.; and Michael B. James: "Evaluation of Seat Back Strength and Seat Belt Effectiveness in Rear End Impacts." SAE 872214, Proceedings of the 31st Stapp Car Crash Conference, New Orleans, Louisiana. October 9-11, 1987.

[Saczalski, 1989]
Saczalski, Kenneth: Petition to National Highway Traffic Safety Administration. Docket 89-20-NO1-001, April 18, 1989.

[SAE J879, 1963]
"Motor Vehicle Seating Systems." 1963 SAE Handbook, Recommended Practice J879. 1963.

[Severy, 1967]
Severy, D.M.; H.M. Brink, J.D. Baird: "Preliminary Findings of Head Support Designs," Proceedings, 11th Stapp Car Crash Conference, (SAE 670921), October 10-11, 1967.

[Stalnaker, 1982]
Stalnaker, R.L.; L.F. Klusmeyer, H.H. Peel, C.D. White, G.R. Smith, H.J. Mertz: "Unrestrained, Front Seat, Child Surrogate Trajectories Produced by Hard Braking," 26th Stapp Car Crash Conference. SAE # 821165, p. 231, 1982.

[Takeda, 1980]
Takeda, Hideo T.; Saburo Kobayashi: "Injuries to Children From Airbag Deployment." 8th ESV, Wolfsburg, Germany, p. 325, October 21-24, 1980.

[Thomas, 1982]
Thomas, C.; G. Faverjon, F. Hartemann, C. Tarriere, A. Patel, C. Got: "Protection Against Rear-End Accidents," 1982. IRCOBI Conference, p. 17, 1982.

[Wilson, 1963]
Wilson, D.: "A Study of Fabric-on-Fabric Dynamic Friction," Journal of the Textile Institute Transactions, p. T143, April, 1963.

APPENDIX:

SEATBACK FRICTION AND RAMPING

The simplest physical model for seatback ramping may be derived from a quasistatic analysis of the forces acting on a seated occupant in a rear impact. The model relates the limiting static friction coefficient, μ, the [assumed rigid] seatback recline angle, ϕ, and the vectorial relationship between horizontal acceleration, a, and the acceleration of gravity, g. It predicts the seat recline angle at which ramping will begin, as a function of limiting static friction coefficient and horizontal seat acceleration. It is important to note that the analysis neglects dynamic effects which substantially encourage initiation of ramping, by developing both upward momentum and relative velocity at the friction surfaces. It also neglects details of seat design, anthropometry, and biomechanical stiffnesses, which can have second-order effects on the initiation of ramping. Frictional test values and graphs of resulting relations are shown in the text in Figures 2 - 4.

Assume that a subject having weight W is at the point of incipient sliding, or ramping up a seatback which is reclined at an angle ϕ behind the vertical as depicted in Figures A-1 and A-2. The force R which tends to potentiate ramping is caused by body inertia, reacting to

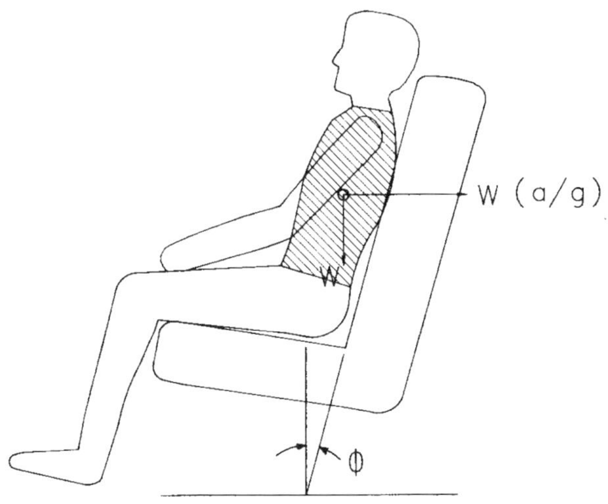

FIGURE A–1: TEST SUBJECT

the forward acceleration, a, of the seat. The net force resisting occupant upward motion is a vector combination of the gravity force W and the frictional force, F, which is itself a product of the friction coefficient, μ, and the

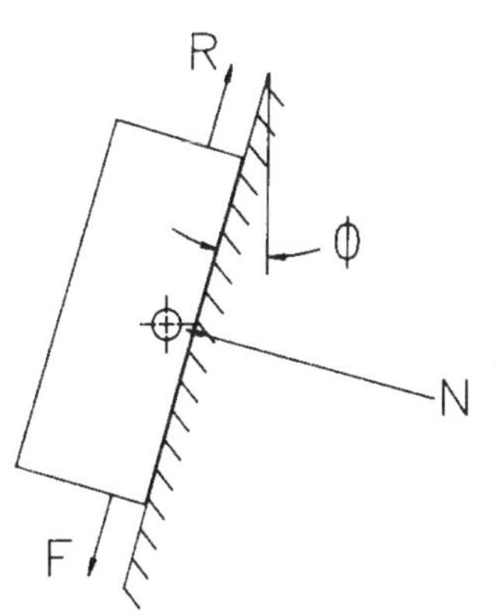

FIGURE A–2: POTENTIAL RAMPING FORCES

normal force, N. The normal force is also a combination of the gravity force and the inertia "force" due to acceleration, hence

$$R = W\,(a/g)\cos(90-\phi) - W\cos\phi$$

$$= W\,[(a/g)\sin\phi - \cos\phi], \quad \text{and}$$

$$N = W\,(a/g)\sin(90-\phi) - W\sin\phi$$

$$= W\,[(a/g)\cos\phi - \sin\phi\,].$$

At incipient ramping, $R \geq F = \mu N$, or

$$\boldsymbol{\phi \geq \arctan\{[1 + \mu(a/g)]/[(a/g) - \mu]\}}$$

This relation is plotted in Figure 4 of the text for normally encountered values of a and μ in rear-end collisions.

872214

Evaluation of Seat Back Strength and Seat Belt Effectiveness in Rear End Impacts

Charles E. Strother and Michael B. James
Collision Safety Engineering, Inc.

ABSTRACT

The issues of front seat energy absorption and seat belt effectiveness are investigated first through the review of prior experimental and analytical studies of rear impact dynamics. These prior studies indicate that the current energy absorption characteristic of seats is a safety benefit. Prior efforts to construct a rigidized seat indicate that such designs are likely to be impractical due to excessive weight and cost. Additionally, these studies indicate that seat belts provide an important safety function in rear impacts. Static tests of production seats were conducted, added to an existing data base, and analyzed to better understand the strength and energy absorbing characteristics of production seats. Crash test results from the New Car Assessment Program as well as earlier test programs were analyzed to describe the response of occupants and seats in rear impact and the protective function of seat belts in such collisions. Finally, accident statistics were reviewed to determine the present level of occupant protection provided by current seats. Field experience indicates that, even with a largely unbelted population, current seats are providing an impressive level of occupant protection in rear impact. A significant potential was observed for increasing rather than decreasing occupant injury potential in rear collisions by obviating the current energy absorbing characteristic of seats.

THE OBJECTIVE OF THE RESEARCH LEADING TO this paper was to determine whether or not there was any merit in the concept of rigidizing seat structures for rear impact and secondly, whether seat belts were of benefit in these types of collisions. The main body of the paper is divided into three sections. In the initial section, prior research on rear impact is reviewed in order to answer a series of fundamental questions concerning the role of seat back yield and seat belts in providing occupant protection. This is followed by a section describing the characteristics of current production seats, as determined from both static and crash test analysis. The concluding section of the paper examines current accident statistics in order to better understand the relative importance of rear impact injuries and the levels of protection currently being realized in this accident mode.

SURVEY OF PRIOR LITERATURE

Automobile seats and their effect on occupant safety in rear end collisions were a concern in the safety literature for an approximately twenty-year period beginning in the mid-1950's. One of the confounding factors in summarizing this literature is the lack of continuity that one typically finds in researching such a safety issue. The literature can be considered to be divided among three basic types of efforts. Perhaps the most interesting category of papers deals with experimental attempts to investigate the various factors involved in rear impact protection. A second group of papers developed mathematical models to simulate the motion of the seat and occupant during a forward acceleration in order to investigate many of these same factors. A third category offers views on various aspects of rear impact protection, but do not present either experimental or analytical results which would support these views. Prior experimental and analytical efforts were examined in order to attempt to answer three questions which were felt to be fundamental to the issue of the protection of occupants in rear collisions.

---

*Numbers in parentheses indicate references at the end of the paper.

These questions are:

(1) In comparing yielding seats to rigid seats, is there any safety advantage to either design?

(2) Is a rigid seat practical?

(3) Are seat belts effective in rear impacts?

<u>Desirability of Rigid Versus Yielding Seats</u> - In 1955, Severy and Mathewson (1)* conducted some of the first rear impact collisions using 1947 Plymouths as the struck vehicle. Both human volunteers and crash dummies were used as subjects for the tests which were conducted at impact speeds of 7 to 19.8 mph. In comparing run 3, conducted at an impact speed of 9.9 mph with run 5, conducted at 19.8 mph, those authors noted the unexpected result that the head accelerations were essentially the same despite the large difference in impact energy levels. This was attributed, after film analysis, to the additional seat deflection in the higher speed test. It is noted that these tests were conducted on seats lacking head restraint.

The experiments begun by Severy et. al. in the mid-fifties were continued up to about 1968 (2-4) to investigate the effects of rigidizing front seats on belted occupant response. The most interesting of these were the series of 13 tests beginning with Experiment X93 and ending with X106. These tests, summarized here in Appendix A, allow a comparison to be made that addresses the issue of the desirability of rigidizing seat structures. Tests X96 and X99 were identical in nearly all respects except for the fact that the seats in the later test were rigidized to prevent permanent seat back deformation. Each involved lap-belt-restrained 95th percentile male dummies with 25-inch and 28-inch high seat backs. In comparing these two tests, the investigators noted that the dummies in the stiffer seats experienced significantly higher whiplash motions.

Martinez et. al., (5-7) developed and exercised a 3-mass 2-dimensional analytical model and conducted a series of animal experiments in an attempt to compare rear impact response with rigid, elastically yielding and plastically yielding seat backs. The analytical model assumed that the occupant moved with the seat back (in effect assuming that the occupant was belted to the seat with an integral belt system). The model predicted disadvantages for the elastically yielding seat, but indicated that both the rigid and the plastically yielding seat were reasonable design options for the fully restrained occupant with a properly positioned and designed head restraint.

In 1967, Mertz and Patrick reported on a series of sled tests using lap-belted dummies, cadavers and human volunteers (8). Among the parameters investigated were the presence or absence of head restraint and seat belt restraint, and rigid versus yielding seat backs. It was concluded that, with or without head restraints, the controlled yielding of the seat back reduced the severity of the impact to the occupants.

Mertz also developed and exercised a 2-mass, 2 degree of freedom planar model as an outgrowth of the above experimental study (9). The math model, which assumes a body rigidly linked to the seat back, indicated that satisfactory performance could be achieved with a rigid seat back under these conditions but allowing the seat back to yield in a controlled fashion reduced both the head articulations and loads when compared to the rigid seat simulations.

Berton reported on a series of sled tests and vehicle crashes in 1968 using belted 50th percentile male dummies (10). In comparing yielding versus non-yielding seat backs, Berton concluded that the yielding seat back reduced the injury exposure of the head.

In 1969, Kihlberg examined the Automotive Crash Injury Research (ACIR) data to determine whether injury exposure could be related to seat deformation (11). He found statistically significant reductions in injury frequency in cases where seat deformation occurred. He noted that this was despite the fact that those cases involving seat deformation were of relatively higher impact severity.

Under contract to the National Highway Traffic Safety Administration (NHTSA), the Highway Safety Research Institute (HSRI) of the University of Michigan developed and tested a rigid seat platform called the Integrated Safety Seat (ISS) in the rear impact mode as part of their efforts to develop a deployable head restraint (12,13). Although this program was limited to work with a rigid seat platform, the analytical and experimental results indicated that ramping problems may be associated with the use of rigid seats by both lap belted and unrestrained occupants.

At the 19th Stapp Conference, investigators at Wayne State University presented the results of a study incorporating both analytical model exercises and three cadaver sled tests (14). The analytical model was a planar 78 degree of freedom model of the articulating head, spine, and pelvic system together with provisions to allow the simulation of a rotating seat back and the movement of an unrestrained upper body relative to the back. The experimental sled tests were used to verify the model.

Comparing model simulations with rigid and yielding seat backs, the authors noted that head angular acceleration and spinal forces and moments were reduced with the yielding seat back. Also noted was the tendency for the unrestrained pelvis to ramp up the seat back. Noting that the upward motion of the pelvis in the yielding seat simulations was greater than that predicted for the rigid seat, the authors suggest that further investigation be made.

Hu et. al., at the 21st Stapp Conference, presented the results of a series of dummy and cadaver sled tests simulating a rear impact with a 16 mph delta-V (15). All tests involved belted subjects and a seat back with the head restraint in the lowest position. Experiments were conducted with both rigid and yielding seat backs. The severity of these impacts were such that all but one of the six cadaveric subjects sustained serious neck injury (the one cadaver without serious neck injury was noted by the authors to have significant anatomical differences). The authors concluded on the basis of these experiments that the deflecting seat is effective in reducing chest and head severity indices. In addition, it was noted that the dynamic responses in those cases involving rigid seats were very sensitive to changes in the impact conditions (including the occupant's position at impact) whereas in the cases involving deflecting seat backs they were not.

Practicality of Rigid Seats - Aside from the issue of whether a significant increase in the strength of current production seats would be a net safety benefit is the issue of whether such increases could be practically achieved. That is, whether rigidized seats meeting consumer expectations could be built without either incurring prohibitive weight and/or cost penalties or compromising comfort and convenience. In this section of the paper, past efforts at building rigidized seats are reviewed.

One of the first attempts known to the authors to build a rigidized seat was by Cox of Watford Ltd. in Great Britain (16). Figure 1, reproduced from Reference 16, is an illustration of the so-called "car seat of the future". Essentially this seat was a gusseted L-shaped structure containing an integral head restraint and four-point active seat belt system. In a later publication, the developers of this seat indicated that it was "too large, clumsy, and costly to be adapted to any British model" (17). It could also be pointed out that given the low usage rates for the relatively simple 2 and 3-point belt systems, usage of the 4-point system, an important part of the concept, would logically be expected to be slight. Also, this seat was lacking in a number of comfort features that occupants currently expect in their vehicle seats. Among these is the ability of the seat

FIGURE 1 – THE COX SAFETY SEAT (1963)
(Figure Reproduced from Reference 16)

to accommodate a wide range of occupant sizes. This requirement is compromised by the inclusion of gussets linking the seat cushion and frame, an important feature when one is trying to rigidize this connection with minimum weight penalty. Another feature which occupants have come to expect in their seats is seat back adjustability. This feature has become increasingly important with the reduction in car size since taller persons would find it next to impossible to occupy many small cars without the ability to recline the seat back. Inserting an adjustment feature in this location of high stress presents a particularly difficult design problem that will significantly increase the weight and cost of the design. No indication was given in this report that the presented design had been installed and tested in an actual vehicle.

Three years later, Hilton of the same Cox group presented another rigidized seat design at the 10th Stapp Car Crash Conference (17). This design, depicted in Figure 2, reproduced from Reference 17, was viewed as a production-engineered improvement on the earlier seat. The seat frame was apparently still configured to be rigid, but the mounting of the seat to the vehicle compartment was to permit controlled yielding so that the seat would pitch forward or rearward upon impact. The newer seat incorporated a integral 3-point rather than a 4-point belt system and retained the gusseted seat back feature of the earlier seat. The results of a dynamic frontal test were briefly discussed in this report, but no mention was made of any installation and test in the rear impact mode.

One year later, Severy of ITTE at the University of California at Los Angeles (UCLA) presented the prototype Capsule Seat designed for the Liberty Mutual Safety Car, see Figure 3 (18). As envisioned by its creator, this swivel seat was, like the Cox designs, equipped with an integrated active seat belt system wherein the belt anchors were attached to the seat. The

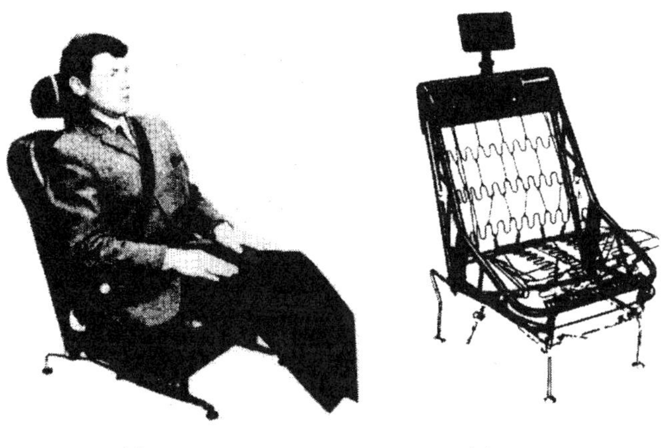

(a) Complete Seat (b) Seat Frame

FIGURE 2 – COX SAFETY SEAT 2 1966
(Figure reproduced from Reference 17)

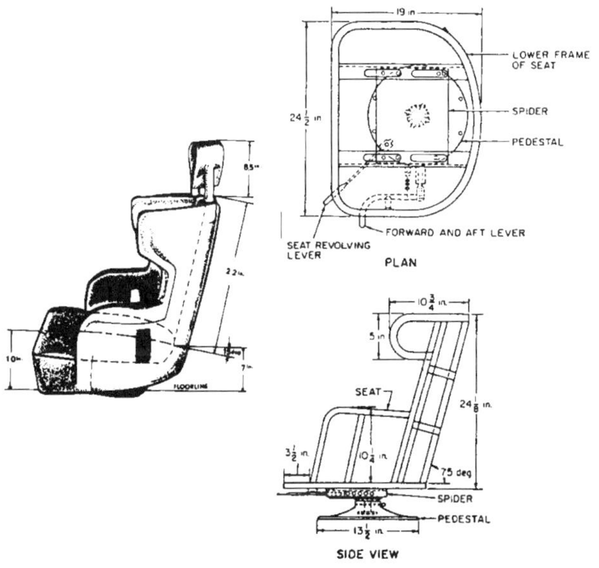

FIGURE 3 – THE LIBERTY MUTUAL CAPSULE SEAT (1967)
(Figure Reproduced from Reference 18)

weight of this seat was about 100 pounds, contrasted with a typical production bucket seat weight of about 25 pounds. This seat, like the Cox designs, lacked comfort and fit features. The seat back could not be adjusted and, for the same engineering reasons, the seat was provided with gussets at the cushion/back junction that would limit its accommodation range. Additionally, the LM Safety Seat was provided with shoulder-level "wings" for lateral impact protection that would raise negative consumer reactions since they restrict shoulder room and complicate entry and egress.

Severy, around 1966 began a series of 13 rear impacts with 1967 Fords as both the struck and striking vehicle (see Appendix A). For the purpose of investigating the effects of seat parameters on occupant response in the struck vehicles, Severy evolved what he termed a Calibrated Seat, designed to sustain rearward forces well beyond the range of production seat designs (2,3). Figure 4, reproduced from Reference 2, depicts this seat of conventional cantilevered seat back design, which contained an adjustment feature at the seat back pivot for strengthening the rigidity of the back-to-cushion link to prescribed moment or force levels. This link was capable of being adjusted to produce rearward moments about the seat cushion plane of 33,000 in-lbf. or more, making it potentially about 8-9 times as strong as production bucket seats. While this seat does not have the apparent disadvantage of cushion-to-back gussets (or "wings") there must be a significant weight penalty due to the necessity of building additional moment-carrying capacity

device was not one that could actually be used in a vehicle since it lacked adjustability and, moreover represented a hazard for rear seat occupants in frontal impacts (19).

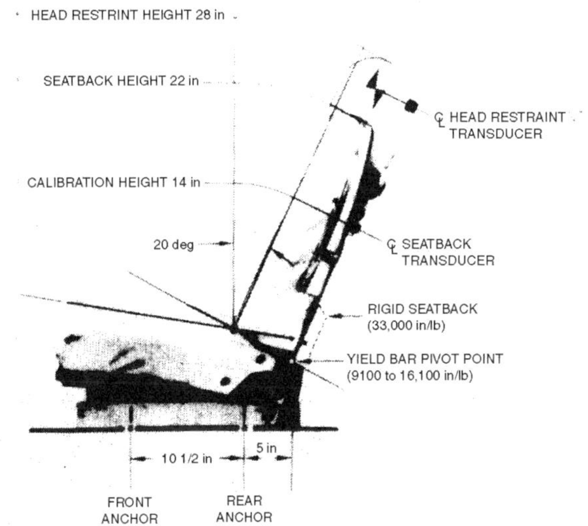

FIGURE 4 – SEVERY "CALIBRATED" SEAT
(Figure Reproduced from Reference 2)

Referring to Appendix A, the experiments reported involve the calibrated seat adjusted to various strength levels and back heights. It is noted that, with the singular exception of Experiment X106, all of the seats in this series of tests experienced dynamic angular deflections. This indicates a potential problem in trying to develop a rigid seat, namely that one may, instead, produce an elastically deforming seat which will store and return a significant percentage of the impact energy to the occupant. This energy can manifest itself in the form of enhanced whiplash motion and/or rebound velocity into frontal structures. In fact, whiplash motions and rebounding were observed to occur in a number of these tests. The only instances where there was no permanent rearward deflection of the seat back angle were in Experiment 97, involving a delta-V of only about 5.5 mph; in Experiments 99 and 102, where there was a delta-V of about 16.5 mph; and in Experiment 106, where a beam was welded between the B-pillars of the vehicle to support the seat backs (delta-V about 30.3 mph). In Experiment 97 this complete restoration of back angle was accomplished with the Calibrated Seat set at a torque of 16,100 in-lbf., which translates to a force of about 1125 lbf. at a level of 14 inches above the plane of the seat cushion. Referring to the section of this report dealing with the static strengths of production seats, it is observed that this seat strength corresponds to the maximum level of current designs. In order to achieve no residual seat back deflection at a delta-V of about 16.5 mph (Experiment 99) it was necessary to have a seat yield strength that was about four times the average of production seats. At a delta-V of just over 30 mph, it was found (X104) that even placing a 4-inch channel beam under the floor to anchor the seats was insufficient and that it was necessary to place a beam behind the seat backs in order to prevent seat back deformation (X106).

A second important observation made from the Severy tests is the importance of having both a high seat back and an effective belt system in place when using a relatively stiff seat. The need for the high seat back is increasingly important to avoid severe hyper-extension or "whiplash" motions given that seat-back-produced torso loads will increase with the stiffening of the seat. The need for an effective seat belt system is increased due to the phenomenon of ramping which occurs with even high backed seats, and also the need to control rebound. This need was observed to increase as the stiffness and impact severity level increased, as demonstrated in Experiment X106. In this test, the lap and shoulder belt system of the Cox seat controlled the occupant's ramping to a much greater degree than did the lap belt used with the Calibrated Seat, which allowed the dummy to ramp about 4 inches up the seat back.

In the early 1970's the University of Michigan, under contract to the National Highway Traffic Safety Administration (NHTSA), built a prototype of what was termed an Integrated Safety Seat (ISS). It consisted of a rigidized seat platform to which, like the LM Capsule Seat, seat belt anchorages were attached (12). This platform, shown in Figure 5, was lacking in adjustment features and was too large for most vehicles. The authors stated that the ISS weighed about 70 lbf., not including the necessary reinforced mounting hardware. The seat was initially dynamically tested on the HSRI sled. Later, the ISS was installed in a production vehicle and crash tested at a severity (delta-V) of about 40 mph. The seat rotated rearward due to floor anchorage problems; similar to the

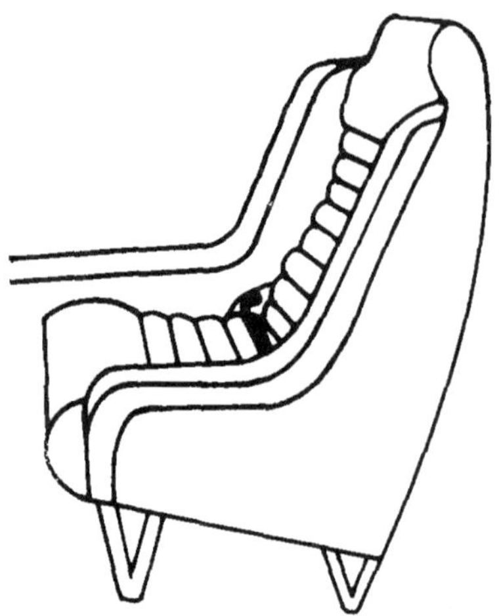

(a) With Upholstery

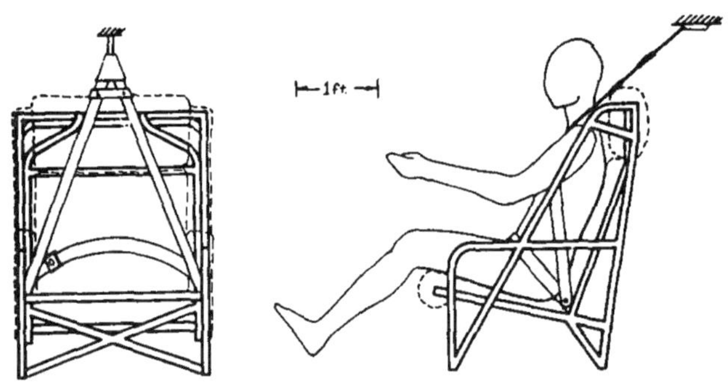

(b) Seat Frame

FIGURE 5 – THE HSRI INTEGRATED SAFETY SEAT (ISS)
(Figure Reproduced from Reference 12)

performance of Severy's Calibrated Seat (20). The observations relative to the practicality of gusseted seats and seats lacking in adjustment features discussed above would also apply to the ISS (as would the caveat that such seats would be fully effective only with a belted occupant population).

At the Fifth Experimental Safety Vehicle Conference in 1974, the Ford Motor Company of Britain presented the results of its program on rear impact protection (21). The test condition chosen for the British Ford design exercise was a 30mph (48 km/hr) impact of the Ford ESV by a rigid mobile barrier weighing nearly 6000 lbf. (2720 kg.). For the front bucket seats of the Cortina ESV, a 20g dummy loading design goal was evolved, which translated to a seat back strength of nearly 2700 lbf. (12 kN) at the torso center of gravity (cg). A figure in the report suggests that the seat was tested and was intended to be used with an active lap belt restraint. When loaded, the seat deflected about 4 inches at the torso cg level which would correspond to a back angular deflection of about 10-11 degrees. Unfortunately, not enough information is given in the report on which to base a judgement of other aspects of the practicality of the seat. Repeating the experience of Severy, the seat was installed in a prototype safety vehicle and tested under the selected design conditions, only to find that the seat rotated rearward about 20 degrees due to yielding of the compartment floor.

The late 70's saw two attempts to strengthen seat backs by attaching the top of the seat back to the roof. In the design of Severy (22), this attachment was made by use of seat-belt type webbing material attached via emergency-locking retractors, see Figure 6.

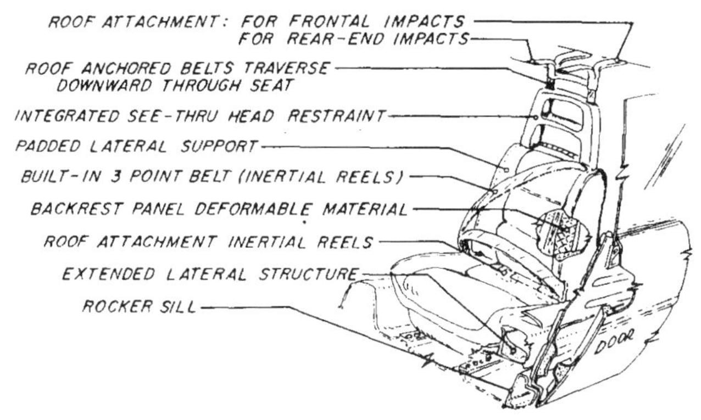

FIGURE 6 – THE INTEGRAL SAFETY SEAT OF SEVERY ET AL
(Figure Reproduced from Reference 22)

This seat was built in prototype form and statically tested, but apparently was not subjected to vehicle testing. The Minicar's Research Safety Vehicle (RSV), also used a roof connection which consisted of a see-through plastic membrane attached to the seat back top via energy-absorbing tension elements, see Figure 7 (23). A prototype of this seat, using sheet metal in place of the plastic membrane, was installed in an RSV test article and impacted at 40 mph by a production vehicle. Seat integrity

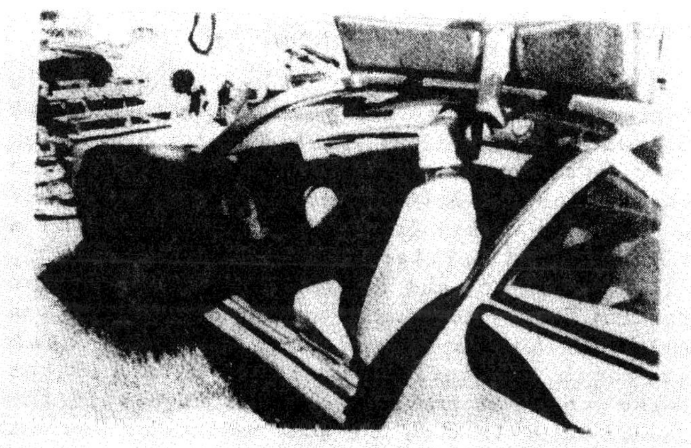

FIGURE 7 – THE MINICARS RSV DRIVER'S SEAT

was maintained in this test and the energy absorbers controlled the rebound motion of the unrestrained dummy reasonably well. For use in two-door vehicles, both of these schemes have problems, even though the use of safety belt retractors in the "belt" design are an advantage. Some sort of force-limiting mechanism is probably necessary in these schemes to avoid the tremendous forces developed in trying to manage longitudinal acceleration with an essentially vertical membrane. In the RSV design, a means of ensuring long-term rearward visibility through the plastic membrane is not presently known to the authors of this paper. In addition, the roof fixtures present impact targets for unrestrained rear occupants in frontal collisions. In summary, when one finishes addressing all of the design problems introduced by such schemes, one is likely to find the result to be complex, and hence prohibitively expensive. In addition, there may be legitimate biomechanical concerns involved in forcing necks into flexion during rearward seat back deflection.

Seat Belt Effectiveness in Rear Impacts - The potential effectiveness of belt restraints in rear impacts is fourfold: 1) the restraint may be able to control or eliminate the phenomenon of ramping up the seat back, 2) the restraint may reduce the velocity of the occupant relative to the vehicle interior and thus reduce injuries resulting from occupant contacts, 3) belts could minimize the potential for occupants to be out-of-position at impact, and 4) belts may be effective in controlling forward rebound. Prior analytical and experimental studies were reviewed in an attempt to determine whether or not any of these potential benefits were demonstrated.

Mertz and Patrick, in 1967, noted that their sled test results (at delta-V's of 9 and 15 mph) indicated that restraints had little effect on occupant response for rigid seats and yielding seats with head restraints, but on yielding seats without head restraints the restraint tended to increase injury severity (8). Berton, on the other hand, found that restraints were useful in preventing ramping at higher speed impacts with yielding seat backs incorporating head restraints (10).

Severy's series of rear impact tests (2-4) were all performed with lap belted dummies. Film observation indicated the effectiveness of the restraints in controlling the degree of ramping and forward rebound.

Melvin and McElhaney, in the course of conducting a series of model simulations of a rigid seat back using the HSRI 2D model, found that the model predicted that larger occupants at high impact severities would tend to ramp up the seat back with or without a lap belt (13).

In 1972, Levine and Campbell published the results of their study of 1966 and 1968 accidents in North Carolina (24). Their analysis of the data indicated that seat belt use decreased the incidence of any injury in a rear collision by about 12 percent and decreased the incidence of serious injury by over 57 percent.

Summary of Findings from the Literature - Relative to the issue of the performance of yielding versus rigid seats, a review of prior analytical and experimental efforts indicates that in general, plastically yielding seat backs produce lower injury exposures. Rigid seat backs may, in fact, be a distinct danger to occupants who may not be in the normal seated position at impact. The use of an integral seat belt system would likely be a requirement of such seats to preclude the ramping of occupants over the seat back in severe rear collisions. A belt system would also be required in order to minimize the probability of an occupant being out of position at impact and to prevent injury upon rebound out of the seat. Furthermore, a potential problem involving the impact of rear seat occupants with rigid front seat backs and/or roof mounting fixtures in frontal collisions has not been adequately addressed by prior investigations.

Incorporating rigidity in a conventionally-configured vehicle seat has been shown to involve significant weight penalties and the sacrifice of a number of comfort and convenience features that consumers have come to expect in their seats. Designing a rigid seat involves more than designing the seat per se, but also requires the rigidizing of the entire load path, including the vehicle floor. Attaching the seat back to the vehicle roof offers the potential to build a rigidized seat of reasonable weight, but introduces new problems, namely the complication of entry and exit, visibility, and the biomechanical consequences of creating forced flexion in severe rear impact situations.

Prior experimental and analytical efforts indicate that seat belts are effective in rear collisions, particularly at the higher accident severities (i.e. at delta-V's above 15 mph). Restraints tend to minimize relative motion of the upper body with respect to the seat back and help control rebound motion. These findings are consistent with more recent published opinions of safety researchers (25,26).

LABORATORY PERFORMANCE OF PRODUCTION BUCKET SEATS

The response of production bucket seats to rearward loading has been investigated in two ways. First, the strength characteristics of seats were investigated through the expansion of a series of static loading tests in a format originally suggested by Severy (2). It should be pointed out that while the Federal Motor Vehicle Safety Standard (FMVSS) 207 compliance testing also involves a static rearward loading of the seat back, data from these tests are not very useful because the tests are terminated whenever the seat force exceeds the required minimum level (27).

Static Tests - The procedure for the static tests is illustrated in Figure 8.

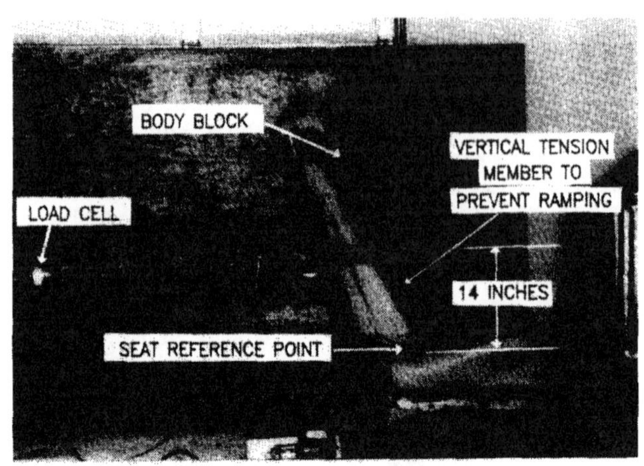

FIGURE 8 – TYPICAL SET-UP FOR STATIC SEAT PULL TESTS

Briefly, a body block is used to distribute seat back loads in a representative manner. A horizontal load is applied to the block through a lateral bar extending beyond the seat back sides 14 inches above the seat reference point. Vertical tension members are attached to this same bar to prevent the block from ramping up the seat back during loading. A load cell is attached to the loading cable and records the applied force while simultaneous measurements of the horizontal displacement of the block and the seat back angle are recorded. The test is stopped when either (1) the seat "breaks loose" (i.e., is no longer capable of resisting force) or (2) the seat back yields to the point that the angle reaches 45 degrees.

The results of these tests are summarized in Table 1. Seat stiffness (column 4), is defined as the initial slope of the force deflection curve. Maximum seat strength (column 5) is the peak horizontal force developed during the test. Absorbed energy (column 6) was calculated from the area under the force deflection curve up to the point where the test was stopped. Inspection of Table 1 indicates that the typical production bucket seat has an initial stiffness of about 150 lbf/in, a peak force capability of about 650 lbf and the ability to absorb about 250 ft-lbf of energy. The maximum and minimum stiffnesses differ from the average by about a factor of 2. This factor is about 3 for peak force and energy absorption.

FMVSS 207 requires that the seat back be capable of withstanding a rearward moment about the seat reference point of 3,300 inch-lbf. At

TABLE 1 - SUMMARY OF STATIC SEAT PULL TESTS

SEAT DESCRIPTION			SEAT PERFORMANCE				TEST CONDUCTED BY:
MODEL YR.	MANU-FACTURER	MODEL	STIFFNESS (lbf/in.)[1]	STRENGTH (lbf.)[2]	ABSORBED ENERGY (ft.lbs)[3]	REMARKS	
'64	MG	-	117.6	430.0	118.6		CSE
'64	AMC	Rambler	85.1	560.0	93.6		CSE
'64	VW	Type I	137.9	620.0	263.7		CSE
'64	Triumph	-	266.7	600.0	174.3		CSE
'64	Pontiac	LeMans	265.0	550.0	89.7		CSE
'65	Renault	Dauphine	73.2	455.0	100.1		Severy
'66	Chevrolet	Chevelle	103.4	505.0	170.5		Severy
'66	Datsun	411	169.3	325.0	174.2		Severy
'66	Dodge	Dart	85.1	660.0	323.2		Severy
'66	Ford	Cortina	105.8	415.0	114.1		Severy
'66	Ford	Falcon Ranchero	106.0	615.0	244.6	One side of split bench seat	Severy
'66	Plymouth	Barracuda	121.2	850.0	233.7	One side of split bench seat	Severy
'66	Saab	95	144.9	425.0	175.4		Severy
'66	VW	Type III	131.6	640.0	227.5		Severy
'66	Volvo	544	160.0	730.0	418.5	Friction clutch at s/back pivot adjusted for maximum torque	Severy
'67	Fiat	850	137.9	475.0	163.6		Severy
'67	Plymouth	Barracuda	105.3	680.0	374.2	Split bench seat, one side	Severy
'67	Renault	R10	66.7	490.0	191.2		Severy
'67	VW	Type I	115.6	610.0	322.5		Severy
'68	Datsun	510	102.6	360.0	87.7[5]		Severy
'68	Ford	Mustang	160.0	787.0	278.4		Severy
'68	Opel	Kadett	88.9	530.0	241.9		Severy
'68	Saab	96	137.9	475.0	277.1		Severy
'68	Volvo	144	103.6	700.0	509.8		Severy
'69	VW	Type I	119.0	825.0	N/A*		CSE
'69	Mercury	Cougar	40.4[5]	715.0	N/A		CSE
'70	Opel	1900	88.9	570.0	N/A		Severy
'70	Renault	R10	52.6	525.0	N/A		Severy
'70	Toyota	Corolla	40.8	310.0	N/A		Severy
'70	Subaru	-	55.5	300.0[5]	N/A		Severy
'70	Datsun	510	142.9	365.0	N/A		Severy
'70	Fiat	850	95.2	873.0	N/A		Severy
'70	Austin	America	222.2	830.0	N/A		Severy
'70	AMC	Gremlin	285.7	1400.0[4]	N/A		Severy
'70	Chevrolet	MonteCarlo	139.1	772.0	480.2		CSE
'70	VW	Type I	210.0	790.0	243.3		CSE
'71	Plymouth	Duster	182.4	830.0	535.9[4]		CSE
'71	Volvo	144	222.2	630.0	373.4	Friction clutch at s/back pivot adjusted for maximum torque	CSE
'72	Chevrolet	Vega	160.0	720.0	N/A		CSE
'73	Ford	Pinto	117.6	950.0	370.2		Severy
'73	VW	S. Beetle	606.1	1055.0	376.5		CSE
'74	Chevrolet	MonteCarlo	285.7	830.0	243.3		CSE
'74	Ford	Maverick	57.1	346.6	145.0		CSE
'76	Chevrolet	MonteCarlo	62.9	508.0	132.5	Swivel Bucket Seat	CSE
'76	Chevrolet	Chevette	144.9	671.0	207.6		CSE
'76	Datsun	B210	169.5	650.0	280.6		CSE
'76	Volvo	242DL	178.6	1113.0	294.2	Seat broke just prior to reaching 45 degrees	CSE
'77	VW	Rabbit	444.4[4]	1180.0	N/A		CSE
Average Values			148.3	642.1	247.0		
[4]Highest Value			444.4	1400.0	535.9		
[5]Lowest Value			40.4	300.0	87.7		

*Not available

[1]Average over first 200 lbfs.

[2]Maximum Force during deflection, stopped at U = 45 degrees

[3]Energy out calculated beyond 0 = 45 degrees

the 14 inch elevation, this corresponds to a force capability of about 235 lbf. Thus, the yield strength of the average bucket seat is about 2 1/2 - 3 times that prescribed by the backrest strength requirement of FMVSS 207.

It is noteworthy that a strength requirement, such as FMVSS 207, does not address the amount of energy that a seat can absorb. The 1966 Plymouth Barracuda seat, for example, has a maximum strength capability of 850 lbf, making it one of the strongest seats tested. Its energy absorbing capacity, however, is just below the average.

Dynamic Tests - In addition to static tests such as those described above, a number of full-scale dynamic (crash) tests involving production seats have been published over the past fifteen years. Those known to the authors are summarized in Appendix B. These tests generally were conducted by impacting either a rigid moving barrier or another production vehicle into the rear of a stationary vehicle, the direction of the impact being straight forward. Inspection of Appendix B indicates that 50 such tests are represented, involving impact severities (delta-V's) from 5.5 mph to about 30 mph. Moreover, it is seen that in every one of these tests, the residual seat back angle differed measurably from the pretest angle. The variation ranged from about 9 degrees for the test involving a delta-V of 5.5 mph to about 30-35 degrees for tests involving delta-V's of about 20 mph or more.

Currently, rear impact crash tests of stationary vehicles by a 4000-pound rigid moving barrier are routinely conducted by the NHTSA as compliance or New Car Assessment Program (NCAP) tests to evaluate the integrity of the fuel system. 50th percentile male dummies, generally lap-and-shoulder-belted, occupy the front seats of these vehicles. Appendix C contains a summary of the results of the NCAP tests involving model year 1979, 1980, and 1981 vehicles. These tests are seen to involve impact severities of about 17.5 to 25.3 mph, depending mainly on the weight of the subject vehicle. Again, even though these tests involve vehicles of substantially later vintage than those in Appendix B, the residual seat back angles in these tests invariably differ from the pretest angles by, typically, 25 degrees or more.

The residual seat back angles of those vehicles of Appendices B and C wherein the angle was either measured or could be estimated from post-test photographs are shown plotted versus accident severity in Figure 9. A band enclosing the vast majority of the data has been constructed on Figure 9 and indicates that standard production bucket seats, when loaded by adult male occupants, typically experience 0.8 to 1.6 degrees of seat back angle change for every mile-per-hour of vehicle delta-V. Thus, at accident severities of about 20-25 mph one would expect to find that the resistive capacity of the seat had been exceeded.

Dynamic seat back angular deflection can be expected to exceed the residual deflection indicated by post-test inspection. Data on the

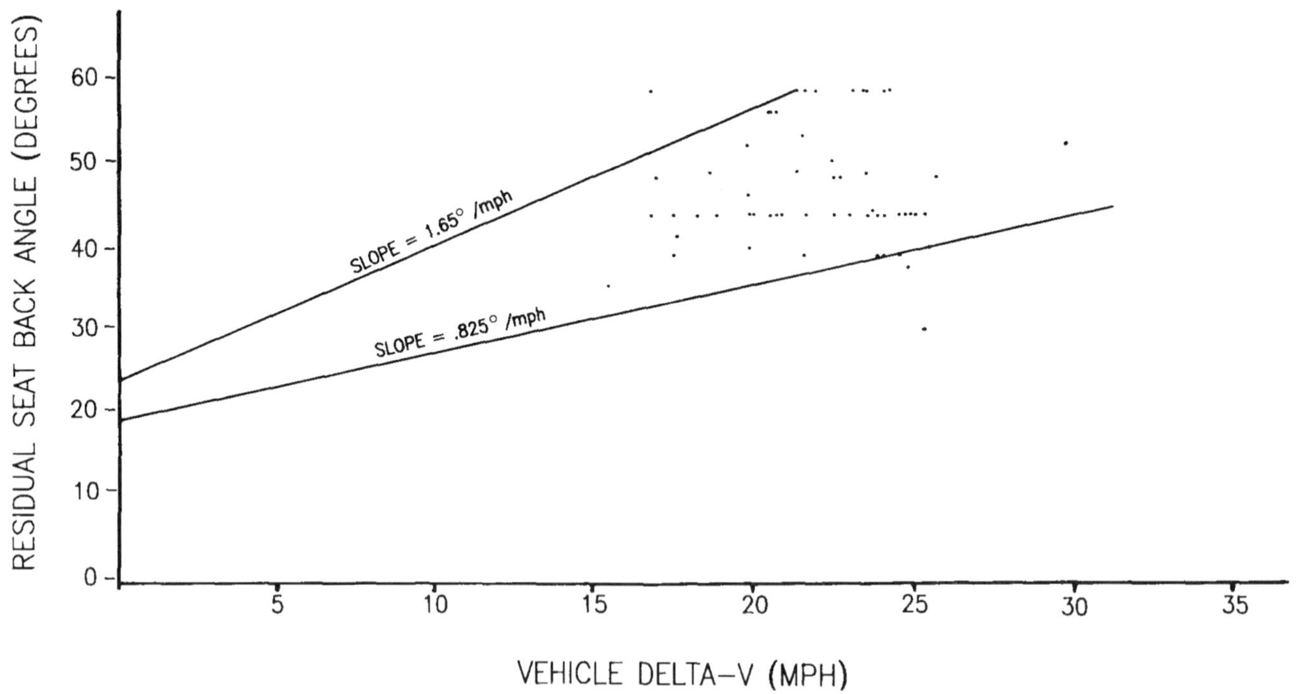

FIGURE 9 — OBSERVED RESIDUAL SEAT BACK ANGLE AS A FUNCTION OF ACCIDENT SEVERITY

amount of seat back angle recovery are very limited, but the existing data seem to indicate that the degree of recovery decreases with accident severity and angular deflection. Data from References 28 and 29 indicate that at delta-V's of 10-15 mph or less, the peak change of dynamic seat back angle can be as much as twice that indicated by post-test observation, while at a delta-V of 20 mph or more angle recovery is about 25% or less. It should be emphasized, however, that these indications are based on very limited data and should not be relied upon in a quantifiable sense.

Occupant Kinematics - An analysis of the motion of unrestrained occupants in severe rear impact collisions such as those of the NCAP series indicates that, typically, very little forward motion of the occupant occurs until the vehicle impact is about one-third over. At about this point in time the compression of the seat back padding and rear surface of the occupant torso has progressed to the stage where the loading of the seat back frame effectively begins. During this initial phase of motion, significant relative velocities can build up between unrestrained, normally-seated occupants and their vehicle compartment. These differences in velocity continue and, in the case of a severe rear impact, typically increase through the remaining duration of the impact. This is because production seats, due to practical considerations, are incapable of producing and sustaining occupant acceleration levels needed to reduce or even maintain these relative velocities. That is, severely rear-impacted vehicles continue to accelerate to their eventual velocity at average rates of perhaps 10-15 G's or more, which they will sustain for the remaining two-thirds of the impact. These vehicles are capable of sustaining these accelerations by virtue of the relatively large distances available in the form of the remaining mutual crush distances of the struck and striking vehicles. By contrast, the applied accelerations to the occupants are limited to those resulting from the approximately 600-700 pound capability of seat backs. Since the effective upper body weight of a 50th percentile male, for example, is about 80-100 lbf, typical seats will produce occupant accelerations of about 6-9 G's, levels comparable to, but typically less than those of the vehicle compartments. Moreover, these seat-produced occupant accelerations can only be effectively applied over the limited distances involved as the seat backs rotate from their initial angles of about 20-25 degrees (with respect to the vertical) to angles where they are not in a position to produce effective horizontal accelerative forces. Thus the period of occupant acceleration is typically relatively short, lasting on the order of about 25 msec, and consequently, results in relatively small changes in the velocities of the occupants.

Also apparent from an examination of occupant kinematics in severe rear impact situations is the fact that unrestrained dummies, at least at the severe levels associated with NCAP testing, do not generally gain elevation as they proceed forward in space and rearward with respect to the vehicle. The phenomenon of "ramping up the seat back" is thus seen to be strictly describing the motion of occupants with respect to the seat backs rather than with respect to either the vehicle or ground coordinate systems. Under even more severe conditions than the NCAP environment, it is probably true that dummies, once seat back rotations have been arrested by virtue of their angulation or stiffness (and possibly the interference of seat backs with rear structures), may begin translating upward along the seat back planes with respect to vehicle and ground coordinates.

It is also important to remember that the above discussion of occupant kinematics involves so-called normally seated occupants. Accidents can be expected to occur wherein occupants are not in this advantageous configuration. If the occupant, for example, is pitched forward at the instant of impact, an even longer period of time will elapse before significant seat back forces are applied. This in turn will increase the relative velocity between occupant and vehicle throughout the event, resulting in an even shorter period of seat back loading and a lower seat back induced occupant velocity change.

Thus far our discussion of occupant kinematics has been concerned with the unrestrained occupant. The differences in trajectory of a belt restrained occupant will be discussed next. It is noted in Appendix C that most of the dummies in the NCAP test series were belt-restrained. However, it appears from inspection of this table that there is little difference in dummy injury measures between the belted dummies and the few unbelted dummies in the experiments. Closer examination of the data in Appendix C and inspection of the associated test reports and films indicate, however, that there are significant effects of the lap belt on dummy kinematics in these tests. First it is noted that where there was both a belted and an unbelted dummy front seat occupant, it is generally the case that the belted dummy received the higher accelerations. This is a consequence of the fact that the belted dummy was more closely linked to his vehicle compartment than was his unbelted partner, i.e. restraint forces were applied. At the severities involved in the NCAP tests, this additional linking was generally of little consequence, but inspection of post-test photographs and test movie film indicates that under more severe conditions this difference could prove critical.

At the severities involved in these NCAP tests, the closer linking of the occupant to the vehicle usually only means that the unbelted dummies strike the rear seat back at a higher elevation than do the heads of the belted dummies. Figure 10, taken from Reference 30, illustrates this difference in head strike locations between the unbelted passenger dummy head and the belted driver dummy head in the NCAP test of the 1979 Volvo 244DL.

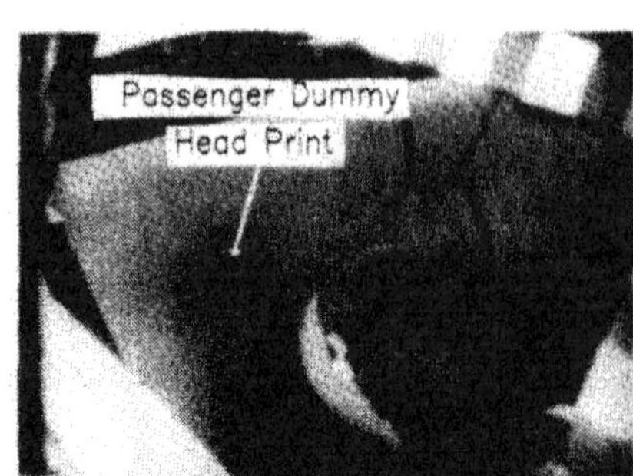

FIGURE 10 – COMPARISON BETWEEN BELTED (DRIVER) AND UNBELTED (PASSENGER) HEAD STRIKE LOCATIONS IN THE NCAP 35 MPH REAR IMPACT TEST OF THE 1979 VOLVO 244DL (Reference 30)

At higher accident severities, two effects of being belt-restrained should be evident. First, the suppression of seat back "ramping" can be expected in some instances to be the difference between striking the relatively friendly padded and proximate rear seat cushion and ramping upward against the arrested seat back to contact the more remote and less compliant surfaces of the vehicle roof and/or rear window surround. In fact, this point is illustrated in the NCAP test involving the 1980 Honda Prelude (31), wherein the unrestrained passenger dummy actually penetrated the rear window opening with his head and struck the window perimeter while the belted driver dummy contacts were limited to the rear seat cushion. Figure 11 is a frame taken from the overhead high-speed camera documenting this test. Secondly, the lap belt forces will reduce the velocity of the occupant with respect to vehicle interior surfaces thereby attenuating impact severities.

FIGURE 11 – UNRESTRAINED PASSENGER DUMMY HEAD LOCATION DURING NCAP REAR IMPACT TEST WITH 1980 HONDA PRELUDE (Reference 31)

The examination of the pictorial data indicates that the belted dummies tend to stay with their seat back to a greater degree than the unbelted dummies. The present authors have examined the films from the rear impact crash tests summarized in Appendices B and C and have yet to discover a case of a belt restrained front seat occupant slipping out of his lap belt. This does not happen because the thighs and knees of front seat occupants jack-knife upward and rearward in response to the forward acceleration of the vehicle, making it virtually impossible for the occupant to thread his way out of the lap belt.

In addition to tying the occupant to the vehicle, the belt restraint system also offers the beneficial effect of tending to keep the occupant in the normally seated position prior to impact. As discussed above in connection with the unrestrained occupant, being out of position will generally reduce the ability of the seat to provide protective, well distributed impact forces.

ACCIDENT STATISTICS

Accident statistics from the Fatal Accident Reporting System (FARS) and the National Crash Severity Study (NCSS) are reviewed to gain a perspective on the relative importance of rear impacts and the relationships between injury and accident severity.

Data from FARS for the years 1970-1985 were examined in order to gain insight into the relative significance of the rear impact mode. Figure 12 presents the FARS data showing the numbers of fatalities for the different clock positions of Principal Direction of Force

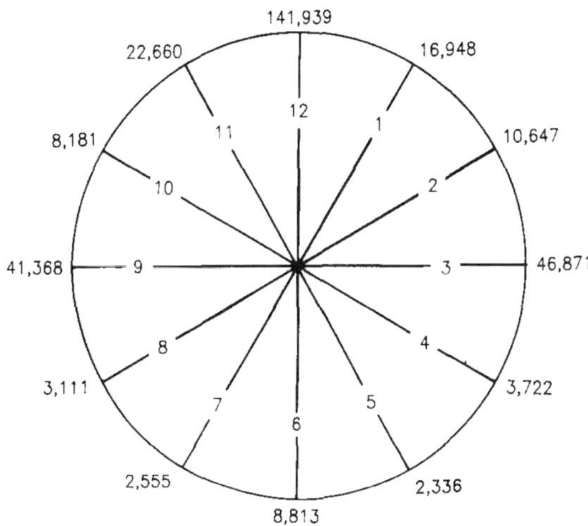

FIGURE 12 – THE DISTRIBUTION OF FATALITIES IN FARS 1970–1985 AS A FUNCTION OF PRINCIPLE DIRECTION OF FORCE

(PDOF). These numbers have been combined to four categories below:

Frontal	(11,12,1 o'clock PDOF)	- 181,547 (51%)
Side	(2,3,4 and 8,9,10 o'clock PDOF)	- 113,900 (32%)
Rear	(5,6,7 o'clock PDOF)	- 13,704 (4%)
Other	(Rollover,etc.)	- 44,126 (13%)

As is evident from Figure 12 and the above summarization, relatively few occupants are killed in rear impacts. This would be expected if it were true that rear impact is a relatively infrequent accident mode. It has been reported a number of times in the literature (32,33,34) that the rear impact mode is the most common accident type. However, when one examines the NCSS data (35), which includes only accidents where at least one of the vehicles was towed away from the scene, it is found that only 5.3 percent of the involved vehicles (with 5.7% of the exposed occupants) were exposed to PDOF's of 5,6, and 7 o'clock . This same file shows that the rear-impacted vehicles accounted for only 1.6% of the fatalities. One conclusion that can be drawn from these data is that a large number of rear impacts apparently occur at low severities. Even at the severities in the NCSS file however, there appears to be factors that render the rear impact situation relatively less injury-producing than the other impact modes.

In order to examine the issue of the distribution of impact severities among the various accident modes, one would need to have a field accident data file wherein accident severities were measured and all severities were investigated. Unfortunately, such a data base does not presently exist. Accident severities were measured in the NCSS data file, however not all accident severities were included. Figure 13, based on data in Reference 35, shows a comparison between the distribution of rear impact accident severities and frontal plus side

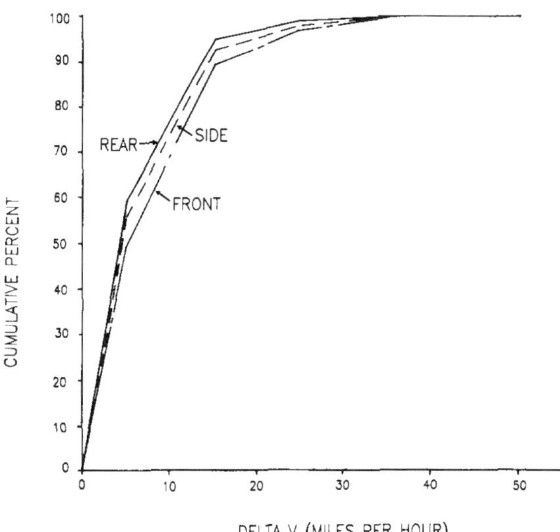

FIGURE 13 – THE DISTRIBUTION OF ACCIDENT-INVOLVED OCCUPANTS STRUCK IN THE REAR, SIDE AND FRONT BY ACCIDENT SEVERITY (DATA TAKEN FROM REFERENCE 35)

impact severities in the NCSS data base. Although the differences are not great, these data do suggest that there are a disproportionately larger percentage of low level rear impacts compared to front and side collisions. Thus the distribution of rear impact severities does seem to be part of the explanation for the relatively few fatal rear collisions.

An additional factor contributing to low occurrences of serious injury in rear impact could very well be that the vehicle protective system in rear impacts is taking advantage of the generally higher tolerance levels of the human body when loaded by well distributed forces in the posterior-anterior direction and thus offers advantages over the other accident modes. NCSS data were examined for differences in the injury severity/accident severity relationships among the various accident modes to gain further insight into this question. Figure 14, based on data presented in Reference 35, is a graph showing the probabilities of sustaining serious (AIS 3) or greater injury as a function of accident severity for the frontal, side, and rear impact modes. As is evident in this Figure, injury rates for the rear mode are significantly lower at all delta-V's compared to the other two modes. These data strongly suggest that a significant factor responsible for rear impacts being a relatively insignificant accident mode is the vehicle protective system (i.e. the characteristics of present vehicle seats) in combination with the higher human tolerance levels when loaded in the posterior-anterior direction.

The NCSS data can also be used to examine the distribution of specific injuries in rear collisions with accident severity. Such distributions are useful in assessing whether a particular injury was understandable in terms of the severity of the accident. Figure 15,

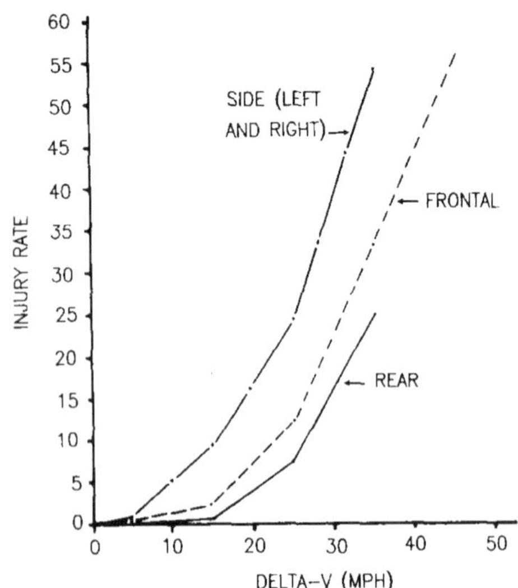

FIGURE 14 - SERIOUS (AIS 3) OR GREATER INJURY RATES TO OCCUPANTS WITH FRONTAL, SIDE, OR BACK DAMAGE IN THE NCSS DATA FILE (Data Obtained from Reference 32)

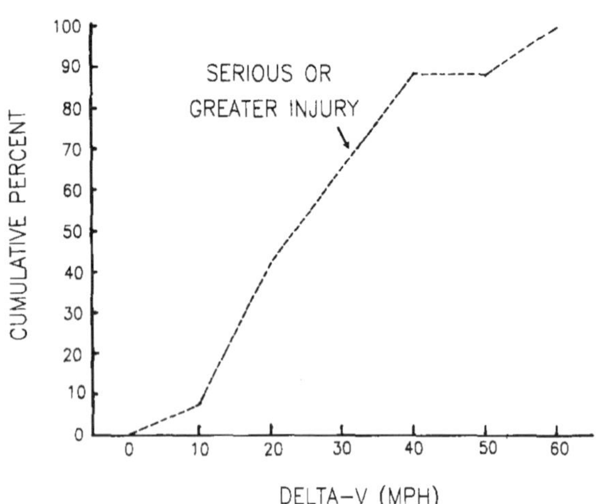

FIGURE 15 - THE DISTRIBUTIONS OF SERIOUS OR GREATER INJURY TO OCCUPANTS OF REAR-IMPACTED VEHICLES AS A FUNCTION OF ACCIDENT SEVERITY (DELTA-V) BASED ON NCSS DATA

reproduced from Reference 35, indicates that 65 percent of all serious or greater injuries occur at a delta-V of 30 mph or less. There are only 6 fatalities in this data base, but they seem to follow the same general pattern found for fatalities as a function of accident severity in the other accident modes (see Figure 16).

SUMMARY AND CONCLUSIONS

Prior experimental and analytical studies of rear impact collisions were reviewed in an attempt to answer three key questions regarding occupant response. With regard to the

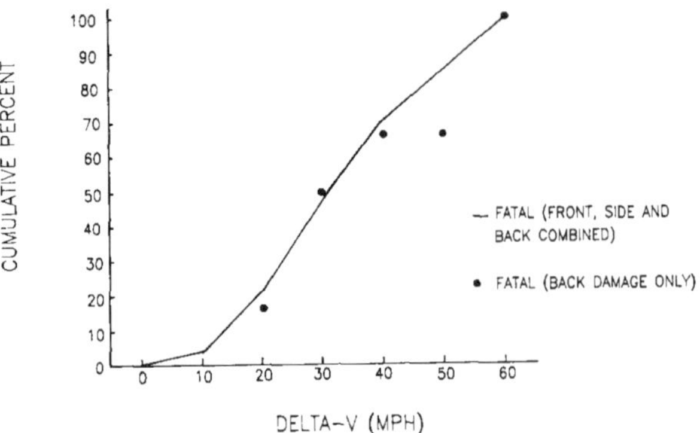

FIGURE 16 - DISTRIBUTIONS OF FATALITIES FOR COMBINED DAMAGE MODES AND BACK DAMAGE ONLY BASED ON NCSS DATA

desirability of a rigidized front seat, it was concluded that such a design would not be likely to decrease occupant injury occurrences and severities, and could possibly increase occupant injury exposure. Prior investigations have not established the feasibility or practicality of a rigid seat and moreover have indicated that attempts at such designs are likely to produce seats with unacceptable compromises of comfort and convenience features and may involve significant weight and cost penalties. Seat belts were found to be advantageous, especially at higher impact severities, by creating a more positive link between the occupant and the vehicle compartment.

Production seats were found to be capable of producing restraining forces significantly beyond the requirements of FMVSS 207. Crash tests indicate that permanent seat back deflection can be expected of virtually all production seats at impact severities (delta-V's) of 15 mph or greater and that at delta-V's about 20-25 mph the energy absorption capacity of all production seats will be exceeded.

Examination of accident statistics for rear collisions indicates that occupant injury exposure in this impact mode is significantly lower than that associated with frontal and side impacts. This field accident experience suggests that any benefits associated with seat design changes for rear impact protection are probably limited and of low priority. Attempts to realize such benefits could possibly increase rather than decrease rear impact protection. Finally, experimental and analytical efforts as well as field accident experience indicate that the most effective and practical means for enhancing rear impact protection is the simple use of available seat belt restraint systems.

REFERENCES

1. D.M. Severy, J. Mathewson, O. Bechtol, "Controlled Automobile Rear-end Collisions, and Investigation of Related Engineering and Medical Phenomena", Canadian Services Medical Journal, XI: 727 - 759, November 1955.

2. D.M. Severy, H.M. Brink, J.D. Baird, "Preliminary Findings of Head Support Designs", Proceedings, 11th Stapp Car Crash Conference, (SAE 670921), October 10-11, 1967.

3. D.M. Severy, H.M. Brink, J.D. Biard, "Backrest and Head Restraint Design for Rear-end Collision Protection", Automotive Engineering Congress, Detroit, (SAE 680080), January 8-12, 1968.

4. D.M. Severy, H.M. Brink, J.D. Baird, "Vehicle Design for Passenger Protection from High-Speed Rear-end Collisions", Proceedings, 12th Stapp Car Crash Conference, (SAE 680774), October 22-23, 1968.

5. J. Martinez, G. Garcia, "A Model for Whiplash", Journal of Biomechanics, Vol.I, June 1967.

6. J.L. Martinez, "Headrest and Seat Back Design Proposals", Proceedings, 12th Stapp Car Crash Conference, (SAE 680775), October 22-23, 1968.

7. J.L. Martinez, "Yielding Seatback is Effective in Protecting Car Occupant from Whiplash if Properly Damped", SAE Journal, December 1969.

8. H.J. Mertz, L.M. Patrick, "Investigation of the Kinematics of Kinetics of Whiplash", Proceedings, 11th Stapp Car Crash Conference (SAE 670919), October 1967.

9. H.J. Mertz, "Car Occupant Response to Rear-end Collisions - A Mathematical Model", 1st International Conference on Vehicle Mechanics, p. 588, July 1968.

10. R.J. Berton, "Whiplash: Tests of the Influential Variables", Automotive Engineering Congress, Detroit, (SAE 680080), January 8-12, 1968.

11. J.K. Kihlberg, "Flexion-Torsion Neck Injury in Rear Impacts", Cornell Aeronautical Laboratory, April 1969.

12. J.W. Melvin, J.H. McElhaney, "Deployable Head Restraints", National Highway Traffic Safety Administration, Contract #FH-11-7612, June 30, 1971.

13. J.W. Melvin, J.H. McElhaney, "Occupant Protection in Rear-end Collisions", Automotive Engineering Congress, Detroit, (SAE 720033), January 10-14, 1972.

14. P. Prasad, N. Mital, A.I. King, and L.M. Patrick, "Dynamic Response of the Spine During +Gx Acceleration", (SAE 751172), Proceedings, 19th Stapp Car Crash Conference, 1975.

15. A. Hu, S. Bean, and R. Zimmerman, "Response of Belted Dummy and Cadaver to Rear Impact", (SAE 770929), Proceedings, 21st Stapp Car Crash Conference, 1977.

16. F.W. Babbs, B.C. Hilton, "The Packaging of Car Occupants--A British Approach to Seat Design", Proceedings, 7th Stapp Car Crash Conference, p. 456, 1963.

17. B.C. Hilton, "Design of Low Cost Seating for Effective Packaging of Vehicle Occupants", Proceedings, 10th Stapp Car Crash Conference, (SAE 660797), November 8-9, 1966.

18. D.M. Severy, H.M. Brink, J.D. Baird, "Collision Performance, LM Safety Car", Mid-year Meeting, Chicago, (SAE 670458), May 15-19, 1967.

19. D.M. Severy, Deposition in Lundgren v. Volkswagen, p. 127, October 14, 1978.

20. J. Hilyard, J. Melvin, J. McElhaney, "Deployable Head Restraints", HSRI, University of Michigan, DOT-HS-800802, January 1973.

21. D. Burland, "Occupant Protection in Rear Impact", 5th Experimental Safety Vehicle Conference, June 4-7, 1974.

22. D.M. Severy, D. Blaisdell, J.F. Kerkhoff, "Designing Safer Seats", Automotive Engineering, Vol. 84, No. 10, October 1976.

23. N. Dinapoli, M. Fitzpatrick, C. Strother, D. Struble, R. Tanner, "Research Safety Vehicle Phase II, Vol. II, Comprehensive Technical Results", Contract #DOT-HS-5-01215, November 1977.

24. D.N. Levine, B.J. Campbell, "Effectiveness of Lap Seat Belts and the Energy-Absorbing Steering System in the Reduction of Injuries" Journal of Safety Research, Vol. 4, No. 3, September 1972.

25. D.M. Severy, D. Blaisdell, L. Horn, "Motorist Head and Body Impact Analysis Methodologies and Reconstruction", 1985 International Congress and Exposition, Detroit, Michigan, (SAE 850097), February 25 - March 1, 1985.

26. G. Nyquist, E. Kennedy, "Accident Victim Interaction with Vehicle Interior: Reconstruction Fundamentals", (SAE 870500), 1987.

27. Current version of Federal Motor Vehicle Safety Standard 207, 49 CFR. 571.207.

28. H. Scheuerman, R. Young, "Automotive Rear-end Collision Tests", National Highway Traffic Administration, NAFEC Report, September 1974.

29. R. Pirtle, "Test and Evaluation of Head Restraints, Seat Backs, and Anchorages in Vehicles Subjected to Rear Impact Collision", National Highway Traffic Safety Administration, Dynamic Science Report, June 1978.

30. W. Levan, B.J. Kelleher, "New Car Assessment and Standards Enforcement Indicant Testing - FMVSS Nos, 212, 219 and 301-75", NHTSA, DOT-HS-8-01938, September 27, 1979.

31. M. Pozzi, "FMVSS 301-75 New Vehicle Assessment and Standards Enforcement Indicant Testing - 1980 Honda Prelude - NHTSA 800553", NHTSA, DOT-HS-9-02274, August 1980.

32. J. O'Day, L.D. Filkins, C.P. Compton, T.E. Lawson, "Rear-Impacted-Vehicle Collisions: Frequencies and Casualty Patterns, Highway Safety Research Institute, July 1975.

33. J.D. States, J.C. Balcerak, J.S. Williams, et. al., "Injury Frequency and Head Restraint Effectiveness in Rear-end Impact Accidents", Proceedings, 16th Stapp Car Crash Conference, (SAE 720967), November 8-10, 1972.

34. B. O'Neill, B. Haddon, Jr., A.B. Kelley and W.W. Sorenson, "Automobile Head Restraints - Frequency of Neck Injury Claims in Relation to the Presence of Head Restraints", American Journal of Public Health, March 1972.

35. L. Ricci, "NCSS Statistics: Passenger Cars", HSRI, University of Michigan, June 1980.

APPENDIX A - SEVERY TESTS WITH CALIBRATED SEATS

TEST CONDITIONS						TEST RESULTS		
EXPERIMENT NO.	VEHICLE DELTA V (MPH)	DUMMY TYPE (AND RESTRAINT)	HEAD OFFSET (INCHES)	CALIBRATED SEAT BACK TORQUE (IN-LBFS) (FORCE AT 14")	SEAT BACK HEIGHT (INCHES)	SEAT BACK RESIDUAL ANGULAR DEFLECTION (DEGREES)	SEAT BACK DYNAMIC ANGULAR DEFLECTION (DEGREES)	REMARKS
x93	16.5	95M (L)	3	9,100 (650 lbf)	22	19	27	Seat back rotated rearward to strike rear dummy knees. Dummy in rigid rear seat experiences head G's twice that of this dummy (24G v. 12G).
x94	11.0	95M (L)	3	9,100 (650 lbf)	22	8	14	Dummy in rigid rear seat experiences head acceleration about three times this dummy (23G v. 8G).
x93	16.5	95M (L)	0	16,100 (1150 lbf)	22	7	16	Dummy in rigid rear seat experiences head acceleration about twice this dummy (23G v. 22G). Seat back may have hit rear dummies knees.
x94	11.0	95M (L)	0	16,100 (1150 lbf)	22	3	10	Dummy in rigid rear seat experiences head acceleration significantly higher than this dummy (50G v. 14G). Head rotated more than in x93 (see above line).
x95	16.5	95M (L)	6	16,100 (1150 lbf)	22	3	10	Dummy in rigid rear seat experiences head acceleration about twice that of this dummy (19G v. 10G).
x95	16.5	95M (L)	12	16,100 (1150 lbf)	22	4	10	Dummy head acceleration higher due to greater offset.
x96	16.5	95M (L)	3	16,100 (1150 lbf)	25	7	16	Dummy in rigid rear seat experiences head acceleration about twice that of this dummy (24G v. 11G).
x96	16.5	95M (L)	3	16,100 (1150 lbf)	28	8	17	Dummy in rigid rear seat experiences head acceleration about twice that of this dummy (21G v. 11G).
x97	5.5	95M (L)	0	16,100 (1150 lbf)	28	0	7	Dummy in rigid rear seat experiences head acceleration about 2.5 times that of this dummy (24G v. 10G).
x98	22.0	95M (L)	3	16,100 (1150 lbf)	28	21	33	Dummy ramped up seat back to get head over head rest. Seat back contacted rear dummy knees. Dummy in rigid rear seat experiences head accelerations about 2.5 times that of this dummy (75G v. 33G).
x100	16.5	95M (L)	6	16,100 (1150 lbf)	25	7	20	Dummy leaned forward about 10 degrees. Severe whiplash motion.
x100	16.5	95M (L)	6	16,100 (1150 lbf)	28	11	23	Dummy leaned forward about 10 degrees. Seat back almost bottomed out. Head rest protected against whiplash motion.
x101	16.5	95M (L)	12	16,100 (1150 lbf)	25	14	22	Ramped up seat with severe whiplash motion. No apparent contact of seat back with rear dummy knees.
x101	16.5	95M (L)	12	16,100 (1150 lbf)	28	18	29	Severe whiplash motion. Strong rebound with chin bottoming on chest.
x102	16.5	50M (L)	0	16,100 (1150 lbf)	25	0	7	Whiplash motion of head. Dummy rebounded into steering wheel.
x99	16.5	95M (L)	0	33,000 (2357 lbf)	25	0	7	Whiplash motion was significantly larger compared to x96.
x99	16.5	95M (L)	0	33,000 (2357 lbf)	28	0	7	Whiplash motion was significantly larger compared to x96.
x102	16.5	50M (L)	0	33,000 (2357 lbf)	25	0	10	Whiplash motion similar to driver (see above). Head bottomed out on chin during rebound.
x103	30.25	50M (L)	0	33,000 (2357 lbf)	25	6	18	Seat back bottomed out against rear seat dummy knees. Floor pan yielded. Driver ramped up seat back with severe whiplash motion. Rebounded into steering wheel.
x103	30.25	95M (L)	3	33,000 (2357 lbf)	18	15	25	Seat yielded, bottomed out against rear dummy knees. Dummy headed for severe whiplash motion when contact with rear dummy chest occurred. Cox seat in left rear yielded back and bottomed out on rear package tray.
x104	30.25	50M (L)	0	33,000 (2357 lbf)	28	8	21	Beam (4" channel) placed under front seat anchor. Head over top of seat back with severe whiplash motion, rebounded into steering wheel.
x104	30.25	95M (L)	3	33,000 (2357 lbf)	28	10	18	Beam in place (see above). Seat bottomed out against rear dummy knees. Severe whiplash motion with severe rebound motion.
x106	30.25	50M (L)	0	97,000 (6929 lbf)	29	0	0	Cox Seat with back against beam welded between b-pillars. Ramping and rebound controlled by integral belt system.
x106	30.25	50M (L)	0	97,000 (6929 lbf)	28	0	0	Calibrated seat with beam against seat back. Dummy ramps 4" up back despite lap belt.

APPENDIX B · REAR IMPACT CRASH TESTS INVOLVING PRODUCTION SEATS 1967-1978

REFERENCE DATA			IMPACT CONDITIONS						RESULTS	
DATE OF REPORT	TITLE OF PUBLICATION (SOURCE DATA)	AUTHOR(S) (CORP. AUTHOR)	TEST NO.	STRIKING VEHICLE OR OBJECT	CASE VEHICLE	CASE VEHICLE DELTA V (MPH)	DUMMY USED (WT)	RESTRAINT	RESIDUAL CHANGE IN SEAT BACK ANG (DEGREES)	REMARKS
10/67	Preliminary Findings of Head Support Designs (11th STAPP)	Severy, Brink & Baird (ITTE, UCLA)	x93 x94	'67 Ford "	'67 Ford "	16.5mph 11.0mph	95M(D) 95M(D)	Lap Lap	19 8	Actually seats were lab. seats. Tests shown with S/B cal. @ 650# (14") · i.e. per prod. seats (9100 in.lbfs)
1/68	Backrest and Head Restraint Design for Rear Impact Protection (SAE 680079)	Severy Brink & Baird (ITTE, UCLA)	x97 x98	'67 Ford "	'67 Ford "	5.5mph 19.9mph	95M(D) 95M(D)	Lap Lap	1.5 12	Production seats used. Except for add. of 6" head restraint.
11/73	Six Rear Moving Barrier Automobile Crash Tests (GEI Rept. A-4650.01) (IIHS Cont. No. 6561)	H.E. Holt (General Environment Corporation)	1	Rigid M.B	'73 Ambassador	15.1mph	50M(2)	??	Fully Reclined	Split bench seat.
			2	"	'73 Ply. Fury	14.7mph	50M(2)	??	"	Fixed bench seat back.
			3	"	'73 Toyota Corona	19.0mph	50M(2)	??	"	Fixed bench seat back.
			4	"	'73 Chev. Vega	19.1mph	50M	??	"	Bucket seats.
			5	"	'73 Ford Pinto	19.1mph (calc.)	50M	??	"	Bucket seats.
			6	"	'73 Opel 1900	19.0mph (calc.)	50M	??	"	Bucket seats.
9/74	Automotive Rear-end Collision Tests (NHTSA Report DOT-HS-801-163)	Scheuerman & Young (NAFEC)	428	Rigid M.B	'68 Ply.	8.5mph	95M(D) 50M(C) 95M(P)	None	3.5	Bench seat.
			433	"	'68 Ply.	9.8mph	Same as above	None	18(L)	Bench seat adjusted. Head rests dropped 2".
			437	"	'68 Ply.	14.3mph	"	None	14.5(L)	Bench seat. Left side hit knees of rear dummy.
			438	"	'68 Ply.	16.2mph	"	None	15.5(R)	Bench seat. No rear seat dummies. Front seat moved back unit r. seat contact.
			441	"	'69 VW I	22.7mph	95M(D) 95M(P)	None None	27 16	50M behind passenger seat Driver seat slipped back to r. seat. Passenger seat hit r. pass. (50M) knees.
			461	"	'71 Pinto	19.8mph	95M(D) 95M(P)	None None	25 --	50M behind p. seat. Driver seat slipped back to hit r. seat. Passenger seat rear legs collapsed.
			462	"	'70 Buick Electra	12.4mph	95M(D) 50M(C) 95M(P)	None None None	19.5(L) 34.5(R)	Bench seat. Front seat impacted r. dummies knees.
			463	"	'69 Ply. Sta.wagon	9.6mph	Same above	None	28(L) --(R)	Bench seat. Left side hit r. dummies knees. Seat broke loose from floor.
			468	"	'70 LTD	14.7mph	Same as above	None	34(L) Note(R)	Bench seat. Dr. dummy head hit r. dummy chest. Pass. dummy head hit rear seat back. Seat broke loose from floor.
			469	"	'68 Ply.	9.3mph	59M(D) 50M(P)	None	7	No rear dummies.
			470	"	'68 Ply.	17.2mph	50M(P)	None	29(L) 34(R)	No rear dummies.
10/73	Six Moderate Speed Front-into Rear Automobile Crash Tests (GEI Report A.-4348.10) (IIHS Contract #6539)	H.E. Holt (GEI)	1	'73 AMC Gremlin	'73 Toy Corona	23.5mph	*M(D) F(P)	L&S "	Yes** Yes**	*Department store manikins filled to proper weight for size with rubber mat.
			2	'73 Datsun 610	'73 Ford Pinto	21.9mph	*M(D) F(P)	" "	Yes** Yes**	Male(M) · 153#, Female (F) · 112#
			3	'73 Ford Gal 500	'73 AMC Ambassador	21.3mph	*M(D) F(P)	" "	Yes** Yes**	**No angles measured Rear seat manikins helped prevent further seat back rotation.
			4	'73 VW Type I	'73 Ply. Fury III	14.1mph	*M(D) F(P)	" "	Yes** Yes**	
			5	'73 Ply. Fury III	'73 Chevy Vega	27.4mph	*M(D) F(P)	" "	Yes** Yes**	
			6	'73 Chevy Impala	'73 Opel 1900	25.9mph	*M(D) F(P)	" "	Yes** Yes**	
4/78	Accident Investigation Studies · Stage Rear Impact Collisions (Dynamic Science Report (3010-78-47)	R. Pirtle (Dynamic Science)	1	'71 Impala	'71 Pinto	17.6mph	50M(D)	L&S	-20	15° Impact angle w/pre-braking (.7g)
			2	'71 Impala	'71 Pinto	19.8mph	"	"	37	0° Impact angle w/pre-braking (.7g)
			3	SAE Bar.	'71 Pinto	21.5mph	"	"	32	
			4	'71 Impala	'72 Vega	25.4mph	"	"	18*	15° Impact angle w/pre-braking (.7g). *Looks like lab fixture in rear stopped seat back rotation.
			5	'71 Impala	'72 Vega	29.7mph	"	"	31	0° Impact angle w/pre-braking (.7g)
	(Seat Data in:) Test and Evaluation of Head Restraints, Seat Backs & Anchorages in Vehicles Subjected to Rear Impact Collision (PB 289 779)	R. Pirtle (Dynamic Science)	6	SAE Bar	'72 Vega	25.7mph	"	"	27	
			7	'71 Impala	'71 Pinto		"	"		No measurements made.
			8	"	'72 Pinto	22.5mph	"	"	27	0° Impact angle w/pre-braking (.4g)
			9	"	"	22.4mph	"	"	29	0° Impact angle offset 14"
			10	"	'72 Vega		"	"		No measurements made
			11	"	'71 Vega	24.8mph	"	"	16	0° Impact angle w/pre-braking(.4g)
			12	"	"	24.7mph	"	"	No data	0° Impact angle offset 11+"
			13	"	'74 Pinto	23.7mph	"	"	23	0° Impact angle w/prebraking (.6g)
			14	"	'71 Impala	18.7mph	"	"	27.5	0° Impact angle w/prebraking (.7g)
			15	"	'74 Pinto	16.7mph	"	"	No data	0° Impact angle
			16	"	'76 Pinto	19.0mph	"	"	Collected	0° Impact angle w/prebraking (.6g)
			17	"	"	14.9mph	"	"	"	0° Impact angle
			18	"	'72 Pinto Sta.wagon	19.6mph	"	"	"	0° Impact angle w/prebraking (.6g)
			19	"	'76 Pinto Sta.wagon	17.4mph	"	"	"	0° Impact angle
11/80	Test and Evaluation of Volkswagen Type I Front Seat Integrity in Rear Impact Collisions (Dynamic Science Report No. 8340-80-192)	M. Pozzi (Dynamic Science)	1	'73 Impala	'72 VW I	20.7mph	50M(D) " (P)	L&S "	-35 -35	No seat damage other than back rest & rear frame yield.
			2	"	'74 VW I	20.4mph	" (D) " (P)	" "	-35 -35	" " " " " " " " " "
			3	"	'72 VW I	20.5mph	" (D) " (P)	" "	-35 -35	" " " " " " " " " "
			4	"	'74 VW I	20.7mph	" (D) 6 yr(P)	None None	-35 7	Driver side "medial hinge" broken.

APPENDIX C.1 - NCAP REAR IMPACT TESTS - 1979 MODEL YEAR VEHICLES

VEHICLE/SEAT DESCRIPTION			TEST DESCRIPTION			DUMMY RESPONSE			SEAT RESPONSE
Model Yr	Manu-facturer	Model	Driver(D) or RFP(P)	Belted(B) Unbelted (U)	ΔV (MPH)	HIC	Pk Chest G's (-3ms)	Residual Seat back Angle (deg's)	Remarks
1979	Chrysler	Cordoba	D P	B B	19 19	177.8 357.5	19.0 30.5	~35 ~35	Split bench seat.
1979	Dodge	Aspen	D P	B U	20 20	258.4 245.4	26.0 24.0	~45 ~45	Seat track broke, seat moved aft ~2".
1979	Dodge	Colt	D P	B B	24.6 24.6	301.7 327.8	51.4 34.8	~40 ~45	Head contacted B-post & RFP head. Head contacted B-post & driver head.
1979	Dodge	Diplomat	D P	B B	18.8 18.8	263.7 469.2	18.3 21.2	~45 ~45	Bench seat moved 2" rearward.
1979	Plymouth	Horizon	D P	B B	22.9 ·22.9	115.4 146.5	17.3 26.6	?? ??	Both dummy heads impacted rear seat cushion during seat back recline.
1979	Ford	Fiesta	D P	B U	24.8 24.8	386.9 359.7	44.0 33.0	~45 ~45	
1979	Ford	LTD Landau	D P	B B	18.7 18.7	140.0 192.3	10.4 19.6	~40 ~40	Both front seat backs bent inward and onto rear seat cushion.
1979	Mercury	Monarch	D P	B U	19.9 19.9	1098.2 585.0	25.6 24.0	~45 ~45	Bench seat reclined allowing both dummy heads to contact rear seat back.
1979	Mercury	Zephyr	D P	U B	22.5 22.5	341.0 419.0	23.0 26.0	~45 ~45	
1979	Ford	Mustang	D P	B B	21.6 21.6	251.6 105.8	33.6 28.8	~40 ~40	Stopped by front seat backs striking rear seat.
1979	Ford	T-Bird	D P	B B	17.5 17.5	52.5 70.9	15.9 13.7	~40 ~45	Both seat backs bent onto rear cushion.
1979	Honda	Civic	D P	B B	25.3 25.3	219.9 245.9	32.9 34.3	~45 ~30	Passenger-side headrest torn off.
1979	Datsun	210	D P	B B	24.1 24.1	445.2 442.1	54.6 68.9	~45 ~40	Driver headrest became detached, both seats moved 4" aft in tracks.
1979	Buick	Electra	D P	B B	18.0 18.0	41.9 94.4	10.74 11.56	~45+ ~45+	Both front seats moved 1" aft.
1979	Buick	Rivera S	D P	B B	18.2 18.2	35.3 113.8	10.4 11.3	~40 ~45	Split bench seat. Driver seat back broke.
1979	Chevrolet	Camaro	D P	B B	19.3 19.3	214.0 311.3	17.5 27.7	~60 ~45	Both seats moved aft (Driver - 1.4" Passenger - 1").
1979	Chevrolet	Chevette	D P	B B	23.06 23.06	137.4 309.9	24.0 29.4	~45 ~45	Passenger seat moved 1" aft in tracks.
1979	Pontiac	Catalina	D P	B B	19.0 19.0	415.0 289.0	14.0 14.0	~45 ~45	Bench seat back broke. Both heads contacted rear seat back.
1979	Pontiac	Gran Prix	D P	B U	19.5 19.5	189.2 100.7	14.0 30.0	~60 ~45	Split bench seat. Both seat backs collapsed. Both heads hit rear seat back.
1979	Pontiac	Sunbird	D P	B B	21.3 21.3	293.5 410.6	17.0 19.	~50 ~60	Both front seats yielded at impact and the and the dummy heads struck rear seat back.
1979	Toyota	Celica	D P	B U	21.9 21.9	210.0 286.0	27.0 30.0	~60 ~60	Both front seat backs yielded on impact. Both dummy heads struck rear seat back.
1979	Toyota	Corolla	D P	U B	23.5 23.5	623.0 1147.0	36.0 42.0	~45 ~45	Passenger head contacted rear seat back support bracket.
1979	VW	Rabbit	D P	B(VWRA) B(VWRA)	23.8 23.8	114.5 102.8	17.0 17.0	~40 ~45	Both bucket seat backs yielded.
1979	Volvo	244DL	D P	B U	20.6 20.6	51.1 210.2	13.4 18.4	~45 ~45	Both seat backs yielded at impact. Both heads struck back of rear seat.
1979	Ford	Pinto	D P	B B	22.3 22.3	288.6 303.2	47.0 76.0	~45+ ~45+	Both seat backs yielded at impact. Both heads struck back of rear seat.
1979	Chrysler	Newport	D P	B B	18.7 18.7	196.0 138.2	13.7 12.9	~60 ~60	Bench seat. Bench seat back yielded, both dummies reclined.

APPENDIX C.2 - NCAP REAR IMPACT TESTS - 1980 MODEL YEAR VEHICLES

VEHICLE/TEST DESCRIPTION			TEST DESCRIPTION			DUMMY RESPONSE		SEAT RESPONSE	
Model Yr	Manu-facturer	Model	Driver(D) or RFP(P)	Belted(B) Unbelted (U)	ΔV (MPH)	HIC	Pk Chest G's (3-ms)	Residual Seat Back Angle (deg's)	Remarks
1980	AMC	Concord	D P	U B	19.9 19.9	78.9 89.1	16.0 12.2	N/A ~45	Driver seat separated from tracks. Both heads struck rear seat back.
1980	Dodge	Mirada	D P	B B	19.6 19.6	-- --	-- --	~45 ~45	Both seats moved 1.5" aft in tracks. Split bench seat.
1980	Mercury	Cougar	D P	B B	20.1 20.1	78.0 57.0	12.9 13.0	~60 ~60	Bench seat moved 3" aft. Track severely buckled. Left track 80% split.
1980	Subaru	GL	D P	B B	23.5 23.5	-- --	-- --	-- --	Both dummy heads contacted rear seat back.
1980	Honda	Civic 1500DX	D P	B B	24.7 24.7			~45 ~45	Left front seat moved 3" aft. Outboard anchor of RFP seat torn from floor.
1980	Chevrolet	Citation	D P	B B	21.4 21.4	171.3 255.8	18.0 31.0	~45 ~45	Bench seat. Seat back reclined.
1980	Honda	Prelude	D P	B U (Belt detached)	23.5 23.5			~50 ~60	Passenger forehead struck backlight. Driver seat moved 4.5" aft., Driver head hit rear seat back.
1980	Datsun	310	D P	B B	24.2 24.2	-- --	-- --	~45 --	
1980	Renault	LeCar	P P	B B	24.7 24.7	-- --	-- --	~45 --	
1980	Datsun	200SX	D P	B B	22.2 22.2	-- --	-- --	?? ??	Both dummies hit rear seat back.
1980	Mazda	626	D P	B B	22.3 22.3	-- --	-- --	~45 ~45	Both seat backs bent rearward onto rear seat
1980	VW	Rabbit Convertible	D P	B B	23.4 23.4	-- --	-- --	~45 ~45	Both dummies hit rear seat back with head. Seat backs bent onto rear seat cushion.
1980	Toyota	Corolla Deluxe	D P	B B	23.5 23.5	-- --	-- --	?? ??	Both dummies hit rear seat back. Passenger dummy head hit B-pillar.
1980	Oldsmobile	Cutlass	D P	B B	19.3 19.3	-- --	-- --	~60 ~60	Split bench seat. Both dummies contacted rear seat back.
1980	Fiat	Strada	D P	B B	23.8 23.8	-- --	-- --	?? ??	Both dummy heads contacted rear seat back.

APPENDIX C.3 - NCAP REAR IMPAC TESTS - 1981 MODEL YEAR VEHICLES

VEHICLE/SEAT DESCRIPTION			TEST DESCRIPTION			DUMMY RESPONSE		SEAT RESPONSE	
Model Yr	Manu-facturer	Model	Driver(D) or RFP(P)	Belted(B) Unbelted (U)	ΔV (MPH)	HIC	Pk Chest G's (-3ms)	Residuel Seat Back Angle (deg's)	Remarks
1981	AMC	Spirit	D P	B B	21.6 21.6	-- --	-- --	~60 ~45	Both dummy heads hit rear seat back.
1981	Chrysler	Imperial Speciality	D P	B B	18.0 18.0	-- --	-- --	~40 ~40	Split bench seat. Both dummy heads contacted rear seat back.
1981	Dodge	Aries	D P	B B	22.8 22.8	Dummies --	Uninstrumented --	~80 ~80	Bench seat came off tracks.
1981	Honda	Civic	D P	B B	24.1 24.1	-- --	-- --	~45 ~60	Both dummy heads contacted rear seat back.
1981	Datsun	810	D P	B B	20.7 20.7	-- --	-- --	~45 ~45	Head restraints became detached.
1981	Renault	18i Deluxe	D P	B B	23.4 23.4	-- --	-- --	~60 ~60	Both seat backs reclined to rear seat cushion. Both heads contacted rear seat.
1981	Mazda	GLC Custom	D P	B B	24.2 24.2	-- --	-- --	~60 ~60	Both dummy heads made contact with the rear seat back.
1981	Toyota	Cressida	D P	B(Passive B(2-pt)	20.8 20.8	-- --	-- --	~45 ~45	Both dummy heads hit rear seat back.
1981	Toyota	Starlet	D P	B B	25.0 25.0	-- --	-- --	~45 ~45	Both seat backs bent onto rear cushion. Both heads hit rear seat back.
1981	VW	Jetta	D P	B(VWRA) B(VWRA)	23.1 23.1	-- --	-- --	~60 ~60	Both dummy heads hit rear seat back.
1981	Chevy	Cavalier	D P	-- --	22.7 22.7	-- --	-- --	?? ??	Both dummies impacted rear seat back cushion.
1981	Plymouth	Sapporo	D P	B B	21.3 21.3	-- --	-- --	~50 ??	Both dummy heads hit rear seat back. Passenger seat pulled loose.
1981	Isuzu	I-Mark	D P	B B	23.6 23.6	-- --	-- --	~45 ~60+	Seat backs bent rearward. Passenger seat twisted inboard. Passenger head hit roof healiner. Driver head hit rear seat back.
1981	Mercury	Lynx	D P	B B	23.7 23.7	-- --	-- --	~45 ~45	Passenger seat right rail separated from track. Both heads contact rear seat back.

760810

AUTOMOTIVE SEAT DESIGN AND COLLISION PERFORMANCE

D. M. Severy, D. M. Blaisdell and J. F. Kerkhoff
Severy, Inc.

ABSTRACT

Eighty-five laboratory full-scale force-deflection tests were conducted on passenger vehicle seats, foreign and domestic, for purposes of evaluating specific resistance to a collision environment and mechanisms of collision induced seat distortion. These tests evaluated seats that span the past thirty years; additionally, seat design studies were conducted evaluating basic features of automotive seating during the past eighty years.

Data from full-scale collision experiments and from a large number of actual accidents facilitated the establishment of seat design criteria for greatly improved collision performance. Evolution of seat and head support standards in the United States and Europe are presented with evaluation of their relative significance to the requirements of automotive seat collision performance.

The foregoing research provided foundation for modification of a production automobile seat into an integral safety seat, based on a design concept that minimizes bending moments during collision. The modified seat was subjected to the same laboratory test procedure applied to the 85 non-modified production seats and results of its performance is given.

Design concepts are presented that would serve to mitigate undesirable seat distortions during collision and thus improve seat restraint capabilities without compromising the important factors of comfort and cost.

ALTHOUGH SAFETY IS A BASIC CONSIDERATION in all aspects of automotive engineering, it is a fact of life that safety advances are not uniformly accomplished. Some essential aspects of motorist protection have in the past, and probably will in the future, continue to be greatly outstripped by advances made in other areas. In some instances, the design challenge simply exceeds current technology; on other occasions an adequate safety standard and concerted industry interest can correct obvious safety deficiencies. For example, although improved tires resulted only after new materials evolved, a greatly improved motorist restraining system resulted from industry initiated research that culminated in a standard requiring installation of combination cross chest lap belts for all outboard front seat locations.

During the past 20 years, new materials and techniques have improved the comfort and wear resistance of automotive seats while simultaneously reducing their weight and cost; however, significant safety related improvement in seat design during this period has not been accomplished. A summary of findings from prior research (1)* published in 1966 classified a structurally redesigned Integral Seat with built-in 3-point belt restraint as critically important for reduction of motorist injuries; this summary also pointed out the lack of safety-seat research up to that time. Although some research has been conducted since that time, the development of a safety seat still has not been accomplished. In the same study special devices, including the air bag, were rated as only of moderate relative importance.

After 10 years of government regulated safety standards and nearly that many years of intensive air bag development sponsored by both government and industry, the consensus of many automotive safety researchers in Europe (2), (3) and elsewhere is to change the emphasis back to development of active restraint systems and to increase effectiveness by means of mandatory use laws. Air bags, if used, would serve a supplementary function in conjunction with active systems. Progress in safer seat design has been impeded throughout a decade of automotive safety standard making, and many lives have been lost that could have been fully protected with a fraction of the inventive genius and funding lost to air bag development. A basic common sense approach to motorist protection from collision trauma calls for special attention to design of the critically important structure nearest the motorist, his seat.

A vital consideration of seat design is its capacity to protect motorists, to the extent practical, from all types of collision injury exposures. This paper provides basic design data for crashworthy automotive seat systems with integral active restraints, as determined from the authors' collision research and laboratory studies as well as experience gained from investigation of relevant accidents.

---

*Number in parentheses designate References at end of paper.

BACKGROUND

The first automotive seat, like the first automobile was an adaptation from the horse-drawn carriage. Springs to absorb road shocks were primitive, effective padding was non-existent and seat adjustability had not yet been considered. Commencing around 1900, motorist safety while travelling over rough roads was improved by development of deeply contoured seats that reduced the likelihood of motorist ejection as the car body pitched and rolled. Generally, however, little consideration was given for the safety of the occupants, Fig. 1.

Front seat fore-and-aft adjustment was not available until about 1929 when adjustable front seats for the driver became a feature of higher priced automobiles. Occupant comfort was given increased attention as engineering problems concerning motor vehicle performance and reliability became more effectively managed. Improvements in seat design continued and by the mid 1930's, seats, tracks and runners closely resembled those of the mid 1960's. During the period between the thirties and the sixties, the only significant innovation in seat design was the introduction of power seats and adjustable reclining backrests during the early 1950's, Table 1. Seatback height reached reasonable levels in the late 1960's but by the mid-seventies the height of backrests on many models had declined to levels less effective than thirty years ago. Seatback strength has not increased significantly over the past thirty years and remains inadequate to resist even moderate collision forces. The remarkable safety improvement facilitated by the backrest head support standard FMVSS 202 has been compromised through failure to upgrade criteria to maintain performance commensurate with intent. Seventy percent of adjustable head restraints are used in the downmost position (4). Little effective protection is afforded the motorist unless the head restraint is positioned behind or slightly above the head and remains in such support position during collision.

SEAT FUNCTION AND CONFIGURATIONS

It is not a simple engineering task to design a good automotive seat; it must provide comfort, style and safety, and yet be sufficiently light weight to facilitate vehicle fuel economy and to minimize collision inertial stresses. Seat designs and materials must be affordable and durable to give acceptable service over the life of the car. In addition to provisions for comfort and position adjustments, a seat also should have adequate structure for housing safety and convenience accessories. Trade-offs are imposed by this complex mix of requirements; however, in seat design we should no longer overlook the requirement for a reasonably safe, collision resistive structure with built-in active restraint system.

SEATING CATEGORIES - Automotive seats for passenger vehicles have a number of basic configurations and sizes depending on intended use and location. Position adjustment depends on factors such as location (front or rear); availability of comfort features, including center armrests, depends on vehicle

Fig. 1 - Driver's seat, horseless carriage

TABLE 1

SEAT DESIGN EVOLUTION

INTRODUCED	ITEM	EXAMPLE
1890 - 1900	Automotive Bench Seats	Philion*
1900 - 1910	Deep Bucket Seats	Thomas*
1910 - 1915	Fold-forward Backrests	Model-T Ford*
1910 - 1915	Console Between Seats	Wescott*
1910 - 1915	Pedestal Seat	Argo Electric*
1915 - 1920	Swivel Seat	Cole*
1920 - 1925	Fold-down Armrest	Dusenberg*
1925 - 1930	Fore-and-aft Adjustment	Viking*
1950 - 1952	Power Seats	Packard
1960 - 1963	Optional Head Restraints	All U.S.
1968	Integrated Head Restraint	Volkswagen
1969	Standard Head Restraints	All U.S.

*Based on authors' study of vehicles made available through cooperation of Harrah's Automotive Museum, Reno, Nevada.

price, type, purpose and country of origin.

BENCH SEAT - SOLID BACK - Standard and compact 4-door sedans are frequently equipped with solid-back bench seats, front and rear. The front seat has a single adjustment mechanism and the entire seat is usually adjusted according to the needs of the driver. A bench seat, if adequately designed for three occupants, must be capable of sustaining a rearward static moment, calculated about the H-point of the seat, equal to three times that of a single seat.

BENCH SEAT - SPLIT BACK - Fold-forward backrests provide access to the rear seat of 2-door vehicles. Since only the upper portion of the seat is divided, the fore-and-aft adjustment can accommodate only one person. Backrest angle, however, can be made adjustable according to the individual needs of driver and passenger.

BUCKET SEATS - Contoured seats for individual occupancy are called bucket seats, although this term originally applied to more deeply contoured racing-type seats. The containment feature of deeply contoured bucket seats has been identified by the authors and others (5),(6),(7) as a consideration in design of a safer seat, owing to improved lateral retention; however, the extent of contouring must be based on consideration of comfort and practicality as well as safety.

Adjustable backrest angle is commonly available and provides improved comfort and reduced muscle fatigue. Direct attachment to the seat of both lap and shoulder belts is a safety and convenience goal that has been identified previously, (1),(5),(6),(7),(8) and is further evaluated in this study.

FOLDING SEATS - These seats require special design considerations to allow them to be folded flat when hauling cargo. These requirements include special sizing, latching devices and flat, durable cargo surfaces. Seatback height and strength as well as provision for isolating cargo are collision safety factors to be considered.

PEDESTAL BUCKET SEATS - These seats are common to buses, vans, motor homes and trucks. In addition to the normal fore-and-aft position adjustments, variable backrest angle and swivelling mechanisms are sometimes provided. Because of their elevation, direct restraint system attachment requires special design considerations.

FIXED, BUS-TYPE BENCH SEATS - Although similar to the rear seats of passenger sedans these seats are elevated and generally provide open space beneath; structural limitations and remoteness of seat from floor increase the difficulty of direct restraint attachment. Seats used in buses that transport children require special considerations, (9),(10).

SEAT SYSTEM COMPONENTS

A brief description of seat components will allow a better understanding of basic seat requirements and the special seat structure necessary for adequate passenger protection from collision trauma, Fig. 2.

STRUCTURAL FRAME MEMBERS - The seat framework is usually constructed of steel that has been formed into tubular configurations or of stamped or rolled sheet metal. Historically, the function of this structural backbone has been limited to providing shape for the cushioning members and support for its own weight and that of its occupant. Redesign and strengthening of the seat framework in conjunction with its anchorages can provide the force resistance necessary for occupant restraint during moderate and severe front, side and rear-end collisions.

NON-STRUCTURAL SEAT MATERIAL - Cushions, springs and upholstery provide the necessary means of load distribution between occupant and seat frame; they also provide contour and geometry necessary for occupant comfort. A redesigned integral seat would include energy absorbing (E-A) padding at the sides, top and back to provide additional force moderation and load distribution during collision. This concept is typified by the Crandell-Liberty Mutual Survival Car II (11).

SEAT ADJUSTMENT MECHANISMS - These mechanisms should be strengthened to withstand collision forces as well as the rigors of everyday usage. Considerations of driver anthropometric differences necessitate a multiplicity of seat adjustments to assure that a given seat provides the multiple functions of driver positioning for correct reach and view, as well as for adequate comfort. Desired seat position may be accomplished by means of longitudinal, vertical and tilt adjustments. Other adjustments include backrest angle, head restraint position and lumbar support stiffness.

Head restraint height adjustment is a common feature of current production seats. Unfortunately, many recent seat designs provide no head support protection from even minor rear-end collisions when the head restraint is in its most commonly used, retracted position. Proposed legislation (12) would modify head restraint requirements to ensure proper utilization by preventing mis-adjustment or complete removal. Backrests with integral head restraints eliminate adjustability deficiencies, but should be designed to protect 95th percentile adult male occupants. Adjustable headrests should extend to 80 cm (32-in) and not allow adjustment below 70 cm (28 in); the headrest in its adjusted elevated position must not collapse from head impact.

SEAT ANCHORAGES - Most vehicle designs depend on seat attachments at the floor or at the sill and tunnel to transmit

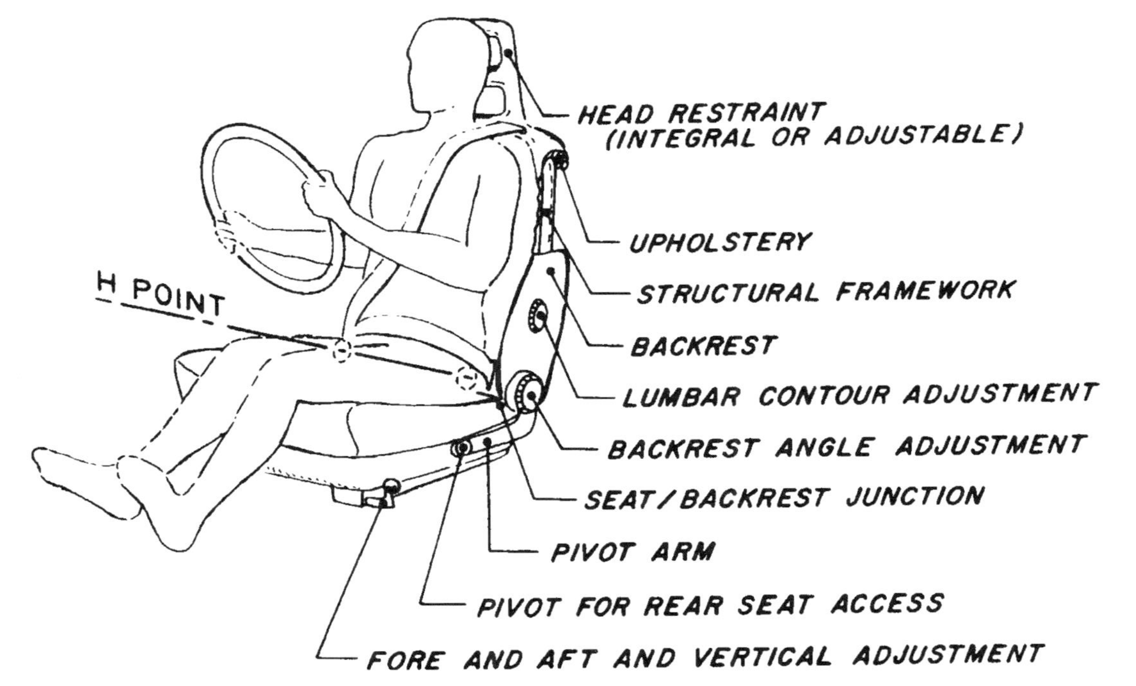

Fig. 2 - Seat nomenclature

forces between the vehicle and the seat. This anchorage is generally interposed between the seat adjustment mechanism and the vehicle floor pan or lower structure.

Under static conditions, the seat anchorage transmits compression, tension and shear forces from the seat to the floor or side structure. In the event of collision or handling accelerations, the resulting forces are transmitted in reverse direction from floor to seat. The seat anchorage structures and attachments require a design of adequate strength to accommodate seat and occupant inertial forces identified specifically in other sections of the paper. Besides the usual floor, sill and tunnel anchorages, other potential anchorage locations for strengthened seat systems include side and roof attachment.

SEAT DESIGN CONCEPTS

Restraint compliance, is the change in restraint geometry induced by motorist's inertial forces during collision. For example, a lap belt may initially course from its floor anchorage toward the lap of the wearer at 45 degrees but as a frontal impact develops, this angle is reduced owing to a combination of factors that result in the occupant's buttocks being compressed downward and forward on his seat as the belt restraining forces increase, (13). Similiarly, variations in support force, restraint angle and resulting backrest pressure modulations occur for the seated occupant as the seat backrest undergoes restraint compliance during a rear-end collision. This is illustrated for restrained and also for inadequately restrained occupants, by the geometric areas within the curves depicting displacements of occupants relative to their vehicle, Fig. 3. Also shown is the added physiological trauma associated with unstructured, inefficient restraint compliance. Variations in occupant collision-induced accelerations (slopes, Fig. 3) depend on the effectiveness of occupant restraints which determines in large measure the trauma during ride-up/ride-down accelerations; the mechanism of transmission of accelerative forces represent the other operative factor for such traumatic exposures. The problem faced by a seat designer is to determine the force resistive levels, and therefore the degree of collision severity, a specific seat design should be able to accommodate before backrest restraint compliance reaches values where occupant forced displacements expose him to serious or critical injuries.

When not faced with the realities of cost and operational practicality, as well as the need to provide design features that favor public acceptance, a simplistic approach is to set safety design criteria at some arbitrarily high level, one considered sufficient to encompass virtually all probable collision exposures. Such arbitrary criteria is insensitive to orders of priority and cost for safety design improvement and disregards the interactive response it may impose on other related, or concurrently important, passenger and operational safety features.

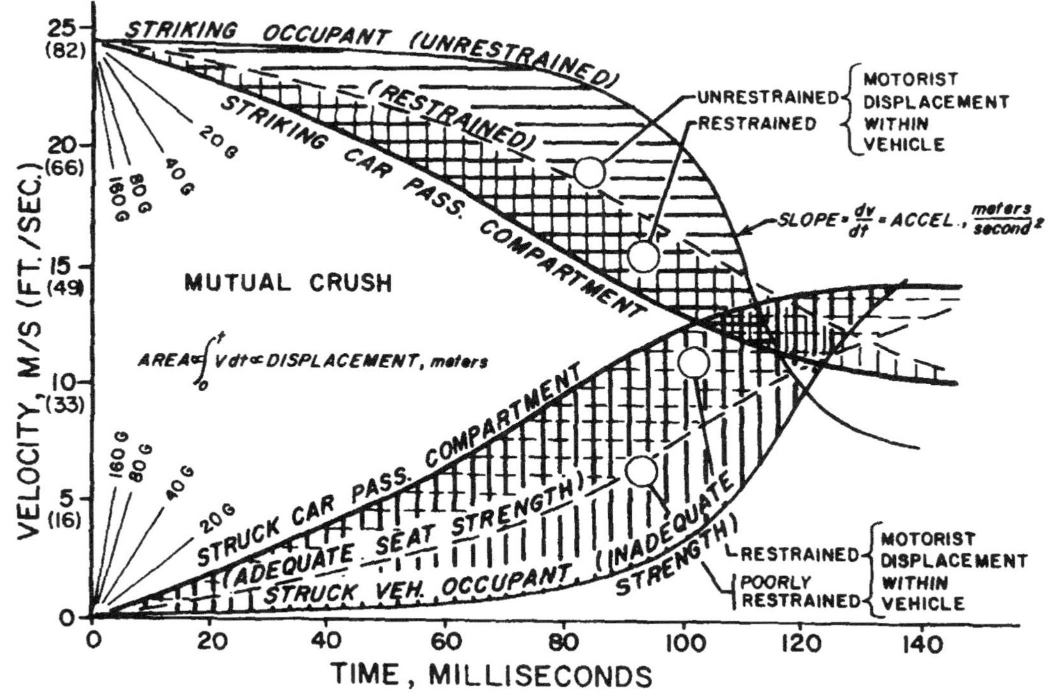

Fig. 3 - Relative abruptness of occupant "Ride-down/Ride-up," according to efficiency of restraint

The front seat backrest and the steering column have in common the engineering design problems that result from dealing with cantilevered structures acted upon by occupant collision forces. Additionally, backrest angles of 15 to 25 degrees increase design difficulties by incorporating angles that allow less efficient crash force moderation.

LABORATORY SEAT TESTS

Passenger vehicle seats of differing make vehicles are as varied structurally as the motor vehicles within which they are mounted. Aside from their apparent commonalities consisting of seat, backrest, head support, covering, padding, springs and frame, the multiplicity of design and size variations account for their significantly differing weights (under 9 kg to over 45 kg or about 20 to over 100 lbs.) height of backrests (from 51 cm to about 76 cm or from 20 to about 30 in) and maximum bending moment before excessive backrest deflection (from 45 m-kg to about 95 m-kg or about 4000 in-lbs to 17,000 in-lbs), Fig. 4. As remarkable as these variations would appear, they represent little difference in motorist protective capacity owing to the fact that none of them are capable of effectively resisting motorist inertial forces for any substantial impact exposure.

Based on laboratory tests of seats from cars large and small, foreign and domestic, and from vehicles 30 years old to near new, backrest strengths were found to be remarkably alike, Fig. 4. Force-deflection performance curves of representative production seats as well as recommended performance for passenger vehicle seats capable of protecting their occupants during severe collisions are shown, Fig. 5.

INTEGRATED SEAT - The concept of an integrated seat for motorists has been considered and evaluated theoretically; experimental seats have been constructed and tested experimentally, (1), (2),(5),(6),(7),(11),(14). In general, the approach has been to design an entirely new seat, inducing the argument that practical aspects of manufacturing, occupant comfort and similar design burdens remain yet unsolved. With this in mind, the authors redesigned a Volvo production bucket seat into an integrated safety seat. This approach should facilitate recognition by automobile manufacturers and the Federal Government's safety standards making authority, that the goal for greatly improved motorist safety is reasonably attainable through development of an integrated front seat.

The Volvo seat was stripped of its upholstery and subjected to an evaluation of modifications considered essential to developing performance specifications for an integrated safety seat, Fig. 6. Next, this production seat frame was modified to provide the added structural support required for the special protective hardware, Fig. 6b.

Seat Modifications - Changes were made in the production seat by additions of components to augment existing

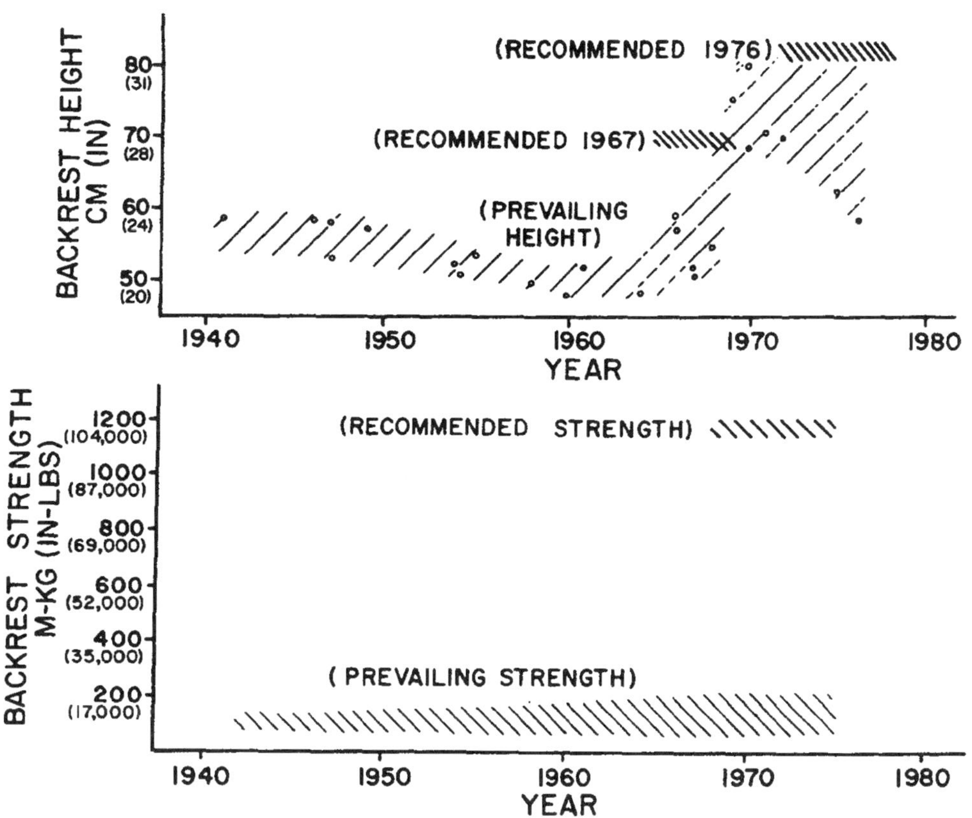

Fig. 4 - Automotive seat strength and backrest height variations over the past thirty years

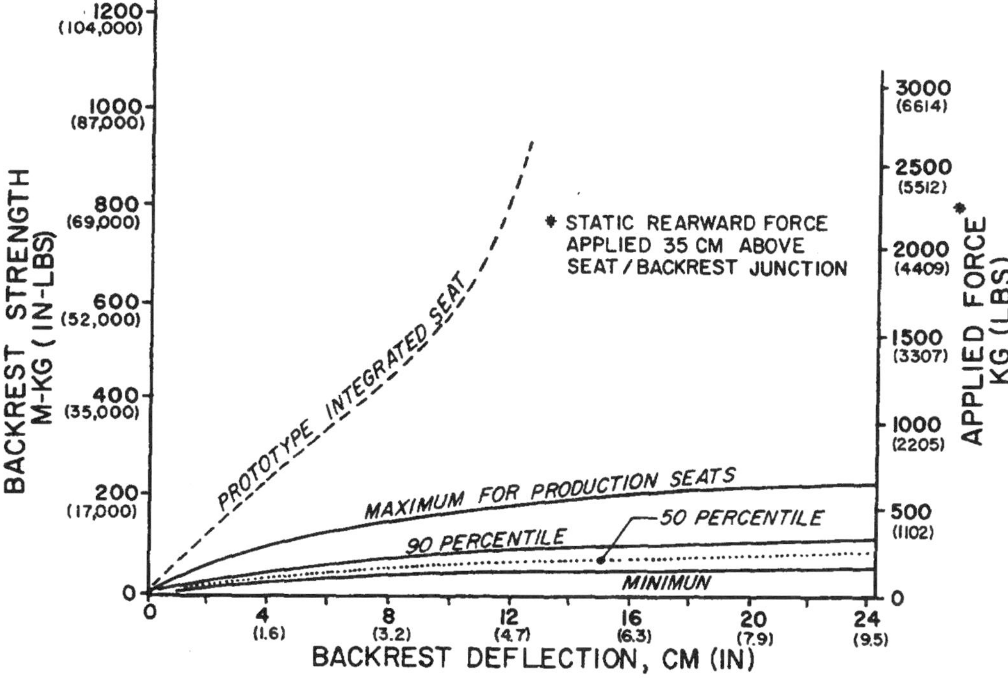

Fig. 5 - Backrest force-deflection data for 85 production automobiles, foreign and domestic, large and small sized vehicles

Fig. 6A - Volvo production passenger car seat frame, modified to high-performance integral safety seat

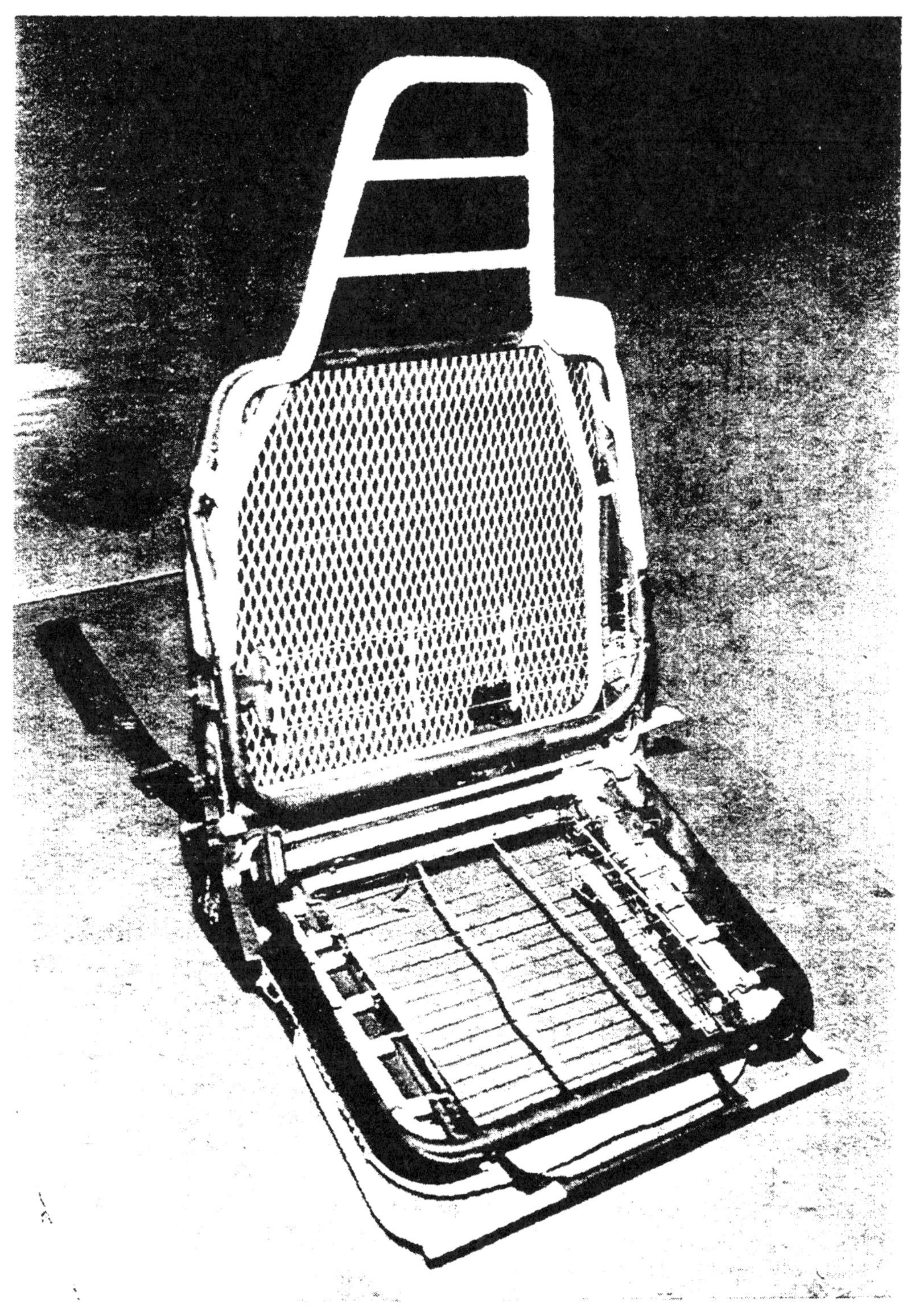

Fig. 6B - Volvo production passenger car seat frame, modified to high-performance integral safety seat

safety and comfort features and to introduce new protective functions as follows:

(a) High section-modulus tubular steel backrest, from seat attachment through head support.

(b) Double seat tracks, for more positive anchorages.

(c) Lap and cross-chest safety belt inertial reels mounted to seat frame for ease of fastening and for optimum occupant restraint for all adjustment positions.

(d) Energy absorbing back panel, to moderate and pocket occupant inertial loads during medium and high-speed rearend collisions and to provide knee pocketing for kinematic control of rear seat occupants thrown forward during frontal collisions.

(e) A pair of seat-mounted inertial reels for belt fabric which connects the base of the seat at its rear to the roof; the belt fabric is passed through the backrest to the roof anchorage. This arrangement allows the seat to be adjusted fore-and-aft and folded for entry accommodation; however during a front or rear-end collision, the dual "safety belts" are capable of immediate seat inertial snubbing.

(f) Deeply contoured bucket seat geometry to assure occupant positive lateral support and to improve comfort.

(g) Retention of the most advanced comfort and safety features of backrest angle adjustment, lumbar spine firmness adjustment and louvered see-through head restraint.

Most prior integrated seat concepts have utilized the backrest cantilever principle which necessitates strengthening floor and seat anchorages along with seat and backrest framing in an effort to combat the extreme bending moments, especially attending front and rear-end collisions, (7),(8),(14). The advantage of this approach is that seat fore-and-aft positioning and backrest folding for access to the rear seat can be accommodated without roof or side attachment; disadvantage is that it is a very inefficient method of resisting large forces owing to the massive increases in structure required to offset extremes in bending moments. The authors found that many production seats could resist between 900 and 1800 kg force (about 2000 to 4000 lbs) horizontally rearward or forward at the seat rail. If these floor anchorages were only slightly improved, and connected by belt webbing to the roof, the cantilever design could be replaced by the basically greater efficiency of the tension semi-hoop design spanning floor to ceiling. The Experimental Safety Vehicle by Opel also uses a form of roof anchorage for its front seat backrests, (2). Encouragement was given to the practicality of this concept by a preliminary test of roof strength of a mid-sixties 2-door hardtop sedan. This unmodified roof structure withstood over 1800 kg (4000 lbs) horizontally rearward at a localized backrest to roof attachment location, without significant downward crush.

The integrated seat was subjected to laboratory tests along the lines previously described as shown by Fig. 8. The improvement in force-displacement over non-modfied production seats is shown by the dotted curve of Fig. 5. Considering the nature of changes necessitated by using an existing seat, it is apparent that a somewhat lighter seat could be made even stronger if it were an original design rather than a retrofit. The composite design factors for a remarkably stronger, attractively contoured seat are shown by Fig. 7. Extensive crash testing of such new concepts are recommended for sound development before they are used in production vehicles.

SEAT COLLISION PERFORMANCE

Relative to vehicle safety, performance connotes a level of attainment, a plateau from which concepts for improvement may be assessed, both as to practicality and possible adverse influence on other safety requirements. Like the occupant that uses it, a seat has rather specific collision force tolerances; exceeding its design criteria allows crash forces to yield or fracture whatever component of the seat is most susceptible. This collision response is common to all seats, as determined from destructive tests to over 85 different automobile seats mounted on their floor pans. Although individual designs determine which components may fail first, their level of susceptibility is remarkably similar. Perhaps the foundation for this similarity in seat strength levels is generic to the basic design requirements that were dictated long before collision forces were identified as factors for consideration. For example, an adult male can apply 90 to 135 kg (200 to 300 lbs) force to the backrest during emergency braking. If passengers also brace for a possible collision, the backrest of a bench seat can be called upon to resist as much as 275 kg (600 lbs) force. It is understandable, therefore, that seats in the forties and fifties had backrest yield resistance of about 275 kg (600 lbs) even though they were not designed in a period when collision safety was a prominant consideration; incomprehensible is the fact that seats in the seventies show no significant improvement.

The basic collision performance criteria that applies to a seat is its resistance to backrest yield and its ability to transmit forces through the seat to its anchorages. Although seat design has in no way kept pace with advances made in other safety related areas, there are some notable exceptions; certain Mercedes and Porsche automobiles, for example, attach lap belts to the seat, rather than using the less satisfactory arrangement of floor anchorages. Volvo and Saab have provided see-through head restraints that improve visibility. The track record of most vehicles, however is not impressive. In general, backrest strength and height has not improved, possibly because the need for stronger, full-support backrests is not fully understood by regulatory agencies and those making design decisions.

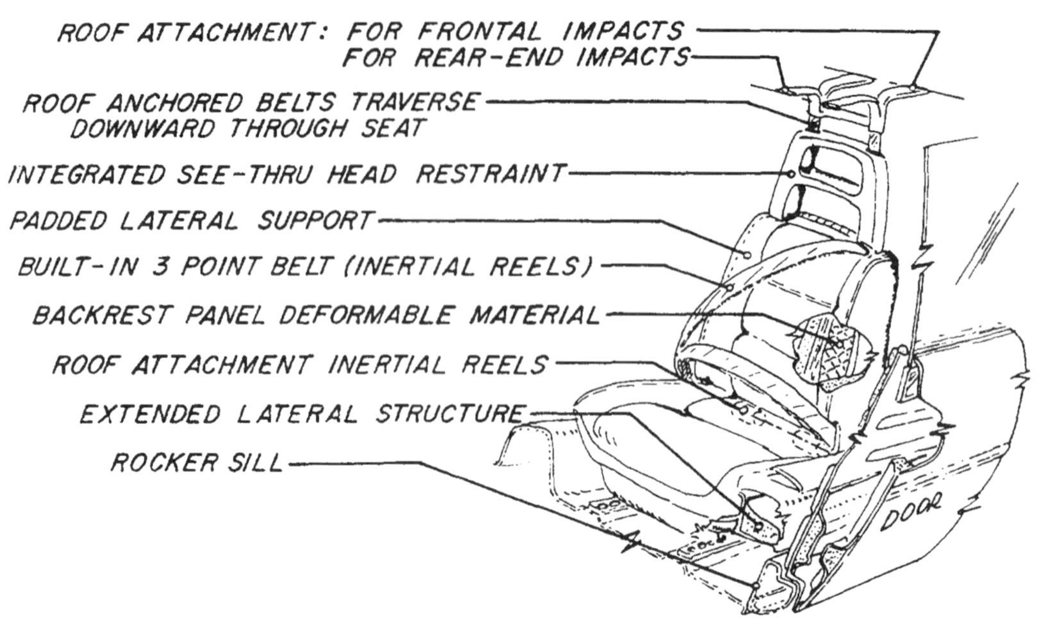

Fig. 7 - The Integral Safety Seat conceptualized and tested by the authors

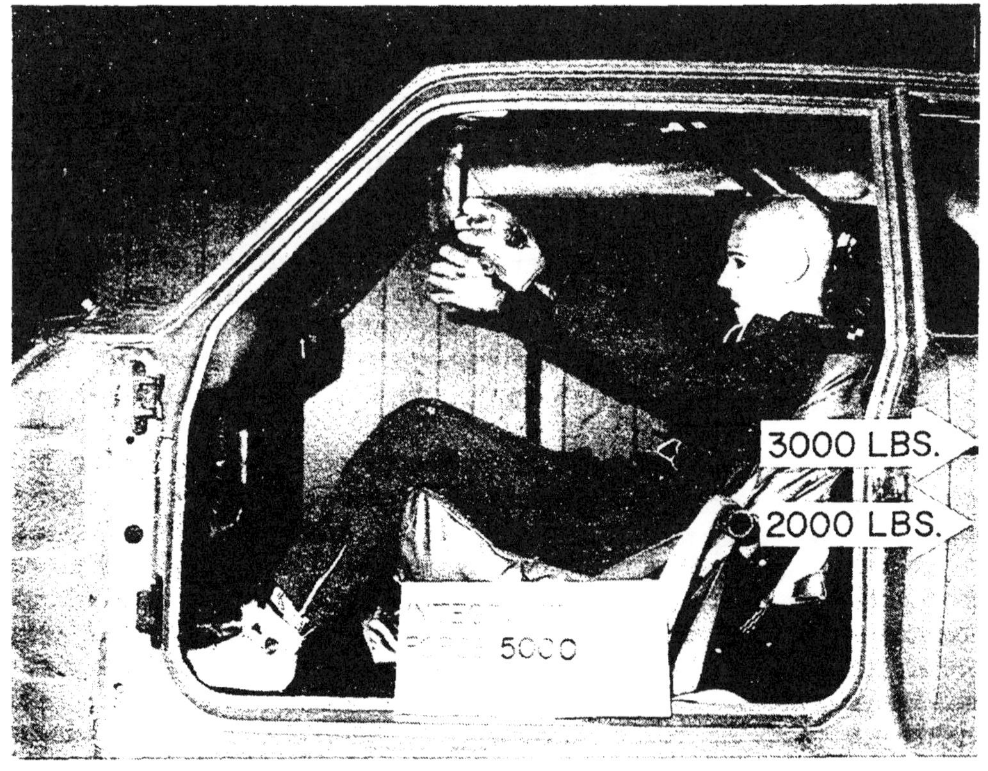

Fig. 8 - Integrated seat during testing; 2268 kg (5000 lb) force applied rearward

SEAT COLLISION PERFORMANCE - FRONTAL COLLISIONS - It is becoming increasingly apparent that passive restraint systems alone cannot protect motorists from the varied exposures encountered during motor vehicle accidents (2),(3). Systems that protect during a barrier-type crash may be completely ineffective if the same frontal collision is accompanied by one or more minor impacts or an upset. The cross-chest lap belt restraint system, on the other hand, provides effective motorist protection during most collisions or combinations of collisions, including upset. If an integral three-point restraint system is available, in conjunction with the strengthened seat to which it should be attached, it can provide levels of protection unattainable with any other single system, however costly or complex. In addition to minimizing or eliminating impact forces to the vehicle interior, the combined restraint system appreciably reduces the forces that would be applied for a lap belt only protective system, (15).

One advantage claimed by air bag proponents is the additional ride-down distance provided as the motorist crushes forward relative to the vehicle interior. Direct attachment of belt restraints to the seat minimizes uncontrolled belt slack and at the same time allows the designer to incorporate belt pre-stressing and other ride-down improvements for the belt and seat anchorages. The belt system can thereby provide maximum protection for front-end collisions and also provide more adequate protection during the wide range of other collision exposures, including moderate to severe rear-end collisions by keeping the occupant in the seat. Even a floor mounted lap belt can provide considerable protection under these conditions, (16).

Rear seat occupants that fail to fasten their belt would obtain some degree of passive restraint from a strengthened front seat backrest. The knee pocketing concept evaluated and described previously (6), will add considerably to rear seat occupants' collision protection by controlling posture during their "second collision." The spacing between rear occupants and front backrest provides a time delay before the rear occupants strike the front backrest; therefore, front seat occupant ride-down would not be compromised significantly from subsequent impact to the backrest by the rear seat occupant. Rear seat occupant protection for all but rear-end collision exposures can best be accomplished by means of active restraint systems.

OCCUPANT RESTRAINT WITH A RECLINED SEAT - The cross-chest lap belt is less effective in front-end impacts where the passenger has reclined his seat; the torso slides forward, shifting the lap belt upward across the viscera. When belt snubbing eventually occurs, forces are transmitted to portions of the body unable to sustain them without exposure to serious injury. Also, the neck or head may strike the cross-chest belt in an ineffective, if not dangerous manner. One solution is to provide an additional pair of straps attached at the front of the seat and passsed upwards through positioning guides. Before reclining the backrest, the hanging ends of the straps are pulled up between the legs and fastened to attachment loops sewn to the lap safety belt. This arrangement maintains the lap belt across the pelvic girdle during collision and, for non-integral seats, reduces mal-alignment of the cross chest strap as the torso flails forward against it.

SEAT COLLISION PERFORMANCE -- REAR-END COLLISIONS - The most frequently occurring type of motor vehicle collision is the rear-end impact. For example, forty-nine percent of the 14.6 million accidents involving motor vehicles in the United States during 1968 were same direction or rear-end type accidents; representing nine percent of the total number of fatal collisions, (17).

Seat backrest performance during rear-end collisions parallels a condition of walking on thin ice; minor torso inertial forces receive adequate support but as impact severity increases, the resistive support approaches negligible values. The mechanism by which this passive restraint is lost is generally unimportant insofar as its effect on motorist kinematics is concerned. Torso inertial movement does not respond differently with differing mechanism of loss of back support; it simply responds inertially to fall-off of resistive force. Thus, whether backrest resistance is subsequently lost because of excessive backrest yield or by combinations of backrest yield and anchorage separations, the motorist inertial forces will act in the same manner.

Backrests are slightly reclined for comfort and to assist with torso support. However, as the seat is forced to recline additionally during rear-end collision, the torso is predisposed to ramping up the plane of the backrest, Fig 9. On reaching backrest angles exceeding 45 degrees, the ramp component falls off according to the function "$\sin\alpha\cos\alpha$," where alpha (α) is the backrest angle rearward from the vertical axis. Simultaneously, the backrest support force, F_{SB}, which accelerates the motorist as the backrest bears against him during a rear-end collision, falls off as a function of backrest yield-angle, likewise depicted by Fig. 9. Obviously, a zero or near zero angle provides the most effective torso support by the backrest during rear-ending and torso support tends to fall off thereafter, as a cosine function of increases in the backrest angle.

These factors are particularly relevant for production seats involved in light to moderate rear-end collisions where backrest strengths generally are adequate for those respective exposures. However, for most moderate to severe rear end collisions, all production seat backrests deflect to angles providing no effective motorist support. This situation is illustrated by considering the relationship of backrest support provided the torso as a function of rear-end collision speed, Fig 10. Sixty-five percent of the body weight may be used as a minimal approximation of the torso and adjacent flailing body components that operate any instant to inertially act on the backrest and may be characterized as the occupant inertial force, F_{OI}. The seat backrest yield point, or point

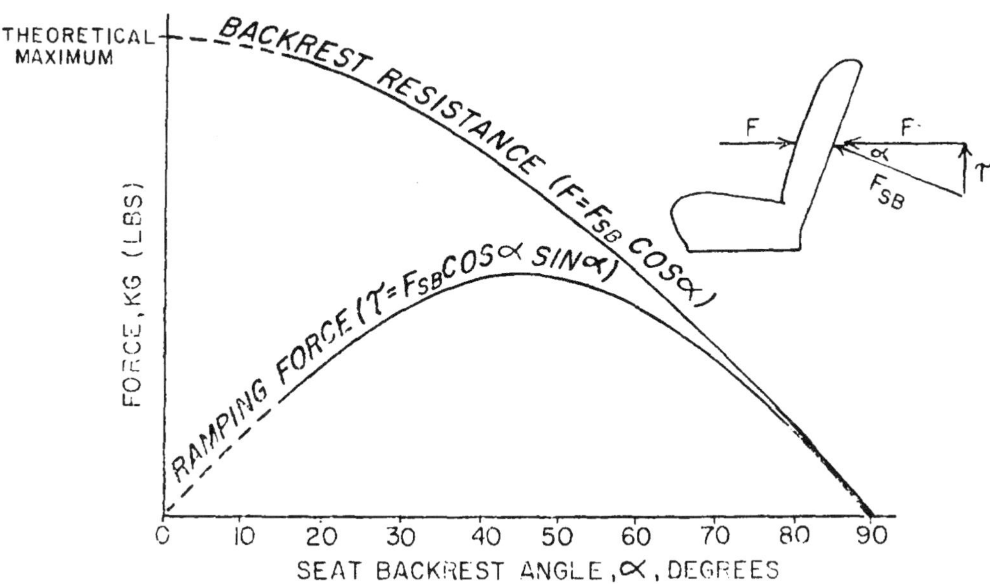

Fig. 9 - Degradation of backrest support as function of collision induced deflection

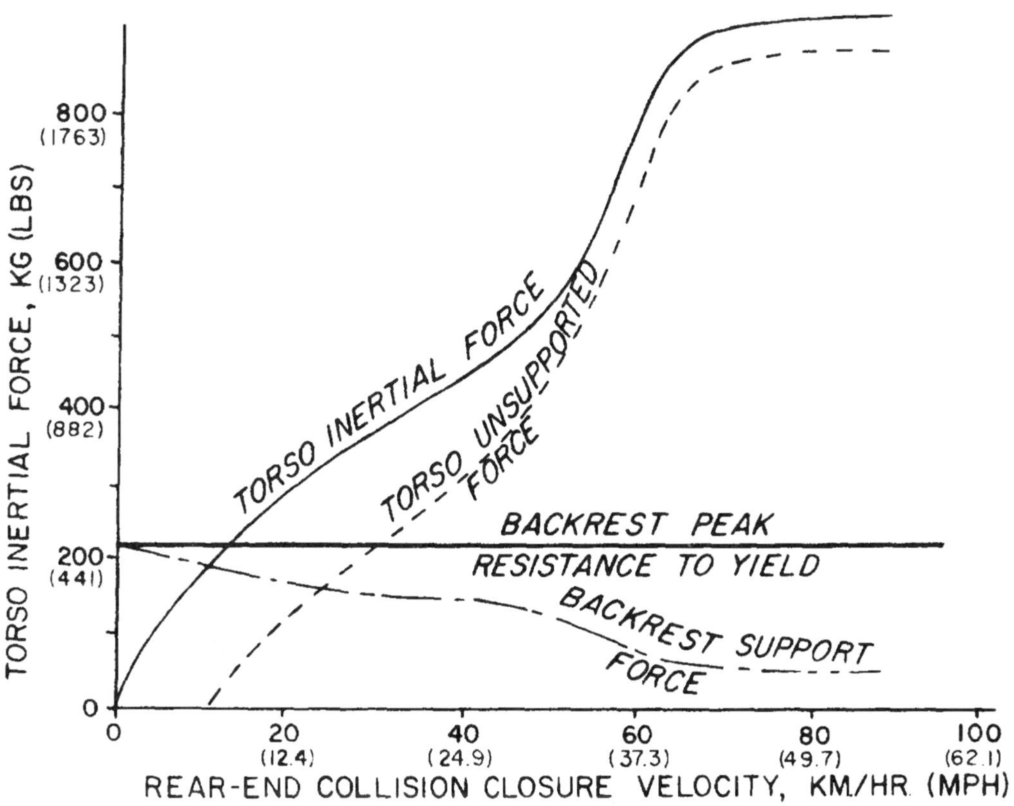

Fig. 10 - The relationship of backrest torso support, as a function of rear-end collision speed

of maximum backrest resistive force, F_R, is reached when the occupant inertial force plus the seat backrest inertia force exceeds F_R as shown by $F_{OI} + F_{SI} = F_R$. Based on collision experiments (14) the torso inertial force increases with collision speed, in the manner shown by Fig. 10. Regardless of speed of impact, the backrest can provide only a given maximum resistance to yield; however, as rear-ending speeds increase, the collision severity is reached where the backrest can at best support solely its own weight, but none of the occupant, Fig. 10. Thus the effective backrest strength is dependent on a combination of backrest resistance to yield and backrest weight, referred to as backrest inertial yield resistive force. The lighter the weight of a seat, relative to a given resistance to deflection, the better able the seat to resist torso displacement during rear-end collision.

Given a sufficiently severe collision, any seat will fail; the type of failure however, depends on design. Rear-end collisions induce five basic types of failure; a particular collision exposure for a particular seat design may result in one or a combination of these five types of failure:

(a) Excessive backrest yield

(b) Track-runner vertical separation

(c) Track-runner longitudinal separation

(d) Compressive crush at seat base

(e) Floor yield at seat attachment

All automotive seats have collision induced failure modes that are caused by moments and forces exceeding one or more of the seat component yield points. Critical reaction requirements can be calculated for various occupant sizes and acceleration exposures using free body analyses of the seat system, Fig. 11.

For rear-end collision exposures, the addition of a roof anchorage for the backrest serves to distribute forces and considerably lessens design requirements for individual components. For example, the extremely difficult design task of providing 75,000 inch-lbs yield resistance at the backrest pivot for a 50% adult male in a 30G seat system can be easily managed by utilization of a roof anchor. If one-third of the 6540 pound horizontal force is applied at the top of the seat and acts 32 inches above the backrest pivot, we now have the following moment at the pivot arm:

$M_1 = 75{,}000 - (32)(2180) = 5300$ in-lbs, a reduction to 7% of the original required bending moment. Most production seats can provide this level of bending resistance without need for modification. With a roof attachment, the maximum bending moment will be at a higher elevation on the seat and in a location of the backrest free of complications of pivots and adjustment mechanisms. Correspondingly, the reduction in reaction strength required at R_R, R_H and R_F, if a roof anchorage is added, provides a requirement of only 2%, 65%, and 2% respectively of the values necessary without a roof attachment.

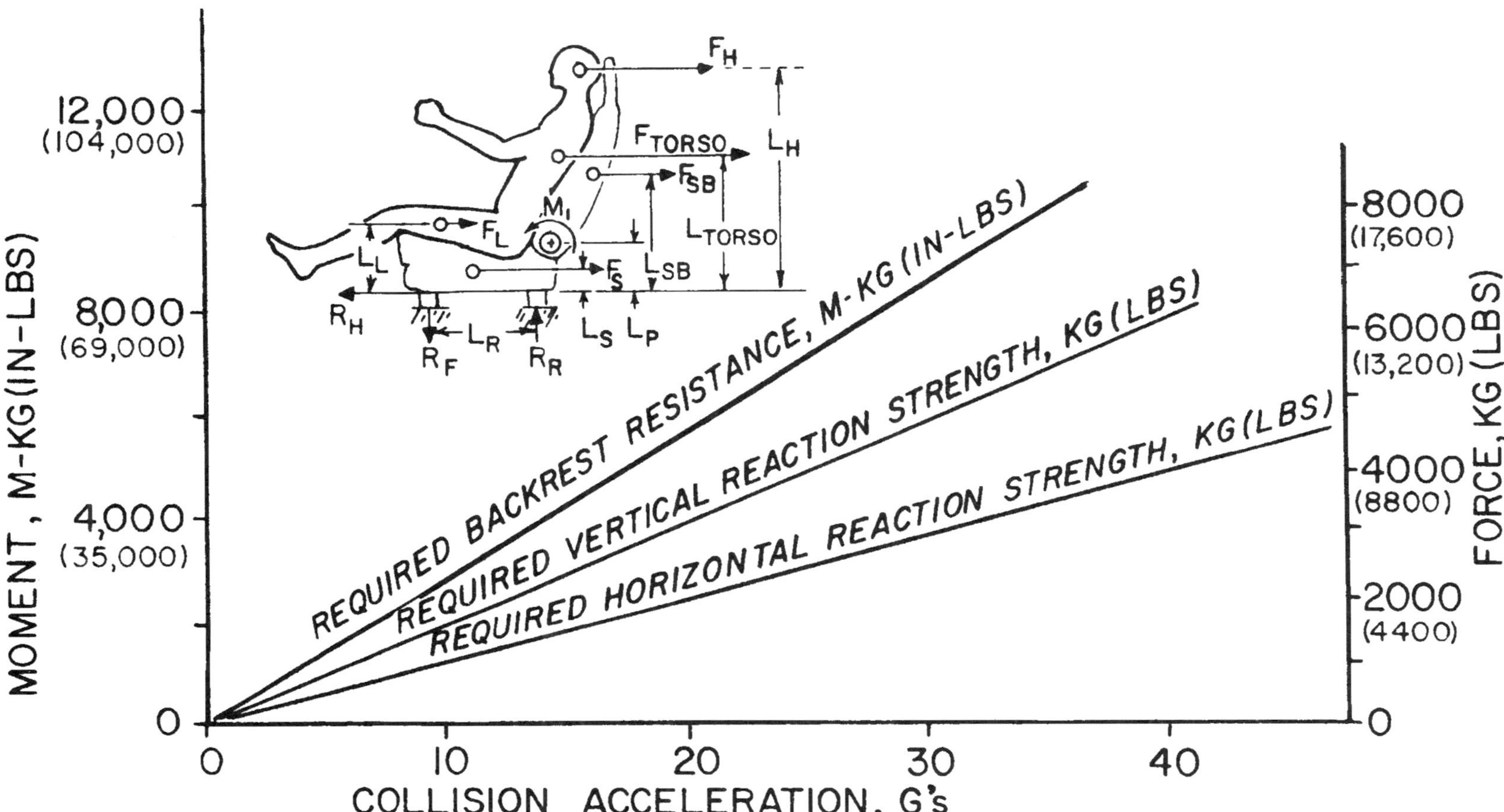

Fig. 11 - Collision induced seat stresses, rear-end collision exposure

SEAT COLLISION PERFORMANCE - SIDE IMPACTS -Direct side impacts into the passenger compartment represent the gravest danger to motorists at a given speed of impact for any type of collision, (18). Usually less than a foot of intervening structure, much of which is non-structured space, separates the motorist on the impacted side from the front end of the impacting vehicle. During collision, the near-side occupant suffers a considerably more severe acceleration than does the vehicle in which he is riding. This impact exposure can be lessened somewhat by increasing the amount of side structure stiffness and thereby minimizing passenger compartment intrusion. However, the point of diminishing return is soon reached as intervening structure is added, since crush distance is still limited, and delayed occupant response serves to magnify the resulting exposure. A considerably simpler solution is to attach the motorist firmly to a deeply contoured seat by means of an integral 3-point belt and to provide intermediate structural members between striking car and seat to cause the seat, with occupant attached thereto, to be accelerated more promptly but less abruptly.

Analysis of the velocity-time profile of a typical side impact collision, Fig. 12, provides further understanding of the nature of accelerative exposure and resulting injury mechanism as it exists for the unrestrained struck motorist. This graph also shows the remarkable reduction in occupant acceleration that can be accomplished if the seat is structured and interlocked with side panels so that it is accelerated more promptly following initial contact by the striking vehicle. This improved side impact acceleration, or "ride-up," is similar to the "ride-down" concept of protection from frontal collisions, identified by prior publication, (19). The more prompt acceleration also reduces the impact abruptness or "Stapp"-factor. Since the combined weight of the seat and the occupant is often less than 90 kg (200 lbs), it is not necessary to provide significantly heavier structure than already present to accomplish the objective of prompt seat acceleration during a collision. It is only necessary to position the structure at the proper elevation and location to allow more direct collision force application to the seat structure to occur before substantial inward crush has taken place. Additional motorist force moderation is provided by the pocketing effect of the integrated seat. This seat side-structure tends to shield the motorist from direct impact by his vehicle side structure as well as to minimize "slack" between the motorist and his restraint system. This shielding also reduces motorist exposure to punctures and shear forces.

ROLLOVER PROTECTION - Injuries and fatalities during rollover accidents occur most frequently as a result of ejection (20),(21); a basic requirement for occupant restraint systems is to retain the occupant within the vehicle during upset. Even if adequate retention is achieved, protection is compromised if there is inadequate supportive roof structure.

Roof structures are sometimes flattened during upset sufficient to dangerously reduce the motorist survival space. This

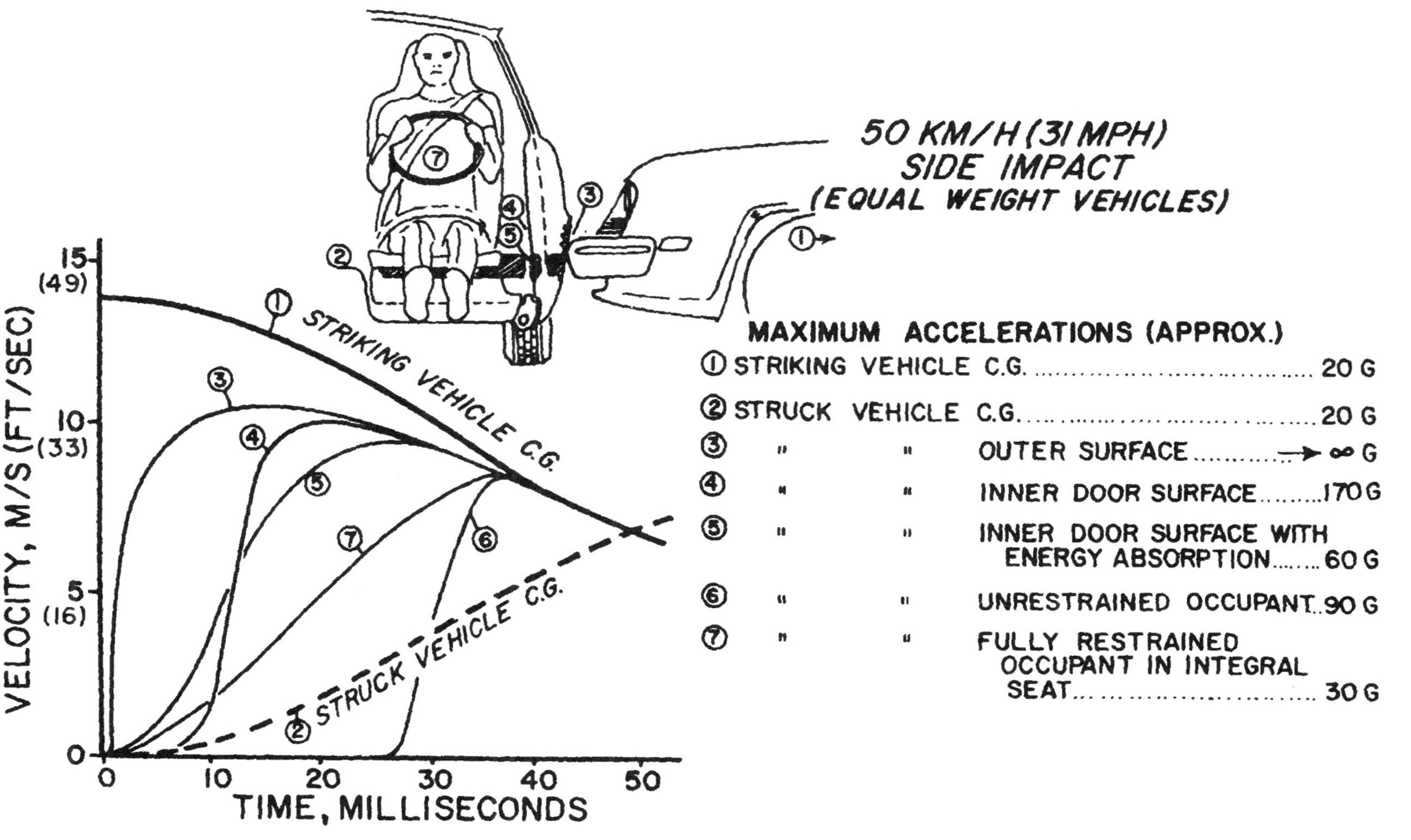

Fig. 12 - Integral seat, side-impact exposure analysis

crush can be reduced considerably by constructing the seat backrests with sufficient strength to act as central pillars for the roof. A strengthened seat with integrated head restraint of a height necessary for adequate head support for taller persons 80 cm or more (32 in) is recommended if central roof collapse resistance is to be increased by backrest structural support. Seat backrest structure of sufficient strength and height to answer the requirements imposed by other collision exposures will also prevent complete collapse of roof structure in nearly all collision exposures.

STANDARDS AND RECOMMENDED PRACTICES

In November 1963, a Society of Automotive Engineers committee report on passenger seats and adjusters was approved. This action, for the first time, extended the concept of recommended practices and performance standards to include automotive seats. Two tests were specified by recommended practice J879, the first of which established minimum strength requirements for horizontal inertial seat loadings for front and rear impacts. This 20G specification corresponds to approximately a 900 kg (about 2000 lbs) seat anchorage loading for a typically large domestic bench seat and serves to evaluate the retention capability of the adjuster mechanism and seat anchorages. The second test prescribed by SAE J879 concerned backrest strength and specified a minimum moment of 4250 in-lbs. Although the scope of the recommended practice included an intent to submit it to a "continuing review" no substantial changes were made until July 1968.

In 1965 the Federal Register published General Services Administration (GSA) Federal Standard 515-6 governing "Anchorage of Seats." This initial seat standard became effective in 1966 for 1967 model automobiles. GSA adopted basically the original SAE recommendations with no major changes. Although this transfer of J879 from an SAE Recommended Practice to a GSA standard was applicable only to government purchased vehicles, it was understood to be the forerunner for a standard directly affecting every passenger car sold in the United States.

In 1966, by act of Congress, the Department of Transportation (DOT) was added to the Federal Government. The 1965 GSA 515-6 Standard, which was actually the SAE Recommended Practice of 1963, was adopted by DOT in January 1968 as Federal Motor Vehicle Safety Standard (FMVSS) 207. The FMVSS 207 established backrest strength for standard passenger vehicles at 3300 in-lb, differing insignificantly from the GSA/SAE 4250 in-lb value, owing to a change in reference systems. FMVSS 207 references the Hip Pivot or H-point (as described by SAE J826b) whereas the SAE J879 and GSA 515-6 used a reference pivot point that was about 20 cm lower (8 in) at the interface between the seat frame and adjuster. Although 3300 in-lbs may appear to be an impressive value, production seats from the nineteen forties and fifties tested by the authors were found to substantially exceed this standard. FMVSS 207 included the requirement that the fold-forward backrest locks withstand a 20G minimum inertial load which represented a modest improvement. In 1972, FMVSS 207 was amended to include multipurpose passenger vehicles, trucks and buses. Additionally, the standard was expanded to include requirements relating to rear seats.

Head restraints, whether separable, or integral with the front seat backrest were regulated by FMVSS 202 in 1969. This standard requires the head restraint to be capable of resisting a 200 pound force applied horizontally to it and actually consists of a more severe loading to the backrest pivot than does FMVSS 207, to reduce the chances of head support failure before backrest failure.

In Europe, seat strength standards applicable to non-exported vehicles were enacted by the Council of European Communities in 1974; these are basically the same as FMVSS 207, except for an increase in the minimum moment resistance of the backrest from 3300 in-lb to about 4700 in-lb (54 m kgf), both with respect to the same H-point reference.

Prior to the advent of the automotive standards making authority by the Federal Government, the automotive industry was largely self-governed, relative to implementation of safety concepts. Slow, methodolical but nevertheless significant progress was made during this self-governed period, as pointed out in a prior publication, (19). The emphasis was on operational and performance safety; accordingly, collision safety advances were limited.

Following the mid-sixties, the responsibility for progress in collision safety was taken over by the Federal Government and assessment of the progress made in this more recent period depends on the specific safety category being considered. In general, commendable improvement in collision safety has occurred during this period of regulation by the National Highway Traffic Safety Administration (NHTSA). However, with respect to automobile seat collision safety, little progress has been made. A near obsession with air bag development regrettably has diverted a tremendous amount of research time and resources from the readily achieveable goal of making the automobile seat crashworthy. As a result, many automobiles being manufactured a decade after the onset of Federal Government standard making still have seats no stronger than they were during the pre-regulatory period. Any real improvements made in seat design have been on the initiative of the manufacturers because the seatback standards relating to strength have remained virtually unchanged. The presence of an inadequate, seldom upgraded standard limits initiative of automotive manufacturers because of the implication that satisfactory conditions prevail.

SUMMARY OF FINDINGS

1. Improved automotive seat systems should provide adequate structure and built-in active restraints to insure a reasonable degree of protection from collision forces. Seats need adequate contour to provide effective support and sufficient adjustments to

meet the varied and changing requirements of their occupants. Redesign and strengthening of the seat framework and anchorages is required for direct attachment of lap and shoulder belts and to provide the force resistance necessary for occupant restraint during moderate and severe front, side and rear-end collisions.

2. All automotive seats develop collision-induced yield or separation modes caused by forces and moments from motorist and seat inertia when such stresses exceed one or more of the seat component yield points. Although the specific component of a seat that fails first may differ, the degree of susceptibility of all seats is remarkably similar. Laboratory tests establish that production seats from cars large and small, foreign and domestic, and from vehicles 30 years old to near new, have backrest strengths remarkably alike. All are incapable of effectively resisting motorist inertial forces for any but light impact exposures without inducing excessive yield and/or component separation.

3. Variations in support force and restraint angle occur for the seated occupant as the seat backrest undergoes restraint compliance during a rear-end collision. The mechanism by which this passive restraint is lost is generally unimportant insofar as its effect on motorist kinematics is concerned. Torso inertial movement does not respond differently with differing mechanism of loss of back support; it simply responds inertially to the degradation in resistive force.

4. Intended headrest height requirements are not met for many seats when the headrest is in its most commonly used, retracted position. Backrests with integral head restraint should protect 95th percentile adult occupants; it is preferred that headrests be contiguous with the backrest and of fixed elevation; if adjustable, headrest adjustment should provide extension to 80 cm (32 in) and not allow adjustment below 70 cm (28 in).

5. Passive restraint systems that effectively control motorist collision dynamics during a barrier crash may be ineffective if the same frontal collision is complicated by multiple impacts or upset. The cross-chest lap belt restraint system, on the other hand, provides effective motorist protection during most collisions or combinations of collisions, including upset.

6. Collision-induced occupant accelerations vary according to the effectiveness of occupant restraints. These variations in "ride-up" and "ride-down" velocity changes determine in large measure the extent of trauma during collision; the mechanism of transmission of accelerative forces represent the other operative factor for such traumatic exposures.

7. The authors designed, constructed and laboratory tested a high performance integral safety seat that was made by structurally modifying an existing production seat design without compromising its advanced design comfort features. The roof-to-floor anchored webbing for backrest crash force attenuation remarkably simplifies

seat structural requirements because bending moments are decreased more than ten-fold. By use of inertial reels this front seat conversion is accomplished without inhibiting rear seat access for two-door vehicles or backrest reclining or seat fore-and-aft positioning. Direct attachment of cross chest, lap belt restraints to the seat minimizes uncontrolled belt slack and allows incorporation of belt pre-stressing and other restraint compliance improvements. Increased costs for this restraint system concept would be nominal considering the benefits, however extensive crash test evaluation of any safety seat design is required prior to production to make certain all safety design features act constructively without degrading other aspects of motorist's safety.

8. The integral safety seat provides the basis for effective protective measures against direct side-impact collisions. With the motorist comfortably restrained by a 3-point belt and seated within a deeply contoured seat, structural members between the seat and outside door surface act, during collision, to accelerate the seat, with occupant attached, during the initial contact and crush phase of the collision and therefore accelerate the motorist less abruptly. Additionally, this design minimizes "slack" between the motorist and his restraint system and shields him from direct puncture and shear forces.

9. The integral safety seat protective capability against frontal impacts is well established, and roof crush during rollover can be considerably reduced by taller and strengthened front seat backrests serving as central pillars for augmentation of roof support; with an integral seat-restraint system, regardless of the direction of impact, levels of motorist protection can be achieved that are unattainable with any other single system, however costly or complex.

10. Commendable improvement in collision safety has occurred during a decade of regulation by the National Highway Traffic Safety Administration. However, with respect to automobile <u>seat</u> collision safety, effective progress has been negligible. A near obsession with air bag development regrettably has diverted a tremendous amount of research time and resources from the readily achievable goal of making the automobile seat crashworthy. Although the 3300 in-lbs backrest resistance feature of the FMVSS 207 sounds impressive, production seats from the nineteen forties and fifties tested by the authors were found to substantially exceed this 207 principal criteria.

11. Judged on any rational basis, whether as to initial cost, maintenance costs, functional reliability, exposure to dangerous malfunction or lack of effectiveness for many types of collision exposures, the air bag comes up as a king-size loser when compared with the collision protection of the integrated seat which includes integral cross-chest lap belt restraint. While the United States endlessly ponders the air bag issue, the rest of the automotive world has reached the conclusion that cross-chest lap belts in combination

with mandatory use laws represents the simplest, least costly and most effective motorist protection measure.

REFERENCES

1. D. M. Severy, "Collision Performance of the Motorist's Compartment," Highway Safety Research Institute, University of Michigan, Ann Arbor, Michigan, April 1967.

2. Anon., "Fifth International Technical Conference on Experimental Safety Vehicles," sponsored by U. S. Department of Transportation, Washington, D.C., hosted by the U.K. Department of the Environment, London, England, 1974.

3. K. Langwieder, "Car Crash Collision Types and Passenger Injuries in Dependency Upon Car Construction," Sixteenth Stapp Conference, SAE No. 720968, Society of Automotive Engineers, Inc., Warrendale, Pennsylvania, 15096, 1972.

4. J. D. States, et al, "Injury Frequency and Head Restraint Effectiveness in Rear-end Impact Accidents," Sixteenth Stapp Conference, SAE No. 720967, Society of Automotive Engineers, Inc., Warrendale, Pennsylvania, 15096, 1972.

5. D. M. Severy, H. M. Brink and J. D. Baird, "Collision Performance L.M. Safety Car," SAE No. 670458, presented at SAE Mid-year Meeting, Chicago, Illinois, May 1967.

6. D. M. Severy, H. M. Brink, J. D. Baird and D. M. Blaisdell, "Safer Seat Design," SAE No. 690812, Proceedings of Thirteenth Stapp Car Crash Conference, Harvard School of Public Health, December 1969.

7. F. W. Babbs and B. C. Hilton, "The Packaging of Car Occupants - A British Approach to Seat Design," Chapter 32, Proceedings of Seventh Stapp Car Crash Conference, University of California, Los Angeles, California, October 1963.

8. D. M. Severy, H. M. Brink, J. D. Baird and D. M. Blaisdell, "Active Versus Passive Motorist Restraints," SAE No. 70024, International Automobile Safety Conference Compendium, Detroit, Michigan, May 1970; Brussels, June 1970.

9. D. M. Severy, H. M. Brink and J. D. Baird, "School Bus Passenger Protection, SAE No. 670040, Automotive Engineering Congress, Detroit, Mchigan, January 1967.

10. A. W. Siegel, A. M. Nahum and D. E. Runge, "Bus Collision Causation and Injury Patterns," SAE No. 710860, Proceedings of Fifteenth Stapp Car Crash Conference, Society of

Automotive Engineers, Inc., Warrendale, Pennsylvania 15096, 1971.

11. C. Clark and C. Blechschmidt, "Human Transportation Fatalities and Protection Against Rear and Side Crash Loads by the Airstop Restraint," Proceedings of Ninth Stapp Car Crash Conference, Society of Automotive Engineers, Inc., Warrendale, Pennsylvania 15096, 1965.

12. National Highway Traffic Safety Administration, "49 CFR Part 571 (Docket No. 74-13; Notice 1), Federal Motor Vehicle Safety Standards," Federal Register, Vol. 39, No. 54, Washington, D.C., March 1974.

13. D. M. Severy, J. H. Mathewson and A. W. Siegel, "Automobile Head-on Collisions, Series II," SAE Transactions, Society of Automotive Engineers, Inc., Warrendale, Pennsylvania 15096, 1959.

14. D. M. Severy, H. M. Brink and J. D. Baird, "Backrest and Head Restraint Design for Rear-end Collision Protection," SAE No. 680079, presented SAE Congress, Detroit, Michigan, January 1968.

15. L. M. Patrick, et al, "Impact Dynamics of Vehicle Occupants," Tenth Stapp Conference, Society of Automotive Engineers, Inc., Warrendale, Pennsylvania 15096, 1966.

16. F. Preston, "A Comparison of Contacts for Unrestrained and Lap Belted Occupants in Automobile Accidents," Seventeenth Conference American Association for Automotive Medicine, 801 Green Bay Road, Lake Bluff, Illinois 60044.

17. J. W. Melvin, J. H. McElhaney, V. L. Roberts and H. D. Portnoy, "Deployable Head Restraints - A Feasibility Study," SAE No. 710853, Fifteenth Stapp Car Crash Conference, Society of Automotive Engineers, Inc., Warrendale, Pennsylvania 15096, 1971.

18. J. H. McElhaney, R. L. Stalnaker, V. L. Roberts and R. G. Snyder, "Door Crashworthiness Criteria," SAE No. 710864, Fifteenth Stapp Conference, Society of Automotive Engineers, Inc., Warrendale, Pennsylvania 15096, 1971

19. D. M. Severy, H. M. Brink and J. D. Baird, "Passenger Protection from Front-end Impacts," SAE No. 690068, International Automotive Engineering Congress, Detroit, Michigan, January 1969.

20. D. F. Huelke, J. C. Marsh, et al, "Injury Causation in Rollover Accidents," Proceedings of Seventeenth Conference of the American Association for Automotive Medicine, Oklahoma City, Oklahoma, 1973.

21. P. V. Hight, A. W. Siegel and A. M. Nahum, "Injury Mechanisms in Rollover Collisions," SAE No. 720966, Sixteenth Stapp Car Crash Conference, Society of Automotive Engineers, Inc., Warrendale, Pennsylvania 15096, 1972.

720967

Injury Frequency and Head Restraint Effectiveness in Rear-End Impact Accidents

John D. States, John C. Balcerak, James S. Williams, Alexander T. Morris, William Babcock, Robert Polvino, Paul Riger, and Raymond E. Dawley
Research Accident Investigation,
University of Rochester

Abstract

All of the rear-end impact accidents occurring in the city of Rochester, New York, in a three-month period were surveyed by tabulation of the police accident reports. Special police information forms, telephone interviews, and mail questionnaires were used for further data acquisition. Vehicle photographs and medical examinations were conducted for approximately every 20th vehicle.

During the data collection period, 691 rear-end impacts occurred. Although a computer program revealed 1371 accidents, defects in the program accounted for the large difference. Whiplash injury frequency based on telephone interview and mail questionnaire data obtained one to seven days after the accident revealed a whiplash injury frequency of 38%, which was approximately twice that determined by on-scene police investigators. Head restraints reduced whiplash frequency by 14% and fixed head restraints appeared to be more effective. Seventy percent of adjustable head restraints were in the downmost position. Women sustained whiplash injury more frequently (51%) but benefited from head restraints more (whiplash injury frequency 38%).

An extensive review of the literature related to whiplash injury and head restraint design and effectiveness is presented.

REAR-END IMPACTS are one of the most common of all accidents, and whiplash injury is the most commonly occurring injury from such an accident. Hyperextension of the cervical spine causes whiplash in most instances, and head restraints have been designed to prevent whiplash by preventing hyperextension in rear-end impact accidents.

These statements are examined in the light of past research and of an epidemiological study of rear-end impacts in an urban area (Rochester, New York).

Literature Review

Rear-End Accident Frequency - The National Safety Council estimates that 24% of all accidents are rear-end impacts. In these totals, it is estimated that 27% of urban accidents and 18% of rural accidents are rear-end impacts (1)*. Consider-

---

Note: The contents of this report reflect the views of the performing organization, which is responsible for the facts and the accuracy of the data presented herein. The contents do not necessarily reflect the official views or policy of the Department of Transportation. This report does not constitute a standard, specification, or regulation.

*Numbers in parentheses designate References at end of paper.

ing only fatal accidents, 4.7% are caused by rear-end impacts. Of these, 4.2% occur in urban areas and 4.8% in rural areas. Locally in Monroe County, which excludes the city of Rochester, 32.7% of all accidents are rear-end impacts (2). Of these, 32.4% resulted in injury and 0.3% caused death. In the city of Rochester, 29% of all accidents were rear-end impacts in 1970 and 31% in 1971.

Based on the National Accident Summary File, Carlson (3) of the Highway Safety Research Institute (HSRI) of the University of Michigan determined in 1971 that 40.1% of multivehicle accidents were rear-end collision and that 33% caused injury, while 0.12% caused death. Rear-end impacts were the second largest group of collision types, the largest being angle collisions, which accounted for 50.8% of all multivehicle collisions.

In March 1970, the Ford Motor Co. Automotive Safety Research Office (5) reported that 7.04 million accidents out of a total of 14.6 million property damage, injury, and fatality accidents in 1968 were rear-end impacts. During that same year, in which over 46,800 fatal accidents occurred, 4500 or 9% were rear-end impacts.

Mackay of the University of Birmingham, United Kingdom (6), reported that 9.8% of a group of 643 accidents occurring in the Birmingham area were rear-end impacts. The highest percentage, 11.3%, occurred in a rural area, compared with 9.4% in a metropolitan area and 5.3% on limited access highways.

New York State statistical data for 1970 (4) reveal that 38.4% of all accidents were rear-end impacts, which constituted the largest single group. Angular impacts accounted for 33.2% of the total.

Nationally, it is estimated that there are four million (7) rear-end impacts each year and that such accidents are more common in urban areas.

Injury Incidence - The incidence of injury occurring in rear-end accidents has never been conclusively determined. In 1963, based on a review of 575 patients with neck injuries, McNab (8) estimated that neck injury occurs in one-fifth of all rear-end impact automobile accidents. Carlson, in his analysis of the National Accident Summary Data (3), cited an injury rate of 33-1/3%, but the basis for the rate was not specified. O'Neill, et al., of the Insurance Institute for Highway Safety (29) in 1971 determined an injury incidence for drivers of 24% for head restraint-equipped cars and 29% for unequipped vehicles. This study was based on a review of 67,000 insurance claims in which 5663 were rear-end impact accidents. Passengers were specifically omitted from the study because seated location data were not available. Kihlberg, from Cornell Aeronautical Laboratory's (CAL) Automobile Crash Injury Research (ACIR) files (45), revealed an overall incidence of approximately 15%. This study used data collected in 1968 before headrests became mandatory equipment. Although not specified in the paper, the time period for determination of injury was restricted to 24 h in past ACIR studies.

Nature of Whiplash - Two factors have made the determination of an injury rate in rear-end impacts difficult. Symptoms of injury resulting from rear-end impacts are often delayed hours or days so that the injury is not evident at the scene of the accident and cannot be reported to police accident investigators (20, 80). Most data on which currently available statistics are based are derived from police accident investigations and do not reflect the true incidence of neck injury. The second factor is the lack of a definition of whiplash injury. Objective clinical evidence of injury does not exist in a typical case, precluding a strict definition of injury.

The clinical entity of whiplash injury has never been strictly defined in the medical literature. Many physicians object to the use of the term "whiplash" because of its lack of objectivity and definition. In 1928, Crowe (9) presented eight cases of neck injuries resulting from traffic accidents and used the term whiplash for the first time. Since then, he has personally regretted the use of the term and has been joined by other physicians (10–13).

Two misleading studies were reported in 1956 and 1959 that received wide attention in the medical and legal communities. Braunstein and Moore (85) reviewed 144 neck injury reports from ACIR and reported that 46 had no objective evidence of injury but were reported to have whiplash injury. The authors concluded that the conscious or unconscious desire for gain on the part of patients may motivate them to exaggerate their injuries and disabilities. Gotten (86) reported a review of 100 patients with neck injuries from rear-end impact automobile accidents in which 88% recovered after litigation was settled. He stated that 85% of the patients had psychosomatic complaints.

These studies were not immediately refuted because the pathological anatomy of whiplash injury was not known at that time. Beginning in 1960, systematic research was undertaken that has clarified the pathology of whiplash by establishing an anatomical basis for the diagnosis. McNab (14, 15) dropped supine anesthetized monkeys down an elevator shaft with their heads unsupported, producing hyperextension of their cervical spines. The animals were dropped from different heights producing injuries of varying severity. Autopsy studies revealed muscle injuries ranging from minor tears in the sternocleidomastoid muscles to partial avulsion of the longus colli. Tears of the longus colli were associated with retropharangeal hematomas and hemorrhages in the muscularis of the esophagus. Separation of the intervertebral disc from the adjacent vertebra was also noted.

Subsequent studies were performed by Wickstrom, Martinez, Rodriguez (16), and others conducted acceleration studies of rabbits and monkeys in which the animals were accelerated in a forward-facing direction ($G + X$) while in a sitting position. Biopsy and autopsy studies revealed that only 8% of the forward-facing and 10.7% of the rearward-facing animals sustained no injury. Injuries reported included multiple ligamentous, muscular, and bony injuries of the neck and cervical spine, hemorrhage and scarring in surviving animals in the brain, and hemorrhage in the thyroid gland and behind the esophagus and larynx in the neck. Electroencephalographic studies of surviving animals revealed multiple abnormalities (16–18).

Luongo (19) in 1969 autopsied an automobile driver who died of a myocardial infarction immediately prior to impact and who was determined to have sustained hyperextension of the cervical spine as evidenced by the vehicle and occupant kinematics and associated injuries. This victim sustained multiple ligamentous and muscular tears in the middle and lower cervical spine. Other human clinical data are lacking because of the rare association of death with whiplash injury.

Central Nervous System Injury in Whiplash - States (20) in a clinical review of 60 rear-end impact accidents reported that concussion occurred in 10 accidents. On the basis of animal studies, Holbourn, Ommaya, Hirsch, and Martinez (21–25) have revealed that significant angular acceleration due to head rotation occurred and caused concussion. Ommaya and Yarnell (26) reported two case histories of patients who sustained subdural hematomas, without impacts, from whiplash injuries. VanGijn and Wintzen reported a case in which a 53-year-old man sustained an extracerebral hematoma confirmed by surgery that was sustained in a side impact accident causing head rotation but not contact (27). An extensive review of the literature concerning the significance of rotational versus linear acceleration in the causation of concussion and other central nervous system injuries is contained in two recent publications (28, 88), and readers are referred to these for a more detailed review.

Other Central Nervous System Injuries - Several authors have reported alteration in the electroencephalogram in clinical and experimental studies. Torres in 1961 reported 21 abnormal EEGs in 45 patients with whiplash injuries (28). Gilbert in 1969 reported that EEGs, if taken within 24 h after injury, usually showed some slowing but the abnormalities cleared within a week (29). An experimental study performed by Hiyama utilizing rabbits revealed that fast-wave changes attributed to the neocortex appeared within a few hours after injury but subsided in two to five days. In contrast, high-amplitude slow waves and spikes occurred in the hypothalamus but did not appear for 7 days and lasted for as long as 45 days (30).

Toglia, et al., reported dizziness and hearing losses in 35 of 50 patients who sustained head trauma or whiplash and had been examined utilizing audiometric and caloric test procedures. One patient was found to have a fracture through the right petrous pyramid and a secondary dislocation of the incus and stapes (31).

Visual disturbances, diplopia, irregularities of pupils, disturbances of visual acuity, and pain symptoms have been reported by several authors. Gibson reported three case studies of patients who had difficulty focusing, diplopia, and ocular pain (32). Roca in 1971 presented a series of 16 patients with many different eye symptoms and signs following whiplash injury (33). Yamaji found a majority of 42 truck drivers who had been struck from behind had defects in dynamic visual acuity and horizontal phoria and width of fusion (34).

Clinical Examination and Findings of Whiplash Injury - The physician caring for a patient with whiplash injury is restricted in his diagnostic armamentarium to those diagnostic methods suitable for a patient with a medical injury of minor severity. The investigative armamentarium includes the medical history, physical examination, and x-rays. In the more prolonged and/or disabling case, special x-rays including cervical myelography and various blood chemistries may be utilized in attempting to make a definitive diagnosis. EEGs are occasionally used when there is evidence of concussion. However, the clinician is severely limited in his ability to diagnose the precise anatomical structure injured in whiplash and revealed by experimental autopsy studies previously cited. Reviews detailing the clinical symptoms, physical examination, and x-ray findings are detailed in publications by DeGravelles and Kelley (35), Goff, Aldes, and Alden (36), States, Korn, and Massengil (20), and Gurdjian (37).

The medical history reveals that the onset of neck, shoulder, or head pain may begin immediately but may be delayed several hours or days following a whiplash-producing accident (80). Location of pain is typically in the back of the neck but may occur in the front and sides of the neck as well as in the shoulder girdle and dorsal spine extending distally to T7. Pain may also occur in the occipital region of the skull frontal region (eyebrows) and throughout the calvarium as a headache of nonspecific nature. Disturbances of perception include visual changes previously cited, disorders of hearing and balance previously cited, difficulty swallowing, as well as stiffness and pain with motion of the neck. Injury to the nerve roots of the cervical spinal cord will cause symptoms of pain, parenthesias, and weakness in the upper extremities, particularly the hands.

Injury of Intervertebral Disc of Cervical Spine - Cervical nerve root irritation may be caused by herniation of the intervertebral discs. Minor disc extrusions or herniations heal spontaneously. More persistent disc extrusions that cause significant disability (weakness in hands, persistent severe pain) require localization by x-ray myelography and surgical excision (38).

Minor neck injuries in which there is no evidence of bone or disc injury may cause persistent and severely disabling symptoms. In a review of 10,000 neck pain problems treated at the Mayo Clinic between 1956 and 1963, James presented 32 who required posterior cervical spine fusion for relief of symptoms. Surgery was deferred an average of 28.3 months after injury. Pathological findings at the time of surgery were marked hypermobility of one or two cervical vertebrae in 13 cases, rupture of the interspinous, superspinous, and ligamentum nuchae and interspinous ligaments, calcification of ligaments, and degenerative changes in the ligaments (39).

Clinically Detectable Severe Musculoskeletal Injuries Due to Hyperextension - Many cases of fracture-dislocations and severe disc herniations resulting in cervical spinal cord injury have been reported (40–43). No large series of rear-end impact accidents has been statistically analyzed with respect to the nature and severity of injuries, however.

Small series are contained in papers by States, et al., (20) and Gurdjian (37). A review of the available medical literature reveals no cases of paraplegia following rear-end impact accidents. Miller and Schultz (44) reported three cases due to hyperextension, but in each instance hyperextension was caused by forehead impact and only one occurred in an automobile accident. The kinematics of this type of hyperextension injury are probably different from simple whiplash caused by rear-end impacts without head contact.

Mechanism of Injury - Neck motion is normally limited by the chest and shoulders, which become anatomical blocks to forward and lateral flexion. The normal range of motion of the neck is reported as follows (46–48):

1. Flexion: 54–67 deg.
2. Extension: 61–93 deg.
3. Lateral Flexion: 41 deg.
4. Rotation: 73–76 deg.

All of the subjects in the above determinations ranged in age from 18–40 years, and the majority were males. Ferlic (50) determined the total range of flexion and extension in a group of subjects ranging in age from 15–64 years. He noted their range decreased from 139 ± 19 deg for age 15–24 to 116 ± 22 deg for ages 55–64. Gadd, Culver, and Nahum (51) in 1971 determined the bending moment of four unembalmed cadavers.. Injury did not occur in any of the cadavers until the necks were extended beyond 70 deg.

The effectiveness of the chest and shoulders as anatomical blocks has not been assessed in the available literature. However, the number of whiplash injuries occurring in head-on and side-impact accidents is small, constituting not more than 30% (20,37,40,49). The implication of these findings is that the neck is usually protected against injury caused by motion exceeding its physiological range by the anatomical blocks provided by the chest and shoulders.

Portnoy, et al., (53), in a clinical review of 55 serious neck injuries caused by automobile accidents, demonstrated that hyperextension neck injuries may occur in head-on automobile accidents. Forehead contact occurs, causing rearward displacement of the neck and extension of the cervical spine.

The majority of whiplash injuries appear to be caused by extension beyond the physiological range of neck motion, in contrast to forward and lateral flexion, which are blocked by the chest and shoulders. There exists no block for extension.

Tuell (52), an orthopedic surgeon, in 1958 allowed himself to be experimentally rear-ended at 20 mph. He was seated in a 1951 Ford that was struck by a 1948 DeSoto sedan. High-speed movies revealed that Tuell's head and neck extended 120 deg. His neck was painful for a few days but he denied any significant disability.

Head Restraint Design - A device to prevent neck extension beyond the physiological range appeared to be the obvious means of preventing whiplash. Ruedemann (54) in 1957 proposed that a padded 6-in fixed headrest be attached to the top of automobile seat backs for neck protection. Severy, et al., (55), after testing a variety of seats, seat backs, and head restraints in crash tests utilizing 1967 full-size Ford sedans, concluded that a 28 in high seat back would protect up to the 95th percentile male occupant. Higher seat backs would protect taller occupants, but rearward and rear corner vision of the driver and forward vision of rear seat occupants would be severely compromised. Severy also demonstrated that head offset of up to 6 in did not significantly reduce the effectiveness of the head restraint. Berton (56) determined the neck extension limiting capability of various head restraint heights in sled tests utilizing anthropomorphic dummies. Twenty-eight inch high head restraints permitted only 20 deg neck extension in spite of a 4 in offset.

FMVSS 202 made head restraints mandatory equipment for passenger cars sold in the United States after December 31, 1968. The majority of headrests installed to meet this standard initially were adjustable. In a study by CAL(57), adjustable head restraints, found in 72.6% of head restraint-equipped cars, were in the up position in only 17.7%. The examination of the heights of headrests in the down position relative to a 95th percentile occupant revealed a significant difference in height relative to the occupant. The lowest head restraint, which extended to the spinous process of the third thoracic vertebra, was installed in the American Motors Matador model. In contrast, the American Motors Ambassador was equipped with a head restraint that in its lowest position extended to the height of C5 spinous process. The highest headrest in the down position was that installed in the Ford Motor Co. Thunderbird, and was at the C3 spinous process level.

In an effort to make head restraints less objectionable, development of deployable headrests using airbag technology has begun. Melvin, et al., (58), using two airbag head restraints and a mechanical sliding head restraint, demonstrated that effective protection could be achieved.

Injuries from Head Restraints - States, et al., (20) demonstrated through case studies that whiplash injury occurred in spite of the presence of head restraints. The exact mechanism of injury is not always apparent. Tall occupants receive injury from neck hyperextension over the top of the adjustable head restraints. Beatty (59) reported an odontoid fracture that occurred with the head restraint in the down position from this mechanism. States also reported the occurrence of whiplash injury in a vehicle equipped with fixed head restraints; the authors hypothesized that hyperextension occurred because of differential rebound of the head and chest from the seat back and restraint that had vastly different spring rates. Severy, et al., (55) noted an alternative mechanism in which dummies slid upward permitting neck hyperextension over the top of the head restraint.

The third injury mechanism was noted in a multidisciplinary accident investigation (MAI) case study RAI 69, University of Rochester (60), in which a torsion injury of the neck occurred. In this rear-end impact accident, a middle-aged male sustained a whiplash injury due to a torsional strain that occurred when he impacted his shoulder belt after rebounding from the seat back and head restraint.

Back seat occupants have sustained injury from impacting head restraints. Fetterman (61) reported that a 14-year-old girl sustained bilateral midshaft clavicle fractures from striking the back of a head restraint in a head-on collision.

A review of the MAI cases automated file of the HSRI in mid-1972 by J. Fell (92) revealed that of a total of 1777 vehicles, 930 were involved in head-on impacts between 10 and 2 o'clock. Eighty of the vehicles were equipped with head restraints and contained rear seat occupants. Eighteen reported head restraint contact. The majority of injuries were head or chest and appeared attributable to the head restraint. Injury severity (Abbreviated Injury Scale) ranged from 1 to 6 and damage severity (last digit of Vehicle Deformation Index) from 2 to 8.

Kinematics and Tolerance for Injury of Human Neck - An early and persistent obstacle in head restraint design is the lack of knowledge of the kinematics and tolerance for injury of the human neck. The complex anatomy of the neck has made investigative approaches difficult. Limited static test data of isolated components of the neck are available (62–65), but provide little useful information because of the complex kinematics and dampening of the human neck.

Far more sophisticated analysis has been undertaken by Mertz and Patrick (66) and Gadd, Culver, and Nahum (51). Mertz and Patrick, utilizing dynamic sled testing, compared the injury tolerance of human volunteers and cadavers and concluded that a 50th percentile male can tolerate a moment of 44 ft-lb about the occipital condyles without injury. Sled accelerations of up to 6.8 G did not cause symptoms, but an acceleration of 9.6 G caused neck and back pain lasting several days. This paper is prefaced by an excellent engineering description of the anatomy of the human neck.

Gadd, Culver, and Nahum (51) dynamically tested four elderly cadavers and determined that neck injury occurred with as little as 160 in-lb moment.

In an earlier study in 1967, Mertz and Patrick (66) compared the kinematics of volunteers, cadavers, and dummies. Neck rotation in cadavers (83 deg) was twice that of volunteers (41 deg), establishing limitations in the use of cadavers in human kinematic research. The authors concluded that neck extension should be limited to 80 deg and preferably to 60 deg to avoid injury. In a recent series of sled tests of airbags by Clarke, et al., (68) utilizing volunteers, it was revealed that neck extension caused by face impact on the bag beyond 68 deg resulted in whiplash injury symptoms, confirming Mertz and Patrick's conclusion.

Mathematical Models of Neck - The advent of volunteer, cadaver, and animal dynamic tests permitted the development of mathematical models of the human cervical spine. These models can be expected to greatly facilitate and accelerate head restraint development and design. Parameters for such models and an injury threshold have been delineated by Mertz and Patrick (66). Martinez (87) developed one of the earliest models, which had an articulated cervical spine. Subsequent models have been developed by Furusho (69), Higuchi, et al., (70), and Pontius (71). Pontius's model is noteworthy because each vertebra is separately considered in the model.

Neck Design in Anthropomorphic Dummies - Most experimental full-scale vehicle crash research in rear-end impact accidents has required the use of anthropomorphic dummies. Although much of our present knowledge of occupant kinematics and head restraint effectiveness is based on this research, dummy neck design remains primitive.

Anthropomorphic dummies were first used as human simulations in aviation design in the early 1920s (72) and for automobile design since the early 1950s (73).

Early dummies were essentially articulated sandbags with dimensions similar to the human head, torso, and extremities. More sophisticated dummy design began in 1949 when Ritterrath of Sierra Engineering Co. (74) undertook the design and manufacture of an anthropomorphic dummy. Other design and construction efforts were undertaken by Mathewson and Severy in 1951 (73), Alderson in 1952 (75), Aldman of Sweden (76) in 1960, and Mira of the United Kingdom in 1972 (77).

Dummy necks most often have been rubber cylinders or stacks of rubber washers tethered by a cable passing through the center anchored to the headform and to the top of the chest. Sierra and General Motors have used a series of ball-and-socket joints with adjustable friction devices and with rubber washers in between for shock absorption.

Although these devices approximate the normal range of motion of the cervical spine, little else in the kinematic performance resembled the human neck. The rubber block or washer types have the problem of a linear load deflection curve in contrast to the human neck, which flexes through its full range of motion with little increase in load until the physiological limit of motion is exceeded, at which time the load rapidly increases (51). Both types lack stretch in the vertical axis and dampening in all directions.

Recently, manufacturers have had difficulty meeting the head acceleration requirements imposed by FMVSS 208, which requires performance testing utilizing a 30 mph barrier impact equivalent for belt restraint systems (79). The high accelerations may be attributable largely to failure of dummy necks to stretch and to absorb energy as occurs in the human being.

Factors Predisposing to Whiplash Injury - The majority of occupants of vehicles struck from behind do not receive neck injury (3,8,20,29). Several factors may modify an occupant's risk of injury:

1. Human or medical factors: sex, age, body build, cervical spine arthritis, seated position in vehicle, and position at moment of impact.
2. Vehicle factors: seat back failure, vehicle crush, and head restraint.

The sex of the vehicle occupant is a relevant factor. O'Neill et al., (29) reported an injury rate for the driver of 37% for females and 24% for males in vehicles without head restraints. Schutt and Dohan (80) reported a much higher neck injury rate for female factory workers than for male factory workers, particularly in urban areas. The rates were 4.8 times the male rate in the urban areas and 1.7 times the male rate in rural areas. Rates for other types of automobile accident injuries were about equal and did not reflect a sex difference. Kihlberg (45) in an analysis of ACIR data in 1969 reported that the female injury rate was twice that of males.

The reasons for this difference in injury frequency are not known. Male necks are larger in circumference than are female necks (81). Using data from military anthropometry studies (81), the ratio of neck cross-sectional area to head volume can be calculated; for the 50th percentile male, it is 1:135 and for the comparable female it is 1:151. The ratio is determined by the formula head circumference^3/ neck circumference^2 = ratio. This ratio quantitates the observation that females have smaller necks relative to head size. It appears reasonable to assume that head weight or mass is directly proportional to head volume because there is little difference in the material structures. Taller occupants may be more vulnerable to whiplash. Kihlberg (45) reported "a faint suggestion of a positive correlation" for the increased risk of neck injury for taller occupants. Seated position is another factor. Front seat occupants more often sustain neck injury than rear seat occupants in his study.

Other factors such as age, body build, presence of arthritis in the cervical spine, muscular tone and conditioning, and occupant position at impact have not been conclusively examined to date. Wiggins (82) reported a prolonged, above-average convalescense of patients with preexisting degenerative disease or arthritic changes following whiplash injury. He reported an average recovery period of 6.1 weeks for 313 patients injured in rear-end collisions who were treated in his office.

Vehicle Factors - Seat back failure appears to protect occupants from neck injury by allowing the torso and shoulders to move rearward with the head (20,29,45). O'Neill, et al., (29) reported that the frequency of neck injuries increased with the introduction of the headrest in 1968 model Volkswagens. A group of 767 rear-impacted VWs were reviewed and 50 seat back failures were identified. These were not a sufficiently large number to explain the low incidence of neck injury (15%) in VW drivers in pre-1968 models.

Vehicle crushability may also be a factor in whiplash injury causation; some vehicles have "softer" rear ends than others and might be expected to protect the occupants from whiplash injury. However, O'Neill, et al., (29), who presented data that classified cars by make, did not establish this. Classification by model and structural type would be necessary to examine this question accurately.

Severy, et al., (55,83), in a series of comprehensive experimental crash tests, reported that occupant compartment accelerations of up to 8 G occurred in rear-end impacts of speeds of less than 10 mph. Little vehicle damage was reported in these impacts.

In an unreported case (84), a 43-year-old athletic male sustained a whiplash injury when he was impacted from behind while driving a 1967 Oldsmobile station wagon equipped with a frame-mounted trailer hitch. There was no damage to the occupant vehicle. He had experienced two previous rear-end impacts without neck injury but in cars without trailer hitches sustaining minor bumper and sheet-metal damage. The rigidity of the trailer hitch apparently resulted in higher occupant space acceleration and protected the vehicle from damage but did not protect the occupant from injury.

Methodology

First-Level Data Acquisition - Police accident reports (Appendix A) of all rear-end impact accidents occurring within the city of Rochester were obtained for a three-month period beginning Jan. 17, 1972. All police reports were reviewed by RAI investigators and those accidents selected in which one vehicle was impacted between five and seven o'clock. (SAE J224a) by another vehicle. A special information form for rear-end impact accidents (Appendix B) was devised and the members of the Rochester Police Bureau were instructed in its use. The form was appended to routine accident reports for rear-end impact accidents.

Second-Level Data Acquisition - One to seven days after case pickup, vehicle occupants were contacted by telephone or postcard questionnaire (Appendixes C and D).

Third-Level Data Acquisition - Approximately one out of 20 cases were randomly selected for vehicle analysis, which was performed by a two-man team. The rear-end impacted vehicle was examined and photographed to record exterior and interior damage. Pertinent measurements were taken and recorded on a special vehicle analysis form (Appendix E). All occupants contacted by telephone were asked to contact the medical office of the senior author for a medical examination and x-ray without cost to the occupant. A special medical history, physical examination, and x-ray form were devised and used for recording of these data (Appendix F).

Injury Determination - Whiplash injury was defined as any injury that caused neck pain and/or stiffness resulting in either disability as evidenced by loss of time at work or significant loss of ability to perform necessary activities of daily living or necessitating medical treatment such as medications, neck support, or physiotherapy. Excluded were injuries that did not cause significant disability and did not require treatment, although the patient may have visited a physician.

Whiplash injury determinations were made on all levels of data acquisition. Police investigators necessarily could use only the criteria of neck pain and/or stiffness because disability could not be evaluated at the accident scene. In second- and third-level data acquisition, treating physicians occasionally were called to corroborate or elaborate on patient statements.

Speed Estimates - No attempt was made to obtain speed estimates in first- and second-level data because such estimates do not routinely appear on police accident report forms in Rochester. Additionally, the 5 mph increments necessary for analysis were considered beyond the expertise of police investigators. Third-level estimates were made using photographs generously loaned by the Insurance Institute for Highway Safety of experimental crash tests of 1972 vehicles that were impacted at speeds of 5, 10, and 15 mph by the front ends of identical 1972 vehicles.

Results

Six hundred and ninety-one accidents that met the study criteria occurred during the data collection period. Three hundred and sixty special information forms (62%) were filled out by police investigators, but not all forms were completed. Two hundred and ninety-seven (38%) occupants in impacted vehicles were contacted, 183 by telephone and 114 by mail. Twenty-two occupants received medical examinations, and 24 cars were examined in depth.

Six hundred and nineteen accidents were two-vehicle collisions, and the remainder involved three or more vehicles (Table 1). A computer program designed to retrieve rear-end impact accidents from the Rochester Police Bureau accident case data bank revealed a total of 1371 accidents. Case-by-case analyses of these

Table 1 - Total No. Accidents, Vehicles, and Occupants

Total No. accidents	691
Two-vehicle collisions	619
Three-vehicle collisions	67
Four-vehicle collisions	4
Five-vehicle collisions	1
Total No. vehicles	769
Total No. occupants in rear-ended impacted vehicles	449

accidents revealed that the 1371 cases included a majority of accidents that did not fit the study criteria. Such accident configurations included side impacts on the rear half of the vehicle, impacts with parked cars, and backward-moving vehicle impacts with fixed objects and parked vehicles.

Injury Frequency - Whiplash injury was the principal injury occurring in the struck vehicle (vehicle 2). Only three injuries (0.7%) other than whiplash occurred in the series. These were of a minor nature.

Occupants of vehicle 2 sustained injury in 16.6% of the police-investigated accidents (level I). In contrast, level II data revealed an injury rate of 38% (Table 2). Symptoms of whiplash injury were delayed in onset in a majority of the cases, accounting for the difference in frequency between level I and level II data.

Head Restraint Effectiveness - Head restraints were effective for both the driver and the right front seat occupant. The whiplash injury frequency (WIF) was 37% for head restraint-equipped vehicles and 43% for unequipped vehicles*. The beneficial effect was more noticeable for the female occupants (38% for head restraint-equipped positions versus 51% for unequipped positions).

Other Factors in WIF - The WIF for the center front and for all rear seat occupants was unusually low (22%), although none of the seats were equipped with head restraints. This may be explained by the fact that these seats were more often

*None of the level II data were found to be statistically significant.

Table 2 - Whiplash Injury Frequency—Level II Data (% in parentheses)

	All Vehicles		Headrest Equipped		No Headrest Driver and Right Front Only	
	Injury	No Injury	Injury	No Injury	Injury	No Injury
All occupants	167(38)	272(62)	78(37)	132(63)	78(43)	101(57)
Drivers	121(41)	176(59)	59(37)	100(63)	62(45)	76(55)
Right front	35(39)	57(61)	19(37)	32(63)	16(39)	25(61)
Center front and rear	11(22)	39(78)	—	—	—	—
Males	92(37)	156(63)	50(35)	92(65)	42(40)	64(60)
Females	70(44)	88(56)	31(38)	50(62)	39(51)	38(49)

Table 3 - Seat Damage versus Injury Occurrence—Level I Data

Type of Damage	Injured	Not Injured
None	46	207
Bent	5	2
Off tracks	2	4
Unknown	53	372
Total	106	585

occupied by younger and smaller occupants whose necks would be protected by seat backs of standard height.

Sex was a significant factor in WIF. The overall WIF for females was 44%, 51% for unequipped vehicles and 38% for equipped vehicles and seated positions. The female WIF approached the male rate of 35% for head restraint-equipped vehicles and seated positions. Age, height, and weight did not exert a detectable influence on WIF nor on head restraint effectiveness.

Head restraint position did not have a statistically demonstratable effect on the WIF based on level I data. The sample size for data in these categories was small and precluded analysis. However, it was determined that 73% of the occupants did not elevate their adjustable head restraints, which were present in 72% of vehicles equipped with head restraints. The remainder of the vehicles had fixed head restraints. Occupants in seats with fixed head restraints appeared to receive additional protection, as evidenced by a WIF of 13% versus 16% overall rate and 21% for vehicles equipped with adjustable head restraints. These percentages are based on level I data.

Seat damage was noted in 13 vehicles (5%) of a sample of 279 vehicles examined by police investigators. Seat backs were bent backward in seven vehicles and pulled off the fore-aft adjustment tracks in six vehicles. This did not have an apparent effect on WIF (Table 3).

Vehicle age, model, and make did not have a statistically significant effect on WIF based on level I data. Surprisingly, the highest WIF was 21% for 1971 model vehicles. This was higher than pre-1969 vehicles, except for the 1963 group, which comprised only 22 out of a 691 vehicle sample.

Level III Data - Level III data acquisition was limited because of the limited capabilities of the research team, which was primarily concerned with level I and II data collection and the failure of occupants to submit to a medical examination. However, examination of 24 vehicles permitted speed estimates, which were made by comparing vehicle damage with an album of photographs of vehicles experimentally impacted at known speeds of 5, 10, and 15 mph (Table 4).

Unfortunately, medical examinations were performed on only 22 occupants. None of these occupants were in the group of vehicles subjected to level III examination. This precluded correlation of vehicle damage and injury severity.

Discussion

The discrepancy in the number of rear-end impact accidents determined by hand-sorting to meet the study criteria of rear-end impacts and those derived from the computer program is a cause for concern. The Rochester Police Bureau uses

Table 4 - Impact Speeds—Level III Data

Impact Speed	No. Vehicles
Up to 5 mph	10
6–10 mph	8
11–15 mph	4
16–20 mph	2

Table 5 - Whiplash Injury, Head Restraint, and Restraint System Use

	Whiplash Injury	No Injury
No head restraint	6	6
Head restraint up	1	0
Head restraint midway	0	1
Head restraint down	3	3
Position unknown	2	1
Seat belt in use	4	0
Seat belt not in use	4	9
Unknown	4	1

the standardized report form based on New York State Department of Motor Vehicle recommendations. These reports are the basis for state and national accident data compilation. This finding cast doubt on the reliability of rear-end impact accident analyses utilizing data derived from police accident reports similar to the Rochester Police Bureau form.

The reliability of injury data derived from level I is doubtful because of the delayed onset of symptoms experienced by a majority of the whiplash-injured occupants. Although the symptom onset delay may appear suspect, it is a common finding in medical practice and is verified by level II data. There were insufficient level III cases to determine the significance of the immediate onset of symptoms. It was not established that these patients had more severe whiplash injury than those who experienced a delayed onset.

The overall WIF determined by level II data compares favorably with O'Neill, et al., (29), who determined an incidence of 28% based on insurance claims. The difference can be explained by the fact that not all injured occupants file insurance claims. This is particularly true of vehicle occupants who are members of the vehicle owner's family.

The head restraint's injury-reducing effect was determined to be 14% (37% WIF for equipped vehicles, 43% WIF for unequipped vehicles). This is comparable to the findings made by O'Neill, et al., (29). Head restraint effectiveness is much less that can be anticipated because most occupants do not raise their head restraints, leaving them in the downmost position. Fixed head restraints appear to be more effective, but whiplash injury still occurred with this type of head restraint. Studies by States (20) and Lange (91) reveal a need for further research in head restraint design with respect to occupant kinematics.

Level III data revealed suggestive evidence that the use of lap belts may increase WIF (Table 5). This can be explained by the alteration in occupant kinematics imposed by the belt, which prevents the occupant's pelvis from moving forward and which would relieve hyperextension of the neck over the top of the seat or head restraint.

Conclusions

1. Seventy-three percent of adjustable head restraints are left in the down position and not adjusted by users.

2. Head restraints reduce the frequency of whiplash injury by 14%. This disappointing finding is attributable in part to the failure of users to adjust their head restraints actively and to less than optimal head restraint design largely caused by inadequate knowledge of human neck kinematics.

3. Fixed head restraints appear to be more effective than adjustable head restraints.

4. Women sustained whiplash injury more frequently than men (44% WIF versus 37% WIF for males).

5. Head restraints decreased WIF for women (51% to 38%) more than for men (40% to 35%).

6. Center and rear seat occupants have an unexpectedly low WIF (22%), possibly because children more often occupy these seated positions.

7. The overall WIF was 38%, 37% for head restraint-equipped seats and 43% for unequipped driver and right front seats.

8. The actual number of rear-end impacts meeting the study criteria was approximately half that retrieved by a computer program from an accident report data bank. Data derived from such computer programs should be used with suitable caution and constraints.

Acknowledgments

The authors are indebted to James C. Fell and Eugene E. Flamboe and to other members of the NHTSA for their patience, their continued assistance in original research design, and for their aid in the literature review and data analysis and interpretation. The authors are also indebted to the Insurance Institute for Highway Safety who generously permitted the senior author to examine their slide files of experimental vehicle impacts and to compose an album of photographs of vehicles impacted at known speeds.

References

1. "Accident Facts 1970." National Safety Council, p. 47.
2. "Annual Report." Monroe County Sheriff's Office, 1968.
3. W. L. Carlson, "AID Analysis of National Accident Summary." HIT Lab Reports, October 1971.
4. New York State Department of Motor Vehicles, "Annual Report." 1970.
5. Ford Motor Co., Automotive Research Office, Report S-70-1, March 1970.
6. G. M. Mackay, "Some Features of Traffic Accidents." British Medical Jrl., Vol. 4 (December 27, 1969), pp. 799–801.
7. "Accident Facts 1971." National Safety Council.
8. I. McNab, Personal Recommendation, 1963.
9. K. Crowe, "Letter to Tribune." Medical Tribune, December 21, 1962.
10. J. D. Ghiardi, "Letter to Tribune." Medical Tribune, October 25, 1965.
11. R. W. Braunstein and J. O. Moore, "The Fallacy of the Term Whiplash." Amer. Jrl. Surgery, Vol. 97 (1959), pp. 522–529.
12. W. L. Ellerbroek, "Whiplash Injuries and Cervical Pain." Headache, Vol. 6 (1966), pp. 73–77.
13. G. C. Manning, "Letter to Editor." Jrl. AMA, Vol. 206 (1968), p. 2320.
14. I. McNab, "Acceleration Injuries of the Cervical Spine." Jrl. Bone and Joint Surgery, Vol. 46A (1964), pp. 1797–1799.

15. I. McNab, "Whiplash Injuries of the Neck." Proceedings Annual Meeting, Amer. Assoc. for Automotive Medicine, 1965, pp. 11-15.

16. J. D. Wickstrom, J. L. Martinez, D. Johnston, and N. C. Tappen, "Acceleration-Deceleration Injuries of the Cervical Spine in Animals." The Seventh Stapp Car Crash Conference—Proceedings. Springfield, Ill.: Charles C Thomas, 1963, pp. 284-301.

17. R. Rodriguez, "Cervical Spine Syndrome in Experimental Acceleration Injuries of the Head and Neck." The Prevention of Highway Injury, HSRI, Ann Arbor: University of Michigan, 1967.

18. R. Rodriguez and D. M. Haines, "Hyperextension and Hyperflexion Injuries of the Head and Neck of Primates." "Neckache and Backache." Springfield, Ill.: Charles C Thomas, 1970, Ch. 10, pp. 108-117.

19. M. A. Luongo, "Pathologist's Sub-Lethal Determinations." Proceedings of Collision Investigation Methodology Symposium, Warrenton, Va. Sponsored by U.S. Department of Transportation, National Highway Safety Bureau, and Automobile Manufacturers Assoc., August 1969.

20. J. D. States, M. W. Korn, and J. M. Massengil, "The Enigma of Whiplash." New York State Jrl. Medicine, Vol. 70, No. 24 (December 15, 1970), pp. 2971-2978.

21. A. H. Holbourn, "Mechanics of Head Injuries." Lancet (October 9, 1943), p. 438-441.

22. A. K. Ommaya and A. E. Hirsch, "Tolerances for Cerebral Concussion from Head Impact and Whiplash in Primates." Jrl. Biomechanics, Vol. 4 (1971), pp. 13-21.

23. P. Yarnell and A. K. Ommaya, "Experimental Cerebral Concussion in the Rhesus Monkey." Bull. New York Academy of Medicine, Vol. 45 (1969), pp. 39-45.

24. A. K. Ommaya, A. E. Hirsch, and J. L. Martinez, "The Role of Whiplash in Cerebral Concussion." Paper 670906, Proceedings of Tenth Stapp Car Crash Conference, P-12. New York: Society of Automotive Engineers, Inc., 1966.

25. A. K. Ommaya, F. Faas, and P. Yarnell, "Whiplash and Brain Damage." Jrl. AMA, Vol. 204 (1968), pp. 285-287.

26. A. K. Ommaya and P. Yarnell, "Subdural Hematoma of the Whiplash Injury." Lancet (August 2, 1969), pp. 237-239.

27. J. VanGijn and A. R. Wintzen, "Whiplash Injury and Subdural Hematoma." Lancet (September 13, 1969), p. 592.

28. W. R. S. Fan, "Internal Head Injury Assessment." Paper 710870, Proceedings of Fifteenth Stapp Car Crash Conference, P-39. Society of Automotive Engineers, Inc., 1972.

29. B. O'Neill, W. Haddon, Jr., A. B. Kelley, and W. W. Sorenson, "Automobile Head Restraints: Frequency of Neck Injury Insurance Claims in Relation to the Presence of Head Restraints." Preliminary report received, September 1971.

30. K. Hiyama, "Alterations in the Electroencephalogram of Rabbits Following Experimental Whiplash Injuries." Jrl. Japanese Orthopedic Assoc., Vol. 42 (1968), pp. 1057-1065.

31. J. U. Toglia, T. E. Rosenberg, and M. L. Ronis, "Vestibular and Audiologic Aspects of Whiplash Injury and Head Trauma." Jrl. Forensic Sciences, Vol. 14 (1969), pp. 219-226.

32. W. J. Gibson, "Ocular Findings in Extension Flexion Injuries of the Cervical Spine." Proceedings Eleventh Annual Meeting, Amer. Assoc. for Automotive Medicine, Springfield, Ill.: Charles C Thomas, 1970, pp. 228-231.

33. P. E. Roca, "Ocular Manifestations of Whiplash Injuries." Proceedings Fifteenth Annual Meeting, Amer. Assoc. for Automotive Medicine, Ann Arbor: University of Michigan, 1971, pp. 308–319.

34. R. Yamaji, Ecta. Soc. Ophthal. JAP, Vol. 75, No. 2 (1971), pp. 1052–1055.

35. W. D. DeGravelles, Jr., and J. H. Kelley, "Injuries Following Rear End Automobile Collisions." Springfield, Ill.: Charles C Thomas, 1969, Ch. 3, pp. 78–109.

36. C. W. Goff, J. L. Alden, and J. A. Aldes, "Traumatic Cervical Syndrome and Whiplash." Philadelphia: J. P. Lippincott Co., 1969, Chapter 1, pp. 13–19.

37. E. S. Gurdjian, "Neckache and Backache." Springfield, Ill.: Charles C Thomas, 1970, Ch. 1, pp. 3–7.

38. W. H. Harris, D. L. Hamplen, and R. G. Ojemann, "Traumatic Disruption of Cervical Intervertebral Disc from Hyperextension Injury." Clinical Orthopedics, Vol. 60 (1967), pp. 163–167.

39. J. M. Jaues and H. Hooshmand, "Extension Flexion Injuries of the Cervical Spine." Mayo Clinic Proceedings, Vol. 40 (1965), pp. 353–369.

40. J. Kulowski, "Crash Injuries." Springfield, Ill.: Charles C Thomas, 1960, pp. 181–191.

41. R. W. Rann and R. H. Crandall, "Central Spinal Cord Syndrome and Hyperextension Injuries of the Cervical Spine." Jrl. Bone and Joint Surgery, Vol. 44A (1962), pp. 1415–1422.

42. A. Morrison, "Hyperextension Injury of the Cervical Spine with Rupture of the Esophagus." Jrl. Bone and Joint Surgery, Vol. 42B (1960), pp. 356–357.

43. W. Gissane, "The Nature, Causes and Prevention of Neck Injuries in Car Occupants." "Accident Pathology," National Highway Safety Bureau, Contract FH-11-6595. Available through Superintendent of Documents, 1968.

44. J. W. Miller and L. R. Schultz, "Paraplegia Associated with Hyperextension Injury to the Cervical Spine." Amer. Jrl. Surgery, Vol. 96 (1958), pp. 618–623.

45. J. K. Kihlberg, "Flexion Torsion Neck Injury in Rear Impacts." Proceedings Thirteenth Annual Meeting, Amer. Assoc. for Automotive Medicine, HSRI, Ann Arbor: University of Michigan (1969), pp. 1–16.

46. C. A. Buck, F. B. Dameron, M. J. Dow, and H. B. Skowlund, "Study of Normal Range of Motion in the Neck Utilizing a Bubble Goniometer." Archives of Physical Medicine and Rehabilitation, Vol. 40 (1959), pp. 390–392.

47. A. D. Glanville and G. Kreezer, "The Maximum Amplitude and Velocity of Human Movements in Normal Male Human Adults." Human Biology, Vol. 9 (1937), pp. 197–211.

48. J. G. Bennett, L. E. Bergmanis, J. K. Carpenter, and H. B. Skowlund, "Range of Motion of the Neck." Jrl. Amer. Physiotherapy Assoc., January 1963, pp. 45–47.

49. R. Jackson, "The Positive Findings in Alleged Neck Injuries." Amer. Jrl. Orthopedics, Vol. 6 (1964), pp. 184–187.

50. D. Ferlic, "The Range of Motion of the Normal Cervical Spine." Hopkins Hospital Bulletin, Vol. 110, 1962.

51. C. W. Gadd, C. C. Culver, and A. M. Nahum, "The Study of Responses and Tolerances of the Neck." Paper 710856, Proceedings of Fifteenth Stapp Car Crash Conference, P-39. New York: Society of Automotive Engineers, Inc., 1971.

52. J. I. Tuell, Seattle, Washington, personal communication.

53. H. D. Portnoy, J. H. McElhaney, J. W. Melvin, and P. D. Croissant,

"Mechanisms of Cervical Spine Injury in Auto Accidents." Proceedings Fifteenth Conference, Amer. Assoc. for Automotive Medicine, HSRI, Ann Arbor: University of Michigan, 1972, pp. 58-83.

54. A. D. Ruedemann, "Automobile Safety Device-Head Rest to Prevent Whiplash Injury." Jrl. AMA, Vol. 164 (1957), p. 1189.

55. D. M. Severy, H. M. Brink, and J. D. Baird, "Backrest and Head Restraint Design for Rear End Collision Protection." Paper 680079 presented at SAE Automotive Engineering Congress, Detroit, January 1968.

56. R. J. Berton, "Whiplash: Tests of the Influential Variables." Paper 680080 presented at SAE Automotive Engineering Congress, Detroit, January 1968.

57. D. F. Morris and J. Garrett, "Performance Evaluation of Automobile Head Restraints." Paper 720034 presented at SAE Automotive Engineering Congress, Detroit, January 1972.

58. J. W. Melvin, J. H. McElhaney, V. L. Roberts, and H. D. Portnoy, "Deployable Head Restraints—A Feasibility Study." SAE Transactions, Vol. 80 (1971), paper 710853.

59. R. A. Beatty, Letters to the Editor, "Fracture of Odontoid Process Related to Automobile Head Rests." Jrl. AMA, Vol. 11 (1970), p. 29.

60. MDAI Case Number RAI 69, University of Rochester, Available from the U.S. Department of Commerce Clearinghouse, Springfield, Va. 22151, 1969.

61. L. E. Fetterman and F. H. DeVan, Letters to the Editor, "Automobile Head Restraint Injury." Jrl. AMA, Vol. 214 (1970), p. 1328.

62. C. F. Simmons and D. N. Herting, "Human Neck Strength—Method of Determination." Technical Report SID 65-1180, North American Aviation, Inc., Space and Information Systems Division, 1965.

63. F. G. Evans, "Stress and Strain in Bones." Springfield, Ill.: Charles C Thomas, 1957.

64. H. Yamada, "Stress Analysis of Individual Tissues." "Accident Pathology." Ed. K. M. Brinkhouse, available from Superintendent of Documents, U.S. Government Printing Office, 1968, pp. 114-116.

65. A. G. Davis, "Tensile Strength of the Anterior Longitudinal Ligament in Relation to the Treatment of 132 Crash Fractures of the Spine." Jrl. Bone and Joint Surgery, Vol. 20 (1938), pp. 429-438.

66. H. J. Mertz and L. M. Patrick, "Strength and Response of the Human Neck." SAE Transactions, Vol. 80 (1971), paper 710855.

67. H. J. Mertz and L. M. Patrick, "Investigation of the Kinematics and Kinetics of Whiplash." SAE Transactions, Vol. 76, paper 670919.

68. P. D. Clarke, C. D. Gragg, J. F. Sprouffske, and E. M. Trout, "Human Head Linear and Angular Accelerations During Impact." Paper 710857, Proceedings of Fifteenth Stapp Car Crash Conference, P-39. New York: Society of Automotive Engineers, Inc., 1971.

69. H. Furusho, K. Yokoya, S. Nishino, and S. Fujiki, "Analysis of Occupants Movements in Rear End Collisions." Bull. Japanese Society of Automotive Engineers, 1971, pp. 150-166.

70. K. Higuchi, M. Morisawa, and T. Sato, "Movement of Automobile Occupants in Collisions." Bull. Japanese Society of Automotive Engineers, 1971, pp. 124-133.

71. U. Pontius, "Computer Analysis of the Mechanics of the Cervical Spine." Computers in Orthopedic Research and Education, Atlanta, March 23, 1972, unpublished.

72. H. T. E. Hertzberger, "Anthropology of Anthropomorphic Dummies." Paper 690805, Proceedings of Thirteenth Stapp Car Crash Conference, P-28. New York: Society of Automotive Engineers, Inc., 1969.

73. D. M. Severy, "Human Simulations for Automotive Research." New York: Society of Automotive Engineers, Inc., SP 266, 1965.

74. S. A. Ritterrath, "Design and Construction of an Anthropomorphic Dummy." U.S.A.F. Technical Report 6365, Air Material Command, Wright Patterson Air Force Base, Dayton, Ohio, February 1951.

75. S. Alderson, Alderson Research Laboratory, Stamford, Connecticut 06902.

76. B. Aldman, "Biomechanic Studies in Impact Protection." A Acta Physiologica Scan., Vol. 56, Suppl. 192, 1962.

77. O. Mira, M50/71, Anthropomorphic Crash Test Dummy, Oggle Design Limited, Birdshill, Letchworth, Hertfordshire, United Kingdom.

78. A. Bloom, W. G. Cichowski, and V. L. Roberts, "Sophisticated Sam—A New Concept in Dummies." Paper 680031 presented at SAE Automotive Engineering Congress, Detroit, January 1968.

79. S. L. Terry, "Identical Crashes Yield Wide Ranges in Dummy Data." Automotive Engineering, Vol. 79, July 1971, pp. 30–32.

80. C. H. Schutt and C. F. Dohan, "Neck Injury to Women in Auto Accidents. Jrl. AMA, Vol. 206 (1968), pp. 2689–2692.

81. Anthropometry of WAF Basic Trainees and Male Basic Trainees. Wright-Patterson Air Force Base, Wright Air Development Center, July 1963.

82. H. E. Wiggins, Jrl. AMA, Vol. 207 (1969), p. 2284.

83. D. M. Severy, J. H. Mathewson, and C. O. Bechtol, "Controlled Automobile Rear End Collisions, An Investigation of Related Engineering and Medical Phenomena." Medical Aspects of Traffic Accidents—Proceedings of the Montreal Conference, Montreal: Traffic Accident Foundation for Medical Research, 1955, pp. 152–184.

85. V. W. Braunstein and J. O. Moore, "Misuse of the Term Whiplash Injury." Amer. Jrl. Surgery, Vol. 97 (1959), pp. 522–529.

86. N. Gotten, "Survey of 100 Cases of Whiplash Injury After Settlement of Litigation." Jrl. AMA, Vol. 162 (1959), pp. 865–867.

87. J. L. Martinez, "Head Rest and Seat Back Design Proposals." Paper 680775, Proceedings of Twelfth Stapp Car Crash Conference, P-26. New York: Society of Automotive Engineers, Inc., 1968.

88. E. S. Gurdjian, W. A. Lange, L. M. Patrick, and L. M. Thomas, "Impact Injury and Crash Protection. Springfield, Ill.: Charles C Thomas, 1970.

89. M. M. Gilbert, Presentation to Amer. Assoc. for the Study of Headache, Medical Tribune, October 16, 1969.

90. J. Fell, Accident Investigation Division, National Highway Traffic Safety Admin., Personal Communication.

91. W. Lange, "Mechanical and Physiological Response of the Human Cervical Vertebral Column to Severe Impacts Applied to the Torso." Max Planck Institute, 46 Dortmund, Germany, 1970.

92. J. Fell, Personal Communication, August 7, 1972.

Whiplash: Tests of the Influential Variables

R. J. Berton

FORD MOTOR COMPANY

WHIPLASH: Tests of the Influential Variables — Seat back rotation, restraint position and collision speed.

INTRODUCTION

The literature (1-10)* abounds with propositions about the cause of injury during rear end collisions. While pointing to extension, flexion, (Figs. 1 and 2) torque and rotation as being the principal agents, it leaves the mechanism of injury undetermined and the selection of influential variables arbitrary.

These tests were undertaken, therefore, to explore variable design parameters to determine whether effective protection can be provided. In contrast to military or space practice, automobile head restraint devices must necessarily accommodate the full range of physical variations found in the general population.

The subjects used in the tests were fifty percentile adult male Sierra dummies; however, the biomechanical relationship of man to these dummies has not been established.

* Numbers in parentheses designate references at the end of the paper.

ABSTRACT

The effect of seat back rotation, head restraint position, and collision speed on crash dummy head acceleration, extension and flexion was determined by tests using an acceleration sled and vehicle collisions. The sled tests were run with a rigid seat and an adjustable back at 10, 20, and 30 mph. Vehicle collisions were conducted with production seats with and without head restraint devices at 10, 15, 20, 25, 30, 40, and 50 mph impact speeds.

Fifty percentile adult male Sierra dummies were used. The head was able to move freely when accelerated backward.

In both sled and vehicle collisions, head restraint devices reduced the measured severity criteria on the crash dummies employed in the tests.

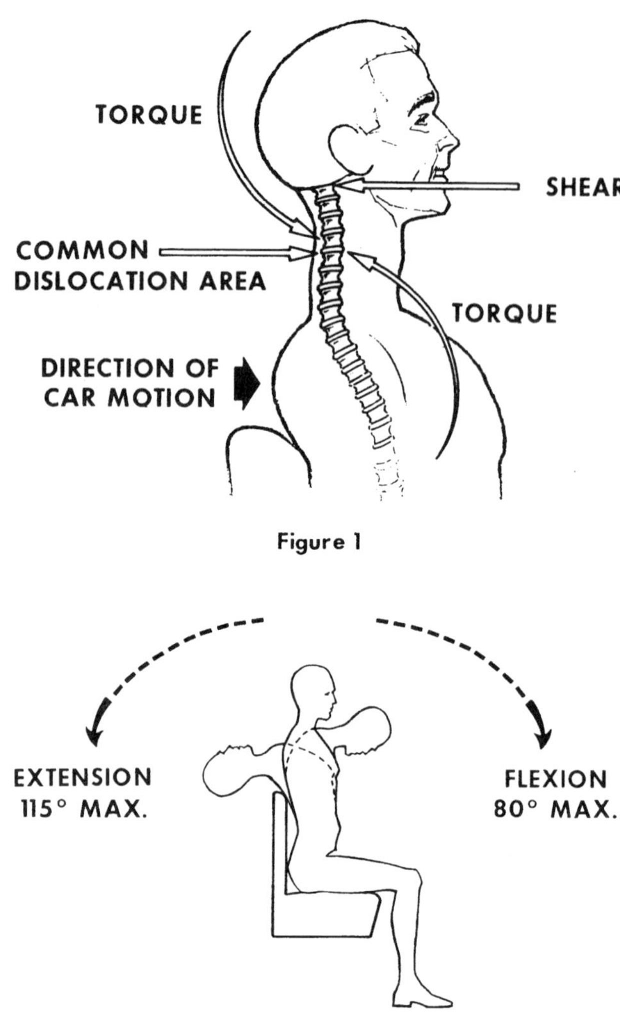

Figure 1

Figure 2

DISCUSSION

Sled Tests with Rigid Seat Back

In these tests, the terminal speed was 10 mph. The seat back height from the "H" point to the top of the cushion was varied from 28 to 22 inches, measured along the torso line. The dummy head in each test was located either 1 or 4 inches away from the seat back cushion, referred to as "horizontal offset."

Accelerations along and normal to the dummy line of sight were recorded continuously for each test. The angular motion of the head relative to the torso line was determined by analysis of film taken at 1000 frames per second.

A multiple regression analysis was conducted to relate the measured acceleration and extension responses to the experimental variables. The response surface obtained indicates that both dummy head acceleration and extension were greater when lower seat backs and larger horizontal offsets were used. In Figure 3, the curved lines represent the various extension angles obtained by varying seat back height and offset. A section taken along line AB gives a more obvious representation of the results by indicating that extension is represented by a curved surface having its highest point with the maximum offset and the lowest seat back.

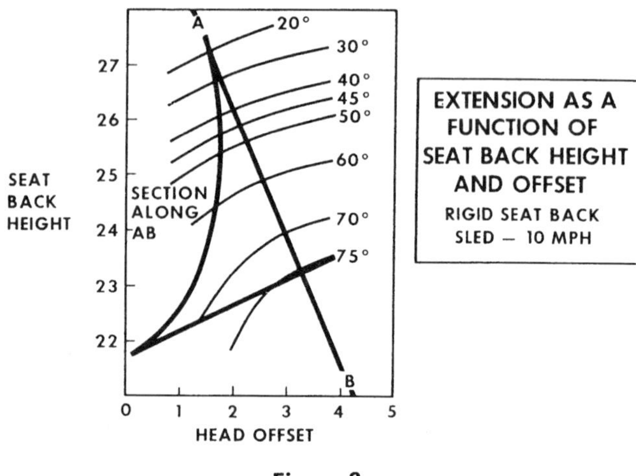

Figure 3

Subjective evaluation of the same tests was conducted by a panel of five Ford employes with backgrounds in engineering and biomechanics. They were instructed to rate the severity of test film sequence according to their own interpretation. As seen in Fig. 4, run #3 was unanimously considered by the panel to be the least severe. In this run, the seat back was as high as the CG of the dummy head and the offset was the 1 inch minimum.

SUBJECTIVE EVALUATION
OF SEVERITY SLED TESTS

RUN #	SEAT BACK HEIGHT	HEAD OFFSET	A	B	C	D	E	AVERAGE
1	22	1	1	1	1	2	1.5	1.27
2	22	4	1	1	1	1.5	1	1.03
3	28	1	5	4	4	4	5.5	4.57
4	28	4	4	3	3	4	4.5	3.72
5	26	1	3	3	2	3	3	2.64
6	26	4	2	2	2	2.5	2.5	2.05
7	25	2.5	2	1	1	1	1.5	1.15
8	24	1	2	1	1	2	1	1.52
9	24	4	1	1	1	1.5	1	1.3
10	25	2.5	1	1	1	1	1.5	1.26

Figure 4

Figures 5 to 9 inclusive represent the results of one test only for each of the variables studied. Individual points should, therefore, not be singled out. Only the general trend of the response should be considered.

10 MPH SLED RUNS
RIGID SEAT BACK ADJUSTABLE IN HEIGHT

RUN #	SEAT BACK HEIGHT	HEAD OFFSET INCHES	NECK PULL 'G'	HEAD/ TORSO ANGLE DEGREES	LONG. ACC. 'G'	SUBJ. EVALUATION
1	22	1	25	65	32	1.27
2	22	4	38	78	45	1.03
3	28	1	15	13	41	4.57 BEST
4	28	4	20	25	47	3.72
5	26	1	23	41	28	2.64
6	26	4	28	58	30	2.05
7	25	2.5	38	67	38	1.15
8	24	1	34	72	40	1.52
9	24	4	38	75	42	1.3
10	25	2.5	35	78	30	1.26

Figure 5

Figure 5 shows that run #3 imposed the lowest normal acceleration (neck pull) and the smallest head/torso angle. Longitudinal acceleration in run #3, however, was among the highest in this series of tests because the head restraint device was high, almost totally preventing head motion. The restraint device absorbs most of the head inertia force.

Figure 6 shows that longitudinal acceleration of the dummy head increases with speed. It is greatest when the head restraint device is highest and the offset is 4 inches.

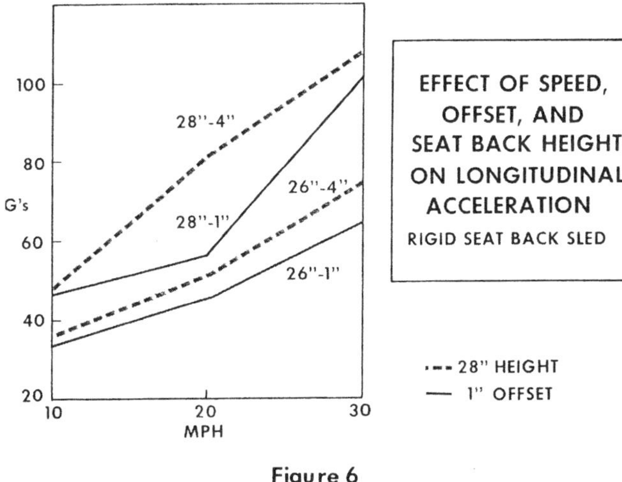

Figure 6

Figure 7 shows that the dummy head/torso angle is smallest when the head restraint device is highest. The offset contributes progressively less to hyperextension with higher seat backs.

Figure 8 is a section through Figure 7 at the 20 mph speed. It indicates the progressive reduction in hyperextension with the increased seat back height.

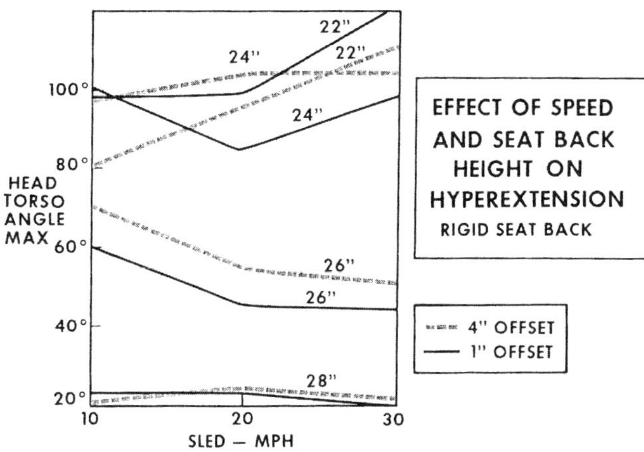

Figure 7

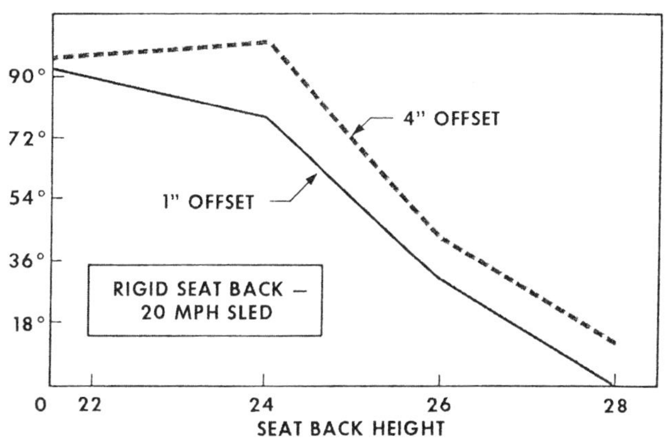

Figure 8

Figure 9. In spite of the crossing over of some of the data, dummy neck pull is generally lower with the higher seat backs.

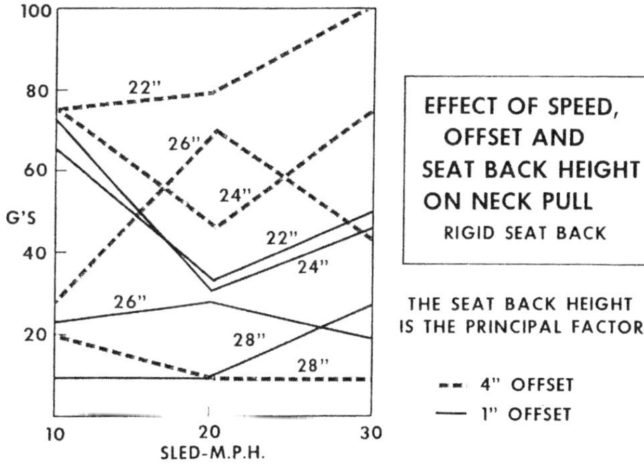

Figure 9

Figure 10 shows that dummy flexion velocity was generally half the extension velocity and that flexion became negligible with 26 and 28 inch seat backs.

EFFECT OF SEAT BACK HEIGHT AND OFFSET ON ANGULAR VELOCITY OF HEAD

RIGID SEAT BACK SLED – 10 MPH

SEAT BACK HEIGHT	4" OFFSET		1" OFFSET	
	EXTENSION (DEGREES/SECOND)	FLEXION	EXTENSION (DEGREES/SECOND)	FLEXION
22	2200	1000	1400 / *1500	900
24	2000	1000	1700	1300
26	1900	0	1100	0
28	1000	0	500 / *550	0

* PRODUCTION SEAT BACK MOVABLE BARRIER/CAR – 30 MPH

Figure 10

Mechanics of the seat back cushion have a significant influence. A 3 inch deflection of the production seat, having a spring rate of 1.5 psi per inch of deflection, can cause a 10 G forward acceleration of the dummy torso. Compression of the cushion allows the dummy to store energy in torso translation. However, when the dummy's head reaches its maximum extension, the seat cushion recovers. This forces the dummy shoulders forward and accentuates the relative torso/head angle and the eventual head extension. (Fig. 11)

EXTENSION AS A FUNCTION OF SEAT BACK CUSHION DEFLECTION TO PRODUCTION SEAT - SLED 10 M.P.H.

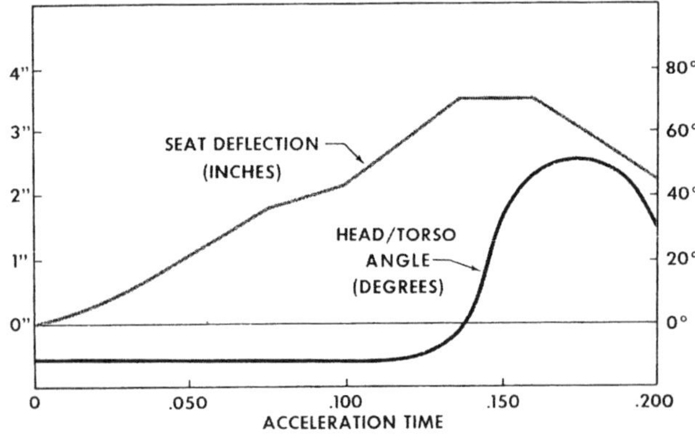

Figure 11

Figure 12 is a graphic representation of a film sequence showing the relative positions of the dummy with 22 and 28 inch seat back heights.

RIGID SEAT BACK

10 M.P.H.

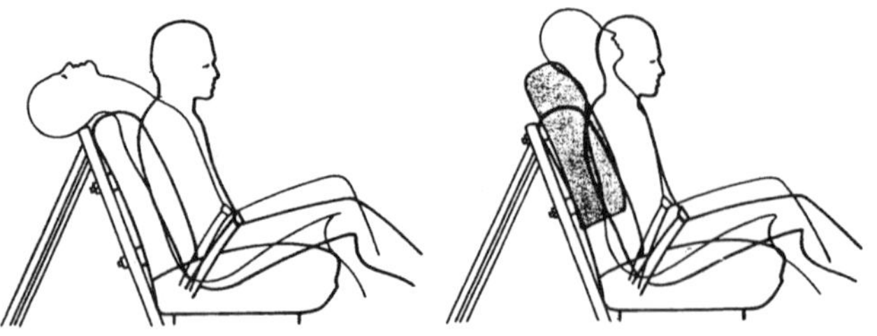

SEAT BACK HEIGHT = 22"
HEAD OFFSET = 4"
HEAD/TORSO ANGLE = 92°

SEAT BACK HEIGHT = 28"
HEAD OFFSET = 1"
HEAD/TORSO ANGLE = 39°

Figure 12

Figure 13a, b shows dummy head response to 10 mph, 10 G pulses ordered from 28 to 22 inch seat back heights and alternated from 1 to 4 inches offset.

Below 26-inch seat back height, there is no meaningful change in the head/torso angle. Dummy vertical head acceleration with rigid seat backs is initially directed upward and is higher in that direction with the 28" high seat back. This condition did not exist with production seat backs which rotated backward under load.

With lower seat back, the longitudinal acceleration becomes choppy, indicative of uncontrolled motion, while the normal acceleration (the dummy neck pull) becomes more severe.

A 28" rigid seat back induces up to 15 G upward acceleration of the dummy head.

Collision Tests with Production Seats

The sled tests were supplemented by vehicle collision tests, in which the impacted vehicle carried four test dummies. The two on the right side were provided with a head restraint device and the two on the left side were without, for comparison. The top surface of the head restraint device was 28 inches above the dummy "H" point, measured along the torso line. The test vehicles were 1965 and 1966 Fords, with roof and doors removed to facilitate high speed photography. They were impacted from the rear by a movable barrier of equal weight, at speeds of 10, 15, 20, 25, 30, 40 and 50 mph. The tests confirmed the effectiveness of the head restraint devices tested in reducing dummy head extension and acceleration. (Fig. 14) They also indicated that when the seat back collapses, the belted dummies sustain negligible neck extension. (Fig. 15)

RIGID SEAT TESTS 10 M.P.H.

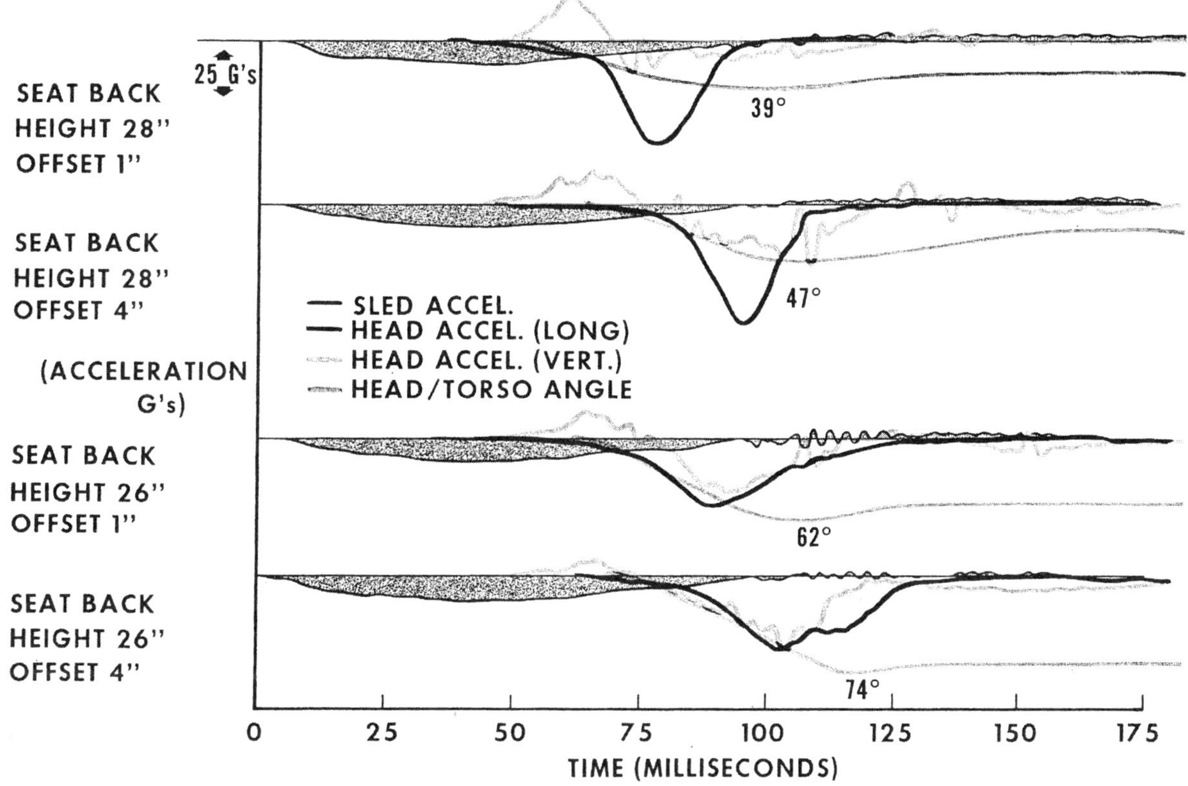

Figure 13a

RIGID SEAT TESTS 10 M.P.H.

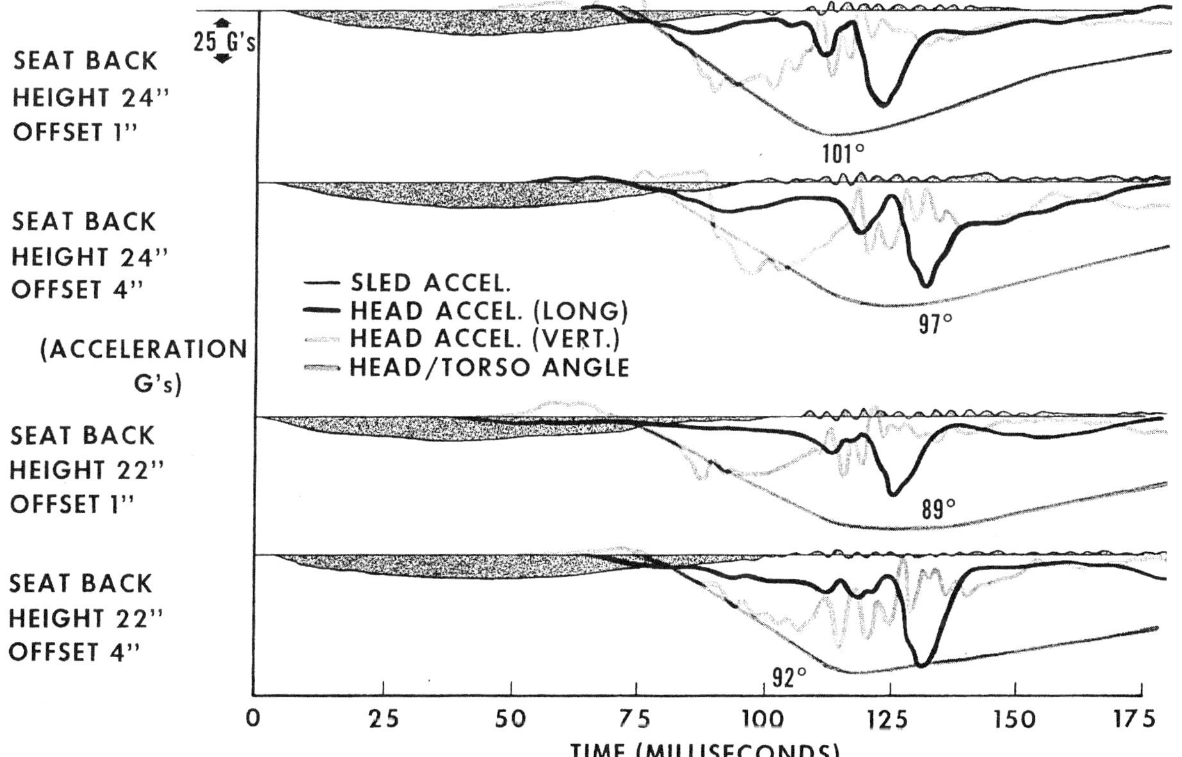

Figure 13b

MOVABLE BARRIER-TO-CAR

20 M.P.H.

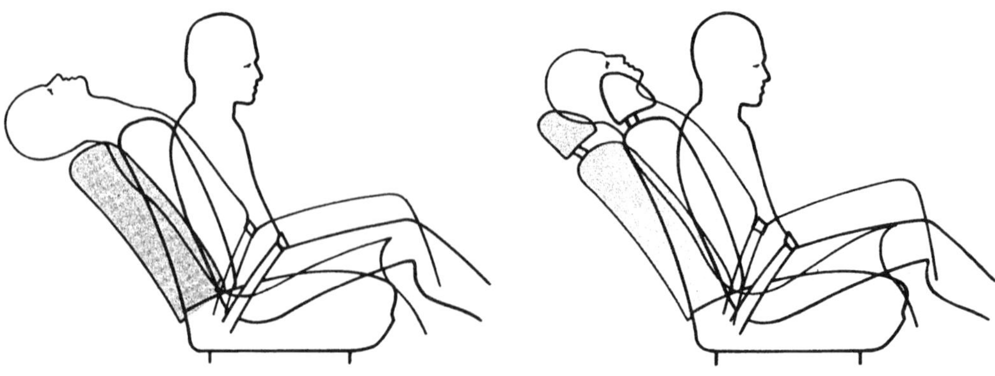

WITHOUT HEADREST	WITH HEADREST
SEAT BACK HEIGHT = 22"	SEAT BACK HEIGHT = 28"
HEAD OFFSET = 1"	HEAD OFFSET = 1"

Figure 14

MOVABLE BARRIER-TO-CAR

40 M.P.H.

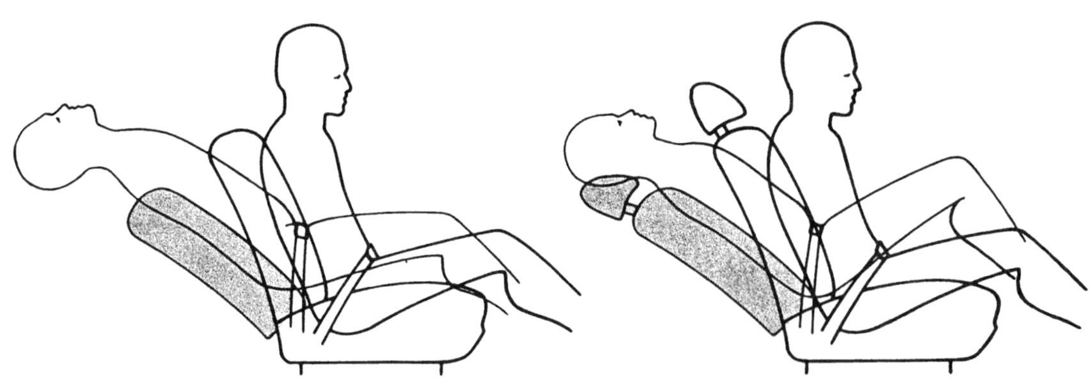

WITHOUT HEADREST	WITH HEADREST
SEAT BACK HEIGHT = 22"	SEAT BACK HEIGHT = 28"
HEAD OFFSET = 1"	HEAD OFFSET = 1"

Figure 15

Figure 16a, b shows the resultant dummy head acceleration and extension as a function of speed, with and without head restraint devices. Extension was halved by the restraint device, but above 30 mph no extension was detected, as the belted dummies impacted the rear seat occupant without a change occurring in the head/torso angle.

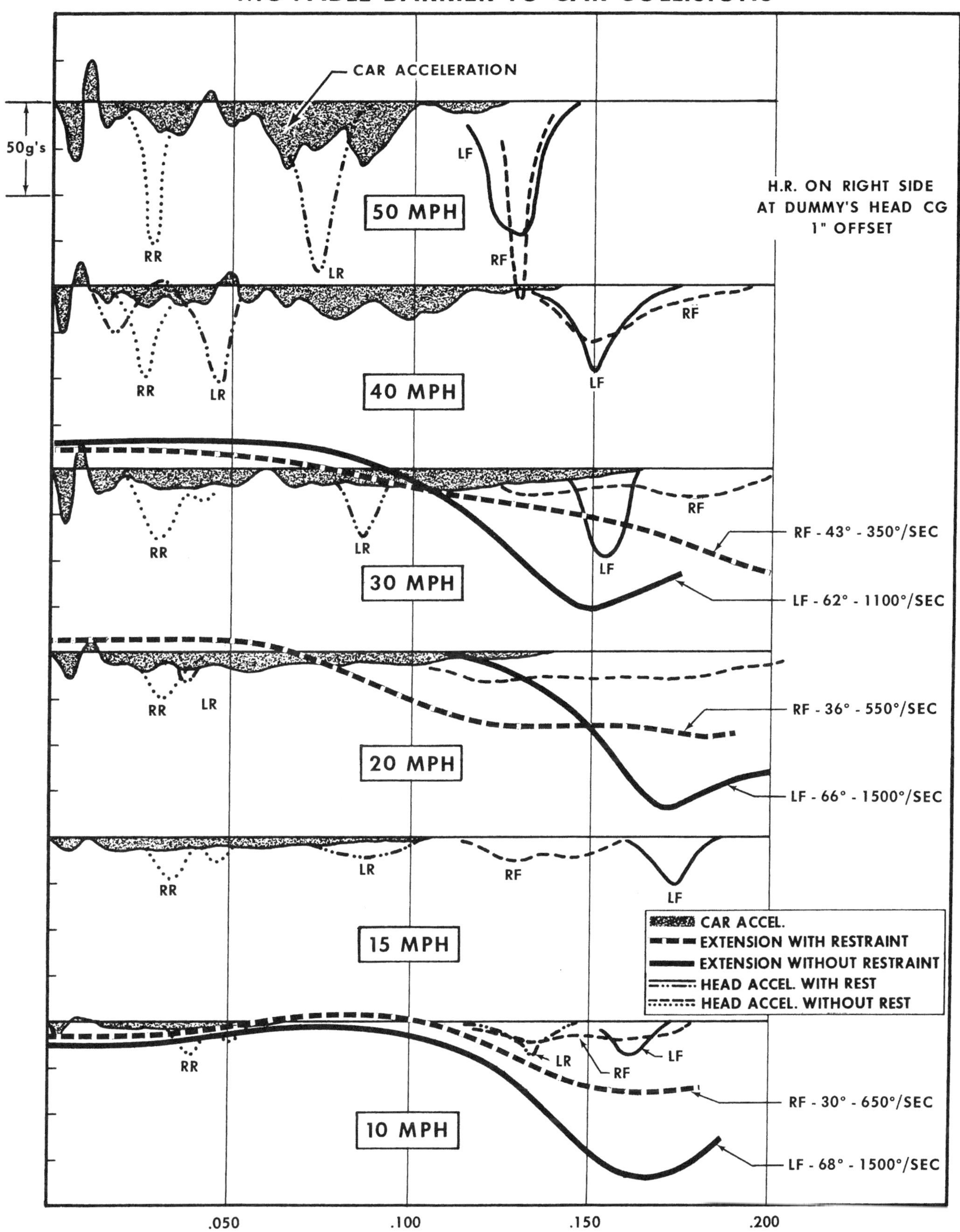

HEAD RESULTANT G's
MOVABLE BARRIER TO CAR COLLISIONS
CAR ACCELERATION
50g's
LF
RR
LR
RF
50 MPH
H.R. ON RIGHT SIDE
AT DUMMY'S HEAD CG
1" OFFSET
40 MPH
30 MPH
RF - 43° - 350°/SEC
LF - 62° - 1100°/SEC
20 MPH
RF - 36° - 550°/SEC
LF - 66° - 1500°/SEC
15 MPH
CAR ACCEL.
EXTENSION WITH RESTRAINT
EXTENSION WITHOUT RESTRAINT
HEAD ACCEL. WITH REST
HEAD ACCEL. WITHOUT REST
10 MPH
RF - 30° - 650°/SEC
LF - 68° - 1500°/SEC
.050
.100
.150
.200

Figure 16a, b

The dummy extension velocity was more than halved by the restraining device.

The resultant head acceleration was also reduced by half and the acceleration pulse length was increased, reducing the slope of the 30 mph acceleration to lower the acceleration rate from 5000 to 1200 G per second.

With the front seat back braced to prevent deflection, flexion was lower than with the unbraced seat when the dummy returned to the normal seated position, indicating that the combined elasticity of the front seat back and seat cushion contributes to flexion.

Figures 17, 18, and 19 indicate that dummy head acceleration changes very little under 30 mph when the dummy head is restrained. The results, however, represent only one point for each variable tested, and individual values should not be singled out for comparison. Only the general trend of the response should be considered.

LONGITUDINAL HEAD ACCELERATION AS A FUNCTION OF SPEED - MOVABLE BARRIER TO CAR

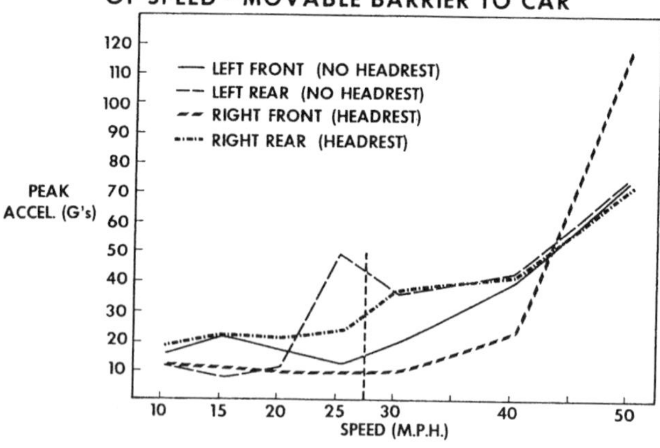

Figure 17

VERTICAL HEAD ACCELERATION AS A FUNCTION OF SPEED - MOVABLE BARRIER TO CAR

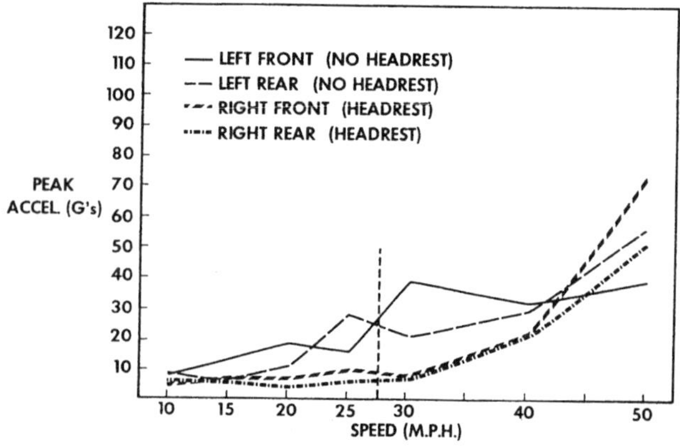

Figure 18

RESULTANT HEAD ACCELERATION AS A FUNCTION OF SPEED - MOVABLE BARRIER TO CAR

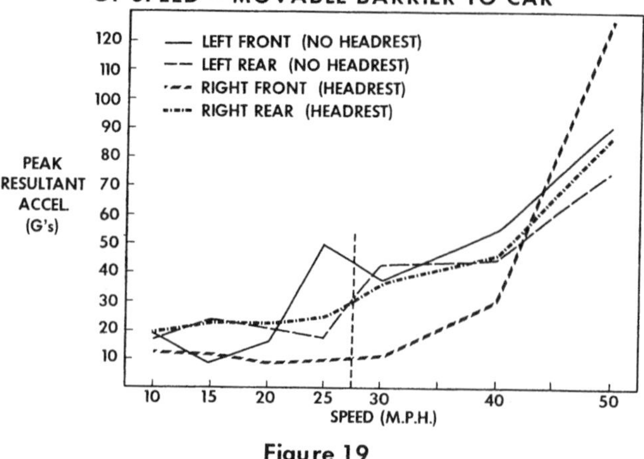

Figure 19

Figure 20 indicates the changes in dummy extension during the vehicle collision tests, showing the reduction achieved by the head restraint device.

HYPEREXTENSION (HEAD/TORSO) AS A FUNCTION OF SPEED MOVABLE BARRIER TO CAR

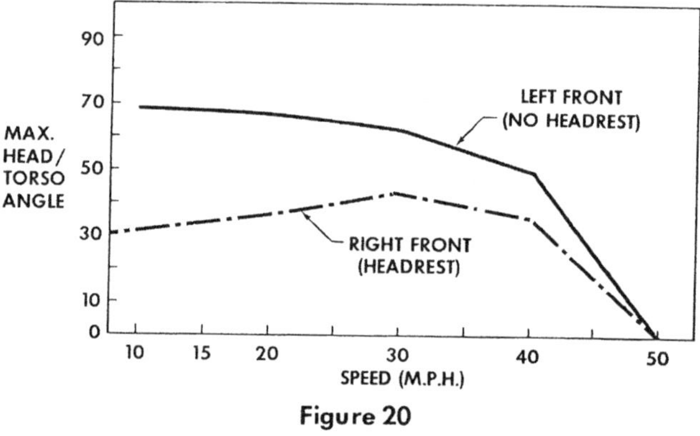

Figure 20

Figure 21 gives a time history of dummy extension during vehicle collision tests, again showing the reduction in extension and rate of extension produced by the head restraint devices.

HYPEREXTENSION TIME HISTORIES MOVABLE BARRIER TO CAR

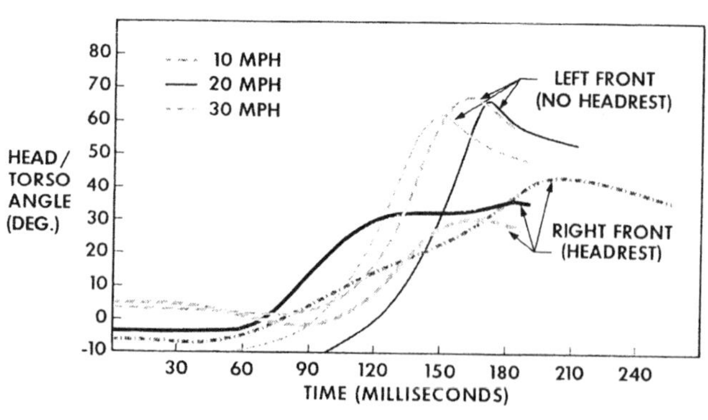

Figure 21

As shown in Figure 22, vehicle structure collapse was directly proportional to speed.

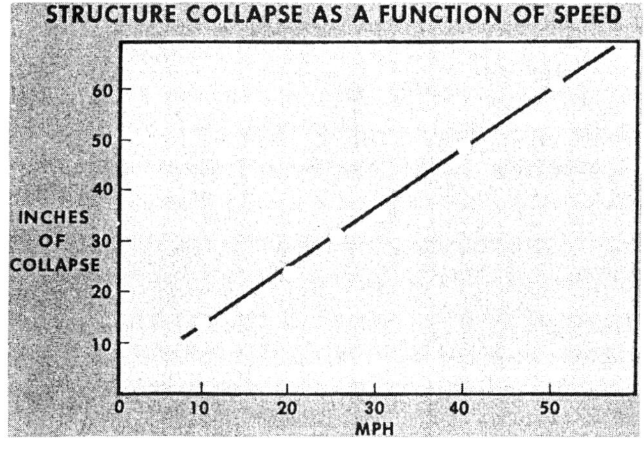

Figure 22

Figure 23 shows vehicle structural collapse as a function of "B" post motion during the tests. Front seat occupants contacted the head restraint after the car structure had completely collapsed and the car was accelerating linearly.

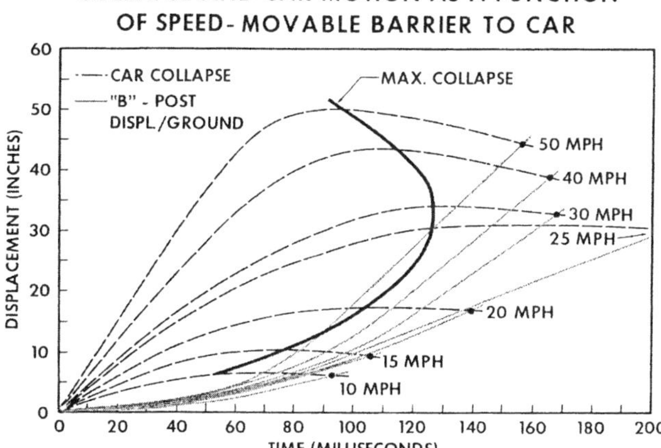

Figure 23

Vision Restrictions

Figures 24, 25, 26, 27, and 28 show the horizontal vision restrictions encountered with head restraint devices on a variety of seat backs in use today. A horizontal line drawn from the top of the restraint to the standard SAE package eyellipse indicates that a 28 inch high restraint of the design tested, as measured along the torso line in packages with a torso line angle of about 26 degrees, obstructs horizontal vision at the 20th percentile level and that a 26 inch restraint imposes no limitations.

T-BIRD vs CAR "X" HEAD RESTRAINT VISION STUDY

Figure 24

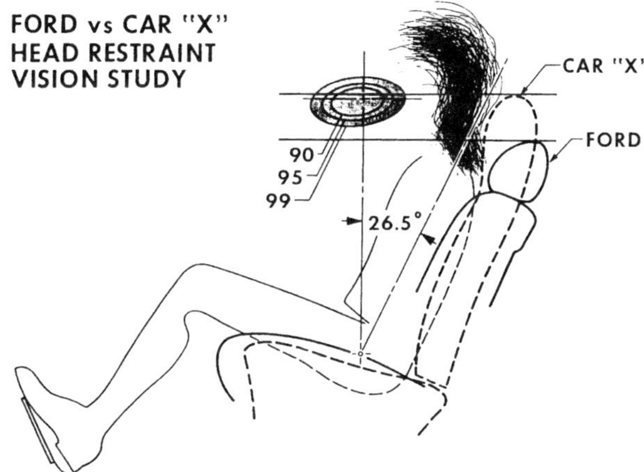

Figure 25

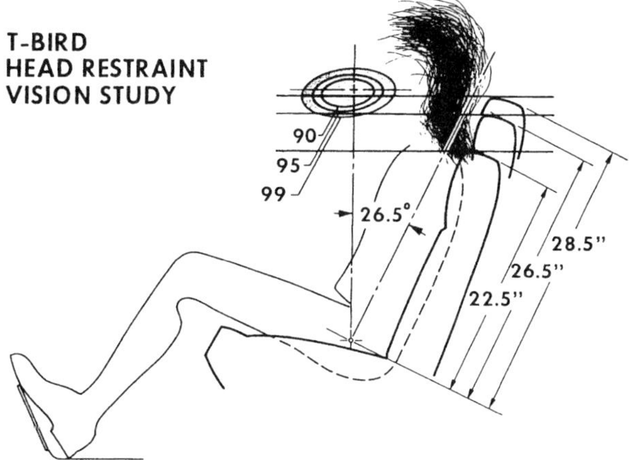

Figure 26

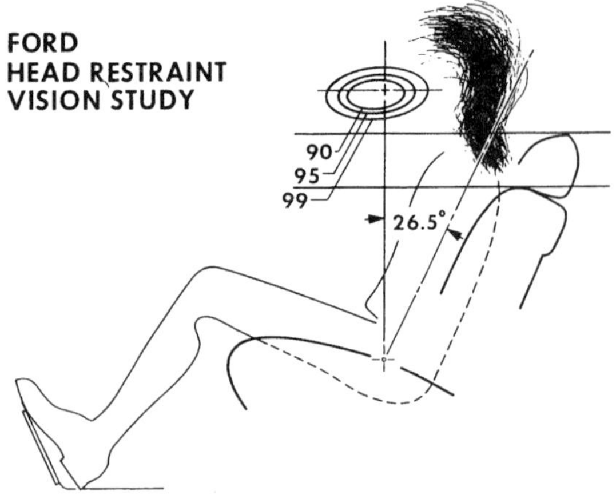

Figure 27

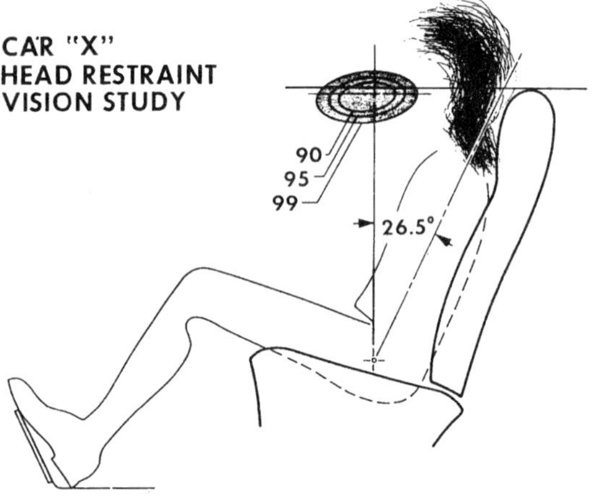

Figure 28

CONCLUSIONS

Sled and vehicle collision tests have shown that head accelerations and neck extensions of the #450 Sierra dummy are lowered when the head restraint device is located near the dummy head CG, with minimum horizontal offset. Restraint devices located 2 inches below the CG of the dummy head, corresponding to 26 inch seat back height, produce accelerations and extensions not very different from those measured with 28 inch high restraint.

Kinematics of the dummy suggest that should a thin head restraint device be located at the base of the dummy skull it might act as a fulcrum for the neck and emphasize neck pull.

Under certain test conditions, the whole dummy is accelerated upward and backward. With rigid seat back structures and high head restraint devices, vertical accelerations are induced, particularly when the head and torso remain aligned. Seat back rotation of 20 degrees would reduce the vertical loading. Breaking the head/torso axis would further reduce it. This could be achieved by designing the restraint devices to deflect 10 degrees more than the seat back. Seat back and head restraint device rotation rearward would also reduce the level of accelerations sustained by the dummy head. When no head restraint device is available, seat back collapse, as occurring in present cars, can reduce extension. With head restraint devices, however, optimum protection requires that the dummy be belted, since at higher speeds it tends to slide up the seat back.

Any seat back structure and cushion deflection should be plastic or be damped to minimize forward acceleration of the dummy torso during neck extension.

Since the biomechanical relationship of man to the dummies used in this study has not been established, the conclusions cannot be applied indiscriminately to human response in rear end collision.

680079

Backrest and Head Restraint Design for Rear-End Collision Protection

Derwyn M. Severy, Harrison M. Brink, and Jack D. Baird
Institute of Transportation and Traffic Engineering, University of California, Los Angeles

IN THIS SERIES OF COLLISION EXPERIMENTS, a stationary 1967 four-door sedan was rear-ended by an identical striking car. The twelve experiments were conducted at five different striking car speeds. Experiment 97, 10 mph; Experiment 94, 20 mph; Experiments 93, 95, 96, 99, 100, 101 and 102, 30 mph; Experiment 98, 40 mph and Experiments 103 and 104, 55 mph. Dual biaxial acceleration data for the heads and chests of the 50th and 95th percentile adult male anthropometric dummies, considered in conjunction with micromotion analyses of body kinematics and related instrumentation, provided information on the relative effectiveness of variations in head restraint and seatback designs. For the rear-ended motorist in a conventional seat (20- to 22-inch seatback), the forces of a col-

ABSTRACT

Scientific methodology and engineering techniques were applied to a series of twelve automobile rear-end collision experiments to provide data relating to seat, seat backrest and head-restraint design. Five speeds of impact, six seatback heights and six seatback strength values were studied. The purpose was to evaluate the relative protective merits and the practicality of various seat designs with respect to the many variables common to rear-end collisions. This research data provides a basic reference system of collision performance for seat designs with respect to occupant size, posture and proximity to injury producing structures.

lision involuntarily place him in an acutely forced adverse posture, representing the primary injury problem; the magnitude of force application is of secondary importance. Micromotion analysis of high-speed motion pictures of rear-ended dummy-motorists provided a basis for correlating relative performances of whiplash restraining systems with respect to the forced movements and recorded accelerations of occupants. Other factors relating to the rear-end collision problem that were evaluated and quantified include performance of fuel tanks, rear-end collapse-action, seat structures, head restraints, seat anchorages, seatback deflection, seatback permanent deformation, lap belts and padding for passenger protection.

Photographic instrumentation included eight high speed cameras mounted at key locations on the collision vehicles to provide close-up, continuous monitoring of occupant movements during these quarter-second collision events. The motion picture cameras mounted on the crashing vehicles were backed up by 19 additional cameras installed on specially constructed towers and at key ground positions circumscribing the collision impact area. Data from photographic, electronic and special instrumentation are presented in this report, using techniques designed to facilitate information-retrieval and presented in a manner that conveys essential findings of this study, even for one having time to make only a cursory review.

BACKGROUND

The UCLA-ITTE Automotive Collision Research Project has conducted five series of rear-end collision experiments. In 1954, the initial series, consisting of five experiments, was conducted for the primary purpose of identifying the forced body movements commonly referred to as "whiplash" and their corresponding force-time patterns as a function of the speed of the striking vehicle (1)*. In 1956, a second series of three rear-end collision experiments was conducted for the principal purpose of determining the extent to which a head restraint, applied to the front seat backrest, would reduce the injury hazard associated with "whiplash" (2). In 1965, a third series of two rear-end collision experiments was conducted for the purpose of determining the reduction in accelerative forces attending more collapsible rear-end structures and to evaluate the protective features of an integrated safety seat (3). In 1966, the highest speed rear-end collision experiment thus far, was conducted with a fully populated school bus for the purpose of studying many different conditions of passenger protection, with special emphasis on school bus safety (4).

These prior four series of rear-end collision experiments, consisting of 11 full-scale crashes, provided a substantial amount of theoretical and practical data lending much confidence to the important role the head restraints concept would play, once adopted for motoring safety. The appropriateness of the concept was well documented and researchers in this field at that time were satisfied that sufficient knowledge had been gained to make further inquiry unwarranted. However, engineering is an exact science and when it became necessary for the advisors to the General Services Administration of the U.S. Department of Commerce and, subsequently, to the National Highway Safety Bureau to provide specifications suitable for automobile safety regulations, voids in the data were quickly identified. Examples were: (1) Strength requirements for seatbacks, as a function of vehicle impact severity and (2) the changes in protective properties associated with a given seatback and head restraint, as a function of size of the motorist and of his head and torso offset from these restraints, to the extent he may be leaning forward at impact. Little information was known about the exposure to whiplash injury or about preventive measures for protecting rear seat passengers. Many other questions of this nature would have to be answered if a practical, reasonable and workable performance standard were to be devised. Accordingly, the UCLA engineers developed an Operations Plan calling for twelve rear-end collision experiments designed to provide the urgently needed design and performance information.

---

*Numbers in parentheses designate References at end of paper.

FACILITIES AND EQUIPMENT

The UCLA-ITTE Field Station for collision research is located adjacent to the U.S. Naval Station, Long Beach, California. The fenced-in facility includes an asphalt surface, mobile shops and buildings as well as several towers to support photographic equipment for observation of collision experiments. The four mobile vans provide research support facilities as follows: Machine Shop, Electronics Shop, Photographic Shop and Project Control Center. An aluminum monorail guide track, installed on level asphalt, provides directional control for the crash vehicles, Fig. 1. Other operational techniques developed and adopted during recent years are described in prior publications by the authors.

The ten passenger cars that were crashed in these collision experiments were donated, along with research funds, by Ford Motor Company for this rear-end collision series. These vehicles were (new) 1967 Ford Custom four-door sedans. Other equipment and facilities required for this research were provided by the University of California, Los Angeles, facilitated by a research grant (No. UI 00032-02) from the U.S. Public Health Service.

EXPERIMENTAL PROCEDURE

The twelve rear-end collision experiments were conducted at five different speeds of impact. Seven of the twelve experiments were conducted at 30 mph to provide a standard collision exposure for evaluating whiplash protection as function of changes in seatback and head restraint design. In addition, one-10, one-20 and one-40 mph impact was conducted to sample whiplash protective equipment performance for these collision exposures; two collision experiments were conducted at 55 mph in order to adequately qualify the relative performance of various motorist protective design features undergoing evaluation.

In order to provide a constant motorist size while evaluating seatback and head restraint design, four identical 95th percentile male anthropometric dummies were used in the struck car for Experiments 93 through 98; in Experi-

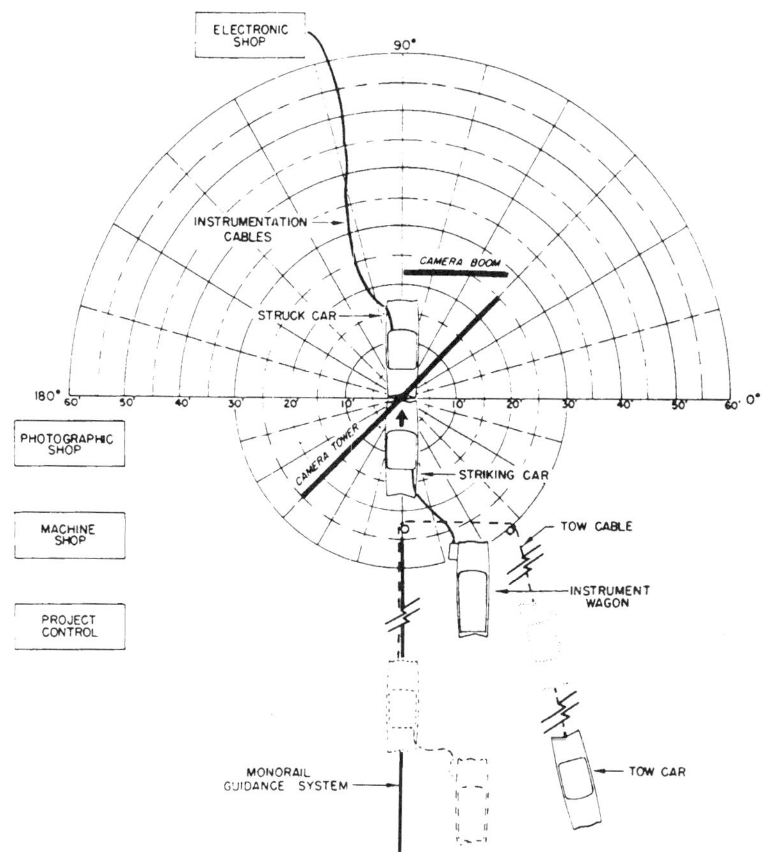

Fig. 1 - Facilities, including vehicle control systems

ments 99 through 102, two 50th percentile male dummies replaced two of the 95th percentile dummies in the struck car and in Experiments 103 and 104 only one 50th percentile dummy was substituted for a 95th percentile dummy in the struck car driver's position.

Referring to Table 1, Schedule of Operations, the first experiment, X-93, was a 30 mph rear-end collision in which occupant seat positions (driver, right front passenger, right rear passenger and

Table 1 - Schedule of Operations

EXPERIMENT NUMBER AND IMPACT SPEED	OCCUPANT LOCATION	SEAT TYPE	SEATBACK HEIGHT-ANGLE	ADJUSTABLE HEAD RESTRAINT HEIGHT-ANGLE	SEATBACK RESISTANCE	ADULT MALE DUMMY SIZE	HEAD OFFSET (SEE FIG. 7)
X-93 30 MPH	DRIVER	BUCKET	22"-20°	–	9,100 IN.-LBS.	95%	3"
	R. FRONT	BUCKET	22"-20°	–	16,100 IN.-LBS.	95%	0"
	R. REAR	BENCH	21"-24°	–	MFG'S SPECS.	95%	12"
	L. REAR	BENCH	21"-24°	–	MFG'S SPECS.	95%	0"
X-94 20 MPH	DRIVER	BUCKET	22"-20°	–	9,100 IN.-LBS.	95%	3"
	R. FRONT	BUCKET	22"-20°	6"-20°	16,100 IN.-LBS.	95%	0"
	R. REAR	BENCH	21"-24°	–	MFG'S SPECS.	95%	12"
	L. REAR	BENCH	21"-24°	–	MFG'S SPECS.	95%	3"
X-95 30 MPH *	DRIVER	BUCKET	22"-20°	6"-20°	16,100 IN.-LBS.	95%	6"
	R. FRONT	BUCKET	22"-20°	6"-20°	16,100 IN.-LBS.	95%	12"
	R. REAR	BENCH	21"-24°	6"-10°	MFG'S SPECS.	95%	12"
	L. REAR	BENCH	21"-24°	–	MFG'S SPECS.	95%	0"
X-96 30 MPH	DRIVER	BUCKET	25"-20°	–	16,100 IN.-LBS.	95%	3"
	R. FRONT	BUCKET	28"-20°	–	16,100 IN.-LBS.	95%	3"
	R. REAR	BENCH	21"-24°	6"-10°	MFG'S SPECS.	95%	3"
	L. REAR	BENCH	21"-24°	–	MFG'S SPECS.	95%	12"
X-97 10 MPH	DRIVER	BUCKET	22"-20°	6"-20°	MFG'S SPECS.	95%	3"
	R. FRONT	BUCKET	28"-20°	–	16,100 IN.-LBS.	95%	0"
	R. REAR	BENCH	21"-24°	–	MFG'S SPECS.	95%	12"
	L. REAR	BENCH	21"-24°	–	MFG'S SPECS.	95%	0"
X-98 40 MPH	DRIVER	BUCKET	22"-20°	6"-20°	MFG'S SPECS.	95%	3"
	R. FRONT	BUCKET	28"-20°	–	16,100 IN.-LBS.	95%	3"
	R. REAR	BENCH	21"-24°	–	MFG'S SPECS.	95%	12"
	L. REAR	BENCH	21"-24°	–	MFG'S SPECS.	95%	3"
X-99 30 MPH	DRIVER	BUCKET	25"-20°	–	33,000 IN.-LBS.	95%	0"
	R. FRONT	BUCKET	28"-20°	–	33,000 IN.-LBS.	95%	0"
	R. REAR	BENCH	21"-24°	–	MFG'S SPECS.	50%	12"
	L. REAR	BENCH	21"-24°	–	MFG'S SPECS.	50%	0"
X-100 30 MPH	DRIVER	BUCKET	25"-20°	–	16,100 IN.-LBS.	95%	6"
	R. FRONT	BUCKET	28"-20°	–	16,100 IN.-LBS.	95%	6"
	R. REAR	BENCH	21"-24°	6"-10°	MFG'S SPECS.	50%	0"
	L. REAR	BENCH	21"-24°	–	MFG'S SPECS.	50%	0"
X-101 30 MPH	DRIVER	BUCKET	25"-20°	–	16,100 IN.-LBS.	95%	12"
	R. FRONT	BUCKET	28"-20°	–	16,100 IN.-LBS.	95%	12"
	R. REAR	BENCH	25"-24°	–	MFG'S SPECS.	50%	0"
	L. REAR	BENCH	25"-24°	–	MFG'S SPECS.	50%	6"
X-102 30 MPH *	DRIVER	BUCKET	25"-20°	–	16,100 IN.-LBS.	50%	0"
	R. FRONT	BUCKET	25"-20°	–	33,000 IN.-LBS.	50%	0"
	R. REAR	BENCH	25"-24°	–	MFG'S SPECS.	95%	3"
	L. REAR	BENCH	25"-24°	–	MFG'S SPECS.	95%	6"
X-103 55 MPH *	DRIVER	BUCKET	25"-20°	–	33,000 IN.-LBS.	50%	0"
	R. FRONT	BUCKET	28"-20°	–	33,000 IN.-LBS.	95%	3"
	R. REAR	BUCKET	28"-20°	–	33,000 IN.-LBS.	95%	0"
	L. REAR	BUCKET	21"-23°	8"-0°	MFG'S SPECS.	95%	0"
X-104 55 MPH	DRIVER	BUCKET	28"-20°	–	33,000 IN.-LBS.	50%	0"
	R. FRONT	BUCKET	28"-20°	–	33,000 IN.-LBS.	95%	3"
	R. REAR	BENCH	21"-24°	6"-10°	MFG'S SPECS.	95%	3"
	L. REAR	BENCH	21"-24°	–	MFG'S SPECS.	95%	3"

* REAR WINDOW AND HEADER REMOVED

left rear passenger) were provided with seats having respectively, 22, 28, 21 and 21 inch seatback heights; other variables are similarly identified in this Schedule of Operations. Rigorous control of variables is essential to a reliable determination of the design data sought. The two front seat positions and the two rear seat positions provide exposure redundancies that allow, within a given rear-end collision experiment, two conditions of motorist exposure to be evaluated for the front and back seat locations.

A discussion of how these experimental techniques fit into the broader plan of the twelve experiments will be deferred to a later section of the paper.

METHODOLOGY

The general experimental techniques identified by prior studies (1), (2), (3) and (4) were employed for this series, augmented and improved as required, to secure the needed data. A brief discussion of these rather standardized experimentation procedures will be followed by the special techniques peculiar to this collision series.

IMPACT CONFIGURATION - The car to be rear-ended was stationary with the face of its rear bumper positioned over the zero-foot marker of a polar grid painted on the pavement; both cars had the same directional heading, and the striking car squarely impacted the stationary car, with no offset. The monorail guide-track system provided directional control of the approaching striking vehicle; speed control was accomplished by use of a tow-cable from the striking car passing around sheaves and on to the towing vehicle. These control systems automatically disengage from the striking car prior to impact, Fig. 1.

VARIABLES UNDER STUDY - Five speeds, 10, 20, 30, 40 and 55 were used in this series; seven of the twelve experiments were conducted at 30 mph, two at 55 mph and one each at 10, 20 and 40 mph. Independent variables such as size of vehicle, degree of collision offset, directional alignment, braking versus nonbraking, etc., were held constant owing to their secondary importance. However, factors such as seatback strength, seatback height, head restraint height, no restraint versus head restraint, motorist size, posture and head offset from seatback typify variables under study, Fig. 2.

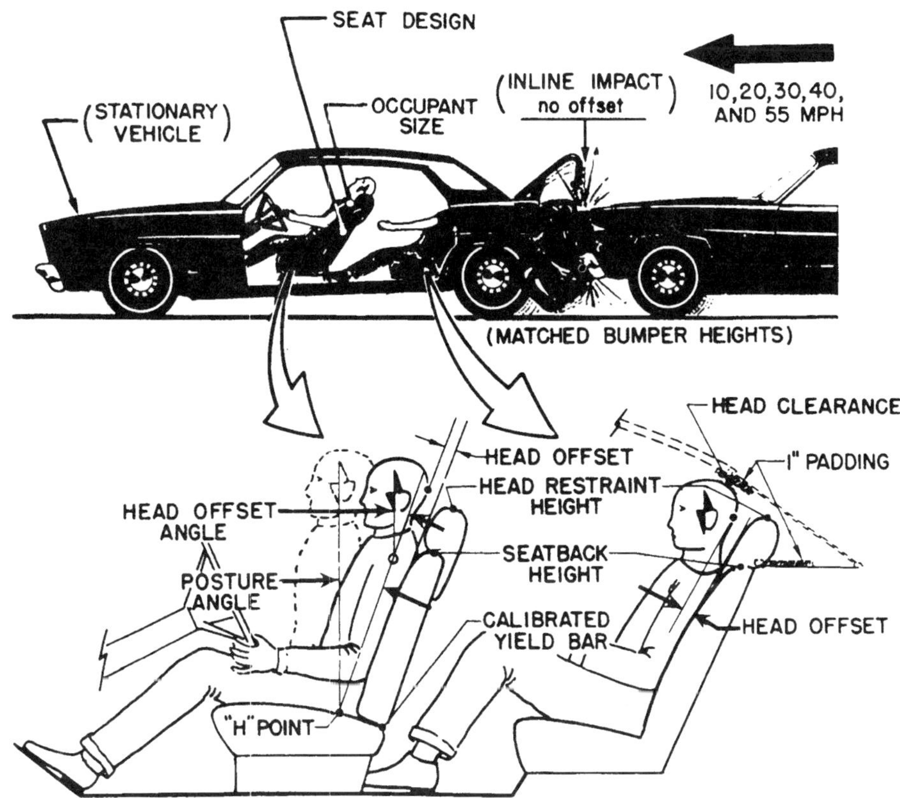

Fig. 2 - Variables under study

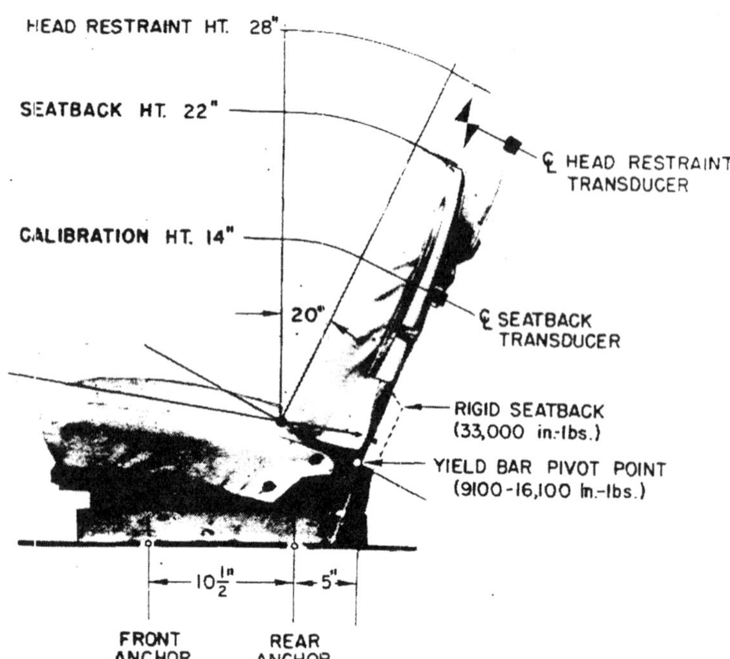

Fig. 3a - Specifications for the UCLA-Modified seat

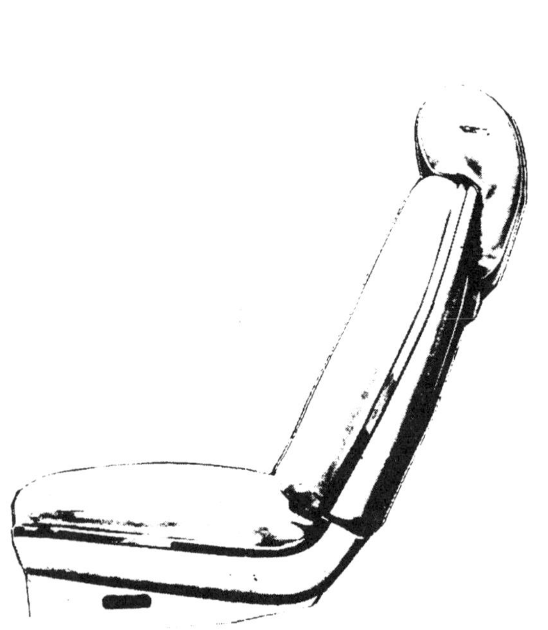

Fig. 3b - Ford production bucket seat with 6 in. head restraint

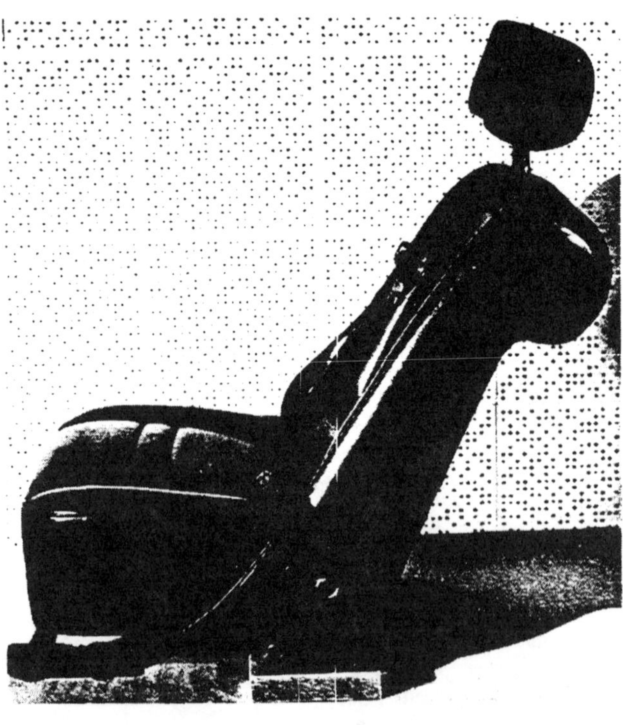

Fig. 3c - Cox production bucket seat with 8 in. head restraint

VEHICLE MODIFICATION - The struck car in these rear-end collision experiments was modified in the following manner:

Seat Arrangements - For all the collision experiments in this series, the front bucket seats were specially designed and constructed for evaluation of their relative protective qualities. These UCLA-modified seats were provided with structural properties that assured adequate performance reliability and a controlled degree of permanent yield, Fig. 3(a). Production seats were used for comparative purposes and were exposed to conditions severe enough to cause seat failure, Figs. 3(b), (c). Findings from static tests applied to seats within the car provided data useful in guiding seat design revisions; these laboratory studies assured the investigators that the seat design developed for these crash studies would not fail or otherwise adversely perform in a manner that would cause loss of valuable data. Additionally, some insight into the relative passenger protective properties of non-yielding seatbacks (rigid design) was gained from the initial several experiments by observing performance of the rigid seatback structure common to rear bench seats, as contrasted with the front seat yielding seatbacks. The especially favorable collision performance of these production rear seats having rigid seatback units provided the basis for designing the UCLA rigid bucket seatback for evaluation in connection with higher speed rear-end collisions. Within the range of inertial forces common to moderately severe rear-end collisions, the UCLA rigid bucket seat limits yield to a nominal amount, corresponding to the conventional concept of rigid structures. For further details on the rigid seat, the reader is referred to the section in the latter pages of this paper entitled "Calibrated Seat Design." Seating arrangement specifications for both the front and rear seats of the struck car for each experiment followed the specifications provided in the Schedule of Operations, Table 1.

Body Revisions - All doors, as well as the upper portion of each doorpost, were removed from the vehicle to be rear-ended to facilitate high-speed photography of occupant kinematics; compression struts were installed to restore structural integrity. The roof and head liner were cut out to improve vehicle interior lighting and to allow photography by overhead cameras. In some experiments the rear window and header were removed to evaluate the possible improvement resulting from increased passenger head clearance.

HUMAN SIMULATION - This entire series of experiments was conducted using anthropometric dummies to provide realistic and reproducible simulation of human response during the rear-end collision; these anthropometric dummies were specially constructed for rear-end collision studies. The dummy was selected as the research tool in preference to the use of cadavers, animals or volunteer subjects for the following reasons:

(a) The dummy allows for selection of precisely equivalent subjects, where two or more subjects of the same size and response characteristics are needed within a given experiment.

(b) The dummy provides exactly repeatable responses without degradation, so far as the exposures contemplated for this series of experiments was concerned.

(c) Prior studies by Stapp, Severy and others have established the reasonably faithful correspondence of forced responses of dummies to their counterparts, human subjects.

(d) Exposures contemplated were hazardous, making impractical the consideration of human volunteers.

(e) The subtle nature of some of the variables (head offset, posture, inertial forces, etc.) made it impractical to use animals having different component weight relations, response characteristics, etc.

Two 50th percentile adult male dummies (165 lbs.) were provided with specially engineered necks for use in these experiments along with four 95th percentile dummies (205 lbs.) also provided with the same special necks, Fig. 4. There are two common procedures used for simulating neck flexion resistance in dummies: The neck resistance to involuntary forced movement is simulated by including relatively high coefficient of friction surfaces between vertebral segments and spacers of simulated discs (ball and socket design) or by use of a spring loaded spinal column (cable-disc design) that resists rearward movement at an adjusted rate, Fig. 4. The UCLA group continued to use the latter because it most closely simulates the human response and because it provides the very important rebound response characteristics for various protective head restraint systems.

Articulation ranges for the dummies used in these experiments are provided by Table 2. The Sierra 292-850 was specifically designed for automobile collision research. The dummy's skeletal structure is enclosed by polyurethane foam flesh and a skin made of vinyl plastisol. The foam flesh and vinyl skin approximate the resilience of human flesh yet give a maximum resistance to tearing or puncturing. Access to the skeletal

Table 2 - Human Simulation Flexion Movement Chart

PARTS MOVED	MOVEMENT	ANGLE OF MOVEMENT	
		50th Percentile Adult Male 292-850	95th Percentile Adult Male 292-95
Head - Relative to Trunk	Ventriflexion (Forward)	90°	90°
	Dorsiflexion (Backward)	90°	90°
	Lateroflexion (To either side)	60°	40°
	Rotation (To either side)	70°	70°
Shoulder	Cone of Rotation	40°	40°
Arm - at Shoulder	Flexion (Forward)	180°	180°
	Extension (Backward)	60°	60°
	Abduction	135°	135°
	Humeral Rotation (To either side)	90°	90°
Forearm - at Elbow Pivot	Flexion	145°	145°
Forearm - w/Hand	Pronation & Supination	90°	90°
Hand at Wrist	Flexion (Forward)	90°	90°
	Extension (Dorsal)	90°	60°
Fingers		PASSIVE FORMABLE	
Thigh - at Hips	Flexion (Forward)	145°	145°
	Extension (Backward)	45°	45°
	Abduction	70°	70°
	Rotation (To either side)	50°	50°
Leg - at Knee	Flexion	135°	135°
Foot - at Ankle	Plantarflexion	75°	75°
	Dorsiflexion	30°	30°
	Med. - Lat. Toe Swing	30°	30° Total
	Med. - Lat. Angle Flexion	30°	30° Total
Torso & Shoulders w/respect to Pelvis	Rotation (To either side)	50°	50°
	Forward Bend	60°	45°
	Backward Bend	30°	25°
	Lateral (To either side)	55°	50°

*Sierra Engineering Company, Sierra Madre, California

structure is simplified by a zipper on each side of the torso. These dummies represent an advancement over the older model dummies. Unusual creativity is typified by the entirely new design of the neck, shoulder, chest, and pelvic areas. However, this Sierra 292-850 design was modified to the full cable-disc neck in order to provide 90 degrees of controlled head rotation and to realistically coincide with the forced limits of head excursion for the human, Fig. 5. Other features of articulation and spinal column design for the 295-95 corresponded essentially with the 292-850 and both designs had identical necks.

Dummy Size To Seatback Height Relationships - The two dummy sizes (50th and 95th percentile) were used with six seatback heights (21, 22, 25, 27, 28, 29). The relationship between these dummy and seatback size variations is illustrated by Figs. 6(a) and 6(b).

Less important than the relative height of the seatback to its occupant is the proximity of the occupant's head and torso to the surface of the seatback. To gain specific knowledge of performance variations according to the extent of head offset, a range of offsets, both less than and more than natural postural relationships, were evaluated under actual collision exposures. The specific head offsets referenced in Table 1 are further clarified by referring to Fig. 7. It was found that the head offset, as measured from the back (occipital) surface of the head perpendicularly to the plane of the seatback, for the values stated in Fig. 7 accounted for torso and head-to-torso angles as stated. The anthropometric dummy did not have the degree of "S-curvature" of the spine normally found with human subjects. When naturally postured, the dummy developed a 3-inch head offset under conditions for which human subjects normally measured 6 inches. However, the dummy head offset was adjusted to cover the full range of 0, 3, 6 and 12 inch offsets for purposes of studying human responses to rear-end collision forces as a function of head location to backrest.

OCCUPANT ASSIGNMENT - Four identical adult 95th percentile dummies having realistically engineered neck articulation and seated, two in the front seats, two in the rear seat, with pre-assigned postures, were used in Experiments 93 through 98. Some 50th percentile adult dummies were used in Experiments 99 through 104, as reflected in Table 1.

DATA COLLECTION - The collision data from transducers in dummies and in the rear-ended vehicle were transmitted via a

Fig. 4 - Neck articulation designed to UCLA specifications by Sierra Engineering

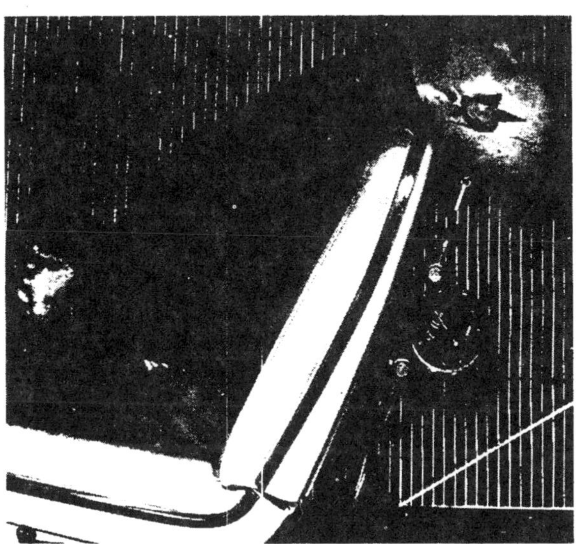

Fig. 5 - Static response characteristics of UCLA special neck

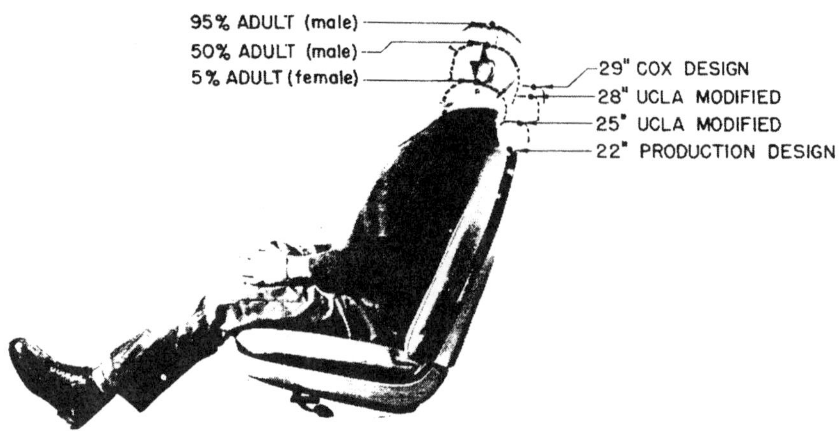

Fig. 6a - Anthropometric dummy and bucket seatback sizes evaluated in these UCLA experiments

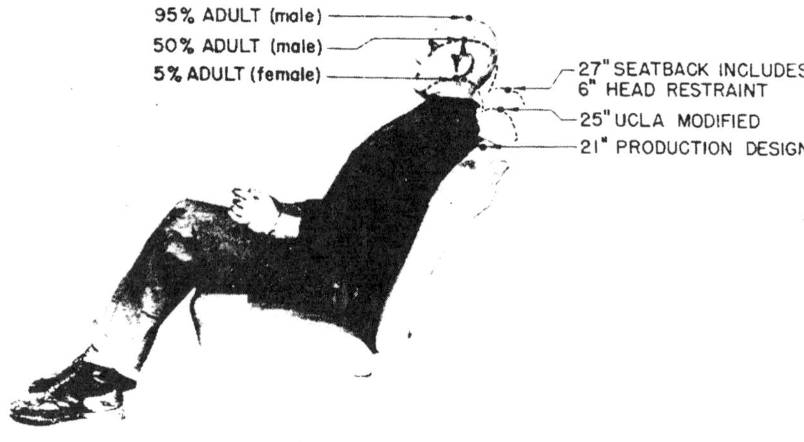

Fig. 6b - Anthropometric dummy and bench seatback sizes evaluated in these UCLA experiments

Fig. 6c - Standing size variations of the near-extremes and intermediate size motorists

100-foot cable to recording oscillographs located in the mobile electronics shop. Photographic instrumentation supporting electronic data collection will be described in a subsequent section.

INSTRUMENTATION

The subtle nature of many of the variables under study for these rear-end collision experiments made it imperative that many exacting measurements be made during the split-second duration of the collision event. Electronic transducers and high speed motion picture cameras represent the principal systems used to obtain data. These and related instrumentation are listed in Summary of Instrumentation, Table 3. Other data collected include pre- and post-crash observations concerning the location of instrumentation, head supports, occupants and their restraints, as well as the condition of vehicle lights and rear windows. Specifics were recorded concerning the collision deformation sustained, nature and location of debris, inferred

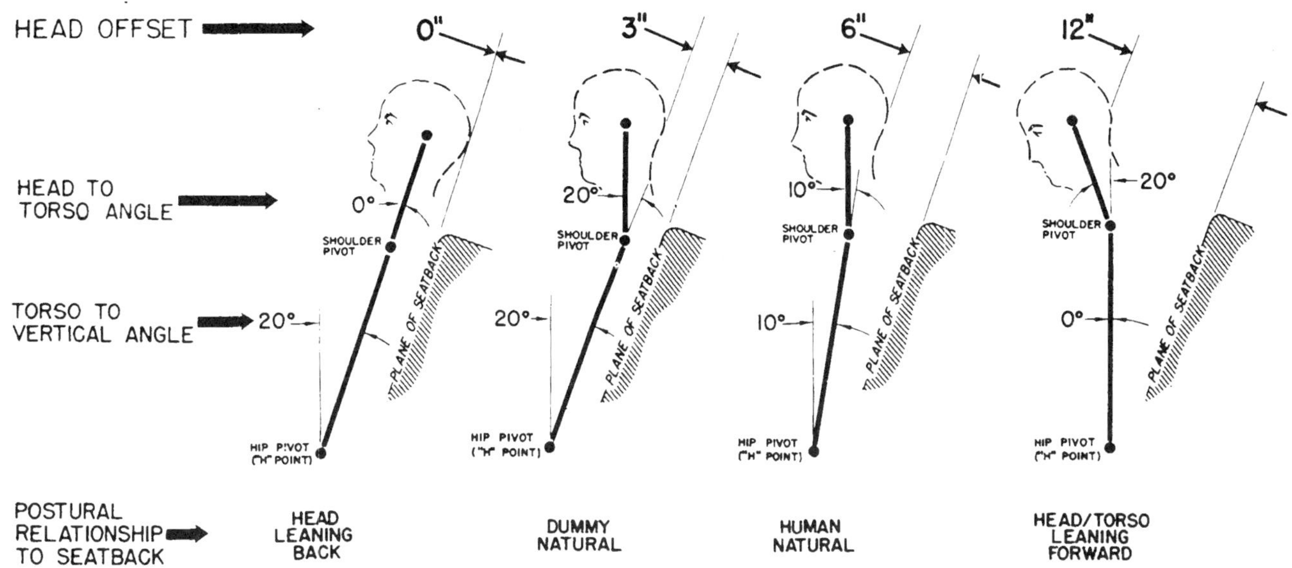

Fig. 7 - Head offsets, and their corresponding neck and back postural relationships to the seatback

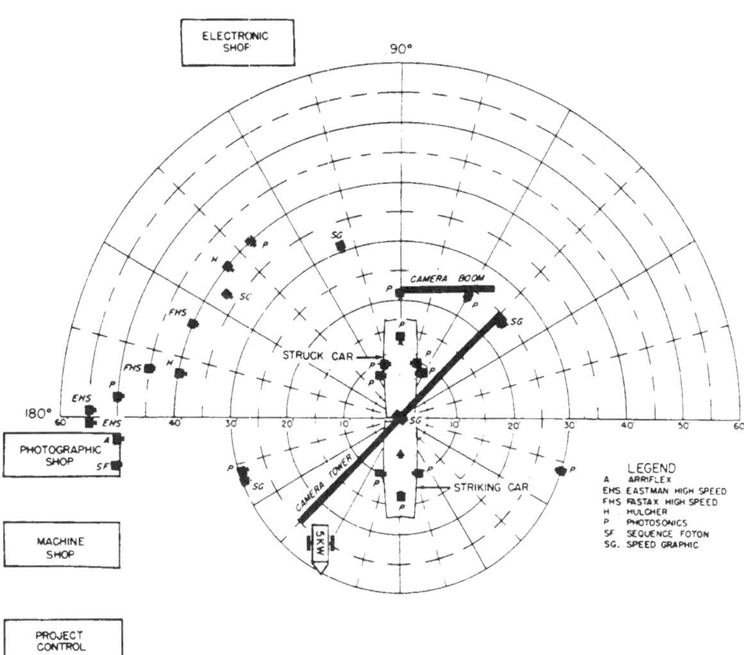

Fig. 8 - Camera assignments

occupant injuries and permanent deformation of seatbacks.

Photographic instrumentation ranged from conventional motion picture cameras to high-G tolerant, moderate-speed motion picture cameras and from automatically triggered, electronically timed, still-cameras to high speed motion picture cameras. The moderate and high speed motion picture cameras provided displacement data of the automobile and seats as well as the corresponding occupant kinematics. Standard speed motion picture cameras were primarily used to provide documentary film coverage. Twenty-seven of the more than thirty photographic devices are shown in Fig. 8.

Dual electronic instrumentation systems were installed in the dummy occupants of the struck car to assure data retrieval and reliability; they included dual biaxial accelerometer units in the heads and chests of all occupants. Other accelerometers were mounted on the car floor pan adjacent to the center doorpost and at several locations on the seats. Displacement transducers measured the rearward deflection of front seatbacks and force transducers monitored the loads applied to the restraint systems. The first four experiments established that passenger seat belt loadings were inconsequential for the rear-end collision exposure; however, it was established that use of a seat belt with a properly designed seatback reduced injury exposure. Other electronic and special instrumentation provided synchronization between photographic and transducer systems as well as other useful information such as the vehicle speed at impact.

Referring to the legend of Fig. 9, the letter designation relates to the structure or device to which transducers were applied and the subscript indicates the number of sensing devices at that location. For example, in some applica-

Table 3 - Summary of Instrumentation

DEVICE	TO PROVIDE	LOCATION	SPECIFICATIONS
High Speed Motion Picture Cameras	Vehicle collision dynamics and passenger kinematics	See impact center camera assignment, prior figure	2-Eastman, High speed; 2-Fastax WF-3; 16mm, 600-1700 frames/sec.
Moderately High-speed Cameras	Vehicle collision dynamics and passenger kinematics	See impact center and vehicle camera assignments prior figures	14-Photosonics, 1B, 1-GSAP MBH 200-16; 200 f/s 16mm
Standard Motion Picture Cameras	General photographic coverage	Adjacent to impact center	1-K100, 16mm; 1-Arriflex, 16mm; 1-Bolex, 16mm; 1-GSAP, 16mm
Special Motion Picture Cameras	Sequential, large format photographs of collision events	Strategic positions at impact center	2-Hulcher, Model 102, 20 f/s 70mm (Camera rotated 90 deg. to permit cine-reduction); 1-Bell & Howell Eyemo, 48 f/s 35mm; 1-Foton 5 frames/sec 35mm Bell & Howell
Still Cameras	Precision-timed photographs	See impact center camera assignment, prior figure	5-Super Speed 4x5" Graflex 1/1000 sec.
Calibrated References	Fixed ground reference points for micro-motion analysis	Near the impact center	Reference markers at five foot intervals and vertical posts calibrated yellow and black at alternate one-foot increments
Reference Targets	Precision photographic references for micromotion analysis	At strategic positions on the vehicles and occupants	Diamond shaped yellow and black targets
Electrical Accelerometers	Acceleration measurements	Head, chest and hips of occupants. Vehicle body and seat structure	B & F-LF 50-50 and LF 100-75-120; Statham A 38a-60-350, A6-100-350, AJ43-100-350, A52-50-300, A5a-100-300, Ca-100-130; CEC 4-202-0001
Seat Belt Tensiometers	Measurement of belt loads	On belt webbing	Dual link and thread-through type, special UCLA construction
Recording Oscillograph	Amplitude-time records of transducer signals	Carried by instruments recording vehicles for striking car and instrumentation trailer for struck car	2-18 Channel Consolidated Type 5-114 P2; 1-50 Channel CEC, Type 5-119, P3; 1-12 Channel CEC, Type 5-118
Electronic Time Delays	Precision timing for still photography	Electric pressure pads near impact center	100, 200 and 500 millisec solid state time delay units built by UCLA
Pulse Generators	Timing for high and moderate speed cameras	Near cameras	Wollensak, 100 cps; 1000 cps and special 100 cps solid state devices
Photographic Oscillographic Synchronization Units	Zero time (vehicle contact); flash bulb for film and pulse for oscillograph	Pressure switches between contacting surfaces of vehicles. Photo-cell and flash bulb on vehicle	SM flash bulb with photocell for oscillograph signal
Flash Bulb Timer	Timing at 0,50,100,150, 200,250 and 500 ms; flash bulb for film and pulse for oscillograph	Near impact center in view of all cameras	Solid state timing device with SM flash bulbs and photocell for oscillograph signal
Auxiliary Timer	Backup timing	Near impact center in view of all cameras	Rotating yellow-black drum; constant speed 1740 rpm motor
Fifth Wheel	Accurate impact velocity	On tow car	Labeco, Model 5101, Weston Scale 0-60, 0-120 mph
Speed Counter	Car speed data; time-displacement	On guide yoke	Induction pickup for oscillograph
Polar Coordinate Grid	Position data for vehicles from point of impact to positions of rest	On asphalt surface at impact center	Yellow traffic marker paint

tions where biaxial accelerometers were installed, a "2" subscript will show. The back-up instrumentation for critical areas not only provided protection against loss of data but substantiated the reproducibility of findings.

Experience has established (2) that even the certified calibration cards accompanying new and reconditioned accelerometers may have errors of substantial magnitude; this statement is based on several inaccurate certifications from more than one manufacturer. Less dramatic errors or variances occur as a result of subtle differences in the location, natural frequency of the instrument and its mounting bracket; the data reduction, analytical and computer treatments leading to the final transducer curve may also provide avenues where errors can occur.

For this study, the transducers were calibrated using recognized dynamic test procedures accounting for the result that most of the back-up instrumentation data became nearly superimposed when plotted to the same scale; where readings showed a significant error, examination of the two parallel data sources provided a method of determining and eliminating the source of error. In cases where minor differences existed but where each transducer cluster was found to be faithfully reporting its input, graphical averaging techniques were used to provide an accelerometer resultant curve representative of the average of the two adjacent cluster locations, Fig. 10.

ANALYSIS AND FINDINGS

Seven of the twelve rear-end collision experiments reported by this paper were at thirty mph; this impact speed was used as a constant in the majority of the experiments to provide a reference system for comparison of the several independent variables under study. Three of the remaining collision experiments were conducted at speeds of 10, 20, and 40 mph and the last two experiments were conducted at 55 mph. In general, the findings for each experiment will include the following information:

(a) Vehicle collision performance and peak acceleration.

(b) Passenger kinematics and peak accelerations.

(c) "Whiplash" protection pro-

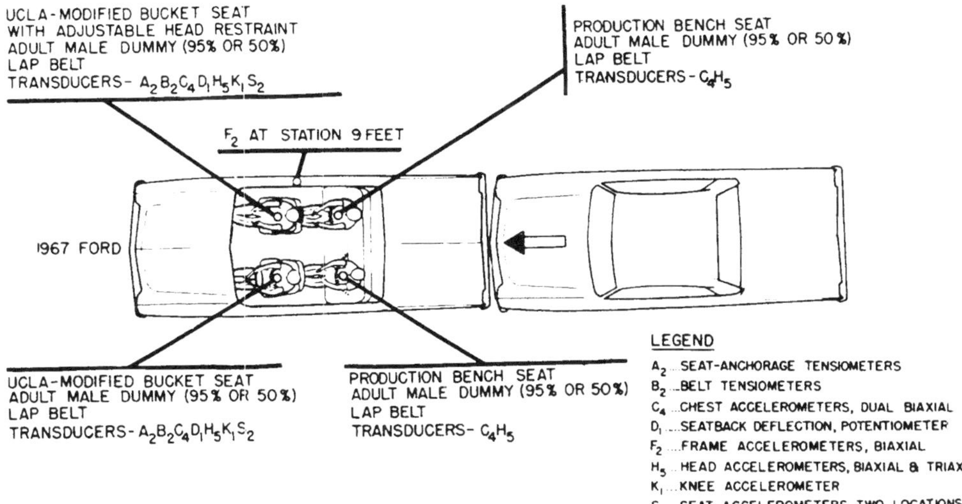

Fig. 9 - Typical instrumentation assignment; example of occupant/seat schedule

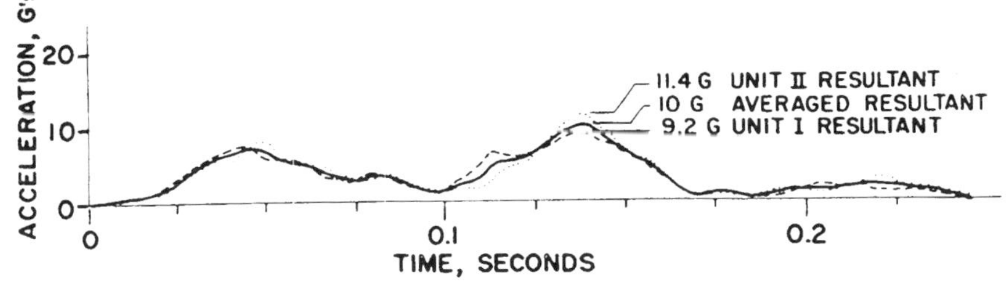

Fig. 10 - Graphical averaging technique, typical resultant curve (Exp. 93, right rear passenger chest)

vided by various seatback heights and strengths as well as related design innovations.

(d) Transducer patterns of occupants, vehicles, restraint systems, head restraints and seat structures.

(e) Description of post-crash observations, including nature and extent of any fuel tank leakage, extent of sheet metal collapse interference with wheel rotation for post-crash roll-out, vehicle path and range following impact and similar observations.

EXPERIMENT 93 - was a 30 mph rear-end collision between two identical 1967 Ford 4-door sedans, Fig. 11. For this report, however, only the findings relating to the rear-ended vehicle will be presented. The two front seat occupants in the struck car provided a contrast in

Fig. 11 - Rear-end collision at 30 mph between two identical 1967 Ford 4-door sedans

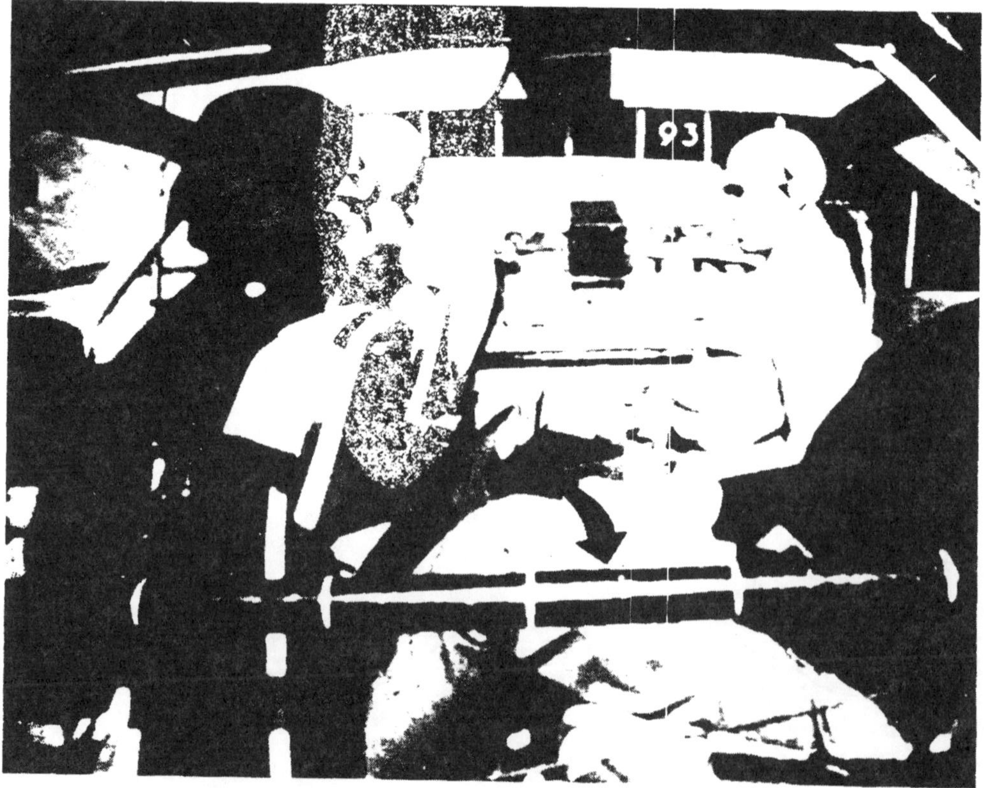

Fig. 12 - Passenger with head restraint; driver's head unrestrained; note compression strut (arrow)

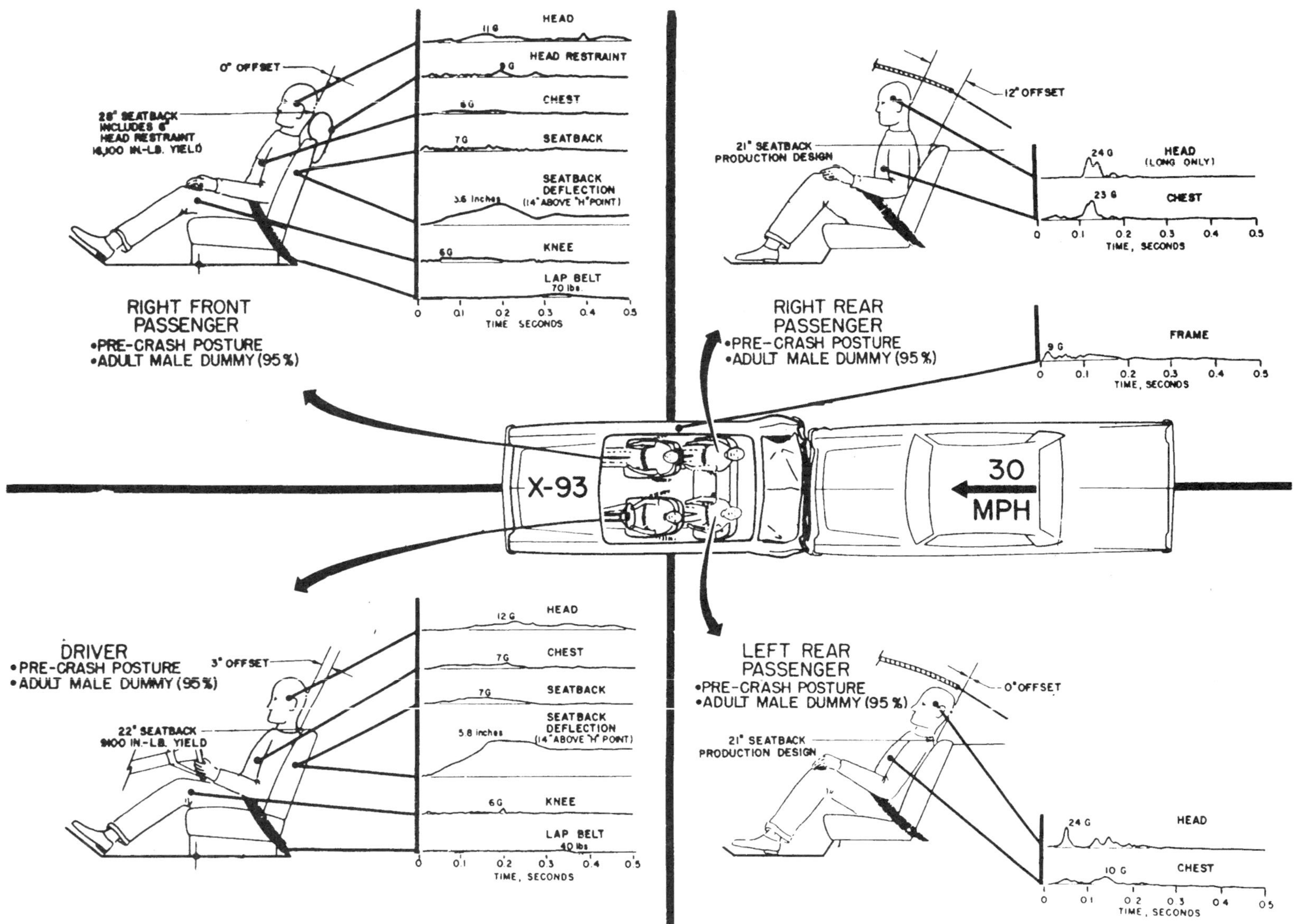

Fig. 13 - Transducer patterns shown with occupant posture and restraint condition at start of impact. Exp. 93

collision injury exposure because a head restraint was provided for the passenger, and the driver's head was unsupported, Fig. 12; each was seated on a bucket seat with a calibrated yield seatback. The two rear seat passengers, although both were seated on a production bench seat with a rigid seatback, responded differently because the left rear passenger was sitting with a zero-inch head offset from the 24 degree seatback, and the right rear passenger was positioned leaning forward, with his head offset 12 inches from the seatback. All four occupants in this struck car were identical anthropometric dummies of the 95th percentile adult male category, each weighing 205 lbs. Their necks were calibrated to resist movement, except for accelerations exceeding 1½ G, corresponding to passive neck muscle tone; each occupant wore a lap belt. The gross weight of the struck car was 4580 lbs. including the four dummy motorists, cameras and outriggers, and 25 gallons of water, colored red to simulate gasoline. This vehicle was struck by an identical vehicle whose weight grossed 4660 lbs. The doors of the struck car were removed and the upper half of the doorpost was cut out to facilitate observation of motorist kinematics. The structural integrity of the passenger compartment was maintained by the installation of horizontal compression struts, Fig. 12.

The Collision - The striking car advanced 1½ feet after the bumpers contacted (through vehicle mutual crushing) without the struck car initiating any advancement. The inertia of the rear-ended car and its rather collapsible rear-end structure account for the fact that this struck vehicle had not commenced to be displaced any significant distance until after its rear section was compressed to the back edges of the rear wheels corresponding to 3 feet of collapse and a peak resultant frame acceleration of 9 G at 15 ms. The peak frame acceleration (9 G at 15 ms) almost coincidental with bumper-to-bumper contact (zero ms) is accounted for because the accelerometer was mounted adjacent to the frame directly in line with the impacting bumpers; additionally, the acceleration build-up did not develop beyond 9 G because of the readily collapsible rear-end structure, but the acceleration again returned to nearly 9 G for a period from 85 ms to 155 ms after bumper contact, Fig. 13. This maximum collapse to the rear-ended vehicle occurred when the striking car had advanced only 4.2 feet beyond the ground reference position of contact between the two colliding vehicles, Fig. 14. Thereafter, the rear-ended vehicle accelerated rapidly and the two cars broke contact with one another after the striking car had advanced 10 feet (350 ms) from zero reference line, Fig. 15.

Before the two cars separated, however, a liberal diffusion of gasoline (simulated) sprayed from the struck car's ruptured fuel tank, Fig. 16.

Driver - A 1967 production bucket seat (22 inch seatback height) was modified in terms of its structural resistance to collision forces and in a manner that localized bending without changing its size and appearance. The seatback strength, calibrated at 14 inches above the junction of the surfaces of the cushion and seatback, was designed to resist 9100 inch-pounds* with localized plastic deflection at the yield bars located at the base of the seatback adjacent to the "H" point, Fig. 3(a). The purpose for redesigning the seatback was to provide an objective basis for comparison of the influence of seatback height on the rear-ended motorist's forced postural changes; this will be discussed in greater detail in the Calibrated Seat Design Section. Because production seatbacks, similar in height to the UCLA-modified 22-inch seatback, provide no head restraint, they are generally less injury-producing during rear-end collisions when the seatback strength is low; however, it is not possible to judge the improvement afforded by adding a head restraint, if the seatback fails rearward before the head support can be brought into full effective performance. In addition, a rear-ended driver in a forced reclined position is in no position to steer and apply brakes, as may be required to avoid a secondary and possibly more serious collision. Ac-

*9,100 in.-lbs. was calculated for 650 lbs. to be applied to the seatback, 14 inches above the plane of the seat, in a rearward direction.

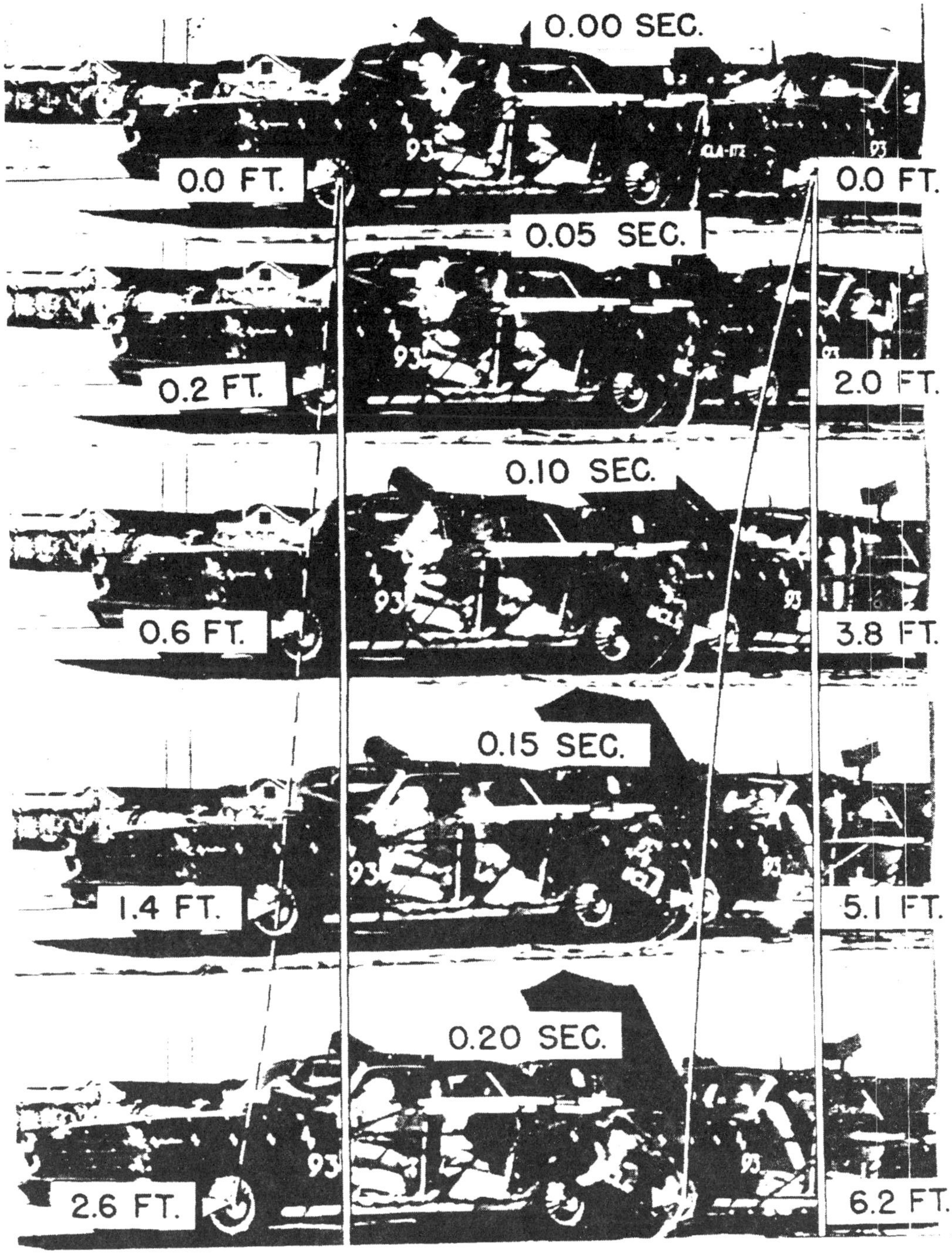

Fig. 14 - Rear-ended car sustains maximum collapse when striking car had advanced 4.2 ft

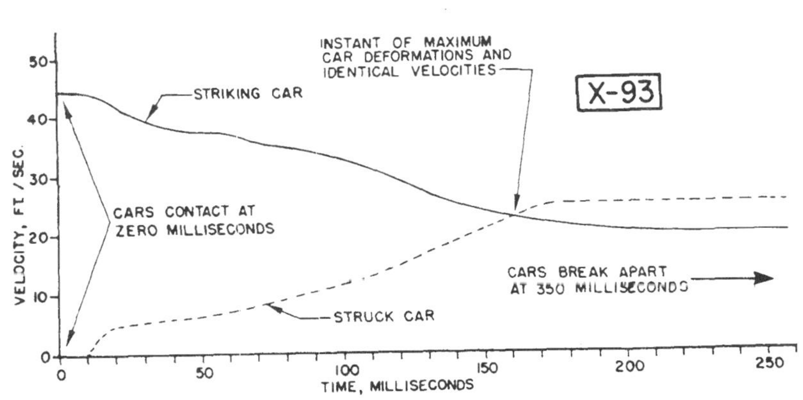

Fig. 15 - Velocity changes, during a 30 mph rear-end collision; cars separate at 350 ms

Fig. 16 - Simulated gasoline sprays from collapsing tank of rear-ended passenger vehicle, Exp. 93

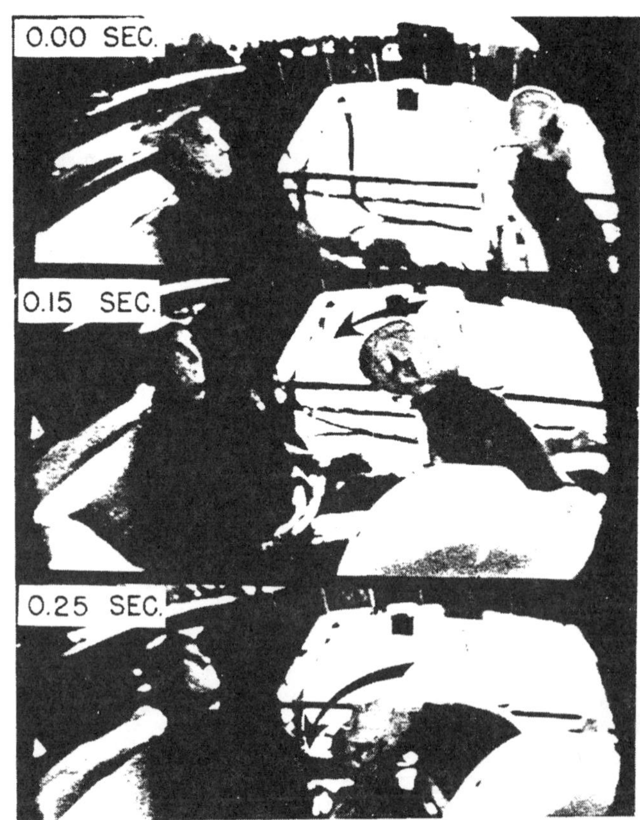

Fig. 17 - Extreme whiplash, no head support, Exp. 93

cordingly, the driver's 9100 in.-lb., 22-inch high seatback is compared in this Experiment 93 with the front seat passenger's 28-inch seatback (including 6-inch head restraint) which was further strengthened to 16,100 in.-lbs. in order to accommodate the increased bending moment caused by the higher seatback.

As the collision progressed to a quarter second after contact, the driver was severely whiplashed, registering a resultant acceleration of 12 G at 220 ms, Fig. 17. At the limit of forced excursion, the driver's head had flailed rearward relative to his shoulders, and rotated backward and downward a total of 120 degrees, or 30 degrees beyond the position at which his eyes would be looking vertically at the roof (78 degrees rearward, relative to torso axis). The 27 degree deflection of the driver's seatback (47 degrees rearward of vertical) forced the accelerometer beam on the seatback, at the 14-inch level, against the tibia of the left rear passenger. Rebound from the acutely dorsi-flexed neck posture was not excessive or sufficient to cause head-to-steering-wheel impact injuries. During the driver's collision acceleration, he sustained a peak chest acceleration of 7 G at 205 ms, a knee acceleration of

6 G at 195 ms and a lap belt load of 40 lbs. at 325 ms. His 9,100 in.-lb., 22 inch seatback, accelerated to a peak of 7 G at 140 ms, sustaining a 5.8 inch rearward deflection at 175 ms, both measured at the 14 inch reference position.

Right Front Seat Passenger - As with the driver, this lap-belted right front passenger was seated in a UCLA-modified bucket seat except that it had a factory installed prototype head restraint. This head restraint installation increased the seatback height from 22 inches to 28 inches. Owing to the increase in bending moment caused by the addition of a passenger's head restraint, the strength of the seatback was increased from 9100 in.-lbs. to 16,100 in.-lbs. with equivalent loading of 1150 lbs. at 14 inches. The head restraint for the right front seat passenger, mounted on the strengthened seatback, provided good resistance to whiplash, Fig. 18. The 16,100 in.-lb. seatback strength resisted excessive rearward inertial deflection accounting for the passenger maintaining an almost normal posture throughout the collision. His head rotated only 65 degrees (31 degrees rearward of torso axis) less than half that of the driver. The maximum head acceleration obtained by the right front passenger was 11 G at 150 ms; his chest and knee both reached a 6 G peak acceleration at 100 and 70 ms, respectively. The seatback, as measured 14 inches above the seat cushion, registered 7 G at 90 ms and the head restraint, 9 G at 190 ms. The seatback deflected rearward 3.6 inches (190 ms) at the 14 inch calibration reference position, representing a total deflection of 16 degrees (post-yield of 7 degrees). Following the forced rearward displacement, the right front passenger's head and torso rebounded forward 18 inches, as measured at his head but was not pitched forceably forward or sufficiently to contact the forward structure.

The Right Rear Passenger - the passengers in the rear seat of the struck car were seated on a conventional bench seat with a standard 21 inch seatback. They were identical 95th percentile anthropometric dummies and the only factor making their exposure different was the fact that the right rear seat passenger was postured leaning with his head offset 12 inches forward of the seatback; the left rear passenger's head was positioned with zero-inch offset. The more erect posture of the right rear pas-

Fig. 18 - Head support in conjunction with strengthened seatback provides good resistance to whiplash during 30 mph collision

senger's 12 inch offset is shown by Fig. 19. The right rear passenger was flailed rearward against the seatback by the rear-ending impact; the top of his head glanced from the header and struck the rear window. Then he sunk into his seat, developing the head clearance that would have missed the header, had he not been sitting so erect at the start of the impact, Fig. 19. He sustained a 24 G head acceleration at 125 ms and a 23 G chest acceleration also at 125 ms. These accelerations were twice the level sustained by the front seat occupants, partly attributed to the closer proximity of rear seat passengers to the position of impact and partly owing to the more rigid seatback construction common to the rear seats of passenger vehicles.

The Left Rear Passenger - The left rear passenger did not sustain a whiplash because his head and torso remained approximately in a normal posture as a result of his head striking the sloping rear window glass, Fig. 17. This wedge-like impact could develop high vertebral compressive forces but the dummy was not instrumented to provide this data. He sustained a 24 G head impact with the sloping rear window glass at 50 ms; his chest acceleration was only 10 G at 140 ms. This relatively low chest acceleration indicates the value of having the torso resting against the seatback for the rear seat position at the start of a rear-end collision. Additionally, the respective time differences of 50 and 140 ms for the head and chest acceleration peaks are explained by the observations that the head initially struck the rear window glass before the chest had crushed all the way into the seatback, Fig. 13.

Observations, Following Collision - The rear-ended vehicle was propelled 140 feet with a final offset to its left from its pre-crash heading of only two feet. Permanent deformation was 3.0 feet for the rear fender-trunk body section above the bumper level; 2.5 feet of permanent collapse was measured at the bumper level. Gasoline (simulated) was liberally sprayed and deposited on the pavement over the entire 140 feet of struck car forced roll-out, except for the initial 10 feet from point of impact. This trail of fluid was also accompanied by a liberal deposit of broken glass, metal parts and other typical collision debris. The driver's seatback sustained a permanent yield of 19 degrees (rearward) as contrasted by the passenger's seatback of 7 degrees, Fig. 20.

EXPERIMENT 94 - was a rear-end collision at 20 mph involving two identical 1967 Ford 4-door sedans, Fig. 21. Except for identifying type, weight, speed and similar factors affecting the rear-ended vehicle, information relating to the striking car will be the subject of a subsequent paper. For the struck car, the dummy assignment and the seat arrangement were identical to conditions for Experiment 93. This arrange-

Fig. 19 - Right rear passenger, Exp. 93, positioned with a 12 in. head offset

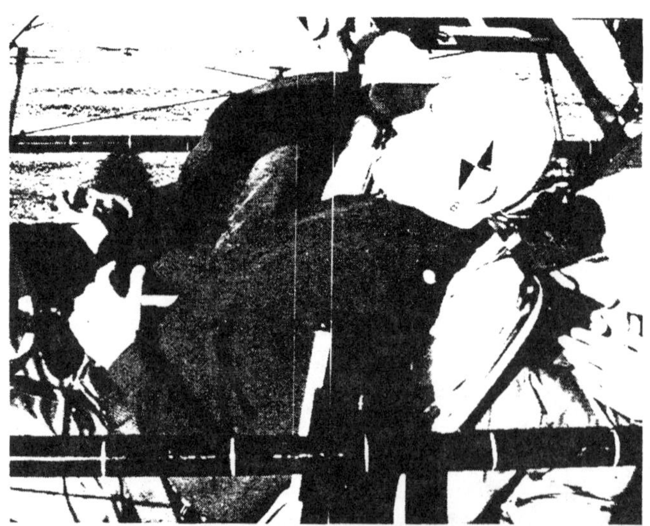

Fig. 20 - Seatback deformation following a 30 mph rear-end collision; driver's seatback displacement was excessive, Exp. 93

ment provided a direct comparison between exposures to rear-end collision injuries by motorists for two of the five speeds of impact. The occupants were four identical, 95th percentile anthropometric dummies (205 lbs. gross), each with special articulated necks that were calibrated to resist movement, except for accelerations exceeding 1½ G, corresponding to passive neck muscle tone; they were positioned two in front and two in the rear seat, Fig. 22. The two front seat occupants provided a contrast in collision injury exposure because the passenger was provided with a head restraint and the driver had no head restraint; each was seated on a bucket seat with a calibrated-yield seatback. The two rear seat passengers, although seated on the same production bench seat with a rigid seatback, responded differently because the right rear was positioned with his head leaning 12 inches forward of the seatback plane, and the left rear was positioned with a 3 inch head offset. All four adult motorist simulations in this struck car had lap belts although the right rear occupant's belt was not fastened at time of collision. The gross weight of the struck car was 4660 lbs., including the four dummy motorists, cameras and outriggers, 25 gallons of water, colored red to simulate gasoline. This vehicle was struck by an identical vehicle whose weight grossed 4700 lbs. The doors of the struck car were removed and the upper half of the doorpost was cut out to facilitate observation of motorist kinematics. The structural integrity of the passenger compartment was maintained by the installation of horizontal compression struts, as previously described.

The Collision - After the bumpers contacted, the striking car advanced 1.2 feet through mutual vehicle crushing without the struck car initiating any advancement. As in the 30 mph impact of Experiment 93, the inertia of the struck car combined with its rather collapsible rear-end structure accounts for the fact that this vehicle had not commenced to be displaced any significant distance until after its rear-end had been compressed 12 inches. A peak resultant frame acceleration of 8 G at 35 ms was recorded, Fig. 22. This observation may be contrasted with a 3 foot rear-end collapse and 9 G at 15 ms, frame acceleration for the 30 mph impact of Experiment 93, struck car. The peak acceleration of 8 G occurred only 35 ms after bumper-to-bumper contact because the accelerometer was mounted adjacent to the frame, in line with the impacting bumpers; additionally, the build-up of acceleration did not develop beyond 8 G because of the readily collapsible rear-end structure. The excellent force modulating properties of this rear-end structure is indicated by the increase of only 1 G peak acceleration for an increase of ten mph impact speed, even though the striking car kinetic energy represented was more than doubled. This maximum collapse to the vehicle rear-end occurred when the striking car had advanced only 4 feet beyond the ground reference position at which contact had

Fig. 21 - Rear-end collision, 20 mph, Exp. 94

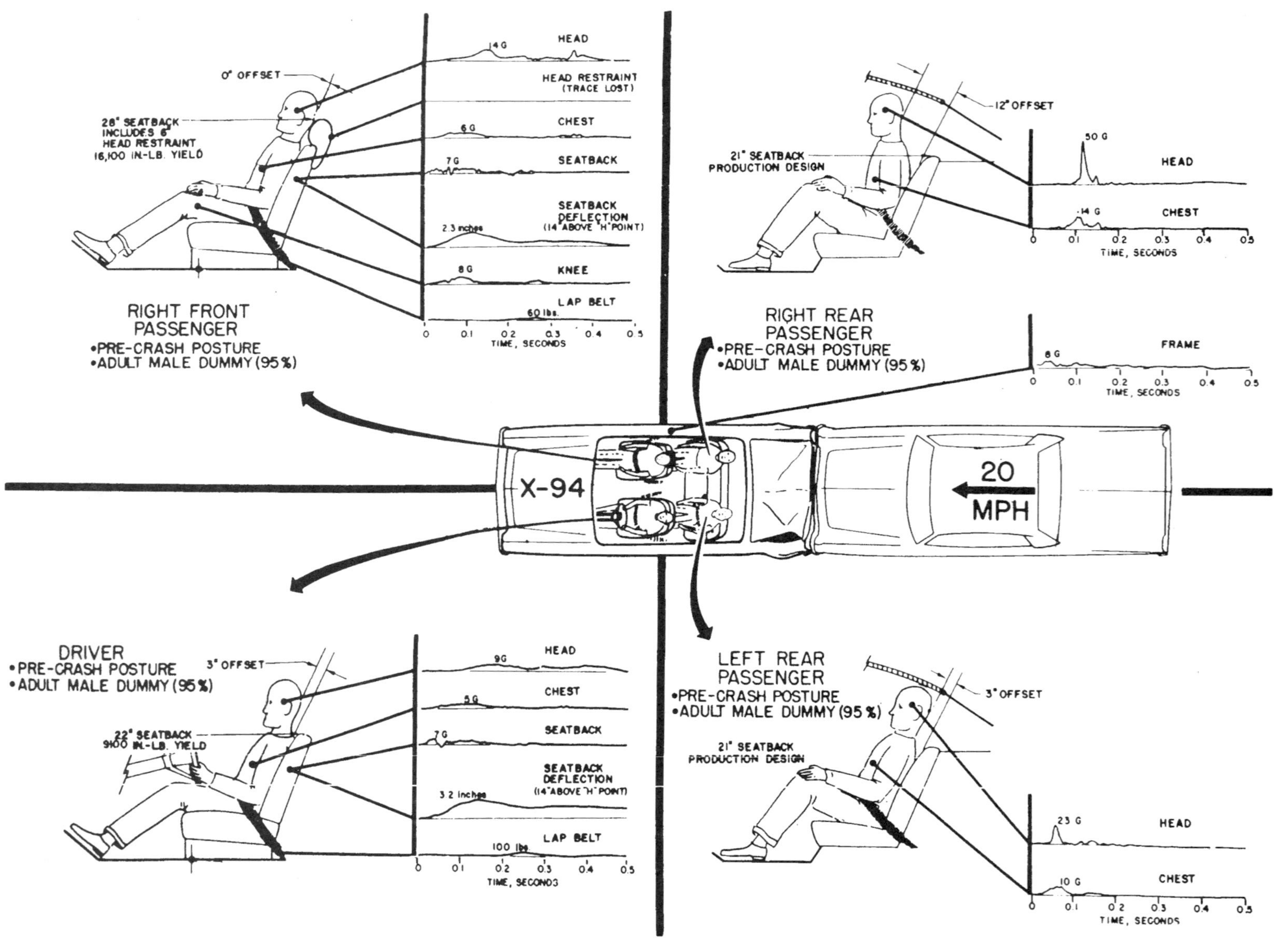

Fig. 22 - Transducer patterns shown with occupant posture and restraint condition at start of impact, Exp. 94

occurred between the two colliding vehicles, Fig. 23. At this instant, the combined or mutual collapse of the two vehicles was 16 inches, Fig. 24. Thereafter, the rear-ended vehicle acclerated rapidly and the two cars broke contact with one another after their contacting surfaces had advanced 6 feet, at 260 ms, Fig. 25. This 20 mph data is in substantial contrast to the 10 foot advance before break-away for the 30 mph rear-ender of Experiment 93.

As the two cars separated, there appeared a liberal diffusion of simulated gasoline sprayed from the struck car's damaged fuel tank, as was also observed in connection with Experiment 93.

Driver - A production 1967 Ford bucket seat (22 inch seatback height) was modified in terms of its structural resistance to collision forces, without changing its size and appearance. The seatback strength, calibrated 14 inches above the hip pivot point, was structured to a 650 lb. loading. The purpose of this structural modification for the seatback was to restrict yield to a predetermined location that allowed objective comparison of the influence of seatback height on the rear-ended motorist's forced postural changes as a function of impact speed. In this 20 mph collision, the driver's 9,100 in.-lb. (650 lbs. at 14 inches), 22 inch seatback is compared with the front seat passenger's 28 inch seatback and head combination, strengthened to 16,100 in.-lb. (1150 lbs. at 14 inches) to accommodate the increased bending moment occasioned by the higher back support. The seatback strength values, refer to their strength as calibrated at the 14 inch level above the junction of the seat cushion and seatback surfaces.

During the first two-tenths second of the collision, the driver was rather severely whiplashed, Fig. 26, registering a resultant head peak acceleration of 9 G at 165 ms, Fig. 22. At the limit of forced excursion, the driver's head had flailed rearward, relative to his shoulders, and rotated backward and downward 105 degrees -- 15 degrees be-

Fig. 23 - Maximum collapse occurred when striking car had advanced only four feet

yond the position at which his eyes would be looking vertically at the roof (74 degrees rearward of torso axis). Seatback deflection of 14 degrees was less than the 27 degrees for Experiment 93 (30 mph) leaving a space of eight

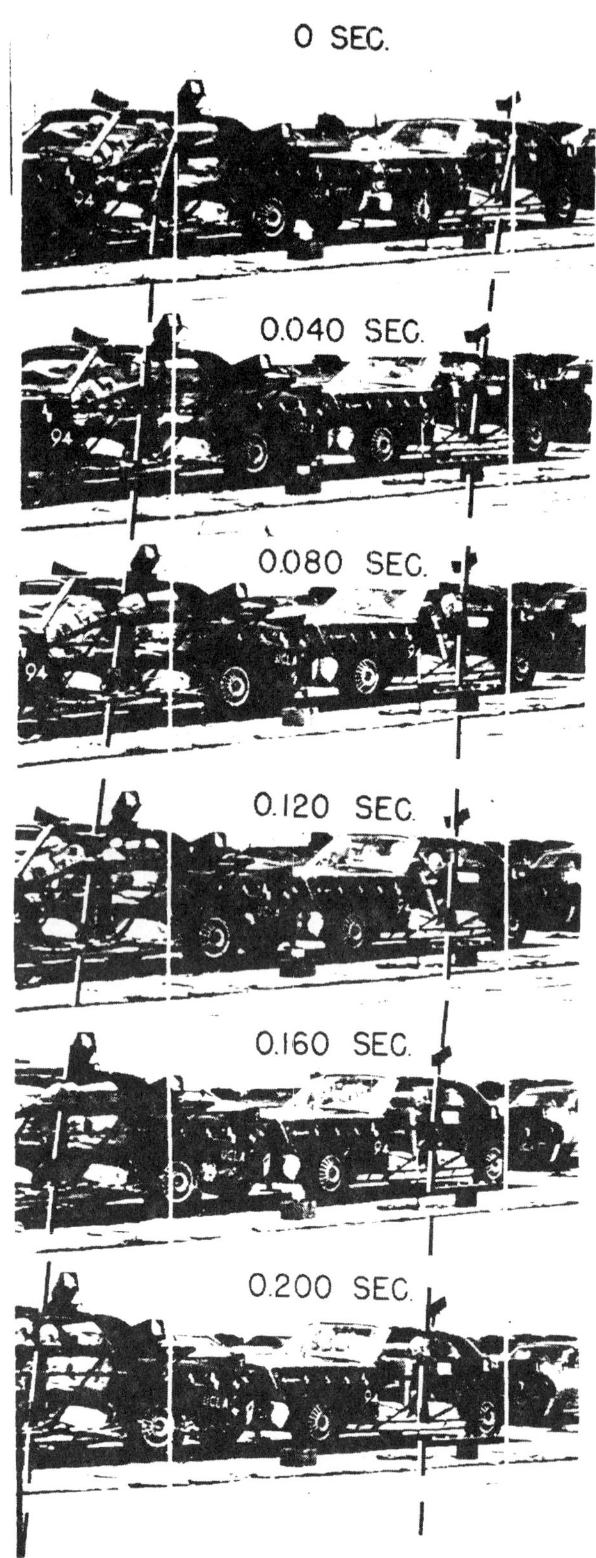

Fig. 24 - Mutual collapse reach 16 in., 20 mph collision

inches between the driver's head and the rear passenger's lap. Rebound from this dorsi-flexed neck posture was not excessive or sufficient to cause head-to-steering-wheel impact injuries. During the driver's collision acceleration, he sustained a peak chest acceleration of 5 G at 90 ms, and a lap belt loading of 100 lbs. at 260 ms. His 22 inch seatback (9,100 in.-lbs. strength), accelerated to a peak of 7 G at 35 ms, sustaining a 3.2 inch rearward deflection at 150 ms, both as measured at the 14 inch reference position.

For this 20 mph collision, the driver did not slide up the plane of the seatback as much as for the 30 mph impact but appeared to have sustained nearly as severe a whiplash injury exposure. The rather small, though possibly significant, differences between the 20 and 30 mph collisions were in the amount of head rotation relative to torso (74 versus 78 degrees) and the head peak acceleration, (9 versus 12 G). Increased seatback yield (from 3.2 to 5.8 inches) and increased mutual collapse of the two vehicles (from 16 to 48 inches) account for the relative constancy of rear-ending forces applied to the driver.

Right Front Seat Passenger - As with the driver, this lap-belted right front passenger was seated in a modified production bucket seat except that it had a factory installed prototype head restraint. The head restraint installation increased the seatback height of the passenger's seat from 22 inches to 28 inches. The added bending moment caused by the passenger's head restraint made it necessary to increase seatback strength from 9,100 in.-lbs. to 16,100 in.-lbs.

The head restraint for the right front seat passenger provided good resistance to whiplash exposure for the 30 mph impact but was slightly less effective for this 20 mph impact. The 16,100 in.-lb. seatback strength resisted rearward inertial displacement accounting for the passenger maintaining a rather normal posture throughout the collision (30 degrees head rotation rearward relative to torso axis) but his head acceleration for 20 mph (14 G at 155 ms) peaked higher than the companion exposure at 30 mph (11 G at 150 ms).

In seeking an explanation for this seeming anomaly, it was observed that although both chest accelerations peaked the same time to 6 G, the higher inertial energy associated with the 30 mph impact was dissipated for the head peak acceleration of 11 G over 180 ms as contrasted with 110 ms associated with the 14 G head peak at 20 mph. Following collision acceleration, his head flailed forward until the neck limitation of motion was reached but the body torso posture remained about the same.

Other information obtained for this right front passenger indicates that his knees sustained 8 G at 85 ms and his belt load reached 60 lbs. at 260 ms. His seat sustained a peak acceleration of 7 G at 65 ms at the 14 inch level above the H-point.

The Right Rear Passenger - The 95th percentile adult passengers in the rear seat of the struck car were on a conventional bench seat with a standard rigid seatback. They were identical anthropometric dummies and the only factor making their exposure different was head offset from the seatback. The right

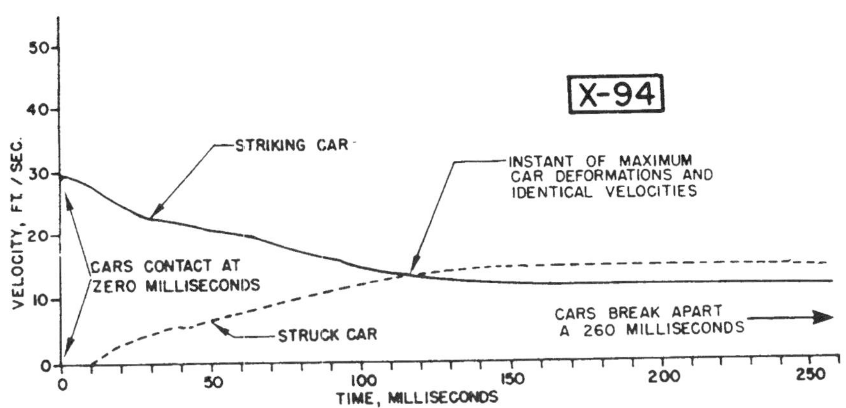

Fig. 25 - Velocity changes during a 20 mph rear-end collision, Exp. 94

Fig. 26 - Driver without head support is severely whiplashed during 20 mph collision, Exp. 94

Fig. 27 - Right rear passenger, 12 in. head offset, Exp. 94

rear passenger was postured with his head offset 12 inches forward of the seatback, Fig. 27; the left rear passenger's head was positioned with a three inch offset. The right rear passenger was slammed rearward against the seatback, his head struck the header, owing to his rather erect seating posture, sustaining a 50 G head acceleration at 115 ms. His chest registered only 14 G at 100 ms indicating the adverse effect of the lack of a head restraint. This right rear passenger maintained approximately the same body posture, with respect to head and torso relationship, as he compressed the seatback padding and this caused him to be displaced rearward until his head struck the upper portion of the rear window header. At this instant, further rearward movement was checked and the passenger continued to maintain about the same postural relationship but started to be displaced forward and go into a slightly bowed posture with his head about 18 inches forward of its pre-crash posture. Thereafter, the effects of this rear-end collision on this passenger became negligible. These accelerations were between two and three times the level sustained by the front seat motorists, partly attributed to the closer proximity of these rear seat passengers to the position of impact and their more rigid seatback but more importantly because of the absence of a head restraint.

As may be seen by reference to Figs. 21 and 26, the requirement for a head restraint isn't as critical for the adult in the rear seat with respect to keeping the head in line with the torso during collision acceleration. The rear seat passengers for this experiment did not "whiplash" because their heads were forced against the close-sloping (30 degree) rear window, Fig. 26. This prevented "whiplash" in its usual form but provided a severe head blow and a potentially dangerous wedge-like, downward, compressive force to the spine. The high head blow (50 G) attests to the conclusion that this design, while sometimes holding the head from whiplashing, is not considered a satisfactory solution to the problem. Human sizes approximating the average adult will whiplash when placed in this rear seat, owing to their shorter torso heights.

The right rear passenger for this 20 mph rear-end collision sustained a 50 G head blow at 115 ms in contrast with the 14 G head blow sustained by the same dummy in the same seat but impacted at 30 mph. This obvious inconsistency is explained by discovery of a technical oversight in which the right rear seat passenger in the first 30 mph collision was slumped lower in his seat and did not strike the rear window header a direct blow. Thus, it may be seen that even though both occupants had the same head offset, a subtle change in posture made a substantial difference in the exposure to injury.

The Left Rear Passenger - did not visibly sustain "whiplash" because his head and trunk remained in an approximately normal posture; this was because his head was positioned within 3 inches of the rear window and it struck the rear window glass, sustaining a head acceleration (23 G at 55 ms) and a 10 G at 70 ms chest acceleration without significant head-to-torso misalignment. He showed no evidence of rebound following this impact with the rear window and the tempered glass did not fracture. As with the right rear passenger, this wedge-like impact caused by the 30 degree slope of the rear window (from horizontal) could develop high vertebral compressive forces but the dummy was not instrumented to provide this data. This relatively low chest acceleration (10 G) was the same value sustained for the 30 mph rear-ender and indicates the value of having the torso resting against the seatback for the rear seat position at the start of a rear-end collision. It also further substantiates the excellent force modulating properties of the rather collapsible rear-end structure of the 1967 Ford in the range of 20 to 30 mph impacts. Additionally, the respective time differences of 55 ms and 70 ms for the head and chest correspond to the observation of the head initially striking the rear window glass before the chest has crushed very far rearward into the seatback.

Observations, Following Collision - The struck car was propelled 139 feet with a final offset to its left from its pre-crash heading of six feet. Gasoline (simulated) was liberally sprayed and deposited on the pavement over the entire 139 feet of struck car forced-movement, except for the intial ten feet from point of impact. This trail of fluid was also accompanied by a liberal deposit of broken glass, metal parts and other typical collision debris. The driver's seatback had a post yield of 8 degrees rearward but the front passenger seat showed only slight evidence of permanent strain (3 degrees).

The relative bumper heights ($21\frac{1}{2}$ inches) for the striking and rear-ended cars were adjusted to provide a good match between the bumpers at contact. The rear frame members buckled ahead of the bumper attachment point. The fuel tank filler pipe pulled free allowing fuel to be dispersed. The taillights were "on" and remained "on" notwithstanding the 20 mph impact to the rear of this car. The trunk lid remained closed throughout the collision. Permanent deformation to the struck car was 11 inches at the top edge of the rear fenders and 13 inches at bumper elevation.

EXPERIMENT 95 - was a 30 mph rear-end collision between two identical Ford 4-door sedans, Fig. 28. Only the information derived from the rear-ended car will be presented at this time. In the front seats of the struck car, the two-205 lb., 95th percentile dummies were seated on identical UCLA-Modified bucket seats, having 22 inch seatbacks, fitted with factory installed (prototype) head restraints that, in effect, increased the seatback height to 28 inches; both of the front seats had calibrated yield seatbacks designed to withstand 16,100 in.-lbs. The only difference in injury exposure relates to the occupant head offsets which were 6 inches for the driver and 12 inches for the front passenger, Fig. 29.

The two rear seat passengers, also 95th percentile anthropometric dummies, were seated on a 21 inch production bench seat, modified for the right rear position to include a head restraint that increased the effective seatback height to 27 inches. The right rear passenger was positioned with a 12 inch head offset from the plane of the seatback; the left rear passenger had no head restraint and zero-inch head offset. Each of the four identical anthropometric dummy occupants had an articulated neck, calibrated to withstand accelerations up to $1\frac{1}{2}$ G, without movement, corresponding

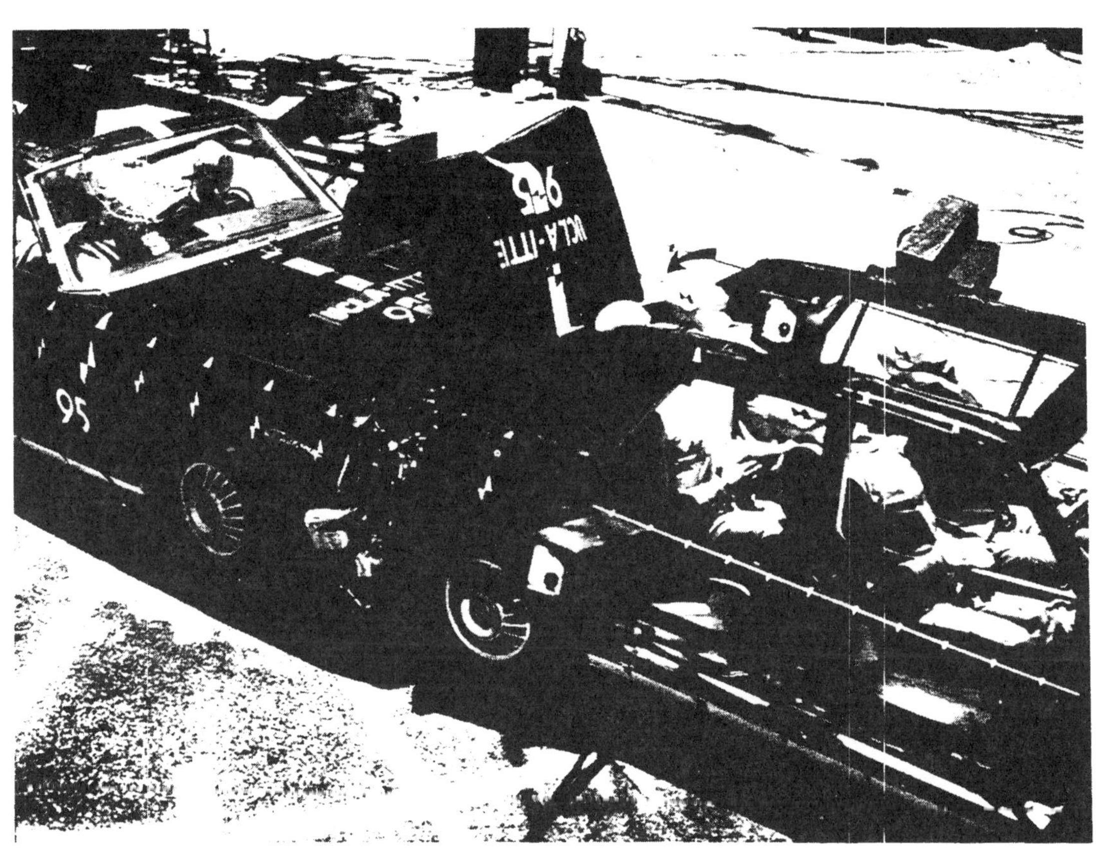

Fig. 28 - 30 mph rear-end collision, Exp. 95

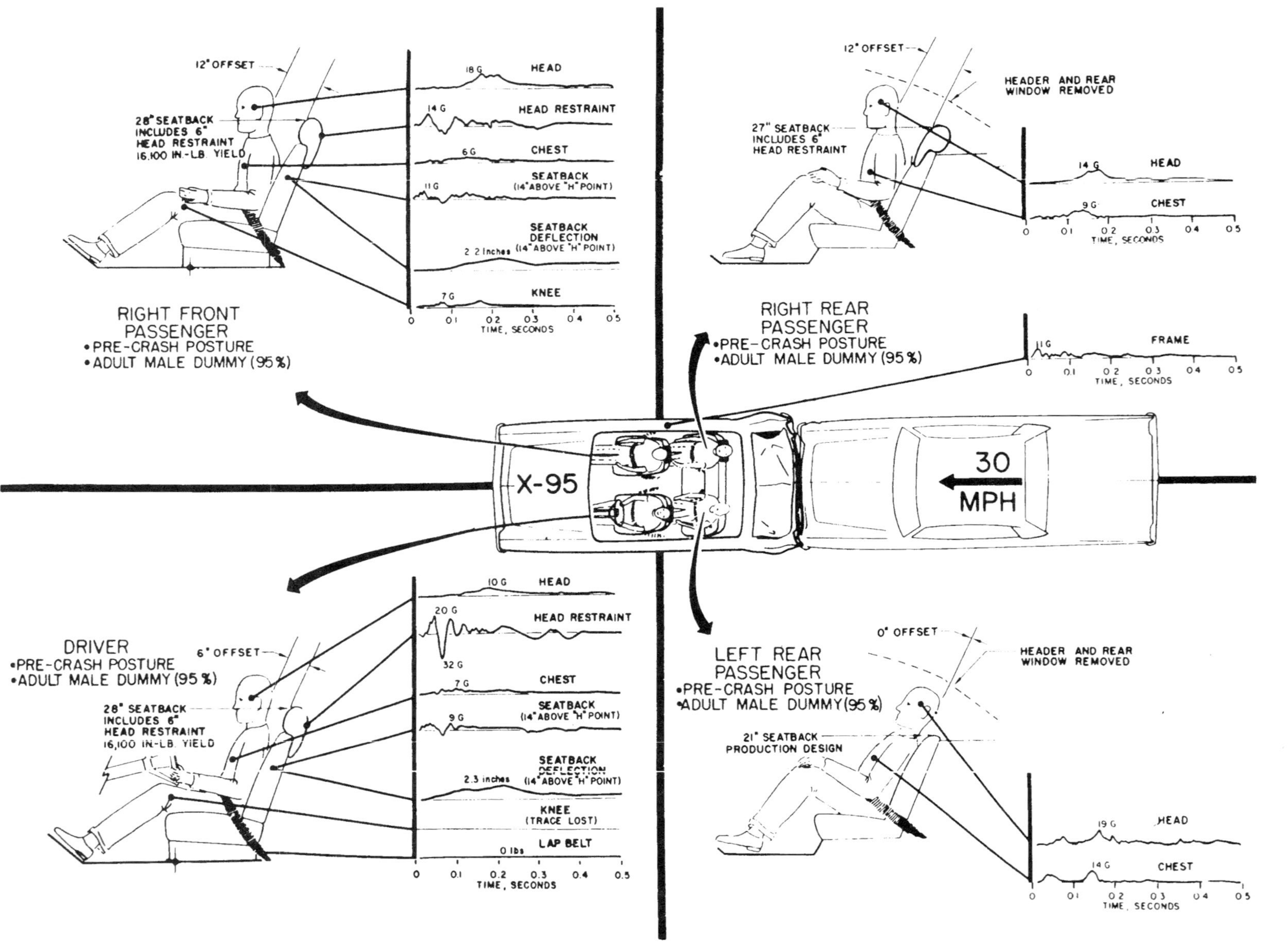

Fig. 29 - Transducer patterns shown with occupant posture and restraint condition at start of crash, Exp. 95

to passive neck muscle tone; each passenger had a lap belt, except for the right front passenger who had no restraint.

The gross weight of the struck car was 4680 lbs. and of the striking car 4700 lbs. For this second 30 mph collision the spare tire was installed to determine possible influence on rear-end collapse of the struck car. The doors of the struck car were removed to facilitate photographic observation of motorist kinematics. The structural integrity of the passenger compartment was restored by installing horizontal compression struts in the door openings, as previously described. The rear window and rear window header were removed from the struck car to evaluate motorist kinematics for vehicle designs having the rear window located more remotely behind the rear passenger's head.

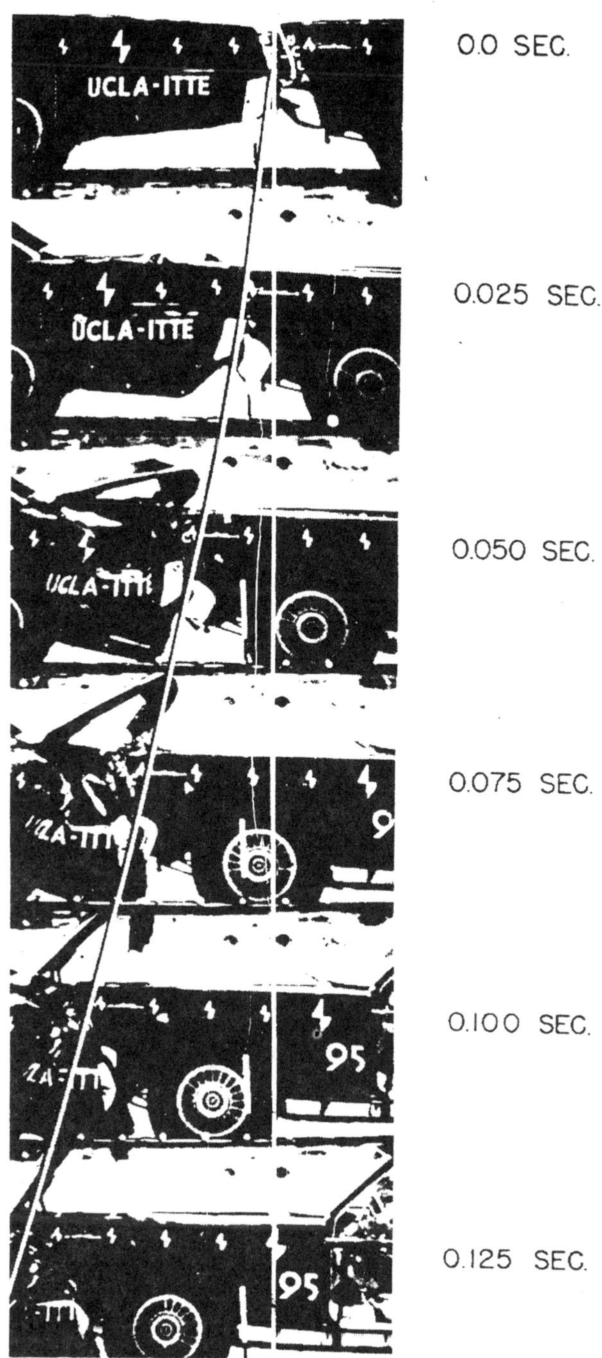

Fig. 30 - Readily collapsible rear-end structure, 30 mph, Exp. 95

The Collision - The striking car advanced one foot after the bumpers contacted (through vehicle mutual crushing) without the struck car initiating any noticeable advancement. As observed with Experiment 93, the inertia of the rear-ended car and its rather collapsible rear-end structure account for the fact that this struck vehicle had not commenced to be advanced any significant distance until after its rear-end was compressed to the rear wheels, representing 3 feet of collapse and a peak resultant frame acceleration of 11 G at 25 ms, Fig. 29. The abrupt acceleration to 11 G, 25 ms after bumper-to-bumper contact is accounted for because the accelerometer was mounted adjacent to the frame directly in line with the impacting bumpers; additionally, the acceleration build-up did not develop beyond 11 G because of the readily collapsible rear-end structure, but the acceleration continued at relatively high values for 100 ms as the rear of the car collapsed forward, Fig. 30. This maximum collapse to the rear-ended vehicle occurred when the striking car had advanced only four feet beyond the ground reference position of contact between the two colliding vehicles. Thereafter, the rear-ended vehicle accelerated rapidly, Fig. 31, and the two cars broke contact with one another after the striking car had advanced 10.5 feet.

Before the cars separated, however, there was a liberal diffusion of simulated gasoline sprayed from the struck car's fuel tank.

Driver - A production bucket seat (22-inch seatback height) was modified in terms of its structural resistance to collision forces and in terms of size by increasing the seatback height to 28 inches with the addition of a head restraint; except for the added head support, the general appearance of the seat was unchanged. The seatback strength was calibrated at 16,100 in.-lbs. As previously mentioned, the purpose for increasing seatback strength was to provide the seat with a designed strength that would allow an objective evaluation of the influence of seatback height on forced postural changes of the rear-ended motorist. Prior studies suggested that a 28-inch seatback would substantially, reduce "whiplash" type injuries even for the very tall adult (e.g., the 95th percentile adult male. The probability of an increased head impact force for the motorist who is leaning somewhat forward at the onset of collision was determined by having the driver positioned with a 6-inch head offset from the seatback plane and the passenger's head positioned with a 12-inch offset. The zero-inch offset situation for the 28-inch seatback height was evaluated in Experiment 93 and Experiment 94, Table 1.

When the driver's chest was accelerated horizontally foward 14 inches

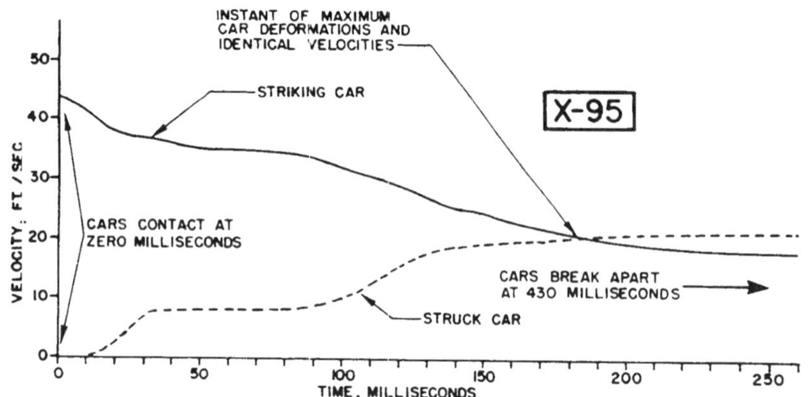

Fig. 31 - Velocity changes, 30 mph rear-end collision; vehicles separate after 10.5 ft advance at 430 ms

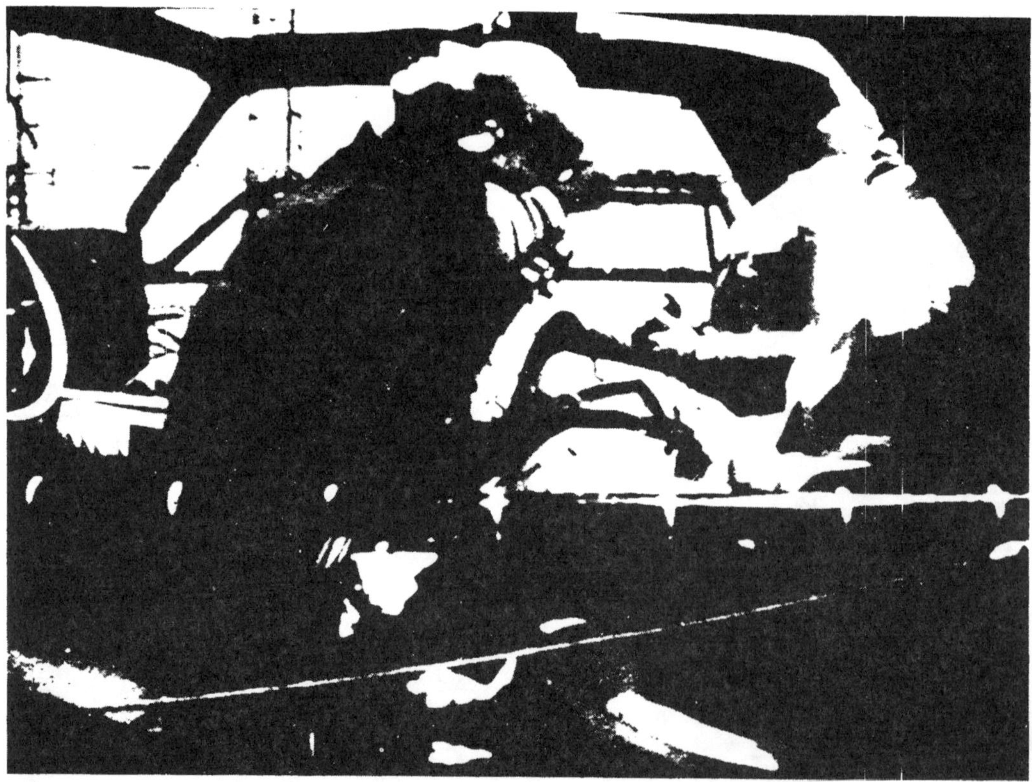

Fig. 32 - Driver's head support provided an effective restraint against excessive rearward rotation of driver's head, Exp. 95; note passenger's comparable head position at 170 ms

relative to a ground reference point, he sustained a head displacement, also in the forward horizontal direction, of only ten inches relative to the same reference point, Fig. 32. The head restraint for the driver provided effective resistance against excessive rearward rotation of the head, limiting its total rotation to 65 degrees rearward of vertical. This driver's collision response was actually only the beginning of "whiplash" because it appeared to be within the physiological limits of natural or voluntary movement for the human neck. The driver's normal posture after closing the pre-set six-inch head offset condition corresponded to the deflected 29 degree seatback slant, rearward of vertical, for chest and head. Therefore, this rearward head rotation of 65 degrees actually was only 36 degrees, relative to the torso, within the average limit of voluntary movement. The driver's head acceleration reached a peak of 10 G at 180 ms and his head restraint registered 20 G at 45 ms. His chest acceleration was 7 G at 95 ms and there was no appreciable load recorded for his lap belt. The seatback, 14 inches above the H-point, registered 9 G at 80 ms, his seatback deflected rearward a maximum of 2.3 inches at 215 ms Rebound from this dorsi-flexed posture was not abrupt and not regarded as injury producing.

The Right Front Seat Passenger - was seated in a UCLA-Modified bucket seat with a 16,100 in.-lb. calibrated seatback and a factory installed prototype head restraint, identical to the driver's seat. The head restraint installation increased the seatback height from 22 inches to 28 inches. The right front passenger was postured leaning forward sufficiently to develop a 12-inch head offset; this posture caused him to be slammed back against the head restraint with greater force than sustained by the driver (6-inch head offset). The passenger's actions appeared synchronized with the driver, Fig. 32, except that his head struck the restraint more severely, registering 18 G at 170 ms. However, his chest sustained only 6 G at 130 ms and his knee (femur) 7 G at 70 ms. The right front seatback at the 14-inch level registered 11 G at 35 ms, his head restraint, 14 G at 45 ms. The seatback deflected rearward 2.2 inches at 215 ms, Fig. 33. As was found for the driver, rebound of the right front passenger was not sufficient

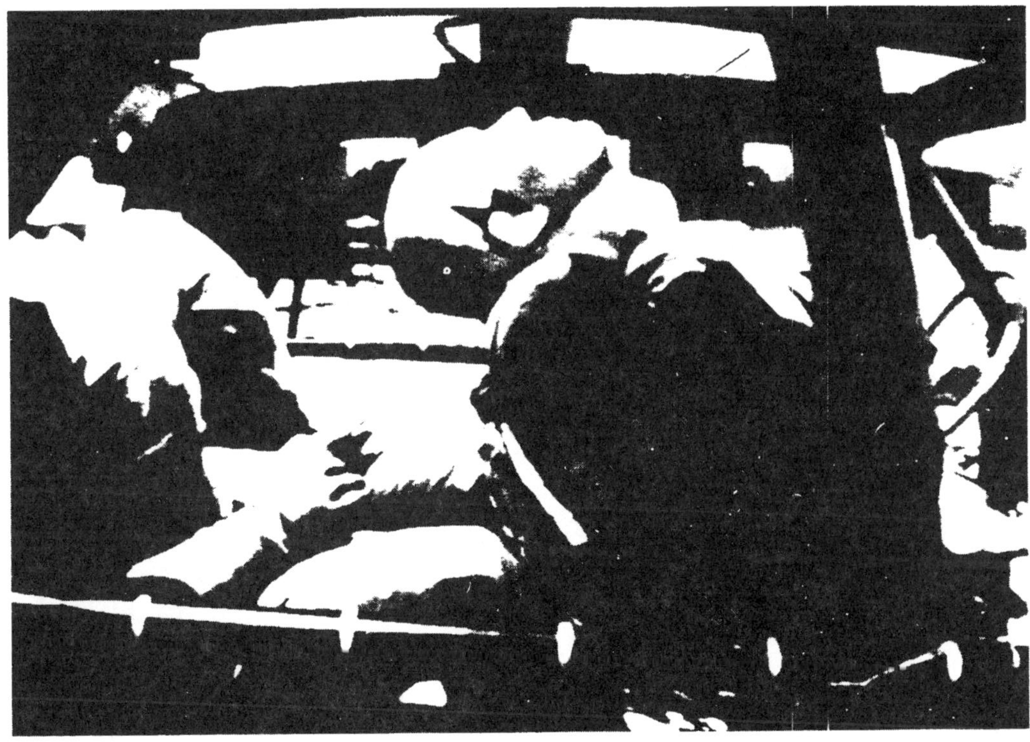

Fig. 33 - Right front passenger deflects seat rearward 2.2 in. at 14 in. seatback elevation (Exp. 94)

to bring him more than slightly beyond a vertical posture and not into contact with any forward structures; his movements against the seatback became negligible as he approached the second cycle. This indicates that the relatively stiff seatback does not provide an adverse feature of springing the motorist against the forward structures following collision acceleration. The head offset of 12 inches for the right front passenger was 6 inches forward of the driver's head and did not result in any significant difference in the linear potentiometer reading of seatback deflections (2.3 in. driver; 2.2 in. passenger, both at 215 ms). As would be expected, the right front passenger "whiplashed" 22 degrees more than the driver (58 vs. 36 degrees).

Right Rear Seat Passenger - Both rear seat passengers were identical 95th percentile anthropometric dummies, seated on a conventional bench seat with a standard 21-inch seatback. The factors making their exposure different were the head restraint and the 12-inch head offset for the right rear seat passenger; his head restraint was the same type used for the front seat, except that it was installed with the padded surface that normally faces rearward turned about to face forward. This head restraint raised the effective seatback height to 27 inches. The right rear occupant appeared to have initiated whiplash-type movements before the head restraint effectively checked his rearward head movement. No adverse posture corresponding to injury producing proportions occurred. His head sustained a 14 G impact with the head restraint at 175 ms, his chest, 9 G at 135 ms.

In addition to the head restraint installation, the rear window and header, were removed; this procedure allowed sufficient clearance for unrestricted head movement, Fig. 34. In Experiment 93, these structures were left in the car and contact was made with the rear window. Rebound for this right rear passenger was mild and did not constitute an injury producing exposure, Fig. 35.

Left Rear Seat Passenger - This 95th percentile dummy was positioned on a conventional 21-inch bench seat without head restraint and with zero-inch head offset. He sustained a severe "whiplash" which could have been moderated considerably if there had been crash padding on

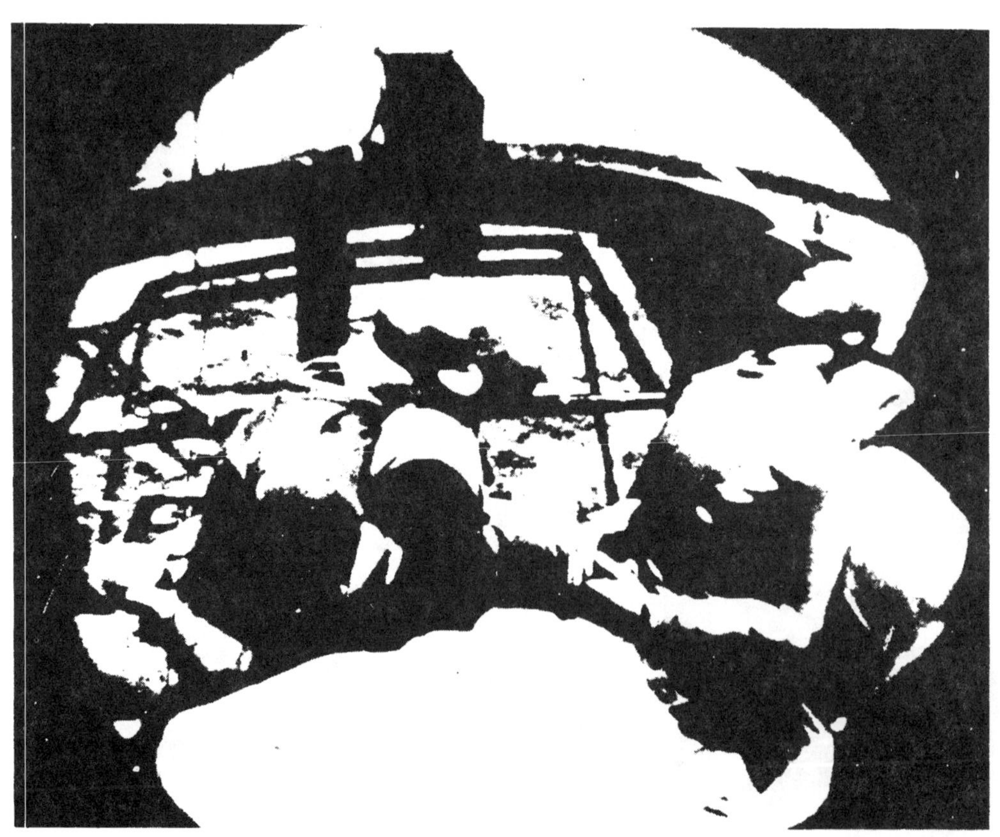

Fig. 34 - Passenger is protected from whiplash by head restraint, right rear seat, Exp. 95; rear window and header removed

the rear window shelf; his head flailed rearward 76 degrees beyond his torso axis until it struck the window shelf, sustaining 19 G at 160 ms, Fig. 36. His chest impact also was comparatively high, 14 G at 145 ms. The value of head restraint for the rear seat passengers is shown by Fig. 37 in which only the chin of the left rear passenger shows while undergoing "whiplash" (no restraint)

Fig. 35 - Passenger rebounds slightly from rear seatback, following impact (Exp. 95)

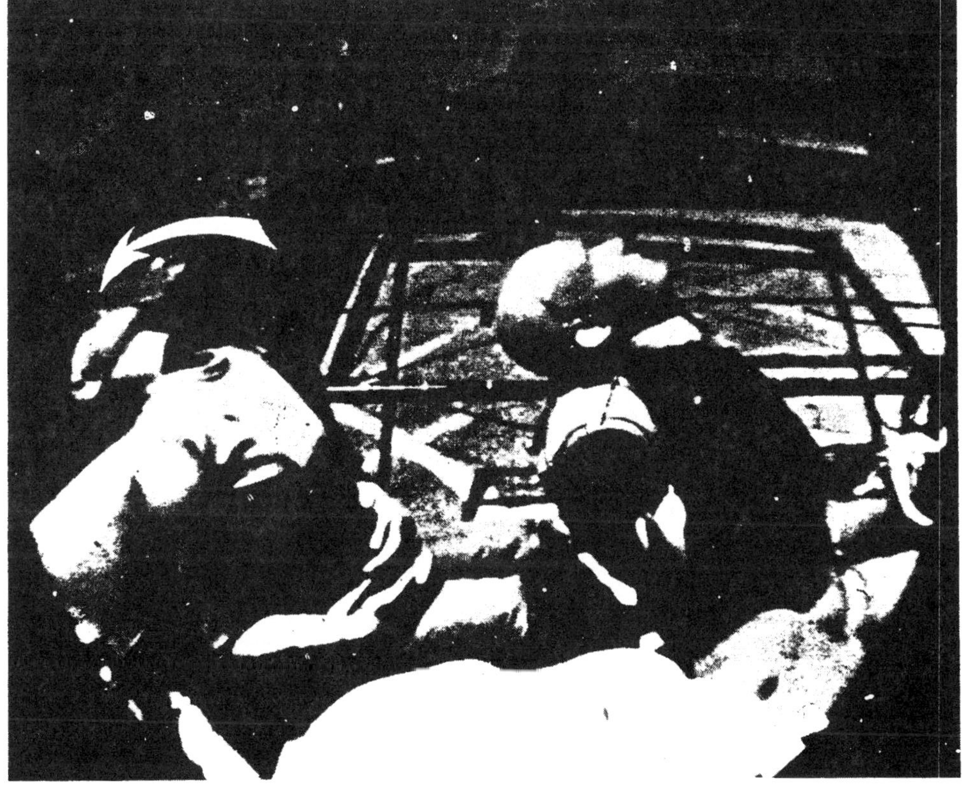

Fig. 36 - Passenger whiplashes head against rear window shelf, Exp. 95

and is contrasted to the right rear passenger's face, still clearly discernible (head restraint), notwithstanding his 12-inch head offset; both rear seat passengers are shown at maximum rearward head movement positions.

<u>Observations, Following Collision</u> - The rear-ended car was propelled 90 feet; it came to rest with a one foot offset to the left of its pre-crash heading. In the initial 30 mph collision (Experiment 93) the struck car was advanced 140 feet. When the two velocity curves for Experiment 93 and Experiment 95 are superimposed, minor variations in velocity changes can be observed but the patterns correlate rather closely, Figs. 15 and 31. Differences in run-out distances for the rear-ended car depend on the extent of collapsed sheet metal involvement with rear wheels, whether or not a tire is punctured and factors of this nature that may vary even for 30 mph repeated collsions. Permanent deformation for the rear fender-trunk section above the bumper level was 3.2 feet, and 2.2 feet at the bumper level. Gasoline (simulated) was liberally sprayed and deposited on the pavement over the entire 90 feet of struck car forced movement, except for the initial ten feet from

Fig. 37 - Value of head restraint for rear seat passenger is shown by this 30 mph collision, Exp. 95

Fig. 38 - Rear-end collision, 30 mph, Exp. 96

point of impact. This trail of fluid was also accompanied by a liberal deposit of broken glass, metal parts and other typical collision debris. The spare tire was installed in the trunk of the struck car for the first time and did not receive observable changes in position as a result of this impact. The shelf was still intact under this tire.

EXPERIMENT 96 - was a 30 mph rear-end collision between two identical 1967 Ford 4-door sedans, Fig. 38. The two front seat occupants in the rear-ended car provided a contrast in collision injury exposure because of differences in seatback heights; the driver had a 25-inch seatback and the right front seat passenger had a 28-inch seatback. Each of the front seat occupants was seated on a UCLA-modified bucket seat with a calibrated-yield seatback of 16,100 inch-pounds; their backs rested against, and parallel with, the 20 degree seatback (slope) and their heads were adjusted to a 3-inch offset from the seatback. The two rear seat passengers, although both seated on a production bench seat with a 21-inch rigid seatback, responded differently because the right rear occupant was sitting with a normal posture (3-inch head offset) resting against the 24 degree seatback and the left rear occupant was positioned leaning somewhat forward to provide a 12-inch head offset, Fig. 39. In addition to posture differences, the right rear seat position was equipped with a head restraint of the "after-market" or retrofit type that increased the seatback height to 27 inches; a similar head restraint was evaluated for the front seat occupants in prior experiments of this series. The left rear passenger seatback was evaluated, unaltered, except for a one-inch crash pad applied to the rear window header, above and rearward of the left rear passenger's head. All four occupants in this struck car were identical anthropometric dummies of the 95th percentile adult male size, each weighing 205 lbs. gross. Their necks were calibrated to resist movement, except for accelerations exceeding 1½ G; this degree of resistance corresponds to passive neck muscle tone. Each occupant wore a lap belt. The gross weight of the struck car, including the four dummy motorists, cameras and outriggers, 25 gallons of water, colored red to simulate gasoline, was 4720 lbs. This vehicle was struck by a similar vehicle whose weight grossed 4660 lbs. The doors of the struck car were removed and the upper half of the doorpost was cut out to facilitate photographic observation of motorist's kinematics. The structural integrity of the passenger compartment was restored by installation of horizontal compression struts, Fig. 40.

The Collision - The striking car advanced one foot after the bumpers contacted (through vehicle mutual crushing) without the struck car initiating any advancement. The inertia of the rear-ended car and the rather collapsible rear structure account for the fact that this struck vehicle had not commenced to be displaced any significant distance until after its rear bumper and trunk compartment were compressed to the back edges of the rear wheels; this represented three feet of collapse and was accompanied by a peak resultant frame acceleration of 9 G at 25 ms; additionally, the car acceleration did not develop beyond 9 G because of the readily collapsible rear-end structure. In the prior Experiment 93, the same 9 G peak frame acceleration occurred at 15 ms, within 1/100 second of the onset of this peak. The frame acceleration for Experiment 96 remained at about half the initial 9 G for a period in excess of 1/10 second after bumper contact, Fig. 39. Maximum collapse of the rear-ended car occurred when the striking car had advanced only four feet beyond the ground reference position of contact between the two colliding vehicles, Fig. 41. Thereafter, the rear-ended vehicle accelerated rapidly and the two cars broke contact with one another at 540 ms, after the striking car advanced ten feet.

Before the two cars separated, however, a liberal diffusion of simulated gasoline was sprayed from the struck car's ruptured fuel tank, Fig. 41. In all four experiments, X-93 through X-96, a separation occurred between the filler spout and the tank; compression of the tank forced fluid from this opening. In addition, the tank was punctured providing other leakage problems.

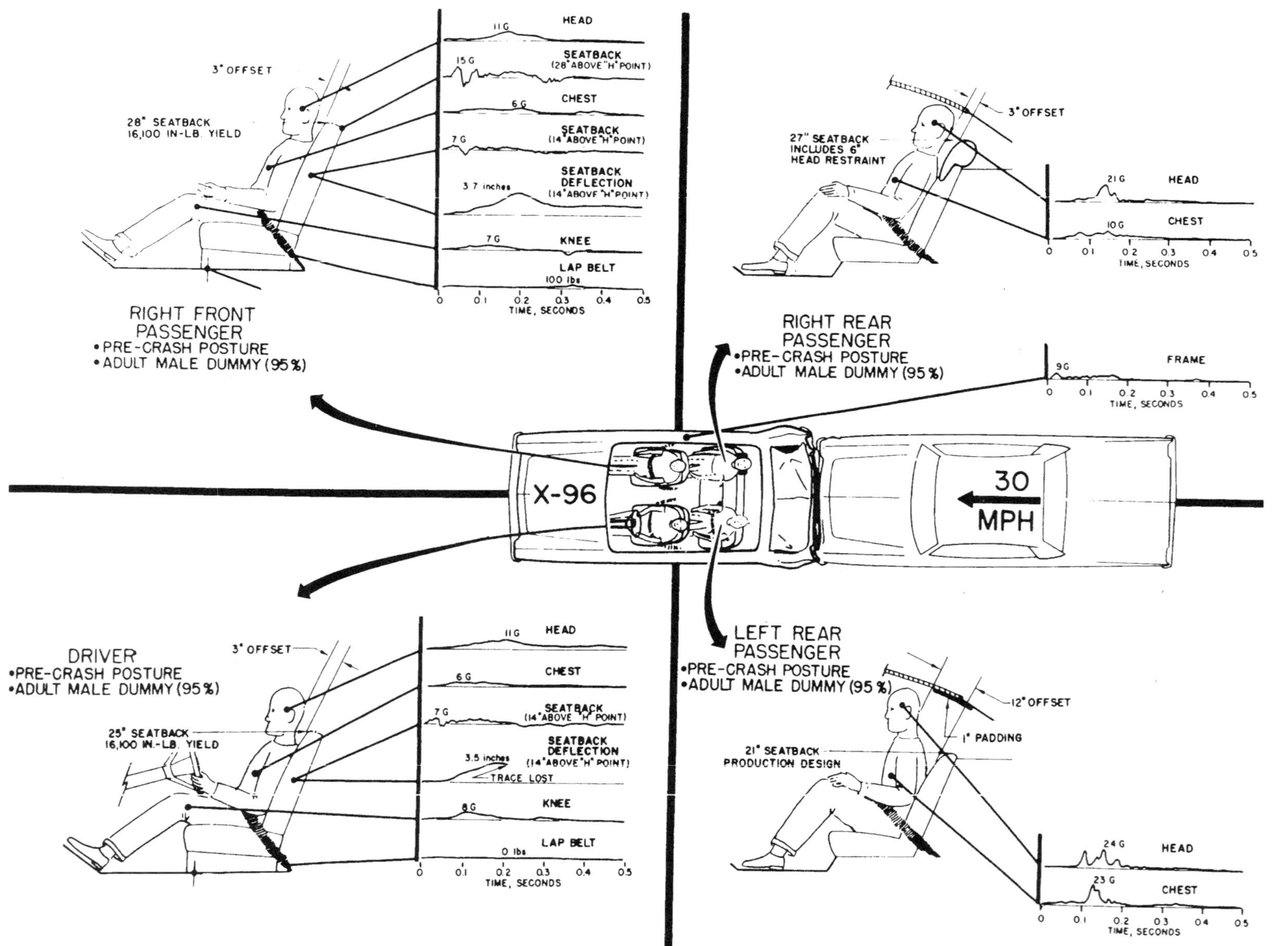

Fig. 39 - Transducer patterns and the postures of occupants and restraint condition at start of collision, Exp. 96

Driver - A 1967 Ford production bucket seat (22 inch seatback height) was modified in terms of its structural resistance to collision forces and by increasing the seatback height to 25 inches, but without otherwise changing its size and appearance. The seatback strength was calibrated to provide 16,100 in.-lbs. resistance to bending. The purpose for increasing the strength of the seatback was to bring the seat into a design category allowing comparison of the influence of seatback height on forced postural changes for the rear-ended motorist. While it is true that current production seat units are generally less injury-producing (during a rear-end collision) if their seatback strength is low, it is not possible to judge the improvement afforded by adding a head restraint if the seatback fails rearward before the head restraint can be brought into full effective performance. In addition, a rear-ended driver, in a forced reclined posture, is in no position to steer and apply brakes as may be required to avoid a second and possibly a more serious collision. This explains the reason a weak seatback is not recognized as an acceptable solution for motorist protection from rear-end

Fig. 40 - Typical 95th percentile dummy seated in modified seat having a 28 in. seatback; note compression struts in door openings (arrows), Exp. 96

collisions. Accordingly, the driver's 16,100 in.-lb., 25 inch seatback is compared in this Experiment 96 with the right front seat passenger's 28 inch seatback, also strengthened to 16,100 in.-lbs. Experiment 96 was the first experiment of this 12 collision series that used a "full height" seatback to accomplish the benefits of a head restraint without having a "tack-on" device.

As collision progressed to a quarter second after contact, the driver's head rotated 90 degrees rearward of vertical. His head registered a peak resultant acceleration of 11 G at 210 ms, Fig. 39. At the limit of forced excursion the driver's head had flailed rearward relative to his torso 54 degrees. Maximum seatback deflection was 16 degrees. Rebound from this dorsiflexed posture was not excessive or sufficient to cause head-to-steering wheel impact injuries. The driver-dummy then returned to an approximately normal posture. During the driver's collision acceleration, he sustained a resultant chest peak of 6 G at 85 ms and a knee acceleration of 8 G at 105 ms; his lap belt loading was negligible. His 16,100 in.-lb. 25-inch seatback accelerated to a peak of 7 G at 35 ms and deflected rearward 3.5 inches at 190 ms, both measured at the 14-inch calibration position.

Right Front Seat Passenger - This lap-belted right front passenger was seated in a UCLA-modified bucket seat. The seatback was increased to 28 inches, and as previously described, the seatback strength was increased to 16,100 in.-lbs.

During the collision acceleration phase of the struck vehicle, this 95th percentile adult pressed firmly against the seatback but continued to maintain a normal posture with the head rotated very slightly rearward. The sharp chrome edge on the upper surface of the seatback had been covered with energy absorbing plastic to prevent the head from directly contacting the metal trim.

The integral head restraint achieved by the extension of the seatback from 22 to 28 inches provided good resistance to whiplash tendencies, Fig. 42. The protective value of a 28-inch seatback is illustrated by the contrast of the two front seat occupants, Fig. 43. The driver's head (right side of picture) is being forced over a 25-inch seatback while the passenger's head has been adequately supported by the 28-inch seatback. The 16,100 in.-lb. seatback strength resists rearward inertial displacement which accounts for the front seat passenger maintaining a nearly normal posture throughout the collision. His head rotated only 16 degrees, relative to torso, less than one-third that of the driver. The maximum head acceleration obtained by the right front passenger was 11 G at 160 ms, his chest reached 6 G at 185 ms and his knee reached a 7 G at 115 ms. A seat belt load of 100 lbs. was measured at 320 ms. The seatback, as measured 14 inches above the seat cushion, recorded 7 G at 35 ms and, as measured at the 28 inch head restraint level, recorded 15 G at 40 ms. The seatback deflected rearward 3.7 inches (190 ms) at the 14-inch calibration reference position, representing a dynamic deflection of 17 de-

Fig. 41 - Sequence showing a 30 mph collision, Exp. 96; note fuel dispersal from ruptured tank, lower two frames

grees and a permanent deflection of eight degrees. Following the forced rearward displacement, the right front passenger's head and torso rebounded forward 18 inches, as measured at his head, his chin pressed against his chest, but he was not pitched forward with sufficient force to contact any forward structure.

Right Rear Passenger - The two passengers in the rear seat of the struck car were seated on a conventional

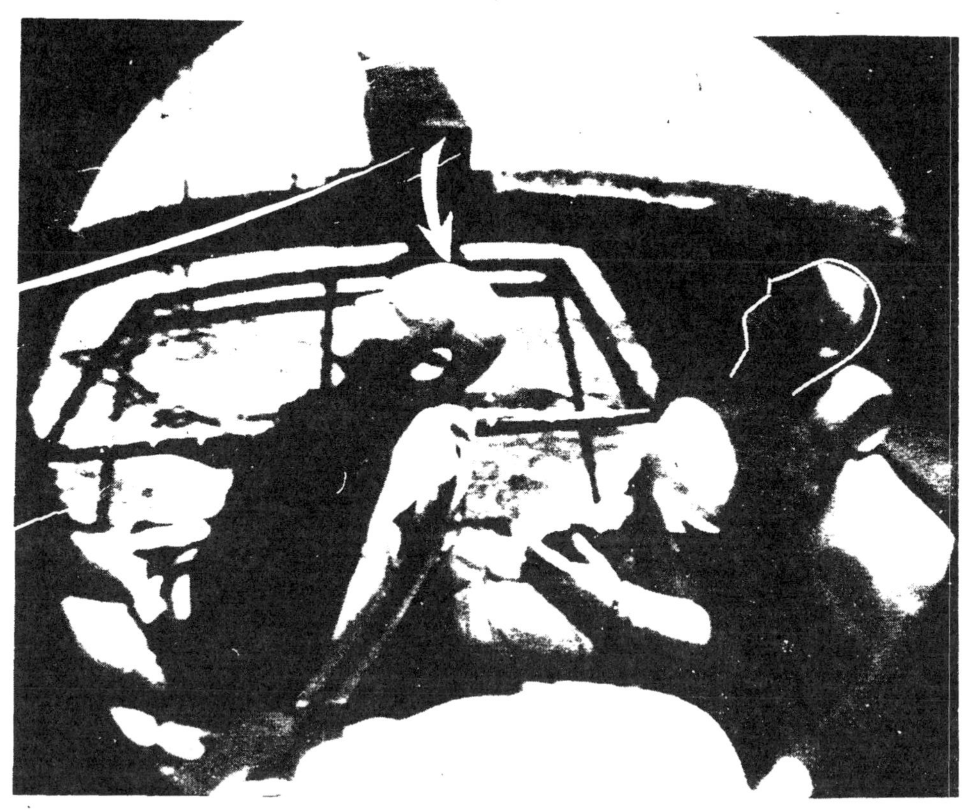

Fig. 42 - No whiplash is evident for the 95th percentile front seat passenger riding in a strengthened seat with a 28 in. seatback, Exp. 96, 30 mph

Fig. 43 - Driver (arrow) with 25 in. seatback has head less supported than passenger with 28 in. seatback; both 95th percentile dummies, 30 mph collision

bench seat with standard 21-inch seatback, except that the right rear passenger was provided with a head restraint of the type used for front seats extending the seatback height to 27

Fig. 44 - The right rear passenger was positioned with a 3 in. head offset from his 27 in. head restraint, Exp. 96

inches. They were identical 95th percentile anthropometric dummies and the only other factor making their exposures different was the posture of the left rear passenger. He was postured leaning forward with his head offset 12 inches ahead of the seatback while the right rear passenger was positioned with a three-inch offset, Fig. 44.

The right rear seat passenger was rather effectively supported throughout the collision event, Fig. 42. The greater rigidity of the rear seatback reduced the rebound of the right rear occupant, as contrasted to the right front occupant, Fig. 45. His head rebounded mildly forward and there was a slight change in position leftward but his shoulders and back maintained contact with the seatback; this motion provided contrast with the right front passenger whose entire upper torso rotated forward approximately 18 inches at the shoulders, pivoting about his hips. The right rear occupant sustained a 21 G head acceleration at 140 ms and a 9 G chest acceleration at 80 ms, followed by a 10 G peak chest acceleration at 145 ms. These accelerations were nearly double the level sustained by the front seat occupant, partly attributed to the rear seat passenger's closer proximity to impact and partly owing to the more rigid seat-

Fig. 45 - The front seat occupants rebound after 30 mph rear-end collision, Exp. 96

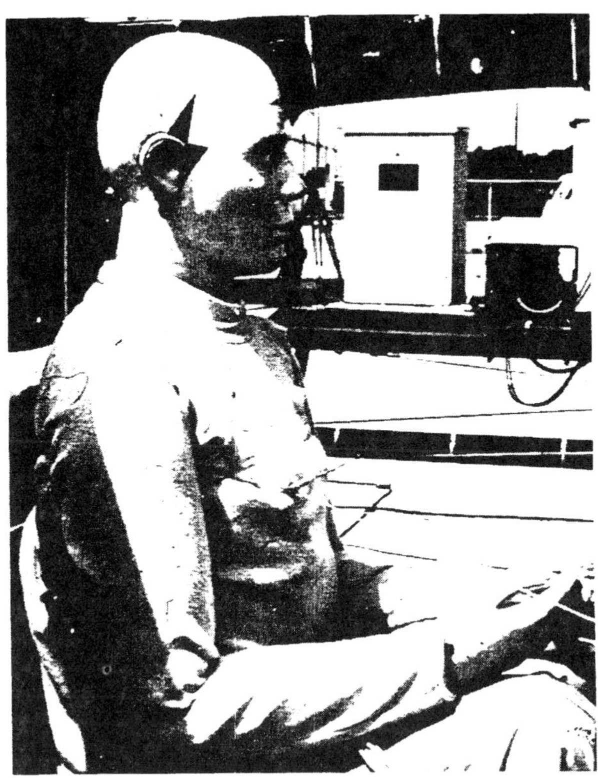

Fig. 46 - Left rear passenger, Exp. 96, positioned with a 12 in. head offset

back construction of rear seats in passenger vehicles.

The Left Rear Passenger - had a 12-inch head offset forward of the seatback plane at the start of the impact, corresponding to a rather erect posture, Fig. 46. As he was forced against the seatback, his head contacted the padded header area, abruptly stopping his head motion as his shoulders continued to crush rearward. This had the effect of flexing his head forward, relative to his torso, forcing his chin against his chest, Fig. 47. Following this action, the dummy was accelerated forward as a result of elastic rebound, and thereafter slumped slightly in his seat, probably as a result of the snubbing effect of the lap belt. Although he did not visibly sustain a whiplash, this wedge-like impact could develop high vertebral compressive forces; the dummy was not instrumented to provide this data. He sustained a 21 G head acceleration at 105 ms and a 24 G head peak at 150 ms with the sloping rear window glass; his chest peak acceleration was 23 G at 125

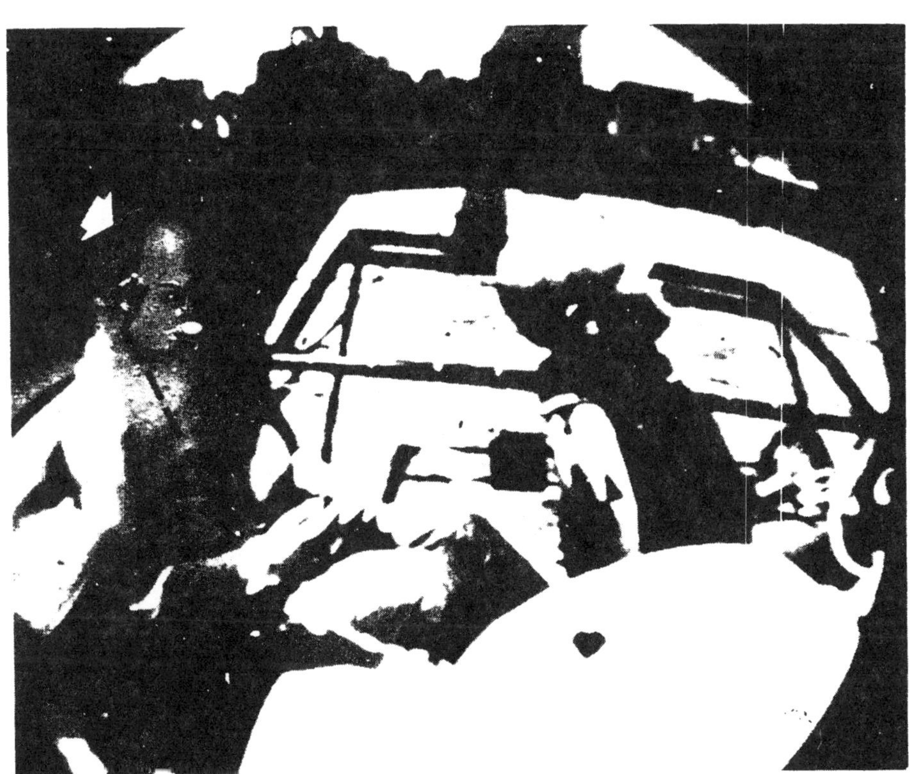

Fig. 47 - Left rear passenger's head is flexed forward, relative to upper torso, as he strikes padded header over rear window, Exp. 96

ms. This chest acceleration is excessive, when contrasted with the left rear occupant's 10 G of Experiment 93. The difference is attributed to the erect posture, 12" head offset for this 30 mph rear-ender, as contrasted with the zero head offset for Experiment 93. The X-93 passenger's back was against his seatback. The advantage of having the torso resting against the seatback for the rear seat position at the start of a rear-end collision was demonstrated; this is attributed primarily to the rigid construction of rear seatbacks.

Observations, Following Collision - The struck car was propelled 77 feet with a 9-foot offset to the right of its pre-crash heading. Permanent deformation for the fender-trunk section above the bumper level was 2.9 feet, and at bumper level, 2.5 feet. Gasoline (simulated) was sprayed at impact and deposited on the pavement over the entire 77 foot run-out of the struck car, except for the initial ten feet. This trail of fluid was also accompanied by a deposit of broken glass, metal parts and other typical collision debris. The front passenger's seatback retained an eight degree rearward deflection and the driver's seat yielded seven degrees, Fig. 48.

When the struck car's trunk collapsed, the sheet metal was pushed against the spare tire, located on the floor pan axle step over the rear axle. Even though the rear of the car was deformed and collapsed to the base of the rear window, Fig. 49, the spare tire was not pushed forward into the rear seatback.

During the course of impact, the fuel tank of the struck car was punctured by the lower transverse stabilizer attachment bolt. Gasoline was sprayed out the left wheel-well when the filler tube disengaged from the tank.

Post-collision observations of the underside of the vehicle indicate that the fuel tank has adequate space and is not fully compressed during impact at 30 miles an hour, Fig. 50. The tank does not release fuel until after the vehicles have completely crushed together; the initial surge of fuel then occurs out the left wheel-well above the axle.

EXPERIMENT 97 - was a 10 mph rear-end collision between two identical 1967 Ford 4-door sedans, Fig. 51. Four identical 95th percentile male anthropometric dummies, Sierra Model 292-95, were seated in the struck car. The driver's seat was a standard Ford production bucket seat with a 6-inch head restraint that elevated the seatback height to 28 inches. This production seat was unmodified as to its yield characteristics. The right front pas-

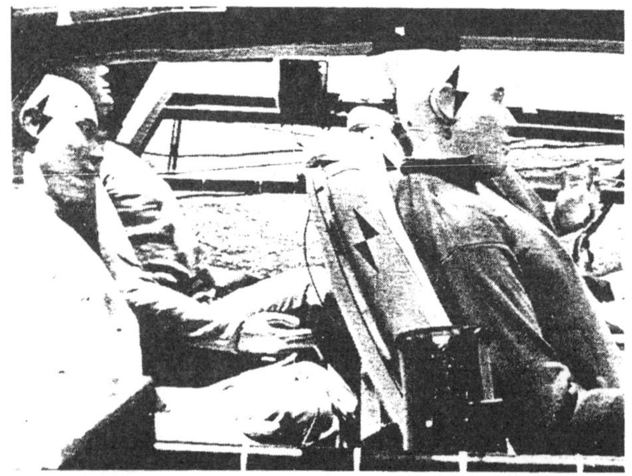

Fig. 48 - The front passenger's 28 in. seatback yielded rearward slightly more than driver's 25 in. seatback, Exp. 96, 30 mph

Fig. 49 - Extreme collapse is evident for this 30 mph rear-end collision, Exp. 96

senger's bucket seat was modified to a 28-inch seatback, designed to yield at 16,100 inch-pounds. The rear seat was a production bench seat having a seatback height of 21 inches. The production rear window and header were unmodified to allow investigation of the influence of head clearance on whiplash exposure, for rear seat passengers.

The right front passenger and the left rear passenger in this experiment were positioned with a zero-inch head offset, the driver was positioned with a three-inch head offset and the right rear

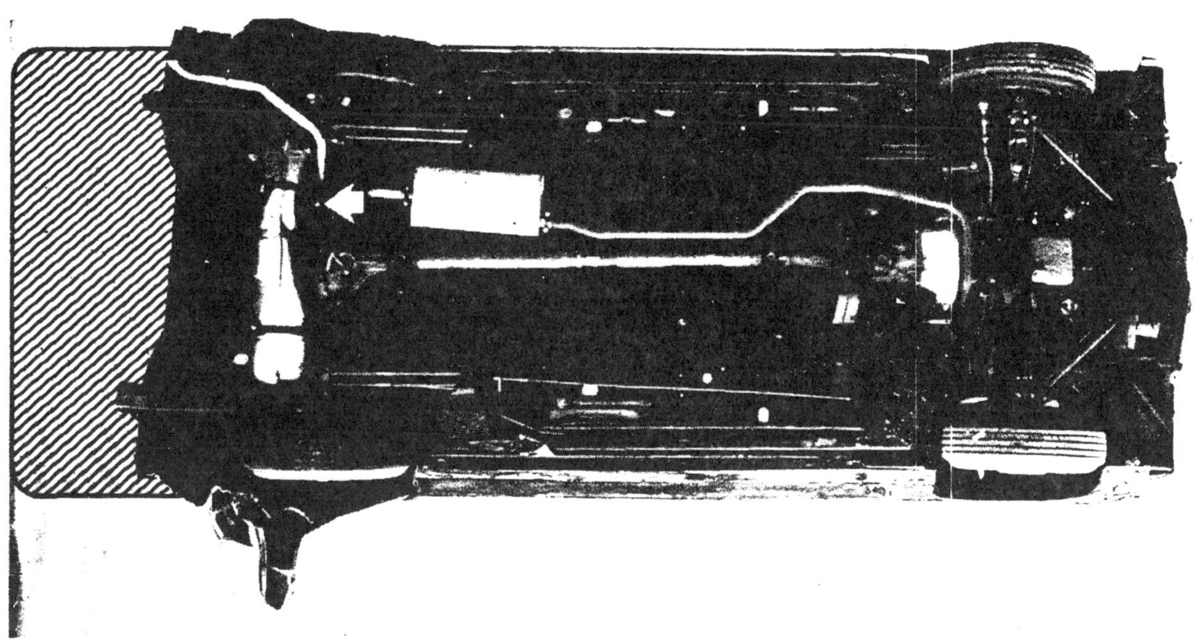

Fig. 50 - Underside of struck car after 30 mph rear-end collision, Exp. 96. Gasoline tank is partially compressed

Fig. 51 - The 10 mph rear-end collision -- only superficial damage, but significant motorist collision forces

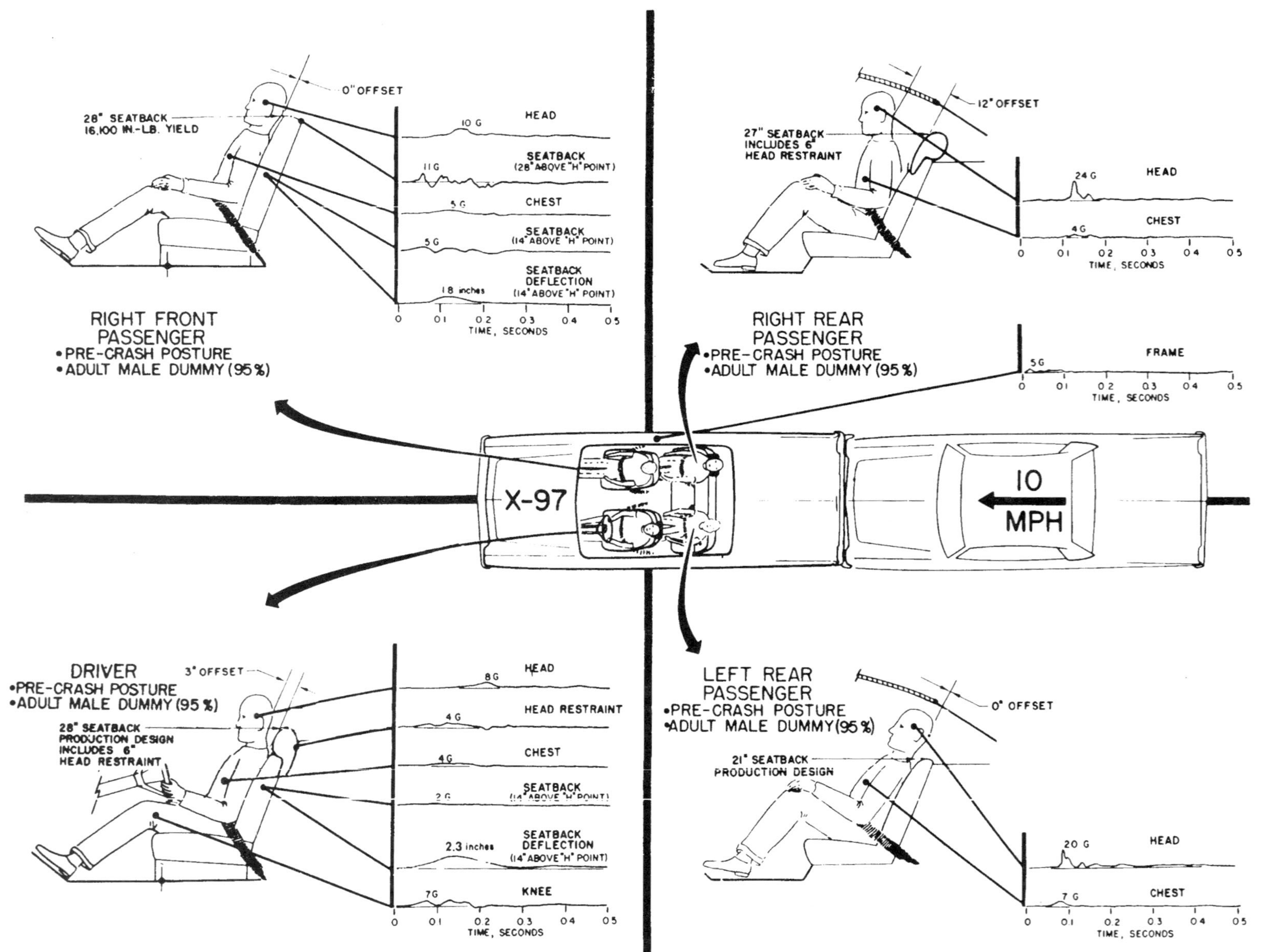

Fig. 52 - Transducer patterns shown with occupant posture and restraint condition at start of impact, Exp. 96

passenger was postured with a 12-inch head offset, Fig. 52. All four occupants were restrained by floor anchored lap belts. Bumper heights were adjusted to bring the rear bumper of the struck car into direct opposition with the front bumper of the striking car to eliminate the variable of bumper misalignment. The gross weight of the struck vehicle was 4690 lbs. including occupants, instrumentation and 25 gallons of fluid simulating gasoline. The striking car weighed 4620 lbs.

The driver's head restraint and the extended seatback for the front seat passenger provided each with an effective back and head restraint height of 28 inches; this height was adequate because their head movements were checked before they could build up to a point characterized as whiplash.

The Collision - The rear-ended vehicle collapsed one inch from this 10 mph collision before its inertia was overcome sufficiently for forward motion of the struck car to be discernible. At this instant (15 ms after the bumpers touched) the body underwent a 5 G acceleration and the frame later peaked at 6 G (50 ms) as the rear-ended vehicle sustained two inches of collapse, Fig. 53. The cars separated one foot beyond the point of impact and the struck car brakes were accidentally applied at a position 4.1 feet beyond impact, Fig. 54. At 3/4 second after impact, a second minor rear-ending action occurred which did not appreciably advance the struck vehicle. The trunk lid and hood were buckled from this impact but they did not open.

Driver - When the driver's head reached maximum acceleration (8 G at 215 ms) his head was flexed rearward over the head restraint at an angle of 53 degrees rearward of vertical (39 degrees rearward of torso). This movement appeared to be within the limit of natural or voluntary head movement and was not considered injury producing, Fig. 55. His chest received its peak resultant acceleration of 4G at 100 ms followed by a maximum seatback deflection of 9 degrees (29 degrees rearward of vertical at 135 ms).

Right Front Passenger - In a somewhat similar manner, the front passenger's head flailed rearward 36 degrees beyond vertical (14 degrees rearward relative to torso axis) sustaining a maximum acceleration of 10 G at 150 ms and did not appear to sustain an injury producing whiplash, Fig. 56. His chest reached a peak acceleration of 5 G at 125 ms after the seatback reached its maximum rearward movement (27 degrees at 115 ms).

Although both front occupants were

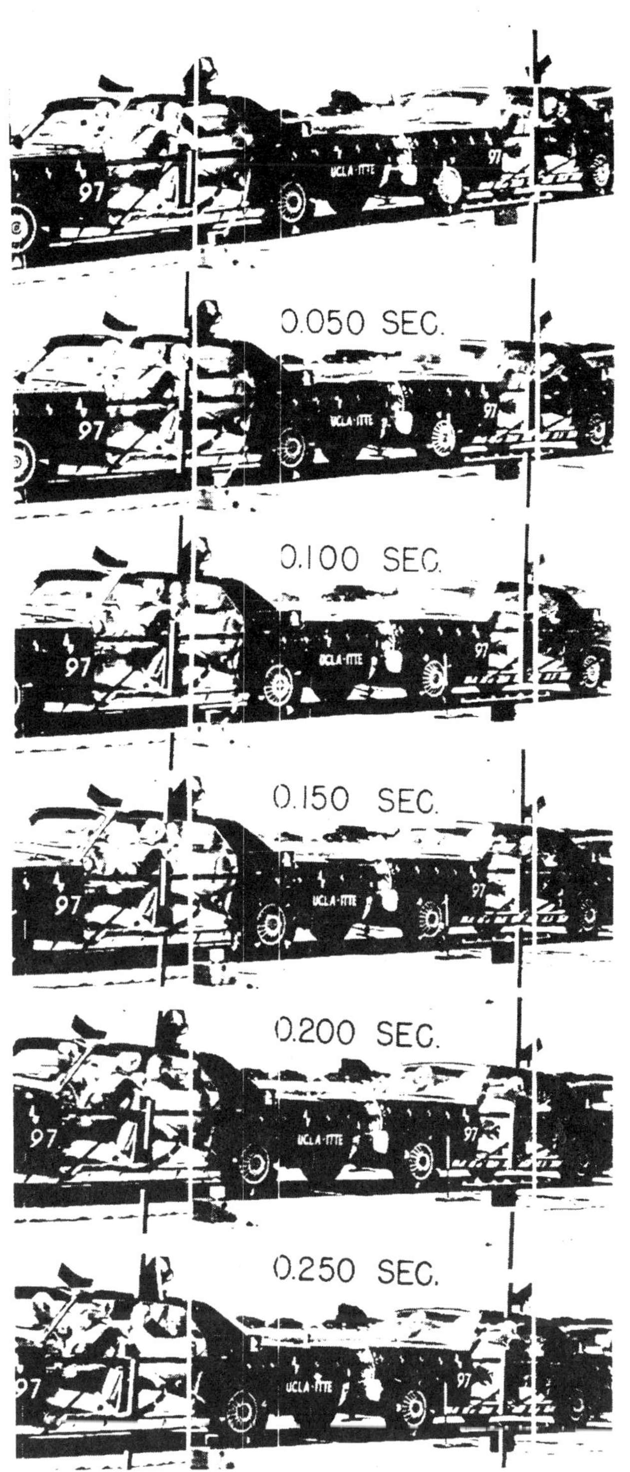

Fig. 53 - 10 mph crash sequence

identical 95th percentile anthropometric dummies, and both were supported by 28-inch seatback heights, the driver seated in a production seat with a factory installed head restraint sustained more head rotation than the passenger with a strengthened UCLA modified seat. This contrast in the relative head and neck protective aspects of these two seating configurations undergoing identical rear-end collision exposures is quite apparent even at an impact speed of 10 mph, Fig. 56 and Fig. 57.

The Right Rear Passenger - (95th percentile adult) was thrown rearward against the backrest from his initial 12-inch head offset position. His head struck the rear window header (24 G peak at 130 ms) after the bumpers contacted, Fig. 58. His head impact stopped his rearward movement, relative to the vehicle. At this same time his chest accelerated to a 4 G peak. Thereafter, his shoulders were pitched forward by seatback-rebound -- having the effect of releasing his head from this wedgelike impact with the sloping rear window header -- rebounding the head slightly forward. Owing to the restitution between the two vehicles and the subsequent application of brakes for the rear-ended car, this struck vehicle was rear-ended a second time; the second impact was minor and only caused the right rear passenger

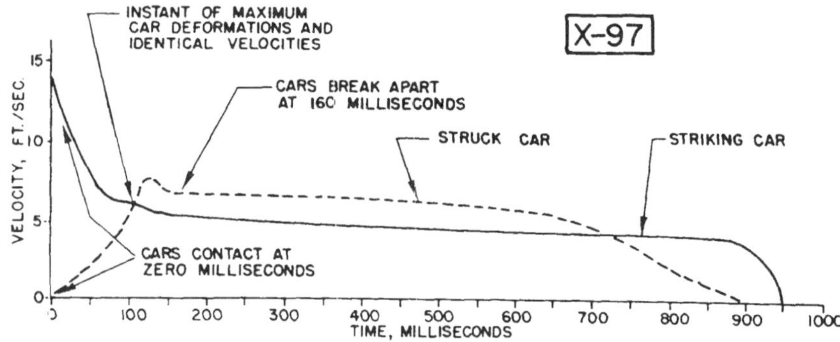

Fig. 54 - Velocity changes during a 10 mph rear-end collision

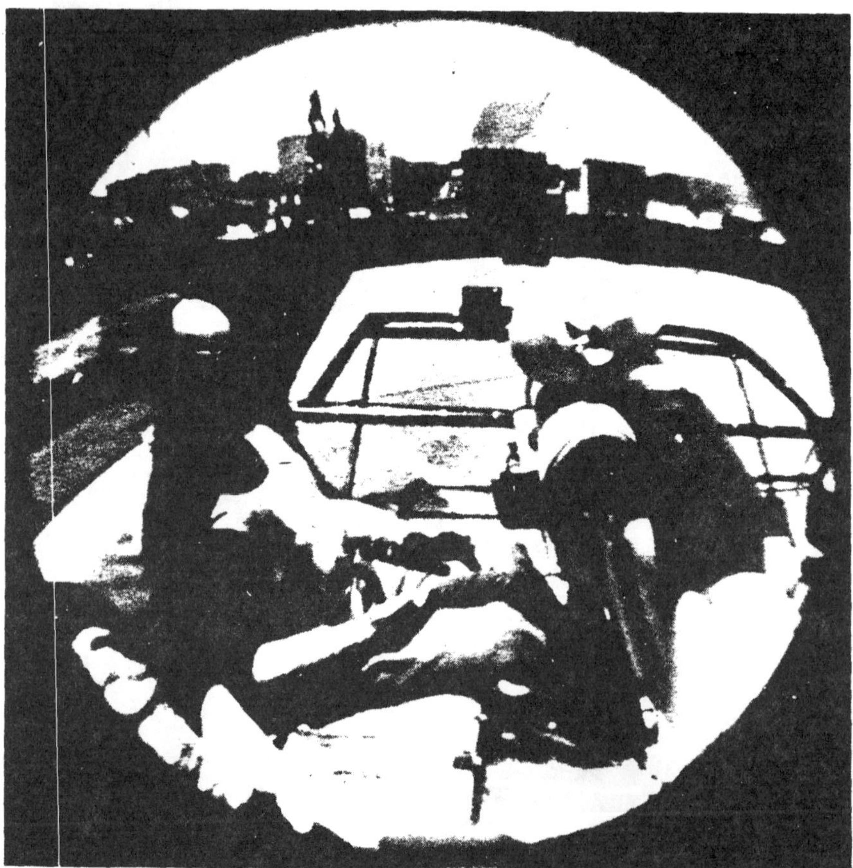

Fig. 55 - Driver's head flexed rearward, within limit of natural movement

Fig. 56 - Right front passenger riding out collision uneventfully

Fig. 57 - Contrast in head restraint; driver (right in picture) mildly whiplashed in a production seat fitted with head restraint -- passenger supported by UCLA calibrated 28 in. seat

to undergo a mild rearward head movement, not sufficient to cause a second head impact.

Left Rear Passenger - The only difference between the exposure conditions for the right and left rear seat passengers was their head offset (12-inch vs zero-inch). The left rear passenger (95th percentile) was positioned on a 21-inch bench seat, with a zero-inch head offset. During this rear-end collision, his head was forced rearward, and sustained an abrupt acceleration (20 G at 90 ms) on contacting the rear window glass; thereafter, the dummy rebounded uneventfully from a forced head-to-torso posture that did not appear to be injury-producing. The left rear seat passenger's impact without head offset is contrasted to the significant wedge-like impact of the right rear passenger, having the 12-inch head offset. This more pronounced collision response of the right rear seat passenger, as compared to the left rear seat passenger, was caused by differences in head posture. The right rear seat passenger's head was positioned 12 inches forward and his torso did not achieve immediate back support. His head developed a far greater relative velocity because of inertial action resulting in a more severe impact. The left rear seat passenger's shoulders were buried into the backrest about the same time his head contacted the glass; therefore, normal posture was maintained without significant differential velocity between the dummy and the car interior.

Observations, Following Collision - This 10 mph impact was not sufficient to cause sheet metal collapse interference with wheel rotation during the roll-out. During impact, the rear-ended car was advanced five feet (5 G at 15 ms, frame). The rear-ended car received collapse action that bowed the central area of the bumper inward approximately 2½ inches, relative to the outer left and right ends of the bumper. There was a slight trunk-lid indentation on the struck car caused by the forward edge of the striking car hood, indicating that the trunk lid had been contacted as well as the bumper. The right rear taillight lens cover was cracked but the left rear taillight was undamaged. There was slight buckling for the rear bumper support brackets. This 10 mph crash did not damage the gasoline tank and spillage did not occur.

EXPERIMENT 98 - The experimental plan for this 40 mph rear-end collision, Fig. 59, was essentially the same as used

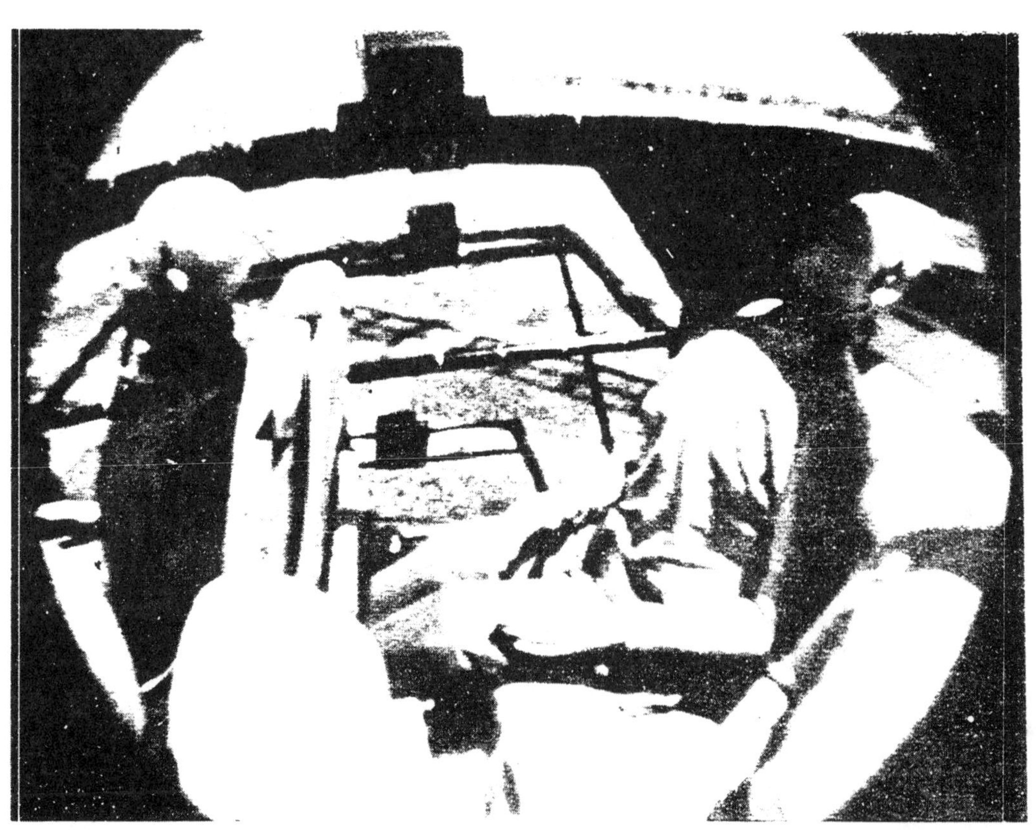

Fig. 58 - Right rear passenger, 95th percentile: 12 in. head offset impacting the rear window header

for the 10 mph impact (X-97). The car type, occupant type and seating arrangement, as well as passenger restraints represent a repeat of Experiment 97, except for the change in impact speed.

A conventional production bucket seat with a factory installed 6-inch head restraint extended the driver's seatback to 28 inches. The right front seat was a UCLA modified bucket seat, having an extended backrest of 28-inch height. This seatback was designed to yield at 16,100 in.-lbs. The rear seat passengers were provided with a conventional bench seat having a seatback height of 21 inches; the right rear passenger had a head offset of 12 inches; the left rear passenger had the same 3-inch head offset as the two front seat occupants, Fig. 60. All four occupants of the struck car were 95th percentile anthropometric dummies weighing 205 lbs; each was restrained with a lap belt. In order to eliminate the variation introduced by different bumper heights, their elevations were matched by applying an appropriate amount of compressed air to the shock absorbers built into the rear suspension systems of both the striking and struck vehicles.

The Collision - The rear-ended vehicle collapsed 2½ feet from this 40 mph collision before it moved forward appreciably. At the time this vehicle was collapsed 3½ feet, 1/10th second after initial bumper contact the frame sustained a peak acceleration of 19 G, Fig. 61. The impact forced the trunk lid open and as the rear-ended car advanced to a position 17 feet from impact, the vehicles separated from one another, Fig. 62. The struck car deviated 7 feet to the left of its original heading and this change in direction was attributed to the restriction of left rear wheel rotation by the collapsing rear-end structures. There was no visible dispersion of gasoline from the filler spout but the tank was ruptured and simulated gasoline commenced to drain onto the pavement 17 feet after the point of impact.

Driver - A 95th percentile dummy in a production, 22-inch, bucket seat was provided with a 6-inch head restraint, having the effect of extending his seatback height to 28 inches; this backrest was unmodified, except for the factory installed head restraint. The acceleration of the car (19 G at 100 ms) forced the driver abruptly against the seatback causing the seat adjustment track to

Fig. 59 - Forty-mph rear-end collision, Exp. 98

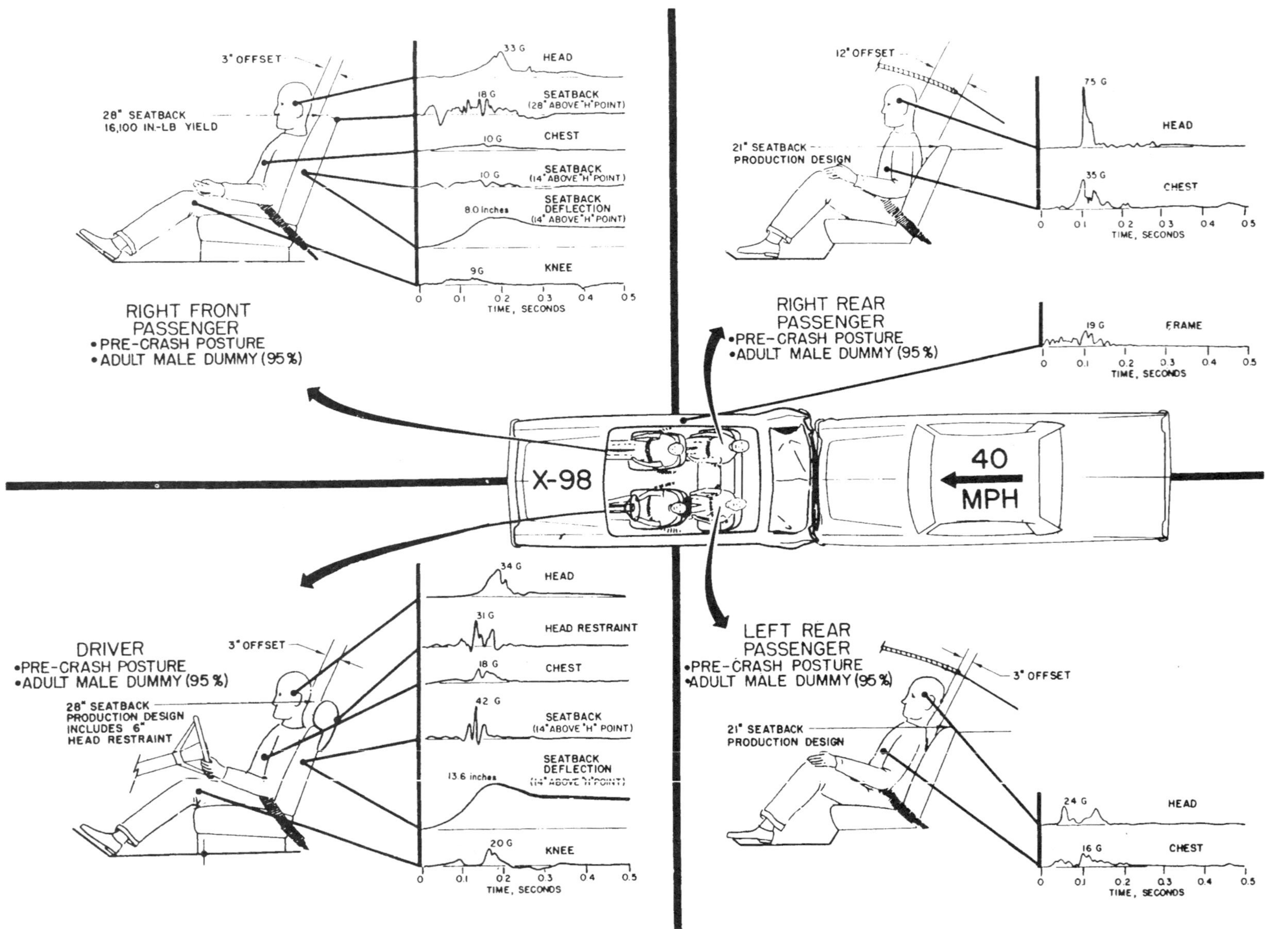

Fig. 60 - Transducer patterns shown with occupant posture and restraint condition at the start of crash, Exp. 98

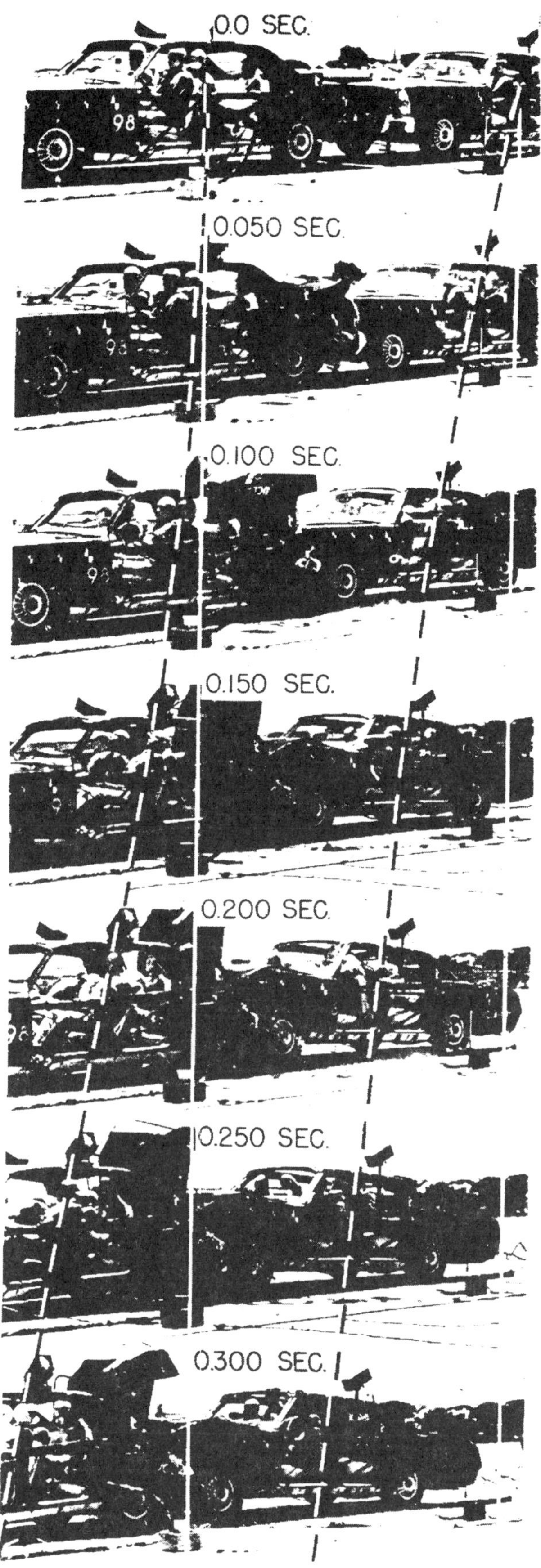

Fig. 61 - Crash sequence, 40 mph, Exp. 98

fail, allowing the seat to displace rearward 6 inches. Thereafter, the seatback yielded 18 degrees (38 degrees rearward of vertical) to a reclined position that was terminated by impact with the knees of the left rear seat passenger; seatback peak accelerations were 31 G at 135 ms for the head restraint and 42 G at 135 ms for the 14-inch elevation. The driver's chest reached its peak acceleration (18 G at 140 ms) and his head continued to rotate rearward and downward for a head peak acceleration of 34 G at 190 ms, Fig. 63. At its maximum rearward movement, the driver's torso had reclined 35 degrees beyond vertical and his head had rotated 70 degrees beyond the axis of his reclined torso. The head was restrained from additional rotation by the head restraint. Thereafter, the driver rebounded forward, lightly contacting the steering wheel, before slumping back into a semi-reclined posture, Fig. 64.

The Right Front Seat Passenger - was provided with a UCLA modified 28-inch seatback having a 16,100 in.-lb yield; he was forced rearward by the extreme impact, yielding the seatback until it contacted the rear passenger's knees. At this time, he recorded a peak chest acceleration of 10 G at 160 ms, Fig. 60. Thereafter, the head of the front passenger continued to rotate rearward and downward until it reached an angle of 79 degrees relative to the semi-reclined torso axis; he sustained a peak head acceleration (33 G at 205 ms) as his head struck the chest of the rear seat passenger, Fig. 65. Notwithstanding the 16,100 in.-lbs. yield bar designed into this seat, the seatback bent rearward approximately 33 degrees (53 degrees beyond vertical) and would have gone further were it not for the seatback interference with the rear seat passenger's knees. If this seatback had been provided with greater strength, the amount of rearward deflection would have been substantially reduced and the torso would not have shifted up the inclined seatback, exposing the neck to the possibility of serious whiplash. As found in prior experiments, stiffer seatbacks assure more effective support of the occupant during rear-end collisions, providing the support is high enough to also resist rearward movement of the head.

The Right Rear Seat Passenger - was

seated on a conventional bench seat with a 21-inch backrest. This severe 40 mph impact forced the torso deep into the seatback upholstery, resulting in a peak chest acceleration of 35 G at 95 ms; the head struck the header over the rear window a severe blow (75 G at 105 ms) and thereafter remained stationary even though the upper torso continued to be forced further into the seatback, thereby positioning the head in a leaning-forward at maximum loading. The 12-inch offset head position for this right rear passenger accounted for the substantially higher head acceleration derived on contacting the header than sustained by the left rear occupant (75 G vs. 24 G). His quick rebound brought him into contact with the rearward movement of the right front passenger accounting for the chest-

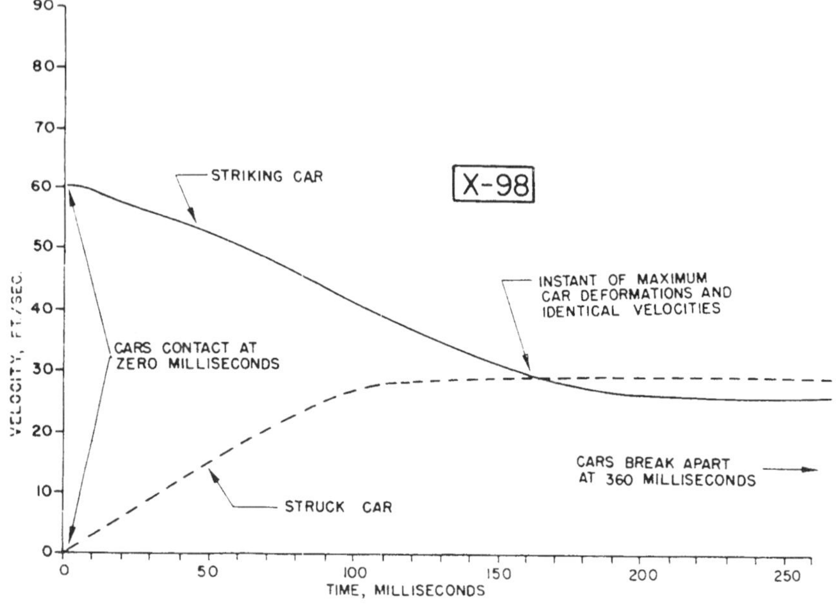

Fig. 62 - Velocity changes, during a 40 mph rear-end collision

Fig. 63 - Peak head acceleration for this rear-ended 95th percentile driver with a tack-on head restraint was 34 G, Exp. 98, 40 mph

to-head contact that served to check whiplash action for the front seat passenger, Fig. 65.

The Left Rear Passenger - was forced rearward in a manner similar to that of the right rear passenger, except that his more natural posture (3-inch head offset) brought his head into contact with the rear window header and glass sooner, resulting in a moderate head acceleration

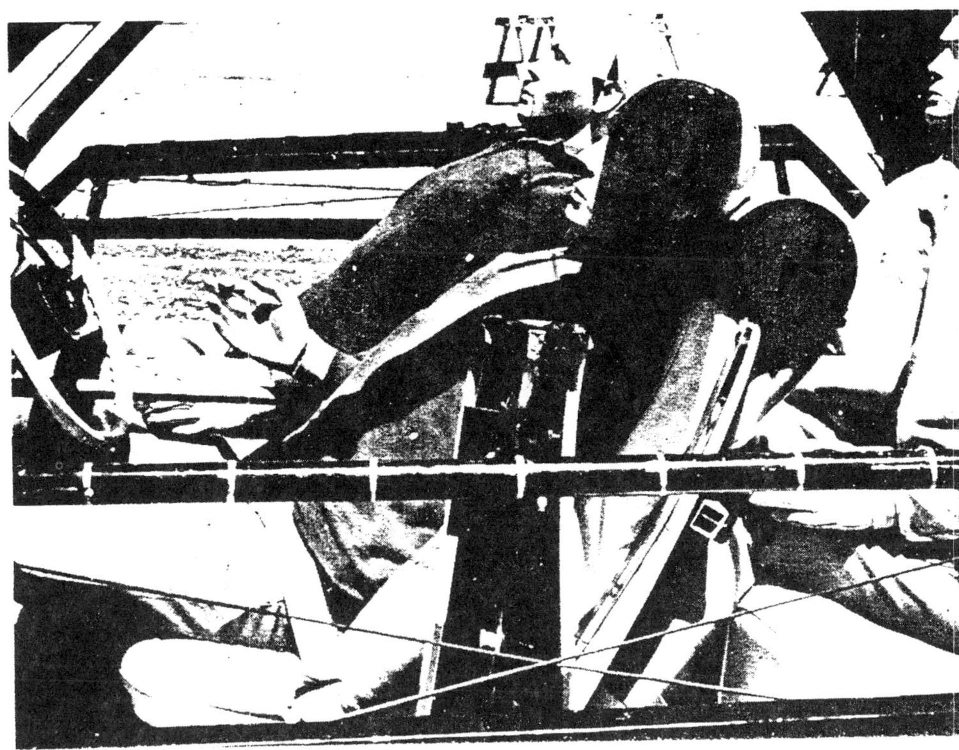

Fig. 64 - The driver, a 95th percentile dummy, rebounded into the steering wheel, then slumped back into his seat, Exp. 98, 40 mph

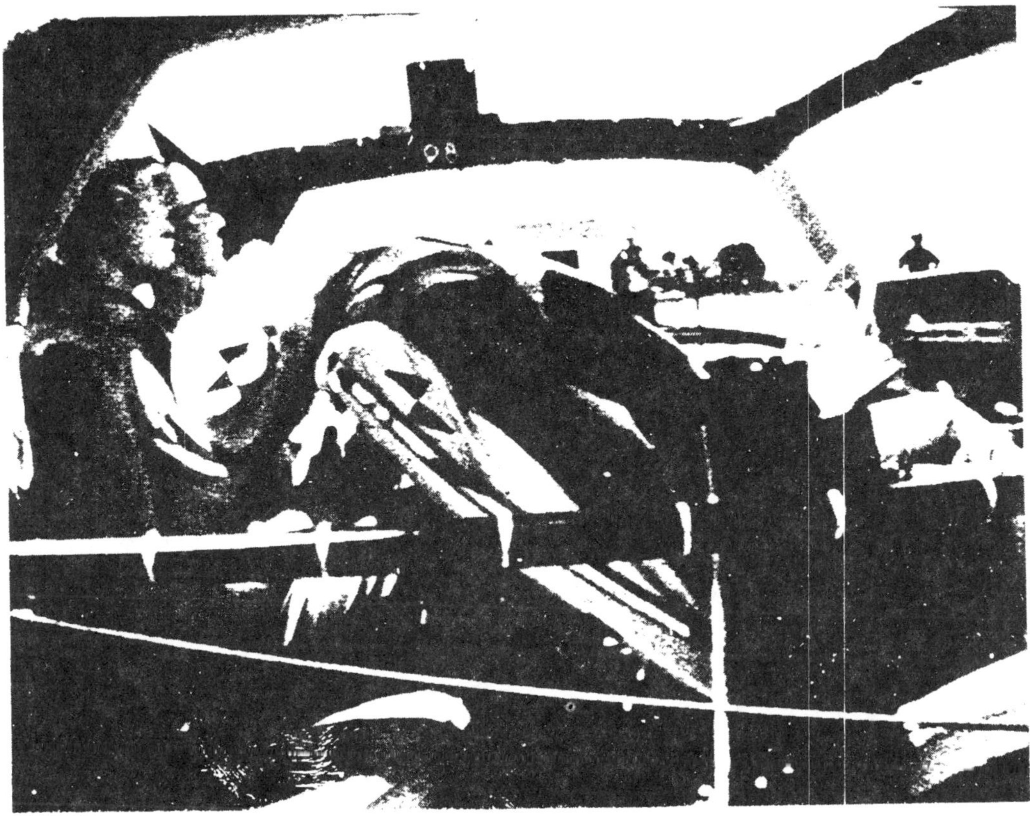

Fig. 65 - Front seat passenger, 95th percentile, whiplashed by 40 mph rear-ender, has head flailed against chest of the rear seat passenger (28 in. modified 16,100 in.-lb. seat)

Fig. 66a - Underside of struck car, following a 40 mph collision, Exp. 98

Fig. 66b - Extreme collapse is evident for this 40 mph collision, Exp. 98

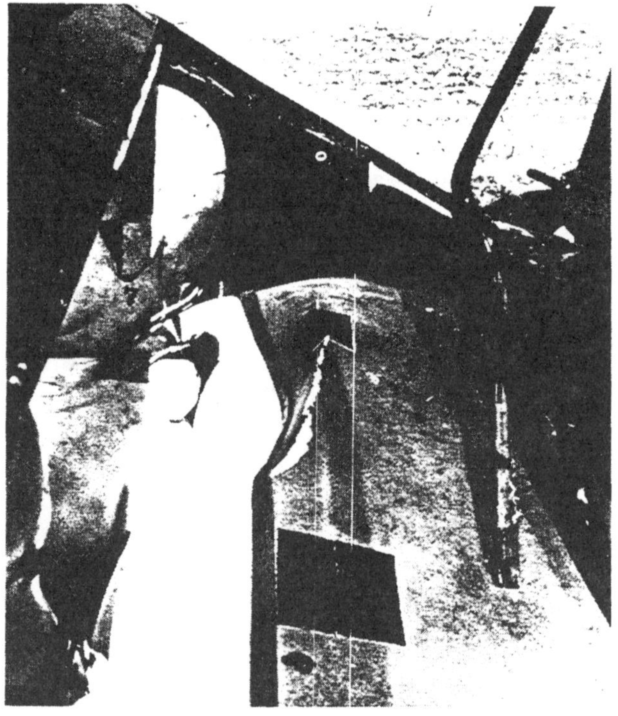

Fig. 67 - Rear seatback deformation, 40 mph collision, Exp. 98

(24 G at 55 ms). Torso acceleration was also lower than the right rear occupant, reaching a maximum of 16 G at 100 ms. Rebound was similar to that of the right rear passenger; both buried their chins into their chests as they returned toward a relatively natural seated posture.

Observations, Following Collision - The rear-ended vehicle was propelled forward 96 feet and deviated to the left 7 feet during the run-out. The rear-end of the struck car collapsed forward against the rear wheels, Fig. 66(a); the trunk was compressed sufficiently to push the spare tire two inches into the rear seat-back panel. Deformation at the bumper level was 3 feet and at the rear tail-light level, 3 feet, 6 inches, Fig. 66(b).

The backrest of the driver's production bucket seat was crushed rearward to within 6 inches of the edge of the rear seat cushion; this close proximity was partly attributed to the reclined nature of the seatback and partly to the failure of the track adjustment mechanism allowing the seat to slip three inches rearward of the rear stop position; this rearward shift of the seat position developed approximately three inches of slack in the lap belt. The seatback hinge mechanism yielded slightly.

The right front seat backrest with 16,100 in.-lb. yield bars was forced rearward to a reclined position within 9 inches of the front edge of the rear seat.

The rear bench seatback was crushed 1½ inches into the rear bulkhead window shelf structure on the right side, Fig. 67. During the collision, the rear window of this struck car was dislodged from its frame but remained partially attached at its base.

The gasoline tank filler spout remained attached to the tank; leakage occurred from a puncture on the front side near the center of the tank.

EXPERIMENT 99 - was a 30 mph rear-end collision between two identical 1967 Ford 4-door sedans, Fig. 68. Two 95th percentile anthropometric dummies were seated in front and two 50th percentile anthropometric dummies in the rear seat, Fig. 6(c). The driver's UCLA modified bucket seat had a 25-inch seatback of rigid design, Fig. 3(a); the right front bucket seat had a UCLA modified 28-inch seatback, also of rigid design. Both rear seat passengers sat on a standard production bench-type seat (21-inch seatback). Each front seat dummy weighed 205 lbs. and each was postured with a zero-inch head offset. The rear seat occupants weighed 165 lbs. each; the left rear passenger was postured with zero-inch head offset and the right rear passenger was postured with a 12-inch head offset. All four dummies for this experiment were restrained by lap belts. The struck vehicle weighed 4640 lbs. including 25 gallons of simulated gasoline; the striking car weighed 4670 lbs.

<u>The Collision</u> - The struck car was collapsed 1.2 feet from this 30 mph collision before visible motion was observed; however, maximum frame acceleration occurred 35 ms after bumpers contacted, reaching 12 G when the rear-ended vehicle had sustained only four-tenths feet of collapse, Fig. 69. During the course of the initial impact, the rear-ended vehicle separated after 12 feet of travel, veered five inches to the right and was subsequently rear-ended again by the following vehicle (after 34 feet of advance) and this impact accentuated the arcing of the rear-ended vehicle further to the right; the rear-ended vehicle was projected 52 feet beyond the point of initial impact and the striking car, 39 feet. No significant amount of fuel was observed to have been forced from the rear-ended vehicle, at least until it had reached its position of rest.

Fig. 68 - Thirty mph rear-end collision, Exp. 99

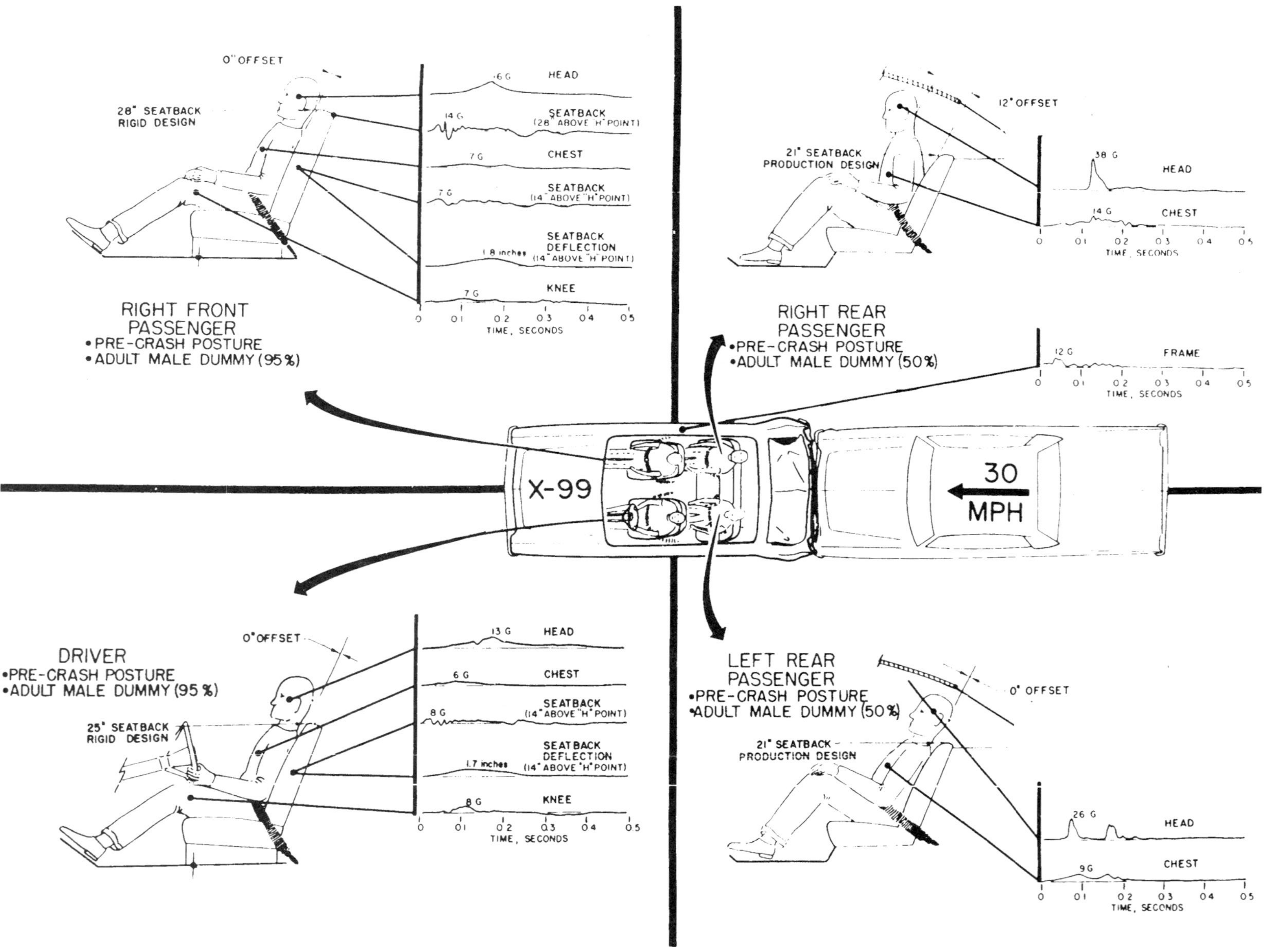

Fig. 69 - Transducer patterns shown with the occupant posture and restraint condition at the start of crash, Exp. 99

The Driver - was a 95th percentile dummy seated on a UCLA modified seat having a rigid 25-inch seatback. This seat provided positive back support but was not of sufficient height to provide adequate head restraint. As a consequence, the driver started whiplash during this 30 mph rear-end collision to the extent that his head rotated rearward and downward 71 degrees beyond vertical and because of the reclining action of the seatback and torso, this amounted to 44 degrees head rotation, as measured with respect to the torso axis, Fig. 70. At this time he sustained a resultant peak head acceleration of 13 G at 175 ms. The chest recorded a 6 G peak acceleration at 80 ms and the seatback (at the 14 inch level) reached a peak acceleration of 8 G at 30 ms.

The Right Front Seat Passenger - during this 30 mph rear-end collision, deflected his 28-inch seatback rearward only 7 degrees, suggesting that the designed seatback yield strength of 2360 lbs. at 14-inch elevation represents an adequately rigid seat for this impact speed. His upper torso was pushed firmly against the seatback (7 G at 105 ms) and his head was flailed rearward receiving 16 G at 165 ms as its motion was checked by the top of the seatback. The seatback top applied a concentrated loading to the base of the skull as his head rotated slightly over its top edge, Fig. 71. Head rotation, measured with respect to his torso axis, was 34 degrees -- 10 degrees less than head rotation for the driver in his 25-inch seatback. Following this impact, restitutional forces repositioned the right front passenger forward to a vertical posture. With all variables constant except for the seatback height, the relative effectiveness of the 28-inch seatback is contrasted with that of the 25-inch seatback in Fig. 72. This figure shows the driver in the initial stages of whiplash before his seat checks further whiplash action; the passenger has had his head movement restrained when his head rotated rearward 34 degrees relative to his torso axis.

Right Rear Seat Passenger - A 50th percentile adult dummy was postured with a 12-inch head offset from the plane of the seatback. On impact, the 21-inch production bench seatback (rigid design) was abruptly accelerated into the back of this dummy causing his head to snap rearward. Owing to this passenger's shorter height, his head missed the header, striking the rear window a glancing blow (38 G at 130 ms). On sliding down the incline of the rear

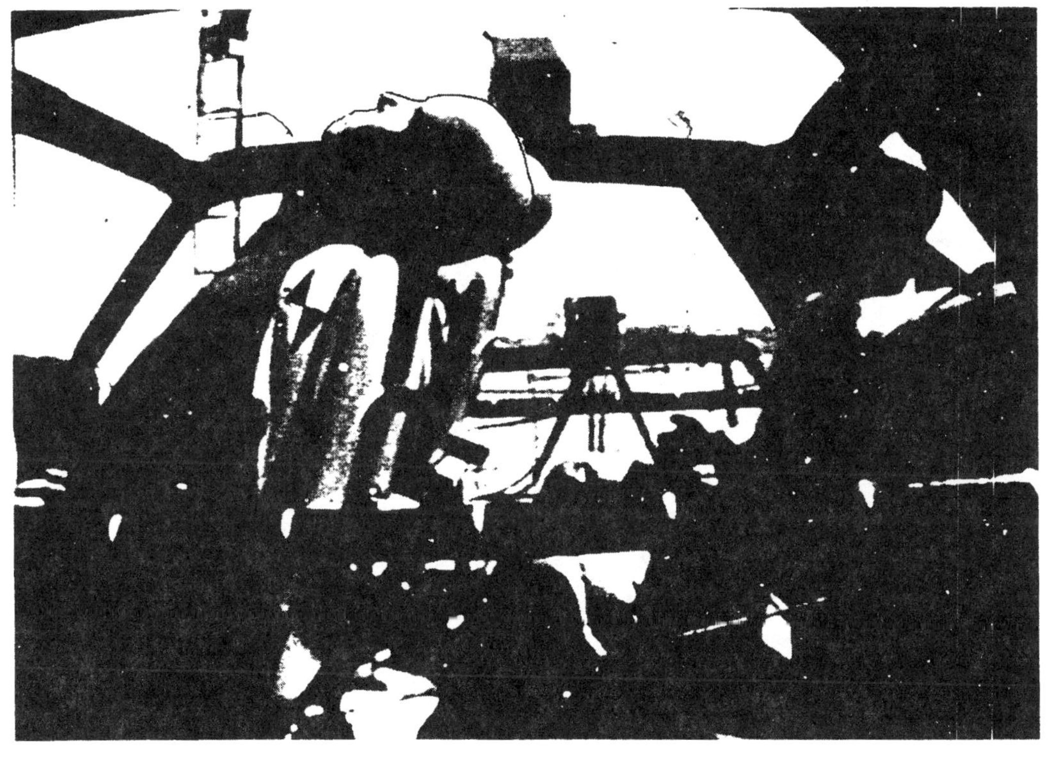

Fig. 70 - Driver (95 percentile) in a UCLA-rigid 25 in. seat, Exp. 99, 30 mph

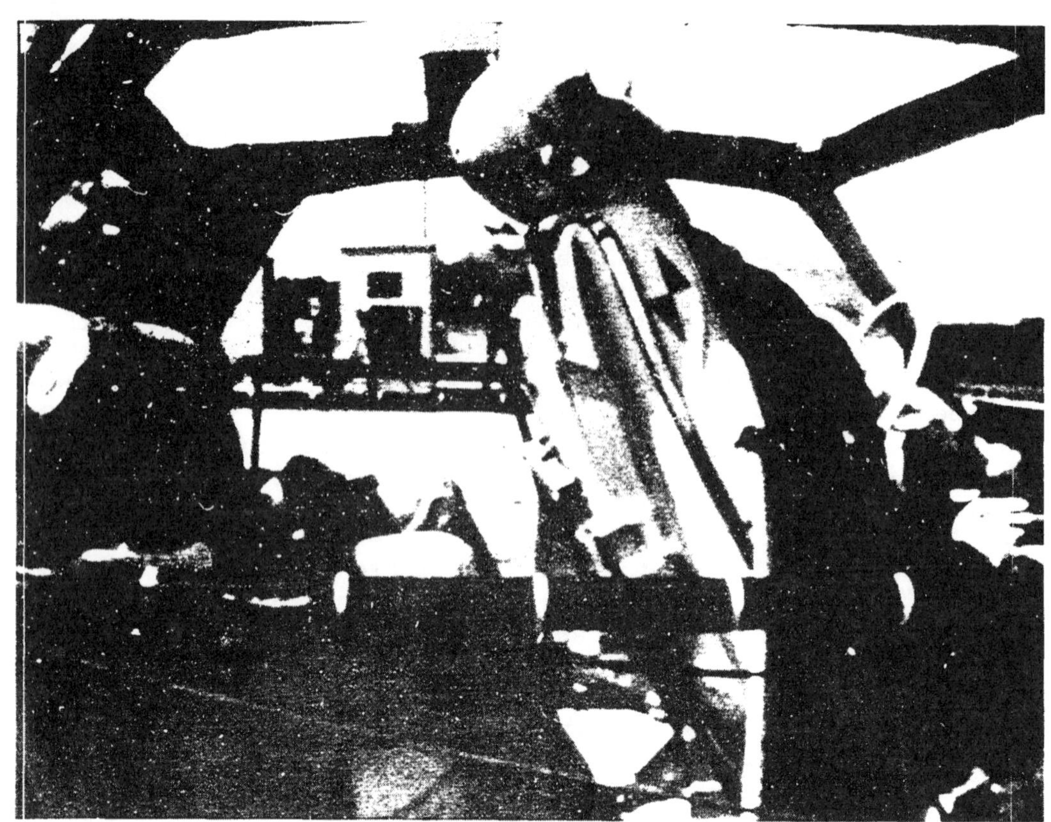

Fig. 71 - Front passenger (95 percentile) in a UCLA-rigid 28 in. seatback, deflected rearward five deg, Exp. 99, 30 mph

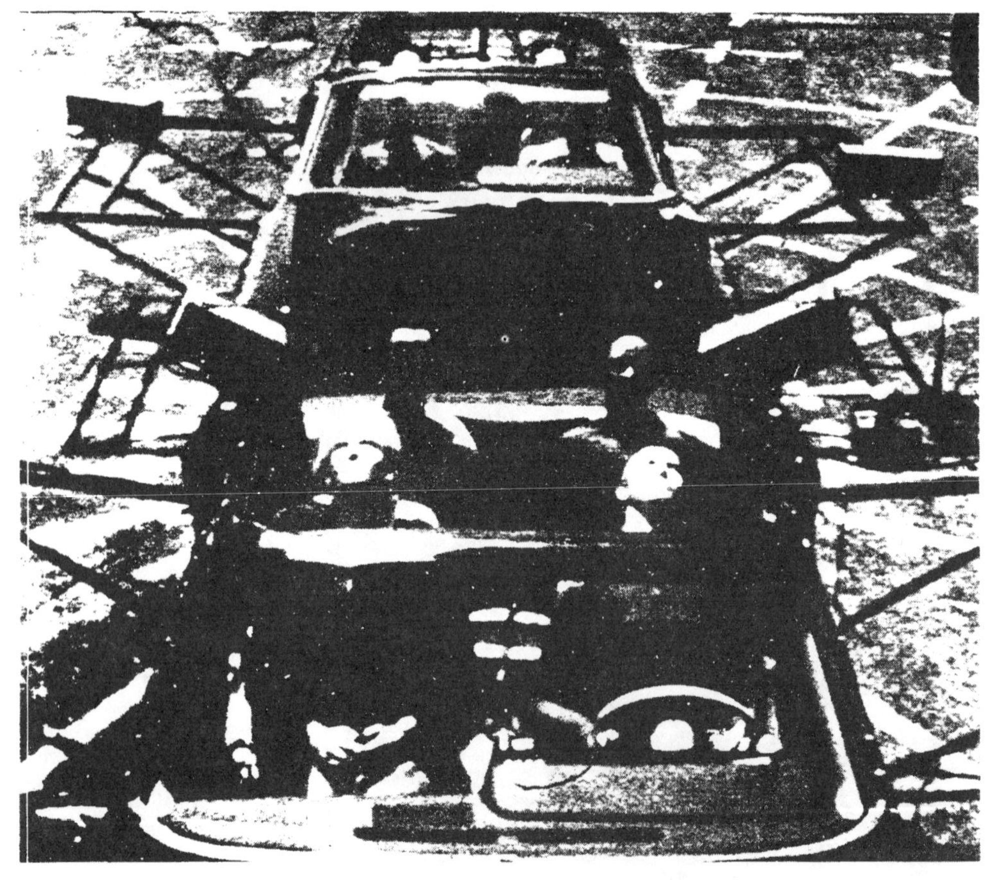

Fig. 72 - Contrast for 30 mph collision between 95 percentile driver in a rigid 25 in. seatback and 95 percentile front passenger in a rigid 28 in. seatback, 30 mph, Exp. 99

window, his head underwent a moderate whiplash movement before rebounding upward, striking the header with a rolling movement. Thereafter, this passenger came to rest in a slumped position. His 14 G peak resultant chest acceleration coincided in time with his head peak-G time (130 ms).

The Left Rear Passenger - was a 50th percentile male adult anthropometric dummy, identical to the right rear dummy. He was seated on the same bench seat with a 21-inch rigid seatback, but had a zero-inch head offset. As the impact developed, his upper torso was crushed against the seatback (9 G at 90 ms) and his head flailed rearward to strike the rear window twice, approximately four to six inches from the header (26 G at 75 ms and 18 G at 165 ms). Thereafter, he rebounded to a position where his head snapped forward but his shoulders were not rebounded from the seatback.

In this experiment, significant observations were: (a) front and rear rigid seat units minimized rebound for all occupants; however, for this tall (95th) dummy the degree of whiplash appeared to be significantly higher for this rigid seat configuration when compared to an exposure that was identical, except for lower seatback strength (X-96: 16,100 in.-lb). Additionally, the rear window serves to limit rearward movement of the head for rear seat occupants at the risk of causing injury producing impacts to the head and the possibility of spinal injury.

Observations, Following Collision - This 30 mph impact propelled the rear-ended car 52 feet beyond its pre-crash position and deviated it approximately 8 feet to the right of its pre-crash heading. The readily crushable rear-end structure collapsed 2 feet, 6 inches, as measured at the trunk lid elevation and 2 feet, 4 inches, as measured at the bumper. The spare tire was impacted by the collapsing sheet metal but did not show evidence of forward displacement. Sheet metal deformation progressed forward from the rear bumper, collapsing the trunk-well to the floor pan axle step and gasoline tank. Both rear tires were punctured by the rear fenders' sheet metal being crushed against the rear edges of the tires; the rear tires also received some circumferential cutting as the roll-out progressed. The collapsing action of the sheet metal around the tires and into the tires caused the rear wheels to skid and these skid marks could be seen approximately four inches forward from their pre-crash position. A typical debris pattern started at approximately 16 feet forward of the struck car's rear bumper pre-crash position and continued to the position of rest for the struck car.

EXPERIMENT 100 - was a 30 mph rear-end collision, Fig. 73, involving two identical 1967 Ford 4-door sedans with two-95th percentile anthropometric dummies in the front seats and two-50th percentile anthropometric dummies in the rear seat. The two front bucket seats were UCLA modified designs with seatback

Fig. 73 - Two 95th percentile front seat dummies, driver with 25 in. and passenger with 28 in. calibrated seatbacks, 30 mph

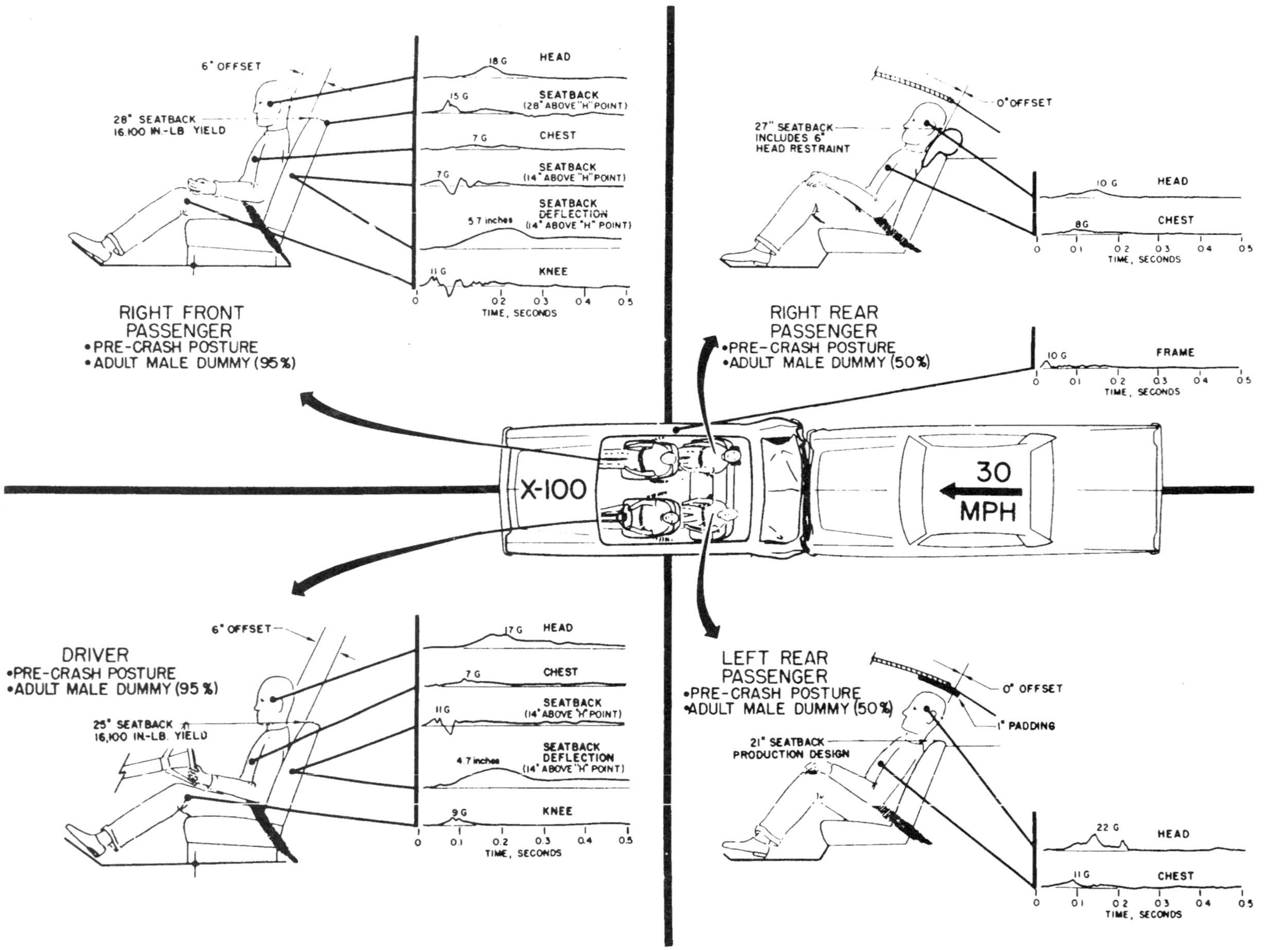

Fig. 74 - Transducer patterns shown with occupant posture and restraint condition at the start of crash, Exp. 100, 30 mph

heights of 25 inches for the driver and 28 inches for the right front passenger; these seatbacks were calibrated to yield at 16,100 in.-lbs. The rear seat was a standard production bench seat having a 21-inch backrest with a 6-inch head restraint for the right rear position. All four occupants were provided with conventional floor-mounted seat belts. The gross weight of the struck car was 4640 lbs. and the striking car was 4670 lbs; both vehicles had 25 gallons of simulated gasoline.

The Collision - The rear-ended car had sustained one foot of collapse from this 30 mph collision before it began significant movement; at this instant (45 ms) the frame acceleration peaked at 10 G, Fig. 74. There was overriding action of the striking car bumper by the time the rear-ended car had been advanced five feet; these vehicles separated at the 15-foot position only to lightly re-collide at the 55-foot position. The entire run-out of the two vehicles was along their original heading.

Driver - a 95th percentile adult male dummy was postured so that his torso axis was approximately 10 degrees forward of the seatback plane and his head was positioned 6 inches forward of the zero head offset position. His UCLA modified bucket seat had a 25-inch backrest with a calibrated yield of 16,100 in.-lbs.

The 10 G passenger compartment acceleration of this 30 mph rear-end collision forced the driver firmly against his backrest and snapped his head rearward beyond the top of the seatback to an extreme whiplash position; his eyes were actually pointed rearward beyond vertical. The seatback, normally inclined rearward 20 degrees beyond vertical, was forced to a position 40 degrees beyond vertical. His head continued its rearward rotation to a position 69 degrees beyond his torso axis. During this extreme whiplash loading condition (17 G at 210 ms) the back of the driver's head was touching the rearward most portion of the top edge of the backrest; the typical shift of the torso up the plane of the seatback as the torso was accelerated forward (7 G at 110 ms) served to intensify whiplash by exposing an additional amount of unrestrained neck, Fig. 75.

The Right Front Seat Passenger - was a 95th percentile anthropometric dummy, identical to the driver in size and weight. The only variable was the 28-inch seatback calibrated to the same yield strength (16,100 in.-lbs.) as the driver's 25-inch seatback. The pas-

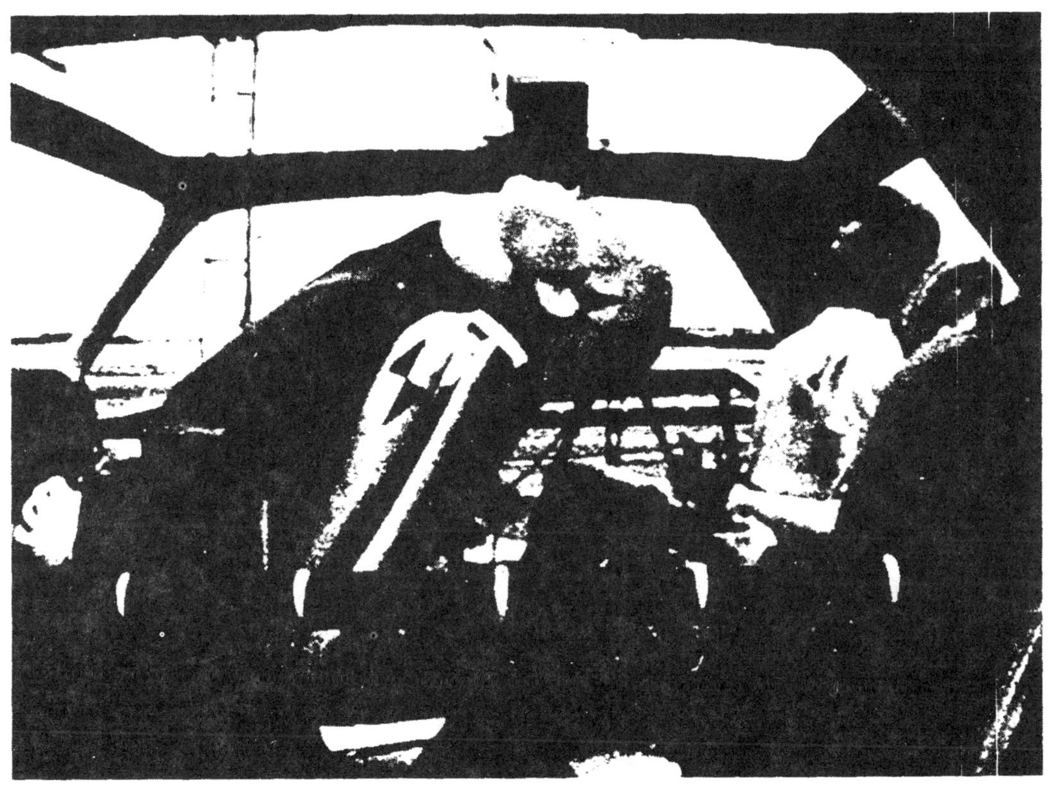

Fig. 75 - Driver, 95th percentile, 25 in. modified seat, 30 mph

senger had the same 6-inch head offset and also the same torso-to-seatback posture of 10 degrees forward of the seatback plane. During the acceleration phase of the collision, the passenger's backrest deflected rearward until it almost touched the right rear seat passenger's knees. Immediately thereafter, the front passenger initiated a whiplash movement over the top edge of the 28-inch seatback; however, the additional head restraint provided by the 28-inch seatback served to check his rearward head displacement. His head sustained a peak resultant acceleration of 18 G at 175 ms. The extended seatback maintained a non-injury producing posture, as shown in Fig. 76. At his maximum postural change, the right front seat passenger's torso and seatback deflected rearward 23 degrees to a position 43 degrees rearward of vertical; at this instant, his head had rotated 66 degrees beyond vertical or 24 degrees beyond the torso axis. Thus, it is seen that the right front seat passenger was adequately protected by the 28-inch seatback as far as whiplash exposure is concerned for this 30 mph rear-end collision, notwithstanding the 6-inch head offset position prior to impact.

Right Rear Seat Passenger - The standard 21-inch production back seat was fitted with a head restraint for the right occupant, elevating the seatback height from 21 inches to 27 inches. This right rear seat passenger was positioned with a zero-inch head offset. Inertial forces thrust his torso against the seatback with an 8 G peak resultant acceleration at 90 ms as his head buried into the head restraint, it grazed the rear window header to strike the glass (10 G peak at 135 ms), Fig. 76. The torso remained in the same relative position to the seatback and his head did not extend above the head restraint which was sufficiently firm to maintain constant alignment of head and torso axes during the accelerative phase of this impact. However, the close proximity of the swept-back rear window to the head position for this 50th percentile adult dummy accounted for the head brushing the window after the seat cushion had been deflected rearward. Rebound was relatively uneventful and the dummy slumped in his seat, slightly leftward,

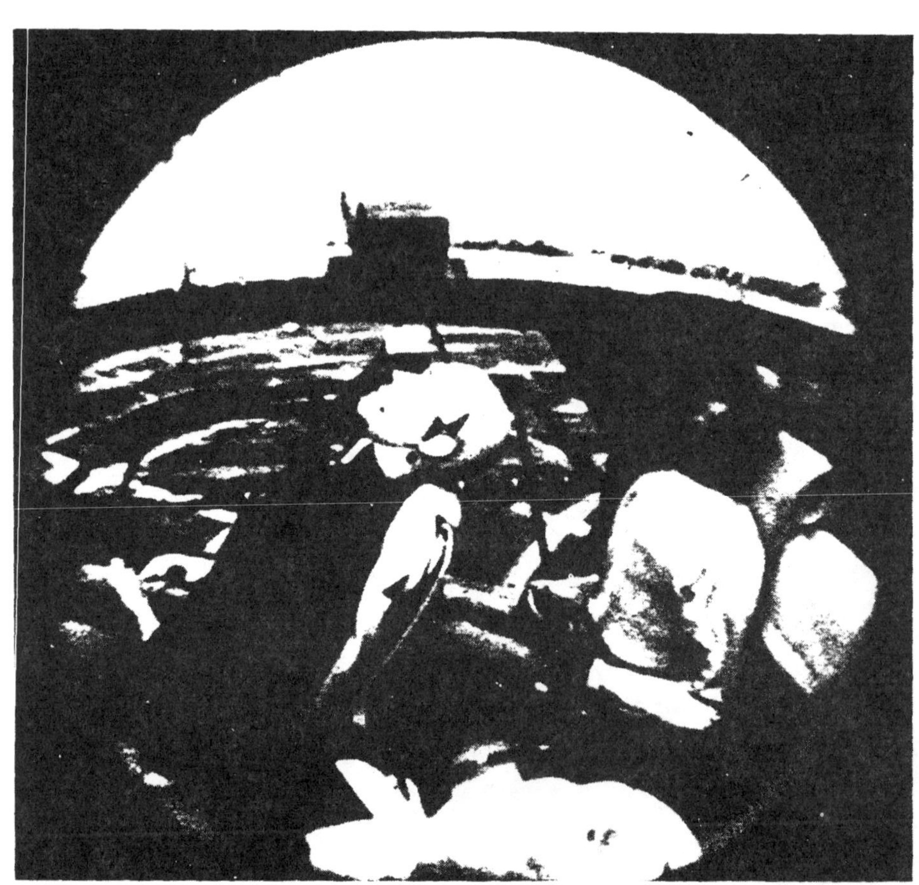

Fig. 76 - Front passenger, 95th percentile, 28 in. modified seat, 30 mph; note head support for right rear dummy

with his back against the backrest.

Left Rear Seat Passenger - A 50th percentile anthropometric male dummy, identical to the right rear passenger, and with the same zero-inch head offset was seated in the left rear seat position. The only variance was the absence of a head restraint, exposing him tothe conditions of a conventional 21-inch bench seatback. During the collision acceleration phase, the inertial loading of the left rear seat passenger forced him against the seatback (11 G at 85 ms) causing his head to lightly brush the rear window and to slide down the plane of the rear window until his head impacted the rear window shelf (22 G at 140 ms); thereafter, he followed a rebound course very similar to the path of his original loading; he finally came to rest in a slightly flexed-forward position, without noticeable torso and shoulder rebound from the backrest. A thin padding was provided over the header and top portion of the rear window, but the head of this 50th percentile dummy passed below the padding to strike the rear window shelf directly. This average size occupant underwent a moderate to serious whiplash, as evidenced by his adverse posture, Fig. 77.

Rebound for each of the four passengers in this struck vehicle was mild and was not of a nature sufficient to develop further exposure to injury.

Observations, Following Collision - Collapse to the rear-ended vehicle amounted to 2 feet, 3 inches at the elevation of the rear bumper and to a maximum of 2 feet, 6 inches at the elevation of the taillights. The taillights were pushed flat against the forward wall of the trunk-well. The accordion-like action of this rear-end folded the trunk lid almost double. The spare tire, forward of this point, remained firmly bolted to the floor pan axle step and did not push forward into the back of the rear bench seat. Sheet metal from the right rear fender was crushed into the trailing edge of the right rear wheel and deflected the tire approximately one inch inward, but was not sufficient to puncture the tire. Sheet metal collapsed around the left rear tire and produced circumferential cuts at the sidewall; the tire remained inflated although wheel rotation was restricted somewhat as evidenced by the presence of light skid marks on the pavement. Improvements made in the fuel-tank filler-spout accounted for a significant reduction in

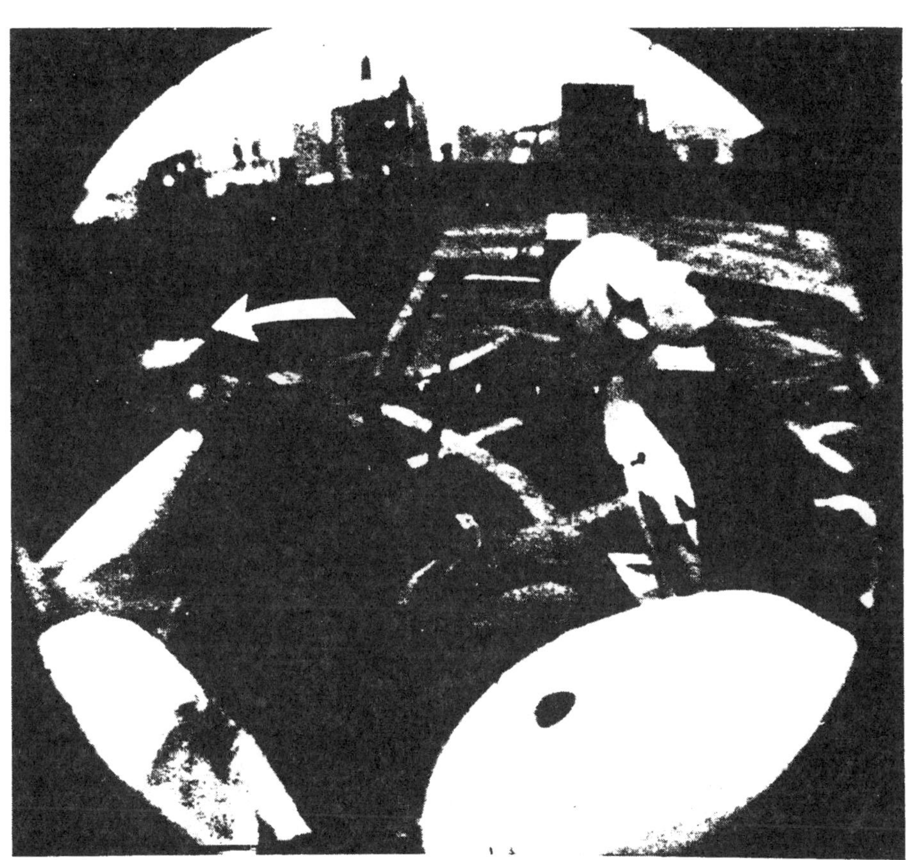

Fig. 77 - Left rear passenger, 50th percentile, whiplashed by 21 in. production bench seatback, Exp. 100, 30 mph, (arrow)

the leakage of gasoline around the dented filler cap, as compared with prior 30 mph crashes.

The driver's 25-inch seatback sustained a permanent rearward yield of 7 degrees and the right front passenger's 28-inch seatback had 11 degrees; both of these modified bucket seats were provided with 16,100 in.-lb. resistant seatbacks.

This rear-end collision advanced the struck car to a position 60 feet forward of its pre-crash position within six inches of the pre-crash axis.

EXPERIMENT 101 - was a 30 mph rear-end collision between two identical 1967 Ford 4-door sedans, each carrying four anthropometric dummy passengers, two 95th percentile dummies in front and two 50th percentile in the rear, Fig. 78. All four occupants wore lap belts. The driver was seated in a UCLA modified bucket seat with a 25-inch seatback having a 16,100 in.-lb. calibrated yield bar. The extended 25-inch seatback was evaluated without a tack-on head restraint and the dummy posture was adjusted to provide a 12-inch head offset. Beside him in the right front seat was another 95th percentile dummy with a UCLA modified 16,100 in.-lb., 28-inch seatback. This front passenger's head offset was also adjusted to 12 inches, Fig. 79. Both rear seat passengers were 50th percentile adult male anthropometric dummies seated in the right rear and left rear positions on a UCLA modified bench seat having a 25-inch seatback, without separate head restraints. The significant difference between the left and right rear seat passenger's exposure was head offset; the right rear dummy was postured with a zero-inch head offset and the left rear dummy was postured with a 6-inch head offset. The total weight of the struck car and occupants, including instrumentation, 25 gallons of simulated gasoline in the fuel tank was 4630 lbs. The striking car weighed 4680 lbs.

The Collision - This 30 mph rear-end collision progressed to 9 inches mutual collapse before the struck car accelerated to a peak of 11 G, 25 ms after the bumpers contacted; the rear-ended car had collapsed 14 inches (50 ms) without significant movement of the passenger compartment. The cars separated at 13 feet beyond the point of initial impact and later re-collided lightly, owing to metal collapse-action braking the struck car speed. The rear-ended vehicle advanced 103 feet before coming to rest and the striking vehicle stopped at 60 feet.

The Driver - This 95th percentile dummy supported by a UCLA modified 25-inch seatback (16,100 in.-lbs.) and postured with a 12-inch head offset, was forced against his backrest during vehicle acceleration that reached 11 G peak at 25 ms for the passenger compartment. His hips pressed firmly against the seatback registering 12 G at 95 ms

Fig. 78 - The 95th percentile driver in a 25 in. modified seat, the 95th percentile front passenger in a 28 in. modified seat, both with 12 in. head offsets, Exp. 101, 30 mph

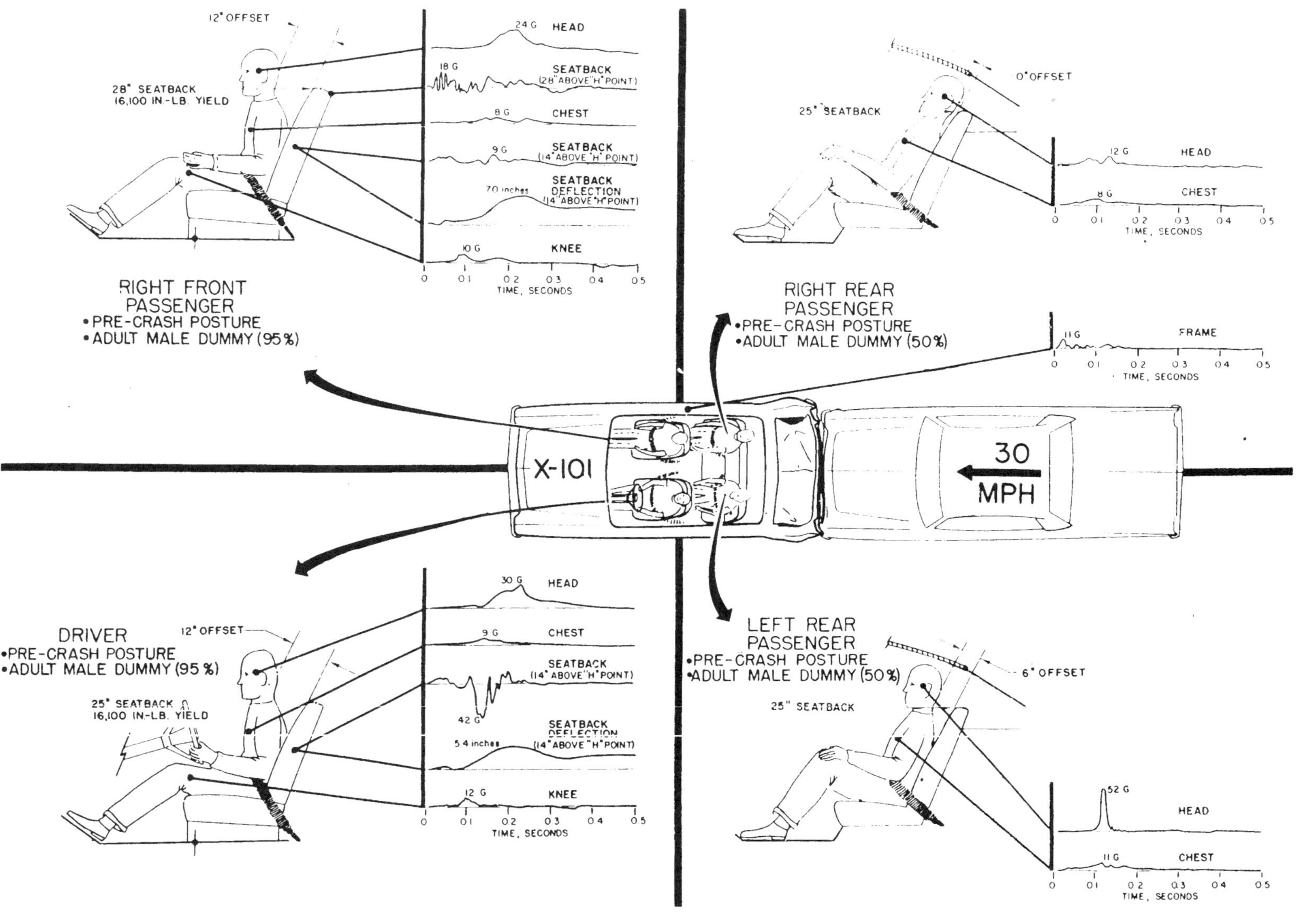

Fig. 79 - Transducer patterns shown with occupant posture and restraint condition at the start of crash, Exp. 101, 30 mph

and his chest sustained 9 G at 135 ms. Thereafter, his unsupported head flexed rearward and downward in an unusually abrupt manner, owing to the initial 12-inch offset. Seatback deflection and the limited support provided by this 25-inch seatback contributed to the elevation of the torso relative to the seatback. At the instant the driver was undergoing a complete whiplash condition, his shoulders were level with the top of his seatback; his head and neck were flexed over the backrest, bringing the base of his skull against the top edge of his seat (30 G at 225 ms), Fig. 80. This exposure constituted a severe whiplash; the driver was actually looking 38 degrees rearward of vertical with an inverted head position occasioned by a maximum dorsi-flexed neck condition at this maximum whiplash, his head had rotated approximately 80 degrees rearward of his torso axis. His torso assumed a 48 degree rearward-of-vertical rotation, indicating that the hips have moved away from the 42-degree seatback slope sufficiently to account for the 6-degree difference between deflected seatback angle and torso angle. Thereafter, the driver rebounded forward off the seat in a rather violent manner, contacting the steering wheel which rebounded him rearward into a reclined but seated posture.

The 16,100 in.-lb. yield design of the driver's seatback was adequate to resist the rearward inertial forces of the occupant and backrest to prevent it from striking the knees of the left rear passenger. In addition to the resistance of this seat, the presence of a smaller size 50th percentile anthropometric dummy in the left rear seat position provided additional space as contrasted with a 95th percentile dummy that had been seated in that position in some of the prior experiments.

The Right Front Seat Passenger - was also a 95th percentile dummy postured with a 12-inch head offset. The only variance from the driver was the seatback height, 28 inches vs. 25 inches; this 28-inch seatback had a 16,100 in.-lb. yield which was identical to the driver's seatback strength.

As the collision progressed, this dummy crushed against the seatback (hips -- 10 G at 90 ms, chest -- 8 G at 165 ms) with the top part of the 28-inch seatback coming into contact about mid-point of his neck. This was a lower position of neck-to-seat contact than

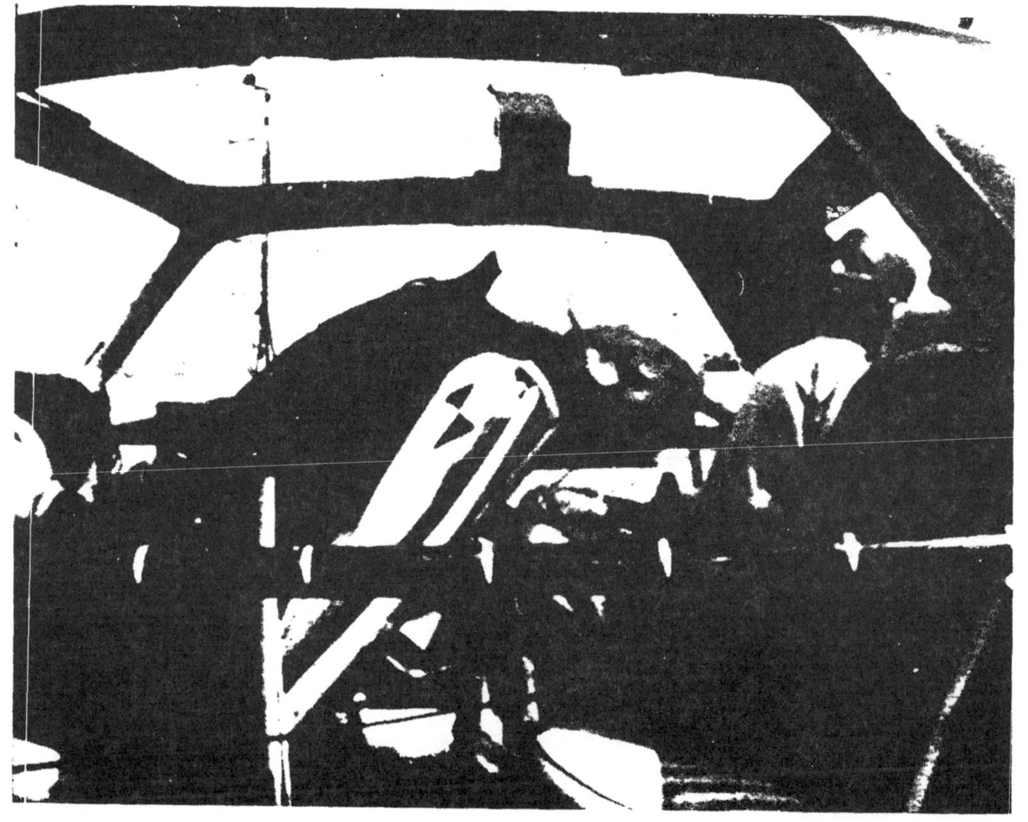

Fig. 80 - During this severe whiplash, the torso elevated sufficiently to bring the base of the skull against the top metal edge of the seat, Exp. 101, 30 mph

this same size occupant experienced when his head had a zero-inch head offset. However, the combination of reclined position of the seat and the inertia of the head offset appeared to overcome much of the advantage of a 28-inch seatback as evidenced by the severe whiplash that resulted, amounting to a 97-degree rearward-of-vertical head rotation about the seatback, or 49 degrees relative to the torso; at this time seatback deflection was 42 degrees rearward of vertical. His peak head acceleration was 24 G at 205 ms, Fig. 81. During whiplash, the midback was seen to arch away from the seat, which accounts for the difference in the torso angle and seatback angle of about 7 degrees. Thereafter, the right front seat passenger was rebounded rather violently forward into a semi-jackknifed posture, coming into close proximity with the instrument panel. However, his head did not strike the instrument panel or forward interior, inasmuch as forward movement was checked by his chin depressed into his chest and by the restraint-action of his seat belt.

Right Rear Seat Passenger - This 50th percentile adult dummy was seated on a UCLA modified bench seat, having a 25-inch seatback; he was positioned with a zero-inch head offset. As his seatback crushed forward against his back, his chest was accelerated forward reaching a peak resultant of 8 G at 85 ms. The 25-inch seatback provided adequate head restraint, accounting for absence of whiplash. The forces applied by the seatback to his head caused a peak acceleration of 12 G at 130 ms, Fig. 79. This average-sized adult supported by the modified 25-inch seatback, underwent this rather severe rear-end collision without his head contacting the header or rear window and without discernible injury exposure, Fig. 81. Thereafter, he rebounded only slightly as his head left the top edge of the seatback but rebound was not sufficient to cause his shoulders to leave the seatback; his rebound was followed by a slumping action as he remained in an almost normal seated posture. This right rear occupant rode out the crash in an uneventful manner and could be judged as not having been subjected to injury-producing forces. In addition, this 25-inch seatback would not interfere with rearward vision for an average-size adult male seated in the rear or front seat of this vehicle.

The Left Rear Passenger - a 50th

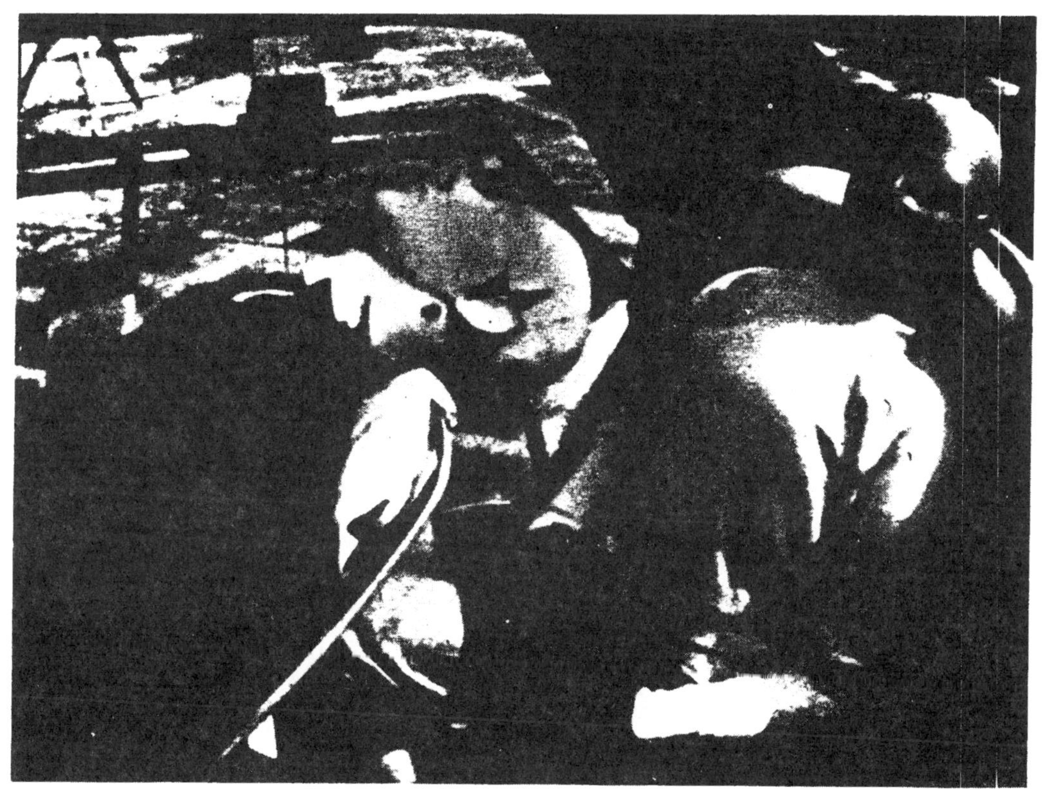

Fig. 81 - A UCLA modified 28 in. front bucket seat, 95th percentile dummy; 50th percentile dummy in modified 25 in. rear bench seat, Exp. 101, 30 mph

percentile adult anthropometric dummy was provided with the same 25-inch bench seatback and was postured with a 6-inch head offset. As the seat crushed forward around him, his torso was accelerated forward (11 G at 115 ms), and his head struck the rear window (52 G at 120 ms), Fig. 82. His head reached its maximum rotation relative to torso of 21 degrees as the torso was rebounded from the seatback. Rebound for the left rear seat passenger was somewhat greater than the right rear, which caused sufficient head rotation forward to bury his chin against the chest.

As has been observed for front seat body kinematics, the head offset serves to partially elevate the torso in the seat, serving to, in effect reduce the seatback level of support. This compromise from a 6-inch head offset was sufficient to cause the 25-inch seatback to be ineffective or less than adequately effective because the head offset caused him to hit the window, accounting for a head blow four times greater than sustained by the right rear passenger, having a zero-inch offset.

Observations, Following Collision - The struck car had advanced about 103 feet and deviated 3½ feet from its original heading when it came to rest. The rear of the struck vehicle at the elevation of the taillight was compressed forward 2 feet, 5 inches, and at the bumper level, 2 feet, 1 inch. The trunk lid was forced forward and upwards, collapsing to a single fold across the width of the trunk lid as shown by Fig. 78. The spare tire was not displaced forward and therefore did not violate the passenger compartment interior. The rear fender area including the taillights were crushed forward to a position corresponding to the forward wall of the trunk-well. Sheet metal was deformed around the right rear tire developing some drag to vehicle roll-out. The two-ply tire, however, was not punctured. An even greater collapse involvement with the left rear tire was observed with a substantially increased amount of gouging of the tire but not sufficient to puncture it. The fuel tank filler spout remained attached to the fuel tank and only a small amount of simulated fuel was purged from the vented filler cap.

EXPERIMENT 102 - A stationary 1967 Ford 4-door sedan weighing 4600 lbs. was rear-ended at 30 mph by an identical 1967 Ford 4-door sedan weighing 4720 lbs.,

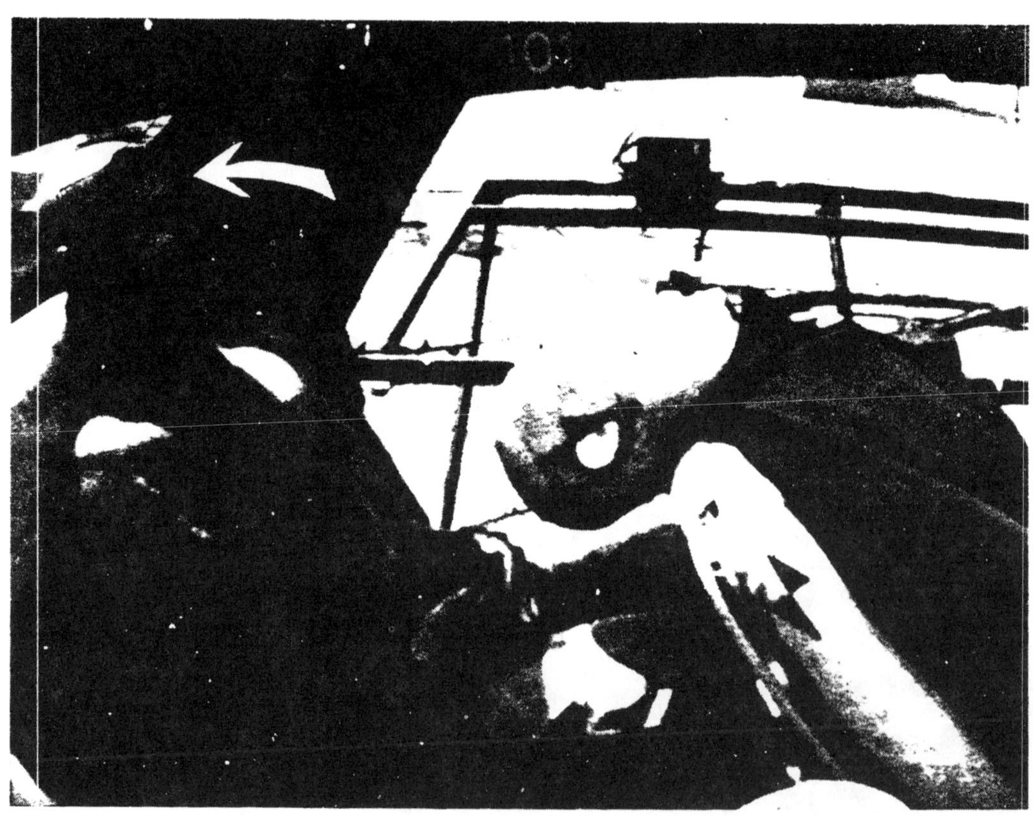

Fig. 82 - Left rear passenger's head (50th percentile) struck the rear window, 25 in. bench seat, Exp. 101, 30 mph

Fig. 83. For this experiment, the rear window and header were removed to evaluat the kinematics of the rear seat dummies without having dummy responses modified by head impacts with the car interior.

All dummies in the struck vehicle had lap belts. Two 50th percentile adult male anthropometric dummies were positioned in the front seats with zero-inch head offset, and two 95th percentile adult male dummies were in the rear seat. The right rear occupant had a 3-inch head offset and the left rear occupant had a 6-inch head offset; both were seated on a bench seat having a backrest extended to 25 inches. The front occupants were positioned on bucket seats with a 25-inch backrest, the only difference was seatback yield strength; the driver's seatback was designed to yield at 16,100 in.-lbs. and the passenger's backrest was rigid, as described in the beginning of this paper in connection with Methodology, Fig. 3(a)

The Collision - The vehicles sustained nearly 12 inches of mutual collapse by the time the rear-ended car registered its peak frame acceleration of 11 G at 25 ms. The trunk lid popped open and the rear-ended vehicle's trunk frame members buckled downward contacting the pavement (9-inches vehicle movement at 100 ms). The cars separated at 14 feet and collided again lightly at 35 feet. Thereafter, the vehicles remained together owing to braking action caused by sheet-metal collapse interference with rear wheels of the struck car; they advanced to a total of 57 feet from initial impact position.

The Driver - This 50th percentile adult dummy was seated in a UCLA-modified bucket seat having a 25-inch seatback with 16,100 in.-lb. yield strength. During collision acceleration he whiplashed over the top of his seat, his head rotating rearward 60 degrees past the vertical, compensated by 27 degrees of rearward seat deflection, resulting in a maximum head rotation of 33 degrees, relative to his torso axis, Fig. 83. His peak resultant chest acceleration (9G at 95 ms) coincided with the maximum seatback deflection and his head resultant peak acceleration occurred slightly later (12 G at 140 ms) Fig 84. On rebound from this whiplash, the driver was thrown sufficiently forward to strike the steering wheel before returning to a normal seated posture. The horn ring and impact pad dislodged from the steering wheel during the abrupt acceleration of the rear-ended vehicle and struck the chest of the driver, thereafter rebounding forward in a projectile like manner.

Right Front Seat Passenger - The experimental conditions encountered by this 50th percentile adult dummy were

Fig. 83 - Rear-end collision, 30 mph, with rear window and header removed, Exp. 102

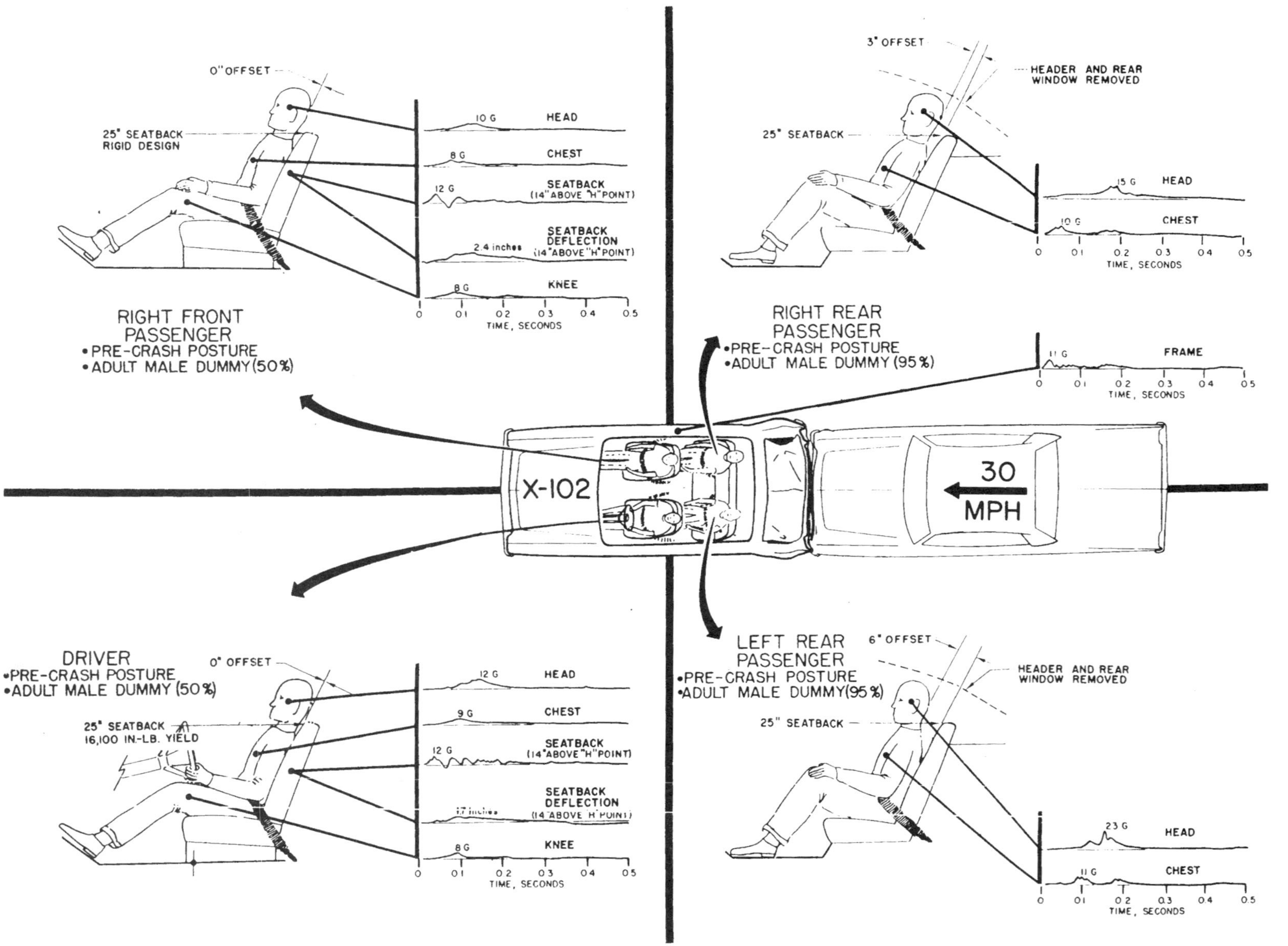

Fig. 84 - Transducer patterns shown with occupant posture and restraint condition at the start of impact, Exp. 102

identical to the driver, also a 50th percentile, except that the right front seatback was a UCLA Rigid design, Fig. 3(a).

During the rear-ended vehicle acceleration, the front passenger's head rotated rearward 29 degrees relative to his torso axis, Fig. 85. The peak resultant head acceleration was actually less than that of the driver (10 G at 135 ms vs. 12 G at 140 ms). His chest peak

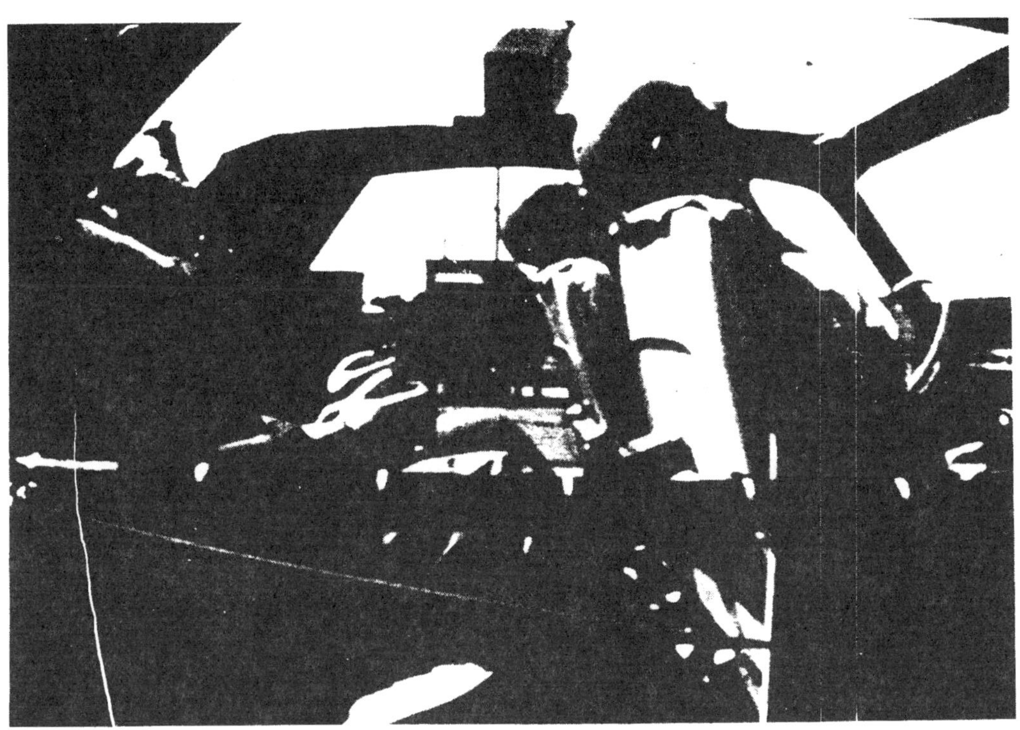

Fig. 85 - Front passenger and driver are 50th percentile dummies with 12 in. head offsets; driver on 25 in. modified seat, passenger in 25 in. rigid seat, Exp. 102, 30 mph

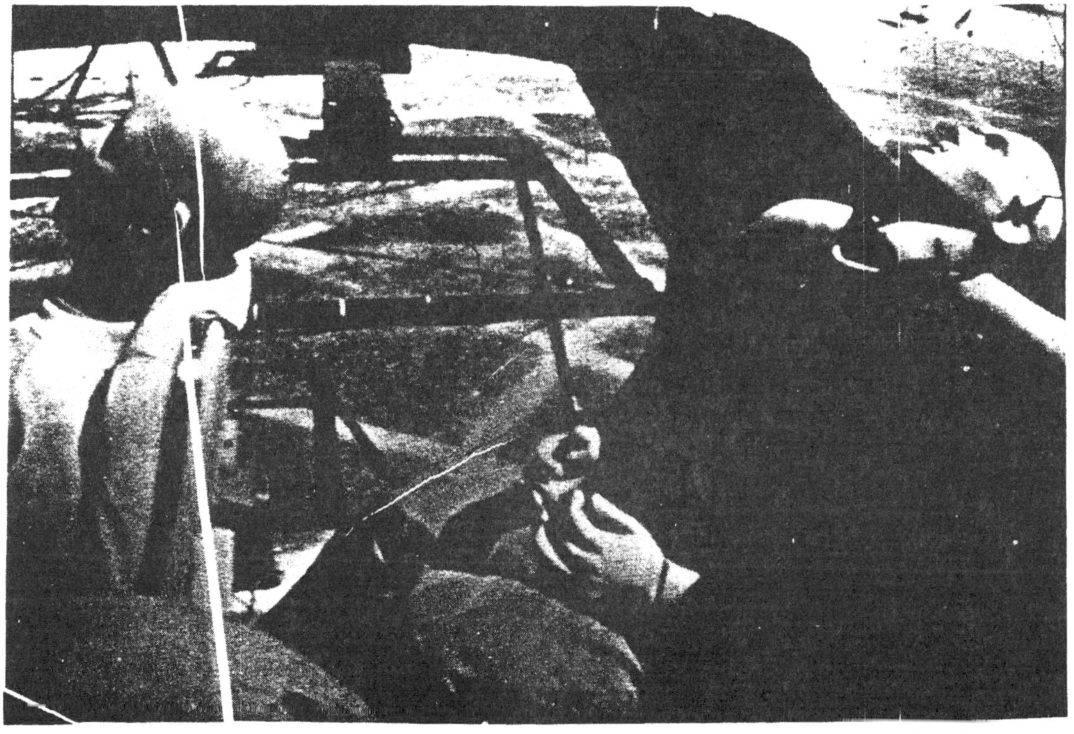

Fig. 86 - Right rear 95th percentile dummy on 25 in. modified bench seat; head acceleration was moderate (15 G) because the rear window and header had been removed for head clearance

acceleration (8 G at 75 ms) was also slightly lower than that of the driver's 9 G peak. His rebound was considered moderate as it only carried him to a posture where his torso was vertical; his head rotated forward, allowing his chin to strike his chest. Thereafter, he slumped rearward to a seated posture leaning to his left.

The Right Rear Passenger - a 95th percentile adult, was seated on a UCLA modified 25-inch bench seat. As the car was slammed violently forward, his chest reached a peak resultant acceleration of 10 G at 55 ms and his head whiplashed over the top of the seat, which was slightly above shoulder level. The removal of the rear window and header allowed his head to flail rearward until it achieved a 45 degree rotation beyond his torso axis (15 G head peak at 190 ms), Fig. 86; thereafter, he rebounded slightly forward with his head passing to fifteen degrees forward of vertical. As this mild rebound occurred, the shoulders pulled away from the seatback approximately four inches before he returned to a normal seated posture. With a rigidly supported 25-inch seatback and with adequate head clearance, notwithstanding the head rotation that occurred, it would appear that this 95th percentile right rear seat passenger could ride out this 30 mile an hour impact without serious injury.

The Left Rear Passenger - a 95th percentile dummy, was exposed to collision conditions identical to the right rear occupant except that the head offset was 6 inches forward of the seatback plane. As the struck car accelerated forward, his head rotated rearward 43 degrees relative to his torso axis, Fig. 87. His chest received a peak acceleration of 11 G at 95 ms, followed by a head acceleration of 23 G at 150 ms. Following this mild whiplash action, he rebounded slightly forward and his torso, not quite reaching a vertical seated position, slumped rearward into the seat in a natural posture.

Observations, Following Collision - This 30 mph rear-end collision propelled the struck car 55 feet forward along the pre-collision axis of the vehicles. Both vehicles came to rest owing to the braking action of sheet metal collapsed about the rear wheels of the struck vehicle. Collision damage at rear bumper level for the rear-end was 2 feet, 4 inches permanent collapse and at the taillight elevation was 2 feet, 11 inches collapse, Fig.

Fig. 87 - Left rear 95th percentile dummy, six-in. head offset for 25 in. modified seat receives mild whiplash, Exp. 102, 30 mph

88. The taillights were compressed forward against the forward edge of the trunk-well and this is the first 30 mph experiment in which the forward wall of the trunk-well was damaged. The longitudinal frame rails were buckled at a position one foot forward of the ends of the frame where the bumper hangers attach. Both frame rails were pushed downward with the left frame rail actually contacting the pavement. The fuel tank was ruptured but fuel was not forced from the filler spout. Sheet metal folding about the rear wheels caused the right rear wheel to sustain a tire failure caused by circumferential cuts. The spare tire remained firmly attached and was not impacted forward into the rear seat compartment. The trunk lid made a partial fold as it was deflected upwards. The horn button and impact pad detached and was found at a distance 47 feet beyond

Fig. 88 - Post crash view of vehicle damage, Exp. 102, 30 mph

Fig. 89 - Rear-end collision, 55 mph, Exp. 103

the position of initial car-to-car contact.

EXPERIMENT 103 - This 55 mph rear-end collision involved two 1967 Ford 4-door sedans, each with four adult dummy occupants, Fig. 89. The rear-ended vehicle, complete with occupants, instrumentation and 25 gallons of simulated gasoline weighed 4610 lbs; the gross weight of the striking car was 4660 lbs. The rear window and header of the struck car were removed to evaluate rear-seat motorist kinematics without the complication of passenger interference with roof and window structures. Additionally, the spare tire was removed from the struck car for this high speed rear-end collision (55 mph) to avoid the possibility of having it crushed forward sufficiently to interfere with the performance of the special bucket seats that were undergoing evaluation at the rear seat locations. Lap belts were worn by the driver, right front and right rear occupants. The left rear occupant wore a three-point belt equipped with an inertia reel. The driver, a 50th percentile adult male, was provided with a UCLA modified 25-inch high seatback of rigid construction. His posture was adjusted so that he had a zero-inch head offset. The right front seat passenger, a 95th percentile adult male, was provided with a UCLA modified 28-inch seatback of rigid construction; his posture was adjusted to correspond to a three-inch head offset. The right rear seat dummy, a 95th percentile adult male, was postured with a zero-inch head offset in a UCLA modified 28-inch seatback of rigid construction. The left rear passenger was a 95th percentile adult male, postured with a zero-inch head offset. He was provided with a Cox safety seat having a head restraint adjusted to the 29-inch height; standard equipment for this Cox safety seat is an integral 3-point belt and inertia reel.

<u>The Collision</u> - The rear-ended vehicle was crushed in 1.8 feet before its forward motion was observable for this 55 mph collision. The frame of the rear-ended vehicle accelerated to 19 G at 85 ms, Fig. 90, at which time the rear-ended car sustained 3.3 feet of collapse. After advancing 17 feet, the cars separated and the trunk lid of the rear-ended vehicle popped open; although the striking car hood was damaged, it remained latched, Fig. 91. The struck car was advanced 110 feet and the striking car 49 feet. As the struck car reached maximum collapse, a large volume of simulated gasoline was seen to erupt from the underside of the vehicle and immediately surged forward under and to both sides of the rear-ended car at a velocity greatly exceeding that of the colliding vehicles. The collision severity and the accompanying sheet metal collapse accounted for the wheels locking up; the striking vehicle's wheels locked up as it reached maximum mutual collapse and the struck car's left-rear wheel rotated 20 degrees after contact and then locked up. During the course of the rear-end collision, the striking vehicle overrode the struck vehicle, at six feet beyond contact and continued this overriding action until it was terminated by rebound. The cars separated 0.45 seconds after initial bumper contact, Fig. 92

<u>The Driver</u> - was a 50th percentile adult male dummy, seated in a rigid 25-inch seatback. During collision acceleration he forced the seatback rearward into a reclining position striking the knees of the rear seat occupant (14 G chest peak at 75 ms). This action was accompanied by the driver's head movement rearward over the top of the seatback and downward to an <u>extreme dorsiflexed</u> condition (35 G head peak at 155 ms) bringing his line of sight rearward of vertical, looking towards the rear window, Fig. 93. At its maximum rearward position the seatback had rotated 38 degrees but the head of the driver had rotated 138 degrees beyond vertical or about 100 degrees beyond his torso line, representing an extreme whiplash. The seatback deflection was largely caused by yielding at the floor anchor points of the seat and not bending at the rigid juncture of the seatback and seat base. Rearward seat movement for this driver was partially checked by action of the rear seat occupant, 95th percentile dummy, when his knees became raised from acceleration forces and contacted the upper part of the driver's seatback at a point about 6 inches from the top edge of the seat; this contact partially resisted the seatback movement in a rearward direction. There-

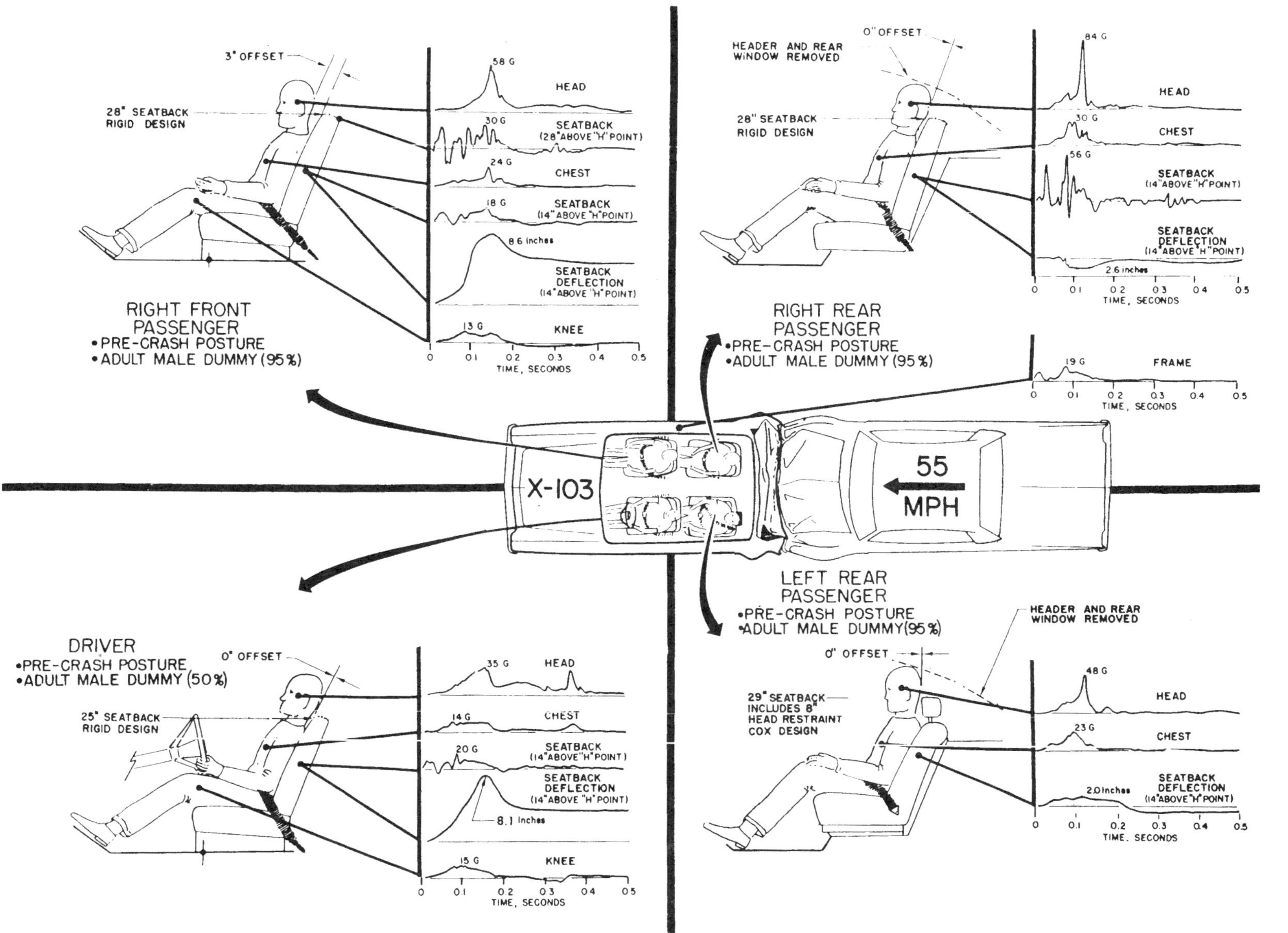

Fig. 90 - Transducer patterns shown with occupant posture and restraint condition at the start of crash, Exp. 103

after rebound accelerated the driver forward against the steering wheel. After striking the steering wheel, even though restrained by a lap belt, the driver flailed to his left and his head struck the compression strut installed to replace the left doors.

Fig. 91 - Crash sequence, 55 mph collision, Exp. 103

The Right Front Seat Passenger - was a 95th percentile adult male dummy in a UCLA rigid 28-inch bucket seat. The inertial forces of his weight and that of the seat during the severe collision acceleration forced his seat rearward until the seatback struck the knees of the occupant behind him. His torso's abrupt acceleration flails his head into a severe dorsi-flexed spinal posture sustaining a 58 G peak at 150 ms; at this same instant, the right rear seat passenger is rebounding forward. Extreme whiplash was checked by contact of the front seat passenger's head with the chest of the rear seat passenger behind him. During the onset of the acceleration, as the seatback was deflected rearward reaching 10 degrees of movement, the passenger commenced to slide up the plane of the seat. By the time the seat attained a rearward deflection of 43 degrees, the passenger's head had slipped 4 inches up the plane of the seat and this facilitated a severe whiplash condition, Fig. 94. On bottoming out of the seatback against the knees of the rear seat passenger, his chest accelerometer attained 24 G at 140 ms. Although positioned with a three-inch head offset, this right front seat passenger obtained a rearward head rotation of 82 degrees, relative to his torso axis, or 125 degrees beyond the vertical reference. Rebound was rather violent, but the lap belt checked his hip movement and his posture did not exceed a vertical position. As contrasted with the driver this right front seat passenger seatback did not rebound appreciably; this additional yielding of the seatback was attributed to floor pan yielding which exceeded the driver's floor pan yielding.

The Right Rear Passenger - was a 95th percentile adult male dummy in a 28-inch UCLA modified bucket seat having a rigid seatback. In order to assess the body kinematics without impact-interference from roof structures, the rear window and header were removed. This 28-inch seatback appeared to reach 2 inches below the center of mass of the dummy's head positioned with a zero-inch head offset and this corresponds to about 8 inches above the shoulder pivot, Fig. 95. The abrupt acceleration forced his head over the top of the seatback approximately 16 degrees rearward of his torso axis. When the rearward movement of his head was checked by impact with the collapsed trunk lid, being folded vertically upwards (84 G at 110 ms), his head was thereafter accelerated forward, Fig. 96. This right rear passenger received a 30 G peak chest acceleration at 95 ms, followed by rebound, facilitated by the seatback being forced forward from collision intrusion; this caused him to reach a vertical posture with his chin buried into his chest. Thereafter, he slumped back into his seat, leaning towards the center of the car.

<u>The Left Rear Passenger</u> - was seated in a Cox safety seat having a seatback angle of 28 degrees with an integral cross-chest and lap belt combination. This 95th percentile adult dummy had a 29 inch head restraint that corresponded to the level of his eyes. The abrupt acceleration caused the top of

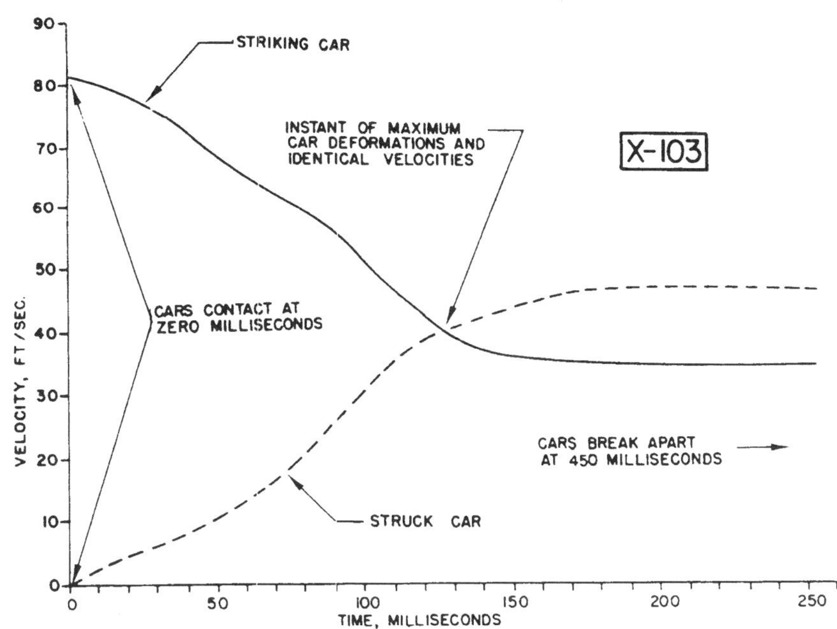

Fig. 92 - Velocity changes during a 55 mph rear-end collision; cars separate at 450 ms

Fig. 93 - Adult male dummy 50th percentile, in an extreme dorsi-flexed spinal posture, 25 in. rigid seatback, Exp. 103, 55 mph

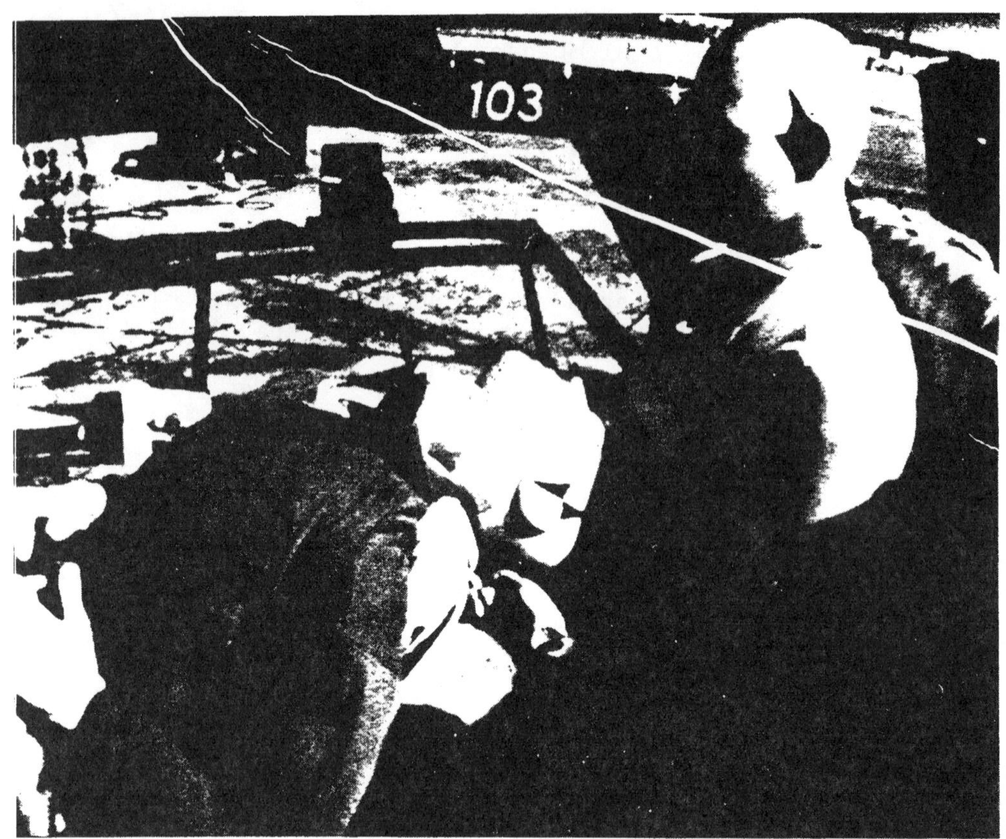

Fig. 94 - Right front passenger's head slipped up the seatback, aggravating a severe whiplash, 55 mph, Exp. 103

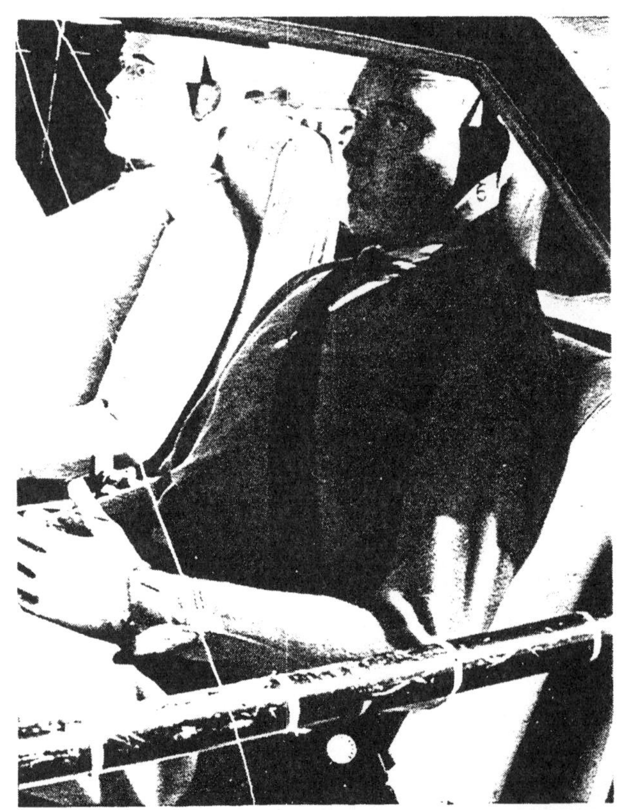

Fig. 95 - Rear seat occupants, pre-crash, Exp. 103

the seatback to be forced rearward approximately two inches before it crushed against the rear window shelf and, at this instant, 90 ms, his resultant peak chest acceleration was 23 G. This left rear occupant's head was supported by a vertical head restraint and as it deflected rearward, it limited head rotation to only 10 degrees, relative to his torso axis; his peak head acceleration was 48 G at 120 ms, Fig. 97. Thereafter, his rebound was uneventful owing to the excellent restraining action of the three-point belt anchored to the seat. Throughout this 19 G collision event, the excellent performance of this Cox safety seat with respect to protection of the 95th percentile passenger was emphasized by the natural posture he maintained, notwithstanding his close proximity to the collapsing rear-end structures.

Observations, Following Collision - The trunk lid had been folded double and rotated to the vertical plane, as the rear-end collapse progressed to the forward part of the trunk well. Deformation to the rear-end at the bumper level was

3 feet and also 3 feet at the taillight elevation, Fig. 98. The rear quarter panel was compressed toward the pavement as the rear frame members folded double and were crushed to the pavement. This gross collapse to the rear-end section buckled the rear window shelf downward; the rear window and header had been

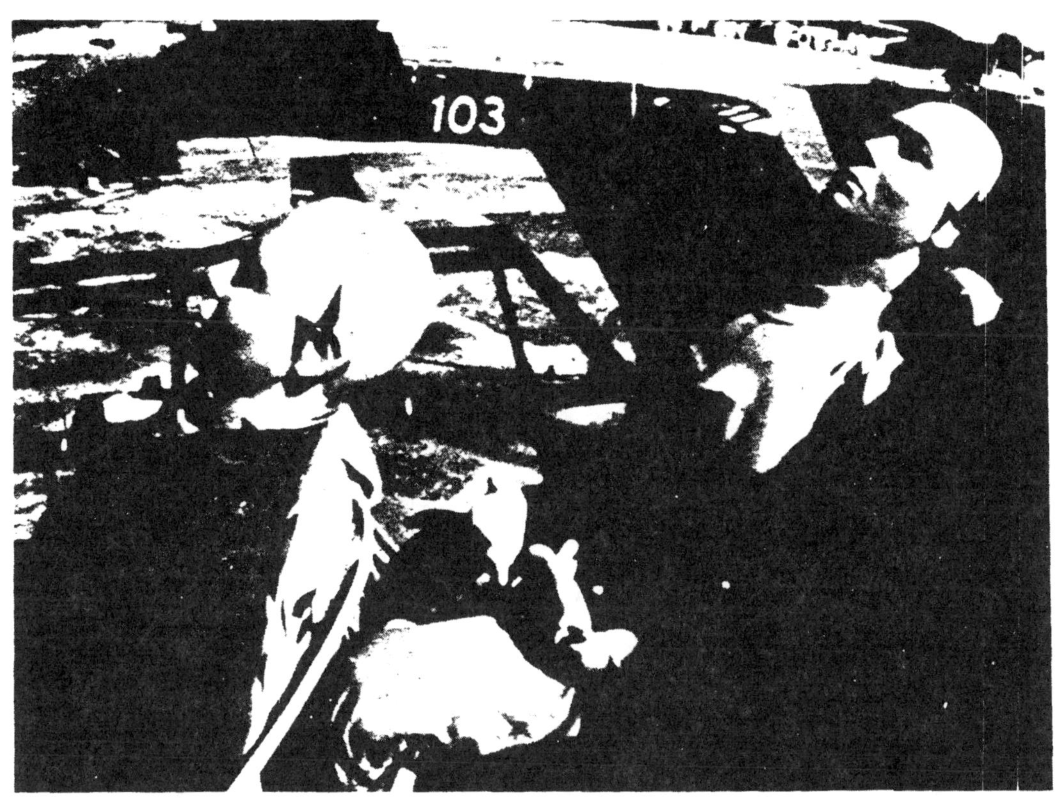

Fig. 96 - Right rear passenger's head "restrained" by collapsing trunk lid, Exp. 103, 55 mph

Fig. 97 - Adult dummy, 95th percentile, head restrained very effectively, seated in a Cox Safety Seat, Exp. 103, 55 mph

removed to give head clearance for the UCLA modified bucket seat installations at the rear seat positions. Slight buckling occurred to the roof side panels and the rear fender sheet metal was crushed approximately one-third the way around the rear wheels; the tires, however, were not punctured. The enveloping sheet metal provided braking action as indicated by the struck car relatively short roll-out of 110 feet; the emergency brake system was not actuated.

Compression to the rear end caused gross splitting of the bottom seam of the fuel tank and liberal dispersal of the simulated fuel over the collision area; this dispersal started at the position five feet forward of point of initial bumper contact and continued for 84 feet, with the preponderance of fuel being dumped in the vicinity of the striking car's position of rest (45 feet from impact point).

Experiment 104 - was a 55 mph rear-end collision between two 1967 Ford 4-door passenger vehicles, Fig. 99. The

Fig. 98 - Underside of struck car after 55 mph rear-end collision, Exp. 103; gasoline tank is compressed

Fig. 99 - Rear-ended at 55 mph, vehicle mutual collapse is 6 ft, Exp. 104

striking vehicle weighed 4620 lbs. and the rear-ended vehicle, 4580 lbs. This latter weight included four adult anthropometric dummies and 41 gallons of water; 23 gallons were colored red to simulate gasoline in the production fuel tank and 18 gallons were colored green for a special fuel tank. The 23-gallon tank was a production tank to which a fuel cell (rubber-like) liner was added. Inside this liner was placed a special installation of low density foam plastic, an open-cell material used for race cars at Indianapolis designed to reduce fire hazards associated with fuel tank rupture. In addition, a second fuel tank, 18 gallons capacity, was installed in the trunk on the floor-pan axle-step over the rear axle for the purpose of evaluating the practicality of this position as a location for fuel tanks. The spare tire was removed from the rear-ended vehicle in order to accommodate this tank. A further modification concerned the transverse installation of a four-inch steel channel to distribute the front seat inertial forces transferred to the forward anchor bolts of the bucket seats.

Both front bucket seats were 28-inch UCLA-Rigid seats (33,000 inch-lbs.); the driver was a 50th percentile dummy with a zero-inch head offset and the front passenger was a 95th percentile dummy with a three-inch head offset.

The rear seat was a 21-inch production bench seat and the rear window and header were left in place; only the right rear passenger was provided with a head restraint and it extended his seatback to 27 inches. Both rear seat occupants were 95th percentile dummies and both were postured with a three-inch head offset. All four occupants wore lap belts.

The Collision - The struck vehicle, rear-ended at 55 mph, was crushed one foot before discernible forward movement occurred (30 ms). As the two cars crushed further into one another, they reached a maximum mutual collapse of six feet, sufficient to lock up their adjacent wheels five feet beyond impact. When the rear-ended car had been knocked 11 feet, its rear frame rails buckled downward and its wheels resumed rotation on advancing 13 feet beyond impact. By 95 ms, the rear-ended car frame reached a peak of 19 G, Fig. 100. The cars became separated after they had advanced 24 feet. A considerable amount of simulated gasoline sprayed from each side of the rear-ended car as it passed the 40 foot position. The striking car came to rest 70 feet beyond impact and the rear-ended car, 94 feet from its position of impact. The rear window was popped out from the right rear seat passenger's head whiplashing against it, 80 ms after contact; then the trunk lid, collapsed forward and fractured the rear window.

The Driver - was a 50th percentile dummy in a 28-inch UCLA-Rigid bucket seat (33,000 in.-lb. seatback), postured with a zero-inch head offset. The front seat mounts were bolted to a steel channel located transversely under the front mounts of the seat, below the floor pan. As the collision progressed, the seat was accelerated forward pressing firmly against the driver's back and head, the seatback appeared to provide good support, Fig. 101(a). Owing to the extreme nature of this 55 mph rear-end collision, acceleration forces and their durations were sufficient to deflect the rigid seatback to a reclined position of 41 degrees rearward of vertical, allowing the center mass of the head to apply its force vector, no longer against the seatback but, over the top of the seatback, developing a head-to-torso angle of 79 degrees. This served to stress the cervical spine and to allow the head to flail over the top of the seatback in a whiplash-like manner, Fig. 101(b). This occured even though the shoulders had not appreciably shifted up the plane of the seatback, an amount sufficient to account for the change in head restraint condition. His knee, accelerated by the seatback through the buttocks, reached a peak of 18 G at 105 ms. Whiplash progressed until the eyes of the driver were actually pointing rearward of vertical, constituting an extreme whiplash (42 G at 140 ms), notwithstanding the protective nature demonstrated for this seat design in experiments at 30 mph and less. His seatback deflected 7.2 inches at 131 ms as measured at the 14-inch level, and sustained 18 G at 140 ms; his chest reached 21 G at 150 ms. Rebound from this whiplash condition was considerably amplified over observations of lesser speed impacts with the same

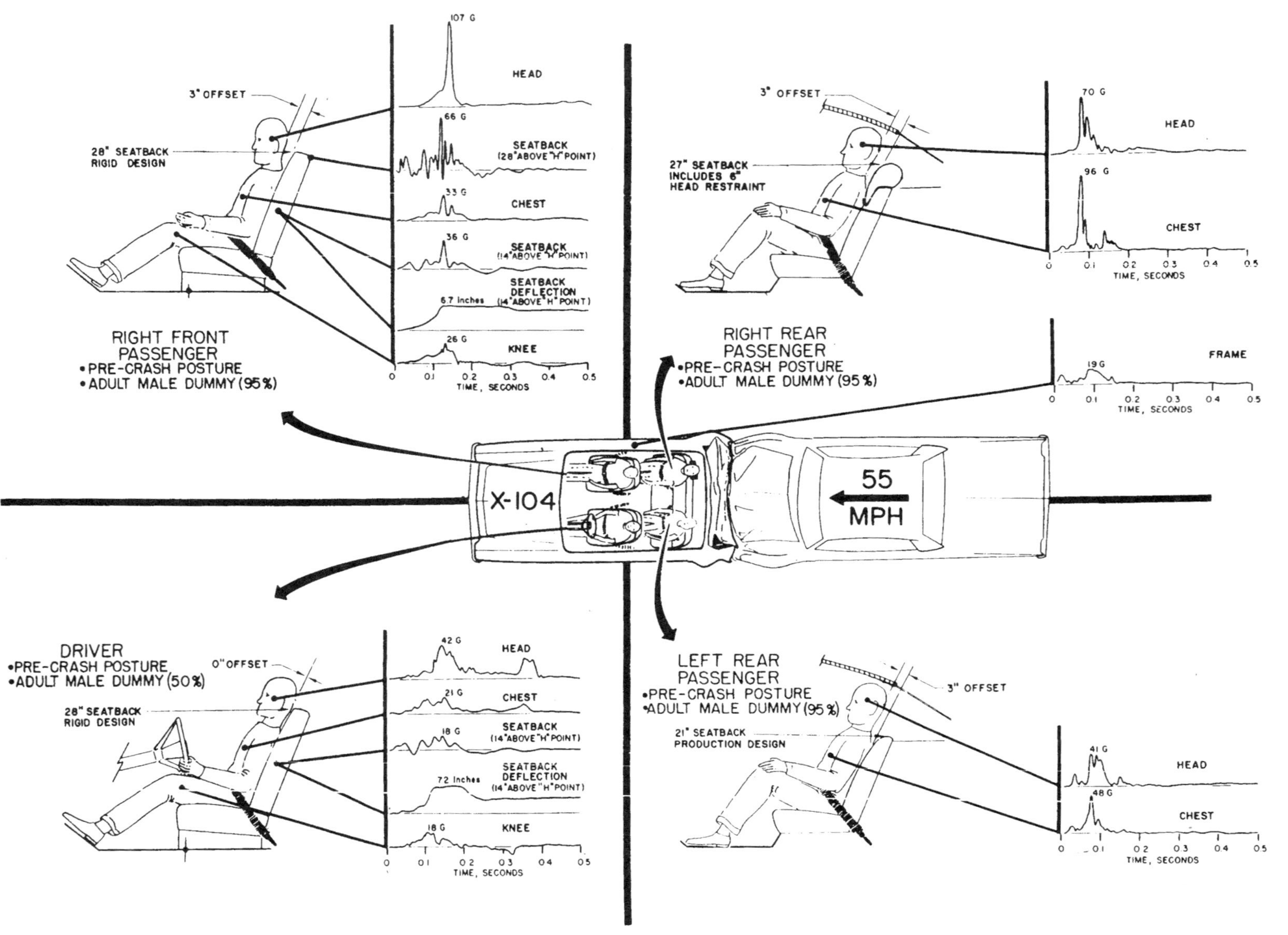

Fig. 100 - Transducer patterns shown with occupant posture and restraint condition at the start of impact, Exp. 104

rigid seat configuration and was of sufficient force to throw the driver against the steering wheel in a forcible manner, producing a 12 G chest and a 24 G head acceleration, both at 350 ms.

The Right Front Seat Passenger - was a 95th percentile dummy in a 28-inch UCLA-Rigid seat, postured with a three-inch head offset. The front seat mounts were bolted to a steel channel located transversely under the floor pan. As previously described for the driver, the inertial forces of the right front seat and passenger were so great that the abrupt acceleration of the seat, forcing against the back of the passenger, caused the seat to be deflected rearward at a rate sufficiently great to keep the unrestrained head of this 95th percentile dummy in line with his torso, Fig. 102(a). The top of the 28-inch seatback registered 66 G at 120 ms. This head-to-torso relationship was maintained until the seatback was almost fully deflected to a position where it bottomed-out against the knees of the right rear seat passenger. His seat had sustained a permanent deflection rearward of approximately 10 degrees with a small amount of this caused by partial failure of the lower bolt anchor of the tension link. The major portion of the seatback rotation was caused by deformation of the floor at the rear anchor of this right front occupant seat. His seatback deflected rearward sufficiently to permit the seatback to strike the knees of the right rear occupant with one inch deformation to the sheet metal cover pan over the back of the front seatback. However, after the crash the seatback remained with approximately one inch clearance between knees of the right rear occupant and the back side of this front seatback. The linear potentiometer recorded 6.7 inch seatback deflection 14 inches above the H-point (125 ms), which represented 12 inches of deflection at the top of the seatback; the dummy's chest registered 33 G also at 125 ms, his seatback at the 14 inch level registered 36 G at 130 ms and simultaneously his knee registered 26 G. For this position, further deflection was not possible and the head had completed a moderate whiplash (head-to-torso axes angle reached 42 degrees and 107 G at 140 ms) bringing the top of the head firmly against the sternum area of the right rear passenger's chest, thereby snubbing further rearward movement, Fig. 102(b). As with the driver, rebound for this right front seat passenger was considerable, as contrasted with other collisions involving the 28-inch UCLA-Rigid seat at lower speeds of impact; rebound was sufficient to throw his head and upper torso against the instrument panel. The windshield had been removed to improve photography through the windshield area and therefore, was not contacted by this dummy.

Fig. 101 (a) and (b) - The excessive accelerations and their lengthened force durations for the 55 mph collision served to modify the well supported (28 in. Rigid Seat) condition of the 50th percentile driver, Fig. (a), to allow him to "whiplash", Fig. (b)

The Right Rear Passenger - a 95th percentile dummy, was postured with a three-inch head offset in a 21-inch production bench seat, provided with a head support which extended the seatback to 27 inches. During collision acceleration the seat crushed forward against the dummy's back reaching a peak chest acceleration of 96 G at 80 ms; his head underwent a moderately severe whiplash (head to torso axis angle of 45 degrees), accompanied by the right rear window fracturing as the result of direct head impact (70 G at 85 ms) with the glass, which forced the right upper corner of the glass from its mount before the glass fractured completely, Fig. 102(b). It was found that his head restraint had a

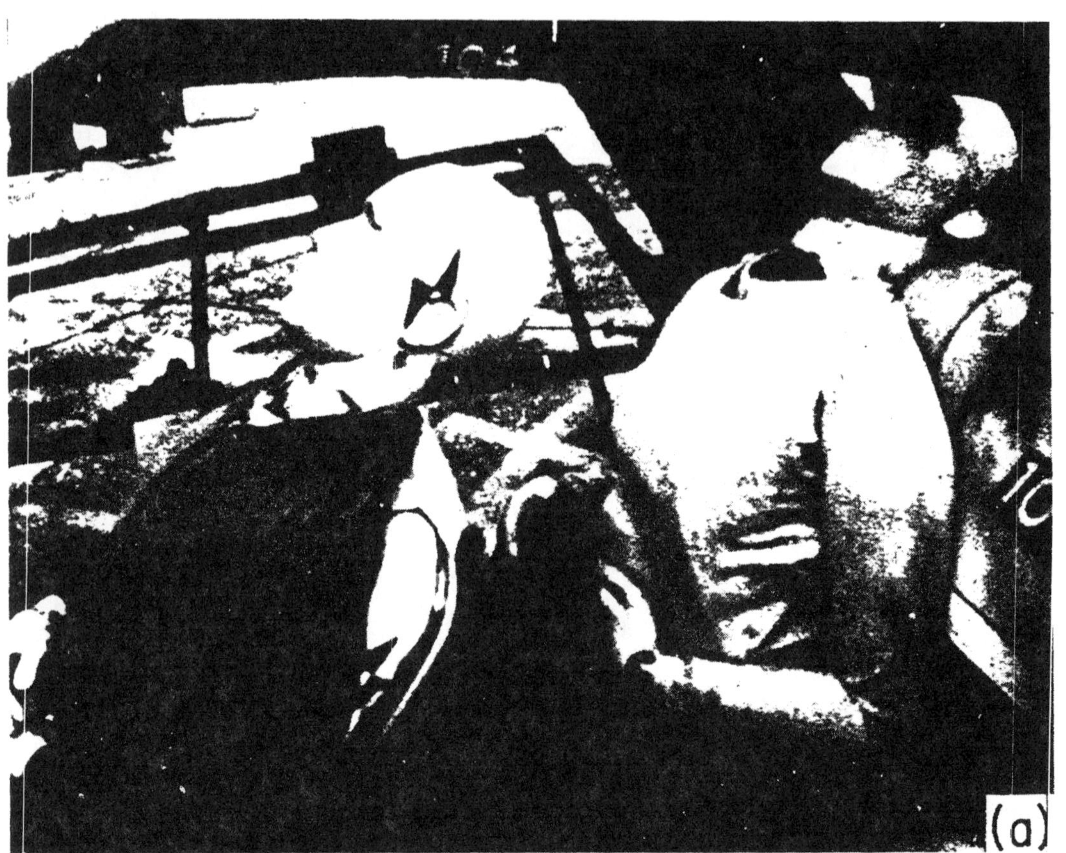

Fig. 102 (a) and (b) - Rigid front and rigid rear seats, 95th percentile dummies, Exp. 104, 55 mph

one-inch deep dent as his head crushed on past it to strike the window. Rebound following whiplash was mild and the dummy thereafter slumped to his left in the seat. The relative lack of rebound for this right rear passenger as contrasted with the right front passenger is attributed to the more rigid seatback structure characteristic of the rear seat production type installations.

The Left Rear Seat Passenger - a 95th percentile dummy in a 21-inch production bench seat without the benefit of a head restraint; he was postured with a three-inch head offset. The only factor for the left rear passenger differentiating him from the right rear passenger was his lack of head restraint. The vehicle carried an extra 20 gallon tank on the floor pan axle step and during the course of collision the tank was displaced forward actually pushing the seatback for the rear seat forward and this effect can be seen (in the motion pictures corresponding to Fig. 101(b)) acting upon the left rear seat passenger. The acceleration and forward displacement of the rear seatback caused the torso of the left rear passenger to be punched forward (chest 48 G at 80 ms); his head pivoted about the header causing a head-to-torso axis modification comparable to a mild whiplash (head 41 G at 80 ms), but the extent of displacement was not sufficient to constitute whiplash (head-to-torso axes reached 46 degrees). Head impact to the rear window header left a depression 1/4 inch deep by 3 inches wide. The main injury exposure was occasioned by the gross acceleration of the seatback owing to its being physically displaced forward by the intervening step fuel tank; rebound for this left rear seat passenger was negligible as was observed also for the right rear seat passenger.

Observations, Following Collision - The vheicle rear-ended at 55 mph was propelled a total of 94 feet forward, coming to rest almost along its original pre-crash heading. Permanent deformation at the bumper elevation for the rear-end was 3 feet, 3 inches and 3 feet, 6½ inches at the taillight elevation. The trunk lid had its characteristic single fold, vertically oriented. The sheet metal was buckled between the forward section of the trunk lid and the rear section of the rear window. The 18-gallon tank on the floor pan axle step (over the rear axle) was collapsed inward approximately two inches, which displaced enough fluid to apply sufficient pressure to blow the filler cap from the filler spout. This was the only leakage sustained by this tank. From this experiment it appears, that if a bellows design were incorporated at each side of a fuel tank, there is generally sufficient space for expansion; the extent of expansion required could be accommodated at the step location for this 55 mph rear-end collision, without loss of any fuel. Inasmuch as the tank would not extend to the sides of the vehicle, it appears that there will be a considerably improved protective area provided for a fuel tank in this location because of the enhanced strength afforded the area opposite the rear wheels by the rear suspension system. Placing the fuel tank in this location puts it in the most preferential, most highly protected area available for a passenger vehicle. The lower tank, a production installation, was modified by the addition of a rubber-like liner bladder and filling the entire tank space with an open cell, low-density, foam material that occupies about 3% of the total volume of the tank. These two devices were designed to reduce the likelihood of fuel drainage from a ruptured tank and flame propagation to within the tank. The tank became punctured sufficiently severely to cause most of the fuel to drain rapidly from the tank. Rear-end collpase progressed forward sufficiently to include the shelf at the base of the rear window, crushing it forward two to three inches in a buckling-like action. The rear wheels were driven forward against the fender-well but both tires remained inflated even though the bumper was pressing against them, particularly the right rear tire. The rear bumper had been crushed forward to the point where it was approximately adjacent to the rear wall of the floor pan axle step. A protective shield of half-inch plywood was placed at the rear side of the special tank installed on the axle step, within the trunk and below the rear window shelf. This was placed in a position to reduce a chance of the collapsing forward structure puncturing the tank. The rear window glass was shattered, subsequently found to be the result of the right rear occupant's head as it punched through the rear window.

This vehicle was impacted in a manner that caused the body to be lifted and stripped forward from the frame to the extent that the frame ahead of the rear wheels was exposed the full depth of the frame section of 4¼ inches and buckling of the frame at this location on the right side was 1-foot ahead of the offset to accommodate the rear wheel well; at the left side, the frame buckled down in the same location and and almost touched the pavement. This buckling of the frame was located directly ahead of the corner-reinforcing gussets at the rear of the rocker panel frame member, Fig. 103. The rear frame support members for the trunk had the typical fold ahead of bumper hanger attachment and these frame rails were bent downward to a point where they were about three to four inches off the ground.

DISCUSSION

The problems associated with automobile rear-end collisions range from the petty annoyance of bumper-tapping contacts to the devastating high speed rear-end collision, frequently accompanied by fire, and usually fatal. Between these extremes, there exists the category of accidents that outnumbers all other types of car-to-car collisions, notwithstanding the absence of accurate frequency data relating to rear-end collisions. The problem is grossly "undersold" when reviewing statistical treatments largely because the majority of these accidents do not result in cuts and lacerations or similar forms of demonstrative injuries. The rear-ended motorist frequently is unaware that he has sustained an injury that may subsequently become disabling; consequently, these accidents are very often excluded from police reports and therefore from the body of statistics.

Regrettably, the tremendous costs of such accidents relate primarily to the painful and disabling injuries and only incidentally to the readily repaired or replaced vehicles; the literature produced by the authors (1), (2), (3), (4) and others typified by (5), (6), (7) attests to the fact that injuries sustained by rear-ended motorists can almost entirely be prevented through proper design of the seats in which they ride. The purpose of this study, as reflected by this report of a series of 12 rear-end collision experiments, is to provide design criteria needed by the automotive manufacturers concerned with providing safer seat designs for passenger vehicles.

Fig. 103 - Underside of struck car after 55 mph rear-end collision, Exp. 104; note buckled frame; the fuel tank is partially collapsed; note front seat anchorage reinforcing channel

VEHICLE SEAT MODIFICATION - The authors initiated this study with an advantage of having very recently conducted two series of rear-end collision experiments directed to the problems of safer-seat design (3), (4). The automobile industry has for many years produced comfortable and inexpensive seat units; the authors considered it would best serve their objective to develop and evaluate safer seats without producing prototypes objectionable and impractical for the automobile manufacturer if their seat design were modified to provide the added features needed for safer seats. Accordingly, production seats were strengthened and seatbacks elevated in conformance with the requirements of a planned series of twelve rear-end collisions.

<u>Production Seat</u> - The Ford produc-

tion bucket seat used in this series of rear-end collision experiments was constructed with a tubular steel frame utilizing flat spring supports for the padded cushion covering, Fig. 104. The seat cushion surface was approximately 9" above the floor of the car and the pressed channel supports were anchored to the floor by four 3/8 inch diameter bolts. On top of the floor supports, the track adjusters provided five inches of front-to-rear movement. The base cushion had five flat springs (8 ga.) aligned longitudinally; they were 2-3/8 inches wide with a 3-inch pitch and a spring rate that deflected 3 inches and 3½ inches, respectively, for the 50th and 95th percentile dummies. This cushion deflection was a compensating factor that positioned the "H" point of the dummies, regardless of size, at approximately the elevation of the unloaded cushion surface. The height of the seatback was also measured from the unloaded cushion surface to provide an easy reference to other types of seats, Fig. 3(a).

The seatback of this production bucket seat had a scissors type hinge with the pivot point one inch above the

Fig. 104 - Ford production bucket seat; note tubular steel frame and flat springs

seat cushion. The 22-inch high seatback had a designed inclination of 20 degrees to the rear from vertical. The seatback frame was one inch O.D. steel tubing, .065 inch wall thickness, with six flat springs (12 ga.), having the geometry of a square wave 1-3/4 inches wide by 1-5/8 inch pitch, supporting the seatback cushion.

In order to have a common basis for comparing different seatback and head restraint conditions, ten laboratory static tests were conducted, Fig. 105, by loading the seatback beyond the yield point to complete failure. Five of these tests determined that the start of seatback yielding occurred when approximately 400 lbs. was applied in a rearward direction at the 14-inch reference elevation above the seat cushion, and excessive yielding occurred at the hinge when a 700 lb. loading was approached, Fig. 106. However, reproducibility was difficult to establish even when the inadequate hinge strength was reinforced; it appeared that the seat would commence to fail at more than one location which made it difficult to pre-determine the areas requiring reinforcement. Therefore, the hinges, supports and track adjustor were bypassed by rigid framing and this allowed loadings in excess of 700 lbs. without failure of the 1-inch tubing of the seatback frame.

UCLA Seat Design - Although the production seat assembly could have been reinforced to provide some degree of reliability, a more versatile design was required. In order to evaluate seatback and head restraint performance during rear-end collisions, it was necessary to design a Calibrated Seat frame with an adaptability for controlled seatback deflections and yield for different collision-induced occupant inertial forces. This Calibrated Seat frame, Fig. 107, accommodated the conventional bucket seat form but isolated the dynamic responses of its seatback to a location adjacent to the occupant "H" point (hip-pivot), at the base of the seatback where most bucket seats are hinged for access to the rear seat. By controlling the seat reaction points, both total deflection and plastic deformation could be calibrated with the added opportunity to localize place-

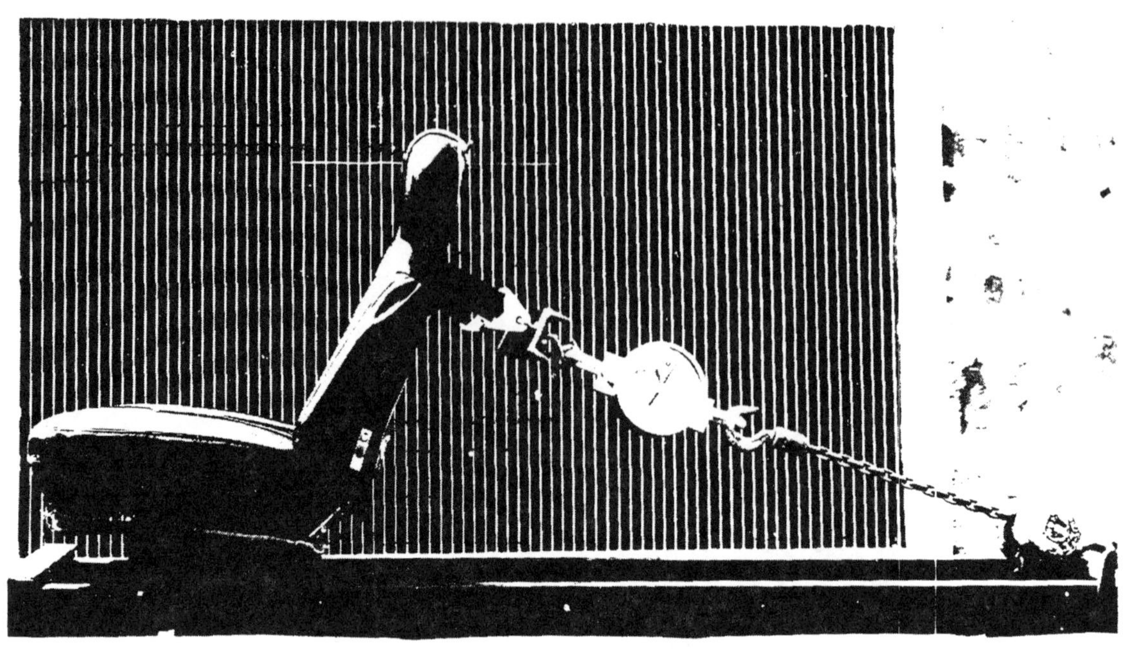

Fig. 105 - Procedure used to conduct static calibrations of seatback strength

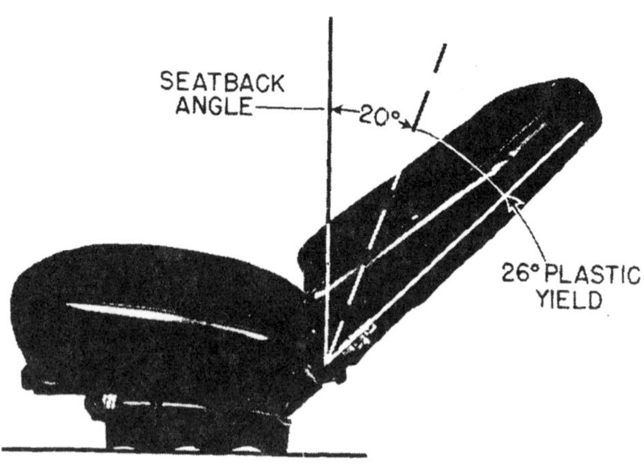

Fig. 106 - Excessive yielding of production bucket seat after 700 lb. load applied to seatback at 14 in. above cushion

Fig. 107 - Calibrated seat frame suitable for a variety of instrumented collision exposures

ment of instrumentation for recording force-time and deflection-time data. An additional design advantage was the interchangeability of the yield brackets which provided seatback strengths of 9,100, 16,100 or 33,000 inch-pounds, Fig. 108. This is an equivalent occupant inertial loading in a rearward direction of 650 lbs., 1150 lbs and 2360 lbs., respectively, at the 14-inch level above the "H" point. This 14-inch height was selected for seatback calibration purposes as the most appropriate location for seatback instrumentation since this approximates the height of the dummy chest accelerometers above their "H" points, Fig. 109. In addition to the calibration insert brackets at the base of the seatback, the calibrated frame accommodates tensiometer anchor

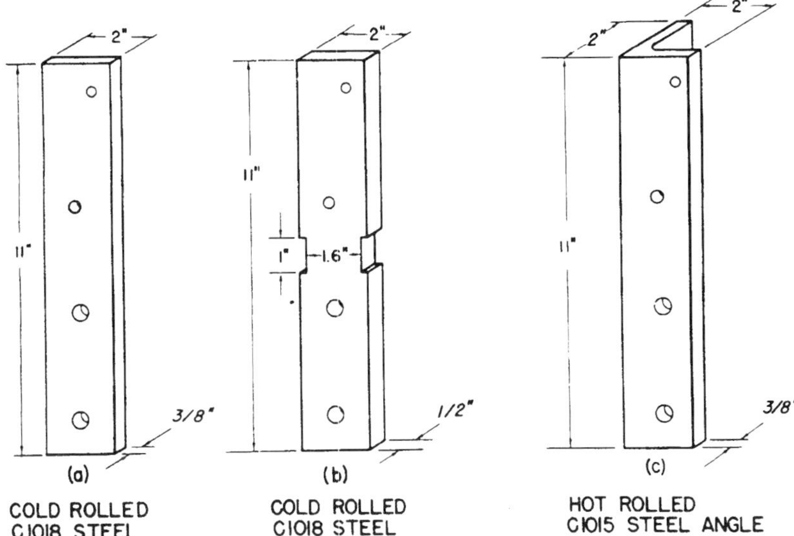

Fig. 108 - Seatback yield localized, by design, to calibrated brackets: two type-(a) brackets provided a seatback strength of 9,100 in.-lbs; two type-(b), 16,100 in. lbs; two type-(c), 33,000 in.-lbs

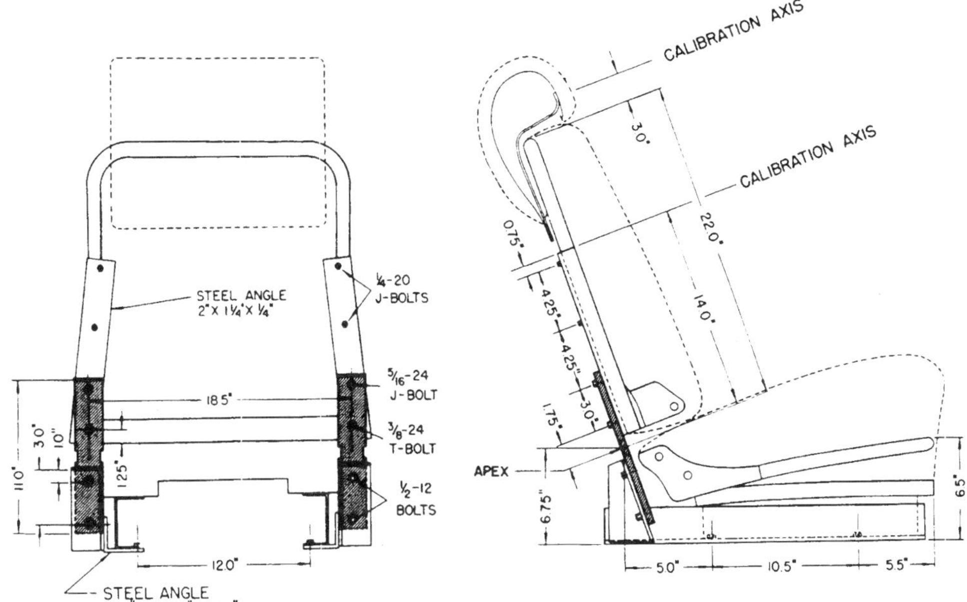

Fig. 109 - UCLA Calibrated Seat Frame

bolts at the front attachment points of the seat with the floor pan. This provides a realistic comparison of dynamic seat loading attributed to the collective inertial forces of seat and occupant, and the stresses of the seat, as determined by static calibration.

Rigid Seat Calibration - Specifications for experiments involving higher energy collisions required a "beefed-up" calibrated seat design, Fig. 110. The seatback supporting brackets were increased in strength to provide double (33,000 in.-lbs.) the bending moment of the former 16,100 in-lb. calibrated seat by removal of the yield brackets and substitution of angle brackets shown by Fig. 108(c). This design was designated "Rigid" because it was adequately structured to resist significant yield until force levels were high enough to commence destruction of the adjacent floor pan. An initial series of static seatback strength tests established that loadings applied by a dummy (via chest strap) produced results comparable to a strap placed directly around the seatback, without dummy, Fig. 111.

Three basic conditions were static tested for the UCLA Rigid seat installation: (a) Two rigid seats were installed in the front seat positions of a 1967 Ford and both seatbacks were subjected to progressively increasing balanced load cycles 14 inches above the H-point (perpendicular to seatback surface) until 1600 pounds (each) was reached. At this point, the floor pan had yielded 7/8 inch and further loading was not applied; floor pan deformation was adjacent to the front and rear anchorages and the seatbacks assumed 15 degree deflections under the balanced load of 1600 lbs. each, with each sustaining a 7 degree permanent deflection following release of their loads. The relationship of seatback loads to seat front anchorage loads are depicted by Fig. 112. (b) Because of the limitations of the floor pan, as demonstrated by the prior test, the seat anchorage at the floor was strengthened by adding a 4-inch transverse channel, 5 feet long to the underside of the floor pan with its webb pressed against the floor. The channel served to distribute the seat front anchor loads over a larger area of the floor pan. Using the same rigid seat design and the same static test procedures, progressive load cycles to 2190 lbs. were applied before the yield point was reached (31,000 in.-lbs.), Fig. 113. This was accompanied by a 7/8 inch depression-deflection of the floor pan at the rear anchorages and a 1/8 inch elevation-deflection of the floor pan at

Fig. 110 - Rigid Seat

Fig. 111 - Initial static seatback strength tests established that loading applied by dummy (via chest strap) produced results comparable to a strap placed around seatback, without dummy

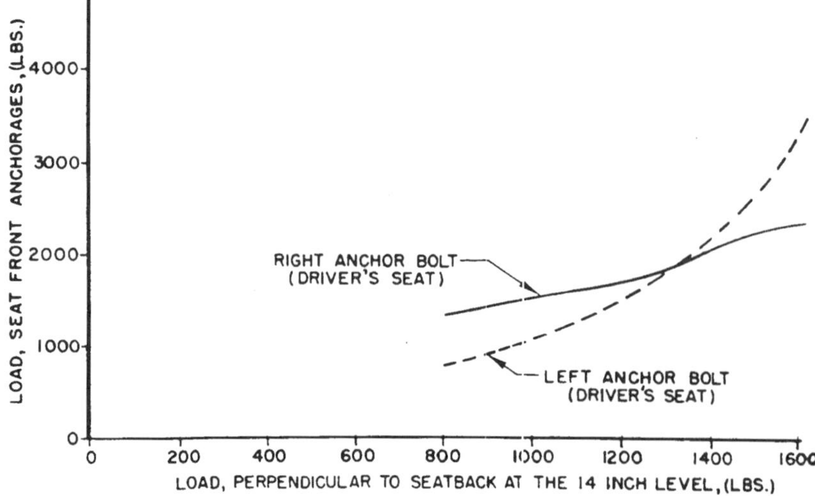

Fig. 112 - Relationship of seatback loads (28 in. rigid seat) to seat front anchorage loads; static calibration (dual seat loading) in the car with production floor pan

the front anchorages, Fig. 114. Additionally, onset of permanent deformation of the rigid seat angle brackets signaled the approach of ultimate strength for the rigid seat. (c) The 28-inch rigid seatback unit was placed on a structural test frame and pulled to destruction, reaching 2360 lbs. or 33,000 inch-lbs. ultimate strength; failure occurred at the rigid angle bracket.

VEHICLE COLLISION PERFORMANCE - The methodology developed for the planned series of twelve experiments was designed to evaluate the practical limits of exposure for rear-ended motorists. Although safer seat design was a primary objective, the realistic experimental procedures adopted provided many unanticipated safety-related observations of value to the automobile designer and to those responsible for the formulation of practical and effective vehicle safety requirements. Of these collision performance observations, the more significant ones refer to the collapsibility of the rear-end structure as related to speed of impact and the presence or absence of luggage in the trunk; the collision performances of the rear window, its header and the fuel tank. The path of vehicle travel and dispersal of debris is depicted in the post collision observation section at the end of each experiment, but is briefly summarized at the conclusion of this section.

Fig. 113 - A balanced static load was applied to both front seat units, simulating rear-end collision front occupants and seat inertial forces

<u>Rear-end Collapsibility of Automobiles</u> - With the current emphasis on unit body construction, some of the unit body collapse characteristics identified a decade ago in connection with head-on collisions (8) and more recently in connection with intersection collisions (9) were also encountered in this series for rear-end collisions. Characteristics encountered are a more readily collapsible structure and lower peak accelerations for speeds of impact up to the energy required to "bottom out" the structure. The 1967 Ford rear-end responds with a comparatively "soft" impact for speeds through 30 mph; the substantial collapse damage at 30 mph suggested that acceleration forces would increase non-linearly for rear-ending speeds much above 30 mph. Data generated for this report relates to the collapsibility of the 1967 Ford; application to other designs of passenger vehicles should be guarded, except where there are comparable rear-end structures. Stiffer rear-end structures will develop higher acceleration values for a given speed of impact and may require correspondingly stiffer seatbacks having yield strengths at higher values, if comparable protection is to be offered. Conversely, seatbacks designed for optimum performance of the 95th percentile male may not necessarily provide optimum protection for the average (50th percentile male and female) motorist owing to their substantially lower inertial forces applied to their seatbacks, for a given speed of rear-end impact. This latter factor is not regarded significant, however, based on findings of this paper indicating that stiffer seatbacks of adequate height provide improved whiplash protection and based on studies with smaller subjects (3) indicating that high-back seats not only provide good protection but in no way contribute to injuries from rear-end collisions.

These experiments were conducted with an empty trunk; a trunk filled with luggage would probably notably stiffen the vehicle's rear-end collapse characteristics and reduce the speed at which the rear seatback is punched forward during impact.

The presence versus absence of the spare tire and wheel was evaluated and for the 30 mph rear-enders, differences were not measurable, except that rear-

Load * (lbs)	X (inches)	X_s (inches)	Y (inches)	Y_S (inches)	Z_F (inches)	Z_R (inches)	α (degrees)	α_s (degrees)	F_L (lbs)	F_R (lbs)
0		0		0				0	320	550
570	1/8		11/32				1 1/2		525	875
0		0		0				0	320	550
1060	1/4		23/32				3°		965	1390
0		0		0				0	320	583
1300	5/16		1 5/16				5°		1500	1550
0		0		0				0	310	615
1620	7/16		2 1/8				7°		2060	1970
0		1/32		5/8				1°	268	875
1740	1/2		2 3/8				9°		2270	2420
0		1/32		13/16				1 1/2°	310	1130
1950	3/8		2 3/4				10°		2600	4110
0		1/16		1				3°	320	712
2040	5/8		3 1/2				12°		2730	6050
0		1/8		1 1/2				5°	375	4440
2150	1 1/16		4 1/8				15°		2980	6400
0		7/16		2 1/8				7°	396	4010
2190	1 3/4		5 1/4				20°		3150	7020
0		1 1/8		3 1/4	1/16 1/8	1/2 7/8		12°	406	4440

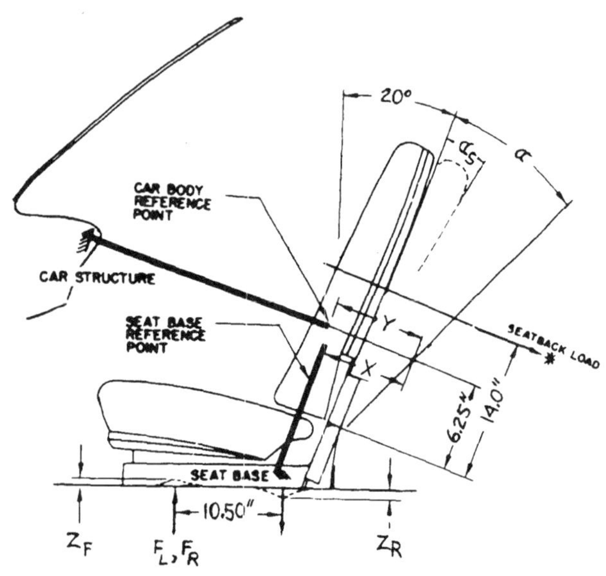

Legend

* - Load applied perpendicular to seatback

F_L - Load at front seat anchorage (left anchor bolt)

F_R - Load at front seat anchorage (right anchor bolt)

X - Seatback deflection relative to rigid seat base

X_s - Seatback permanent set, cumulative, relative to rigid seat base

Y - Seatback deflection relative to vehicle structure

Y_S - Seatback permanent set, cumulative, relative to vehicle structure

α - Seatback angular rotation relative to rear seat anchorage

α_S - Seatback angular set, cumulative, relative to rear seat anchorage

Z_F - Vehicle floor deformation at seat front anchorage

Z_R - Vehicle floor deformation at seat rear anchorage

Fig. 114 - Typical findings of static tests UCLA 28 in. seatback, yield point 2190 lbs (31,000 in.-lbs)

end collapse did reach a point of starting to force the spare tire and wheel forward toward the rear seatback. At 40 mph, the spare tire bolt did not tear loose from the floor pan step; the tire was forced forward about two inches, not sufficient to be significant, except if there were a center-rear seat passenger. No spare tire was included for the two 55 mph crashes because of the possibility of the tire interfering with the more important observations of special seat performances and the evaluation of the special fuel tank location.

Speed of Impact - During the initial phases of this study, it appeared that the gross collapse attending the 30 mph rear-end collisions, so far as the rear-ended car was concerned (crushing the rear-end forward close to the two rear tires, 2¼ to 3 feet collapse at 30 mph) represented a close approach to a bottoming-out condition; this characteristic was found true with the readily crushable unit-body construction vehicles when impacted head-on at speeds approaching 40 mph (8). However, although the bottoming-out condition did occur as anticipated, the 40 and 55 mph rear-end collision impacts registered some rather remarkably low passenger compartment accelerations for the struck car. The more collapse-resistant rear-end structure characteristics associated with speeds of above 30 mph serve to transfer sufficient stress to cause gross collapse of the striking car's front-end structures; this unanticipated situation maintains the constancy of relatively-low peak accelerations in the rear-ended passenger compartments for speeds through 55 mph, Fig. 115.

Analysis of the data of Fig. 115 for the maximum mutual collapse, which includes plastic (permanent) and elastic (rebound) deformation of both cars, shows a rather uniform, almost linear, increase in maximum mutual collapse* as a function of increases in speed of impact. The mechanism of collapse may be further understood by referring to the two lower curves of Fig. 115, depicting the permanent collapse of the striking and struck vehicles, as a function of collision speed. The permanent collapse of the struck car increases abruptly from 10 to 30 mph and then levels off on approaching 40 mph, remaining constant through 55 mph. This bottoming-out of rear-ended structure provides sufficient collapse resistance to cause a sharp in-

*Maximum Mutual Collapse, the maximum amount of closure from initial contact, measured in feet, of the non-collapsing portions of two cars undergoing collision.

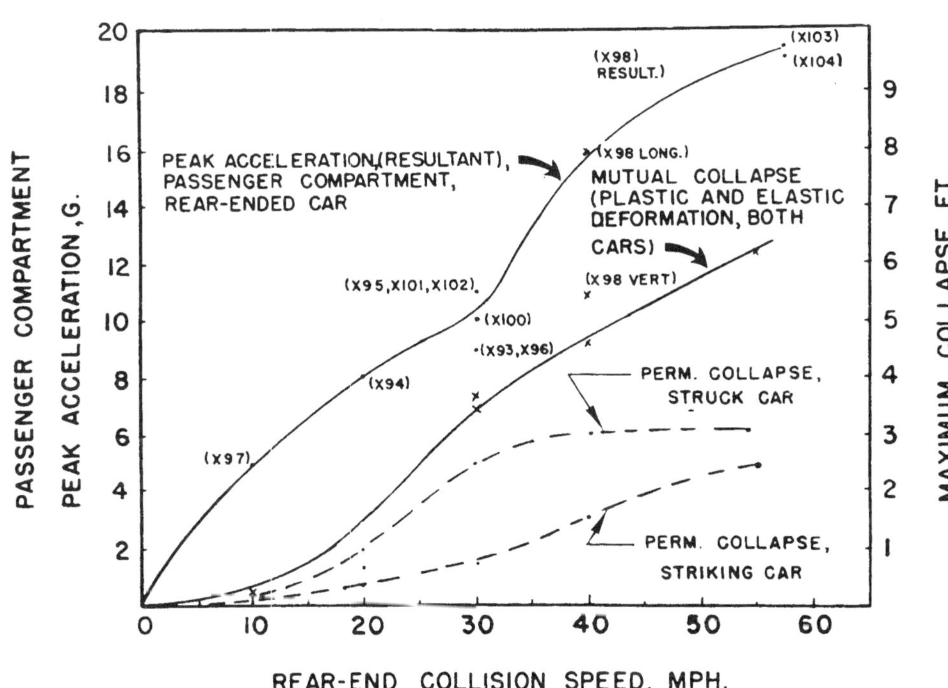

Fig. 115 - Passenger compartment accelerations and maximum collapse

crease in the rate of collapse of the striking car for speeds above 30 mph.

The rather linear change in rear-ended passenger compartment peak acceleration follows the pattern of maximum mutal collapse, as would be expected. During a collision, short of passenger compartment intrusion, it makes little difference whether it is solely the striking car, solely the struck car or a combination of the two vehicles that collapse, so far as the passenger compartment peak accelerations are concerned. It is for that reason that the rear-ended car peak acceleration varied rather uniformly with increases in the collision speed; the singular exception was the 40 mph collision and a closer examination of the component peak accelerations making up the resultant acceleration shows a 16 G longitudinal peak augmented by an 11 G vertical peak. In this collision, both longitudinal and vertical peaks phased together while in the other crashes they were asynchronous and therefore did not augment to provide excessive resultants. The rather high longitudinal peak G is accounted for by observing that the striking car front-end collapsed 1½ feet, sufficient to bottom-out against the engine and front wheels. This caused a very abrupt acceleration, comparatively late in the crash. The additional kinetic energy of the 55 mph impact speed was sufficient to collapse the striking car front end in a gross manner (almost twice the collapse of 40 mph) accounting for the comparatively low peak accelerations of the rear-ended car passenger compartment for the 55 mph collisions.

Rear Window Header - The low profile designs for automobiles of the past decade have markedly increased their stability against upset. One of the trade-offs for this safety improvement is the restricted head clearance dimensions of the passenger compartment occasioned by the lower silhouettes of passenger vehicles. Very little roof collapse space is avilable for upsets that do occur and it was found that the 95th percentile male strikes his head against the rigid header or molding over the rear window during the course of a rear-end collision. This contact occurs because of the sharp slope of the car top starting above the heads of the rear seat passengers and terminating at the trunk deck; in some cars, this entire section is glass. This rigid unpadded structure consistently provided excessive head impact accelerations. The header area should be padded or contoured to reduce passenger injuries. Inasmuch as rear window-header contact is associated only with male motorists approximating the 95th percentile size and considering further that passengers only infrequently occupy rear seats, the authors do not consider alteration of header elevation warranted, solely on this basis.

There is, however, a factor associated with this problem that should be further researched. The rear window and adjacent metal structures provide a wedge-like surface for the head to strike. As the head contacts this inclined plane (30 degrees), assuming no contact with a header projection, the peak acceleration is produced by the wedge-like application of relatively excessive accelerative forces to the head and this may cause excessive compressive forces for the vertebral column. This exposure is not restricted to only the tallest males; even average height males strike the sloping rear window with high-G head blows at 30 mph.

Fuel Tanks - Statistics on the incidence of fire following passenger vehicle collisions is vague but there is no question as to the serious aggravation occasioned by post-crash fire. Much progress in recent years has been made at racing events, like Indianapolis, and similar reductions in passenger car post-crash fires will occur when studies are made and findings used to guide future design of fuel systems. With respect to the 1967 Fords rear-ended in this series, the initial four developed gross fuel tank spillage arising out of the detachment of the filler spout from the rubber type flexible connector at the tank. The rear fender sheet metal crushed forward and pulled the filler spout from the tank. The longer filler spouts (a production change) used on vehicles in the last six experiments, eliminated this disassociation and the associated spillage. Many of the tanks also sustained direct punctures; in all but the 10 mph collision, the tanks

measure of protection from whiplash injuries with an extended seatback that is slightly lower than the top of a tack-on head restraint. For example, in Experiment 95, the driver's head rotation to 36 degrees relative to torso axis and the 28-inch height of the passenger's head restraint restricted the passenger to 58 degrees. In Experiment 96, tack-on head restraints were not used; the driver's seatback was increased from 22 to 25 inches and this extended seatback restricted head rotation to 54 degrees relative to torso axis for the driver. Also in Experiment 96, the passenger's seatback was increased from 22 to 28 inches providing one of the most effective restraining actions evaluated, 16 degrees of dorsi-flexion or rearward head rotation relative to semi-reclined torso axis.

The explanation for this unexpected improvement concerns the relative lack of yield of the higher seatback as contrasted with the lower seatback that carried a tacked-on head restraint. Consequently, a seatback designed sufficiently high to provide adequate protection from whiplash type injuries will not have to extend as high as a seatback carrying a separate head restraint.

Head Offset - One of the problems facing automotive designers for head restraints is the relationship of head restraint effectiveness as a function of the distance the head is postured from the support at the time of impact. Experience gained from catapult-launching of aircraft from cruisers and from aircraft carriers indicates that abrupt accelerations least affect the pilot if his head and back are firmly pressed rearward against their respective supports. This precaution limits the inertial impact forces as the supports are accelerated against the head and back of the pilot. This analogy is especially apropos in view of the fact that the nature and magnitude of forces for 30 mph rear-end collisions involving relatively crushable 1967 Fords are nearly comparable to the upper limits of catapult launches. It was not unexpected, therefore, for rear seated passengers having a conventional rigid seatback design, that head and chest accelerations were substantially higher when postured leaning forward and subsequently slammed rearward against the seatback as contrasted to those occupants postured against the seatback at the time of the collision.

Unlike rigid pilot seats, the front seats for the passenger vehicle have yielding seatbacks. The readily compressible upholstery and flat springs bottom out without much resistance to displacement, followed by seatback tilt that can be controlled by design. The UCLA calibrated seatback provided a rate of yield that greatly moderated the chest and head accelerations. Because of this feature, the added inertial force caused by the head and torso, when positioned forward of the seatback, is compensated by additional deflection, but not necessarily accompanied by additional yield of the seatback. This feature suggests that front seat posture variations common to motor vehicle operation do not pose a serious problem for the automotive designer providing the seatback is designed with realistic yield strengths and has a full support system (28-inch seatback appears adequate).

Seat Belts in Relation to Rear-End Collisions - As was identified in connection with the School Bus study (3) the authors recognize that use of seat belts with the low seatback design, especially the front seat, can contribute to the severity of injuries for the whiplashed motorist, in those instances where seatback failure does not occur. This phenomenon occurs because of the anchor-like action a belt may apply to the hips. The strapping-down of the hips intensifies the head and shoulder inertial forces that are applied to the spinal column and that are resisted by the fulcrum-like action of the seatback. The seatback fulcrum applies force to the upper mid-back of the average adult. The possible adverse influence of a seat belt in use at the time a motorist is rear-ended was originally considered by the authors to be of greater importance than it actually was, as evidenced by this series of collision experiments. Instrumentation for seat belt tension during rear-end collisions indicated that for 10 through 30 mph impacts, belt loads did not exceed 100 lbs. In seeking an explanation for this observation, it was found that during the onset of a rear-end collision, substantial belt slack developed not only from the seat and floor pan crushing forward against the hips but also because elevation of the hips in the rearward direction tends to follow the arc described by the seat belt buckle rotating upwards and rearward without extension of belt length. These factors account for what was confirmed in this series of experiments that body kinematics and forces sustained are not generally significantly altered by the presence or absence of a seat belt. Notwithstanding, the authors recognize that this adverse feature is more than offset by the decided safety advantage the seat belt provides for other collision exposures. If seatbacks have adequate strength and height (at least 28 inches), the seat belt tends to reduce the torso sliding up the inclined plane of the seatback during rear-end collision acceleration. A motorist whose position has shifted up the plane of the seatback will have more of his head and upper torso unsupported and will be correspondingly exposed to more serious whiplash injuries. Seatback yield is primarily the cause of torso shift up the plane of the seatback; the contour and fabric of the seatback represent other minor factors, similar in importance with the use of a seat belt.

<u>Load-deflection Characteristics of Seatback Cushion</u> - The rear-end collision experiments reported in this paper establish the importance of the head-offset from the head restraint at time of collision. The reference point for head-offset is the distance from the back of the motorist's head perpendicular to the head restraint surface, or the plane representing an extension of the seatback surface, where there is no opposing head restraint. This information is significant because a motorist may have zero-inch offset but if his head is resting against a very low-density thick padding, he may receive no appreciable protection until the head restraint has crushed forward perhaps six inches, corresponding in performance to more than a normal three-inch offset.

To obtain this data, the test seat was installed on a load-calibration frame; bracing was applied to the seatback to prevent it from deflecting when loads were applied to the sternum or the dummy's chest, as illustrated by Fig. 116. Standard laboratory techniques were followed in which sequentially increasing loads were applied to the dummy and corresponding cushion deflections were noted, separated by the zero load-deflection readings illustrated by Fig.117.

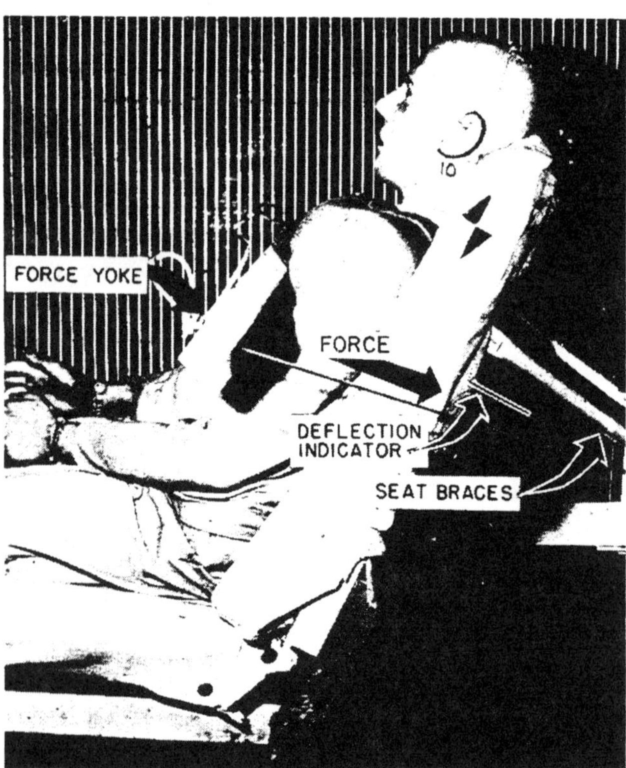

Fig. 116 - Upholstery stress/strain equipment, 50th percentile dummy, 25 in. UCLA modified (33,000 in.-lbs) seat. Shown as 215 lbs and 3 3/8 in. deflection

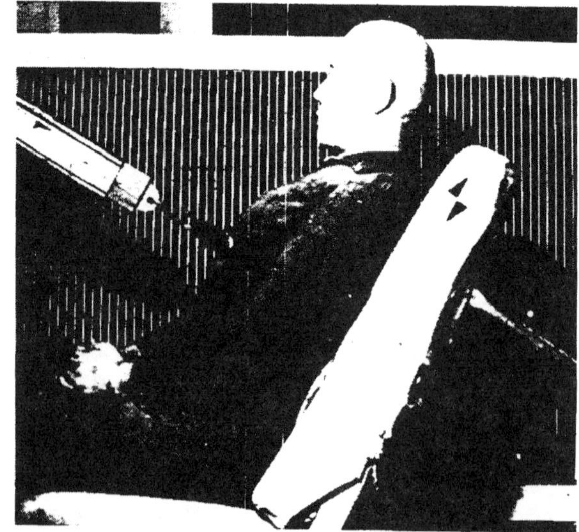
Fig. 117 - Cancelling out 95th percentile dummy's weight between progressive load cycles for 28 in. seatback upholstery calibration

The seatback cushion deflection relationships for the two sizes of dummies (50th and 95th percentile) and for the three principal UCLA seatback modifications are provided by Fig. 118. Higher loads (simulating dummy inertial loads) are required for the 95th percentile dummy because his back and head surface area understandably exceeds those counterparts of the 50th percentile dummy. The ratio of inertial mass attributable to a 95th percentile dummy is 205/165 or 1.24 times as much as that of a 50th percentile dummy; while this ratio is less than the relative 95th/50th stress ratios for a given deflection, the accelerative forces are still sufficient for speeds of impact above 10 mph to bottom-out these seatback cushion designs regardless of occupant size.

Accordingly, the cusion load deflection data indicates that "slack" built into a seatback restraint system, as far as rear-end collision exposures are concerned, compromises motorist protection. A comparable observation was made with respect to "slack" for a seat belt which describes the ineffective yield characteristics of lap belt restraint systems, as identified in Reference 8.

From this analysis, it may be concluded that seatback conventional upholstery and spring construction is relatively ineffective for providing consistent load/ deflection. Therefore, its use should be kept to a minimum, consistent with comfort, in order to provide the motorist with the greatest measure of head and back restraint possible.

Seatback Collision Deflection - Referring to Table 4 (Calibrated Seatback Data) only the front seat units were calibrated; rear seat units are typically rigid and therefore do not require calibration. There was no evidence of rear seatback permanent yield or failure for impact speeds through 30 mph and relatively little collapse for speeds through 55 mph. Three heights of front seatbacks evaluated were 22, 25 and 28 inches. Seatback strengths were 9,100 in.-lbs. for the 22-inch seatback, 16,100 in.-lbs. for the 25 and 28 inch seatbacks and 33,000 in.-lbs. for the 25 and 28 inch rigid seatbacks. Still referring to Table 4, as would be expected for the same size (95th percentile) occupant, seatback deflection was greatest for the weakest seatback (See Experiment 93 and also Experiment 94) and for the higher speed of impact (See Experiment 93 at 30 mph, as contrasted to Experiment 94 at 20 mph).

By considering only 30 mph impacts and the 28-inch seatbacks*, each with

*Experiment 96 used a head restraint that was integral with its seatback, Experiment 93 and Experiment 95 used prototype head restraint or tack-on installations.

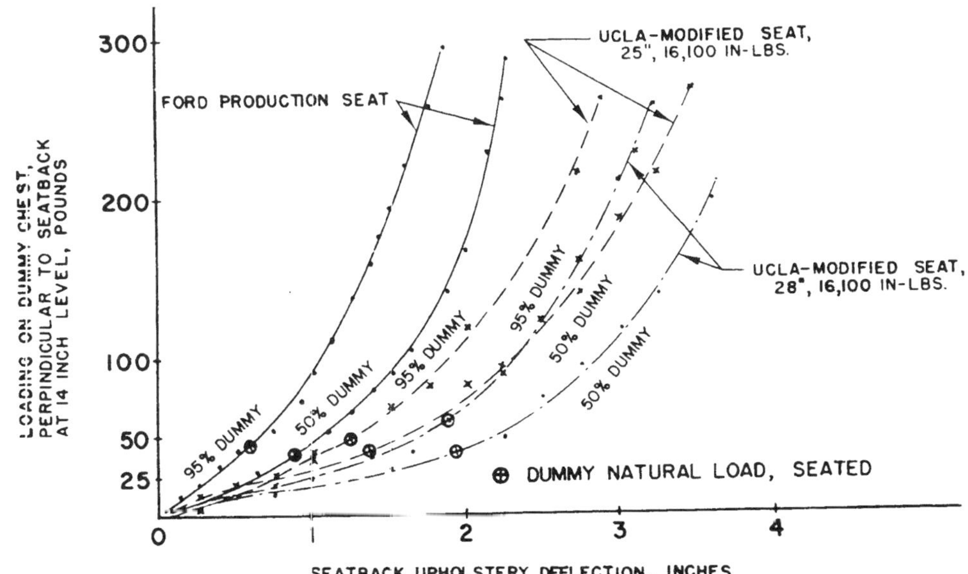

Fig. 118 - Seatback cushion stress/strain characteristics

Table 4 - Calibrated Seatback Data (Front Seat Only)

	30 mph x 93		20 mph x 94		30 mph x 95		30 mph x 96		10 mph x 97		40 mph x 98	
SEATBACK	LEFT	RIGHT	LEFT	RIGHT	LEFT	RIGHT	LEFT	RIGHT	LEFT	RIGHT	LEFT	RIGHT
Height	22"	28"w/hs	22"	28"w/hs	28"w/hs	28"w/hs	25"	28"	22"w/hs	28"	22"w/hs	28"
Strength (in.-lbs.)*	9,100	16,100	9,100	16,100	16,100	16,100	16,100	16,100	Mfg's Specs	16,100	Mfg's Specs	16,100
Peak Defl. (deg/ms)	27°/170	16°/190	14°/150	10°/125	10°/215	10°/215	16°/190	17°/190	9°/135	7°/115	18°/184	33°/215
Post Yield (deg)	19°	7°	8°	3°	3°	4°	7°	8°	1.5°	0	12°	21°
G/ms @ 14"	7/140	7/60	7/35	7/65	9/80	11/55	7/35	7/35	2/65	5/60	42/135	10/155
G/ms (Head Restraint)	–	9/190	–	Trace Lost	32/65	14/45	–	15/40	4/125	11/55	31/135	18/150
FRONT OCCUPANTS												
Head Offset	3"	0"	3"	0"	6"	12"	3"	3"	3"	0"	3"	3"
Head Rotation (deg.)**	78°	31°	74°	30°	36°	58°	54°	16°	39°	14°	70°	79°
G/ms Head	12/220	11/150	9/165	14/155	10/180	18/170	11/210	11/160	8/215	10/150	34/190	33/205
G/ms Chest	7/205	6/100	5/90	6/65	7/65	6/130	6/85	6/185	4/100	5/125	18/140	10/160
G/ms Knee	6/195	6/70	–	8/85	Trace Lost	7/70	8/105	7/115	7/75	–	20/165	9/140
Lap Belt lbs/ms	40#/325	–	100#/260	–	0#/6	–	0#/6	–	***	–	–	–
VEHICLE												
G/ms Frame	9/15		8/35		11/55		9/25		5/15		19/100	

	30 mph x 99		30 mph x 100		30 mph x 101		30 mph x 102		55 mph x 103		55 mph x 104	
SEATBACK	LEFT	RIGHT	LEFT	RIGHT	LEFT	RIGHT	LEFT	RIGHT	LEFT	RIGHT	LEFT	RIGHT
Height	25"	28"	25"	28"	25"	28"	25"	25"	25"	28"	28"	28"
Strength (in.-lbs.)*	33,000	33,000	16,100	16,100	16,100	16,100	16,100	33,000	33,000	33,000	33,000	33,000
Peak Defl. (deg/ms)	7°/88	7°/125	20°/180	23°/208	22°/227	29°/230	7°/90	10°/126	18°/158	25°/144	21°/131	18°/125
Post Yield (deg)	0	0	9°	12°	14°	18°	0	0	6°	15°	8°	10°
G/ms @ 14"	8/30	7/35	11/45	7/60	42/135	9/160	12/30	12/35	20/90	18/140	18/140	36/130
G/ms (Head Restraint)	–	14/60	–	15/80	–	18/55	–	–	–	30/135	–	66/120
FRONT OCCUPANTS												
Head Offset	0"	0"	6"	6"	12"	12"	0"	0"	0"	3"	0"	3"
Head Rotation (deg.)**	44°	34°	69°	24°	80°	49°	33°	29°	100°	82°	79°	42°
G/ms Head	13/75	16/165	17/210	18/175	30/225	24/205	12/140	10/135	35/155	58/150	42/140	107/140
G/ms Chest	6/80	7/105	7/110	7/130	9/135	8/165	9/95	8/75	14/75	24/140	21/150	33/125
G/ms Knee	8/110	7/65	9/85	11/140	12/65	10/90	8/90	8/80	15/90	13/85	18/105	26/130
Lap Belt lbs/ms	–	–	–	–	–	–	–	–	–	–	–	–
VEHICLE												
G/ms Frame	12/35		10/25		11/25		11/25		19/85		19/65	

*Strength in lbs., as measured at the 14-inch level above the "H" point (650, 1150 lbs or 2360 lbs)
**Maximum head rotation rearward relative to torso axis at time corresponding to peak head acceleration
***Lap belt instrumentation discontinued, load values insignificant

16,100 in.-lbs. strength, but with respect to offsets of zero-inches (Experiment 93, right front), 3-inches (Experiment 96, right front), 6-inches (Experiment 95, left front) and 12-inches (Experiment 95, right front), the following observations were suggested, with respect to the amount of head offset, Table 4:

(a) Head peak-acceleration remained constant for head offsets from zero to 6 inches (11, 11, 10 G's) but increased substantially for head offsets of 12 inches (18 G).

(b) Chest and knee accelerations as well as seat belt force did not vary significantly for changes in head offset.

(c) Deflection, the maximum rearward seatback displacement during the collision, remained constant (16 and 17 degrees) for 0" and 3" offsets but was signficantly reduced (10 and 10 degrees) for 6 and 12 inch offsets. Similarly, post-yield, the permanent rearward strain of the seatback, is constant (7 and 8 degrees) for 0 and 3 inch head offsets but likewise reduces (3 and 4 degrees) for the 6 and 12 inch offsets.

Observations (a) and (b) are straightforward; item (c) appears to be varying in the wrong direction, assuming the increased inertial-force of the velocity difference between the head and head restraint developed as the head slammed against the head support, it would be expected that head impacts of increased intensity would be a function of greater head offsets. This departure from predictability is explained by considering the dynamic responses of both the head and its head restraint. When the occupant's head is close to (3 inches) or against (zero-inch offset) the head restraint, the perform as a single mass, applying their collective inertial forces against the seatback in approximate synchrony. This combined mass produces the largest deflections. When the head is sufficiently remote (6 to 12 inches) from the head restraint they respond initially independently with the head restraint and much of the seatback below it deflecting from their collective inertial forces. This develops counter elastic forces that apply to diminish the head and upper torso inertial forces when the rebounding or near-rebounding seatback contacts the head and back of the occupant. Seatback asynchrony with occupant head and torso inertial forces reduces the extent of seatback deflection. On the basis of this explanation, there may be a posture further forward of the seatback than 12 inches that would reestablish synchrony and result in an excessive deflection. Such motorists' postures are not often assumed and should not, therefore, pose a highly significant problem.

For the 1967 Ford rear-ended by an identical car, comparable frame accelerations were obtained for 20 mph (8 G at 35 ms) as for 30 mph impacts (9 G at 15 ms, 11 G at 25 ms and 9 G at 25 ms). Occupants in comparable seats (e.g., 28 inch and zero-offset) sustained rather comparable head, chest and knee accelerations.

From this planned series of experiments, it is concluded that front seat motorists are adequately protected from rear-end collision injuries for striking car speeds through 30 mph if they have an adequately structured seatback with a 28-inch seatback, capable of sustaining without failure

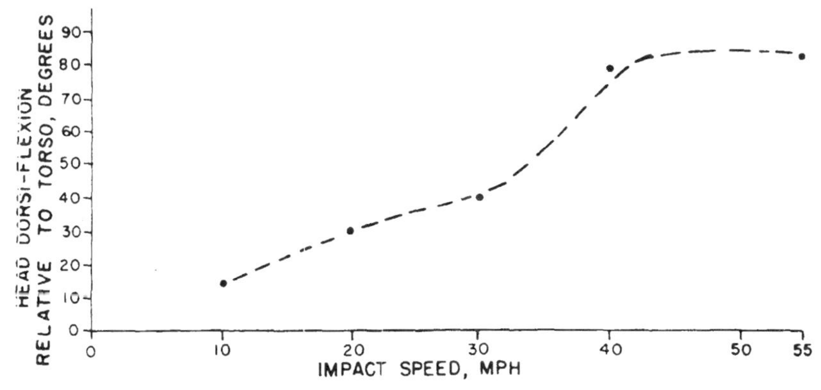

Fig. 119 - Whiplash exposure, as a function of vehicle speed

a 16,100 in.-lbs. inertial force; that head-offset may result in variations in seatback deflections and head peak accelerations, but that both factors can be safely accommodated in the majority of exposures. By analyzing other data in Table 4, the reader may derive conclusions in a similar manner.

Whiplash - During the whiplash phase of a rear-end collision, the extent to which the head axis remains aligned with the torso axis is a common measure of whiplash severity; other factors such as the head peak acceleration attained, the effective elevation of the seatback relative to occupant seated height (potential for neck "karate-chop") and the presence of injury producing surfaces for the head to be flailed against are discussed in other sections of this paper. These experiments have verified the importance of keeping the head and torso axis aligned by a properly designed seatback and head restraint; this data has also established that every whiplash-protective device evaluated demonstrates some measure of degradation of performance as the relative speed increases between the rear-ending cars. This feature is illustrated by Fig. 119, which presents collision data on the extent to which the 28-inch head restraint becomes progressively less effective as a device for keeping the head and torso axes aligned, as the speed of rear-end impact increases. This degradation is primarily attributed to the increased seatback yield caused by increased body and seatback inertial forces attending high speed impacts; this greater seatback yield positions the torso in a more reclined posture which facilitates its movement up the inclined seatback, exposing the neck to a more extensive degree of whiplash injury. Some tendency of this nature occurred even without seatback yield because of the natural 20 degree rearward slope of most seatbacks.

In addition to the variable of increases in speed of impact, as a contribution to whiplash severity, the height of a seatback is an obviously singularly important variable. To illustrate this factor, the number of degrees of misalignment of the head and torso axes, serving as an indicator of "whiplash", is plotted against seatback height, based on 30 mph collision exposures by the 95th percentile dummy, Fig. 120. Included are separate curves for head offsets of 3, 6 and 12 inches indicating also the degradation in protection occasioned by head offset. Thus, for a 3-inch (rather natural) head offset, a 22-inch seatback results in extreme whiplash (78 degrees) and the same 3-inch offset with a 28-inch head restraint reduces whiplash to an incidental 16 degrees. However, if this same 95th percentile dummy with the same 28-inch seatback is positioned with a 12 instead of 3-inch head offset, he whiplashes 49 instead of 16 degrees. A twelve-inch head offset as compared with a 3-inch offset degrades the protection of a head restraint by an amount approximately equivalent to diminishing the head restraint height three inches.

Whiplash Injury Index - The authors recognized that the term "whiplash" has been challenged as to its appropriateness in describing the motorist's involuntarily assumed head-to-torso postures occasioned by automobile collision and other forms of trauma.

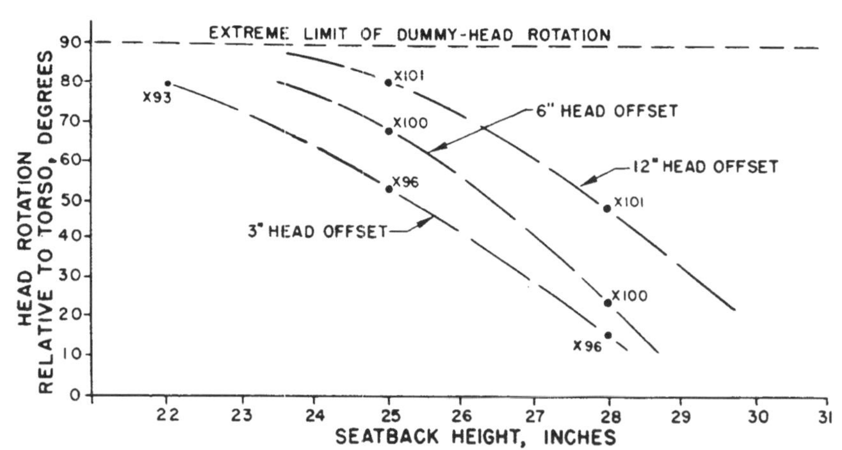

Fig. 120 - Whiplash protection (95th percentile dummy, 16,100 in.-lb seatback) of seatback height, 30 mph collision

Its use in this paper is limited to the forced dorsi-flexed posture of the motorist's head relative to torso characteristic of rear-end collisions. With the production bench seat common to most vehicles on the highway, the average seatback height is approximately 21 inches, with a few vehicles being equipped with head restraints that may extend seatback heights to about 26 inches. This average seatback height is so low that even the 5th percentile adult female driver (among the smallest) can be whiplashed seated in the production 21-inch seat.

With the probability now of rapid improvement in this situation, it becomes important to consider the effect of higher seatbacks that are not sufficiently high to provide positive head restraint. This consideration becomes even more important with the finding that relatively rigid seatbacks (ranging between 16,000 and 33,000 in.-lbs.) are required to derive the greatest whiplash protection for the more frequently occurring, higher speeds of rear-end impact. This is because combinations of motorist size and seatback heights can be assembled that provide rather firm support to the neck, one or more vertebra below the base of the skull. This type of support, while closely approaching full effective support can be injury producing owing to the "rabbit-type chop" given to the back of the neck as the inertial forces of the head come into play, Fig. 101(b).

The special "whiplash-neck" built by Sierra Enginering to UCLA specifications represents an engineering tool for assessing this problem, Fig. 121. While it is recognized that the articulation and related simulation is a crude representation of the human neck, it does provide a device having repeatable characteristics accurately portraying relative performances as variation in seatback heights are introduced.

To provide a reliable gage of effective seatback height (absolute-rigid), five 1/2 inch sheets of plywood were cut to corresponding seatback heights that placed their top-edges opposite (in separate tests) each of the five cervical vertebra indicated in Fig. 121. The chest of the dummy was strapped to the rigid seatback. The head was rotated rearward at 5 degree increments by applying a load in a rearward direction through the center of mass of the dummy's head as shown by Fig. 122.

These data are presented in Fig. 123, illustrating the relationship of simulated inertial head loads to neck-flexion for specific seatback heights.

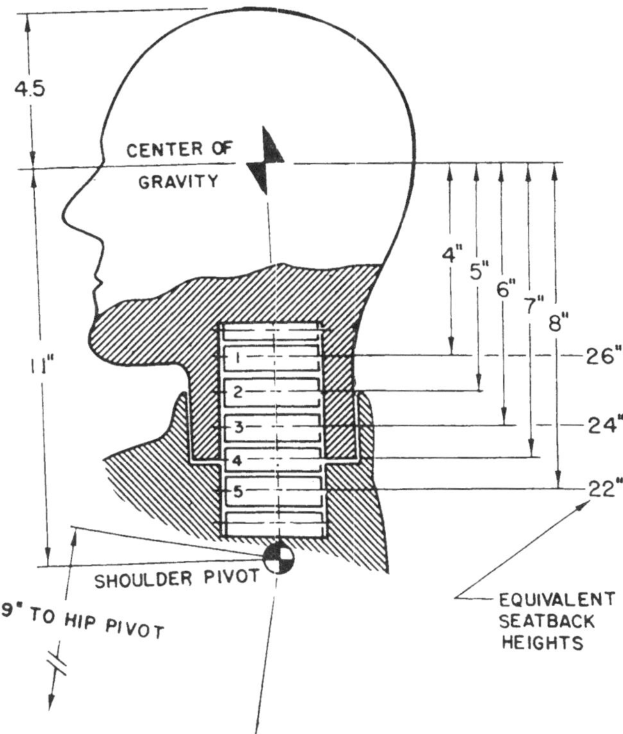

Fig. 121 - The cervical vertebra and their relationship to height of seatback restraint

Fig. 122 - Head stress neck-flexing calibration of articulated 95th percentile dummy; seatback height was 25 in., head rotation 90 deg rearward relative to torso axis for 146 lb loading

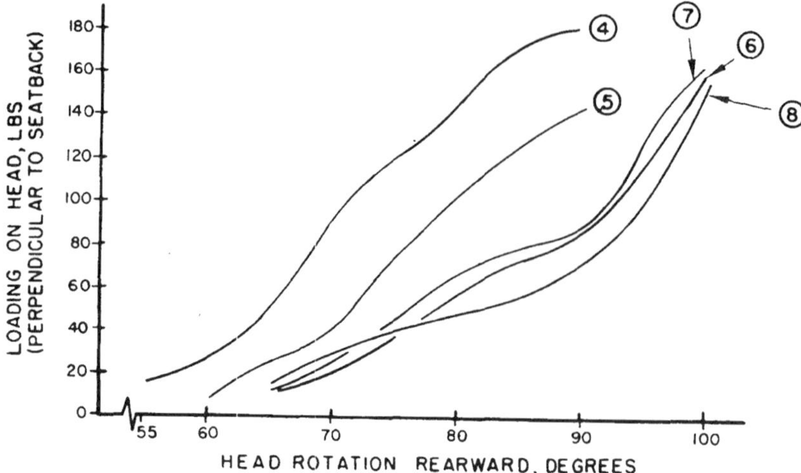

Fig. 123 - Head loading/head rotation as a function of effective seatback height for the 95th percentile adult male dummy

This graph illustrates what might be intuitively concluded, namely that high inertial forces may act through the center of mass of the unrestrained head and neck combination during a whiplash exposure and that the "whiplash" potential injury, per se, may not be clearly evident because the seatback height was sufficient to apply a "rabbit-chop" fulcrum at some level of the neck that reduces limits of voluntary neck excursion, and at the same time increases injury producing shear and bending stresses.

CONCLUSIONS

In the conclusions that follow, the authors wish to point out that these statements are based on specific observations, the majority of which should not be over-interpreted to form generalized conclusions. However, because of the wide variety of conditions evaluated, certain conclusions are broad, not because of a specific observation, but because of a multiplicity of observations that are correlative and that reinforce a specific conclusion, thereby providing a foundation for some degree of generalization.

A. CONCLUSIONS - METHODOLOGICAL - The foundation of scientific inquiry is its methodology, procedures devised for evolving information not commonly available and not readily verifiable. The reader's confidence in data subsequently developed depends on the methodology, the comprehensiveness of the studies and the reputation of the investigators.

1. The twelve rear-end collision experiments reported by this paper are representative of the majority of collision exposures, particularly as related to severity. Passenger vehicles travel in a stream of traffic columnated by lane guide-lines; it is not surprising, therefore that errors in speed control result in rear-end collisions that are either single or multiple and that most frequently occur with no appreciable offset or differences in vehicle heading at impact. The speeds evaluated, 10 through 55 mph represent the majority of injury producing exposures.

2. These collision experiments provide conditions sufficiently realistic and comprehensive to adequately evaluate the relative merits of various passenger vehicle rear-end collsion protective devices and to identify related injury producing factors. Severe whiplash injuries were inferred from these rear-end collisions; these results indicate that a practical level of collision force was obtained for evaluating relative performances of different seatback and head restraint designs.

3. These experiments were planned with a primary objective of developing reference material needed by engineers designing seatback and head restraints for passenger vehicles. For this reason, many variations in design and manufacturing techniques were not included in the interests of holding the experiments to a manageable number.

4. The use of both 50th percentile and 95th percentile adult male anthropometric dummies provided a practical evaluation of the effect of the average and the nearly largest passenger sizes on the various collision protective de-

vices evaluated. Prior studies (4) established that smaller sized subjects, including children are well protected riding in seats adequately protecting adults.

5. On a selective basis, variations in seat types, seat strengths, seatback heights and related protective devices were evaluated under realistic conditions. Selection of a practical variety of protective devices was made to provide objective data useful in making performance judgements concerning units not specifically evaluated in these studies.

6. Comprehensive instrumentation was used in these studies to provide detailed specifics required for safety performance evaluation of the rear-ended passenger vehicle. The use of properly articulated anthropometric dummies, 55 transducers and 29 photographic units represented the most comprehensively instrumented passenger vehicle collision study conducted to date. The extensive photographic coverage of each human simulation during the entire collision event provided new insight into injury causation.

7. The utilization of instrumented full size vehicles providing realistic collision conditions and instrumented trauma-indicating anthropometric dummies simulating passenger collision, their induced movements, and the use of high speed motion picture color photography overlapping all collision movements represents a most practical and reliable method for determining passenger vehicle rear-end collision performance. An evaluation was made of the several procedures in use by research groups in the U.S. and abroad concerned with accidental trauma; the consensus of opinions concurred with this methodological approach. While it is true that fatal injury trauma varies greatly from individual to individual and this range is not understood with any great degree of precision, nevertheless, the procedures used in this study provided an exacting basis for determining the relative performances of safety devices. It is more important to learn which passenger environment provides the most practical and effective improvement to passenger protection than to delve into refinements concerned with the specific traumatic conditions required for permanent injury or death. Progress with the safe transportation of humans will always be a relative matter - there will never be a practical arrangement for transportation that guarantees no injuries.

8. The presence of double or higher order variables was controlled through appropriate methodology to ensure reliability of findings. In this study, the techniques of redundancy in instrumentation and absolute constancy of factors involved, except for the single variable under study, were used as control devices. For example, the seating side-by-side of two identical dummies with identical restraints in identical seats and identical postures except for specific head offset variation provided a means of establishing the precise role minor variations in head-offset may play in motorist protection.

B. CONCLUSIONS - COLLISION PERFORMANCE - The collapse resistance of a rear-ended vehicle as a function of impact speed, the susceptibility of the vehicle to passenger compartment encroachment, the comparative resultant accelerations, the influence of vehicle components on injury causation, the preservation of passenger compartment integrity while undergoing moderate levels of impact acceleration are examples of vehicle collision responses characterizing collision performance. In the ordinary use of a vehicle, these deficiencies do not usually manifest themselves and frequently escape observation by accident investigators owing to the transient nature of these deficiencies or the investigator's lack of familiarity with levels of collision performance. Adequate collision performance provides the passenger with a protective shield from the crashing structures of the primary impact; adequate passenger compartment safety protects the passenger from the injury producing forces of the secondary impact, the one in which the passenger may be hurled against the compartment interior or ejected. This section relates to the former and the section to follow, to the latter.

1. The rather linearly varying passenger compartment peak accelerations sustained for speeds of rear-end impact between 10 and 55 mph attest to the excellent force modulating properties of

7. Seatback yield represents a condition that may reduce the amount of head adverse posturing relative to torso (whiplash) during a rear-end collision. However, it is not practical to consider seatback yield as a protective design feature owing to the wide range of performances needed for a controlled-yield. The many variables having to do with the severity of rear-end collision forces makes this approach impractical.

8. Plastic deformation of high seatbacks (front seats) reduce passenger's rebound toward the windshield in rear-end collisions but greatly increase chances of injury for any rear seat passengers thrown against them. During moderate speed rear-end collision experiments, high-backed seats offered adequate support for the head and torso and when the seatback yielded rearward approximately 20 degrees, rebound was diminished. This rebound hazard would be intensified if a front-end impact occurred, following the rear-end collision. A disadvantage however, of the seatback being forced into a semi-reclined position is that the passenger to the rear is more likely to strike it on rebound. Additionally, the more reclined angle facilitates sliding by the motorist up the seatback, thereby extending his head further beyond the end of the head restraint.

9. Elastic rebound of seatbacks increases the chances of passengers sustaining multiple impact injuries. Increasing seatback rigidity through designs that allow for only minor controlled yield with only nominal elastic action reduces rebound of motorists following peak accelerative forces of a rear-end collision; this consistent observation is explained on the basis that the elastic energy that can be stored by the upholstery padding and springs as the seat crushes forward into the backs of the motorists is negligible when contrasted with the metal seatback frame, anchorages and floor pan elastic energy for designs allowing significant elastic rebound.

10. Rear seatback yield was insignificant for speeds of impact through 30 mph and very slight (to 1½ inches) for speeds of impact through 55 mph, attesting to the rigid nature of rear seatback structures.

11. Even at 30 mph a 28-inch seatback (95th adult) should have a deformable or energy absorbing top edge in order to conform to neck arch during the rear-end loading.

12. Front seat passenger protection against the injury producing forces of rear-end collisions using the current design technique of seatback failure is unsatisfactory. Not only is the passenger subjected to the random chance of critical injuries sustained from striking the rear surfaces of the car interior or the rear seat passengers, the driver is so adversely positioned that he loses all opportunity of regaining control of his vehicle in time to avert potentially more serious secondary collisions. This explains the reason a weak seatback is not recognized as an acceptable solution for motorist protection from rear-end collisions.

13. Rigid seatbacks assure more effective support of the occupant during rear-end collisions, providing the seatback support is high enough to also resist rearward movement of the head. Conversely, a seat that yields appreciably rearward (e.g. more than one foot rearward displacement, as measured at the head restraint elevation) places the motorist in a semi-reclined posture that may serve to attenuate some of the injury producing forces but at the same time adversely displaces the motorist to higher elevations relative to the seatback, thereby reducing the measure of support that may be derived.

14. The more rigid the seatback, the less the tendency of head and torso displacement up the plane of the seatback during a rear-end collision; to the extent that a forward tilt design is incorporated in a seatback at its upper limits, this adverse torso shift is less likely to occur since this design feature straightens the inclined plane of the seatback where it's unneeded but encountered during torso shift.

15. Increasing seatback rigidity reduces rebound of motorists following peak accelerative forces of a rear-end collision; this consistent observation is explained on the basis that the elastic energy that can be stored by the upholstery padding and springs as the seat crushes forward into the backs of the motorists is negligible when con-

trasted with the metal seatback frame elastic energy for designs allowing significant elastic yield.

16. Seat assemblies and anchorages should not fail from collision accelerations under 20 G from rear-end impacts. The fact that most seat anchorages held during the UCLA Series V rear-end collision experiments is attributed to the following factors: most of the seats were special units and required individualized techniques to anchor them. The UCLA engineers made certain these anchorages would sustain the contemplated impacts so that evaluations of protective restraints, etc., would not be compromised by hardware failure.

17. As a restraining material, the relative ineffectiveness of upholstery, having conventional stress/strain characteristics, indicates why its thickness should be kept to a minimum, consistent with comfort, in order to provide the motorist with the greater measure of head and back restraint possible with the extended seatback metal structure.

18. The use of rigid metal trim across the top of bucket seatbacks exposes the rear-ended front seat motorists as well as the front-ended rear seat motorist to head impacts with injuries amplified by these non-yielding sharp surfaces. Padding such trim in the position of most probable impact is ineffective unless the padding is dense.

19. Seatbacks and arm rests should be designed using well padded, broad surfaced metal frames designed to provide the required strength and to attenuate head impact forces. The seat requires a strong frame to prevent seat inertial forces and passenger inertial (impact) forces from excessively deflecting it and breaking it free from its mounts. This strength must be designed into the seat so that small surface areas, and rigid structures are not encountered during "bottoming-out" type head impacts, occasioned by rebound from rear-enders, or direct impact by rear seat passengers thrown into front seatbacks during head-on impacts.

20. In the high-back seat, a contour forward feature for upper third of the seatback provides restraint against torso sliding up the plane of the seat. Relatively small vertical downward forces are required to neutralize the torso movement up the plane of the seatback; a properly designed shoulder harness, integral with the seatback, would accomplish this objective.

21. A seatback having differential strength with least bending moment at top would reduce the dangerous fulcrum-like action of head restraints that do not reach high enough to support the head mass. This matter deserves further research and the preferred solution, based on current information, is to provide vehicles with head restraints high enough to obviate the fulcrum-like action problem.

E. CONCLUSIONS - HEAD RESTRAINTS - Head restraints are devices that restrain the rearward movement of the head relative to the seatback so as to maintain the head and torso axes in a natural relationship during a rear-end collision.

1. Rear-end collisions are one of the most common types of accidents; even low speed impacts can be crippling. Head restraints, designed to function as a part of the seatback, represent a satisfactory, and the best known, solution to this problem. Head restraints should not be optional equipment because that status should be reserved for items not involving motorist safety.

2. Head restraints are as important to the motorist involved in rear-end collisions as the safety belt is to the motorist involved in a front-end impact; the safety belt provides "brakes" for motorists in front-end impacts and the properly designed seatback with head restraint provides "brakes" for the rear-ended motorist's head and torso. The

3. The head restraint for front seat units should be designed as an extenion of the seatback and preferably not as an attachment or an adjustable unit. Additionally, because of the somewhat critical nature of head restraint position relative to the head and its response to seatback inertial and head impact forces, the concept of attaching the head restraint to the roof structure, independent of the seatback, is regarded as objectionable on the basis that it compromises motorist protection and can be a source of injury during upset and other types of collision exposure.

4. The closer the head restraint

to the motorist's head, the better his protection. Head offsets from the head restraint to six inches do not greatly increase a motorist's exposure to injury.

5. Static tests for head restraint performance yield generally exceed the corresponding dynamic yield value owing to force augmentation attributable to seatback inertia not present with static tests. This condition depicts the capacity of the restraint to resist adverse head movement but does not include the requirement for resisting the restraint's own inertial force.

6. Properly devised laboratory static tests of seatbacks with head restraints represents a satisfactory procedure for evaluating head restraint performance, where correlative data is available from dynamic full-scale studies. This conclusion assumes experimental sophistication in terms of including sufficient realism in the laboratory procedure to avoid such serious possible omissions as floor pan yield, seat anchorage bolt bending and yield, and the possible interaction of vehicle fixed interior surfaces with seatback yield and occupant displacement.

7. The head restraint should not be weaker than the seatback. If the seatback yields or fails, the head restraint should maintain the same relationship.

Failure of the head restraint may expose injury producing structures to the occupant flailed against it. Seatback failure, while an undesirable condition, does not generally represent serious injury exposure unless accompanied by failure of the head restraint with respect to the seatback.

8. For the 10 mph impact, a 95th percentile adult male dummy on a rear (bench) seat receives some measure of head restraint from the sloping (30 degree horizontal) rear window. The quality or protective aspects of this wedge-like impact type of restraint is suspect, not only because it cannot be counted on for variations in posture and dummy heights, but also because the restraining force is accompanied by a vertically downward vector that may at times reach sufficient magnitude to cause injury to the spinal processes. Rear seat positions should be provided with head clearance sufficient to prevent significant contact with vehicle interior by a 95th percentile adult male undergoing a rear-end collision by an equivalent striking car traveling between 10 and 40 mph faster than the rear-ended car at impact.

9. Head restraints should be an integral part of the vehicle, preferably requiring no adjustment, if they are to provide the most consistent and effective protection against rear-end collision injuries. The width of the seatback near the upper (head restraint) can be reduced to accommodate increased visibility, without significantly reducing the overall protection afforded by the head restraint.

10. Seatbacks not designed to accommodate the added stress of properly designed and constructed retrofitted head restraints will in general, nevertheless, provide greatly improved passenger protection when so equipped. This

modification, when competently managed, provides the motorist with a means for maintaining a normal head-to-torso axes alignment; the probability of gross seatback deflection or failure is viewed as an undesirable condition but one which would occur anyway and a condition for which structural revision would probably be too expensive to be practical.

11. Seatbacks extended sufficiently to provide effective head restraint need not interfere with driver rear-view direct vision or driver see-through-car-ahead vision, providing the head restraint position of the seatback does not exceed the recommended width of 15 inches. This dimension for driver and right front seat passenger allows adequte see through vision, even for small cars, and a motorist has no problem seeing around it when looking to the rear.

12. Head restraints whether an integral extension of the seatback or a tack-on unit should be 15 inches wide but not less than 10 inches wide. Head restraints less than 10 inches wide, while perhaps adequate for square-on rear-end collisions, pre-supposes perfect postural alignment at the time of impact; owing to the natural lateral shifts in posture of the motorist and the possible occurrence of oblique rear-end collisions, a bias toward the 15-inch dimension will provide protection

for a wider range of rear-end collision exposures. Head restraints wider than 15 inches represent a problem for drivers attempting to see through cars ahead when traveling at highway speeds under congested conditions.

13. As a practical matter of comfort, the head restraint should be no closer than one inch from the back of the head when the motorist is in a natural seated position.

F. CONCLUSIONS - OCCUPANT PROTECTION - During the whiplash phase of a rear-end collision, the extent to which the head axis remains aligned with the torso axis is a common measure of whiplash severity; other factors such as the head peak acceleration attained, the effective elevation of the seatback (potential for neck "rabbit-chop") and the presense of dangerous interior surfaces that the head may strike, represent other mechanisms by which motorists are injured from rear-end collisions.

1. "Whiplash" as used by the authors connotates the motorist's involuntarily assumed, injury producing, torso and head-to-torso postures occasioned by automobile collisions and other forms of trauma.

2. In addition to the extent of "whiplash" injury exposure, as characterized by the angle the head axis assumed rearward of the torso axis, the location of back and neck force applications and their magnitude are also important indicators of potential injury exposure. For this reason, those exposures that do not manifest a critical approach to the voluntary limits of head excursion relative to the torso axis and yet are accompanied by significant whiplash or head impact with the vehicle interior should also be regarded as unacceptable solutions to the protection of motorists from rear-end collision injuries.

3. A 95th percentile dummy in a 16,100 in.-lb. 28-inch high front seat will sustain during a 30 mph rear-ender nearly half the head and chest acceleration of a 95th percentile dummy in the rear seat, extended also to 28 inches. This observation is attributed in part to the closer proximity of the rear seat to impact and also the more rigid seatback construction of rear seats in passenger vehicles.

4. With respect to head offset for the 28-inch seatback, the 95th percentile dummy and the 30 mph rear-end collision exposure, the following conclusions can be made:

(a) Head impact remained relatively constant for 0, 3 and 6 inch offsets (e.g., 11, 11, 10 G) but increased substantially for the 12-inch offset (e.g., 18 G).

(b) Chest and knee accelerations as well as seat belt force did not vary with changes in head offset.

(c) Maximum deflection and permanent rearward deflection of seatbacks remained constant for 0 and 3-inch offsets but was significantly reduced for 6 and 12-inch head offsets. This seeming anomaly is explained by considering the dynamic responses of both head and its head support; they perform as a single mass for 0 and 3-inch offsets and their inertial forces became asynchronous for the larger 6 and 12-inch offsets. Seatback asynchrony with occupant head and torso inertial forces reduces the extent of seatback deflection.

5. Increases in head offset increases tendency for whiplash; normal posture variations of twelve inches were found to cause only slight modifications in whiplash as contrasted with the more dominant variables of seatback height and strength.

6. A twelve-inch head offset for a rear seat passenger in a standard bench seat will more than double chest acceleration as contrasted with a zero-inch head offset during a rear-end collision; this is attributed to the exceptionally rigid construction of seatbacks for the rear seat position.

7. A twelve-inch head offset as compared with a 3-inch offset degrades the protection of a head restraint by an amount approximately equivalent to diminishing the head restraint height by three inches. Design criteria associated with protection devices for the motorist represents a combination of compromises. The problem of unusual head offset at the time of impact is compatible with the problem of excessive seatback yield common to higher speed rear-end collisions; both conditions are obviated by higher seatbacks but higher seatbacks impose additional restrictions to driver's and motorist's visibility. This identifies the compromised con-

dition governing the authors' recommendations for a 28-inch seatback.

8. In addition to "head-offset," it is useful to consider the extent to which readily compressible upholstery further separates the motorist's head from his head restraint. To this end, one can refer to the "Effective Head-Offset," the combination of head-offset and the compressible depth of the upholstery before bottoming-out occurs.

9. Seatback and head restraint upholstery, depending on its thickness, account for a significant amount of the "effective head-offset." Upholstery characteristically offers low force resistance to deflection. To the extent that upholstery is unnecessarily deep for the comfort purpose intended, its depth serves to unnecessarily compromise the motorist by allowing higher differential velocities between the seatback and his head and torso.

10. Owing to variations in the effective elevation of the head restraint, the extent of whiplash injury exposure is not simply a matter of the difference in head axis to torso axis angles because lesser angles are required to physiologically exceed spinal voluntary limits of articulation as a function of the effective elevation of the top of the back restraint. These variations have to do with motorists' seated height variations and with the yield performance characteristics of the backrest undergoing collision. Where head restraint occurs below the skull level, some cervical vertebra may be restrained against flexion, but those vertebra above the restrained segment are free to flex beyond voluntary limits and owing to the reduction in articulation units, the extent of flexion for injury is correspondingly reduced.

11. The seat belt does not contribute significantly to passenger protection from rear-end collision exposures:

(a) The floor anchored lap belt, and to a lesser extent the seat anchored lap belt pivot about their attachments as the motorist is accelerated. This action increases seat belt slack allowing the motorist to slide up the plane fo the seatback, thereby resulting in a reduced level of back support as the whiplash forces reach their maximum values.

(b) The low-back seats (less than 25-inch seatbacks) become even more hazardous during a rear-end collision if a seat belt is being worn because forward movement of hips is restricted thereby increasing the bending moment sustained by the spinal column. The seatback acts as a fulcrum with the lap belt lower limb inertial forces acting at at the base of the spine and the unrestrained head and shoulder inertial forces acting at the upper end of the spine.

(c) Seatbacks found to be of adequate height for one level of collision exposure will not necessarily provide adequate head restraint at higher levels owing to a combination of increased seatback deflection and tendency for head and torso displacement up the inclined plane of the seatback. The higher forces associated with more severe impacts, positions the head inertial force vector operating through the center of mass of the head aimed rearward over the top of the backrest serving to pull the head rearward over the head restraint to a posture more likely to result in whiplash.

(d) The more rigid the seatback, the less the tendency of head and torso displacement up the plane of the seatback during a rear-end collision; to the extent that a forward tilt design is incorporated in a seatback at its upper limits, this adverse torso shift is less likely to occur since this design feature straightens the inclined plane of the seatback where it's unneeded but encountered during torso shift.

G. CONCLUSIONS - GENERAL - These experiments have verified the importance of keeping the head and torso axes aligned by a properly designed seatback; the experiments have also established that every whiplash-protective device evaluated demonstrates some measure of degradation of performance as the relative speed increases between the rear-end collision vehicles. The degradation of seatback performance for protection of the motorist from whiplash as a function of speed increased seatback yield caused by increased body and seatback inertial forces attending high speed impacts that position the torso in a more reclined posture thereby facilitating its movement up the plane of the seat. Some tendency of this nature occurs even without seatback yield because of the natural 20 degree slope of seatbacks.

1. For the moderately severe collision exposures reported in this paper, (to 55 mph), it was established that a well designed safety seat would protect most passengers from sustaining significant rear-end collision injuries. It is apparent that far safer seats can be provided on the basis of performance guidelines established by this paper. The higher initial investment that provides greatly improved safety and comfort is money well spent.

An adequately designed, properly structured and anchored high-back contoured seat (28" or higher, well padded backrest) provided with well padded armrests, harness or a lap belt that is built into the seat-unit with retractable, inertial-lock mechanism represents the essential features of a safety seat that provides sufficient protection for a motorist to sustain, with probably no more than minor injuries, rear-end collisions through 55 mph. The crash performance of seats designed as safety seats represents a decided improvement over conventional seats. This was established in a prior series of experiments designed to evaluate the Liberty Mutual safe-seat configuration, Reference 3. As demonstrated by the Cox contour safety seat with head support and built-in cross-chest lap-belt restraint, the average motorist from child to adult size can ride out a severe rear-end collision (e.g. up to the 55 mph exposure) without significant injury.

2. Seatback strength should include allowance for passengers thrown forward against the backrest. Even though passenger vehicles are provided with lap belts, not all passengers will be sensible or knowledgeable enough to use them. Additionally, lap-belted taller persons can flail their heads and chest against the seatbacks ahead of them during front-end impacts, if not on rebound from rear-end collisions.

3. Front seat motorists are adequately protected from rear-end collision injuries for striking car speeds through 30 mph if they have an adequately structured 28-inch seatback capable of sustaining without failure a 16,100 in.-lbs. inertial force; head offset may cause variations in seatback deflections and head peak accelerations but both factors can be safely accommodated in the majority of exposures.

4. The importance of considering the head restraint protection for the 95th percentile dummy is not so much a special concern for the welfare of the minority-population of "king-sized" males as it is a recognition of the fact that seatback heights, satisfactory for the average adult at low-speed rear-end collisions, must be significantly higher for equivalent protection at higher collision speeds. Seatback heights and strengths providing satisfactory protection for the 95th percentile adult male for 30 mph rear-enders will provide satisfactory protection for the average height motorist for speeds through 30 and substantially above 30 mph.

5. These experiments were structured to be comprehensive but there were observations made during the experiments that suggest further avenues of profitable inquiry:

(a) The top surface of a seatback may provide improved head restraining properties during a rear-end collision if it extends rearward as a ten to twelve inch padded shelf. This design may serve to reduce the minimum elevation required for reasonable protection from whiplash.

(b) The upholstery material and texture used opposite the shoulders for a high-back seat may significantly influence the tendency for torso shift up the plane of the seatback.

(c) The rear window and adjacent metal structures provide a wedge-like surface for the head to strike. As the head contacts this inclined plane (30 degrees), assuming no other hazardous contact such as with a projection, the peak acceleration is produced by the wedge-like application of accelerative forces to the head and this may cause excessive compressive forces of the vertebral column. The magnitude of these forces was not instrumented but they may be physiologically significant and both the 95th and 50th percentile male dummies sustained relatively high head impacts with the rear window.

6. Properly designed high back seats provide an inner protective

shield around the passengers while also compartmentizing them to reduce the possibilities of their interacting with each other during all but the most devastating of collisions. In general, seats when structured for collision safety, represent the most important single lifesaving device available to the motorist.

ACKNOWLEDGMENTS

The authors express their deep appreciation for the financial support provided by the U.S. Public Health Service under Research Grant No. UI 00033-02 administered by the National Center for Urban and Industrial Health.

Special acknowledgment is made to the Ford Motor Company for their research grant, for the donation of ten (10) vehicles and for their valued assistance with our mutual interest in advancing the knowledge relating to motorist protection from rear-end collisions, and to the City of Los Angles for making the Test Site and facilities available.

Acknowledgment and appreciation is expressed for the dedication and high degree of professionalism rendered by Hans Jakob and David Blaisdell, our technical staff supervisory engineers, along with the excellent support of the entire Research Project Staff.

Acknowledgment is also made of the valued service of Bernard McGuire, Edmund Cababa and John Kerkhoff for photography and engineering illustrations; Louise Kirkpatrick and Lillian Nelson for report preparation; Jean Belsha for data reduction; technicians, Virgil Sibell, Ross Sater, Bernard Tai and Stanley Sax for their special attention to preparations for experiments.

Finally, acknowledgment is made of the devoted service of the part-time student workers of the University of California, Los Angeles.

REFERENCES

1. Severy, D. M., J. H. Mathewson, and C. O. Bechtol, "Controlled Automobile Rear-End Collisions - An Investigation of Related Engineering and Medical Phenomena, Series I," Canadian Service Medical Journal, Vol. XI, Nov. 1955, pp. 727-759 (ITTE Reprint No. 40).

2. Severy, D. M., and J. H. Mathewson, "Automobile Barrier and Rear-End Collision Performance," SAE Preprint 62C, a paper presented at the Society of Automotive Engineers Summer Meeting, Atlantic City, New Jersey, June 8-14, 1958.

3. Severy, D.M., H. M. Brink and J. D. Baird, "Collision Performance LM Safety Car," SAE Preprint No. 670458, May 15-19, 1967, 79 pp. Presented at the Society of Automotive Engineers, Chicago, Illinois, May, 1967.

4. Severy, D. M., H. M. Brink and J. D. Baird, "School Bus Passenger Protection," SAE Preprint No. 670040, January, 1967, 165 pp. Presented at the Society of Automotive Engineering Congress, Detroit, Michigan, January 1967.

5. Gay, J. R. and K. H. Abbott, "Common Whiplash Injuries of the Neck," J. Am. Med. Assoc. 152:1698, 1953.

6. Ommaya, Ayub K., Arthur E. Hirsch and John L. Martinez, "The Role of 'Whiplash' in Cerebral Concussion," Proceedings of the Tenth Stapp Car Crash Conference, Paper No. 660804; published by the Society of Automotive Engineers, Inc., November 1966.

7. Wickstrom, J., J. L. Martinez, D. Johnston and N. C. Tappen, "Acceleration-Deceleration Injuries of the Cervical Spine in Animals," Proceedings of the Seventh Stapp Car Crash Conference, D. M. Severy, Ed., Charles C. Thomas, Publisher, 1965; presented at Los Angeles Nov. 1963.

8. Severy, D. M., J. H. Mathewson and A. W. Siegel, "Automobile Head-On Collisions, Series II," SAE Transactions, Vol. 67, 1959, pp. 238-262.

9. Severy, D. M. and H. M. Brink, "Safety Glass Breakage by Motorists During Collisions," Proceedings of the Tenth Stapp Car Crash Conference, published by the Nolte Center for Continuing Education of the General Extension Division, University of Minnesota, Minneapolis, M. K. Cragun, Editor, 1966; pp. 205-236.

10. Severy, D. M., "Human Simulation for Automotive Research" SAE SP-266, January 1965, presented to the Society of Automotive Engineers, Automotive Congress and Exposition, Detroit, Michigan, January 13-17, 1964.

Appendix A

Rear-End Collision Transducer Data (Peak Values - Struck Car)

EXPERIMENT NUMBER IMPACT SPEED	X-93 30 MPH	X-94 20 MPH	X-95 30 MPH	X-96 30 MPH	X-97 10 MPH	X-98 40 MPH	X-99 30 MPH	X-100 30 MPH	X-101 30 MPH	X-102 30 MPH	X-103 55 MPH	X-104 55 MPH
VEHICLE FRAME*	9G/ 15ms	8G/ 35ms	11G/ 25ms	9G/ 25ms	5G/ 15ms	19G/ 100ms	12G/ 35ms	10G/ 25ms	11G/ 25ms	11G/ 25ms	19G/ 85ms	19G/ 95ms
DRIVER												
Head	12G/ 220ms	9G/ 165ms	10G/ 180ms	11G/ 210ms	8G/ 215ms	34G/ 190ms	13G/ 175ms	17G/ 210ms	30G/ 225ms	12G/ 140ms	35G/ 155ms	42G/ 140ms
Head Restraint (28")**	–	–	32G/ 65ms	–	4G/ 125ms	31G/ 135ms	–	–	–	–	–	–
Chest	7G/ 205ms	5G/ 90ms	7G/ 95ms	6G/ 85ms	4G/ 100ms	18G/ 140ms	6G/ 80ms	7G/ 110ms	9G/ 135ms	9G/ 95ms	14G/ 75ms	21G/ 150ms
Seatback (14")	7G/ 140ms	7G/ 35ms	9G/ 80ms	7G/ 35ms	2G/ 95ms	42G/ 135ms	8G/ 30ms	11G/ 45ms	42G/ 135ms	12G/ 30ms	20G/ 90ms	18G/ 140ms
Seatback Deflection (14")	5.8"/ 170ms	3.2"/ 150ms	2.3"/ 215ms	3.5"/ 200ms	2.3"/ 135ms	13.6"/ 184ms	1.7"/ 85ms	4.7"/ 180ms	5.4"/ 235ms	1.7"/ 90ms	8.1"/ 158ms	7.2"/ 131ms
Knee	6G 195ms	–	Trace Lost	8G/ 105ms	7G/ 75ms	20G/ 165ms	8G/ 110ms	9G/ 85ms	12G/ 95ms	8G/ 90ms	15G/ 90ms	18G/ 105ms
Lap Belt	40#/ 325ms	100#/ 260ms	0#/ 0ms	0#/ 0ms	–	–	–	–	–	–	–	–
RIGHT FRONT PASSENGER												
Head	11G/ 150ms	14G/ 155ms	18G/ 170ms	11G/ 160ms	10G/ 150ms	33G/ 205ms	16G/ 165ms	18G/ 175ms	24G/ 205ms	10G/ 135ms	58G/ 150ms	107G/ 140ms
Head Restraint (28")	9G/ 190ms	Trace Lost	14G/ 45ms	15G/* 40ms	11G/ 55ms	18G/ 150ms	14G/ 50ms	15G/ 80ms	18G/ 35ms	–	30G/ 135ms	66G/ 120ms
Chest	6G/ 100ms	6G/ 95ms	6G/ 130ms	6G/ 185ms	5G/ 125ms	10G/ 160ms	7G/ 105ms	7G/ 130ms	8G/ 165ms	3G/ 75ms	24G/ 140ms	33G/ 125ms
Seatback (14")	7G/ 90ms	7G/ 65ms	11G/ 35ms	7G/ 35ms	5G/ 60ms	10G/ 155ms	7G/ 35ms	7G/ 50ms	9G/ 160ms	12G/ 35ms	18G/ 140ms	36G/ 130ms
Seatback Deflection (14")	3.6"/ 190ms	2.3"/ 125ms	2.2"/ 215ms	3.7"/ 190ms	1.8"/ 115ms	8.0"/ 215ms	1.8"/ 125ms	5.7"/ 210ms	7.0"/ 230ms	2.2"/ 130ms	8.6"/ 144ms	6.7"/ 125ms
Knee	6G/ 70ms	8G/ 85ms	7G/ 70ms	7G/ 115ms	–	9G/ 140ms	7G/ 95ms	11G/ 40ms	10G/ 90ms	3G/ 30ms	13G/ 85ms	26G/ 130ms
Lapbelt	70#/ 297ms	60#/ 260ms	–	100#/ 320ms	–	–	–	–	–	–	–	–
RIGHT REAR PASSENGER												
Head	24G/ 125ms	50G/ 115ms	14G/ 175ms	21G/ 140ms	24G/ 130ms	75G/ 105ms	38G/ 130ms	10G/ 135ms	12G/ 130ms	15G/ 190ms	84G/ 110ms	70G/ 85ms
Chest	23G/ 125ms	14G/ 100ms	9G/ 135ms	10G/ 145ms	4G/ 130ms	35G/ 95ms	14G/ 130ms	8G/ 90ms	8G/ 85ms	10G/ 55ms	30G/ 95ms	96G/ 80ms
Seatback (14")	–	–	–	–	–	–	–	–	–	–	56G/ 75ms	–
Seatback Deflection (14")	–	–	–	–	–	–	–	–	–	–	2.6"/ 85ms	–
LEFT REAR PASSENGER												
Head	24G/ 50ms	23G/ 55ms	19G/ 160ms	24G/ 150ms	20G/ 90ms	24G/ 55ms	26G/ 75ms	22G/ 140ms	52G/ 120ms	23G/ 150ms	48G/ 120ms	41G/ 80ms
Chest	10G/ 135ms	10G/ 70ms	14G/ 145ms	23G/ 125ms	7G/ 75ms	16G/ 100ms	9G/ 90ms	11G/ 85ms	11G/ 115ms	11G/ 95ms	23G/ 90ms	48G/ 80ms
Seatback Deflection (14")	–	–	–	–	–	–	–	–	–	–	2.0"/ 110ms	–

* Station 9, on door sill, adjacent to right door post

** Transducer located on seatback at 28 inches above dummy "H" point

Appendix B

Anthropometric Adult Dummy Specifications

50th Percentile Sierra 292-850	95th Percentile Sierra 292-95*	
MEASUREMENTS (inches)		LOCATION
14.5	15.47	Top of head to shoulder pivot
18.4	19.10	Shoulder pivot to hip pivot
16.8	17.68	Hip pivot to knee pivot
16.0	17.35	Knee pivot to ankle pivot
3.3	3.50	Ankle pivot to bottom of foot
32.9	34.57	Top of head to hip pivot ("H" point)
35.7	37.75	Top of head to base of buttocks (sitting)
33.1	35.70	Crotch to heel
17.8	19.40	Width between shoulders
6.1	6.40	Transverse diameter of skull
14.0	15.25	Transverse diameter between shoulder pivots
12.0	13.40	Transverse diameter of chest
13.2	14.40	Transverse diameter of hips
7.9	8.80	Transverse diameter of knee
11.5	12.00	Shoulder pivot to elbow pivot
19.0	19.30	Elbow pivot to tip of fingers
3.5	3.74	Transverse diameter of 4 lateral fingers compressed together
9.0	10.40	Longitudinal diameter of chest
10.4	11.30	Long diameter of foot (heel to toe)
68.7	73.10	Height - inches
16.5	205.00	Weight - pounds

*50th and 95th percentile dummy with full disc and cable articulated neck.

670458

Collision Performance, LM Safety Car

Derwyn M. Severy, Harrison M. Brink, and Jack D. Baird*
Institute of Transportation and Traffic Engineering, University of California

THE PURPOSE of conducting a series of collision experiments involving the Liberty Mutual Survival Car II (LM car) was to determine its collision performance for representative exposures. The two types selected were intersection and rear-end collisions; these exposures collectively represent the most frequently occurring types and, in addition, they include an opportunity for evaluating the relative protective performance of the LM vehicles undergoing front-end impact. These data are available because for every rear-end collision and every intersection type collision, there is a striking vehicle subjected to front-end impact. Consequently, the selection of these two types of impacts provided a reasonably complete evaluation of the potential protective qualities and innovations embodied in the LM car.

Full-scale crash evaluation of the LM car, in terms of its collision performance, was selected as the only reliable approach for obtaining performance data under the multiplicity of conditions and variables that exist during an actual collision. Laboratory simulations of the contemplated exposures are valuable for assisting the design engineer in refining his concepts. Experience has shown that the many interrelated and often unanticipated factors occurring during a collision cannot be accounted for at the design level, with any reasonable confidence. Consequently, it would be a disservice to the public, as well as to those responsible for manufacturing safe motor vehicles, to have a proposed safety car design developed that does not include full-scale crash evaluations of the prototype in order to avoid release of statements not based on fact.

On the face of it, one may conclude that some innovations are so obviously advantageous that they need no evaluation before placing them into operation. Experience has shown, however, that this position is erroneous, even with respect to the simplest of devices. To be an improvement worthy of use, a safety innovation not only must perform the protective feature intended by its designer but simultaneously must not diminish the level of safety achieved by other innovations that may be even more important.

This series of collision experiments brought to 92 the number of instrumented, full-scale, collisions conducted during the past 18 years at UCLA. Each of these 92 engineering experiments has provided objective findings concerning

*The authors respectfully dedicate this paper to the late Frank J. Crandell, Vice-President, Engineering, Liberty Mutual Insurance Co., for his contribution to the cause of improved motorist safety during a period when such efforts were unpopular; particularly for his far-sighted conviction that a properly designed motorist seat represents the greatest single advance to the cause of reducing injuries from vehicular accidents.

ABSTRACT

Engineering evaluations of the collision performance for the Liberty Mutual Safety Car and the 1966 Chevrolet sedan were made, consisting of two 30 mph rear-end and two 40 mph intersection collision experiments. Methodology provided comparative analysis of functional characteristics for five types of seats, each studied for different exposure conditions. Five conditions of restraint were included for four sizes of occupants; the anthropometric dummies, their restraints, seats, and crashing cars all carried appropriate transducers. Seven high G-tolerant, high-speed cameras, carried by the crashing cars provided close-up continuous monitoring of these quarter-second collision sequences, supported by many more tower and ground level special photographic units arranged about the collision scene. Data from photographic, electronic and related instrumentation are presented using photographic and graphical techniques designed to facilitate comprehension.

the injury-producing and related medical engineering factors attending automobile collisions. The identification and quantification of such research findings provide valuable guideposts to designers of automobiles as well as to designers of highways concerned with protective barriers and accessories of this nature. Detailed engineering evaluation of the complex multivariant phenomena attending automobile collisions requires full-scale experimentation. Other forms of inquiry, laboratory studies, case accident studies, statistical treatment of accident data, and so forth, are equally important to the basic objectives of reducing highway traffic deaths and to the frequency and severity of motorists' injuries. These latter approaches, however, will not provide insight to all of the mechanisms by which motorists' injuries occur and are of a less objective form, having a lower reliability factor concerning contributions to design of improved safety systems.

FACILITIES AND EQUIPMENT

The ITTE-UCLA Field Station for collision research is located adjacent to the U. S. Naval Station, Long Beach, Calif. The fenced-in facility includes an asphalt surface, trailers and building for supporting activities, and several towers for photographic observation of the collisions. Four mobile vans provide research support facilities as follows: machine shop, electronics shop, photographic shop, and project control center. An aluminum monorail guide track, installed on the level asphalt runway, provides the means for controlling the direction of crash vehicles. Other operational techniques, incorporated in recent years, are reported in prior publications. (1-3)* Two experimental cars were provided for evaluation by Liberty Mutual Insurance Co., Boston, Mass. These vehicles were 1960 Chevrolet Bel Air 4-door sedans, modified as indicated in a subsequent section on the Liberty Mutual Survival Car II. To provide a third vehicle of the same make, model and year for correlation with these LM Chevrolets, a 1960 Chevrolet 4-door sedan, was purchased. In addition, two 1966 Chevrolet Bel Air 4-door sedans were used in this series. The manner in which these vehicles were selected for the two types of impacts is explained in a subsequent section, "Automobile Assignment Scheduling."

EXPERIMENTAL PROCEDURE

Full-scale collision operations require months of planning and preparation; it is therefore essential that each experiment is successfully completed, as planned. The crash cars are controlled directionally by a front bumper-mounted phenolic shoe that slides in an aluminum monorail guide track secured to the asphalt pavement. The speeds of the crash cars and the synchronization of their positions at impact are accomplished by a tow car connected to the crash cars by a steel cable through appropriately arranged sheaves, as shown in Fig. 1. Adjustment of the tow cable lengths assure perfect synchronization for the type of impact in-

*Numbers in parentheses designate References at end of paper.

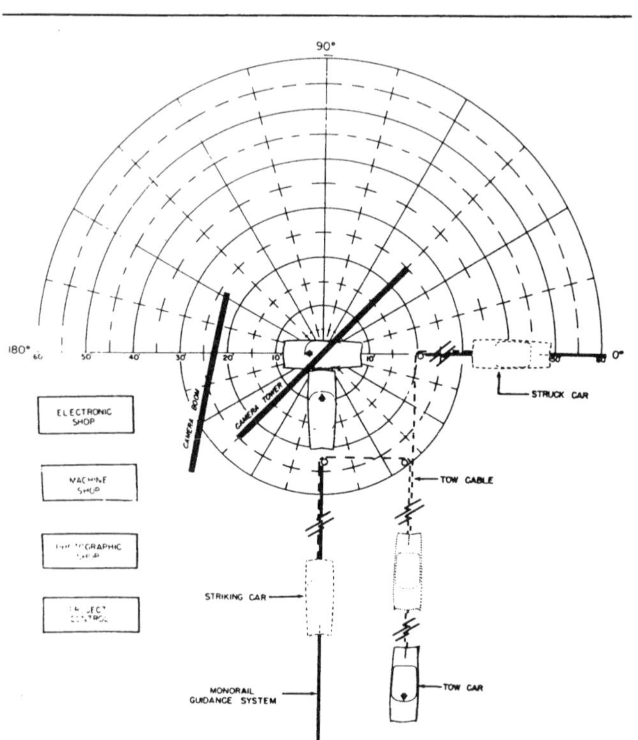

Fig. 1 - Vehicle control systems, side-impact collisions

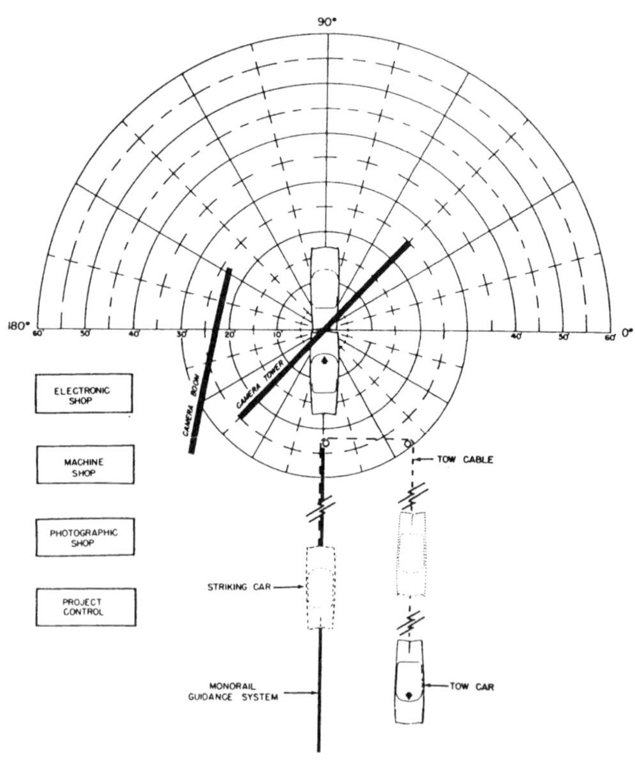

Fig. 2 - Vehicle control systems, rear-end collisions

volved. The tow car accelerates to the predetermined velocity and holds that velocity until the tow cables automatically detach from the crash cars twenty feet before impact. During the last split second before impact, the crash vehicles progress without directional or speed constraints and crash in a manner that corresponds to real accidental collisions. The methodology for rear-end collisions is simplified because the struck car is stationary and the striking car is accelerated to its predetermined impact speed in a manner previously described (Fig. 2). Operational procedures for the controller, the two instrumentation pace vehicles, and the two crash vehicles are described in Ref. 1.

METHODOLOGY

The Liberty Mutual Survival Car II represents a safety-customized 1960 Chevrolet 4-door sedan. In order to properly evaluate the extent to which motorist protection had been improved it was necessary not only to crash the Liberty Mutual cars but also to crash non-modified cars to provide a basis for comparison. Logically, the 1960 LM Chevrolets should be compared with 1960 production Chevrolets. This would have been the procedure for the UCLA evaluation if the LM cars could have been obtained in the early 1960's. Perhaps, understandably, the LM engineers were reluctant to subject their cars to destructive tests before the public had seen the product of their nine years of research and development. The LM cars were placed on exhibit throughout most of the United States and after serving this purpose, the cars were released to UCLA in 1965. Rather than compare these 1960 LM vehicles with 1960 production models that were now regarded as old cars, the UCLA engineers elected to match the 1960 LM cars against the latest production models then available, namely, 1966 Chevrolet Bel Air 4-door sedans.

This series of experiments consisted of two rear-end collisions at 30 mph and two center side-impact collisions, each car traveling 40 mph. In the rear-end collisions a 1966 production vehicle was used to strike the rear of a 1960 LM car in Experiment 88; a 1966 production vehicle was struck in the rear by another 1966 production Chevrolet in Experiment 89. This procedure provided the basis for comparison between the motorist collision protection of the LM vehicle, as contrasted with the motorist protection of the same type of vehicle, but six years newer. Variables under study for rear-end collision Experiments 88 and 89 are shown in Fig. 3.

The first of the two intersection collisions of this series, Experiment 91, involved two LM cars while the second experiment consisted of a 1966 production vehicle struck by a 1960 production model. These two experiments provided a comparison of the relative protective qualities of an LM design versus a new 1966 car when they are struck at the driver's side. In addition, the relative protection provided for the striking car when it is a 1960 LM survival car as contrasted with a striking car that is a 1960 production Chevrolet was evaluated, for the 40 mph side impact, Experiment 91 and Experiment 92, respectively. Variables investigated for these two side-impact experiments are shown in Fig. 4.

Although other impact configurations could be devised to further evaluate advantages and possible disadvantages accrued from the various modifications made in the Liberty Mutual Survival Car, it was felt that this experimental plan provided sufficient data for an objective appraisal of the merits or deficiency of the modifications made by Liberty Mutual.

Automobile Assignment Scheduling - The purpose of developing an automobile assignment schedule is to determine those experiments that allow the use of repaired vehicles and those which require one or both of the vehicles not to have been previously crashed. Assuming that a collision could produce changes in the collision response characteristics of a car that would not be entirely corrected by repairs, such influence may be circumvented by making certain that the repaired section is not directly involved in a subsequent collision. In referring to Fig. 5, it may be seen that Vehicles 1, 2, and 3 were used in two different collision experiments; each of these vehicles was repaired before using it under conditions that brought no previously impacted and repaired structures into contact.

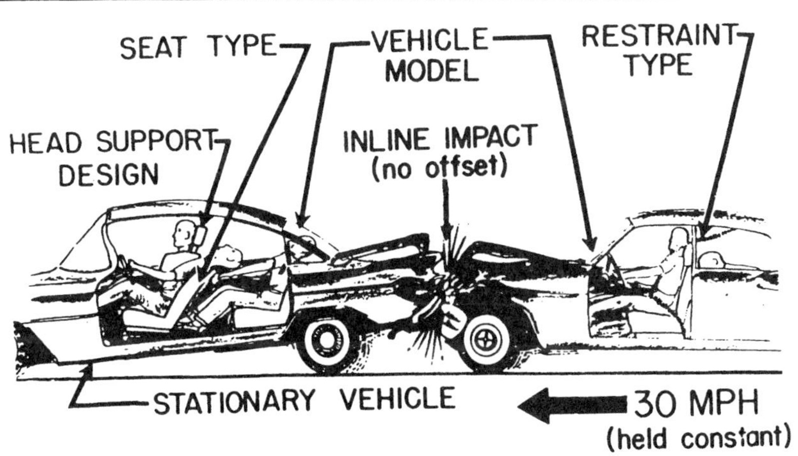

Fig. 3 - Variables under study, rear-end collisions (Experiments 88 and 89)

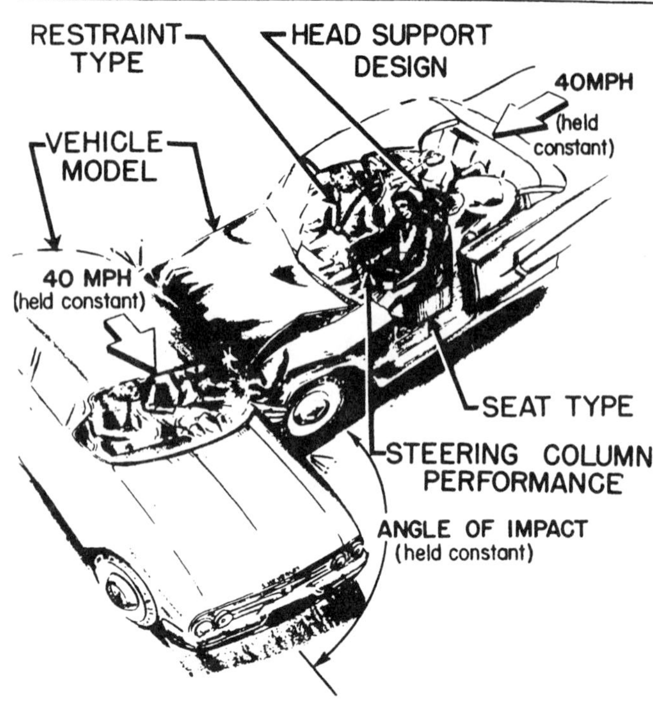

Fig. 4 - Variables under study, side-impact collisions (Experiments 91 and 92)

Vehicle Seat Assignment - It is logical to consider the properly designed seat as one form of motorist restraint. To facilitate use of seat and chest belts, these restraints should be installed in the seat and this combination makes the seat the most important single contribution to motorist protection that may be provided in a car. The relative performance of these more complex integrated seat designs, as typified by the LM capsule chair and the Cox seat as contrasted with conventional production seat designs, was the most important aspect of this study.

Five seat types representing three basic designs were evaluated; the design features are described and illustrated in a later section. Seat types evaluated were:

1. The Liberty Mutual capsule chair (Fig. 6(a)).
2. The General Motors bucket seat with head support (Fig. 6(b)).
3. The Cox safety seat (Fig. 6(c)).
4. A production bench seat (Fig. 6(d)).
5. A Ford bucket seat (Fig. 6(e)).

Except for the LM and Cox seats, the seats were of conventional structural design.

The two 30 mph rear-end collisions both utilized 1966 Chevrolet sedans as striking cars; they were equipped front

EXPERIMENT	IMPACT CONFIGURATION	REMARKS
X-88 Rear-End (30 MPH)	Stationary 1960 Chevrolet (LM Design)	Struck Car: Modified Liberty Mutual Survival Car
	30 MPH 1966 Chevrolet (Production)	Striking Car: Current Production Sedan with Modified Seats
X-89 Rear-End (30 MPH)	Stationary 1966 Chevrolet (Production)	Struck Car: Production Sedan Repaired from X-88 Crash with LM Seat Installed
	30 MPH 1966 Chevrolet (Production)	Striking Car: New Production Sedan with Modified Seats
X-91 Side-Impact (40 MPH)	40 MPH 1960 Chevrolet (LM Design)	Struck Car: Modified Liberty Mutual Survival Car
	40 MPH 1960 Chevrolet (Production)	Striking Car: Liberty Mutual Design, Repaired from X-88 Crash
X-92 Side-Impact (40 MPH)	40 MPH 1966 Chevrolet (Production)	Struck Car: Production Sedan Repaired from X-89 Crash with LM Seat Installed
	40 MPH 1960 Chevrolet (Production)	Striking Car: Production Sedan with Modified Seats

Fig. 5 - Automobile assignment schedule

and rear with production bench seats, with front seat adults restrained by the roof anchored cross-neck, cross-chest lap belt combinations, and the rear seat child dummies unrestrained.

The 1960 LM Chevrolet rear-ended in Experiment 88 was equipped with a 1963 Ford bucket seat for the driver and an LM seat for the front seat passenger (Fig. 7(a)). The rear seat was a conventional bench seat, retrofitted with a webbing head support (left rear) and a shelf mounted head-impact pad (right rear).

The 1966 Chevrolet rear-ended in Experiment 89 was equipped the same, except that the driver's seat was a 1966 GM Astro bucket seat with head support (Fig. 7(b)).

The seat types identified in Figs. 6 (a-e) were also evaluated for the 40 mph intersection collisions, Experiments 91 and 92. The striking car was in each experiment a 1960 Chevrolet; the struck car was a 1960 Chevrolet for Experiment 91 and a 1966 Chevrolet for Experiment 92.

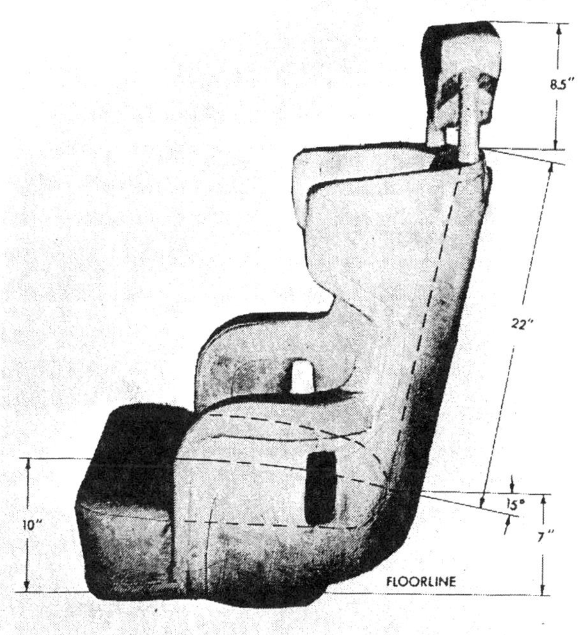

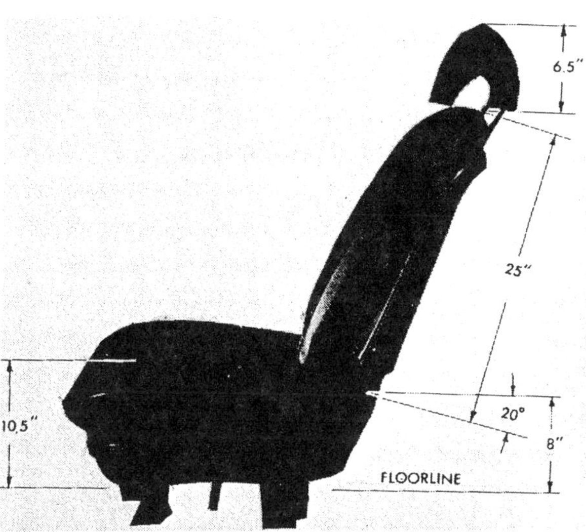

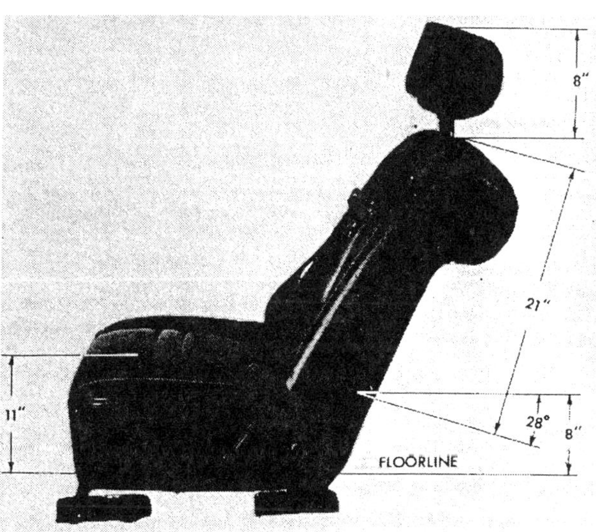

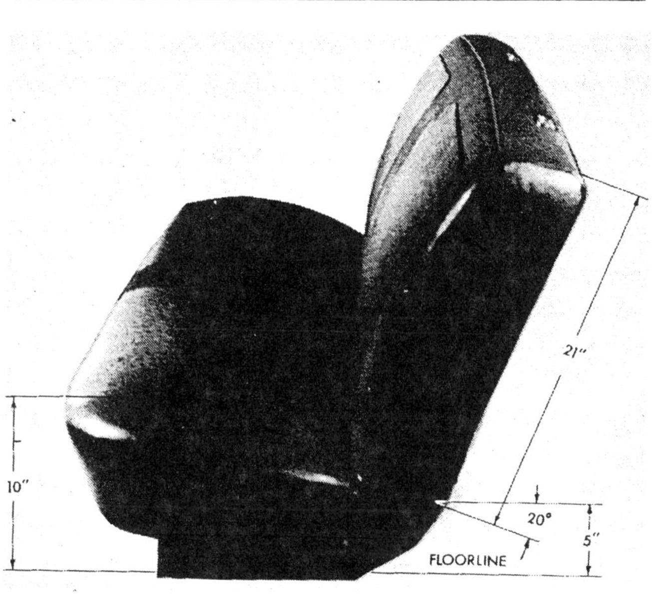

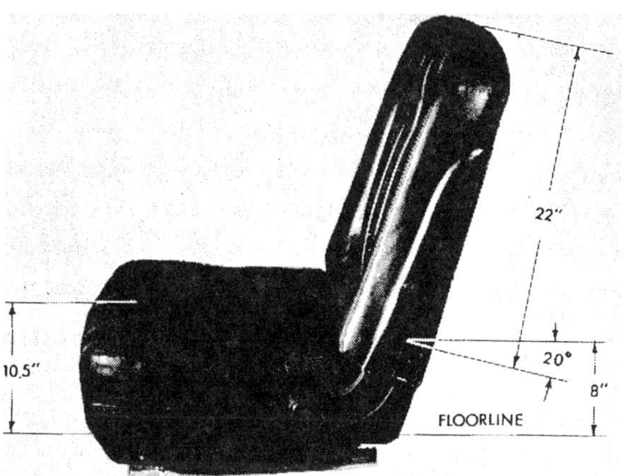

Fig. 6 - Vehicle seat types evaluated: A, Liberty Mutual capsule seat; B, General Motors bucket seat; C, Cox integrated seat; D, Conventional bench seat; E, Ford bucket seat

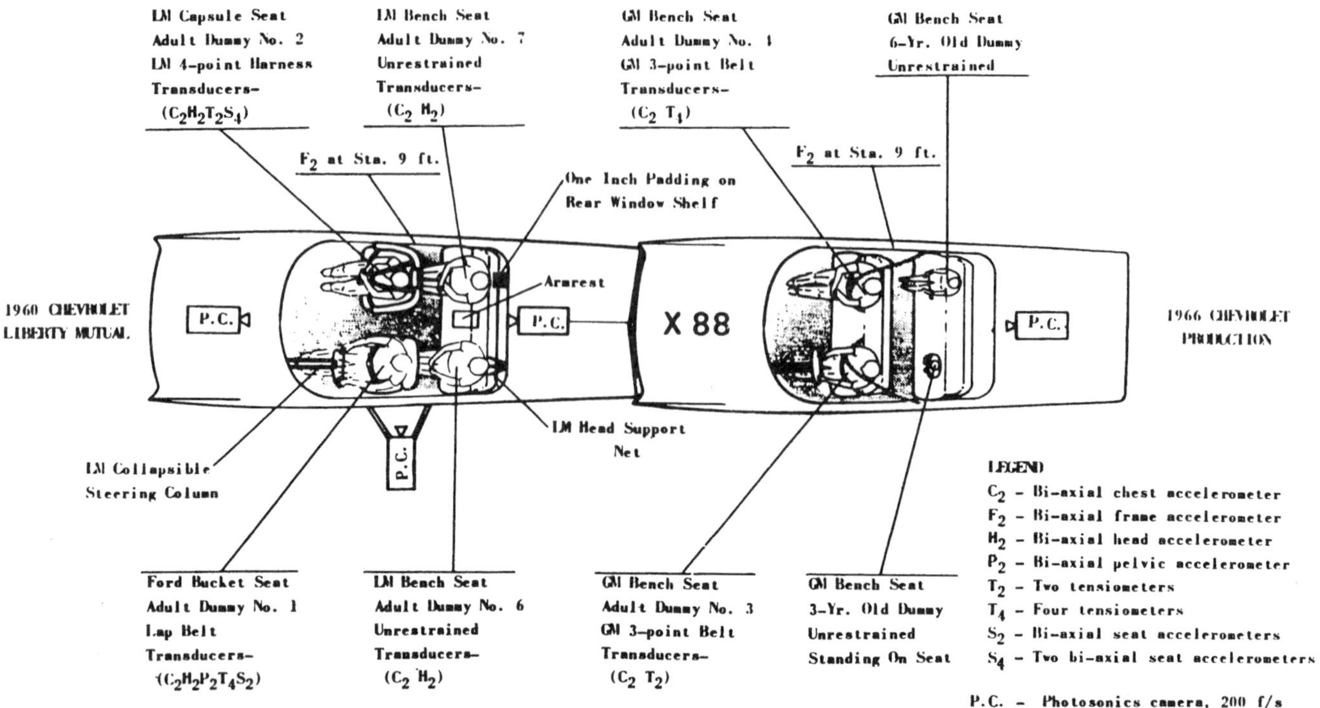

Fig. 7 (a) - Vehicle seat assignment and instrumentation, rear-end collision, 30 mph (Experiment 88)

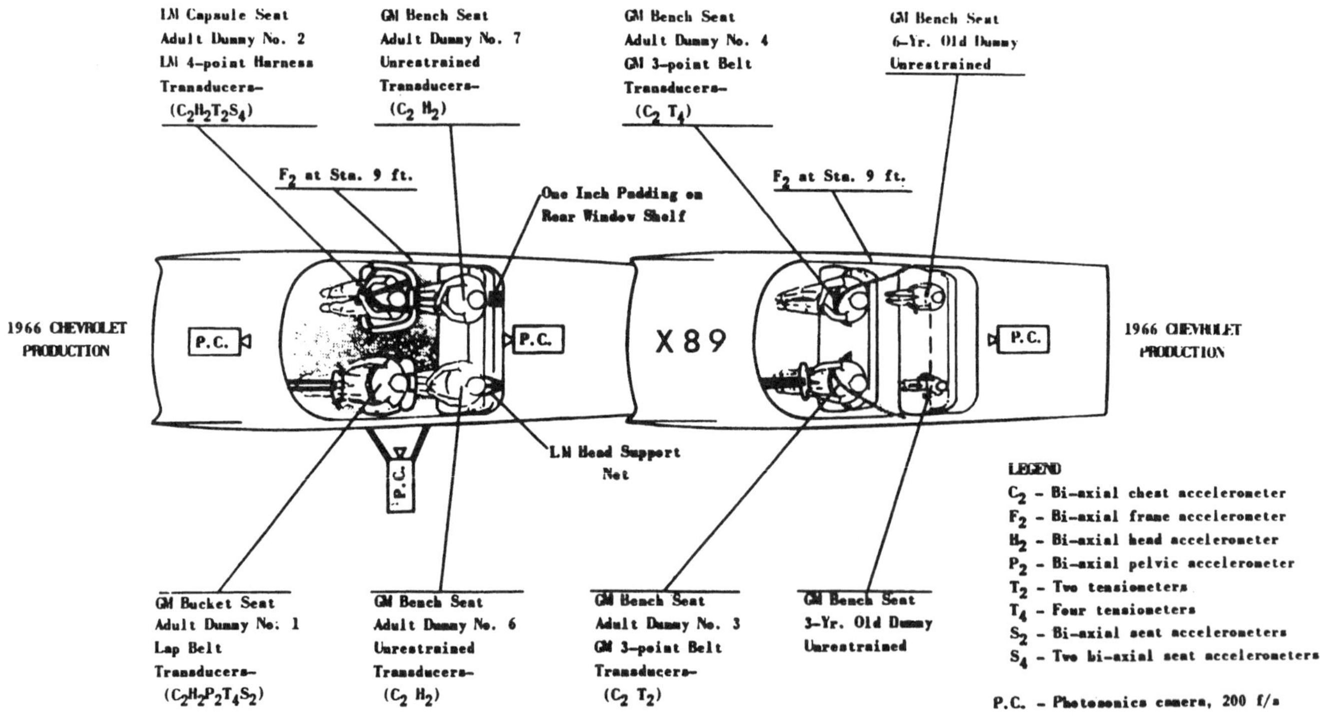

Fig. 7 (b) - Vehicle seat assignment and instrumentation, rear-end collision, 30 mph (Experiment 89)

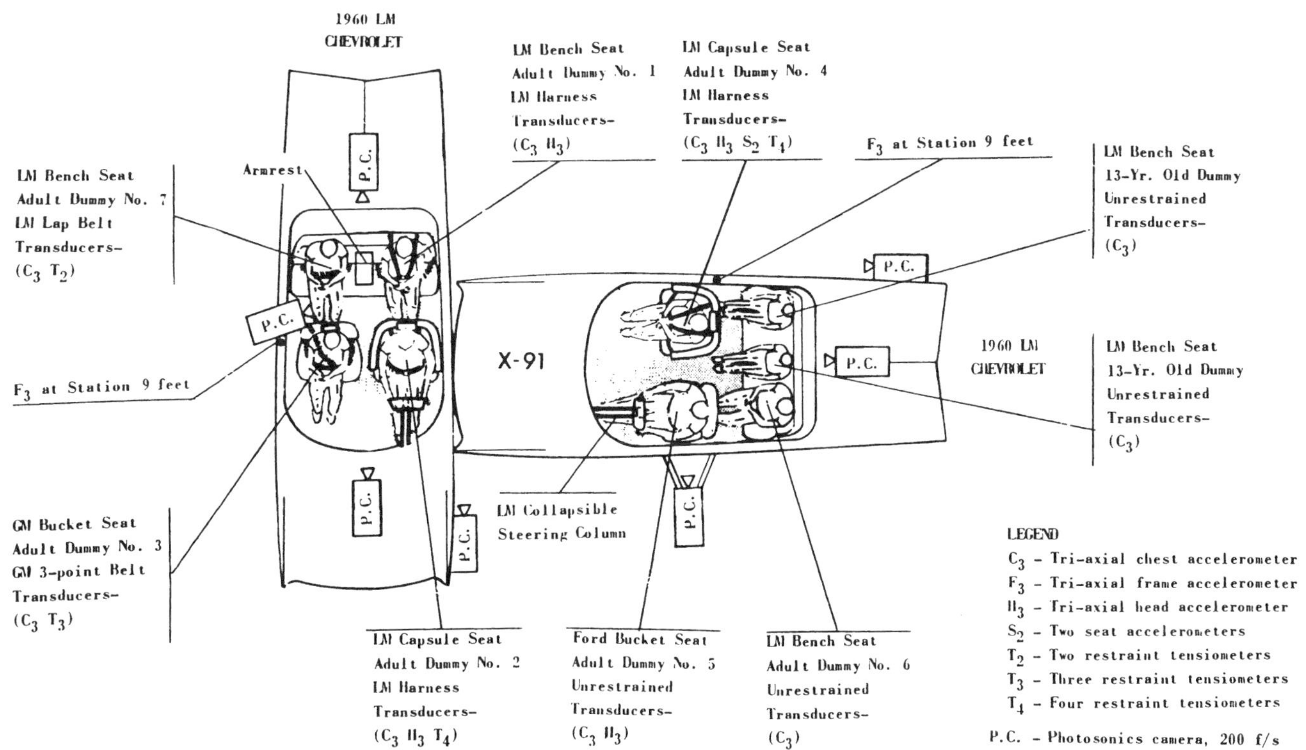

Fig. 8 (a) - Vehicle seat assignment and instrumentation, side-impact collision, 40 mph (Experiment 91)

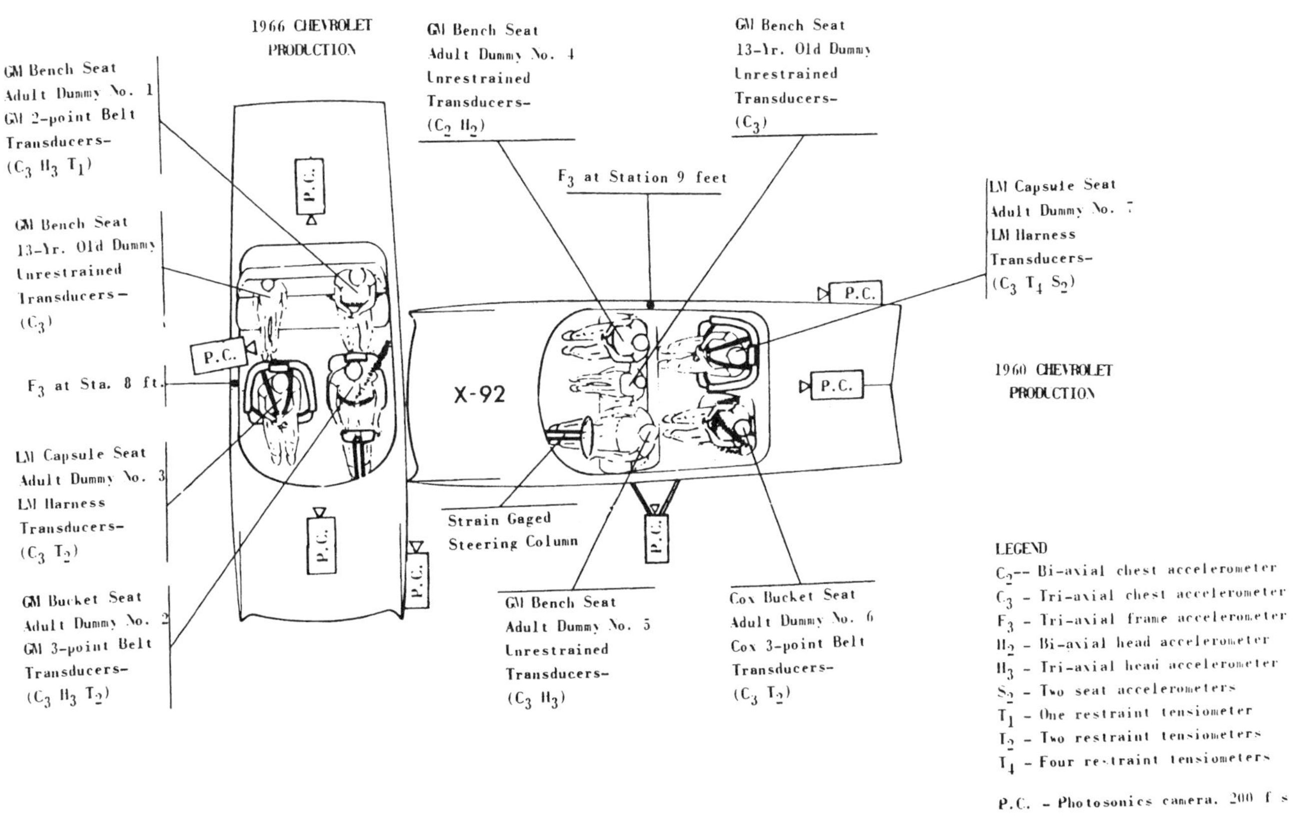

Fig. 8 (b) - Vehicle seat assignment and instrumentation, side-impact collision, 40 mph (Experiment 92)

The seat assignment for Experiment 91 is presented by Fig. 8 (a). Not every seat could be evaluated for every exposure position but it was possible to develop sufficient exposure data with these two experiments and with prior experiments involving the bench and Cox seats to provide a performance evaluation of these units. The seat arrangement for Experiment 92 is shown by Fig. 8 (b) and a comparison of this figure with the prior figure indicates the opportunities these seating assignments allow for their evaluation.

INSTRUMENTATION

General - Regardless of the care and thought applied to developing an effective experimental plan and methodology, the collision event is over in a quarter second and this brief period is too short for meaningful direct observation, even by the trained observer. The value of a successfully conducted series of collision experiments directly relates to the quality and comprehensiveness of its instrumentation.

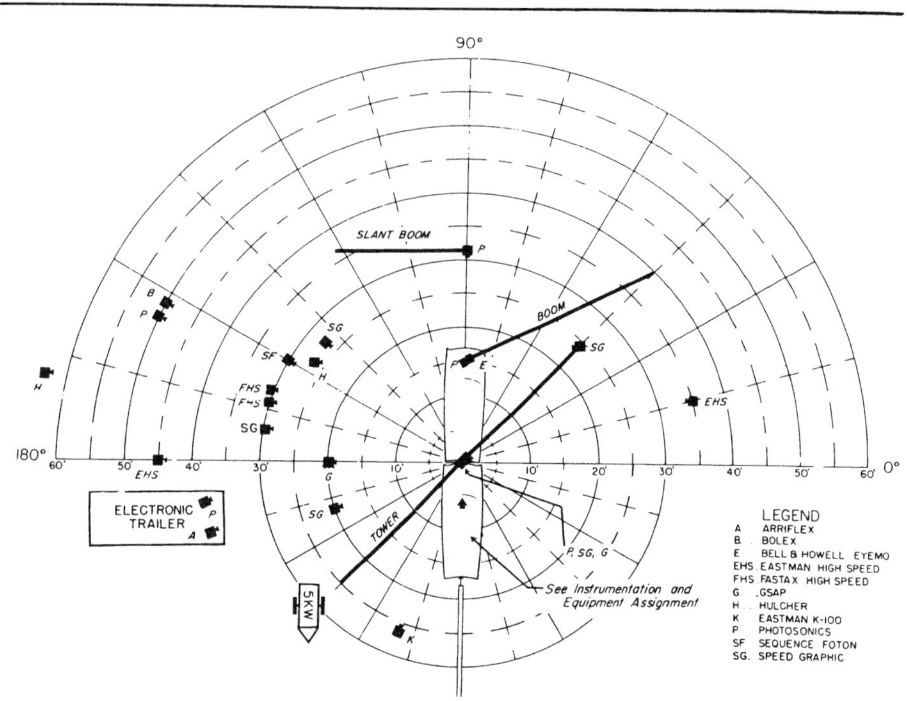

Fig. 9 - Camera assignment, rear-end collision

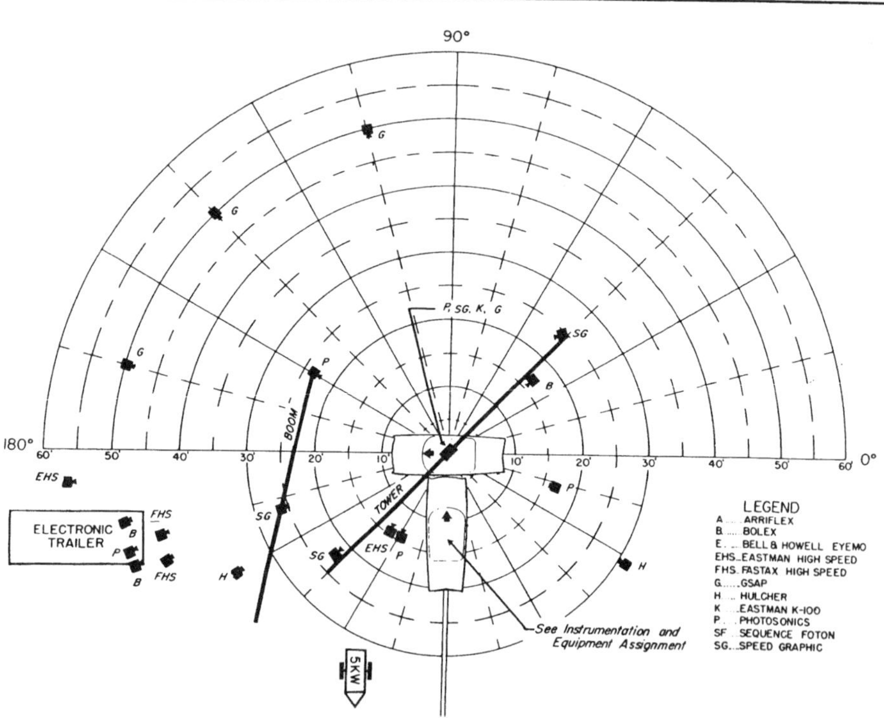

Fig. 10 - Camera assignment, side-impact collision

Table 1 - Summary of Instrumentation

Device	To Provide	Location	Specification
High Speed Motion Picture Cameras	Vehicle collision dynamics and passenger kinematics	See impact center camera assignment, Fig. 10	2-Eastman, High speed; 2-Fastax WF-3; 16mm, 600-1700 frames/sec
Moderately High Speed Motion Picture Cameras	Vehicle collision dynamics and passenger kinematics	See impact center and vehicle camera assignments, Figs. 9 and 10	10-Photosonics, 1B; 1-GSAP MBH 200-16; 200 f/s 16mm
Standard Motion Picture Cameras	General photographic coverage	Adjacent to impact center	1-K100, 16mm; 1-Arriflex, 16mm; 1-Bolex, 16mm; 1-GSAP, 16mm
Special Motion Picture Cameras	Sequential, large format photographs of collision events	Strategic positions at impact center	2-Hulcher, Model 102, 20 f/s 70mm (camera rotated 90 deg to permit cine-reduction); 1-Bell & Howell Eyemo, 48 f/s 35mm; 1-Foton 5 f/s 35mm Bell & Howell
Still Cameras	Precision-timed photographs	See impact center camera assignment, Fig. 10	5-Super Speed 4×5" Graflex 1/1000 sec
Calibrated References	Fixed ground reference points for micro-motion analysis	Near the impact center	8-ft longitudinal reference boards and 4-ft vertical posts calibrated yellow and black at alternate 1-ft increments
Reference Targets	Precision photographic references for micro-motion analysis	At strategic positions on the vehicles and occupants	Diamond shaped yellow and black targets
Electrical Accelerometers	Acceleration measurements	Head, chest and hips of occupants. Vehicle body and seat structure	B & F-LF 50-50 and LF 100-75-120; Statham A 38a-60-350, A6-100-350, AJ43-100-350
Seat Belt Tensiometers	Measurement of belt loads	On belt webbing	Thread-through type, special UCLA construction
Recording Oscillograph	Amplitude-time records of transducer signals	Carried by instrument recording vehicles for striking car and in instrumentation trailer for struck car	2-18 Channel Consolidated Type 5-114 P4; 1-50 Channel CEC, Type 5-119, P3
Electronic Time Delays	Precision timing for still photography	Electric pressure pads near impact center	100, 200, and 500 ms solid state time delay units built by UCLA
Pulse Generators	Timing for high and moderate speed cameras	Near cameras	Wollensak, 100 cps, 1000 cps and special 100 cps solid state devices

(cont'd)

Table 1 (con't)

Device	To Provide	Location	Specification
Photographic Oscillographic Synchronization Units	Zero time (vehicle contact); flash bulb for film and pulse for oscillograph	Pressure switches between contacting surfaces of vehicles. Photocell and flash bulb on vehicle	SM flash bulb with photocell for oscillograph signal
Flash Bulb Timer	Timing at 0, 50, 100, 150, 200, 250 and 500 ms; flash bulb for film and pulse for oscillograph	Near impact center in view of all cameras	Solid state timing device with SM flash bulbs and photocell for oscillograph signal
Auxiliary Timer	Backup timing	Near impact center in view of all cameras	Rotating yellow-black drum; constant speed 1740 rpm motor
Speed Counter	Car speed data; time-displacement of vehicle	On guide yoke	Induction pickup for oscillograph
Polar Coordinate Grid	Position data for vehicles from point of impact to positions of rest	On asphalt surface at impact center	Yellow traffic marker paint

Electronic transducers and photographic equipment represent the two principal data gathering systems applied to these experiments. Pre- and post-crash observations, in terms of locations of instrumentation, condition of lights, windows, door locks, seat adjustments, belt adjustments as well as collision deformation, skid pattern, debris location, glass damage, among others, represent another category of information important to this study.

Photographic instrumentation ranges from conventional motion picture cameras to high G-tolerant, moderate-speed motion picture cameras, and from automatically fired still cameras to high-speed motion picture cameras. Depending on requirements, as many as a dozen different types of cameras and a total of 40 photographic units have been used for a given experiment. In the two types of experiments presented, Fig. 9 provides the camera assignment for rear-end collisions and Fig. 10 presents camera assignments for the intersection collision experiments. Camera specifications and details on other types of instrumentation are given in Table 1.

The Rear-End Collisions - In addition to the camera assignments about the crash scene, several were positioned on the crash cars; these camera locations as well as specific dummy locations, positioning of transducers both within the car and within dummies, the seat types and where they were installed, as well as description of other accessories, are identified in Figs. 7 (a, b).

The Intersection Collision - As may be expected, the impacting vehicles and dummy occupants carry the majority of instrumentation used in this series. Camera, accelerometer, tensiometer, dummy and seat assignments are depicted in Figs. 8 (a, b), along with appropriate collateral information.

THE LIBERTY MUTUAL SURVIVAL CAR II

During the past two decades several individuals from the engineering and medical professions who were involved with work relating to vehicular collisions have attempted to design and build safety modifications into the motor vehicle; in some cases these designs represented changes that would assure reduction in injuries during motor vehicle collisions. For example, several years ago, Roy C. Lininger, M.D., Chairman, Colorado Safety Car Committee, Littleton, Colo., modified a passenger vehicle so as to provide improved mirrors for rear vision, padded dashboards, a safety steering wheel, special restraint harness and similar features, some of which are still in the discussion phase. Other interested medical professionals have manifested their frustrations in wanting a safer automobile through modification of their own passenger vehicles and by issuing brochures to patients they treat to improve their knowledge of safety devices. In a similar manner, engineers have at times modified their personal vehicles in an effort to evaluate innovations they consider represent improvements in motorist safety. Unquestionably, many other individuals and groups have attempted to accomplish this same objective using unique or ingenious techniques that were less publicized and therefore not of common knowledge.

The late Frank J. Crandell, vice-president and chief

engineer for Liberty Mutual Insurance Co., his associate chief engineers, Arthur Gordon and Frank Burkett, and project engineer Charles Sparrell, developed the design for Survival Car II. The Liberty Mutual Survival Car II represents the result of more than nine years of research and experimentation relating to the basic principles of "packaging the passenger." The LM engineers recognized their responsibility to produce more than just a conceptual design and took the necessary precautions to calibrate, in a realistic manner, the seat strength and similar factors to be reasonably certain that their performance would correspond to the safety assertions accompanying the publicity of this vehicle. Studies have shown that for the majority of cars operating on both urban and rural streets, there is only a driver present and that where only two motorists are riding in a given car, almost without exception, the passenger is seated in the right front seat. These observations point up the importance of front seat units as compared to rear seat units and the degree of protection that may be afforded motorists through proper design of these seat units. The completed LM Survival Car II (4) is shown by Fig. 11. In viewing this figure, the most conspicuous changes to this passenger vehicle are the front seat units; without question a properly designed integrated safety seat represents the greatest single contribution to passenger protection possible for current automobile design.

Referring to Fig. 11, the front seat units, 1, were referred to by the LM engineers as capsule chairs designed to restrain the driver and passenger while undergoing 30 G deceleration forces, whether from the front, the rear, or from the side. The rear seat passenger's armrest, 2, was designed to resist lateral movement without injury-producing forces being concentrated at the passenger hips or waist. Still referring to Fig. 11, restraining lap belts and shoulder harnesses, 3, were installed for all seats. A flexible steering shaft, 4, capable of transmitting the necessary torsional forces for steering and yet one that would buckle under the axial loads of a forward collision, was installed between the firewall and the steering gear box. A slip-type steering tube, 5, was installed between the steering wheel and the instrument panel; this tube had a telescoping capacity of 8 in. during a crash. Head restraints, 6, were applied to all seat positions. A rectangular steering wheel, 7, was installed for the stated purpose of preventing breaking or bruising of knee caps; this wheel size and shape was very similar to the 1957 models of Chrysler steering wheels. The LM steering wheel, 8, had a reduced diameter for better visibility by short drivers. Improved maneuverability was also an objective

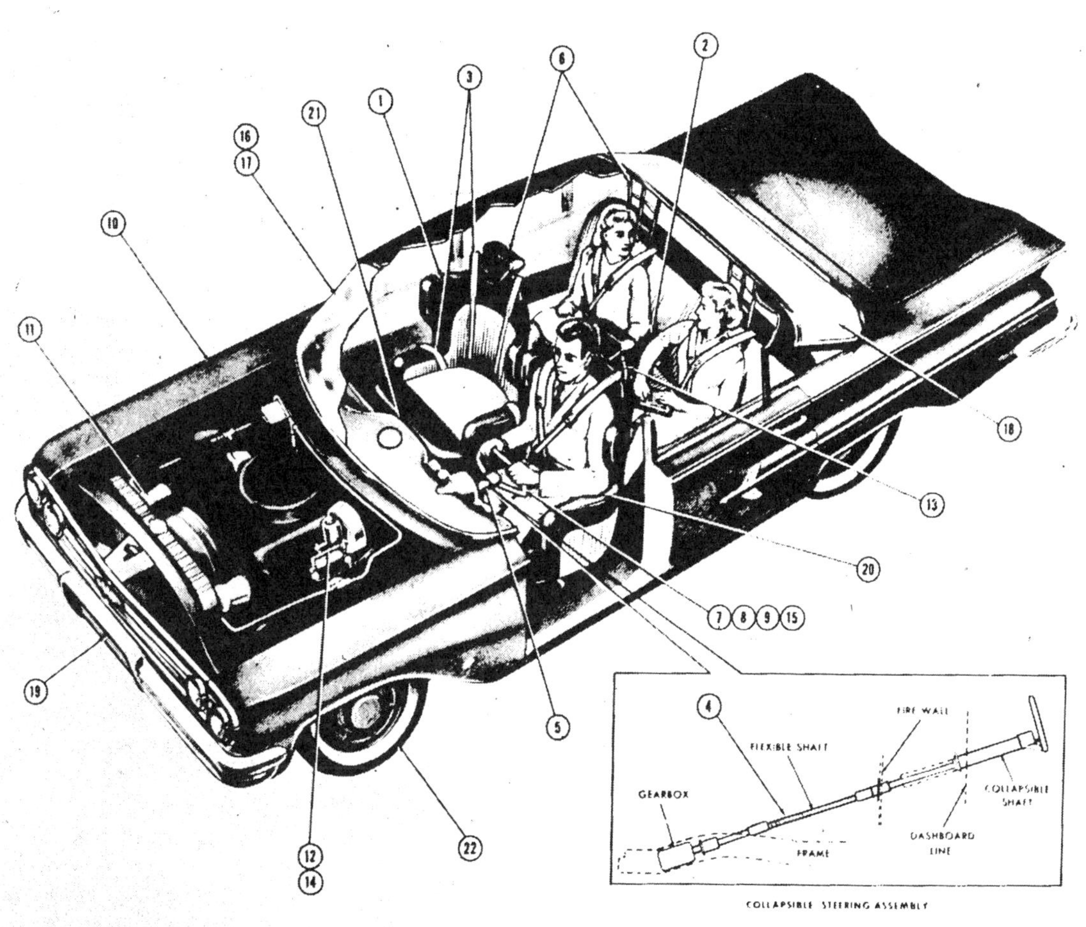

Fig. 11 - LM Survival Car II

of the LM engineers through the use of the smaller and squarish steering wheel, 9. Unit body construction listed by 10 is an error but could have contributed greater collapsibility -- on the theory that passengers would sustain lower peak acceleration during collision. An automatic fire control system, 11, utilizing carbon dioxide with appropriate dispersal locations, was included in this vehicle. A safety hydraulic brake device, 12, was installed that reportedly prevents appreciable loss of fluid during brake failure; a warning light, mounted on the instrument panel is electrically connected with this device to advise the driver of partial brake failure. Included with the capsule chairs are rollover bars, 13, to reduce the likelihood of head impacts during rollovers. Power brakes, 14, and power steering, 15, were included. Recognizing the limitations of the safety glass in laminated windshields, the LM engineers fabricated a windshield having a double thickness of the polyvinyl butyral interlayer, 16. This change was made to provide greater resistance to penetration, without necessarily increasing the forces applied to the human head during impact. Windshield transmission of ultraviolet rays was reportedly reduced 95% by the use of "Saflex" interlayer, 17. The other windows of the car were tinted, 18, for the purpose of reducing the heat load, reportedly by 30%. The vehicle was provided with reflective license plates, 19, for greater viewability at night. Armrests, 20, were provided for the driver and passengers to reduce fatigue. An Alert-O-Matic signal system, 21, was installed for the purpose of awakening the driver, should he fall asleep; if the driver does not awaken the vehicle is brought to a stop by this device. The tires were modified to reportedly improve the resistance to skidding on wet, icy and snow-covered roads by a technique known as "Micro-siping," 22. In addition, side reflection mirrors were installed for greater driver viewability and a smooth hood over the engine was a design consideration to reduce the injury hazards to pedestrians. With respect to the brake safety device, it functioned so there would always be at least two-wheel brakes in operation, should a hydraulic tubing fail. The 1967 domestic vehicles use the added protection of dual master cylinders but at the time that this LM device was installed, seven years ago, it represented a considerable advance over domestic braking systems then available on production cars.

THE LM CAPSULE CHAIR

Two basic concepts of passenger protection are:

1. To restrain the motorist by at least a lap belt and position him within a large passenger compartment, one in which the interior surfaces are sufficiently remote that he cannot be flailed against them, while restrained by the lap belt unit.

2. To ascertain that the restrained motorist is seated within a passenger compartment that has its interior surfaces developed very close to the driver, especially from his shoulder level downward, and to include a head support; that the interior surfaces in close proximity are modified in form and energy absorbing properties to reduce the likelihood of injury when the motorist is crushed against it.

From the current studies a third concept should be added, that of a capsule chair, in which the immediate proximity of the vehicle interior is of secondary importance because the motorist is rather effectively protected within the confines of the capsule chair or motorist protective envelope. This was the motorist protective concept selected by the LM engineers because it was not possible to secure sufficient body clearance within the medium size automobile to evaluate the maximum space concept. The capsule chair includes the concept of the integrated protective motorist seat but improves motorist safety by including a roll bar and side impact "wings" that in no small measure increases the chances of survival for motorist impacted at these vulnerable points. A feature of the capsule chair is that it may swing like a swivel chair to facilitate entrance and exit. In addition to the shoulder "wings" the arms of the chair provide additional protection against the collision forces of a side impact. The capsule chair also has a requirement to move fore-and-aft to accommodate motorists of different heights. The armrests and shoulder protection for side impacts is best portrayed by reference to Fig. 12 depicting the supporting structure of the capsule chair; titanium alloy tubing, having an ultimate fiber stress of 125,000 psi, was selected by the LM engineers to provide high strength for relatively lightweight conditions. Although this tubing of itself would not be capable of resisting the side-impact intrusion forces at 40 mph, the motorist sits within this protective shield and the seat frame is sufficiently rigid to displace the motorist out of range of the intruding forces of a striking car while partially resisting direct penetration. This concept is sound and merits serious consideration in future designs.

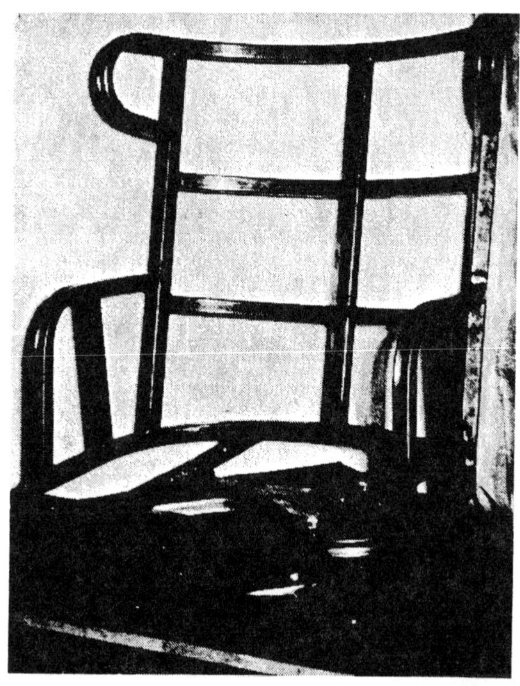

Fig. 12 - Skeleton of LM capsule chair

Referring to Ref. 3, "During the bus side-impact experiment, it was observed that armrests provided a significant improvement in passenger safety by first, preventing (unrestrained) individuals from being ejected from their seats laterally to strike passengers across the aisle, and second, preventing the larger passenger from crushing against a smaller passenger who may be seated in his path." In addition, evaluation of the Cox-Hilton (England) integrated safety seat (3) showed that even the relatively subtle contouring of the bucket seat configuration provided substantial resistance to the fully belted dummy before the lateral forces peaked on the belt tensiometers. The exceptional effectiveness of the armrests and shoulder wings for the LM seat will be described in a subsequent section. The basic design of the LM seat is further illustrated by Fig. 13.

SEAT DESIGN

The concept employed by the LM capsule chair for packaging the passenger consisted of developing a nominally nonpenetrable protective structure, in the form of a seat enveloping the motorist from the sides and rear, and including a built-in full harness restraint for frontal collision forces. The head support was also structured to resist roof collapse. Padding was added for comfort and to assure acceptable distribution of collision forces to the stronger areas of the motorist's body during collision. The Cox integrated safety seat was designed to sustain the added forces of a motorist belted to it during collision without failure to either seat or anchorage, but, unlike the LM capsule chair, the Cox seat design did not include the occupant packaging concept and the nonpenetrability concept designed into the LM seat.

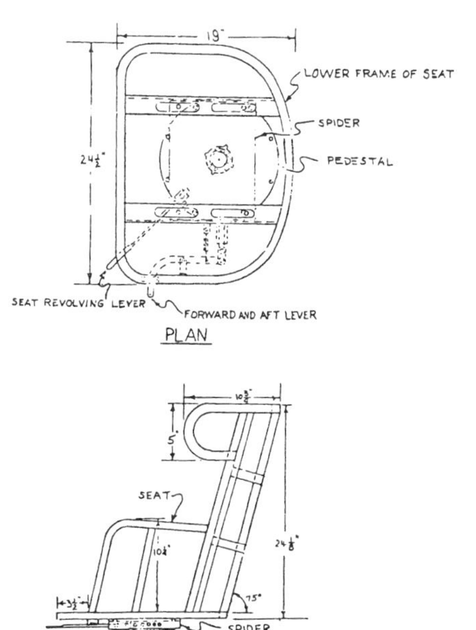

Fig. 13 - LM capsule chair details

In the previous section, the LM design was presented; the Cox seat (Fig. 14) is similar in many respects. (6) This design rather effectively prevents lateral displacement of the occupant relative to the seat during side impacts (3) but does not provide the same measure of protection or degree of resistance to intrusion afforded by the LM design. The structural design of the Cox seat, Fig. 15, (5) represents a great improvement over conventional bucket seat design.

For the most part, only engineers responsible for seat design appreciate the difficulty of constructing a seat suitable for the great variety of body sizes and forms -- a seat that also limits the onset of fatigue, to the extent possible, and one that provides the body support and adjustability needed for attentive driving. Seat design is challenging and the preoccupation with this difficult task has probably diverted much needed attention from the vital factor a car seat should include, namely, motorist protection from collision forces.

Structurally, the motor vehicle bucket seat corresponds approximately to a lightweight, padded living room chair. It has a readily compressible, bundled-wire substructure support attached to a lightweight tubular frame (Fig. 16) (6) that is difficult to anchor effectively and that provides inadequate body support from any direction. Unable to support its own weight during collision deceleration, the conventional car seat design cannot be used for attaching protective restraints; as a result

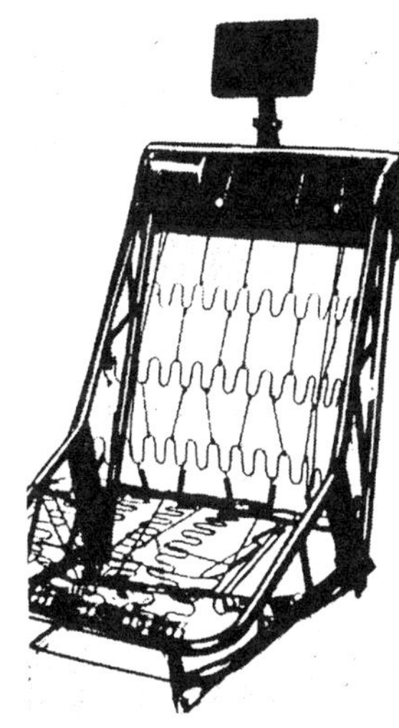

Fig. 14 - Cox seat frame

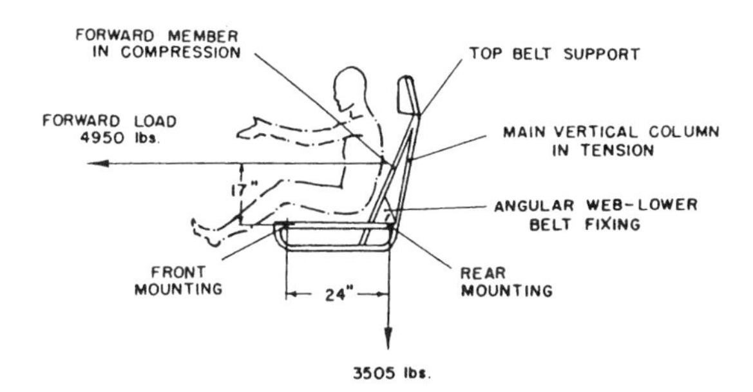

Fig. 15 - Cox seat structural design

of this deficiency, seat belts have to be extended to the floor and shoulder belts to roofs, doorposts, rear seat or floor points for anchorage. These nonappealing, troublesome locations make their use by motorists less probable and provide compromises in protection for those determined to avail themselves of whatever protection is at hand. What the authors are suggesting is that the time has long since passed when gross changes in seat design were obviously needed; a car seat which does not act as a motorist's inner protective shield against collision forces is failing in its most vital role.

ANALYSIS AND FINDINGS

Of the four LM crashes reported by this paper, the two 30 mph rear-end collisions will be analyzed first, followed by the analysis and findings for the two 40 mph intersection collisions. For each experiment the struck vehicle will be analyzed first, followed by analysis of the striking vehicle, with reference to other findings deferred until the section "Conclusions and Recommendations." In general, the analysis and findings for each experiment will include the following topics:

Vehicle collision performance and peak accelerations.
Passenger kinematics and peak accelerations.
Protection provided by various restraint systems.
Acceleration curves of occupants, vehicles and seat structure.
Vehicle collision dynamics including charts of positions of impact and rest.
Description of post-crash observations.

There were very significant differences observed between the collision performance of the 1960 and 1966 Chevrolets, between the capsule chair and the conventional bench seat, and between the integrated harness and the cross-chest lap belt combination harness; these observations will be made in connection with the section "Conclusions and Recommendations."

FINDINGS - REAR-END COLLISIONS

EXPERIMENT 88 - A 1960 LM Chevrolet 4-door sedan (Liberty Mutual configuration) was stationary when rear-ended by a 1966 production Chevrolet sedan; the impact was an "in-line" rear-end collision with the rear of the LM Chevrolet being directly opposed by the front end of the striking 1966 Chevrolet. Passenger vehicle designs have bumpers almost flush with front-end and rear-end body sheet metal panels; consequently, speeds of impact as low as 10 mph will damage body sheet metal sections, even though initially involving only bumper-to-bumper contact. Bumpers as presently designed are little more than parking guards and provide very little resistance to impact, even at moderate speeds; considering the grossly inadequate height of seatbacks common to most domestic vehicles, these "soft" bumpers do serve to reduce the injury-producing forces transmitted to the improperly supported motorist during rear-end collisions. This example of interrelated safety factors shows how changes to improve one factor may be detrimental to other motorist protection features. Domestic cars should have more collapse resistant bumpers, but such improvement should follow, not precede, other more important modifications that would ensure motorist protection from rear-end collision whiplash.

The 1966 Chevrolet traveling 30 mph, crushed into the rear of the stationary 1960 LM Chevrolet, advancing it 2 ft in the initial 150 milliseconds. The striking 1966 Chevrolet front-end at this instant sustained more collapse than the 1960 Chevrolet rear-end (Fig. 17), even though the opposite relation is more frequently identified for rear-end collisions. This difference is partially attributed to the fact that the struck 1960 Chevrolet has a full frame extending to the rear bumper, while the 1966 Chevrolet design is based on collapse principles of unit body construction using a perimeter frame. Almost simultaneous with bumper contact, sheet metal contact at substantially higher elevations presents resistance above the car center of mass elevation; as the collision collapse action advances, the rear-end of the struck

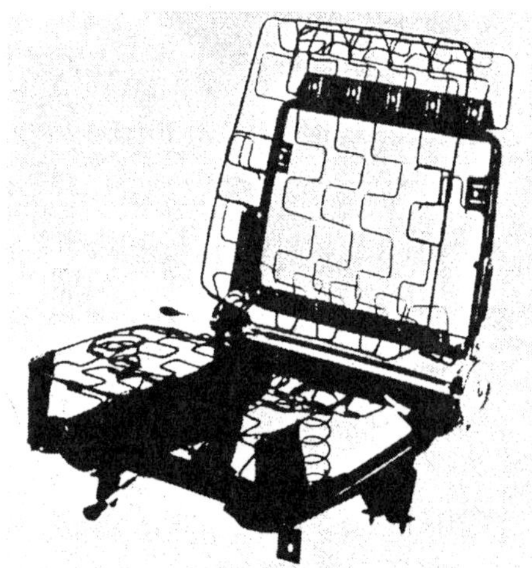

Fig. 16 - Conventional bucket seat frame

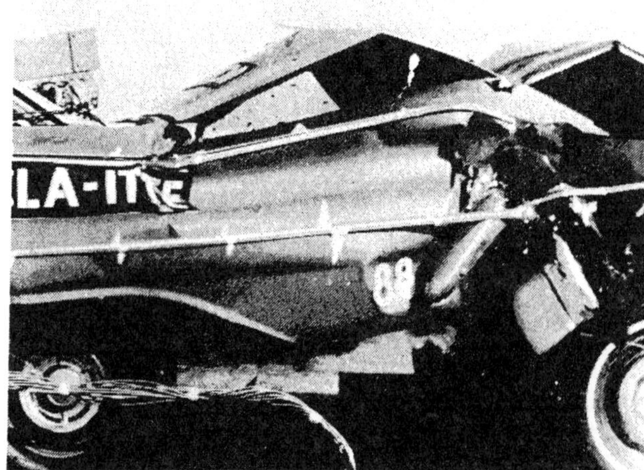

Fig. 17 - 1960 rear-ended Chevrolet receives less damage than 1966 striking Chevrolet (Experiment 88)

Fig. 18 - 1960 Chevrolet, 150 ms, being rear-ended by 1966 Chevrolet (Experiment 88)

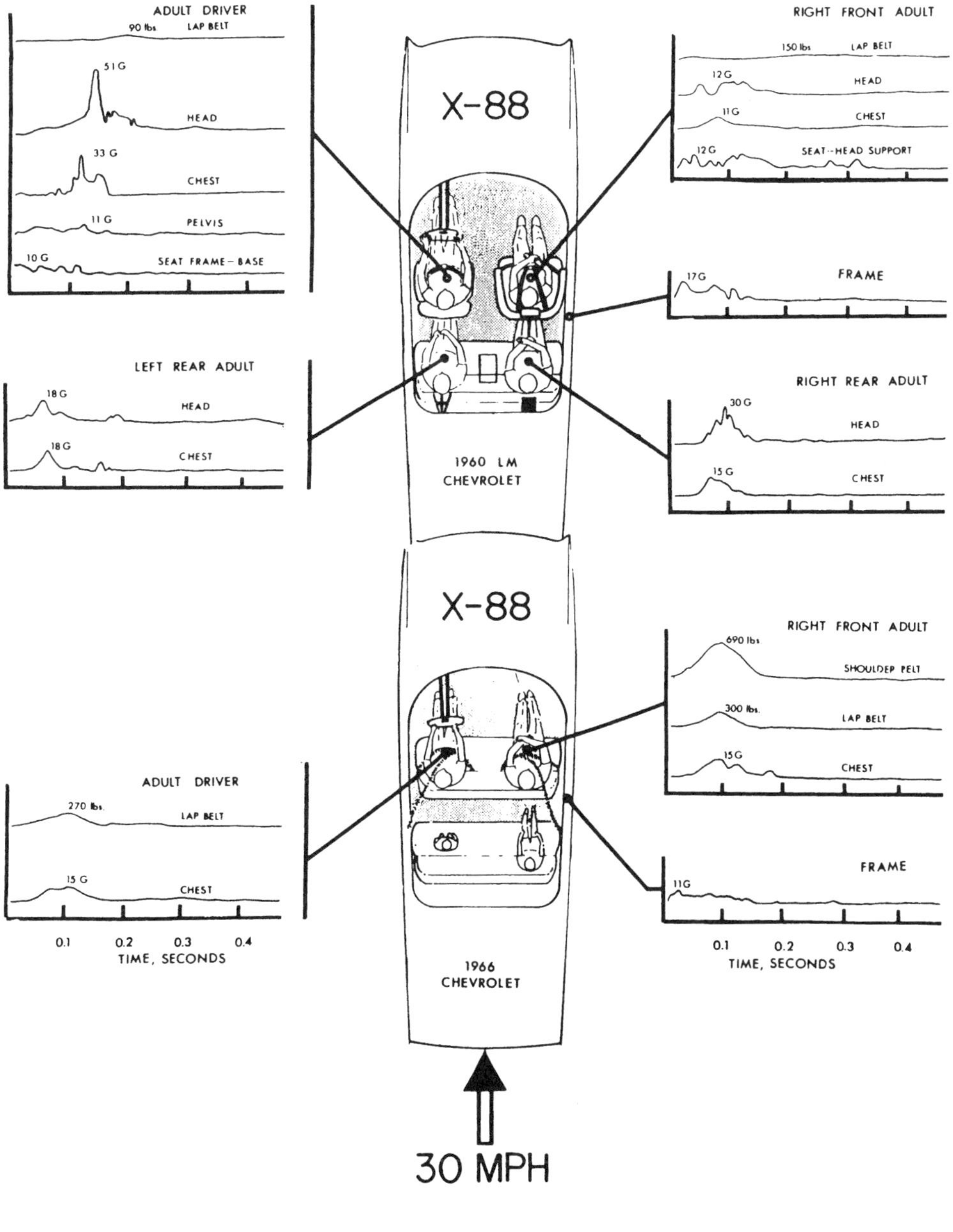

ADULT DRIVER
90 lbs. LAP BELT
51 G HEAD
33 G CHEST
11 G PELVIS
10 G SEAT FRAME - BASE

X-88

RIGHT FRONT ADULT
150 lbs LAP BELT
12 G HEAD
11 G CHEST
12 G SEAT - HEAD SUPPORT

17 G FRAME

LEFT REAR ADULT
18 G HEAD
18 G CHEST

RIGHT REAR ADULT
30 G HEAD
15 G CHEST

1960 LM
CHEVROLET

X-88

RIGHT FRONT ADULT
690 lbs. SHOULDER BELT
300 lbs. LAP BELT
15 G CHEST

ADULT DRIVER
270 lbs. LAP BELT
15 G CHEST

0.1 0.2 0.3 0.4
TIME, SECONDS

11 G FRAME

0.1 0.2 0.3 0.4
TIME, SECONDS

1966
CHEVROLET

30 MPH

Fig. 19 - Transducer patterns, 30 mph rear-end collisions (Experiment 88)

car and front-end of the striking car become a toggle with inertial forces operating to buckle the collapsing vehicle combination upwards, the direction of least constraint (Fig. 18).

Struck Car Occupants, Experiment 88 - The driver's capsule chair in the LM car was replaced by a 1963 Ford bucket seat, having dimensions and design representative of the average domestic single passenger automobile seat. The driver's head and upper torso (Fig. 18) have forced his seat into a reclined position, primarily as a result of failure of the forward anchorages. He was still in a noninjury-producing posture as his seatback impacted the knees of the passenger behind him, causing a 33 G chest acceleration at 120 ms. His head then slammed into the unyielding lap of the left rear adult seated behind him causing a peak head acceleration of 51 G at 145 ms (Fig. 19). This high head acceleration was partially attributed to the unnaturally hard composition of his head. Elastic rebound of the driver-seat combination caused him to undergo whiplash (Fig. 20), and his adverse posture is readily contrasted with that of the right front seat passenger in the LM seat (Fig. 21).

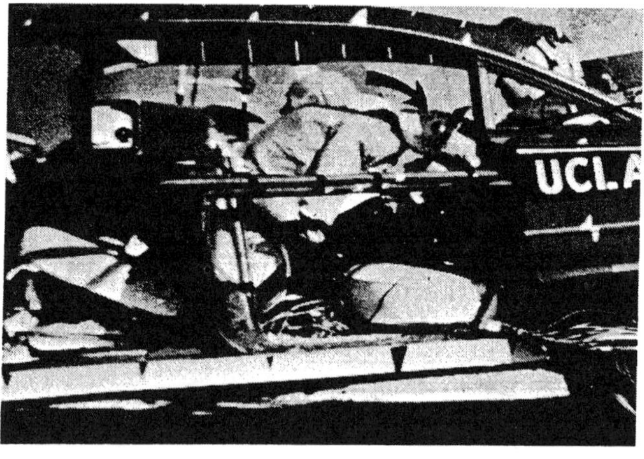

Fig. 20 - Driver of struck car in Ford seat undergoing whiplash, 170 ms (Experiment 88)

Referring to the motion picture analysis for the occupants of this struck 1960 Chevrolet, the driver kinematics were described in connection with Figs. 18 and 20. During this time, the right front seat passenger seated in an LM capsule chair showed no evidence of exposure to injury, because of the protection derived from this contour seat. The seat pedestal base flexed the floor pan but the seated passenger maintained a normal posture throughout the impact. His chest sustained a peak acceleration of 11 G at 70 ms before his head recorded 12 G at 100 ms (Fig. 19). The LM seat and seatback reclined 30 deg rearward at its maximum rotation, bringing the back of the seat touching the knees of the right rear passenger (Fig. 22). His posture remained normal because the entire seat, including the base, rotated backwards in an energy absorbing manner. Also, because the floor deformed, the springback of the seat was nominal and the occupant did not show a rebound inertia effect that would be common for a condition in which the floor pan was not allowed to yield.

The right rear passenger received a 15 G chest acceleration at 65 ms while undergoing whiplash (Fig. 19). His neck design had limited articulation and prematurely reached its maximum excursion (30 G head peak at 90 ms) before his head could rotate 90 deg and strike a 1 in. thick pad of plastic installed on the rear window shelf for evaluation. The pad and the webbing net are shown in Fig. 23. Extension of padding from the rear seatback with a smooth transition to the rear window shelf is a proposed method for minimizing head and neck injuries resulting from rear-end collisions.

The head of the left rear passenger impacted the head support net installed by Liberty Mutual with sufficient force to partially tear it from its roof anchor (Fig. 24). Because of this effective restraining action, his head received only 18 G at 60 ms and his chest also received 18 G at 70 ms (Fig. 19); whiplash action was prevented as the net arrested the head before it was displaced significantly rearward.

Striking Car Occupants, Experiment 88 - This 1966 Chevrolet sedan with production bench seats, was occupied by two adult dummies in the front seat, a 6-year-old sitting

Fig. 21 - Passenger of rear-ended car rides out crash maintaining normal posture in LM chair (Experiment 88)

in the right rear position, and a 3-year-old standing on the seat in the left rear position.

The driver was restrained by GM installed cross-chest lap belt combination installed in the 1966 Chevrolet to GM specifications. Analysis of moderately high speed film revealed that his upper torso and head were pitched forward slightly before being snubbed by the diagonal strap; the increased wedge action, caused by the more remote roof anchoring of the diagonal strap behind the doorpost (Fig. 25), amplified the submarining effect and forced the driver deeper into his seat. At this time the lap belt part of the diagonal chest combination recorded 270 lb at 105 ms. Simultaneously, his chest sustained a 12 G peak deceleration (Fig. 26).

The right front adult passenger, restrained by an identical 3-point GM belt, responded to the collision forces in a manner similar to the driver. His shoulder strap loaded to 690 lb at 95 ms and at this same time his lap belt tension peaked at 300 lb. His chest accelerometers simultaneously reached their peak of 15 G (Fig. 19).

The belt system for the front seat occupants is effective for restraining a motorist from striking his head and chest against the front surfaces of the vehicle interior; however, the application of forces and abrasive action across the neck appears to represent a condition that may cause serious injury. Further study of this belt system appears indicated.

The rear seat child dummies were thrown forward against the front seatback but not over it; they then rebounded with the right rear 6-year-old being pitched onto his seat in a normal seated posture while the left rear 3-year-old landed against the front seatback. As the left rear standee was thrown forward against the front seatback, it was moving rearward, owing to the rebound effect of the front seat occupants from their diagonal belts, and this rearward force caused the 3-year-old's hips to be rebounded rearward as his head firmly struck the center cross-bow of the roof. During the time the right rear 6-year-old was pressing against the front seatback, he flailed his head over it and struck his head firmly against the right front passenger's head who

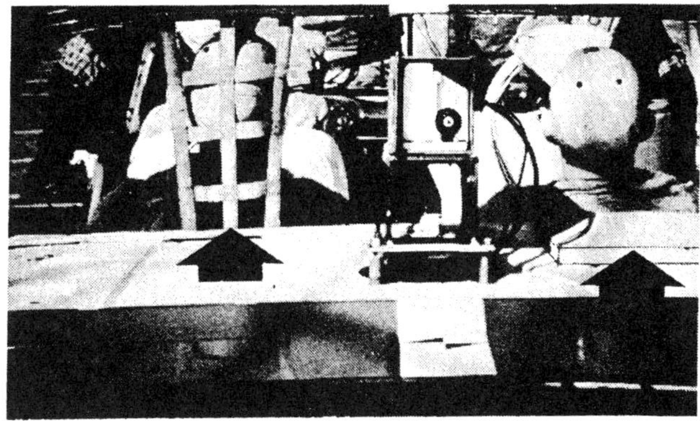

Fig. 23 - Rear seat passengers' whiplash protection, shelf pad (right) and LM head support net (left) (Experiment 88)

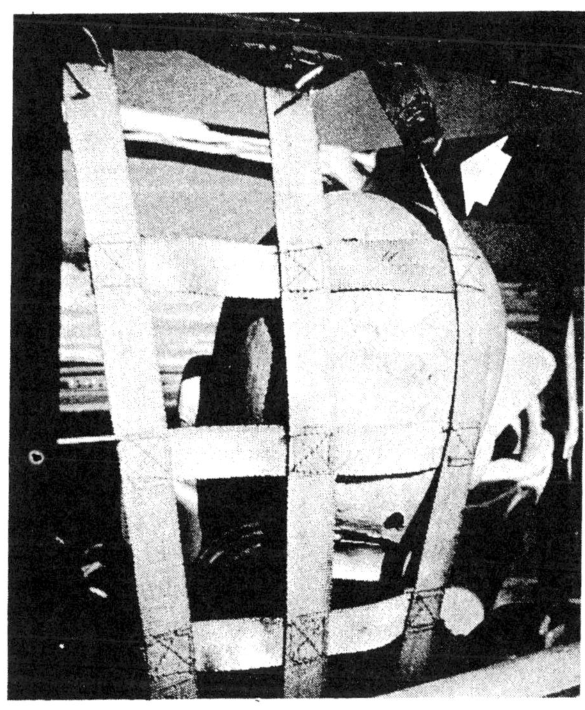

Fig. 24 - LM head support net partially torn from arresting whiplashing head

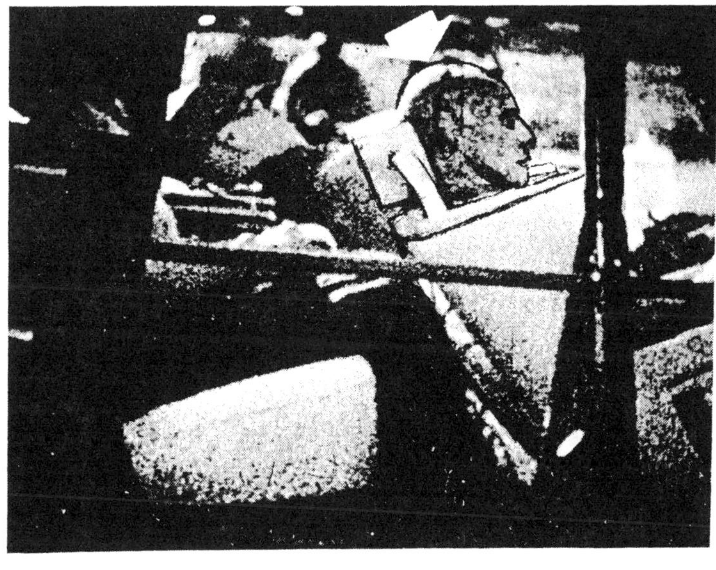

Fig. 22 - LM capsule chair rotated 30 deg rearward at its base during rear-ending (Experiment 88)

was rebounding from the elastic energy stored in the diagonal belt.

Post-Collision Observations, Experiment 88 - Following collision, an inspection was made for purposes of noting significant changes. The struck 1960 LM Chevrolet was propelled 180 ft approximately straight ahead; it deviated to the left 27 ft during this 180-ft travel. The left front Ford bucket seat had a residual separation at the forward anchors of 4 in.; the right front LM capsule chair had a permanent rotation of 15 deg rearward about the pedestal base. Collision damage of this 1960 LM Chevrolet portrayed by Fig. 27 (a) shows a moderate damage as contrasted to the 1966 Chevrolet rear-ended under identical conditions in Experiment 89 (Fig. 27 (b)). However, frame peak acceleration for the stiffer frame collapse of the 1960 Chevrolet was almost 50% higher than the corresponding peak of the newer vehicle.

The striking car, a 1966 Chevrolet, decelerated to rest 104 ft past the point of impact; it traveled a nearly straight course, deviating only 5 ft to the right during the 104-ft runout. The 1960 LM Chevrolet was stationary when rear-ended by the 1966 Chevrolet; the struck car abruptly accelerated and the 30 mph striking car abruptly decelerated so that 1/10 sec after they contacted, they had crushed together a maximum amount and had a common velocity (Fig. 28). Thereafter, restitution took place and 500 ms later, or 600 ms after the cars contacted, they separated as they continued to roll to their positions of rest.

Collision damage to the front of the 1966 Chevrolet is shown in Fig. 29. The left side of the windshield was cracked

Fig. 25 - Roof-anchored cross-chest belt wedges driver of striking car into submarined posture (Experiment 88).

Fig. 26 - Driver's head and upper torso are restrained by roof-anchored 3-point cross-chest belt (Experiment 88)

Fig. 27 (a) - Collision damage, 1960 Chevrolet rear-ended by a 30 mph 1966 Chevrolet (Experiment 88)

Fig. 27 (b) - Collision damage, 1966 Chevrolet rear-ended under same conditions as Experiment 88

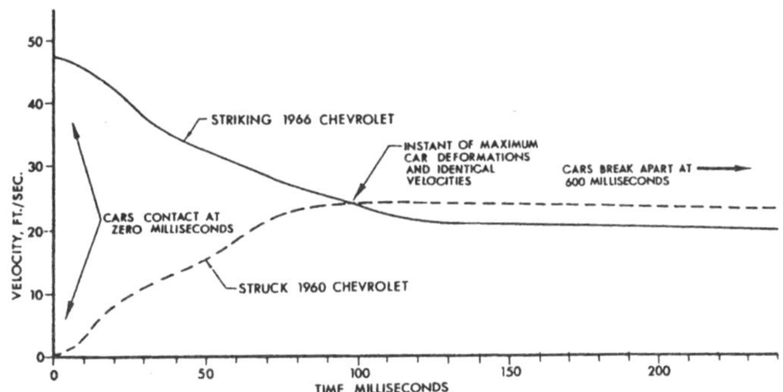

Fig. 28 - Velocity changes for 30 mph rear-end collision (Experiment 88)

Fig. 29 - Front-end damage to 1966 Chevrolet in 30 mph rear-end collision (Experiment 88)

Fig. 30 - Rear-end collision, 30 mph, 1/ 10 sec after contact (Experiment 89)

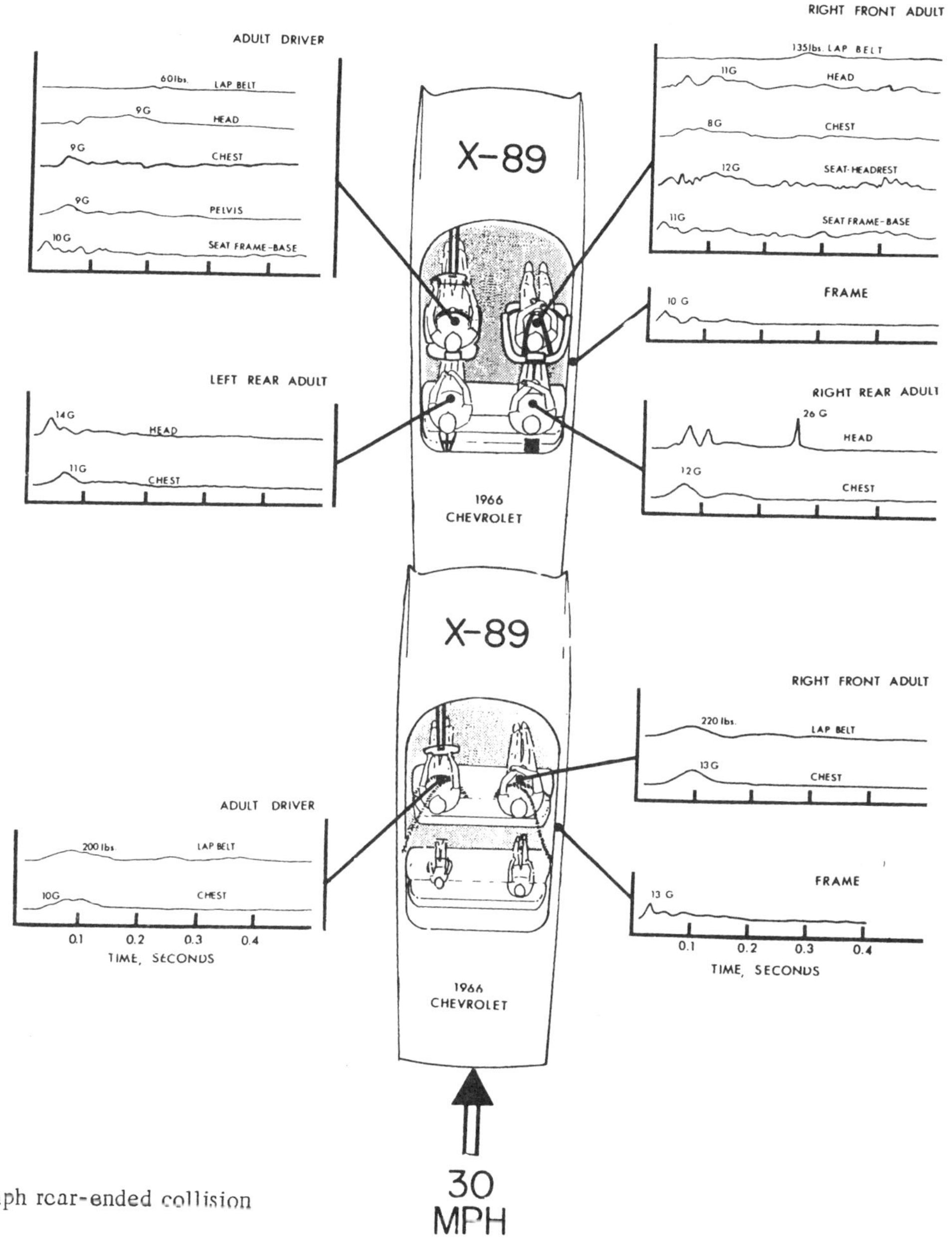

Fig. 31 - Transducer patterns, 30 mph rear-ended collision (Experiment 89)

by rearward hood displacement. The collapse of the right fender applied forces to the windshield sufficient to crack it. The right front door was displaced rearward 3/16 in., deforming the center doorpost. Similar displacement occurred for the left side, even though the left front door had been removed and a compression strut installed in its place. The debris pattern for this 30 mph rear-end collision extended from 15-45 ft past the point of impact along the path of the struck car. Neither vehicle had brakes applied and the transmissions of both vehicles had been placed in neutral.

EXPERIMENT 89 - A 1966 production Chevrolet 4-door sedan was stationary when rear-ended by an identical 1966 Chevrolet, traveling 30 mph at impact (Fig. 30). As with Experiment 88, this collision was an in-line rear-end collision with the rear of the struck car completely overlapped by the front of the striking car. Principal differences between Experiments 88 and 89, as reflected in Figs. 7 (a) and 7 (b), were that in Experiment 88, the struck car was a 1960 LM Chevrolet and, in Experiment 89, the struck car was a 1966 Chevrolet. In Experiment 88, the struck car driver was seated in a 1963 Ford bucket seat and, in Experiment 89, the struck car driver was seated in a 1966 GM bucket seat, with a head support. The LM head support net was installed in the 1966 Chevrolet for the left rear passenger and a 1 in. thick pad was installed on the rear window shelf for the right rear passenger. In Experiment 89, the rear-ended 1966 Chevrolet sustained a frame peak acceleration of 10 G, 30 ms after the bumpers contacted; the striking 1966 Chevrolet registered a frame peak deceleration of 13 G, also at 30 ms (Fig. 31). The 10 G frame acceleration for the 1966 Chevrolet represents a significant difference for Experiment 89 as contrasted to the 17 G frame peak for the more rigid fram design of the 1960 LM Chevrolet (Fig. 19).

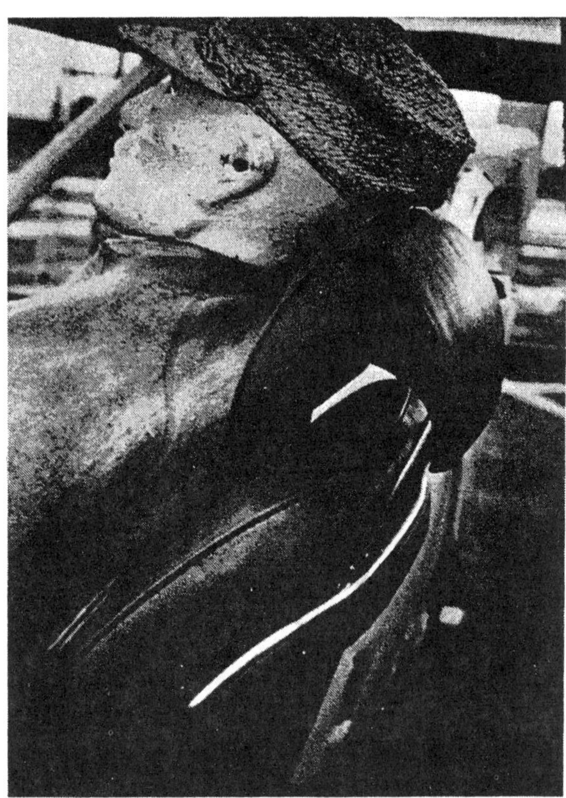

Fig. 32 - Driver in GM bucket seat with head support adjusted to 6-1/2 in. above top of backrest, providing an overall back and head support height of 31-1/2 in. (Experiment 89)

Struck Car Occupants, Experiment 89 - The driver was seated in a GM bucket seat with an adjustable headrest (Fig 32). During the initial phase of the collision his chest reacl a peak acceleration of 9 G at 60 ms and simultaneously his pelvis also recorded 9 G (Fig. 31). These peaks occurred before the seatback deflected rearward appreciably. The inertial forces of his torso displaced the backrest into a sem reclined position (Fig. 33). Yielding of the seatback forme an inclined plane (Fig. 34) which allowed the driver to slid beyond the protective limits of the head support. This condition produced a whiplash, notwithstanding the apparent head-supported condition shown in Fig. 32.

The head of the driver reached a peak acceleration of 9 G at 165 ms, which occurred after he was forced into a whiplash condition. This whiplash action was snubbed by a combination of factors: The seatback struck the knees of the rear seat and occupant; the driver's head was restrained from further rearward rotation by the head support. Thereafter, seatback restitution forces were sufficiently develope to rebound the driver against the steering wheel, causing him to flail his head over the wheel but not against the windshield.

The right front seat passenger was seated in a Liberty Mutual capsule chair and was observed to tilt the chair, as measured at the headrest, 10 in. rearward. The passenger rode through the entire collision in a completely satisfactor manner, sustaining an 8 G chest acceleration at 85 ms and an 11 G head acceleration at 100 ms. The seat and backrest unit was sufficiently rigid to resist bending; the rearward rotation occurred between the base of the seat and the floor pan. Immediately after contact (20 ms), the seat frame

Fig. 33 - Driver's GM backrest rotated rearward 30 deg beyond normal 1/10 sec after contact (Experiment 89)

registered its peak acceleration of 11 G. At 110 ms, the headrest peaked with 12 G. The maximum rearward rotation of the LM capsule seat was 16 deg.

At 65 ms after vehicle contact the unrestrained right rear adult passenger sustained a 12 G peak chest acceleration (Fig. 31). Thereafter, his head flailed rearward, striking the rear window shelf pad; as he rebounded forward he struck the header and received a 26 G blow to the front top of his head, 265 ms after the cars contacted one another.

In contrast to the violent head movement and 26 G head acceleration of the right rear occupant, the left rear adult sustained only a 14 G head acceleration because his head pocketed into the LM head support net at 40 ms; his chest accelerometers reached a peak of 11 G at 65 ms. Through-

Fig. 34 - Whiplashed driver, 150 ms after contact; head support protection diminished by excessive seatback yielding (Experiment 89)

Fig. 35 - Left rear passenger remains in normal seated posture without whiplashing owing to performance of LM head support net (Experiment 89)

Fig. 36 - Front seat occupants of striking car 100 ms after contact (Experiment 89)

out the collision event he remained in a normal seated posture (Fig. 35), and his head movement relative to torso was negligible owing to the effective performance of the head support net.

Striking Car Occupants, Experiment 89 - The striking car for Experiment 89, as with Experiment 88, was a 1966 Chevrolet production vehicle, with standard bench seats. The driver and front seat passenger each were restrained by a roof-rail anchored cross-chest lap belt combination; both front seat occupants underwent submarining with the belts lacerating their faces and necks, indicative of injuries likely to be sustained using the cross-neck design (Fig. 36). A peak tension force of 610 lb was measured on the shoulder strap of the passenger's belt at 90 ms, while the lap portion of the passenger's and driver's belts measured 220 lb and 200 lb, respectively. Notwithstanding the adverse aspects of this cross-neck design, the belts did serve to prevent the front seat occupants from being pitched against the steering column and forward interior structures. The rear bench seat of this striking car was occupied by a 6-year-old on the right side and a 3-year-old on the left. Both children were seated and unrestrained. They were thrown forward from their seats during the collision with the shorter 3-year-old striking the seatback and rebounding to his seat while the taller 6-year-old on the right partially flexed over the seatback before rebounding to his seat.

Post-Collision Observations, Experiment 89 - The struck car was propelled 48 ft directly forward to its position of rest, still interlocked with the striking vehicle (Fig. 37). The relatively short run-out of these vehicles after collision, as contrasted to Experiment 88, was caused by sheet metal that had crushed against the left rear wheel. The gas tank on the struck car was torn free from its anchor straps and was found 25 ft from the impact point.

The rear portion of the struck car was crushed forward 22 in. (Fig. 27(b)). The pressed channel frame members, supporting the trunk floor pan and bumper, buckled at the

Fig. 37 - Interlocked vehicles, both 1966 Chevrolets, traveled 48 ft to their positions of rest

Fig. 39 - Collision damage to 1966 Chevrolet striking car (Experiment 89)

Fig. 38 - Permanent deformation, 30 mph rear-end collision (Experiment 89)

frame attachment point of the bumper support (Fig. 38 (arrows)). However, the major portion of forward movement for the collapsing rear end section occurred in the frame transition member over the rear axle. This accordion action of the frame at the wheel-well area accounts for body metal encroachment, restricting rear wheel rotation. The result was braking action during the runout phase of the collision.

The front-end damage of the striking car amounted to only 4 in. of permanent deformation (Fig. 39). This was comparatively minor for a 30 mph impact but is consistent with the relatively "soft" structure encountered by the striking 1966 Chevrolet.

FINDINGS - SIDE-IMPACT COLLISIONS

EXPERIMENT 91 - This experiment was an intersection type collision in which two 1960 LM Chevrolets impacted while traveling 40 mph (Fig. 40). The collision was a 90 deg side impact, with the striking car directing its force to the driver's side of the struck car at the doorpost; both vehicles received peak decelerations of 11 G at 25 ms after the start of this impact (Fig. 41). The LM chair resisted collision intrusion and served to moderate the forces applied to the driver. Following the car-to-car impact, the struck car spun out counterclockwise, rolling toward its right side as it rotated 180 deg; it then rocked toward its left side bringing the left side into contact twice with the pavement. During these rocker panel contacts, the left door, having been crushed by the impacting car, popped open (Fig. 42). Details concerning dummy occupant assignments, seat types, and instrumentation were presented in Methodology Section (see Fig. 8(a)); for ease of reference, some of this information has been included with the transducer patterns (Fig. 43).

Struck Car Occupants, Experiment 91 - The driver of the struck car was protected by an LM capsule chair. Because of its rigid construction and encapsulating design, the directly impacted chair displaced the driver to his right

Fig. 41 - LM chair resists intrusion and buffer collision forces at 40 mph side impact directed to driver's door

Fig. 40 - Intersection collision of 1960 LM Chevrolet traveling 40 mph, struck at center doorpost by 1960 LM Chevrolet, also traveling 40 mph

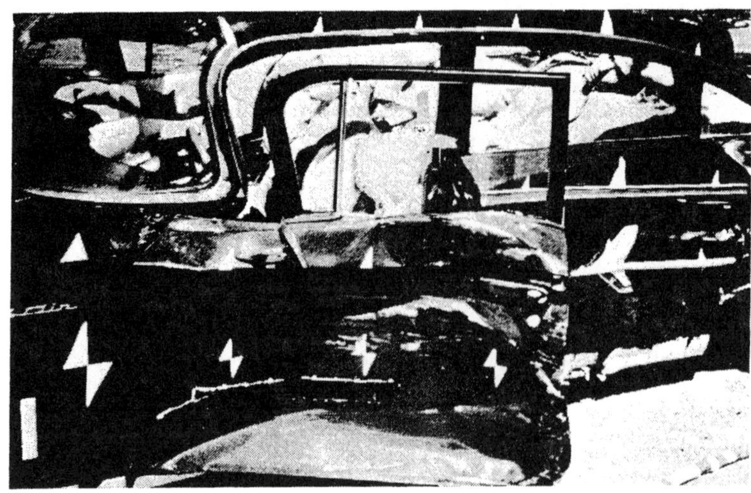

Fig. 42 - "Inner protective shield" (LM capsule se prevented driver from being ejected as impacted d popped open

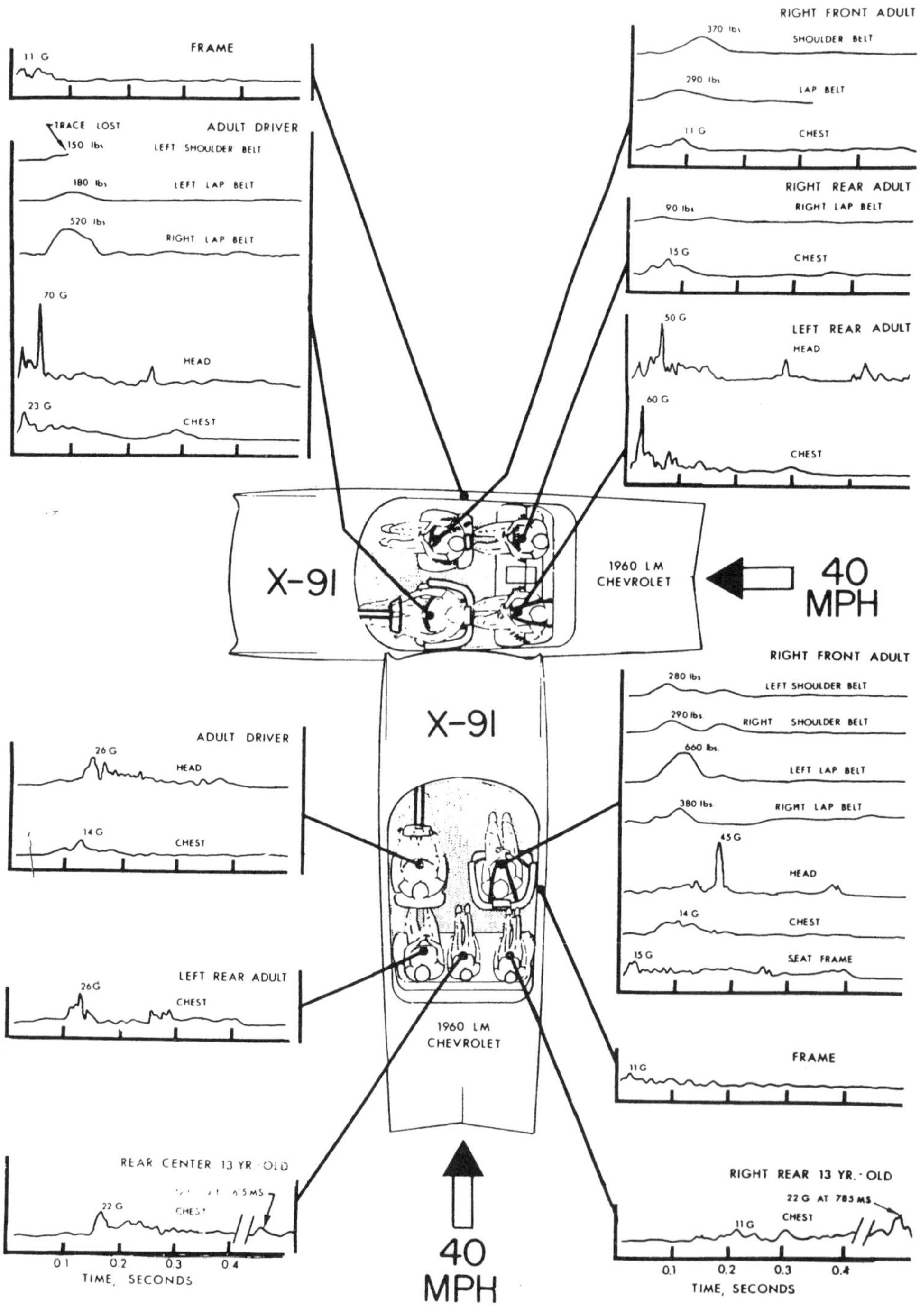

FRAME
11 G

ADULT DRIVER
TRACE LOST
150 lbs LEFT SHOULDER BELT
180 lbs LEFT LAP BELT
520 lbs RIGHT LAP BELT
70 G HEAD
23 G CHEST

RIGHT FRONT ADULT
370 lbs SHOULDER BELT
290 lbs LAP BELT
11 G CHEST

RIGHT REAR ADULT
90 lbs RIGHT LAP BELT
15 G CHEST

LEFT REAR ADULT
50 G HEAD
60 G CHEST

X-91

1960 LM CHEVROLET

40 MPH

RIGHT FRONT ADULT
280 lbs LEFT SHOULDER BELT
290 lbs RIGHT SHOULDER BELT
660 lbs. LEFT LAP BELT
380 lbs RIGHT LAP BELT
45 G HEAD
14 G CHEST
15 G SEAT FRAME

X-91

ADULT DRIVER
26 G HEAD
14 G CHEST

LEFT REAR ADULT
26G CHEST

1960 LM CHEVROLET

FRAME
11G

REAR CENTER 13 YR OLD
22 G CHEST

RIGHT REAR 13 YR. OLD
22 G AT 785 MS
11 G CHEST

0.1 0.2 0.3 0.4
TIME, SECONDS

40 MPH

Fig. 43 - Transducer patterns, 40 mph intersection collision (Experiment 91)

causing a peak chest acceleration of 23 G at 20 ms in a manner not likely to cause serious injury. The "packaged" driver was rather effectively isolated from this extremely dangerous collision exposure, as shown for the struck car driver in Fig. 40.

During impact, the motion picture films of the struck vehicle showed the driver, seated in his LM capsule chair, being forced leftward, causing his head to flail abruptly against the side glass. At the instant of maximum head-to-glass impact he sustained a 70 G blow at 45 ms and developed a 3 in. diameter head-pressure area (Fig. 44). The striking car pushed the driver's door and side glass inward several inches accounting for the contact made by the driver's head with the side glass; if this glass had not displaced inward the adequately restrained driver probably would not have flailed his head far enough to contact the side glass. Notwithstanding the severity of the head impact, it was not hard enough to break the 1/4 in. tempered glass. The present rigidity, resistance to breakage, and weight of side window glass allows it to deliver severe head impacts; the reduced inertia and increased flexibility of thinner glass that is at least as strong represents a change expected to reduce motorist injuries.

The rigid LM driver's seat was displaced 5 in. to the right and forced the adult driver to his right, away from the intruding striking car. The LM seat isolates the driver from the direct blow of the striking car and moderates the forces applied to his body during impact. Most of the lateral force applied to the driver was by the armrest and shoulder wing of the capsule chair instead of by the restraining harness. This was evident by the relatively minor belt loads registered for the driver: his shoulder strap, 150 lb at 86 ms; his left lap belt segment, 180 lb at 101 ms; his right lap belt segment sustained 520 lb at 95 ms. The driver sustained a puncture injury on his left upper arm from a sharp corner of the capsule chair frame. This normally fatal exposure position (40 mph striking driver's door) appears to have been improved to perhaps no more serious than a moderate injury exposure condition because of the LM capsule seat (Fig. 45).

The right front seat passenger (Fig. 46(a)), was thrown to his left, loading the lap portion of the GM combination belt to 290 lb at 85 ms and sustained an 11 G chest peak at 95 ms (Fig. 46(b)). The shoulder strap registered 370 lb at 130 ms as he flailed his head to the left striking the wing

Fig. 44 - Driver's side impacted, head flails against side window (Experiment 91)

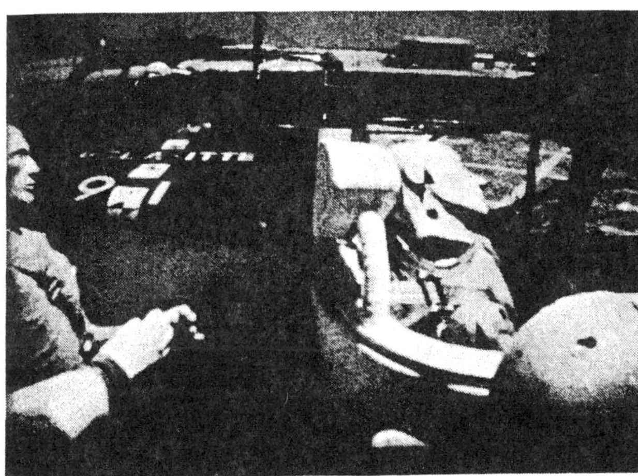

Fig. 45 - Striking car encroachment partially resisted by rigid LM capsule chair

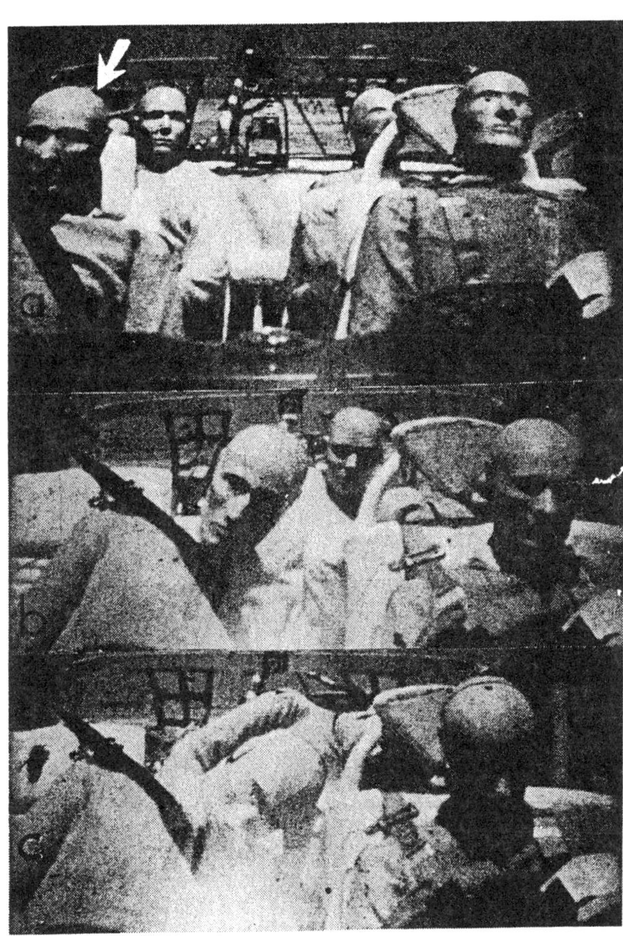

Fig. 46 - Struck car front passenger: a, pre-crash; b, 95 ms; c, head impact (Experiment 91)

edge of the LM seat a significant blow (Fig. 46(c)). This right front passenger restraining action is also shown in Figs. 40 and 41.

The right rear passenger was thrown against the right shoulder of the left rear passenger but his head did not reach the side glass. The lap belt and armrest provided acceptable protection for his particular position of exposure. His right lap belt segment registered only 90 lb at 65 ms (armrest contributed to restraint) and his chest sustained a 15 G acceleration at 75 ms. The magnitude of his head impact was not instrumented, except that motion picture coverage indicated it was not severe.

The left rear seat passenger was thrown to his left, striking his head against the rear doorpost with a 50 G blow at 65 ms; his head rolled forward from the doorpost cracking the glass 6 in. ahead of the rear edge of the window (Fig. 47). The left rear passenger's chest registered 60 G at 30 ms, notwithstanding his full harness restraint. This observation is particularly significant because it provides a controlled comparison of relative collision protection afforded, using conventional seating and an LM full harness restraining system (left rear passenger chest, 60 G at 30 ms) and an LM capsule chair with LM full harness restraining system (left front driver's chest, 23 G at 20 ms). A condition that makes this contrast in protection even more conspicuous is the fact that the striking car was directed at the driver and that while slippage allowed it to work rearward relative to the struck car (Fig. 41), the maximum penetration was associated with the driver's door. The marked difference in collision performance shown by this illustration is attributed to the most significant variable, the protective action of the LM capsule chair.

Striking Car Occupants, Experiment 91 - Contact between the striking and struck cars occurred, as planned, with the striking car centered on the center doorpost, 8-1/2 ft to the rear of the front bumper. As the vehicles crushed together the struck car, owing to its relative lack of penetration, slipped on by and the striking car locked into the struck car at the 10-1/2 ft position (Fig. 41).

The unrestrained driver of the striking 1960 LM Chevrolet was thrown forward against the Liberty Mutual steering wheel. The steering wheel is square shaped with a retractable column. The driver pushed the column inward without encountering significant resistance until his knees struck the instrument panel; his chest thereafter flexed forward against the retracted steering wheel and adjacent instrument panel sustaining a 14 G peak deceleration at 130 ms (Fig. 48). Twenty milliseconds later, his head struck the windshield header sustaining a 26 G blow. A split second later the driver's head, continuing its rightward movement, made a 4 in. scuff mark on the header and was badly slashed by the rear-view mirror bracket (Fig. 49). The driver continued his rightward trajectory until he was flung across the lap of the right front seat passenger. As the driver was pitched against the forward interior structure, his tilt-forward seatback, Ford bucket seat, flailed forward, striking him in the back. The right front passenger, seated in a Liberty Mutual capsule chair with full shoulder and lap restraints, rode out the crash in a relatively uneventful manner (Fig. 50). The LM seat decelerated, reaching a peak of 15 G at 30 ms. The right front passenger, following impact, flailed his head to his right, sustaining a 45 G blow;* his chext peak acceleration was 14 G at 105 ms. His right shoulder strap registered 290 lb; his right lap belt, 380 lb at 105 ms, and the left portion of his lap belt 660 lb at 115 ms. These dynamic

*45 G blow: This right front seat passenger in the striking car, except for the LM chair, would have struck his head against the side glass. However, the side glass was removed by the elimination of the door; motion picture analysis indicates the dummy either hit the LM chair shoulder wing, or reached the limits of neck articulation, or both, to account for the rather high peak head acceleration he sustained.

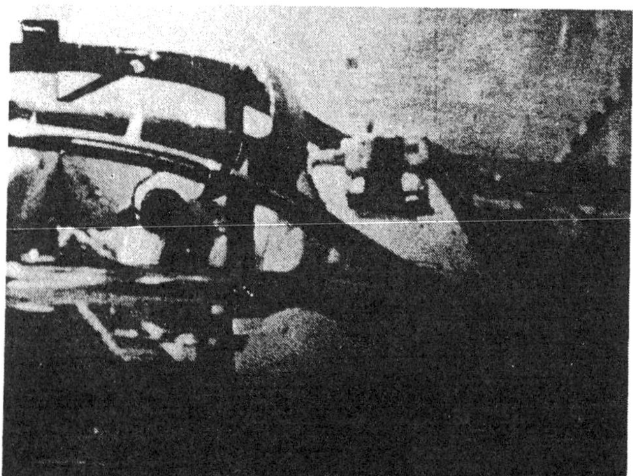

Fig. 47 - Left rear passenger of struck car flails head against side glass in action similar to driver's (Experiment 91)

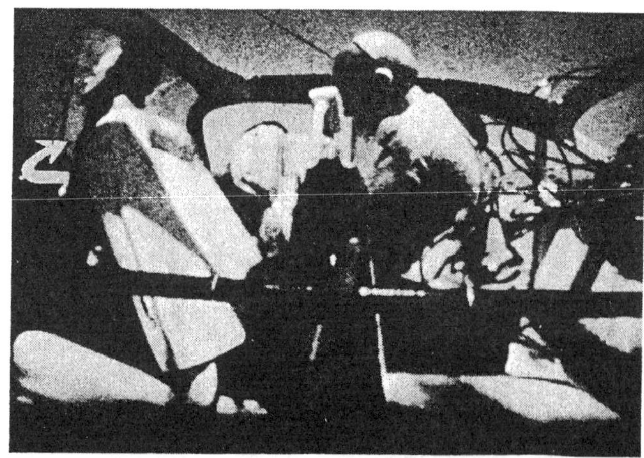

Fig. 48 - Driver in striking car is thrown forward and to his right (Experiment 91)

observations led credibility to laboratory (static) findings that oblique loadings to lap belts do not apply equal forces to the two anchorages.

A 13-year-old, unrestrained in the right rear seat, was ejected through the right rear door opening to forcibly strike the pavement (Fig. 51); the right side doors had been replaced by a compression strut for better viewing of occupants (Fig. 52). Impact with the pavement was primarily with the 13-year-old's head, accounting for the low chest deceleration of 22 G at 3/4 sec after the cars contacted. The rear seat bench cushion pulled free during collision and shifted forward 1 ft.

The other unrestrained 13-year-old in the rear seat, center position, was pitched forward and to his right, striking his shoulder against the left rear of the LM seatback and his head against the metal frame of the head support on this right front LM seat. He sustained a 22 G chest blow at 170 ms.

The left rear adult passenger received his 26 G chest peak deceleration at 130 ms on being thrown initially forward, striking his knees against the already flailed forward backrest of the driver's seat. The subsequent impact with the left front padded edge of the LM seat resulted in a subsequent lower peak deceleration. The adverse action of tilt-seatbacks is also shown in Fig. 40 as the rear seat passenger, for the striking car, is being pitched against the driver's seatback, already slamming forward against the driver's back. Although the left rear adult was thrown forward and to his right, he remained in the rear floor area at the right side.

Post-Collision Observations, Experiment 91 - Following the initial impact, the upper edge of the windshield was

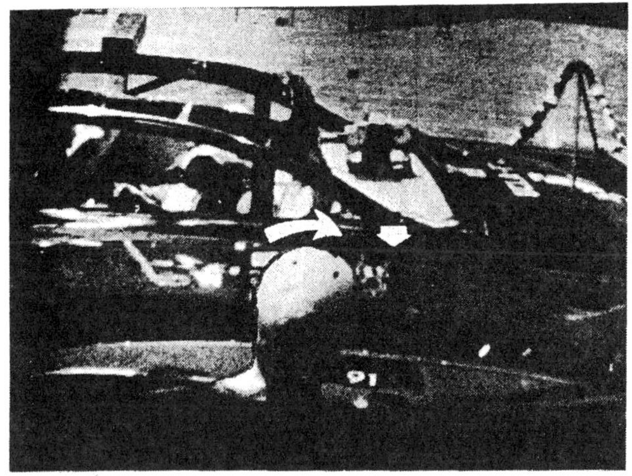

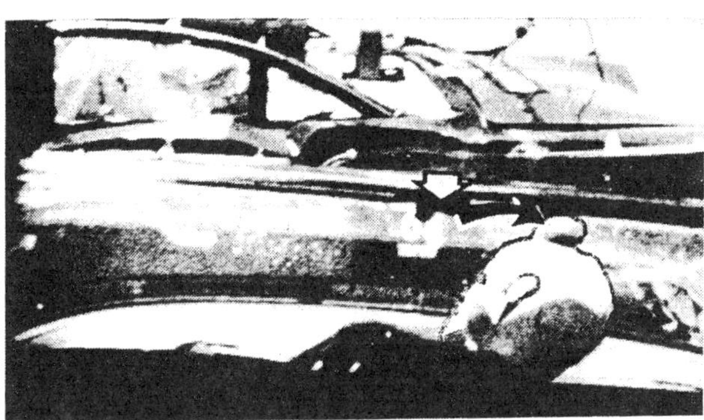

Fig. 49 - Driver of striking car is hurled to his right across rear-view mirror bracket, slashing his scalp (Experiment 91)

Fig. 51 - Thirteen-year-old from right rear seat position ejected from striking car (Experiment 91)

Fig. 50 - Front seat passenger of 40 mph striking car rode out crash uneventfully, securely restrained by his LM capsule chair harness

Fig. 52 - Ejected 13-year-old sustained a relatively low chest deceleration because he struck his head first

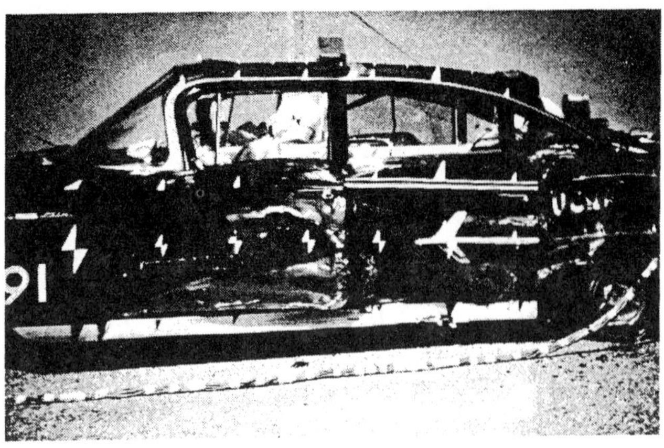

Fig. 53 - Damage to left side was relatively light, considering 40 mph impacting speed (Experiment 91)

Fig. 54 - 1960 Plymouth traveling 40 mph was struck at center doorpost by another Plymouth traveling 40 mph; permanent deformation exceeded 1 ft (Experiment 60, conducted in 1961)

forced outward 3 in. and remained there as the car spun to its position of rest. The struck car came to rest in positions shown in the next section dealing with Experiment 92. The left front door latch failed, the left rear window was cracked from a head impact, and the left rear tire blew allowing the wheel rim to gouge the asphalt. The permanent intrusion to the impacted left side did not exceed 6 in. (Fig. 53). This is in marked contrast to the at least doubled penetration for the struck 1960 Plymouth of prior Experiment 60, owing to the inadequate lower attachment of the doorpost associated with this semi-unit body design (Fig. 54).

The striking car, following collision, came to rest as shown in the collision dynamics chart included in the next section dealing with Experiment 92. Collision deformation to the 1960 LM Chevrolet front end was 11 in. (Fig. 55).

Following the collision, the cars were rolled onto their sides to allow photographs to be taken of their underbodies as a means for illustrating the permanent deformation (Figs. 56(a) and 56(b)).

Fig. 55 - Front-end damage to 1960 LM Chevrolet in 40 mph side-impact collision (Experiment 91)

Fig. 56 (a) - Struck 1960 LM Chevrolet underbody, showing permanent deformation to driver's side (Experiment 91)

EXPERIMENT 92 - This experiment was an intersection collision and, like prior Experiment 91, the striking car impacted the left center doorpost of the struck car. Both vehicles were traveling 40 mph at the time of impact. In place of the two 1960 LM Chevrolets of Experiment 91, the striking car was a production 1960 Chevrolet and the struck car was a production 1966 Chevrolet (Fig. 57). Details concerning dummy occupant assignment, seat types, and instrumentation were presented in the section on Methodology (see Fig. 8 (b)).

On impact, the struck car slid by the contact position 15 in. before positive lock-in of the two vehicles occurred; this observation was contrasted with the 24 in. of slippage for the LM car of Experiment 91. The impacted car's right side leaned low as it spun counterclockwise; this brought the sill into contact with the pavement. In addition, the right rear wheel rim made a gouging contact with the pavement. Maximum penetration of the struck car was 12 in., attended by a frame peak acceleration of 16 G, 20 ms after contact (Fig. 58). The 12 in. penetration (16 G frame peak) for Experiment 92 (Fig. 59) is twice the 6 in. penetration (11 G frame peak) for the companion Experiment 91. Superficially, one would expect the higher penetration to result in a lower frame acceleration. Three out of four of the two pairs of cars involved in the 40 mph intersection collisions (Experiments 91 and 92) were 1960 Chevrolets, each with a frame peak acceleration of 11 G; the fourth (struck car, Experiment 92) was a 1966 Chevrolet that registered a frame peak acceleration of 16 G.

Triaxial accelerometers were consistently mounted on the rocker panel adjacent to the center doorpost. However, this placed them on sheet metal structure for the three "X" frame 1960 Chevrolets, and located them adjacent to the perimeter frame for the 1966 Chevrolet. The force attenuating properties of sheet metal as contrasted with the force transmitting property of the perimeter frame account for the high frame reading associated with the 1966 Chevrolet. Similarly, seats anchored to sheet metal (for example, Experiment 89, right front passenger, 1966 Chevrolet struck car, with perimeter frame and seat attached to sheet metal within frame border) registered 11 G as contrasted to the same experimental plan for Experiment 88, except that the struck car was a 1960 Chevrolet with its "X" frame passing under the right front seat anchorage and the seat so anchored registered 17 G. Consequently, the transducer readings are valid and are representative of frame accelerations even though additional considerations are required for an understanding of the results. The frame peak accelerations are indicated for this right angle impact in Fig. 60.

Struck Car Occupants, Experiment 92 - The driver's GM bucket seat was displaced 11 of the 12 in. of intrusion. This action exposed the driver to the full impact force of the intruding vehicle because his seat did not provide lateral intrusion protection. His chest registered a reduced peak

Fig. 57 - In intersection collision, a 40 mph 1960 Chevrolet is impacted at driver's door by a 40 mph 1960 Chevrolet (Experiment 92)

Fig. 56 (b) - Striking 1960 LM Chevrolet underbody, showing permanent deformation (Experiment 91)

acceleration of 12 G at 45 ms for this exposure because the lateral trace was lost. The driver's diagonal chest belt combination was anchored at the roof rail behind the center doorpost. It coursed downward and forward, across the side of his head and neck, and around his neck forward and downward diagonally across his chest. This driver dummy received a substantial laceration directly across his neck in a manner suggesting a serious injury-producing exposure. His lap belt registered 570 lb at 105 ms. His head struck the side glass a 93 G blow at 40 ms. He appeared to submarine in his seat, ending up in the posture shown in Fig. 57.

The right front passenger of the struck car riding in the safety seat flailed leftward as he loaded his harness. His left shoulder strap registered 255 lb at 120 ms, and his right shoulder strap registered 265 lb at 130 ms. His LM seat was bent slightly to the left but his head did not strike the wing, intended as a protection from side impact, as did the head of the occupant beside the seat for Experiment 91. His chest

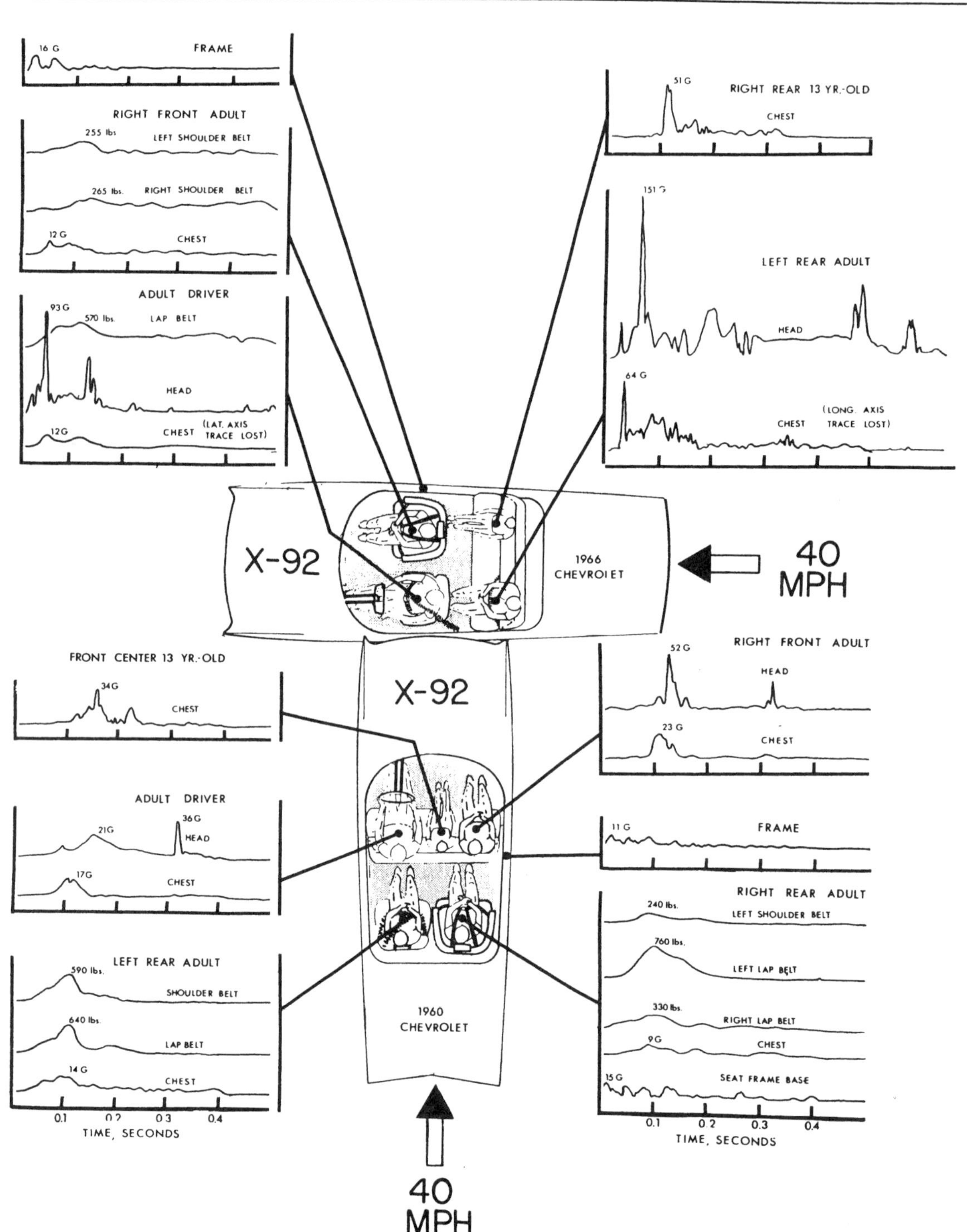

Fig. 58 - Transducer patterns, 40 mph intersection collision (Experiment 92)

registered a 12 G peak at 50 ms, significantly less than the 16 G frame deceleration registered beside his seat. This observation points out the force attentuation afforded the protective "inner shield" for the motorist in an integrated capsule seat, as contrasted with the passenger compartment decelerations representative of the motorist "outer protective shield."

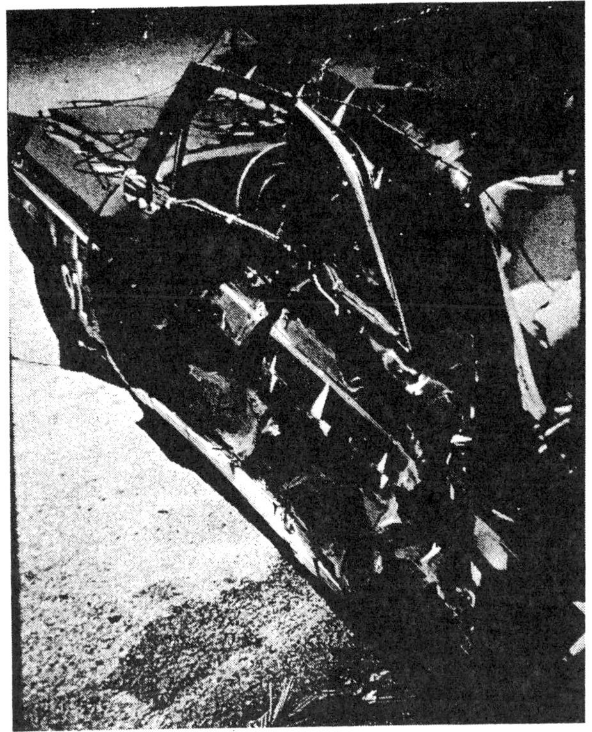

Fig. 59 - In Experiment 92, side intrusion for this 1966 Chevrolet was twice the amount sustained by the 1960 Chevrolet in Experiment 91

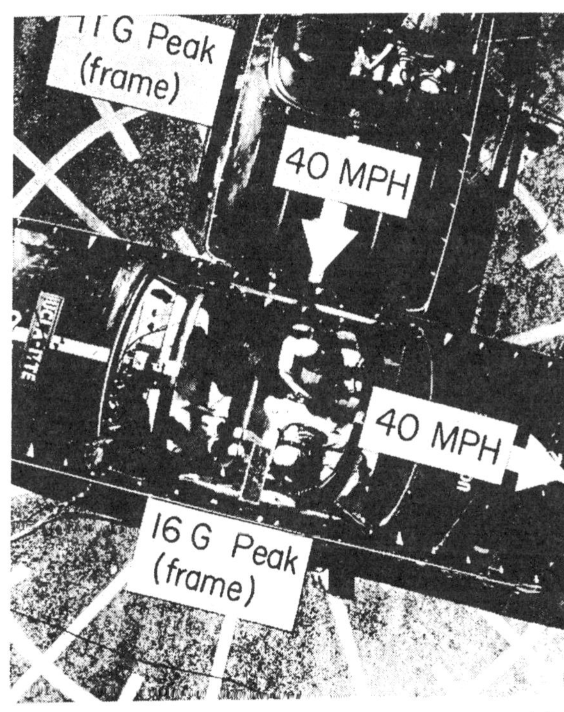

Fig. 60 - Frame peak accelerometer values attained for Experiment 92

The right rear passenger was thrown to the left against the left rear passenger, striking his torso forcibly (chest, 51 G at 110 ms) before rebounding (Fig. 61). This 13-year-old was then thrown to the right rear where he rotated counterclockwise before returning to his own seated position, facing rearward, as though he were looking out the back window (Fig. 62).

The chest of the left rear seat passenger of the struck car sustained 64 G at 40 ms; his head flailed leftward to strike the roof rail a critical blow, 151 G at 70 ms, owing to the height and sloping design of the roof rail behind the rear door window.

Striking Car Occupants, Experiment 92 - For the striking 1960 Chevrolet, the driver was unbelted to allow him to be thrown forward against the steering wheel so as to load the column, instrumented with strain gages.* As the driver of the striking car was thrown forward against the steering wheel, the lower portion of the rim caught him in the ab-

* This data will be included in a subsequent paper along with steering column performance data currently being generated by the 1967 UCLA Rear-End Collision Experiments.

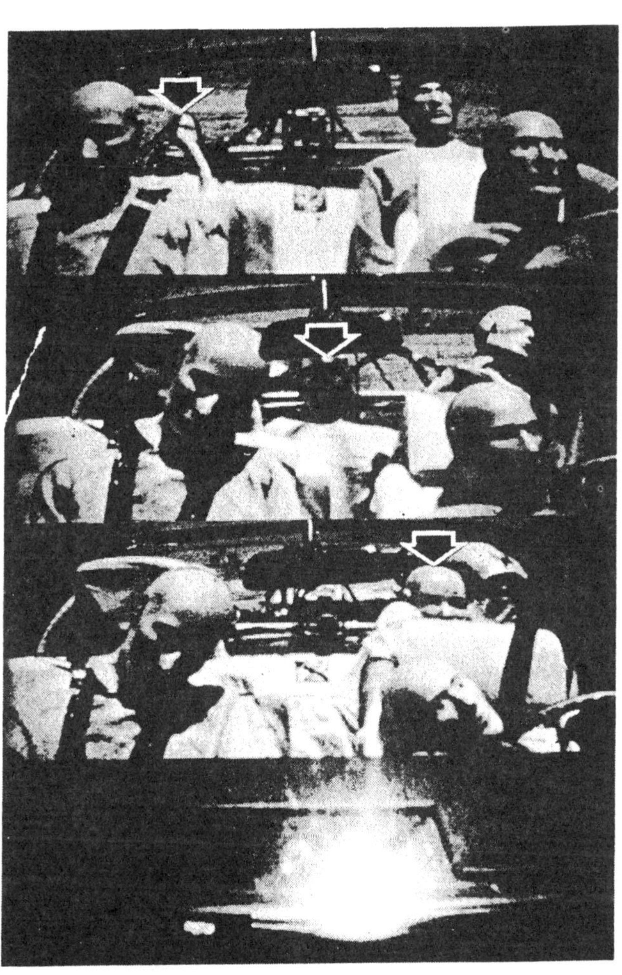

Fig. 61 - Right rear passenger thrown to his left against left rear passenger (Experiment 92)

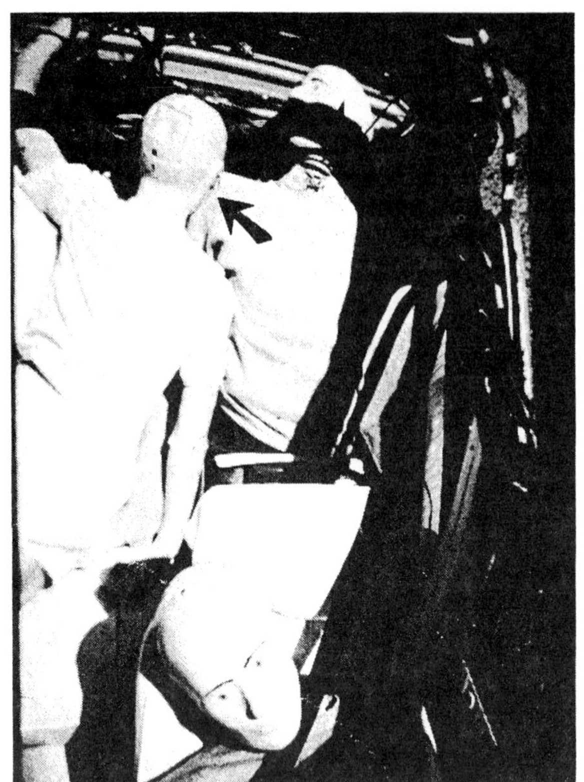

Fig. 62 - Right rear 13-year-old passenger thrown back into his seat position, except facing rearward (Experiment 92)

dominal area (chest, 17 G at 100 ms) tilting the upper section of the rim against the bridge of his nose and across his eyes in a forcible manner (head, 21 G at 150 ms) (Fig. 63). The driver was thrown violently to his right, crushing again the right front seat passenger and sustaining a second head blow of 36 G at 320 ms.

The center front 13-year-old passenger was thrown again the right windshield post and struck his head obliquely, buckling his neck as a result of the head impact and his torso inertia. He received a 34 G blow to his chest at 155 ms; his head did not have electronic instrumentation.

The right front adult passenger was unrestrained, and was forced to his right against the door sustaining a 23 G chest acceleration at 105 ms, and striking his head against the side glass sustaining a 52 G blow at 125 ms. The door remained closed but the glass cracked from the shoulder impact distortion of the window sill.

A Cox seat was installed for the left rear position and an LM capsule chair for the right rear position of the striking car (see Fig. 8 (b)). With respect to this LM chair, the frame of the chair registered 15 G at 10 ms. The adult passenger sustained only a 9 G chest acceleration at 95 ms, typical of the excellent force attenuating performance of this LM capsule chair and restraint system. His left shoulder belt registered 240 lb at 85 ms; his right lap belt, 330 lb at 100 ms, and his left lap belt 760 lb at 100 ms. Motion picture film,

Fig. 63 - Upper rim of steering wheel is deflected into driver's eyes (Experiment 92)

coupled with transducer data, indicated that this passenger rode out the crash under noninjury-producing conditions.

The left rear adult passenger, wearing a Cox diagonal-chest lap belt combination integral with the seat, appeared to be restrained as effectively as the double shoulder harness system performed, throughout the primary impact. However, during the subsequent spin-out of the striking car, the shoulder strap dropped from his shoulder, partially across his chest and near his elbow (Fig. 64(a)), as contrasted to the harness remaining in place for the LM safety seat (Fig. 64(b)). He sustained a 14 G chest acceleration at 95 ms, 640-lb lap belt load at 110 ms, and a 590-lb shoulder belt load at 115 ms. The advantage of the LM system is that, should there be a subsequent collision as this car is spun out to its position of rest, as for example, hitting a fixed object, the LM passenger would be in a position to sustain additional impact forces without injuries likely to occur. In contrast, the left rear passenger in the Cox seat, with a diagonal-chest belt, would have his upper torso and head thrown forward, striking the car interior at this point, owing to the diagonal strap dropping from his shoulder.

Post-Collision Observations, Experiment 92 - The struck 1966 Chevrolet sustained penetration to 12 in. for the left doors, as contrasted to 6 in. for the 1960 Chevrolet, Experiment 91. The doorpost did not tear from the rocker panel. The left rear tire was flat as a result of a puncture by the fender metal. The gas tank anchor strap for the right side of the tank tore loose, dropping the tank to the pavement and causing rupture.

The striking 1960 Chevrolet was compressed at its front 10 in.; the right front tire was flat from bumper penetration (Fig. 65); the right front side glass was broken from impact by the right front passenger's shoulder, and the steering wheel was deformed from the driver being thrown against it.

Collision Dynamics - These two intersection experiments (91 and 92) both involved Chevrolets impacting the center side, with all vehicles traveling 40 mph at impact. They differed only in the type of cars; all cars were 1960 Chevrolets except the struck car in Experiment 92, a 1966 Chevrolet. With these similarities (frame type construction), it is not surprising that the collision dynamics for Experiment 91 (Fig. 66(a)) correspond closely to Experiment 92 (Fig. 66 (b)). When the design of the car is changed (semi-unit body construction) but other conditions are kept constant,

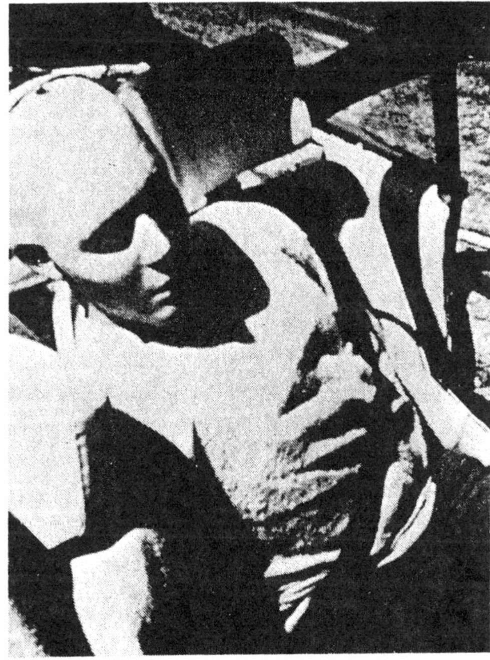

Fig. 64 (a) - Cox cross-chest strap slips from shoulder after impact (Experiment 92)

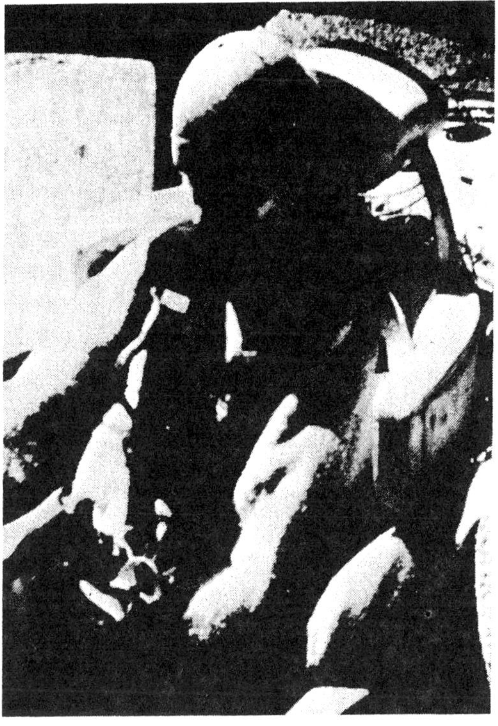

Fig. 64 (b) - Passenger in LM harness, effectively restrained throughout collision (Experiment 92)

Fig. 65 - Front-end damage sustained by 1960 Chevrolet in 40 mph side-impact collision (Experiment 92)

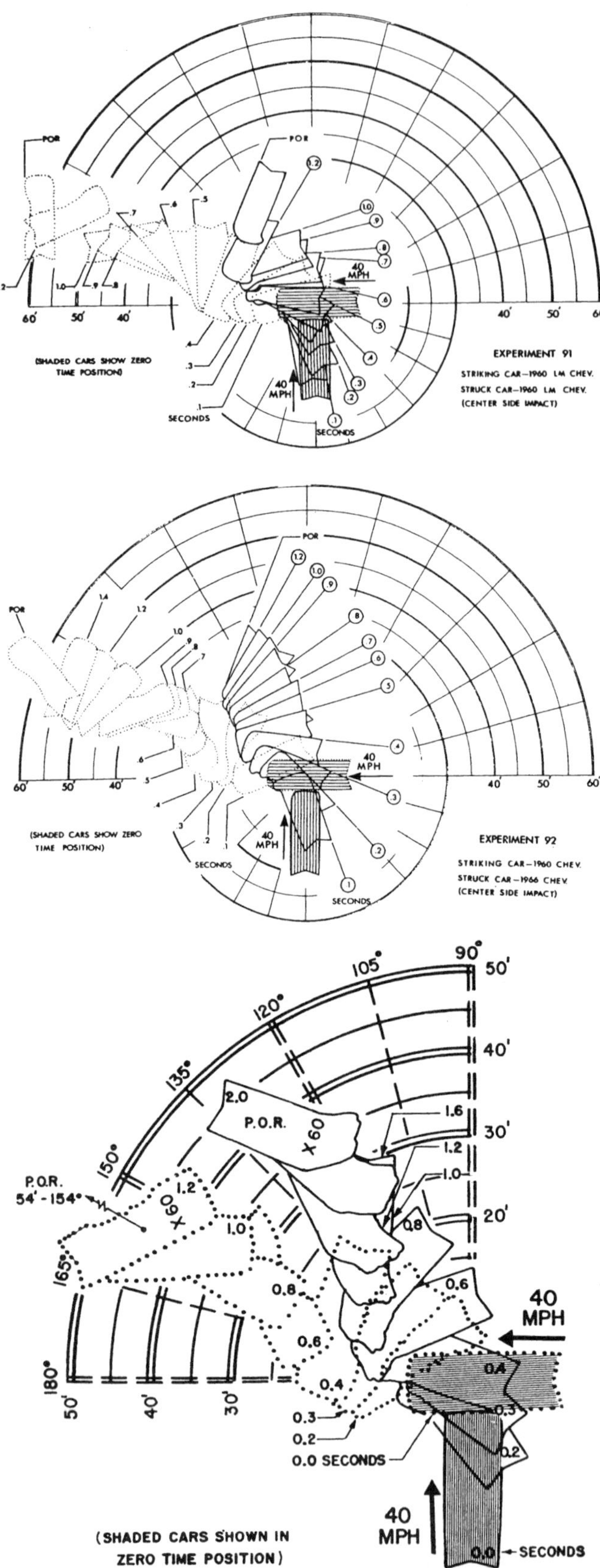

Fig. 66 - Collision dynamics, center-side impact at 40 mph: A, 1960 LM Chevrolet striking a 1960 LM Chevrolet (Experiment 91); B, 1960 Chevrolet striking a 1966 Chevrolet (Experiment 92); C, 1960 Plymouth striking a 1960 Plymouth (Experiment 60)

we find that the collision dynamics correspond rather closely to Fig. 66 (c), where both vehicles remained in the locked-in condition longer owing to the greater penetration of the struck car. The reason for the bias in the pre-crash direction of the struck vehicles, following contact between the cars for this accident configuration (center-side impact), relates to the effective resistance the struck car tires apply to the impacting car's force tending to side-slip the struck car. Contrariwise, the striking car wheels are not being forced sideways and it tends to track around to its left, locked into the side of the struck car until it breaks away and spins out. For this reason the struck car usually travels greater distances than the striking car after contact, except when lock-in time is increased, both vehicles tend to act as one and they approach similar spin-out patterns while expending their kinetic energies.

CONCLUSIONS AND RECOMMENDATIONS

1. The prototype for a safer car design built in 1960 may include concepts and devices that represent substantially improved motorist protection over conventional designs of that year; to be of interest to present-day car designers, such 1960 safety-customized vehicles must be collision evaluated with equivalent current production vehicles. This procedure accounts for the extent, if any, the safety innovations of yesteryear have been embodied within present-day production models.

2. The selection of rear-end and side-impact accident types provided a reasonably complete evaluation of the potential protective qualities of the LM car; in addition, the striking vehicle for these two-car collisions allowed opportunity for evaluating exposures for front-end type impacts.

3. So serious are motorist collision injuries that all reasonable avenues of endeavor should be considered; one such approach has been the evolvement of safety or safer car type specifications. The static tests included in the Liberty Mutual report (4) provide reasonable validation for the important aspects of the safety specifications of the LM capsule seat. Generalized observations, such as "better door latches will reduce ejections," may not require qualifications but specific performance criteria for safer cars should include the specific sources of evidence on which assertions are based. This approach fulfills the responsibility of engineers to establish the basis of credibility of design criteria instead of shifting the burden of proof or rejection to manufacturers who ultimately are expected to transpose a heterogeneous mixture of pure assumptions, partially presumptuous data, and fully factual data into a roadworthy vehicle, enthusiastically acceptable to the motoring public and one that also is crashworthy under a multiplicity of collision exposures.

4. Laboratory simulations of contemplated exposures provide valuable assistance for refining concepts of the design engineer; however, based on experience of researchers of this field, it has been established that many interrelated factors occurring during collision are not anticipated at the

preliminary design level or observed during laboratory tests. In matters as vital as motor vehicle safety, the public interest is best represented by reserving publicity on the anticipated merits of vehicle safety devices or designs until such innovations have been validated through full-scale collision studies.

5. Automotive design engineers, engaged directly or indirectly with the automotive industry, who are intent on serving the public interest by reducing motorist collision injuries should prepare their specifications, with scientific documentation, indicating those areas where they are unable to find or develop studies for reference and the basis, if any, for estimates they include in lieu of valid data.

6. Some safety devices are apparently so advantageous as to appear to need no evaluation before reaching a decision on their use; experience has shown, however, that this position is erroneous, even with respect to the simplest of devices. To represent an improvement worthy of installation, a safety innovation must not, while performing its function, simultaneously diminish the level of safety achieved by other design features, regardless of whether they are more or less important.

7. Interrelated safety factors (ISF) must be considered when making changes in vehicle design to avoid the mistake of improving one factor while deteriorating another. Examples of ISF were presented in Ref. 3, and a typical finding of this nature arising out of this study is given by Conclusion 11.

8. Although preliminary in nature, it appears that steering columns that are not provided with energy absorption performance would be less dangerous if retractable rather than rigid. The telescoped (at the passenger compartment side of the firewall) column, allowed the knees, arms, chest, and head to receive a more distributed impact with the forward structure than is possible when the rigid column acts as a spear-like rearward thrusting, or, at best, rigid projection.

9. The rear-ended 1966 Chevrolet was substantially more collapsed than the rear-ended 1960 Chevrolet impacted under identical conditions; however, the frame peak acceleration for the more rigid 1960 Chevrolet was nearly 50% higher than the corresponding peak acceleration of the newer vehicle. Where increased collapse reduces collision acceleration without violating the integrity of the passenger compartment, reductions in collision injuries will occur.

10. Acceleration measurements for the passenger compartment are influenced by the proximity of transducers to the position of impact and by the type of structure used for anchorage. Accordingly, for purposes of recording general acceleration values for the passenger compartment of the struck car in side impacts, the transducers are placed on the non-struck side of the passenger compartment; additionally, significant differences in passenger compartment acceleration values were observed when transducers were frame mounted versus sheet metal anchored (rocker panel).

Consideration of such potential variations is necessary when contrasting values that may be at variance because of the structure to which the transducers are attached as well as the independent variable under consideration.

11. Bumpers, as presently designed, are little more than parking guards and provide negligible resistance to vehicle impact, even when traveling at moderate speeds. However, the present grossly inadequate height of backrests common to most domestic vehicle seats presents a reason for not strengthening bumpers or, for that matter, seat backrests, until backrests are increased in height and strength to provide full back and head support.

12. For intersection collisions involving the same weight cars impacting at the same speed, the struck car usually travels, after contact, greater distances than the striking car. An exception is that when lock-in time is prolonged both vehicles tend to act as one and expend their kinetic energy in similar spin-out patterns. The position-of-impact to position-of-rest direction is biased in the pre-crash direction of the struck car owing to its more effective resistance to tire side-slip during the initial phases of collision.

13. The LM capsule chair resists collision intrusion (side-impact encroachment) by isolating the motorist from the direct blow of the striking car, and in addition moderates the passenger compartment forces reaching the motorist.

14. The LM capsule chair does not entirely protect the motorist from side glass impacts for heads flailed against the glass, a contact often facilitated by side glass being pushed inward from side impacts. This deficiency is regarded as relatively minor, considering the exceptional overall protective qualities of the LM capsule chair.

15. Thinner, more flexible, side window glass that is not more readily fractured will reduce the impact injuries sustained from side window glass head blows.

16. The conventional car seat, unable to support its own weight during collision deceleration, cannot be used for attaching protective restraints; as a result of this deficiency, seat belts have to be extended to the floor, and shoulder belts to roofs, doorposts, rear seat, or floor points for anchorages. These nonappealing, troublesome locations make their use by motorists less probable and provide compromises in protection for those determined to avail themselves of whatever protection is at hand.

17. Occupant transducer loads of these experiments substantiated findings of earlier studies (1-3) that unrestrained rear seat passengers during collision not only become injured from being hurled about the car's interior, or ejected, but also cause injuries to others they dangerously impact during their forced involuntary movements.

18. The LM head-arresting net for rear seat passengers had excellent performance in preventing whiplash injuries, but the 1 in. webbing somewhat obscures vision through the rear window; a practical compromise may be represented by a lightweight netting that allows better viewability (thin nylon cord).

19. For motorists whiplashed by a rear-end collision, the diminished head clearance for rear seat passengers above and to their rear presents a significant head impact hazard and wedge-like force. Higher rear seat backrest pads ex-

tending onto the rear window shelf would reduce whiplash injuries, providing sufficient clearance exists for rearward rotation of the head without impact to the rear window header.

20. The seat anchored cross-shoulder belt is an effective upper torso restraint; however, during some collisions the strap may drop from the shoulder across the chest thereby depriving the motorist of this protection, should a subsequent impact occur. The LM double shoulder harness does not have this performance problem.

21. Side impacts produce significantly different belt loadings for the left and the right segments of the harness and of the lap belt units. Motorist-to-lap belt friction and body kinematics account for these differences.

22. Most of the lateral force applied to the driver was from his contact with the seat's padded armrest and shoulder wing with relatively little restraint for this direction achieved through the shoulder harness and lap belt.

23. The cross-neck, cross-chest lap belt combination restrains the head and upper torso from striking the steering wheel or other forward structure; however, application of forces and abrasive action across the neck appears to represent a condition that may cause serious injury; additionally, the dummy motorist, restrained in this manner, submarined, and crushed deeper into the seat during collision because of the wedge-like action this belt applied as a result of its anchor point being located above the window level and behind the center doorpost. Based on this limited evaluation, this belt is not recommended for use by the public.

24. Design concepts for protection of motorists from collision injuries are:

a. Restrain the motorist, at least with a seat belt, and position him within a large passenger compartment, one in which the interior surfaces are sufficiently remote to assure the motorist does not strike them, even if only restrained at the hips.

b. Restrain the motorist who is seated within a minimum free volume passenger compartment, portions of which have been hinged to swing aside for access and egress but having surfaces normally very close about the seated motorist, including a head support, but especially from his shoulder level downward. The interior surfaces adjacent to the motorists are contoured and provide energy absorbing properties that reduce the likelihood of injury for motorists crushed against them.

c. A modified capsule or integrated seat for which the immediate proximity of the vehicle interior is of secondary importance owing to the rather excellent protection afforded the motorist within the protective confines of the integrated semi-enveloping chair.

Concept a is restricted to large vehicles; concept b, while unquestionably effective, is inconvenient and congested; and concept c has neither of these objections and is readily adaptable to current vehicular designs. This was the concept pursued by the LM engineers because it provided a nominally nonpenetrable protective structure in the form of a seat enveloping the motorist from the sides and rear, including a built-in full harness restraint for frontal collision forces. The head support was also structured to resist roof collapse.

25. The LM seat occupant of the rear-ended LM car showed no evidence of exposure to injury because of the protection derived from his contoured full back seat with head support; the pedestal base flexed at its juncture with the floor pan tilting the LM chair about 30 deg rearward, but the seated dummy maintained normal posture throughout the impact.

26. The conventional automotive seat corresponds approximately to a lightweight, padded living room slipper-chair; it has a readily compressible, bundled-wire substructure support, attached to a lightweight tubular frame that is difficult to anchor effectively and that provides inadequate body support in any direction.

27. The inadequate front seat anchorage strength and backrest strength tend to reduce the severity of whiplash from rear-end collisions. This condition does not represent a consistent or satisfactory or comprehensive solution to the rear-end collision injury problem because: low speed impacts do not provide sufficient force to break the anchorage or backrest, resulting in serious whiplash; high speed impacts may force the front seat passenger up the plane of the backrest to whiplash him or break the seatback, releasing him to the rear seat area or even out the rear door, notwithstanding the use of a lap belt; neither of these extremes represent acceptable or satisfactory solutions to a serious and very frequently occurring type of accident.

28. Interoccupant impact injuries may occur when the front seatback yields excessively and allows the occupant to be flailed rearward to strike the rear seat passenger during rear-end collisions.

29. Backrest rebound of a rear-ended, front seat motorist may pitch him forcibly against the forward structure of the car interior; injuries from this situation can be prevented if seat or backrest yield is designed to respond plastically, rather than elastically. Good plastic deformation occurred in the floor pan attachment for the rear-ended LM capsule chair.

30. Tilt-forward backrest designs serving to accommodate rear passenger entry and exit provide serious impacts to the mid-back of front seat occupants; such action diminishes the restraint afforded rear seat passengers hurled forward against tilt backrests.

31. A rigid seat frame, designed along the lines of the LM capsule chair, while not capable of resisting the side impact intrusion forces at 40 mph, does provide a protective shield for the motorist that partially resists collision intrusion and that otherwise becomes displaced sideways, carrying the encradled motorist with it away from the intruding vehicle. This concept is sound and merits serious consideration with respect to future designs.

32. A contoured bucket-type seat provides resistance to lateral forces; prior studies (3) as well as this study have established the value of well-padded but strong armrests and the value of backrest contouring (shoulder wings) as a protection against lateral collision forces.

33. The LM capsule chair concept provides a nominally

nondeformable structure that encircles the significant areas for motorist protection, providing a within-the-car protective shield for the motorist from excessive passenger compartment forces and distortions brought about by extreme outside collision action.

34. In addition to prevention of direct intrusion force from reaching the motorist, the LM capsule chair reduced the peak acceleration transmitted to the motorist to nearly one-third the amount sustained under the same conditions for a motorist with a full harness restraint but seated on a conventional seat.

35. The Cox safety seat embodies many of the advances in motorist protection afforded the LM capsule chair (for example, shoulder and lap retractable inertial reel units attached to a strong, firmly anchored, contoured seat frame, full backrest with head support) but does not provide as much lateral support or resistance from collision intrusion as does the LM chair. The diagonal chest strap of the Cox seat can slip from the shoulders but the LM harness remains correctly placed during and following collision.

36. Better padding is needed for the armrest and shoulder wings of the LM chair owing to the important role they play in side impacts and the high forces they apply to the motorist during such exposures. The shoulder wing top surface requires special attention to minimize injuries from heads flailed sideways onto that structure.

37. The time has long passed by when the need was clearly apparent and the design criteria adequately available for an integrated safer car seat to become a production item for the automobile, a car seat that provides an inner protective shield about the motorist, functioning in a manner similar to the outer protective shield afforded by the motorist compartment.

38. An integrated safety seat includes the following design criteria:

a. A tubular metal seat frame or unit body constructed seat that includes a full height backrest and head support (a 1967 series of UCLA rear-end collision experiments will provide seatback dimensions) capable of resisting permanent seat deformation not exceeding 2 in. for collision forces of 9000 lb in the forward direction, 7000 lb for the side impact direction, and 5000 lb for the rear-end direction; these seat design loads provide for a 100-lb integrated seat, and a 200-lb motorist harnessed to the integrated seat undergoing moderate speed collision exposures (approximately 30 mph).

b. Cross-chest belt or double shoulder harnesses with upper anchorage installed at shoulder level within the seatback and provided with an inertia snubbing, automatic retroacting device; similar devices and design will be applied to accommodate the lap belt portion for the 3- or 4-point harness.

c. A tubular seat frame that includes side structure for support of padded armrests and for resisting collision intrusion below the armrests to the base of the seat; shoulder-wing structure should also be provided that extends 5 in. forward of the backrest.

d. Padding arranged for comfort and, to the extent possible, for distributing collision forces applied to the motorist.

e. An anchorage system for the seat that does not allow appreciable yield, one that collectively, with the seat yield, limits deflection to 6 in. for 9000-lb force in the forward direction operating through the combined motorist/seat center of mass, to 3 in. for 7000 lb applied laterally, and 4 in. for 5000 lb applied in the rearward direction.

f. Fixed integral head supports capable of sustaining 2000 lb vertically downward as an assist for reducing roof collapse during rollovers.

39. The majority of vehicles operating, urban and rural, carry only a driver; where two occupants are present the passenger, for a majority of the vehicles, is sitting in the right front seat position. The front left and the front right seat positions are the most important in the car. For practical purposes, the remaining seats in the automobile could continue to be of conventional design, with improved anchorages and equipped at least with lap belts.

40. The automobile seat is a restraining device. A poorly designed seat and seat anchorage system becomes an injury-producing agency during collision; contrariwise, a properly designed integrated seat system represents the most important safety feature that may be provided for the motorist. At the time of this writing (1967), the first integrated safety seat for a production car has yet to be produced, domestic or foreign.

ACKNOWLEDGMENTS

The authors express their deep appreciation to the U. S. Public Health Service for the substantial financial support of the project through a research grant to the University of California; to Liberty Mutual Insurance Co. and General Motors Corp. for donating the collision vehicles; to the city of Los Angeles for making the Test Site and facilities available.

Acknowledgment is made for the valued services of the entire project staff, particularly the exceptional engineering services of Hans Jakob and David Blaisdell; William Archibald, John Kerkhoff and Bernard McGuire for data reduction, photography and engineering illustrations; mechanicians James Currey, Virgil Sibell and Ross Sater for a wide range of experience; and especially Louise Kirkpatrick for the excellent secretarial service.

Acknowledgment is also made of the highly competent and exhaustive efforts of the UCLA Engineering Reports Group, under the leadership of Estelle Dorsey Ratner. Finally, and not least, acknowledgment is made of the devoted services of the part time University student workers.

REFERENCES

1. D. M. Severy, J. H. Mathewson, and A. W. Siegel, "Automobile Head-On Collisions, Series II." SAE Transactions, Vol. 67 (1959), pp. 238-262.

2. D. M. Severy, J. H. Mathewson, and A. W. Siegel, "Automobile Side-Impact Collisions, Series II." SAE SP-232, 1962.

3. D. M. Severy, H. M. Brink, and J. D. Baird, "School Bus Passenger Protection." Presented at SAE Automotive

Engineering Congress, Detroit, January 1967, paper 670040.

4. F. J. Crandell, A. Gordon, F. E. Burkett, and C. F. Sparrell, "Packaging the Passenger." Liberty Mutual Insurance Co., 1965.

5. F. W. Babbs and B. C. Hilton, "The Packaging of Car Occupants - A British Approach to Seat Design." Proceedings, Seventh Stapp Car Crash Conference, Springfield, Ill., Charles C Thomas (1965), pp. 456-464.

6. B. C. Hilton, "The Development of Safety Seating for Injury Prevention." Proceedings, 10th Stapp Car Crash Conference. New York: Society of Automotive Engineers, Inc. (1966), pp. 124-131.

APPENDIX A

ANTHROPOMETRIC DUMMY PASSENGER SPECIFICATIONS

Ref. No.	Age Group	Ht., in.	Wt., lb	Manufacturer	Model No.
1	Adult	72	195	Sierra Eng. Co.	120
2	Adult	68	200	Sierra Eng. Co.	15
3	Adult	68	200	Sierra Eng. Co.	15
4	Adult	68	175	Sierra Eng. Co.	135
5	Adult	69	170	Sierra Eng. Co.	262
6	Adult	72	195	Sierra Eng. Co.	292
7	Adult	72	195	Sierra Eng. Co.	292
1	13-yr	61	115	UCLA	13-SP*
1	6-yr	46	42	Sierra Eng. Co.	492-06
1	3-yr	38	32	Sierra -UCLA	492-01

*13-SP, 13-year-old, Solid-Pour construction, anthropometric dummy

APPENDIX B
TRANSDUCER DATA (PEAK VALUES)

Transducer Location	Rear-End Collision Experiments				Side-Impact Collision Experiments			
	Experiment 88		Experiment 89		Experiment 91		Experiment 92	
	Striking	Struck	Striking	Struck	Striking	Struck	Striking	Struck
	1966 Chev.	1960 LM Chev.	1966 Chev.	1966 Chev.	1960 LM Chev.	1960 LM Chev.	1960 Chev.	1966 Chev.
Vehicle Frame	11G	17G	13G	10G	11G	11G	11G	16G
Driver								
Head	--	51G	--	9G	26G	70G	36G	93G
Chest	12G	33G	10G	9G	14G	23G	17G	**12G
Pelvis	--	11G	--	9G	--	--	--	--
Shoulder Strap	--	--	--	--	--	*150/-lb	--	--
Lap Belt	270 lb	90 lb	200 lb	60 lb	--	*180/520 lb	--	570 lb
Seat Frame-Base	--	10G	--	10G	--	--	--	--
Front Center Pass. (13-yr)								
Chest	--	--	--	--	--	--	34G	--
Right Front Pass.								
Head	--	12G	--	11G	45G	--	52G	--
Chest	15G	11G	13G	8G	14G	11G	23G	12G
Shoulder Strap	690 lb	--	--	--	*280/290 lb	370 lb	--	*255/265 lb
Lap Belt	300 lb	150 lb	220 lb	135 lb	*660/380 lb	290 lb	--	--
Seat Head Support	--	12G	--	12G	--	--	--	--
Seat Frame-Base	--	--	--	11G	15G	--	--	--
Right Rear Pass.								
Head	--	30G	--	26G	--	--	--	--
Chest	--	15G	--	12G	22G	15G	9G	51G
Lap Belt	--	--	--	--	--	*-/90 lb	*760/330 lb	--
Shoulder Strap	--	--	--	--	--	--	*240/-lb	--
Seat Frame-Base	--	--	--	--	--	--	15G	--
Rear Center Pass. (13-yr)								
Chest	--	--	--	--	22G	--	--	--
Left Rear Pass.								
Head	--	18G	--	14G	--	50G	--	151G
Chest	--	18G	--	11G	26G	60G	14G	64G
Shoulder Strap	--	--	--	--	--	--	590 lb	--
Lap Belt	--	--	--	--	--	--	640 lb	--

*Left side/Right side
**Incomplete value (trace lost)

W. G. CICHOWSKI
General Motors Proving Ground, General Motors Corp.

THE AUTHORS of this paper should be complimented on their very interesting presentation. In addition, compliments should be extended to the U.S. Bureau of Public Health, Liberty Mutual Insurance Co., and the University of California for providing funds, equipment, and man power for this test series. The films and pictures used by the authors of this paper give vivid portrayals of what is happening during impact in these types of collisions. Some of the physical testing techniques that the research engineers of the Institute of Transportation and Traffic Engineering of the University of California use have been followed by the people in the industry that are in the accident simulation field.

I would like to expand upon several areas mentioned by the authors. A similar series of rear end impact tests have been conducted at the General Motors Proving Ground using car-to-car impacts and the SAE J972 Standard moving barrier as the bullet car.

This series of tests were conducted at various speeds from 2.6 mph to over 26 mph to evaluate a typical vehicle in terms of the acceleration pulse that it feels during this type of an impact. The impact into the rear end of the vehicle puts a force impulse into it. The occupants then respond to this pulse.

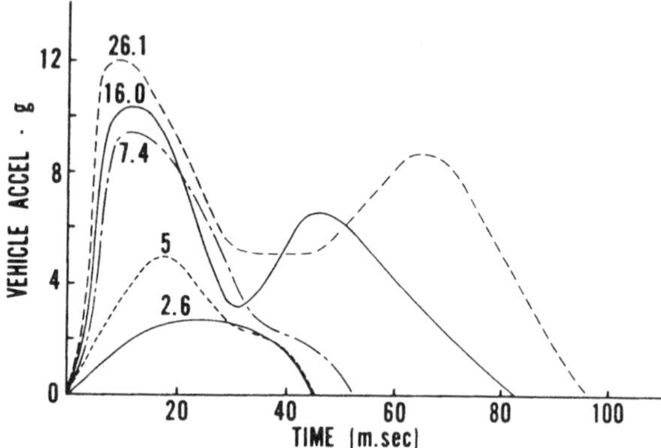

Fig. A - Relative velocity, mph

Fig. A shows a series of faired curves that were generated from some of the tests conducted at the Proving Ground. As the relative velocity of the impact increases, the acceleration trace takes on a typical pattern. The initial peak rises and the time duration of the impact is extended. As the vehicle is subjected to a greater relative velocity impact, the secondary acceleration peak appears.

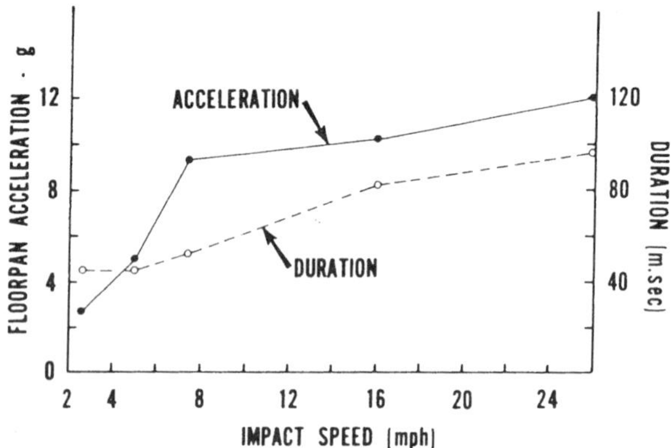

Fig. B - Peak acceleration points versus impact speed and time duration

By plotting the peak acceleration points versus the impact speed and the time duration as shown in Fig. B for these impacts, we begin to notice a pattern. At the lower velocity impacts, most of the reaction between the vehicles is an elastic impact. As the impact speed approaches 7-1/2 mph, the plastic deformation of the impacted vehicle begins. The acceleration experienced then by the impacted vehicle increases at a much lower rate with the increase in relative speed because of this plastic deformation. By the establishment of these test parameters, it was then possible for us to reproduce these typical rear end impacts on the impact sled at the General Motors Proving Ground.

Fig. C is a view of the General Motors Impact Sled with the vehicle oriented in a position to simulate a forward impact. Rear impacts are simulated by simply reversing the vehicle upon the sled.

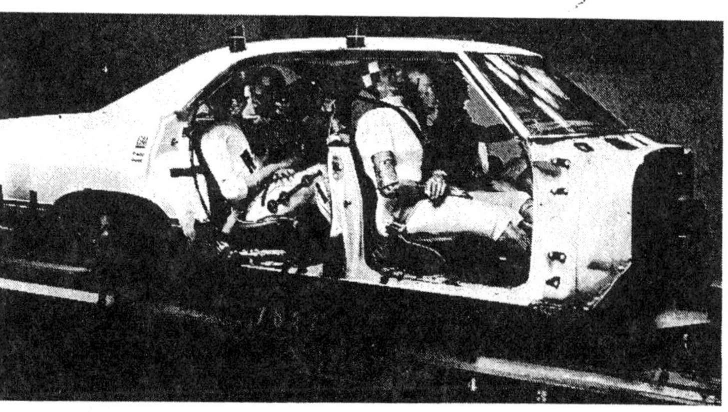

Fig. C - GM impact sled

sion can extend past 60 deg aft of the torso line. With some head rotation, either to the right or left, the limitations of voluntary motion may reduce this limit to around 45 deg. On this basis, our dummies are calibrated to begin a linear stiffness at a greater rate at approximately 30 deg as shown by the lower curve on this figure.

Fig. H is a view of a dummy seated on a production seat on the impact sled. You will note the bar attached to the dummy chest which is used as a reference plane parallel to his torso line. Fig. I is a view of this dummy after being subjected to a simulated rear impact. These before and after shots do not really reveal what happens in between. Fig. J, taken during the peak excursion of the dummy, shows what the action of the dummy was during this rather high g level acceleration pulse.

Let's study some of the parameters of "whiplash" during a series of tests which we conducted that very closely duplicate the acceleration levels described in the UCLA paper.

The sequence of Figs. K-N are of a 95th percentile dummy subjected to two variables:

1. Restrained or unrestrained.
2. Headrest up and down.

The first, Fig. K, portrays the dummy motion during the peak head extension of a 95th percentile dummy on a production 1966 GM seat with the headrest down. He was unrestrained during this test.

Fig. L is a repeat of the same test but with the dummy now restrained by a lap belt. The dynamic motion of the dummy is now restricted somewhat and, as I will explain later, the head rotation angle is increased.

Fig. M is a repeat of the same test series with the headrest now extended and the dummy again unrestrained. Note more closely confine the torso and the head rotates over the top of the headrest ass

With the headrest up and adding a lap belt restraint, we more closely confine the torso and head rotates over the top of the headrest as shown in Fig. N.

At first appraisal, this data boils down to a "scattergram" Fig. O. This diagram includes more tests than I have shown you and there were other variables such as different seat back heights and different bending moment stiffness of the seat back.

By filtering out specific factors, it is possible to evaluate

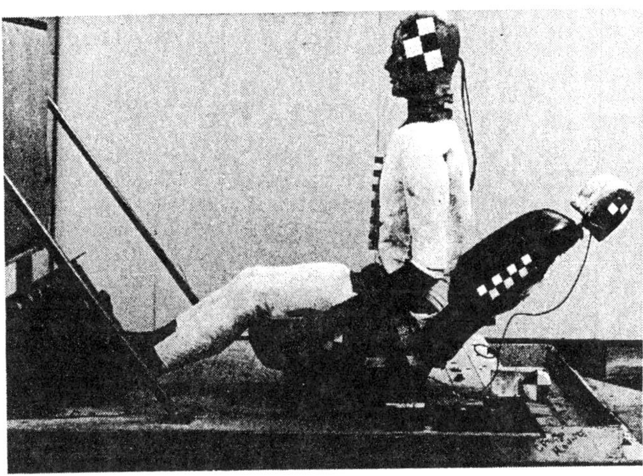

Fig. I - View of dummy after simulated rear impact

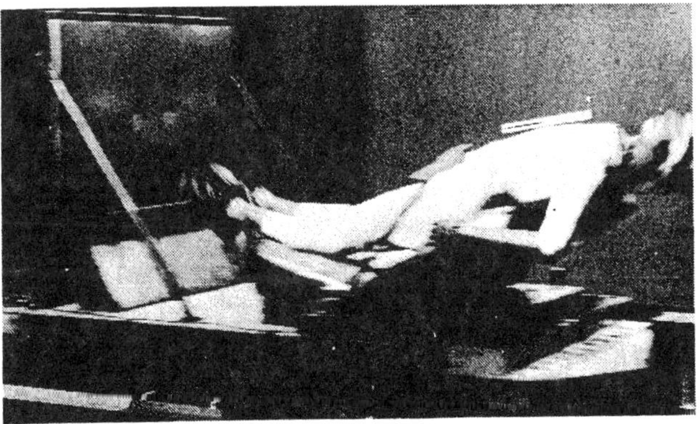

Fig. J - Peak excursion of dummy shown in Figs. H and I

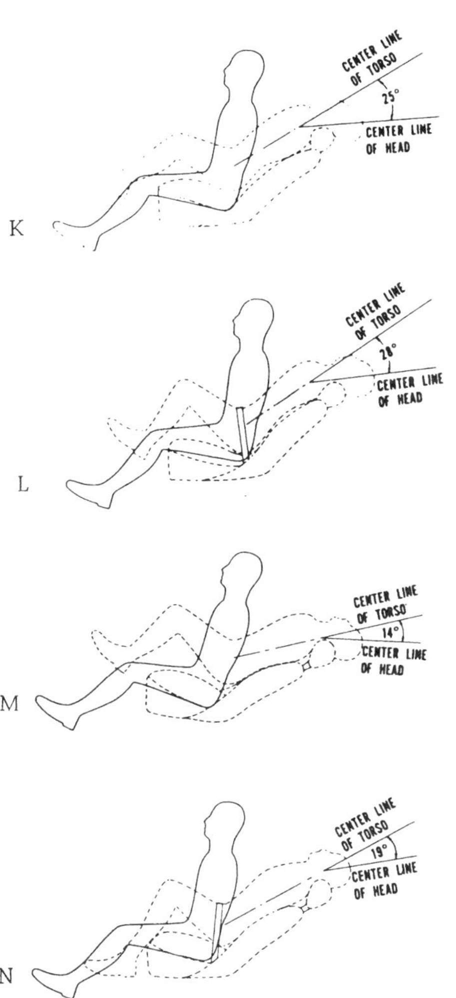

Figs. K, L, M, and N - 95th percentile dummy subjected to two variables: Restrained or unrestrained; headrest up or down

The anthropomorphic dummies used for whiplash studies in our testing are calibrated for neck motion. Figs. D and E are portrayals of how this calibration is done statically. A force is applied through the center of gravity of the head and the head motion or angle of rotation is related to degrees aft of the torso line. Notice the two levels placed upon the dummy to measure their angle.

Complicated mechanisms, manufactured by the Alderson Research Labs., Inc., as shown in Fig. F, are available to simulate a human neck. These necks enable us to more closely simulate the action of a human neck during rear impact.

One of the more difficult tasks during studies of human simulation is the calibration of our test equipment. These calibration curves (Fig. G) of various dummy necks that have been used show the variation that is possible using different necks or different settings on a particular neck.

Our studies indicate that voluntary motion in hyperexten-

Fig. D - Anthropomorphic dummy used for whiplash studies

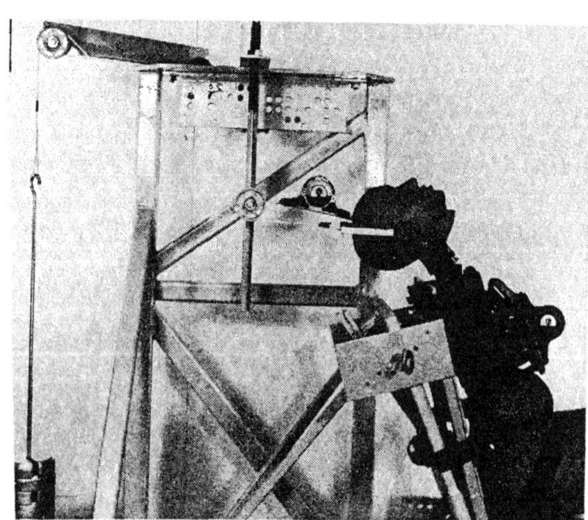

Fig. E - Closer view of Fig. D showing calibration for neck motion

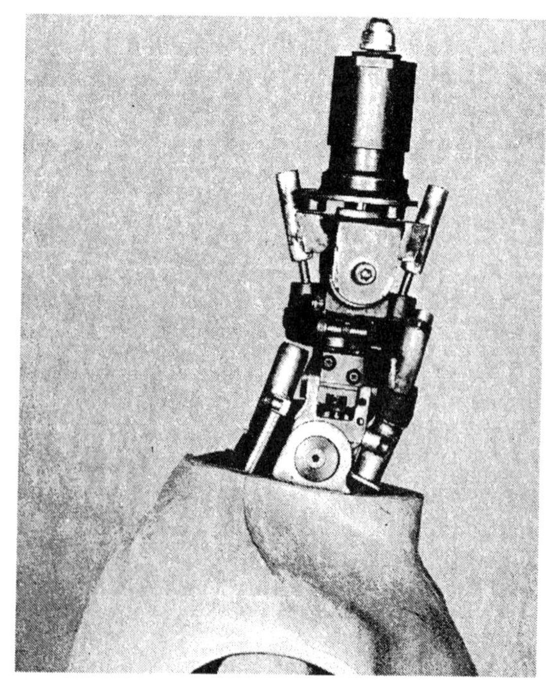

Fig. F - Mechanism simulating human neck

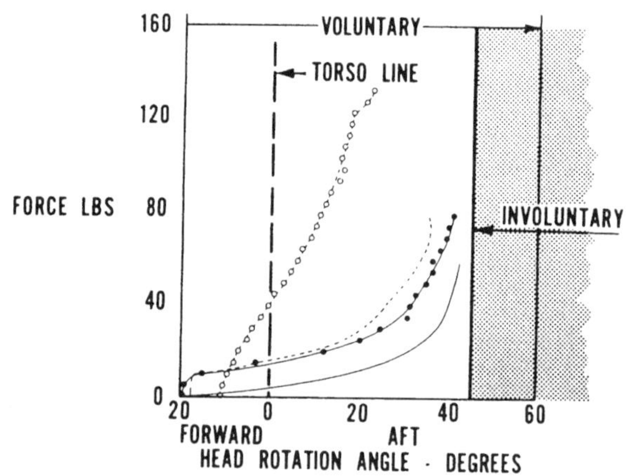

Fig. G - Calibration curves of various dummy necks

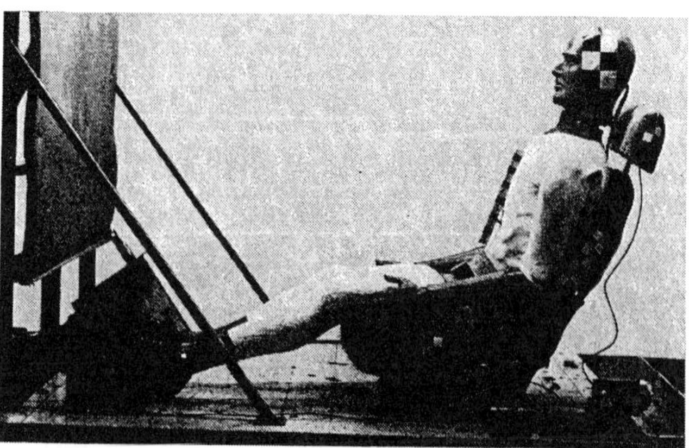

Fig. H - Dummy seated on production seat on impact sled

the various parameters during these tests. Fig. P shows that the use of a lap belt restraint tends to increase the rearward head rotation angle aft of the torso line with the same seat back height.

We also found (Fig. Q) that the more deformable seat back increased the extension. This is caused by the dummy sliding up the seat back. You will note that the lines pass through the abscissa. This indicates that the opposite or flexion of the neck can occur if a headrest is too high for the occupant. I would like to point out that in none of these tests portrayed did we even approach the limits of voluntary head motion.

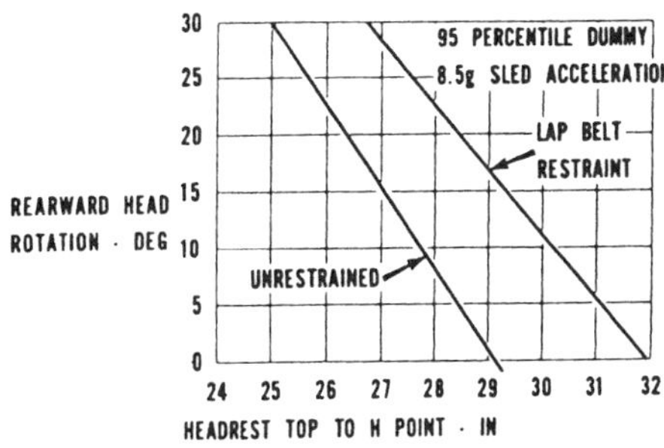

Fig. P - Effect of lap belt restraint

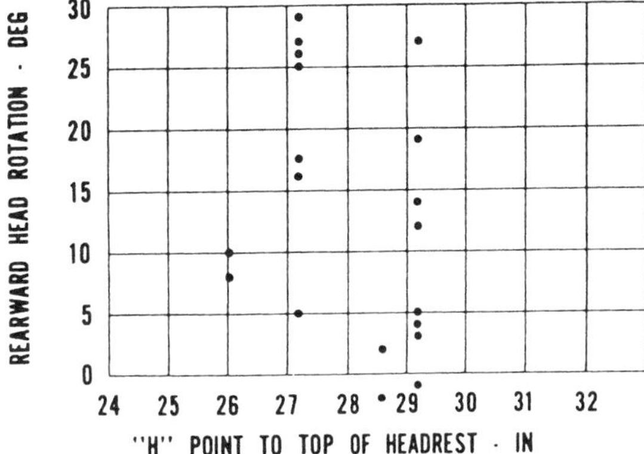

Fig. O - Headrest height evaluation

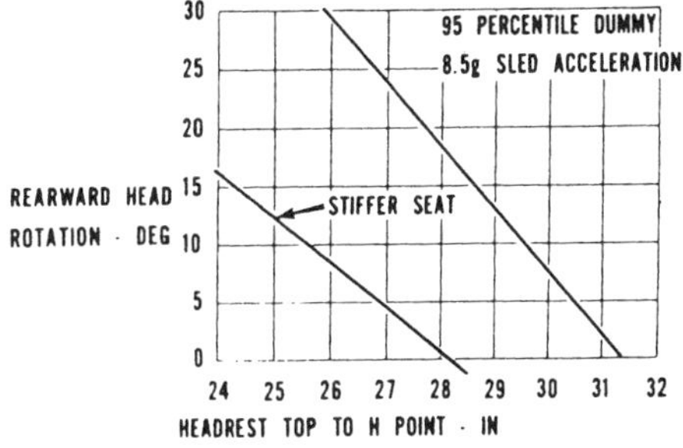

Fig. Q - Effect of seat back stiffness

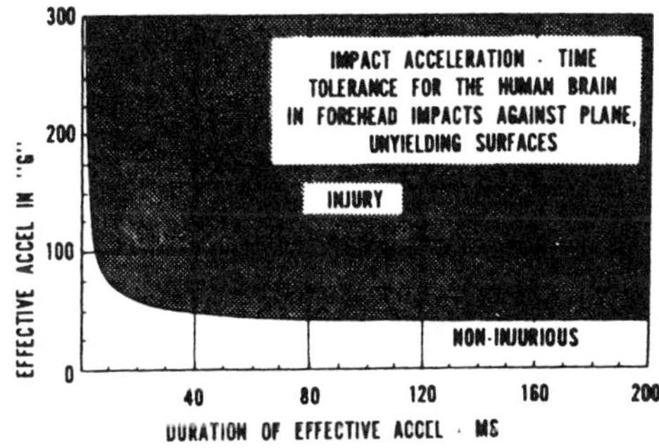

Fig. R - Data for head impacts on flat, unyielding surface

Occupant injury during any impact cannot be evaluated by one parameter alone. The investigator must consider all applicable parameters. He must look at the load applied and its duration. He must also analyze the distribution of force or the area of contact. Another parameter to be considered is the rate of onset or jerk. In the case of head "whiplash," this parameter could be the rate of change of angular acceleration the head experiences during its excursion.

Head injury has been shown to be related to these parameters by the curve shown in Fig. R. This curve was derived by a series of tests conducted on cadaver skulls at the Wayne State University. Data for this curve were obtained from head impacts on a flat unyielding surface.

By application of a mathematical formula,

$$(1) = \int_{t_1}^{t_2} a^n \, dt$$

where:

a = Acceleration

t_1 = Beginning of pulse

t_2 = End of pulse

n = Arbitrary exponent

a number or severity index can be computed for head injury. This concept developed by GM is related to the previously shown graph. This mathematical formula is essentially a means of relating the peak acceleration on a higher emphasis along with the time duration that the force impulse is applied.

The application of this mathematical formula with the

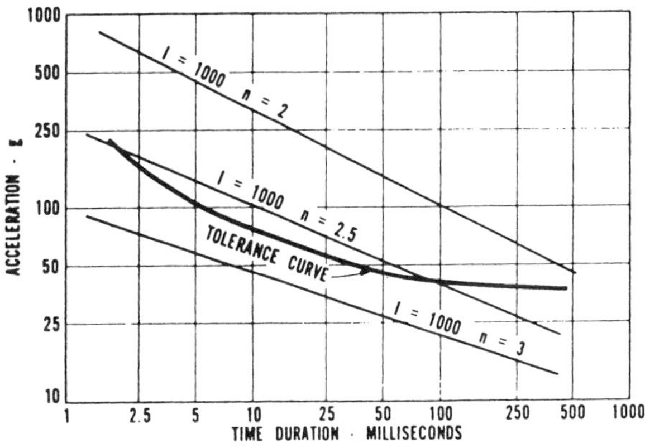

Fig. S - Severity index versus tolerance data

various exponents "n" can be shown by Fig. S. The exponent "n" in the formula is a function of the slope of the line plotted on log graph paper. When "n" is equal to approximately 2.5, the slope of the line closely approximates the previously shown impact tolerance curve.

When this criterion of severity index was applied to the rear impact data that were obtained during the UCLA test, none of the values calculated approached any level which may be considered injurious to life. The values calculated from the UCLA data were approximately 18.

Table A - 1967 GM Safety Features

Energy Absorption	Instrument panel and padding steering wheel and steering column
Minimizing Injury Potential	Improved instrument panel control knobs and gearshift lever and knobs Soft vinyl sun visor hooks, coat hooks, and low profile window regulator knobs Inside rearview morror with vinyl clad breakaway base and vinyl edge bezel
Structural Improvements	Safety door latches and hinges Passenger-guard door locks Stronger seat structure and anchorage Stronger fuel tank retention Folding seat back latches
Operational Features	Four-way hazard warning flasher Dual master cylinder brakes with warning light Larger tires for single inflation pressure Push-button seat belt buckles with front retractors Day-night mirror Lane change feature in direction signal controls

We have made several surveys of the available data trying to evaluate the various criteria for "whiplash" injuries. We have found no significant statistics other than an indication that the right front seated occupant is more susceptible to this type of injury.

It is interesting to make a comparison of the Liberty Mutual car features with those that are now available in the 1967 Chevrolet (Table A). Fuel tank retention and location and construction have been discussed at great lengths. At the General Motors Proving Ground, during our full scale impact tests, which we conduct with the fuel tank filled with gasoline and the engine running and all electrical systems on, we have had only one fire occurrence. This resulted from a very severe impact. This fire, however, was confined to the carburetor air cleaner and did not proceed any further. Our experimental data are substantiated by ACIR data to the effect that vehicle fires, as a result of impact, are of very low incident. In fact, vehicle fires presently occur in less than 1/2 of 1% of all vehicle accidents reported and a good share of the fires are electrical in nature.

The Liberty Mutual Safety Car II was basically a 1960 Chevrolet sedan. The door latches in the 1966 Chevrolet used in this test series had a greatly improved door latch mechanism, approximately two times stronger in push-out strength than the 1960 model. The interlock or longitudinal strength of this new latch is about 22-1/2 times stronger than the 1960 version. This accounts for the greatly increased performance of the door latch assemblies during side impacts as verified by the UCLA side impact test series.

The laws of physics are irrevocable. The basic formula, $V^2 = 2\,gh$, as it applies to velocity and acceleration, can be thought of in terms of jumping from the top of a tall building.

A free-fall from a building 107 ft tall will end in a ground velocity at impact of 56.5 mph; this is the vectorial sum of two elastic colliding bodies each traveling at 40 mph at 90

Fig. T - Dummies used during impact test series (authors' experiment 88)

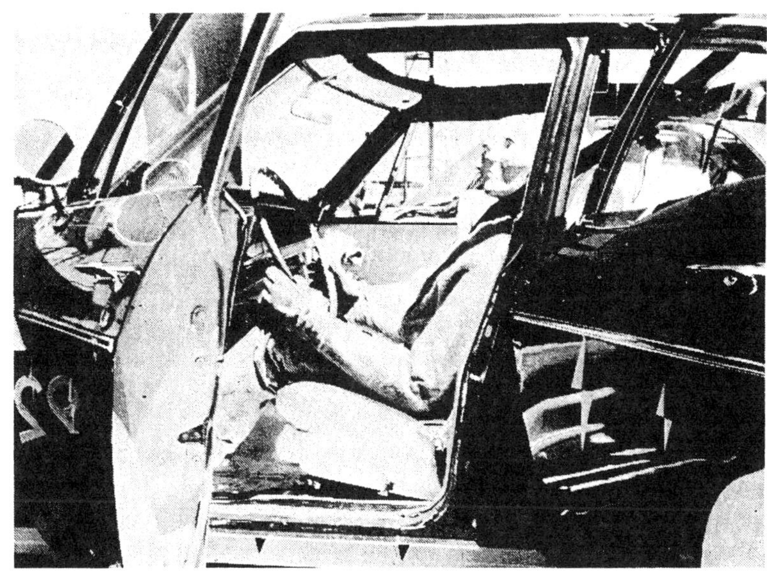

Fig. U - Dummy used during side impact collision (authors' experiment 92)

deg to each other upon impact. To continue this discussion a bit further, surveys of free-falls into fire nets during emergency situations indicate that very few rescues have been successful in height of falls over 70 ft. If one wanted to really be brave and consider that the mattress used to absorb his free-fall was 100% efficient and that it would resist him with a square-wave impulse of 47 g, it would only have to be 2 feet, 3 inches thick. Forty-seven g was used in this calculation because this is the peak acceleration that was experienced by Colonel Stapp's sled during his famous test at Holloman Air Force Base. Obviously, the construction of an automobile cannot be 100% efficient in energy absorption, whether it be hard on the outside and soft on the inside, or hard on the inside and soft on the outside, or any combination of these factors. The best that the automotive designer can design for the occupants in the presence of these conditions is protection against localized penetration by the striking vehicle and presentation of contact area for the occupant.

I would like to ask the authors several questions. The first is about the dummies that were used during the impact test series (Fig. T). How realistically were their necks calibrated to simulate human response during rear impacts?

The second question relates to the driver dummy that was used during the side impact collision test. As you will note in Fig. U, his elbow was nearly touching the front seat cushion. The dimensions of this dummy indicate that his torso length approximates that of the first percentile male driver. This creates a rather adverse appearing shoulder harness belt attitude when the seat is placed in the full rear position. My question is, how realistic was this driver simulation?

AUTHORS' CLOSURE TO DISCUSSION

IN ANSWER TO Mr. Cichowski's first question, we are not primarily concerned with direct head impact in whiplash accidents (usually rear end). An effort is made to limit the posture of the body in a rear end collision.

As for the second and third questions, dummies are compromises. The torso was shortened (represents first percentile male driver) on the dummy used for the side impact collision in an effort to get the dummy to slouch. Neck injuries are likely to occur when using a shoulder harness. Some improvement was obtained when the roof anchor point was lowered. We feel there is still room for improvement in the selection of shoulder harness anchor points.

We have an excellent calibration system, to crank whatever resistive forces are desired into the necks of the dummies. A human being usually exerts just a little more than a 1 g force to hold his neck in a neutral position. When a car occupant is forewarned of an impending crash, his stiffened neck muscles can resist greater accelerations before the head is whiplashed uncontrollably in any direction. In the UCLA tests, neck structures of the dummies were adjusted to simulate an unwarned crash victim.

We also have some comments to make on severe rear end collisions (40-50 mph) which result in fire. Although reports indicate that less than 1% of today's accidents result in fire, we are quite concerned over the oftentimes tragic cremations that do result. We feel that gas tank location and structure, among others, are particularly important facets, and industry is already working on improvements. Car occupants must be able to get out quickly.

We use a tow system to bring test cars into the accident situation because we cannot afford a fire in which they might lose their equipment. We agree it is a better idea to have the engines running, as in the GM tests.

ORAL DISCUSSION

Q. We at Firestone would like to know if a range of sizes would be required with the LM encapsuled seat? What degree of inconvenience is experienced in every day ordinary use of a car?

Mr. Brink: Entry is somewhat difficult with a 10 in. wing. A proposal has been made to reduce this to 5 in. It might be possible to make part of the seat integral with the door. Actually, there is considerable latitude of movement for the occupant. The seat could rotate. However, it isn't easy to design a seat which will be totally comfortable for the various sized people expected to use same.

Q. In accident situations, what is the effect of car weight differentials (that is, Cadillac versus Volkswagen)?

Mr. Severy: Deceleration rates were comparable when the Ford Motor Co. crashed a Falcon, a Mercury, and a Continental into a barrier at like speeds. Obviously in "two car" accidents the heavier car has the advantage.

Mr. McCockney, Buick Motor Div., GMC - Regarding the effect of a six year age differential (1960 Chevrolet versus 1966 Chevrolet), we question your confidence in the use of one accelerometer. Could you clarify the curve describing the velocities of the struck and striking vehicle (rear end collision)?

Mr. Severy: Aging (rust) would normally have an effect, but the 1960 vehicle was not rusted as it had been a showroom vehicle all this time. The variance in crushability of the 1960 versus the 1966 Chevrolet illustrated the difference in their structures. The greater deformation of the 1966 vehicle is advantageous, provided the passenger compartment is not infringed upon.

Concerning the rear end collision, it took 100 ms before both cars attained the same velocity. Unless locked together, the cars usually part within 600 ms. The struck car usually moves away about 2 mph faster than the striking car.

By no means do we depend on one transducer to measure deceleration. Usually they are clustered in sets of three at various locations on the car. As many as 70 transducers are employed in a single test. The paper itself explains how data of this type (which appeared doubtful) were cross checked and confirmed or eliminated.

Q. What effect would shoulder harnesses have on forward direction whiplash?

Mr. Cichowski: Whiplash injuries are not well defined. The chin would bottom out on the chest at a neck angle of approximately 45 deg. The threshold of neck injuries has not been defined.

Mr. Severy: The chin provides a limit of motion in the forward direction. Whiplash injuries can also be sustained in a sideward direction.

Mr. Margolin, International Harvester Co.: Would not high "G" accelerations result when the Liberty Mutual chair moved the occupant sideward in a side type collision? Perhaps the human is crushable and this would result in a broken shoulder (as an example) rather than killing him.

Mr. Brink: There is a significant improvement (reduction) in "G" forces imposed upon the seat occupant when he is moved away from the direct impact area.

Mr. Severy: The human body is not a good energy absorber. Many injuries are not reversible; for example, a cracked kneecap could result in a crippled walk, etc. Thus, it is essential to minimize use of the body as an energy absorber. Forces should be transmitted to the hips or shoulders, through belts because these are the strongest parts of the body.

SUMMARY OF SAFETY ISSUES RELATED TO FMVSS NO. 207, SEATING SYSTEMS

Prepared by:
The National Highway Traffic Safety Administration
September 1992

Table of Contents

Summary of Safety Issues Related To FMVSS No. **207,** Seating Systems

I. INTRODUCTION

The purpose of this report is to summarize recent work on the safety issues related to the Federal Motor Vehicle Safety Standard (FMVSS) No. 207, "Seating Systems." This technical report **has** been prepared in support of the notice published in the Federal Register by the National Highway Traffic Safety Administration (NHTSA) concerning FMVSS No. 207. This report first reviews the background and requirements of the present standard, identifies seating system performance issues fur safety **goals,** summarizes current NHTSA evaluations, research, and laboratory testing, and finally reviews the technical literature on seating system safety.

II. FMVSS **NO.** 207, SEATING SYSTEMS

A. The Current Standard

Federal Motor Vehicle Safety Standard (FMVSS) No. 207, "Seating Systems," first went **into** effect on January 1, 1968, for passenger cars only. The 1968 standard was basically adopted from the requirement of the Society of Automotive Engineers Recommended Practice **J879,** "Passenger Car Front Seat and Seat Adjuster," November 1963.

On January 1, 1972, the standard was extended to multipurpose passenger vehicles, trucks, and buses. The amended FMVSS **No.** 207, "Seating Systems,@@ adopted the test procedures in the revised SAE 5879. The SAE 5879 was revised in July 1968 and was renamed **"Motor** Vehicle Seating Systems **-** SAE **J879b."**

On March 19, 1974, NHTSA proposed a modification of **FMVSS No.** 207, but the rulemaking action was terminated in 1978. Consequently, the standard has not been upgraded since January 1, 1972.

B. Recent Petitions to Upgrade the Standard

On March 3, 1988, Edward **J.** Hurkey petitioned NHTSA to look into the **"slingshot"** effect on restrained occupants during rear impacts and to amend the requirements **for safety** belt retractors in FMVSS No. 208, "Occupant Crash Protection," and **FMVSS No.** 209, **"Seat** Belt Assemblies." The **"slingshot"** effect is **a rebound** effect whereby an occupant is propelled forward from a deformed seat due to the recovery of elastic energy. The petitioner **believed that** the emergency lock retractor (ELR) on **some** safety **belts unlocked during the occupant's rebounding and** therefore **could not prevent the "slingshot" effect during an** impact. **On** July 24, 1989, NHTSA notified **Mr.** Horkey that his petition was granted. The petition **was** entered into **NHTSA's** Docket No. **89-20-N01-001.**

On April 18, 1989, Kenneth J. Saczalski petitioned NHTSA to reexamine the general performance requirements of FMVSS No. 207. In particular, the petitioner suggested upgrading the seat back requirements for rear impacts. On July 24, 1989, NHTSA notified Dr. *Saczalski* that his petition was granted. The petition was also entered into NHTSA's Docket No. 89-20-N01-001 and Docket No. PRM-207-001.

On October 4, 1989, NHTSA published a Request for Comments notice and sought comments on the Horkey and Saczalski petitions. Based on responses to the notice (Docket No. 89-20-N01) and agency review and analysis, on October 17, 1990, NHTSA published a Termination of Rulemaking notice on the Horkey petition. The termination notice stated that NHTSA was unable to establish that amending the requirements for safety belt retractors would provide any significant safety benefits.

On December 28, 1989, Alan Cantor petitioned NHTSA to amend FMVSS No. 207 to eliminate "ramping" along a collapsed seat back during rear impacts. "Ramping" is movement of an occupant rearward and upward along the seat back during a rear impact. On February 28, 1990, NHTSA notified Mr. Cantor that his petition was granted. The petition was entered into NHTSA's Docket No. PRM-207-002.

C. Test Requirements of the Current Standard

Test requirements of the current FMVSS No. 207 are summarized in the following paragraphs.

Each occupant seat, other than a side-facing seat or a passenger seat on a bus shall "withstand" the applied forces.

o Force: A force equal to 20 times the weight of the seat is applied through the center of gravity (c.g.) of the seat in a forward and in a rearward longitudinal direction. If a seat belt assembly is attached to the seat, an additional force (FMVSS *No.* 210, "Seat Belt Assembly Anchorages," S4.2's requirement) is applied to the forward direction only for forward-facing seats and rearward only for rearward-facing seats.

o Moment: A force that produces a 3,300 inch-pound moment about the seating reference point is applied to the seat back for *each* designated seating position that the seat provides. The force is applied to the rearward direction only for forward-facing seats and forward only for rearward-facing seats.

o Seat back lock or seat back restraint device: For a forward-facing seat, a forward longitudinal force equal to 20 times the weight of the seat back is applied through the c.g. of the seat back. Similarly, for a rearward-facing seat, a rearward force equal to 8 times the weight of the seat back is applied. In addition, the restraining device is required to **"not release"** when it is subjected to an acceleration of 20g in the direction opposite to that in which the seat folds.

D. Comparison of U.S. and Foreign Seating System Safety' Requirements

The following table shows a comparison of test requirements on seating systems between the U.S. and some other countries. Note that the test requirements are essentially the same except that the ECE Regulation **#17** requires a higher test moment (4,690 in-lb vs. 3,300 in-lb).

Comparison of U.S. and Other Country's Seating Systems Test Requirements

Country	Test Force on Seating System C.G. of Seat	Test Moment on Seating Reference **Pt.** (Per Seat Occupant)	Test Force on Seat Back Lock C.G.-Seat Back
U.S.A.	**20g**	3,300 in-lb	**20g**
Canada	**20g**	3,300 in-lb	**20g**
ECE Reg. **#17** Rev. 3	20g	4,690 in-lb (53 **daNm)**	**20g**
Japan	**20g**	3,300 in-lb **(Approx.)** (38 Kg-m)	20g
Sweden	**20g**	3,300 in-lb **(Approx.)** (38 Kg-m)	20g
Brazil	**20g**	3,300 in-lb	**20g**
Australia	20g	3,300 in-lb	**20g**

XII. ANALYSES **OF** SAFETY PROBLEM

A. Seating System Performance Issues

Based on the Saczalski and Cantor petitions, the comments submitted in response to the October 4, 1989 notice, and agency research, the agency has determined that there are four categories of performance issues which need to be addressed as part of the consideration of any upgrade of Standard No. 207.

The first category is seating system integrity. Seating system integrity refers to the ability of the seat and its anchorage to the vehicle to withstand crash forces without failure. Examples of failure of the seating system would include: breakage of the seat adjusters, breakage of the folding seatback locks and supports, or separation of the anchorage from the vehicle.

The second category is the energy absorbing capability of a seat. The energy absorbing capability of a seat includes the manner in which the seat and its attachment components absorb energy, and the manner in which the seat and its attachment components release energy.

The third category is compatibility of a seat and its head restraint. The concern in this category is that any change in seat back energy absorbing capability could exacerbate head or neck injuries if the geometry and energy absorbing capability of the head restraint is not also changed.

The fourth category is the safety belt restraint system. A seating system and its safety belt restraint system must complement each other to prevent injury. Several manufacturers are considering integrated seats, i.e., seats which have the safety belt attached to their seat structure to increase the compatibility of these systems.

B. Seating System Safety Concerns

Most of the concerns raised in the rulemaking petitions, in comments submitted in response to the October 4, 1989 Request for Comments, and in the literature relate to the energy absorbing characteristics of the seating system. Specifically, they concern how to achieve a proper "balance" in stiffness. Concern has been expressed by commenters and in the literature that if a seating system is too stiff, injuries could be increased *in* a rear impact collision because of the exacerbation of several problems: occupant rebound off the seat back into the frontal components, ramping of the occupant into the roof of the vehicle, direct contact with the seat back, and phasing problems between the neck/back body regions contacting the head restraint and the seat back. On the other hand, concern has been expressed that if the seating system appears to bend too far backward when the vehicle is struck in the rear, injuries to front seat occupants could be increased by the exacerbation of several other problems: ramping toward the rear components, contact with the rear seat and/or rear seat occupants, and loss of vehicle control.

Further, there could be an increase in injuries fur rear seat occupants also.

In an attempt to define the proper **"balance"** related to this energy absorbing **or** stiffness characteristic plus **a better** understanding **of the** other issues discussed above, a number of efforts have been undertaken. The first effort was to publish a notice of Request fur Comments in the Federal Register requesting comments on the subject petitions and information on seat back performance characteristics. Partial results of this effort were published in another notice, a Termination of Rulemaking, in the Federal Register and are discussed further in subsequent sections of this report.

The remaining part of this section is concentrated on seating **system** related data analysis and review. Analyses were undertaken of NHTSA accident data files including both an exploratory data analysis and a hard copy investigation of selected **cases.** A review of NHTSA tested seating **system** performance data was conducted including **FMVSS** No. 207 and **FMVSS** No. 301 rear impact tests, and New Car Assessment Program (NCAP) rear and frontal impact tests. Also, the agency's recent defect investigation files related to this subject were examined.

C. Accident Data

1. National Accident Sampling System - Exploratory Analysis of Computerized Database

The 1988 to 1990 National Accident Sampling System (NASS) data describe seat type and seat performance for occupants of light passenger vehicles that were towed from the scene because of **damage** received in the crash. NASS also collects data describing the crash circumstances, vehicle damage, occupant factors, and resulting injuries. The NASS sites were randomly selected to represent the national accident experience and cases were selected randomly in each site, **so the** weighted NASS data produce national estimates of light vehicle tuwaway accidents. The complete analysis of the NASS data is presented in **a** detailed report, **"Seat** Damage and Occupant Injury in Passenger Car **Towaway** Crashes," Susan C. Partyka, Office of Vehicle Safety Standards, Rulemaking, National Highway Traffic Safety Administration, June 8, 1992. This report can be found in **NHTSA's** Docket No. **89-20-** N03.

In interpreting the results, the report states that **"[t]his** report describes an exploratory **data analysis of seat damage and** occupant injury performed to provide insight into injury mechanisms, to suggest questions for further research, and **to** help establish the safety priority of these questions. Occupant involvements and injuries were estimated from a statistical survey of towaway crashes, and these estimates are subject to sampling and nonsampling error. Estimates of sampling variability are beyond the scope of the present effort, and no claims of statistical significance are implied by the comparisons made here."

Ten percent of occupied front-outboard passenger car seats (the driver and right-front seats) were deformed (by occupant contact or intrusion) or their hardware was damaged in the crash (the seat adjusters, folding locks, tracks, or anchors were broken).

While most of the interest related to seating system performance has been focused on rear impacts, it should be stressed that seat damage occurs in all crash modes. Rear impacts account for a third of seat damage, and seat damage was much more common in rear impacts than in other towaway impact types. In recent towaway crashes,

38 percent of seats in rear impacts were damaged,
19 percent of seats in near-side impacts were damaged,
9 percent of seats in rollover crashes were damaged,
7 percent of seats in far-side impacts were damaged, and
5 percent of seats in frontal impacts were damaged.

The type of seating system damage differed among the different crash modes. Deformation from occupant loading (which relates to the energy absorption issue) accounts for 71 percent of the damage types in rear impacts. The seat damage in frontal impacts and rollover crashes relate about equally to broken hardware (which relates to the structural integrity issue) and occupant loading. For side impacts, seat damage primarily relates to deformation from vehicle intrusion, which is not being considered within this effort. It appears that damage in frontal crashes is associated with more severe crashes compared to that seen in rear impacts. For crashes with known severity as measured by delta-v, 41% of the seats damaged in frontal impacts occur with a delta-v of at least 20 mph, while in rear impacts, 24% occur with a delta-v of at least 20 mph. The energy absorption concern appears to be more an issue in rear impacts, where the indication of structural integrity problems appears to be much less than those associated with occupant loading.

In examining the accident data for information linking the performance of seating systems with injury causation it was found that occupants in damaged seats tended to be injured more severely than occupants in undamaged seats, largely because seat damage indicated a high-severity crash. Thus, in order to further explore the relationship between seat damage and injury, several methods were utilized to control for crash severity. These methods are fully discussed in the detailed paper on this analysis as presented in the previously referenced report.

Overall the results were unclear and did not consistently show a pattern of increased likelihood of injury fur damaged seats.

For the injury rate comparisons that were made within ranges of crash severity, in frontal crashes, occupants of damaged seats had higher injury rates than did those in undamaged seats fur each 10 mph range of delta-v. It is not **clear** from these data whether stronger seats would have prevented injury, because crashes in which seats are damaged may differ in important ways from those in which seats are nut damaged, even within a range of crash severities. It also should be stressed that the type of damage in frontal crashes are mixed -- the damage types consists of what was defined earlier as structural integrity and energy absorption characteristics. Also, it should be repeated that only a small percentage of seats sustain damage in frontal impacts. In any case, there dues appear to be some indication that the likelihood of injury to occupants is greater fur damaged seats.

Fur rear impact crashes, the data are very limited after -controlling for crash severity. However, the results are mixed, with occupants of damaged seats having higher injury rates than those in undamaged seats in the lower delta-v ranges, while in the higher delta-v ranges the occupants of undamaged seats had higher injury rates. Fur rear impacts this damage was primarily related to the energy absorption characteristic.

The previously referenced report examines many other variables related to seat damage that would need to be controlled for in isolating the effect of seat damage on injury causation. For example, the analysis concluded that the likelihood of seat damage increases with increasing occupant weight and increasing age of the vehicle. Also, belt use rates were higher in undamaged seats, possibly because belt users tended to be involved in less severe crashes, **or** because belt use reduces occupant loading of the seat.

The exploratory NASS analysis **also examined** any similarities **and/or** differences in injury patterns for occupants of damaged and undamaged seats. The analysis found that unbelted occupants of damaged seats in rollover, frontal impacts, and side impacts were **more** likely to be ejected than were unbelted occupants of undamaged seats. In large part, the differences in ejection rates reflect differences in crash severity -- severe crashes were more likely to result in seat damage and to involve occupant ejection and injury. There are very few ejections from cars with damage to the rear, but the available data do nut suggest that occupants in damaged seats were more likely to be ejected than were occupants in undamaged seats.

As to the distribution of body regions injured and the source of these injuries, the previously referenced report presents the results of this analysis for all crash modes. In this summary, specific attention is focused on rear crashes. For these impacts, the head/neck area generally accounted for 42% of the moderate or greater injuries (AIS 2+), while the torso accounted for 57% of serious and more severe injuries (AIS 3+). Seat backs were cited as the source of moderate and more severe injury in a high percentage of cases:

13 percent for those unbelted in undamaged seats,
6 percent for those belted in undamaged seats,
26 percent for those unbelted in damaged seats, and
24 percent for those belted in damaged seats.

Occupants of seats damaged in rear impacts receive more of their injuries from seat back contact than did occupants in rear impacts with undamaged seats. This may reflect, in part, the severity of the crash that resulted in heavy occupant loading of the seat back.

For occupants with damaged and undamaged seats in rear impacts, a high percentage -- approximately 70% for undamaged seat injury sources and approximately 30% for damaged seat injury sources -- of the injuries were attributed to passenger compartment front components; this suggests that the occupant may have rebounded from the seat, forward, or contacted frontal components as the seat back rotated backward, or contacted the frontal components in a subsequent frontal impact.

There were very few moderate and more severe injuries attributed to a contact that could be determined to have been made behind a normally-seated front-outboard occupant -- none for occupants in undamaged seats and approximately 3% for damaged seats. As discussed in the detailed report, defining rear contacts is difficult from the automated file. Injury sources that could be defined unambiguously as rear include the rear header, backlight area, and pillars rearward of the B-pillar. However, many of the identified injury sources were components that extended both forward and rearward of the occupant -- for example, the roof and side rails; the B-pillar was also classified as an ambiguous source.

The likelihood of injury and the pattern of injuries for damaged and undamaged seating systems especially in rear impacts does not provide clear evidence as to the direction for upgrading the stiffness of seat backs. Possible injuries due to ramping or contacting the rear components are limited. There are a great number of frontal contacts which could be due to rebound and must be considered in any improvement. The seat back is cited as a source of injury in many cases for damaged and undamaged seats. While this percentage is greater for damaged seats, this may be a result of the greater crash severities associated with seat back damage. In any case, because of the high frequency of serious torso injuries and seat back contacts, any change in energy absorption characteristic must consider this direct interaction. The high percentage of head/neck injury also demonstrates the need to consider the head restraint and occupant height along with any **modifications** to the seating system.

2. Hard Copy Studies

To further evaluate how injuries occur and their relation to seat damage in frontal and rear impacts, hard copy cases from the NASS 1988-90 files were selectively reviewed. This work was conducted as part of an agency contractual research project, and results of the study are presented in a report, "Upgrade Seating - Patents, Literature Search, and Accident Analysis," Kennerly Diggs and John Morris, University of Virginia, Prepared fur the National Highway Traffic Safety Administration, September 1, 1992. This NHTSA sponsored research report can be found in Docket No. **89-20-** N03. Three sets of cases were selected as follows.

Set	Impact	No. of Cases	Case Selection Criteria
1	Rear	18	30-39 mph delta-v
2	Rear	31	**AIS 2+** or seat damage
3	Frontal	6	Seat damage

A general discussion of the reviewed cases is presented. Since the cases are unweighted NASS data, the discussion is anecdotal.

a. Eighteen Rear Impact Cases

The 18 rear impact review set encompass all cases in 1988-90 NASS with a rear damage delta-v 30 to 39 mph. Actual impact speed of those cases would be much higher than the delta-v range. For example, **NHTSA's** New Car Assessment Program conducted 35 mph rear impact tests on 1979 to 1981 model vehicles, the delta-v range was 18-26 mph.

The distribution of occupant injuries based on the maximum **AIS** was as follows.

	Front Occupants	Rear Occupants
AIS 0	3	1
AIS 1	12	3
AIS 2	6	
AIS 3	2	
AIS 4-5	0	
AIS 6	1 (Burn)	
AIS 7 (Unk Inj)	1	

Rear impact caused injuries were at low levels for these severe impacts.

The most frequently injured body part was the head and face, and the highest severity injuries were inflicted to the head and chest. In terms of the probable injury source, steering wheel (9) was the most frequent, followed by the seat back (8), headrest (5), and flying glass (4). However, the number of noncontact injuries (9) was equal to the number of injuries caused by the steering wheel.

In this small sample, 19 of the 29 occupants were restrained. The injury rate for AIS 2+ for restrained occupants is lower than the unrestrained occupants (26% vs. 50% based on unweighted data).

In addition, 11 of 25 front seat occupants suffered injuries from contacts with frontal compartments. Two of these frontal injury cases involved only rearward crash forces.

For seat damage effect, ramping of the occupants was difficult to identify in all the cases. Ramping was clearly not relevant to injuries in 21 of 29 occupant cases. Seat deformation may have contributed to only two of the occupant injuries. In one case, ramping may have contributed to the head injury (AIS 2) from the rear header, and an associated AIS 2 abdominal injury from the safety belt. The other case's relevant injury was at a minor severity level.

On the other hand, rebound may have contributed to injuries in 12 of 25 cases. In many cases, a frontal impact followed the rear impact, thereby adding to the rebound velocity. This is also consistent with the observation that the steering wheel is the most frequent cause of injury. For the most part, the rebound injury was minor.

b. Forty-Nine Rear Impact Cases

The 31 additionally selected rear impact cases had one group with two cases of delta-v greater than 40 mph and the other group with restrained occupants suffering AIS 2+ injuries at any delta-v. To provide a more detailed review of the seat performance, the 31 cases are combined with the already examined 18 cases to become a review of 49 rear impact cases.

Three cases of seat adjuster failure were noted and all were associated with seats deformed by occupant impact. Yet no **AIS 2+** injuries from seat adjuster failure could be positively identified from the rear impact cases examined.

Fourteen cases of folding lock failures were reported. Five of the occupants may have suffered **AIS** 2 injuries that were exacerbated by the failed locks. In four of five cases, the **source** of the **AIS** 2 injury is unknown. Thus, the evidence of seat luck failure contribution to injury is unclear.

Two cases of seat anchorage failures were reported. The maximum injury fur the occupants in both of these cases were **AIS** 1. It appears that seat anchorage failures were not associated with any serious rear impact injuries.

To further examine issues concerning occupant injury vs. seat damage, the analysis concentrated on the seat back yielding issue. The combined 49 cases were divided into two groups **-** no permanent seat back yielding and permanent seat back yielding. In addition, only injuries of **AIS 2+** were included. Thirty-five occupants in the 49 cases met the criteria fur inclusion in this analysis.

In the **"no** permanent seat back yielding" group, 16 of 17 occupants were restrained. No injuries were identified which resulted from ramping **or** seat related deformation. However, frontal injuries were present in 8 of the cases. This suggests that rebound may **be** frequent and the possibility that rebound **may** have contributed to some injuries.

In the **"permanent** seat back yielding" group, 16 of 18 occupants were restrained. Seat deformation may have contributed to three of the injuries, and ramping may have contributed to one. However, rebound is also frequently present. Seven of 16 occupants had injuries from frontal sources. Four of the cases were single event rear impacts.

This analysis also examined 8 cases in which vehicles contained rear seat occupants. Some minor **AIS** 1 injuries could be attributed to the front seat deformation. However, it is not possible from the data to assess the degree to which front **seat** deformation contributes to the injuries of rear seat occupants.

c. Six Frontal Impact Cases

Six frontal impact cases were reviewed. They involve three seats which are reported to have track/anchorage failures, two which deformed under occupant loading and one which moved forward under impact. From the previously referenced University of Virginia report: "For the seats which *were* deformed by occupant impact, no contribution of the seat to the injury was evident. However, for the seat that moved forward during a frontal impact and the three track/anchorage failures, seat contribution to the injury severity is possible. In all of these cases, occupants experienced injuries higher than expected for the crash severity. The higher injuries were consistent with those expected from undesirable seat loading of the occupant."

d. Summary of Hard Copy Studies

The agency contractual report summarizes the findings as follows:

> "This preliminary analysis suggests that improvements in seat performance is a more complex matter than simply increasing the strength *of* the seat back. Legitimate concerns exist over the potential increase in neck injuries and rebound injuries which might accompany strengthened seats.
>
> Harm analysis by Malliaris [Reference: Data Link, 1990] provides insights of injury frequency and severity in rear impacts. His analysis shows that noncontact neck injuries constitute more than 20% of the Harm to restrained occupants. The head restraint is the largest source of contact Harm (17%). The role of head restraints in noncontact neck injuries and contact head injuries needs to be studied in conjunction with any seating system modifications.
>
> Foret-Bruno (91) found a significant increase in head restraint effectiveness as seat back strength in Renault cars was increased to meet the EEC standard. He suggested that the lower-strength, prestandard seats deformed at a force level below that which induces noncontact neck injury. He concludes that strengthened seats are likely to increase the demand on head restraints to mitigate the neck injury risks.
>
> Our data analysis did not permit the quantification of *neck* injury risks for deformed versus nondeformed seats. In the accident cases we analyzed, we found three noncontact neck fractures in seats which did not deform. No noncontact neck fractures *were* observed in seats which deformed.

'Rebound' type injuries occur frequently in crashes which involve rear impacts. Malliaris found that 16% of the Harm in rear impact **NASS** cases was from frontal contacts. In many of the cases, the rear impact is followed by a frontal impact, either in a line of stopped **traffic**, or by being accelerated into a fixed object. In these cases, and in cases where no subsequent impact occurs, injuries from impact with frontal components of the vehicle are frequently observed. Fur the data set of **AIS 2+** injuries in selected rear impacts, injuries from frontal components were believed to be presented in 15 of 35 occupants. It is not possible to determine how many of these injuries were related to the elastic response of the seat, or from other phenomena. However, **some** seat induced rebound phenomena can be observed in FMVSS **301** rear impact tests. It is evident that the rebound phenomena needs to be researched in conjunction with future seat **improvements."**

b. Seating System Performance in NHTSA Conducted Tests

1. FMVSS No. 207 Compliance Tests

From the **FMVSS** No. 207 compliance tests that covered fiscal year (FY) 1972 to 1986, seats in 6 out of 169 (4%) tested vehicle models failed the requirements. For FY 1991, no seats in the 10 tested vehicle models failed the requirements. The compliance test results indicate that FMVSS No. **207's** requirements have achieved the intended goals for maintaining a certain minimum performance level for the integrity of vehicle seats. More information about the six seat failure cases is presented as follows*

Model **Yr/Make**	Model	Seat Failure Description
1972 Toyota	Corona	Front seat back/rear seat back cushion
1972 Ford	Mark IV	Front seat tracks
1974 Ford	Monterey	Rear seat back anchorages
1975 Ford	Mustang II	Rear seat cushion
1982 **GM**	s-15	Track
1982 GM	c-10	Track

2. **FMVSS** No. 301 Rear Impact Compliance Tests

To study the performance of production vehicle seat backs in rear impacts, FMVSS No. 301, **"Fuel** System Integrity," compliance tests were reviewed. **FMVSS** No. **301's** rear impact test requires a 4,000 lb. flat-face rigid barrier crash at 30 mph. It also specifies Part 572 **50th-percentile** restrained test dummies (uninstrumented) at each front outboard designated seating position.

A total of 54 test reports (FY 1987 to FY 1991) were reviewed. Seat back deflection remaining after the test was measured in terms of degrees of backward rotation from the pre-test orientation. The rotations were measured from post-test photographs in the reports. The following list shows the seat back deflection from those tests. It is noted that significant seat backward rotation in tested vehicle were very frequent.

FMVSS No. 301 Rear Impact Report Review on
Vehicle Seat Back Deflection (Degrees of Rotation)

FY	0 - 10	10+ - 20	20+ - 30	30+	Total Number
87	15.4%	53.8%	23.1%	7.7%	13
88	11.1%	11.1%	44.5%	33.3%	9
89	--	--	60.0%	40.0%	5
90	10.0%	40.0%	10.0%	40.0%	10
91	11.8%	23.5%	47.1%	17.6%	17

To examine occupant's ramping and rebound effects during rear impacts, 12 recent FMVSS No. 301 rear impact test films were reviewed (two-1991 and 10-1992 vehicles). The observed results are presented as follows. From this review, no ramping was observed due to the 30 mph impact and the belt system appears to prevent or reduce the degree of rebound.

FMVSS No. 301 Rear Impact Test Film Review on
Occupant Ramping and Rebound Effects

Model Yr/Make	Model	Ramping	Belt Restrained Dummy Rebound
1991 Ford	Explorer	No	Dummies did not contact belts
1991 VW	Jetta	No	Yes
1992 Mitsubishi	Expo	No	Yes
1992 Toyota	Camry	No	Yes
1992 Ford	C. Victoria	No	? Dummies out of camera's view
1992 Oldsmobile	Royal 88	No	? Same as above
1992 Buick	Road Master	No	? Same as above
1992 Hyundai	Elantra	No	? Same as above
1992 Plymouth	Acclaim	No	? Same as above
1992 Toyota	Paseo	No	? Same as above
1992 Acura	Vigor	No	? Same as above
1992 Plymouth	Voyager	?	? The van has tinted windows

? = Unknown

In summary, from reviewing FMVSS No. 301's 30 mph rear impacts, it appears that front seat backs frequently deform to a high degree but no apparent ramping effects were observed. In addition, it appears that rebound effects were minimized by the use of belt systems.

3. New Car Assessment Program Test

NHTSA conducted 55 FMVSS No. 301 rear impact tests at 35 mph under the New Car Assessment Program (NCAP) on 1979 to 1982 model year vehicles. The tests show that, under the 35 mph test, it was estimated that **all** the front seat **backs rotated** backward permanently for more than 30 degrees (13 seat backs rotated 30 to **40** degrees and 80 seat backs rotated 45 to 80 degrees) and must of the front seat backs touched the rear seat back. The legs of **àll** the driver dummies hit the steering wheel, consequently, the driver dummies mainly rotated with the deformed seats. To further study the performance of production vehicle seat backs in rear impacts, films and reports of three NCAP tests that sustained seat back rotation of 60 degrees **or** more were reviewed. Results are listed as follows.

FMVSS No. 301 Rear Impact NCAP Tests on Occupant Ramping and Rebound Effects

No.	Model **Yr/Make**	Model	Ramping		Rebound	
(1)	1982 Chevrolet	Cavalier	**No (D)**	**No (P)**	No (D)	Some (P)
(2)	1981 AMC	Spirit	**No (D)**	**No (P)**	No (D)	Some (P)
(3)	1981 **Isuzu**	I-Mark	No (D)	Some (P)	No (D)	Some (P)

D = Driver Dummy P **=** Passenger Dummy
Some = Possible occurrence but nut apparent, decision was based on contact marks on dummy and film observation.

(1) Both dummies' heads contacted the rear seat back. The passenger dummy has a little contact mark on the chin and there is contact mark on the shoulder belt.

(2) The driver dummy's head contacted the rear seat back. The passenger dummy has a line of facial contact mark probably due to contact with the shoulder belt. The rear seat back was pushed up during the impact to meet the dummies.

(3) The driver dummy's head contacted the rear seat back. The passenger dummy's head contacted the roof area above the rear window. There are sume scratch marks on the dummy's face **and** a little contact mark **on** the shoulder belt.

NHTSA has conducted 35 mph frontal impacts on vehicles since 1979. Although the focus of this present effort is mainly related to rear impacts, recent NCAP frontal impacts were also reviewed as to seating system performance. From the 1987 to 1991 tested vehicle models, three were identified fur further study from those having seating system damage. The three test results are reviewed and summarized **as** follows:

FMVSS No. 208 Frontal Impact NCAP Tests on Seating System Damage

No.	Model Yr/Make	Model	Seating System Damage
(1)	1990 BMW	325i	Both seat height adjusters released
(2)	1991 Buick	Century	Both seats shifted forward
(3)	1991 GM	Saturn SL2	Driver seat shifted full forward

NCAP tests on FMVSS No. 208 have instrumented dummies. Responses of the dummies for the three tests are listed as follows:

No.	Head Injury (HIC)		Chest Deceleration		Ave. Femur Loads	
	Driver	Passenger	Driver	Passenger	Driver	Passenger
(1)	1,036	No Data	56g	No Data	859lb.	No Data
(2)	815	1,144	47g	40g	1,177lb.	247lb.
(3)	918	1,018	44g	46g	1,168lb.	1,076lb.

All the dummies had knee contacts with the instrument panel.

(1) The driver dummy's head contacted the deployed airbag and the passenger dummy's head contacted the instrument panel.
(2) The driver dummy's head and chest contacted the steering wheel and the passenger dummy's head contacted the instrument panel.
(3) The driver dummy's head contacted the steering wheel.

In summary, from reviewing the NCAP's 35 mph test impact data, it appears that seats routinely deformed to a high degree in rear impacts. Yet, only limited ramping effects were observed and sume possible minor rebounding was observed. It is also interesting to note the leg contact made with the steering wheel due to the rotation of the dummy in rear impacts. This may account for some of the frontal contacts seen in the accident data. Relatively infrequent seating system damage was observed in frontal impact tests.

E. Defect Investigation Data

NHTSA is responsible for investigating safety related defects on in-use motor vehicles. The agency has received many consumer complaints on seating systems. Between FY 1985 and FY 1992 (August), the agency has initiated 55 investigations due to possible seating system defects. The number of vehicles affected by these investigations amount to 17.5 million vehicles and the vehicle model years are from 1981 to 1992.

The 55 vehicle seat defect investigations relate to 13 cases on seat backs, 11 seat track or anchorage failures, 2 seat track and anchorage failures, and 29 other seat failures. In addition, at least 15 of the 55 cases have indicated that the defective seating system may have resulted in loss **of** vehicle control. Possible occupant injuries related to the investigated cases are **159** nonfatal injuries and 5 fatalities.

The 55 investigated cases resulted in 9 safety recalls which affected about 3.2 million vehicles and 74 nonfatal occupant injuries. The following two tables **show** a **summary** of the 9 safety recall cases which include numbers of complaints received **by** NHTSA and the affected manufacturers. Attachment 1 provides more details about **the 9 safety recalls.**

In summary, the agency vehicle seat defect investigations indicate that some seats lost integrity **or** failed while the vehicles were in operation without any impact. This could indicate an additional seat failure mode due to initial design **or** construction inadequacies or component fatigue, a failure mode which is not covered in the current standard's requirements.

NHTSA Vehicle Seat Defect Investigations
Safety Recalls Between FY 1985 And FY 1992 (August)

No.	NHTSA Recall Campaign No.	Manufac	Model Yr.	Vehicle Population	Seat Defects
1	89V011000	Chrysler	1985	60,000	Seat Frame (1)
2	86V082000	Ford	1984	195,732	Seat Back
3	87V079000	General Motors	1983 to 1984	1,136,407	Track and/or Anchorage
4	89V170000	Ford	1985 to 1987	1,375,500	Track
5	88V061000	Chrysler	1985 to 1986	1,500	Seat Frame/ Seat Back
6	88V060000	Chrysler (AMC)	1983 to 1984	149,000	Seat Back
7	91V036000	Ford	1988 to 1989	278,000	Seat Fire (Power Seat)
8	88V058000	Utili-master Corp.	1988	10	Track (2)
9	88V147000	Mack Trucks	1982 to 1988	11,000	Seat Tether (2)

(1) Fatigue Failure (2) Faulty Installation

Number of Reported Complaints and Injuries

Manufacturer;	01	02	03	04	05	06	07	08	09	Total
					Complaints					
NHTSA	33	25	25	18	29	18	8	0	1	157
Manufac	83	115	136	50	99	43	50	0	0	576
	116	140	161	68	**128**	61	58	0	1	733
					Nonfatal Iniuries					
NHTSA	0	6	10	2	1	1	0	0	1	21
Manufac	3	29	16	1	3	1	0	0	0	53
	3	35	26	3	4	2	0	0	1	74

IV. PETITIONS AND REQUEST FOR COMMENTS

A. Petitions

Three recent petitions requested NHTSA to amend **FMVSS** No. 207, "Seating Systems." All the three petitions related to rear impacts. Edward J. Horkey petitioned the agency on March 3, 1988, to look into the "slingshot" effect on restrained occupants. The petitioner suggested that NHTSA amend the requirements for safety belt retractors in **FMVSS** Nos. 208 and 209. Kenneth J. Saczalski petitioned the agency on July 24, 1989, to reexamine the general performance requirements of FMVSS No. 207, in particular, to upgrade the seat back requirements. Alan Cantor petitioned the agency on December 28, 1989, to amend FMVSS No. 207 to eliminate **"ramping"** along a collapsed seat .

The following table summarizes the relevance of the requested amendments from the three petitions to the four previously defined seating system performance issues.' The first issue concerns seating **system** structural integrity, the second issue refers to the energy absorbing capability of the seat, the third issue relates to the compatibility of the seat back with respect to the head restraint, and the fourth issue relates to the safety belt restraint system..

Petition	Issue 1	Issue 2	Issue 3	Issue 4
Horkey	No	**Some**	No	Yes
Saczalski	Yes	Yes	Some	Some
Cantor	Yes	Yes	Yes	Yes

NHTSA granted the three petitions and started the rulemaking actions. The agency published a Request for Comments on

October 4, 1989, seeking comments on the Horkey and Saczalski petitions. The Cantor petition was not in time to be included in the notice, but is included in this review.

NHTSA terminated the rulemaking action of the Horkey petition on October 17, 1990, based on responses to the Request for Comments notice and agency review and analysis. The agency stated in the termination notice that NHTSA was unable to establish that amending the requirements for safety belt retractors would provide any significant safety benefits. However, in the notice, the agency indicated it would continue to examine this issue. Rulemakings on the Saczalski and Cantor petitions are still ongoing.

B. Request for Comments

The Request for Comments notice encompasses all seating system issues raised in the petitions. In terms of the Horkey and Saczalski petitions, the notice asked a series of questions contained in six issues.

- Seat Back Stiffness in Rear Impact
- Dual-Mode Sensing Emergency Locking Retractors
- Costs of Requiring Dual-Mode Emergency Locking Retractors
- Consumer Acceptance of Dual-Mode Emergency Locking Retractors
- Other Seat Back Performance Requirements
- Seat Performance Measurement and Test Requirements

As of March 1, 1992, there were twenty-two (22) entries in Docket No. 89-20, Notice 1 responding to the Request for Comments. Ten (10) entries are from automobile manufacturers (Manufac), six (6) from vehicle safety consultants (Consult), two (2) affiliated with university accident research teams (Univ), one (1) from a contractual report prepared for Transport Canada (Report), and one (1) from a safety belt association (Asso). The remaining two (2) entries are the copies of the two petitions and an NHTSA sponsored research report (Other).

Analyzed Entries in Docket No. 89-20, Notice 1

No.	Doc. Date	Page	Group	Name
002	11/01/89	1	Manufac	Navistar International Trans. Corp
003	11/29/89	6	Manufac	Ford Motor Company
004	11/28/89	2	Manufac	Volvo North Am. & Volvo Car Corps.
005	11/28/89	6	Manufac	Chrysler Motors
008	11/30/89	2	Manufac	Thomas Built Buses, Inc.
009	12/04/89	6	Manufac	General Motors Corporation
013	12/01/89	1	Manufac	Fiat Auto U.S.A., Inc.
014	12/07/89	5	Manufac	Mercedes-Benz of North Am., Inc.
015	12/06/89	4	Manufac	Volkswagen of America,' Inc.
016	12/08/89	4	Manufac	Toyota Motor Corp. Services of Am.
007	11/27/89	7	Consult	Horkey, Horkey & Associates Inc.
011	11/29/89	95	Consult	Saczalski, Environ. Res. & Safety
012	11/30/89	3	Consult	Hoar, Ralph Hoar & Associates
017	12/01/89	2	Consult	Warner, Collision Safety Engr. Inc
022	12/03/91	46	Consult	Shaw, Shaw Research & Consulting
006	11/27/89	1	Univ	Gorski, Accident Research, UWO
019	01/03/90	24	Univ	Macnabb, Accident Research, UBC
010	12/04/89	3	Asso	AORC, Auto. Occ. Restraint Council
018	01/23/90	45	Report	TES Limited, Ontario, Canada

The agency interpretation and analysis of the comments is conducted according to the previously defined four seating system performance issues. They are Issue 1: Seating System Integrity, Issue 2: Energy Absorbing Capability of Seat, Issue 3: Compatibility of Seat Back and Head Restraint, and Issue 4: Compatibility of Seat Back and Safety Belt Restraint.

The comments of the above listed nineteen docket entries are analyzed in terms of the four seating issues and the results are summarized in the following tables. Original entries of the comments are in Docket No. 89-20, Notice 1.

Analysis of Comments
Docket No, 89-20, Notice I, 54 FR 40896

Issue I: Seating System Integrity

No,	Group	Name	Seat Damage Occurs	Damage Causes Occupant Inj.	Upgrade FMVSS 207
002	Manufac	Navistar	--	--	--
003	Manufac	Ford	?	No	No
004	Manufac	Volvo	--	--	No
005	Manufac	Chrysler	?	No	No
008	Manufac	Thomas	No	No	No
009	Manufac	GM	Yes	No	No
013	Manufac	Fiat	--	--	No
014	Manufac	Mercedes	No	No	Yes
015	Manufac	VW	--	--	No
016	Manufac	Toyota	--	--	No
007	Consult	Horkey	--	--	--
011	Consult	Saczalski	Yes	Yes	Yes
012	Consult	Hoar	Yes	Yes	Yes
017	Consult	Warner	?	?	No
022	Consult	Shaw	Yes	Yes	Yes
006	Univ	Gorski	--	--	--
019	Univ	Macnabb	Yes	Yes	Yes
010	Asso	AORC	--	--	--
018	Report	TES Ltd	Yes	Yes	Yes

Notes: -- = No response ? = Doubts expressed by commenter

Analysis of Comments
Docket No. 89-20, Notice 1, 54 FR 40896

Issue 2: Energy Absorbing Capability of Seat

No.	Group	Name .	Current Seating Sys. EA Capabi	Ramping	Seat Back Stiffness	Upgrade 3300in-lb
002	Manufac	Navistar	--	--	--	No
003	Manufac	Ford	O.K.	--	O.K.	No
004	Manufac	Volvo	--	--	--	No
005	Manufac	Chrysler	O.K.	?	O.K.	No
008	Manufac	Thomas	O.K.	No	O.K.	No
009	Manufac	GM	?	Some	?	No
013	Manufac	Fiat	--	--	--	Yes
014	Manufac	Mercedes	O.K.	No	Low	Yes
015	Manufac	VW	O.K.	--	O.K.	No
016	Manufac	Toyota	--	--	--	No
007	Consult	Horkey	--	--	--	--
011	Consult	Saczalski	N.G.	Yes	Low	Yes
012	Consult	Hoar	N.G.	Yes	Low	Yes
017	Consult	Warner	--	--	--	No
022	Consult	Shaw	N.G.	?	Low	Yes
006	Univ	Gorski	--	--	--	--
019	Univ	Macnabb	N.G.	Yes	Low	Yes
010	Asso	AORC	--	--	--	--
018	Report	TES Ltd	N.G.	Yes	Low	Yes

Notes: -- = No response ? = Doubts expressed by commenter
O.K. = Adequate N.G. = Inadequate
Low = Stiffness of current seat back is low
High = Stiffness of current seat back is high

Analysis of Comments
Docket No. 89-20, Notice 1, 54 FR 40896

Issue 3: Compatibility of Seat Back and Head Restraint

No,	Group	Name	Current Head Restraint Sym Problem	Source	Causes Inj	Seat Back Interaction
002	Manufac	Navistar	--	--	--	--
003	Manufac	Ford	No	--	No	--
004	Manufac	Volvo	--	--	--	--
005	Manufac	Chrysler	--	--	--	--
008	Manufac	Thomas	--	--	--	--
009	Manufac	GM	--	--	--	--
013	Manufac	Fiat	--	--	--	--
014	Manufac	Mercedes	--	--	--	--
015	Manufac	VW	--	--	--	--
016	Manufac	Toyota	--	--	--	--
007	Consult	Horkey	--	--	--	--
011	Consult	Saczalski	Yes	?	Yes	Yes
012	Consult	Hoar	--	--	--	--
017	Consult	Warner	--	--	--	--
022	Consult	Shaw	?	?	?	Yes
006	Univ	Gorski	--	--	--	--
019	Univ	Macnabb	Yes	Design	Yes	Yes
010	Asso	AORC	--	--	--	--
018	Report	TES Ltd	Yes	Design	Yes	Yes

Notes: -- = No response ? = Doubts expressed by commenter
Design = The source of current problem is in design

Analysis of Comments
Docket No. 89-20, Notice 1, 54 FR 40896

Issue 4: Compatibility of Seat Back and Safety Belt Restraint

No.	Group	Name	Rebound Effect Rear	Rebound Effect Frontal	Dual-Mode **ELRs**	Seat Back Interaction
002	Manufac	Navistar	--	--	--	--
003	Manufac	Ford	Some	--	No	--
004	Manufac	Volvo	--	--	Yes	--
005	Manufac	Chrysler	No	No	Yes	--
008	Manufac	Thomas	No	No	Yes	--
009	Manufac	GM	Some	--	No	--
013	Manufac	Fiat	Some	--	No	--
014	Manufac	Mercedes	No	--	Yes	Yes
015	Manufac	VW	--	--	Yes	Yes
016	Manufac	Toyota	No	--	No	--
007	Consult	Horkey	Yes	--	Yes	Yes
011	Consult	Saczalski	--	--	--	--
012	Consult	Hoar	Yes	--	--	Yes
017	Consult	Warner	?	--	--	Yes
022	Consult	Shaw	Yes	--	--	Yes
006	Univ	Gorski	--	--	--	--
019	Univ	Macnabb	Yes	Yes	--	Yes
010	**Asso**	AORC	--	--	No	--
018	Report	TES Ltd	Yes	--	--	Yes

Notes: -- = No response **?** = Doubts **expressed** by **commenter**

V. LITERATURE REVIEW ON SEATING SYSTEMS

A. Seating System and Occupant Protection

The modern vehicle seat is generally recognized as a fundamental portion of the total occupant crash protection system. However, the design of belt restraint and head restraint systems have largely progressed independently from the seating systems. Although in recent years vehicle designers and vehicle safety researchers have looked into the interaction of the three systems* the emphasis and the results are not reflected upon the production and occupant injury reduction.

A review of the literature on seating systems in relation to occupant protection resulted in identifying numerous reports and papers. Many of the technological and innovative designs have been or are being considered for incorporation into production seats. For example, the concept of an integrated seat and restraint system was studied in the early 1900s and a patent was awarded in 1903 on the design of an integrated seat with a lap belt and two shoulder belts. Today, General Motors, Ford, Chrysler, and others are all conducting research on cars with seats integrated with restraints. Mercedes-Benz and BMW both offer integrated seats in their production vehicles.

Some of the earliest automotive safety research sponsored by the Federal government was directed to improve seating systems. For example, the Public Health Service sponsored a research program at the University of California at Los Angeles by Severy, et al. The research resulted in an SAE paper, entitled "Backrest and Head Restraint Design for Rear-Collision Protection," SAE Paper #680079 (Starting from 1955, Severy et al, published a series of papers on seating system safety - see bibliography in the previously referenced University of Virginia report (Docket No. 89-20-N03). In the mid 1960s, the Federal Highway Administration funded research at the Cornell Aeronautical Laboratory and at the Highway Safety Research Institute in the University of Michigan, etc. (These are summarized within the above cited report.)

B. Literature Review on Seating System Safety

Through the years, numerous reports and papers have been published on seating system safety. The NHTSA sponsored report ("Upgrade Seating - Patents, Literature Search, and Accident Analysis," Kennerly Digges and John Morris, University of Virginia, September 1, 1992) has a detailed discussion and list of literature on seating system safety. A paper submitted into the NHTSA docket also contains a bibliography on seating system safety publications (Docket No. 89-20-N01-022, by L. M. Shaw).

Because of limitations in time and resources, this literature review on seating system safety only includes some more recent publications. The reviewed publications are summarized as follows.

1. **"Response** of Belted Dummy and Cadaver to Rear Impact," **Final Report,** DOT-HS-5-01201, July 1976 and **"Response** of Belted Dummy and Cadaver to Rear Impact", A. S. Hu, S. P. Bean, and R. M. Zimmerman, **Final** Report, DOT HS-805-792, December 1980.

In the late **1970s,** the agency sponsored research which examined the reaction of dummies and cadavers seated in rigid and yielding seatbacks during simulated rear impact crashes. This research is reported in two documents as listed above.

Sled impact tests were conducted to simulate the motion of a standard size car at rest impacted from the rear by a second car of equal weight travelling at 32 mph. The test subjects were anthropomorphic dummies and unembalmed cadavers. They were seated in a bench seat and were belted and unbelted. In one test mode the **seatback** was held rigid and in a second test mode the **seatback** rotated rearward in response to the test subject's loading.

In general, the results indicate lower dummy injury criteria values for the head and chest for the deformable seat back. The head and chest severity indices **for** the rigid seat back subjects were more severe than those for the deflecting **seatback** subjects.

The cadavers suffered neck injuries in all except one case -- the one case was a rigid seat back. However, it appears that these injuries are more associated with the placement of the headrest rather than the stiffness of the seat back. The authors indicate that results of the cadaver tests are inconclusive to reach conclusions regarding the relative risk of injuries for deforming vs. rigid seats.

In interpreting these results the characteristics of the rigid and yielding seatbacks should be explained. The yielding **seatback** utilized a standard bench seat apparently based on a typical Model Year 1975 vehicle. This seat itself was modified to accommodate the installation of the extensive instrumentation utilized in the study. Test data indicated the **seatback** deflected backward to approximately 40 degrees from the vertical during the sled experiments. The rigid **seatback** configuration used extensive bracing which resulted in **no seatback** deflection during the tests. Also, the headrest was placed at the lowest position in most of the test.

The results of this study support the concept of optimizing the **seatback** energy absorption or deflection characteristic and avoiding designing seatbacks that are too stiff. However, as indicated above, the deflecting seat back utilized in these tests did not show as high a deflection angle as those seen in higher severity crashes. Also, the study demonstrates the need to properly consider the integration of the head restraint and seat back geometry and energy absorption characteristics.

2. "The Minicars Research Safety Vehicle Program Phase III, Volume I - Technical Final Report," V. K. Ausherman, A. V. Khadikar, S. R. Syson, C. E. Strother, and D. E. Struble, Final Report, DOT HS-806-213, September 1981.

This is a report on the NHTSA funded Research Safety Vehicle (RSV) program. The RSV program was to design, build, and test prototypes vehicles that would exhibit advanced safety performance. Minicars, Inc. was one contractor for the RSV program and Calspan, Inc. was another contractor.

In the Minicars RSV, "[t]he front seats are constructed from modified Dodge van seats (1971 to 1976 model year) and are adjustable to accommodate all occupant sizes between a 5th percentile female and 95th percentile male. The seat frame backs carry a thin sheetmetal panel to resist intrusion by the knees of back seat occupants in rear impacts. Each seat frame top is narrowed and attached to a 0.06 inch (1.5mm) thick clear Lexan sheet. The Lexan sheet, in turn, is connected to the roof, which substantially improves the seat's structural strength in rear impacts. The Lexan attachment to the seat frame incorporates mild steel tape force limiters which provides 700 pound (320 kg) load limiting. The foam seat cushions are also narrowed, then built up with additional foam to form a more desirable contour. All four seats, frontal and rear, are covered with standard automobile vinyl."

The modified Dodge van front seats use Volvo seat tracks for fore and aft adjustment. There are no belt restraints for the frontal seats. "Sled and crash tests have demonstrated that this seat not only is extraordinarily crashworthy, but also correctly and comfortably positions drivers ranging from 5th percentile females to 95th percentile males." The test results include measurements of dummy responses (head and chest accelerations, knee contact forces) and calculation of HIC and CSI.

3. "A Preliminary Evaluation of Seat Back Locks - For Two-Door Passenger Cars With Folding Front Seatbacks," Charles J. Kahane, NHTSA Technical Report, DOT HS-807-067, February 1987.

This is an effectiveness evaluation report on part of FMVSS No. 207, "Seating Systems." The evaluation is limited to only the part of the standard that contains requirements concerning seat back locks (Section S4.3. Restraining device for hinged or folding seats or seat backs). In particular, the evaluation determines if seat back locks are effective in reducing deaths or injuries **and** measures the actual costs of the locks.

The evaluation is based on statistical analyses of three States, **FARS,** Multidisciplinary Accident Investigation **data, sled test** analyses, and **a** cost study of production lock assemblies. The evaluation indicates **that seat back locks hold seat backs in** place in crashes when the **back seat is unoccupied, but locks or other seat** components often separate at moderate crash speeds when **there are unrestrained back** seat **occupants. The** report also presents data from a series of sled tests and accident data that indicate that in frontal impacts there is a high percentage of seat hardware and/or anchorage breakage.

The report concludes that there are **no** statistically significant injury or fatality reductions found **for seat back locks from the** studies. The locks add about $14 (1985) to the lifetime cost of owning and operating a car. However, the report cautions that in using statistical analyses it is easy to prove definitively **that a** safety device is effective but difficult to prove that it is nut. Hence the report is **labeled as a "preliminary"** evaluation.

4. **"Evaluation** of Seat Back Strength and Seat Belt Effectiveness in Rear End **Impacts,"** Charles E. Strother and Michael B. James, Collision Safety Engineering, Inc., SAE Paper **#872214,** Proceedings of the **31st** Stapp Car Crash Conference, New Orleans, LA, October **9-11,** 1987.

This paper is to determine whether or not there is any merit in the concept of **"rigidizing"** (stiffening) **seat** structures for rear impact and, whether **seat** belts are of benefit in these collisions. To achieve the objectives, the authors first review prior research on rear impact, then evaluate data of laboratory tests of current production seats, and finally examine current accident statistics.

For prior research on seat safety, the authors reviewed literature covered 25 years, beginning in the mid-1950s. They divide the literature into three basic types: experimental, mathematical, **and** other. They examine only the first two types **and try** to answer **three** questions: (1) In comparing yielding seats to rigid seats, is there any safety advantage to either design? (2) Is a rigid seat practical?, .and (3) Are seat belts effective in rear **impacts?**

The authors conclude **that occupants in rigidized seats experience** significantly higher injuries than yielding seats with **"controlled** yielding" of the seat back. The rigid seats result in more ramping and whiplash related occupant injuries with or without **head restraints** and lap belted or unrestrained. **In addition, an** occupant's response in rigid seats is very sensitive to **impact** conditions (occupant's position at impact, etc.).

From the review of the history of rigid seat development and test data, the authors conclude that rigid seats are impractical due to excessive weight and cost. Rigid seats would lead to technical and biomechanical complications and could be effective only with a belted occupant population and adequate head restraints.

The authors conclude that seat belts are effective in rear impacts, particularly when delta-v exceeds 15 mph. In addition, restraints tend to minimize the motion of the upper body relative to the seat back and help control rebound motion.

From conducting static tests on production seats, the authors conclude that production seats are capable of producing restraining forces significantly above the requirements of FMVSS No. 207. From reviewing NHTSA conducted NCAP rear impact tests, the authors state that permanent seat back deflection can be expected of virtually all production seats at impact severities of delta-v of 15 mph or greater.

From examining accident statistics for rear impacts, the authors indicate that occupant injury resulting from this impact mode is significantly less than that associated with frontal and side impacts. Finally, they state that from all indications the most effective and practical means to reduce rear impact injury is the use of available seat belt restraint systems.

This paper provides results of static seat tests on production bucket seats conducted by *Severy* et al and the authors (Table 1). *In* addition, this paper provides results of rear impact tests on production vehicles for examining dynamic characteristics of seats (Appendices A, B, and C). Those crash tests were conducted by Severy et al, NHTSA, and others. Accident data collected from *FARS* and NCSS are presented.

5. "Accidents Involving Seat Back Failures," TES Limited, Kanata, Ontario, Prepared for the Ministry of Transport Canada, Report No. C1322/2, December 1989.

This is a contractor's report conducted by TES Limited for the Canadian government. This report was submitted to NHTSA Docket No. 89-20-N01-018. The study involves the examination of 23 real-world case reviews in which passenger cars have experienced seat back failures. Many of the cases involve very severe crash severities and involve vehicles that experienced considerable rotation and multiple crashes during the impact. However, there were a number of cases that were examined where the seat back was damaged during normal operation without any crash event.

This study indicates that seat backs fail mostly due to the occupant's weight on the seat back in rear impacts and some are due to the occupant's weight while the vehicle is stationary or traveling at a constant speed. The two most common types of failure are from the seat back support system -- locks and hinges and the deformation of the seat back frame.

The study specifically addressed the question of the effect of seat back failure on passenger ejection. Of the twenty-three cases examined, eleven of these resulted in one or more of the passengers being ejected from the vehicle. Most of these cases involved multiple impacts and rotation and appear to be of a high severity. About half the occupants that were ejected were using their restraint system. The report indicates: "It is, therefore, apparent that if seat back failure occurs, the use of the restraint system may not prevent the occupant from being ejected from the vehicle". The report indicates that in several of the cases the ejection would probably have been avoided if the seat back had remained in an upright position when loaded by the occupant during the collision.

The report indicates that in three cases in which major injuries were sustained, there were two occupants in the vehicle in which one seat back failed but the other did not. In all three cases the occupant of the failed seat suffered major or fatal injuries, while the occupant of the other seat suffered only minor injuries or no injuries at all.

Three cases involve rear seat passenger injuries due to the collapse of front seats or head restraints. In one case, the left rear passenger was fatally injured by the detached head restraint of the driver's seat, another case involved the right rear passenger who suffered minor injuries when the right front passenger's seat back collapsed, and the third case involved the right rear passenger who was fatally injured and was probably impacted by the driver during his ejection from the driver's seat.

6. "Current Issues of Occupant Protection in Car Rear Impacts," Technical Memorandum, DTRS-57-87-C-00117 (Task #6), prepared by Data Link, Inc., February 1990

This NHTSA funded report was prepared by Data Link, Inc., in February 1990, as part of the agency's investigation into the safety problem related to rear impacts.

This report utilized the agency's NASS 1979-1986 data files and analyzed rear impacts in both an aggregate and clinical method.

As to seat back "failure" or "collapse," the study found that it is quite frequent in rear impacts.

In examining the NASS data the report indicates the following:

"The harm proportion assigned to occupant contacts with the rear compartment is higher in rear impacts than it is in all other impacts. Although this is not unexpected, we observe that the proportion in question is small and rather insignificant by comparison to other sources of injury shown in Table 13." Table 13 assigns 2.8% of harm to rear injury sources compared to 42.2% for noncontact and 16.1% for frontal injury sources.

In examining ejection, the report presents a table that shows that injury from ejection in rear impacts account for about 3% of the total harm. This compares to about 19% for all other modes. "It is evident in the results shown in this table that ejections are not a particularly significant problem in rear impacts, per se or in comparison with ejections in other impacts."

In specifically examining selected cases -- high injury and/or damage in rear impacts -- the report presents a series of tables and concludes the following: "The dominant source of injury is "noncontact", and there are substantial proportions assigned to seatbacks and head restraints, as expected." It continues: "In the case of front seat occupants, irrespective of restraint status, about 15% of the harm is assigned to contacts with frontal interior surfaces and components. A part of this harm might be associated with the 'rebound effect'."

In regard to safety belt effectiveness in rear impacts, the report concludes that the effectiveness is about 40% in the reduction of harm.

7. "Occupant Protection in Rear-End Collisions: I. Safety Priorities and Seat Belt Effectiveness," Michael E. James, Charles E. Strother, Charles Y. Warner, Robin L. Decker, and Thomas R. Perl, Collision Safety Engineering, Inc., SAE Paper #912913, Proceedings of the 35th Stapp Car Crash Conference, San Diego, CA, November 18-20, 1991.

This paper reports the results using four published accident data reports which examined injuries associated with rear impacts. The four reports are (1) a December 1987 report, "Fatalities in Rear-Impacted Small Cars from 1982 through 1987," Susan C. Partyka, one of the four papers in the report, "Papers *on* Vehicle Size -- Cars and Trucks," Page 125, DOT HS-807-294, NHTSA Technical Report, June 1988; (2) a 1989 Data Link report, "Car Crash Outcomes in Rear Impacts," NHTSA supported research report, July 1989; (3) another Data Link report, "Current Issues of Occupant Protection in Car Rear Impacts," NHTSA supported research report, 1990 (also has been reviewed separately in this section); (4) a 1990 Peugeot/Renault supported report, "Risk of Cervical Lesions *in* Real-World and Simulated Collisions," Foret-Bruno, J. Y., et al., the 34th AAAM Conference Proceedings, Scottsdale, Arizona, Page 373, October 1990.

These accident data studies show that rear impacts do not account for a significant portion of automobile injuries; current production seat backs, provide a high level of protection in rear impacts; and the injury mechanisms which may be addressed by rigid seats make up a minuscule proportion of rear impact injuries.

These accident.data also provide additional information on the effectiveness of seat belts in rear impacts. Reductions in Societal Harm or injury of 37% to 47% are indicated.

8. "Occupant Protection in Rear-End Collisions: II. The Rule of Seat Back Deformation in Injury Reduction," Charles Y. Warner, Charles E. Strother, Michael B. James, and Robin L. Decker, Collision Safety Engineering, Inc., SAE Paper X912914, Proceedings of the 35th Stapp Car Crash Conference, San Diego, CA, November 18-20, 1991.

The authors indicate that this paper is partly in response to NHTSA's Docket No. 89-20, in relation to FMVSS No. 207, "Seating Systems." They intend to arrive at an appropriate seat design philosophy by comparing the appropriateness of a rigid seat design and a controlled yielding seat design.

The approach is to review the legislative history of seating systems standards, production seat characteristics, and rigid seat back concerns in real-world rear impacts - ramping and rebound effects.

The paper states that the rulemaking history of "seat design" standards indicates that the concept of yielding, energy absorbing seat backs has always been considered the appropriate design approach for passenger cars. Research findings associated with the rulemaking efforts confirm that rear impacts do not represent a substantial contribution to occupant injuries, and that yielding seat backs can provide occupant protection.

The paper reviews static test data of 61 production seats. Forty-eight of them are from previous tests (SAE Paper #872214 which has been reviewed in this section.) and 13 new tests. They state that , the static tests indicate that the strength of production passenger car seats has not substantially changed over the past three decades.

The paper discusses three most important concerns related to rigid seat backs - ramping, rebound, and out-of-position occupants. The authors indicate that since frictional coefficients between occupants and seat backs decrease dramatically with increasing pressure on the seat backs, ramping will occur on rigid seat backs at lower rotation angle from vertical. They indicate that rigid seat backs have to be built on seats with adjustment features due to practical considerations. During rear impacts, adjustable rigid seats will store more elastic energy than non-adjustable seats, therefore, will cause more rebound. Finally, they indicate that "[i]mpacts with a yielding seat back structure by an out-of-position occupant would be expected to be of reduced severity because of the energy absorbing properties of the yielding seat."

Based on the reviews, the authors conclude that rigid seat backs have the potential to increase injury exposure in real-world impacts due to the above discussed three major concerns:

(1) Rigid seats have the potential of exposing unrestrained occupants to impacts with the roof structure in severe rear impacts. Ramping can be expected with seat back angles as low as 25 degrees from the vertical.

(2) Non-yielding seats, particularly those with adjustable seat backs, can increase occupant rebound because all of the seat deflection will represent elastic energy which will be returned to occupants in the form of rebound velocity. Injuries associated with rebound include whiplash and contacts with frontal vehicle components.

(3) Rigid seats are potentially dangerous *to* occupants not in the "Normal Seated Position," especially unrestrained occupants. A majority of rear-impacted vehicles experience pre-impact changes in momentum which could cause occupants to be out-of-position at impact.

9. "Influence of the Seat Belt and Head Rest Stiffness on the Risk of Cervical Injuries in Rear Impact," J. Y. Foret-Bruno, F. Dauvilliers, and C. Tarriere, Report No. 91-S8-W-19, the 13th ESV Conference, Paris, France, November 4-7, 1991.

Rear impacts are statistically less frequent and severe than other collision modes based on an analysis utilizing French accident data files. Cervical injuries are the most frequent injury -- 27% of the cases. Cervical injuries are mostly minor (AIS 1). This is the case in 99% of the cases. In 1% of the cases these injuries are severe, including fractures of the cervical vertebrae.

Recent studies by Renault indicate that head restraints are 30% effective in reducing neck injuries. This effectiveness was calculated at up to 60% for newer vehicles.

Information indicates that newer vehicles have a lower frequency of seat back breakage. At above 25 km/h, 1971 to 1976 vehicles have a breakage frequency of 62%, whereas 1977 and newer models have a breakage frequency of 55%. It is indicated that the head rest has become more effective because seats have become stiffer. Thus, they conclude that the systematic installation of head rests in all vehicles was a good initiative.

Their accident data indicate that the risk of cervical injury in current seat types is highest without head rests mainly when the seat back does not break. For seats that had head rests, the risk of cervical injury appears similar for seats with and **without** breakage. This is true throughout the different vehicle age categories.

A series of experimental sled **tests** utilizing the Hybrid III dummy were performed. Tests simulated a 15 km/h impact. Parameters studied included variations in the effect of head rest and seat back stiffness. The results indicate "that the head rest, **no** matter what its position in relation to the head, and no matter whether it held under the force exerted by the head, reduces the must predictive criteria of cervical injury".

The report indicates that the most predictive criteria are shear force and **flexion** torque. The tension force (pull on the neck) and the extension torque are also given.

For seat back stiffness differences, it was indicated that utilizing the **"experimental"** head rest, the most predictive criteria were similar for both the stiff and normal seat backs. The pull force (tension force) was reduced with the stiff seat back.

10. "Upgrade Seating - Patents, Literature Search, and Accident Analysis," Kennerly Digges and John Morris, University of Virginia, prepared for the National Highway Traffic Safety Administration, September 1, 1992.

This report describes current research sponsored by the agency. The report is included in **NHTSA's** Docket No. **89-20-N03.** It also reviews selected literature, including some of the earliest studies performed on this subject by Severy and the Cornell Aeronautics Laboratories.

VI. SUMMARY

The primary safety concerns being raised regarding seating system performance relate to the possibility of increased risk of injury when the seat back is damaged in rear impacts. The damage to seating systems in real-world collisions has been characterized as seat **"failure"** and the solution proposed by some of the petitioners and other commenters on this subject relates to increasing the strength of the seat back. The analyses presented in this report indicates that improvements in seating system performance is more complex than simply increasing the strength of the seat back. Legitimate concerns exist over the potential increase in certain types of injuries that might be associated with strengthened seats. As indicated in several of the reports evaluated, a proper **"balance"** in stiffness and compatible interaction with head and **belt** restraint systems must **be** obtained to ensure optimal injury mitigation.

To assist *in* sorting out the information on this complex issue, specific seating system performance issues were identified from the literature review presented in this report. The first category is concerned with seating system integrity. This refers to the structural integrity of the seat and its anchorage to the vehicle structure. The second category of issues refers to the *energy* absorbing capability of the seat itself. The energy absorbing capability is characterized by the stiffness of the seat back. It includes not only the manner in which the seat back absorbs energy but also the manner in which it releases energy. It is the structural integrity and energy absorbing issues that are the primary focus of the rulemaking petitions and public interest.

The third category of issues relates to the compatibility of the seat back with respect to the head restraint, and the fourth category relates to the compatibility of the seating system and the safety belt restraint system. These two issues may relate to FMVSS No. 202 on the head restraint and FMVSS Nos. 208 and 209 on the safety belt restraint system.

Also, it should be stressed that while the current concern has been focused on rear impacts, the seating system must function properly in all crash modes. Thus, the four performance issues discussed must be evaluated not *only* for rear impacts but, also, frontal, side, and rollover crashes.

The agency's National Accident Sampling System (NASS) was utilized to examine the performance of seating systems in real-world collisions. Analyses were conducted to identify any differences in injury risk between occupants of damaged and undamaged seats. Since both the likelihood of seat damage and injury increase with crash severity, the only way to examine a possible causal relationship between seat damage and injury is to control for crash severity. In interpreting the results of these analyses, it should be cautioned that some of the results may not be statistically significant *since* the sample size is fairly small especially after controlling for crash severity. The analyses were conducted separately for each crash mode. For rear impacts, the results were mixed. For low crash severity crashes, occupants of damaged seats had higher injury rates than occupants of undamaged seats. However, for the higher crash severities, the occupants of damaged seats had lower injury rates. *In* frontal crashes, the results were more consistent. In all ranges of crash severity the occupants of damaged seats had higher injury rates.

The real-world collision data utilizing NASS also does not indicate a greater likelihood of ejection with or without seat damage. However, from information and reports presented in the docket, based *on* analyses of selected individual collisions with seat damage, there are indications of ejection associated with seat damage in rear impacts. The data do not indicate that occupant **"ramping"** up a damaged seat is a major concern vis-a-vis exposure to rear components. There are data to support the concept of **"rebound"** into the frontal components after rear impacts. However, the evidence of this phenomenon does not suggest a **significant** safety problem and is based primarily on minor injuries received from contacting frontal components. But it does indicate that any stiffening of the seat back must consider this phenomenon.

The real-world collision data do not support any concern related to the operation of safety belt systems. Further, it should be cautioned that many of these frontal contacts are not due to rebound off the seat, but could **be** attributable to secondary frontal impacts or contact with frontal components as the occupant is rotating back in response to rear impact forces. These analyses also reinforce the need to evaluate head restraint' performance along with any changes in seat back stiffness.

The laboratory data evaluated in this report demonstrate that a proper balance in stiffness must be achieved to obtain optimal safety results. Again, the laboratory data are inconclusive in indicating any evidence that seating systems are not performing appropriately. Most of the laboratory efforts have utilized dummies which were not developed to simulate rear impact kinematics. The research does support the need to address both the head restraint and seat back performance together.

The agency's compliance and New Car Assessment Program rear and frontal crash tests are analyzed as to seating system performance. The tests indicate frequent seat damage in rear impacts due to dummy loading. The production vehicle seats all rotated backward permanently and most of them touched the rear seat backs when subjected to 35 mph rigid barrier rear impacts. The results of this review did not indicate a problem with ramping, but did indicate potential rebound problems. No problems were noted with the operation of the safety belt in the rear impact tests. Damage in frontal impacts was rare and involved seat hardware or anchorage breakage. The frontal NCAP tests indicate that the loss of seat integrity exacerbates the dummies' injuries.

A review of the agency's consumer complaint and recall files indicated several cases of seat back collapse under normal driving conditions without any crash event. Some of the seat defects are not crash related, but they may cause the driver to lose control of the vehicle while the vehicle is in operation.

In reviewing the responses to the agency's public request for information on this subject area, the responses were mixed. Debate continues as to whether today's production seats provide appropriate performance characteristics to minimize injuries. In general, the vehicle manufacturers indicate that the seating system is performing adequately and that the standard dues not need to be upgraded. They indicate that a safety need has not been demonstrated, and that stiffening seat backs may create additional problems. Several accident investigators and researchers have presented case study information on cases involving occupant injuries in vehicles with damaged seats. Most of the arguments utilize the same available accident files, tests, *and* reports to arrive at different conclusions. The agency cannot clearly conclude the issue based on the comments.

From the literature review, the difference in judgment on the performance of seating systems continues the debate. There is the view that seat backs should be designed to have a "controlled yielding" for energy absorption. However, the literature does not have practical designs for "controlled yielding" seat backs. There is another view that seat backs should be designed rigid enough to resist rotating backward and to prevent ramping. Both design ideas are with merit with appropriate limitations and is why many researchers indicate designs must strive for a proper "balance" of these characteristics. The characteristics refer to a consideration of the four issues discussed in the previous paragraphs on structural integrity, energy absorption, head restraint compatibility, and safety belt restraint compatibility.

D:\SEATING\207-035.JJL

About the Editor

Dr. Viano is a specialist in injury biomechanics and impact protection in automotive crashes, sport impacts, and defense/law-enforcement actions. He is Adjunct Professor of Traffic Medicine at Chalmers University of Technology and Biomedical Engineering at Wayne State University, where he serves as the Director of the Sport Biomechanics Laboratory. He has advised and graduated five doctoral students. He is a Fellow of SAE, ASME, AIMBE and AAAM, and serves on the Board of Directors of several scientific, medical and charitable organizations. Dr. Viano is Editor-in-Chief of *Traffic Injury Prevention.* He has published over 200 papers on injury biomechanics and impact protection; and he recently published the book, *Role of the Seat in Rear Crash Safety* with SAE. He has more than a dozen patents, most notably the first commercial active head restraint system to prevent whiplash and the high retention seat for rear impact safety. He is recently retired as Principal Scientist from General Motors Corporation, and now serves as a consultant to the National Football League in their efforts to understand and prevent concussion. He works through his company ProBiomechanics LLC. Dr. Viano served on the National Academy of Science committee that wrote *Injury in America*, and received the Award of Engineering Excellence from NHTSA/DOT as well as numerous SAE awards. Dr. Viano received his PhD in Applied Mechanics from the California Institute of Technology, Pasadena, California and Dr. med. from the Karolinska Institute and Medical University, Stockholm, Sweden.